The Embryologic Basis of Craniofacial Structure

Michael H. Carstens
Editor

The Embryologic Basis of Craniofacial Structure

Developmental Anatomy, Evolutionary Design, and Clinical Applications

Volume II

Springer

Editor
Michael H. Carstens
Plastic and Reconstructive Surgery
Saint Louis University Plastic and Reconstructive Surgery
Falls Church, VA, USA

ISBN 978-3-031-15635-9 ISBN 978-3-031-15636-6 (eBook)
https://doi.org/10.1007/978-3-031-15636-6

This Springer imprint is published by the registered company Springer Nature Switzerland AG
The registered company address is: Gewerbestrasse 11, 6330 Cham, Switzerland

Paper in this product is recyclable.

This work is dedicated with love to my wife, Socorro Gross Galiano, and to our family: Momo, Lily, Alana, Xavi and Josh.

Foreword

It is with both pride and respect that I express appreciation to Dr. Michael Carstens for the privilege of "story telling" about him personally and providing an admittedly biased commentary on *The Embryologic Basis of Craniofacial Structure: Developmental Anatomy, Evolutionary Design, and Clinical Applications*. Simply said, this book is like none I have ever encountered. What makes this work unique?

That the ***content*** is 100% innovation-embedded will be obvious to anyone. Author, surgeon, "different thinker," Dr. Michael Carstens is a person possessed with an endless, open-minded curiosity and a commitment to basic science principles who, over the past three plus decades, has observed, studied, and recorded his observations and clinical management notes on multiple facial cleft patients. Having been such an individual himself, one who has benefited from excellent surgical care both in infancy and with later revisions, he knows at a gut level what plastic surgery can do for patients. Dr. Carstens has provided a "give back" account of a life committed to advancing our knowledge of how exactly the structures of the head and neck are based on an unorthodox mélange of scientific disciplines, from basic molecular embryology to advances in epigenetics. This treatise is clearly personal and even a "love story" about a passion for understanding coupled with relentless persistence.

But, from my perspective, it is the ***context*** of this work, its relationship to its author, that I find most striking. Behind the scientific and intellectual offerings of every piece of literature there is always a "back story" which can make reading or owning a book or manuscript even more poignant and either highly enjoyable or thoroughly detested. I would like to share with you my take on what this "back story" looks like.

My storytelling about the author begins when Michael first came to us from general surgery residency in Boston to begin training in Plastic and Maxillofacial Surgery at the University of Pittsburgh Medical Center (UPMC). He was initially a Resident in plastic surgery and subsequently our first Fellow in cranio-maxillofacial surgery. Why did we choose him to come to Pitt? As a training program committed to developing dedicated future leaders in both research and clinical "doctoring," we learned the value of selectively recruiting individuals for enthusiasm and work ethic and providing an ambience for trainees to "be themselves". In particular, we were always looking for those with ambition to make a difference, possibly while being different themselves. As a Stanford undergraduate Michael had his own very special mentor, Dr. Donald Laub, chief of Stanford's program in plastic and reconstructive surgery. Don Laub's legacy to our specialty is unorthodox, as his Interplast Foundation (now Resurg) has demonstrated global reach for providing surgical care and teaching for underserved countries, plastic surgery as a way of "giving back" to the world. He is an acknowledged role model for countless residents and colleagues. After phone calls and personal discussions, Laub assured me that Michael was just the kind of person we needed. His endorsement, in pure "Laubese," was exceptionally strong, "this kid has a *stickity-stick-to-itiveness* that just won't quit." From the very start of his residency, Michael did not disappoint. During his time at Pitt, besides intelligence and work ethic, I found myself driven to uncover just what might be locked up in his character. I can recall repeatedly challenging Michael to be "really remarkable." That single phrase encompasses what this book is all about. Despite the overwhelming volume of scientific detail, it has a remarkably personal tone, as Michael adds color to each chapter by his own "get it done" approach to research and problem solving.

This book has had a long gestation but there is no doubt in my mind that the seeds of its author's "out of the box" thinking were sewn during his time at Pitt. The origins began with early discoveries with anatomic dissections, and the author gives worthy acknowledgment to his professors, especially George Sotereanos, DMD, for his support and early work done together on vascular anatomy of the head and buccal mucosa, and the pivotal role periosteum has played as a "biosynthetic envelope" encasing stromal and regenerative stem cells. This work gave rise to the publication of two new flaps, the subgaleal fascia, and with co-resident Guy Stofman, the buccinator, which have contributed significantly to our reconstructive armamentarium. These projects also proved in the long run to be a source of important clinical information regarding the developing fetus and subsequently in cleft surgery. I have subsequently followed Michael's trajectory for over 30 years, with great interest, and even awe. So, years later, it comes as no surprise to me that he should have produced this work. It was all there at the start.

As stated previously, this is a work that cannot be separated from the life experiences that produced it. Perhaps motivated by the early work by Ralph Millard during the Korean war, perhaps under the spell of Don Laub to go back to Latin America after residency and fellowship, Michael's career took a wide detour to conflict-ridden Central America, as a surgical consultant for the Pan American Health Organization in Nicaragua. I thought he was nuts at the time, but something also told me that working under those conditions might give him lessons in both surgery and life that would mold his character and define him as a plastic surgeon. Much of Michael's life has been subsequently spent in less developed countries where cleft reconstruction surgery was, and is even today, less available, particularly for the poor. Those patients, who have contributed by their own fate of having facial cleft pathologies, and who have benefitted from Michael's skill and expertise, are "his people." I am sure that much of what he has been able to accomplish is due to these nameless contributors … from whom he has learned and to whom he feels indebted.

Those who have volunteered overseas know that the gain routinely is far more than the contribution. Michael's service taught him invaluable lessons, learned under duress. In areas of the world where the infrastructure is rudimentary and where ancillary components for proper cleft care do not exist, one must create the "system" on his or her own, right on the spot. This led to the establishment of a residency in plastic surgery at the National University of Nicaragua, an innovative program of high quality, supported by the Interplast Foundation, the University of Wisconsin, and the A.O. Foundation from Switzerland. Burn care was a challenging problem as well. In this regard, among mentors who have been especially meaningful in Michael's life are Carlos and Vivian Pellas, significant entrepreneurs and burn survivors after an airplane crash. Their suffering and recovery led them as philanthropists to create—with Michael's input—a plastic surgery and burn foundation APROQUEN which remains self-sufficient to this day, three decades later! Supporting such a national center for burn care has taken capital and resources. Accordingly, the Pellas wisdom and generosity has created a physical infrastructure in the form of the Hospital Vivian Pellas (Joint International Commission-certified!) in Managua. Working in collaboration with multiple others, the work of these innovators created a "system" for care and learning from scratch. This book by Dr. Carstens might be considered a physical representation of those experiences.

So, let's return to the work at hand. Our specialty has benefited from constant evolution in terms of techniques and innovation. Yet in terms of understanding basic developmental mechanisms the field of congenital anomalies has lain rather fallow. Plastic surgery for these patients has been one of evolving surgical designs, each one hotly contested, yet without significant change in our underlying understanding of the problem itself. Around the turn of the twentieth century, the prominent French anatomist and plastic surgeon Victor Veau was critical of this approach, stating presciently that "cleft surgery is applied embryology." Carstens takes Veau's challenge into the context of molecular biology and to the next level by laying out the role for regenerative medicine and cell therapy as the future of craniofacial surgery in the twenty-first century.

As one approaches this book, perhaps the advice from Henry David Thoreau, written in the 1850s, should be that "the price of anything is the amount of life you exchange for it." For those who commit to these two volumes, the price will be an investment of your own time in "deep learning," to appreciate the principles and mechanisms of development that creates final configuration of our own adult anatomy, as described by Gray. The book is not an easy read, nor is it intended to be. The benefit today for readers from multiple disciplines is new learning and likely a better understanding of the threads woven and completed by nature. The effort made by the author to "connect the dots," to make understandable the complex, is one of devotion and, yes, even love for the hidden story behind our structure. "Developmental Field Theory," born from simple discoveries at Pittsburgh, has morphed into a complex narrative that the author always strives to make comprehensible. As he states, "Anatomy is a knowable truth."

For me, the most riveting and provocative part is Chap. 20 on "Biologics." Expected topics such as the physiology of osteoconduction and osteoinduction are modified by the impact of morphogens but there is also a much wider array of concepts. Chapter 20 is itself a crystal ball which envisions cell therapies as the building blocks of regenerative medicine. Here, the work of Professor Arnold Caplan, PhD, from Case Western Reserve University, is appropriately credited for the identification in 1991 of a construct of cells within human bone marrow that can be harvested, and culture-expanded in vitro. Such cells were found to have the capacity to form multiple mesodermal phenotypes such as cartilage, bone, nerve, etc. leading Caplan to coin the term mesenchymal stem cell (MSC). Thus, an avalanche of research on the properties of "stem cells" was launched. Thirty years later we now know their properties much better. Although these cells do not themselves regenerate missing tissue, they do "home" to sites of injury where they produce a host of molecules resulting in healing process involving reprogramming in situ endogenous stem cells. Caplan has re-conceived MSCs as "mesenchymal signaling cells." And from work in Pittsburgh in the 1990s and beyond it is recognized that these cells also exist in fat, arising as pericytes from the microvasculature.

Over the last decade, Michael's work has focused on the clinical applications of these stromal vascular fraction cells (SVF) derived from adipose tissue. From his faculty position at the Wake Forest University Institute of Regenerative Medicine and as Professor of Surgery at the National University of Nicaragua, he has done pioneering work in proof-of-concept studies with SVF cells in diabetic wound healing and neuropathy, pathologies involving scarring and fibrosis as in post-COVID lung patients, inflammatory conditions such as the aggressive and routine fatal kidney disease known as Mesoamerican nephropathy, and even an application for brain pathology, in particular Parkinson's disease. These studies have been performed under Ministry of Health and national IRB approval. Michael credits this work as being a final lynchpin for him to understand the mechanism of tissue failure underlying craniofacial clefts. This clinical research involving continuing data collection and practice follow-up has proven to be a productive "detour" in a long-range thought process resulting in a model of tissue formation based on normal versus pathologic stem cell function, a "squaring of the circle," as it were. For me, that is what Chap. 20 is all about.

In sum: I believe Michael's work to be of profound and of lasting value. What its final contribution will be … only time will tell. But if I could gaze into a crystal ball to converse with a book reviewer in the future, perhaps writing a century from now, say in 2123, I could envision the following words: "This is clearly a unique offering, now recognized even as a true masterpiece Magnum Opus, written 100 years ago by Professor Michael Carstens, MD. The passage of time and collateral supporting information allow one to say now that the author was extremely prescient in his dissertation about how to 'connect the dots of life' and the need for 'silos of medical disciplines' to interrelate through craniofacial embryology, and how specifically the biology of regenerative medicine and cell therapies has evolved." The commentator might continue: "My assessment is that in 2023, when the treatise was first published, it was likely received with skepticism and even suspicion by some. It is now known and accepted that Dr. Carstens, by rejecting much of the dogma of his time and through empirical, open-minded observations, together with an emphasis of principles and process in biologic research, demonstrated that to think creatively and with a vision of the future, a great work can be created, and our scientific-healthcare world has benefitted from his efforts."

In closing, I would like to add a very personal comment, one which Michael might find embarrassing, given his unassuming nature. But this is my chance, and I am going to take it. As one who knows him well, I believe only Michael Carstens could have written this book. My assessment of life-long evidence suggests that he is a true innovator, perhaps at the McArthur genius level, as he has demonstrated exceptional creativity, a track record of significant accomplishments, and promise of a collaborative spirit to foster continued learning and teaching. The AMA (American Medical Association) has recently defined certain characteristics that are most desirable in "doctors of character," and healthcare "system-citizens." These individuals possess critical thinking skills to treat without dogma individual patients and entire populations, while seeing patterns and synthesizing observations as a "system thinker." Culled from both clinical practice and research, they are humble, yet determined, while constantly striving to exceed expectations. They relate to others as trusted team members who are not simply analytical on-lookers but problem-solvers and mentors. In sum, this book, the result of a 30-year pilgrimage, exemplifies both the McArthur Foundation "characteristics" and the AMA Standards of Excellence for "doctors of character" who are also "characters" in their own right. And Dr. Michael Carstens is, indeed, one of these.

So, with my back story in mind, as you read this book—the front story—best of luck and here's a promise: although the onslaught of new thinking awaiting you may seem daunting, take heart! In the end, your persistence will be richly rewarded. Anatomy is, indeed, knowable. You will come away seeing and understanding the human head and neck anatomy in a deeper and more satisfying way than you ever thought possible.

Respectfully,

J. William Futrell, MD

Professor of Plastic Surgery, Emeritus
The University of Pittsburgh
Pittsburgh, Pennsylvania

Foreword

The Open Sesame for Craniofacial Biology

Dr. Michael Carstens' book *The Embryologic Basis of Craniofacial Structure: Developmental Anatomy, Evolutionary Design, and Clinical Applications* presents the students of craniofacial biology with an "open sesame" to reveal the treasures that the field possesses in numberless profusion. Here, the reader finds the vertebrate biology of embryology merged with what is known of the molecular biology accompanying these cellular migrations. It is richly illustrated. The bibliographies are exhaustive. At first glance, one might think it is an entire life's work. In fact, it is the product of many, many lives whose efforts have been diligently searched out and integrated into this comprehensive display of the vastness of the modern craniofacial biology.

The Enlightenment was significant for many things, among them Denis Diderot and his effort to record all useful information in the *Encyclopedia*. Dr. Michael Carstens is clearly our century's Diderot in craniofacial biology! Hopefully like Diderot, he and colleagues will update this magnum opus to periodically keep us readers up to date. The effort will be of lasting value.

There are several reasons to study this massive production. The first is to absorb as much knowledge as you can, or you need. There is no other single resource like it in the world. It represents a carefully curated synthesis of knowledge from many academic fields and their experts. It also references the history of our attempts to correct congenital defects. The collected reference lists are of great value in themselves.

Second, for anyone, whether student or teacher, the illustrations of this text will prove priceless for the preparation of didactic lectures and as background to illustrate the research of the day. These illustrations will also be available online. Generations of students and scholars will be in debt to Dr. Carstens. The amount of professional time saved by lecturers preparing to teach this subject in the future will simply be immense.

Third, anyone working in the field of craniofacial biology—and especially those striving to grow bone where needed—will find this text to be the starting point for his or her lab. Students, residents, and postdocs joining labs will find this to be their Bible for acquiring the knowledge base to build the work ahead of them.

This is not a text that has been dumbed down for easy consumption by casual readers or beginning students. The information seems limitless, but the reader may have to work to get it at times. A diligent reader will be richly rewarded. Given the way the teaching of anatomy and embryology has been steadily reduced in medical schools over recent years, medical personnel of all levels will find this presentation to be far beyond anything they have previously encountered. Not only will this book inform clinical and research work going forward, but it will also provide illustrations and information that may help better inform patients and families of the biology of the defect at hand.

Seldom in one's professional life does one see such a major production of such importance to so many. Michael's effort is truly epic in scope and depth. Many will be in his debt for many years to come. For those of us reading this effort, nothing would please its author more than for his learning that one of us has used this information to advance the field, in a basic science or clinical effort of benefit to patients. For that is where it all began!

Steve Jobs gave the memorable Stanford University graduation address in 2005 famously quoting the parting words of editor Stewart Brand on the back of the last edition of the Whole Earth Catalog: "Stay Hungry; Stay Foolish." Anyone studying *The Embryological Basis of Craniofacial Structure* will be impressed at the level of commitment, time, and work required to prepare and share this opus. He or she will ask, "How could anyone have possibly produced this work while practicing surgery and doing research?" The answer is simply, "Stay Hungry, Stay Foolish." Thank you, Dr. Carstens, for sharing a large part of your life and passion.

Professor of Plastic Surgery Emeritus, The University of Pittsburgh
Pittsburgh, Pennsylvania

Ernest K Manders

Adjunct Clinical Professor of Plastic Surgery
The Ohio State University
Columbus, Ohio

Foreword

The year was 1997, and the meeting involved our hospital's Medical Executive Committee. Of concern was a proposal to initiate an innovative, markedly different approach to congenital facial anomalies in children. In this case, treatment of cleft deformities which had never been previously described. The surgical procedure was intended to maximize cosmetic, but most important, functional outcome. As Surgeon-in-Chief at that time, the concern of some non-surgical leadership of the medical staff was voiced with little understanding of the anatomy, embryology, and science involved. Navigating the "arbitration" of that process was arduous. As surgeons, we are in a dynamic, ever-changing field. Innovation with new procedures and augmenting the established ones is a most important component of surgical advancement. It was early in Dr. Carstens' career at our institution. He had based his theory and proposed procedure on detailed review and inclusion of his work in human fetal embryology and facial development. Ultimately, the new repair was instituted at our hospital; it went on to become the basis for developmental field theory and ultimately gave rise to this book.

In 1999–2000, Michael and our maxillofacial surgeon, Dr. Martin Chin, were the first to implement rhBMP-2 in craniofacial reconstruction under a special FDA-approved compassionate use protocol. This case was the beginning of a fruitful collaboration between Carstens and Chin as they subsequently pioneered in the implementation of morphogen-based bone reconstruction, a concept that is now achieving a place in the clinical armamentarium of craniofacial surgery.

Over the past 3 decades, he has refined long accepted procedures, as well as bringing new ideas and techniques, sharing them with colleagues through juried journal publications. These three volumes reflect his dedication to treating these children and are a memorialization of his life's work. All who practice in this noble calling of pediatric craniofacial surgery will appreciate this important contribution to the discipline. It is a privilege to introduce my colleague, mentor, and friend's seminal work.

Professor and Chief of Surgery for the UCSF
Benioff Children's Hospital Oakland
San Francisco, CA, USA

James Betts, MD

Foreword

Understanding how daily natural processes work is an essential mission of science. This is a special challenge for the clinician-scientist treating children with developmental craniofacial disorders. Explanations of human developmental anatomy taught in medical education failed to provide a framework upon which improved surgical procedures could be engineered. The results of conventional surgical repairs were disappointing, particularly as the child grew into an adult, and it became apparent that the early intervention impaired future growth.

Establishing a complete and accurate model of how an embryo assembles itself has been the objective of scientists and philosophers for centuries. The evolution of these models is gradual. Progress reflects thoughtful observers' extension of science and intuition. The accepted premise is that abnormalities in development result in anatomic deficiencies related to deviations of the normal process of development. It is therefore essential to understand how the process of normal development works. Crafting effective treatments for craniofacial disorders demands the clinician-scientist engineer surgery that complements the underlying biology that gave rise to the problem in the first place. Conceptualizing developmental anatomy is challenged by rapidly changing three-dimensional relationships, changes in tissue volume, and differentiation of cells for specialized function. Simultaneous events occur at the macroscopic, cellular, and molecular levels.

Simple descriptions of how unexpected biologic development results in clinical craniofacial disorders are suitable for basic biology students. When surgeons assume trust of an anxious parent of an infant who has a facial deficiency, it is essential to base treatment on the best model science can deliver.

Michael Carstens spent decades conceptualizing an improved model of developmental anatomy. The mission of this work is to document for students, scientists, and clinicians the complex synthesis of anatomy, emerging biologic science, and intuition required to understand how the embryo assembles itself. This description provides an opportunity for clinicians and scientists to engineer procedures and treatments that are more effective and less invasive. Carstens' dual role of scientist and clinician affords a unique perspective on both immediate surgical outcomes and the result after years of growth. Thoughtful observation of how variations in surgical design can produce more-or-less desirable outcomes as children grow takes time, patience, and will.

It has been an honor to work with Michael for over 30 years. Medical science evolved during that time, allowing us to advance treatments in ways we could not have fully appreciated as students. The route to success must incorporate the difficult and sometimes controversial adoption of new ideas. In the end, we all want to do the best thing for the child. It is our calling as doctors.

Martin Chin, DDS

Attending Surgeon, California Pacific Medical Center
San Francisco, California

Founder, Beyond Faces Foundation

Foreword

Michael Carstens, MD moved to Los Angeles in 2003 to be the craniofacial fellow at Children's Hospital Los Angeles. Dr. Carstens was unique as he didn't fit the mold of more typical post-resident fellows. He had been in solo private practice pediatric plastic surgery for many years at the Children's Hospital Northern California in Oakland. His intense intellectual and surgical curiosity were stifled. Eventually, he decided to follow his drive for a more academic environment by applying for the craniofacial fellowship at CHLA. He left his family in Oakland and rented a small apartment near Children's Hospital. Thus began his academic journey. His time in Los Angeles with me was crucial, as it allowed him to formulate and develop a profound understanding of craniofacial embryogenesis, one that led to many innovations in surgical technique.

This book is the culmination of almost two decades of relentless exploration and mapping of craniofacial embryological development. Dr. Carstens' intention is to offer surgeons a rationale and philosophy for how to think about surgical repair for a wide variety of congenital defects. He presents in complex detail a model to help surgeons in their reconstructive efforts to move closer to nature's intended mechanics.

Dr. Carstens' contributions in this book are embryologically sophisticated and original. They are invaluable to anyone with a deep interest in cleft formation and repair.

I have enjoyed knowing Michael for almost 20 years. He has an optimistic and enthusiastic approach to life. He engages deeply with whatever captures his attention. For example, although Mike was born in Iowa, he developed a deep interest in Latin American culture. Having spent time in Central America, he spoke Spanish like a native speaker. He was so immersed in linguistic fluency, that I would see him warmly trading stories in Spanish with the janitorial crew at the hospital. Over the course of the year, Michael became an integral part of CHLA. He was well-liked by all.

Michael is also unique in that he himself was born with a cleft lip and palate. Perhaps his personal experience is the nidus of his passionate interest in craniofacial development and treatment.

I am honored to witness the fruition of his work.

Los Angeles, CA, USA
October 13, 2022

John F. Reinisch

Foreword

A Unified Theory of Developmental Field Repair

It is an honor and a privilege to be asked to write a foreword to the book written by my friend and colleague Prof. Michael Carstens *The Embryologic Basis of Craniofacial Structure: Developmental Anatomy, Evolutionary Design, and Clinical Applications.*

Let me begin with a short story. I discovered Michael on a sleepy afternoon in the comfort of a couch in the library of Bombay Hospital and Institute of Medical Sciences. The year was 2005, I was browsing through "Clinics in Plastic Surgery" and stumbled upon his article on clefts. Within minutes I was wide awake and fascinated by his completely new approach. It was a delightful and scholarly change from the endless discussions about "Rotation advancement" *vs.* "Triangular flap," etc. for Michael dived deep and tried to bring into focus where and how it all starts and how we have got it wrong all these years. I was then the President-elect of the Indian Society of Cleft lip Palate and Craniofacial Anomalies. I felt a great urge to get this man to come to India and address our members. He came to the annual meeting in Guwahati, Assam State of India and the audience was held spellbound by his analysis.

Michael came often to address various meetings. During the last two decades, Michael has kept working on his theories and worked across disciplines to try and get a coherent whole. We kept corresponding. It was through this association that he has honored me thus and I am grateful.

The book consists of 20 chapters, and I must admit it was a task to read it. Mainly due to my vast ignorance of the various areas of scholarship involved. Once I got the hang of it, the fascination was rekindled, and the pieces of the jigsaw started falling in place.

I want to quote a few snippets from Chap. 1, they are not continuous in the text, but they create a whole as far as the philosophy behind the book is concerned and give you, the reader, the essence of the approach.

1. "Understanding the sequence of field deficiency has bi-directional consequences: it leads us further inward to work out the genetic sequences that produce the field; in the first place; and it leads us outward towards a more developmentally-based and innovative therapeutics."
2. "The well-known dictum of DeMeyer in 1964, 'the face predicts the brain,' can be inverted to 'the brain predicts the face,' as mechanisms of induction are now better understood."
3. "Despite advances in presurgical orthodontics and operative techniques we continue to be faced with results that deteriorate over time; most of our patients requiring multiple secondary interventions. Such a model cannot be correct."

Read in sequence they reflect the essence of the book and the novelty in its approach.

This is a book for the anatomist, vertebrate biologist, and also for surgeons involved in treating craniofacial clefts. For the latter, it ought to be required reading before they start independent surgery as Fellows/Senior residents. Whether they and their mentors adopt the thinking or not, they need to be aware of it to make informed choices.

There are at least four disciplines involved in understanding embryology: paleontology (going back in phylogeny), genetics, molecular biology (gene signaling) and, of course, anatomy, which is influenced by the preceding disciplines in a predictable manner as Michael shows so cogently.

As he freely admits, it is by standing on the shoulders of previous giants that Michael has been able to look farther, and his references give ample and due credit to scores of people whose work he has relied upon. The difference is that he has digested this enormous amount of scholarly material and distilled a holistic narrative from it.

It is nice to note the credit given to Tessier, Talmant, and Delaire among others. The fact that Tessier accurately described the pathology 50 years ago without the knowledge of a lot of later material in gene signaling and regulators like BMP4 is uncanny to say the least and testament to the meticulous observation and surgical mastery of the great man. Prof. Talmant told me he was trained in philosophy by Delaire and in Surgery by Tessier, which made him the best equipped to deal with clefts. It is important to synthesize multiple sources and come to your final technique.

The first 13 chapters lay the ground for the last 7 which are largely clinical with recommendations for rational incisions and rational repair. They are logical, and I hope, despite the enormous influence Prof. Millard still has over the cleft world, people synthesize his ideas with the rationale of Michael so that a newer type of repair evolves.

Finally, this book is for the intrepid reader. It is very heavy going for those only interested in technique. However, without the background in newer sciences and their findings; I think the clinical case for a technical shift cannot be made.

In the end, I want to quote one more nugget from Michael:

"But the iron fact of the matter is this: without a detailed understanding of the developmental anatomy of the face based on modern developmental biology, genetics, comparative anatomy and neuroembryology, paediatric plastic surgery is a collection of techniques in search of a science."

In my view, this is a very clear call for a rethink and reappraisal.

The great physicist Albert Einstein noted the fundamental forces of gravitation, electromagnetism, and the strong and weak nuclear forces, and during the last decades of his life in Princeton tried very hard to bring them all into a single theory which he called the "unified field theory." Unfortunately, he did not succeed and that struggle in theoretical physics goes on.

Dare I say Michael has succeeded in creating a unified field theory of cleft origins, pathology, and anatomy and that the "Developmental Field Reassignment repair' will fill the lacuna long present in our understanding and treatment of clefts. I hope it does. I wish him the best and commend this book to you the reader.

Mumbai, India
August 28th, 2022

Mukund R. Thatte

Foreword

All surgeons involved with cleft care know fully well the frustration of seeing well-executed repairs in infancy transform into a predictable sequence of secondary deformities requiring further correction. Even in the best of hands, re-operation rates may reach as high as 85%. What exists here is not failure of technique but, an inadequate biologic model of the problem in the first place.

"If the pathologic anatomy of the cleft site pivots on a deficiency state in a specific developmental field, and if the surgical correction of the cleft does not include reconstitution of that defective field such that it will grow normally over time, and such that it will cease to perturb the growth of its neighboring fields, then all forms of cleft surgery are condemned over time to varying degrees of relapse." Michael Carstens

Most facial anomalies represent defects in specific *developmental fields*. The success or failure of surgical manipulations permits a more accurate understanding of just exactly where these fields exist and how they act.

When a deficient developmental field is released from normal surrounding fields, subsequent facial growth can be anticipated to be more normal.

We have long wanted a publication on the subject. Michael Carstens needs no introduction to those who have followed this literature. We have recognized the outstanding qualities which made him the logical choice to write this complete research ever since we watched his brilliant and devoted work.

He is introducing a new and clinically relevant model of craniofacial pathogenesis based on concepts of developmental biology and neuroembryology that are yet little known in medicine.

Without a detailed understanding of the developmental anatomy of the face based on modern developmental biology, genetics, comparative anatomy, and neuroembryology, pediatric plastic surgery is a collection of techniques in search of a science.

The clinical significance of the neuromeric organization is that it enables us to map out the anatomic site of origin for all zones of ectoderm and mesoderm supplied by a given zone of the nervous system.

Evolution of new concepts in flap surgery based on expanding knowledge of skin circulation increased understanding of the biology of wound healing and tissue transplantation.

The craniofacial/plastic surgeon can learn from his research in basic sciences, thus enlarging the body of knowledge of the specialty.

Dr. Carstens has left his imprint and injected his own philosophy of diagnosis and treatment throughout this book.

He has offered talent, time, and energy to make possible the completion of this extensive three volumes edition making an original scientific contribution.

We can confidently predict the progressive growth of this area, a specialty with unlimited possibilities engendered by the fertile imagination of this plastic surgeon and hope he is deservedly recognized for his titanic effort.

Post graduate professor, School of Medicine, Nat. Univ. of Buenos Aires Ricardo Bennun
Director Cleft Lip/palate and Craniofacial Program
Asociación PIEL, Buenos Aires, Argentina

Foreword

The work of Dr. Jean-Claude Talmant, pediatric plastic surgeon in Nantes, France, will perhaps not be fully appreciated for another generation. He has single-handedly explored the relationship between fetal breathing mechanics facial anatomy, using facial clefts as his model. Pressures exerted normally within the nasopharynx and nasal cavity during development play an undeniable role in shaping the morphology of the lip-nose complex at birth. Dr. Talmant documented these effects of fetal ultrasound. But he did not stop with merely proving an anatomic point; he carefully and methodically developed techniques to address these effects during primary surgical repair with a resulting restoration of not only breathing mechanics but of their cerebral control. Jean-Claude's life work also demonstrates a principle that is very much in the tradition of French anatomy: the careful observation and recording of natural events over time. In this regard, his contributions are a surgical version of Marcel Proust's revolution 7-volume masterpiece, *À la Recherche du Temps Perdus* (Remembrance of Things Past). Like Proust, he has lovingly captured the passage of time as reflected in the visage of each one of his patients. Although his case series is small, it has been recorded with the utmost precision and care. No one has done it better and his attention to detail remains a model for surgeons to emulate, both now and in the future.

Commentary—Jean-Claude Talmant

In March 2006, Mukund Thatte, president of the Indian Society of Cleft Lip & Palate had invited us, before the Guwahati congress, to visit the white rhinoceros sanctuary of Kaziranga Reserve. Our convoy crossed the green hills of Assam, covered with groves and tea plantations. I was in the back of the land rover with my wife, Odile. To the left of the driver, an American passenger had dozed off during the trip. From time to time he awoke and seemed to emerge from a bubbling and uninterrupted reflection: It was Michael Carstens. The thirst for understanding and the urgency to transmit from his enthusiastic nature which is his character, even when resting. Both of us, during this congress, felt the convergence of our approaches to cleft lip and palate. He sent me his work revealing the connections between the embryological evolution of facial vasculature and the clinical classification of facial clefts by Paul Tessier. He read the CME lecture (later published by Mukund Thatte in the Indian Journal of Plastic Surgery), following up on many of the arguments that I had put forward in my presentations.

A few years passed, interspersed with shipments of chapters from Michael, as he was immersed in the writing of his work. We came from two different worlds. Michael, through investigations, meetings, and research, has gathered an encyclopedic knowledge of craniofacial embryology. His ability to synthesize and his talent as a storyteller make us forget the aridity of the subject. He has the ability to make developmental anatomy come alive and outline its principles without it being necessary to memorize all the details. To think of anatomy as biologic and conceptual, rather than a collection of unrelated facts is truly a paradigm shift. Michael writes clearly and patiently to explain how each structure can be identified by its neuromeric origin and can be tracked along its pathway from the primitive stages of the embryo into its final form, following an algorithm of steps triggered by molecular signals. He deduced from this a fundamental principal for surgical therapeutics—to restore the deformed and displaced elements to their rightful place and to thereby reconstruct the missing embryological field. Defects of embryogenesis leave behind a scrambled map of mismatched tissues incapable of growing harmoniously. By unscrambling these mismatched fields, normal growth can be restored. The principal of Developmental Field Reassignment is virtually identical to that proposed by Victor Veau almost a century before, but this time is substantiated by modern molecular embryology.

For my part, I learned a lot from Paul Tessier, who clarified the anatomy of facial clefts, exploring them through precise and extensive dissections in the only sub-periosteal and subperichondrial planes which would respect the structures (fields), until he could envision them as distinct entities, thus making it possible to identify and reposition them. Jean Delaire, by entrusting me with the sequelae of the many cleft lip and palate cases that he was following, positioned me in a unique to observe and understand the anatomy of previously operated clefts. Thus, I became able to detect the poor outcomes arising from different protocols. I paid all my attention to certain common themes in these deformities, in particular, that of the nostril whose alar cartilage had obviously abnormal relations with the myrtiformis muscle. Michael saw the logic of my analysis and found that it could be perfectly explained as an application of reassignment of displaced fields; he thus integrated it into the DFR model.

I am also indebted to my brother, Jacques Talmant, who was the first to show me a video of fetal ventilation captured shortly before the year 2000 by his wife, sonographer Claude Talmant. It is thus better understood that the facial growth deficit in utero is due to a perfectly normal growth potential locked up in an abnormal anatomical context.

Since the beginning of the 2010s, our meetings have multiplied. Michael came to see us in Nantes with David Matthews, one of Paul Tessier's three musketeers. Michael and David know each other well and like each other. They attended operations and multidisciplinary consultations with my consultant orthodontist, Jean Pierre Lumineau who, since the mid-1980s, has performed anterior maxillary expansion by quad helix in our 4-year-old patients to reconstruct the sector missing from the lateral incisor by gingivoperiosteoplasty and iliac bone graft before the age of 5.5 years. Michael returned for a week in 2012 with Jyotsna Murthy and I remember a passionate lesson, where at the end of our surgical intervention, he drew his famous non-Philtral Prolabium flap (NPP) with a surgical marker right on the operating field drapes!

Michael wanted that we continue to make an impression at the congresses with our complementary presentations. He invited me to Santiago, Chile to visit his friend Luis Monasterio in 2014. We presented again with Ricardo Bennun of Buenos Aires for the Congress of the South American Cleft-Craniofacial Society in 2016 in Salvador de Bahia. We met in 2017 in Chennai, at the invitation of Jyotsna Murthy at Sri Ramachandran Medical Center. These meetings gave us the opportunity to discuss our views and our doubts, addressing our differences frankly. After re-reading Michael's Chaps. 18 and 19 where the description of my protocol is presented, identical in uni- and bilateral clefts, I think a few comments are warranted to clarify what I consider the most important components of a DFR-based cleft lip/nose/palate protocol.

The first operation, done at 6 months (without orthopedic preparation), reconstructs the lip and seeks a complete correction of the nose to introduce the child to nasal ventilation. At the same time, the velum is closed with an intravelar veloplasty according to Sommerlad (a point to which we will return). What is the rationale for the velar closure?

Prior to operation, the child ventilates through his palate cleft; thus, he has not had the opportunity to develop in his brain a cortical representation of the nostril. By separating the nasal and oral passages, the child is forced to choose a nasal mode of ventilation for the first time. At 6 months, the hyoid bone is still positioned so high that there is no space to lower the tongue: Nasal ventilation is the only option at rest and in sleep. An immediately patent nostril, one that is neither too small in its dimensions nor readily collapsible due to the mal-positioning of the nasalis muscle, will allow *nasal ventilation* and *bilabial contact*. This is striking in bilateral clefts where the premaxillary deformity that initially prevents labial occlusion corrects spontaneously in 4 months. *No orthopedic treatment can do this*! Forcing the child to depend upon oral ventilation at this early age is against nature. In the absence of adequate nasal ventilation, the initial programming of the breathing process produces an imprint on the cerebral cortex that experience has shown me is difficult to erase.

In the months following the intervelar veloplasty, the cleft of the hard palate shrinks to such an extent that it can be closed in two planes, without a raw surface, almost always between 14 and 18 months. Fistulas are rare; and they are often so small that locating them is more difficult than closing them.

Around age 4, if necessary, orthopedic expansion of the anterior maxilla by the quad helix device restores intercanine width at least 4 mm greater than its mandibular counterpart to reconstruct the alveolar cleft before age 5.5.

The choice of the procedure implemented for the closing of the soft palate deserves discussion because its implications are considerable.

Let us first review the fetal development of the velum. As soon as its fusion is acquired, nasal ventilation begins. With each exhalation, the bony palate is pulled forward by the projection of the tip of the nose, while the muscular structure of the veil is retained by its continuity with the velopharyngeal sphincter. It is even driven back by the reflux of amniotic fluid into the reservoir of the nasal fossae which follows each exhalation. Thus, the aponeurosis of the tensor veli muscle stretches between the hard palate and the pharynx. It is at birth that it is in proportion the longest. It will then be invaded by the muscles which remain at a distance from the posterior edge of the palatal plates. Only the tensor inserts into the proximal 1/3, the zone of the aponeurosis. The remainder of the velar sling is situated in the middle 1/3 of the velum. In the event of a cleft, the palatine aponeurosis, whatever its primary hypoplasia, is not subject to any distension and remains retracted on the pterygoid process. The palato-pharyngeal and levator veli muscles converge anteriorly into a single muscle body in the cleft. They insert aberrantly into the aponeurosis without inserting directly on the bone. Thus, if the repair does not take into account the pathologic antero-posterior orientation of the muscles, they will remain unable to move back and raise the veil due to this abnormal anchoring.

At the beginning of my career, I sutured these muscles without intravelar dissection and lengthened the veil by suturing the posterior pillars, thus creating a flange limiting the ascent of the veil and ventilation. These soft palates remained motionless and short in their functional part in front of the uvula. For this first generation of my patients with non-syndromic clefts, 50% required pharyngoplasty. My evolution since 1999 toward the posterior transposition of the velar muscles after extensive dissection, rotation of nearly 90° and suturing in tension with overlapping has completely transformed the morphologic outcome. If the muscles are of good quality, the velum lengthens under the effect of the correction of the vector of the muscular action. For non-syndromic clefts, the Sommerlad veloplasty is well-suited to this analysis and achieves the desired anatomic and physiologic objectives.

If I happened to revise by the same technique (Sommerlad) some of my previous soft palate cases in association with a successful lipofilling of the pharynx, I found that I no longer had to do pharyngoplasties in these non-syndromic cases.

Things are different when it comes to syndromic forms. *Whatever the operation, the result will be disappointing if the muscles are pathological.* Pharyngoplasties and sphincteroplasties are then resorted to, which are dangerous due to their impact on nasal ventilation. It is then that the int*erposition flap of the cheek mucosa carried by the buccinator muscle* becomes a recommendable option for reconstructing the retracted sector of the palatine aponeurosis because it respects nasal ventilation. I read the work of Michael and that of Robert Mann, both of whom, having developed this technique for 30 years, deserves consideration and respect.

I remember a conference where Michael proposed the buccal flap as the primary treatment for velar clefts in 2016. It had been very severely criticized, and that is not surprising given the context at the time. The cheek mucosa has little to do embryologically with the velopharyngeal structure, especially if one is a follower of Victor Veau. This was a real paradigm shift, even more so than my premise of insisting that the nasal deformity must be repaired in the primary surgery.

These surgical practices that go against everything that has been promoted over the past 50 years are beginning to prove themselves! Robert Mann's idea of dressing up his concept in a more respectable way by speaking of it in terms of modern embryology can be understood, insofar as the foreshortening of the palatine aponeurosis settles in utero, but one should not be misled: Retroposition of the velum en masse by interposing a cheek flap with the vascular supply of the buccinator muscle is an opportunistic technique that is more akin to the repair of tumors than that of malformations.

In syndromic clefts, the soft tissue structures present in the sidewalls of the pharynx and soft palate are not normal, and their repositioning is disappointing. The situation is akin to that of a trauma, where the principle to be implemented is to replace the deficient or missing pathological structure—here, the palatine aponeurosis—by the best equivalent available. On the basis of Robert Mann's 30-year experience, the benefit of soft palate and soft palate scar reduction that accompanies his technique, and the mediocrity of most other solutions, ultimately give the buccal flap a prominent status. It will soon be one of the good answers to the velar insufficiency sequelae of many cleft veloplasties, if not the best, especially if there is a minimum of remaining muscle tone which is still functional. Michael, embryology expert, is not dogmatic. He is pragmatic. He knows that the efficiency of a technique and its harmlessness are its first justifications. Common sense takes precedence over blind respect for repairs with embryologically similar tissue. *If a field is deficient or otherwise absent, it must be recognized and replaced.*

For 50 years, more than two generations of practitioners, entrapped by dogma, have explored very opposing options in the treatment of cleft lip and palate. Only Victor Veau's philosophy of restoring displaced structures to their normal anatomic position and function has stood the test of time. The most effective protocol pursues this logic throughout the course of treatment: *Restore without compromise, with ambition and despite the technical demands, the lip, the nostril, the septum, the alveolar arch, the dental occlusion, the palate, the veil by installing nasal ventilation from the first operation and preserving it.*

Correction of the deformation of the alar cartilage and the septum and reconstruction of the missing embryological field of the lateral incisor are therefore aesthetic and functional prerequisites. The nose, forgotten in recent decades, whose subtle functions and interactions are better understood, will finally take its rightful place: that of the great organizer of facial growth. This is a very demanding standard that only a coherent protocol can satisfy at a cost. Rigorous learning of the underlying biology on the part of all specialties involved is required to achieve a shared awareness, a consensus about what the problem and the principles of its correction.

In pediatric plastic surgery, understanding the nature of a deformity and knowing the normal, going beyond mere appearances to perceive all the functional implications of a repair is the goal of the science of morphology, still in its infancy and hardly taught. "Function creates form," a well-known aphorism by Louis Sullivan, architect of the first skyscraper on the eve of the twentieth century, becomes medical evidence when one devotes one's life to repairing cleft lip and palate. And yet, ignorance of the role of nasal ventilation remains preponderant in the medical profession, as does the prohibition of all infant nasal surgery. Paul Tessier, although little suspected of lacking interest and curiosity for craniofacial anatomy, lamented in an essay later found by Tony Wolfe in his archives, his "ignorance at all times of embryology, functions and functional interactions in the child and adult." He regretted not having been able to really remedy this shortcoming due to lack of time, and no doubt also, because of the difficulties of access to this knowledge. Perhaps for this reason, although he and Michael never met one another, he welcomed two binders of writings and drawings that Michael sent to him from California. These were proof positive that his systematization of craniofacial clefts had a solid basis in developmental anatomy.

It is not certain that anatomy and embryology are much better considered and taught today's curricula and especially if there exists a will to recognizing their ongoing importance.

Fortunately, and we discover it in this book, born of a driving initiative, motivated by the sole ambition for a radical advance in the treatment of malformations. Such initiatives come from practitioners whose experience has sharpened their capacity for judgment without taking away their originality, creativity and, above all, the constancy essential to the pursuit of work that requires rigorous and uncompromising evaluation. True innovators take advantage of their independence to escape from the uniformity of worn-out thought and to question knowledge that is stagnant and wanting in depth. The fundamental and practical research pursued by innovators, whatever their favorite fields, while respectful of medical ethics, is most often done with a curiosity and an open-mindedness that leads them to explore new paths without fear of controversy or confrontation. When like minds meet, they take advantage of their convergences as well as their differences to progress and rise together. They blow a breath of fresh air that invigorates a sleepy world. May this breath, symbolized by the work you hold in your hands, inspire you and many other colleagues, to move forward and innovate for yourself, for in this pathway lies the promise of a better life for your patients.

Jean-Claude Talmant, MD

Cleft Lip and Palate Center
Pays de la Loire, Clinique Jules Verne
Nantes, France

Foreword

This is one of the few books that can be truly called a magnum opus. My first connection with it was in the early 2000s, when Michael first contacted me with some embryological questions. Since then, he and I have been in regular contact. My role has been principally listening to and commenting on a stream of ideas that have put together a remarkable edifice that contains an integration of fundamental principles of fields as diverse as paleontology, anatomy, and reconstructive surgery, with embryology serving as a connecting link. This book represents the culmination of at least two decades of original thought that has resulted in a system for understanding the fundamental relationships that underlie the incredibly complex anatomy of the adult human head and neck and putting that understanding into practical use in reconstructive surgery.

The basis for this system is a recognition of the highly segmental organization of the human body. In the trunk and abdomen, a fundamental segmental organization is evident even to the moderately attuned observer, but it is not so apparent in the craniofacial region. Starting with the neuromeric organization of the central nervous system, Dr. Carstens has extrapolated this fundamental pattern to more peripheral tissues and has been able to produce a system that allows one to coherently organize many complex facets of craniofacial anatomy into understandable segments. In some cases, actual laboratory investigations have not been performed to confirm extrapolations from animal models to humans, but the way that the information is presented in this book allows researchers to devise experiments that would test relationships and mechanisms hypothesized in the book. For students of anatomy, there is no better way to understand the complex anatomy of the head and neck than organizing the myriad of structures on the basis of their embryological organization.

Starting with some early personal research on vascular fields in the head, the author has gone into some rarely cited embryological literature on the vasculature of human embryos. As a result, he has laid out a basis for understanding the basis of adult vasculature, especially of the face, that not only sheds considerable light on the anatomical basis of many adult structures, but more importantly mechanisms underlying the genesis of many facial clefts. In this domain, he has produced a new understanding of the developmental basis for the widely recognized Tessier system for classifying craniofacial clefts.

As he was developing his embryological models for craniofacial development, Dr. Carstens began to dive into the increasingly rich paleontological literature on the phylogenetic development of the head. This enriched his way of viewing the vital role of segmentation as an organizing principle in both phylogeny and ontogeny. I am aware of no book that juxtaposes the anatomy of fossil fishes with discussions of congenital malformations of the craniofacial region and their surgical treatment.

Because of the vast amount of detail, much of which will probably be unfamiliar to many readers, this is not an easy book to read. Nevertheless, with the help of the many illustrations, the reader finally has access to what will be for many years the definitive source of information about both the embryonic and phylogenetic development of the craniofacial region. More importantly, for the practicing surgeon, understanding the embryonic field relationships in the developing head can lead to devising reconstructive procedures for congenital malformations that are notorious for requiring one or many follow-up surgeries.

Prefessor of Anatomy, Emeritus University of Michigan
Ann Arbor, MI, USA

Bruce M. Carlson, MD, PhD

Preface

What Brings You to Open This Book?

A preface is a contradiction in terms. Although positioned at the beginning of a book, it represents the last act of its creation and a final reflection upon its value. A preface poses a question which can only be answered by you, the reader …: Why go forward? Of all subjects in biology, anatomy seems supremely impenetrable, a "known world" of facts … not much to get excited about here. Could there be more to it than meets the eye? Is it worth your time and effort to seek beneath the surface of cut-and-dried facts to find something else? The answer is *yes*. Anatomy is a Rosetta stone that permits you to understand the hidden language of development. To decipher it requires three keys: a willing suspension of preconceptions, patience, and persistence. So, not unlike an archaeologist unearthing past treasures, your search for the story of development will become its own reward.

The Open Sesame of Structure

It all began with a skull. I was a resident in plastic surgery at the University of Pittsburgh; the year was 1987. On my basement workbench was a model skull, a drill, colored pipe cleaners, Moore's Anatomy, and Grant's Atlas. The model lacked many of the small orifices and fissures as shown on the pages of the atlas. Convinced that these all had an explanation, I set out to find them, drilling out those that were not present. The pipe cleaners served as markers for the exit and entry sites of nerves and vessels; obviously, soft tissue development took place prior to that of the bones. It occurred to me that, with growth, the skull bones would advance toward one another like drifting continents until making contact, and the orifices, foramina, and sutures represented boundary zones between these fields. Thus, vessels and nerves did not penetrate bone fields. They would remain as markers at their peripheries, thus identifying a system of primitive embryonic fields. I never forgot that moment. I was convinced that *anatomy had to make biologic sense*. Hidden in plain sight, within each structure of the face and skull was a secret story, written in code, of how it came to be, one which would explain the reason for its ultimate shape, size, location, and function. If only there were a way to decipher it …!

Curtains and a Surgical Innovation

Fast-forward to 1996 … residency, fellowship, and tour of duty with the WHO in Central America were behind me. As I was preparing for board exams at home in Berkeley, California, I found myself thinking about secondary cleft patients I had cared for previously in Nicaragua and puzzling about curtains. The glass wall of the second-floor bedroom looked westward to the Golden Gate Bridge. Late-afternoon light flooded in; there were no curtains. I found myself imagining a pleated curtain attached to the wall on the right. As I pulled it leftward, the pleats, of course, became asymmetrical. I remembered the remarkable similarity of the previously operated patients … and how, on the side of their repairs, the soft tissues of the face always seemed stretched taut, like a curtain pulled medially from a lateral bony anchorage point. Then I thought, "What if the curtain were detached from the wall, and allowed to move freely to the midline?" Would not the pleats stay symmetrical? And what if, at the time of surgery, the soft-tissue envelope were to be detached from the maxilla and allowed to move forward into the midline? Would not a tension-free closure be achieved? If so, the face could regain its natural symmetry …. The key to this insight was to think backward in time, like a videotape in reverse, to the beginnings of facial development. Little did I know how relevant paleontology would become.

The 4Ds

Subsequent surgical cases proved this prediction to be true. Using a developmental model for cleft repair, with a dissection radically different from the status quo, the on-table results were quite striking. Working in a different plane to shift the tissue into the center, I saw the tension melt away. By moving the soft-tissue "curtain" laden with stem cells into the center, further deposition of bone could take place where it was intended. But I could not understand neither how nor the why this should work … my thoughts kept returning to the image of tectonic plates. It was as if the cleft side of the face had drifted away from the midline.

Instead of using cookie-cutter designs to rearrange the tissues as we saw them on the table, should we not be attempting to put into reverse a pathology in four dimensions stemming backward in time to the cleft event itself? In this model, an unknown site of **d**eficiency, causing a **d**ivision of tissue, would lead to **d**isplacement of the cleft side of the face as it drifted away from the midline, like a tectonic plate. And of course, with the explosive growth of the facial envelope during the embryo-to-fetus transition, neighboring structures (such as the nose) would become **d**istorted. This process theory made sense—but *what was the original problem*? I felt myself on the edge of a paradigm shift, searching for answers.

Sojourn in Basic Science

Reviewing all three volumes of Millard's *Cleft Craft* and reading its myriad of references produced nothing. If I were going to get *anywhere*, I would have to immerse myself in basic science—several to be exact: embryology, developmental biology, genetics, neuroscience, comparative anatomy, and evolutionary biology. Little did I know that this process would take 20 years. Over time, many colleagues helped me along the way. I first learned about homeobox genes and their potential mapping of the pharyngeal arches from the fetal pathologist Geoffrey Machin at Kaiser Oakland (we were doing vascular injection studies). Two years later, at John Rubenstein's neuroscience lab at UCSF, I found that the homeotic gene map had been extended forward to encompass the entire CNS. Moreover, Michael Depew, Rubenstein's fellow, introduced me to the *Dlx* system of genes, one which further mapped out the pharyngeal arches. He also showed me how to relate comparative anatomy to skull evolution. By 2004, at Children's Hospital Los Angeles, with John Reinisch, I worked out how the developmental field map of the skull could explain synostoses. But it was not until 2006, when I stumbled on the work of Dorcas Paget, medical artist and amateur neuroembryologist at the Carnegie Institution in the 1940s and 1950s, that the actual neurovascular basis for these fields became clear. By 2009–2010, at Saint Louis University, I began writing out these ideas as a series of essays and notes to myself as it were ….

Stem Cell Stimulus

The jump start for this book really came about when I transitioned from surgery into stem cell biology and clinical research with adipose-derived stromal vascular fraction (SVF) cells. The intense focus on stem cell biology made possible a coherent assembly of the developmental story of craniofacial anatomy based on the neuromeric model. It also made it possible to consider regional anatomic organizations such as the pharyngeal arches or the orbit, map them out, and describe their functional role in the formation of the head and neck. I believed that this vision could be a great value and felt compelled to bring it to life. And so, it was in 2017 that the concept of this book was presented to, and approved by, Springer-Verlag.

Qualifications

I suppose that this is as good a time as any to introduce myself. The facts of my professional life are available elsewhere (as in the back of the book), which is where they deserve to stay … since they have nothing to do with why you and I are meeting here in these pages.

So how in the world did I ever get to this place, 35 years in the making? The ideas in this book span a wide spectrum of disciplines. To amalgamate these into something relevant required curiosity, persistence, a whole lot of patience to let ideas mature, and a steely conviction that somehow it would all be worth it in the end. In retrospect, the process by which this happened began in 1972 at the office of Dr. Donald Laub at the Division of Plastic Surgery at Stanford. I had returned to campus after doing thesis research for a year in Ecuador. I had with me a copy of *Where There Is No Doctor* by David Werner, a primary care manual in Spanish for rural communities (it later earned a Guggenheim prize). I had met Werner, and we agreed that I would edit the language during my senior year and carry out a field test in Sinaloa, Mexico. I thought I should sign up for an anatomy course, and Laub, being a friend of Werner's, was my contact at the Medical School. While waiting, I was transfixed by the photos of cleft children from Latin America repaired by him through his brainchild, the Interplast Foundation. My heart was pounding. As I handed him a note of introduction from Werner, written on a 3 × 5 card, Laub looked at me intently [disclosure: I have a repaired cleft lip and palate]. Then, pointing a finger at me, he said with utmost assurance, "You, Michael, *you* are going to be a plastic surgeon!" I felt he had seen right through me, it felt like a hot knife slicing down to the bone, to something undiscovered. It was a challenge, that is for sure. I made up my mind right then and there to take him up on it. So that was it for me, for the rest of my life.

My surgical career morphed into a search for a new model for cleft biology, a way to do things better. Perhaps it was the strange admixture of interests with degrees in Latin American affairs and chemistry that enabled me to appreciate the value of merging seemingly disparate disciplines, like interlocking Venn diagrams, into something distinct. Perhaps it was visualization. From childhood, I loved drawing. I saw anatomy as something dynamic, in four dimensions. Somehow, I was able to create a mental videotape of embryogenesis that would let me move around the conceptus in my mind's eye from different angles at different stages to imagine how development takes place. No matter what the factors, I acquired through experience a core belief: that *nature makes sense*. Beneath its awesome complexity lies an unsurpassed beauty and an ultimate simplicity—something worth loving, worth pursuing, and worth sharing.

In summary, the why of this book is that the beauty of anatomy, its hidden story, has *always* been with us; it is a story that needs to be told. I have been acting through the years as its custodian, but it can no longer remain with me … it is not mine. It is yours, to have, to enjoy, to develop further, and to share with others.

Why Should You Read This Book?

Research requires a leap of faith, an investment of your time in the hope of discovering the unexpected. The contemplation of anatomy is a true case in point … will teach you something far beyond what you imagined. It will speak to you in accordance with your interests. For medical students, residents, and clinicians, it offers insights permitting translation of the *why* of structure into the *how* of therapeutic innovation. For graduate students and for scientists, it will add to, and perhaps challenge, the "known world" of structural biology.

This book can be read in whatever sequence you like, as all its parts are interconnected. The figures and accompanying legends are designed to help you visualize the developmental process. Take your time with them. Develop your own videotape. Push yourself to see familiar structures in a new light. Make this knowledge your own and, by thinking creatively, drive yourself past dogma to find innovative clinical solutions for your patients. Above all, may the beauty of development speak to you. *Open sesame* … let your own search begin.

Falls Church, VA, USA Michael H. Carstens

Acknowledgments

This book and the ideas it contains have its own evolutionary history; it spans a lifetime. So many, both living and departed, have contributed … these comments are incomplete and inadequate. The only way I can capture this moment is by use of a timeline:

My parents (deceased), Keith and Virginia Carstens, and sisters, Sue and Lynnea.

Dr. Eduardo Cornejo and Rosemarie Doring (my family in Ecuador and a lifelong connection with Latin America).

Robert Kiekel, Oregon State University, and Alicia de Ferraresi, Stanford (my professors of Spanish and Portuguese).

Casa de la Cultura Ecuatoriana, Quito, Ecuador (sponsorship of *La Novela Ecuatoriana: La Generación de 1930*).

Drs. Rodney D. Chamberlain and Kathleen Conyers Chamberlain (my Palo Alto family who instilled in me the ideal of medicine).

David B. Werner, author of *Where There Is No Doctor* (health care for the poor).

Donald R. Laub, Stanford, plastic surgery and founder of Interplast (my inspiration for plastic surgery and lifetime mentor).

Robert A. Chase, Stanford (for my love affair with anatomy).

Mark Gorney and Ed Falces, St. Francis Hospital Plastic Surgery and the Reconstructive Surgery Foundation, San Francisco (you showed me in the most personal way the transformative power of plastic surgery, supported the initial work in Central America, and lived the values of our profession).

David E. Marcello, Erwin F Hirsch, and Gene A Grindlinger, Boston University (my mentor in general surgery).

J. William Futrell, George F Sotereanos, and Wolf Losken, University of Pittsburgh (for my education in plastic surgery and craniomaxillofacial surgery; anatomy and technique of the subgaleal fascia flap and buccinator flap).

Joachim Prein and Klaus Honigmann, Kantonsspital Basel and the AO Foundation (for unleashing creativity and a lifetime of service to Nicaragua).

Carlos Pellas and Vivian Pellas; the Fundación APROQUEN (philanthropy for burns and plastic surgery), Hospital Vivian Pellas (excellence in medical care), and innovators for regenerative medicine.

Enrique Ochoa, Antonio Fuente del Campo, Luis Monasterio and Ricardo Bennun - masters of Latin American cleft/craniofacial surgery, mentors, and friends.

Dr. Rigoberto Sampson (deceased), the Nicaplast Foundation, the National Autonomous University of Nicaragua – León, Fundación Nicaplast (the foundation of modern reconstructive surgery in Nicaragua).

Jim Betts, Children's Hospital Oakland (creation of the CHO pediatric plastic surgery service), with support from Richard Rowe and Michael Austin.

Martin Chin, Children's Hospital Oakland (innovation in cleft surgery and pioneer with rhBMP-2 reconstruction).

Jerold Z. Kaplan, Alta Bates Hospital Berkeley Burn Center (reconstruction with Integra® dermal matrix).

Alexandra Cabri and Jo Wolters, medical artists (for bringing surgical concepts to life).

Robert Hardesty, Loma Linda University plastic surgery (for the first academic support for the work and for your poolhouse).

John F. Reinisch, Children's Hospital Los Angeles (mentoring pediatric and craniofacial surgery, master craftsman, and friend—for saving my career).

Christian E. Paletta and Robert Johnson, Saint Louis University (chairs at SLU plastic surgery and general surgery—you made innovation possible at Cardinal Glennon Children's Hospital).

Debbie Watters, RN, Cardinal Glennon Children's Hospital, Saint Louis University (for leadership of the cleft-craniofacial team at Glennon).

David C. Matthews, Carolinas Medical Center (true mastery of craniofacial surgery, out-of-the-box thinker, and lifelong friend).

Jean-Claude Talmant, Jean Delaire, Paul Tessier (contributors in the tradition of French anatomy).

Harvey B. Sarnat and Laura Flores Sarnat (beginnings of neuroembryology).

Mukund Thatte and Karoon Agrawal (introduction to the plastic surgery tradition of India).

S.M. Balaji, Chennai (you brought morphogen-based surgery to India and driving force for innovation).

Rolf Ewers, University of Vienna (the anatomic evidence for the Tessier cleft system).

Bruce Carlson, University of Michigan, and William Bemis, Cornell University (mentors in developmental anatomy and evolutionary biology).

Ernie Manders, University of Pittsburgh (for your inspiration and support for work and its place at Pitt Plastic Surgery).

Sonia Castro, Carlos Saenz, and Carlos Cruz, and Ministry of Health of Nicaragua (for you support for innovation in regenerative medicine and for your commitment to health care for all).

Socorro Gross Galiano—the beginning and the end of this work—I owe it all to you.

Contents

Volume I

Volume II

About the Contributors

Karoon Agrawal For over four decades, Dr. Karoon Agrawal, Professor of Plastic Surgery at the National Heart Institute of New Delhi, India, has been a thought leader with extensive contributions in both cleft lip/palate and hypospadias. A tireless academician, he has also proven himself to be a glutton for punishment on two accounts: first, he is the editor of a 6-volume treatise published with Thieme covering the entire specialty; and second, he willingly committed to review this book from cover to cover, contributing careful and in-depth comments. For the time and effort this required, I sincerely hope his wife will forgive me. Karoon cuts a fine-featured, slender and scholarly figure, accompanied by a warm smile and restless intellect. He is also a first-class debater, which I discovered to my discomfiture. We were summoned by the Indian Society for Cleft Lip and Palate to debate on the relative merits of alveolar extension palatoplasty versus the traditional concept of cleft palate repair. Right off, I knew it was a dangerous proposition…no mortal could expect to approximate Karoon's surgical experience. But I had not counted on his sense of humor. There, up on the screen, with his very first slide, in front of the entire audience, was a picture of a crouching King Kong in the posture of a sumo wrestler with my face pasted on top! I suppose it was intended as a back-handed compliment about AEP but at the time I wanted to crawl in a hole. Anyway, the discussion was lively and the whole audience loved it. Later on, as the journal editor for the ISCLP, Karoon invited me to contribute to a special issue to present the developmental science of palatoplasty. Although it was like pulling teeth, he was patient with me and the result was two articles that eventually formed the blueprint for Chaps. 14–17. Karoon's gentle but persistent questioning forced me to focus on mechanism and good writing. I can say, in retrospect, that the consent chapters are as much his thinking as my own. As both colleague and friend, Dr. Karoon Agrawal represents the very best of Indian plastic surgery, a gift for which I will always be indebted.

S. M. Balaji From the vantage point of the Balaji Dental and Craniofacial Hospital in Chennai (formerly Madras), Tamil Nadu, India, Prof. Dr. SM Balaji cuts an imposing figure with an impossibly long list of accomplishments. As his full name is also impossibly long, I shall refer to him as Bala; and he is a true force of nature. A relentless perfectionist, academician, optimist, and general over-achiever, his energy (seemingly boundless) is somehow contained within a sizeable and powerful frame. However, behind the booming voice and irrepressible smile is a restless and innovative mind, rather unencumbered by dogma and quick to seek clinical implementation for concepts he considers of value. Given this constellation of attributes, it is not surprising that Bala has been the driving force behind the introduction of rhBMP-2 to reconstructive surgery for the Indian subcontinent and Southeast Asia. We started out doing cleft cases together, but he quickly found a way to innovate with bone grafts with an eye toward regenerative applications in the future. In the operating theater, I found him to be a master technician and a never-ending source of new ideas. Away from the hospital, Bala and his wife, Sachin, were the most gracious of hosts. But our cooperation did not stop there, for Bala is a tireless organizer and connector—his conferences in the Seychelles and Maldives gathered together like minds with the results that were eye-opening. Bala showed me the promise of a much larger world. I am convinced that innovation in medicine for the twenty-first century will come about as clinicians and scientists from India harness the incredible power of their clinical experience and produce studies that will change the direction of our thinking and techniques. As this story unfolds, I am sure that Dr. SM Balaji will be in the forefront; I hope to follow along to see its denouement.

William Bemis at Cornell likes to go by "Willy" and that it is totally appropriate. Although I have never had the pleasure of walking into his office, the exuberance of his personality is matched only by the length of his beard. He is the quintessential college professor you never forget and with whom you know the office door will always be open for you. But he is also the author of the single book that changed my professional life and made the one in your hands become a reality. *Comparative Anatomy of the Vertebrates: An Evolutionary Perspective* drove home an important lesson: one cannot understand human adult anatomy without knowing both its development and its prehistory. The final form of our structures and systems must be understood in the context of other vertebrate species extent, and of their historical precursors through the passage of time. Consider the humble ossification centers of any bone (the occipital bone, for example), described in small print in Gray's Anatomy. Do these not represent separate, previously autonomous bone fields which have been melded together in the distant past to create the structure before our eyes today? Bemis text was riveting; each chapter brought new insights into structures that I thought I knew…only to find something different more relevant. Although he wears multiple hats, Willy Bemis lives and breathes his subject, be it paleontology, evolutionary biology, genetics, or comparative anatomy. When I discussed this project with him, very much aware of my own ignorance, he straightaway sent me all the slides from his book and threw open his mind (and his time) to help me out. As a scientist he would be better seen in khakis and a butterfly net than with a lab coat. But I am convinced that any medical curriculum with a serious interest in the developmental biology of the twenty-first century and beyond would do well to make Willy Bemis' text required reading for all premedical curricula. For it is this approach that will make the final interface with genetics and molecular biology. Without Willy Bemis, Chap. 8 (indeed all of them) could never have been imagined. My only regret is not to have had Dr. Bemis as my professor or, nowadays, as that eccentric but delightful neighbor just around the corner.

Ricardo Bennun Alveolar extension palatoplasty as a concept is uniquely South American, both its design and in its surgical verification. AEP, in concept the palatal version of subperiosteal tissue transfer for cleft lip repair, was drawn out on a paper napkin during an airplane flight to Ecuador. Although the operation was described in 1999, it was subsequently picked up by Dr. Luis Monasterio, distinguished cleft surgeon in Santiago, Chile. Lucho invited me to do some cases with him at the Fundación Ganz. Since the incisions required to reposition the entire embryonic field of the hard palate were made on the alveolar ridge, the effect of these on dental eruption was an issue of debate. Dr. Monasterio did his own series, following dental development for several years. He determined that eruption was unaffected (which he subsequently reported to the 12th American Cleft Palate Association meeting). He also introduced me to Dr. Ricardo Bennun from the University of Buenos Aires and director of the Fundación Piel. Ricardo subsequently took the operation to the next level beginning a case series which now extends to over a decade. Ricardo is a true biologic surgeon, reflecting his long-time commitment to burn care; he keeps on asking questions and seeking answers. I had the opportunity to contribute two chapters for his 2015 work, *Cleft Lip and Palate Management: A Comprehensive Atlas*; in the process his attention to detail and insistence of quality forced me to think about the issues more deeply than before. It is his influence that really pushed me over the edge, daring me to write this book, a task that seemed overwhelming to me at the time. True to form, his atlas of AEP cases faithfully recorded herein will stand the test of time as surgical proof that developmental biology can win out over dogma for better patient outcomes. For this, all cleft surgeons will be grateful.

James Betts Some people are larger-than-life. Jim Betts, pediatric surgeon and urologist, Professor and Chief of Surgery for UCSF Benioff Children's Hospital Oakland (formerly Children's Hospital Oakland) and volunteer EMT and fire fighter in Big Sur, has been serving the community for four decades. His modesty is the real deal, but it belies a life of devotion to patients, colleagues, and friends, with a long trajectory from his roots in small town Vermont. He is a true mensch who never fails to answer the call of the moment. In 1989, amidst the chaos and destruction of the Loma Prieta earthquake, Jim Betts was at the scene, rescuing the entrapped, even performing an amputation of a leg to save a boy's life. At 73, he still works on the weekends at the fire station in Big Sur (where he has a home). But there is another side of Jim Betts that underscores his character, one which literally made my own life work possible, because as the phrase goes, "justice, justice shall you pursue." Dr. Jim Betts lives by this…he puts his principles first. When he believes in something, he stands up for it. I had started a hospital-based pediatric plastic surgery service at CHO to meet the needs of our largely disadvantaged community and thus saw Jim in the pediatric emergency room both day and night. When brought my ideas to him about embryologic cleft surgery, he because a stanch ally, despite the controversy that ensued. Unbeknownst to me Jim Betts advocated for the possibility for a new form of cleft surgery, one that he thought had merit…. Time and again, in the board room, at the directors' meeting and in front of the CHO IRB, Jim Betts held the line. The work we were trying to push forward could have been so quietly and very adroitly shut down and swept into the dustheap but Jim believed in what we were doing; he never backed down. In the end, with his support, embryologic cleft surgery and morphogen-driven bone reconstruction became realities, Martin Chin and I kept our heads down, published the work, and never looked back. Now, many years later, the very existence of this book is a tribute to Jim Betts, to his vision for achieving better outcomes for children, and, most of all, to his unshakeable integrity.

Bruce M. Carlson You cannot separate Bruce Carlson, MD, PhD, from his fish and the pond at his cabin in the backwoods of Michigan. As Professor Emeritus at the Department of Anatomy from the University of Michigan, Dr. Carlson is, of course, the author of the do-no-pass-go textbook *Human Embryology and Developmental Biology*. He also holds a MS in ichthyology—a touchstone for understanding vertebrate evolution This single book (which I got my hands on in 1997) proved to be the very beginning for my own work. Of all the contributors to the book, Bruce stands alone as those who thought it should be done from the very beginning. By making me an honorary member of the department Bruce enabled me to work with the University of Michigan library to amass the thousands of references required for this research. Although I am sure his wife Jean must have questioned his sanity, Bruce did actually go through the manuscript time and again, from its crudest interaction. And though I am just an MD without the benefit of doctoral work in anatomy, Bruce kept track of the good ideas, allowing me to get out from my errors gracefully, while supporting the concept of embryologically based surgery. So, as this book takes shape and can one day sit on your shelf, consider it my tribute back to you, Bruce Carlson, the professor I never had but can always count on.

Michael H. Carstens Dr. Michael Carstens received his AB in Latin American Studies with honors and Phi Beta Kappa from Stanford University. He subsequently completed a BS in chemistry with honors from Colorado State University. He received his MD from Stanford. He completed residency training in general surgery at Boston University and in plastic surgery at the University of Pittsburgh. He holds two fellowships in craniofacial surgery from the University of Pittsburgh and from the University of Southern California. He is currently working with the Wake Forest Institute of Regenerative Medicine at Wake Forest University and teaches in several medical schools around the world. His academic interests are craniofacial embryology, developmental origin of stem cells, and surgical management of facial clefts.

Dr. Carstens speaks six languages; this makes him a well-respected ambassador for craniofacial surgery around the world. His career has been remarkable for social involvement. In 1990–1992, a consultancy for the World Health Organization brought Dr. Carstens to Nicaragua; he became a co-founder of the renowned APROQUEN foundation for burn care in that country. Currently, he works for WFIRM and serves as a scientific consultant in regenerative medicine for the Ministry of Health of Nicaragua where he continues doing research on developmental embryology and clinical applications of stromal vascular fraction (SVF) cells for complex wounds and chronic inflammatory states.

Martin Chin In 1992, at Children's Hospital Oakland, I met a man who would become my intellectual partner in crime, the inspiration for innovative surgical procedures, and over the years an unforgettable friend. Oral-maxillofacial surgeon Martin Chin, DDS, is the personification of a "thinking man's surgeon," a consummate professional and humanitarian who looks at problems in a unique way, always just a little different; his first concern is to define the problem from first principles, putting goals and concepts first, then adapting technique to suit the case. When it comes to a device or an alteration, nothing gets in his way. A mechanical genius, ensconced in his garage, Martin can literally make anything, thus enabling him to pioneer in alveolar distraction osteogenesis. Martin combines these talents with a gift for biologic thinking. When we say a little girl in the ICU with a complex set of facial clefts, it was if we both could see the same map, but Martin had the inspiration to ask if we could engineer the bone using a morphogen, rhBMP-2, and thus fill zones she was missing in a completely novel way. In his own quiet way, Martin always got around. He made a point of going to Loma Linda University to collaborate with oral surgeon Dr. Phil Boyne in grafting mandibular defects. But Martin's vision was to create a structure that did not exist. So armed with this wild idea, and against the opposition of other colleagues at CHO, with the support of our Chief of Surgery Dr. Jim Betts, we somehow secured permission from the FDA, treated the child and succeeded far beyond expectation. Martin never stopped going forward; our treatment of alveolar defects, considered by many to be beyond the pale, has now been fully integrated into the surgical armamentarium. In the long run though, Martin Chin is a model of a surgical biologist who sees just a little farther than the rest of us. I strain to keep up with him…but, even if he pushes me past my comfort zone, I know his heart is in the right place and, well, it's good for me…sort of like avocados. All this is to say that developmental field reassignment cleft surgery started at Children's Hospital Oakland and with Martin Chin. Anyone who ever benefits from this, be they patients or professionals, owes him a debt of gratitude. Of course, in his quiet way, he would be the last to say so…this book is my chance to set the record straight.

Rolf Ewers Over the years, Dr. Rolf Ewers has been not only a friend and trusted colleague, but a key contributor to this work, although he himself is unaware of the fact. When I visited his unit in Vienna, Rolf was (and is) an outstanding thought leader in craniomaxillofacial surgery. At the time I was working out the vascular embryology of the face as a means of understanding the Tessier classification of rare facial clefts. Rolf and Hildegund welcomed me to their home and subsequently directed me over to Federal Pathological-Anatomical Museum of the University of Vienna. This collection is housed in the so-called Narrenturm, a former hospital for the mentally ill founded in 1784. In addition to many anatomical oddities collected from all over Europe over the centuries are skeletal clefts of all types and configurations. Here, before my eyes, was proof positive that Tessier's system was based on focal neurovascular abnormalities. I returned from Vienna convinced that Dorcas Padget's observations were correct; Chaps. 6 and 7 describing the vascular embryology of both the intracranial and extracranial arterial systems are the result. I am indebted to Rolf for his seminal contribution; these chapters are but a very inadequate way to express my appreciation, and to acknowledge the historical role of Vienna in the development of science and medicine.

J. William Futrell Pittsburgh can be bitterly cold in the winter. In 1985, as a third-year resident in general surgery, I found myself trudging through the snow up the hill to Scaife Hall to interview with Dr. J William Futrell for a position at the University of Pittsburgh plastic surgery program, wondering what I was doing there in the first place. A convert to California (from Iowa), the years spent in Boston only reminded me of how much I wanted to return. There was a residency spot waiting for me with David Furnas at UC Irvine. But the storied Pitt program, built up by Bill Futrell, could not be overlooked so I went to check it out. Folksy, quirky, and professorial, he walked in with a novel in his lab coat and reading glasses hanging from one ear. On the wall behind him was a photo of Futrell in the Duke backfield, the quintessential football coach. What I remember from that hour, years later, is the sense of excitement and innovation he projected. Pitt was awesome and challenging, overshadowing the competition. For reasons I have yet to understand, he decided to take me. When his letter arrived, I accepted right away, the snowdrifts be damned. Futrell gave me a career and a way of thinking, encapsulated by "good enough, isn't." All Pitt graduates know the essence of Futrellism, as he kept pushing everyone's buttons, searching to find the way to get the best out of us. Futrell's thinking was a Rorshach: "You want to take a job in Central America? You're crazy! But if you do, you will find out something about clefts that you will *never* learn here." "You say you found a new flap with the puny buccinator muscle? It'll *never* work! But you should do it anyway…" And so it was…for all of us. But he and Annie made a home for all of us, each in a very personal way. With the passing years their door always remained open. I think Bill Futrell had his own ideas about what each of us could do, about how we would (and should) contribute to plastic surgery. As such, the choices I made must have driven him nuts, but he was always there, the perennial coach, and an unwavering advocate for commitment and creativity. So a quarter of a century later, this book is a fitting tribute to J William Futrell. I could sum it up thus: "To whom much is given, much is expected." Bill Futrell gave me his very best. This book is intended as payback, in full measure.

Ernest K. Manders The early life of Ernie Manders was a true hejira, from a small town in British Columbia, to Alabama, back to the logging town of Coos Bay on the rugged Oregon coast, and thence forward into a surgical career of unceasing innovation. In 1997, when Bill Futrell recruited him for Pitt Plastic Surgery from Penn State, he had already made his mark by engineering a three-dimensional tissue expander for breast reconstruction, a real advance for cancer survivors. This was just the beginning, as his interests translated into tissue biology and bone biology. But the most striking feature of Ernie Manders' career is his combination of imagination, humanity, and service intertwined as a gifted professor and mentor for everyone who sought him out. In practice and struggling with ideas that seemed to not fit any known model, I turned to Ernie, knowing that I would find encouragement and new ideas. He read the drafts of all my early cleft papers and made them better. When it came to tissue engineering with rh-BMP2, his support for what was a yet unproven concept helped Martin Chin and me get FDA authorization for our compassionate-use protocol at Children's Hospital Oakland. Ernie Manders has never forgotten where he came from. He gave back to Coos Bay by establishing a plastic surgery residency rotation there to encourage surgeons to serve in rural communities. Perhaps for this reason he always stood behind my commitment to work in Central America. Finally, Ernie's contributions to this book deserve recognition. Over the years, his criticisms were always on-point, but positive…and eloquently delivered. The last chapter, on biologics, is my tribute back to him.

Robert Mann For cleft surgeons worldwide, the work of Dr. Robert Mann needs no introduction. For over three decades he has relentlessly explored and refined the concepts of interposition palatoplasty using vascularized buccal tissue, all the while patiently and thoroughly documenting his results. Our collaboration represents a convergence of thought and technique, a phenomenon not uncommon in science. In this case, Robert's work dramatically illustrates the impact for patients when embryologic principles are applied to a clinical problem. It is also a testimony to his unflagging persistence in the face of controversy and to his constant willingness to give of himself to share these new ideas with the coming generation of surgeons. His thinking and the cases herein presented stake out a definitive new standard in the treatment of congenital palate defects.

David C. Matthews Known to but a fortunate few, the lifework of Dr. David Matthews, pediatric plastic and craniofacial surgeon, can be summarized in two words: Semper Fidelis. A survivor of Vietnam, he transformed the scars of that experience into a lifetime of commitment to children and the underserved. David was the youngest of Paul Tessier's fellows; their emotional bond was strong. Possessed of great technical skills, he never forgot his mission, to provide service to all. He moved to Charlotte, North Carolina. When his practice group there demanded that he change his priorities to augment the bottom line, David went to set up a solo practice of reconstructive plastic service to the poorest of the poor. Year after year, he took the hits for the trauma service at Carolinas Medical

Center, supporting his dedicated office staff, and living on a shoestring in an apartment above his office. The constancy of his service to the community was unmatched. Despite the human toll exacted by these choices, David's commitment never wavered, nor did his creativity. David's insights into cleft care and bone biology are the best in the world. When we met, as Tessier's disciple, he immediately grasped the significance of development field theory. We share common values as well. I no longer felt alone outside the box and we became friends. With David and his wife Anna I discovered friendship in another dimension, one that has remained constant through the years. With him it is Semper Fi, regardless of the circumstances. When the chips are down, David is there. That's just the way he is. But David Matthews is not done. His thinking has pushed forward into wholistic medicine, acupuncture, nutrition, and electrical properties of cells. He remains ahead of the curve. The rest of us can only try to keep up.

John F. Reinisch is best known for his pioneering work with ear reconstruction. In 2003–2004 I was fortunate to serve as his fellow in pediatric plastic surgery and craniofacial surgery at Children's Hospital Los Angeles. A true master of soft tissue closure, the techniques I learned from him revolutionized my appreciation of wound healing. A true craftsman, John's open-mindedness encouraged me to go further than I had thought possible to define the developmental fields of the face, recognize their abnormal position in clefts, and relocate them to fulfill their physiologic roles. Working with Dr. Mark Urata, this was the beginning of developmental field reassignment. John's tutelage proved to be the resuscitation of my surgical career. For this, and for the many kindnesses John and his wife Nan showed me during my hejira far from home, I will always be grateful. DFR and JFR will always be inseparable.

Harvey B. Sarnat In 1998, at Children's Hospital Oakland, Dr. Martin Chin and I were working though the concept of in situ osteogenesis, using the morphogen BMP-2 to convert stem cells in the periosteum into bone-forming cells. In this process we were in contact with two pioneers in biologic surgery, maxillofacial surgeon Dr. Phil Boyne at Loma Linda and craniofacial surgeon and bone biologist Dr. Bernard Sarnat at UCLA. I thus met Dr. Harvey Sarnat, a pediatric neurologist and neuropathologist also at UCLA, and the nephew of the latter. Harvey had written a book about developmental neuroanatomy which timed with the appearance of Butler and Hodos work on comparative vertebrate neuroanatomy. The sources together formed a basis to understand developmental field from the standpoint of Hox genes and segmentation. Later on, during my fellowship year at CHLA with John Reinisch, I got to know Harvey and his wife Laura, also a distinguished pediatric neurologist. Our collaboration grew into two chapters on a new model of craniofacial clefts based on neuroembryology. Over time Harvey's insights into the system became the structural ribcage for developmental fields, welding together the origins of craniofacial and CNS tissues. Despite my ignorance of the many critical details Dr. Sarnat gave generously of his time and talent, making Chap. 5 accessible for all readers, regardless of specialty, to have an internal look into how the neuromeric model really works. Without Harvey's contributions, this book would not have been possible.

Jean-Claude Talmant The work of Jean-Claude Talmant will perhaps not be fully appreciated for another generation. He has single-handedly explored the relationship between fetal breathing mechanics and fetal development using facial clefts as his model. Pressures exerted normally within the nasopharynx and nasal cavity during development play an undeniable role in shaping the morphology of the lip-nose complex at birth. Dr. Talmant documented these effects using fetal ultrasound. But he did not stop with merely proving an anatomic point; he carefully and methodically developed techniques to address these effects during primary surgical repair with a resulting restoration of not only breathing mechanics but of their cerebral control. Jean-Claude's life work also demonstrates a principle that is very much in the tradition of French anatomy: the careful observation and recording of natural events over time. In this regard, his contributions represent a surgeon's version of Marcel Proust's revolutionary 7-volume masterpiece, *À la Recherche du Temps Perdus* (Remembrance of Things Past). Like Proust, he has lovingly captured the passage of time as reflected in the visage of each one of his patients. Although his case series is small, it has been recorded with the utmost precision and care. No one has done it better and his attention to detail remains a model for surgeons to emulate, both now and in the future.

Mukund Thatte Finding Mukund, and finding India, was an extraordinary event that almost did not happen. Dr. Thatte had been looking for me (unsuccessfully) for over a year to speak at an upcoming meeting in Guwahati, Assam, in March 23–25, 2006, for the Indian Cleft Lip and Palate Association. I had been in transit between Children's Hospital Oakland and Saint Louis University; somehow the contact was lost. But, to my surprise and good fortune, we eventually connected and, of course, I would attend. India is a natural place to go for innovation in plastic surgery…the specialty was literally born there from the pioneering work of Suśhruta, author of the world's first treatise of reconstructive surgery, the "Suśhruta Samhita" (Suśhruta's Compendium), written in ancient Sanskrit. The journey was arduous. Assam being tucked into India's northeast corner abuts to the south with Bangladesh. At the end of the road awaited the Kaziranga wild animal preserve, home of the rare white rhinoceros and, within the shelter of a tea plantation, the Borgos Hotel, at the doorway of which stood Dr. Thatte…as striking as were his surroundings. Stately, tall, impeccable in both manner and speech, Mukund cut a figure of charm and gravitas, admixed with warmth and a sly sense of humor. As one of the true doyens of Indian plastic surgery I was stunned to find that he had literally picked me out from a 2004 journal. As I have discovered over the years, Mukund is possessed of an unbounded curiosity and an intellect curiously allergic to dogma. I am not sure that he knows what the inside of a box looks like. At any rate, I owe to him the introduction to the collection of world-class surgeons and, with his wife

Urmila, remarkable friendships. Over the years Mukund has been a staunch advocate of the value of my basic science approach to plastic surgery. With his support, developmental field theory assumed an honored place in the specialty among Indian colleagues with open minds. For this India will always be special…for it was here, in the tradition of Suśhruta, that something utterly new could come safely into existence.

The Orbit

13

Michael H. Carstens

> Intellectual excellence lies in having faith in the observation of apparently nontranscendental and unimportant facts. To observe an anatomic element calmly, with an open, analytical spirit, and with a spiritual freedom, can lead to an explosive vortex of new knowledge.—Miguel Orticochea, M.D.

Introduction

Craniofacial dermoid cysts are a common management problem comprising 60% of all facial cysts [1]. Their distribution is remarkably consistent. The majority are *periorbital*, relating to the lateral eyebrow (63%) and medial eyebrow (26%). The embryologic mechanism of *nasofrontal* dermoid cysts involving the medial dorsal nose is different; these lesions are uncommon (4%) [2]. These three sites are associated with three sutures, all relating to the frontal bone: frontozygomatic, frontoethmoidal, and frontonasal. These pedestrian observations raise a number of questions, all of which are related.

- What are the tissues that make up a dermoid and where do they come from? How can we explain their location?
- What are the three sources of neural crest that contribute to the orbit?
- What is the neurovascular basis of developmental fields?
- Why does each frontal bone have two ossification centers?
- What is the significance of eyebrows?
- Why in anencephaly and exencephaly are the orbital rims preserved, despite the absence of frontal bone?
- What is the paleontology of the orbit?
- What is the relationship between the neuroangiosomes of the orbit and Tessier zones 9–13?
- What is the origin of the stapedial artery?
- What are the unique characteristics of Tessier zone 9?

In our search to understand the biology of the simple dermoid we must take a long but necessary segue into the developmental anatomy and evolution of the orbit itself. Having examined this material, we will certainly be equipped to answer these questions about a seeming minor pathology but, even more importantly, we can consider a wide range of craniofacial problems ranging from clefts, to orbital malpositions, to synostoses. So this, then, is the point of this chapter, to set the stage for a deeper appreciation of clinical problems involving the orbit, thereby setting the stage for innovative ways to improve their correction.

The histology of dermoids is well known and consists of a stratified squamous lining that is keratinized with adnexal contents, including glands. They are considered to arise at ectodermal lines of fusion or from an underlying suture. The traditional concept of mechanism (dating back to 1910) as proposed by Pratt [3] and later by Posnick (1994) holds that a projection of dura is included in a suture line, comes into contact with skin, and is eventually pinched off. This model, as in the case of midline dermoids, is equally applicable to the embryogenesis of gliomas and encephalocoeles.

These concepts are not supported by contemporary neuroembryology. Let us be clear that no mesoderm in involved in these lesions whatsoever. With the exception of the (1) basioccipital bone complex of the cranial base and (2) the striated muscles of the eye, pharyngeal arches, and tongue, all craniofacial mesenchymal structures, be they bone, cartilage, fascia, fat, or dura, are of exclusively neural crest derivation. So too are the contents of periorbital dermoid, the tissue that tracks potentially through a suture the solid mass of intracranial neural crest. It is of the same derivation as dura but is more superficial, hence it is not itself in contact with neural tissue – it is *not* a hollow sac. If *full-thickness dura and leptomeninges* are involved, the inclusion of underlying neural tissue becomes anatomically possible.

M. H. Carstens (✉)
Wake Forest Institute of Regenerative Medicine, Wake Forest University, Winston-Salem, NC, USA
e-mail: mcarsten@wakehealth.edu

M. H. Carstens (ed.), *The Embryologic Basis of Craniofacial Structure*, https://doi.org/10.1007/978-3-031-15636-6_13

How to Use This Chapter

We shall approach the orbit from the ground up. First, let us take a look at its building blocks, the mesenchymal tissues for which it is constructed and their embryologic origins. We then examine how these tissues are assembled. Periorbital anatomy in the human cannot be understood without an appreciation of its overevolutionary history. The phylogeny of the individual bones was previously described in Chap. 8. This time we recount how they are rearranged to produce the primate orbit. Theories of the evolutionary changes relating stereoscopic vision and primate evolution are presented emphasizing the role of orbital anatomy in the process. The vascular supply of the orbit is the next topic. For those of you familiar with Chaps. 6 and 7 this is a review. For those coming de novo to this chapter the discussion will set the stage for understanding orbital clefts. The chapter concludes with a consideration of selected pathologies of the orbit from the standpoint of developmental fields.

Developmental Fields and the Neuromeric Map

Big picture idea: The orbit is composed of nine neurovascular units: developmental fields.

The orbit contains structures from three different tissue types. The eye represents an evagination of the brain from prosomere hp2, specifically alar plate of terminal hypothalamus. Extraocular muscles arise from paraxial mesoderm (PAM) of somitomeres 1–3 and 5. All remaining structures of the orbit are NC derivatives: fat, tendon, fascia, and bone. Neural crest mesenchyme involved in the construction of the orbit arises from three distinct sites along the neuraxis, migrates in three distinct patterns, produces three specific tissue structures, and is supplied by three separate vascular systems.

The orbit is thus a collection of *developmental fields*, composites of tissue supplied by a unique arterial axis accompanied by a sensory nerve. The growth cone of nerve produces vascular endothelial growth factor (VEGF), inducing the artery. At the same time, the growth cone of the artery produces nerve growth factor (NGF), inducing the nerve. Thus, *neuroangiosomes* form the building blocks of the craniofacial anlage. Using simple concepts of neuroanatomy, it is possible to map out the neural crest source for each orbital bone and trace it back to its site of origin along the neural folds.

Embryonic developmental fields are blocks of tissue that interact in a programmed fashion to produce the topology of the face. Like tectonic plates, these units possess autonomous innervation and blood supply and move along developmental "slippage planes" according to their mass, velocity, directionality, and timing. Facial development can therefore be considered as the *formation*, *migration*, *coalescence*, and subsequent *interaction* of separate genetically based fields. Adjacent tissue units required to form a structure (such as the nostril) can be termed *partner fields*. Their interface is a *field boundary*. Any process that disrupts the physical integrity of a field or field boundary will mechanically disrupt normal field interactions leading to *field mismatch*.

Paul Tessier enumerated distinct biological zones in which craniofacial clefts occur [4]. These were subsequently identified as neuroangiosomes supplied by individual branches of the extracranial stapedial system [5]. Our purpose here is to define how these arise and what structures are included in each Tessier zone. We will start with some basic terminology of neural crest mapping, describe the formation of membranous bone, and then relate these findings presented by dermoids as evidence of an ancestral arrangement of bone fields (Figs. 13.1 and 13.2).

Big picture idea: All mesenchymal tissues in the orbit and the optic anlage except the extraocular muscles are derived from neural crest originating from the two sites: midbrain (m1–m2) and rostral hindbrain (r0–r1–r2).

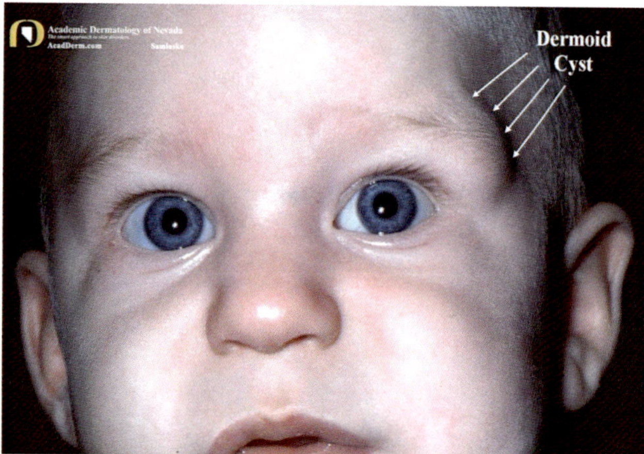

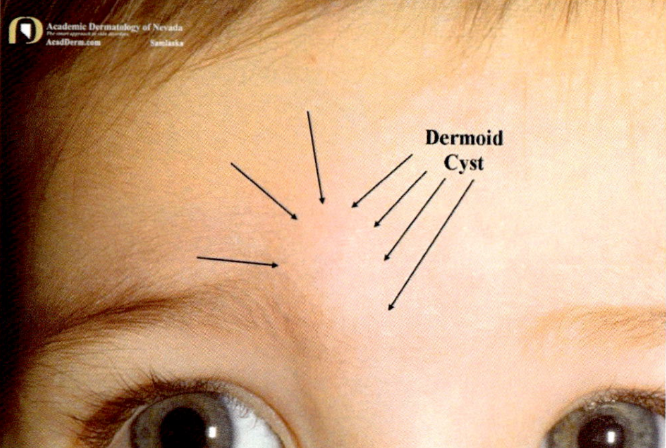

Fig. 13.1 Dermoids occupy specific exit sites on either side of the eyebrows. [Courtesy of Dr Curt Samlaska, MD, FACP, FAAD]

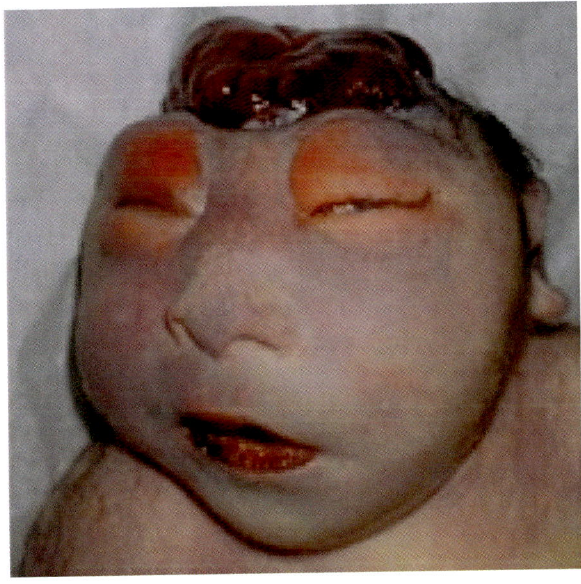

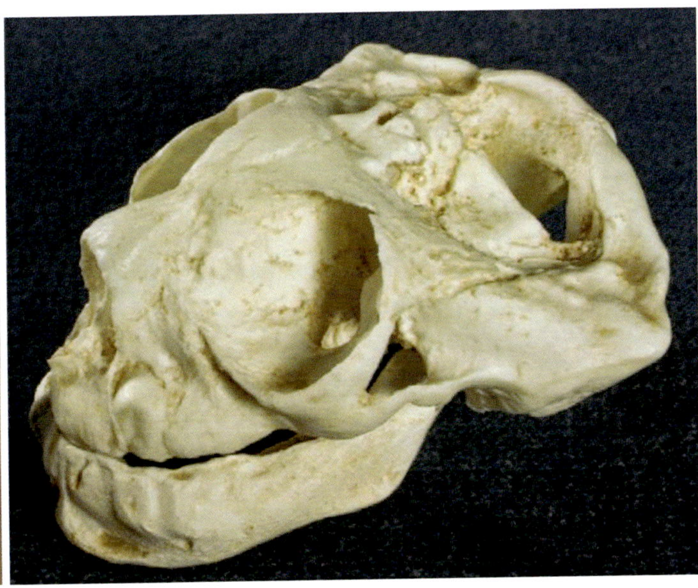

Fig. 13.2 Exencephaly. In absence of the membranous calvarium, orbital roof remains intact. Note prefrontal (PrF) and postfrontal (PF) bone fields. In mammals these are absorbed into the frontal bone. Periorbital serves as the program for the inner lamina of orbital roof. Frontal bone attaches above. Medial and lateral dermoids exploit sutures at the boundaries of PrF and PF. [Reprinted from ECLAMC Centro Anomalias de Malformaciones Congenital, Professor Ignacio Zarante, Institutode Genetic Humana, Potifica Universidad Javeriana Bogotá, Colombia.]

By way of review, at stage 7 the process of *gastrulation* produces a trilaminar embryo with three germ layers. In the center of the upper layer (ectoderm) a neural plate extends in the midline from head to tail. Running along the sides of the neural plate are neural folds separating the neural ectoderm of the CNS from the non-neural ectoderm of the future skin. In the subsequent process of *neurulation* at stage 8, with the neural folds acting as hinge points, the neural plate rolls up upon itself like a cigar, and the neural folds fuse to create a *neural tube*. Neural crest cells arise all along the folds and (in mammals) immediately migrate into the face and pharyngeal arches even before closure is completed. Eventually NC is dispersed throughout the body.

How can we describe which neural crest population comes from which part of the brain? We need a convenient mapping system. Some unavoidable terminology is needed. Fusion of the neural folds results in closure of the neural tube and establishment of the embryonic brain. Initially it has three parts: *prosencephalon* (forebrain), *mesencephalon* (midbrain), and *rhombencephalon* (hindbrain). Later a five-part structure emerges. Hindbrain divides into *metencephalon* (isthmus/pons) and *mylencephalon* (medulla). Midbrain stays the same. Forebrain subdivides into a posterior *diencephalon* and an anterior *secondary prosencephalon* which further divides into *diencephalon* and *secondary prosencephalon*.

Much like a submarine is divided into water-tight compartment, the embryonic brain is divided into 19 developmental units called *neuromeres*; these are located cranial along the neuraxis in their order of development. Each neuromere is a six-sided box based on the neural plate and running vertically to include all segments of the neural tube: floor plate, basal plate, alar plate, and roof plate. The anatomic boundaries of each neuromere are defined by expression patterns of genes of the homeobox family. Each neuromere has its own unique hox "barcode." For any given level, e.g., r2, the right and left boxes have the same hox code but may, through variations in other genes, have morphologic differences, explaining why pathologies can be unilateral. Alterations in the hox code can result in more drastic defects, such as partial deletion or complete loss, or even the additional segment, as in a hemi-vertebra leading to scoliosis.

Neuromeric zones allow the structures of the brain to fit into a genetic "map" as summarized below. For a detailed description of the neuromeres, see Chap. 5 (Fig. 13.3).

- Hindbrain consists of 12 *rhombomeres* (r0–r11) and develops from r0 backward. The basal plate has motor neurons and the alar plate has sensory neurons.
  - Rhombomere r0, the isthmus, is a narrow zone separating hindbrain from midbrain and is an important signaling center. It contains the trochlear nucleus.
  - Rhombomeres r2–r11 contain the cranial nerves for all five pharyngeal arches (two rhombomeres per arch) with abducens located in r5.

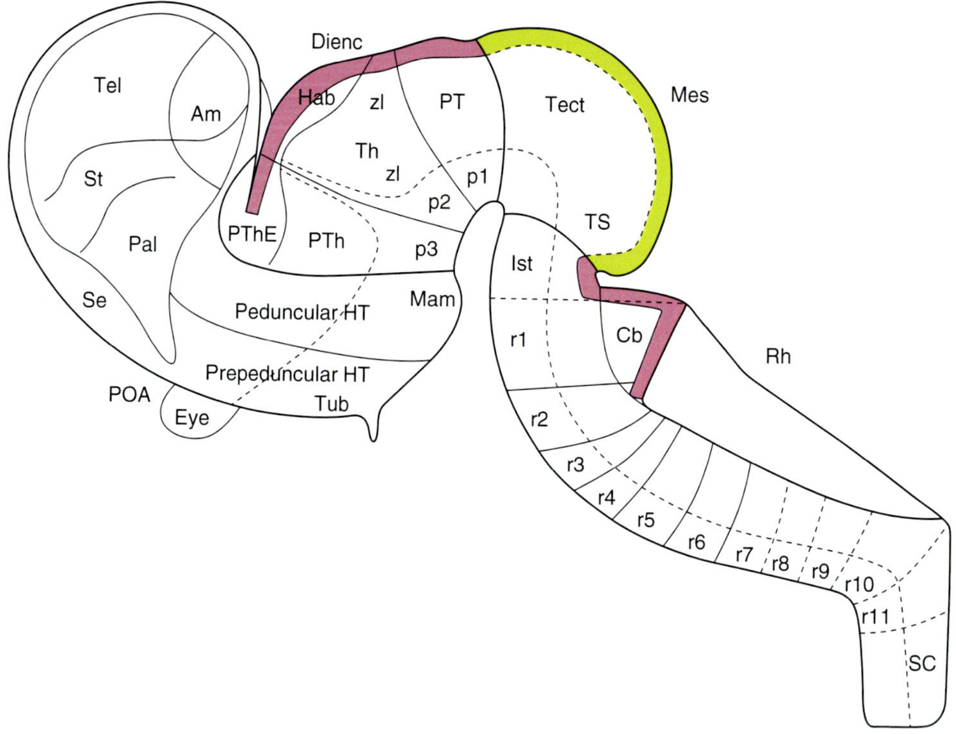

Fig. 13.3 CNS neuromeres. Neural crest from r0 to r1 (magenta) migrates first to the forebrain and future frontonasal tissues. Midbrain is very large in the early CNS development. Migration from m1 to m2 (yellow) is directed to the eye. Because r0–r1 and m1–m2 co-mingle to populate the same zones and because they are all innervated by V1 we refer to these neuromeres collectively as midbrain neural crest (MNC). Derivatives of r1 are dura, calvarial bone, and dermis. Hindbrain neural crest proper from r2 to r11 takes two routes. Hypaxial RNC populates the five pharyngeal arches with two rhombomeres per arch. Epaxial RNC from r2 and r3 flows forward to complete coverage over the forebrain, thereby creating dura, calvarial bone, and dermis. Prosomeres p1–p3 (pink) migrate forward under the epithelium of hp1 and hp2. Frontonasal dermis comes from p1 to p3 neural crest while frontonasal epidermis comes from the non-neural ectoderm of hp1 and hp2. [Reprinted from Puelles L, Harrison M, Paxinos G, Watson C. A developmental ontology for the mammalian brain based on the prosomeric model. Trends in Neurosciences 2013;36(10): 570–578. With permission from Elsevier]

- Midbrain develops from r0 forward. It starts out singular and then subdivides into two *mesomeres*. Midbrain basal plate is tegmentum and alar plate is tectum.
  - Mesomere m1 is larger and contains the oculomotor nuclei and superior colliculi associated with the visual system.
  - Mesomere m2 is smaller and intercalated between m1 and the isthmus. It contains the inferior colliculi of the auditory system.
- Forebrain, caudal (diencephalon) has three prosomeres:
  - Prosomere p1: pretectum
  - Prosomere p2: thalamus (formerly the dorsal thalamus)
  - Prosomere p3: prethalamus (formerly the ventral thalamus)
- Forebrain, rostral (secondary prosencephalon) has two hypothalamic prosomeres:
  - Hypothalmic prosomere hp1
    Peducular hypothalamus (ventral)
    Telencephalon (dorsal)
  - Hypothalamic prosomere hp2
    Terminal hypothalamus (ventral)
    Eye and optic fields (dorsal)

The neural folds form the median boundary of the subjacent neuromere; they share the same homeotic barcode. Thus, the NC cells above any given neuromere retain its hox code like a tattoo: no matter where they travel, they always related back to the sector of CNS from whence they originated. This is clinically useful. If we know the neuroanatomy of a neural crest structure, we can tell where it came from. In sum: all that is required to map out the orbit is neuroanatomy and some basic concepts of neural crest organization.

Fortunately for us, each neuromere also has a *specific anatomic content* which allows us to recognize it. Thus, m1 contains the oculomotor nuclei, sensory trigeminal V2 resides in r2, etc. And each of the five pharyngeal arches receives neural crest from paired rhombomeres. Thus, the first arch is supplied by r2 and r3. These contain, respectively, the nuclei of V2 and V3. Thus, the neural crest origin of any structure can be determined from its innervation. All the bones of the orbit, except alisphenoid and the zygomatico-maxillary complex are derived from r1 neural crest (our generic term for midbrain neural crest; Table 13.1).

Orbital Mesenchyme: Neural Crest

Big picture idea: The neural crest entering the face comes from *three different populations* with *three different migration patterns*. NC makes everything but striated muscle (Figs. 13.4, 13.5, and 13.6).

Table 13.1 Derivatives of pharyngeal arches. Blood supply: (1) stapedial, (2) external carotid

Pharyngeal arch	Rhombomere/ nerve	Derivatives
First (mandibular	r2–r3/V2, V3	Maxilla complex, mandible
First (mandibular)	r2–r3/V3	Non-jaws, mastication
Second (hyoid)	r4–r5/VII	Mastication, animation, glands
Third (glossopharyngeal)	r6–r7/IX, X	Palate, pharynx, glands
Fourth (pharyngolaryngeal)	r8–r9/X	Constrictors, larynx, glands
Fifth (internal laryngeal)	r10–r11/X	Arytenoids

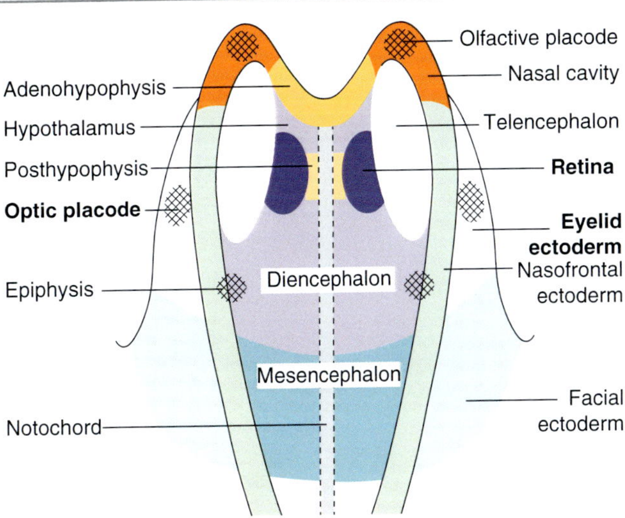

Fig. 13.4 Cephalic ectoderm fate map. The neuroectodermal territory fated to give rise to the retina (violet) lies in the diencephalic floor plate (pink), interposed between the presumptive telencephalic neuroepithelium (yellow) and the posthypophysis (beige). The optic placode (grey lattice) at the origin of the lens is lateral to the anterior diencephalic non-crest cell producing neural fold and is surrounded by the presumptive eyelid ectoderm (white). [Reprinted from Ezin M, Barembaum M, Bronner ME. Stage-dependent plasticity of the anterior neural folds to form neural crest. Differentiation. 2014;88(2-3):42-50. With permission from Elsevier.]

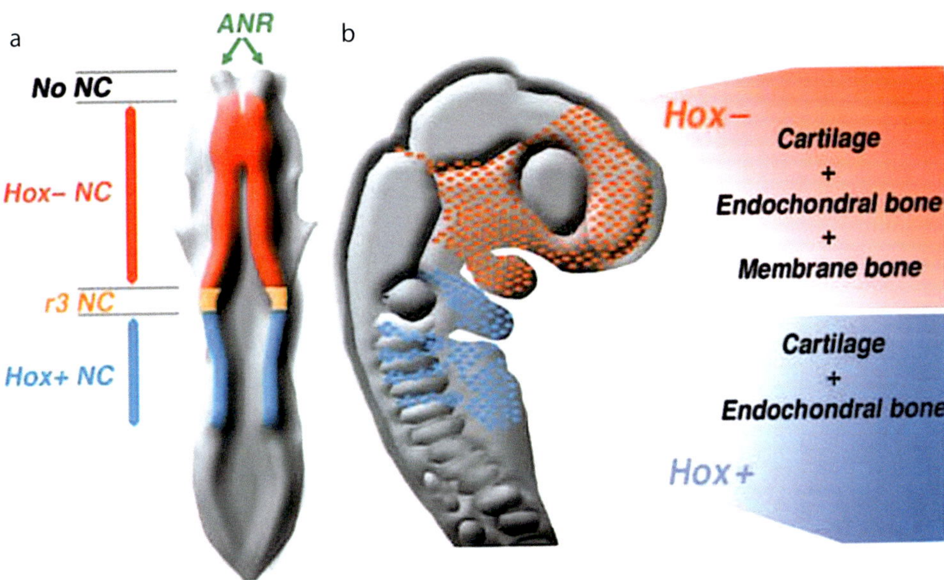

Fig. 13.5 Cephalic neural crest: Hox gene expression and skeletogenic properties. In a 5ss chick embryo, the cephalic NC is divided into an anterior Hox-negative domain (in red) and a posterior, Hox-positive domain (in blue). The transition between these two domains corresponds to r3 (in orange). The neural fold (NF) rostral to the mid-diencephalon does not produce NCCs. It contains specialized non-neural ectoderm (NNE) that will become frontonasal epidermis

- Hox-negative NCCs (in red) yield cartilages as well as endochondral and dermal bones of the entire upper face and jaws. Neuromeres anterior to r3 are specified by non-Hox homeotic genes.
- Hox-positive NCCs (in blue), by contrast, have skeletogenic properties limited to chondrogenesis and endochondral ossification in the hyoid structure.

[Reprinted from Creuzet S, Couly G, Le Douarin NM. Patterning the neural crest derivatives during development of the vertebrate head: insights from avian studies. J Anat 2005; 207:447–459. With permission from John Wiley & Sons.]

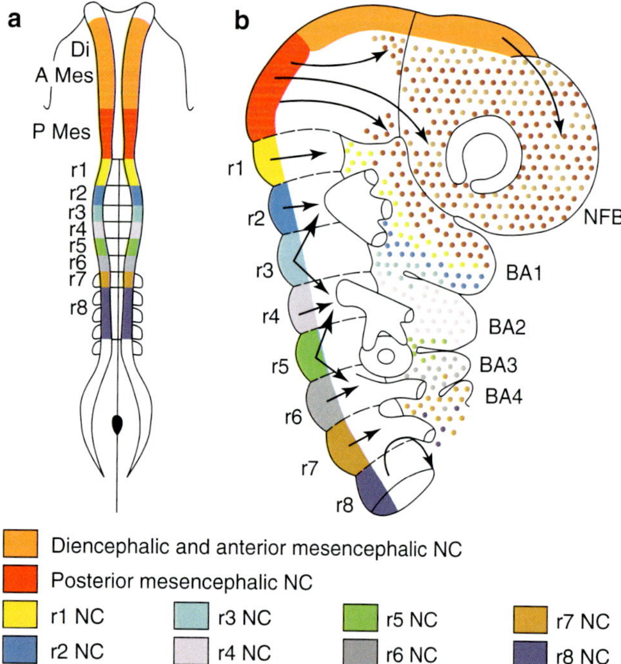

Fig. 13.6 Cephalic neural crest: cell migration streams and fate map

- Presumptive diencephalic, mesencephalic, and rhombencephalic territories of the NF in the 5 somite stage (HH 8.5, 30 hours) avian embryo. The neuromeric map and the relative sequence of NC migrations is the same for mammals.
- Migration map of cephalic NCCs in the avian embryo. The origin of NCCs found in the nasofrontal, periocular regions, and BAs is color-coded as in A. NCCs arising from the posterior diencephalon and mesencephalon populate the nasofrontal and periocular region.
  - In the chick posterior mesencephalon also participates in BA1.
  - The mammalian model locates all r0-r1 neural crest in the anterior cranial fossa and frontonasal mesenchyme
- NCCs from r2–r3 migrates in two directions. Epaxial r2-r3 completes the coverages of the forebrain to produce dura and membranous bone. Hypaxial r2-r3 supplies the 1st pharyngeal arch (BA1).
- NCC from r4-r11 migration is exclusively hypaxial in the remaining pharyngeal arches. Although some populations become confluent with others, the rule of thumb is two rhombomeres per pharyngeal arch.

[Reprinted from Creuzet S, Couly G, Le Douarin NM. Patterning the neural crest derivatives during development of the vertebrate head: insights from avian studies. J Anat 2005; 207:447–459. With permission from John Wiley & Sons.]

Forebrain: *Prosencephalic Neural Crest* (PNC) > Fronto-orbito-nasal Skin

- The neural folds above the forebrain have two different zones, anterior and posterior. Tissues from these sites create the skin of the forehead, nose, and upper eyelid. Epidermis arises from *non-neural ectoderm* (NNE) of the anterior zone.
- Dermis arises from the *neural crest* of the posterior zone (PNC). PNC migrates forward underneath NNE as a *single sheet* with distinct genetic zones for prosomeres, p1–p2–p3. These correspond to Tessier cleft zones 10–13. FNO skin is unique; *frontonasal dysplasias* seen in the upper face are a consequence of its development.

Midbrain: *Mesencephalic Neural Crest* (MNC) > Dura, Orbit, and Upper Face

MNC consists of four populations that travel forward in two distinct *streams* over the lateral aspect of the forebrain to reach the midline.

- The first stream to migrate is neural crest from the rostral hindbrain, r0–r1. These cephalic rhombomeres do not enter the pharyngeal arch system. Recall that in extinct agnathic fishes these r0–r1 were assigned to a putative *premandibular arch* supplied by the nervus terminalis (cranial nerve 0) and a placodal sensory nerve that becomes V1. Later in evolution, with gnathostomes, r0–r1 became formally reassigned to brain coverage and frontonasal mesenchyme. Innervation of these tissues is strictly V1.
  - Derivatives of r0–r1 are dura, frontonasal bones (the sphenoid complex (except alisphenoid), the ethmoid complex, the lacrimal bone, the frontal bone, and the membranous bones of the orbital series), and all V1-innervated nonorbital soft tissues.
- The second stream to migrate consists of the two mesomeres immediately forward from r0–r1. Midbrain neural crest sensu stricto develops from m1–m2. These populations are assigned to the eye and orbit. They are innervated by V1.
  - Derivatives of m1–m2 are all orbital soft tissues (fat, fascia, intraocular muscles), excluding the extraocular muscles.
  - When the optic cup evaginates from diencephalon it becomes coated with MNC.
- Thus, four neuromeres produce mesenchyme supplied by V1. Although midbrain neural crest sensu stricto refers to m1–m2 we shall refer to MNC sensu lato as being comprised of r0–r1 and m1–m2. All are supplied by V1-induced branches of stapedial ophthalmic axis, StV1 neuroangiosomes.

Hindbrain: *Rhombencephalic Neural Crest* (RNC) > Midface

RNC develops from r2–r11 in the hindbrain and migrates into the five pharyngeal arches. Each arch is supplied by a pair of rhombomeres. Blood supply for all these derivatives is external carotid. In the first arch a separate population of RNC from r2 and r3 is responsible for synthesizing the mandible, zygomatico-maxillary complex, and alisphenoid. These *non-pharyngeal arch jaw fields* are supplied by V2- and V3-induced branches of the stapedial maxillo-mandibular axis. These neuroangiosomes are referred to as StV2 and StV3.

Orbital Neural Crest: How Does It Get There?

Quail-chick mapping by Creuzet [6] demonstrated three distinct streams of neural crest migration to the optic vesicle (OV). These findings are germane to our discussion of mesenchymal derivatives. They are organized by Carnegie stage as the equivalent to chick embryonic day (E stage). Note: *mapping uncertain for MNC derivatives (Figs. 13.7 and 13.8).

- Diencephalic NC (p1–p3)
  - Covers the prosencephalic vesicle and frontonasal process
  - Derivatives: dermis and activation of placodes, source of intraocular neural crest for retina and ciliary muscles
- Mesencephalic NC (m1–m2, r0–r1)
  - Populates optic vesicle and future orbit
  - Derivatives of m1–m2*: sclera, fascia, fat
  - Derivatives of r0–r1*: orbital bone
- Rhombencephalic NC (r2–r3)
  - Carnegie stage 13 (E2.5)
  - Midfacial tissues (V2–V3 innervated)

The timeline and destination of neural crest cells to the eye and orbit are summarized below. **Note:** For a more inclusive timeline of eye development vide infra.

Stage 8

- Neural folds appear
- Mesoderm organized 18 somitomeres

Stage 9

- Mesencephalic neural crest: MNC = r1 then m1
- First pharyngeal arch

Stage 10

- MNC migrates to optic primordium
- Rhombencephalic neural crest, rostral: RNC_R = r2–r5
- Second pharyngeal arch

Stage 11

- Rhombencephalic neural crest, caudal: RNC_C = r6–r11
- RNC_R migrates to brain and face (first and second arches have fused)
  - First and second arches have fused
- MNC completely distributed around the eyeball:
  - Migrates along wall of brain to cover the optic stalk
  - Fills space between optic neurectoderm and hp2 ectoderm
- Third pharyngeal arch

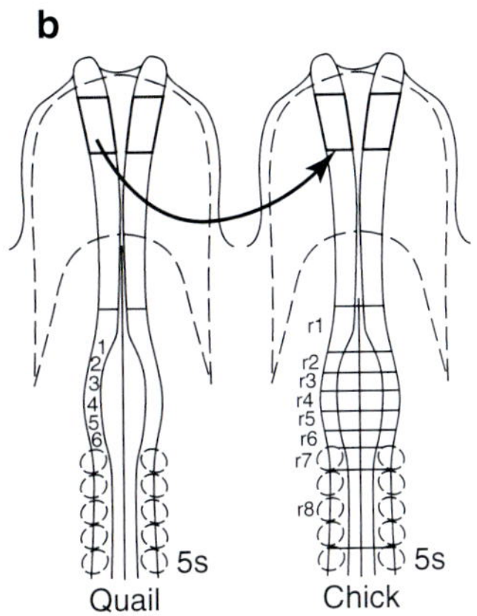

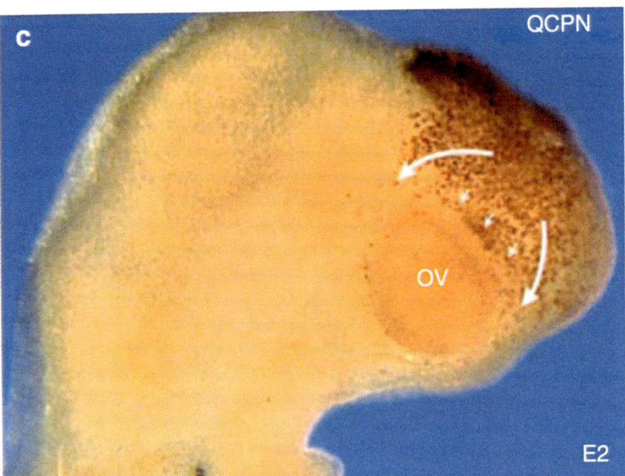

Fig. 13.7 Cephalic neural crest: cell migration streams and fate map

- Presumptive diencephalic, mesencephalic and rhombencephalic territories of the NF in the 5 somite stage (HH 8.5, 30 hours) avian embryo. The neuromeric map and the relative sequence of NC migrations in the same for mammals.
- Migration map of cephalic NCCs in the avian embryo. The origin of NCCs found in the nasofrontal, periocular regions and in BAs is colour-coded as in A. NCCs arising from the posterior diencephalon and mesencephalon populate the nasofrontal and periocular region.
  - In the chick posterior mesencephalon also participates in BA1.
  - The mammalian model locates all r0–r1 neural crest in the anterior cranial fossa and frontonasal mesenchyme
- NCCs from r2–r3 migrates in two directions. Epaxial r2–r3 completes the coverages of the forebrain to produce dura and membranous bone. Hypaxial r2–r3 supplies the first pharyngeal arch (BA1).
- NCC from r4–r11 migration is exclusively hypaxial in the remaining pharyngeal arches. Although some populations become confluent with others, the rule of thumb is two rhombomeres per pharyngeal arch.

[Reprinted from Creuzet S, Couly G, Le Douarin NM. Patterning the neural crest derivatives during development of the vertebrate head: insights from avian studies. J Anat 2005; 207:447–459. With permission from John Wiley & Sons.]

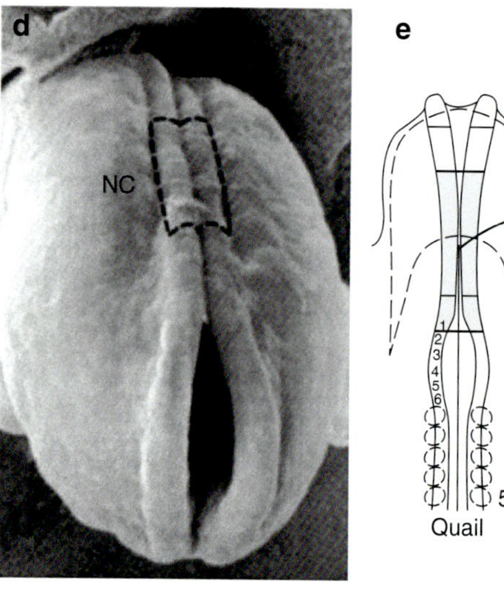

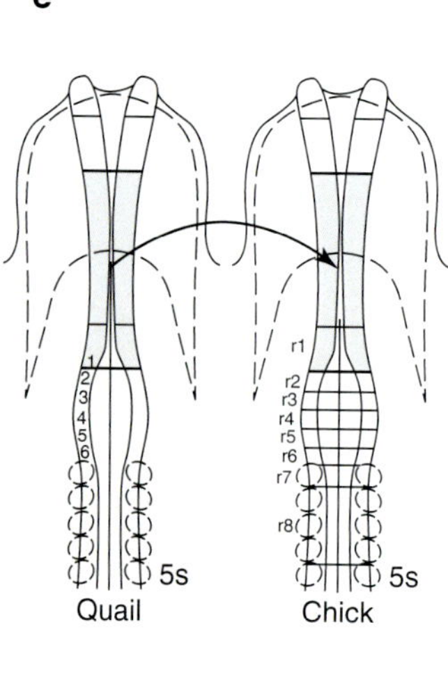

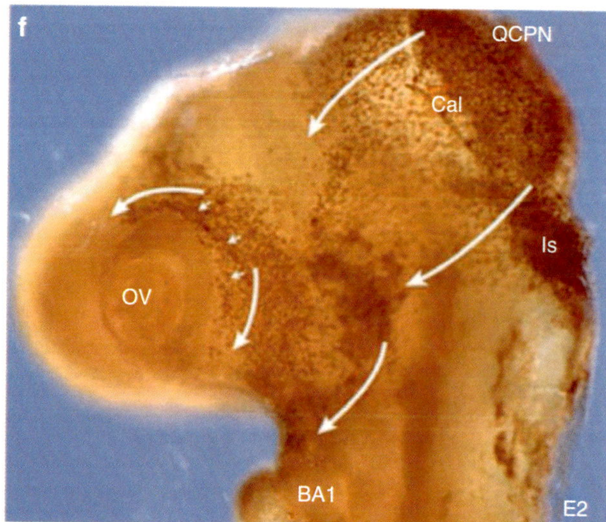

Fig. 13.8 Migration of r0–r1 and m1–m2 neural crest to the orbit

- SEM as in A in which the mesencephalic and anterior metencephalic neural fold is delineated (dotted lines).
- Scheme representing the interspecific transfer of this territory from a 5 somite stage (HH 8.5, 30 hours) quail donor embryo to a stage-matched chick host.
- Whole-mount detection of the quail neural crest cells that have spread from the graft into the host environment at E2.5; laterally migrating neural crest-derived cells have massively moved (arrows) around the ocular region by encompassing the posterior three quarters of the optic vesicle (OV). Note that a large part of graft-derived cells which are destined to form the calvarium (Cal) remains dorsally located thus covering the posterior mesencephalon and the isthmus (Is). This translates to (1) **dura**, (2) **calvarium**, and (3) **periocular mesenchyme**.

[Reproduced with the permission of UPV/EHU Press from Cruezet S. et al. (2005) Int. J. Dev. Biol. 49: 161-171.]

Stage 12

- Optic neural crest at its maximum: "frightened hedgehog"
- Fourth pharyngeal arch

Stage 13

- NC accumulates at optic cup margin (interface of ORE and IRE)
  - Optic epithelium now double layer of outer and inner retina
  - Pigmented layer of the retina appears
- In anterior segment NC forms ciliary body
- Cornea is double: outer epithelium (r1) and inner endothelium (NC)

Neural Crest Zones of the Sclera

There are three distinct zones affecting the globe. These "embryonic hemispheres" are significant as they relate to (1) the insertion of extraocular muscles; and (2) the anatomy of embryonic arteries to the eye with primitive dorsal ophthalmic (DOA) developing prior to the primitive ventral ophthalmic (VOA).

An alternative hypothesis is that the orbit has a single population of MNC which is regionalized in genetic expression patterns emanating from the globe itself. As we shall see, opposing gradients of SHH and BMP4 divide the retina into quadrants. These signals may interact with the surrounding MNC population. This is in keeping with the concept of the role of the eye as "master coordinator of the orbit."

- Dorsolateral (r1)
- Ventromedial (m1)
- Lateral (r2)

Orbital Mesenchyme: Mesoderm (Second String to Neural Crest)

Compared with its "big brother," neural crest orbital mesoderm is a "bit player" with a limited repertoire: it makes striated extraocular muscles. Excluded from this "gang of seven" are Müller's muscle, the ciliary muscles, and the intraocular muscles that control the pupil. Mesodermal mapping demonstrates the origins of the extraocular muscles (Figs. 13.9, 13.10, and 13.11).

Prechordal Mesoderm Versus Somitomeres: Where Do the EOMs Come From?

Confusion exists regarding the source of mesoderm for the EOMs and the mechanism by which it develops. Consider the following:

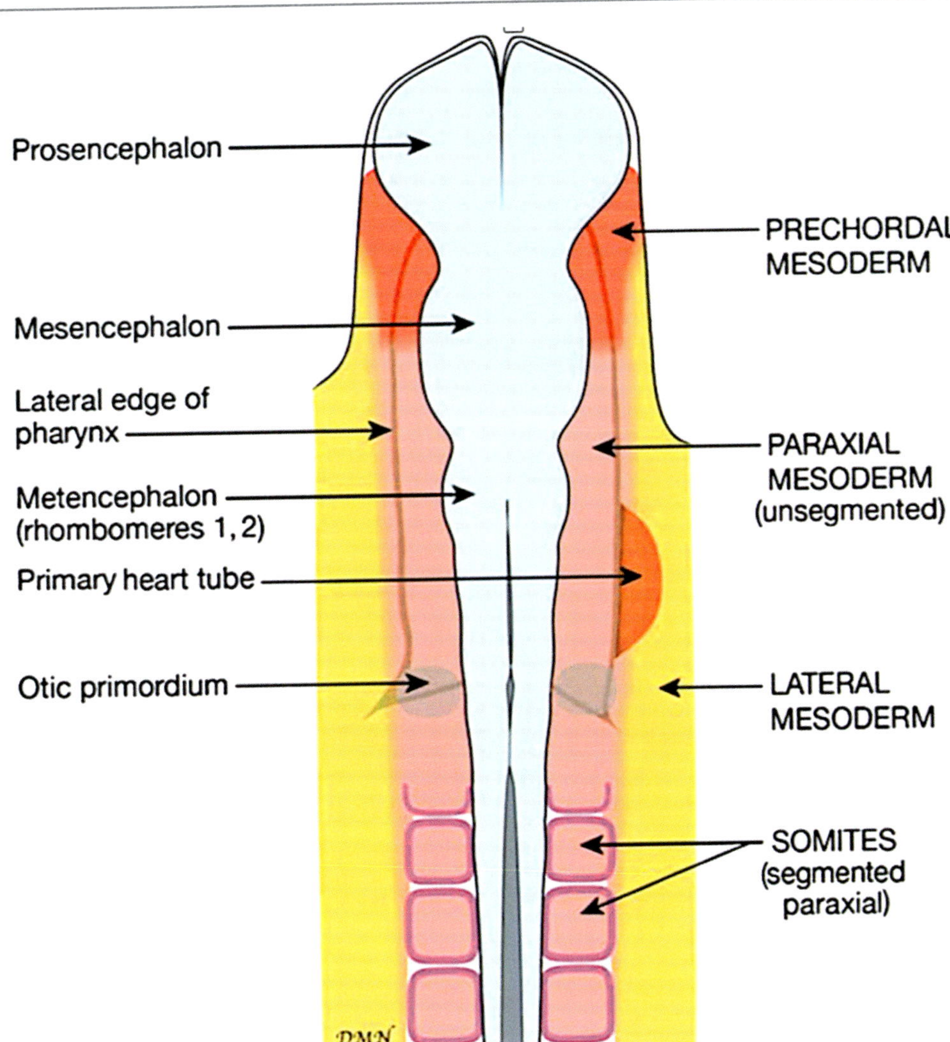

Fig. 13.9 Prechordal plate mesoderm. PCM is produced at stage immediately upon arrival of the primitive node at its future r0 position. PCM flows forward to surround the buccopharyngeal membrane and reach the cardiac mesoderm. Notochord, at the start of gastrulation, slides forward in the midline beneath PCM and above the endoderm. It will program the floor plate of the forebrain. As gastrulation continues, primitive nodes retreat backwards, laying down additional notochord in front of it. Thus notochord from r0 forward is produced caudal-cranial and from r0 backward is produced cranial-caudal. PCM, red; mesoderm, pink. [Reprinted from Shoenwolf GC, Bleyl SB, Brauer PR, Frances-West PH (eds). Larsen's Human Embryology, 5th ed. Philadelphia, PA: Elsevier; 2015. With permission from Elsevier.]

- Prechordal mesoderm (PCM) has long been considered as the source for EOMs except lateral rectus but its origins and anatomic organization are poorly appreciated [7].
- We all know that cephalic mesoderm is organized into seven somitomeres.
- We also know that the extraocular muscles have been fate-mapped to four distinct somitomeres as follows: **Sm1** = SR, MR, and IR; **Sm2** = SR, LPS; **Sm3** = SO; and **Sm5** = LR [140].
- Of these four somitomeres, only Sm5 is physically associated with the hindbrain.
- Sm1–Sm3 develop in proximity to r1 and the midbrain, in a region called *prechordal mesoderm* (PCM).
  - PCM is in genetic register with r1, and m1–m2 develop in a unique way.

Prechordal Plate Mesoderm Develops from r0–r1

The birth of PCM is the first act of gastrulation. As soon as the primitive node has arrived in position over neuromeric level r0, at stage 6b its dorsal sector produces a column of cells that pass directly forward beneath the ectoderm and spread out to fill up the space between the node and the cardiac mesoderm. This zone is known as *prechordal mesoderm*. At stage 7 primitive node emits a midline rod, the notochord, that plows through the center of PCM and stops when it reaches the genetic signal of hp2.

Prechordal Plate Reorganizes into Three Somitomeres

Prechordal mesoderm has long been considered a source for the extraocular muscles [8]. Mapping in the avian model by Noden demonstrates location of muscles corresponding to somitomeres 1–3. Sm1 contains SR and IR. Sm2 bears MR and IO (they come from a common anlage). Sm3 has SO. These somitomeres mature in cranial-caudal order. Note: *Development of PCM is simultaneous with that of PAM*. For this reason, superior rectus from Sm1 PCM appears at the same time as lateral rectus from Sm5 PAM. This fact underlies the insertion sequence of extraocular muscles, vide infra.

Other derivatives from PCM are not associated with somitomeres. Because of its location along the first aortic arches,

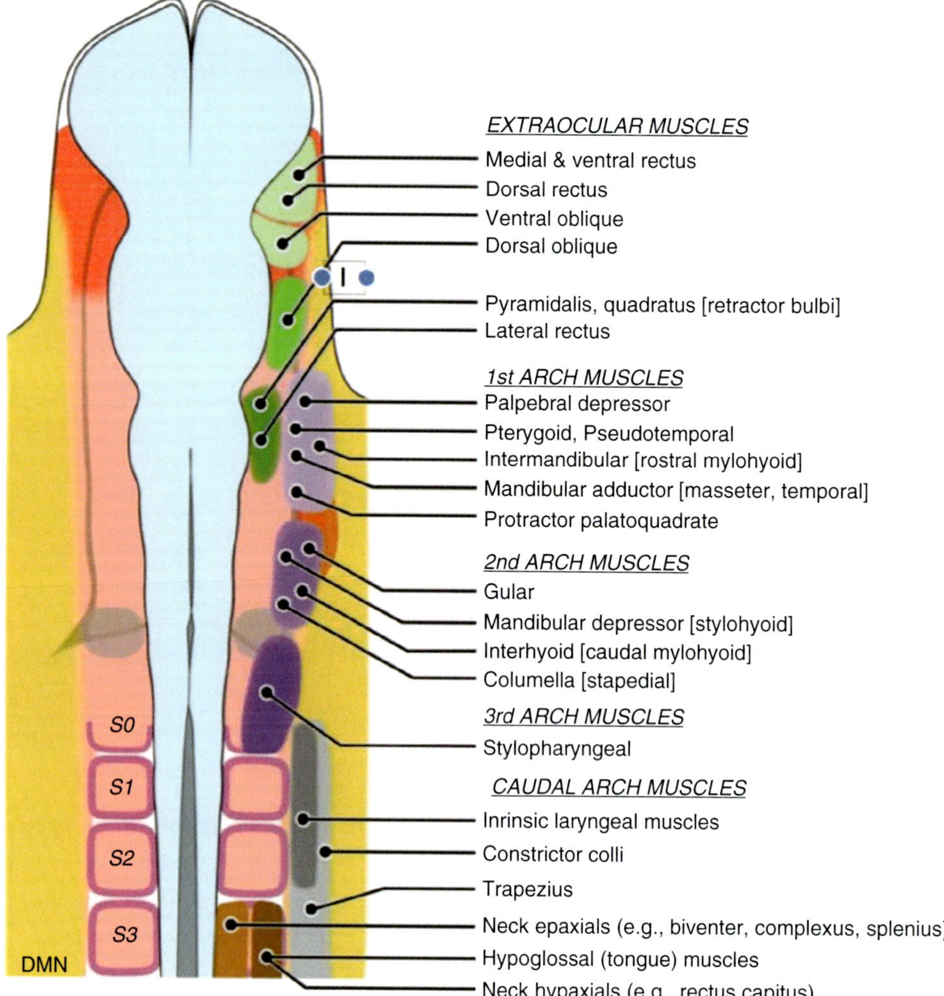

Fig. 13.10 Primordia of cranial muscles (mesoderm) versus neurocranium (neural crest). Map is based on transplantation and in situ labeling studies in avian model. The relative positions of each primordium or group are based in part on frequency of co-labeling; the boundaries shown are approximate. The medio-lateral positions indicated for branchial arch progenitors distinguish paraxial (medial) from lateral mesoderm origins. [Mammalian homologues] *contributions by somites 1 and 2 have been reported. Notes: (1) Extraocular muscles are very accurate with mammals. (2) Sm4 and Sm5 are superimposed but abducens nerve is caudal to trigeminal. (3) Palatal muscles and constrictors not mapped (Sm7, third arch). Revised from [140]. (4) Constrictors (purple) map to r6–r8; these are considered to be "lateral plate" derivatives along with SCM and trapezius. (5) Sternocleidomastoid and trapezius in mammals are enclosed by neural crest fasciae that originate from c1–3 and from c2–c6 respectively. These muscles correspond to the lateral motor column. In the spinal cord this is the continuation of the branchiomeric nucleus ambiguus. Therefore SCM and trapezius develop not from lateral plate mesoderm but from laterally displaced head mesoderm which has extended backwards. This mesoderm represents, in turn, the source material for ancient branchial arches BA6–BA7–BA8 as can be seen in some modern-day sharks. (6) Fronto-orbital bones cluster around neuromeres r0–r1 and m1–m2, collectively referred to as r1 or as MNC. Parietal bone maps clearly to r2–r3. [Reprinted from [139]. With permission from John Wiley & Sons.]

PCM becomes repositioned by virtue of embryonic folding as a source for the posterior mesodermal part of basisphenoid. (Figs. 13.9, 13.10, 13.11, 13.12, and 13.13).

An interesting role for mesoderm in anterior chamber development has been described. During lens development apoptosis occurs. Mesodermal cells insert themselves between the surface ectoderm and the lens to clean up the mess. This allows for deposition of a primary corneal epithelium, the prerequisite for NC penetration into the subcorneal space. NC–mesodermal interaction takes place to create vasculature both within the choroid and externally.

Paraxial Mesoderm Supplies the Brain and Orbit

Conventional gastrulation involves egress of mesoderm via primitive streak (which is posterior to primitive node). This process, from level r1 backward, produces distinct populations of paraxial mesoderm that contribute vascular supply for the brain and extraocular muscles.

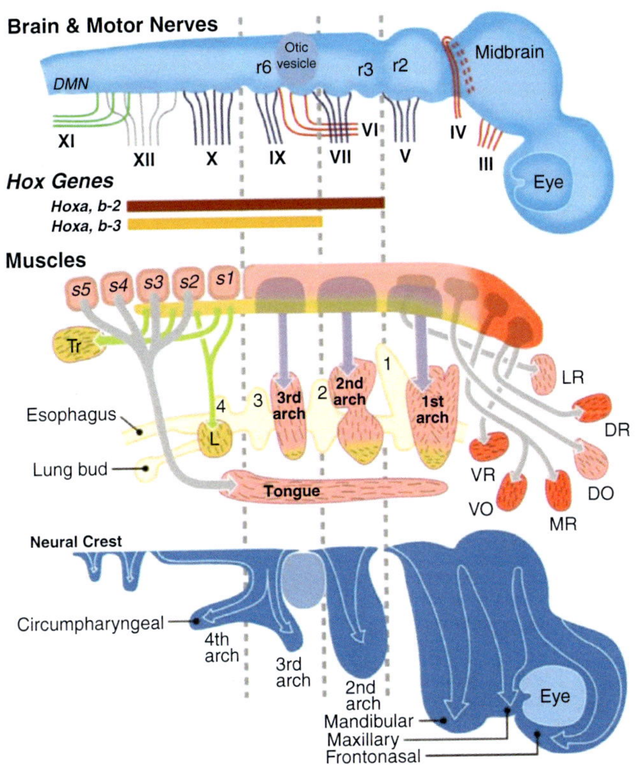

Fig. 13.11 Spatial relations among motor axons, myoblasts, and neural crest cells. Cohorts that populate branchial arches arise at matching axial locations and maintain registration as they move ventrally to establish the arches. In contrast, EOMs, their motor nerves, and periocular neural crest cells arise at disparate axial levels and do not come together until they approach reach their terminal periocular niches. Tongue muscle precursors from several somites form a cohesive hypoglossal cord that extends beneath the caudal pharynx and then grows rostrally, where it penetrates connective tissues derived from first- and second-arch crest cells. Laryngeal (L) and trapezius (Tr) muscles arise primarily within lateral mesoderm located beside the occipital (first 4–5) somites. Note: Palate muscles not mapped (Sm7, third arch). [Adapted from [140] With permission from John Wiley & Sons.]

- Brain: Epaxial displacement of PAM arising from the primitive streak at levels r1–r3 is immediately dragged epaxially over the developing forebrain to provide endothelial cells for the primitive head plexus.
- Lateral rectus and rectractor bulbi develop from Sm4 and Sm5, in register with r4–r5. Both of these muscles are supplied by abducens nerve which has nuclei in both rhombomeres.

Insertion Sequence of Extraocular Muscles: The Spiral of Tillaux

As discussed previously in the neuromuscular chapter, development and insertion of the extraocular muscles follow a choreographed spatio-temporal sequence (the suprascript refers to the Carnegie stage): **LR**^{13} > **SO**^{13} > **SR**^{14} > **MR**^{16} > **IO**^{16} > **IR**^{17} > **LPS**$^{19-21}$. Note that superior oblique (Sm3) appears before superior rectus (Sm1) and inferior rectus (Sm1) develops before inferior oblique (Sm2). Note that within PCM, Sm1 monopolizes the recti. Note that the order of innervation is CN VI, CN IV, and CN III.

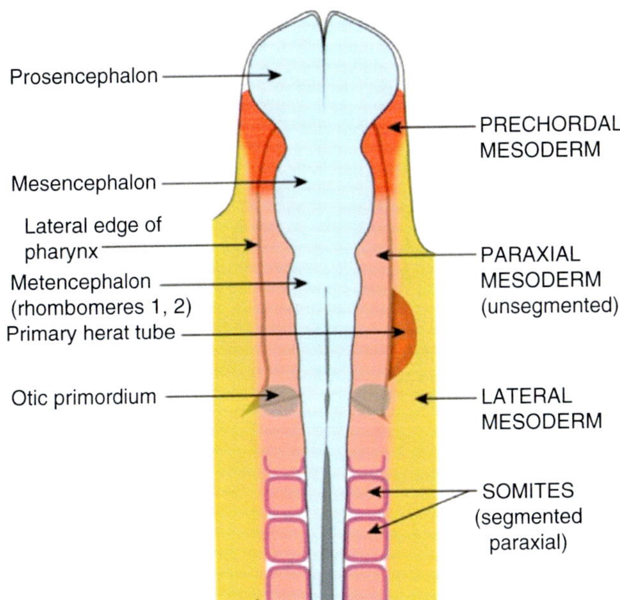

Fig. 13.12 The locations of mesodermal populations that contribute to head musculature. Prechordal mesoderm is identified based on sites of Pitx2 expression and tracing of cell movements in avian embryos. Note the location of the primary heart tube, which is the precursor of atrial and left ventricular compartments, at the boundary where first and second branchial arches will subsequently develop. Note: The embryo has folded by this time (stage 9). Some researchers define the first somite as the partially epithelialized caudal boundary of unsegmented head paraxial mesoderm; in this article the term is used for the first fully epithelial somite. We use the former definition for human mesoderm: the first somite is fused cranially with Sm7 but epithelized caudally, being distinct from the second somite. Prechordal mesoderm refers to Sm1–Sm3 and provides six of seven extraocular muscles. [Reprinted from [139]. With permission from John Wiley & Sons.]

Note: These muscle relationships must be interpreted based on the original position of the globe at 90 degrees from the neuraxis. These are easily recognized by the spatial location in the primitive state of prDOA territory (caudal-dorsal) and prVOA (cranio-ventral). When the orbit relocates, the territories of DOA becomes temporal-superior while that of VOA becomes nasal-inferior. Don't get stuck on the labels, the globe is basically two opposing L-shaped zones.

The developmental sequence exactly follows the insertion pattern defined by French surgeon Paul Jules Tillaux, who is better known for his description of the Salter Harris type III fracture of the tibia. He observed that a line connecting the insertions of the recti followed a spiral MR > IR > LR > SR and that each insertion was more progressively more poste-

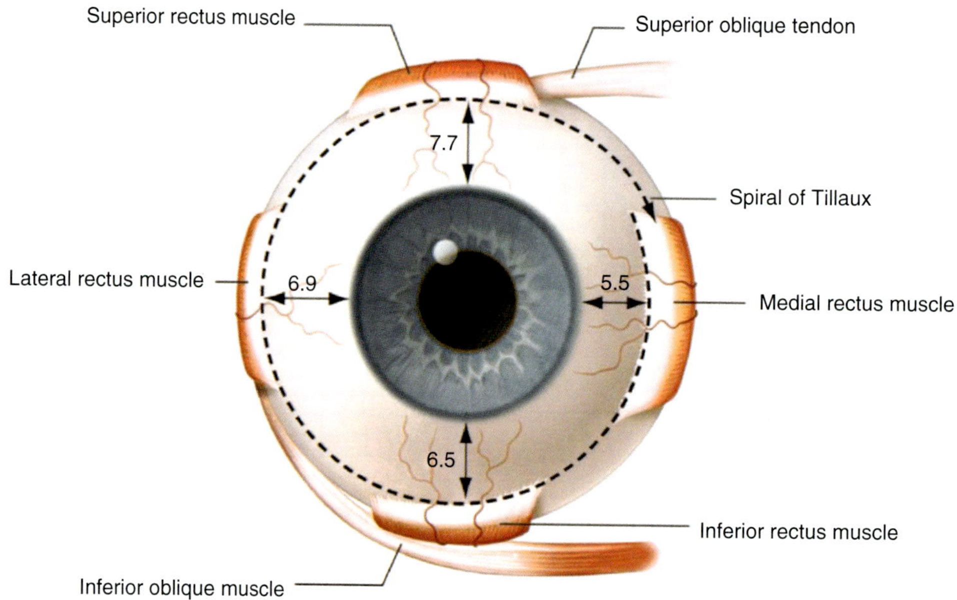

Fig. 13.13 Spiral of Tillaux. Reproduction of a spatiotemporal relationship regarding the territories of embryonic vascular supply to the developing eye and insertion of the four rectus muscles. The term developing eye refers to the initial position of the globe at 90 degrees from the axis of the neural tube. The embryonic territory of prDOA is caudal and superior. This sector of sclera matures first. It receives the insertions of LR[13] and SR[13] at stage 13 with SO[14] inserting posteriorly at stage 14. The embryonic territory of prVOA is rostral and inferior. This sector matures later. It receives the insertions of MR[16] before IR[16] both at stage 16. Inferior oblique inserts at stage 17. In both situations the developmental order is x-axis, y-axis, and z-axis. [Reprinted with permission from American Academy of Opthalmology. Retrieved from: https://www.aao.org/image/new-mediabeacon-item-7.]

rior from the limbus. The spiral of Tillaux has a neuromeric rationale (Fig. 13.13).

- The territory of prDOA is temporal-dorsal and it "fills up" with muscles in sequence. LR inserts first, at 6.9 mm from the limbus. SR inserts third, at 7.1 mm from the limbus. SO inserts second, behind the insertion of SR. Because SR tracks back directly to gain its secondary insertion into the annulus of Zinn, SO cannot compete. It is forced laterally where, via the trochlea, it ricochets backward to join the annulus, but outside the ring formed by the four recti. SO is not a member of the "Gang of Four." [A reference to the leadership of the Chinese Cultural Revolution, led by Jian Quing, the last wife of Chairman Mao Zedong].
- The territory of prVOA is nasal-ventral; it also contains muscles that develop in sequence. Medial rectus inserts at 5.5 mm from the limbus, inferior oblique inserts posteriorly, and inferior rectus inserts at 6.5 mm.
- In both territories the developmental order is geometric: x-axis first, z-axis, followed by the y-axis.

The Mechanism of Primary Insertions

Neural crest arrives in the orbit between stage 11 and 12 so by stage 13 the sclera is ready to receive the insertion. The globe in its original form projects laterally at 90 degrees to the neural axia. The first somitomere is the most complex, having three groups of myoblasts. Within Sm1, SR is caudal with MR and IR being cranial. We further assume that Sm2 is biologically more mature than Sm3 (they are innervated, respectively, by cranial nerve III and cranial nerve IV). So myoblasts from Sm2 will be ready to migrate prior to those from Sm3 (Fig. 13.14).

- At stage 13, prDOA sclera is available. Sm1 and Sm5 are both ready to go. They both send out a muscle, blastema muscle.
  - The x-axis requires the closest blastema which comes from Sm5 due to its location: it becomes lateral rectus.
  - The y-axis attracts the blastema from Sm1, which is farther away from the prDOA insertion zone. The most caudal population of Sm1 takes the superior insertion as superior rectus.
  - The z-axis is the final insertion site. It attracts a population from Sm3, "the girl next door," which becomes superior oblique.
- At stage 16, prVOA sclera becomes available. By now, there are only two somitomeres available (Sm3 and Sm5 having been expended). Sm1 being biologically more mature than Sm2, it will obviously have priority. Thus, inferior oblique is condemned to be last.

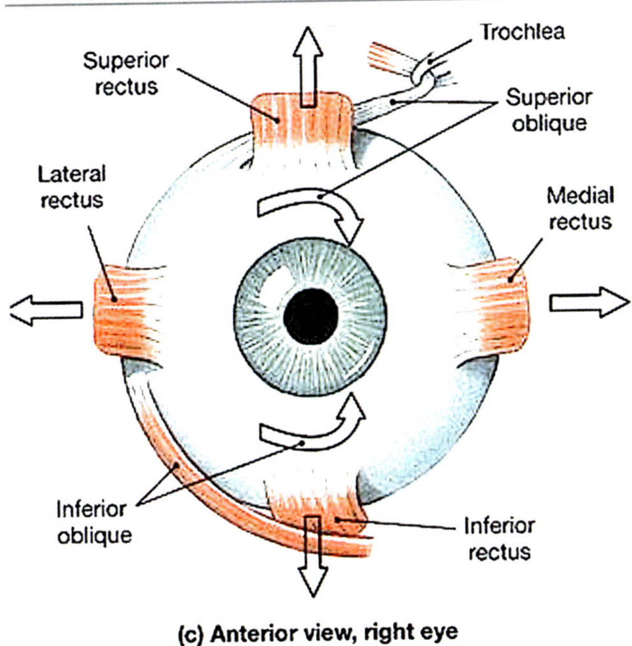

Fig. 13.14 Primary insertions, the 3-axes model. x-axis: lateral and medial recti; y-axis: dorsal and ventral recti; z-axis: dorsal oblique (anterior) and ventral oblique (posterior). [Reprinted with permission from American Academy of Opthalmology.]

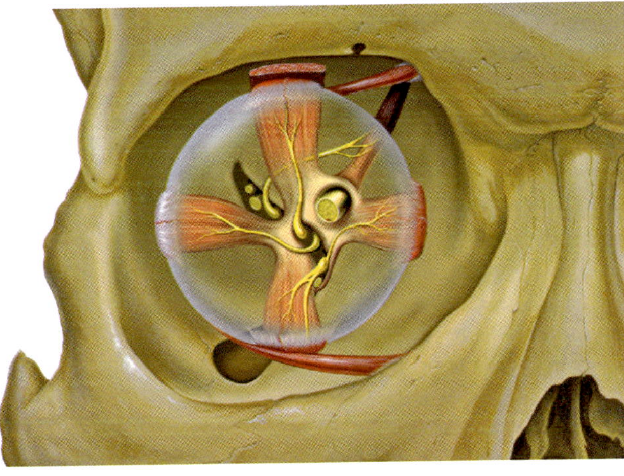

Fig. 13.15 Annulus of Zinn. Superior oblique from Sm3 inserts behind superior rectus and is forced to ricochet from the trochlea back to the annulus but outside the ring. Accessory lateral rectus/retractor bulbi will be internal to the recti. Do these follow the axes of the muscular arteries? [Reprinted from Wikipedia. Retrieved from: https://commons.wikimedia.org/wiki/File:Eye_orbit_anterior.jpg. With permission from Creative Commons License 2.5: https://creativecommons.org/licenses/by/2.5/deed.en.]

- The x-axis attracts the nearest blastema from the most proximal population within Sm1; this will become medial rectus.
- The y-axis gets the final remaining myoblasts from the extreme cranial zone of Sm1. These will insert as inferior rectus.
- Finally, the z-axis receives the sole population from Sm2 as inferior oblique.

The Mechanism of Secondary Insertions

Extraocular muscles migrate using the same pattern we see in the limbs. The myoblasts travel distantly to their targets with sclera being the primary insertion. They then retrace their steps retrograde to make secondary insertions into the annulus of Zinn or into bone. Zinn is defined by the four recti. The obliques behave in similar fashion. Both make a beeline for bone: SO supero-medially and IO infero-medially. Of the two SO is able to return to Zinn, perhaps because it shared a genetic cue with the recti that prevents attachment at the trochlea and directs it to the annulus. IO is "Zinnophobic" so it contents itself to bind to maxilla (Fig. 13.15).

Lateral rectus presents many clinical correlations. The close association of abducens nucleus with facial nucleus explains the condition of bilateral lateral rectus palsy with Moebius syndrome. In Duane syndrome, lateral rectus nucleus is absent with consequent innervation of LR by inferior division of III. Adduction leads to retraction of the globe and narrowing of the palpebral fissure whereas abduction causes widening of the palpebral fissure. In this regard it is useful to consider the *retractor bulbi* muscle, rarely present in man but interesting. This muscle exists within the cone and sends slips outward to the other four recti. Rectractor bulbi is found only in synapsids (mammals) but not in the haplorhine/simiiforme/cattarhine primate line (Old World monkeys, apes, man). In most mammals, when present, retractor bulbi runs in parallel between superior rectus and lateral rectus. The primary insertion is lateral and its secondary insertion is into the annulus of Zinn. Its innervation is usually abducens but can be oculomotor, superior branch. This makes eminent sense since it belongs to the prDOA sector and could arise from either Sm5 or Sm1. Human cases have been documented in which retractor bulbi is internal and sends a connecting slip to all four of the rectus muscles [9]. Accessory abducens nuclei are described; these can be correlated with double proximal heads of lateral rectus [10] (Figs. 13.16 and 13.17).

Orbital Mesoderm: How Does It Get There?

Mesodermal organization takes place immediately upon gastrulation and is complete for the head certainly by stage 9. Optic evagination takes place at stage 11 and the vesicle is ensheathed by neural crest. It erupts through overlying mesoderm of somitomeres Sm1–Sm3. Slightly later in time the vesicle will be in proximity to Sm5 (containing lateral rectus). Differentiation of muscles is programmed within the somitomeres and muscle migration follows a strict timeline with insertions into the sclera determined by its neuromeric zones (Fig. 13.18).

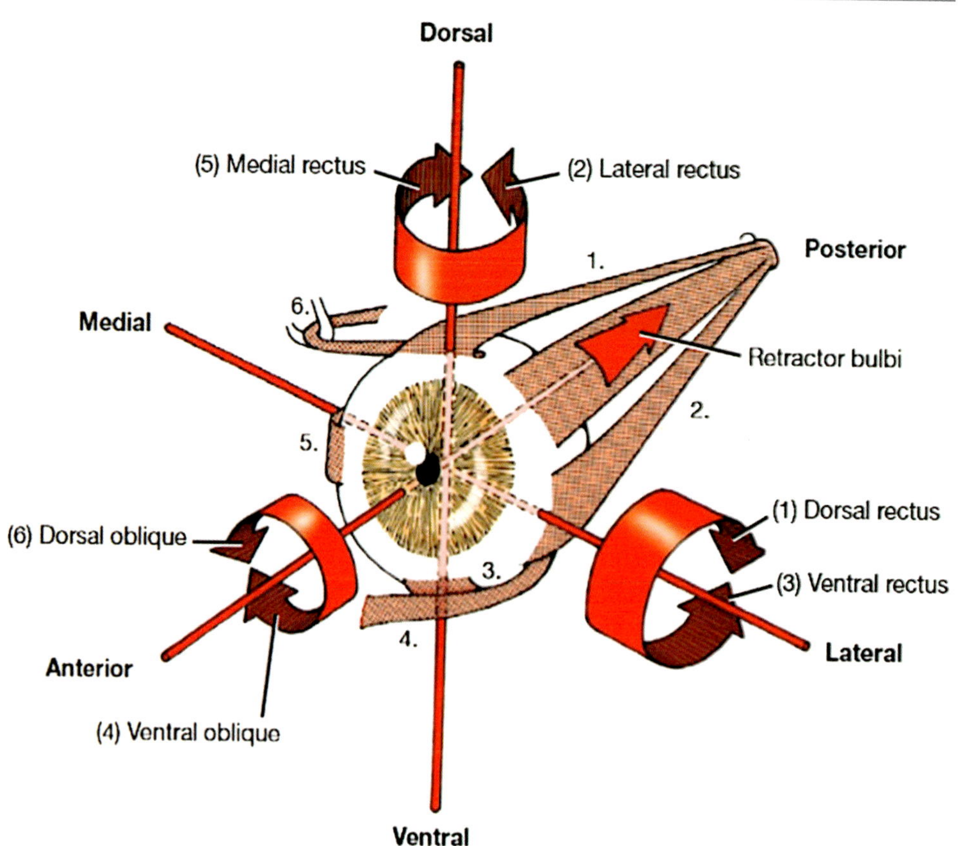

Fig. 13.16 Retractor bulbi. This muscle is often located intermediate between superior rectus and lateral rectus. For this reason, it has been found to be innervated by either oculomotor or abducens nerve. Found only in synapsids and non-hominid primates, the muscle was lost in the haplorhine/simiiforme/catarrhine (Old World monkeys, apes, humans). [Reprinted from Evans HE, Lahunta A. Guide to the dissection of the dog, 7th ed. Missouri, MO: Saunders Elsevier; 2010. With permission from Elsevier.]

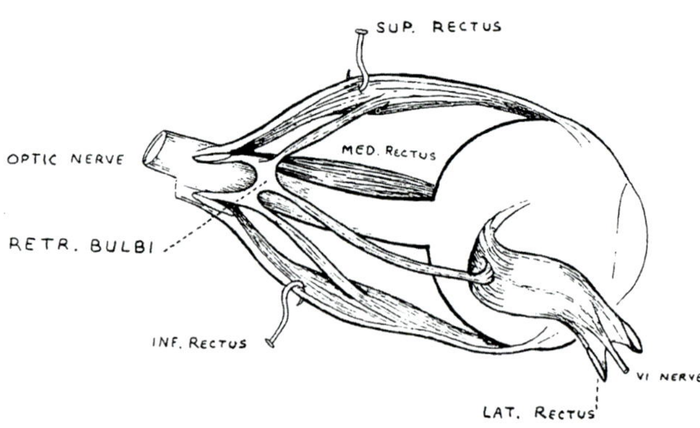

Fig. 13.17 Retractor bulbi muscle in the right orbit. The lateral rectus muscle has been cut from its two heads of origin and turned forwards, thus exposing the tendon of the retractor bulbi, with its four muscular slips passing to the four rectus muscles. The third and fourth nerves passed between the sheath of the optic nerve and the tendon of the retractor bulbi, the latter separating these nerves from the sixth. [Reprinted from Whitnall SE. An instance of the retractor bulbi muscle in man. J Anat Physiol 1911; 46(Pt1):36-40.]

Phylogeny of Extraocular Muscles

Extraocular muscles are a unique component of the vertebrate head. Our knowledge of the agnathic condition is limited to the Myxinoidea (hagfishes) and Petromyzontida (lamprey). Hagfishes, being mud-burrowing scavengers, do not have EOMs (perhaps for good reason) but lampreys as such demonstrate a full complement of six EOMS with the same basic distribution as gnathostomes but at different locations in the orbit and with different innervations. Kuratani's group attempted to define the ancestral state of extraocular muscles using a comparison between the basal agnathan, lamprey *Lethenteron camtschaticum* with a basal gnathostome, cloudy catshark, *Scyliorhinus torazame*. In both species EOMs arose from three locations of the head mesoderm termed premandibular, mandibular, and hyoid. These correspond of course to Sm1–Sm2 supplied from m1 by oculomotor nerve, Sm3 supplied from r0 by trochlear nerve, and Sm5 supplied from r5 by abducens nerve. This innervation pattern remains invariant in vertebrate evolution (Fig. 13.19).

Lamprey EOMs demonstrate several anatomic differences with higher vertebrates. CN III in lamprey only supplies three EOMs, not four. Trochlear nerve supplies caudal oblique muscle that is 180 degrees opposite to the position of superior oblique in gnathostomes. Abducens supplies two distinct rectus muscles, ventral rectus and caudal rectus. To keep the phylogenetic relationships straight, consider the chart below (Table 13.2).

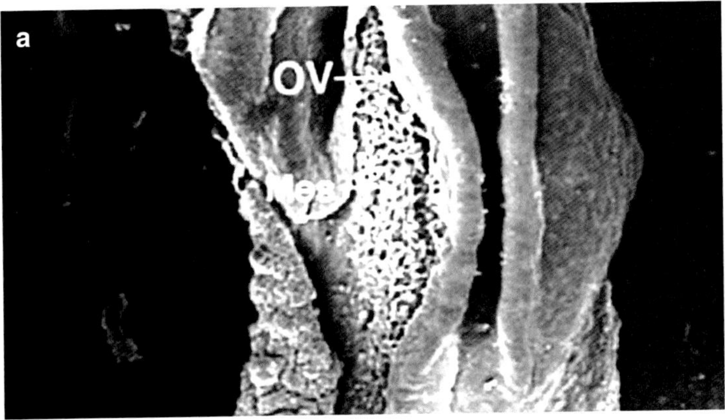

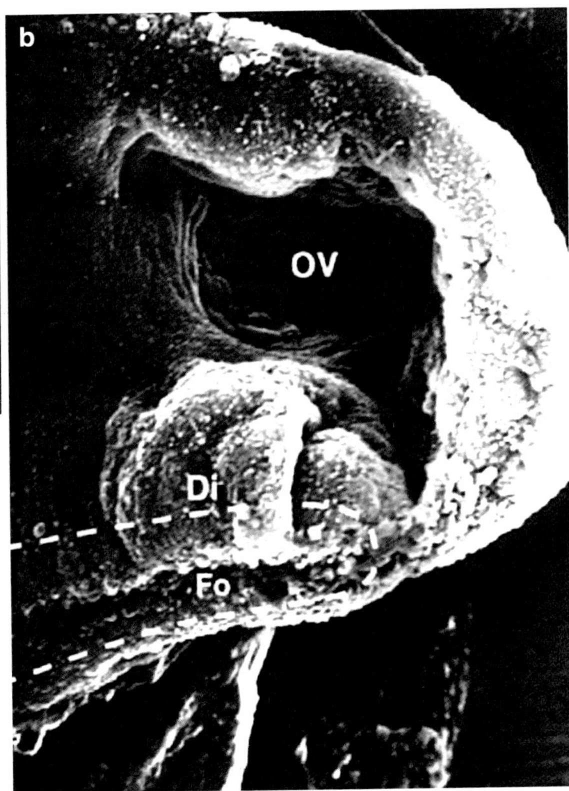

Fig. 13.18 Mesenchymal covering of the optic vesicle at chick equivalent of stage 11.

- Dorsal view (**a**) of a scanning electron micrograph (SEM) of the 3 somite stage (HH stage 8) chick head in which the lateral ectoderm has been extirpated thus showing the lateral aspect of the outgrowing optic vesicle (OV) surrounded by the cephalic mesoderm.
- Inner aspect of the left-hand half of a 7 somite stage (HH stage 9) chick embryo sagittally cut (**b**); at this stage, the neuroectoderm of the optic vesicle expands laterally at the transverse level of the diencephalic floor plate (Di) to reach and contact the ectodermal anlage of the optic placode. This corresponds to Carnegie stage 11 at which time the vesicle is surrounded by cephalic mesoderm and m1–m2 neural crest. Note the position of the foregut endoderm (Fo; dotted line).

[Reproduced with the permission of UPV/EHU Press from Cruezet S. et al. (2005) Int. J. Dev. Biol. 49: 161-171.]

In assessing the changes from the agnathic state to gnathosomes the one muscle we have to account for is *medial rectus*. This muscle arises by a duplication from a preexisting blastema. This event takes two directions. Remember that cartilaginous fishes and bony fishes diverge from a common ancestor. In elasmobranchs SR duplicates. In our ancestral line IR duplicates. That SR is spared had later significance. The evolution of the mobile upper eyelid required a separate muscle, levator palpebrae superioris. In mammals SR was available to duplicate and form LPS.

Clinical note: inferior and medial rectus are susceptible to pathologies.

In Grave's disease, thyroid ophthalmopathy causes fatty infiltration of the extraocular muscles. This likely represents a reassignment of resident stem cells (frequently listed as fibroblasts) to an adipose line. An inflammatory cascade is well described [11] and the specific response of resident mesenchymal stem cells (pericytes) to inflammation can involve cellular transformation [12]. Although all muscles are involved, the order is very specific: Inferior rectus is most common, followed by medial, superior, levator, and lateral rectus [13]. The anatomic distortion of the muscles affected the results in ocular misalignment, diplopia: inability to look up when the eye is adducted, the so-called "double elevator palsy." The peculiar susceptibility of IR and MR may be due to their origin from a common blastema descended from the cyclostome anterior rectus. Both muscles belong to the axis supplied by the primitive ventral ophthalmic artery. Recall the prVOA develops one stage later than prDOA so if a vascular insult hits the central axis, the inferior and medial sec-

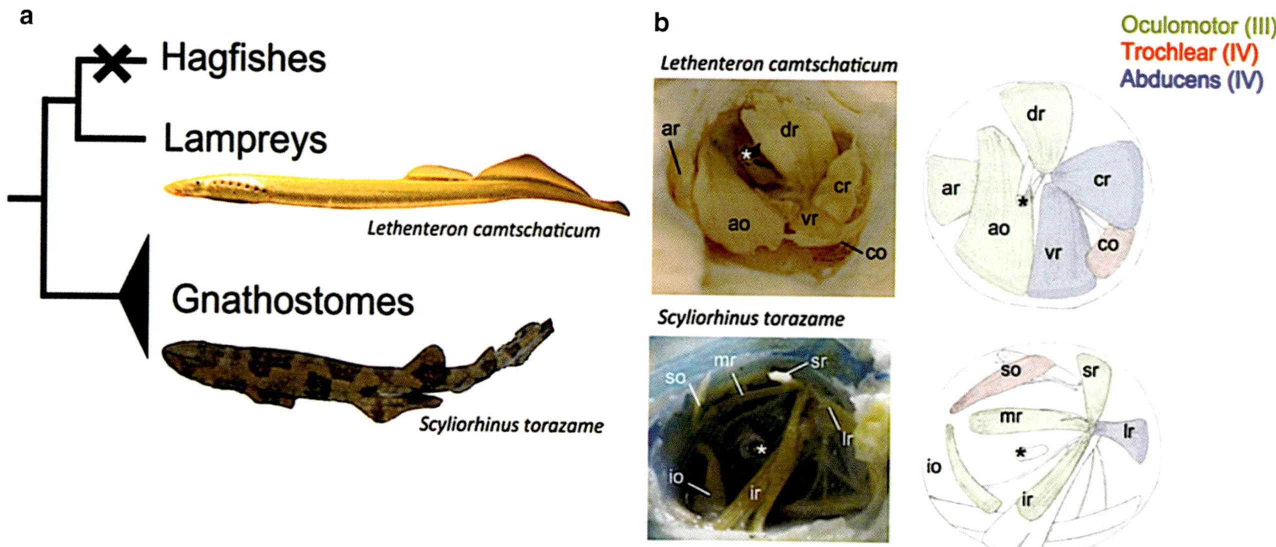

Fig. 13.19 Phylogenetic relationships of extraocular muscles. (**a**) Phylogenetic tree of the vertebrates. The hagfishes lack extraocular muscles. (**b**) Gross anatomy of the extraocular muscles of the lamprey (*Lethenteron camtschaticum*) and shark (*Scyliorhinus torazame*). Asterisks indicate the optic nerve. [Reprinted from Suzuki DG. Fukomoto Y, Yoshimura M, Yamazaki Y, Kosaka J, Kuratani S. Comparative morphology ad development of extraocular muscles in the lamprey and gnathostomes reveal the ancestral state and developmental patterns of the vertebrate head. Zoological Letters 2016; 2:10-24. With permission from Creative Commons License 4.0: https://creativecommons.org/licenses/by/4.0/.]

Table 13.2 Evolution of extraocular muscles

	Cyclostomes	Gnathostomes	
	Lamprey	Chondrichthyans	Osteichthyans
Oculomotor (III)			
Dorsal branch	Dorsal rectus	Superior rectus Medial rectus	Superior rectus
Ventral branch	Anterior rectus	Inferior rectus	Medial rectus Inferior rectus
	Anterior oblique	Inferior oblique	
Trochlear (IV)	Caudal oblique	Superior oblique	
Abducens (VI)	Caudal rectus	Lateral rectus	
	Ventral rectus	Retractor bulbi	

tors of the globe are affected first. This same pattern pertains to the spectrum of holoprosencephaly. Finally, the reported cases of congenital absence of the extraocular muscles are skewed toward inferior rectus [14] (Fig. 13.20).

Smooth Muscles from Neural Crest (Surprise, Surprise…)

NC accesses the iridial angle later in development once ingression of the lens has taken place. Because of this anterior entry the NC cells are more likely to originate from the diencephalic population (p1–p3) than from midbrain (m1–m2) but this has not been sorted out. The derivatives include pigment cells, stroma, and the unique muscles of the iris and lens, controlled through the sympathetic autonomic system, itself an exclusively neural crest derivative (Fig. 13.21).

The iris consists of three layers: an outer stromal layer, a superficial layer of pigmented epithelium, and a deep layer of unpigmented epithelium. Neural crest cells migrate first into the stroma, superficial to the outer pigmented layer.

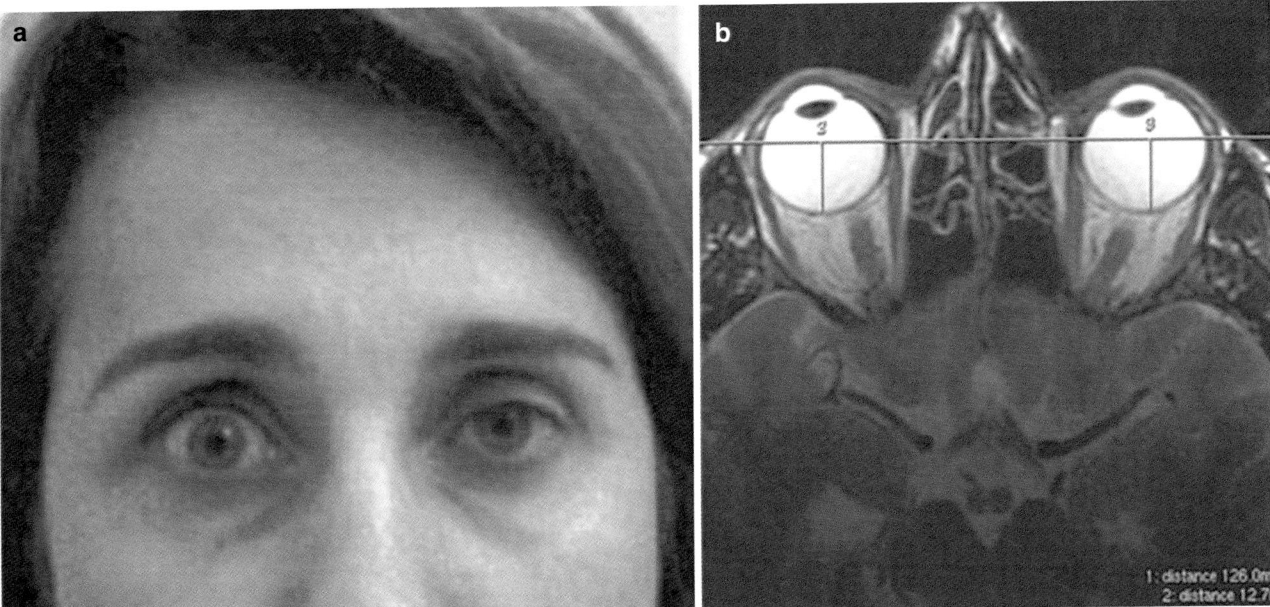

Fig. 13.20 Grave's disease. Eosinophilic round cell infiltrate invades extraocular muscles, particularly IR and MR. The susceptibility of these muscles makes an interesting correlation with phylogeny. [Reprinted from Machado KFS, Garcia MM. Oftalmopatia tireoidea revisitada. Radiol Bras. 2009;42(4):261–266. With permission from Brazilian College of Radiology and Diagnostic Imaging.]

They then penetrate into the iris and there, within its stroma, they form *sphincter pupillae* and *dilator pupillae* muscles. Neural crest remaining in the stroma forms *ciliary muscles* that adjust tension on the suspensory ligaments of the lens and thereby control the musculature.

The derivation of the ciliary muscle sheds light on the origin and clinical significance of Müller's muscle (MM). Long considered to be an independent muscle, its proximal insertion is from the posterior layer of aponeurosis of levator palpebrae superioris (LPS); it is inserted distally into superior border of tarsal plate. Orbital dissections by Kakizaki [15] demonstrated bilateral extension of MM into the peribulbar network including pulleys of medial and lateral rectus muscles. In pursuing its lateral course, fibers from Müller's pass through and contribute to the lacrimal fascia of the palpebral lobe. As such, Müller's represents a smooth muscle derivative of neural crest, likely from midbrain. It is not derived from paraxial mesoderm (some studies describe an admixture of striated fibers as well).

The anatomic and clinical behavior of smooth muscle MM is at variance with that of its striated extraocular *confrères*. (1) Unlike the extraocular muscles, contraction of Müller's muscle has no directionality. (2) The innervation of MM is strictly sympathetic. (3) Surgical specimens from cases of congenital ptosis show MM to be normal, without atrophy. (4) In Graves's ophthalmopathy, MM remains un-infiltrated. (5) MM as a smooth muscle retractor is analogous to the inferior palpebral muscle of the lower eyelid, also a part of the peribulbar smooth muscle network [16].

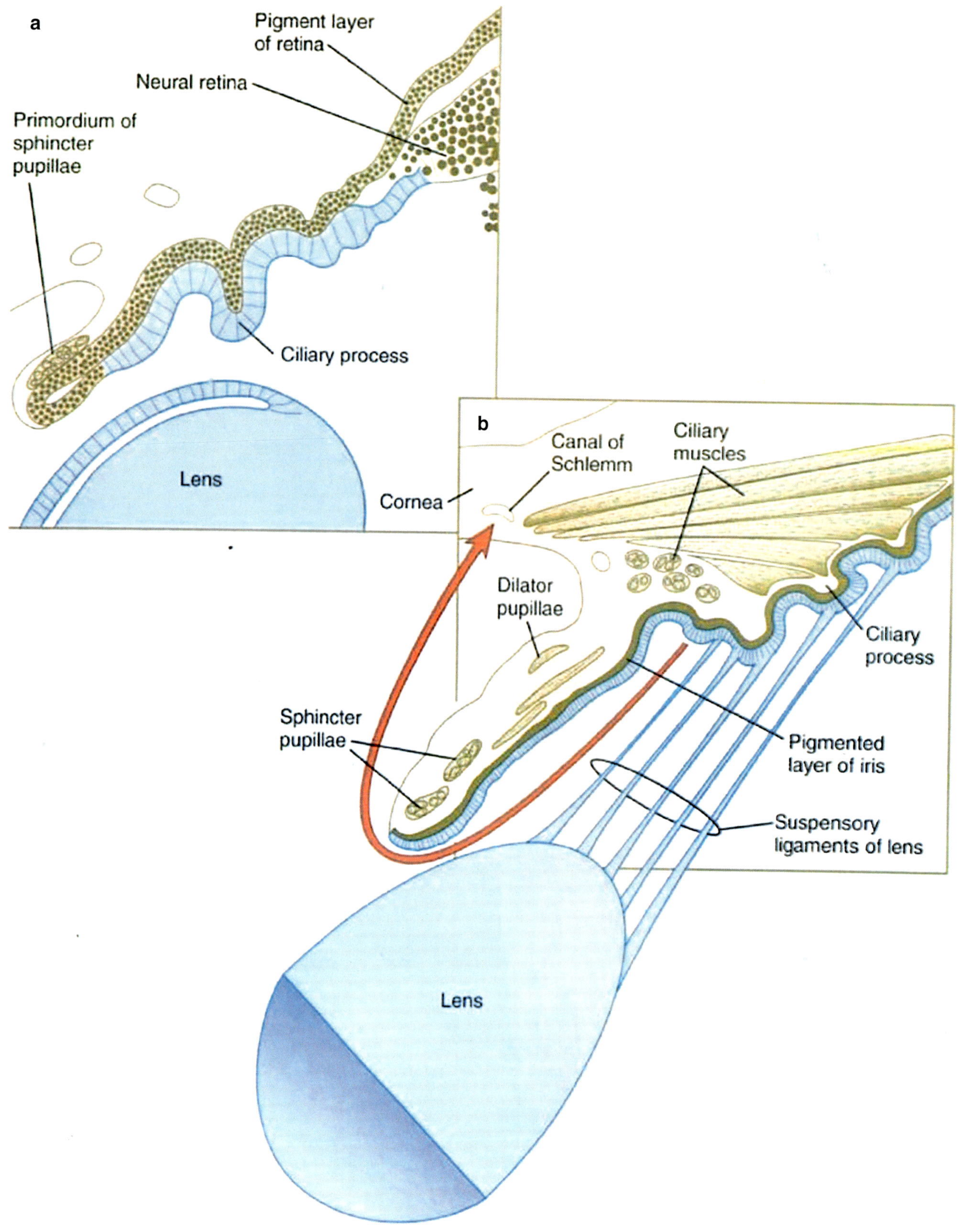

Fig. 13.21 Neuroectodermal smooth muscles of the eye. Note the three layers of iris. Neural crest remaining in the stroma forms the ciliary muscles to control the shape of the lens. Neural crest that penetrates into the iris forms dilatator and sphincter muscles to control the amount of light entering the eye. [Reprinted from Carlson BM. Human Embryology and Developmental Biology, 6th edition. St. Louis, MO: Elsevier; 2019. With permission from Elsevier.]

Construction of the Orbit: Component Parts

A-B-C's of the Orbit

Big picture idea: Developmental fields can be lumped into three groups: A + B + C. The face is constructed from three blocks of fields on either side of the midline. We can classify them by their blood supply and innervation. This requires understanding of the stapedial arterial system.

A field neuroangiosomes consist of an *FNO skin coverage* and *bone fields of the anterior cranial fossa and medial-superior orbital walls*. The epidermis contains *nasal and optic placodes*, islands of specialized non-neural ectoderm that form the nasal cavity and complete the development of the globe. The MNC bone fields are sphenoid (except alisphenoid), the ethmoid complex, the frontal bone, and the orbital series (lacrimal, prefrontal, and postfrontal). These are innervated by V1. The blood supply is V1-induced branches of the stapedial ophthalmic axis.

B field neuroangiosomes are ***non****-pharyngeal arch bone fields* of the jaws and supporting bones: mandible, vomer, premaxilla, maxilla, palatine, zygoma, and alisphenoid. These are innervated by V2. The blood supply is V2-induced branches of the stapedial maxillo-mandibular axis. Mandible, supplied by StV3, is not included in our discussion.

C field neuroangiosomes: *pharyngeal arch tissues* (except the jaws). These are innervated by V2. The blood supply is from the branches of external carotid (Table 13.3).

Mesenchymal Derivatives Summarized

To get oriented, let's list right off the bat the various structures of the eye and orbit and their mesenchymal derivation. This is based on the quail-chick mapping work of Sophie Creuzet. Bone derivatives have been worked out by Drew Noden (Table 13.4).

Quick Summary of Cranial Neural Crest Derivatives in the Skull

The mesenchymal map has been worked out by Drew Noden (Fig. 13.22).

- With the exception of the cranial base, the orbitosphenoid, and the parietal cartilage, all craniofacial bones are derivatives of neural crest, whether they develop directly in membrane or proceed via a cartilage intermediate.
- The four occipital somites fuse together to form basisphenoid, exoccipital, and basioccipital bone below the transverse sinus.
- Above the transverse sinus the interparietal bone fields are membranous.
- Recall that the postparietals are neural crest while the tabulars are PAM.

Table 13.3 Neuroangiosome classification

Field	NCrest	Neuromeres	Blood supply	Derivatives
A	PNC	p3–p1	StV1	Fronto-naso-orbital skin
A	MNC	m1–m2, r0–r1	StV1	FNO bone fields
B	RNC	r2–r3	StV2/StV3	Jaws and supporting bone
C	RNC	r2–r11	Ext carotid	All other pharyngeal arch structures

Table 13.4 Embryonic origins of orbital tissues

Superficial epithelium	Mesenchyme		Neuroepithelium
Ectoderm	Neural crest	Mesoderm	Neural plate
Corneal epithelium	Corneal stroma and endothelium	Blood vessel endothelium	Iridial epithelium and lamella
Lens	Iridial stroma and muscles	Myofibers of the EOMs	Pigmental and sensory retina
Eyelid epithelia	Ciliary muscles		
	Ciliary corpus mesenchyme		
	Wall of canal of Schlemm		
	Sclero-corneal limbus		
	Choroid membrane		
	Sclerotic cartilages		
	Eyelid mesenchyme		
	Eyelid muscle??		
	Lacrimal gland mesenchyme		
	Fascia of the EOMs		
	Pericytes of ocular blood vessels		
	Periorbital bones		

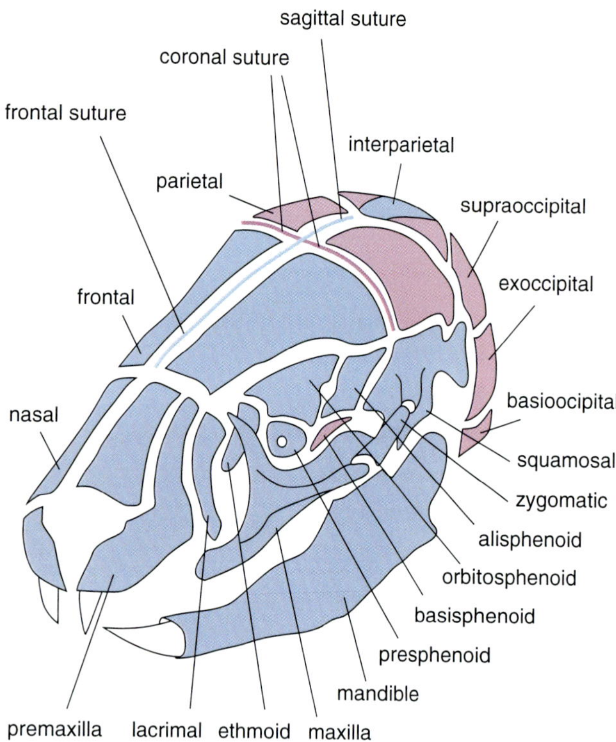

Fig. 13.22 Mesenchymal sources of skull bones. This is a murine model differing in topology with *Homo sapiens* but the bone fields are the same. Note that parietal (PAM) interfaces with a midline strip of r1 neural crest, the sagittal suture, in two zones: anterior (r2 vs. r1) and posterior (r3 vs. r1). Interparietal is likely r6–r7 neural crest; its boundary with parietal is the lambdoid suture. Alisphenoid is r2 and forms spheno-parietal suture. Thus, all five periparietal sutures represent the opposition between a PAM bone field and neural crest bone field. Neural crest, blue; paraxial mesoderm, red. [Reprinted from Mishina Y, Snider NT. Neural crest cell signaling pathways critical to cranial bone development and pathology. Exp Cell Res 2014; 325(2):138-147. With permission from Elsevier.]

- All neural crest craniofacial bones develop from stem cells in the supratentorial dura. Those bones form within an opposing external neural crest as well. When these two laminae are from different embryonic sources, the resulting membranous bones are bilaminar, with a potential space for bone marrow or a sinus.

Orbital Fat: Observations

Like all white fat, orbital fat is a derivative of r1 pericytes (see Chap. 11). The relationship between pericytes and neural crest is protean. We can consider white fat to be a neural crest derivative.

Within the orbit, fat is dispersed throughout without distinct compartments. However, it can be described as intraconal versus extraconal with the four recti dividing the fat up like a "four-leaf clover." Histologic differences define two zones. Retrobulbar fat has large lobules septae. All remaining fat has small, densely packed lobular and robust septae.

Post-aponeurotic fat (the so-called central fat pads) is yellow in color and sits directly behind the septum orbitale. Their medial boundaries are passively determined by neighboring structures. The upper lid central fat pad sits lateral to trochlea and abuts lacrimal gland, from which it is separated by the thin lacrimal capsule demonstrating that the two tissues are embryologically distinct. In the lower lid, central fat pad is lateral to inferior oblique muscle. Medially, both upper and lower lids have medial fat pads which are white in color.

Contrary to some reports, color distinctions do not represent brown fat versus white fat. The medial fat pads are both supplied by small branches from the StV1 nasociliary axis whereas the supply for the central fat pads is StV1 supraorbital. The arterial branches to the medial pads are capable of retraction back into the orbit thus requiring careful hemostasis so as not to create a focus for hemorrhage within the orbit. The supraorbital vessels, on the other hand, are connected sidebranches to the extraocular muscles and being thus tethered are less vulnerable to retraction and easier to control. Thus the central fat pads are developmentally related to the recti and obliques whereas the nasociliary fat relates to the orbital plate of the ethmoid. These zones could have different homeotic coding and therefore different neuromeric origin (such as m1 versus m2).

Postaponeurotic fat pads are defined passively in a similar manner and have a substantial literature with excellent reviews [17, 18]. The preaponeurotic fat pad of the upper lid is of embryologic interest. This r1 structure has a thin capsule with blood vessels and can be dissected with care as a unit. Laterally it extends around the lacrimal gland and is *continuous with the fat of the orbit* [19, 20]. As such it betrays the migration of r1 neural crest forward from the orbit and into the upper lid where it also forms the dermis. A similar preaponeurotic fat pad exists for the lower eyelid. It too is encapsulated but in continuity with the face, not the orbit [21] (Fig. 13.23).

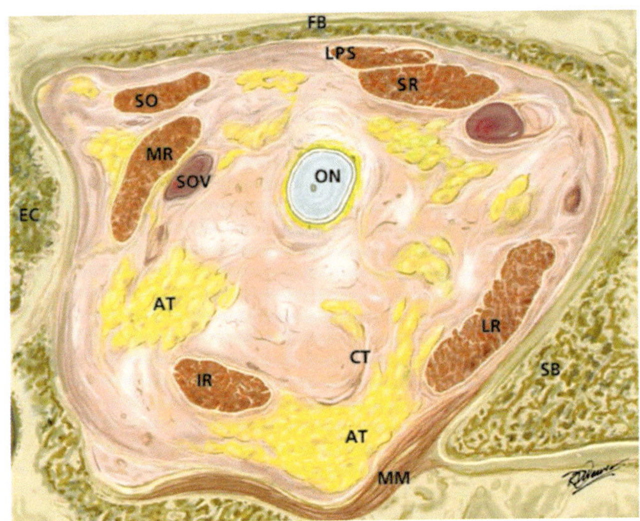

Fig. 13.23 Orbital fat. White fat is derivative of neural crest-related pericyte cell line. Fat positioned front of the lateral two-thirds of septum orbitale is septated. Preaponeurotic (central) fat pad is distinct from medial superior and inferior fat pads. These are white in color and extend back into the orbit with a small, but important arterial pedicle from the anterior ethmoid axis. Left: [Reprinted from Tawfik HA, Dutton J. Embryologic and fetal development of the human orbit. Ophthal Plast Reconstr Surg 2018; 34(5):405-421. With permission from Wolters Kluwer Health.]. Right: [Reprinted from Amrith S, Min Yeug S. Chapter 2 Anatomy in: Amrith S, Sundar G, Min Yeug S. Orbital Adnexal Lesions: A Clinical, Radiological and Pathological Correlation. Singapore: Springer Nature; 2019: 3-11. With permission from Springer Nature.]

Construction of the Orbit: Timetable of Developmental Events

Summary of Events (see Chap. 4)

Event #1: *Deposition of Mesoderm*

Prechordal mesoderm (PCM) is present immediately after gastrulation in stage 7. It may form the basipostsphenoid pre-somitic mesoderm at stage 8 and will give rise to the extraocular muscles. Basipostsphenoid cartilage and midline cartilage form PCM in the anterior cranial fossa.

Event #2: *Migration of Neural Crest Mesenchyme*

Beginning at stage 9, streams of MNC cover the entire forebrain and the anterior cranial fossa. All cartilages anterior to the pituitary are of MNC origin. These include the trabeculae, the alae orbitalis, and the alae temporalis. PNC travels forward and downward like a sheet under the FNO epithelium to complete the skin cover. RNC fills up the five pharyngeal arches during stages 10–15. A portion of RNC from r2 and r3 will be set aside to create the jaws and all supporting bone structures. These will include the palatine bone, the floor, and the lateral walls of the orbit.

Event #3: *Closure of the Neural Tube, Forebrain Growth and Epithelial Coverage with Non-neural Ectoderm*

Neural tube closure begins at stage 10 with the rostral neuropore closed at stage 11. Under normal circumstances, the optic placodes (sites of exit for the eyes) are 180° apart. The nasal placodes are 120° apart.

Event #4: *Approximation of Tissue Fields*

Frontoethmoid MNC undergoes *apoptosis* which brings the nasal fields into the midline. First and second arches merge in stage 12 to form the B–C complex which makes physical contact with the A complex. Tissues from upper inner sector of the B–C complex (the future maxillary tissues) flow medially below base of the forebrain to create the lining of the future oral cavity. Failure of apoptosis causes hypertelorism. Inadequate MNC causes hypotelorism.

Event #5: *Formation of Nasal Passages*

Nasal placode epithelium invaginates into the underlying MNC forming tunnels directed backwards toward the fused adenohyphyseal placodes that make up Rathke's pouch located at the roof of the future nasopharynx. Maxillary and mandibular processes are physically in place. Fusion of neurovascular fields takes place.

Event #6: *Development of the Globe*

Contact between the optic cup and the optic placode forms the ophthalmic anlage. The globe is surrounded by MNC that forms sclera. Programming from sclera and overlying dura or from lateral nasal chamber epithelium creates orbital bones.

Event #7: *Completion of Facial Fields*

The naso-orbital A-fields form the letter "T" in the midline with B–C complexes located on either side of the vertical limb. The inferior walls of the nasal chambers, both medial and lateral, are in place. External to the B field bones, soft tissue coverage from C fills out the midface, which is defined by the orbital process of palatine bone. Bones originating from r2 neural crest are (1) palatine, at the postero-medial angle of the orbital floor, (2) maxilla, (3) zygoma, and (4) alisphenoid.

Carnegie Stages of Eye and Orbit Development

For vascular development of the eye see Chap. 6 (Fig. 13.24). Stage 9 (19–21 days), 1–3 somites)

- First pharyngeal arch
- Neural crest from r0–r1 in place for dura

Stage 10 (22–23 days, 4–12 somites)

- Second pharyngeal arch
- Neural crest from m1–m2 in place: source material for orbital fat
- Optic primordia appear @ 8 somites
- Optic primordia meet in the midline to form the optic chiasm

Stage 11 (23–26 days,13–20 somites)

- Third pharyngeal arch.
- Fusion of first and second arches, repositioning of PA1–PA2 in front of the of third arch, buccopharyngeal membrane separates them.
- Optic evagination @ 14 somites
  - Lateral wall contacts first and is developmentally more primitive; this will be the site of primitive dorsal ophthalmic artery.
- Neural crest separates the evagination from the hp1 preplacodal ectoderm.
  - Definitive r1–r2 ectoderm for the eyelids is not yet in place.
- Optic vesicle forms @ 17–19 somites
  - Optic vesicle is defined from the forebrain by *caudal limiting sulcus.*

Stage 12 (26–30 days, 21–29 somites)

- Fourth pharyngeal arch.
- Optic neural crest is at the maximum extent.
  - Optic vesicle is completely ensheather: "frightened hedgehog."

Stage 13 (28–32 days, 4–6 mm, 30+ somites)

- Fifth pharyngeal arch.
- Optic vesicle lies beneath the PPE (preplacodal ectoderm) from prosomer hp2.
- Two basement membranes develop:
  - Optic vesicle BM.
  - Preplacodal ectoderm BM.
- PPE responds to optic vesicle
  - Optic (lens) placode.
- Lens placode in contact with the retinal disc.

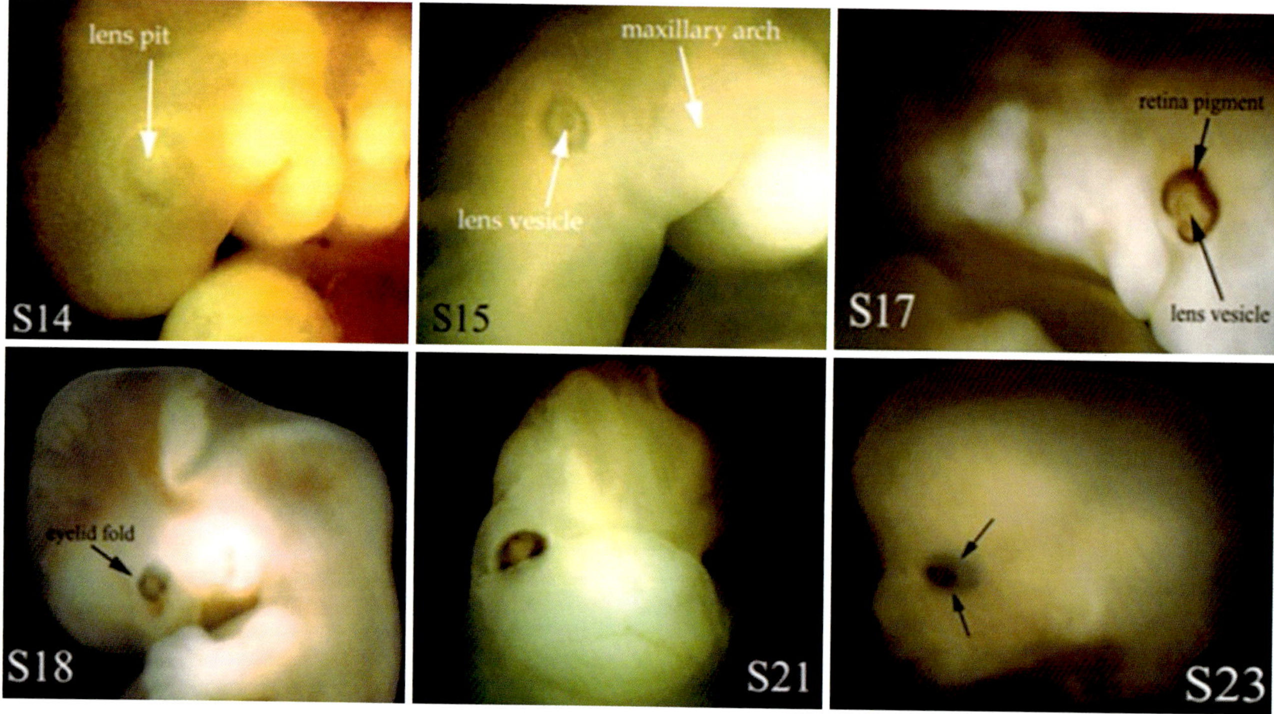

Fig. 13.24 Eyelid development Carnegie stages. Upper eyelid fold forms at stage 18 and lower fold at stage 19. Eyelid fusion is lateral first, then medial, reflecting the lateral to medial deposition of r1 and r2 neural crest mesenchyme. Recall that r1 migration takes place prior to that of r1. [Reprinted from Byun TH, Kim JT, Park HW, Kim WK. Timetable for upper eyelid development in staged human embryos and fetuses. Anat Rec 2011; 294:789–796. With permission from John Wiley & Sons.]

- Optic (neural) retina, pigment layer of retina, optic stalk extends backwards.
- Primitive dorsal ophthalmic artery appears.
- prDOA zone
  - Lateral rectus.

Stage 14 (31–35 days, 5–7 mm)

- Lens placode indented as lens pit.
  - Communicates with the surface via a lens pore.
- Retinal fissure (choroid fissure).
- Two-layer retina inverts to form the optic cup.
  - Non-neural retina has five layers:
  - (1) external limiting membrane, (2) mitotic zone, (3) intermitotic zone, (4) marginal zone, (5) internal limiting membrane.
  - Neural retina has same layers as cerebral wall.
- Sclera develops.
- Superior rectus/superior oblique from common primordium (prDOA).
  - SO inserts posterior to SR on the z-axis.
- Primitive ventral ophthalmic artery appears.
- Blood vessels develop in the uveal tract on both temporal and nasal sides.
- Cranial nerve III appears (SR is innervated before SO).

Stage 15 (35–38 mm, 7–9 mm)

- Lens pit closes over to form lens vesicle.
- Lens vesicle and optic cup lie just beneath, and in contact with, the surface.
- Initial epithelium in front of the lens is PPE.
- Retinal pigment appears from local neural crest.
- Vitreous body develops behind the lens.
- Hyaloid artery develops in the retinal fissure.
- Surface preplacodal ectoderm is now the anterior epithelium of the cornea, complete with basement membrane.
- Cranial nerves IV and VI appear.

Stage 16 (37–42 days, 8–11 mm)

- Arrival of r1 ectoderm and r2 ectoderm.
- Eyelid grooves form prior to lids: first upper, then lower.
- Retinal pigment.
- Lens is D-shaped.
- Ciliary ganglion.
- Medial rectus (prVOA zone).
- Inferior rectus/inferior oblique from a common primordium (prVOA zone)
  - IO inserts behind IR along the z-axis and the tracks downward and medially to insert on maxilla.

Stage 17 (42–44 days, 11–14 mm)

- Eyelid grooves deepen.
- Nasolacrimal groove.
- Inferior rectus insertion.

Stage 18 (44–48 days, 13–17 mm)

- Upper eyelid fold.
- Palpebral primordium.
- Neural crest mesenchyme from r1–r2 invades between r1–r2 epithelium and the hp1 anterior cornea epithelium.

Stage 19 (48–51 days, 16–18 mm)

- Lower eyelid fold.
- Lateral canthus.
- Superior conjunctival fornix thickens.
- Posterior cornea epithelium.

Stage 20 (51–53 days, 18–22 mm)

- Medial canthus.
- Epithelium organizes into nodules. Mesenchyme surrounding this zone in the fornix assumes an oval shape.
- Nerve fibers from retina reach the brain.

Stage 21 (53–54 days, 22–24 mm)

- Epithelial buds further condense into an identifiable fornix. Invagination begins.
- Levator palpebrae superioris.

Stage 22 (54–56 days, 23–28 mm)

- Eyelids fully formed but not closed
- Scleral condensation

Stage 23 (56–60 days, 27–31 mm)

- Eyelids closed.
- Retina: 10 layers established, from external (facing the vitreous) to internal:
  - Inner limiting membrane – Müller cell footplates.
  - Nerve fiber layer – retinal ganglion axons eventually the optic nerve.
  - Ganglion cell layer – neuronal cell bodies of retinal ganglion cells, their axons form the nerve fiber layer and eventually the optic nerve.
  - Inner plexiform layer – another layer of neuronal processes.

- Inner nuclear layer – neuronal cell bodies.
- Outer plexiform layer – another layer of neuronal processes.
- Outer nuclear layer – neuronal cell bodies.
- External limiting membrane – layer separating inner segment portions of photoreceptors from their cell nuclei.
- Photoreceptor layer – rods and cones that convert light into signals.
- Retinal pigment epithelium.

Phylogeny of the Orbit

We shall now recount the story of how the orbital bone fields came into being. The orbit is composed of nine bone fields, not seven as is commonly thought. These are

- Prefrontal bone of pars orbitalis
- Postfrontal bone of pars orbitalis
- Lacrimal bone
- Ethmoid bone, lamina papyracea
- Palatine bone, orbital process
- Maxilla, orbital process
- Zygoma, orbital process
- Orbitosphenoid (lesser wing)
- Alisphenoid (greater wing)

In reviewing the phylogeny of the human orbit, we shall follow a straight line through evolution, concentrating, first, on the bone fields defining its peripheral surround, and, later, on how those changed resulted in frontalization and the development of stereoscopic vision. The story involves two phases. The first involves the development of initial bone periorbital bone fields and their positioning to protect the laterally-directed eye, concentrating primarily on the zygoma. The second phase involves changes as we ascend through the synapsid line to primates in which the orbit becomes radially repositioned and a new back wall constructed from alisphenoid. We shall see orbital anatomy as dynamic, responding to evolutionary advances in skull design, mastication, and palate stability.

The Early Periorbital Structure

The dawn of vertebrate history is difficult to reconstruct because primordial vertebrate did not have mineralized tissues. Recent finds from China dating back to the early Cambrian give us a start. *Haikouella* and *Yunnanzoon* represent crown protovertebrates. They had a body plan like Amphioxus with all relevant chordate features (notochord, pharyngeal slits, thyroid gland/endostyle, a dorsal hollow nerve cord, and a post anal tail). They also had lateralized eyes but lacked an ear capsule and had no skull. Slightly later (and also from China) *Haikouichthys* and *Myllokunmingia* had similar features but were true vertebrates. Their eyes were larger but they still lacked a skull (Fig. 13.25).

The first manifestation of cranial bone was seen in the late Cambrian with *conodonts*, tiny vertebrates with a number of important synapomorphies: a definite cranium (without a skull) having paired sense organs, eyes controlled by extrinsic eye muscles, and a sclerotic eye capsule. They had a primitive tooth development program, the *odontode* [22]. Histologic samples show calcified cartilage, enamel, and dentin produced by odontoblasts, themselves neural crest derivatives. The discovery of conodont tooth-like structures was the basis of the now-refuted "inside-out" hypothesis that teeth evolved independently of the vertebrate skeleton and prior to the invention of jaws. This was replaced by the "outside-in" hypothesis that teeth evolved as an extension of osteogenic competence from the outer dermis to the inner oral epithelium very shortly after the invention of jaws in placoderms (Duncan 2013) (Fig. 13.26).

Conodonts were followed in the Ordovician by armored jawless fishes, now with *micromeric* dermal bone made of dentin, capped with enamel, and having pulp cavities in the center. *Astraspis* (Gk "star shield'), found in Bolivia, had a massive bony head shield. Along the sides of the body were eight branchial openings corresponding to the nine branchial

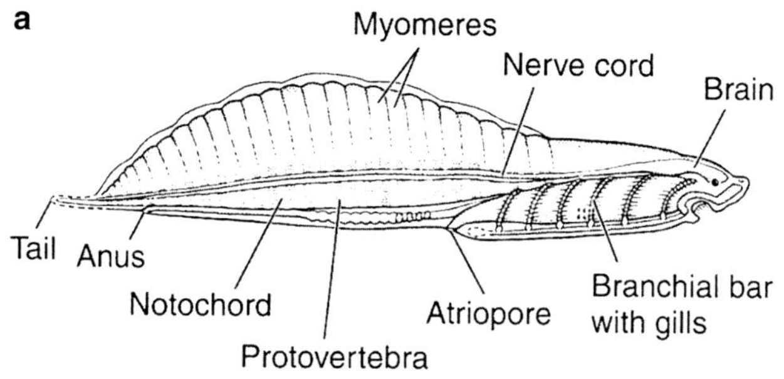

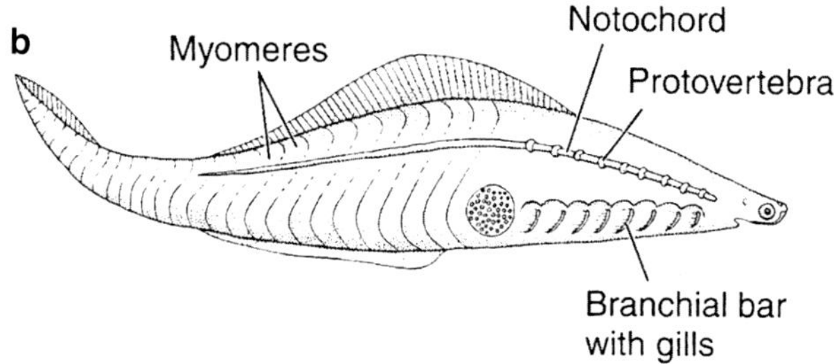

Fig. 13.25 *Haikouella lanceolata.*Crown protovertebrate *Haikouella* had lateralized eyes but no extraocular muscles and no skull. This arrangement continued in first true vertebrates, *Haikouichthys* and *Myllokunmingia*. [Reprinted from Mallet J. Fossil sister group of craniates: predicted and found. J Morphol 2003; 258(1): 1-31. With permission from John Wiley & Sons]

Fig. 13.26 Conodont revolution. These tiny agnathans resembling eels were very common. They define stage 2 of the Cambrian period. They show the first manifestations of neural crest, had a skull without bone, and lateralized eyes with extraocular muscles. Conodont teeth represent the first evidence of mineralized tissue in vertebrates. Left: conodont teeth; Right: reconstruction of conodont large lateralized eyes and branchial slits. [Reprinted from Duncan JEM, Dong X-P, Repetski JE, et al. The origin of conodonts and vertebrate mineralized structures. Nature 2013; 502(7472):546–549. With permission from Springer Nature.]

arches. These fishes evolved a lateral line system with neuromasts, a cerebellum, and a second vertical set of semicircular canals. The eyes of *Astrapsis* were surrounded by a myriad of tiny bones, the tubercles of which were diamond-shaped (Fig. 13.27). The next significant step in orbital evolution took place in the Ordovician, 470 million years ago when the agnathic Ostracoderms consolidated the micromeric bones into larger, regular units.

The first gnathostomes arose in the ensuing Silurian and Devonian periods. These fishes represented by placoderms had large bone plates. Despite their body armor they had a head–trunk joint which permitted elevation of the head for prey capture. This joint prefigured the future innovation of a neck. Basal placoderm *Dunkleosteus* was non-maxillate but had an identifiable jugal bone located in the suborbital position. It covered the cheek and made contact with the opercular bone which covers the gills (Fig. 13.28). *Qilinyu rostrata* bridges the unknown development of the gnathal (lower jaw only) and later maxillate (upper jaw) conditions for extinct fish. Researchers propose that the maxilla, premaxilla, and dentary are homologous to the gnathal plates of placoderms and that all belong to the same dental-arched group. The gnathal-maxillate transformation occurred concurrently in upper and lower jaws, predating the addition of infradentary (a serially homologous group ventral to the dentary) bones to the lower jaw. This condition is fully developed in the crown placoderm *Entelognathus primordialis* (primordial "complete jaw"): premaxilla, maxilla, mandible, and lacrimal are all present. Jugal is interposed between lacrimal in front and maxilla below (Figs. 13.28, 13.29, and 13.30).

At this juncture, we take leave of the cartilaginous fishes and move directly to Osteichthyes. (We are primate-centric,

Fig. 13.27 Astraspis reconstruction. Jawless and fin-less, Astraspis lived in Bolivia during the Ordovician and had nine branchial arches separated by eight branchial pores (three are depicted here). [Reprinted from Wikimedia. Retrieved from: https://commons.wikimedia.org/wiki/File:Astraspis_desiderata.jpg. With permission from Creative Commons License 2.5: https://creativecommons.org/licenses/by-sa/2.5/deed.en.]

Fig. 13.28 Basal placoderms are represented by the Gogo fish, *Mcnamaraspis kaprios,* the fossil emblem of Australia. This arthrodire placoderm shows bone plates beginning to consolidate. It has a large orbit and an immobile eye surrounded by large bone plates. It demonstrates a ring of bones around the snout. As a conventional placoderm it lacks a maxilla. Jugal field is below the orbit. [Courtesy of Prof. John Long, College of Science and Engineering, Flinders University, Adelaide, Australia.]

are we not?) These have important characteristics: endochondral bone, teeth are now embedded in the jaws, gular plates in the floor of the mouth, identifiable skull bones and scales and lungs as well as gills. Osteichthyes are divided into two clades: the spiny-finned *actinopterygians* leading to derived fishes and the fleshy-finned *sarcopterygians* (our exclusive focus here) leading to tetrapods. The sarcopterygians are characterized by paired lobed fins with an internal skeleton and muscles which extend into the fins themselves. They subdivide into (1) *coelacanths*, primitive "living fossils" living in the depths of the Indian Ocean represented today by *Latimeria*, (2) *dipnoi* the lungfishes, and (3) our ancestors, the *rhipidistians.*

Let's follow the lineage. Stem sarcopterygian *Guiyu* demonstrates a more extensive jugal that extends backward to articulate with preopercular. Postorbital remains an independent field present. Another early form is the Onychodontid *Strunius*, which flourished in the early Devonian. Scleral ossicles persist. The orbit is surrounded by four bones in

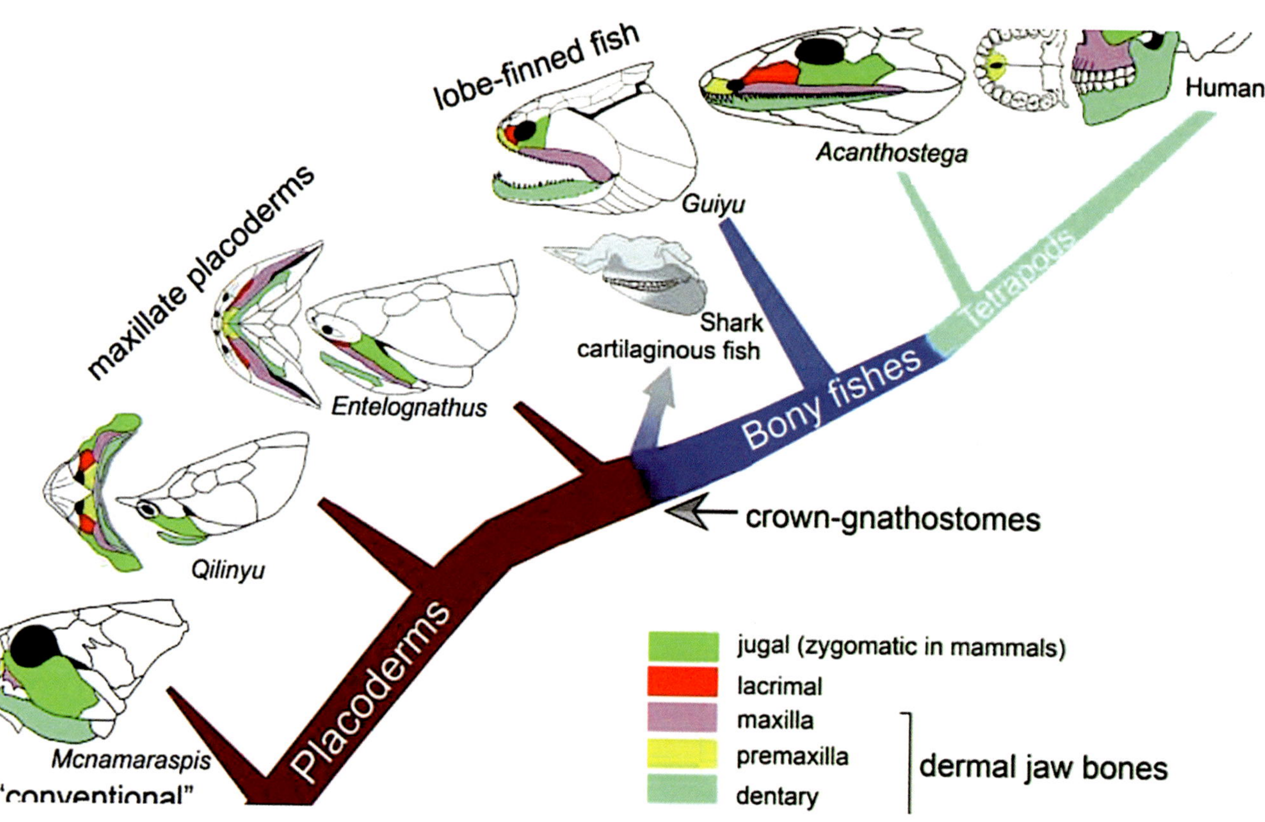

Fig. 13.29 Phylogeny of the jugal. Jugal is a key component to understand the orbit. In placoderms suborbital = jugal; anterior supragnathal = premaxilla; posterior supragnathal = maxilla; infragnathal = dentary. Basal placoderms *Dunkleosteus* and *Mcnamaraspis* were non-maxillate: they had dentary only. *Qilinyu* is the first maxillate placoderm. Note that jugal is displaced for the orbit in *Entelognathus* but this condition is temporary. [Reprinted from Zhu M, Ahlberg PE, Pan Z, et al. A Silurian maxillate placoderm illuminates jaw evolution. Science 2016;354(6310): 334-336. With permission from The American Association for the Advancement of Science.]

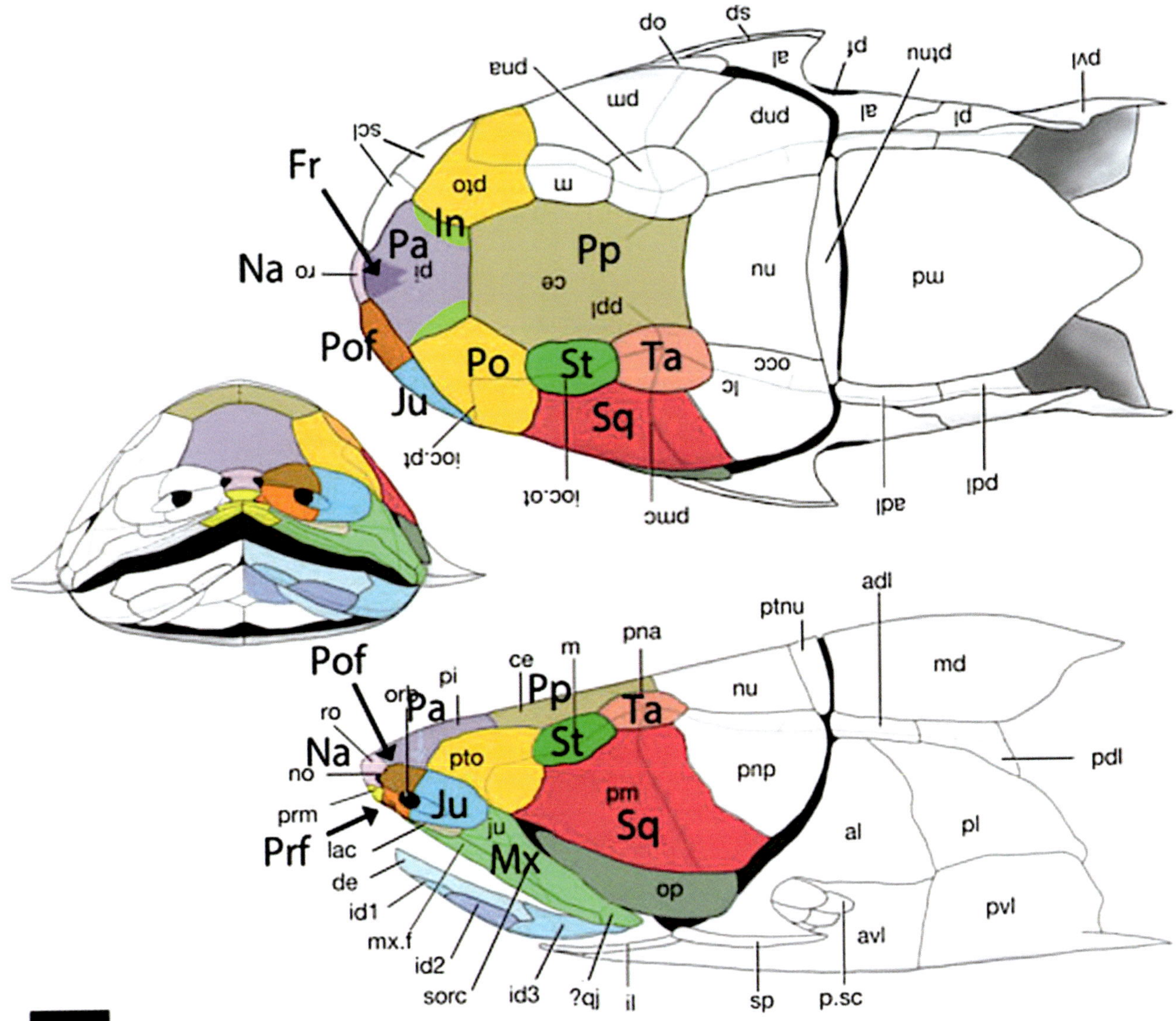

Fig. 13.30 Orbit of *Entelognathus principalis*. *Entelognathus* represents the crown placoderm state. For the first time, we have homologies between placoderm bone and the zoologic bone fields recognized in fishes and higher vertebrates. The abbreviations below are universal. You have seen or will see them elsewhere in this text. *Key:* adl, anterior dorsolateral plate; al, anterior lateral plate; amv, anterior medioventral plate; av, anteroventral plate; avl, anterioventrolateral plate; ce, central plate; de, dentary; gu, principal gular, id1–id3, first to third infradentary; il, interlateral plate; ioc otic, otic branch of infraorbital line groove; loc pt, postorbital branch of infraorbital line groove; ju, jugal; lac, lacrimal; lc, main lateral line groove; m, marginal plate; mand, mandibular line groove; md, median dorsal plate; mx f, facial lamina of maxilla; mx p, palatal lamina of maxilla; no, nostril; nu, nuchal plate; occ, occipital cross commissure; op, opercular; orb, orbital fenestra; pdl, posterior dorsolateral plate; pf, pectoral fenestra; pi, pineal plate; pl, posterolateral plate; pm, postmarginal plate; pmc, postmarginal line groove; pmx f, facial lamina of premaxilla; pmx p, palatal lamina of maxilla; pna, anterior paranuchal plate; pnp, posterior paranuchal plate; ppl, posterior pitline; prm, premedian; ptnu, postnuchal plate; pto, postorbital plate; pvl, posterior ventrolateral plate; p sc, pectoral fin scales; qj, quadratojugal; ro, rostral plate; sbm a, anterior submandibular; sbm p, posterior submandibular; scl, sclerotic plate; sorc, suproral line groove; sp, spinal plate. [Reprinted from Zhu M, Yu X, Ahlberg PE, et al. A Silurian placoderm with osteichthyan-like marginal jaw bones. Nature 2013;502(7470): 188-93. With permission from Springer Nature.]

roughly four quadrants. The anterior superior rim consists of supraorbital; postorbital lies behind it. The inferior rim contains premaxilla anteriorly and jugal posteriorly (the maxilla is excluded). Opercular, preopercular, subopercular, and the extrascapulars are in place (Figs. 13.31 and 13.32).

Osteolepiforms radiated in the late Devonian and early Permian. *Eusthenopteron* shows subdivision of supraorbital into two fields: prefrontal and postfrontal. Postorbital and jugal remain the same. Although the three opercular bone fields remain, extrascapulars are lost, being replaced by supracleithrum, anocleithrum, and cleithrum. Crown sarcopterygians are represented by the Elpistostegalian *Panderichthys*. Here, for the first time, postorbital inserts itself into the orbital rim creating a ring of five bones. This innovation permitted expansion of orbital volume. In addition to the extrascapulars, the opercular bones are lost as well. The head is flattened and the snout is elongated (Figs. 13.33 and 13.34).

The discovery of *Tiktaalik* by Shubin marked a real breakthrough in the understanding of the tetrapod transition [23]. Postfrontal does not extend past the orbit; postorbital and parietal are in contact. Postfrontal forms a brow ridge. Remarkable changes in the position of the orbit were of great adaptive value as described by McIver. Large eyes predated terrestriality. Eye socket volume in digited aquatic tetrapod increased 1.42× that of finned aquatic tetrapods. This correlated with larger pupils. These factors plus the change in optics through air versus water led to a total visually surveyed volume increase of over 1 million times the aquatic volume…certainly favorable for predation and rewarding movement onto land (Figs. 13.35, 13.36, and 13.37).

We see next the tetrapod transition. To quote Prof. Jennifer Clack [24], a tetrapod is "a sarcopterygian with hands and feet." The ancestors made their debut in two forms, *Ichthyostega* and the later *Acanthostega*. We shall concentrate on the latter species, so-named (Gk "spine armor") for the prong-like "horns" projecting backward from its tabular bone. The most obvious features of *Acanthostega* can be seen in the comparative evolution of the pectoral girdle. The total number of orbital bones drops to four as lacrimal is excluded. Sclerotic bones around the eye persisted. It is also significant the jugal makes contact with quadratojugal...an association it will maintain throughout evolution. In so doing squamosal is separated permanently from maxilla (Figs. 13.38, 13.39, and 13.40).

The earliest amniote, *Paleothryis*, demonstrates a very large orbit surrounded by six bones: prefrontal, postfrontal, postorbital jugal, plus two new-comers. Lacrimal extends all the way to the nostril. This is in keeping with the need for an efficient hydration system for the cornea. Below lacrimal, *maxilla* comes into contact with the inferior orbital rim for the first time. During the Carboniferous, amniotes split into two fundamental groups based on the configuration of the

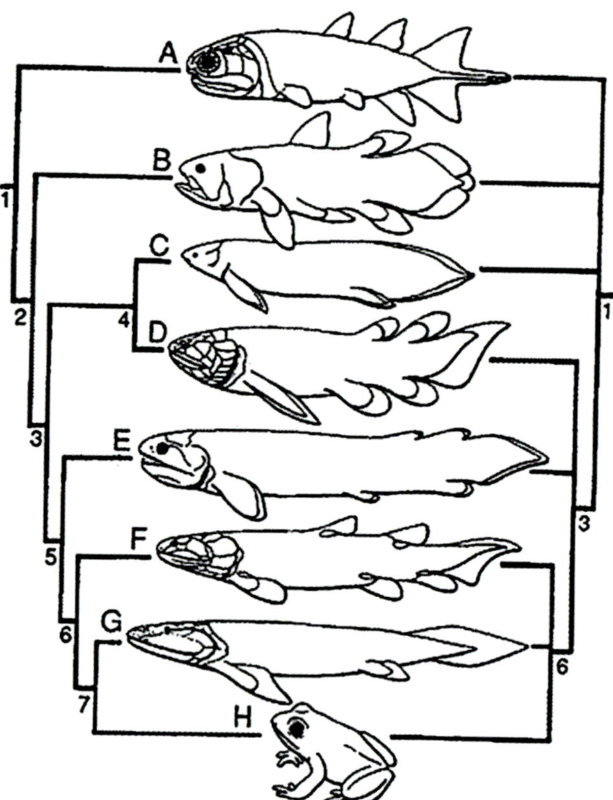

Fig. 13.31 Sarcopterygian phylogeny. (**a**) Onychodontiformes (*Strunius*) – early to late Devonian. (**b**) Actinistia – middle Devonian to Recent. (**c**) Dipnoi – late Silurian or early Devonian to Recent. These are the coelacanths. They have multiple periocular scleral bone fields. (**d**) Porolepiformes – (*Onychodus* possibly) early to late Devonian. (**e**) Rhizodontiformes – late Devonian to Carboniferous. (**f**) Osteolepiformes – (*Eusthenopteron*) middle Devonian to early Permian. (**g**) Elpistostegalia – (*Panderichthys*) ?middle to late Devonian. (**h**) Tetrapoda – late Devonian to Recent. [Reprinted from Ahlberg P, Johanson Z. Osteolepiforms and the ancestry of tetrapods. Nature 1998;395, 792–794. With permission from Springer Nature.]

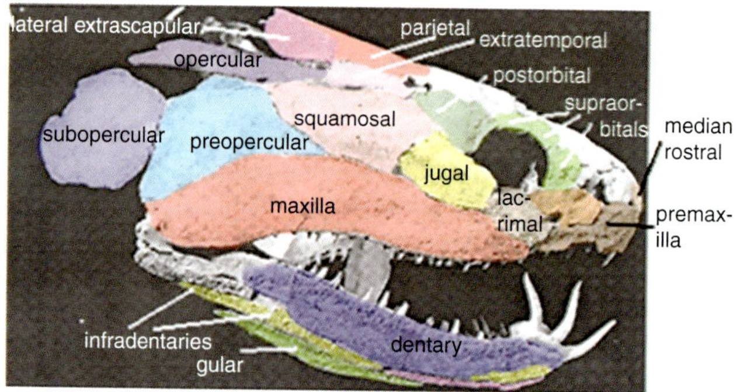

Fig. 13.32 *Onychodus*. An Onychodontiforms, *Onychodus* was a primitive sarcoptyergian. It shows identifiable bone fields around the orbit, but prefrontal and postfrontal are not yet distinct. Behind the opercular and subopercular (not fully shown) are the extrascapular series. [Reprinted with permission from Palaeos.com. Retrieved from: http://palaeos.com/vertebrates/sarcopterygii/onychodontiformes.html#Onychodontiformes]

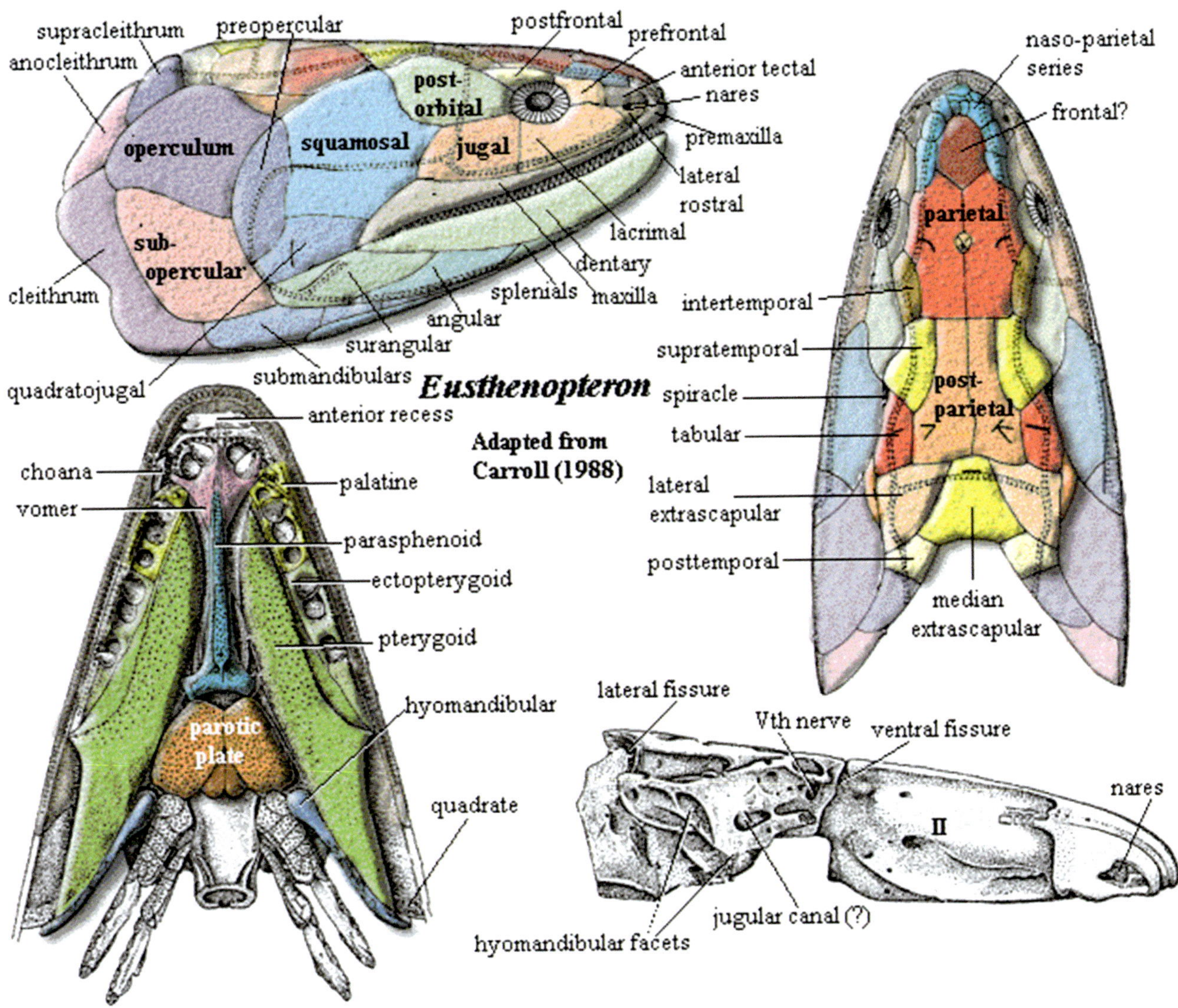

Fig. 13.33 *Eusthenopteron*. First definition of the supraorbital roof into prefrontal and postfrontal bone fields. Postorbital distinct from jugal. Dorsal view shows frontals excluded from the orbit. They are positioned in the midline and forward. Lateral line canals indicated with hatching. Palatal view shows ectopterygoid as the terminal field of the palatine series. Intracranial joint remains but now lost from the external skull. Note: Kinetic skull required support between the palate and skull base provided by epipterygoid. [Reprinted with permission from Palaeos.com. Retrieved from http://palaeos.com/vertebrates/sarcopterygii/eusthenopteron.html.]

temporal fossa: diapsida and synapsida. The former line is characterized by two temporal fossae and includes reptiles, dinosaurs, and their descendants, the birds. Synapsids (our lineage) have a single fossa and our lineage (abbreviated) consists of pelycosaurs, therapsids, cynodonts (directly ancestral to mammals), and mammals (Fig. 13.41).

Orbital changes are seen in the basal pelycosaur *Edaphosaurus*. Its diapsid skull shows invasion of the prefrontal and postfrontal bone fields by frontal leading to their expropriation and creating a thickened supraorbital ridge. Postorbital persists but its posterior process is reduced. The crown pelycosaur *Dimetrodon* has a diapsid skull in which lacrimal is retracted back away from the naris. What is glaringly obvious is that there is no bone cover for the lateral brain (Figs. 13.42 and 13.43).

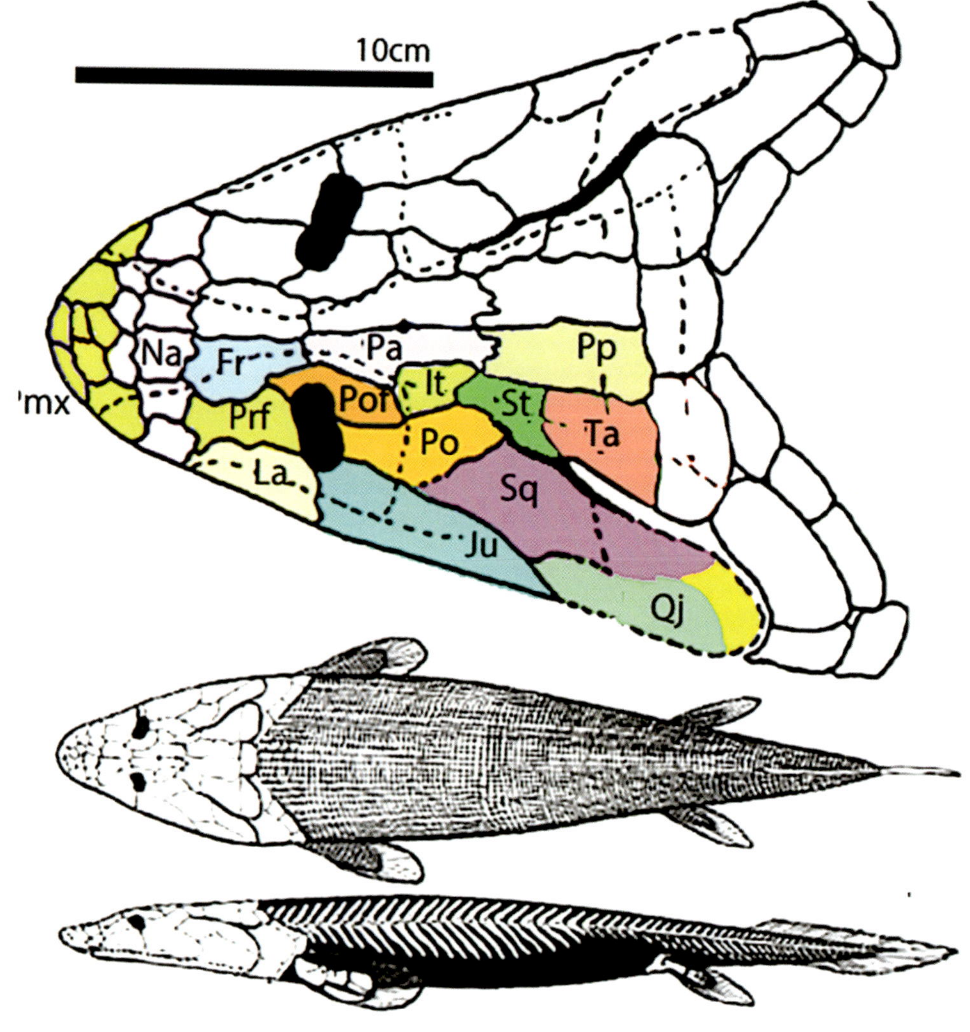

Fig. 13.34 *Panderichthys*. The orbit now has a fifth bone included in its rim, postorbital. Frontals moving backward and are now positioned just above the prefrontal (Prf). [Reprinted with permission from Palaeos.com. Retrieved from: http://palaeos.com/vertebrates/sarcopterygii/panderichthys.html#Panderichthys.]

Making the Postorbital Wall

At this point, we switch gears. All the circumorbital bones are in position. We now consider the changes of the bone fields of the palate and temporal fossa, changes that will set the stage for the final configuration of the orbit in anthropoid primates. Here is a quick synopsis of the key issues. You are encouraged to review Chap. 8 on bones for additional backup information. We will then pick up the rest of the story regarding the outer bone fields. We are now going to trace the development of the alisphenoid, a definitive. To accomplish this goal, we must consider the fate of four bones that make up the primitive hard palate: palatine, ectopterygoid, pterygoid, and epipterygoid. The original three-dimensional anatomic relationships can be appreciated in a baseline amniote, the iguana (Figs. 13.44, 13.45, 13.46, 13.47, 13.48, 13.49, and 13.50).

The terms ectopterygoid is a misnomer. Let's clear up the confusion. Ectopterygoid was originally the fourth and last of a series of palatine bone fields (Pl1–Pl4) lined up in a row just inside maxilla. All four fields were tooth-bearing. In the cynodont transition, the palatine fields Pl1–Pl3, being closely related to maxilla, produce *perpendicular plate* which extends a toe into the orbit, and *horizontal plate* which tracks along the backside of the maxillary shelf. The most posterior field Pl4 maintains its allegiance to the pterygoid bone, becoming *ectopterygoid*.

Lying internal to the palatine series were large *pterygoid* bones, also tooth-bearing. In course of cynodont evolution, the pterygoid bone fields get reduced and fuse with basisphenoid. Ectopterygoid follows along with them. These remnants are seen in marsupial *Adelobasileus* as medial pterygoid process and intermedial pterygoid process. These form, respectively, the *medial pterygoid process* and *lateral pterygoid process*, which are suspended from the junction of the sphenoid body and the alisphenoid.

Epipterygoid is also a misnomer. During evolution several dermal bones developed on either side of the palatoquadrate cartilage, forming external and internal laminae. Epipterygoid (Epi) is programmed from by the most posterior sector of the palatoquadrate bone field. It is a true PQ

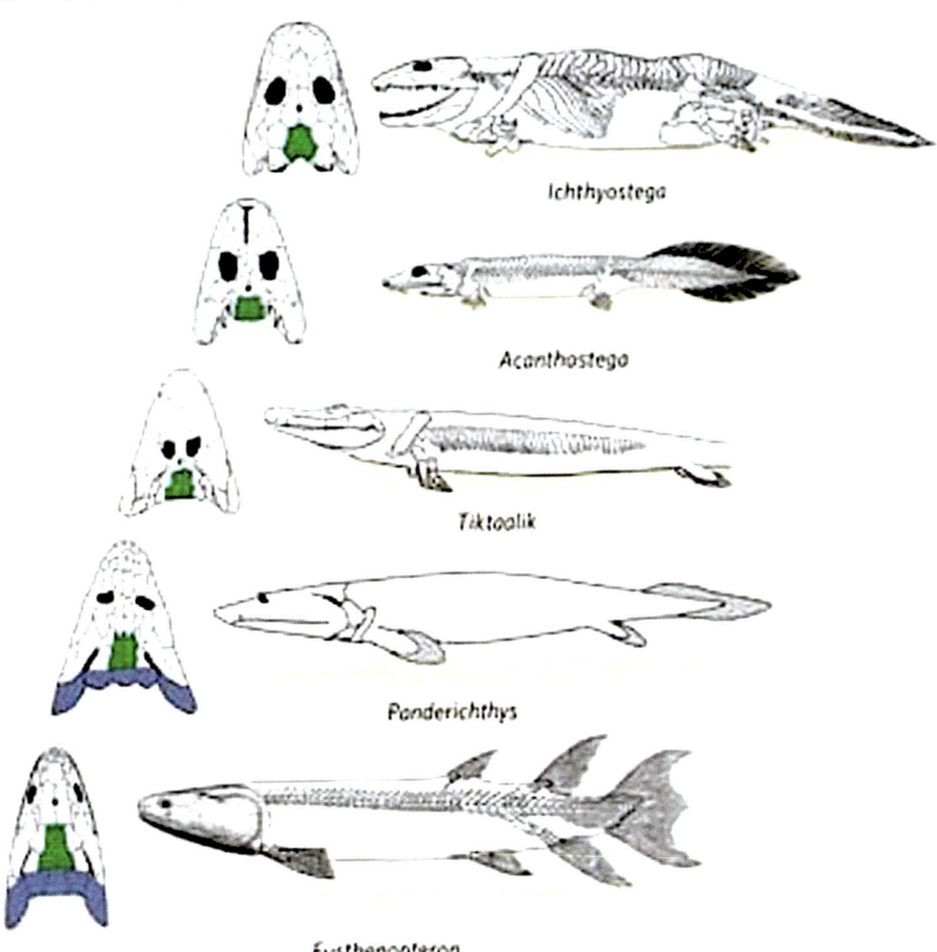

Fig. 13.35 *Tiktaalik*: transitional to tetrapods. Upper extremities developed compared to lower extremities. Radical change in the skull with dorsalized orbital position, making it ideal for hunting at the water's edge. This advantage was not retained by the early tetrapods. [Reprinted from MacIver MA, Schmitz L, Muganc U, Murphey TD, Mobley CD. Massive increase in visual range preceded the origin of terrestrial vertebrates. Proc Nat Acad Sci 114; (12): E2375-E2384. With permission from Proceedings of the National Academy of Sciences of the United States of America.]

Fig. 13.36 *Tiktaalik* orbits. Meet the first tetrapod! Note the dorsalization of the eyes. [Reprinted from Academy of Natural Sciences, Drexel University. Retrieved from: http://ansp.org/exhibits/online-exhibits/stories/meet-your-sister/.]

Fig. 13.37 Dorsalized eyes: modern-day advantages for prey-capture. The following link demonstrates the application of dorsalization for predation: https://www.youtube.com/watch?v=eFnnyWbj97U. [Reprinted from Fandom. Retrieved from http://spec-evo.wikia.com/wiki/File:Crocodile_skull_cast_mugger_crocodile.jpg. With permission from Creative Commons License 3.0: https://creativecommons.org/licenses/by-sa/3.0/.]

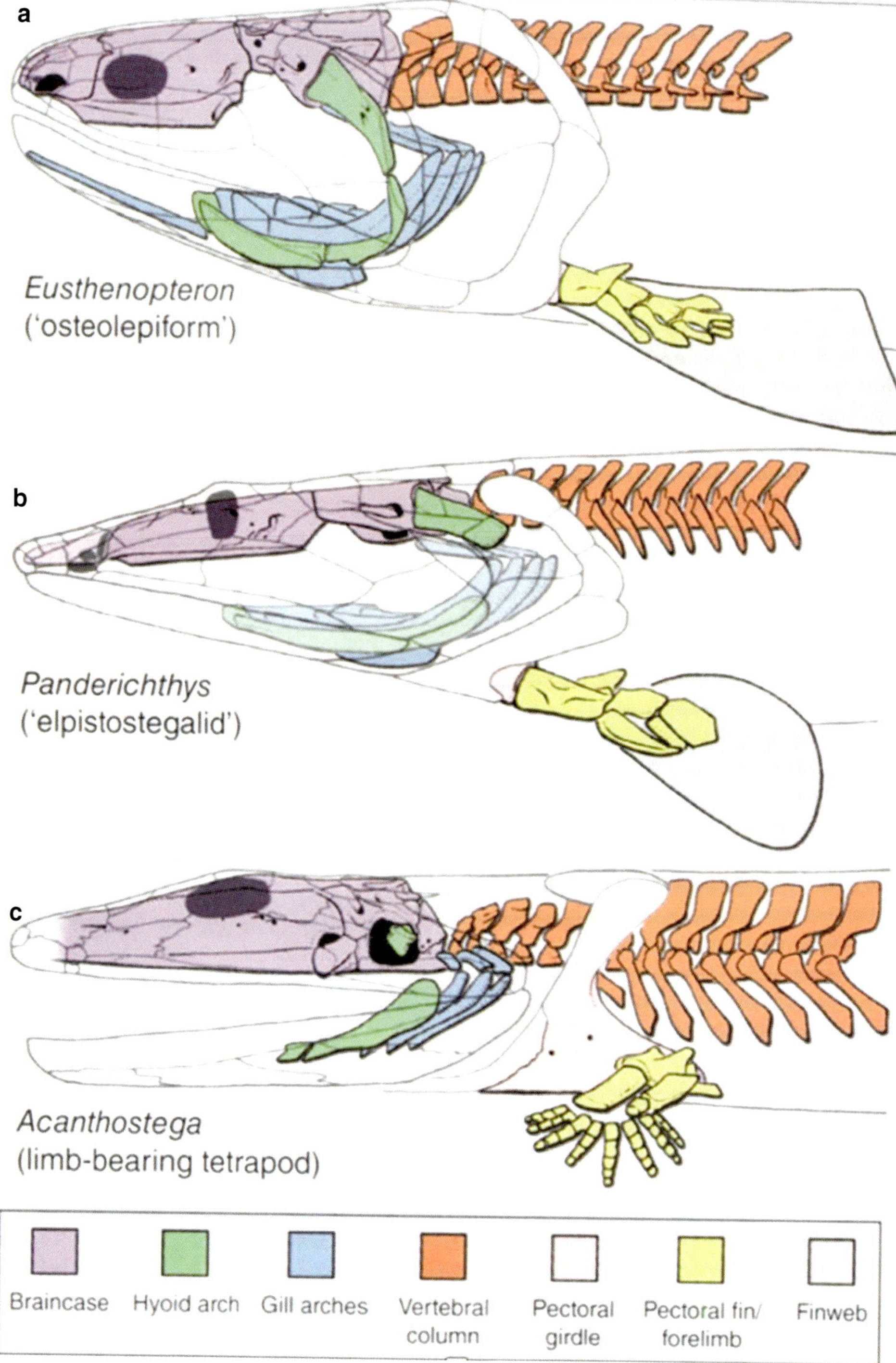

Fig. 13.38 Transition from tritichopterid (osteolepiform) fish to panderichthid (elpistostegalid) fish to tetrapod. *Eusthenopteron* loses the extrascapulars and gains a shoulder girdle which remains attached. Pandericthys and certainly Tiktaalik achieve separation when the operculars are lost. The beginnings of a pectoral girdle are present. The pectoral girdle stabilizes with *Acanthostega*, having five main identifiable bones: anocleithrum (attached to tabular), cleithrum, clavicle, interclavicle, and scapulocoracoid. [Reprinted from Benton MJ. Vertebrate Paleontology, 3rd ed. Oxford, UK: Wiley-Blackwell, 2005. With permission from John Wiley & Sons.]

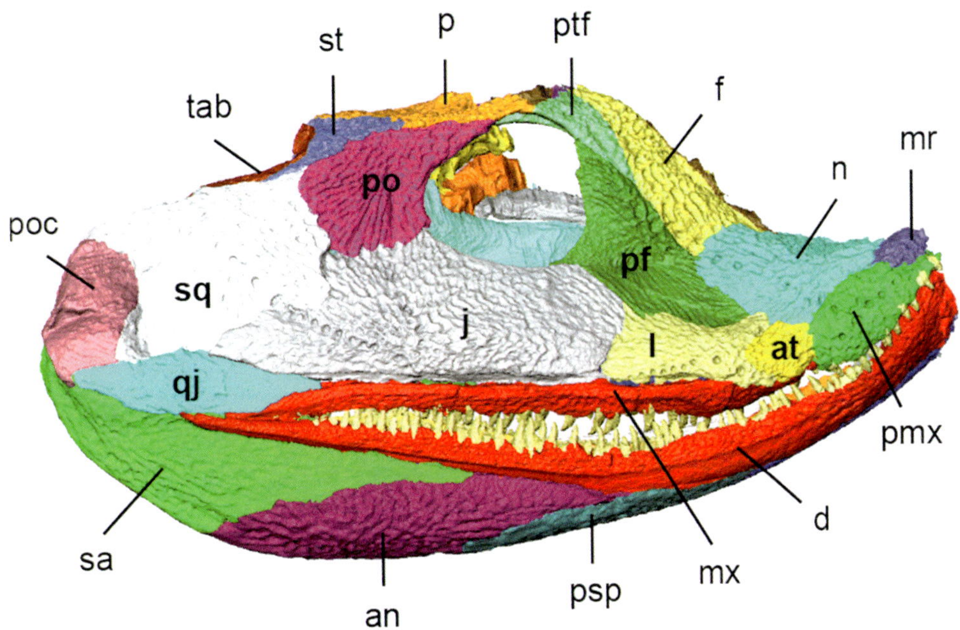

Fig. 13.39 Basal tetrapod *Acanthostega gunnari*. Note appearance for the first time of prefrontal (pf, dark green) and postfrontal (ptf, pale green). Basisphenoid (bs, orange) can be seen. Jugal (j, gray) and postorbital (po, magenta) are joined. Key: Individual bones are shown in various colors. Anatomical abbreviations: ad, adsymphysial; an, angular; ar, articular; at, anterior tectal; bo, basioccipital; bs, basisphenoid; co1, coronoid 1; d, dentary; ect, ectopterygoid; f, frontal; j, jugal; l, lacrimal; mr, median rostral; mx, maxilla; n, nasal; p, parietal; pa, prearticular; pf, prefrontal; pl, palatine; pmx, premaxilla; po, postorbital; poc, preopercular; ps, parasphenoid; psp, postsplenial; pt, pterygoid; ptf, postfrontal; ptp, postparietal; q, quadrate; qj, quadratojugal; sa, surangular; sp, splenial; sq, squamosal; st, supratemporal; tab, tabular; v, vomer. [Reprinted from Porro LB, Rayfield EJ, Clack JA. Descriptive anatomy and three-dimensional reconstruction of the skull of the early tetrapod *Acanthostega gunnari* Jarvik, 1952. *PLOS ONE* 2015;10(3): e0118882. With permission from Creative Commons License 4.0: https://creativecommons.org/licenses/by/4.0/.]

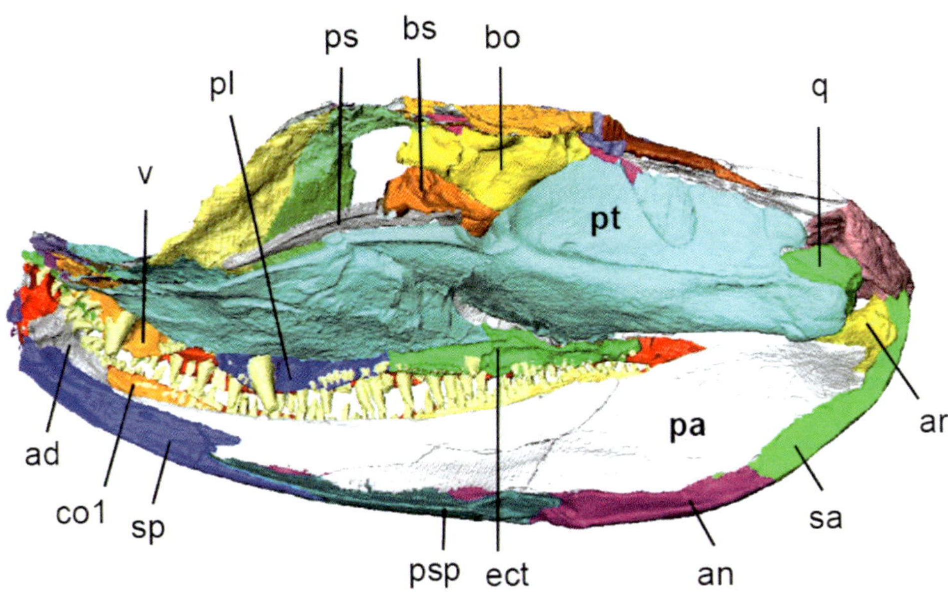

Fig. 13.40 Basal tetrapod *Acanthostega gunnari*. Interior of the skull is seen in sagittal view (nose to the left). Note: Presphenoid (ps, gray) runs in the midline from basisphenoid (bs, orange) to frontal (f, yellow). Palatine fields (pl, blue) continue backward as palatine #4, ectopterygoid (ect, bright green). Ectopterygoid (the future alisphenoid) connecting pterygoid (pt, aqua) to frontal is not seen. Key: Individual bones are shown in various colors. Anatomical abbreviations: ad, adsymphysial; an, angular; ar, articular; at, anterior tectal; bo, basioccipital; bs, basisphenoid; co1, coronoid 1; d, dentary; ect, ectopterygoid; f, frontal; j, jugal; l, lacrimal; mr, median rostral; mx, maxilla; n, nasal; p, parietal; pa, prearticular; pf, prefrontal; pl, palatine; pmx, premaxilla; po, postorbital; poc, preopercular; ps, parasphenoid; psp, postsplenial; pt, pterygoid; ptf, postfrontal; ptp, postparietal; q, quadrate; qj, quadratojugal; sa, surangular; sp, splenial; sq, squamosal; st, supratemporal; tab, tabular; v, vomer. [Reprinted from Porro LB, Rayfield EJ, Clack JA. Descriptive anatomy and three-dimensional reconstruction of the skull of the early tetrapod *Acanthostega gunnari* Jarvik, 1952. *PLOS ONE* 2015;10(3): e0118882. With permission from Creative Commons License 4.0: https://creativecommons.org/licenses/by/4.0/.]

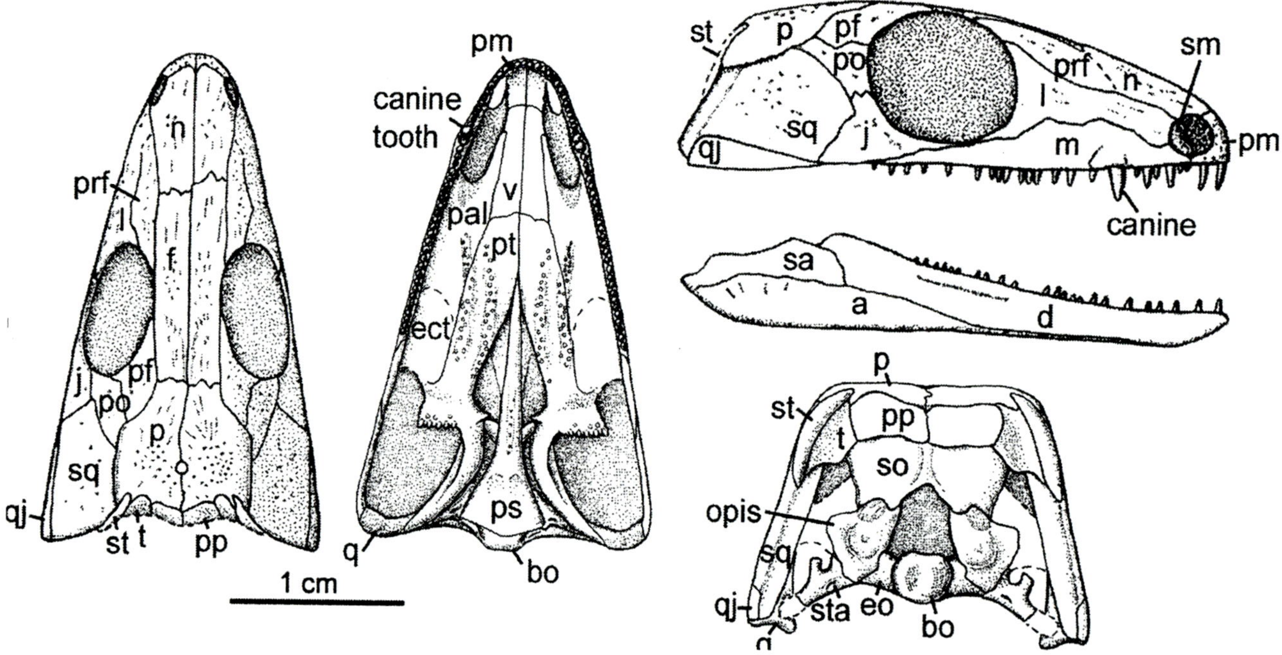

Fig. 13.41 *Paleothyris acadiana.* This basal amniote has an anapsid skull. It shows definitive changes to the external orbit. Six bones surround the orbit, with lacrimal re-included and maxilla ascending to the orbital rim. Epipterygoid is present internal to maxilla. Note that frontal remains midline. It only becomes part of the orbit by expropriating prefrontal and postfrontal. Note that canines appear for the first time with anapsids and then continue into synapsids. [Reprinted from Carroll RL. A middle Pennsylvanian captorhinomorph, and the interrelationships of primitive reptiles Journal of Paleontology 1969; 43: 151. With permission from PALAEOS.]

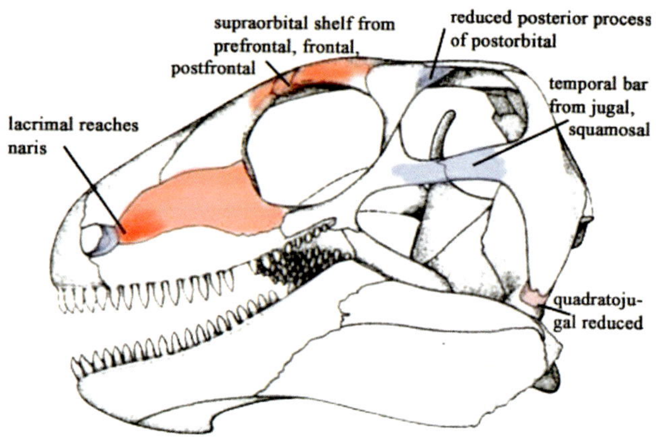

Fig. 13.42 Basal pelycosaur Edaphosaurus. Lacrimal now in the orbit but not maxilla. Supraorbital rim is now complete, as frontal absorbs the prefrontal and postfrontal bone fields. Postorbital is becoming an exclusively orbital bone as its posterior process retreats away from temporal roof. At the same time zygomatic arch is formed. Note the single temporal fossa, indicating the synapsid line. Jaw articulation is *below* the tooth row. QJ is getting reduced. [Reprinted from Palaeos.com. Retrieved from: http://palaeos.com/vertebrates/synapsida/edaphosauridae.html. With permission from the American Palaeontological Association.]

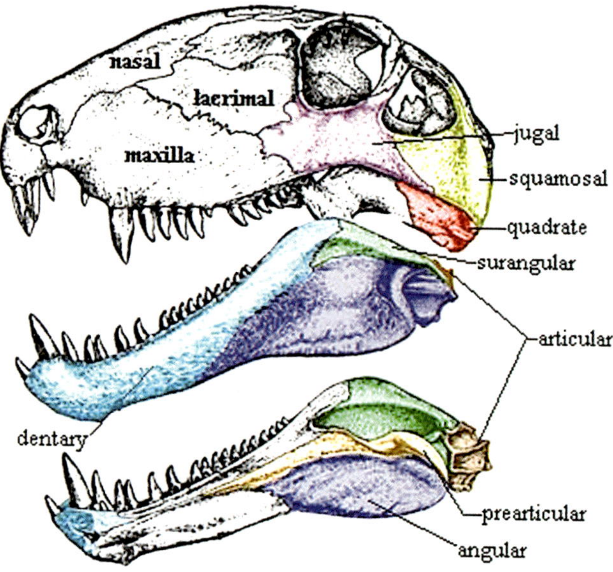

Fig. 13.43 Crown pelycosaur Dimetrodon. Changes in the mandible and jaw joint prefigure expansion of the temporal fossa with consequences for the orbit. Coronoid appears for change in muscle insertion. Inner view of angular (purple) shows its postdentary rod, the source of the tympanic bone. Articular (brown) and quadrate (red) form malleus and incus. Surangular (green) will produce ramus. [Reprinted with permission from Palaeos.com.]

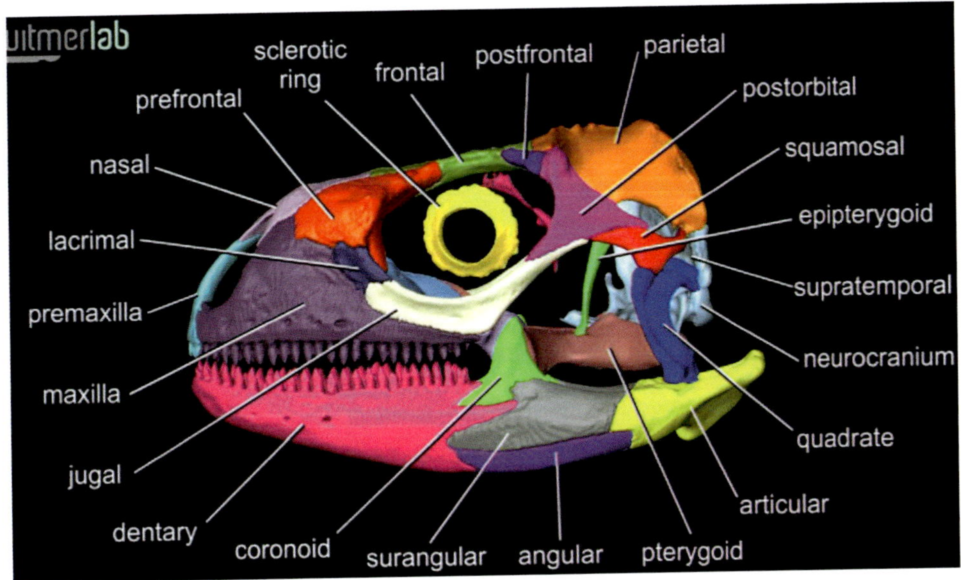

Fig. 13.44 Iguana lateral epipterygoid. The lizard, being a diapsid, retains important features basic to the amniote skull and in common with primitive synapsids. Note articulation between postorbital, postfrontal, and parietal. Epipterygoid (medium green) lies internal to postorbital and squamosal. Its footplate is on pterygoid. In synapsids, when hard palate develops the pterygoid involutes forward to become medial pterygoid plate. This frees Epi as it becomes alisphenoid. As parietal bone is expanded outward, it drags Epi with it. Alisphenoid thus becomes externalized; it is intercalated between the external plates of squamosal and postorbital. [Courtesy of Lawrence M. Witmer, PhD. Retrieved from: https://people.ohio.edu/witmerl/3D_iguana/WitmerLab_VisInt-Iguana_OUVC10667_harlequin-skull_1920x1080.mov. With permission from Ohio University.]

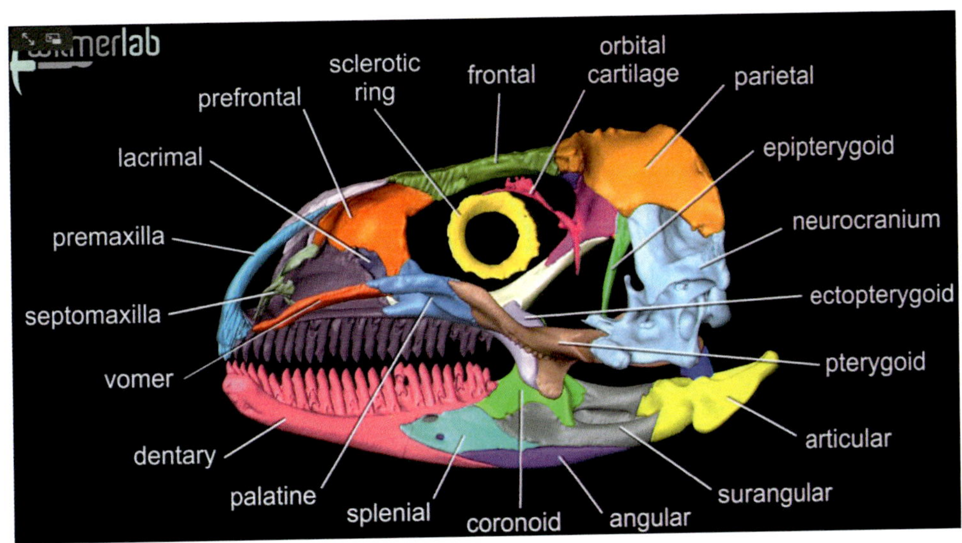

Fig. 13.45 Iguana sagittal: epipterygoid. Palatine bone has four fields: palatine proper (pl1–pl3) and ectopterygoid (p4). Palatine (medium blue) and ectopterygoid (lilac) are dermal bones synthesized internal to maxilla (violet). Note ascending process of epipterygoid (medium green) articulates with otic capsule (light blue) and with parietal (orange). Note that when pterygoid (brown) involutes forward it carries ectopterygoid (lilac) with it. Pterygoid will form the medial pterygoid plate and ectopterygoid will hang down as lateral pterygoid plate. [Courtesy of Lawrence M. Witmer, PhD. Retrieved from: https://people.ohio.edu/witmerl/3D_iguana/WitmerLab_VisInt-Iguana_OUVC10667_harlequin-skull_1920x1080.mov. With permission from Ohio University.]

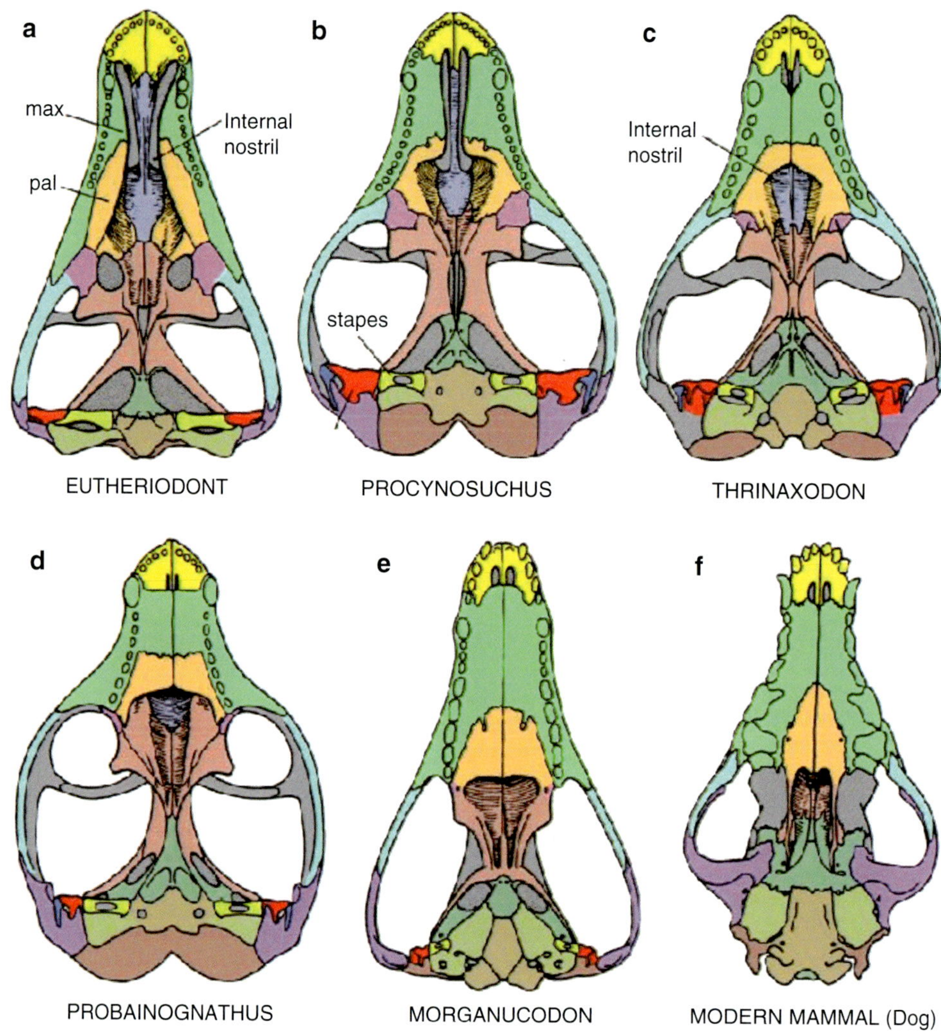

Fig. 13.46 Development of the secondary palate in cynodonts. The palatine bone fields close inward and from front to back. In primates the anteroposterior dimension reduces. Changes in the pterygoid disconnect it from epipterygoid, which remains in contact with parietal and becomes externalized. When the transition to mammals takes place, epipterygoid will become alisphenoid. Color code: premaxilla, yellow; vomer, gray; parasphenoid, lilac; maxilla, green; palatine, tan; pterygoid (medial plate), flesh; ectopterygoid (hamulus), magenta; articular (malleus), blue; quadrate (incus), red; columella (stapes), lemon; prootic/opisthotic (petrosal), purple. [Reprinted from Hopson JA. Systematics of the non-mammalian Synapsida and implications for patterns of evolution in synapsids. In: Controversial Views on the Origin of Higher Categories of Vertebrates (Ed. by H. P. Schultze & L. Trueb), Ithaca: Cornell University Press; 1991. With permission from Cornell University Press.]

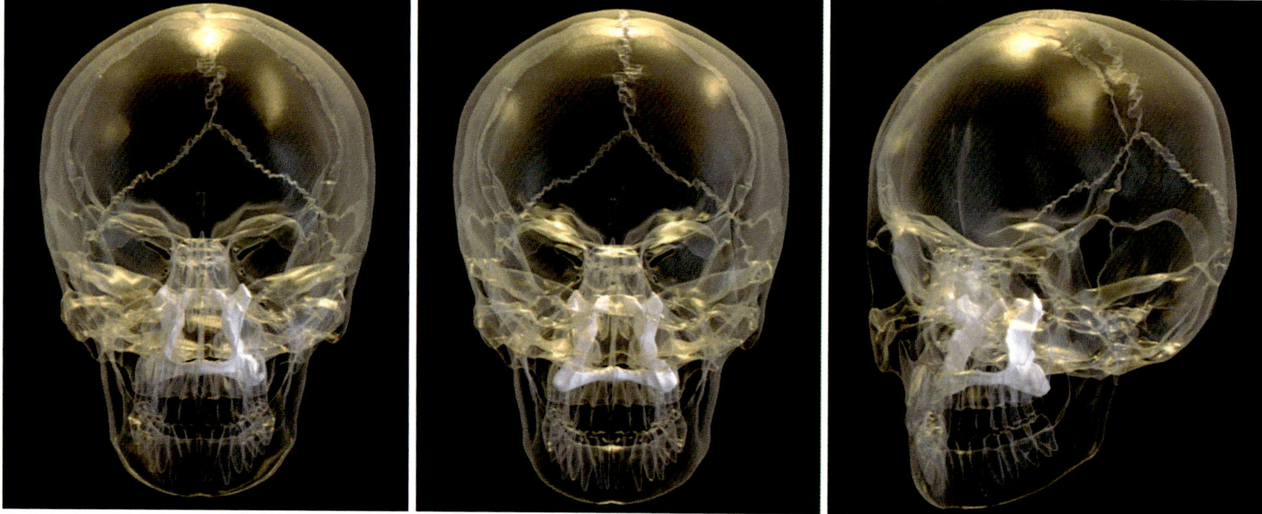

Fig. 13.47 Palatine bone fields in situ. The palatine bone fields develop in contact with and posterior to the maxilla. One lamina follows the secondary hard palate shelf to become transverse palatine shelf. The other population of palatine mesenchyme follows the backwall of the maxilla upward into the orbit and is thus interposed between it and the lateral pterygoid plate. It achieves its position in the orbit very early, at least by the time of the synapsids. [Adapted from Wikimedia. Retrieved from: https://commons.wikimedia.org/wiki/File:Palatine_bone_-_animation_02.gif#file. With permission from Creative Commons License 2.1: https://creativecommons.org/licenses/by-sa/2.1/jp/deed.en]

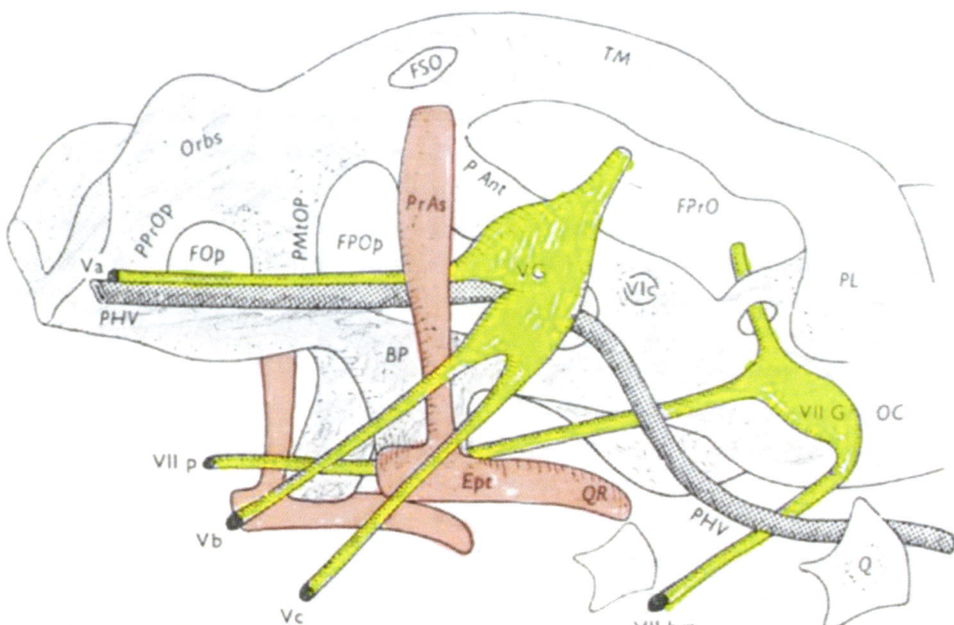

Fig. 13.48 Alisphenoid. *Processus ascendens* respects the vascular pedicle to the orbit and stays lateral to it as greater wing. Note *pila metoptica* (PMtOP) forming orbitosphenoid (lesser wing). It forms a U-shaped fusion with *pila preotica* (PPrOp) to create optic foramen. Note that pila antoptica is anterior to pila metoptica. They fuse together to create orbitosphenoid in mammals. Key: BP, basitrabecular process; Ept, epipterygoid; FOp, foramen opticus; FPOp, fenestra postoptic; FPrO, fenestra prootica; FSO, fenestra supraorbitale; OC, otic capsule; Orbs, orbitosphenoid; Pant, Pila antotica; PL, parietal lamina; PHV, primary head vein; PMtOp, pila metoptica; PPrOp, pila preoptica; PrAs, processus ascendens of epipterygoid; Q, quadrate; QR, quadrate ramus of epipterygoid; TM, taenia marginalis; VG, trigeminal ganglion; Vic, abducens canal; VIIG, facial ganglion; VII hm, facial nerve; VIIp, palatine (vidian) nerve. [Reprinted from Presley R, Steel FLD. On the homology of the alisphenoid. *J Anat* 1976; 121(3):441-459. With permission from John Wiley & Sons.]

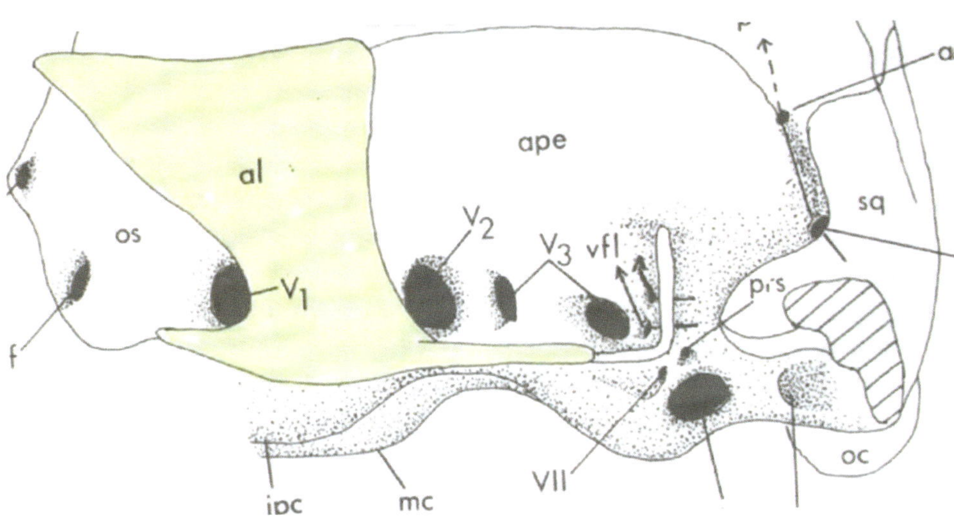

Fig. 13.49 Epipterygoid > alisphenoid in *Adelobasileus*. Note the persistence of the quadrate ramus of epipterygoid which is now disconnected from the quadrate. Note: median pterygoid crest is the remnant of pterygoid; it will form medial pterygoid plate. Intermediate pterygoid crest (ipc) is the ectopterygoid; it forms lateral pterygoid plate. Alisphenoid and lateral pterygoid plate fuse. Blood supply to AS is STV1 and StV2 while that to lateral pterygoid plate is lateral pterygoid artery from internal maxillary. Key: **ad**, foramen for ascending branch of arteria diploetica magna; **al**, alisphenoid; **ape**, anterior lamina of petrosal; **etf**, ethmoid-temporal foramen; **ipc**, intermediate pterygoid crest; **lc**, lambdoidal crest; **mc**, median pterygoid crest; **oc**, occipital condyle; **of**, optic foramen; **os**, orbitosphenoid; **p**, parietal; **prs**, preotic sinus canal; **ptc**, post-temporal canal; **sc**, sagittal crest; **sq**, squamosal; **vfl**, vascular foramen of lateral flange. [Reprinted from Lucas SG, Luo Z. *Adelobasileus* from the upper Triassic of west Texas: the oldest mammal. J Vert Paleontol 1993; 13(3):309-334. With permission from Taylor & Francis.]

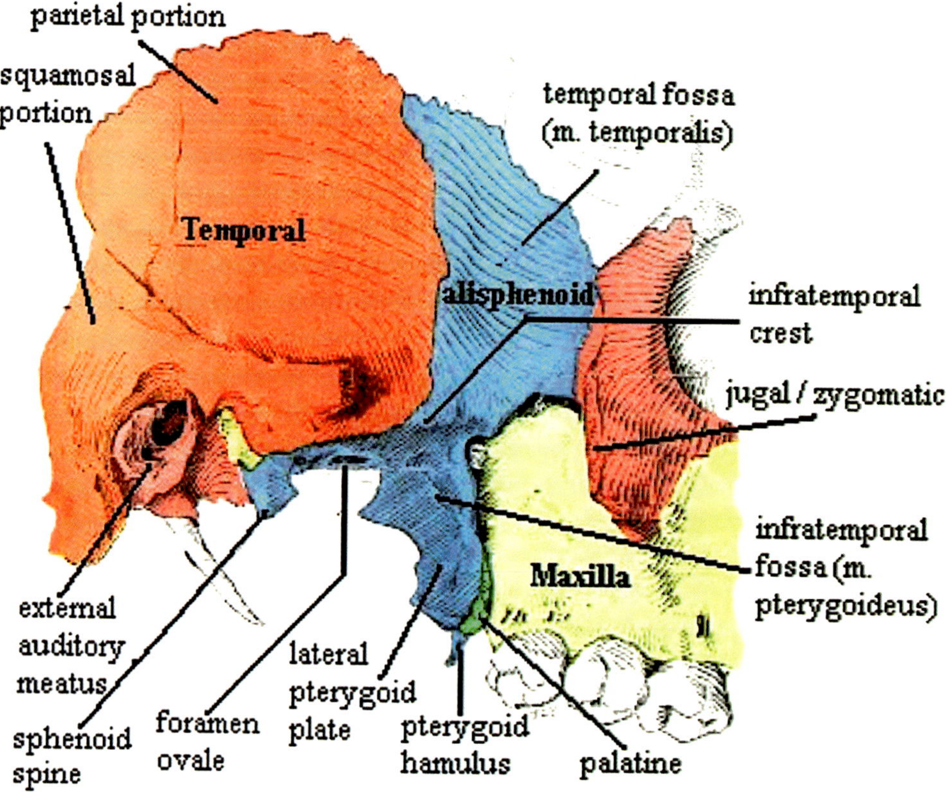

Fig. 13.50 Alisphenoid and ectopterygoid in *Homo sapiens*. *Processus ascendens* forms greater wing with an exteriorization which unites with orbitosphenoid, orbital plate, lateral zone (corresponding to postfrontal bone field), and postorbital field of zygoma. Ectopterygoid forms lateral pterygoid plate which is fused at the level of basisphenoid with medial pterygoid plate, the sole surviving remnant of the original pterygoid bone. [Reprinted from Boileau JC. Grant Atlas of Human Anatomy Philadelphia: Williams & Wilkins 1943.]

derivative. Its baseplate sits on pterygoid bone (therefore its name). It has no developmental relationship with pterygoid. Epipterygoid tracks upward as *processus ascendens* to connect with the skull base, thus suspending the palate via a joint. Via *processus posterioris*, epiterygoid connects with the otic capsule and parietal bone. As we shall see, each one of these four attachments is lost during evolution, permitting the epipterygoid in mammals, now as alisphenoid, to seek out insertions with new bone fields.

Let's follow this process in cynodonts, the final step in synapsida before mammals. The cynodonts presented many innovations including (1) closure of the secondary hard palate, (2) new muscles of mastication, (3) a new form of jaw suspension, (4) double occipital condyle, and (5) massive expansion of the brain. Synapsids were the first to entirely fill up their endocranial cavity with brain. In other words, a rapid expansion of the brain placed outward stress on the surrounding calvarium. The parietal bone pushes outward. Given the context of a developing jaw joint and the construction of a three-ossicle ear the posterior skull was a crowded place indeed. Basal cynodonts such as *Procynosuchus*, like the therapsids, used a reptilian mechanism to close their jaws. Adductor mandibularis was housed in a space between *inner surface of parietal bone* and the braincase and inserted into the coronoid process and the inside of the mandible. More room for muscle was needed (Figs. 13.51, 13.52, 13.53, and 13.54).

Intermediate cynodont *Thrinaxodon* resolved this dilemma by expanding the zygomatic arch outward. This scooped-out, convex configuration allowed for the attachment of greater muscle mass. Parietal bone grew downward and adductor attached outside to it, giving more room for expansion. The zygomatic bar, now a bowed-out zygomatic arch, presented an entirely new surface for attachment. Adductor split up, creating temporalis and a masseter. Crown cynodont *Probainognathus* went one step further to split the masseter into two opposing directions, permitting more sophisticated mastication. The resulting changes in force vectors would exert itself in the back wall of the orbit, with the need for additional insertion surface (Figs. 13.55 and 13.56).

Facial change and orbital are seen in cynodonts as well. *Procynosuchus* has full forward-facing orbits with likely stereoscopic vision. The skull shows nasal bone and lacrimal bone in contact. Lacrimal separates prefrontal from maxilla. Frontal bone remains excluded from the orbital rim. Significant changes occur in the epipterygoid bone. It now shifts forward to attach to frontal bone. Epipterygoid articulates with prootic at two sites: superior-posterior process is overlapped by prootic. Along its base, epipterygoid presents a horizontal quadrate ramus, directed backward. It also makes contact with the lateral flange of prootic. In *Thrinaxodon* midline closure of the secondary palate, albeit short, is achieved. The infraorbital rim remains jugal and the back wall of the orbit is open. Important: Frontal bone completely absorbs the prefrontal and postfrontal fields in *Probainognathus*.

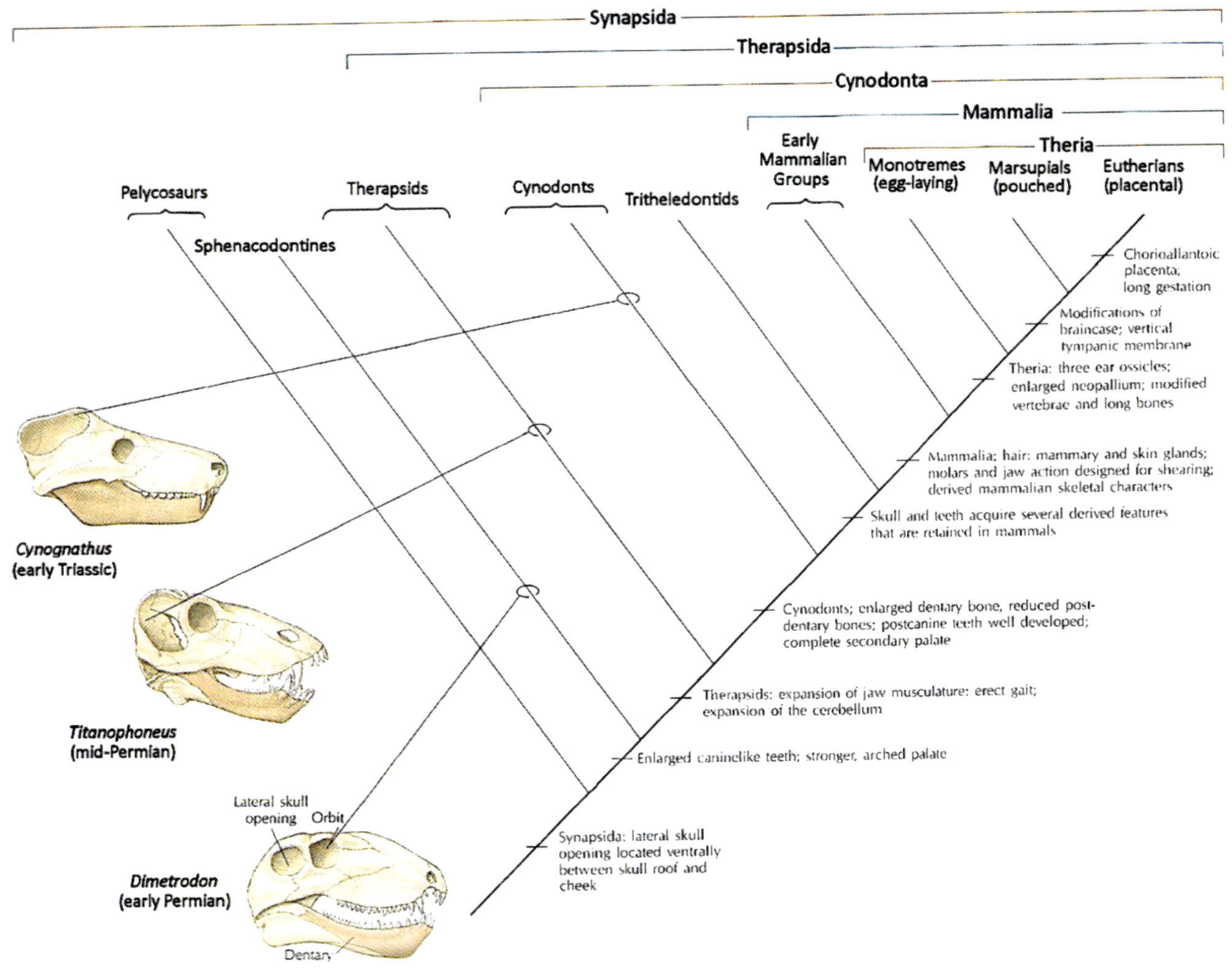

Fig. 13.51 Evolution of the synapsid line from pelycosaurs to placental mammals. [Courtesy of Prof. Stephen M. Carr, Department of Biology Memorial University of Newfoundland, St. John's NF, Canada.]

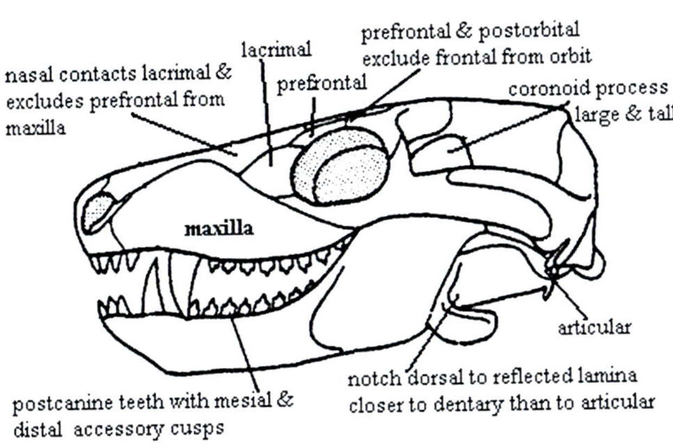

Fig. 13.52 *Procynosuchus*: characteristics of the cynodont skull. Frontal and maxilla remain excluded from the orbit. Specialization of postcanine teeth will lead to modifications for complex chewing to complement the invention of the TMJ. [Reprinted with permission from Palaeos.com. Retrieved from: http://palaeos.com/vertebrates/cynodontia/overview.html.]

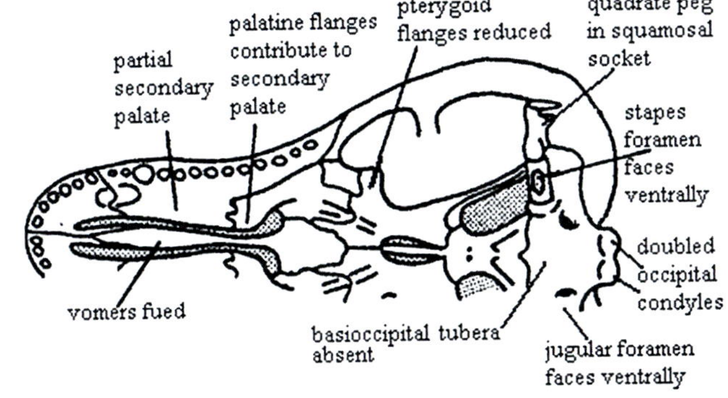

Fig. 13.53 *Procynosuchus*: characteristics of the cynodont palate. Palatal view shows horizontal shelves of palatine bone (posterior zone, Pl 3) forming horizontal plate. When pterygoid shifts forward, the flanges will hang down as the internal pterygoid plates. Quadrate is closely aligned with squamosal. In the mammalian transition this contact leads to its incorporation into the middle ear as the malleus. Note double condyles for definitive craniovertebral joint. [Reprinted with permission from Palaeos.com.]

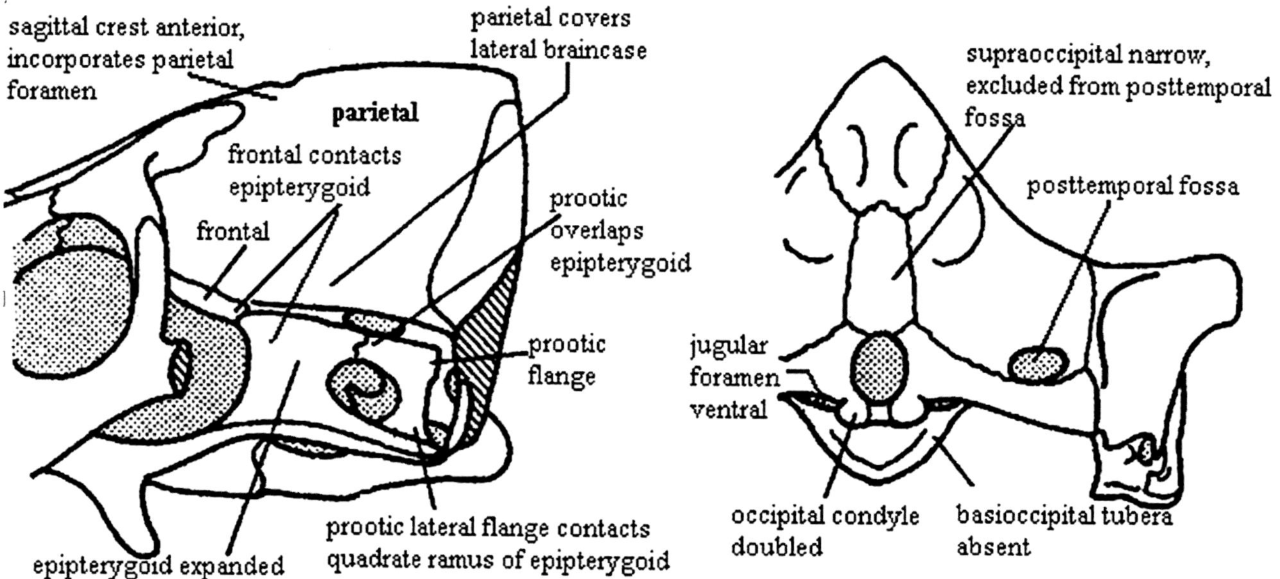

Fig. 13.54 *Procynosuchus*: characteristics of the cynodont occiput. Imagine epipterygoid sliding forward until it reaches the orbit. In so doing, it breaks with pro-otic but retains contact with the frontal. Parietal expands downward to cover the lateral braincase. It unites with epipterygoid. Posteriorly parietal unites with squamosal (not shown here). [Reprinted with permission from Palaeos.com.]

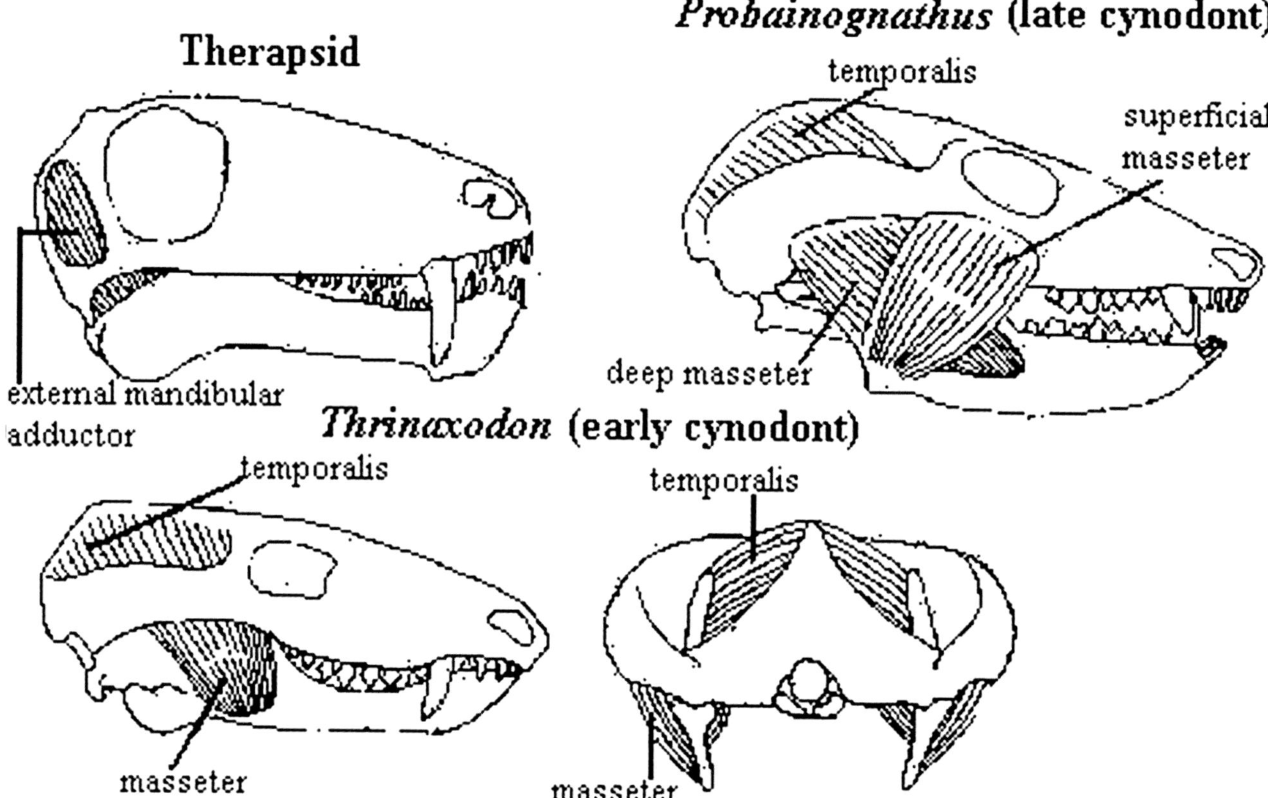

Fig. 13.55 Switch of masseteric jaw muscles. Changes in jaw musculature affect the jugal bone and place stress on the orbit. Note in *Probainognathus* the postdentary rod projecting backward at the same time as a new articulation between dentary (via the surangular) and the squamosal as the future TMJ. [Reprinted with permission from Palaeos.com. Retrieved from: http://palaeos.com/vertebrates/cynodontia/overview.html]

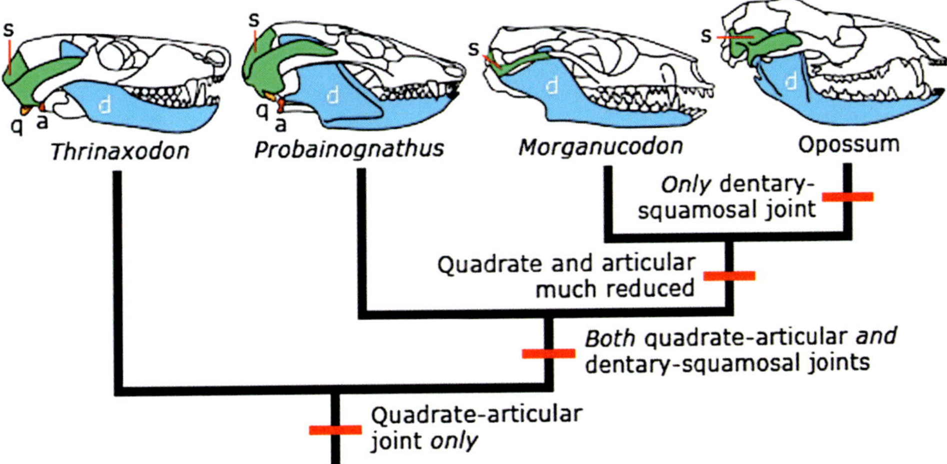

Fig. 13.56 Transition of the jaw joint. Procynosuchus and Thrinaxodon had an original reptilian quadrate-articular joint. Transition forward in mammals permitted quadrate and articular to enter the middle ear as malleus and incus. It also involved change in jaw muscles: temporalis moved forward and deep masseter split to form the pterygoids. Changes in jaw musculature affect the jugal bone and place stress on the orbit. Key: a, articular; d, dentary; q, quadrate, s, squamosal. [Copyright © 2020 by The University of California Museum of Paleontology, Berkeley, and the Regents of the University of California. Reprinted from Understanding Evolution. Jaws to ears in the ancestors of mammals. May 13 2020: https://evolution.berkeley.edu/evolibrary/article/evograms_05.]

Primate Innovations

The basal mammal, *Adelobasileus*, shows the transformation of the epipterygoid into alisphenoid. Events in the palate liberate Epi for reassignment. Closure of the hard palate eliminated the need for suspension. Furthermore, pterygoid, upon which Epi was based involved forward. Finally, the lateral wall of the skull was dynamic with changes in parietal. As a result of these factors, newly minted alisphenoid remained frontalized and was incorporated into the anterolateral skull wall. These factors remained attached to frontal bone, but posteriorly it is sutured to the newly created lateral wall of the skull, squamosal. Alisphenoid added additional external bone surface capable of bearing muscle insertion. Ahead of it a new bone field orbitosphenoid forms the medial back wall of the orbit. It is in contact with frontal bone and the small orbital surface of palatine bone. Note that in all mammalian orbits, save those of haplorhinid primates, alisphenoid is *not* incorporated into the backwall of the orbit (cf. Fig. 13.42).

One of the defining features of the primate orbit is the role of the zygoma. Here we must backtrack to review the development of zygoma with respect to the orbit and temporal fossa. This is depicted in Figs. 13.57 and 13.58.

Primates are divided into two clades. *Plesiadapiforms* are a small group of 11 families that lived in the Paleocene and Eocene. They were arboreal and squirrel-like with large eyes set sideways so they did *not* have stereoscopic vision. All remaining primates are members of clade *Euprimates* and have forward-facing orbits. They are divided into two groups. *Strepsirrhini* (G. *strepsis* "inward, curved" + *rhinos* "nose") are the lorises and lemurs of Madagascar. They have wet, comma-shaped noses, relatively smaller brains with large olfactory lobes and large eye sockets. *Haplorhini* (Gk *haplos* "simple" nose) are divided into the tarsiers and anthropoidea. They possess dry noses with intact nostrils, larger brains with reduced olfactory lobes and enlarged optic lobes (Fig. 13.59).

Let's make a few quick comparisons between these two groups of euprimates

- Strepsirrhini: lemurs and lorises
  - Small brain/body mass ratio
  - Primary sense olfaction
  - Nocturnal with *tapedum lucidum* for greater light capture but no fovea
  - Stereoscopic vision but orbits not completely frontalized
  - Postorbital bar between zygoma and frontal, no post-orbital plate
  - Alisphenoid contributes to temporal fossa
  - Upper lip relatively immobile
  - Rhinarium, upper lip, and alveolus tethered by philtrum
  - Diastema between incisors
  - Specialized tarsal joints enable complex rotation (inversion) of ankle
- Haplorhini: tarsiers and simians (anthropoidea)
  - Large brain/body mass ratio
  - Primary sense vision
  - Diurnal, no tapetum, have fovea
  - Stereoscopic vision, full frontalization of orbits
  - Postorbital bar and post orbital plate

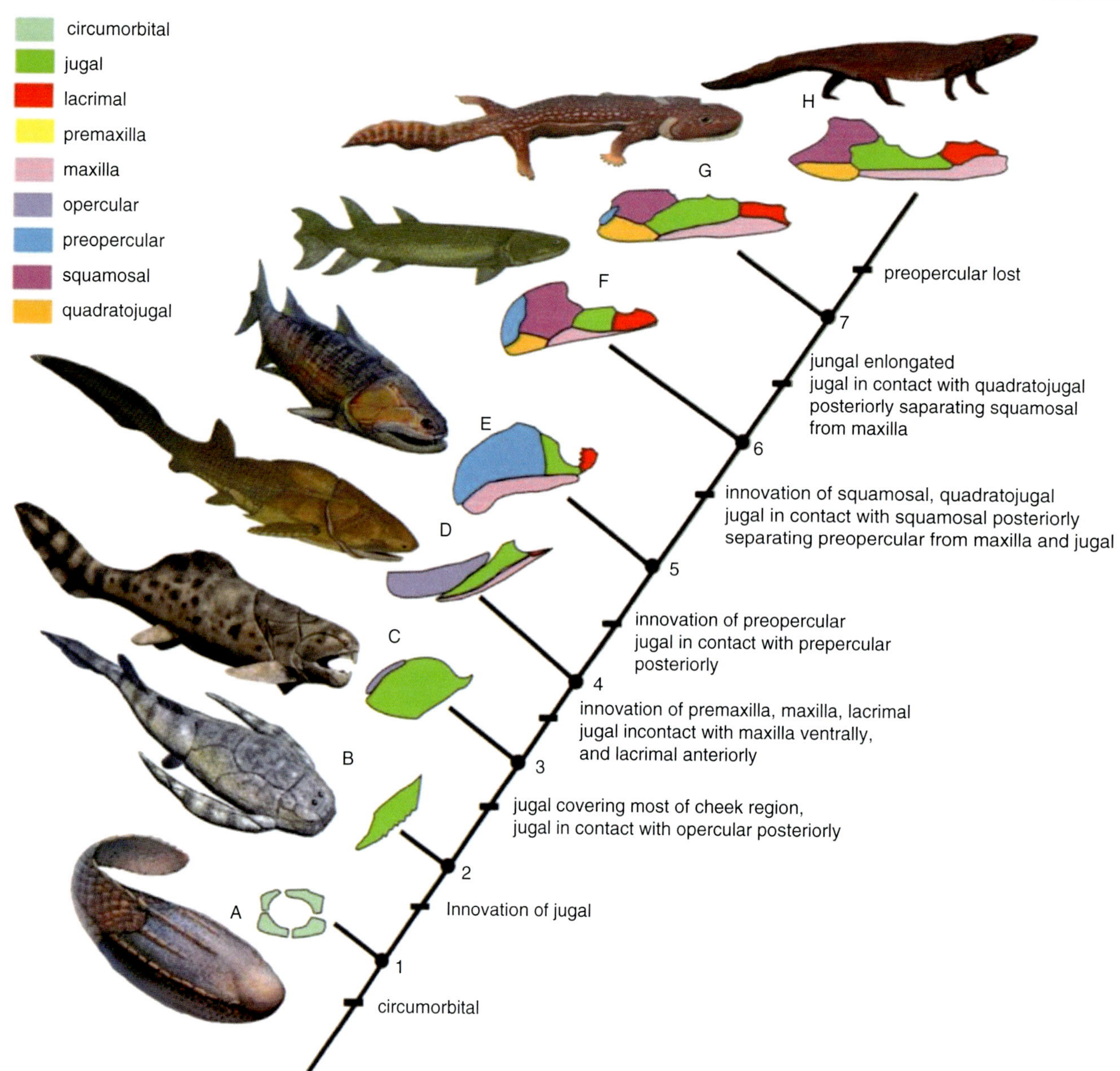

Fig. 13.57 Evolution of the jugal bone (not postorbital). Key: A. *Astraspis* (armored agnathan); B. Bothriolepis (antiarch placoderm, premaxillate); C. *Dunkleosteus* (basal placoderm); D. *Entelognathus* (crown placoderm); E. *Guiyu* (basal sarcopterygian); F. *Eusthenopteron* (crown sarcopterygian tetrapodomorph); G. Acanthostega/Ichthyostega (basal tetrapod); H. *Dendrerpeton* (temnospondyl amphibian in pre-amniote line). [Reprinted from Gal Z, Yu X, Zhu M. The evolution of the zygomatic bone from agnatha to tetrapoda. Anat Rec 2017; 300:16-29. With permission from John Wiley & Sons.]

- Alisphenoid makes up orbital plate (back wall of the orbit)
- Upper lip is mobile
- Philtrum does not extend to nose
- No diastema, philtrum is free from gum (perhaps the frenulum is a remnant)

In non-primate terrestrial mammals, direct continuity exists between the orbit and temporal fossa. There is no "back wall." In many species (dogs, rabbits) postorbital is simply absent. A defining feature of euprimates is the presence of a postorbital bar between the postorbital field of zygoma and the frontal bone. Recall that zygoma = jugal + postorbital. In strepsirrhines the orbital bar is a narrow band conforming to the lateral aspect of the globe. Zygomatic process of frontal is very prominent. The contribution of alisphenoid to the strepsirrhine orbit is minimal.

Anthropoids present another case altogether. Zygomatic process of frontal is reduced. The postorbital field makes up a much greater proportion of the lateral orbital rim reaching

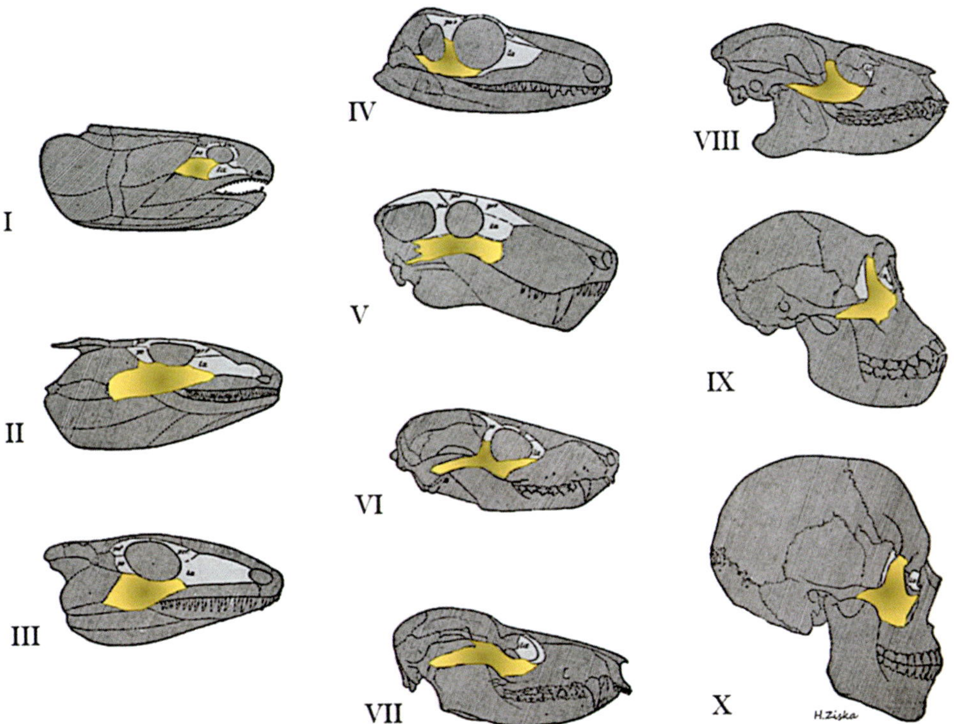

Fig. 13.58 Evolution of the zygoma. The jugal field is at all times part of the orbit. Consolidation with postorbital takes place with cynodonts. Maxilla is excluded from the rim until mammals. Key: I. Sarcopterygian, Devonian age; II. Early tetrapod, Lower Carboniferous; III. Primitive cotylosaurian, Permo-Carboniferous; IV. Primitive theromorph, Permo-Carboniferous; V. Gorgonopsian, Permian; VI. Cynodont, Triassic; VII. Marsupial, Cretaceous; VIII. Primitive primate, Eocene; IX. Anthropoid, chimpanzee; X. [Reprinted from Richtsmeier JT, Schonenebeck JJ, Schwartz T, Heuzé Y,Kawasaki K, Developmental and evolutionary significance of the zygomatic bone in *Homo sapiens*. Anat Rec 2016; 299(12):1616-1630. With permission from Creative Commons License 4.0: https://creativecommons.org/licenses/by/4.0/.]

Fig. 13.59 Strepsirrhines. Greek "curved nose" refers to downward curvature from the nose into the upper lip, the rhinarium. Strepsirrhines consist of two groups: (1) lemurs (Madagascar) and (2) lorises (India, Southeast Asia). Characteristics: "Wet nose" that is connected to the upper lip; small brain with large accessory olfactory system (pheromones), i.e., the primary sense is smell; periocular bone ring; nocturnal vision with a *tapidum lucidum* (reflective layer of the retina for enhanced light capture). The strepsirrhine uterus is bicornate. This lemur can be seen licking its fingers. Haplorhines. Gk. "simple nose," referring to absence of rhinarium. Haplorhines consist of two groups: (1) tarsiers and (2) simians, divided into platyrrhines (New World monkeys) and catarrhines (Old World monkeys and apes). Characteristics: Dry nose. The upper lip replaces the rhinarium; it is disconnected from the nose allowing for expression; large brain; primary sense is vision; tarsier uterus is bicornate and simians uterus is single chamber. Infant Lar Gibbon enjoying some grass. [Courtesy of Duke Lemur Center.]

the level of the coronal suture. The frontozygomatic suture lengthens medially to contact alisphenoid. Alisphenoid presents a laterally directed lamina in the coronal plane, the *orbital plate* separating the globe from the temporal fossa. A small flange of alisphenoid projects outward from the skull to articulate with the posterior lamina of postorbital bone. Temporalis inserts into this cavity. The body of zygoma (jugal bone field) is rotated medially and projects outward.

The root cause of these changes lies in the expansion of frontal and temporal lobes. In the middle cranial fossa, alisphenoid develops as membranous bone in contact with temporal lobe dura. On its orbital side intraorbital fat is interposed between alisphenoid and the sclera. Thus, with only one source of stem cells, the greater wing is a simple laminar bone. In functional terms, alisphenoid may serve to insulate the foveate eye from vision-disturbing contact with temporalis muscle during mastication [25].

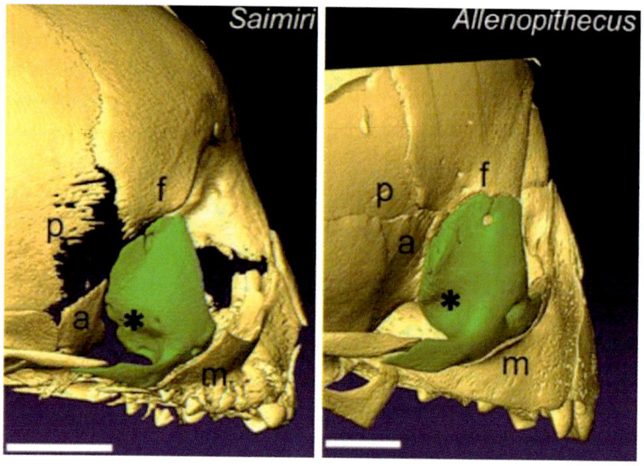

Fig. 13.60 Alisphenoid height changes with brain expansion. Relative contributions of alisphenoid to temporal fossa and posterior orbital wall in platyrrhine monkey *Saimiri* and catarrhine monkey *Allenopithecus*. [Reprinted from DeLeon VB, Smith TD, Rosenberger AL. Ontogeny of the postorbital region in tarsiers and other primates. Anat Rec 2016; 299:1631-1645. With permission from John Wiley & Sons.]

Comparison of skulls shows the expansion in the height of alisphenoid between the platyrrhine squirrel monkey *Saimiri sciureus* and the catarrhine monkey *Allenopithecus*. Articulations of alisphenoid also differ (Figs. 13.60 and 13.61).

Let's summarize the contributions of individual bone fields to the orbit

- Ethmoid – constant
- Jugal—constant
- Prefrontal—constant but incorporated by frontal
- Postfrontal—constant but incorporated by frontal
- Postorbital—inserts in *Panderichthys*
- Palatine – likely into the floor with cynodonts
- Lacrimal—incorporated with mammals
- Maxilla—excluded until primates
- Orbitosphenoid
  - Fusion of pila metoptica with pila antoptica
  - Forms lateral margin of optic canal
- Alisphenoid—transformation from epipterygoid occurs with mammals

The Tetrapod Eye: Innovations

The tetrapod eye faced several challenges compared to the fish eye. Since the eye is mostly water, fish do not have a refractive index for light coming through an aqueous

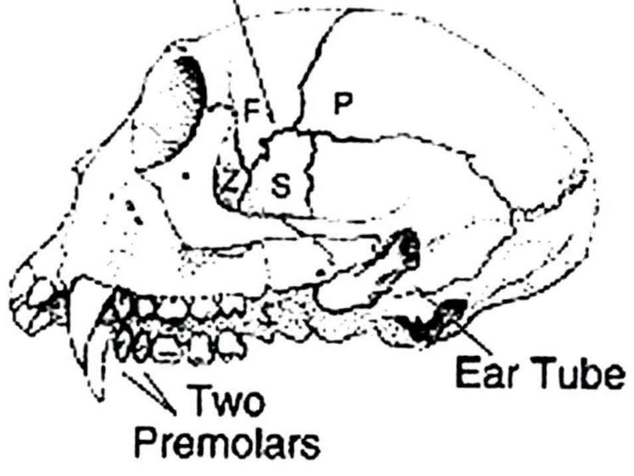

Fig. 13.61 Platyrrhine vs. catarrhine calvaria. Platyrrhines (Gk. "flat nose") are New World monkeys. Catarrhines (Gk. "hook nose") are Old World monkeys, apes, and human. Note parietal (P) excludes alisphenoid (S) from contact with frontal. Alisphenoid has contact with three bones. Back wall of orbit is almost entirely a posterior extension from zygoma (Z). In catarrhines alisphenoid is incorporated and orbital wall has greater projection. In so doing it excludes parietal. Alisphenoid has contact with four bones adding to possibilities for calvarial expansion. Note reduction of premolars. [Reprinted with permission from Palaeos.com. Retrieved from: http://palaeos.com/vertebrates/primates/anthropoidea.html]

medium. Light reaching the tetrapod eye through air and then changing to water has a significant refractive index. The fish lens is a non-compressible sphere. Fishes achieve accommodation by moving the position of the lens within the eye. Amphibians have a spherical lens as well but evolve protractor and retractor muscles.

The amniote cornea acts as an additional source of refraction. The amniote lens is compressible. Most amniotes squeeze the lens using ciliary muscles action along its equator. Because the thickened cornea of amniotes has refractive power itself, the lens is located further back in the eye, behind the iris allowing for the pupil to constrict as well.

We have made reference previously to the presence of sclerotic bones surrounding the eyeball. These bones in lobe-finned fishes and early tetrapods were flat but in amniotes such as lizards and birds they become concave to fit the surface of the globe. Muscles along the inner surface contract and compress the lens; at the same time the bony ring resists the augmented intraocular pressure that results.

Tetrapod eyes are protected against dessication by lacrimal glands with a drainage system into the nose. Eyelids and levator palpebrae superioris appear with tetrapods.

Tetrapods evolved another unique eye muscle, *retractor bulbi*. This muscle, absent in birds and higher primates, is supplied by abducens nerve and acts to prevent hydrostatic congestion of the eyeball in animals that maintain a head-down posture. VI supplies muscles in two somitomeres. Since lateral rectus has been identified in Sm5 it is likely that retractor bulbi comes from Sm4 (Figs. 13.16 and 13.17).

The Orbit and Primate Evolution

Stereoscopic Vision in a Nutshell

What, if any, are the evolutionary implications of the human orbit? Does frontalization have an effect at a deeper level? The primate orbit achieves an alignment in three axes fundamental for vision. The optic axis is the axis of symmetry through the cornea and the lens. The visual axis is the line running through the point of fixation and the fovea centralis. The orbital axis is the line of symmetry of the orbit. Changes in orbital anatomy are required to binocular fields, and the mechanism of stereoscopic vision. This topic has an extensive literature and is well reviewed [26].

Binocular Vision

First off, we start with a disclaimer: binocular vision is not the only game in town. Mammals with divergent orbits and panoramic vision have mechanisms to judge depth of field, irrespective of orbital orientation. Interposition refers to the recognition of one object placed in front of another serving as a depth cue. A man standing in front of a tree, partially obscuring it, serves as a depth cue indicating the distance of the tree. Neither eye can see the part of the tree that is obscured. Perspective lines are used in painting to indicate a third dimension. Motion parallax refers to differences in perceived velocity between objects close to the subject and those farther away. When one turns one's head, closer objects projected on the retina appear to move more quickly that those further away. Near-field depth cues can be obtained through the process of accommodation by the lens for a proximal object. Vergence eye movements required to direct the eyes to focus on a near-field object provide additional information.

Frontalization is responsible for the unparalleled degree of visual field overlap enjoyed by primates. Binocularity has three main advantages:

- Enhanced light sensitivity: The greater the degree of field overlap, the greater the probability of light capture: a 2× increase. Ambient light available on a moonless night can be 100×10^6 less available. Nocturnal primates thus have an advantage over nocturnal species with panoramic vision.
- Contrast discrimination: Differences in luminescence between objects or among parts of the same object help to identify it and differentiate it from its surroundings. Because the eyes generate two separate images for each object, the physiologic information is doubled and can be summated in the cortex.
- Contrast differences: The disparities between an object and its surroundings, between meaningful information and "noise," can translate to the perception of an edge.

Stereopsis

For our purposes, stereopsis is all about depth perception based on binocular vision. This results in an object being perceived from two different retinas, thus creating two slightly different images that are integrated in the visual cortex level to produce a single image.

It is a common misconception that stereoscopic vision is somehow a crown achievement of primates, enabling our ancestors to swing through the trees. In fact, stereopsis is found, to a greater or lesser degree, throughout the animal kingdom [27, 28]. It has even been experimentally tested in the invertebrate praying mantis. 3-D perception does not require two functional eyes (although it helps). Wiley Post, the first aviator to circumnavigate the globe, was monocular!

Stereopsis exists in different forms in nature [29] and is very ancient.

McIver has demonstrated the tremendous advantage obtained by first terrestrial vertebrates such as *Tiktaalik* when their eyes were projected cranially, with some degree of binocularity and a 1 million-fold increase in the available visual field for hunting. Thereafter, stereopsis, either for prey capture or to avoid that fate, was a strong selection factor. Stereopsis takes two forms; this will have bearing on our ensuring discussion.

- Coarse stereopsis
  - Depth-of-field regarding objects in motion in peripheral field
  - Orientation in space while moving through one's surroundings
  - Primary in infancy; may guide vergence movements necessary to develop fine stereopsis
- Fine stereopsis
  - Depth-of-field of static objects in the central visual field
  - Required for fine motor tasks
  - Requires both eyes

Discovery of Stereoscopic Depth Perception

Understanding the relationship between the eyes and the brain has challenged anatomists for centuries. Although the Galenic teaching that a visual "spirit" diffused outward to the eyes from the ventricles was debunked by Renaissance anatomists, René Descartes in *L'homme et un traitté de la formation du foetus du mesme autheur* [30] was the first to conceptualize nerve fibers from corresponding sites of the retinas converging via uncrossed pathways to form a mirror image in…the pineal gland, where they were interpreted. Although written in 1633, Traité was not published during his lifetime. Descartes was aware of what happened to Galileo and was afraid of the Inquisition. Isaac Newton in 1704 discovered the crossover of optic nerve fibers at the optic chiasm. In 1881 the visual cortex was finally localized by Hermann Munk to the occipital lobe (Fig.13.62).

Mapping the visual cortex was carried out by two main groups. David Hubel and Torsten Weisel at Harvard found in cats that a high proportion of cells of the visual cortex received input from both eyes. From 1958 their work over 25 years culminated in the Nobel Prize in medicine emphasizing the discovery that dominance columns in the cortex had a critical period for development requiring bilateral input, thereby explaining deprivation amblyopia and the requirements for binocular vision. They did not consider the possibility that visual cortex cells could also be selective for retinal disparity (processing two different images simultaneously) and thereby interpret 3-dimensional space. The neurobiology of stereopsis remained a black box until Jack Pettigrew at Queensland University, working with birds, cats, and monkeys, showed the existence of neurons capable of detecting *binocular disparity*. Moreover, he found that monocular deprivation (blocking vision from one eye) during development had profound effects on ocular dominance across species, suggesting that binocular visual experience is important for the development of higher visual centers across species [31–34].

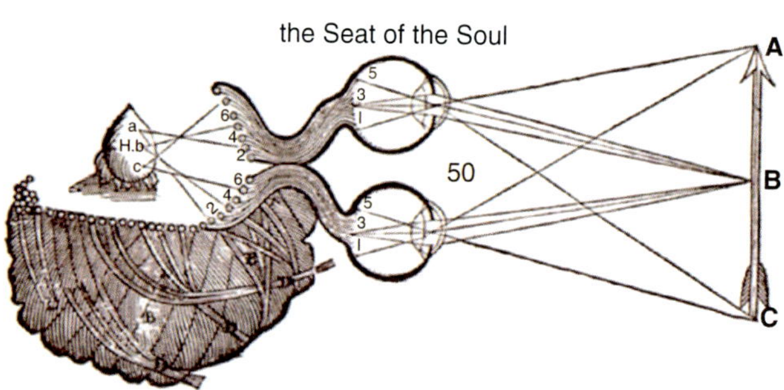

Fig. 13.62 Descartes and binocular vision. His diagram has ipsilateral projections from each eye referred to the pineal gland. [Reprinted from Descartes R. *L'homme et un traitté de la formation du foetus du mesme autheur*. Paris: C Angot; 1664.]

Mechanisms of Stereoscopic Vision

Parallax (Gk *parallaxis* difference) is the difference in the apparent position of an object viewed from two different lines of sight. Recall that the visual cortex has cells which respond to discrepancies.

Binocular disparity: Depth of field is calculated based on where an image appears on the retina relative to an ideal fixation point when the visual axis is directed straight ahead. If the object is close, the image will project temporally. If it exceeds the fixation point, it will project nasally. The disparity between the ideal and the recorded image gives an angle *alpha* from each retina that is processed in the visual cortex (Fig. 13.63).

- Object N in line with the visual axis: alpha Rt = alpha Lt
  - Discrepancy (delta) for N is near zero
- Object not in line with the visual axis: alpha Rt not = alpha Lt
  - Discrepancy (delta) for point N = X which is detected by visual cortex

Size of binocular visual field (VF): Enlarging the VF increases the potential number of points that can be seen by both eyes, thereby giving more stereoscopy (Fig. 13.64).

Da Vinci stereopsis: When an object is close to the viewer it blocks the background for one eye but the other eye remains capable of seeing the background. Partly occluded areas are monocular zones. The right eye sees the monocular zone on the left and the left eye sees the monocular zone on the right. This generates information as to whether the components we see are part of a whole or represent separate objects: *depth ordering* detects camouflage (Fig. 13.65).

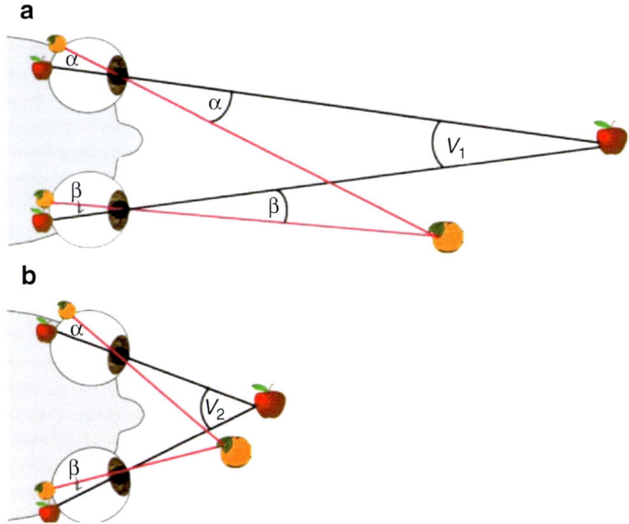

Fig. 13.63 Binocular disparity: stereopsis in mobile eyes. In both (**a**, **b**), the apple is imaged at the fovea while the orange is to the left of the fovea in both eyes, by an angle α in the left eye and β in the right. The retinal disparity is therefore the same in both cases: the absolute disparity of the apple is 0, and the absolute disparity of the orange is $\alpha - \beta$, which is also the relative disparity between the two objects. However, the different positions of the eyes (less converged in (**a**), strongly converged in (**b**) means that the locations of the objects in space are very different in the two cases. In both cases, the fact that the orange is closer can be deduced from the relative disparity, but to deduce the absolute distance to either object requires knowledge of the vergence angle ($V1$, $V2$). [Reprinted from Niyananda V, Read JCA. Stereopsis in animals: evolution, function and mechanisms. J Exp Biol 217; 220(14):2502-2512. With permission from Company of Biologists.]

Review of Visual Pathways

Now that we have localized the where and how of stereoscopic visual processing, let's get oriented to the post-retinal pathways that provide the brain with visual information. After this review we will turn our attention back to the orbit: how its position in space affects the visual fields, what relationship many obtain between the orbit and the ipsilateral retinal projection (IRP), the role of fine motor control and grasping, stereoscopic vision, and a possible sequence to explain the role of the orbit, both active and passive, in primate evolution (Fig. 13.66).

Visual Afferent Pathways

- Optic nerves contain three types of neurons
  - 90% diencephalon thalamus (p2): lateral geniculate ganglion orders somatotopic processing
  - 10% midbrain (m1): superior colliculus to coordinate eye movements, head-turning, and shifts in attention
  - Retinohypothalamic tract
    - Diencephalon pretectum (p1): pupillary reflex
    - Suprachiasmatic nucleus (hp2): biologic clock
    - Preoptic nucleus, ventrolateral: (hp2) sleep regulation
- Optic chiasm—alar plate of hp2, just above hypothalamus
  - 50% fibers ipsilateral/50% fibers contralateral

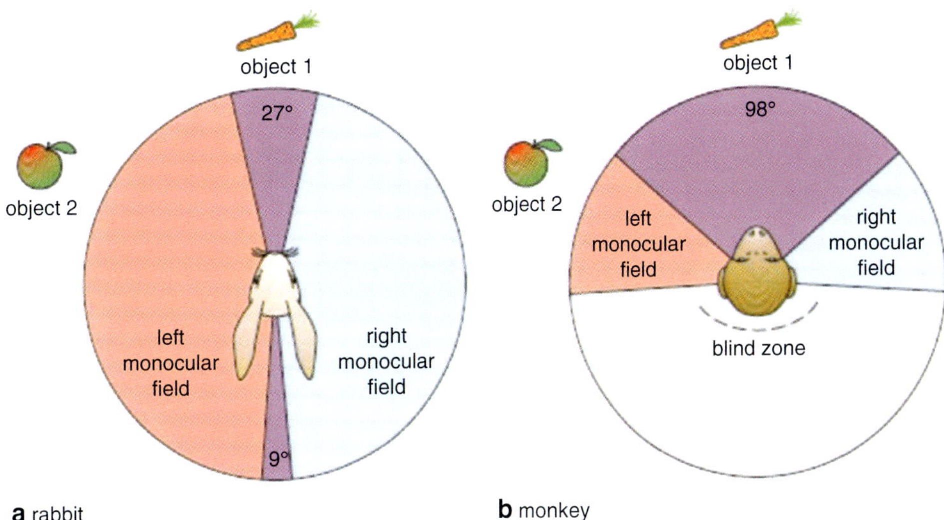

Fig. 13.64 Binocularity. Visual field overlap and binocularity: relationship to orbital position. Rodents has 2–3% ipsilateral retinal projection neurons. Primates have 45–50% ipsilateral retinal projection neurons. [Reprinted from Bradbury JW, Vehrencamp SL (eds). *Principles of Animal Communication*. Sinauer Associates, Inc; 1998. Reproduced with permission of the Licensor through PLSclear.]

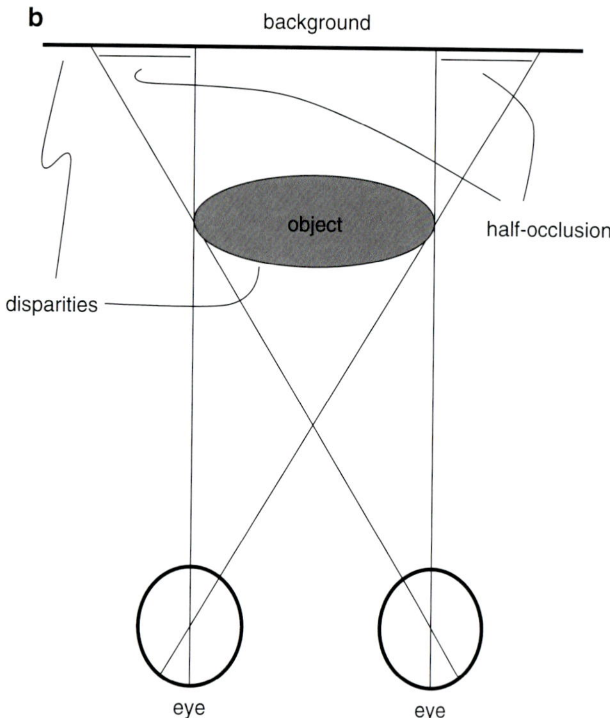

Fig. 13.65 Da Vinci stereopsis. (**a**) Insect on a leaf can be detected. (**b**) The nearer object occludes the background surface, which creates features visible to only one eye. The partially occluded regions are called here as "half-occlusions" or monocular zones. The portions of the object and background visible to both eyes contribute binocular disparity cues for stereoscopic perception, whereas the monocular regions are free of disparity cues. Stereopsis and monocular zones combine to generate the perception of object "pop out," which rapidly draws attention to a target of interest. Basic and da Vinci stereopsis facilitate object detection and object ordering. (**a**) [Reprinted from Wikimedia. Retrieved from: https://en.wikipedia.org/wiki/File:Bristol.zoo.dead.leaf.mantis.arp.jpg]. (**b**) [35] [Reprinted from Asssee A, Qian N. Solving da Vinci stereopsis with depth-edge-selective V2 cells. Vision Research 2007; 47:2585–2602. With permission from Elsevier.]

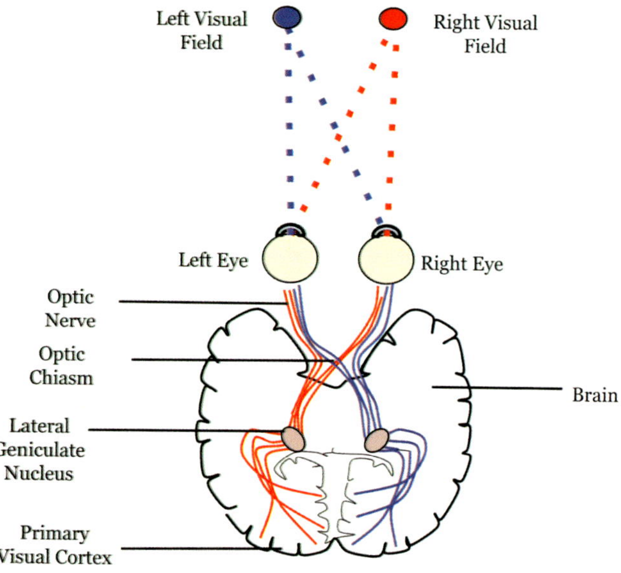

Fig. 13.66 Visual pathways. Note how primary visual cortex receives input from both the ipsilateral temporal retina and contralateral nasal retina. [Reprinted from Wikimedia. Retrieved from https://commons.wikimedia.org/wiki/File:Neural_pathway_diagram.svg. With permission from Creative Commons License 4.0: https://creativecommons.org/licenses/by-sa/4.0/deed.en]

- Optic tract: short pathway containing fibers from front temporal (lateral) retina of ipsilateral eye and fibers from nasal (medial) retina of contralateral eye
- Lateral geniculate nucleus
  - Function: determine the range and velocity of objects in the field
  - Somatotopic organization in six layers
    - Layers 1, 4, 6 contralateral
    - Layers 2, 3, 5 ipsilateral
  - Processing center: receives information from cerebral cortex and visual cortex
  - Relay center: primary visual center in calcarine sulcus of occipital lobe
- Optic radiation
  - Separates fibers dorsal-ventral
- Visual cortex
  - V1 spatial organization, edge detection
  - V2 determines depth, distinguishes foreground (breaks camouflage)
    - Relays forward to V1, V3–V5
    - Relays via pulvinar V1 to control visual attention
  - V3 determines speed and directionality of objects
    - Relays to inferior temporal lobe
  - V4 simple shape analysis

- V5 motion analysis: self vs. environment
- V6 motion analysis: motion of objects in the environment
  - Topographic map
- Other important areas:
  - Inferior temporal gyrus
    - Complex shape analysis, face recognition
    - Visual memories
  - Pretectal area (p1)
    - Anterior nucleus: blocks pain (interesting new therapies)
    - Medial nucleus: accommodation
    - Posterior nucleus: REM sleep
  - Suprachiasmatic nucleus: responds to morning light by blocking melatonin

Visual Efferent Pathways [36, 37] (Fig. 13.67)

- Dorsal stream
  - Parietal lobe
    - Map of the visual field
    - Analysis of objects: 3-D awareness
  - "Where" function: recognizes position of objects in space
  - "How" function: directs action toward those objects (how to grasp it)
- Ventral stream
  - Temporal lobe
    - Object recognition
    - Creating form
  - "What" function: what sort of object is it? Do I recognize it? How do I feel about it?
    - Temporal lobe, medial: long-term memory
    - Limbic system: emotions

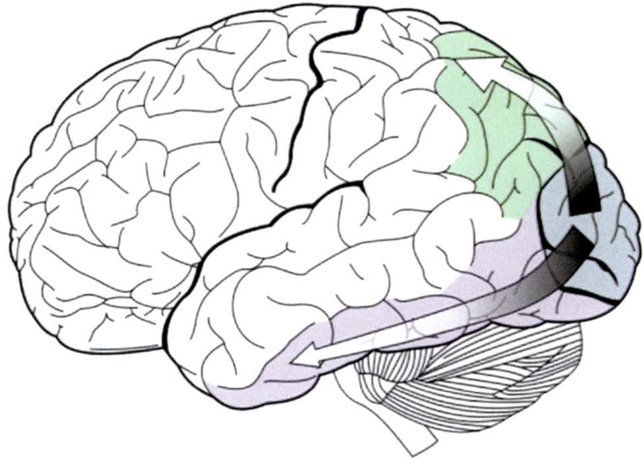

Fig. 13.67 Dual visual systems. [Reprinted from Wikimedia. Retrieved from https://commons.wikimedia.org/wiki/File:Ventral-dorsal_streams.svg. With permission from Creative Commons License 3.0: https://creativecommons.org/licenses/by-sa/3.0/deed.en.]

The Orbital Axis, Visual Fields, and the Optic Chiasm

From the time of Newton we have known that the degree of stereoscopic vision is directly proportional to the percentage of retinal ganglion cells (RGC) that remain uncrossed at the optic chiasm. This was formalized in 1942 by GF Walls as the *Law of Newton-Miller-Gudden* (NMG). This in turn depends upon the geometry of the orbit. In general, species that are subject to predation such as ungulates (hoofed animals), equids (zebras, tapirs), and lagomorphs (rabbits) have lateral-directed eyes and *panoramic vision* consisting of excellent peripheral fields for predator detection and minimal binocular overlap. The percent of ipsilateral retinal projection (IRP) fibers at the chiasm in rodents is 2%. For these animals with poor binocular vision the number of ipsilaterally projecting RGCs in the temporal retina is low and the pathway of contralateral retinal projection (CRP) fibers directed to the opposite hemisphere is considerably longer. Predators have increasing degrees of frontalization to give them a stereoscopic advantage. The degree of uncrossed fibers in cats is 30% while in humans the IRP ratio is between 45 and 50% [38–40]. This implies an increase in the number of temporal RGCs and a shortened retina-to-visual cortex distance (Fig. 13.64).

Selective pressure in primate evolution has resulted in a significant shift in the IRP/CRP ratio at the optic chiasm (OC). We know that optic nerve fibers are organized somatotopically. Note that in mice, with only 3% IRP fibers, almost all of them are directed to the ventral temporal crescent via Meyer's loop, thus indicating an origin from the dorsal temporal retina. What factors are operative at the optic chiasm to control the direction of optic nerve fibers? (Fig. 13.68).

The optic chiasm is described as developing from ventral diencephalon. This is incorrect. It is found in the secondary prosencephalon hp2 just above the alar plate of terminal hypothalamus. We also know the entire secondary prosencephalon is patterned by notochord not, as previously thought, by prechordal mesoderm. PCM produces secondary signals but is subordinate to the notochord and floorplate.

The behavior of RPG neurons when they reach the optic chiasm is a "battle of the Ephrins." The decision whether or not to cross the midline depends upon whether or not RPG neuron venturing outward from the retina expresses guidance receptor Ephrin B1 (EphB1). EphB1 has a sworn enemy, its confrère Ephrin B2 (EphB2), which is expressed in the midline and will repel those axons that are sensitive to it. Neurons expressing EphB1 have receptors for EphB2 (Figs. 13.69 and 13.70).

<u>In sum</u>, and admittedly simplifying:

- Eph1+ fibers bind Eph-B2 and cannot cross the midline.
- Eph1– fibers will track across OC to the contralateral pathway.
- EphB2+ fibers are required to cross the midline.

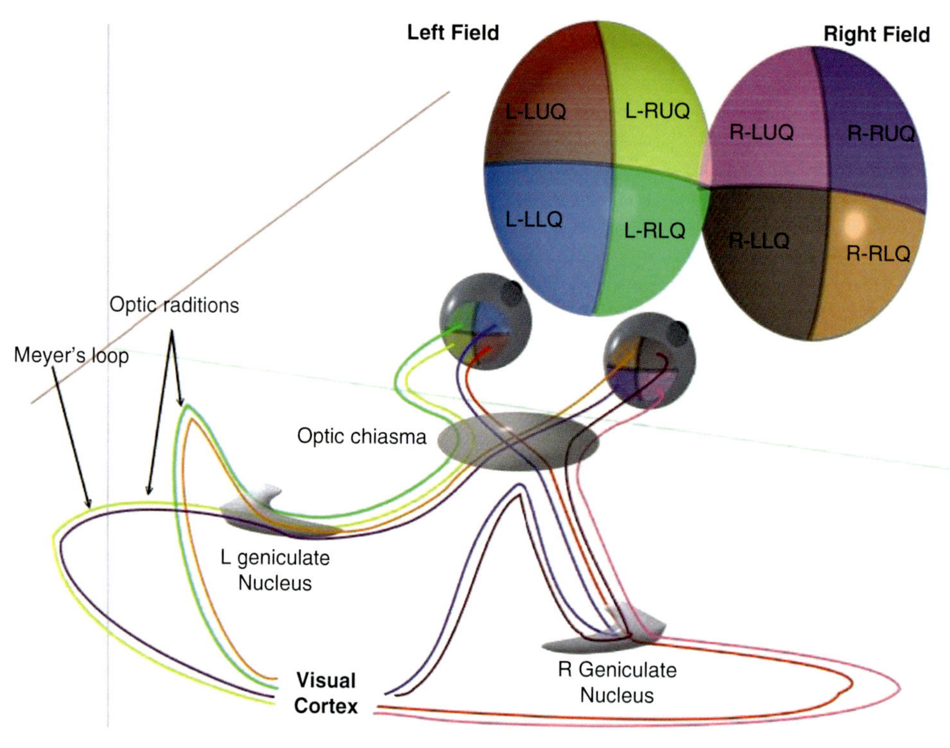

Fig. 13.68 Somatotopic relationship of retina to the visual cortex. Note the ipsilateral RGCs come from the temporal ventral (lateral-inferior) quadrant of the retina. Inferior radiations proceed to visual cortex via Meyer's loop; superior radiations proceed to visual cortex as the optic radiations. [Reprinted from Wikimedia. Retrieved from: https://commons.wikimedia.org/wiki/File:ERP_-_optic_cabling.jpg. With permission from Creative Commons License 3.0: https://creativecommons.org/licenses/by-sa/3.0/deed.en.]

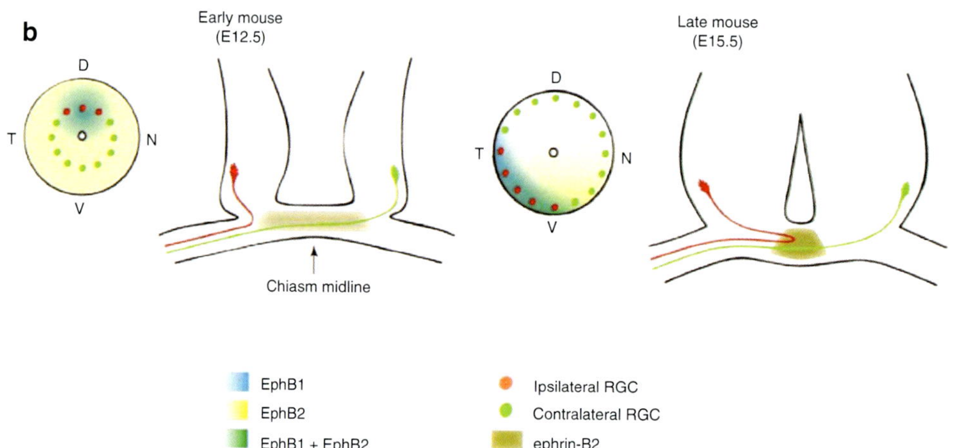

Fig. 13.69 IRP neurons localize in the VT retina. In the mouse, at E12.5 (Carnegie stage 13) the early projection from the retina includes both IRP and CRP neurons. Ephrin-B2 is expressed in the chiasm during this period. The pattern of receptors shows that EphB2 is broadly expressed initially, whereas EphB1 is found in a population of RGCs in dorsocentral retina, a region that gives rise to a small, transient population of IRP axons. At E15.5 (Carnegie stage 19), EphB2 forms a ventral gradient, whereas EphB1 becomes confined to ventrotemporal retina. VT axons express the highest levels of EphB receptors, are the most sensitive to ephrin-Bs, and are therefore repelled from the chiasm midline. [Reprinted from Williams SE, Mason CA, Herrera e. The optic chiasm as a midline choice point. Curr Opin Neurobiology 2004; 14(1):51-60. With permission from Elsevier.]

Mapping retinal neurons destined to form ipsilateral and contralateral pathways using receptors for EphB1 and EphB2 reveals a developmental sequence for mammals.

- At E12.5 (Carnegie stage 13) the early projection from the retina includes both IRP and CRP neurons. Ephrin B2 expression begins in the chiasm during this period. The receptor pattern shows that EphB2 is broadly expressed at first, whereas EphB1 is found in a population of RGCs in the dorsocentral retina, a region that gives rise to a small, transient population of IRP axons.
- At E15.5 (Carnegie stage 19), EphB2 forms a ventral gradient, whereas EphB1 becomes confined to ventrotemporal retina. VT axons express the highest levels of EphB receptors, are the most sensitive to ephrin Bs, and are therefore repelled from the chiasm midline. The nasal retina expresses EphB2+ which is required to cross the midline.

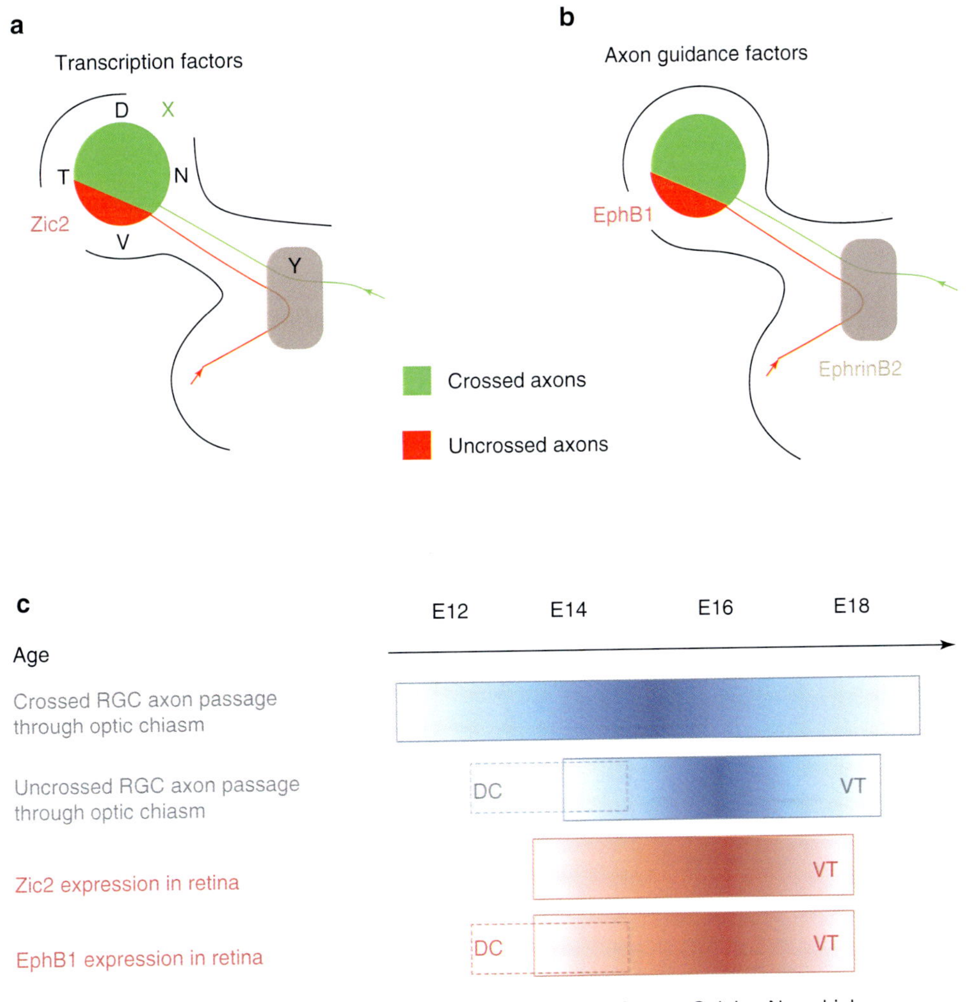

Fig. 13.70 Molecular control of uncrossed projections at the optic chiasm. Key: mouse E stage (days post coitus) versus human Carnegie stage: E12 (CS 11.5); E14 (CS 15.5); E16 (CS 19.5); E18 (CS 23+). (**a**) Expression pattern of the transcription factor Zic2 in the RGCs of VT retina (red), which give rise to the uncrossed retinal projection. Different transcription factor(s) may be expressed only in crossed RGCs (green) (X). It is also probable that a set of transcription factors are localized at the ventral diencephalon (Y) that are important for the specification of the cell/molecular cues of the chiasm midline and crucial to retinal axon divergence. (**b**) Summary diagram of the distribution of EphB1 (red) and ephrin-B2 (brown) proteins in RGCs and the chiasmatic midline. Uncrossed RGC axons from VT retina turn away from ephrin-B2-expressing midline glia near the midline, whereas crossing RGCs (green) traverse the ephrin-B2 zone. Whether the crossing axons actively overcome the inhibitory cues or use an entirely different molecular mechanism to cross the midline is not known. (**c**) The expression of the transcription factor Zic2 and guidance receptor EphB1 closely overlap in time, during the outgrowth of the permanent uncrossed projection through the optic chiasm, and space, as both Zic2 and EphB1 are expressed in postmitotic cells in VT retina. [Reprinted from Williams SE, Mason CA, Herrera e. The optic chiasm as a midline choice point. Curr Opin Neurobiology 2004; 14(1):51-60. With permission from Elsevier.]

What causes this population of EphB1 neurons to localize in the ventrotemporal retina? Zinc finger transcription factor Zic2, previously identified with neural patterning, is found in the VT retina, specifically in the ipsilateral retinal projection neurons. It is spatially co-localized with EphB1. In fact, the timing of ZIC2 expression at E13.5 (Carnegie stage 15) immediately precedes the inception of IRP migration across the optic chiasm at E14 (Carnegie stage 15.5). In sum, we have the following timeline (Table 13.5).

In Summation

The binocular visual system of primates is characterized by an unusually high number of retinal projecting neurons from the temporal retina that fail to cross to the contralateral side. This system is under control of simple genetic cues between the retina and the optic chiasm. We also observe that the overall embryonic mapping of the retina is defined by the two sequential developments of two primitive ophthalmic arterial axes: first, prDOA and later on, prVOA. So, it is no

Table 13.5 Development of the visual commissure by stage

Mouse stage	Carnegie stage	Events
12	12	EphB2 is expressed in DC retina. EphB2 begins contralateral migration from nasal retina
12.5	13	EphB1 now co-expressed with EphB2 in DC retina
13	13.5	EphB1 from DC retina begins ipsilateral migration
13.5	14.5	EphB1 and Zic2 co-expressed in the VT retina
14	15.5	EphB1 from VT retina begins ipsilateral migration
15.5	18.5	EphB2 divides into quadrants
18	23	Uncrossed IRP pathway complete
18.5	23+	Crossed CRP pathway complete

surprise that retinal expression of ephrins is dorsal-ventral. The insertion of the lateral rectus muscle gives additional evidence of timing. LR is first to appear at stage 13, is supplied in part by lacrimal artery, and occupies an insertion site into the ventrolateral sclera outside the domains of the other. Once again, the lateral (or caudal) globe is the more primitive, developing first.

Orbital Anatomy and Theories of Primate Evolution

Primate are diagnosed by 30 characteristics, the most significant of which are (1) locomotor agility in the trees, (2) large brains, (3) acute daylight vision, and (4) parental care of the young. A postorbital bar (the strut separating the orbit from the temporal fossa in related mammals) is uniformly present as is an expanded otic capsule (the petrosal) to house the auditory system. It has been suggested that primates were successful because they took advantage of a new food source of fruits and flowers, the primate-angiosperm theory [41].

Let's review the adaptive complex for early primates that characterizes the primate visual armamentarium. These are (1) large orbits; (2) expansion of neocortex in the occipital and temporal lobes—and in humans, the parietal lobe as well; (3) frontally directed eyes; (4) an area of central retina with high visual resolution; (5) binocular overlap of 90 degrees; (6) ipsilateral projection of retinal neurons to the ipsilateral geniculate nucleus and optic tectum, i.e., the eye-forelimb advantage; (7) expansion of the visual field representation in three centers—lateral geniculate nucleus of thalamus, colliculi of the midbrain, more complex cortex; (8) laminated LGN with inputs from both eyes to separate layers; (9) optic tectum with the representation of the contralateral half of the visual field; (10) expanded primary visual cortex; (11) extrastriate visual center is expanded; (12) foramen magnum changes from posterior and to ventral position for upright posture; (13) prehensile thumbs; (14) eye–hand coordination; (15) accommodative focusing; and (16) fine-grain stereopsis [42].

Arboreal Hypothesis: A Fall Can Be Fatal (Figs. 13.71 and 13.72)

Our remote ancestors were arboreal and herbivorous. Of the four clades of placental mammals, we belong to Euarchontoglires (also called supraprimates) which emerged about 95 mya and divided into two groups: (1) *Glires* (rodents and lagomorphs, i.e., rabbits) were terrestrial, (2) *Euarchonta* (treeshrews and primatomorphs) were arboreal and non-predatory. Just as the non-avian dinosaurs were dying out at the beginning of the Paleocene 65 mya "pre-monkey" primates emerged, the Plesiadiforms. *Plesiadapsis* had large eyes that faced sideways and a long snout: although it had some convergence it probably did not have binocular vision. The alisphenoid remained temporal. This is the first species to have a fingernail, albeit on the great toe alone.

The concept that primates developed their vision as an adaption for an arboreal existence, i.e., for leaping from branch to branch is misleading. Many arboreal mammals, such as squirrels, have sophisticated acrobatics without a fatality despite laterally directed orbits. Most animals use *optic flow perception* (visual information travelling across the retina) to calculate heading, velocity, and time-to-impact. Data from birds show that binocularity is not necessary to make use of optic flow information during flight.

Visual Predation Hypothesis: You Can't Eat What You Can't See

The *visual predation hypothesis* [43] postulated the visually directed predation was the key ecological specialization leading to primate development. Smell as a primary sense declined with primates, making them visually dependent, a trend that increases from strepsirrhines to haplorhines. The "follow and grab" model postulated that predation on insects was combined with grasping extremities capable of pursuing and plucking lunch from slender twigs and stems. In primates, both wide visual field overlap and orbital convergence are maximized (the bony orbits face the same direction, anthropoids being more convergent than strepsirrhines).

This model was heavily influenced by the nocturnality of early primates. Heesy found in 1300 specimens of predatory mammals that nocturnal species had the highest orbital convergence, exceeded only by diurnal primates. Frontalization proved to be advantageous adaption to low-light conditions.

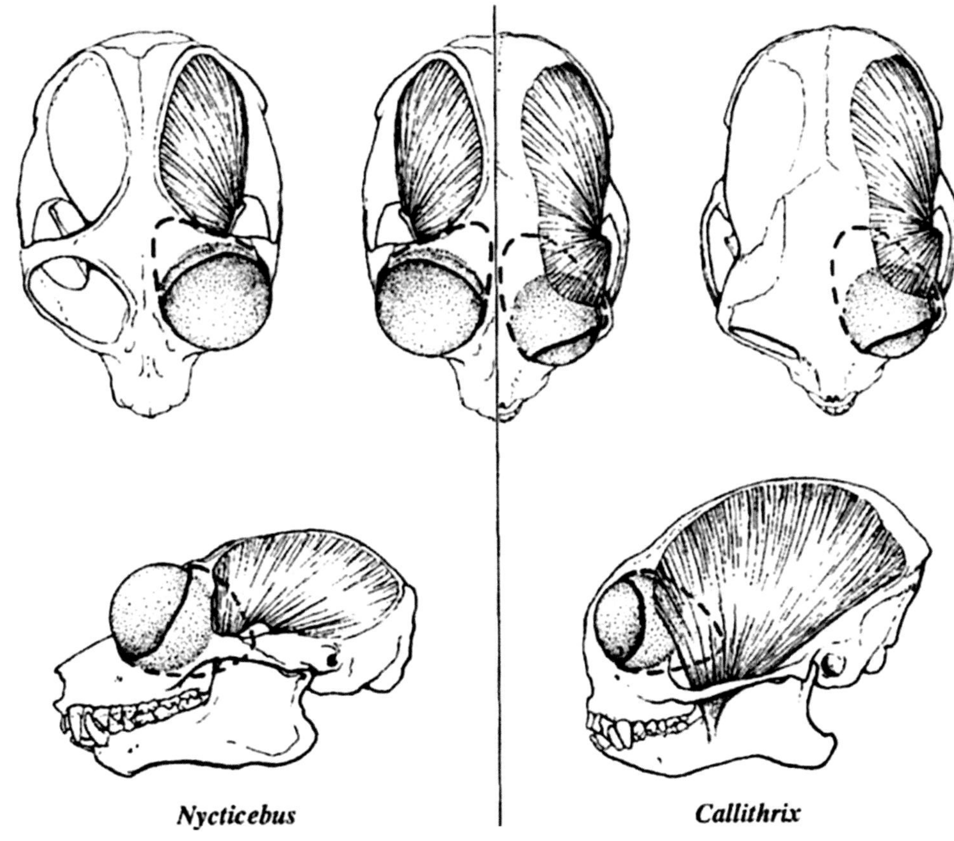

Fig. 13.71 Significance of the postorbital bar for anthropoids. Comparison of temporalis-orbital relationships in a lorid *Nycticebus* and an anthropoid *Callithrix*. The vertical eye requires a post-orbital septum to keep temporalis lateral to the eye and prevent distortion of the globe. [Reprinted from Ross C. Muscular and osseous anatomy of the primate anterior temporal fossa and the functions of the post-orbital septum. Am J Phys Anthro 1995; 98(3): 275-306. With permission from John Wiley & Sons.]

- Convergent optic axis: Binocularity increases light capture by a factor of 1.25–2. In animals with divergent optic axes, when they have to look forward at an object light enters through the most curved part of the lens. This produces *spherical aberration*…and a reduction in image quality.
- Compensation for pupil size: Nocturnal primates require large pupils. The best solution, for diurnal species is a small pupil which forces light to travel through the center of the lens, avoiding spherical aberration. Cornea sizes were evaluated in 55 primate species by Kirk [44] with significant differences documented between strepsirrhines and anthropoids. While the relative cornea sizes of the former are comparable to other nocturnal non-primate mammals, anthropoid pupils are significantly smaller. This change has to do with genes determining the construction of the iris and neural crest ciliary muscles. But this solution is problematic for nocturnal animals as low-light conditions do not permit them to reduce their pupil size. Stem primates were both nocturnal and arboreal: selection pressure favored frontalization, a simple solution permitting retention of large pupils.
- Centralization of the fovea centralis: In order to take advantage of low-light conditions, "best image" should be concentrated at the point of maximum sensitivity. Alignment of the visual axis favors a more centralized fovea.

Visual predation theory has several drawbacks. First, many ocular changes had already occurred in the plesiadapiforms, the last common euprimate ancestor prior to streptorrhini and haplorhini (tarsiers and anthropoids). Second, primatomorphs were vegetarian, with no need for "strike capture" of a moving target. Third, the evolution of the hand preceded that of the orbit. The currently extent pen-tailed treeshrew *Ptilocercus* (L. *ptilo* "feather" + Gk. Kerkos "tail") is basal to primates. It demonstrates a functional carpometacarpal joint (CMC) and an opposable thumb. Thus, grasping capabilities emerges prior to changes in the orbit. As we shall see this finding becomes very important for the evolution of the visual system (Fig. 13.73).

Nocturnal Restriction Hypothesis: Neural Adaptation to Night

Vega-Zúniga et al. [45] evaluated anatomic changes in the visual pathways of two virtually identical species of living octodontine rodents found in Chile, their only difference being ecological: *Octodon lunatus* being nocturnal and *Octodon degus* being diurnal. The species are closely related, being separated in the Eocene about 2 million years ago. *Octodon degus* had 50 degree of visual field overlap; *O. lunatus* had 100 degrees of overlap. *Octodon lunatus* showed a reduction in retinal ganglion cells of 40% but its fovea centralis was repositioned more centrally (less temporal).

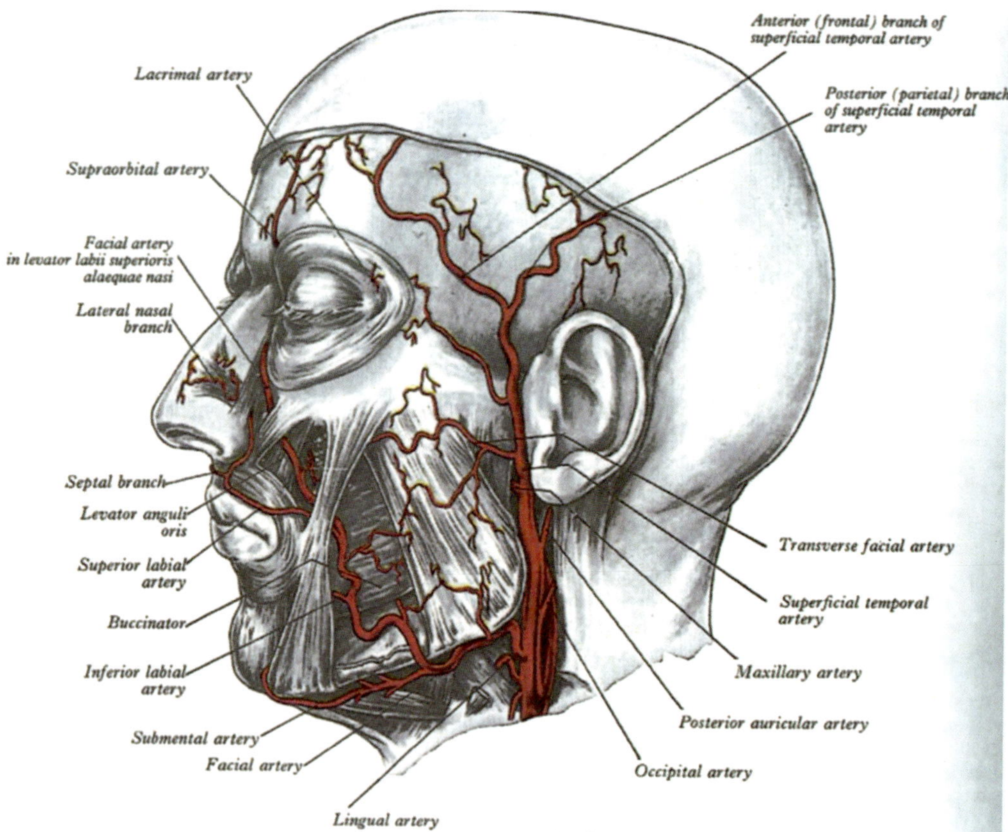

Fig. 13.72 Zone 8: the postorbital bone field. Zone 8 SIF plane involves the superficial plane of the postorbital field of zygoma. It has a dual arterial axis from StV2 zygomatico-temporal branch of zygomatic artery and its innervation is V2 and from StV3 superficial temporal artery. The latter has an immediate postorbital branch to orbicularis oculis. Cranially a larger frontal branch courses toward the lateral zone of the frontalis muscle. Note that both orbicularis and frontalis are contained within the SMAS fascia. Zone 8 overlies the postorbital temple. Immediately medial to zone 8 lies zone 9, the greater wing of sphenoid. The arterial axis of alisphenoid is StV1 meningeal branch (recurrent meningeal) which communicates embryonically with lacrimal and middle meningeal. In most cases, recurrent meningeal involutes. There are three keys to understanding zones 8 and 9 of the lateral orbit: (1) the separate nature of postorbital and jugal bones, both being r2 derivatives; (2) the origin of AS from r1; and (3) the lacrimal gland. [Reprinted from Williams P (ed). Gray's Anatomy, 38th ed. New York, NY: Churchill Livingstone; 1995. With permission from Elsevier.]

Fig. 13.73 *Ptilocercus*, the pen-tailed treeshrew. The short fingers of *Ptilocercus* demonstrate the ability to spread so widely that digits 1 and 5 oppose one another by 180 degrees, a character unheard of in mammals other than bats. [Reprinted from Le Gros-Clark WE. On the Anatomy of the Pen-tailed Tree-Shrew (*Ptilocercus lowii.*) Proceedings of the Zoological Society of London 1926; 96: 1179-1309. With permission from John Wiley & Sons.]

Volumes of lateral geniculate nucleus (LGN) and superior colliculi (SC) were 15% in *O. degus*. The most striking finding was the ipsilateral projections in *O. lunatus* to LGN and SC: 500% larger, thus indicating a much more developed IRP through the optic chiasm.

The authors interpreted their findings as showing that nocturnality was selective for neuroanatomic expansion of the binocular visual field pathways; they proposed a more comprehensive model. The *nocturnal restriction hypothesis* holds that low-light conditions favored increased scotopic photoreceptors (rods) and decreased numbers of more efficient ganglion cells combined with a large iris and cornea. Recall that field size relates to the ratio of corneal diameter to axial eye diameter…and if the ocular axis is lateral, a larger cornea will throw the temporal retina out of focus [31]. In mammals the fovea centralis always targets the center of the binocular field. When the orbits are more convergent the fovea centralis is more centralized. Thus, the temporal retina enlarges, as does its ipsilateral retinal projection fibers to the optic chiasm.

Eye–Forelimb Hypothesis: Fine-Tuning Fine Motor Control

This hypothesis builds on the observation that increasing degrees of frontalization correlate with an increase in the percentage of retinal neuron remaining ipsilateral at the optic chiasm [38–40]. E-F theory postulates that *increased ipsilateral retinal projection is functionally important for animals that use forelimbs for complex movements*. IRP neurons from the temporal retina receive images from the central nasal zone, i.e., the zone of maximum stereoscopic acuity and convey that information via short pathways to the somatosensory and motor zones involved in the control of the contralateral limb. For example, the left temporal retina perceives movements of the right hand (which is controlled from the left hemisphere) and communicates with the ipsilateral cortex responsible for control of that hand. E-F model considers visual feedback from the upper limbs to be highly adaptive for the creation of neural circuits capable of directing precise and complex motor activities (Figs. 13.74, 13.75, and 13.76).

Central to the E-F model is a relationship between the CNS and visual pathways. As previously stated, retinal ganglia cells follow a series of guidance molecules expressed at each anatomic level: optic disc, optic stalk, optic commissure, and optic tract. These include ephrins in the retina, Pax-2 in the stalk, and at the commissure the EphB1 repellent, Eph-B2, acting in concert with the growth cone attractor, Netrin-1, and Shh acting as a block to axon growth. Contralateral EphB2+ neurons home to the ventromedial hypothalamus (HTvm). In a feedback loop, glial cells in the HTvm produce ephrinB2 and act backward at OC to repel any neurons bearing the ephrinB1 receptor (Fig. 13.77).

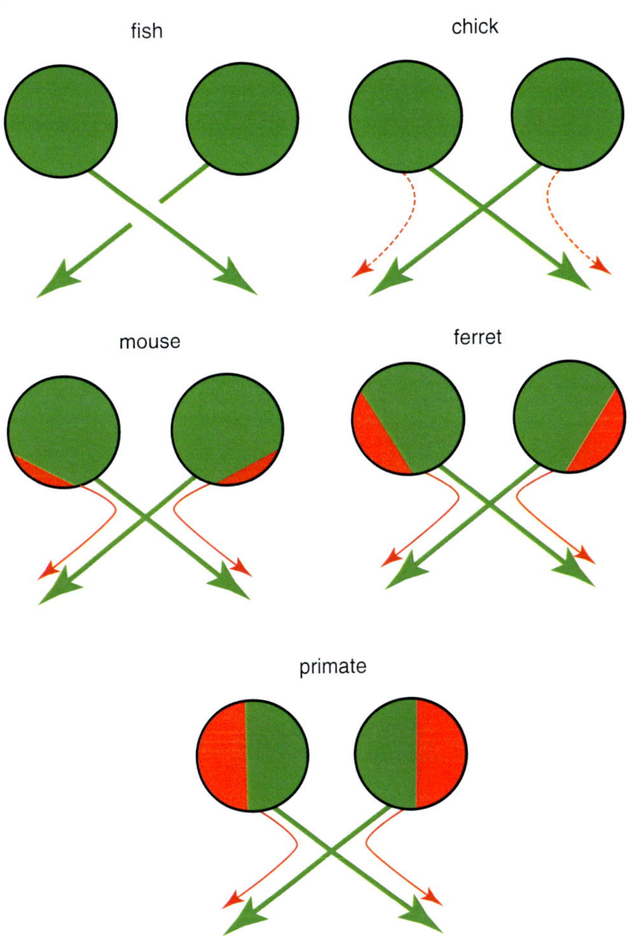

Fig. 13.74 In fish, except cyclostomes, all retinal ganglial cells project contralaterally and retinal axons do not intermingle at the chiasm. In birds, retinal axons project contralaterally; however, there is a minor transient ipsilateral projection (red dashed lines) that disappears at hatching. Binocular vision is partially developed in mouse and ferret with most retinal axons crossing the midline (green), but they also possess ipsilateral retinal projections (red). Binocular vision is fully developed in mammals with frontally located eyes. Line thickness indicates the proportions of fibers that project to ipsilateral or contralateral targets. [Reprinted from Herrera E, Mason CA. The evolution of crossed and uncrossed retinal pathways in mammals. In: Krubitzer I, Kaas J. *Evolution of the Nervous System*. Oxford, UK: Elsevier; 2007, pp. 307-317. With permission from Elsevier.]

There is good reason to expect that high percentage of IRP in primates will lead to more precise motor control. Pettigrew's demonstration of "binocular neurons" in primates helps to explain the ability of primates with frontalized orbits to identify food under difficult conditions, i.e., camouflage. When slightly different images of an object (a moth on a leaf) are perceived, the resolution of these disparity cues generates information about edges and object solidity vis à vis the background. Furthermore, cells of the visual cortex receive information from both sides and relay it forward to the parietal cortex via the dorsal stream. Thus, if the right hand is required to reach out to pluck a berry from a

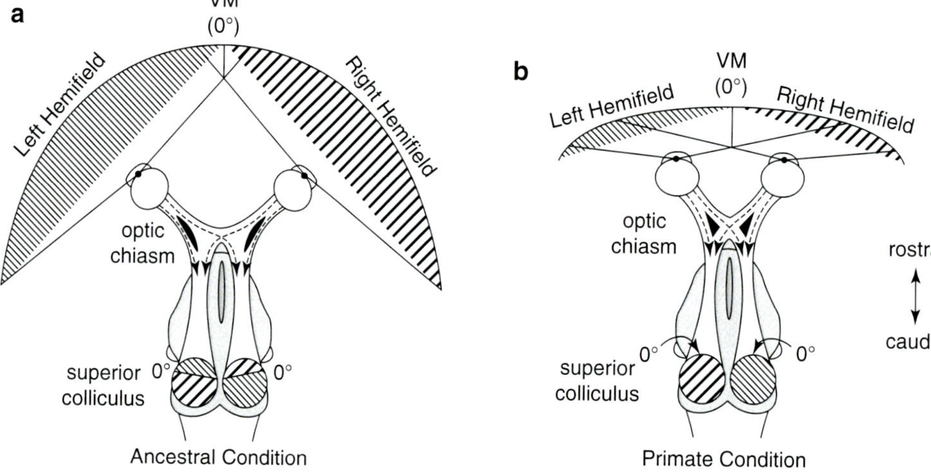

Fig. 13.75 Eye–forelimb model. The ancestral condition in cyclostomes has IRP. Most fishes (except primitive sturgeons) have no IRP. In squamates (reptiles) that lose their limbs secondarily, the percentage of IRP remains higher than for limbed animals. This reflects the need for them to control both sides of the body. [Reprinted from Larsson ML. Binocular vision and ipsilateral retinal projections in relation to the eye and forelimb coordination. Bran Behav Evol 2011; 77(4):219-230. Copyright © 2011, © 2011 S. Karger AG, Basel.]

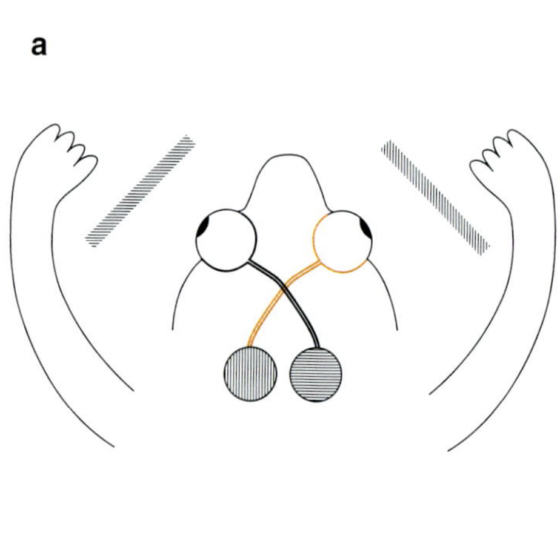

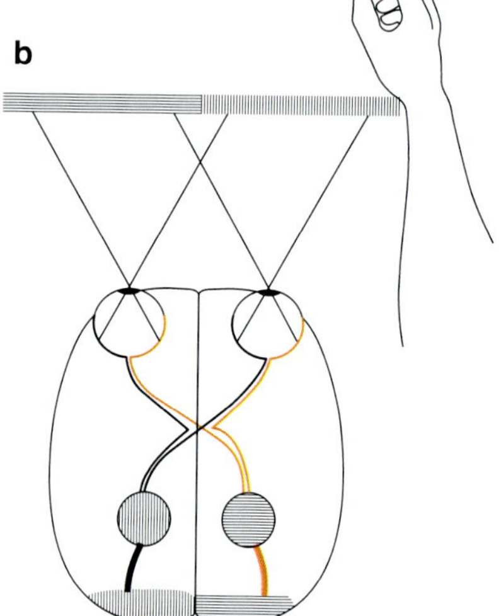

Fig. 13.76 Eye–forelimb model. (**a**) In animals with laterally placed forelimbs, visual, motor, tactile, and proprioceptive information concerning the forelimb are processed in the contralateral hemisphere, thus emphasizing the nasal retina. An evolutionary change in the OC took place causing a reduction of IRP compared with cyclostome-like ancestors. (**b**) Increased IRP from the temporal retina provides second source of information to the visual cortex for processing and secondarily to the sensorimotor cortex responsible for limb control. Cerebral hemispheres receive visual information solely from the contralateral visual hemifield. [Reprinted from Larsson M. The optic chasm: a turning point in the evolution of eye/hand coordination. Front Zool 2013; 10:41. With permission from Creative Commons License 2.0: https://creativecommons.org/licenses/by/2.0/.]

branch, the motor cortex is guided by simultaneous information from both left temporal retina and the right nasal retina.

The eye–forelimb hypothesis fits well with the observed inclusion of grasping and the opposable thumb. By way of review recall that the autopod (hand and wrist) first appeared in the late Devonian. The original tetrapod hand had eight symmetrical digits, much like the flippers of dolphins and whale. By the beginnings of the Carboniferous 350 mya, evolution settled on five digits. The carpometacarpal joint dates to the dinosaurs. Opposition appeared in Euarchonta about 65 mya and took two forms. In Scandentia (treeshrews) it appears in the hand whereas in the primatomorph line (the last known species before primates) *Plesiadapis* has a grasping great toe [46] (Fig. 13.78a).

Euprimates arose in the Eocene 55 mya from an unknown ancestor and divided into two (now extinct) lines: Adapiforms (stem-strepsirrhines) and Omomyiforms (stem-haplorhines). The adapiform *Notharctus tenebrosus* ("false bear" + "dark")

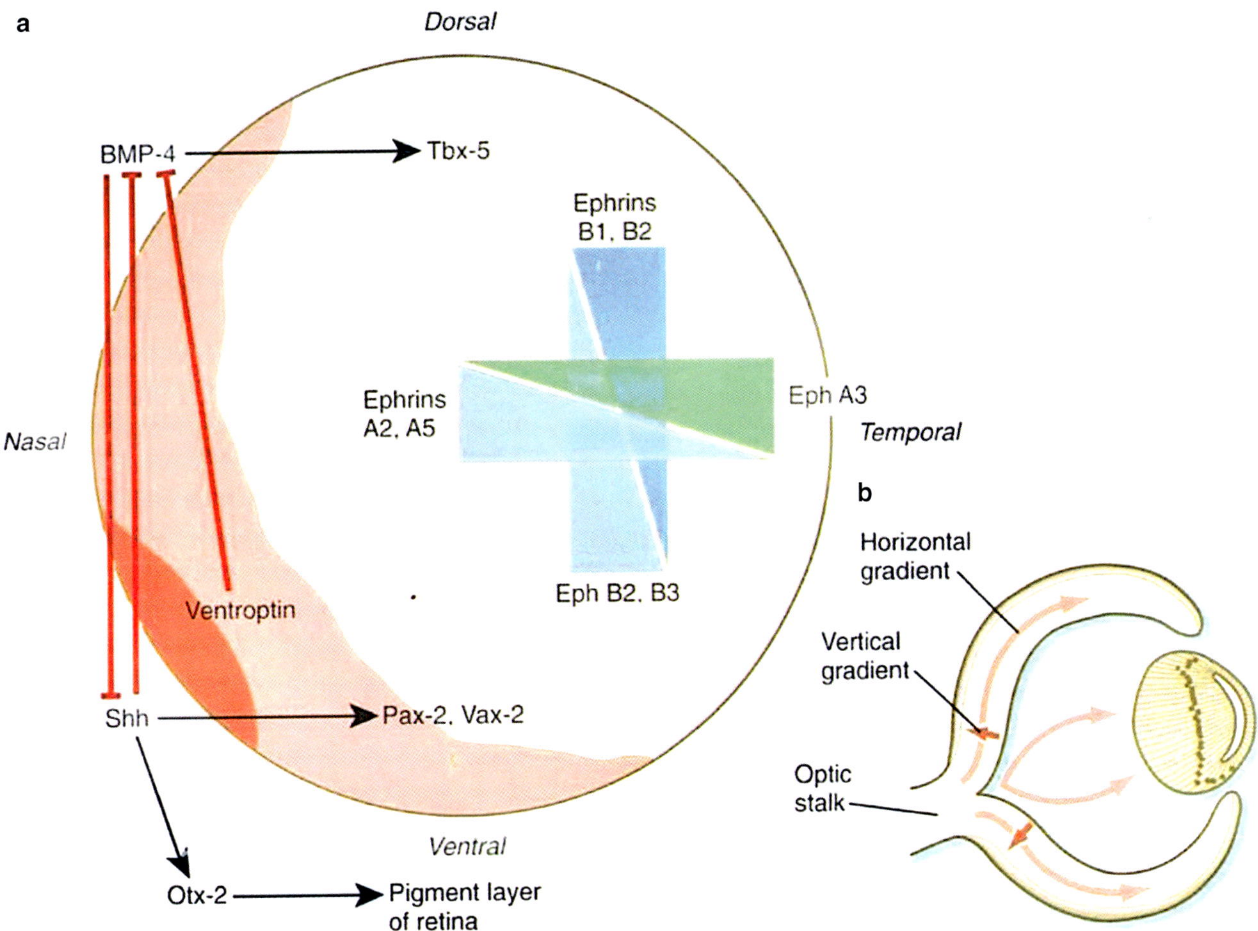

Fig. 13.77 Molecular map of the retina establishes quadrants. Original molecular drivers: (1) ventral, shh; (2) dorsal DMP-4. These are supplemented by dorso-ventral and naso-temporal gradients, indicating that the most primitive zone of retina and sclera is the distribution of primitive dorsal ophthalmic artery (prDOA). Note that inferolateral quadrant, ventrotemporal quadrant is the most recent, corresponds in the retina to the ipsilateral retinal projection neurons and in the sclera to insertion of lateral rectus muscle. [Reprinted from Carlson BM. Human Embryology and Developmental Biology, 6th edition. St. Louis, MO: Elsevier; 2019. With permission from Elsevier.]

found in Wyoming had both opposable thumbs and toes. Also from Wyoming at the same time was the omomyid *Tetonus homunculus*. Not only did it have full opposition but its skull shows the first contact between alisphenoid and the zygoma (postorbital field). This demonstrates the complex thumb of haplorhine and the frontalization of the orbit with the complete backwall were now synchronous processes (Fig. 13.78b and Table 13.6).

Napier further refined the concept of true opposability as (1) ability to bring the thumb completely into the palm such that its ventral surface is 180 degrees against that of the digits and (2) ability to sweep the thumb across the palm all the way to the base of the fifth finger. This requires flexion, abduction, and medial rotation at the metacarpal-trapezium (MTz) joint. Only catarrhines can perform this motion due to (1) a unique saddle morphology of the trapezium and (2) a laxity or "misfit" at the MTz joint which allows for additional motion. This saddle MTz joint is a prerequisite for precision pulp-to-pulp grip. It is not found in other primates. Therefore neither the strepsirrhine line nor the platyrrhines within haplorhine line could ever develop a human grasp pattern. We see this pattern as well in the orbit as full frontalization is achieved only in the haplorhine line.

Fine motor coordination and left–right spatial localization are all parts of an increasingly sophisticated primate approach to the environment. The E-F model seems the most inclusive yet none of the theories to date concern themselves with the fundamental requirement for the whole system to work: a frontalized orbit. Notice we are not focusing on orbital size, as that is dependent on interactions between globe and its bony surround. Our point of emphasis is on the orbital axis, extending from optic foramen forward to the centric point of the orbital aperture, when this is redirected in evolution from 180 degrees to 45 degrees from the midline. The eye in forward position is half this angle from the midline, or 22.5 degrees. What factors are involved for the bony orbit to make this shift?

Fig. 13.78 Plesiadapiform versus Euprimate. (**a**) Plesiadapiform *Plesiadapis cookei* from Late Paleocene of Wyoming and (**b**) Euprimate adapsid *Notharcthus tenebrosus* from Early Eocene of Wyoming. Plesiadapiforms were considered for long as ancestors of primates. Studies of nearly complete skeletons of plesiadapiforms and Early Eocene primates suggest that they are sister groups or that their resemblance results from a convergent evolution. Euprimates have eyes in front with postorbital bar, nails, opposable thumb (hallux) and big toe (pollex), and elongate tarsals (calcaneum and astragalus). The adapiforms were stem to strepsirrhines and did not have contact between alisphenoid and zygoma. Plesiadapiforms do not present these characters and have lateral eyes and claws. (**a**) Source: Smith T. Contributions of Asia to the evolution and paleobiology of the earliest modern mammals. Bulletin des séances-Académie royale des sciences d'outre-mer. 2011; 57(2-4-2011):293-305. https://www.researchgate.net/publication/265171523. (**b**) [Reprinted from Wikimedia. Retrieved from: https://commons.wikimedia.org/wiki/File:Notharctus_tenebrosus_paris.JPG. With permission from Creative Commons License 3.0: https://creativecommons.org/licenses/by-sa/3.0/deed.en.]

Table 13.6 Primate classification based on thumb anatomy [47]

Thumb type		Class	Brain size
Non-opposable but grasping		Tarsiers	Small
Pseudo-opposable	Thumb to index	Strepsirrhini	Medium
Opposable	Thumb to all digits	Haplorhini	Large

Developmental Fields and the Primate Orbit

We have dedicated some effort to familiarizing ourselves with the developmental fields of membranous bone that make up its walls and constitutes its neighbors. When we compare the plesiadapiform and primate orbits something is glaringly obvious: the primate orbit has a backwall. Two fields are responsible for this: (1) the postorbital field and (2) the greater wing from alisphenoid. Postorbital (PO) emerges once again, having been lost in mammalian evolution. No longer is it in contact with parietal or squamous. It simply spans the gap from jugal to frontal. PO starts out as a thin strut but thickens with the haplorhines and sends out a medially directed lamina that joins with a projection of alisphenoid. Temporalis muscle extends forward to insert into this fossa. In various surgical approaches to the orbit it must be stripped away for access. Recall that alisphenoid (AS) is incomplete in strepsirrhines but it expands dramatically with haplorhines as the unilaminar greater wing of sphenoid (GWS). Most of GWS is intracranial but there is a small but sturdy lamina that fuses with PO. Thus the primary contribution to the orbit of postorbital is extracranial while alisphenoid makes up the intracranial separation of orbit from temporal fossa and defines the lower border of superior orbital fissure (Fig. 13.79).

PO and AS are not the only actors. Jugal and its posteriorly directed temporal process have flared out. The prefrontal and postfrontal fields that make up the orbital roof, albeit expropriated by frontal are larger and face forward. In haplorhines, maxilla takes up a parking space in the inferior orbital floor, thereby assisting in the lateralization of jugal. In this process a number of factors can be identified as drivers of change:

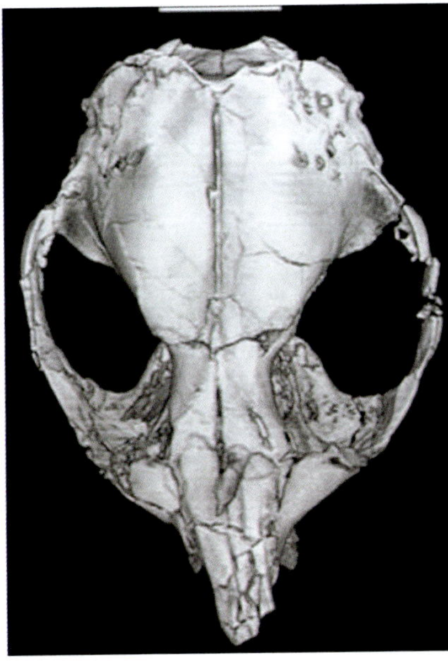
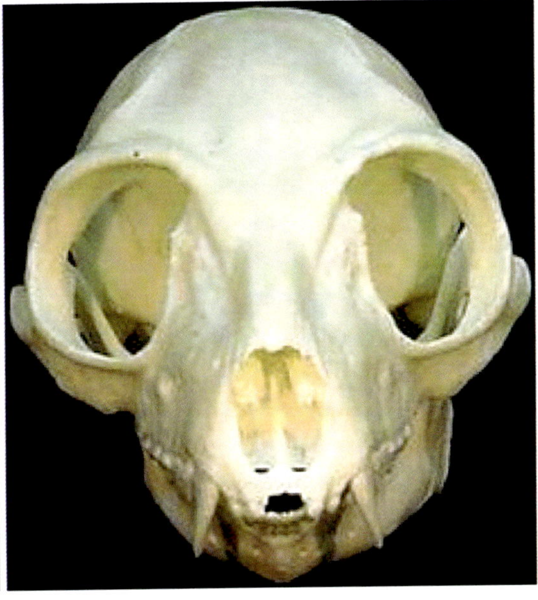

Fig. 13.79 Skull characteristics of euprimate transition. Plesiadapiform *Ignacius grabullianus* has no postorbital bar. The early Euprimate, the lemur-like adaptiform *Lemur catta*, is distinguished by frontalization. Postorbital is more well developed

- Repositioning of parietal/squamosal bones secondary to parieto-occipital lobe
  - Accommodate parieto-occipital growth
  - Provide for expanded muscle insertion
- Flaring of the zygomatic arch
  - New masseter insertions placing stress on the arch
  - Increased muscle volume forcing filling of the temporal fossa
- Hard palate closure (in cynodonts) – frees the base of epipterygoid
- Displacement of epipterygoid into the external calvarial wall as alisphenoid
  - Previous connection with frontal is maintained in the lateral roof of the orbit
- Expansion of greater wing secondary to growth of the temporal lobe

Of all these factors probably the most important is brain growth. Parietal expansion drives liberation of alisphenoid from an intracranial to extracranial position at the cynodont-mammal transition. Moreover, beginning precisely with the euprimates, greater wing expands upward into the orbit. It retains contact with frontal and drags it out laterally. The development of postorbital may have to do with the approximation of temporalis muscle, a source of mesenchymal stem cells. As PO is resynthesized, local BMPs could stimulate migration of MSCs from temporalis into the immediate post-orbital area, thereby erecting the posterolateral wall of the orbit (Figs. 13.71 and 13.72).

Temporal Lobe Growth and the Alisphenoid: A Speculation

Development of the primate hand is asynchronous with that of the orbit. Inferotemporal retinal neurons monitoring hand activities refer ipsilateral information to visual cortex layer V3 having to do with spatial direction of objects. V3 in turn refers to medial temporal lobe, precisely the source for EphB2 and IRP directionality. This could set up a positive feedback look which would (1) progressively reinforce the IRP pathways through the optic chiasm until 50% uncrossed status is achieved; and (2) stimulate the expansion of temporal lobe as it is specifically tasked with visual memory. In this case, temporal growth would power alisphenoid growth to achieve closure of the orbital backwall.

Does Color Have a Role in Catarrhine Evolution?

Anthropoids have expanded neocortex and visual cortex. The back wall of the orbit is separated from the adductor muscles of the temporal fossa, perhaps explaining the extent of frontalization. Their corneas are small with a longer focal length and better visual acuity, features adapted to a diurnal existence. There are two groups of anthropoids, both of which have color vision. New World monkeys, the platyrrhines ("flat or broad nose") have nostrils broadly spaced and prehensile tails. Old World moneys, apes and humans are referred to as catarrhines (hook nose) and have narrow snouts and non-prehensile tails. Both groups have color vision but catarrhines have three types of cones.

This change is intriguing in that molecular phylogenetic studies show four of the five vertebrate opsin genes existed in the early Cambrian. Fossil rods, cones, and melanin have been found in the eye of one of the last common ancestors of gnathostomes, the acanthodian fish *Acanthodes bridgei* dating 300 mya. The rod/cone ratio indicates that *A. bridgei* was diurnal and probably had color vision [48]. But mammals lost full color vision. They survived the challenge of the Cretaceous–Paleocene extinction crisis 66 mya by virtue of being small and developing a burrowing existence. In so doing, they lost the opsin necessary for distinguishing red and green, thereby becoming dichromats. Marsupials were an exception, retaining primary color vision. Taking advantage of the dispersion of angiosperm plants primates successfully exploited this niche by becoming arboreal. It is thought that evolutionary pressure favored their ability to discern colored fruits amidst the green vegetation [49]. This probably occurs by duplication of an opsin gene [50].

Catarrhines have larger brains than platyrrhines. Their orbits show considerably more zygomatic projection as well. The trichromat state for catarrhines likely presents an increased volume and complexity of information to the visual cortex. How does that translate to the volume of output via the lower pathway to temporal lobe? Could this, in combination with the sensorimotor data derived from their hands, also play an indirect role in the cerebral cortex growth, expansion of the temporal dura, and spatially directed membranous osteosynthesis?

Blood Supply of the Orbit

The Orbit Can Be Understood as a Museum of Paleontology

The blood supply to the orbit is a rather simple-minded affair. Recall that the orbit is a five-sided box composed of two L-shaped struts and a back wall, with a palatine finger in the corner. Each bone field is supplied by a separate pedicle, according to the neuromeric source of its neural crest mesenchyme.

- Medial wall (ethmoid) and roof (prefrontal, postfrontal): intracranial StV1.
- Backwall (orbitosphenoid, alisphenoid greater wing): intracranial StV1.
  - Note that temporal plate of alisphenoid incorporated into calvarium is supplied by intracranial StV2.
- Floor and lateral wall (maxilla, orbital process of palatine, jugal, postorbital): extracranial StV2

Why does this make so much evolutionary sense? Previously, we spent considerable time following the meandering history of the orbit. Recall that the stapedial system came into being with gnathostomes. Early orbits were simple in design. They had a backwall consisting of a sphenethmoid complex, outside of which were a collection of bone fields. The superior orbital rim was the dynamic duo of prefrontal and postfrontal, with frontal positioned above them but excluded from the orbit. Thus, V1 stapedial in fishes has to perfuse only two sets of fields: an internal branch to ethmoid (what we now call nasociliary) and an external branch to the orbital bone field (supraorbital) which could then continue upward to frontal. The postorbital bone field was distinct, being located initially outside the orbit; it was supplied by a branch of StV2 distinct from its neighbors. Since the orbit looked outward on the world from the side of the head, no lateral zone existed.

Tetrapods invented a lacrimal system to protect the eye. The gland in turn required protection. The lacrimal functional unit was constructed from four sources of mesenchyme: (1) periocular r1 epithelium for ducts, (2) orbital MNC neural crest for stroma, (3) r4 neural crest for secretory elements, and (4) r2 jugal bone and postorbital bone for protection. This unique anatomy explains why the lacrimal artery appears as a lateral "afterthought," distinct from the primary stem and of smaller caliber. It is of interest that each rectus muscle (with the glaring exception of LR) carries with it two anterior ciliary arteries forward to enter the sclera at the muscle insertion. Lateral rectus, on the other hand, conveys only one anterior ciliary artery. These vessels pass anteriorly to the episclera and supply the anterior segment of the eye, including the sclera, limbus, and conjunctiva. Lacrimal artery has an unusual relationship with lateral muscle. All muscular arteries exit the ophthalmic stem within the cone and penetrate the rectus muscles proximally, along their undersurface. Lacrimal arises outside the annulus of Zinn and runs alongside lateral rectus on its upper border. Despite this cozy arrangement, lacrimal artery is generally *not* involved in supplying the muscle. This fits with the appearance of the other extraocular muscles early in evolution, long before any would-be tetrapod ancestor dared to venture out of the water.

Palatine bone is intimately related to the posterior border of maxilla. Recall its original position as a complex of four tooth-bearing fields programmed inside maxilla (the fourth palatine field becomes ectopterygoid). With the vertical growth of maxilla in mammals and the consequent expansion of the maxillary sinus, the vertical plate of palatine, being dead-level with the maxilla, ascends to articulate with the orbital floor in front and the sphenoid behind [51].

The Stapedial Artery System is the Rosetta Stone of Facial Clefts

The blood supply of the <u>globe</u> is provided exclusively by the primitive ophthalmic artery, only the second branch of the internal carotid system (hypophyseal being the first; Fig. 13.80). In contrast, the extraocular structures of the <u>orbit</u> are supplied by stapedial arterial system. The developmental anatomy of these systems is discussed in detail with figures in Chap. 8 (Fig. 13.81 orbital arteries).

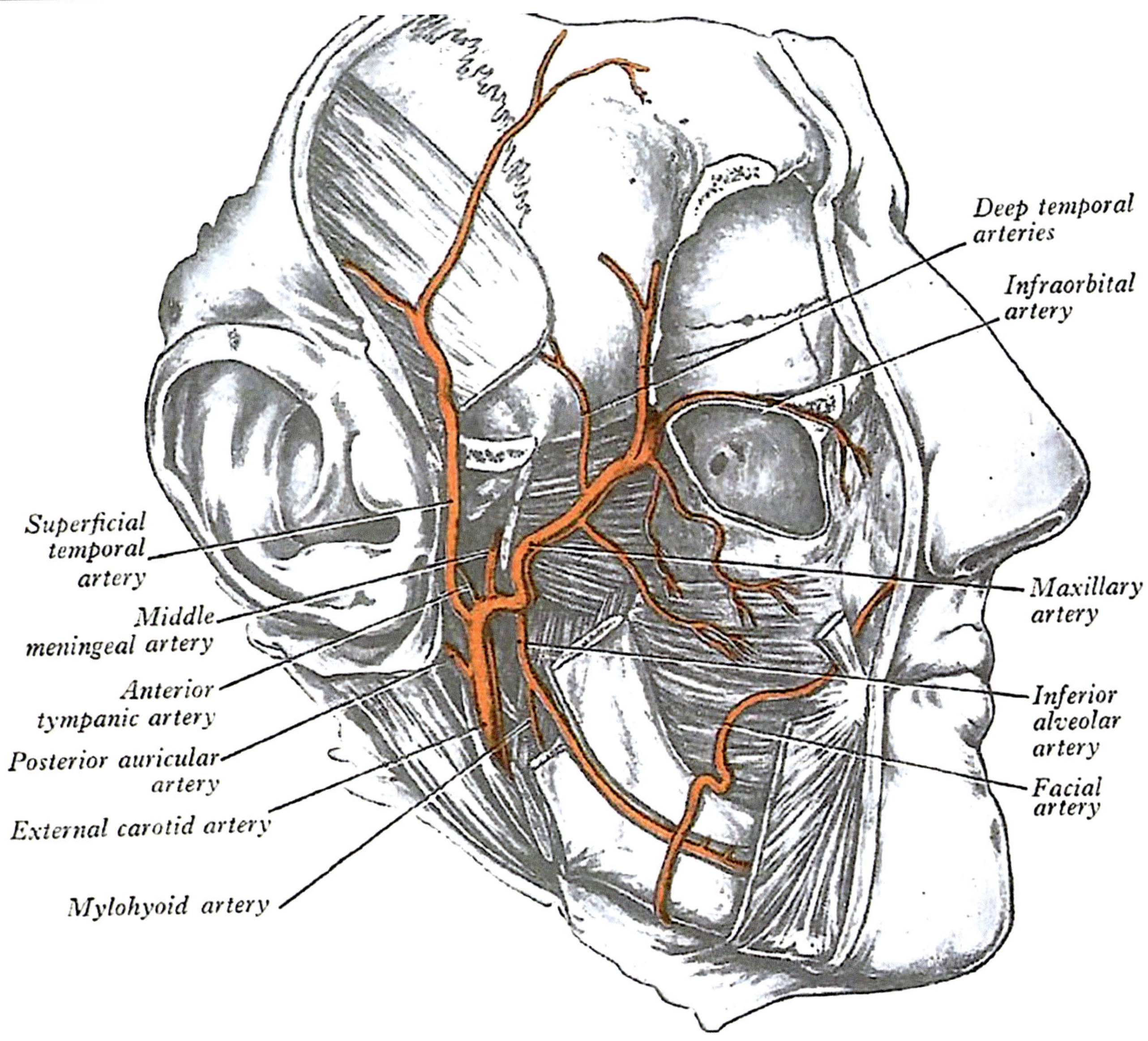

Fig. 13.80 Zone 8 DIF plane. Temporalis being a muscle of mastication is supplied by two deep branches from internal maxillary. Anterior deep temporal artery underlies zone 8, supplying the osseous component of the extracranial posterior orbital wall. [Reprinted from Williams P (ed). Gray's Anatomy, 38th ed. New York, NY: Churchill Livingstone; 1995. With permission from Elsevier.]

- StV1 intracranial via superior orbital fissure is anastomotic to ophthalmic. It gives off major branch that subdivides to supply the orbit and a minor branch (lacrimal) found only in tetrapods.
  - Ethmoid (nasociliary): zones 13 and 12
  - Frontal (supraorbital): zones 11 and 10
  - Lacrimal: zone 9
- StV1 intracranial via meningo-orbital foramen to lateral orbit (usually involutes)
- StV2 extracranial:
  - Infraorbital: zones 4–5–6
  - zygomatic artery: zones 7 and 8

The *stapedial artery system* is one of the great innovations of evolution, dating back to the invention of jaws. Its various branches are programmed by the trigeminal nerve complex. The distribution of stapedial is vast, but very specific. It encompasses all fronto-nasal-orbital structures, the jaws, and the dura. Intracranial stapedial has an anterior branch that extends to the posterior border of anterior cranial fossa. Here, one branch follows V1 into the orbit while the other branch remains with V2 in the floor of the anterior middle cranial fossa. As we know, this primitive meningeal system gets supercharged by an ascending anastomosis with the internal maxillary axis, the middle meningeal artery. At

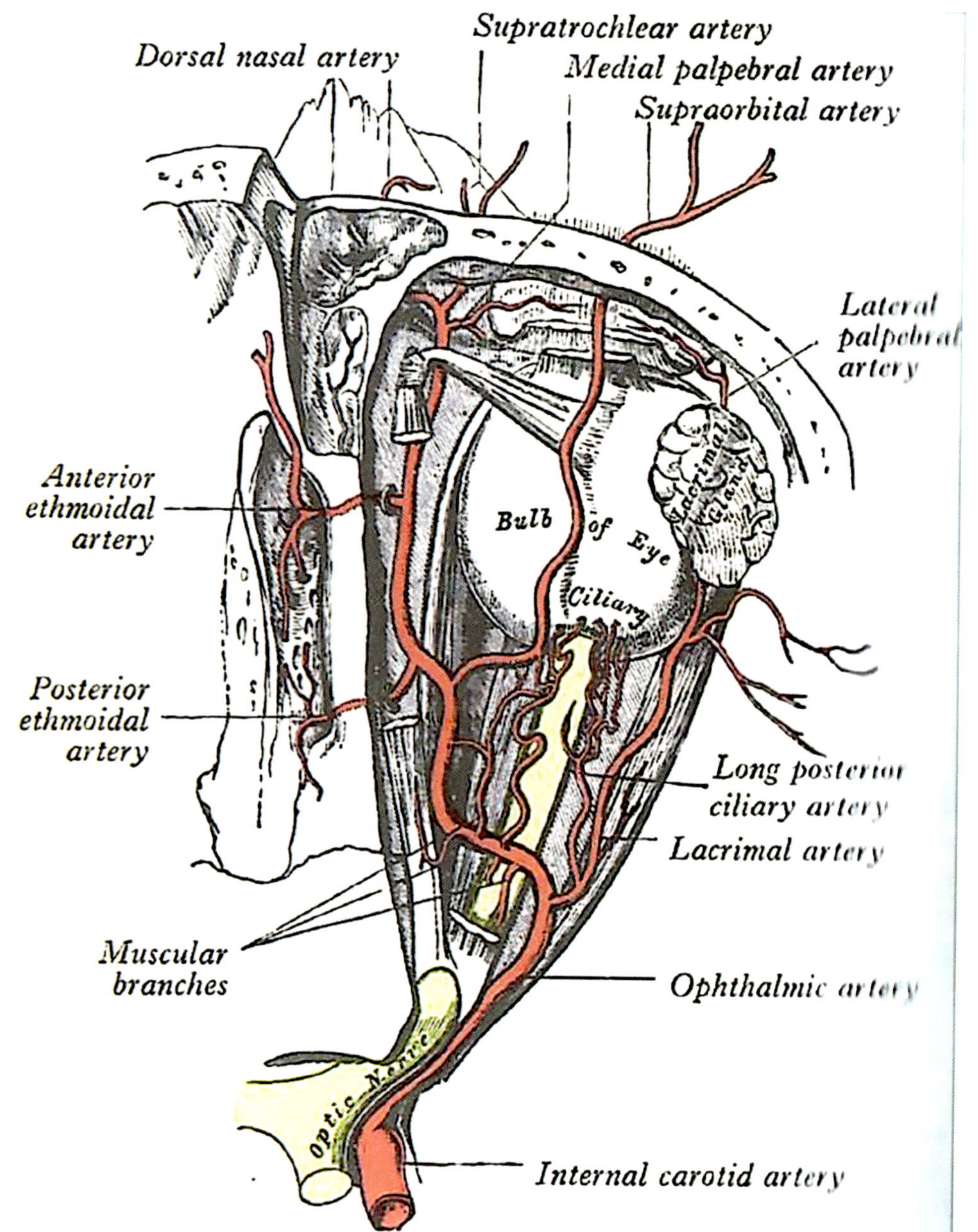

Fig. 13.81 Non-ocular orbital arterial system. Ophthalmic stem anastomoses with ophthalmic stem from which the central retinal artery and posterior ciliary arteries are given off to both temporal and nasal choroid; these represent the distributions of prDOA and prVOA. Thereafter stapedial proceeds as a common trunk which splits. (1) Nasociliary branch supplies zones 12 and 13 which include the ethmoid complex, V1 dura and, via supratrochlear, the glabellar frontal bone. The nose is supplied by branches anterior and posterior ethmoid branches, while the infratrochlear artery supplies the soft tissues of the dorsum and lateral nose. (2) Supraorbital branch supplies zones 11 and 10, the orbital plate. (3) Lacrimal branch is an evolutionary neo-morph; it evolves in tetrapods along with the gland. Ocular arterial supply. These vessels originate from four sources: primitive dorsal ophthalmic artery (prDOA), primitive ventral ophthalmic artery ((prVOA), internal ophthalmic artery (IOA), and the stem artery, external ophthalmic artery (EOA). Recall that the origin of EOA from internal carotid moves proximal once prVOA is established. This sets the stage for anastomosis between EOA and the StV1 stem accompanying V1 into the orbit.

- The lateral muscular branch of the ophthalmic artery is dedicated to the prDOA zone. It supplies the lateral rectus, superior rectus, and superior oblique muscles.
- The medial muscular branch is dedicated to the prVOA zone. It supplies the inferior rectus, medial rectus, and inferior oblique muscles.
- Muscular arteries travel within the cone and enter the muscles on their internal aspect.
- Seven anterior ciliary vessels arise from the muscular branches, which travel with the four rectus muscles to provide circulation for the anterior segment of the eye. These vessels pass anteriorly to the episclera and supply the anterior segment of the eye, including the sclera, limbus, and conjunctiva.
  - Each rectus muscle has two anterior ciliary arteries, except for the lateral rectus muscle, which has one vessel.
  - Anterior ciliary vessels penetrate the episclera at the four-muscle insertion. They then (1) supply the anterior segment of the eye, including the sclera, limbus, and conjunctiva; and (2) anastomose within the choroid with the posterior ciliary arteries.
- Short ciliary arteries supply the choroid and ciliary processes

[Reprinted from Williams P (ed). Gray's Anatomy, 38th ed. New York, NY: Churchill Livingstone; 1995. With permission from Elsevier.]

this point intracranial stapedial supply from the tympanic cavity disappears. Flow changes force StV1 to annex ophthalmic, causing additional flow reversal. Orbital StV1 connection to the middle meningeal system is usually lost. StV2 persists as a forward branch of the neomorphic "middle meningeal" system, middle meningeal.

Understanding the stapedial system is critical: it represents the Rosetta Stone of craniofacial cleft states. What follows is a brief account of this developmental sequence and how the final vascular anatomy of the orbit is achieved. Appended below is a quick timeline for reference (Figs. 13.82, 13.83, and 13.84).

Stage 13: hyoid stem, primitive maxillary supplies eye

Stage 14: primitive dorsal (temporal) ophthalmic from distal ICA, stapedial is born

Stage 15: primitive ventral (nasal) ophthalmic from more proximal ICA

Stage 16: ventral pharyngeal artery (precursor of ECA) arrives at V3: IMMA

Stage 17: prDOA and prVOA unite, StV3 middle meningeal

Stage 18: tympanic vessels visualized

Stage 19: stapedial enters orbit via superior orbital fissure and meningo-orbital foramen

Stage 20: anastomosis StV1 to ophthalmic

Stage 21: involution of stapedial system complete

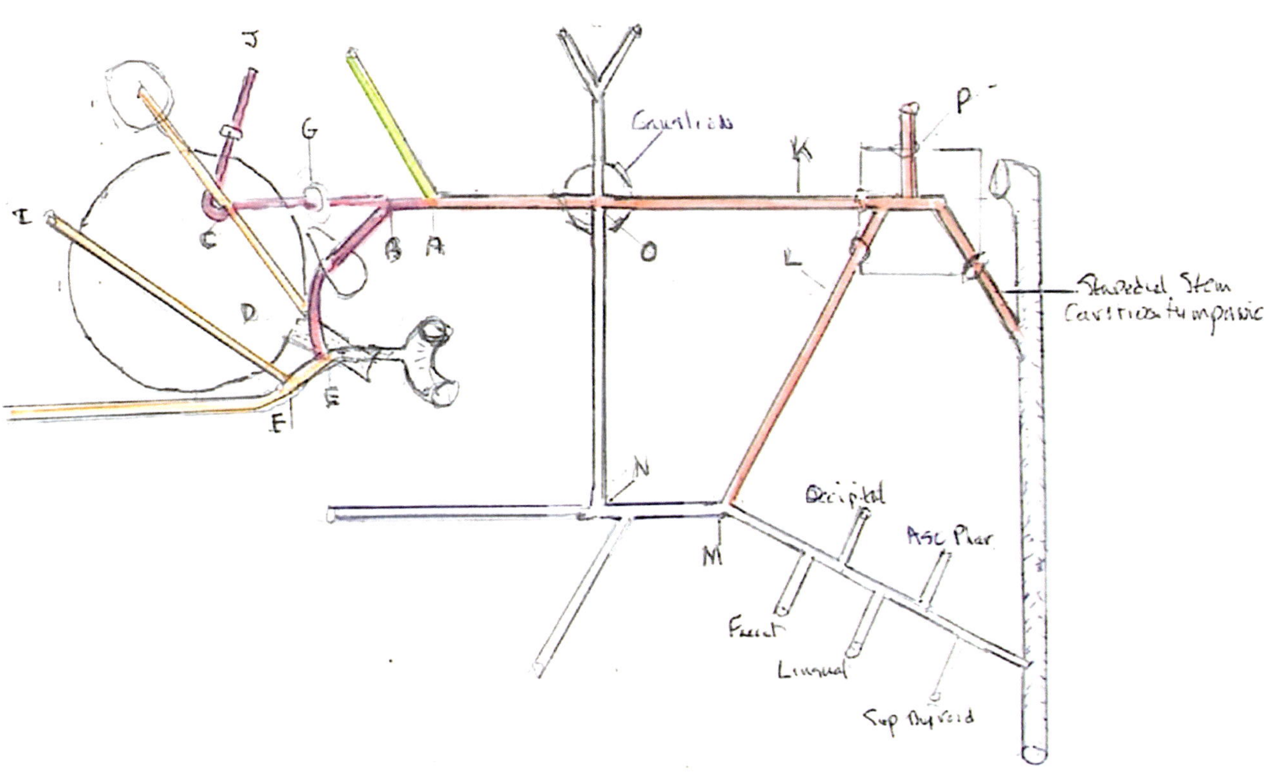

Fig. 13.82 Stapedial artery development. Stapedial stem comes from derivative of AA3 common carotid, into tympanic cavity. When the embryonic stapedial system involutes, this persists as the caroticotympanic artery.

Stage 16: Stapedial stem (pink) comes from derivative of AA3 common carotid, into tympanic cavity (T)

- TP: Upper division, posterior (pink) is programmed by branches of VII; supplies posterior dural mesenchyme.
- TKA: Upper division, anterior (pink) is programmed by VII (greater petrosal) until it reaches V ganglion. It supplies anterior dural mesenchyme.
  - Although terminology is outdated we will refer to anterior branch of upper division as the supraorbital artery (pink).
- TL is lower division. It is approached by ventral pharyngeal artery.

Stage 17: Supraorbital approaches the orbit. Anastomosis of VPA with lower division at M forms internal maxillo-mandibular MN which gives

- StV2 to pterygopalatine fossa, StV3 to mandible, and StV3 to middle meningeal which re-enters the cranial cavity with recurrent meningeal nerve.

Stage 18: Supraorbital splits at point A to give off StV2 meningeal branch (yellow) to anterior temporal fossa and common StV1 AB (magenta) to orbit. Middle meningeal is united with supraorbital at O.

Stage 19: StV1 enters the orbit: meningeal branch BC and orbital branch BD. Three orbital branches (orange) EF and EH develop; they are still supplied from behind by supraorbital. StV3 developing to the dura (frontal branch and parietal branch) causing proximal involution of supraorbital OK. MN will persist as anterior tympanic.

Stage 20: Anastomosis between StV1 system and ophthalmic artery.

Stage 21: Supraorbital AB involutes.

Establishment of anastomosis between the lower division of the stapedial artery (future IMAX) and the ventral pharyngeal artery.

[Courtesy of Michael Carstens, MD]

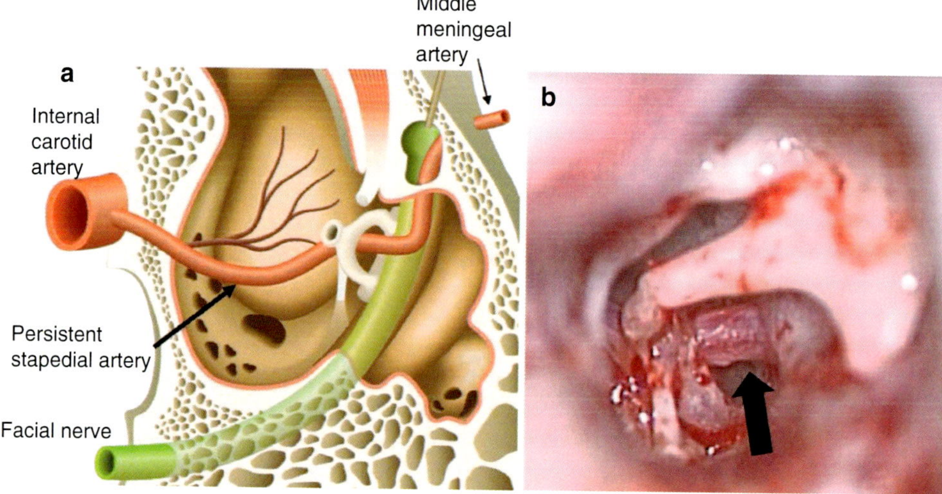

Fig. 13.83 Persistent embryonic stapedial stem in tympanic cavity. Stapedial stem comes from derivative of AA3 common carotid, into tympanic cavity. When the embryonic stepedial system involutes, this persists as the carotico-tympanic artery. Here it is seen passing out of tympanic cavity to connect with middle meningeal. The more distal anterior supraorbital system has involuted. [Reprinted from Hisashi Sugimoto H, Ito M, Hatano M, Yoshizaki T. Persistent stapedial artery with stapes ankylosis. Aurus Nasus Larynx 2014; 41(6):582-585. With permission from Elsevier.]

Aortic Arch Arteries

Human embryos possess five pharyngeal arches (PAs) with which craniofacial structures are assembled. Running through the core of each arch is a vascular axis, *aortic arch artery* (AA), that spans from the cardiac outflow tract tucked below the pharynx upward to paired dorsal aortae. The first four pharyngeal arches are important. PA5 is diminutive, producing the arytenoids, cricoid, and associated muscles. For this reason, the fifth aortic arch artery involutes almost immediately, making PA5 dependent upon the blood supply of PA4 for its survival. *There is no sixth pharyngeal arch.* Instead, the fate of the sixth and final aortic arch artery is to form the pulmonary circulation.

Note: Textbook illustrations of the aortic arches are misleading. We can easily understand how they work by means of a little *embryonic origami*. Fact: the first aortic arch artery does *not* "sprout forth" from the cardiac outflow tract. At stage 8 the embryo is still a flat trilaminar disc. The embryonic heart field is located in front of the future brain and is connected to the two dorsal aortae that run passively backwards along the length of the embryo. At stage 9 explosive growth of the forebrain forces the embryonic head to flex downward almost 150°. The hinge point, located at the midbrain, is called the *mesencephalic flexure*. The heart is now positioned under the neuraxis, specifically, directly below posterior pharynx. Because the dorsal aortae are attached to the ventricles of the heart, in the process of folding they become stretched downward and backward, forming bilateral arcs connecting the heart below with the embryo above. They are now termed the *first aortic arch arteries*. The remaining five aortic arch arteries develop one per stage. Each arch represents the union of two distinct stems. From above, the dorsal aortae send out ventrally directed stems while, from below, the outflow tract produces paired dorsally directed stems. The two sets of stems unite to produce the aortic arch artery. Failure of this anastomotic process explains (1) the failure of AA5 and (2) the breakdown and reorganization of the system.

The six aortic arches appear in craniocaudal sequence in stages 9–14. The pharyngeal arch becomes visible one stage later. After their formation in stages 10–11, the first and second pharyngeal arches merge together in stages 12–13, thus creating the tissue masses that will complete the construction of the face. In the process their aortic arch arteries disintegrate, leaving behind arterial remnants dangling from the dorsal aortae. *Hyoid artery*, the dorsal remnant of AA2, will become the source of the stapedial system.

By the end of the pharyngeal arch period (stage 14), AA3 is transformed into the common carotid; it gives off the external carotid artery and then continues onward as the extracranial internal carotid. AA4 produces paired subclavians and, on the left side, the aortic arch. The two dorsal aortae into which AA4 inserts have merged into a single descending aorta. As a result of embryonic growth and spatial rearrangement of the pharyngeal arches the hyoid artery becomes the *first branch of the extracranial internal carotid artery* just below the otic capsule. Hyoid artery will morph into the stem of stapedial system.

The reader will note that, at this point, the skull has not yet developed. Osteogenic mesenchyme subsequently condenses around pre-existing soft tissue structures. The various fissures and foramina of the skull are a result of this process.

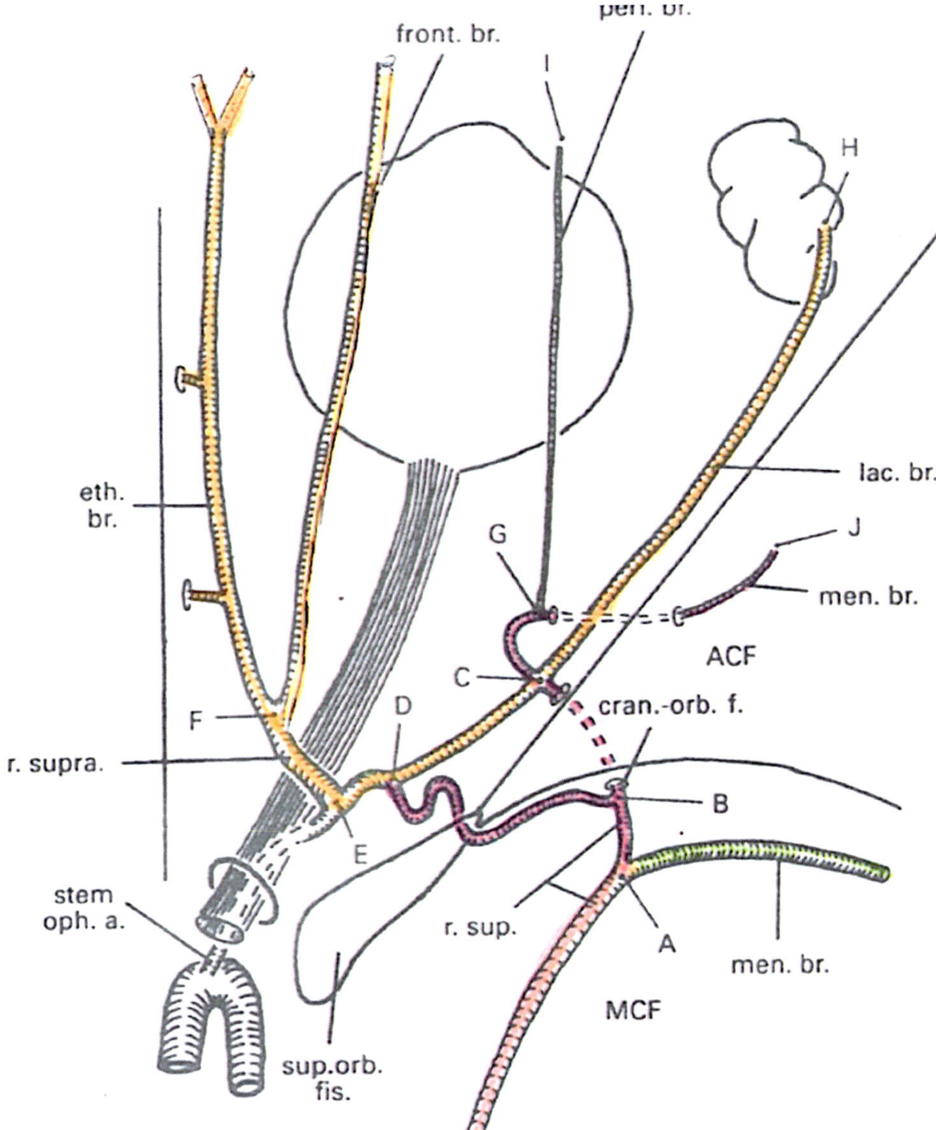

Fig. 13.84 Embryonic stapedial system to the orbit. Stapedial stem given off by internal carotid via the tympanic cavity; it then divides into (1) extracranial stapedial (StE) to join the external carotid system and (2) intracranial stapedial (StI) which supplies the entire dura, the individual meningeal branches being programmed sensory branches of trigeminal nerve. Anterior branch of StI (pink) proceeds from trigeminal ganglion through middle cranial fossa (MCF) toward anterior cranial fossa.

- At point A, StI divides: (1) AB follows V1 toward the orbit—this is StV1 (magenta); (2) the remainder of StI follows V2 branches to the dura of anterior middle cranial fossa. It also supplies the calvarial lamina of alisphenoid.
- At point B, StV1 divides into two V1-programmed branches: (1) medial orbital branch BDE through great orbital fissure into the orbit and (2) lateral meningeal branch BCGJ passes through meningo-orbital foramen in alisphenoid to enter the lateral orbit; it then exits via an un-named foramen to enter the floor of the anterior cranial fossa where it supplies the dura.
- At point D, cranial nerve V1 divides into three sensory: the large common branch continues medially, then splits to become ethmoid n. and frontal n.; a small lateral branch proceeds laterally as lacrimal nerve.
- At point E, StV1 makes an anastomosis with stem ophthalmic. It then follows V1 sensory nerves. StV1 proceeds medially as EF to supply zones 13–12 via the ethmoid branch (nasociliary) and zones 11–10 via the frontal branch (supraorbital). StV1 proceeds laterally as EDCH to supply zone 9 via lacrimal artery.
- Involution of embryonic stapedial supply has alternative consequences:
  - ABD involutes leaving the intraorbital stapedial system connected to ophthalmic. This is the most common situation.
  - BD can involute with ABC intact with recurrent meningeal supplying the intraorbital stapedial system.
  - Minor variants have persistent ABD ± BC.

[Reprinted from Diamond MK. Homologies of the stapedial artery in humans, with a reconstruction of the primitive stapedial artery configuration of euprimates. American Journal of Physical Anthropology 2005; 84(4): 433-462. With permission from John Wiley & Sons]

Big picture idea: Sensory nerves of the trigeminal system accompany and program the individual arteries of the stapedial and external carotid systems. The neuroangiosomes that result delineate the migratory pathways of neural crest to the face and dura.

How the Trigeminal Nerve Innervates the Dura and Programs the Stapedial System

During stage 14 the cranial nerves develop rapidly. They (especially trigeminal and facial) determine the trajectories of the stapedial system. The innervation pattern is all-important for programming the branches of the stapedial system. *Of particular importance is the anatomic pathway for trigeminal supply to the dura.* After leaving the trigeminal ganglion V1 and V2 travel forward in the lateral wall of the cavernous sinus while V3 descends directly out of the skull. Note: ***the most proximal sensory branch of each part of trigeminal is dedicated to the dura***.

The following facts about trigeminal are relevant to understand how these nerves conduct the blood supply to the orbit:

- V1, upon exiting the cavernous sinus and just prior to entering the orbit, gives rise to the ***anterior meningeal nerve*** supplying the dura of the anterior cranial fossa. It then enters the superior orbital fissure. Once in the orbit, V1 again supplies the dura via a separate *dural branch of anterior ethmoid nerve* that re-enters the cranial cavity via the cribriform fossa.
- V2, before exiting the skull via foramen rotundum, gives off the ***middle meningeal nerve*** to the postorbital dura of anterolateral frontal lobe behind alisphenoid**.**
- V3, after exiting the skull via foramen ovale, gives off the ***recurrent meningeal nerve*** supplying the temporo-parietal-occipital lobes. This requires that it tracks upward with middle meningeal artery via foramen spinosum.

The Stapedial Stem Divides Inside Tympanic Cavity

The stem of stapedial is the remnant of the primitive *hyoid artery* to the second pharyngeal arch. When the second aortic arch to PA2 disintegrates at stage 12 it leaves behind a dorsal remnant, the hyoid artery, dangling from the dorsal aorta. With growth of the embryo the hyoid is repositioned backward toward the otic capsule where it eventually is identified as the sole extracranial branch of ICA.

At stage 14, the definitive circulation of the internal carotid is established. The hyoid artery remains extracranial but it gives off the stapedial stem which immediately tracks upward into the tympanic cavity where stapes forms around it (henceforth the name). Directly beyond stapes, the artery bifurcates. The upper division exits tympanic cavity and proceeds forward under the direction of VII toward trigeminal ganglion. The lower division follows chorda tympani out from the tympanic cavity into the face via the pterotympanic fissure. It then picks up V3 sensory nerve just below foramen ovale (Figs. 13.82, 13.83, and 13.84).

Fate of the Intracranial Embryonic Stapedial

Upper Division Stapedial: Forward to the Orbit and Dura (r1–r2)

Intracranial stapedial anterior (StA) follows greater petrosal nerve forward to the trigeminal ganglion. Here, it takes leave of VII and seeks out cranial nerves V1 and V2 which are running forward together in the lateral wall of cavernous sinus. Immediately upon leaving the sinus, they are picked up by STA and run forward together. In the anterior temporal fossa, and just behind the orbit, V1 and V2 part company. StA also divides.

- One branch continues as StV2 to supply the dura of the lateral floor and side wall of temporal fossa.
- The other branch continues as StV1. It runs a short distance until just behind the orbit where it makes a critical bifurcation, one that we must be aware of, anterolateral prior to exiting the skull.
  - *Orbital branch of StV1* goes medial to enter the superior orbital fissure at its extreme superolateral corner. In the orbit it (1) follows the three sensory branches of V1 and (2) anastomoses with stem ophthalmic artery.
  - *Meningeal branch of STV1* continues on straight ahead to enter the orbit via its own meningo-orbital foramen. There, it anastomoses with lacrimal artery and then exits the orbit via an un-named foramen to access dura of the anterior cranial fossa above the orbital roof.

StV1 becomes identifiable at stage 18, enters the orbit during stage 19 and forms an anastomosis with the *secondary ophthalmic stem* at stage 20. (Recall from Chap. 6 that the *primary ophthalmic stem* translocates backward along internal carotid at about stage 18.) The pressure from ICA into the intraorbital stapedial system results in a *reversal of flow* in the proximal extraorbital segment of StV1 leading back to the junction with StV1.

- Orbital StV1 involutes but occasionally persists (Fig. 13.85).

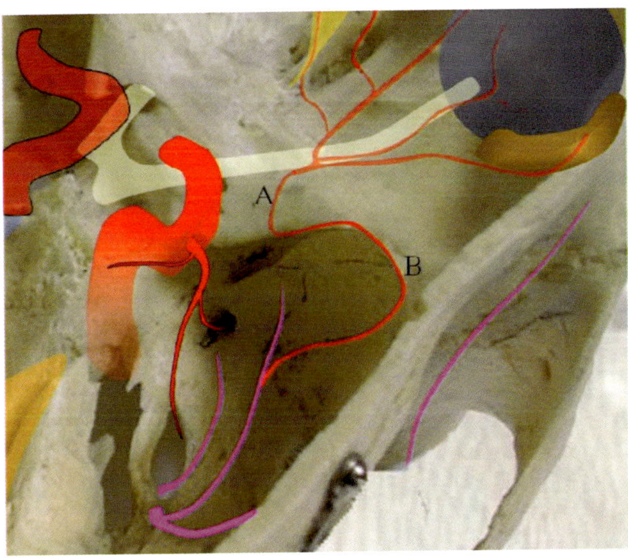

Fig. 13.85 Persistent embryonic StV1 supplying the orbital arteries. Angiogram shows ophthalmic artery disconnected from the system. The source of orbital circulation is now via StV2 branch of middle meningeal. [Reprinted from neuroangio.org, used with permission. Retrieved from: http://neuroangio.org/anatomy-and-variants/middle-meningeal-artery/.]

- Meningeal StV1 involutes as well, but it also can persist, creating a connection between middle meningeal artery and lacrimal artery. In this situation meningeal StV1 has acquired two confusing names: *recurrent meningeal artery* or *meningo-orbital artery*.
  - In rare cases, orbital STV1 fails to develop, leaving meningeal StV1 as the sole source of blood supply for the orbital arteries.

Thus, *the final product, "ophthalmic artery," is a hybrid system*. Primitive ophthalmic, being derived from ICA, supplies CNS tissue exclusively, i.e., the optic nerve and globe. All remaining tissues of the orbit innervated by orbital StV1 (or by meningeal StV1) are supplied by branches from the derivatives of the original stapedial system.

Recall that at stage 17 dura mater is beginning to organize at this time. At stage 18 an anastomosis between V3, which had exited to skull to reach infratemporal fossa, sends a recurrent dural branch upward, the middle meningeal nerve. It is accompanied by middle meningeal artery from the newly constituted maxillary division of internal maxilla-mandibular artery, vide infra. By stage 19 high-pressure flow and complete trigeminal innervation set the stage for the definitive meningeal arterial system following the branches of V1–V2–V3. Once again, a reversal of flow occurs and the proximal segment from middle meningeal backward to the tympanic cavity becomes irrelevant and involutes. Thereafter, all meningeal arteries are supplied by MMA (principally) with contributions from other branches of the external carotid system.

Lower Division Stapedial: Forward to the Jaws and Upward to the Dura (r3)

As soon as chorda tympani departs from the tympanic cavity, it makes a beeline for the sensory root of V3, where it seeks out lingual nerve by which to convey itself to the tongue. Lower division stapedial tracks along with the nerve. At the same time, the poorly named maxillary branch from external carotid tracks forward toward V3. At stages 18–19 an anastomosis between lower division stapedial and ECA creates the hybrid *maxillo-mandibular artery* (MMA).

MMA has two functions and three distinct zones. Branches associated with the external carotid system supply all original structures of the first arch, such as the muscles of mastication, fat, glands, and endoderm. Branches associated with the stapedial system supply those neural crest structures *reassigned* from the first arch: jaws, the suspensory bones of the maxilla, and the dura. The zones of MMA are as follows:

- *Proximal (mandibular) zone* gives off stapedial derivatives that re-enter the skull to supply the tympanic cavity and the dura of the middle and posterior cranial fossa. It also sends StV3 inferior alveolar downward to supply the mandible. The sole ECA branch of the proximal zone supplies mylohyoid.
- *Middle (infratemporal) zone* branches are distributed exclusively to muscles of mastication (from somitomere 4).
- *Distal (pterygopalatine) zone* gives off StV2 branches in the pterygopalatine fossa; these are subsequently distributed to the maxilla and its suspensory bones.

Big picture idea: Anastomoses to the stapedial system are responsible for its dissolution.

Reunification and Disappearance of the Stapedial System (See Figs. 13.82, 13.83, 13.84, and 13.85)

Toward the end of the embryonic period the anatomy of the stapedial system is drastically altered, making it virtually unrecognizable. Three anastomoses are responsible for these changes.

1. At stage 20, within the orbit, StV1 is annexed by ophthalmic. Its proximal segment dies back to the bifurcation with StV2.
2. Also at stages 18–19, inferior division stapedial joins external carotid just lateral to the sensory root of V3 immediately below its exit from the skull. This anastomosis creates the hybrid *maxillo-mandibular artery* (MMA).

3. From MMA, middle meningeal artery follows sensory V3 middle meningeal nerve upward to supply the dura. At stages 19–20 middle meningeal annexes StV2. It does not annex StV1 proximal to the orbit because this segment has already undergone involution. Thus, intracranial middle meningeal becomes a hybrid system: its anterior branches arise from the original StV1 and StV2 and its posterior branch arises from StV3.

The common denominator of these anastomoses is the *exposure of the distal vessel to higher flow*; the proximal segment therefore involutes. StV1 dies backward from the superior orbital fissure to its bifurcation with StV2. Superior division of stapedial is eliminated completely back to stapes. Inferior division of stapedial distal to stapes persists as the *anterior tympanic artery*. Proximal to stapes, the stapedial stem persists as the *caroticotympanic artery*.

In stem: Anastomoses to the stapedial system cause hemodynamic changes that lead to its demise.

Fate of the Stapedial System

StV1 Arteries

Ethmoidal artery (zones 13–12)

- Intraorbital supplies orbital plate of ethmoid
  - Anterior ethmoidal artery: passes through the anterior ethmoid canal and subsequently supplies the anterior and middle ethmoidal cells, frontal sinus, lateral wall nose, and nasal septum. Sends branches up to cribriform plate.
  - Posterior ethmoid artery: passes through the posterior ethmoidal canal, supplying the air cells of the posterior ethmoid sinus. Sends branches up to cribriform plate.
- Extraorbital
  - Supratrochlear artery: Leaves the orbit at its medial angle above the trochlea and supplies the forehead and scalp creating an anastomosis with the supraorbital artery terminal branches.
  - Infratrochlear artery

    Nasal artery: Superior lacrimal sac and nasal skin.

    Medial palpebral artery: Terminal branches include superior and inferior medial palpebral arteries. These vessels supply the lacrimal sac and eyelids creating an anastomosis with the two lateral palpebral branches from the lacrimal artery.

Frontal (supraorbital) artery (zones 11–10)

- Intraorbital supplies orbital plate of prefrontal and postfrontal bone fields.
- Extraorbital passes through the supraorbital foramen and its terminal branches supply the eyebrow and forehead.

Lacrimal artery (zone 9)

- Runs along the lateral wall of the orbit and supplies the lacrimal gland.
- Anastomosis with zygomatic artery to make zygomaticotemporal artery.
- Terminal branches of the lacrimal artery include the superior and inferior lateral palpebral arteries which supply the lateral upper and lower eyelids and conjunctiva (a 100% r1 derivative).

StV2 Arteries: The Posterolateral Orbit

The development of the extracranial StV2 system is discussed in detail in Chaps. 6 and 10. Branches that emanate from the pterygopalatine fossa are the basis for maxillary craniofacial clefts which we shall discuss later, vide infra.

The posterior wall of the orbit is comprised of two units of the sphenoid complex: orbitosphenoid and greater wing of alisphenoid, both of which are neural crest cartilages via cartilage intermediates (the calvarial lamina of alisphenoid ossified in membrane). OS or lesser wing has one ossification per side and is supplied by internal carotid. AS greater wing has a single ossification center and is perfused by StV1. It is perfused by StV1 lacrimal and/or "recurrent meningeal" (sic) and represents zone 9. Field deficiency here can lead to enlargement of the superior orbital fissure with encephalocoele into the orbit.

Greater wing is only part of the alisphenoid. It also forms the anterolateral corner of temporal fossa and it contributes a lamina to the lateral wall of the skull. These zones are perfused by intracranial StV2. Externally, lateral lamina AS is supplied by periorbital branches of transverse facial. Defective lateral lamina can lead to frontosphenoid synostosis, vide infra. Note: Anatomy pictures aside, lateral pterygoid plate is *not* part of alisphenoid. It really represents a derivative of ectopterygoid. Its ossification is strictly membranous and it is supplied by ECA lateral pterygoid branches to the muscle and bone.

Blood supply to the lateral wall of the orbit is more complex. Recall that in mammals the maxilla and jugal form a developmental complex consisting of three bones: maxilla proper, jugal, and postorbital. These are supplied by a common vascular stem, StV2 zygomatico-maxillary branch (ZMA). Just before entering infraorbital canal ZMA gives off posterior superior alveolar artery to zone 6. After entering the canal ZMA has two more branches: the common zygomatic artery (ZA) supplies both the jugal and the postorbital fields; anterior superior alveolar artery (ASAA) continues forward to supply the rest of the maxilla in zones 4 and 5.

Zygomatic nerve (ZN) follows the floor of the orbit laterally to give off *zygomatico-facial nerve* (ZF) which exits the jugal through the ZF foramen. ZN then goes upward as *zygomatico-temporal nerve* (ZTN) along lateral orbital wall. At the lacrimal gland it bifurcates sending a medial branch to the gland and a lateral branch to zone 8. Thus, ZTN brings PANS to lacrimal to supply the gland; then it darts off laterally to exit the postorbital bone through the ZT foramen. It is a matter of nomenclature whether ascending branch of V2 is termed zygomatic nerve or zygomatico-temporal nerve.

In the meantime, arterial supply from common zygomatic artery follows zygomatic nerve laterally inside the orbital rim until it encounters ZF nerve; it then gives off a branch which escapes through the malar eminence as *ZF artery* to **zone 7**. It now continues upward with the nerve as zygomatico-temporal artery (ZTA) until it reaches the lacrimal gland, which it enters to make an anastomosis with lacrimal. At the gland or before it zygomatico-temporal artery is given off to **zone 8**. This makes embryologic sense because the secretory elements of lacrimal gland are supplied by superior salivatory nucleus from r4–r5. These cells must come from r4 neural crest in the first arch. Their blood supply must come from the extracranial stapedial system.

- It should be noted that lacrimal artery anastomoses with ZFA in the gland. It can therefore provide blood flow to ZTA as well. There are instances where it may be the dominant supply for zone 8. *This does not mean that zone 8 is r1 territory*. Craniofacial fields are always defined by the sensory nerve because that indicates the path of neural crest migration. The key is the neuroanatomy. The ZF and ZT nerves do the programming…the artery, departing from its lacrimal blood source, simply follows orders.

StV3 Arteries: Bone Fields of the Mandible and Middle Ear

To complete our picture, V3 stapedial supplies two functionally related zones of neural crest derivatives. Inferior alveolar supplies the dental units and the ramus whereas anterior tympanic refers backwards to supply the eardrum, malleus, and incus. In so doing, these reflect the complex paleohistory of the mandibular bone complex in which its proximal components such as articular were brought backwards into the skull as structures of hearing.

The Evolutionary Rationale of the Stapedial System

The earliest vertebrates were jawless fishes, represented today by the lamprey and the hagfish. These had eight (or more) aortic arches and a very simple circulatory system consisting of two pairs of aortae, dorsal and ventral. In the lamprey/hagfish model, the heart pumps oxygen-poor blood through the ventral aortae from whence it passes through gills, picks up oxygen, and flows into the dorsal aortae, and from there is distributed to the body.

Jawed fishes (gnathostomes) immediately reduced the total number of arches to six. This gave rise to two variations. (1) Cartilaginous fishes (chondrichthyans) such as sharks lose the first branchial arch, the tissues being reassigned from respiration to the creation of jaws. Sharks supply the upper jaw, eye, and face from a stapedial artery originating from the dorsal aorta at its junction with the second branchial arch. The lower jaw is supplied by an external carotid originating from the ventral part of the first collector loop, located at the junction between second and third arches. Thus, sharks have five gills. (2) Bony fishes (osteichthyces) reassign both first and second branchial arches as a composite unit; they have four gills. External carotid extends to tissues of the upper jaw as well where it interacts with the stapedial system.

A subgroup of fishes, the dipnoi, represented by the lungfish, developed an alternative system for breathing air and were precursors to mammals. This converted the sixth arch to a pulmonic circulation. In addition, the third and fourth aortic arches lost their gills, forming a complete connection between ventral aortae and dorsal aortae, the pressure drop between the circulations was eliminated.

Tetrapods lost the fifth aortic arch artery and gained a complete atrial septum. In adult amniotes, the internal carotid arteries now arise from the third aortic arch and short segments from the dorsal aortae and the ventral outflow tract. The external carotids now annex the more primitive extracranial stapedials to form a common maxillo-mandibular system. The final configuration of MMA reaches its most complex expression in mammals with the reassignment of the bulk of second arch musculature to muscles of facial expression.

The Lacrimal Gland: Key to Understanding the Lateral Orbit

The structure of the lacrimal gland is complex and has provocative implications. Its epithelial components, the ductules, represent invaginations of r1 ectoderm associated with the development of the upper eyelids. General sensory afferents from the gland are carried by V1. Zone 9 of the orbit, which is inhabited by lacrimal gland is constructed with r2 neural crest. V2 ascending from the pterygopalatine fossa as zygomatic nerve transports parasympathetic motor efferent fibers from the facial nucleus. Thus, the parenchyma of the lacrimal gland arises from r2 neural crest as well. If so, this gives rise to a single unifying hypothesis regarding the origin of salivary glands, all of which arise from either r2 or r3 neural crest. The penetration of each gland by ductal epithelium occurs by physical proximity. Common affectation of lacrimal and salivary glands as is seen in Sjögren's syndrome therefore has a common embryological basis (Figs. 13.86, 13.87, and 13.88).

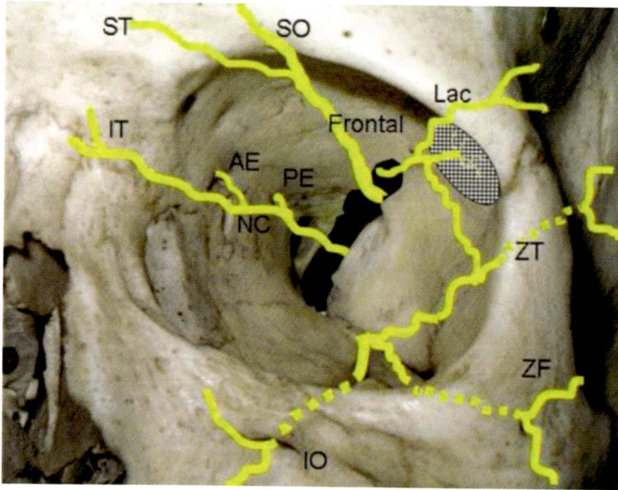

Fig. 13.86 Sensory nerve supply to the orbit. Note: Zygomatic nerve exits directly as ZF and then climbs up the lateral wall, where it divides into ZT and a lacrimal branch Zlac. Lacrimal artery will follow Zlac out to the zygomatico-temporal foramen, exiting as zygomatico-temporal artery. Note that the long posterior ciliary nerves and the sensory root to the ciliary ganglion are not depicted. The hatched structure is the lacrimal gland. Key: ST, supratrochlear nerve; SO, supraorbital nerve; Lac, lacrimal nerve; ZT, zygomatico-temporal nerve; ZF, zygomatico-facial nerve; IO, infraorbital nerve; NC, nasociliary nerve; AE, anterior ethmoidal nerve; PE, posterior ethmoidal nerve; IT, infratrochlear nerve. [Reprinted from René C. Update on orbital anatomy. Eye 2006;20:1119-1129. With permission from Springer Nature.]

Neurovascular Anatomy

Lacrimal gland arterial supply can be classified as follows [52]:

- Type I: Lacrimal artery is a branch of ophthalmic and is the sole supply for the gland. This is the most common anatomy.
- Type II: Lacrimal artery is a branch of middle meningeal via a persistent StV2 meningo-orbital artery penetrating great wing of sphenoid via Hyrtl canal.
- Type III: Lacrimal from ophthalmic and meningo-orbital from middle meningeal co-exist as separate vessels, both of which supply the gland and anastomose with each other therein.

The innervation of lacrimal gland by zygomatico-temporal nerve (ZTN) and lacrimal nerve (LN) has three variations as well. (1) Type I: ZTN and LN enter the gland separately, with ZTN always more lateral. (2) Type II: ZTN and LN communicate prior to both entering the gland separately. (3) Type III: ZTN anastomoses with LN prior to the gland, passes off its fibers but then exits without penetrating the gland [53].

In sum:

- LN continues to pass through the substance of the gland to supply the skin below the lateral eyebrow. In some cases, it can carry PANS fibers from ciliary ganglion.
- ZTN always gives off PANS to the gland, either by itself alone, in partnership with LN, or strictly indirectly (relying on LN to distribute the fibers). The final destination of ZTN is GSA to zone 8.

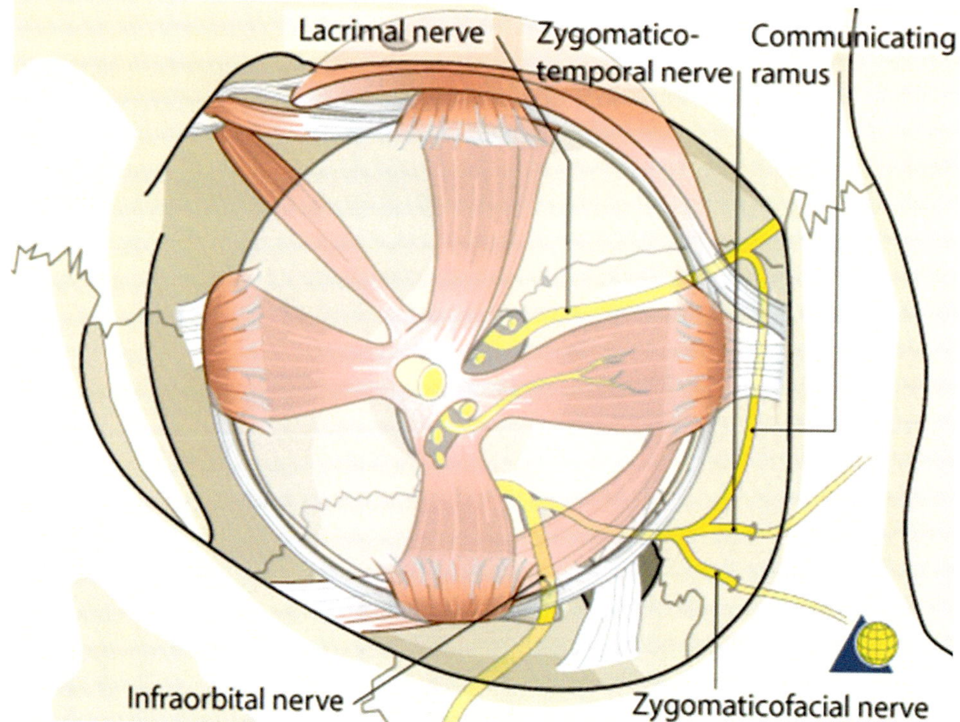

Fig. 13.87 Zygomatic and lacrimal nerves, relations to extraocular muscles. Lacrimal nerve runs along the upper border of lateral rectus. Oculomotor nerve lies internal to the other recti. V2 zygomatic nerve supplies both jugal and postorbital fields. Recall that postorbital was originally part of the calvarium, posterior–superior to the orbit. ZN courses external to internal oblique. It proceeds directly to the cheek while the ZN proper ascends along the postorbital wall. Upon reaching to lacrimal gland, zygomatic nerve anastomoses with lacrimal nerve (thus bringing PANS fibers to lacrimal gland). It then exits as ZT through a separate foramen. [Copyright by AO Foundation, Switzerland.]

The cranial nerve supply to lacrimal gland is likewise diverse. Tracing studies in a primate model shows the following sources [54] (Fig. 13.89):

- PANS from superior salivatory ganglion (r4–r5) supplying first arch neural crest
  - Synapse in pterygopalatine ganglion
  - Access lacrimal gland via zygomatic nerve
  - Minor contribution from contralateral pterygopalatine ganglion
- PANS from ciliary ganglion (m1–m2) supplying midbrain neural crest (minor)
  - Direct innervation via V1 lacrimal nerve
  - Secretomotor (minor)

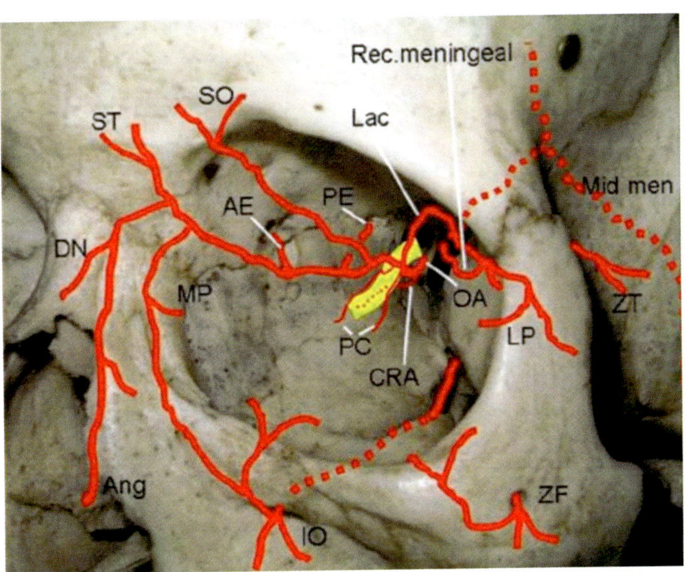

Fig. 13.88 Arterial supply to the orbit. Note the anastomoses between external and internal carotid supplies. The muscular branches, short posterior ciliary, long posterior ciliary, and anterior ciliary arteries are not depicted. The yellow structure is the optic nerve. Key: OA, ophthalmic artery; CRA, central retinal artery; PC, posterior ciliary artery; Lac, lacrimal artery; LP, lateral palpebral artery; IO, infraorbital artery; ZT, zygomatico-temporal artery; ZF, zygomatico-facial artery; Mid men, middle meningeal artery; Rec meningeal, recurrent meningeal artery; Ang, angular artery; MP, medial palpebral artery; DN, dorsal nasal artery; ST, supratrochlear artery; SO, supraorbital artery; AE, anterior ethmoidal artery; PE, posterior ethmoidal artery. [Reprinted from René C. Update on orbital anatomy. Eye 2006;20:1119-1129. With permission from Springer Nature.]

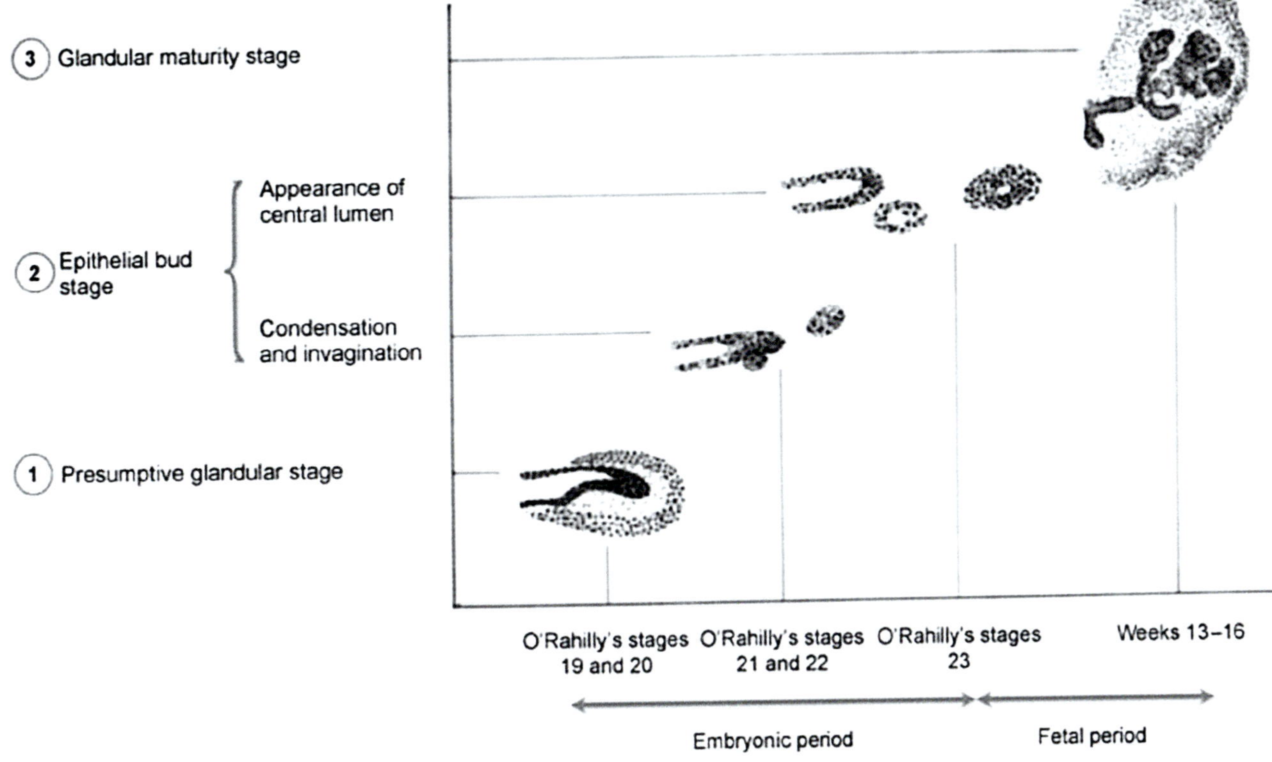

Fig. 13.89 Lacrimal development. Lacrimal morphogenesis has three phases. (1) The glandular stage (Carnegie stage 18–20 refers to the interactions between a designated population r4 neural crest superolateral to the globe being surrounded and intermingled by m1 neural crest. The bud stage (Carnegie 21–22) takes place once the upper lid sulcus is established, giving an opportunity for r1 epithelium to migrate inward, bringing ductal elements into the gland. Mesenchyme, The glandular maturity stage involves epithelial–mesenchymal interactions that induce acini and the supportive stroma. [Reprinted from de la Cuadra-Blanco C, Peces-Peña MD, Merida-Velasco JR. Morphogenesis of the human lacrimal gland. J Anat 2003; 203:531-536. With permission from John Wiley & Sons.]

- SANS from superior cervical ganglion (neuromeres c1–c3)
- GSA from lacrimal gland via lacrimal to V1 trigeminal (r1)
- GSA via zygomatic to V2 trigeminal (r2)

The Lacrimal Functional Unit: Applied Neurology

Evolution has provided elaborate defenses for the protection of vision from a hostile environment. All of us are aware of the sensations caused by foreign body or a corneal abrasion. In the dry eye syndrome corneal sensations are normally described as gritty, sandy, or itchy. These complaints are usually accompanied with a pathophysiological state that indicates a chronic state of inflammation and a disadvantageous change in tear film composition. Infiltrating inflammatory cells within the ocular surface tissues have been reported. Let's review a bit of the neurology so essential for defense of the eye (Fig. 13.90).

Subconscious stimulation of the free nerve endings within the cornea generates afferent nerve impulses through V1 to the midbrain (pons).

- Afferent signals are integrated in the midbrain and then travel via the efferent branch through the pterygopalatine ganglion.
- Targets: main (Wolfring) and accessory (Krause) lacrimal glands.
  - This pathway also controls meibomian glands and conjunctival goblet cells.

Proper function of the LFU supports homeostasis on the ocular surface by controlling secretion of the three major tear film components (mucin, aqueous, and lipid) to maintain the optimal quantity and quality of tear fluid; however, dysfunction of the LFU

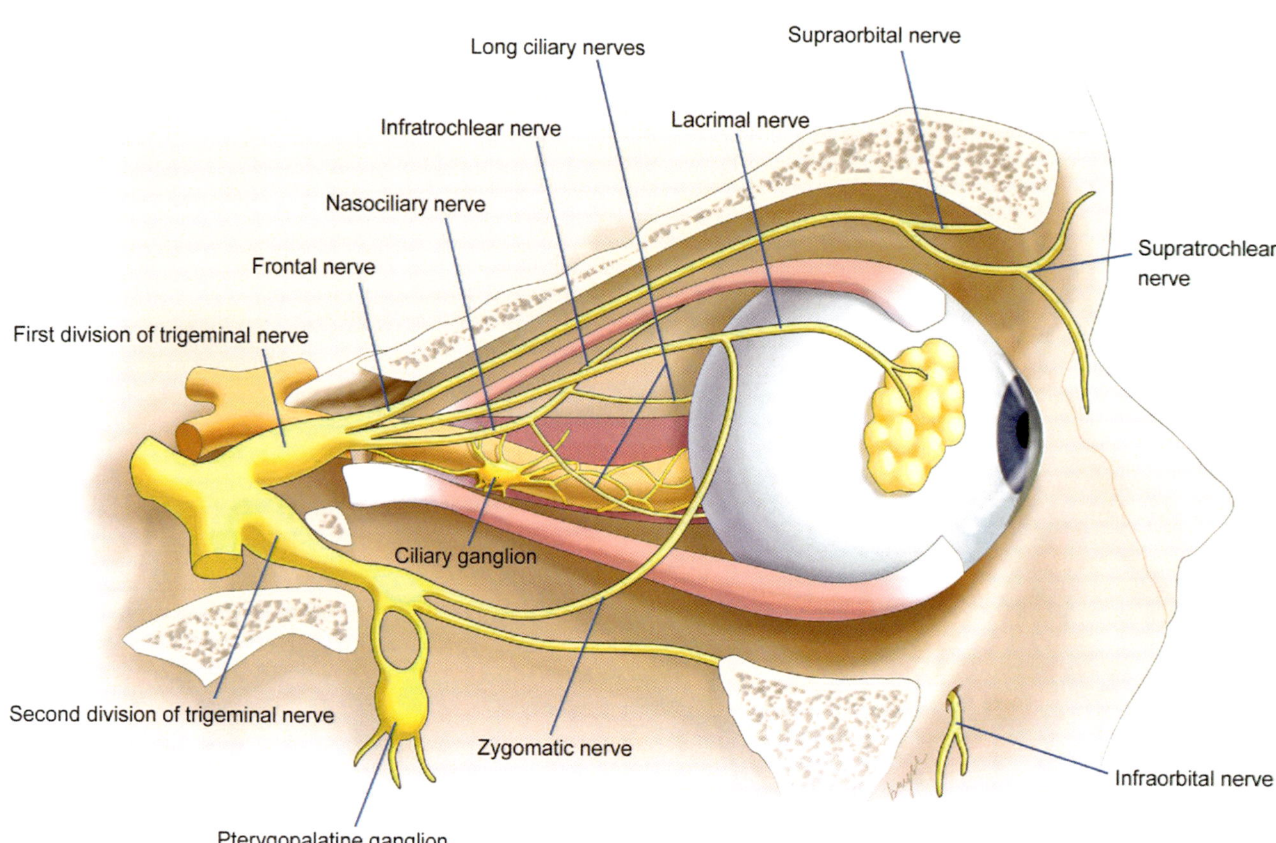

Fig. 13.90 Innervation of the lacrimal gland. Three systems are involved. Somatic sensation from the cornea is conveyed backward from the anterior ciliary nerves and nerves of the immediately adjacent conjunctiva to the first division of trigeminal. Parasympathetic secretomotor supply for lacrimation travels from the superior salivatory nucleus in the greater petrosal nerve. Sympathetic supply originates from superior cervical ganglion and passes via the internal carotid plexus to form the lesser petrosal nerve. PANS and SANS fibers join together as the short nerve of the pterygoid canal (vidian nerve) to reach the pterygopalatine ganglion. Here secretomotor fibers ascend along V2 zygomatic nerve to connect with the V1 lacrimal nerve. [Reprinted from Amrith S, Min Yeug S. Chapter 2 Anatomy in: Amrith S, Sundar G, Min Yeug S. Orbital Adnexal Lesions: A Clinical, Radiological and Pathological Correlation. Singapore: Springer Nature; 2019: 3-11. With permission from Springer Nature.]

- Altered tear film composition > dry eye disease.

Requirements of tear layer for optimal optical performance: (1) stable, (2) anti-infective, and (3) epithelial supportive.

- Ocular surface sensory nerves that supply continuous input into the CNS regarding change in ocular surface milieu.
- The brain then sends signals to the specialized support tissues, for example, lacrimal and meibomian glands to optimize tear quantity and composition.

Corneal nerve stimulation initiates normal tears.

- Unconscious process with many stimuli.
- Identify dry spot formation.
- Out of necessity for survival, evolution cornea is the most densely innervated epithelial surface in the body.
- Conduction of pain originates from myelinated and unmyelinated nerves that terminate in the cornea, limbus, and conjunctival epithelium.

Neural receptors in the cornea are free nerve endings that terminate in all of the corneal epithelial layers and are protected from direct irritation by zonula occludens and the tear mucin gel.

- Afferent (ocular surface to the brain) nerve traffic through V1 ophthalmic refers to midbrain and the para spinal SANS.
  - Signals are integrated with cortical and other inputs.
- Transmitted to the efferent (brain to ocular surface tissues and associated glands) secretomotor impulses resulting in secretion of the homeostatic tear-film.

Tear secretion by the lacrimal gland also occurs in response to neural stimulation.

- PANS: The acini, ducts, and blood vessels of the lacrimal gland are innervated by parasympathetic, sympathetic, and sensory nerves.
  - Initial signal is PANS cholinergic nerves via acetylcholine.
  - ACh binds to M3 muscarinic ACh receptors of secretory epithelia.
- VIP: Vasoactive intestinal peptide (VIP) binds to VIPergic receptors.
- Norepinephrine binds to a1- and p-adrenergic receptors.
- Substance P: Innervation of the accessory lacrimal gland?
- Sensory, sympathetic, and parasympathetic neuropeptides are present in the ocular surface tissues and associated glands. Conjunctival goblet cells have M3-muscarinic receptors and a1a- and p3-adrenergic receptors. Meibomian glands have both types of receptors as well. Thus autonomic nerves can regulate lipid secretion onto the ocular surface.

Morphogenesis of the Eyelids and Lacrimal Gland by Carnegie Stage

The human lacrimal gland consists of two lobes separated by the tendon of levator palpebrae superioris. Orbital lobe is invaded by five to six epithelial buds from the upper fornix. Development is complete at stage 23 (8 weeks). Palpebral lobe development lags behind, being initiated at stage 23. The following landmarks should be compared with the developmental stage of the eye and orbit previously described [55, 56] (Fig. 13.91)

Stage 9

- First pharyngeal arch
- Neural crest from r0–r1 in place for dura

Stage 10

- Second pharyngeal arch
- Neural crest from m1–m2 in place: source material for the future lacrimal capsule
- Optic primordia appear

Stage 11

- Third pharyngeal arch
- Fusion of first and second arches, repositioning in front of third arch and buccopharyngeal membrane: r4 neural crest mesenchyme now in position as source material for future lacrimal gland elements

Stage 12

- Fourth pharyngeal arch

Stage 13

- Fifth pharyngeal arch.
- Optic vesicle lies beneath the PPE (preplacodal ectoderm) from prosomere hp1.
- PPE thickens in response to form optic (lens) placode.

Stage 14

- Lens placode indented as lens pit

Stage 15

- Lens pit closes over to form lens vesicle.

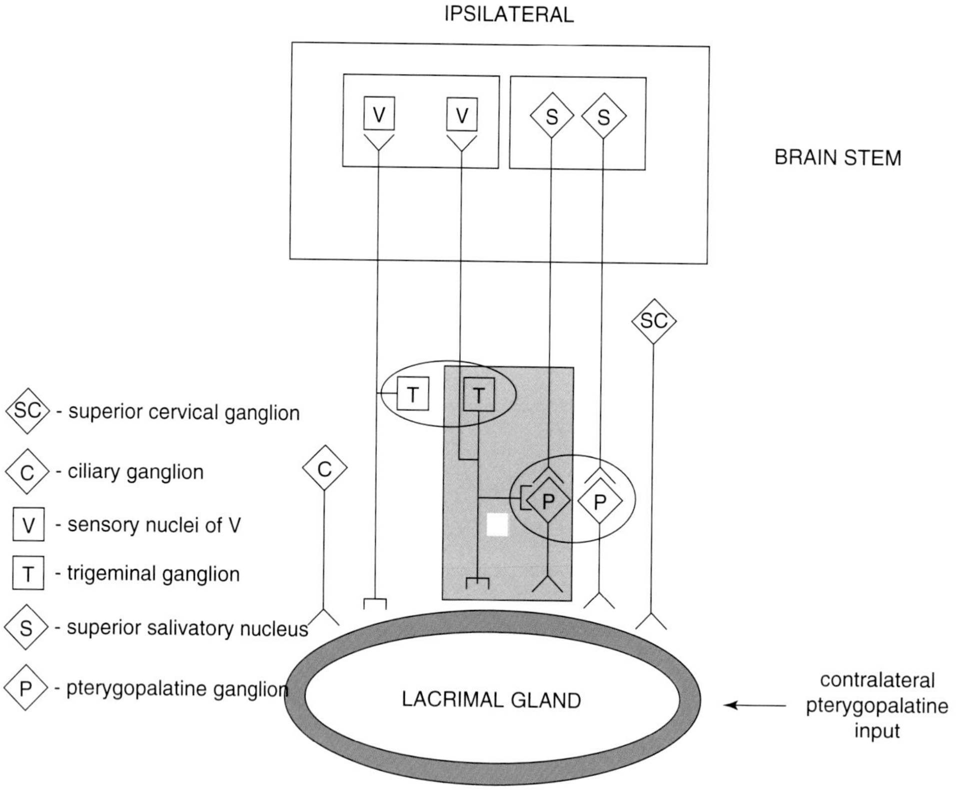

Fig. 13.91 Neuroanatomy of the primate lacrimal gland innervating following tracers. Lacrimal functional unit has three components: (1) afferent sensory, (2) efferent PANS, (3) efferent SANS

- Subconscious stimulation of the free nerve endings within the cornea generates afferent nerve impulses through V1 to the midbrain (pons). The afferent signals are integrated in the midbrain and then travel via the efferent branch through the pterygopalatine ganglion, terminating in the main (Wolfring) and accessory (Krause) lacrimal glands. Evidence suggests that this pathway also controls secretion from meibomian glands and conjunctival goblet cells.
- Proper function of the LFU supports homeostasis on the ocular surface by controlling secretion of the three major tear film components (mucin, aqueous, and lipid) to maintain the optimal quantity and quality of tear fluid; however, dysfunction of the LFU may lead to altered tear film composition and dry eye disease.
- Function of the LFU is to control secretion of tear constituents that help sustain a stable, anti-infective, and epithelial supportive tear layer essential for optimal optical performance. Signals emanating from ocular surface sensory nerves supply continuous input into the CNS that tells the brain what changes are occurring within the ocular surface milieu. The brain then sends signals to the specialized support tissues, for example, lacrimal and meibomian glands that are programmed to secrete the optimal tear quantity and composition.
- The process by which normal tears are secreted is initiated following corneal nerve stimulation. The process occurs unconsciously and in response to many stimuli; however, environmentally induced dry spot formation is thought to be among the most common. Out of necessity for survival, evolution has shaped the cornea to become the most densely sensory nerve innervated epithelial surface in the body. Conduction of pain originates from myelinated and unmyelinated nerves that terminate in the cornea, limbus, and conjunctival epithelium. Neural receptors in the cornea are free nerve endings that terminate in all of the corneal epithelial layers and are protected from direct irritation by zonula occludens and the tear mucin gel. Afferent (ocular surface to the brain) nerve traffic through V1 ophthalmic enters the central nervous system in the area of the pons (midbrain) and the para spinal sympathetic tract. These signals are integrated with cortical and other inputs and are then transmitted to the efferent (brain to ocular surface tissues and associated glands) secretomotor impulses resulting in secretion of the homeostatic tear-film components.
- Tear secretion by the lacrimal gland also occurs in response to neural stimulation. The acini, ducts, and blood vessels of the lacrimal gland are innervated by parasympathetic, sympathetic, and sensory nerves. The initial signal originates from the parasympathetic cholinergic nerves via acetylcholine release, which then binds to M3 muscarinic acetylcholine receptors on the basolateral cell membrane of secretory epithelia. At the same time, vasoactive intestinal peptide (VIP) binds to VIPergic receptors, and norepinephrine, a sympathetic neurotransmitter binds to a1- and p-adrenergic receptors. Neural innervation of the accessory lacrimal glands has also been reported and fibers positive for CGRP and substance P are associated with secretory tubules, interlobular and excretory ducts, and blood vessels. However, the degree of neural influence over accessory lacrimal glands is still being elucidated.
- Sensory, sympathetic, and parasympathetic neuropeptides are present in the ocular surface tissues and associated glands. Conjunctival goblet cells have a secretory response to the parasympathetic cholinergic muscarinic output from the pterygopalatine ganglion. Goblet cells express M3-muscarinic receptors on their membranes. The M1 and M2 receptors are located throughout the conjunctiva. The presence of a1A- and p3-adrenergic receptors on conjunctival goblet cells suggests the presence of sympathetic innervation. In addition, transmission electron microscopy of meibomian glands revealed the presence of unmyelinated axons with granular and agranular vesicles. Substance

P- and CGRP-positive axons have also been identified, but their function is uncertain, as these neurological peptides would be expected to conduct information into, rather than away from, the CNS. It is predicted that parasympathetic fibers innervating the meibomian glands are indeed present at higher levels. Parasympathetic neurotransmitters neuropeptide Y and VIP have been found around the meibomian glands, as well as tyrosine hydroxylase in sympathetic axons, implicating that both types of autonomic nerves may play an important role in stimulating lipid secretion onto the ocular surface.

Patients with LKC commonly complain of constant corneal sensations normally described as a gritty, sandy, or itchy. These complaints usually accompany with a pathophysiological state that indicates a chronic state of inflammation and a disadvantageous change in tear film composition. Infiltrating inflammatory cells within the ocular surface tissues have been reported in dry eye. These inflammatory cells, in addition to ganglioside-specific antibodies and other neural proteins, may result in regional degeneration of small diameter axons and their terminals. Chronic dysfunction of the LFU results in a shift toward inflammation and persistent psychological distress.

[Reprinted from Van der Werf F, Baljet B, Prins M, Otto JA. Innervation of the lacrimal gland in the cynomolgous monkey: a retrograde tracing study. J Anat 1996; 188:591-601. With permission from John Wiley & Sons.]

- Lens vesicle and optic cup lie just beneath, and in contact with, the surface.
- Initial epithelium in front of the lens is PPE.

Stage 16

- Eyelid grooves form prior to lids: first upper, then lower.
- Retinal pigment.

Stage 17

- Eyelid grooves deepen.
- Nasolacrimal groove.

Stage 18

- Upper eyelid fold
- Palpebral primordium

Stage 19

- Lower eyelid fold.
- Lateral canthus.
- Superior conjunctival fornix thickens.

Stage 20 (51–53 days, 18–22 mm)

- Medial canthus.
- Epithelium organizes into nodules. Mesenchyme surrounding this zone in the fornix assumes an oval shape.

Stage 21 (53–54 days, 22–24 mm)

- Epithelial buds further condense into an identifiable fornix. Invagination begins.

Stage 22

- Eyelids fully formed but not closed.
- Neurovascular supply arrives at the gland. The epithelial formations become isolated within the mesenchyme. Arrival of lacrimal nerve at the posterior medial aspect of orbital lobe of the gland. No lumen formation.

Stage 23

- Eyelids closed.
- Ducts appear in the glandular elements. Upper rectus muscle inserts in the sclera.

Fetus 9–12 weeks

- 9 weeks: levator palpebrae superioris appears, upper lid shows tarsal plate, eyelash follicles, and orbicularis oculi, orbital part identified.
- 10 weeks (52 mm): LVP divides the gland into two lobes.
- 12 weeks: acini of glands appear. Anastomosis between zygomatic nerve and lacrimal nerve. Septum orbitale of upper lid appears as an extension of supraorbital periosteum.

Fetus 13–14 weeks

- Arborization of parenchyma. Intraglandular anastomosis between V2 lacrimal and V2 zygomatic nerves.
- 14 weeks: Upper lid sebaceous glands. Central fat pad and Müller's muscle identified.

Fetus 15–16 weeks

- Stromal condensation, formation of lobes
- By 16 weeks both lobes vascularized

Fetus 17–18 weeks

- Upper lid: preseptal fat pad

Fetus 19–20 weeks

- Medial fat pad with white fat
- Müller's muscle and LPS joined

In sum: Origins of lacrimal gland mesenchyme

The mesenchymal composition of lacrimal gland represents the target cell populations of the nerves. It consists of three principle mesenchymal components: (1) a surrounding capsule and stroma, (2) tear-producing glands, and (3) ducts that convey the tears to the surface of the eye. Intraorbital MNC provides the capsule of lacrimal gland and its stromal fibers. The glandular elements come from the first and second pharyngeal arch complex, which is a mixture of r2–r5. The topographic organization of all pharyngeal arches is determined by the *Dlx* code. When the first and second arches meld together their composite structure respects this system. Thus, the maxillary "upper deck" contains first arch r2 fields paired with second arch r4 fields. The mandibular "lower deck" pairs up r3 with r5. Lacrimal gland belongs not to the orbit sensu stricto but to the maxillary-zygomatic complex. We therefore hypothesize that genetic profile of glandular elements matches that of rhombomere 4 and that neural crest glandular cells originate at that site. The ductal epithelium originates from r1 ectoderm on the inner surface of lid and penetrates into the substance of the gland.

Selected Clinical Pathologies of the Orbit

Periorbital Dermoids Arise on Either Side of Ancient Bone Fields

The histology of dermoids is well known and consists of a stratified squamous lining that is keratinized with adnexal contents, including glands. They are considered to arise at ectodermal lines of fusion or from an underlying suture. The traditional concept of mechanism (dating back to 1910) as proposed by Pratt and later by Posnick holds that a projection of dura is included in a suture line, comes into contact with skin, and is eventually pinched off. Posnick considers this theory, in the case of midline dermoids, as equally applicable to the embryogenesis of gliomas and encephalocoeles [57, 58] (Figs. 13.92 and 13.93).

These concepts are not supported by contemporary neuroembryology. Let us be clear that no mesoderm is involved in these lesions whatsoever. With the exception of the (1) basioccipital bone complex of the cranial base and (2) the striated muscles of the eye, pharyngeal arches, and tongue, all

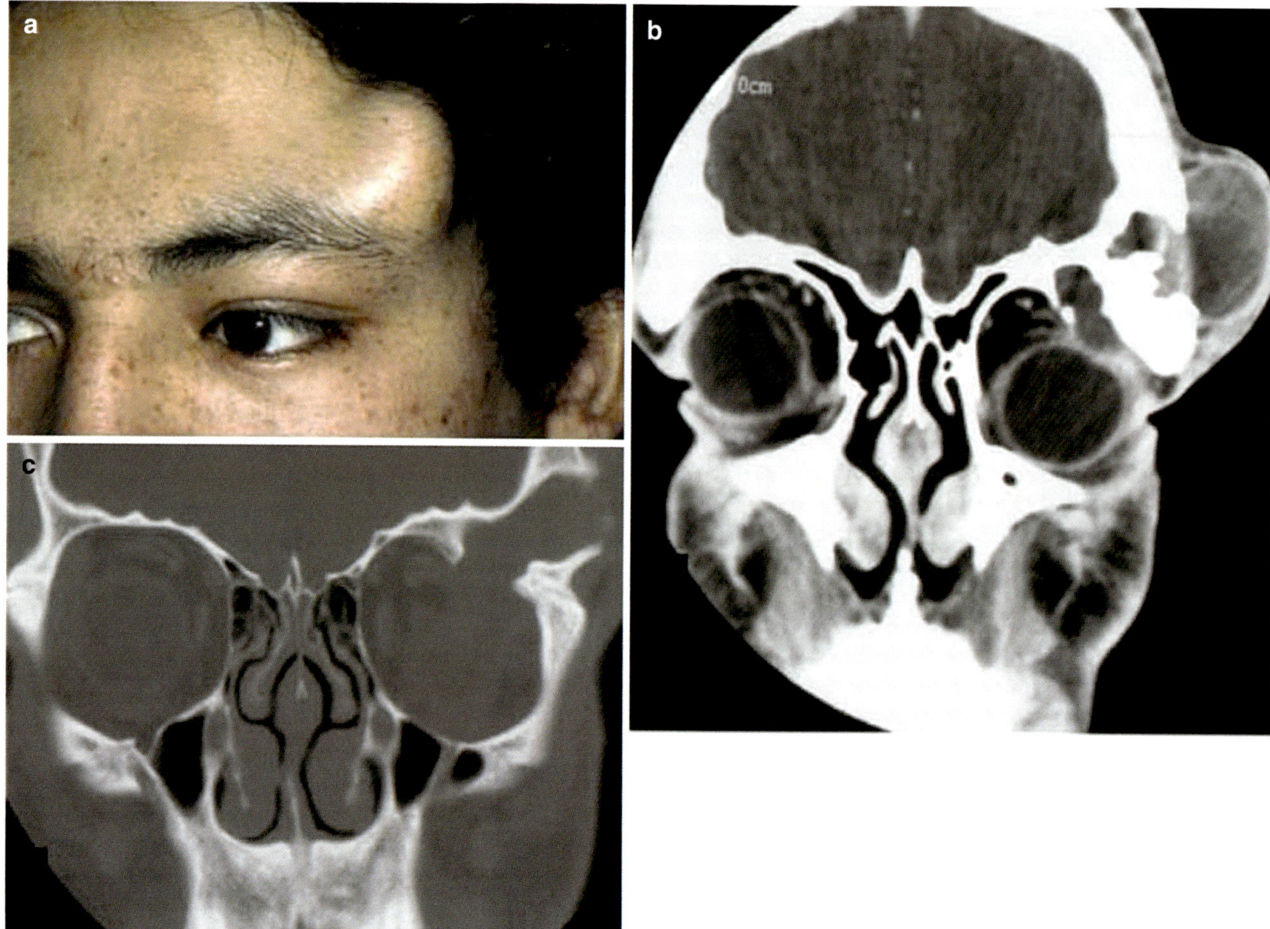

Fig. 13.92 Dermoid zones 9. Note defect in lateral orbital roof in postfrontal zone bordering with alisphenoid. Mass effect causes depression of globe at zone 10, where the orbital plate is generally unilaminar. [Courtesy of Michael Carstens, MD]

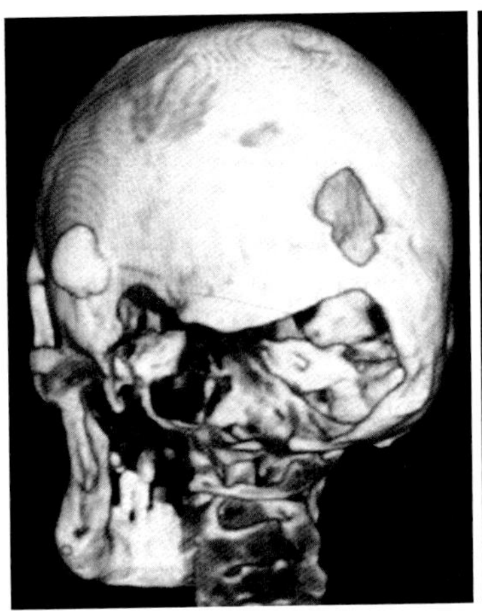

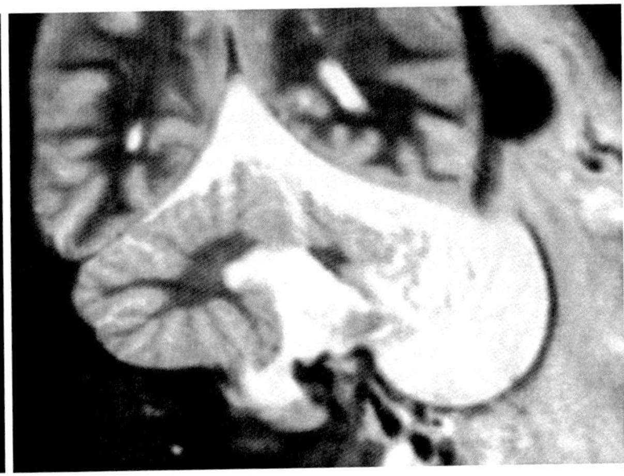

Fig. 13.93 Zone 9 encephalocoele caused by loss of great sphenoid wing. Infratemporal fossa sphenoid via defect in petrous margin.

craniofacial mesenchymal structures, be they bone, cartilage, fascia, fat, or dura, are of exclusively neural crest derivation. So too are the contents of periorbital dermoids. This neural crest tissue of the same derivation as the dura. As mentioned earlier in this chapter dermoids cannot contain neural tissue. However, if full-thickness dura/leptomeninges exit through an unsealed border between adjacent cranial bone fields, the resultant hernia can contain neural tissue, i.e. either a glioma or an encephalocoele.

In general, neural crest bone undergoes membranous ossification. It can also pass through a cartilaginous intermediate, as in the presphenoid or orbitosphenoid, or it can produce chondral bone as in parts of clavicle and scapula. Membranous bone represents a response to an epithelial signal, be it from dura, dermis, or submucosa. When a responder population is programmed from two distinct sources, the bone is *bilaminar* and can contain marrow, as in the membranous bones of the calvarium. If invaded by oronasal epithelium, the two walls will grow away from one another, forming a sinus cavity. *All sinuses represent field separation planes.*

The frontal bones are paired, each one having two ossification centers. The primary ossification center appears at the end of the second month of fetal life and is located just above the supraorbital notch. This represents the vascularization of the osteogenic matrix. With time, ossification spreads upwards, tracking as expected along the axis of the supraorbital neurovascular axis. A second ossification center appears later rostrally. Frontal bone is programmed from r1 dura and from FNO skin and, with orbital roof, by r1 dura above and sclera below. This explains the frontal sinus and its transverse extension laterally into the orbital plate.

Returning to the subject of dermoids one notes that the eyebrows define a zone running the entire span of the superior orbital rim. This indicates the presence of underlying, independent developmental fields, bilaterally situated on either side of the supraorbital axis. In cases of anencephaly or exencephaly, with near total loss of the frontal bone, the orbital rims and eyebrows are preserved. The fossil record provides a clear-cut explanation for these phenomena. In discussing these bone fields, we will use standard zoologic terminology.

In paleontology, Herring's law stipulated that multiple bone fields tend to coalesce and simplify with time. The parietal originally included bones such as interparietal and tabular as evolutionary hold-overs. Review of the anatomic literature reveals many reports of so-called "ectopic" bones in the membranous occipital complex (see Chap. 8).

In the ancestral tetrapod model the orbit was surrounded by a series of membranous bones. The *orbital series* consisted of lacrimal (L), prefrontal (Prf), postfrontal (Pf), postorbital (PO), and jugal (J) [59]. Lacrimal was much larger and was in direct contact with jugal, thereby excluding maxilla. Medialization of the orbits along with function changes in jaw suspension simplified the situation. Separation of jugal from lacrimal permits maxilla to be incorporated into the orbital floor. The modern zygoma is now a combination of jugal and postorbital. Prefrontal and postfrontal in humans are fused to frontal. They can appear in isolation in cases of anencephaly [60, 61] (Fig. 13.2).

In the medial back wall of the orbit the tetrapod *spheneth-moid* complex split into a membranous presphenoid and a chondral basisphenoid. The *epipterygoid* bone served as a mobile articulation between the skull base and palate. It subsequently expanded superolaterally and fused to the basisphenoid to create the modern *alisphenoid* (AS). Since palate is a non-RNC field supplied by StV2, it is not surprising that the blood supply for AS retains its ancient allegiance to the oral cavity. Separation of jugal from lacrimal permitted maxilla to become incorporated into the anterior wall and medial floor.

Embryonic developmental fields correspond to blocks of tissue that interact in a programmed fashion to produce the topology of the face. Like tectonic plates, these units possess autonomous innervation and blood supply and move along developmental "slippage planes" according to their mass, velocity, directionality, and timing. Facial development can therefore be considered as the *formation, migration, coalescence*, and *interaction* of separate genetically based fields. Adjacent tissue units required to form a structure (such as the nostril) can be termed *partner fields*. Their interface is a *field boundary*. Any process that disrupts the physical integrity of a field or field boundary will mechanically disrupt normal field interactions leading to *field mismatch*.

Mechanism of Dermoid Formation

In the case of the orbit, although prefrontal and postfrontal are no longer distinct, the genetic "idea" behind these fields is still present during development. Prf and Pf are served by the same neuroangiosome that supplies the overlying frontal bone. Thus, during the embryonic fusion process interposition of intracranial MNC between the postfrontal and postorbital or between prefrontal and lacrimal creates the conditions for a dermoid cyst at those locations.

To complete the story, recall that r1 MNC is the source of dura for the frontal lobes and cerebellum. Primitive dura has two layers, a deep layer in contact with arachnoid and is separated by a *dural limiting membrane* that constitutes a barrier between it and the CNS beneath. Hence follows a *loose layer* containing subdural veins. The outer so-called *skeletogenous layer* is the source for frontal bone. When mature, the outermost dura is thick and tough with dural arteries running along its surface. A final superficial layer, the deep periosteum, permits reflection of the dura from the calvarium in a virtually bloodless plane. Thus, the mesenchyme of a dermoid cannot contain CNS tissues. The etiology of glioma or encephalocoele involves a protrusion of all the layers. This requires a more substantial escape route, such as in the foramen cecum. For this reason, these more complex masses can be found to penetrate the frontonasal junction passing above nasal bone or escaping beneath nasal bone. The common denominator for these lesions is widening of crista galli of the ethmoid complex. This failure to close may be the primary problem, analogous to an encephalocoele, or it may be secondary to the interposition of dural mesenchyme.

In conclusion, dermoids may well constitute a *forme fruste* of craniofacial cleft. The lateral dermoid is at the interface between zone 8 postorbital bone and zone 9 alisphenoid. The medial dermoid occurs at the interface of zones 11 and 12. Midline dermoids represent a fusion failure due to inadequate apoptosis of ethmoid mesenchyme. Other locations can occur wherever field failure is present (Fig. 13.93).

Encephalocoeles: Field Failure and Escape Routes

Encephalocoeles are a variation on the same theme as dermoids; both represent masses that exploit a defect in a cranial bone field to occupy a pathologic position. They are typically found extracranially but they can occur within the skull itself, invading the membranous labyrinth through a defect in the superior semicircular canal, or into the transverse sinus [62, 63]. Encephalocoeles present in two forms. *Meningocoele* refers to a sac containing meninges and cerebrospinal fluid but no CNS tissue. *Meningoencephalocoele* indicates the sac presence of neural tissue as well. Extracranial encephalocoeles are covered by either skin or (within the nose or pharynx) by mucosa (Figs. 13.93, 13.94, and 13.95).

Classification of encephalocoeles is commonly by location of the bony defect but adds little information as to pathogenesis. Distribution of encephalocoele is geographic, with nasofrontal being common in Southeast Asia while occipital is more frequent in Africa [64].

Frontoethmoidal: Exit anterior to cribriform, can present externally

- Nasofrontal
  - Can track along nasal dorsum
- Nasoethmoidal: anterior to cribriform plate into nose
- Naso-orbital

Basal: Exit within or posterior to cribriform plate, presents internally

- Transethmoidal: cribriform plate into nasal cavity
  - Field 12 defect lateral
  - Field 13 defect medial
- Spheno-ethmoid: more posterior and into nasal cavity
- Spheno-sphenoid
  - Pre-sphenoid-postsphenoid suture
  - Penetrates sphenoid sinus or nose
- Spheno-orbital
  - Via superior orbital fissure
  - Dystopia
- Sphenomaxillary: via orbitosphenoid/alisphenoid into pterygoid fossa

Occipital

- Most common (75–80%)
- Bone fields of interparietal
- Can reach very large size [65]

Cranial vault: through membranous bones of calvarium

- Interfrontal
- Interparietal (includes fontanelles)
- Parieto-temporal

Intracranial

- Middle cranial fossa

Contrary to dogma, encephalocoele does not represent a neural tube defect. It is a strictly neuromeric issue involving an inadequate field boundary. In most cases, these boundaries are identified by sutures but some cranial bone fields represent an incorporation of one or more pre-existing fields. This is the case of zygoma, in which jugal fuses with postorbital with no marker other than the zygomatico-facial foramen. Prefrontal and postfrontal are snatched up by frontal during evolution. We can separate them in the midline by the axis of supraorbital nerve and artery. The emergence of encephalocoeles serves as a marker of their original medial and lateral boundaries.

Encephalocoeles can occur under secondary circumstances due to trauma or tumor. Pulsatile exophthalmos is a good example of the former. Destructive lesions such as plexiform neurofibroma can create a passage for encephalocoele, as reported in a case that destroyed the left lower occipital field and neighboring petrosa field [66] (Fig. 13.93).

Encephalocoeles are not lytic, they are opportunistic. What strikes one about the various presentations of congenital cranial encephalocoele is that many involve r1 fields (frontal, ethmoid, sphenoid). The remainder such as temporal, interparietal, or via the fontanelles involve r2–r3 neural crest membranous bones. And in every case a field boundary can be identified. But if we dig deeper we can learn lessons about skull embryology from presentations that are a bit less pedestrian (Fig. 13.94).

Unusual Calvarial and Intracranial Encephalocoeles

The parietal bone presents multiple boundaries. At the parietal squamosal suture, nine intradiplöic encephalocoeles have been reported [67]. An atretic form of parietal encephalocoele has been described at the parietal–interparietal boundary, i.e., the lambdoid suture. These lesions consist of a stalk containing arachnoid tissue tracking from the dura through a bone defect and attaining a subcutaneous position. The lack of representation along the bi-parietal suture is notable, with encephalocoeles confined (quite logically) to the fontanelles.

Inside the skull, patency of the developmental boundary between the prootic and opisthotic fields can lead to invasion of encephalocoele into the middle ear or disruption at the superior semicircular canal [63]. A common presentation is meningitis secondary to otitis media. Contact between the encephalocoele and the ossicular chain can cause pulsatile tinnitus. Petrous apex involves the boundary between the Sm5 prootic and basioccipital fields. In Mulcahy's report all three patients had congenital defects in Meckel's cave [68]. This represents a weakness or absence of tegmen tympani [69]. Additional etiologies for encephalocoelic herniation into the temporal bone include infectious (otitis) and post-traumatic (or iatrogenic). Temporal lobe encephalocoeles can track forward in defects between V1 and V2 to enter the sphenoid sinus [70]. In other cases they have been associated

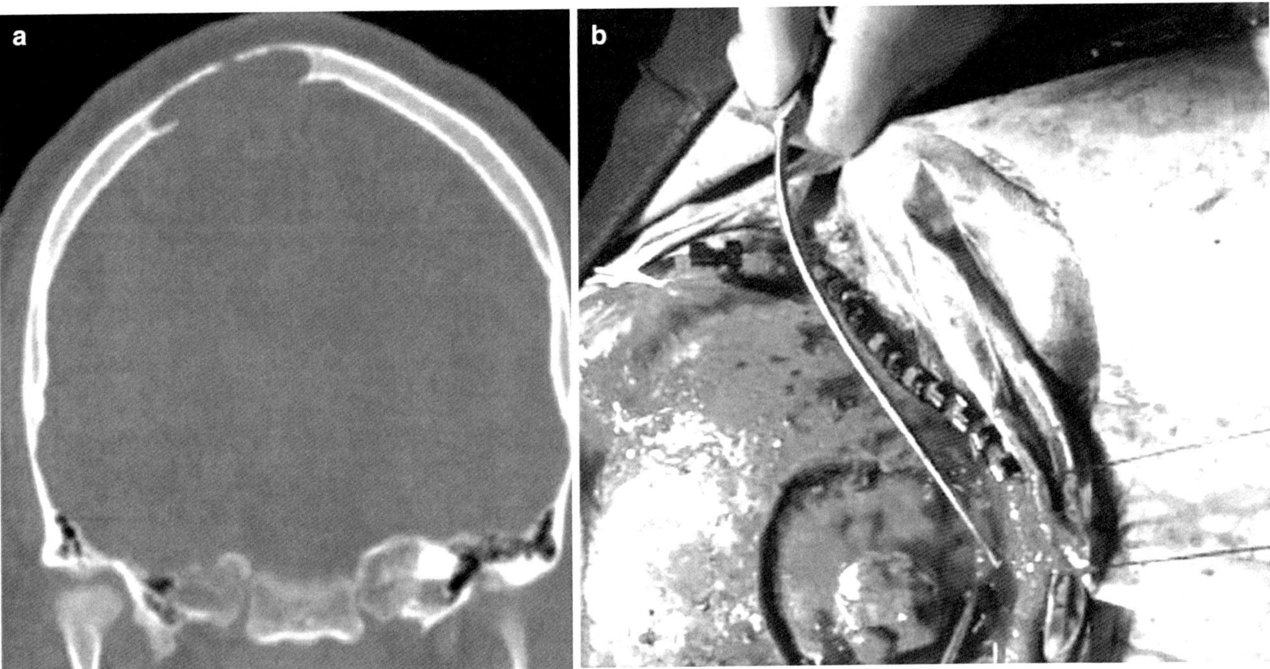

Fig. 13.94 Parietal diplöic encephalocoele. Defect on CT with mass seen at surgery (defect encircled). [Reprinted from Shi C, Flores B, Fisher S, Barnett SL. Symptomatic parietal intradiplöic encephalocele: case report and literature review. J Neurol Surg Rep 2017;78:e43–e48. With permission from Thieme.]

with temporal lobe epilepsy refractory to medical therapy. Surgical intervention in such cases is promising [71].

Basal Encephalocoeles

Let's return to r1 bone fields once more. The many developmental fields of the orbito-spheno-ethmoid complex present opportunities for encephalocoele development. Recall that cerebellum develops from r1. A clival defect through the boundary between r1 basisphenoid and somitic basioccipital permitted encephalocoele to enter the nasopharynx. The cerebellum was hypoplastic and the pons was kinked, being dragged downward to the boundary along with the polyp. The brainstem was likewise dysplastic, pointing toward a more global problem affecting the homeotic fields of multiple rhombomeres [72].

It should come as no surprise that encephalocoeles involving the sphenoid are associated with a high incidence of facial anomalies. The presence of an encephalocoele bespeaks an embryologic event: "the brain predicts the face." And herein lies a lesson. As we have previously discussed, the developing forebrain is almost immediately surrounded by neural crest cells. These synthesize the meninges covering the forebrain; they also assist in the formation of blood vessels, in the form of pericytes. From this layer will come the bony protection for the brain. It is likely that all events leading to encephalocoele take place around the same time, with one exception…midline closure of the r1 fields. If sphenethmoid fields achieve union beneath the forebrain, no matter if an encephalocoele can form, it will not be capable of blocking normal facial approximation. But early defects in the midline can forestall this process leading to gross widening of the face or even intrauterine demise.

The Australian Craniofacial Unit review of basal encephalocoeles found clinical distinctions between the transethmoid (TEE) and sphenoid-related presentations. TEE can be difficult to diagnose but it is sometimes accompanied by pulsatile exophthalmos. By mass effect TEE can present as hypertelorism with widening of the midline but an otherwise normal brain. Sphenoid variants, on the other hand, always involve a drastic defect in r1 neural crest migration or faulty performance in situ. These relate to midline CNS structures such as corpus callosum, abnormalities of the optic disc and tract, and intracranial lipomas representing reassignment of neural crest tissue to fat. Midline facial manifestations include median cleft nose with a flattened root, loss of premaxilla, and vomer (median cleft lip). Defects of the sella, as in sphenomaxillary encephalocoeles, may be accompanied by endocrine dysfunction as they involve the hypothalamus and pituitary in prosomere hp2 [73].

Encephalocoeles and the Orbit

Key concept: Craniofacial clefts, congenital defects in four distinct neurovascular fields permit encephalocoeles to gain access to the orbit, causing dystopia. We shall return to the cleft zones subsequently (vide infra) (Fig. 13.95).

Herniation of cranial tissue into the orbit can take place from only two sites: via the roof and through the back door.

The orbital roof is thin r1 membranous bone consisting of three fields.

- Zone 9 is supplied by StV1 lacrimal artery and extends forward from the fronto-alisphenoid suture to form a triangle in the lateral one-third of the orbital floor. Zone 9 is generally unilaminar.

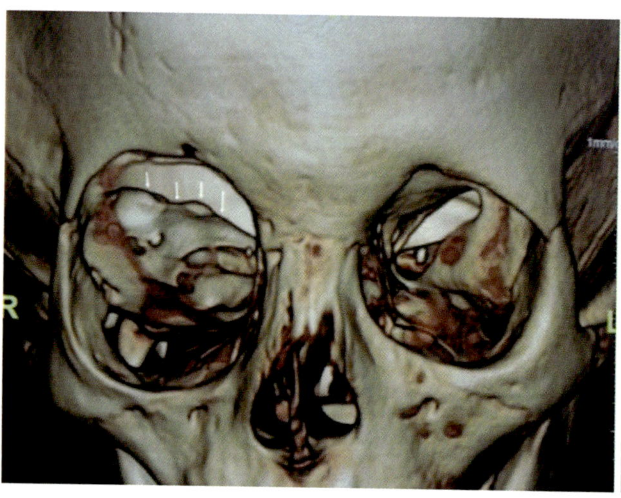

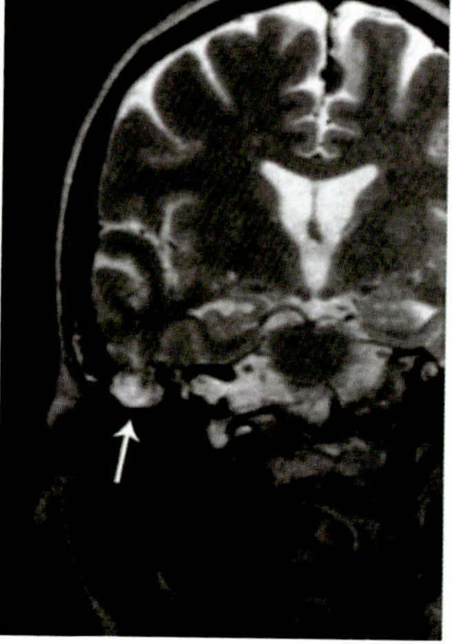

Fig. 13.95 Orbital and petrosal encephalocoeles reveal field boundaries. Left: Zone 9 encephalocoele caused by loss of great sphenoid wing. Right: Infratemporal fossa sphenoid via defect in petrous margin. Left: [Reprinted from Asii K, Gunduz Y, Yaldiz C, Aksoy YE. Intraorbital encephalocele presenting with exophthalmos and orbital dystopia: CT and MRI findings. J Korean Neurosurg Soc 2015; 57(1):58-60. With permission from the Korean Neurosurgical Society.]. Right: [Reprinted from Sanna M, Fois P, Russo A, Falcioni M. Management of meningoencephalic herniation of the temporal bone: personal experience and review of the literature. Laryngoscope. 2009 Aug;119(8):1579-85. doi: 10.1002/lary.20510. Review. Erratum in: Laryngoscope. 2010 Jan;120(1):217. With permission from John Wiley & Sons.]

- Zones 10 (post-frontal) and 11 (prefrontal) lie on either side of the StV1 supraorbital axis. They form a quadrangular zone defined posteriorly by the frontal-orbitosphenoid suture. Recall that the transverse extension of frontal sinus occupies zones 11 and (to a lesser extent) 10; this part of the orbital roof is bilaminar. For this reason defects in the supraorbital fields favor encephalocoele formation laterally, where the bone is a single layer.
- Abutting against zone 11 is the anteroposterior border with the ethmoid complex. The arterial supply is StV1 nasociliary. The frontoethmoid suture running longitudinally forward from lesser wing to crista galli constitutes a potential entry point for encephalocoele into the medial orbit.

The back door of the orbit is alisphenoid. Its arterial supply is the StV1 meningo-orbital artery (otherwise called the recurrent meningeal artery). Recall that embryologically RMA ran forward into the orbit to anastomose with lacrimal but in most cases, by stage 23 this connection falls apart with intraorbital StV1 lacrimal picking up the slack. This unilaminar membranous bone field stretches like a sheet from the greater orbital fissure downward until it reaches the middle temporal fossa, where it promptly does two things: (1) laterally, it forms a suture with a horizontal sheet from the squamous temporal bone field; and (2) medially, it sends a vertical lamina down into the infratemporal fossa – the lateral pterygoid plate. If the bone field is synthesized from below upward, any mesenchymal deficit will affect the width of the fissure, thereby allowing entry of the encephalocoele.

Craniosynostosis: Orbital Deformation

It is obvious that any distortion of surrounding fields can deform the orbit in different directions. Fields can be too small as in synostosis, or too large, as in neurofibromatosis. Craniostosis is an obvious place to start. While we are not going into extensive detail let's set out some principles. Ignoring the more pedestrian forms of synostosis, we will see how these principles are represented by three pathologies: (1) trigonocephaly for its effects on the orbital roof, (2) frontosphenoid synostoses, in which the alisphenoid field distorts the back wall of the orbits, and (3) squamosal synostosis, a poorly appreciated condition that reinforces the model of parietal developmental fields.

The Pathogenesis of Calvarial Craniosynostosis

Parietal bone, as we have previously discovered, is more complex than it looks. This seemingly bland neural crest bone played a decisive role in mammalian evolution when it expanded outward, providing a new system for chewing muscles and the invention of the temporomandibular joint. We also appreciate parietal as being the only membranous calvarial bone composed predominantly of Sm4 paraxial mesoderm (admixed with r2–r3 neural crest). This mesenchymal mixture divided parietal into *four distinct quadrants* of neural crest, making the boundaries it shares with its purely neural crest neighbors (frontal, squamosal, and interparietal) all subject to deformation secondary to a premature synostosis. These partner fields of distinct neuromeric composition are frontal (r1), squamosal (r2–r3), and interparietal (Sm7 ?admixed with cervical PAM) (Fig. 13.96).

Mastoid does not count. This is a purely PAM bone from Sm7 and firmly attached to the cranial base.

Alisphenoid does not count. AS and anterior inferior parietal have identical r2 populations at their common border and therefore cannot form a premature synostosis. The anterior zone of parietal and the entirety of alisphenoid are synthesized by r2 neural crest. Both bones are covered with Sm4 muscles (temporalis and lateral pterygoid). Both zones receive blood supply exteriorly, via anterior deep temporal and pterygoid branches of internal maxillary, and interiorly, by the StV2 branches of the middle meningeal system: recur-

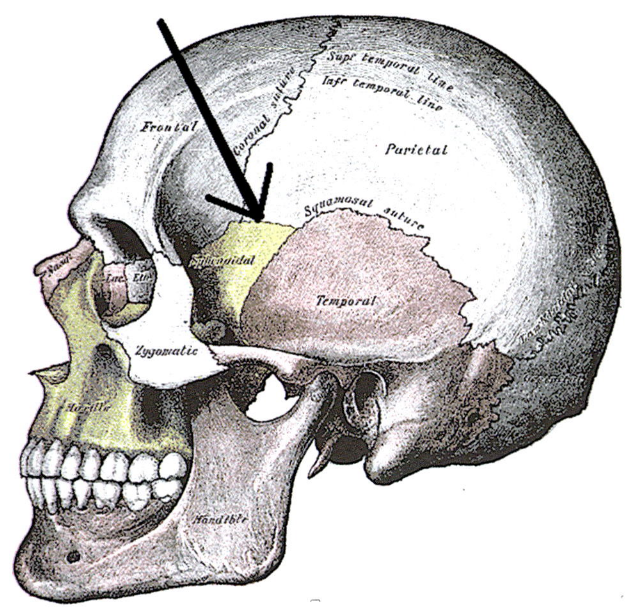

Fig. 13.96 Calvarial synostosis involves four parietal bone populations. Parietal bone consists of four distinct populations. It is surrounded by four4 boundaries involving bones of unlike neuromeric composition: frontal anteriorly and along the median sagittal strip, interparietal; and squamosal. Neuromeric fields of adjacent alisphenoid and parietal have the same composition; therefore parieto-sphenoidal synostosis does not occur. Neuromeric mismatch exists between frontal with alisphenoid and/or parietal. Thus, synostosis can take three forms: frontosphenoidal (FSS), frontoparietal (FPS), and with both (FSS + FPS). [Reprinted from Lewis, Warren H (ed). Gray's Anatomy of the Human Body, 20th American Edition. Philadelphia, PA: Lea & Febiger, 1918.]

rent meningeal artery, previously mentioned in the principle blood supply of alisphenoid.

- There are *no* cases of isolated alisphenoid-parietal synostosis reported in the literature.

Alisphenoid and squamosal also share similar populations and blood supply and cannot form a premature synostosis.

- There are *no* cases of isolated alisphenoid-squamosal synostosis reported in the literature (although its theoretical existence has been postulated by Sakamoto).

Alisphenoid is very dissimilar with frontal bone (r1) and with postorbital bone (r2) but with different blood supply.

- Synostosis between the alisphenoid with frontal bone and between alisphenoid and the postorbital (zygoma) is well documented.

In Summation

Craniosynostosis has sponsored an extensive literature, much of which has been dedicated to ascertaining its cause. Although multiple gene defects have been reported, potential correlations regarding mesenchymal mapping, rhombomeric homeotic labels, and gene defects have received little attention. Nonetheless, undaunted, let's set out three neuromeric principles, chosen for their future accessibility to experimental verification.

Principle 1. Craniosynostosis involves membranous bones of dissimilar neuromeric composition.

Principle 2. Potential sites of craniosynostosis:

- Parietal (all four borders): frontal, parietal, lambdoid, squamosal
  - exception: parieto-alisphenoid suture
  - exception: parieto-mastoid suture
- Alisphenoid (anterior border)
  - Anterior border
    - Frontal bone in orbit
    - Zygoma (postorbital field) backwall of orbit
  - Exception: superior border with parietal = unaffected
  - Exception: posterior border with squamosal = unaffected

Principle 3. Pathogenesis of craniosynostosis

- An abnormal mesenchymal field that responds to normal signals from the dura in an abnormal way
- An abnormal signaling zone of dura that sends faulty signal to the neural crest responder field

Trigonocephaly: An Embryologic Misnomer

Trigonocephaly refers to a prominence of the midline metopic suture in conjunction with a spectrum of clinical findings. In its mildest form, midline ridging is seen accompanied by a deficiency in projection of the lateral brow. The degree of suture prominence is proportional to the severity of the orbital plate deficiency (see below). It takes three forms:

1. Calvarial and extracranial
   - Stapedial system: StV1
   - Small anterior cranial fossa
   - Triangular orbital roof
   - Frontal bone normal, restricted growth laterally, above zone #9
   - Hypotelorism: transverse mesenchymal deficiency in the ethmoid complex, causing approximation of the orbits; when severe, associated with CNS
   - Depressed pterion: StV2 deficiency reduced size of temporal plate of alisphenoid
   - High-arched palate: vertical mesenchymal deficiency in the perpendicular plate of the ethmoid pulling the palatal shelves of the maxilla upward—no cleft
2. CNS
   - Internal carotid system: r1 anterior cerebral distribution
   - Midline structures: corpus callosum
   - Dropout of white matter
   - Cerebellar hypoplasia, including the vermis
   - Increased ventricular size
   - Developmental delays
3. Syndromic
   - Say–Meyer: short stature, skeletal abnormalities have been reported [74, 75]
   - Opitz C syndrome: dysmorphic facies, "gonadal mosaicism" [76]

Trigonocephaly, misconstrued as metopic synostosis, does not bear that label well. It is now established that physiologic closure of this suture takes place as early as 3 months. In virtually all patients the suture is closed by 9 months of age [77]. Classic theories of bone growth and synostosis cannot explain how a vertical midline suture can affect the growth of the lateral orbital roof. Finally, such a midline synostosis should ipso facto cause bilateral deficits but such is not the case as unilateral cases of trigonocephaly have been documented (Figs. 13.97, 13.98, 13.99, 13.100, 13.101, and 13.102).

The metopic suture and its histology have been explored in the past in a fruitless attempt to find commonality with more pedestrian forms of synostosis (e.g., coronal) [78]. In

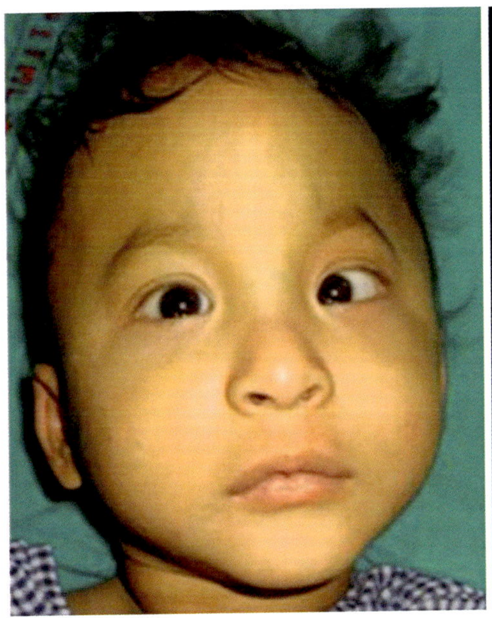

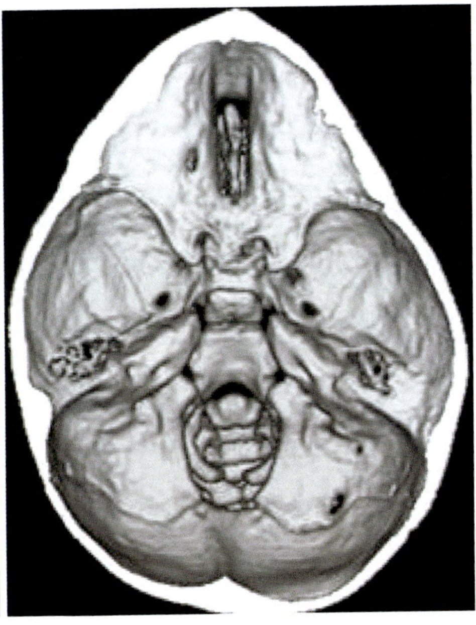

Fig. 13.97 Trigonocephaly. Note the frontal keel, the constriction of the frontal lobes and, from the vertex view of anterior cranial fossa, modes of retraction of lateral supraorbital rims with severe deficits of surface area in the lateral frontal fields. Left: [Reprinted from Rath G, Dash HH. Anaesthesia for neurosurgical procedures in paediatric patients. Iindian J Anesthesia 2012: 5(5):502-510. With permission from Wolters Kluwer Medknow Publications.]. Right: [Reprinted from Pictorial essay: Khanna PC, Thapa MM, Iyer RS, Shashank RS Prasad SS. The many faces of craniosynostosis. Indian Journal of Radiology and Imaging 2011;21(1): 49-56. With permission from Wolters Kluwer Medknow Publications.]

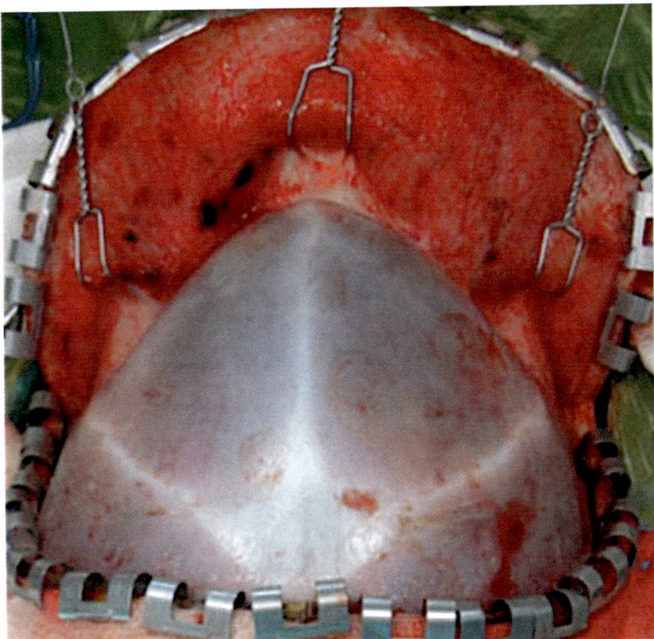

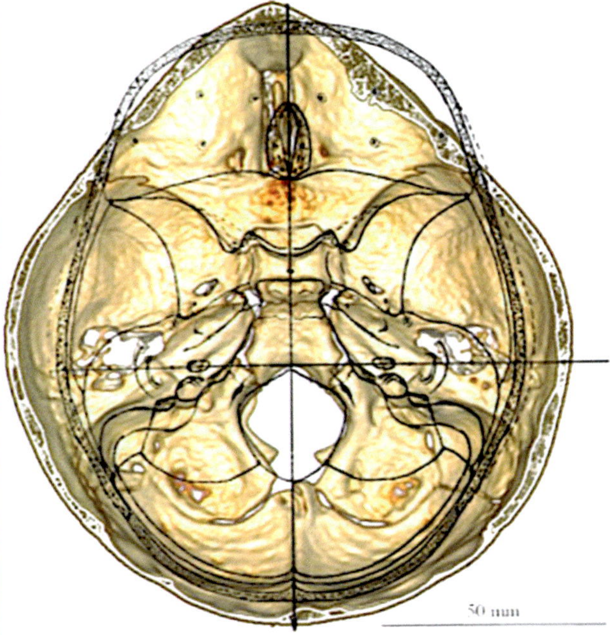

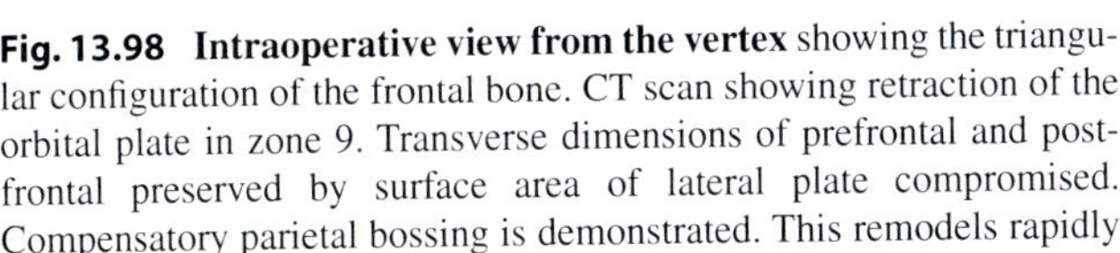

Fig. 13.98 **Intraoperative view from the vertex** showing the triangular configuration of the frontal bone. CT scan showing retraction of the orbital plate in zone 9. Transverse dimensions of prefrontal and post-frontal preserved by surface area of lateral plate compromised. Compensatory parietal bossing is demonstrated. This remodels rapidly after surgery. [Reprinted from Pellerin P, Vinchon M, Dhellemes P, Wolber A, Guerreschi P. Trigonocephaly: Lille's surgical technique. Childs Nerv Syst (2013) 29:2183–2188. With permission from Springer Nature.]

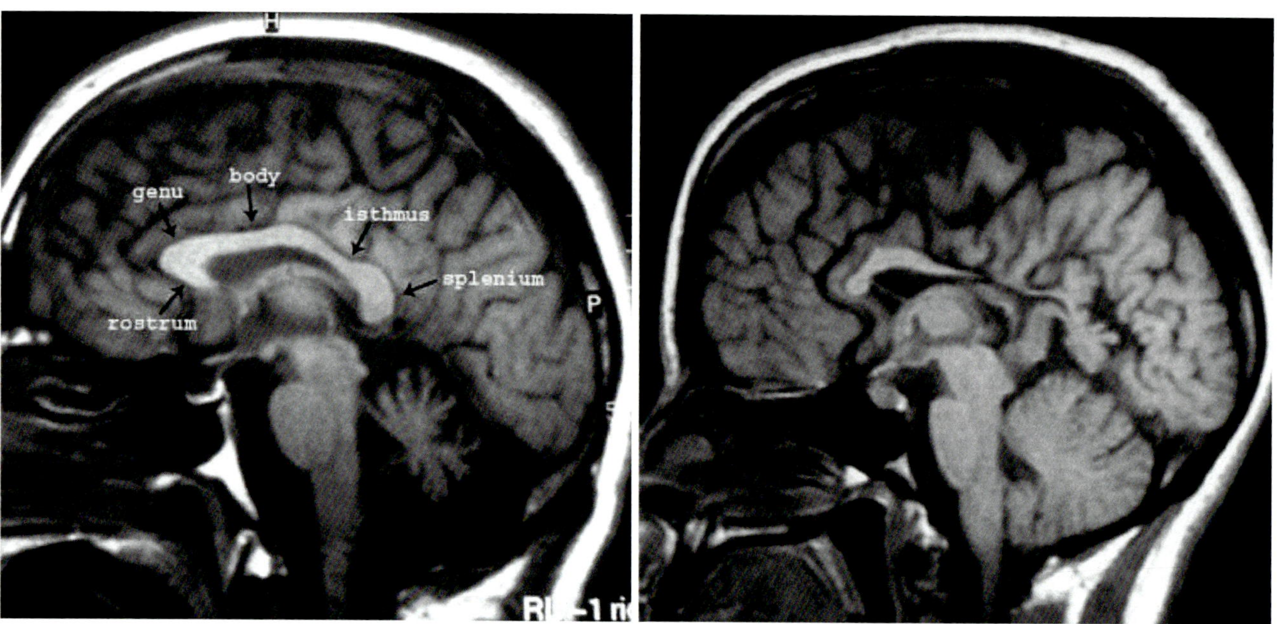

Fig. 13.99 Corpus callosum dysgenesis/agenesis in trigonocephaly. Midline defects are associated with severe trigonocephaly. Normal anatomy shown left. Agenesis of the corpus callosum follows a distal-to-proximal gradient with progressively more proximal involvement of anterior cerebral artery neuroangiosomes. Partial agenesis shows genu and anterior body of the corpus callosum visualized, whereas the posterior body, splenium, and rostrum are absent. [Image reproduced with permission from Medscape Drugs & Diseases (https://emedicine.medscape.com/), Imaging in Agenesis of the Corpus Callosum, 2018, available at: https://emedicine.medscape.com/article/407730-overview.]

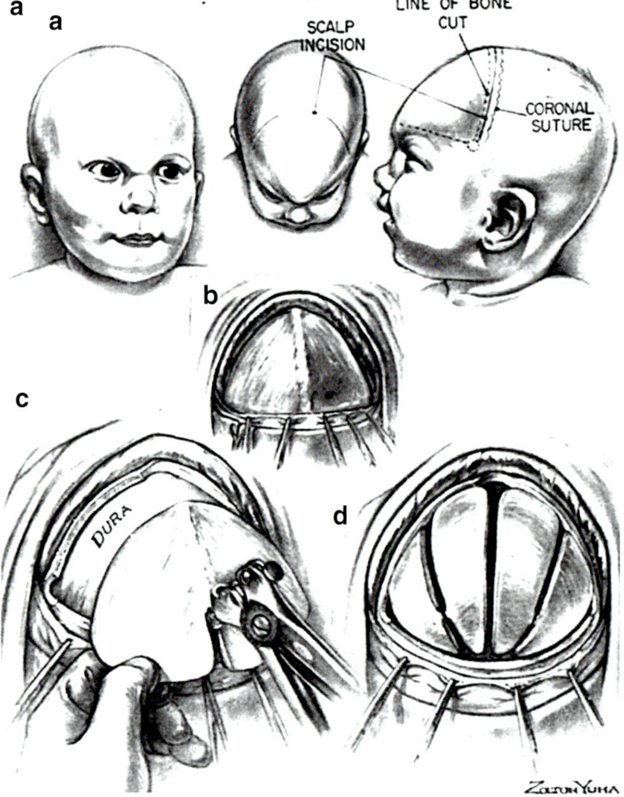

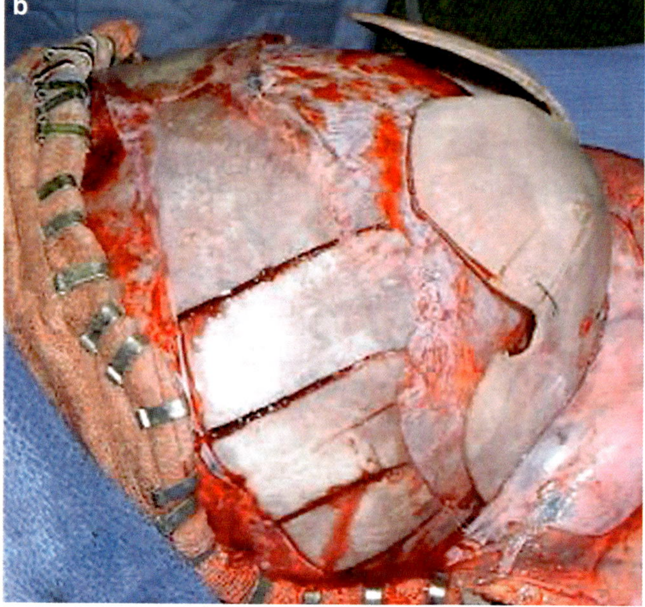

Fig. 13.100 **Original technique for correction** was focused on the metopic suture and the forehead. This ignored the deficiency state of the orbit. The constriction effect of the frontal bone on the parietal was not appreciated. At the time, fronto-orbital advancement was unknown. Contemporary techniques expand the parietal. Note the extensive mesenchymal deficit of the frontal bone fields. The zone 9 pathology affects all remaining fields 10–11–12–13 with the <u>deficit worsening as one ascends toward the vertex</u>. This indicates how effective the restriction of the upward flow of mesenchyme is from the orbital rim upward. (**a**) [Reprinted from Anderson FM, Gwinn JL, Todt JC. Trigonocephaly: identification and surgical treatment. J Neurosurg 1962; 19-723-730. With permission from Journal of Neurosurgery Publishing.]. (**b**) [Reprinted from Pellerin P, Vinchon M, Dhellemmes P, Wolber A, Guerreschi P. Trigonocephaly: Lille's surgical technique. Childs Nerv Syst 2012; 29:2183-2188. With permission from Springer Nature.]

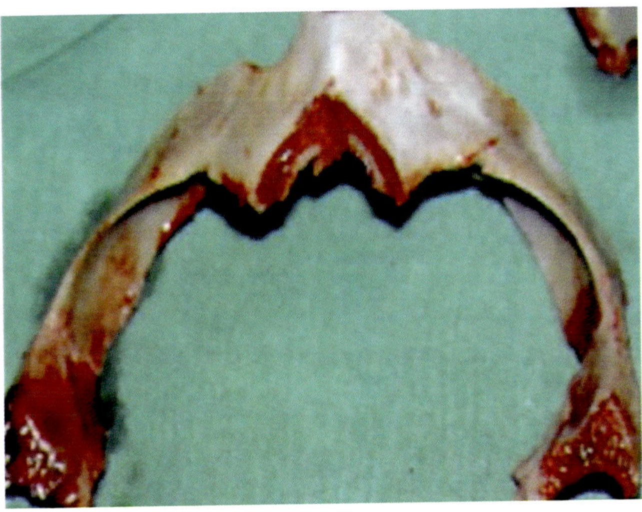

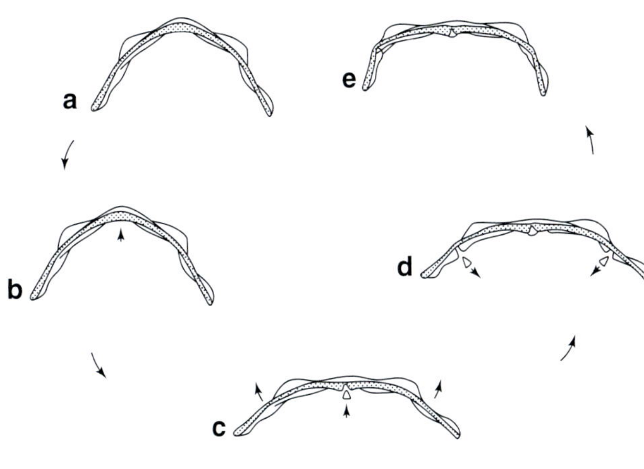

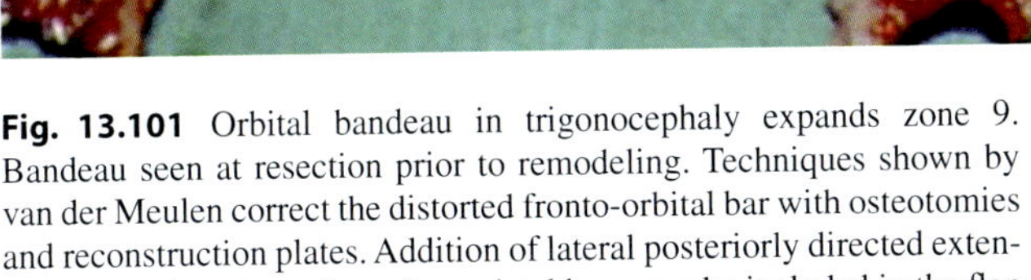

Fig. 13.101 Orbital bandeau in trigonocephaly expands zone 9. Bandeau seen at resection prior to remodeling. Techniques shown by van der Meulen correct the distorted fronto-orbital bar with osteotomies and reconstruction plates. Addition of lateral posteriorly directed extensions (not shown) cut from the parietal bone can be included in the flap and slid forward to assist in stabilization. [Reprinted from Philippe Pellerin P, Vinchon M, Dhellemmes P, Wolber A, Guerreschi P. Trigonocephaly: Lille's surgical technique. Childs Nerv Syst 2012; 29:2183-2188. With permission from Springer Nature.]

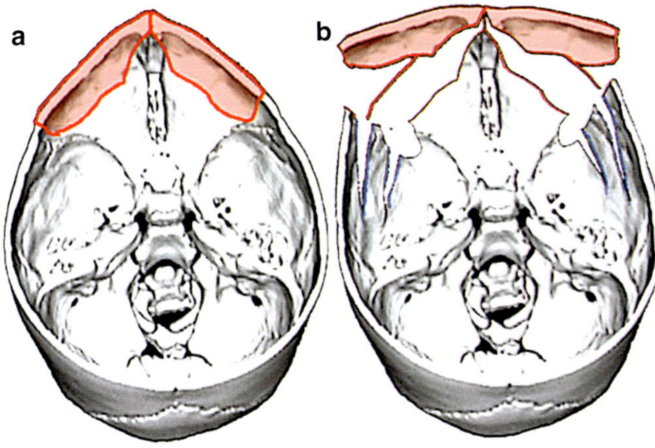

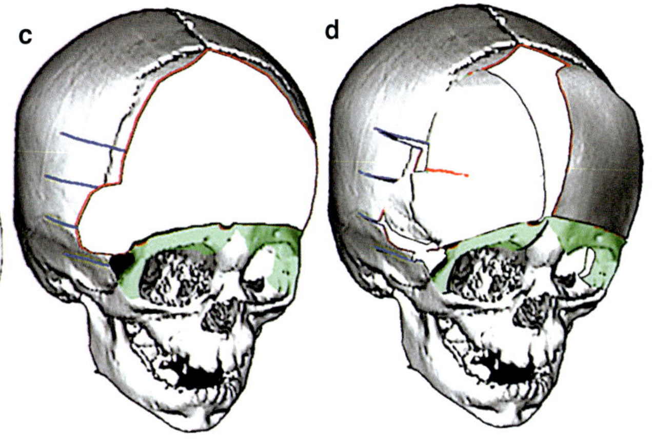

Fig. 13.102 Surgical correction of trigonocephaly. Correction at two levels. Fronto-orbital bar (bandeau) opens up the anterior cranial fossa. Note lateral releases into the temporal zone of parietal. A variation of this is a tongue-in-groove advancement with horizontal struts in continuity with the frontal bandeau that are cut from the parietal and slid forward for additional stability. Frontal correction involves expansion of surface areas with (1) osteotomies or (2) dividing the frontal bone flap vertically, rotating each piece 90 degrees and then transposing them. [Reprinted from Philippe Pellerin P, Vinchon M, Dhellemmes P, Wolber A, Guerreschi P. Trigonocephaly: Lille's surgical technique. Childs Nerv Syst 2012; 29:2183-2188. With permission from Springer Nature.]

reality, the pathology of trigonocephaly is a field defect in zone 9, the lateral orbital plate. The orbital plate is made up of two bone fields: prefrontal (zone 11) and postfrontal bone (zone 10). The fields develop their mesenchyme from posterior-anterior and medial-lateral pattern reflecting the way MNC arrives and is distributed. The frontal bone fields sit above PrF and PF; they develop in a medial-lateral and ventral-dorsal sequence. These relationships are not obvious because the fields are fused and appear confluent. The lateral defect changes the stress vectors across the frontal fields caused by the constraint at the orbital rim. Midline hypertrophy of the suture results. The greater the degree of lateral deficit and constraint, the more prominent the suture becomes. Alternatively, hypertrophy could represent an accumulation or "piling up" of frontal mesenchyme which cannot distribute itself normally. In any case, trigonocephaly allows us to visualize how the sequence of bone fields interacts to create the forehead…particularly, because it exists as a spectrum.

Extracranial trigonocephaly involves the angiosomes of the StV1 arterial system to the orbit. Lacrimal is always involved as it is the supply deficit in the #9 zone of the orbital plate.

Supraorbital, lateral branch to the #10 zone, could be affected in severe cases but it is generally uninvolved. Nasociliary supplies the ethmoid complex and explains the hypotelorism and arched palate. StV2 to alisphenoid can explain the frontosphenoidal synostosis and temporal pinching.

Intracranial trigonocephaly is more sinister. The hits can come from two directions. (1) Recall that the stapedial system provides feeder vessels to the dura, and thence to the outer cortex of the frontal lobe. Thus, trigonocephaly may represent a reduction of blood supply to this part of the brain. (2) Branches of the axis of anterior cerebral are affected in a distal to proximal gradient. This anatomy was reviewed in our discussion of holoprosencephaly (Chap. 8). As a rule-of-thumb, intracranial distribution of V1-innervated dura is a good guide to blood supply as well. The most distal branches of the pericallosal artery provide collaterals to cerebellum. Callosomarginal artery sends ascending branches along the internal walls of the cortex all the way to the surface. Single-photon emission CT scans taken at one year in 54 patients with mild trigonocephaly and developmental delays showed 76% with a reduction in blood flow to the frontal lobes [79].

Trigonocephaly commonly gets lumped into syndromes in association with typical forms for craniosynostosis such as Baller-Gerold, Muenke, and Saethre–Chotzen. We have previously discussed forms of craniosynostosis derived from field defects of the parietal bone. Metopic suture synostosis is a myth. Frontal bone embryology is utterly different, involving neurovascular axes of the StV1 arteries supplying the bone fields and underlying dura. Unlike the synostoses, trigonocephaly can be associated with pathologies of midline CNS structures supplied by the anterior cerebral artery. In syndromes associated with trigonocephaly neuropathology is always present.

Syndromic forms of trigonocephaly demonstrate multiple effects [80]. These include additional StV1 and StV2 angiosomes supplying the face but many other developmental fields are involved as well. The CNS component of syndromic trigonocephaly is not any more profound. Once the intracerebral fields are hypoplastic the damage has already been wrought.

Surgical correction for this mild trigonocephaly is straightforward and intellectual deficits are not common. With increasing degrees of severity, more of the orbital roof is compromised as is brain development. Like all bone fields the lateral third of the orbit has a gradient of osteosynthesis which follows from the pattern of mesenchymal deposition. If the lacrimal zone is a square development proceeds from medial-lateral and posterior-anterior. This means a progressive reduction in the lateral zone of the supraorbital artery neuroangiosome causes the bone fields to become progressively triangular. Timing of surgery should generally be before six months of age in case strabismus is present, as early intervention is more favorable for visual development [81].

Although trigonocephaly was defined by Welker in 1882 its radiologic features and surgical management were worked out in the early 1960s [82, 83]. The initial approach was to simply remove the forehead and open it up with multiple osteotomies like barrel staves. The frontal bar was not addressed because no one had attempted such combined osteotomies. It would remain for Tessier and Marchac to design the combined orbital-frontal bone flaps now used to correct trigonocephaly. These are nicely illustrated by the Lille group [84]. The orbital plate entrapment is corrected by reshaping and advancing the orbital bar. Forehead remodeling can take a variety of forms, all of which reveal the developmental deficit in the frontal fields occasioned by their respective embryonic field defects.

Frontoparietal Synostosis (FSS)

Coronal synostosis, in its non-syndromic form, can be unilateral or bilateral; it results in plagiocephaly, the features of which are well described. Orbit volume remains unchanged. The anatomic findings of importance to the orbit are listed (with comments) (Figs. 13.103, 13.104, 13.105, 13.106, 13.107, 13.108, 13.109, and 13.110).

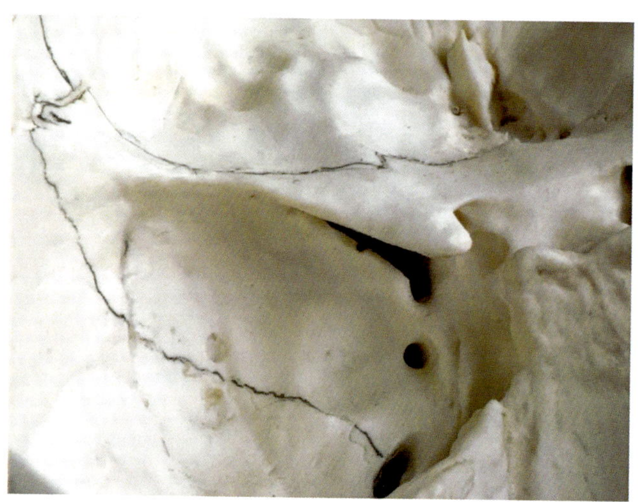

Fig. 13.103 Alisphenoid in zone 9. Alisphenoid developed in conjunction with maxilla; its mesenchyme comes from first arch. Blood supply to this zone is StV2 intracranial. It is traversed by StV1 meningeal branch en route to anterior cranial fossa via the orbit but that does not affect its composition. Defects in AS are distal, affecting its contact zone with orbitosphenoid and frontal. This causes downward traction on the greater orbital fissure and backward traction on the frontal and zygomatic fields of lateral orbital wall. Zone 9 is complex, combining deficits due to the StV1 lacrimal axis with deficits of the StV2 axis. [Courtesy of Michael Carstens, MD.]

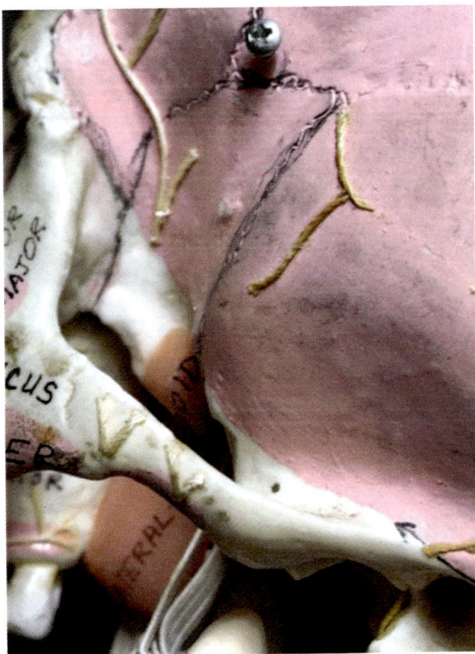

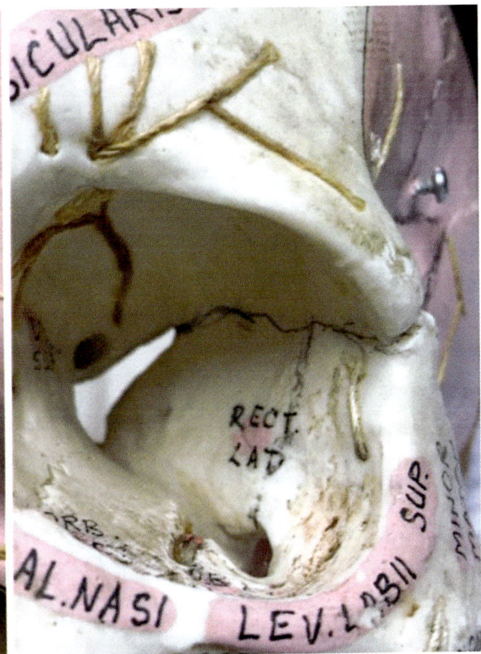

Fig. 13.104 Alisphenoid in situ. Alisphenoid calvarial lamina shares neuromeric populations with r2 infero-anterior quadrant of parietal, squamosal, and zygoma. These sutures do not develop synostosis. The frontosphenoid suture is an r1/r2 boundary. Alisphenoid is synthesized from the cranial base upward. Deficiencies affect its distal population, which is dorsal. This zone abuts against frontal. [Courtesy of Michael Carstens, MD.]

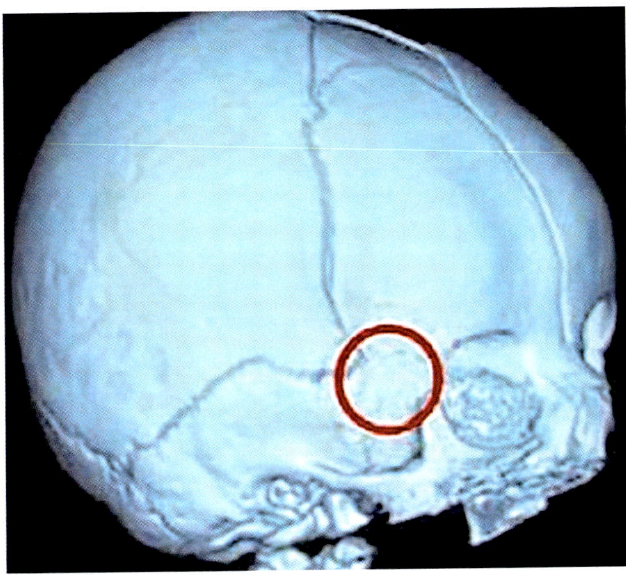

Fig. 13.105 Alisphenoid synostosis develops in only one zone. Note in the case of isolated frontosphenoidal synostosis that the alisphenoid sutures with both the squamosal and parietal bone fields remain patent. These fields share a common r2 neural crest population. The suture with the frontal bone is incomplete. [Reprinted from de Ribaupierre S, Czory A, Pittet B, Jacques B, Rilliet B. Frontosphenoidal synostosis: a rare cause of unilateral anterior plagiocephaly. Childs Nerv Syst (2007) 23:1431–1438. With permission from Springer Nature.]

<u>Craniofacial</u>

- Frontal bone
  - Forehead deficiency with ipsilateral retraction of roof and supraorbital rim
    - Floor is OK
  - Orbital roof deficiency: not localized to zones 9, 10, 11
    - Smaller volume
- Anterior cranial base: ipsilateral deviation 7 degrees [85]
  - Angle between midpalatal suture and posterior cranial base with locus just behind anterior clinoids, i.e., at presphenoid/basisphenoid border
- Ethmoid complex: A-P length normal; normal nasal envelope
- Parietal bone
  - Anterior zones deficient, both A-P and vertical
  - Pulls frontal backward and alisphenoid upward
- Sphenoid (assuming a normal alisphenoid)
  - Elevation of orbitosphenoid (lesser wing) due to frontal traction
  - Elevation of alisphenoid (greater wing) due to parietal vertical deficiency
    - Harlequin sign
    - Narrowing of pterion (alisphenoid reduced in volume or compressed)
    - Ipsilateral pterygoid plate displaced ventral
- Temporal bone: intertemporal
  - Ipsilateral temporal widening of middle cranial fossa
  - Bowing of squamosal temporal
  - Forward displacement of petrous temporal (Sm5–Sm6) following sphenoid
    - Ear follow petrous forward and upward
    - Mastoid position: torticollis in 14%

<u>Orbit</u> [86, 87]

- Altered orbital height/width and volume

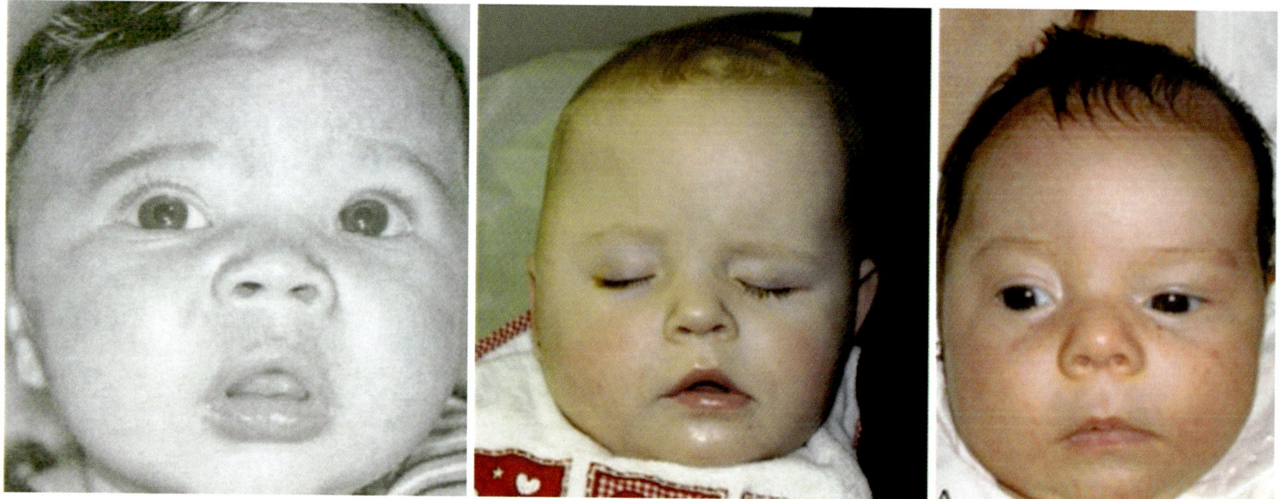

Fig. 13.106 Frontosphenoid synostosis (FSS) vs. frontoparietal synostosis (FPS) vs. combination of FSS+FPS. FSS (left): Coronal suture normal bilaterally. Note the "reverse Harlequin" appearance of the right orbit. The nasal root deviates away from the lesion. Ipsilateral temporal hollowing seen. The right frontal zone 9 is reduced and pulled downward in its lower zone. Frontal height difference (bregma) is minimal compared with normal side. No ipsilateral parietal bossing. FPS (right): Note "Harlequin" orbit, more severe ipsilateral frontal deficiency, especially in the vertical dimension, affecting upper projection. The nasal root deviates toward the lesion. The right frontal zone 9 is reduced posteriorly and pulled upward. Frontal height difference (bregma) is reduced compared with normal side. Parietal bossing ipsilateral. FSS + FPS (center): Coronal effect is stronger. Ipsilateral orbit is less deformed. Nose still deviated ipsilateral but less so. Alisphenoid deficiency retracts greater wing downward at the same time as parietal deficiency pulls it upward. The two Harlequins cancel each other out. Left: [Reprinted from Rogers GF, Proctor MR, Mulliken JB. Unilateral fusion of the frontosphenoidal suture: a rare cause of synostotic frontal plagiocephaly. Plast Reconstr Surg 2002; 110(4): 1011-1021. With permission from Wolters Kluwer Health, Inc.]. Right: [Reprinted from Dundulis BS, Becker DB, Grovier DP, Marsh J, Kane AA. Coronal ring involvement in patients treated for unilateral coronal synostosis. Plast Reconstr Surg 2004; 114(7):1695-1703. With permission from Wolters Kluwer Health, Inc.]

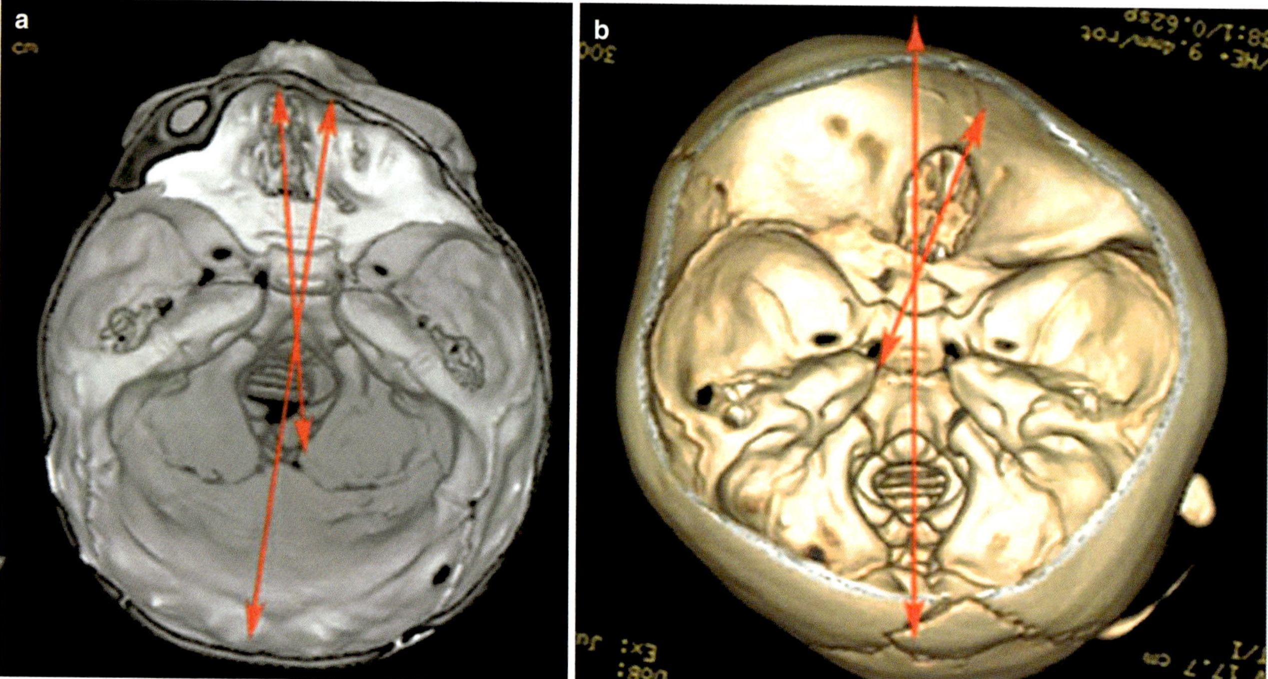

Fig. 13.107 Cranial base angulation in isolated frontosphenoid versus isolated frontoparietal synostoses. Endocranial base view showing that the angulation of the anterior cranial base in FSS with respect to the posterior cranial base is 9 degrees *away from* the synostotic side. Endocranial angulation in right unilateral FPS is 11 degrees *toward* the synostosis. Note: The combination of FPS with FSS ameliorates the FPS form of cranial base angulation, i.e., anterior cranial base remains angulated toward the synostosis, but to a lesser degree. [Reprinted from [88]. With permission from Creative Commons License 4.0: https://creativecommons.org/licenses/by-nc-sa/4.0/.]

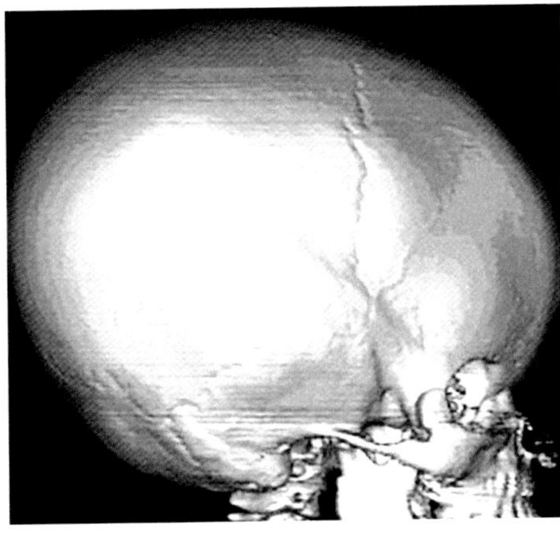

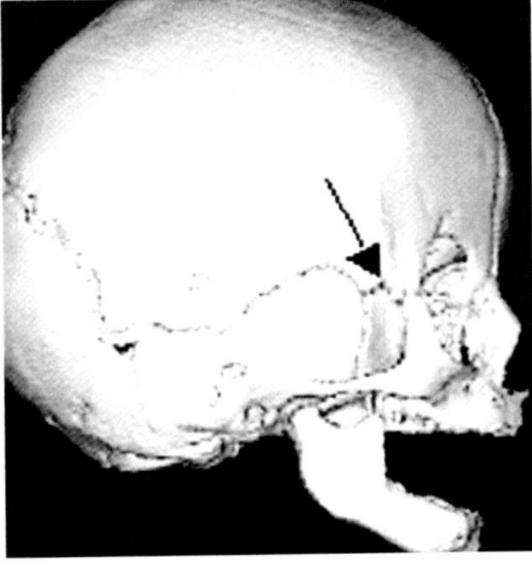

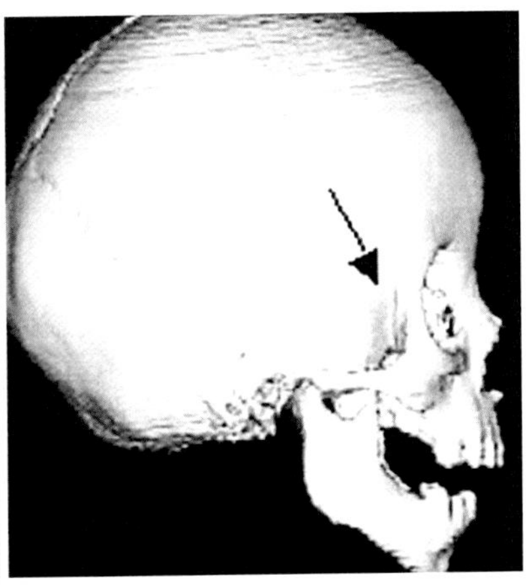

Fig. 13.108 Comparison of post-temporal skull (right side). Left: FSS, isolated; Center: FPS, isolated; Right: FPS+FSS. Note: FPS + FSS has more severe post-orbital deficiency; the axis is centered at the parieto-sphenoid suture (which remains patent). Left: [Reprinted from Hubbard BA, GorskinJL, Muzaffar AR. Unilateral frontosphenoidal craniosynostosis with achondroplasia: a case report. Cleft Palate–Craniofacial Journal 2011; 48(5):631-635. With Sage Publishing.]. Right: [Reprinted from Dundulis JA, Becker DB, Govier DP, Marsh JL, Kane AA. Coronal Ring Involvement in Patients Treated for Unilateral Coronal Craniosynostosis. Plast Reconstr Surg 2004; 114(7):1695-1703. With permission from Wolters Kluwer Health, Inc.]

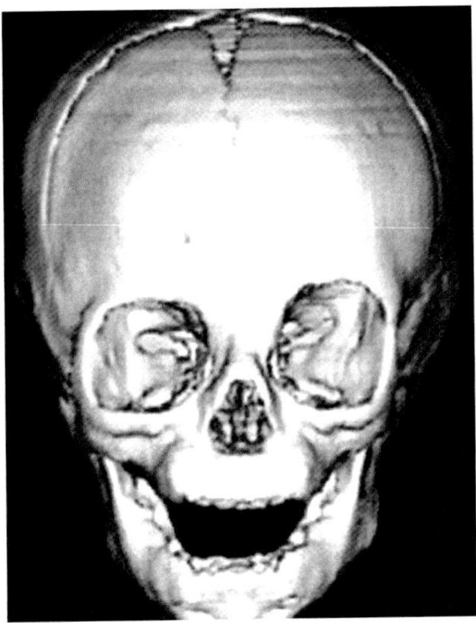

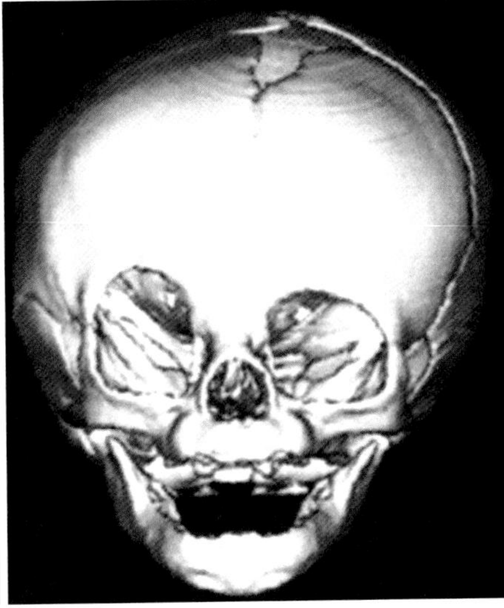

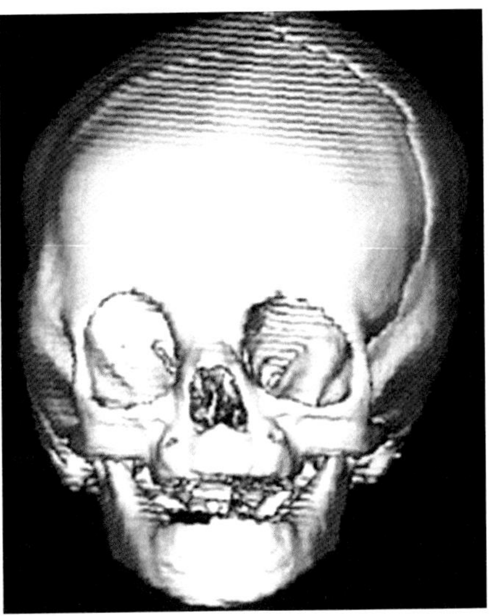

Fig. 13.109 Comparison of orbits (right-side synostosis). Left: FSS, isolated; center: FPS, isolated; right: FPS + FSS combined. Upper left: FSS, isolated; center: FPS + FSS; upper right: FPS, isolated. Note: Combination has a mitigating effect on orbital shape. Coronal effect is stronger. Ipsilateral orbit is less deformed. Nose still deviated ipsilateral but less so. Alisphenoid deficiency retracts greater wing downward at the same time as parietal deficiency pulls it upward. In "the battle of the Harlequins," the effects tend to cancel each other. Parietal boss is mitigated. [Reprinted from Dundulis JA, Becker DB, Govier DP, Marsh JL, Kane AA. Coronal Ring Involvement in Patients Treated for Unilateral Coronal Craniosynostosis. Plast Reconstr Surg 2004; 114(7):1695-1703. With permission from Wolters Kluwer Health, Inc.]

- Contralateral orbit is mildly abnormal
- Orbital dimensions [89]:
  - Length (A-P distance) = contralateral = normal
  - Height: 114% of normal
  - Width: no difference/slight deficiency from normal
  - Volume: 94% of normal

Eye

- More anterior globe in a smaller and more vertically elongated orbit
- Globe dimensions [90]
  - Volume of globe: 93% of normal
  - Projection of globe: 27% > normal
- Strabismus 37% patients
  - V-pattern most common
- Eyelids: decreased transverse distance

Nose

- Root of nose deviated toward deficiency
- Soft tissue envelope adequate size

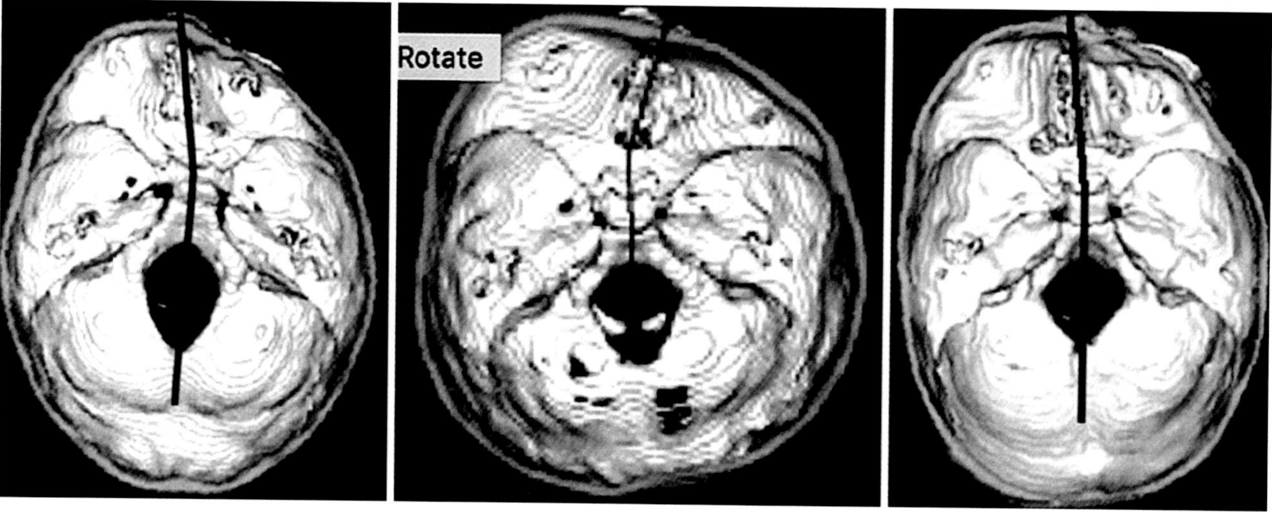

Fig. 13.110 Comparison of cranial base configuration (right-side synostoses). Left: FSS, isolated; center: FPS, isolated; right: FPS + FSS. Note: Combination has a mitigating effect on orbital shape: "battle of the Harlequins." Parietal boss is mitigated. [Reprinted from Dundulis JA, Becker DB, Govier DP, Marsh JL, Kane AA. Coronal Ring Involvement in Patients Treated for Unilateral Coronal Craniosynostosis. Plast Reconstr Surg 2004; 114(7):1695-1703. With permission from Wolters Kluwer Health, Inc.]

Maxilla-zygoma complex

- Maxilla-zygoma volume not different than normal [86]
  - Normal (quasi) development depends on release
- Dental crowding occurs with time [91]
  - In unoperated cases, worse with permanent dentition
  - Missing teeth 45% between 1 and 5 dental units
  - Flaring of incisors—patients cannot chew with them

Neuro

- No neurological effects and normal intelligence

Extracraniofacial

- Cervical spine fusion: less common in Crouzon than Apert
  - C2–C3 in 50% Crouzon (somites S6–S7)
  - C5–C6 normal

Genetic

- Non-syndromic: coronal synostosis is sporadic (no known causative gene)
  - Incidence in first-degree probands 1%, lowest of all single-suture synostoses
- In FCS cases, the following genes confer increased risk of recurrence in first-generation offspring
  - *FGFR3* gene, P250R mutation causes gain-of-function with increased [FGF] Muenke syndrome (coronal synostosis, low-set ears, hearing loss)
  - *TWIST1* loss-of-function > abnormal transcription
  - *TGF12* loss-of-function leads to *TWIST1* mutation

In sum:

- Frontoparietal synostosis (FPS) affects anterior (r2) quadrants of parietal bone: it affects the shape of the orbit without causing a significant reduction in volume.
- Alisphenoid in FPS can be reduced in volume but is not synostotic.
- No intrinsic r1 deficiency.
  - Peripheral effects on r1 fields (frontal, ethmoid) and r4–r5 (petrous temporal) are due to overall reduction in parietal length and height.
- Coronal synostosis, non-syndromic, has no gene identified with pathology.

Frontosphenoidal Synostosis (FSS)

In 1995 Francel published a case report of frontoplagiocephaly and an ipsilateral synostosis of the frontosphenoidal suture. FSS is a rare entity. By 2015 only 26 cases had been reported [92]. The presentation of FSS is initially subtle and the diagnosis is frequently missed. FSS initially appears like a routine anterior deformational plagiocephaly with mild flattening of the ipsilateral orbital rim and contralateral bossing. Of note: ear position is nearly always normal. The initial diagnosis is positional plagiocephaly and helmet therapy is often implemented but with no success. With time (often by six months) the plagiocephaly worsens.

- Physical findings are follows: (1) Forehead flattening is disproportionate to contralateral occiput (different that positional). (2) Temporal hollowing presents. (3) Maxilla and chin remain normal. (4) Eye–ear distance unchanged symmetrical. (5) No torticollis.

- Radiologic features of FSS are quite distinct from frontoparietal (coronal) synostosis (FPS): (1) No elevation of sphenoid wing – no "Harlequin" sign is present. The lateral superior orbit may have a "teardrop" shape, inclining *downward*, giving a "*reverse Harlequin*" shape to the orbital rim. (2) Anterior cranial fossa is smaller but, unlike FPS, the tip of the nose is normal or toward the flattening. (3) Angulation of the cranial base also is normal or slightly contralateral. (4) The sagittal suture remains in the midline.
- Surgical treatment uses a conventional coronal approach with care taken to extend the osteotomy all the way along the frontal border of alisphenoid [92–95].

Anatomy of the Alisphenoid Field: Tessier Zone 9 (Read This with a Skull in Your Hand)

Alisphenoid (AS) is complex structure. Medially it is attached to presphenoid: the apertures of foramen rotundum and foramen ovale serve as landmarks for the fusion plane (you can observe this for yourself in the skull). From there AS trifurcates. It sends downward a lower laminar field, the lateral pterygoid plate that partners with medial pterygoid plate (the remnant of the ancient pterygoid bone). This zone is richly supplied by StV2 descending palatine and external carotid deep temporal branches. In continuity with the pterygoid plate AS sends an upward lamina as well, which forms the lower lateral corner of the calvarium covering the middle fossa. Blood supply here is dual as well with StV2 recurrent meningeal on the inside and deep temporal on the outside. This finger-like lamina unites from below with postorbital zygoma and terminates with frontal bone. Intracranially the greater wing extends laterally in parallel with OS lesser wing until it contacts frontal bone and terminates in the lateral orbital roof. Once again, the blood supply is dual: StV2 recurrent meningeal on the backside (temporal fossa) and StV1 lacrimal on the orbital side (Figs. 13.103 and 13.104).

Note in the skull photographs the medial extensions of the sutures on either side of alisphenoid. Posterior suture runs transversely into middle cranial fossa in rough parallel with the transverse axis of the orbit. Anterior suture disappears at the inferior border of alisphenoid and then reappears inside the lateral orbital wall where it separates greater wing of sphenoid from postorbital, the lateral orbital wall.

Pathologies of the Alisphenoid

Hypoplasia/Aplasia/Dysplasia

Alisphenoid pathology produces (1) *zone-specific form of hypoplasia* or (2) *synostosis*. The "flow" of r2 neural crest that produces alisphenoid causes it to develop in these same three vectors. The most vulnerable zones are distal. A defect in alisphenoid will affect in the upper deck, first, its contact with frontal bone and, second, its contact with postorbital. The lower deck affects the distal lateral plate, extending proximally. Medial defects, where AS meets the central column of presphenoid, open the spaces of cranial nerve foramina, providing escape routes for temporal encephalocoeles into maxilla or the infratemporal fossa.

Field defects of the alisphenoid affect the greater wing and can lead to massive orbital encephalocoele. Note that this zone of AS is the most distal to its arterial supply, the orbital communication between StV2 middle meningeal and StV1 lacrimal, known as the *recurrent meningeal artery*. Even though the communicating branch is absent in the fetal state, the position of the artery is marked by the cranio-orbital foramen. The *forme fruste* of zone 9 could be a dermoid projecting into the temporal fossa or middle cranial fossa [1]. With progressive deficit, a knock-out at the superolateral orbital margin would appear progressing on the complete loss of the greater wing. In the sphenoid complex, medial pterygoid is a separate field which may or may not be affected.

The rarity of zone 9 clefts, per se, is likely due to the redundancy of arterial supply to the area. An isolated defect in the medial aspect of the great sphenoid wing has been reported. Rare cases exist of *lacrimal gland aplasia or hypoplasia*. Such cases manifest a distal deficiency state of the lacrimal artery or recurrent meningeal artery. More common pathologies are *sphenoid dysplasia due to neurofibromatosis type 1*. *Meningiomas* of the anterior cranial fossa tend to localize in the sphenoid wing.

Frontosphenoid Synostosis

These findings of unilateral FSS are caused by foreshortening of the upper lateral orbital reflecting a contraction of zone 9 at the superolateral margin of AS, at its junction with frontal. The sphenoid wing is flattened, not elevated, causing less orbital height with compensatory orbital width. This causes the abnormal shape of the external orbit. It is transversely ellipsoid with a lateral reduction in height giving it a "teardrop" or "reverse Harlequin" deformity. Depressions of the lateral orbit in isolated FSS are noted in multiple reports [95].

Unlike FPS, the cranial base angulation in FFS is normal or even slightly contralateral. This is because diffuse involvement of the frontal bone with the stenosed frontoparietal suture torques the entire frontal bone complex toward the lesion. Given the attachments of frontal and ethmoid fields a deforming force is exerted. This is simply not operative in FSS as alisphenoid has minimal contact with frontal (save in the orbital backwall).

Embryologic Explanation of the Coronal Ring: FSS, FPS, or FSS + FPs

In 1985 Al Burdi defined the coronal ring as a continuous set of cranial articulations, beginning with the (frontoparietal suture) per se and continuing downward on the both sides of the alisphenoid, posteriorly with sphenosquamosal suture and anteriorly with the sphenofrontal suture [96]. His model continued the posterior suture across the floor separating the frontal and middle cranial fossae and the anterior suture transversely to separate frontal and ethmoids from the orbitosphenoid (Fig. 13.111).

Coronal synostosis literature lumps together pathologic presentations of what we now know to be two distinct embryologic fields: frontosphenoid (FS) and frontoparietal (FP). These can be clearly distinguished. The "Harlequin sign" of unilateral frontoparietal synostosis (FPS) results from an upward cant of the superior orbital fissure and the superolateral corner of the orbit. In isolated FSS this sign is not present. When the two synostoses occur in combination the vectors of frontosphenoid counteracted those of frontoparietal. Thompson's group evaluated the orbital shape in three groups: isolated FS, isolated FP, and FS + FP using a modified orbital index, MOI, to measure the severity of the Harlequin deformity. CD represents the greatest diameter of the orbit perpendicular to a line AB indicating the longest diameter of the orbit. Thus: line CD/line AB always < 1. Decreasing ratios indicated the increasing degrees of ellipsoid shape. FPS was 0.84, worse than 0.92 in FSS, the intergroup comparison being $p = 0.024$. When combined FPS + FSS was even worse (0.78) it suggested that for isolated FP synostosis, the presence of an open FS suture provides some degree of vertical compensation which cannot occur in the combination [97].

In sum:

- Isolated frontosphenoid synostosis is a field malfunction in zone 9.
- The clinical manifestations of FSS are distinct from FPS.
- Coronal ring synostosis has three forms: FSS, FPS, or FSS+FPS.
- When they occur together the clinical picture of FPS predominates but the degree of orbital deformity is worse.

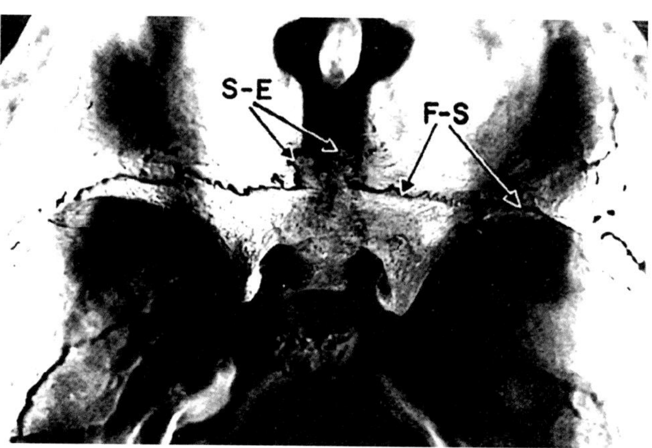

Fig. 13.111 Coronal ring of Burdi. The continuity of sutures between ethmoid, sphenoid, frontal, and parietal extends around the skull. The relationships between additional borders of ethmoid are central for understanding the fasciosynostoses of Crouzon, Apert, and Pfeiffer. The ethmoid complex is a six-sided box. Pathology at one of its borders can affect the development of otherwise normal bone fields. The boundaries are as follows: (1) upper lateral border of ethmoid with frontal, (2) lower lateral border of ethmoid with maxilla, (3) anterior lateral border with frontal and nasal bones, (4) perpendicular plate inferior border with vomer, and (5) perpendicular plate anterior border with septum. [Reprinted from Burdi A, Kusnetz AB, Vernes JL, Gebarski SS. The natural history and pathogenesis of the cranial coronal ring. Cleft Palate Craniofac J 1986; 23(1):28-39. With permission from Sage Publishing.]

Uncommon Synostoses

Craniosynostosis in its most common forms represents a form of mismatch between the parietal bone and its neighbors. Parietal mesenchyme is unique in being of paraxial r2 and r3 mesoderm composition (admixed with neural crest) whereas its partner fields are purely neural crest derivatives. Parietal is a collection of four developmental fields arranged essentially as quadrants: dysfunction in any of these can lead to synostosis. Furthermore synostosis is not a blind genetic condition. *FGFR2*, *FGFR3*, *FGFR1*, *TWIST1*, and *EFNB1* (all associated with coronal synostosis) have zonal actions depending upon the developmental field in which they are expressed. Biologic confirmation of this principle can be seen in rare synostoses of the squamosal suture and lambdoid suture.

Unilateral Squamosal Synostosis

Craniosynostosis between the squamosal and parietal bones has received little attention in the literature. Isolated involvement was being reported by Ranger [98]. A bilateral presentation has since been described in occurrence with Muenke syndrome by Doumit in 2014. The clinical findings in the unilateral form are quite distinct with palpable ridging of the suture, a prominent temporal bulge, and a "reverse Harlequin eye." Barlett's retrospective study produced 14 patients having unilateral or bilateral squamosal synostosis, 4 of which were syndromic: two with multiple suture involvement (Saethre–Chotzen, Pfeiffer), and two with craniofacial microsomia. Seven patients had an additional suture leaving three with isolated parieto-squamosal synostosis, of which one was unilateral. Clinical findings were ipsilateral supraorbital retrusion, forehead flattening, temporal flattening, mild orbital hypoplasia, mandibular retrusion. Jaw position is potentially related to effects by the neural crest membranous components of the temporal bone complex on the position of the articular eminence. Contralateral occipital bossing was present. Murphy et al. identified 26 patients of whom 3 were isolated, including 1 unilateral. None of the three patients had signs of elevated intracranial pressure. Helmet therapy was carried out but one of the bilateral cases required posterior remodeling. This was similar to Ranger's report of a unilateral case which was refractory to molding and underwent posterior vault expansion (Fig. 13.112).

In sum:

- Squamosal synostosis is caused by a pathologic field along the inferior border of parietal.
- Findings relate to overall reduction in the height and, to a lesser extent, the surface area of squamosal.
- The relationship between the squamosal field and tympanic field may influence TMJ position.
- Isolated cases affect the occiput and may require posterior remodeling.

Unilateral Lambdoid Synostosis

Posterior plagiocephaly because a significant surgical problem after the 1992 when the American Academy of Pediatrics, in an attempt to combat sudden infant death syndrome, instituted a "back to sleep" campaign encouraging parents to keep their infants sleeping on the back. This led to a 10-fold rise in *positional plagiocephaly* (PP) cases, constituting up to 22% of infants at 7 weeks [99]. The constant differential diagnoses in this situation are unilateral lambdoid synostosis versus torticollis versus PP. Assessment criteria for diagnosis have been worked out and the algorithm is well known [100–102] (Fig. 13.113).

Lambdoid synostosis is uncommon to start with, being only 1–5% of all non-syndromic craniosynostoses. Within that subset, the bilateral form is equally uncommon (15%) but striking, with a characteristic "Mercedes Benz" shape and a high incidence (67%) of a Chiari I malformation [104]. Clinical findings in the unilateral variant are (1) palpable suture, (2) ipsilateral mastoid boss, (3) ipsilateral skull base displacement downward, dragging the ear with it, (4) right parietal boss, and (5) cervical spine compensation. Recall that parietal has two distinct fields in contact with each interparietal bone complex. Synostosis can be induced for either one or both.

In sum:

- Unilateral lambdoid synostosis is caused by a defect in the posterior zones of parietal bone.
- ULS can present as a lateral, medial, or generalized defect depending on the parietal field involved.
- CT and MRI analysis of the posterior fossa is required to rule out congenital defects predisposing the cerebellar tonsil herniation.

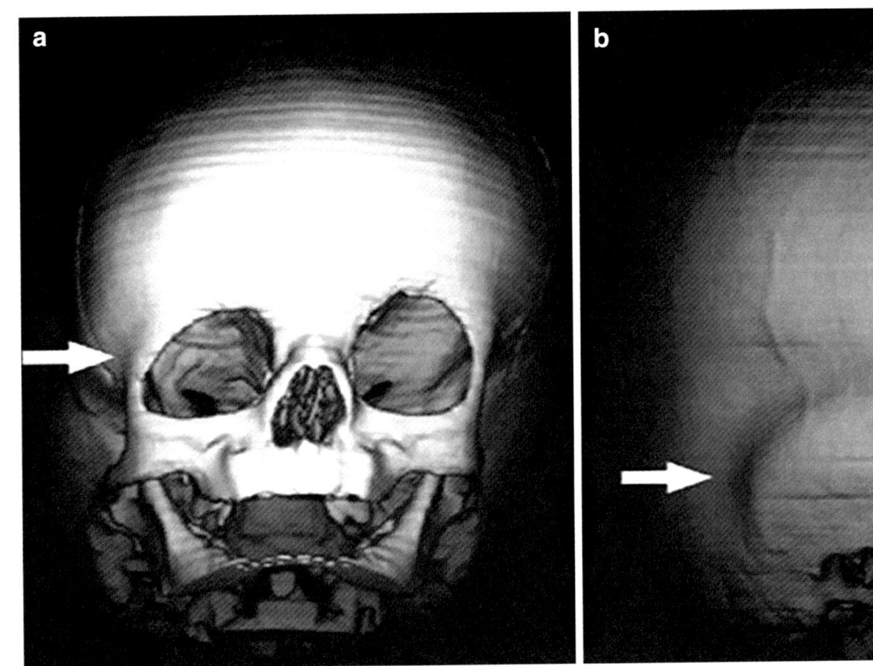

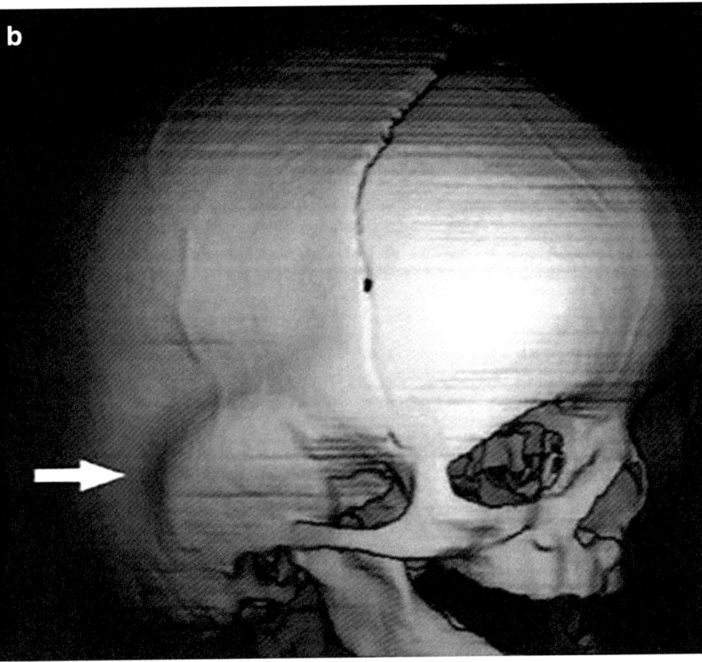

Fig. 13.112 Squamosal suture synostosis. Traction exerted on the alisphenoid causes right temporal plagiocephaly (yielding a diamond-shaped skull, when viewed from above), as well as an associated orbital deformity with a "reverse Harlequin eye," more exaggerated than with frontosphenoid synostosis alone [configuration on the side of the affected suture]. Note prominent bossing of the squamosal. Note that like FSS, nasal root deviates away from the pathology. [Reprinted from [98]]

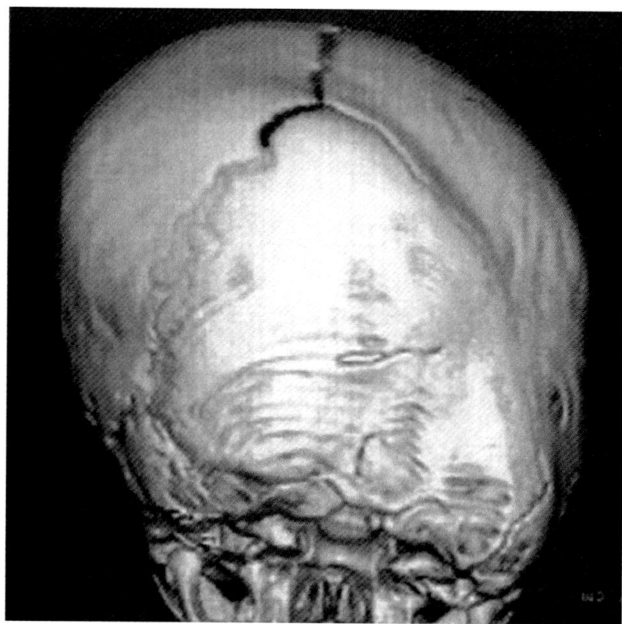

Fig. 13.113 Lambdoid suture synostosis, unilateral. Interparietal bones compensate with ipsilateral bossing to the mastoid. This clearly demonstrates involvement of both posterior quadrants of parietal, lateral more the medial. This is consistent with parietal mesenchyme flowing upward. Reduction in the height of the posterior skull shows upper posterior quadrant is affected more than the lower posterior quadrant. [Reprinted from [103]]

Fasciosynostoses

Craniofaciosynostoses: Apert, Crouzon, and Pfeiffer Syndromes

Certain conditions arising from the intrinsic walls of the orbit can result in changes in its volume as well. The well-known syndromes of Apert and Crouzon, as well as the rarer Pfeiffer syndrome, share a common appearance of midface retraction and bulging eyes with retracted lids. The orbits are similar, being shallow both from the standpoint of the orbital roof and the maxillary floor, causing an A-P deficiency-based proptosis. The orbital angles are widened, causing a lateral deficiency-based exorbitism. The ethmoid fields are hypoplastic. The airway is compromised, both from the size of the nasopharynx and the posteriorly positioned maxilla. All three forms have synostosis, with coronal being practically universal, although other sutures can be involved, more so in Apert, with Pfeiffer syndrome being multi- or pan-synostostic. All three syndromes share a common mutation to fibroblast growth factor receptor 2 (*FGFR2*) on chromosome 10. Despite such commonalities, these syndromes differ in important ways which are best understood embryologically. Neuromeric analysis leads to a number of provocative observations (Figs. 13.111, 13.114, 13.115, 13.116, 13.117, 13.118, 13.119, and 13.120).

Crouzon Syndrome

Craniofacial

- Coronal synostosis: associated synostoses rare
  - Acrocephaly
  - Compensatory transverse widening
    Traction on alisphenoid = Harlequin sign
- Cranial base angles unchanged [86]
- Frontal bone
  - Anterior cranial fossa foreshortened (A-P) due to ethmoid entrapment
- Ethmoid complex
  - A-P foreshortened
- Sphenoid complex: volume not different from normal [86]
- Alisphenoid
  - Elevation and forward bowing of alisphenoid
  - Duplication of sphenoid ridge
- Temporal bone: Intertemporal
  - Intertemporal widening (expansion of middle fossa)
  - Bowing of squamosal temporal
  - Forward displacement of petrous temporal (Sm5–Sm6) following sphenoid

Orbit [86, 87]

- Wide shallow orbit
  - Ethmoid wall foreshortened (A-P reduction)
  - Zygomatic wall foreshortened (?) or just angulated (?)
  - Increased zygoma–zygoma distance = widening
  - Widened (more obtuse) lateral orbital angle
  - Greater wing elevation (Harlequin sign)
  - Transverse expansion = hypertelorism not due to ethmoid
- Orbital dimensions—*less than Apert, **more than Apert
  - Length (A-P distance): 17% < normal**
  - Height: 7% > normal*
  - Width: no difference from normal
  - Volume: 23% < normal**

Eye

- Globe dimensions [86]
  - Volume: 36% > normal
  - Projection: 119% > normal
  - Volume outside the orbit: 179% >
  - Volume inside the orbit: 28% < normal
  - Soft tissue volume: 29% < normal

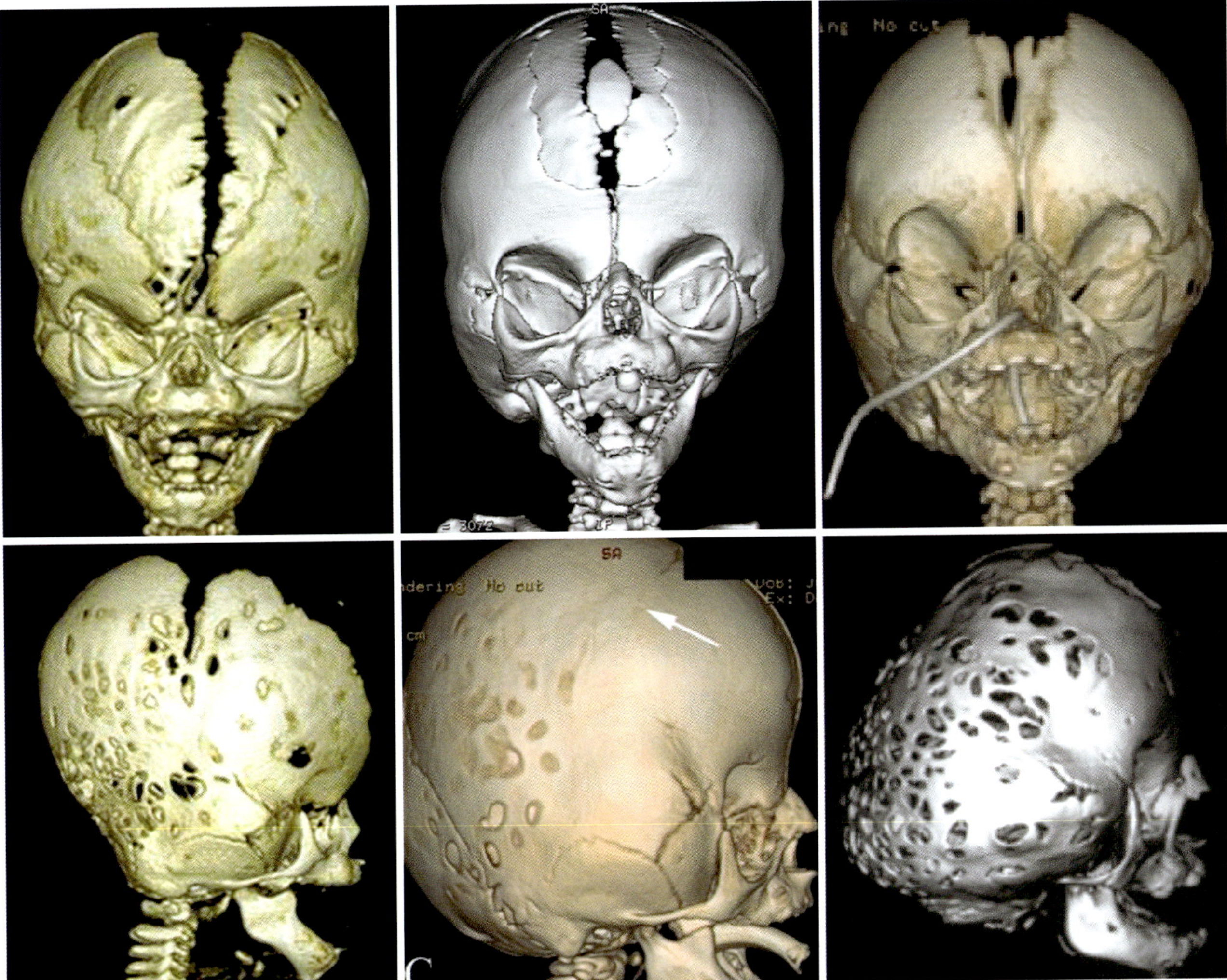

Fig. 13.114 Crouzon (left, top, and bottom), Apert (middle, top, and bottom), and Pfeffer (right, top, and bottom) have common ethmoid and coronal field defects with increasing degrees of severity as seen in the orbits and in additional synostoses. Findings are shown for Crouzon. Left (top and bottom): [Reprinted from Norgaard P, Hagen CP, Hove H, et al. Crouzon Syndrome associated with acanthosis nigricans: prenatal 2D and 3D ultrasound findings and postnatal 3D CT findings. Acta Radiologica Short Reports 2012; 1: 15-19. With permission from SAGE Publishing.]. Middle (top and bottom): [Reprinted from Khanna PC, Thappa MM, Iyer RS, Prasad SS. Pictorial essay: The many faces of craniosynostosis. Indian J Radiol Imaging 2010; 21(1): 49-56. With permission from Wolters Kluwer Medknow Publications.]. Right (top and bottom): [Reprinted from Rai R, Iwanga J, Dupont G, Ouskian RJ, Loukas M, Oakles WJ, Tubbs RS. Pfeiffer type 2 syndrome: Review with update on its genetics and molecular biology. Child's Nervous System (2019) 35:1451–1455. With permission from Springer Nature.]

- Eye (*FGFR-2* related)
  - Cornea: megalocornea, microcornea
  - Iris: aniridia, coloboma
  - Lens: cataract, ectopia lentis

Nose

- Bulbous, "parrot-beak"
- Foreshortened nasal bones (r1)

Maxilla-zygoma complex

- Maxilla-zygoma volume not different than normal [86]
- Dental crowding [91]
  - Missing teeth 45% between 1 and 5 dental units
  - Flaring of incisors—patients cannot chew with them

Neuro

- Least affected neurologically: normal intelligence
- In contrast with Apert, Crouzon ventricles usually normal (perhaps up to 25%)

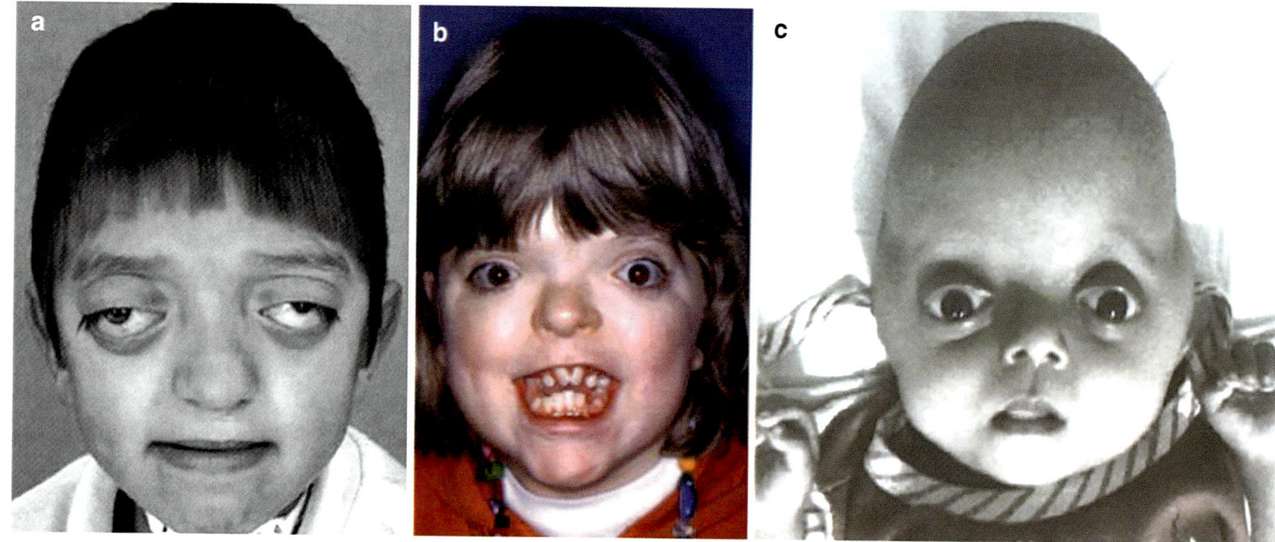

Fig. 13.115 Facies of Crouzon (**a**), Apert (**b**), and Pfeiffer (**c**) differ in degree. Crouzon tends to have ptosis and a well-formed nose. Apert is more severe with a small bulbous nose. Pfeiffer can be mild like the others or have Kleeblattschädel with extreme degrees of proptosis. (**a**) [Reprinted from Derderian C, Seaward J. Syndromic craniosynostosis. Sem Plast Surg 2012 May; 26(2): 64–75. With permission from Georg Thieme Verlag KG.]. (**b**) [Reprinted from Hohoff A, Joos U, Meyer U, et al. The spectrum of Apert syndrome: phenotype, particularities in orthodontic treatment and characteristics or orthodontic surgery. Head & Face Med 2007; 3; 10. With permission from Creative Commons License 2.0: https://creativecommons.org/licenses/by/2.0/.]. (**c**) [Reprinted from Singh RK et al. Common primary fibroblastic growth factor receptor-related craniosynostosis syndromes: A pictorial review. J Pediatr Neurosci 2010; 5(1):72-75. With permission from Creative Commons License 4.0: https://creativecommons.org/licenses/by-nc-sa/4.0/.]

Extracraniofacial

- Cervical spine fusion: less common in Crouzon than Apert
  - C2–C3 in 50% Crouzon (somites S6–S7)
  - C5–C6 normal

Genetic

- FGFR2 mutation
  - Parietal bone, anterior border
  - Ethmoid
- Autosomal dominant

Apert Syndrome

Craniofacial

- Coronal synostosis: additional synostoses common
  - Squamosal = towering
  - Posterior sagittal, lambdoid = reduced volume posterior fossa

Orbit [86]: *less than Crouzon, ** greater than Crouzon

- Length (A-P distance): 12% < normal*
- Height: 14% > normal**
- Width: no difference from normal
- Volume: 21% < normal*
- Globe dimensions
  - Volume: 15% > normal*
  - Projection: 99% > normal*
  - Volume outside the orbit: 179% > normal
  - Volume inside the orbit: 28% < normal
  - Soft tissue volume: 19% < normal*

Eye (*FGFR-2* related)

- Glaucoma (in some)
- Retina: pigment alterations
- Iris: aniridia, coloboma
- Lens: cataract, ectopia lentis

Nose

- Decreased r1 mesenchyme: "parrot-beak" similar to Crouzon

Maxilla-zygoma complex

- Maxilla-zygoma volume same as Crouzon
- Dental crowding [91]
  - Dental agenesis: maxillary canines
  - Ectopic maxillary molars
  - Dental crowding
  - Enamel dysplasia 45% between 1 and 5 dental units
  - Flaring of incisors—patients cannot chew with them
- V-shaped palatal arch: foreshortened perpendicular ethmoid
- Soft palate cleft 44%

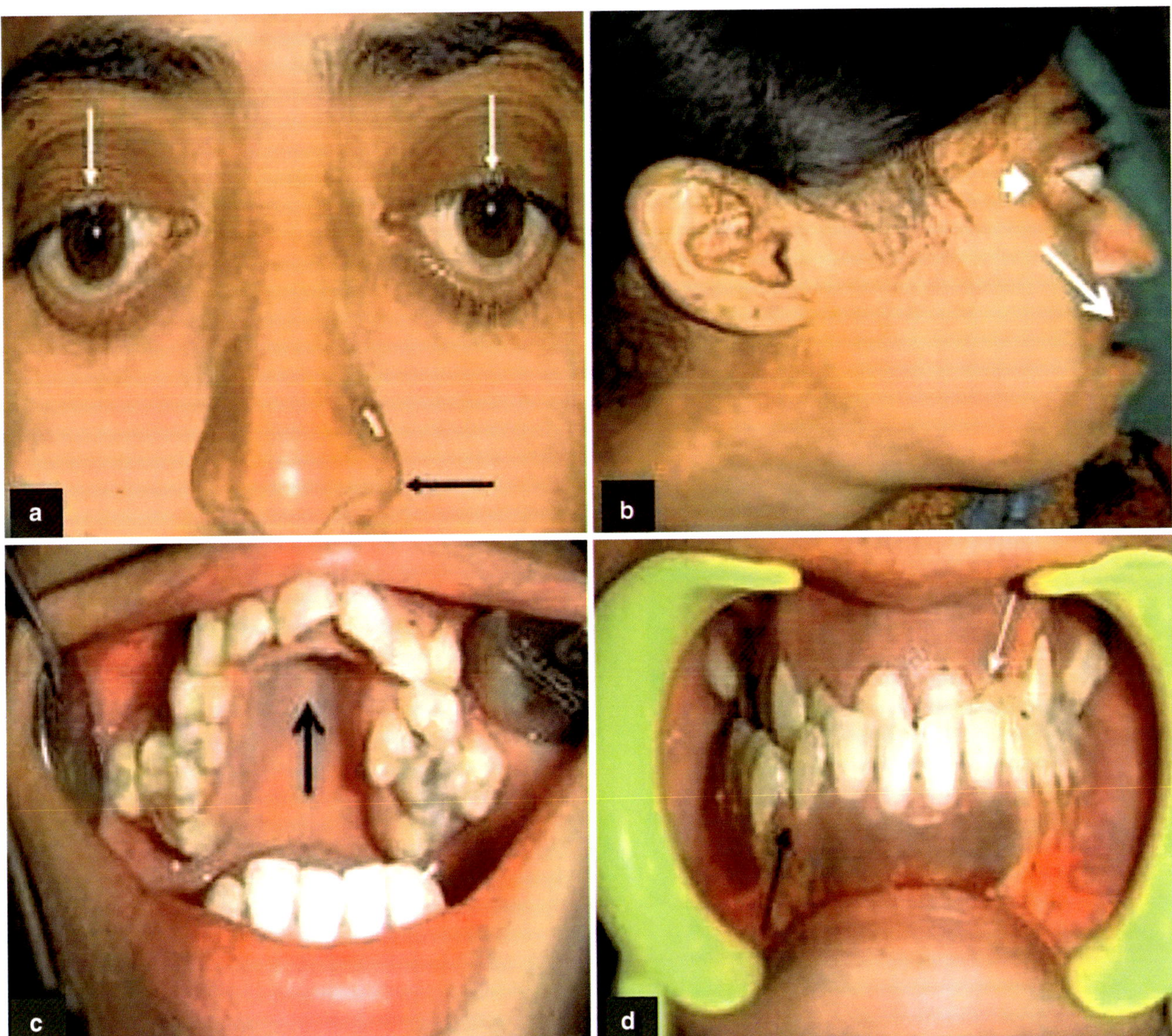

Fig. 13.116 Clinical features of the fasciosynostoses are similar between Crouzon, Apert, and Pfeiffer. Proptosis, ptosis, class III dento-skeletal relationship, and high-arched palate. Central to these pathologies is the combination field defects involving the ethmoid complex and the parietal fields, causing bicoronal synostosis in Crouzon, with increasing involvement of other sutures in Apert and Pfeiffer. Increasing degrees of synostosis. [Reprinted from Mohan RS, Vemanna NS, Verma S, et al. Crouzon syndrome: clinico-radiological illustration of a case. J Clin Imaging Sci 2012;2:70. With permission from the Journal of Clinical Imaging Science.]

<u>Neuro</u>

- CNS abnormalities are common with Apert
  - Corpus callosum and septum pellucidum
  - Limbic system leading to hippocampus in the temporal lobe
  - Abnormal gyri, especially temporal lobe
  - Optic nerves and optic chiasm stretched
- Learning disabilities 44%

<u>Extracraniofacial</u>

- Thumb radial deviation, syndactylism of the hands
- Cervical spine fusion: 2× greater in Apert than Crouzon
  - C2–C3: normal in Apert
  - C5–C6: two-thirds of Apert patients (correlates with thumb, somites S9–S10)

<u>Genetic</u>

- *FGFR2* mis-sense mutations cause increased FGF–ligand binding
  - Parietal bone: all zones
  - Ethmoid
- Autosomal dominant, paternal

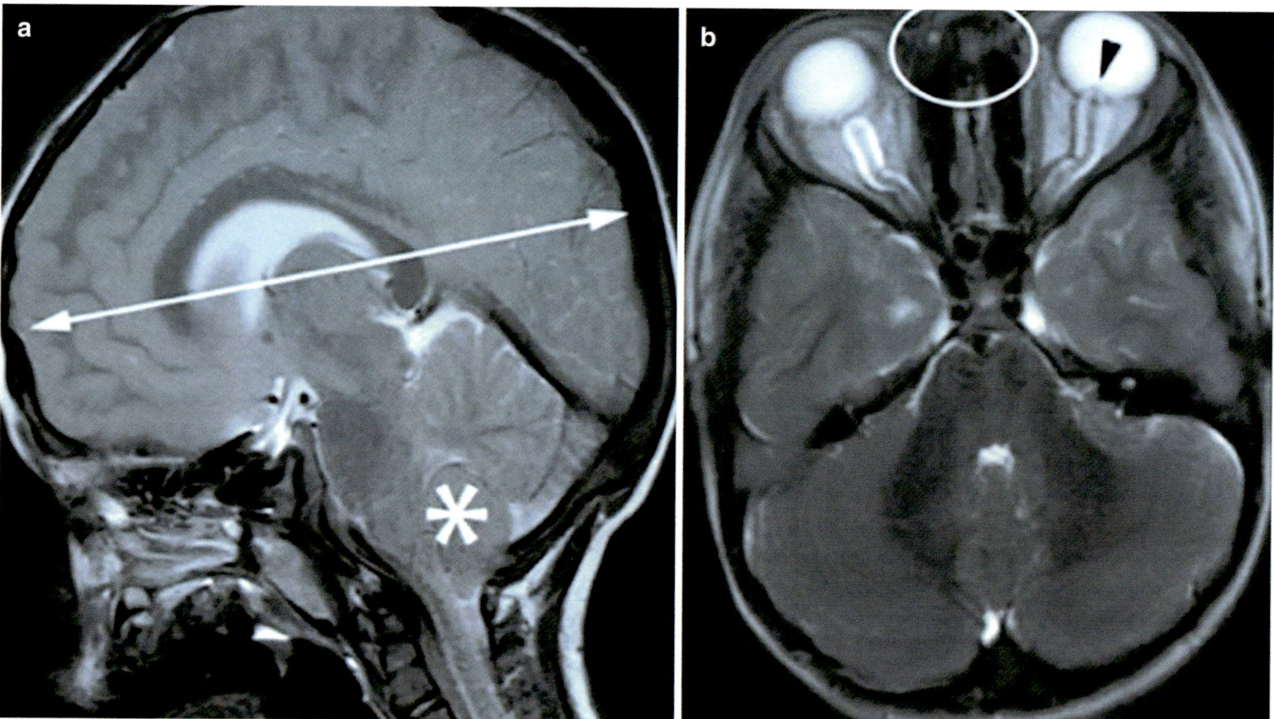

Fig. 13.117 Crouzon's syndrome. Anterosuperior 3DCT volume rendered image (**a**), sagittal T2W MRI (**b**), and axial T2W MRI show premature fusion of the coronal sutures (arrows) and brachycephaly (double-headed arrow). Note also the crowded posterior fossa with tonsillar herniation through the foramen magnum (*). There is also hypertelorism (oval). Note the cerebrospinal fluid (CSF)-distended optic nerve sheaths bilaterally and bulging left optic disc (arrowhead) indicating raised intracranial pressure. These findings are mechanical in nature. [Reprinted from Khanna PC, Thapa MM, Iyer RS, Prasad SS. Pictorial essay: The many faces of craniosynostosis. Indian J Radiol Imaging. 2011; 21(1):49-56. With permission from Wolters Kluwer Medknow Publications.]

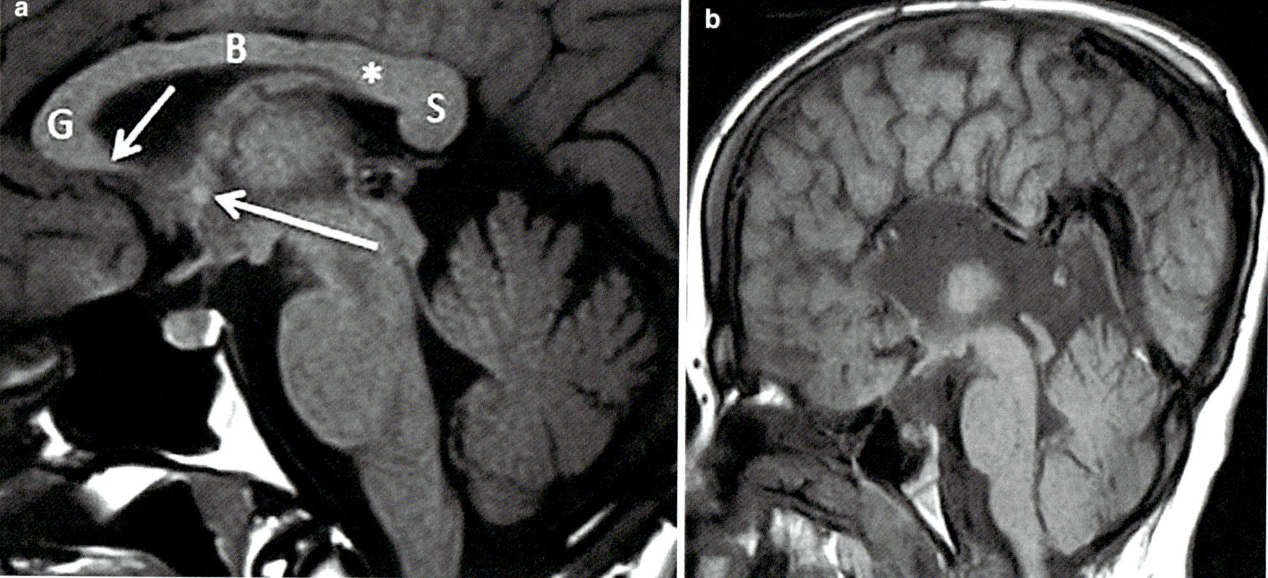

Fig. 13.118 Apert MRI – CNS pathology in the midline. The corpus callosum comprises of four main segments, the rostrum (short arrow), genu (G), body (B), and splenium (S), all of which are expected to be formed by 20 weeks of gestation. Columns of the fornix (long arrow) run medially and meet in the midline, just beneath the isthmus (asterisk) of the corpus callosum. T1-weighted image confirms the complete absence of midline corpus callosum and cingulate gyrus. Note the characteristic "spoke-like" appearance of the sulci on the medial surface of the cerebral hemisphere. Concomitant lambdoid synostosis in Apert's can lead to reduction in the posterior cranial fossa and herniation of the cerebellar tonsils. [Reprinted from Tan AP, Mankad K. Apert syndrome: magnetic resonance imaging (MRI) of associated intracranial anomalies. Child's Nerv Syst 2018; 34 (2):205-216. With permission from Springer Nature.]

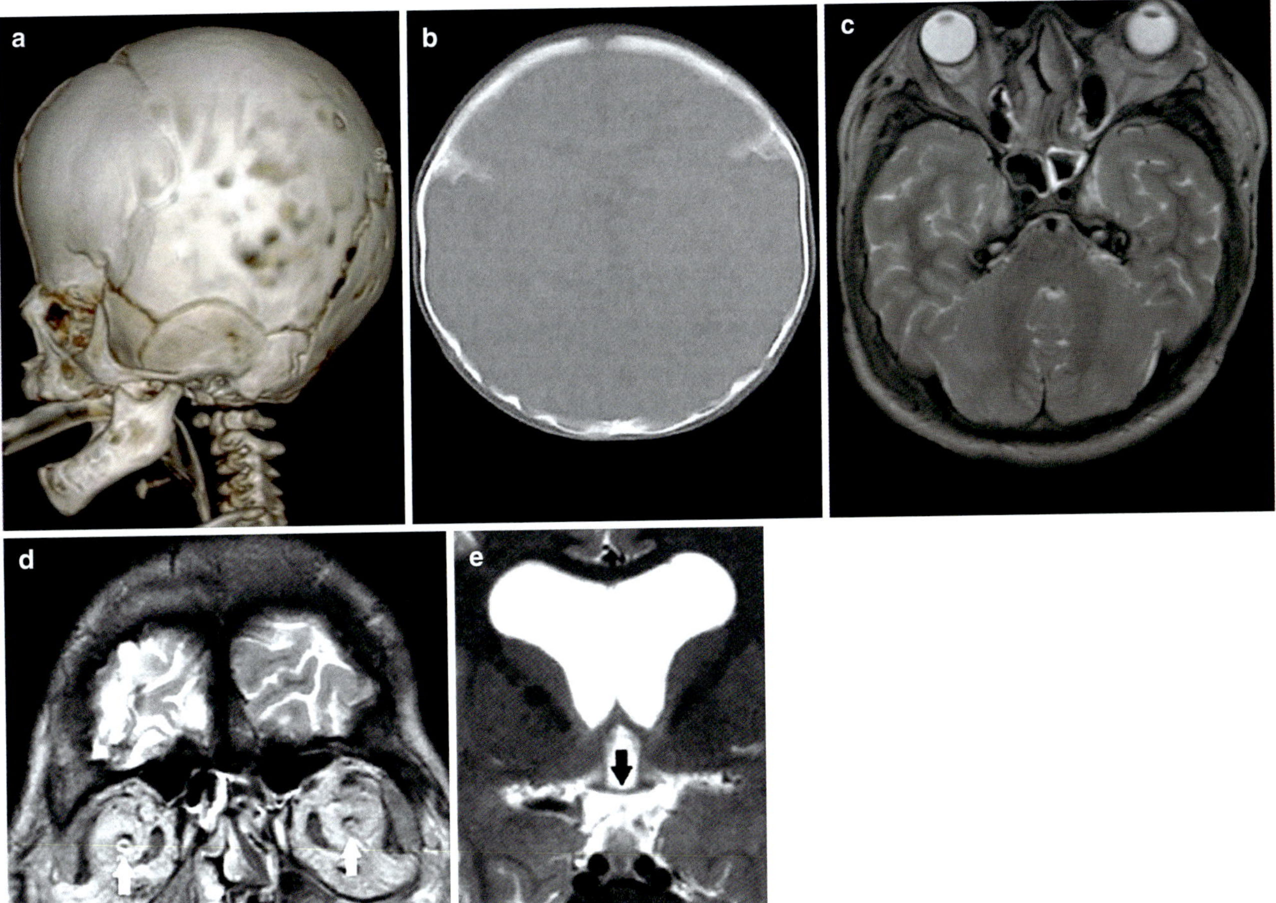

Fig. 13.119 Synostosis of the coronal and squamosal sutures. Increasing the number of sutures has physical consequences for the CNS. (**a**) 3-D CT image shows prominent convolutional markings. (**b**) Axial CT image (bone window) shows brachycephaly with a short and wide anterior cranial fossa. The metopic suture remains patent. (**c**) Axial T2-weighted image demonstrates foreshortened ethmoid complex, hypertelorism, shallow orbits, and bilateral proptosis. (**d**, **e**) Coronal T2-weighted images show small optic nerves (white arrows in **d**) and optic chiasm (black arrow in **e**). Ventriculomegaly and absence of the septum pellucidum are also seen. [Reprinted from Tan AP, Mankad K. Apert syndrome: magnetic resonance imaging (MRI) of associated intracranial anomalies. Child's Nerv Syst 2018; 34 (2):205-216. With permission from Springer Nature.]

Pfeiffer Syndrome

All parietal fields are involved giving multiple synostoses, up to and including *Kleebattschädel*, the coronal being closed bilaterally. In distinction with Apert, Pfeiffer has both upper and lower extremity deformities, including broad thumbs and toes and radio-ulnar synostosis.

There is partial or complete midline failure of frontal bone closure, in the latter case down to glabella. The squamosal fields bulge in compensation but are otherwise normal. The orbits are shallow with an anti-mongoloid slant of the palpebral fissures. Ptosis is present. The nose is short and bulbous. The airway is reduced transversely and the perpendicular plate is short causing a high-arched palate which sometimes has a midline cleft.

<u>Craniofacial</u>

- Polysynostosis with consequent CNS effects
- Petrous bone: 50% hearing loss

<u>Extracranial</u>

- Abnormalities of both upper and lower extremities
- Wide thumb and great toe, short fingers, pollux and hallux varus (digits bend in the opposite direction) (Hox reversal of programming at the joint?)

<u>Genetic</u>

- Nosology [105]
  - Type 1: autosomal dominant (advanced paternal age), normal intelligence
  - Type 2: involvement of all parietal borders: *Kleebattschädel*, sporadic, severe neuropathology, can be salvaged
  - Type 3: same as type 2 but more severe proptosis, early demise

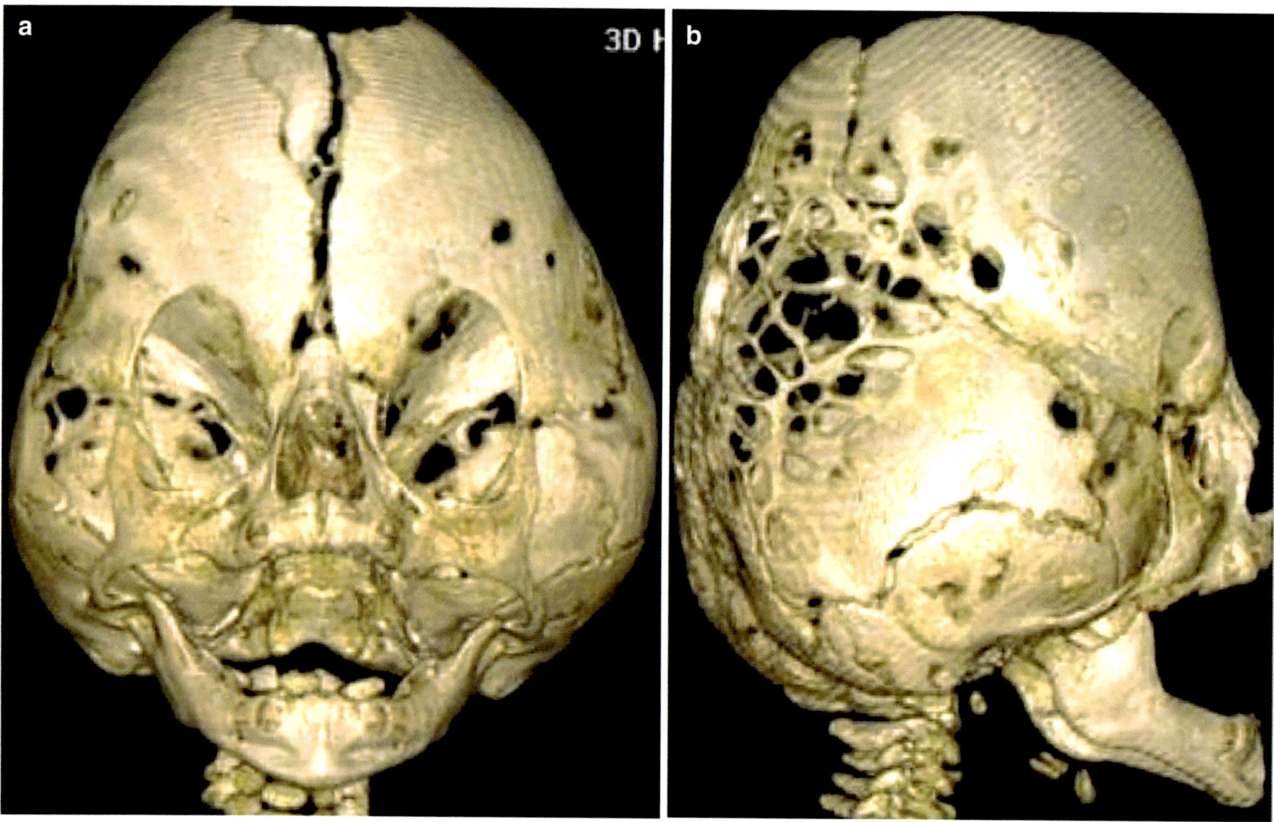

Fig. 13.120 Kleeblattschädel. "Cloverleaf" skull resulting from pansynostosis such as in Pfeiffer syndrome. Note the extreme distortion of the orbits. [Reprinted from Khanna PC, Thapa MM, Iyer RS, Prasad SS. Pictorial essay: The many faces of craniosynostosis. Indian J Radiol Imaging. 2011; 21(1):49-56. With permission from Wolters Kluwer Medknow Publications.]

- Genetic link to Crouzon syndrome [106]
  - *FGFR2* on chromosome 10
  - *FGFR1* on chromosome 8 [107, 108]

Neuromeric Field Analysis of Faciosynostoses: A Spectrum

- Unknown factors lead to abnormal parietal/sphenoid fields in non-syndromic FCS.
- Heterozygous mutations of *FGFR2* lead to syndromic parietal fields in Crouzon, Apert, and Pfeiffer syndromes.
- FGFR2 adds an r1 component by reducing ethmoid mesenchymal volume.
  - A-P deficit: foreshortened ethmoid sinus affects forward growth at two levels: (1) orbital plate of frontal bone is restrained at the frontoethmoid suture just lateral to cribriform plate and (2) and the maxillary field is restrained at its suture with the orbital plate of ethmoid, i.e., the lateral wall of the ethmoid sinuses.
  - D-V ethmoid deficit: vertical shortening of perpendicular plate.
  - Transverse ethmoid deficit: choanal atresia, < interorbital distance.
- Increased binding of FGFR2 worsens the pathology: Crouzon < Apert < Pfeiffer.
  - Increasing involvement of parietal fields = more synostoses.
  - Increasing loss of orbital volume.
  - Abnormalities of the extremities.
  - Concomitant neuropathology.

Neuromeric Differential Diagnosis of Faciosynostoses

	r1	r2 parietal	Intracranial	Orbit vol	Gene	Extremity
FCS	–	AI, AS	–	95% nl	None	–
Crouzon	+	AI, AS	+	67% nl	*FGFR2*	hands
Apert	+	AI, AS, PS	++	62% nl	*FGFR2*	hands/feet
Pfeiffer	+	AI, AS, PS, PI	+++	?	*FGFR2* + pm	hands/feet

Parietal field key: AI, anterior inferior; AS, anterior superior; PI, posterior inferior; PS, posterior superior.
Coronal = AI + AS; sagittal = AS + PS; lambdoid = PI + PS; squamosal = AI + PI.
pm = three-point mutations in codon 290, three different forms of Pfeiffer syndrome [106].
Note: FGF2, FGF3 encode *mis-sense proteins* with gain-of-function: increased proliferation, osteoblast activation, premature differentiation.

Craniofacial Clefts and the Orbit

Neuromeric Model of the Tessier System

Over the course of previous chapters, we have worked together to create a model of craniofacial development based on neurovascular fields, blocks of tissue that are ultimately related back to the central nervous system and their site of origin. Now it is time for us to put these concepts into practice as we examine craniofacial clefts, nature's gift to all those to understand head and neck anatomy. Why make such a bold statement? Because these experiments of nature give us an insight as to the identity of these building blocks…and how they are put together (Figs. 13.121 and 13.122).

First off, cleft is a misnomer. It leads us to suppose an event, a destruction that never happened. Let's set the record straight: **A cleft is a pathologic state of tissue deficiency or excess within the context of its neurovascular axis**.

Chapters 6 and 7 describe the sequential development of arterial systems to supply the CNS, face, skull, and neck. The stapedial system received particular emphasis which was reinforced in our discussions of the meninges and the orbit. Study the following chart with care as we shall use this as our point of departure to describe the pathologies which befall tissues in these neurangiosomes. We'll go through the anatomic and clinical details of each zone. Once we have the Tessier cleft system under our collective belts we will be able to take on subsequent topics of facial development affecting the pharyngeal arches. And so, to begin.

You may find a journeyman's approach to this subject in the following reference. It is a succinct review of the background material you have been reading and is available online [5]:

http://www.ijps.org/article.asp?issn=0970-0358;year=2009;volume=42;issue=3;spage=19;epage=34;aulast=Ewings

Successful treatment of congenital craniofacial defects relies on a thorough understanding of the embryologic processes that lead to their formation.

In 1964 DeMyer, in his landmark study of holoprosencephaly, understood that developmental processes of the face and brain were somehow linked. As a result, he proposed the concept "***the face predicts the brain***" meaning that worsening facial appearance indicated advanced neuropathology [109]. His phrase was catchy, and remains widely cited, but, in reality, the reverse is true. Central to our study of congenital facial clefts is the concept "***the brain predicts the face***," the inverse of DeMyer's original principle. In the neuromeric model, the embryonic nervous system constitutes a roadmap by means of which all facial tissues can be traced back to their site(s) of origin. The elucidation of this neuroanatomy and the pathways in which craniofacial tissues arise are translated into position and differentiate the foundation on which we can classify clefts, observe patterns and syndromes, and predict patterns of growth or relapse.

The human embryo has its own neuroanatomy. The embryonic central nervous system develops from discrete segmental units of the neural tube called neuromeres named in cranio-caudal order according their origin in the three-part embryonic brain. The anatomic boundaries of neuromeres are defined by the expression pattern of a unique series of homeotic barcodes. There are 5 prosomeres (hp2–hp1 and p3–p1), 2 mesomeres (m2–m1), and 12 rhombomeres (r0–r11), each one with its own homeotic "barcode." The neural crest cells found outside the neural tube maintain genetic markers in common with their neuromere of origin. Further, neural crest cells from a given neuromeric level supply specific zones of ectoderm, mesoderm, and endoderm with the same barcode. As such, the embryologic nervous system can be seen as the master integrative agent of development.

The clinical observations made by Paul Tessier [4, 110, 111] regarding patterns of craniofacial cleft formation were made from empiric observations, but they closely match known patterns of neural crest migration. By applying the neuromeric model to these pathologies, the Tessier system proves to have a solid basis in neuroembryology, rather than superficial topography. The developmental fields involved in the areas where the numbered Tessier clefts fall have been mapped out (Table 13.1). Abnormalities in the functional matrices giving rise to these fields, or the neurovascular sup-

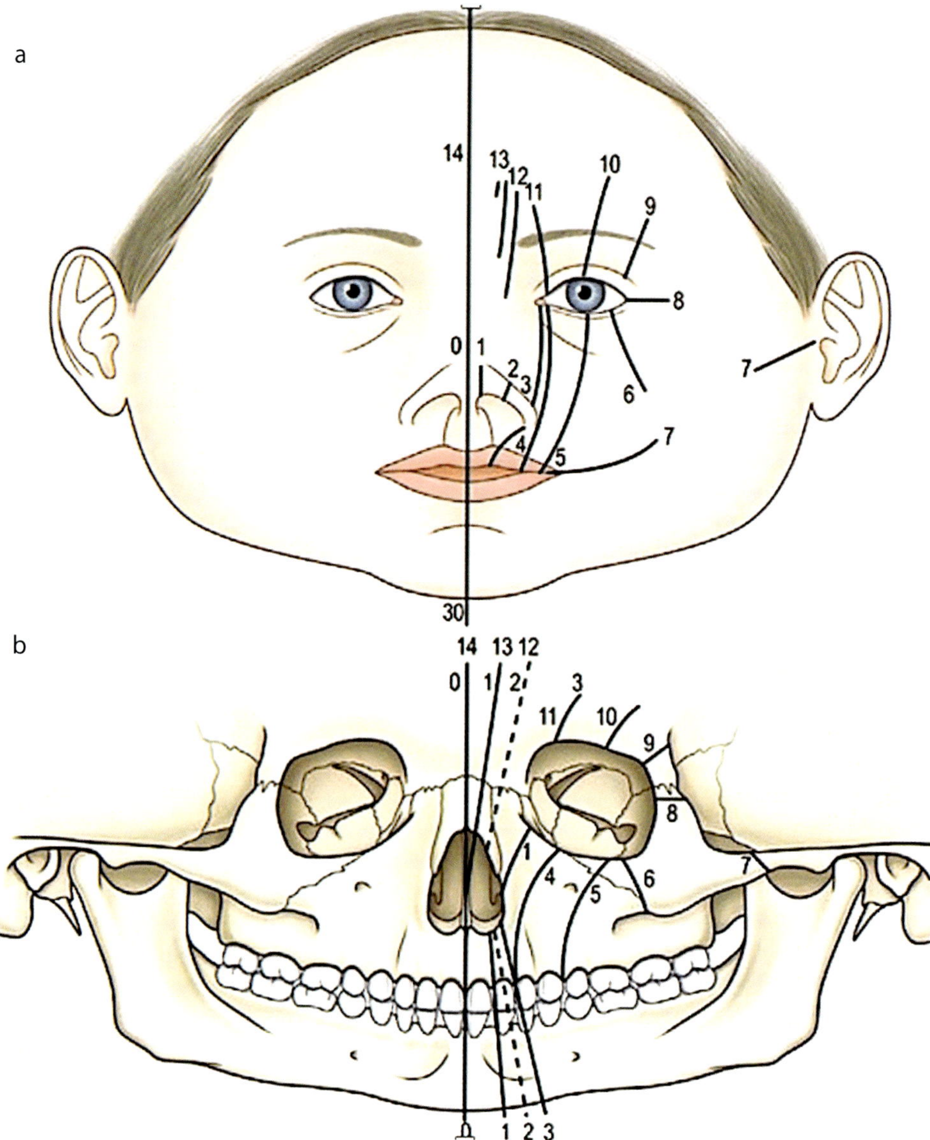

Fig. 13.121 Tessier numerical classification of craniofacial clefts. Skeletal clefts organized by stapedial branches: orbital series zones 10–13, maxillary series zones 1–9. Zone 14 does not exist. Midline mesenchymal are due to deficiency or excess in zone 13. Failure of both halves of the face would locate to the midline but not in a field defect sensu stricto. Soft tissue clefts represent the effect on the fusion of embryonic fields of inadequate amounts of BMP4 produced by the corresponding bone field(s). They fail to repress Shh in the soft tissues thus preventing fusion between adjacent soft tissue fields. Clinical appearance of soft tissue clefts follow gradients established from below. Piriform fossa deficiency in zone 2 can result in a lip cleft that extends from caudal to cranial and from superficial to deep depending on the quantitative amount of BMP4 available to the tissues. [Reprinted from Sari E. Tessier #30 facial cleft: A rare maxillofacial anomaly. Turk J Plast Surg 2018; 26: 12-19. With permission from Turkish Journal of Plastic Surgery.]

ply supporting them, may lead to the very same clinical malformations observed by Tessier.

Each cleft zone of the Tessier system will be explored individually. Several common themes underlie the collective series of clefts. First, all Tessier cleft zones faithfully follow the neurovascular anatomy of the stapedial system. Second, several of the topographically numbered clefts of the original Tessier classification may at first seem to overlap the same developmental field. When examined in more detail, each zone has a unique neurovascular anatomy. Furthermore, several zones can be grouped together according to their embryologic beginnings.

- Zones 4–9 are simplistic, marked by single bones with clearly identified fields (maxilla, zygoma).
- Zones 10–11 are laminated, in that they are derived from two epithelial layers (dura-sclera, dermis) which interact with an intervening layer of mesenchyme. This causes a split into two laminae, with the resultant formation of sinus cavities (frontal, ethmoid sinuses).
- Zones 12–13 may be seen as "stacked": r1 (ethmoid) fields lie internal to p5 (frontal) fields.
- Zones 1–2 clefts manifest pathology in zones 12–13.
- Zone 0 does not exist as a distinct cleft. Rather, the absence of midline structures results from insult to the internally situated precursor, the ethmoid. Therefore, midline hypotelorism is an extension of pathology in zone 13, while midline "cleft" hypertelorism results from insults to an external midline approximation mechanism, rather than intrinsic flaws in the tissues (Table 13.7).

Anatomic Description of Cleft Zones

Zone 13: This zone consists of medial ethmoid fields (cribriform plates and crista galli) and medial glabellar forehead. It

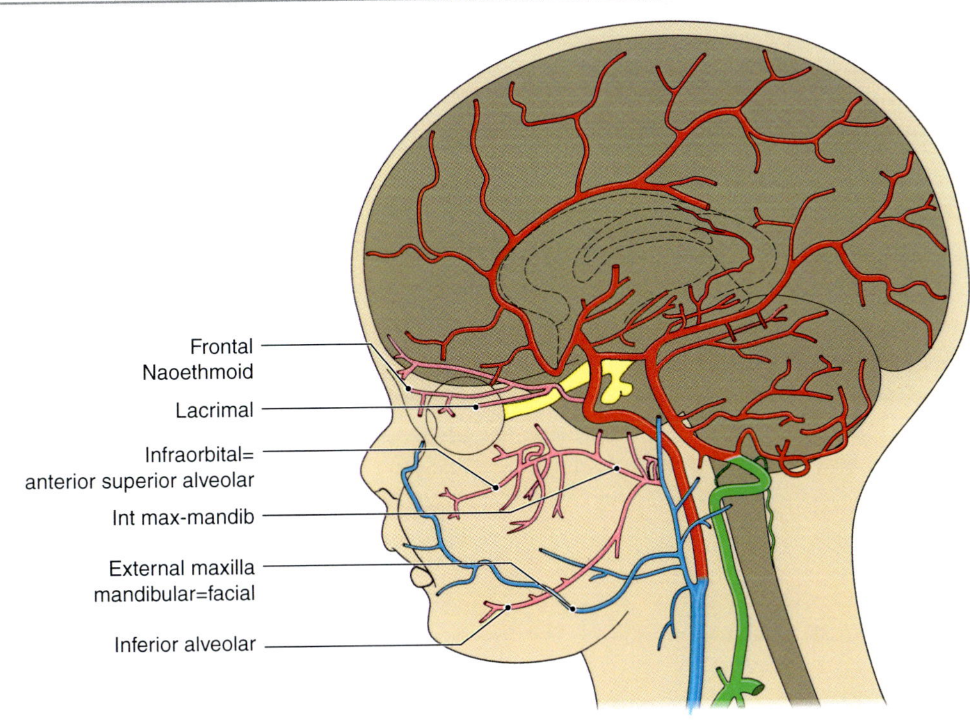

Fig. 13.122 **Stapedial system** originates at stage 17 and produces (1) meningeal aa (anast with external carotid), (2) StV1 extraocular ophthalmic aa (anast with primitive ophthalmic), and (3) StV2 and StV3 sphenopalatine aa (anast with external carotid). Tessier cleft zones correspond to individual knock-outs of the stapedial system. [Courtesy of Michael Carstens, MD.]

Table 13.7 Comparison of Tessier cleft zones. ***Big picture idea***: All Tessier clefts represent knock-outs, either partial or complete of a specific neurovascular axis

Tessier zone	Bone derivatives	Neurovascular supply
Neural crest derivatives from r1		**Artery: V1 stapedial**
13	Ethmoid cribriform	Frontal
12	Ethmoid labyrinth	Frontal
13–12	Medial forehead, p3	Frontal
13–2	Nasal bones	Frontal
11	Orbitosphenoid	
11	Prefrontal, lacrimal	Supraorbital
10	Postfrontal	Supraorbital
11–10	Lateral forehead, p2	Supraorbital
9	Lateral brow p1	Lacrimal
Midbrain/hindbrain neural crest r1 and r2		**Artery V1–V2 stapedial, ECA**
9	Alisphenoid	Lacrimal, StV1 Meningeal (StV2) Anterior deep temporal (ECA)
Hindbrain neural crest from r2		**Artery: V2 stapedial**
8	Postorbital	Zygomatico-temporal[a]
7	Jugal	Zygomatico-facial[a]
6	Buttress, posterior maxilla, molars	Posterior superior Alveolar
5	Lateral anterior maxilla, premolars	Anterior superior alveolar, lateral br.
4	Medial anterior maxilla, frontal process, lateral incisor, canine	Ant. superior alveolar, medial br.
3	Ascending process palatine	Descending palatine
3	Palatine hard palate	Lesser palatine
3	Maxillary hard palate	Greater palatine
3	Inferior turbinate	Lateral nasopalatine
2	Premaxilla, incisors	Medial nasopalatine
1	Vomer	Medial nasopalatine

[a]See description of neuroanatomy vide supra and in Chaps. 8 and 10

abuts the nasal bone. The blood supply is from the StV1 axis of nasociliary artery. (1) Anterior and posterior ethmoid arches track intracranial to supply the ethmoid plate. (2) Supratrochlear, medial branch supplies medial glabella (Figs. 13.123, 13.124, and 13.125).

Bone

- Hyperplasia/hypoplasia causes field deficits of medial ethmoid fields (cribriform, crista gall) or medial glabella.
- Cleft proceeds through olfactory groove.
- Encephalocoeles can result from defective sutures between ethmoid and nasal bone allowing subcutaneous descent into the nose or between lateral margin of cribriform and frontal giving access to the orbit.
- If tissues clefted or expanded: hypertelorism.
- Hypoplasia of medial ethmoid fields: hypotelorism.
- Nasal bone eroded or absent.

Soft tissuesv

- Eyebrow is spared
- Nasal cleft through the intermediate crus of the alar cartilage. This leaves alar base free; it does not rotate.

Combination

Zone 12

This zone consists of lateral ethmoid fields (both sinuses and transethmoid intracranial), the internal and external soft tissues of the nose, and the lateral glabellar forehead. It abuts the frontal process of maxilla. The blood supply is from the StV1 axis of nasociliary artery. (1) Anterior and posterior ethmoid branches supply the ethmoid sinuses and then descend into nose to supply lateral nasal vestibular lining and skin. (2) Infratrochlear branch supplies upper lateral nasal skin, medial to lacrimal duct. (3) Supratrochlear, lateral branch supplies lateral glabella (Figs. 13.126, 13.127, and 13.128).

Bone

- Dysplasia: cleft between nasal bone and medial canthus traversing lateral fields of ethmoid, i.e., the sinuses. Lateral displacement of tissue fields causes telecanthus. Upward displacement of maxillary frontal process and foreshortened piriform fossa.
- Hyperplasia: Enlargement of the ethmoid labyrinthine and/or frontal sinus leading to lateral displacement of the orbits and hypertelorism.
- Hypoplasia of lateral ethmoid fields: hypotelorism.

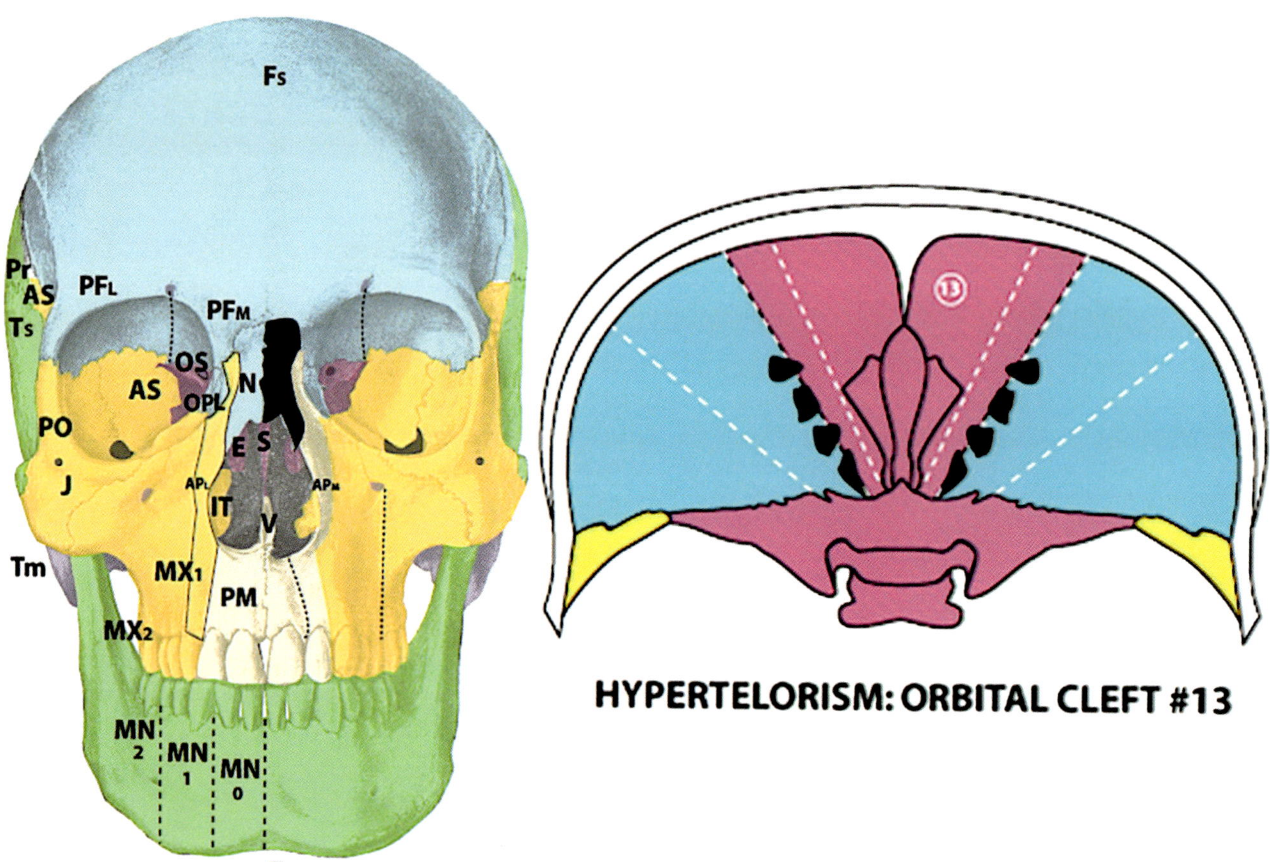

Fig. 13.123 Zone 13. Ethmoid axis: (1) intracranial via anterior and posterior ethmoids via sinuses and through cribriform plate to crista galli; (2) extracranial: via infratrochlear to nasal skin envelope, and via supratrochlear to median glabella. [Reprinted from Ewings E, Carstens MH. Neuroembryology and functional anatomy of craniofacial clefts. Indian J Plast Surg 2009; 42(Suppl):S19-S34. With permission from Creative Commons Attribution License.]

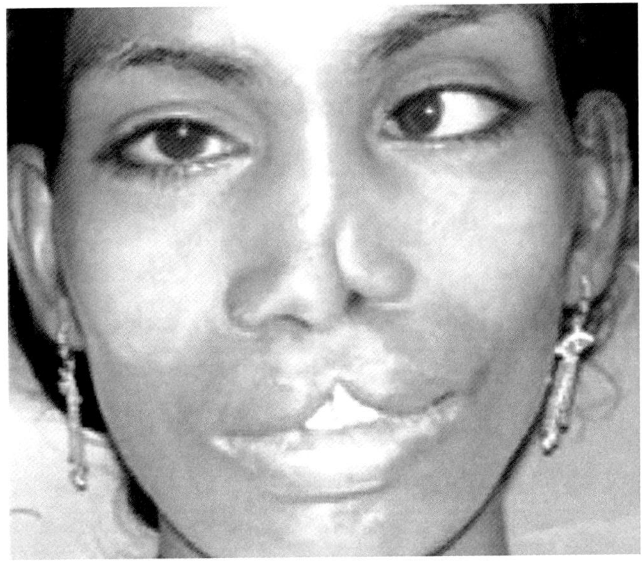

Fig. 13.124 Zone 13 nasal effects. Cleft in the zone 13 leading to ethmoid hyperplasia and left-sided hypertelorism. This is commonly considered a no. 2 cleft but its angiosome belongs to StV1. [Reprinted from Ewings E, Carstens MH. Neuroembryology and functional anatomy of craniofacial clefts. Indian J Plast Surg 2009; 42(Suppl):S19-S34. With permission from Creative Commons Attribution License.]

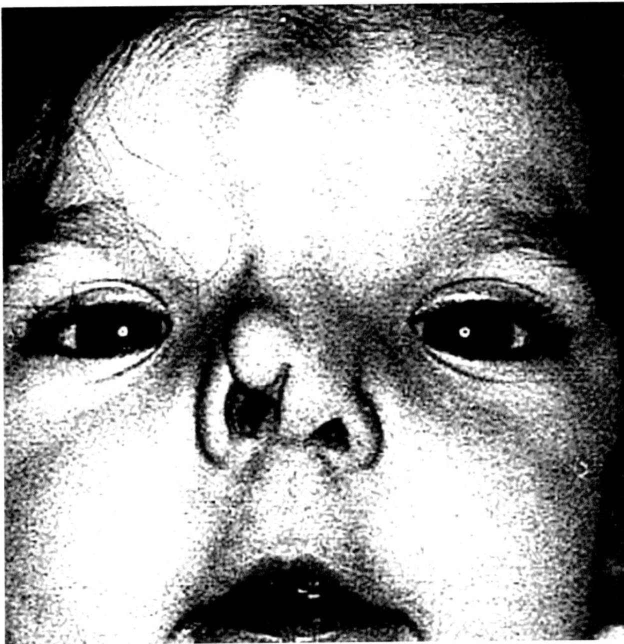

Fig. 13.125 Zone 13 frontal mass. Zone 13 defect with continuity from the nasal tip to loss of the nasal bone to a frontal dermoid mass. [Reprinted from Ortiz Monasterio F, Fuente del Campo A, Dimpoulus A. Nasal clefts. Ann Plast Surg 1987; 18(5):377-397. With permission from Springer Nature.]

Soft tissues

- Eyelid uninvolved. Coloboma at medial root of eyebrow.
- Nasal defects located just lateral to the alar cartilage with lateral ala rotated upward.

Combination

- Cleft through ala lateral to alar cartilage; is associated with zone 2 cleft through the ipsilateral philtral column.

Zone 11

This zone refers to the bone fields of medial orbit and nonglabellar forehead. These are defined by the medial branch of StV1 supraorbital artery and include bone fields of both the orbit and forehead: (1) The prefrontal bone forms the medial half of the orbital roof from the supraorbital axis to the orbital plate of the ethmoid and the medial supraorbital rim. (2) Medial forehead is defined from the vertical axis of the supraorbital neurovascular pedicle medial to a vertical line extending from the frontoethmoid upward to the vertex (Fig. 13.129).

Bone

- Notch is present at medial supraorbital margin.
- Encephalocoeles are much less common because medial orbital roof of prefrontal is bilaminar due to the presence of transverse frontal sinus. A defect of frontoethmoid suture can permit entry and medial margin encephalocoele, causing dystopia. Frontal encephalocoele and hyperplasia can occur.

Soft tissue

- Coloboma of the medial third of the upper lid may be observed, along with brow and hairline distortion in the same region.

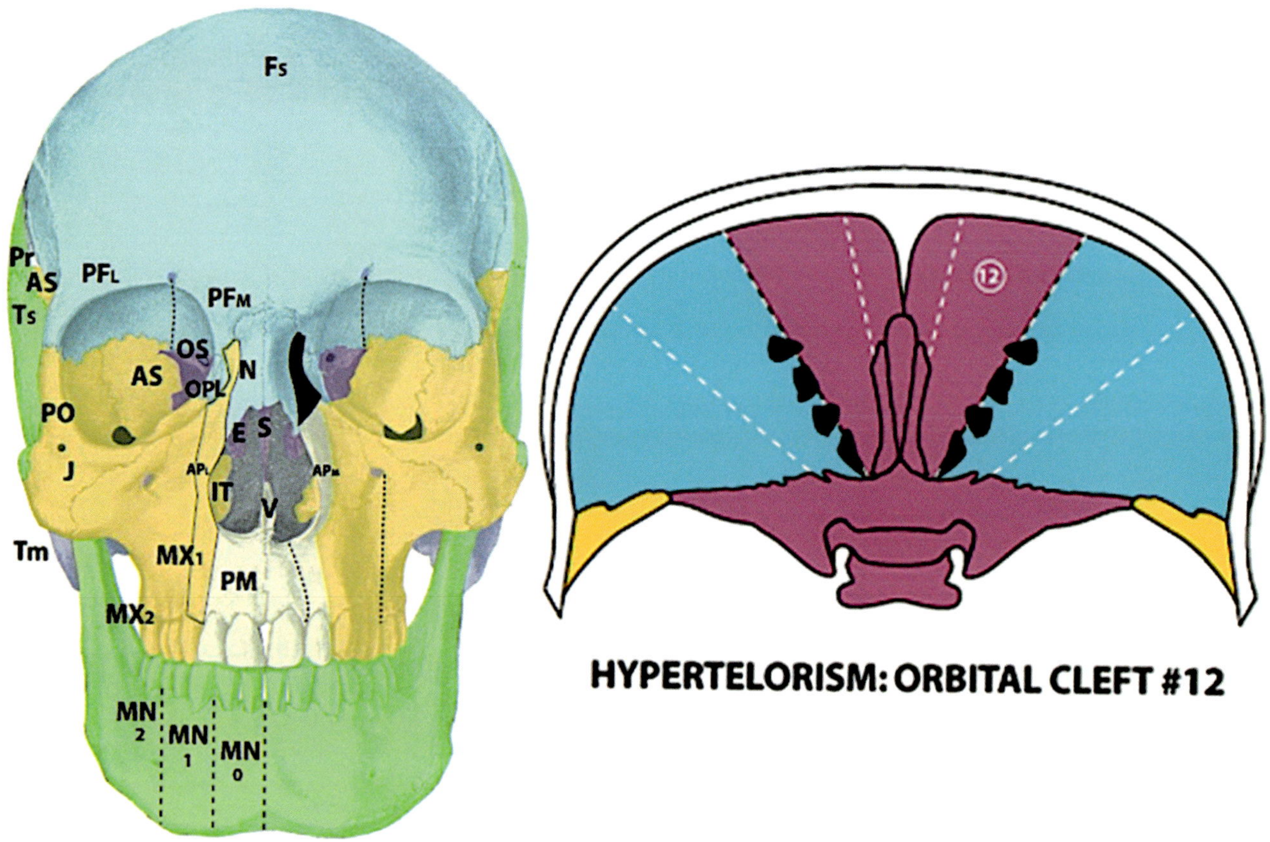

Fig. 13.126 Zone 12. Blood supply is based on the branches of StV1 ethmoid: (1) intracranial via the anterior/posterior ethmoid angiosomes to the sinuses and to ethmoid labyrinth; (2) extracranial via supratrochlear axis. Pathologies range from hyperplasia, expansion of the frontal sinuses, or encephalocoele. [Reprinted from Ewings E, Carstens MH. Neuroembryology and functional anatomy of craniofacial clefts. Indian J Plast Surg 2009; 42(Suppl):S19-S34. With permission from Creative Commons Attribution License.]

- A more subtle form of the zone 11 cleft is the isolated lacrimal bone deficiency. Here the neurovascular axis is the supratrochlear and dorsal nasal arteries and nerves. Lacrimal stenosis may occur when distorted orbicularis insertions affect the lacrimal pump mechanism. Zone 11 deficiency can pull the posterior ala upward.
- The sclera, which arises from r1 neural crest supplied by the nasociliary nerve (V1) can produce epibulbar dermoids.

Combination (zone 3)

- Medial lower eyelid dermis can also have epibulbar dermoids.
- Thus, 3–11 clefts involving nasal displacement above the medial canthi demonstrate the relationship to the p6/p5 cribriform plate.

Zone 10

This zone includes bone fields of both the orbit and forehead. These are defined by the lateral branch of StV1 supraorbital artery. (1) The post-frontal bone forms the lateral half of the orbital roof minus the lacrimal zone corresponding to post-frontal bone and lateral half of the supraorbital rim. (2) Lateral forehead is defined from the curvilinear fascial band separating frontalis from temporalis medial to the vertical axis of the supraorbital neurovascular pedicle (Figs. 13.130 and 13.131).

Bone

- Zone 10 defects can allow encephalocoeles into the orbit, where they will displace the globe inferolaterally, distorting the anterior cranial fossa. Forehead encephalocoeles can also occur. Mesenchymal excess states can cause downward dystopia as well. Mesenchymal deficit can present with more severe forms of trigonocephaly.

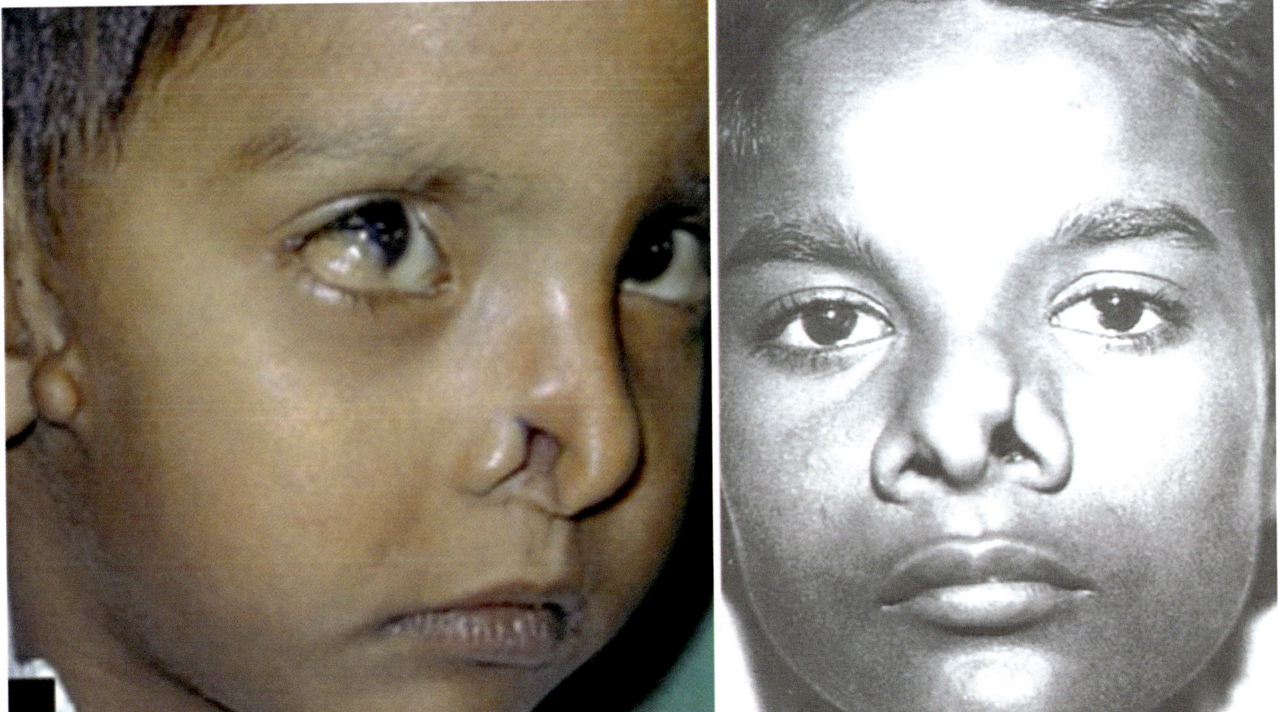

Fig. 13.127 Zone 12, nasal angiosomes from StV1. Left: Lateral nasal dorsal artery from anterior ethmoid perfuses lateral nasal skin. Neural crest part of the ala is cleft at junction of lateral crus and sesamoid cartilages. Right: Bilateral nasal clefts with masses (widened, thinned-out frontal sinuses) in zone 12 of the forehead (left > right). Left medial canthus displaced laterally 4 mm compared to the right side. Widened ethmoid sinuses give hypertelorism (left sided). Left: [Reprinted from Deraje V, Ahuja RB. A rare combination of Tessier 2 cleft with Goldenhar syndrome. J Cleft Lip Palate Craniofac Anomalies. 2016; 3:100-102. With permission from Wolters Kluwer Medknow Publications.]. Right: [Reprinted from Tiwari P, Bhnagar SK, Kalra GS. Tessier number 2 cleft, a variation. J Cranio-Maxillofac Surg 1991; 19(8)346-347. With permission from Elsevier.]

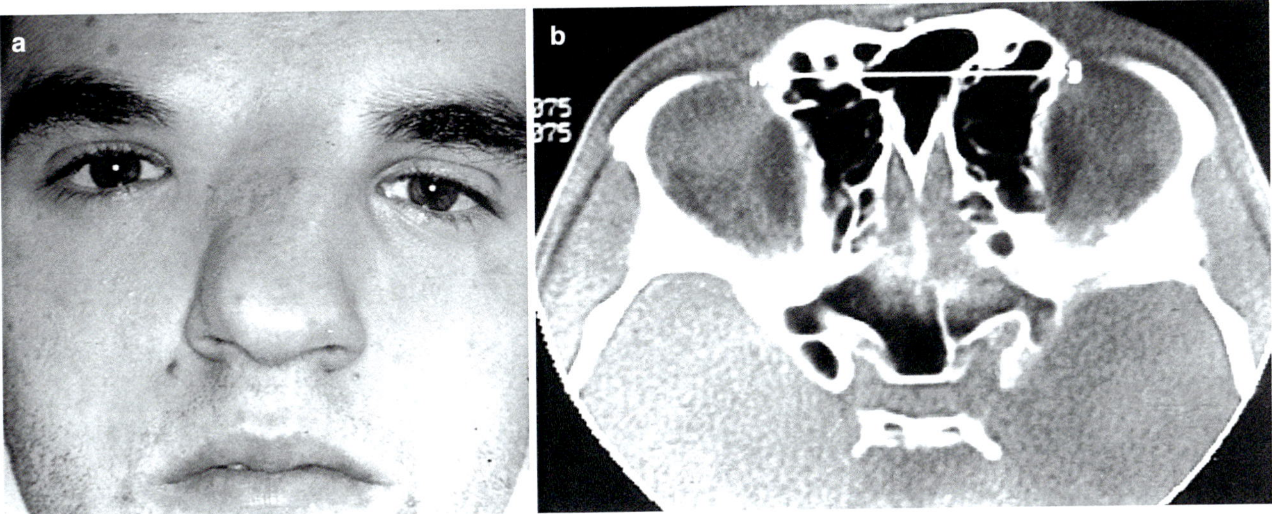

Fig. 13.128 Zone 12. Loss of right nasal bone with widened ethmoid labyrinth causing hypertelorism. [Reprinted from Ozuk C, Gundogan H, Bilkay U, et al. Rare craniofacial anomaly: Tessier no 2 cleft. J Craniofac Surg 2001; 12(4):355-361. With permission from Wolters Kluwer Health, Inc.]

Fig. 13.129 Zone 11. Frontal (supraorbital), medial supplies prefrontal field of the orbital plate, medial one-third of eyebrow, and medial forehead. Shortened dorsal nasal field causes upward rotation of r2 ala. [Reprinted from Ewings E, Carstens MH. Neuroembryology and functional anatomy of craniofacial clefts. Indian J Plast Surg 2009; 42(Suppl):S19-S34. With permission from Creative Commons Attribution License.]

Fig. 13.130 Zone 10. Frontal (supraorbital), lateral supplies postfrontal field of the orbital plate, central one-third of eyelid/eyebrow, and lateral forehead. [Reprinted from Ewings E, Carstens MH. Neuroembryology and functional anatomy of craniofacial clefts. Indian J Plast Surg 2009; 42(Suppl):S19-S34. With permission from Creative Commons Attribution License.]

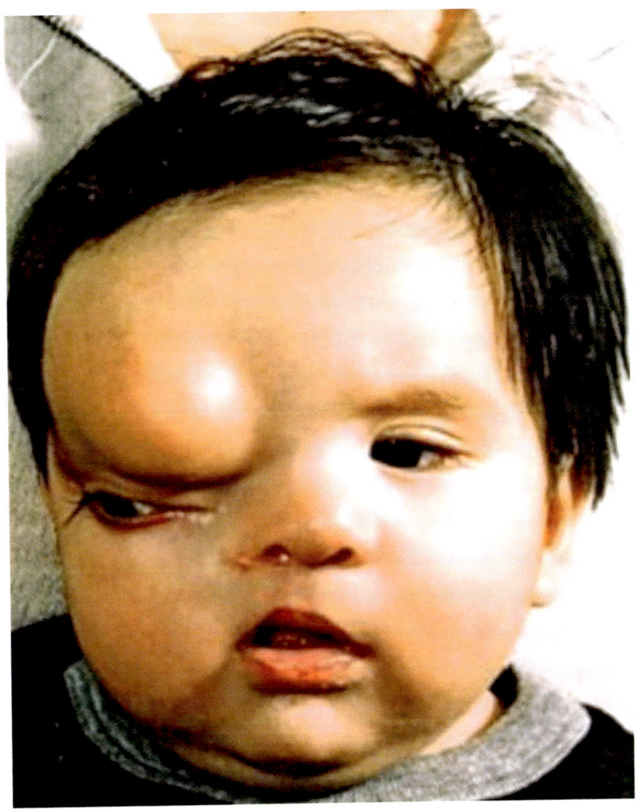

Fig. 13.131 Zone 10 encephalocoele. Defects in frontal zones 10–13 permit herniation of tissue. [Reprinted from ECLAMC Estudio Cooperativo Latinoamerericano de Anomalías y de Malformaciones Congénitas, Professor Ignacio Zarante, Institutode Genetic Humana, Potifica Universidad Javeriana Bogotá, Colombia.]

Soft tissue

- Middle third of the upper eyelid coloboma defects are common along with distortion of the middle eyebrow and downward displacement of the frontal hairline towards the brow.

Combination

- No reference cases found: in theory would include soft and/or hard tissue involvement in the neuroangiosome corresponding to the medial branch of infraorbital nerve and/or medial branch of superior alveolar nerve.

Zone 9

Zone 9 defects involve the lateral one-third of the orbital plate and the alisphenoid. The blood supply is dual: StV1 lacrimal artery and (sometimes) StV2 recurrent meningeal are described as quite rare (not seen by Tessier himself) but, with an embryologic definition, they become easier to understand and are probably quite common! The blood supply to alisphenoid (AS), as previously described, was originally from the StV2 recurrent meningeal artery which generally involutes, leaving the unilaminar bone supported from StV2 dural branches to anterior temporal fossa and the StV1 lacrimal on the other side. As previously described, the inferior extension of AS, lateral pterygoid plate, is perfused from the pterygoid branch of internal maxilla-mandibular (Figs. 13.132, 13.133, 13.134, and 13.135).

Bone

- Recall that zone 9 includes anteriorly the lateral one-third of the orbit and posteriorly the alisphenoid. The most common defect is trigonocephaly which can cause distortion of the neighboring frontal, and to a minor degree the parietal bone fields.
- Synostosis of frontozygomatic suture is observed, as is a hypoplastic lateral pterygoid plate. Note: FS is sometimes referred to (erroneously) as the lower coronal suture.
- Aplasia of greater sphenoid wing leads to intraorbital encephalocoele.

Soft tissues

- Lateral third of the eyelid can be disrupted by coloboma. The lateral eyebrow can also be dysplastic or missing. Additionally, the hairline is distorted with upward displacement of the brow and forward displacement of the temporal hairline and sideburn. This occurs due to mesenchymal deficiency of the r2-derived dermis in this region; reduction in this skin pulls the sideburn forward.
- Frontalis and upper orbicularis palsies can be present.

Combination: Zone 9 can be associated with a zone 5 defect.

Zone 8

Zone 8 refers to the postorbital bone field, supplied by StV2 zygomatico-temporal artery, the "origin" of which, in most cases, is lacrimal, although the original embryonic supply from recurrent meningeal artery sometimes remains present (Figs. 13.136, 13.137, and 13.138).

Bone

- The bony defect observed in the zone 8 cleft involves the posterior extension from zygoma which forms a suture with greater wing of the sphenoid. Recall that this suture is *extracranial*. A minor defect at the sutural level can allow the postorbital escape of a dermoid cyst and lateral orbital wall with downward deformation of the lateral orbital rim. The jugal bone field is unaffected, as expected.

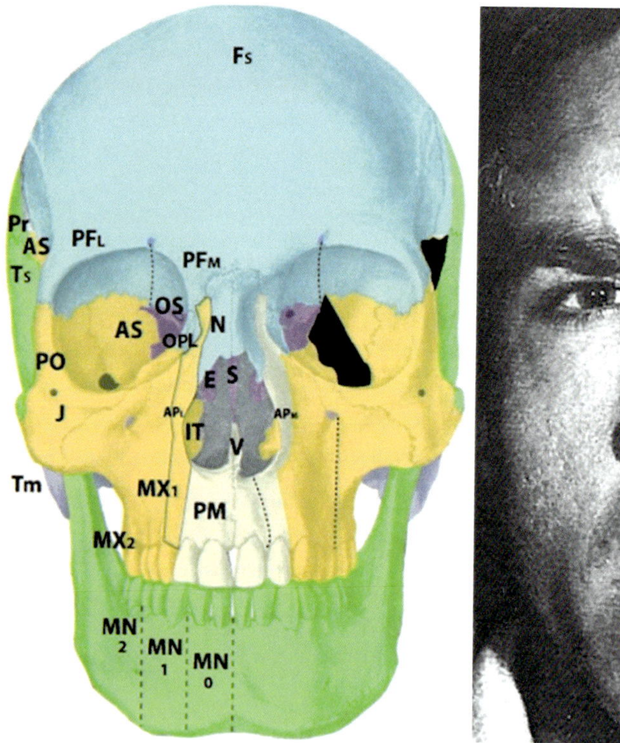

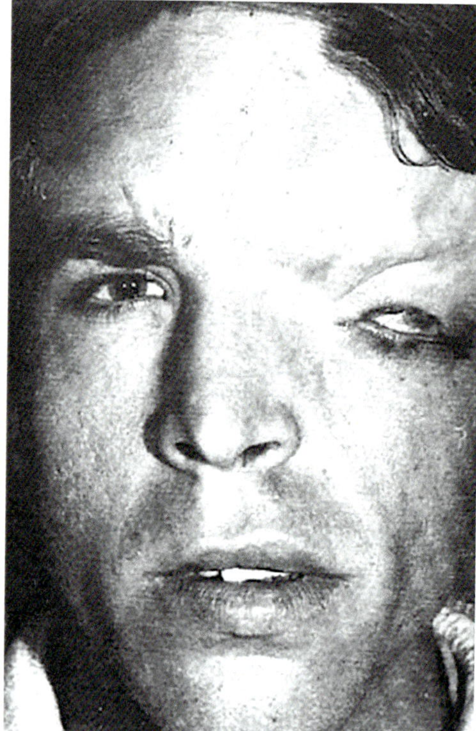

Fig. 13.132 Zone 9 is complex. StV1 lacrimal deficit affects the lateral corner of the orbital plate, specifically the lateral supraorbital rim, and can contribute to trigonocephaly. StV2 to alisphenoid affects superior orbital fissure. Soft tissue effects in zone 9 include the lateral brow, lateral canthus. Left: [Reprinted from Tessier P. Plastic Surgery of the Orbit and Eyelids (trans Wolf SA) Paris, France: Mosby, 1981. With permission from Elsevier.] Right: [Reprinted from Ewings E, Carstens MH. Neuroembryology and functional anatomy of craniofacial clefts. Indian J Plast Surg 2009; 42(Suppl):S19-S34. With permission from Creative Commons Attribution License.]

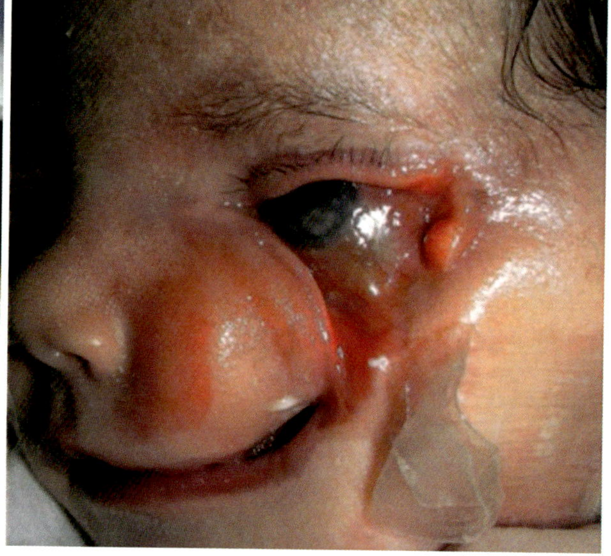

Fig. 13.133 (Left) Zone 9 + zone 2. Zone 9 cleft involving lateral one-third of eyebrow, lateral orbital rim, greater wing of sphenoid, and adjacent squamous temporal bone. Lateral canthal defect permits displacement of globe. The right ala is notched with displaced chondrocutaneous mass above the notch, possibly the missing intermediate crus. (Right) Zone 9 + zone 5. Theoretical combination of 5 and 9 is extremely rare. Here the upper crus of lateral canthus is missing. Upper eyelid pulled upward and lateral. The 5 cleft presented as a defect in the premolar zone extending lateral to the infraorbital neurovascular bundle. Left: [Reprinted from Bubanale SC, Kurbet SB, De Piedade Sequeira LMG. A rare case of cleft number nine associated with atypical cleft number two. Indian Journal of Ophthalmology 2017; 65(7):610-612. With permission from Wolters Kluwer Medknow Publications.]. Right: [Reprinted from Pereira FJ, Milbratz GH, Cruz AAV, Vasconcelos JJA. Opthalmic considerations in the management of Tessier cleft 5/9. Ophthal Plast Reconstr Surg 2010;26(6):450–453. With permission from Wolters Kluwer Health, Inc.]

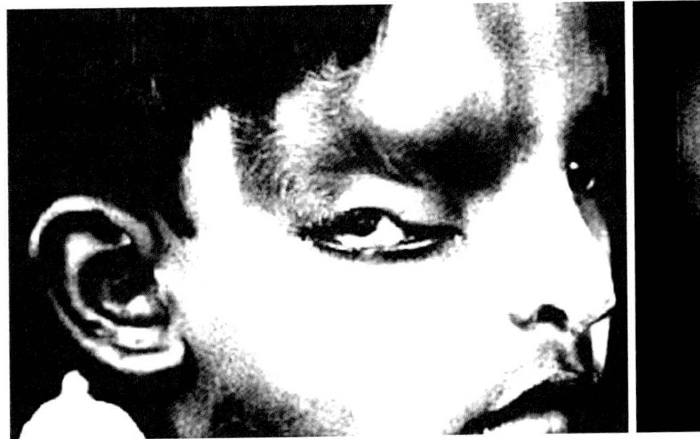

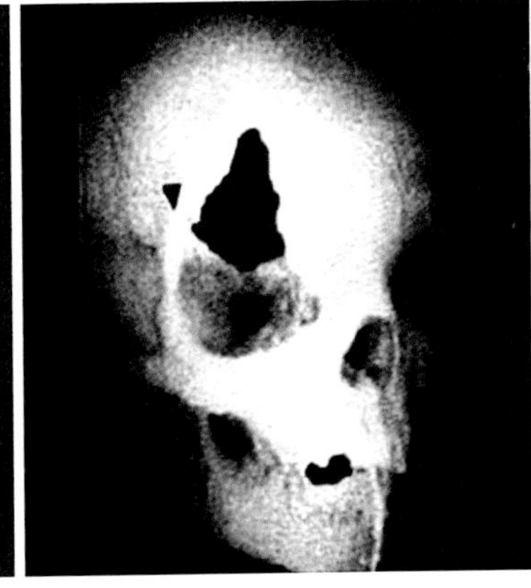

Fig. 13.134 Zones 9 + 10. Patient reported by Dr David with a combination of 10 and 9 clefts. The zone 10 cleft presents with encephalocoele with zone 9 characterized by hair distortion. CT scan shows distinct notching of the lateral orbital margin that is separate from zone 10. At surgery a microform cleft of the superolateral margin was confirmed. [Reprinted from David DJ, Moore MH, Cooter RD, Chow S-K. The Tessier number 9 cleft. Plast Reconstr Surg 1989; 83(3):520-525. With permission from Wolters Kluwer Health, Inc.]

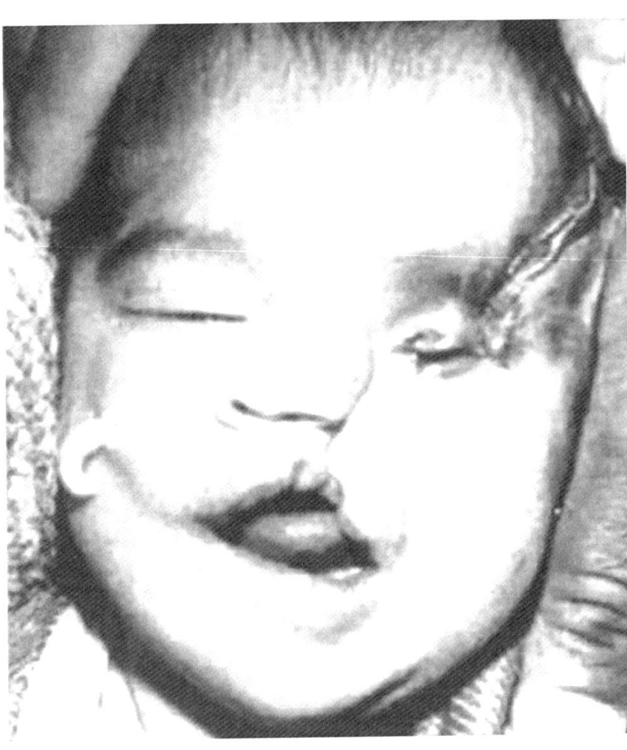

Fig. 13.135 Zones 9 + 3. Note dramatic separation between r1 frontonasal mesenchymal fields and r2, r4 composite fields from the combined first and second arches. [Reprinted from Darzi MA, Chodri NA. Oblique facial clefts: a report of Tessier numbers 3, 4, 5, and 9 cleft. Cleft Palate Craniofac J 1993; 30(4):414-415. With permission from Sage Publishing.]

Soft tissues

- An absent lateral canthus with coloboma and epibulbar dermoid formation and loss of v globe support.
- The zone 8 cleft may again be observed as a component of several craniofacial syndromic conditions. Soft tissue deficiencies predominate in Goldenhar's syndrome whereas osseous defects predominate in Treacher-Collins syndrome.

Zone 7

This zone refers to the jugal bone field. This is supplied by the StV2 zygomatico-facial artery. Recall that StV2 to the orbit anastomoses with lacrimal. Soft tissues of upper lateral orbit are supplied from superficial temporal via lateral facial, zygomatico-orbital branch (Figs. 13.139 and 13.140).

Bone

- The isolated zone 7 cleft demonstrates a deficiency in the jugal bone, or caudal zygoma. The maxilla may show retrusion as a restrained neighboring field, with Angle Class III occlusion.
- As compared to the zone 6 cleft, in which the maxillary buttress is absent but the zygoma is present, a zone 7 cleft shows *presence of the zygomatico-maxillary buttress* but possible involvement of the zygomatic arch.

Soft tissues

- The lateral canthus is displaced downward.
- Clefting from the inferior orbital fissure toward the commissure itself. Again, the zone 7 cleft is seen as a part of several syndromic conditions, but may exist in an isolated form. The soft tissue cleft represents a communication upward toward the surface from the underlying r2/r4 bone fields such that soft tissues cannot migrate across the defect.
- This is *not* to be confused with the laterofacial cleft extending from commissure to the ear. This so-called "number 7 cleft" represents a divide between the r2/r4 fields of the midface and the r3/r5 fields of the lower face. We will discuss this separately.

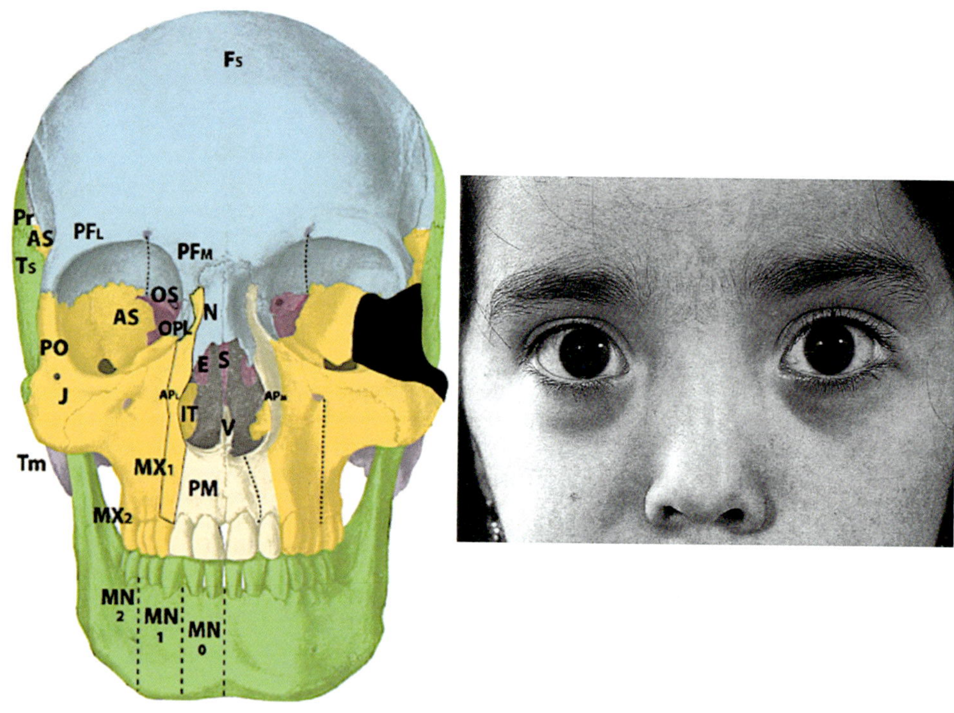

Fig. 13.136 **Zone 8** is the postorbital bone field of zygoma. Arterial axis is zygomatico-temporal. Bony defects more common in Treacher-Collins–Francheschetti. The apex of the defect is the frontozygomatic suture. As the mesenchymal deficiency worsens, the postorbital field recedes. Soft tissue defects are more common in Goldenhar syndrome and include lateral canthus distortion. [Reprinted from Ewings E, Carstens MH. Neuroembryology and functional anatomy of craniofacial clefts. Indian J Plast Surg 2009; 42(Suppl):S19-S34. With permission from Creative Commons Attribution License.]

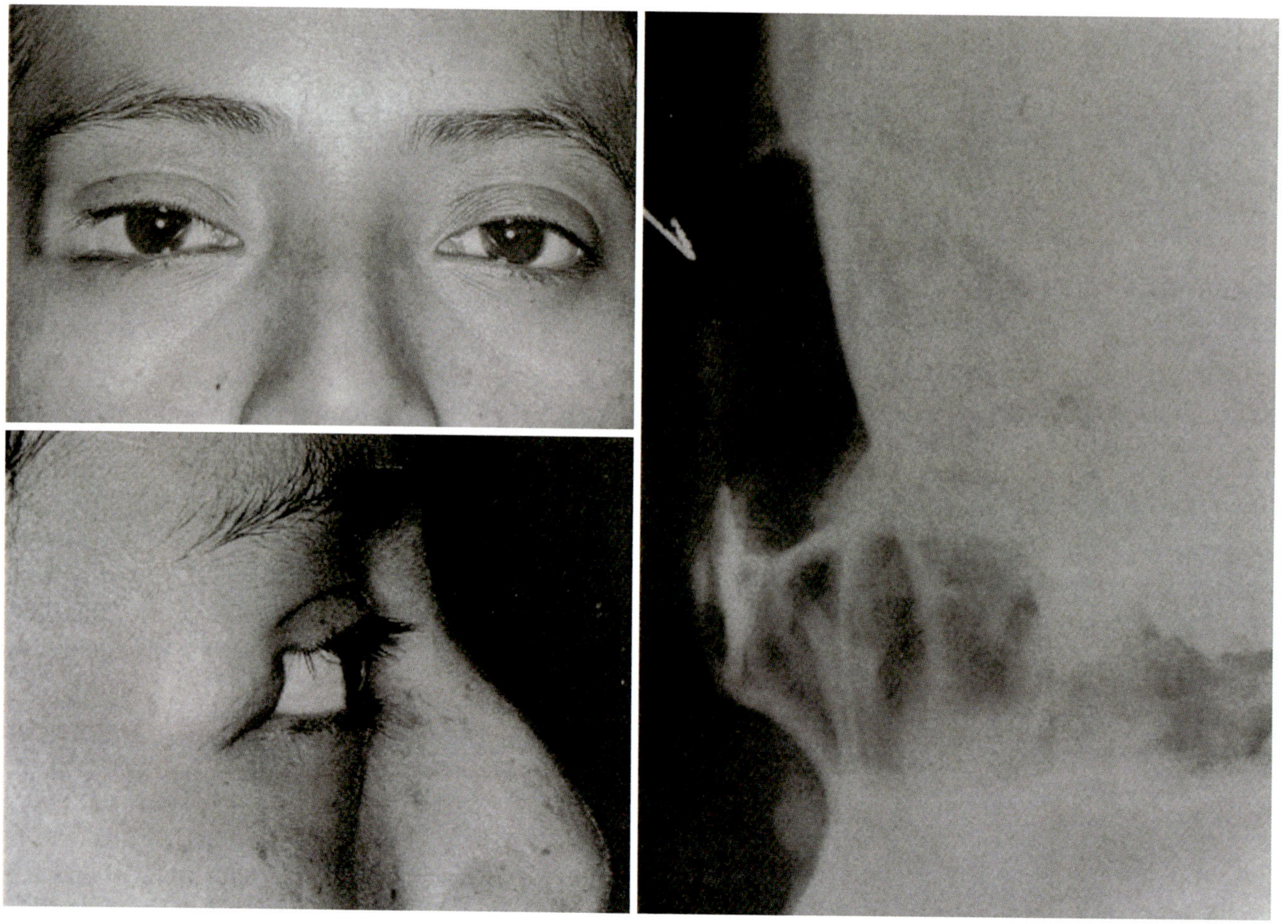

Fig. 13.137 Zone 8 cleft, isolated. Note absence of normal commissure; absent canthus permits eyelids to spread apart. Radiograph shows superior orbital rim and inferior orbital rim with malar field: arrow indicates missing postorbital field. The great wing positioned just behind it is normal. [Reprinted from Fuente del Campo A. Surgical correction of Tessier number 9 cleft. Plast Reconst Surg 1989; 86(4):658-661. With permission from Wolters Kluwer Health, Inc.]

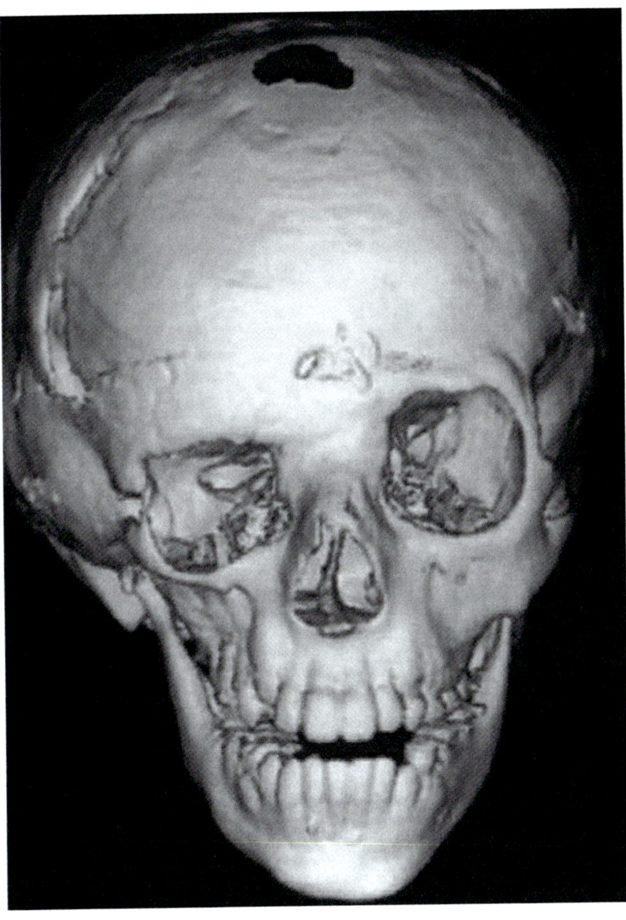

Fig. 13.138 **Zone 8 cleft** of the right postorbital field in a case of right-sided craniofacial microsomia. [Reprinted from Cassi D, Magnifico M, Gandolfini M, et al. Early Orthopaedic treatment of hemifacial microsomia. Case Reports in Dentistry Volume 2017, Article ID 7318715. With permission from Creative Commons License 4.0: https://creativecommons.org/licenses/by/4.0/.]

Zone 6

Defects in zone 6 represent deficiency of the lateral maxillary wall, the so-called "buttress" leading up to zygoma. The neurovascular supply to this zone comes from the StV2 posterior superior alveolar that penetrates maxilla via foramina, with collaterals from the ascending pharyngeal artery and nerve (Figs. 13.141 and 13.142).

Bone and dental

- Vertical shortening of the posterior maxilla and maxillary buttress; because the suture between zygoma and maxilla can extend all the way to the inferior orbital fissure it can also be affected.
- Hypoplastic molars with an alveolar "crease."
- The deficiency state curves around the posterior part of maxilla to affect palatine bone, causing high vaulting of the posterior palate with posterior choanal atresia.
- Zone 6 does not affect the jugal (malar) field and the zygomatic arch is not involved.

Soft tissues

- Colobomas of the lateral one-third of the lower eyelid with downward displacement of the lateral palpebral fissure and lateral canthus; vertical furrowing of the lateral eyelid to the commissure in cases of incomplete clefting. Zones 6–8 are macrostomia, the mechanism of which is very different (vide infra).

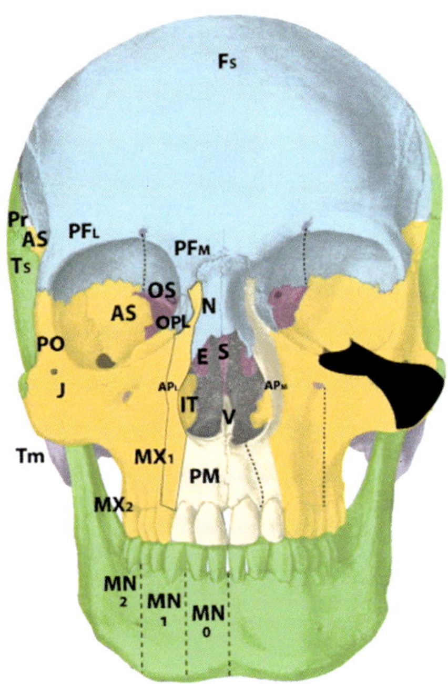

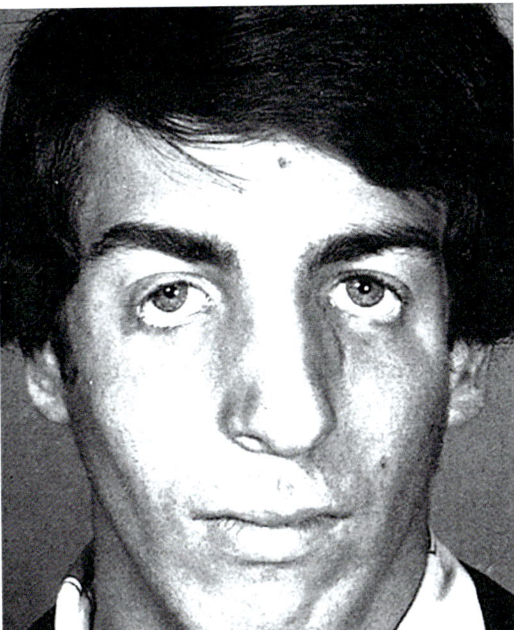

Fig. 13.139 Zone 7. The axis is zygomatico-facial artery. If isolated, the zone 6 maxillary buttress is intact and the molars are unaffected. [Reprinted from Ewings E, Carstens MH. Neuroembryology and functional anatomy of craniofacial clefts. Indian J Plast Surg 2009; 42(Suppl):S19-S34. With permission from Creative Commons Attribution License.]

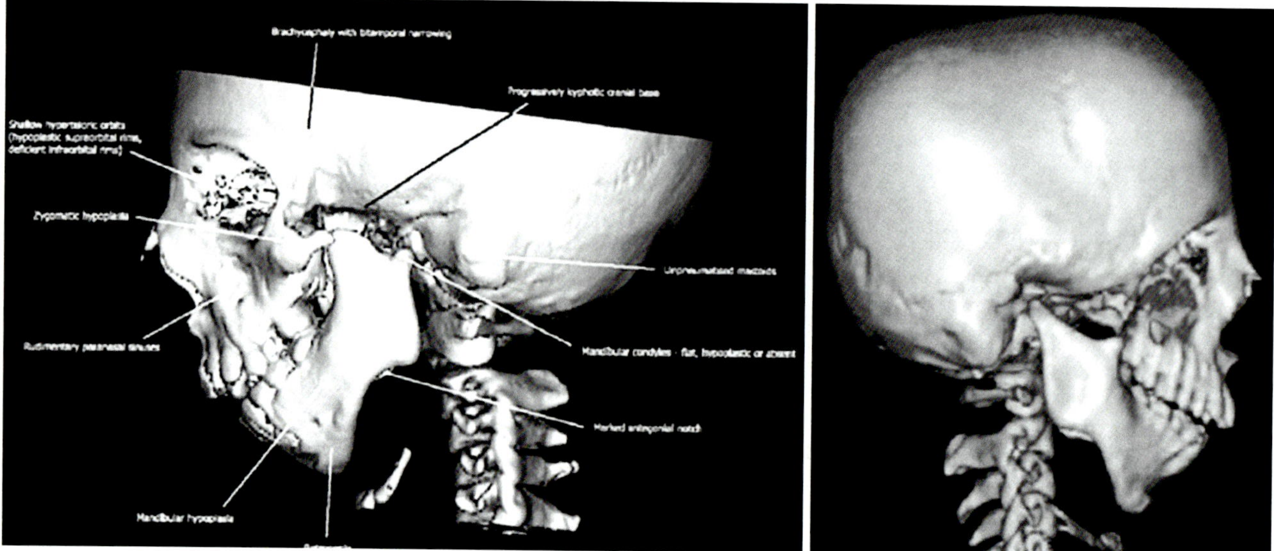

Fig. 13.140 Zone 7 clefts are characteristic of Treacher-Collins–Franscheschetti syndrome. (Left) Complete absence of the jugal field. Maxilla is fused to postorbital. (Right) Partial absence of jugal field with missing arch (from the squamosal?) and reduction in volume of the jugal field. Postorbital is intact. Left: [Reprinted from Cobb ARM, Green B, Gill D, Ayliffe P, Lloyd TW, Bulstrode N, Dunaway DJ. The surgical management of Treacher Collins syndrome. Br J Oral Maxillofac Surg 2014; 52(7):581-589. With permission from Elsevier.] Right: [Reprinted from Shete P, Tupkari J, Benjamin T, Singh A. Treacher Collins syndrome. J Oral Maxillofac Pathol. 2011 Sep;15(3):348-51. With permission from Wolters Kluwer Medknow Publications.]

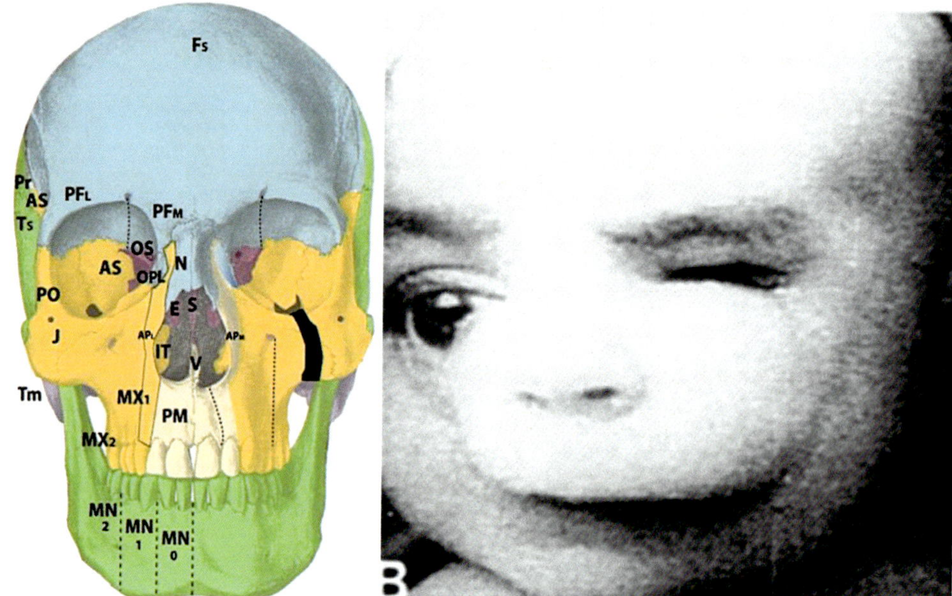

Fig. 13.141 Zone 6. The axis is posterior superior alveolar artery. The 6 cleft is a maxillary field defect with forme fruste at the zygomaxillary suture (maxillary buttress). Left: [Reprinted from Ewings E, Carstens MH. Neuroembryology and functional anatomy of craniofacial clefts. Indian J Plast Surg 2009; 42(Suppl):S19-S34. With permission from Creative Commons Attribution License.]. Right: [Reprinted from Bilkay U, Gundogan H, Ozek C, Gurter T, Akin Y. A rare craniofacial cleft: bilateral Tessier no. 5 cleft accompanied by no. 1 and no. 6 clefts. Ann Plast Surg 2000; 45(6):654-657. With permission from Wolters Kluwer Health, Inc.]

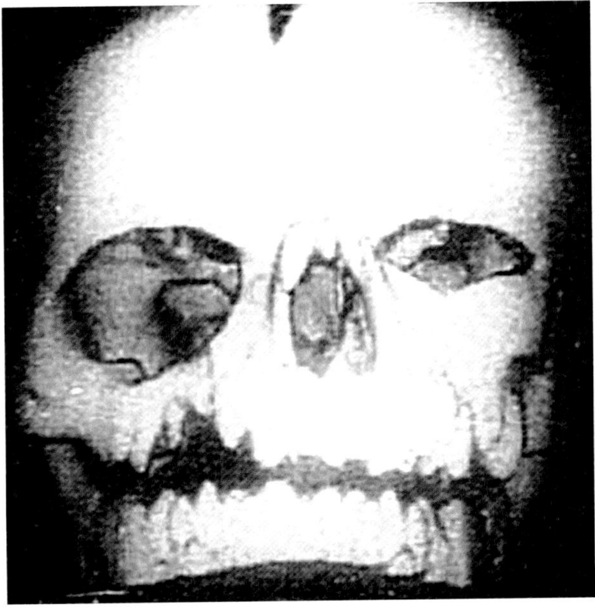
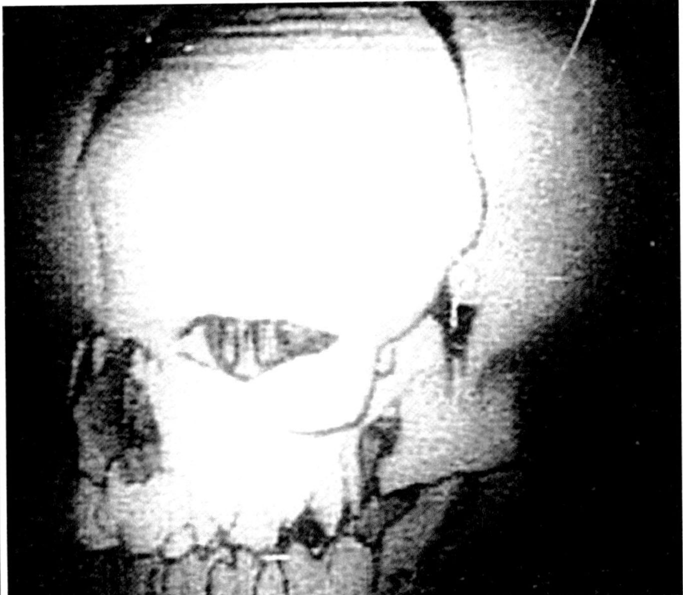

Fig. 13.142 Isolated zone 6 cleft is very rare. Here we can see it ascending through the molar zone along the zygomatico-maxillary buttress with a partial thickness soft tissue defect tracking superficial to the suture and leading to the lateral orbit. Right-sided no. 5 cleft ascends to but does not disrupt the orbital rim. The left orbit is anophthalmic and therefore smaller. [Reprinted from Bilkay U, Gundogan H, Ozek C, Gurter T, Akin Y. A rare craniofacial cleft: bilateral Tessier no. 5 cleft accompanied by no. 1 and no. 6 clefts. Ann Plast Surg 2000; 45(6):654-657. With permission from Wolters Kluwer Health, Inc.]

Zone 5

Defects in this zone involve the maxillary wall from the infraorbital nerve lateral to the buttress. The neurovascular axis supplying this zone is StV2 anterior superior alveolar which gives off, just prior to traversing infraorbital foramen, a lateral branch to the alveolar bone containing the premolars. Then, after emerging from the foramen, it sends a lateral branch to periosteum and soft tissues. Superficial tissues of muscle fat and skin are supplied by transverse facial artery via perforators supplying the facial musculature (Figs. 13.143 and 13.144).

<u>Bone and dental</u>

- This is a narrow zone so the maxillary sinus is spared, but there is a secondary deformity of zygoma. It affects the premolars (the canine belongs to zone 4).

<u>Soft tissues</u>

- Middle third of the lower eyelid and can involve lateral canthal dystopia.
- A cleft of the upper lip is present just medial to the commissure but lateral to the typical cleft lip. The commissure is privileged territory where r2 and r3 meet. Orbicularis from the r5 lower zone criss-crosses that from the r4 upper zone. In fact, in cleft zones 4–6, the axis of the lip defect mirrors that of the lower eyelid.

Zone 4

This zone spans the maxillary wall from the lacrimal groove to the inferior orbital foramen. This tissue is derived from the mesenchyme of r2. The neurovascular axis supplying this zone is StV2 anterior superior alveolar which gives off, just prior to exiting the foramen, a medial branch to the alveolar bone containing maxillary lateral incisor and canine. It then exits the infraorbital foramen whereupon it sends a medial branch to periosteum and soft tissues. Superficial tissues of muscle sygomaticus major and fat and skin are supplied by the facial system, specifically the nasolabial axis (Figs. 13.145 and 13.146).

<u>Bone and dental</u>

- The defect involves the medial wall of maxilla and medial orbital rim resulting in maxillary sinus extrophy, as well as loss of the medial third of the orbital rim and floor. Medial globe prolapses.
- The canine is included in the cleft zone and is missing.

<u>Soft tissues</u>

- Infraorbital orbicularis musculature is disrupted (*meloschisis*) secondary to failure of myoblast migration through the cleft in the developing embryo.
- The punctum, which remained present in the zone 3 cleft, is involved in the zone 4 cleft.

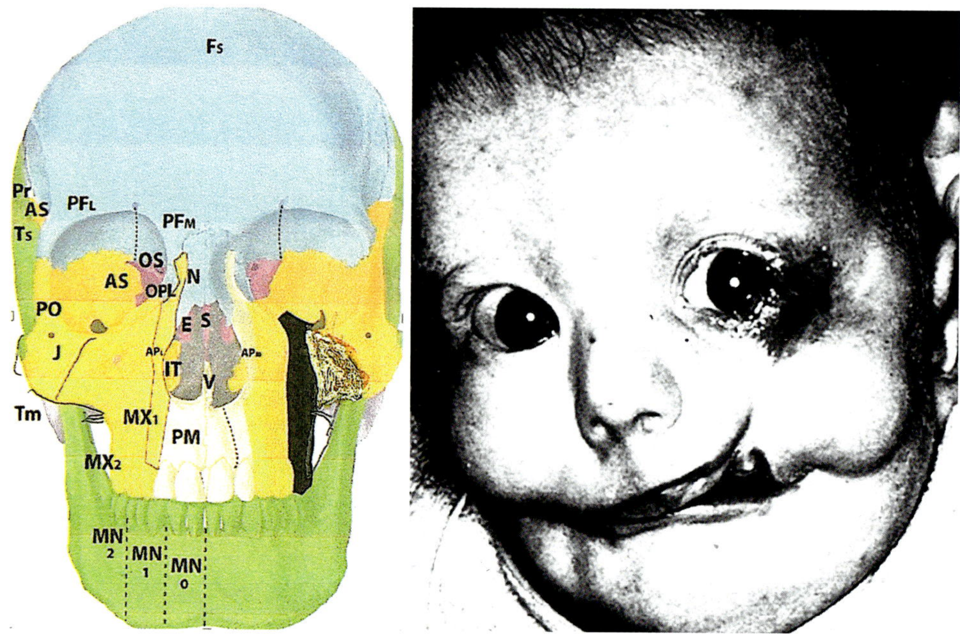

Fig. 13.143 Zone 5. The axis is anterior superior alveolar, lateral branch. Alveolar defect is in the premolar zone. It ascends to the orbital rim lateral to the nerve. Coloboma of the lower lid is common. Medial canthus is normal but the lateral canthus is dystopic. Associations with zones 4 and 3 denote the mesial maxillary side of StV2 (anterior superior alveolar + sphenopalatine) versus the zygomatic. Soft tissue cleft recognizes descent of neural crest mesenchyme superomedially around the eye medial to zone 4. Tissues lateral to zone 5 come superolaterally and laterally. These zones recognize the angiosomes. Note on diagram that zone 6 (molars and buttress) is preserved. [Reprinted from Ewings E, Carstens MH. Neuroembryology and functional anatomy of craniofacial clefts. Indian J Plast Surg 2009; 42(Suppl):S19-S34. With permission from Creative Commons Attribution License.]

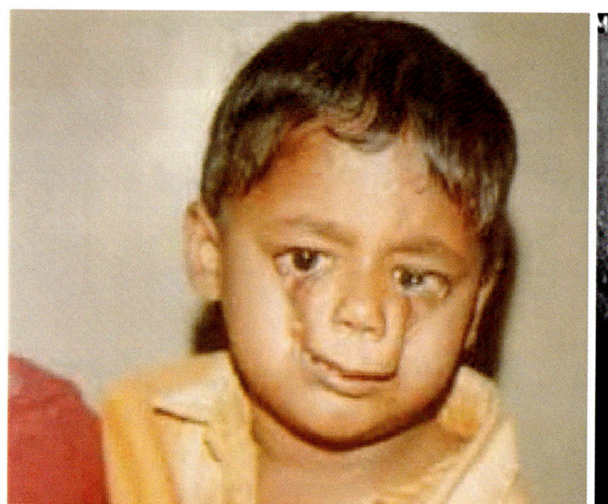

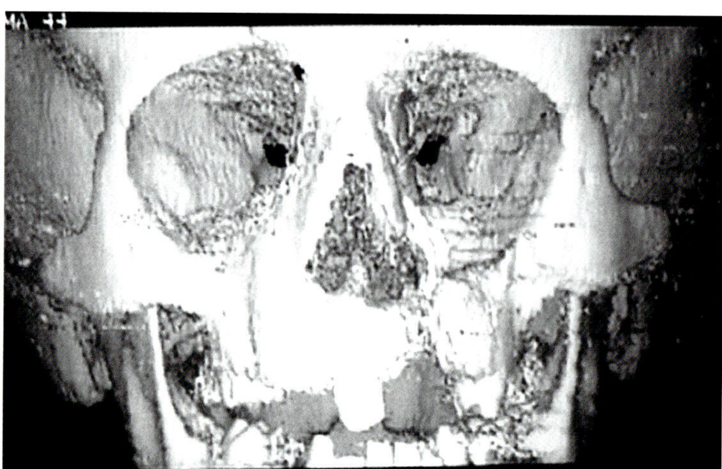

Fig. 13.144 **Bilateral zone 5 clefts** showing soft tissue pattern. Zone 5 (right) and zone 3 (left) on CT with involvement of the right premolar zone. Left: [Reprinted from Chattopadhay D, Murmu MB, Gupta S, et al. Bilateral Tessier number 5 facial cleft with limb constriction ring: The first case report with an update of literature review. J Oral Maxillofac Surg, Med Pathol 2013; 25(1);43-45. With permission from Elsevier.]. Right: [Reprinted from da Silva-Freitas R, Alonso N, Shin J. The Tessier number 5 facial cleft: surgical strategies and outcomes in six patients. Cleft Palate–Craniofacial Journal, 46(2):179-86. With permission from Sage Publishing.]

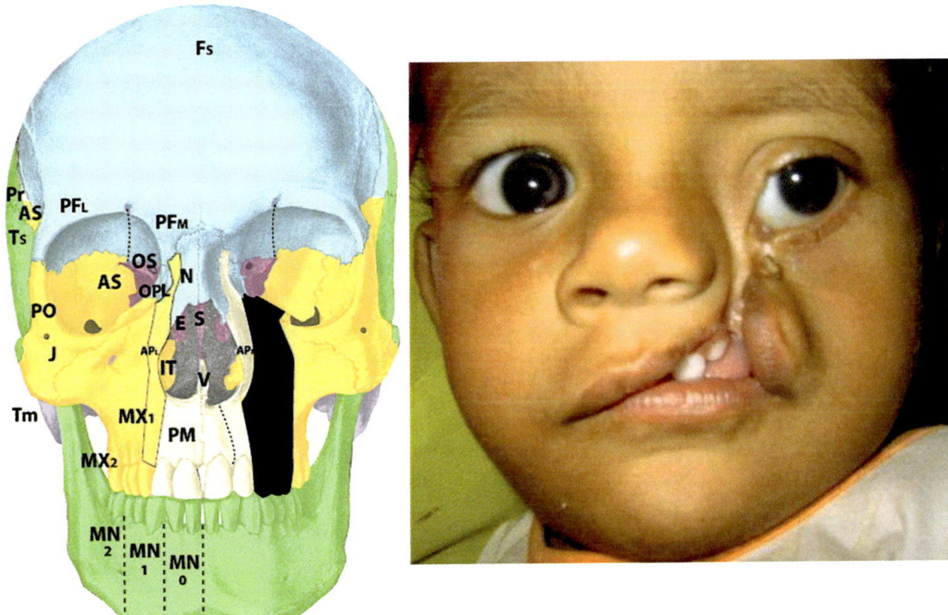

Fig. 13.145 Zone 4. The axis is anterior superior alveolar, medial branch. Note that a small branch supplies the frontal process of the maxilla and the outer lamina of the piriform fossa. The alveolar defect is in the canine and maxillary incisor zone. This dental unit is often termed "ectopic" but really is a part of zone 4. Osseous defect passes lateral to the piriform fossa and terminates medial to the nerve. Maxillary sinus extrophy can be present. The many complex features of this cleft are summarized by Kawamoto. Left: [Reprinted from Ewings E, Carstens MH. Neuroembryology and functional anatomy of craniofacial clefts. Indian J Plast Surg 2009; 42(Suppl):S19-S34. With permission from Creative Commons Attribution License.]. Right: [Reprinted from Resnick NI, Kawamoto HK. Rare craniofacial clefts: Tessier no. 4 clefts. Plast Reconst Surg 1990; 85(6):845-849. With permission from Wolters Kluwer Health, Inc.]

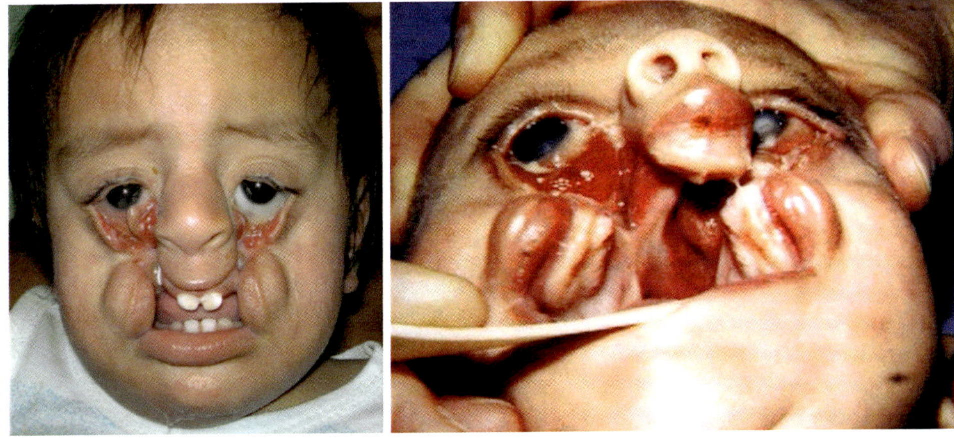

Fig. 13.146 Bilateral no. 4 clefts. Skeletal defect at the canine; the maxillary lateral incisors are missing in the cleft or never developed. Premaxillary lateral incisors intact. Nasolacrimal sac and duct are intact but the inferior canaliculus is dysplastic or absent. Soft tissues around the philtrum are in continuity with the no. 11 infra-trochlear angiosome, indicating that these neural crest tissues pass superomedially to the eye and descend to the level of the lip. Left: [Reprinted from Alonso A, da Silva-Freitas R, Azor de Oliveira G, Goldenberg D, Tolazzi ARD. Tessier no. 4 facial cleft: evolution of surgical treatment in a large series of patients. Plast Reconstr Surg 2008; 122(5):1505-1513. With permission from Wolters Kluwer Health, Inc.]. Right: [Reprinted from Laure B, Picard A, Bonin-Goga B, et al. Tessier number 4 bilateral orbito-facial cleft: a 26-year follow-up. J Craniofac Surg 2010; 38(4):245-7. With permission from Elsevier.]

- The nasolacrimal groove, which follows a line from the punctum to the lateral crus of the nasal alar cartilage and is involved in the zone 3 cleft, is medial to the zone 4 cleft and thus is spared. And unlike zone 3, the lacrimal sac is normal but dilated.
- Deformation of the surrounding normal fields occurs through disconnection of the normal force vectors between them: the premaxilla becomes protrusive, the lateral nasal walls become retrusive, and the pterygoid plate is displaced forward.
- Overall, this lack of support from tissues lateral to the nose causes a compression of the nasal passages, leading to the clinical appearance of choanal atresia.
- The zone 4 cleft does not include the piriform fossa or the medial maxillary sinus wall, and the lateral incisor remains intact.

Zone 3

The inferior lateral nasal wall (medial maxillary wall) is a complex zone consisting of four distinct fields: palatine bone (vertical plate and horizontal plate), palatal plate of the maxilla, and inferior turbinate. Positioned in front of this wall, anterior piriform margin consists of a bilateral lamination consisting of frontal process of zone premaxilla and zone 4 frontal process of maxilla. PMxF is supplied by medial nasopalatine artery. This zone is supplied by two StV2 neurovascular axes. (1) Descending palatine artery (DPA) supplies the vertical plate and then divides into lesser palatine (LPA) directed backward to horizontal palatine shelf and greater palatine (GPA) directed forward along the oral side of the maxillary palatal process (plate). (2) Lateral nasopalatine (LNPA) runs in parallel with greater palatine to supply the nasal side of the palatal process and the inferior turbinate.

Above zone 3, zone 11, the superior lateral nasal wall, is composed of the middle and superior turbinate processes of ethmoid complex. These are all supplied by StV1 lateral nasal branches from the StV1 nasociliary axis in the orbit. Because of the many structures involved, pathologies of zone 3 are varied. Complete clefts in the zone are notoriously difficult to manage (Figs. 13.147 and 13.148).

Bone

- Development of the DPA and the LNPA is independent. The maxillary palate and palatine bones form at about the same time; these may predate the inferior turbinate. Thus, in zone 3 cleft, the inferior turbinate, which develops later, can be deficient while the palatine bone is present. The palatal cleft observed in zone 3 begins in the palatine bone and spreads posteriorly. This results in a "horseshoe" shaped cleft, with the lateral palatine bone present but falling away medially and posteriorly. The footplate of the lacrimal bone rests on the inferior turbinate, and

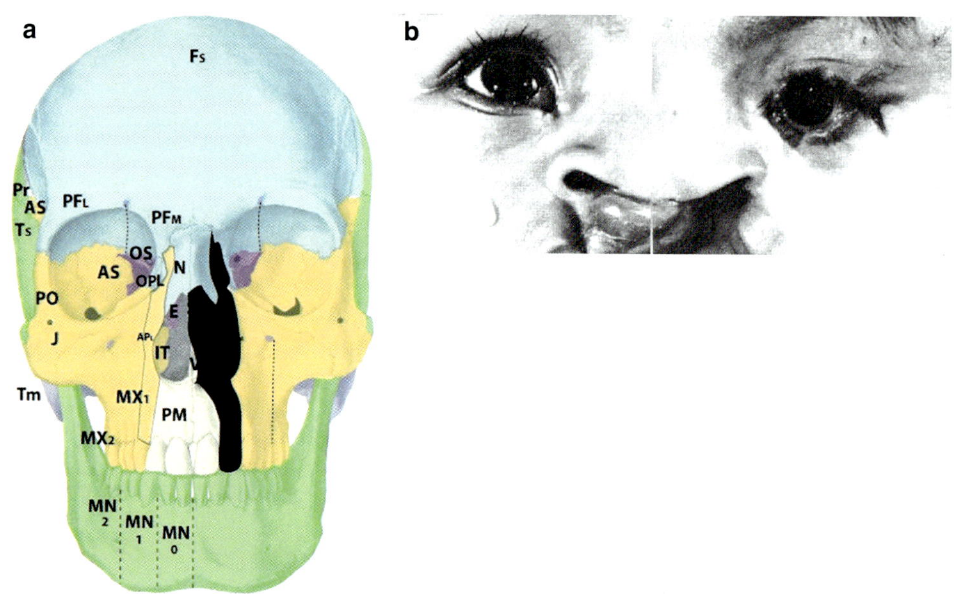

Fig. 13.147 Zone 3. Axis is lateral nasopalatine to the inferolateral nasal wall, including the inner lamina of piriform fossa. This is one of the most complex clefts to repair. As the soft tissue defect occurs along the nasolacrimal groove the lacrimal system is disrupted. Tessier 3 clefts are reviewed in Allam KA, Lim AA, Elsherbiny A, Kawamoto HK. The Tessier number 3 cleft: A report of 10 cases and review of literature. Journal of Plastic, Reconstructive & Aesthetic Surgery 2014; 67(8):1055-1062. (**a**) [Reprinted from Ewings E, Carstens MH. Neuroembryology and functional anatomy of craniofacial clefts. Indian J Plast Surg 2009; 42(Suppl):S19-S34. With permission from Creative Commons Attribution License.]. (**b**) [Reprinted from Allam KA, Lim AA, Elsherbiny A, Kawamoto HK. The Tessier number 3 cleft: A report of 10 cases and review of literature. Journal of Plastic, Reconstructive & Aesthetic Surgery 2014; 67(8):1055-1062. With permission from Elsevier.]

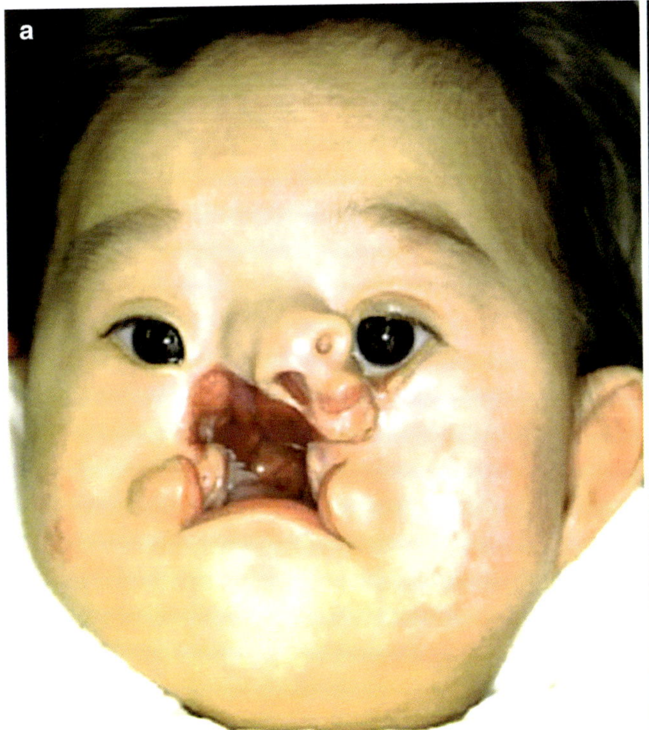

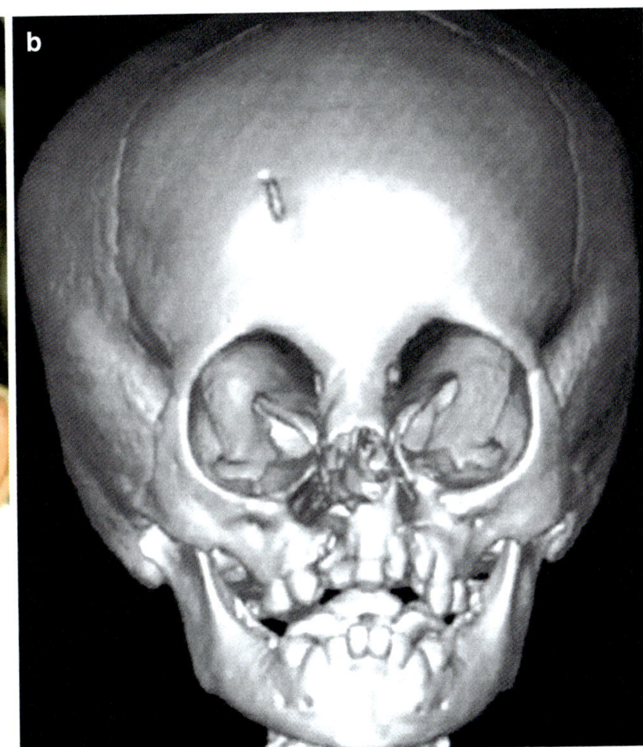

Fig. 13.148 Right no. 3 cleft between central incisor and canine (lateral incisor did not form). Lateral wall of the nose was absent, exposing the maxillary sinus. The frontal process of maxilla is disrupted as well (because it is laminated on the frontal process of premaxilla). Right orbital rim remains intact. On left side the no. 4 cleft extended as expected from between the left lateral premaxillary incisor and the maxillary incisor. [Reprinted from Allam KA, Lim AA, Elsherbiny A, Kawamoto HK. The Tessier number 3 cleft: A report of 10 cases and review of literature. Journal of Plastic, Reconstructive & Aesthetic Surgery 2014; 67(8):1055-1062. With permission from Elsevier.]

absence of the turbinate can cause disruption of the lacrimal system.

- Frontal process of premaxilla is wiped out. The zone 4 process of maxilla bearing the lateral maxillary incisor (variable) and canine is lateralized but is disconnected from its frontal process which is preserved by the infratrochlear axis. This explains the preservation of the medial canthus vide infra. The incisor-bearing frontal process of maxilla is lateralized. Lateral incisor zone of premaxilla is absent as well. Inferior turbinate is missing. This combination indicates that the stem of the nasopalatine artery is completely disrupted prior to its bifurcation into MNPA and LNPA. The maxillary sinus is wide open into the nose. Interruption of MNPA affects vomer and can cause cleft palate even assuming a normal width of the maxillary shelf. Additional pathology of the descending palatine artery may affect soft palate (LPA) or hard palate (GPA).

Soft tissue

- The nasolacrimal system is disrupted beginning at its exit point beneath the missing inferior turbinate and running upward to the punctum, which remains intact. Lacrimal sac is laid open with tear drainage onto the cheek and propensity to infection. If zone 11 is involved as well (affecting the upper duct) the entire lacrimal system is destroyed.
- Medial canthus is dystrophic but intact. The lower lid medial to the punctum can have a coloboma.
- Skin medial to the lacrimal system is PNC supplied from above by infratrochlear, dorsal nasal axis (assuming zone 11 is normal).
- The nasal ala from PNC is deformed, disconnected from the lip and cheek, and retracted upward.

Zone 2

This zone is the site of two of the most common labiomaxillary clefts, involving the premaxilla and/or the vomer. The arterial supply is medial nasopalatine artery (MNPA). Premaxilla may be divided into three distinct developmental fields involving the (1) central incisor, (2) lateral incisor, and (3) frontal process. During embryological development, the mesenchymal neural crest populates these areas from medial to lateral. The premaxillary fields actually straddles two neuromeric (numeric) Tessier zones, with the central incisor belonging to zone 1 while the lateral incisor and frontal processes belonging to zone 2 (Fig. 13.149).

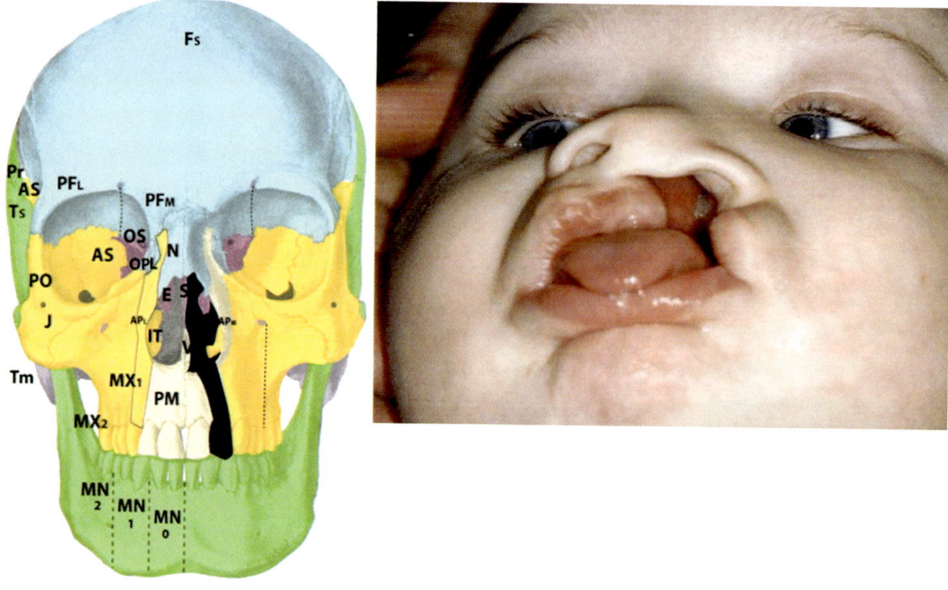

Fig. 13.149 Zone 2. The axis is medial nasopalatine. If unilateral, collateral flow from the opposite angiosome supports the premaxilla (partially). Defect begins in the frontal process of premaxilla and descends to lateral incisor and its alveolar housing. Occasionally, the entire hemipremaxilla is deficient but vomer is also involved, this transitions over to the midline pathologies such as holoprosencephaly. [Reprinted from Ewings E, Carstens MH. Neuroembryology and functional anatomy of craniofacial clefts. Indian J Plast Surg 2009; 42(Suppl):S19-S34. With permission from Creative Commons Attribution License.]

Bone

- Osseous deficit is variable depending on the extent of the vascular defect: thus, the order of involvement (least-affected to most-affected) is PMxF < PMxB < PMxA.
- Bone defects cause a decrease in [BMP4] leading to decreased repression of sonic hedgehog (Shh) and the pathological retention of epithelial integrity with consequent failure of fusion.
  - PMxF: Loss of the ascending frontal process of the premaxilla results in the reduction of the lateral nasal wall lining directly in front of the inferior turbinate. This results in an overall airway reduction of 30–40%. The piriform aperture is always involved in defects of the frontal process.
  - PMxB: Loss of the lateral premaxillary segment results in loss of the lateral incisor, an alveolar cleft, and frequently (by mechanical displacement of vomer with remaining premaxilla), secondary palatal clefts.
  - PMxA: When the defect extends to the central incisor the central portion of the entire hemi-premaxillary is lost. This condition is rare and occurs with cases of holoprosencephaly.
- The osseous defects observed in the minimal cleft with normal lip and cleft stigmata in the nose indicate that the actual site of the defect resides in the lateral piriform fossa, specifically the frontal process of the premaxilla.

Soft tissues

- Lip defects run through the ipsilateral philtral column and follow the [BMP4] concentration gradient: deep-to-superficial and caudal-to-cranial. Minimal defects begin at the vermilion border and ascend. Skin involvement begins as just a crease but deepens to become full thickness.
- Nasal deformity is complex and caused by (1) bony distortion of the piriform and (2) disruption of normal insertion pattern of nasalis complex [112].

Combination

- Clefts of the nasal tip are always manifestation of a concomitant pathology in 12, as this is the source of the blood supply to the deep tissue of the nasal ala.
  - Soft tissue defect in the 2–12 extends through the ala lateral to the lower lateral cartilage. Recall that zone 11 defects affect the alar crease.

Zone 1

The central zone refers to the vomer. Recall the vomer is a bilaminar structure supplied by paired medial nasopalatine pedicles. Deficiency of one side can affect ipsilateral closure of the secondary hard palate. Premaxilla can remain viable either with reduced flow or, in the case of a unilateral vomerine knock-out, from the contralateral side. Pathologies depend on unilateral versus bilateral involvement of the MNPAs (Figs. 13.150 and 13.151).

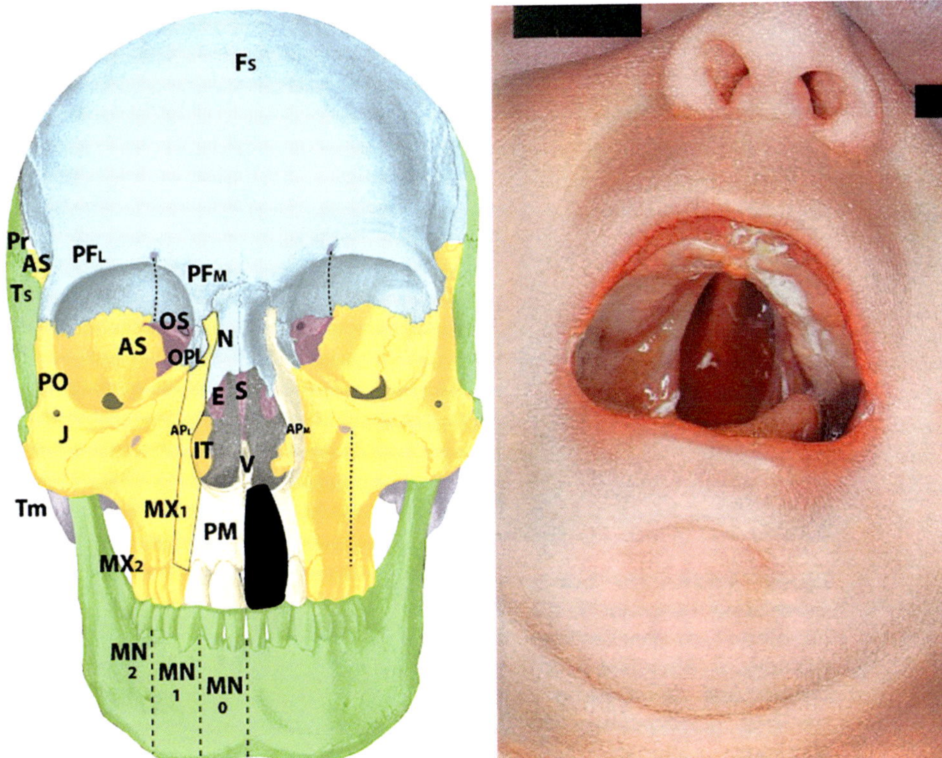

Fig. 13.150 Zone 1. The axis of the cleft in the maxilla in medial nasopalatine. This cleft is described as causing a "typical" cleft lip involving the alveolar bone, selectively knocking out the central incisor. It moves into zone 13 anterior ethmoid territory causing a crease in the ala through the intermedial crus. Splitting between nasal bone and frontal process of maxilla occurs in continuity with zone 13 creating a crease in the ala that extends through the intermediate crus. Isolated zone 1 clefts may underlie simply off of the incisor. Right: Medial nasopalatine involved in deficiency of the vomer and a midline cleft palate with an intact alveolar arch. [Reprinted from Ewings E, Carstens MH. Neuroembryology and functional anatomy of craniofacial clefts. Indian J Plast Surg 2009; 42(Suppl):S19-S34. With permission from Creative Commons Attribution License.]

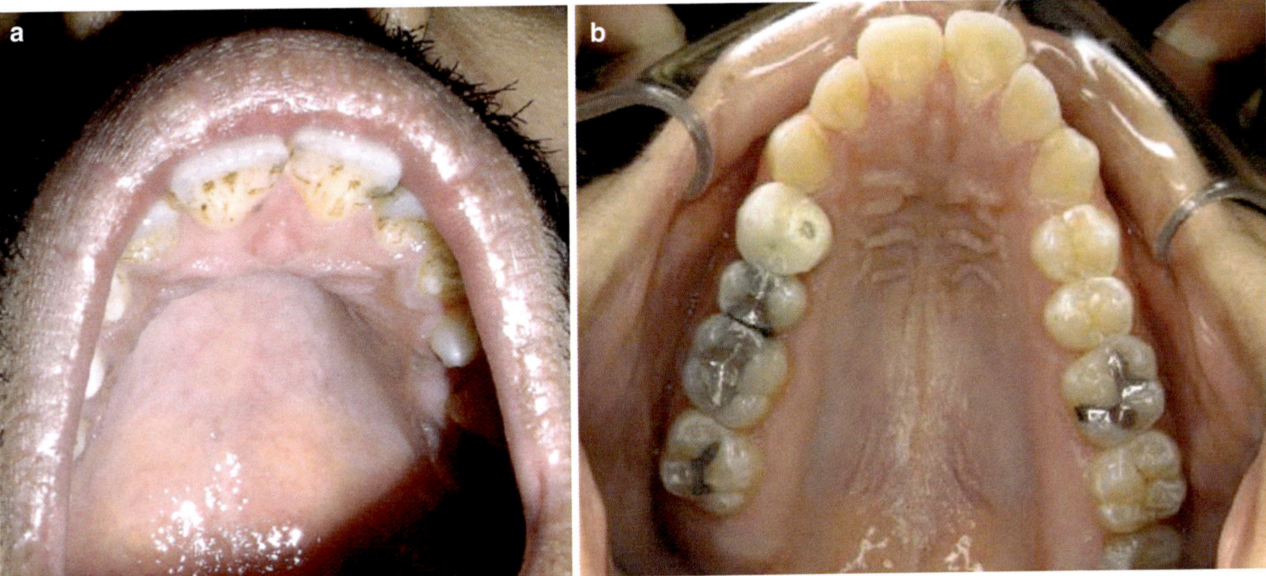

Fig. 13.151 Zone 1 palate. When the vomer is vertically foreshortened but it connects with the hard palatal shelves a high-arched palate results. The vomer, being a bilateral bone, may be too small on one side to achieve fusion but attach to the ipsilateral hard palate causing a complete cleft on one side of the midline. If neither shelf can make contact a bilateral midline cleft results. Left: High-arched palate. Right: Complete cleft of secondary hard palate, bilateral. Left: [Reprinted from Yung-Chuan Liu S, Guyilleminault C, Huon L-K, Yoon A., Distraction osteogenesis maxillary expansion (DOMER) for adult obstructive sleep apnea patients with high arched palate. Otoloryngol Head Neck Surg 2017; 157(2): 345-348. With permission from SAGE Publishing.]. Right: [Courtesy Dr. Michael Carstens]

Bone

- Vertical dysplasia of vomer can cause high-arched palate.
- Cleft secondary palate with normal width of palatal shelf and normal vertical height of vomer but fusion failure.
- Crease but not a frank cleft between lateral incisor and central incisor.
  - Decreased supply from ipsilateral MNPA.

Soft tissue
Combination

- Clefts of the nasal tip are always manifestation of a concomitant pathology in 12, as this is the source of the blood supply to the deep tissues of the nasal tip.
  - Soft tissue defect in the 1–13 extends through the ala lateral to the lateral crus of alar. Ala unaffected.

Laterofacial Microsomias and the Orbit

Field defects involving the first two pharyngeal arches can affect the size, shape, and position of the orbit. A number of these are embedded in the literature and share many features in common: craniofacial microsomia (otomandibular syndrome), Goldenhar syndrome (oculoauricular dysplasia), and Treacher-Collins–Franceschetti syndrome for a spectrum. Elaborate attempts were made to separate them by their phenotypic features (see the chart by Tessier) but in point of fact, they are more easily understood as variations on the theme of rhombomere-specific lesions. Structures developing from r1–r11 can be mixed and matched into varying combinations, either unilaterally or on both sides of the face. Let's consider these three syndromes (plus Parry–Romberg syndrome), in regard to their effects on the orbit (Figs. 13.152, 13.153, 13.154, 13.155, 13.156, 13.157, 13.158, 13.159, 13.160, 13.161, 13.162, 13.163, and 13.164).

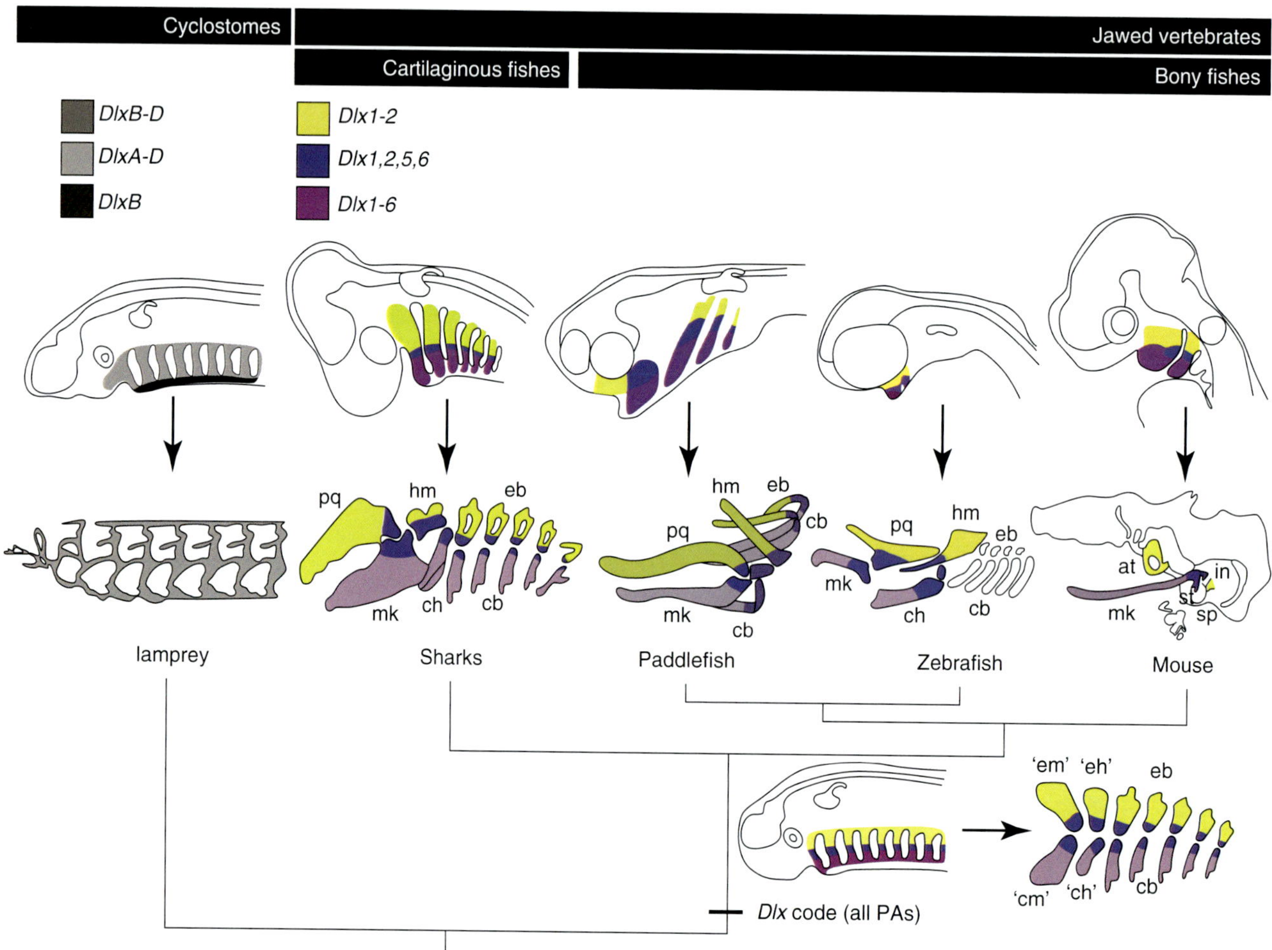

Fig. 13.152 Craniofacial microsomia and number 8 cleft Craniofacial microsomia. Neuromeric formula is r2–r3–r4–r5 with emphasis on the mandible and lower face (r3, r5) sectors. Here, a defect in the postorbital field is seen as a Tessier 6 cleft. Bone deficits are worse than soft tissues; Goldenhar is the opposite. [Reprinted from Gillis J, Modrell M, Baker C. Developmental evidence for serial homology of the vertebrate jaw and gill arch skeleton. Nat Commun 2013;4:1436. With permission from Springer Nature.]

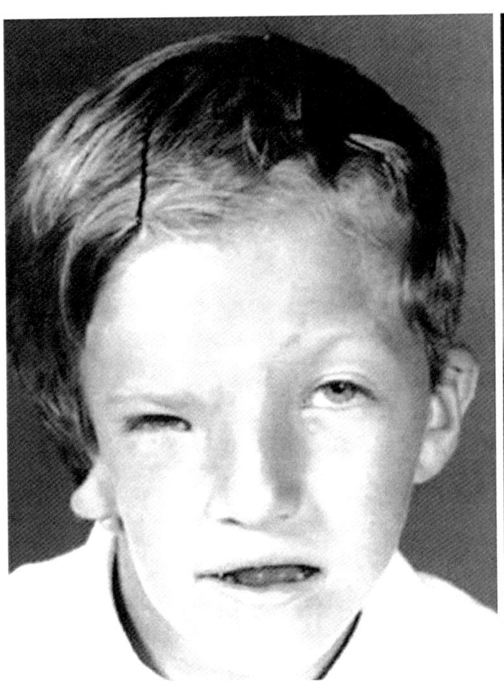
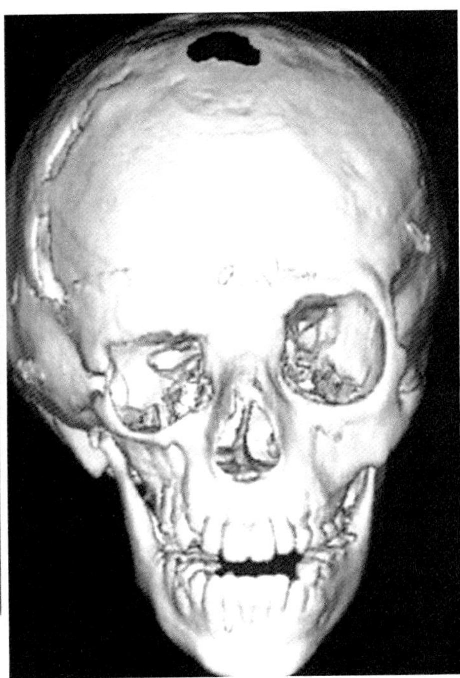

Fig. 13.153 Craniofacial microsomia with more severe involvement of the mandible (loss of the ramus/condyle) and absent zygomatic arch or zygomatic process of the squamosal. Squamosal is probably an r3 derivative and is an underrated contributor to the loss of the TMJ. [Reprinted from Radiopaedia. Retrieved from: https://radiopaedia.org/cases/hemifacial-microsomia. With permission from Creative Commons License 3.0: https://creativecommons.org/licenses/by-nc-sa/3.0/.]

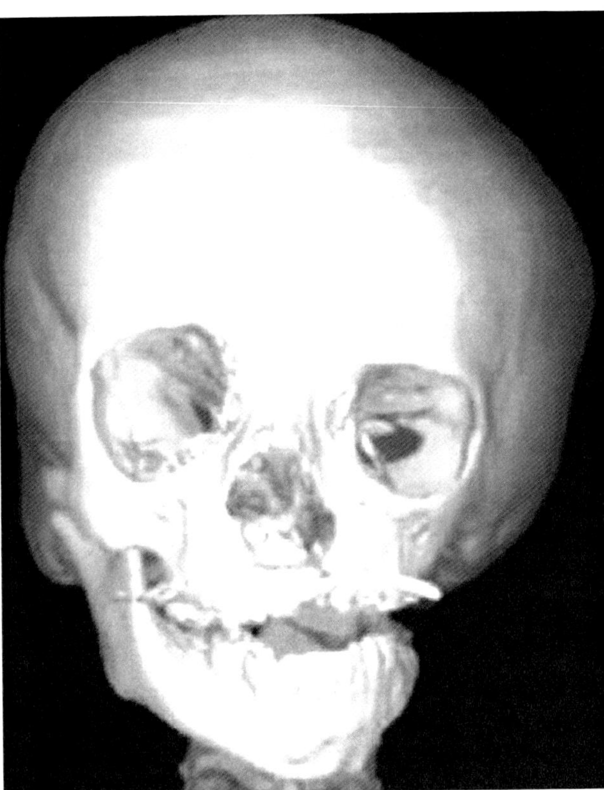
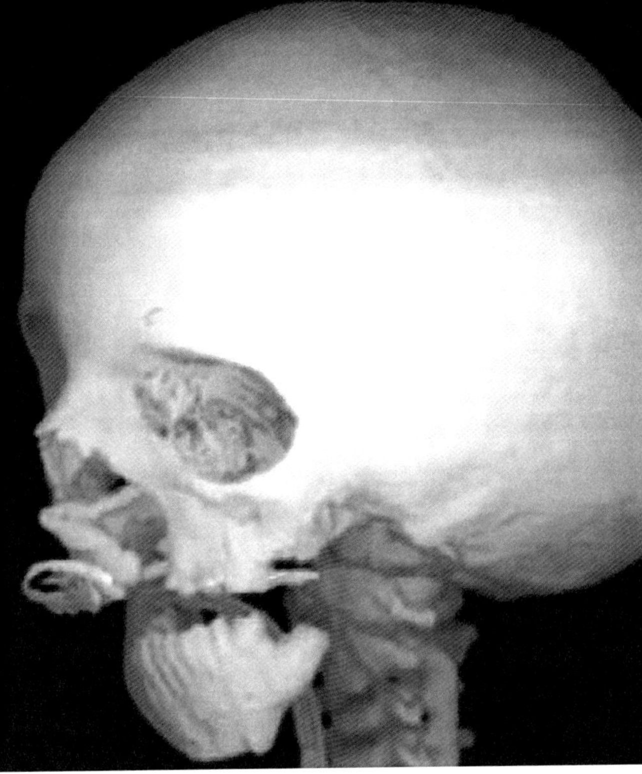

Fig. 13.154 Goldenhar syndrome and lateral facial cleft. Ocular dermoids are a key part of the diagnosis. Greater soft tissue involvement, including cleft soft palate. This presentation is commonly called number 7 but it has nothing to do with the stapedial system. Rather it represents a boundary defect between the upper first arch (f2, r4) and the lower first arch (r3, r5). [Reprinted from Chauhan DS, Grurprsad Y. Goldenhar Syndrome with Tessier's 7 Cleft: Report of a Case J Maxillofac Oral Surg 2015;14(Suppl 1):42-6. With permission from Springer Nature.]

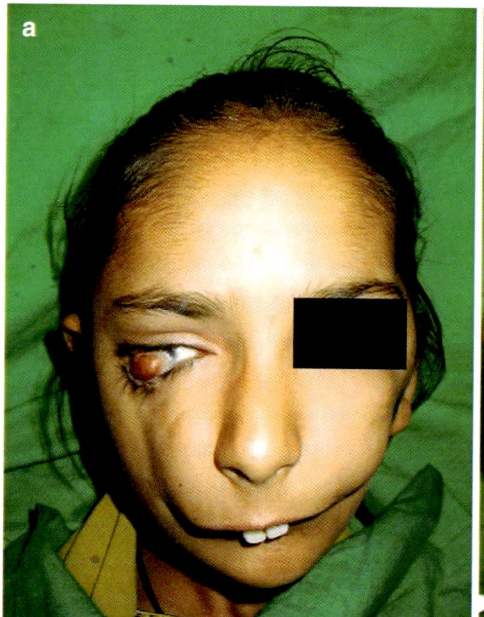

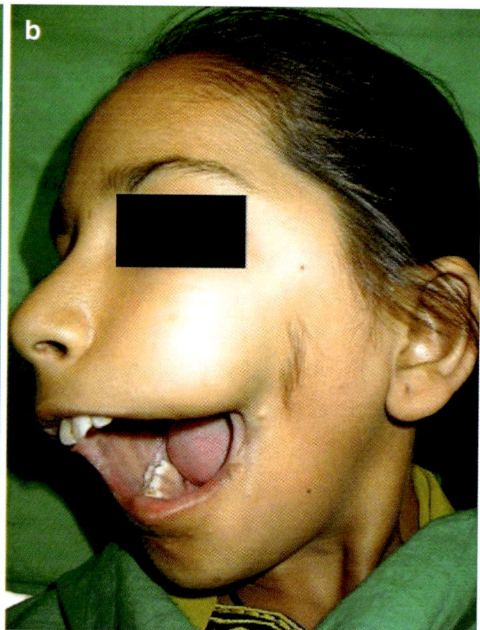

Fig. 13.155 (**a**) Pharyngeal arches: anteroposterior definition from the Hox code. Role of Dlx genes in proximodistal pharyngeal arch patterning. Diagram of a generalized gnathostome with neurocranium (Nc) and pharyngeal arches. The identity of pharyngeal arches along the anterior–posterior axis is regulated by Hox and Otx genes, which are expressed in a nested pattern. Dlx genes have a similar nested expression (and function) along the proximal–distal axis. Bb, basibranchial; Cb, ceratobranchial; Ep, epibranchial; Hb, hypobranchial; MC, Meckel's cartilage; Pb, pharyngeobranchial; PQ, palatoquadrate. Pharyngeal arches are programmed from paired neuromeres. Each has a distinctive homeotic code. In the case of the first arch and second arch, when they merge at stage 11, all four constituent neuromeres (r2–r5) are clustered together. Lateral facial clefts occur along the midline between r2–r4 and r3–r5; this is indicated by the blue zone. (**b**) Pharyngeal arches: proximal–distal definition for the Dlx code. The Dlx code arose along the gnathostome stem and was primitively deployed in all pharyngeal arches. The evolutionary relationship between the Dlx code of gnathostomes and the nested expression of DlxA-D in the pharyngeal arches of lamprey remains unclear. Dorsal (Dlx1-2-expressing) and ventral (Dlx1-6-expressing) domains of the Dlx code would have primitively given rise to dorsal "epimandibular," "epihyal," and "epibranchial" elements and ventral "ceratomandibular," "ceratohyal," and "ceratobranchial" elements (in the mandibular, hyoid, and gill arches, respectively), while intermediate (Dlx1-, 2-, 5- and 6-expressing) domains would have given rise to the region of articulation between these elements. The primitive role for the Dlx code in patterning the mandibular, hyoid, and gill arch endoskeletal segments has been conserved in elasmobranches, and presumably in non-teleost actinopterygians (for example, paddlefish, a stem bony fish prior to the teleosts), while post-hyoid arch expression of the Dlx code has been modified or obscured in amniotes (for example, mouse) and possibly in teleosts. at, ala temporalis; cb, ceratobranchials; 'ch', hypothetical ceratohyal; ch, ceratohyal; 'cm', hypothetical ceratomandibula; eb, epibranchials; 'eh', hypothetical epihyal; 'em', hypothetical epimandibula; hm, hyomandibula; in, incus; mk, Meckel's cartilage; pq, palatoquadrate; sp, styloid process; st, stapes. Left: [Reprinted from Depew MJ, Lufkin T, Rubenstein JLR. Specification of jaw subdivisions by Dlx genes. Science 2002; 298, 381–385. With permission from The American Association for the Advancement of Science.]. Right: [Reprinted from Gillis JA, Modrell MS, Baker CVH. Developmental evidence for serial homology of the vertebrate jaw and gill arch skeleton. Nature Communications 2013: 4:Article 1436. With permission from Springer Nature.]

Lamination of the Pharyngeal Arches

By way of review, recall that the five pharyngeal arches are produced between stages 9 and 13. Each one is initially supplied by its own aortic arch artery connecting the outflow tract up through the arch to the dorsal aorta. The aortic arch arteries form from opposing buds from the same neuromeric levels, r2 dorsal aorta sends a ventral bud and r2 outflow tract sends a dorsal bud; they connect in the middle of the pharyngeal arch. A special zone of cardiac neural crest arises from the posterior neuromeres and sweeps forward to encounter the aortic arch arteries. These NC cells provide a means by which NC cells track downward along the arteries to enter the heart.

As the arches develop they laminate within each other. The most dramatic case is that of first and second arches. By stage 11 they are merging with second arch pursuing a massive subcutaneous migration around the entire head being blocked only when it encounters the foundaries of cervical somite dermis (see Chap. 11 on skin and fascia). The third, fourth, and fifth arches are successively inside one another like Russian dolls. In point of fact, all pharyngeal arches share a common topologic organization. A nested set of *Dlx* (Distal-less) genes divides each arch into sectors with morphogen gradients (posterior-anterior, ventral-dorsal, and medial-lateral). This system, described in a seminal paper in 2005 by DePew, allows for overlap of homologous sectors. Thus first arch is divided longitudinally into r2 and r3 zones.

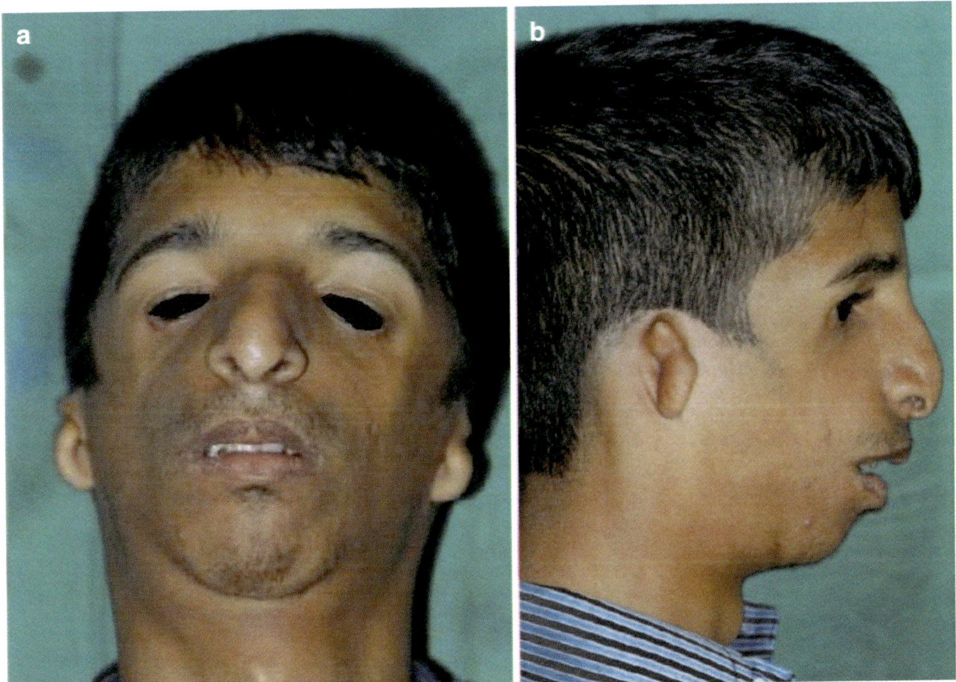

Fig. 13.156 Treacher-Collins–Franceschetti syndrome. (**a**) Antimongoloid facies, malar defects, colobomata. (**b**) Fish-like facies, preauricular hair displacement forward, auricular deformities. TCF can have familial inheritance. It also affects neuromeric fields across the board in pharyngeal arches 1–5: facial deformities as above are first arch defects. (Soft palate and auditory apparatus stem from second and third arch defects. These can present as muscle weakness of the palate and pharynx. Long-term hearing issues have been improved with bone-assisted hearing aid (BAHA). Patients have difficult airway management and may require tracheostomy. These issues reflect defects in arches 4 and 5. [Reprinted from Shete P, Tupkari J, Benjamin T, Singh A. Treacher Collins syndrome. J Oral Maxillofac Pathol. 2011 Sep;15(3):348-51. With permission from Wolters Kluwer Medknow Publications.]

When second arch becomes sandwiched into the center of the first arch the r2–r3 zones are lined up with the corresponding r4–r5 zones. We can thus explain the bifurcation of VII into upper and lower motor branches. Similarly, a developmental failure along the axis of PA1 and PA2 can lead to a partial or full-thickness soft tissue defect extending from the oral commissure to the external auditory canal (Fig. 13.152).

The third, fourth, and fifth pharyngeal arches share the same system which enables them to interact and they laminate within one another. However, these latter arches are of little consequence to the orbit, except as they contribute to the laterofacial microsomias. What we can observe is that these pathologies result from insults to homeotic genes shared among one or more neuromeres in a coordinated way.

Let's compare the findings in these selected pathologies from a neuromeric perspective. We will then compare and contrast them.

Craniofacial Microsomia (Otomandibular Syndrome): r2–r5

Craniofacial microsomia (CFM) refers to a spectrum of mesenchymal deficiencies affecting multiple developmental fields of the first and second arches. The more common term, hemifacial microsomia, is incorrect because CFM can have bilateral manifestations of varying severities of facial asymmetry seen on either side. CFM in the first arch involves both r3 and r2 structures (r3 more severely); in the second arch the pattern is the same with r5 affected more than r4. The axis of pathology runs from the oral commissure to the external auditory canal. Orbital findings (r1) are present as well. The OMENS-Plus classification (orbit, mandible-maxilla, ear, nerve, soft tissue, plus extracraniofacial manifestation) is comprehensive and clinically useful [113, 114] (Figs. 13.153 and 13.154).

<u>Orbit</u>

- About 15–43% (average 25%) of patients have either a smaller orbit or dystopia.
- Deficiency of either maxilla or zygoma will reduce orbital volume.
- Maxillary insufficiency can displace the orbit downward.
- Squamosal deficiency can displace orbit backward.

<u>Orbital soft tissues</u>

- Eye is normal: microopthalmia is rare.
  - Literature reports may confuse CFM with Goldenhar [115].
  - CNS associations noted [116, 117].

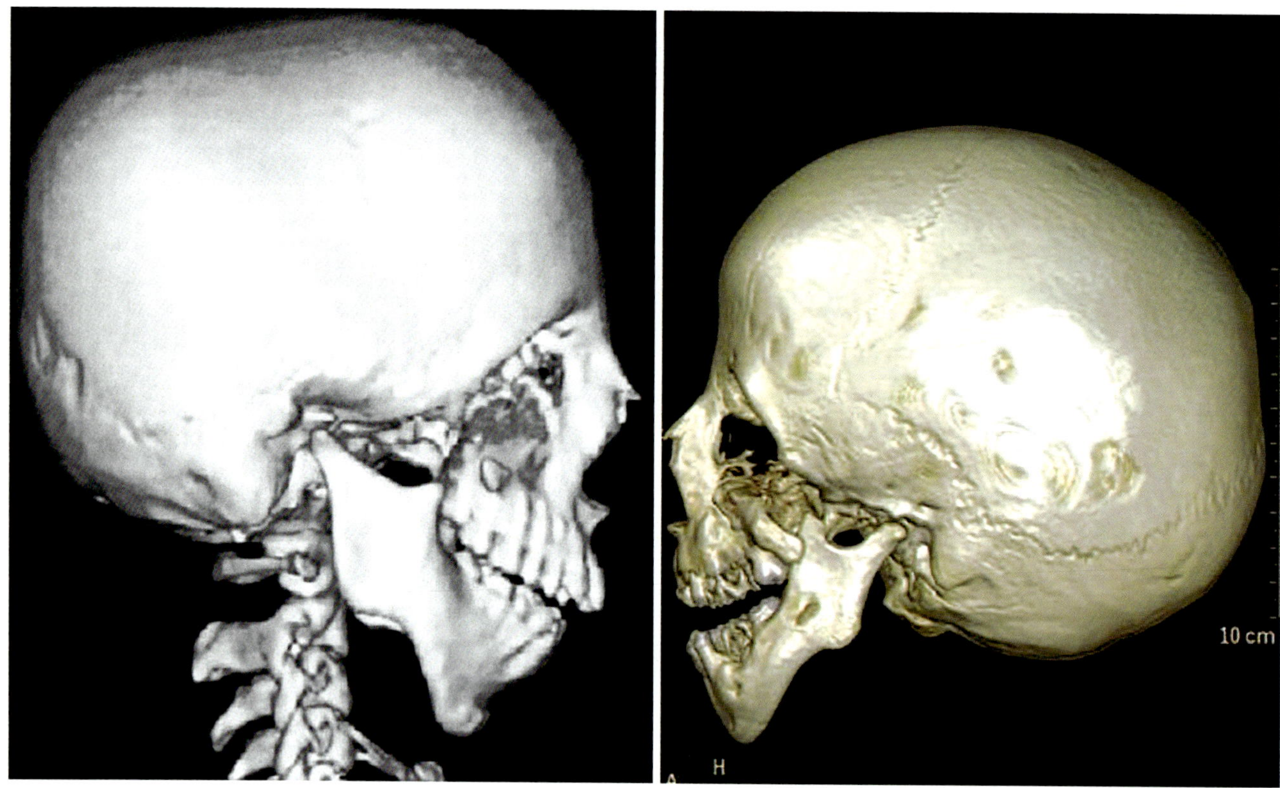

Fig. 13.157 Treacher-Collins Alisphenoid. Radiographic examination revealed underdeveloped condylar and coronoid processes, hypoplastic zygomatic arches, bilateral prominent antegonial notch on mandibular angle, and short rami with resultant crowding of teeth [Figs. 5 and 6]. A noncontrast high-resolution (0.6) CT scan report of skull and face clearly demonstrated maxillofacial deformity with more prominent mandibular hypoplasia. However, the cranial vault and skull base appeared normal. It also revealed bilateral hypoplasia of external pinnae, marked hypoplasia/aplasia of bony external auditory canals, hypoplasia of middle ear cavity on either side with absence of ear ossicles [Fig. 7]. No paranasal sinus pathology was evident. Neuromeric analysis of TCF is r2–r3–r4–r5 with zygoma more affected than mandible (but the latter can have a small condyle). Orbital effects exophthalmia and strabismus due to changes in orbit shape affecting extraocular muscles. Left: [Reprinted from Shete P, Tupkari J V, Benjamin T, Singh A. Treacher Collins syndrome. J Oral Maxillofac Pathol 2011;15:348-51. With permission from Wolters Kluwer MedKnow Publications.]. Right: [Reprinted from Radiopaedia. Retrieved from: https://radiopaedia.org/articles/treacher-collins-syndrome. With permission from Creative Commons License 3.0: https://creativecommons.org/licenses/by-nc-sa/3.0/.]

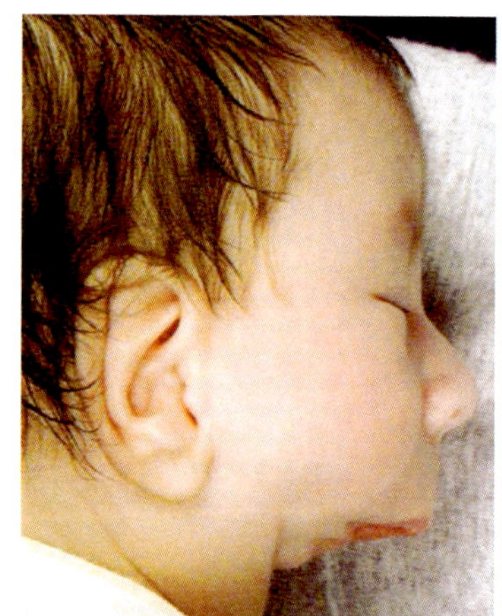

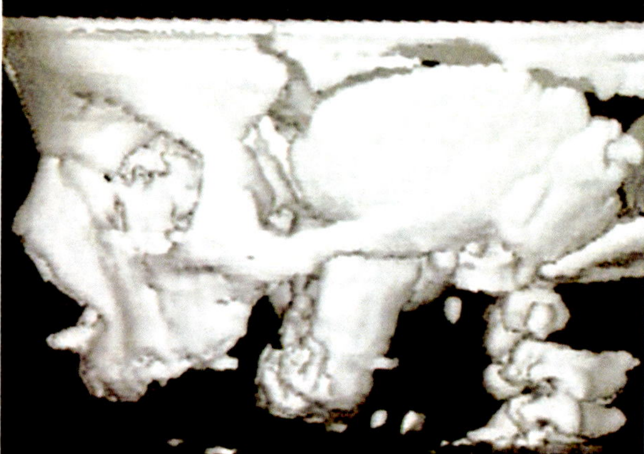

Fig. 13.158 Nager alisphenoid. Note extreme foreshortening of ramus and angulation. Position of mandible and hyoid may be a manifestation of global ventral pharyngeal arch problem with foreshortened pretracheal fascia. [Reprinted from Hunt JA, Hobar PC. Common craniofacial anomalies the facial dysostoses. Plast Reconstr Surg 2002; 110(7):1714-1725. With permission from Wolters Kluwer Health, Inc.]

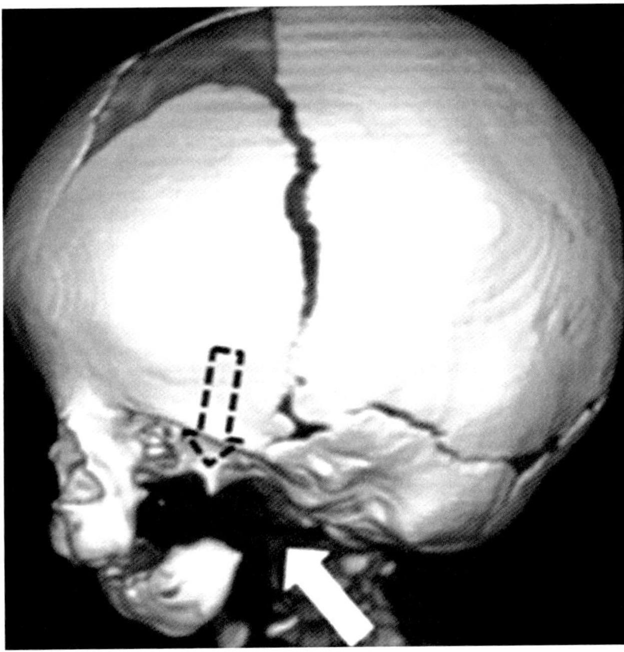

Fig. 13.159 Nager, severe. 3D reformatted image demonstrates severe maxillary and zygomatic hypoplasia (black open dashed arrow) and severe micrognathia and retrognathia (white block arrow). Alisphenoid and hyoid are missing. Axis of deficiency centers around the postdentary mandible (ramus) and corresponding squamosal support. [Reprinted from Weaver KN, Watt KEN, Hufnagel RB, et al. Acrofacial Dysostosis, Cincinnati Type, a Mandibulofacial Dysostosis Syndrome with Limb Anomalies, Is Caused by POLR1A Dysfunction. The American Journal of Human Genetics 2015;96(5): 765-774. With permission from Elsevier.]

- Epibulbar dermoids *not* common.
- Lower eyelid: dermolipoma but no coloboma.
- Lateral canthus dysplastic but intact (zygomatic insertion preserved).

Nose

- If unilateral, nasal bone and soft tissue deviation to affected side; if bilateral, the more involved side.
  - Deviation worse in vomer—it follows the maxilla.
- Ala lateral deviation and elevation.

Maxilla–zygoma complex

- Under-appreciated feature, maxilla is vertically deficient, more so in Tessier zone 6, and posteriorly rotated.
  - Downward displacement of orbit—vertical dystopia.
- Alisphenoid retracted back by squamosal: diastasis with maxillary tuberosity.
- Zygoma develops as an "outrigger" to maxilla but can be affected.
  - Squamosal (r3) contains the condylar fossa. Deficiency in squamosal field can torque zygoma backward, causing lateral dystopia.
  - Deficiency in either jugal or postorbital fields can reduce orbital volume.
- Zygomatic arch: severe hypoplasia, even aplasia.
  - Lateral canthus displacement.

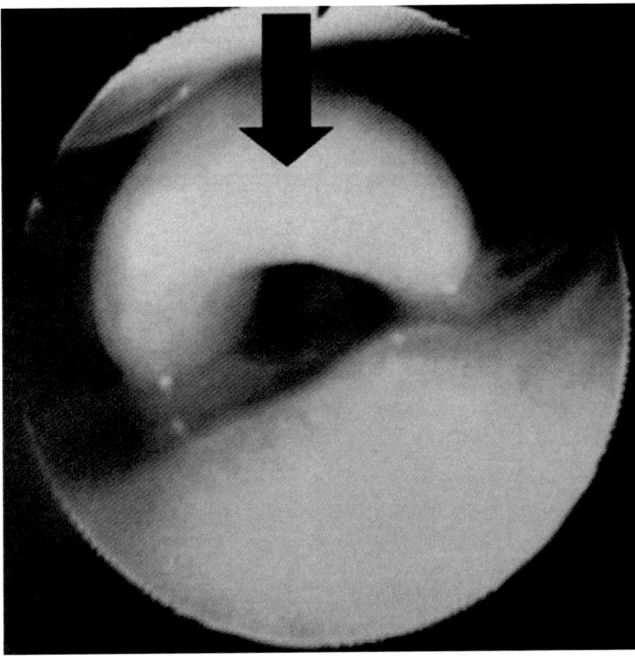

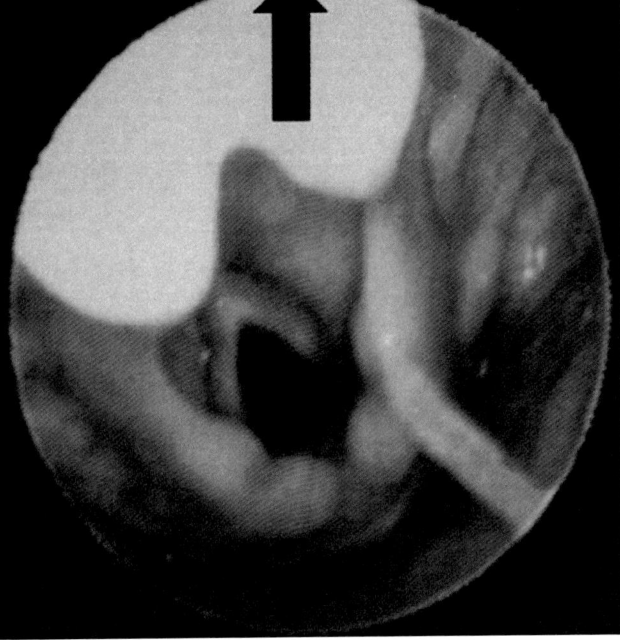

Fig. 13.160 Airway reconstruction in Nager syndrome. The Nager airway is characterized by patient retrognathia and retropositioned hyoid with tongue collapse and collapsed epiglottis. Distraction genioplasty and hyoid advancement combined using fascial slings to the mandible change the airway by improving the hyoid and projecting the collapsed epiglottis. The intervention is a success. [Reprinted from Heller JB, Gabbay JS, Kwan D, et al. Genioplasty Distraction Osteogenesis and Hyoid Advancement for Correction of Upper Airway Obstruction in Patients with Treacher Collins and Nager Syndromes. last Reconstr Surg. 2006 Jun;117(7):2389-98.]

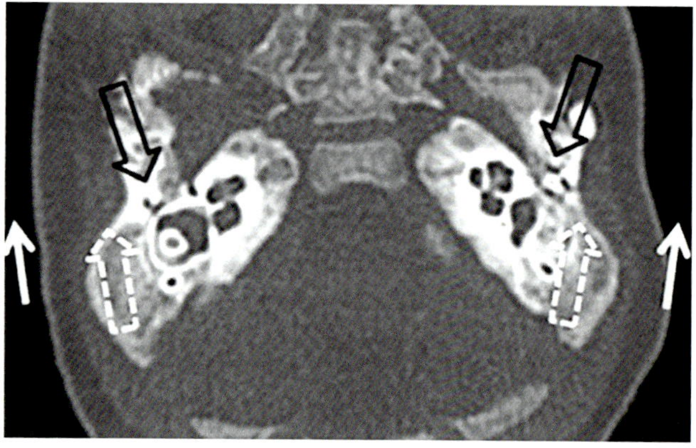

Fig. 13.161 Nager temporal bone. Axial CT of the temporal bones in Nager syndrome demonstrates severe microtia with absent pinnae (white arrows), external auditory atresia (white open dashed arrows), and severe middle-ear hypoplasia and ossicular dysplasia (black open arrows). These findings implicate rhombomeres r2–r3 for the ossicles and r4–r5 for the petrous complex, synthesized from Sm5–Sm6–Sm7. Sm7 may belong to mastoid; the mapping is undetermined. [Reprinted from Herrmann BW, Karzon R, Molter DW. Otologic and audiologic features of Nager acrofacial dysostosis. International Journal of Pediatric Otorhinolaryngology 2005;69(8): 1053-1059. With permission from Elsevier.]

- Temporal depression:
  - Absent temporalis.
  - Persistence of ancient temporomandibularis.

Mandible

- Mandibular hypoplasia resides in the ramus-condyle fields and is a defining feature of CFM (49–100%). Recall that condyle in mammals arises as a posterior outgrowth, the *postdentary rod*, from the surangular field of the mandible in the pre-mammal cynodont *Thrinaxodon*. Surangular becomes incorporated into dentary bone as the future ramus. The mandible in CFM has a normal complement of teeth: the body of mandible (dentary field) remains relatively unaffected. Dental crowding and malocclusion are not intrinsic to CFM.
- The retrognathia seen in CFM is simple. Like most retrognathias the pathology lies in the ramus. Surgical corrections use either (1) sagittal split osteotomy to separate the ramus and body, permitting advancement of the normal dentary field forward away from the deficiency site; or (2) osteotomy through the ramus itself to expand the deficiency site itself.

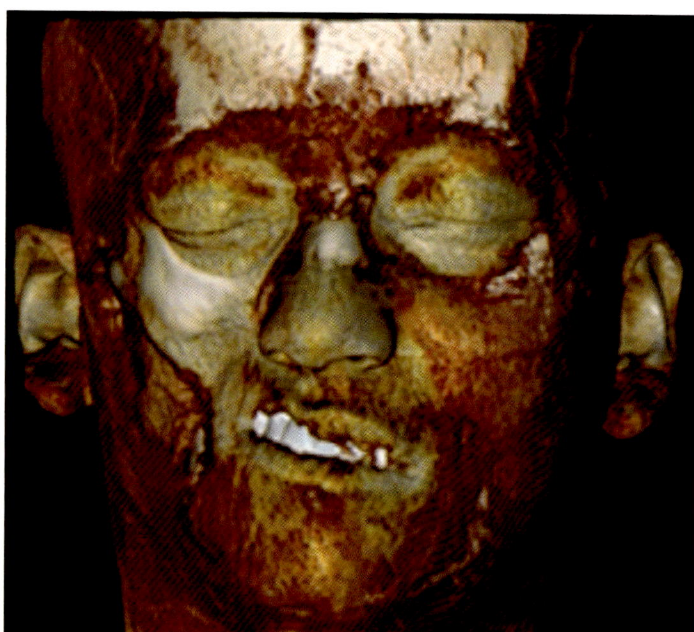

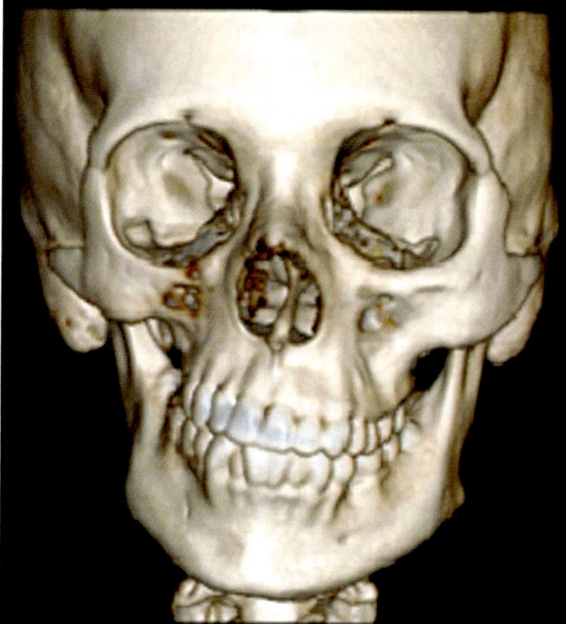

Fig. 13.162 Parry–Romberg syndrome. Deficiency site centers around zone 6 of the maxilla and jugal field causing rotation of the dental arch. Resorption of bone takes place, indicating reduction in a poorly understood trophic mechanism between overlying soft tissues and bone. [Left: Reprinted from Wikimedia. Retrieved from: https://commons.wikimedia.org/wiki/File:Parry_Romberg_syndrome_CT_reconstruction,_soft_tissues.jpg. With permission from Creative Commons License 3.0: https://creativecommons.org/licenses/by-sa/3.0/deed.en.]. [Right: Reprinted from Wikimedia. Retrieved from: https://commons.wikimedia.org/wiki/File:Parry_Romberg_syndrome_CT_reconstruction,_bone.jpg. With permission from Creative Commons License 3.0: https://creativecommons.org/licenses/by-sa/3.0/deed.en.]

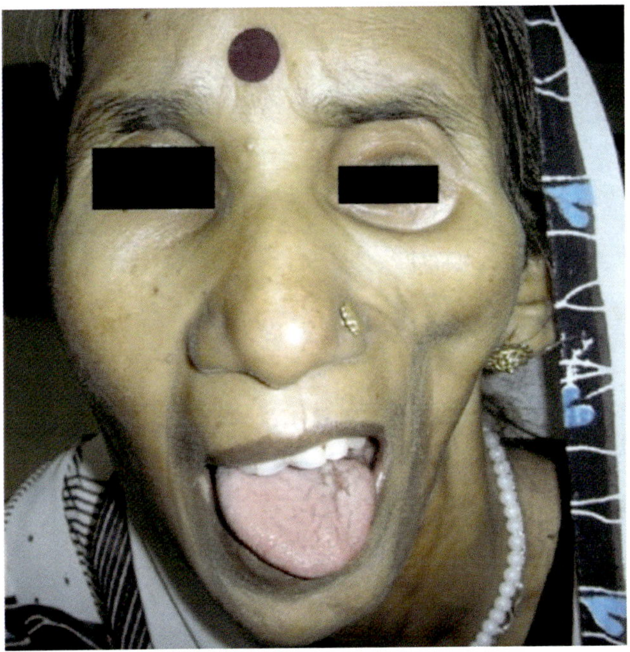

Fig. 13.163 Parry–Romberg tongue. Fascia and septae of the tongue are neural crest and may regulate the trophic state of the tongue muscles. [Reprinted from Verma R, Ram H, Gupta M, Vidhate MR. A case of extensive left-sided atrophy of Romberg. Natl J Maxillofac Surg 2013; 4(1):77-80. With permission from Wolters Kluwer Medknow Publications.]

- Glenoid fossa is frequently malformed (24–27%). Recall that the pre-mammal articular-quadrate joint is lost with a new articulation between post-dentary rod and squamosal, also an r3 derivative. These fields lie on either side of the axis of pathology.
- In CFM the field defects have a spectrum of severity defined by Figueroa and Pruzansky.
  - Type I: Mild hypoplasia of ramus. Condyle and glenoid fossa are normal.
  - Type II: The condyle is small and malformed. Glenoid fossa is converted into a flat surface. The coronoid field can be absent. Although unrelated to condyle, the two fields develop at the same time.
    Type IIA: Glenoid fossa still in anatomic position.
    Type 2B: Glenoid fossa anterior, medial, condyle severely dysplastic.
  - Type III: TMJ is gone. Ramus is reduced or absent.
- Coronoid process deficiency: absent temporalis.

<u>Ear</u> (r3–r4 pinna and ossicles)

- Microtia/anotia follows a well-known sequence.
  - Stage 1 is generalized hypoplasia of both first arch and second arch hillocks with the more cephalic ones being affected first.

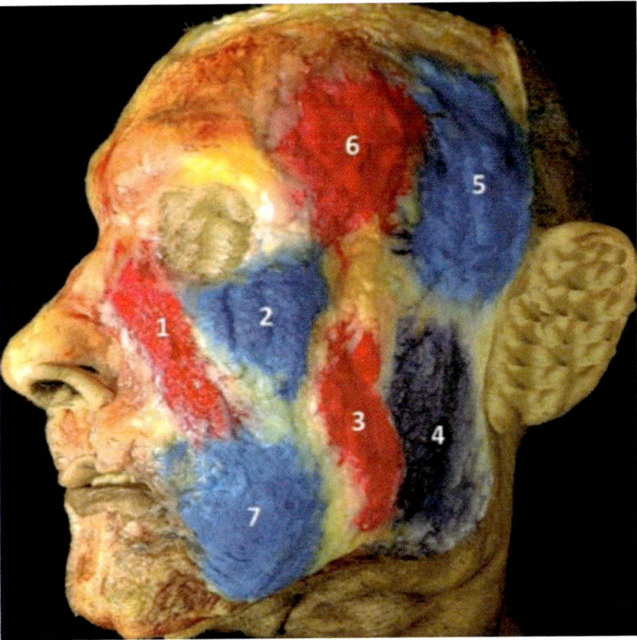

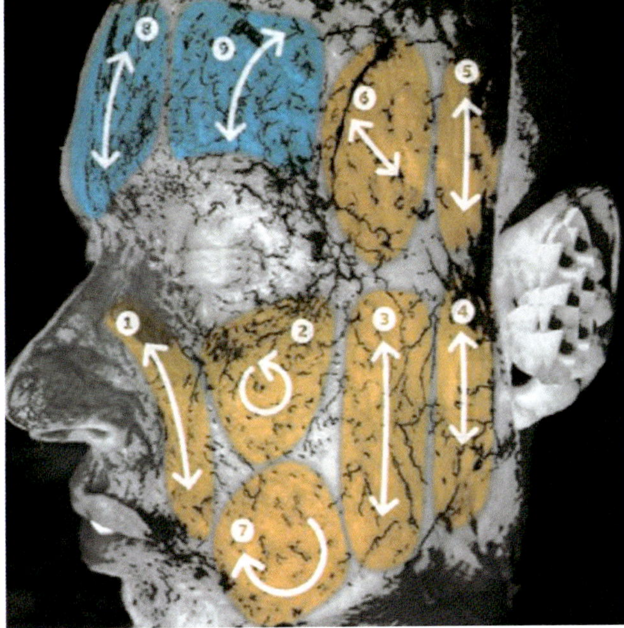

Fig. 13.164 Developmental fields, adipose. Fat populations (left) are shown with their vascular supply (right). Important mapping studies of facial fat demonstrate regional zones associated with either upper deck maxillary structures or lower deck structures. Eventual mapping of fat is likely to show these populations differ in neuromeric composition. [Reprinted from Schenck TL, Koban K, Sykes J, Tagosinski S, Earlbacher K, Cotofana S. The functional anatomy of the superficial fat compartments of the face: a detailed imaging study. Plast Reconstr Surg 2018; 141(6):1351-1359. With permission from Wolters Kluwer Health, Inc.]

  - Stage 2: loss of external auditory canal from tympanic temporal bone field in association with squamosal temporal field dysplasia of the condylar fossa.
  - Stage 3: auricle absent; lobule persists because it contains somitic mesenchyme from S6 to S7, i.e., C2 to C3.
- Aural atresia: hypoplasias of external auditory canal and TMJ related through tympanic versus squamosal fields.
- Hearing loss due to defects in r4–r5:
  - Ear deformity in CFM 80% chance of hearing loss.
  - Mechanism: conductive 73%, sensorineural 1%; therefore problems stem with the function of the ossicular chain: malleus and incus (r2), stapes (r3) with stapes most affected. Ossicular findings could represent a forme fruste of the disease as 8% of children with ipsilateral findings had concomitant hearing loss on the contralateral side [118].

Nerve

- Arterial axis of second arch is eccentrically placed: posterior (r5) zones of second arch develop prior to anterior (4) zones.
- Lower division of VII (r5) develops prior to upper division of VII (r4), hence temporal branch is more vulnerable to late insult.
- The severity of facial nerve involvement is in parallel to the degree of ear deformity (r3). This makes sense since first arch pathology affects r3 derivatives more than r2 structures.

Soft tissues

- Nose deviates to affected size.
  - Insufficient maxillary base, piriform deformity.
- Subcutaneous fat and skin are affected along a line extending from the oral commissure to the external auditory canal.
  - Mechanism: inadequate production of [BMP4] by the defective bone fields leading to reduced repression (and persistent function) of Shh at the epithelial margins, preventing fusion.
  - Mechanism: localized growth factors inducing
- High-arched palate secondary to deficit of ethmoid perpendicular.
- Soft palate cleft.

Plus (extra-craniofacial) findings

- Homeotic genes in common with the same homeotic "barcode" of the first and second arches lead to concomitant pathology at distant sites.
- Cardiac neural crest (inadequate population, pathologic migration)
- Vertebral abnormalities are strictly homeotic from identifiable somites.
  - Hemi-vertebrae and/or vertebral fusion with consequent scoliosis.
  - Rib anomalies.
- Renal: kidney development follows a strictly neuromeric and homeotic pattern [119].
- Brain malformations
  - Brazilian series of 46 patients reports epaxial r1 involvement: anophthalmia (1), lacrimal gland dysplasia (1), septum pellucidum (1), and corpus callosum (3).

In sum: Neuromeric diagnosis is hypaxial r2–r5, posterior and proximal zones of first and second arches are dominant.

- Craniofacial microsomia represents dysplastic and/or dysfunctional neural crest populations arising from rhombomeres r2–r5. Mandible is dominant over maxilla. Zygoma normal but distorted. Involvement of r1 is rare but may be associated with increase morbidity. Necropsy case reports co-existence of microphthalmia with cerebellar hypoplasia (cerebellum arising from r1) (Pegler 2016, [116]).

Goldenhar Syndrome (Oculoauriculodysplasia): r1–r5

Oculo-auricular-vertebral syndrome and its differentiation remain a dilemma within the context of physical diagnosis criteria, with some studies considering OAV as a mere subset of CFM [114]. The distinguishing features cited for the diagnosis of Goldenhar versus CFM are (1) epibulbar dermoids, (2) vertebral anomalies, and (3) bilaterality. Nowadays since CFM is no longer referred to as "hemifacial microsomia," the distinction of bilaterality is no longer important. Using the OMENS system, the findings in Goldenhar syndrome can be compared to CFM. As we shall see, certain features of Goldenhar have neuromeric significance and warrant comment [120, 121] (Fig. 13.155).

Orbit

- Less bony involvement than in CFM
- Orbital rim: lateral depression but no supraorbital dystopia
- Hypertelorism r1 excess (ipsilateral to cleft lip?)

Orbital soft tissues

- Micro-ophthalmia more common than in CFM
- Sclera (r1) presents with epibulbar dermoids

- Coloboma of upper eyelid, iris, choroid, retina
- Upper eyelid: coloboma, sometimes a peculiar aberrant eyelid is seen (with lashes) in the superior temporal quadrat
- Lower eyelid and fornix: dermolipoma*
- No canthus disruption—malar field defect is mild
- Miscellaneous ocular: strabismus [121] – lateral orbital floor deficit?

Nose

- Nasal bone and soft tissue deviation to affected side

Maxilla–zygoma complex

- Less hypoplasia
- Rotation/retrusion less than CFM
- Malar flattening mild, no frontonasal diastasis
- Zygomatic arch hypoplasia minimal

Mandible

- TMJ can be hypoplastic but never absent
- Ascending minimal to moderate hypoplasias but never absent
- Body (dentary) minimal hypoplasia
- Coronoid intact
- Malocclusion mild
- Rotation (chin deviation) less than CFM

Ear

- Microtia sequence severe but not as marked as CFM
- Hearing loss, conductive, due atresia of tympanic cavity and ossicles
  - Temporal bone abnormalities in one-third of Goldenhar patients include cochlear hypoplasia, 25–43% dysplasia of semicircular canals [122, 123]
  - Stria vascularis and semicircular canal involvement [124]
  - Findings implicate more profound involvement of r4–r5
- Abnormal semicircular canal and vestibular dysplasia r4–r5, Sm5–Sm6

Nerve

- Facial nerve weakness (not paralysis) involvement similar to CFM

Soft tissues

- Lacrimal apparatus OK
- Cleft lip/palate is common but wide variation in reports (5–50%) [125]
- High-arched palate (intact): deficient r1 perpendicular plate of ethmoid
- Macrostomia but less severe than CFM
- Temporal region OK (intact temporalis), epaxial r3 deficiency less marked

Plus (extracranial)

- Associated peripheral malformations more common than CFM
- CNS developmental delay

In sum: Neuromeric diagnosis hypaxial r1–r5, all quadrants of first and second arches more equally represented. Involvement of r1 neural crest is significant.

- Goldenhar syndrome represents dysplastic and/or dysfunctional neural crest populations arising from rhombomeres r2–r5.
- Mandible is still dominant over maxilla, but less so.
- Zygoma normal but distorted. Involvement of r1 is rare but may be associated with increase morbidity.
- Necropsy case reports co-existence of microphthalmia with cerebellar hypoplasia (cerebellum arising from r1) (Pegler 2016, [116]).
- Note: human homeobox gene *NKX5-3* abnormalities have been reported in three consanguineous family members having the newly described variant of **oculo-auriculo syndrome** presenting with microphthalmia, colobomas, anterior segment dysgenesis, abnormalities of retinal pigment epithelium, and rod–cone dystrophy (asterisks indicate identifiable abnormalities of *prosencephalic neural crest, **r1 neural crest). Gene knock-outs *nkx5.3* in *Dario rubrio* (zebrafish) using morpholina oligonucleotides show induced microphthalmia and retina dysplasia [126, 127].

Treacher-Collins Franceschetti Syndrome

This complex syndrome is predicated on severe hypoplasia or absence of the zygomatic fields, both jugal and postorbital. TCF is always bilateral. The same OMENS findings give a very different neuromeric interpretation (Figs. 13.156 and 13.157).

Orbit

- The orbital floor is missing to the zygomatic field defect.
  - Floor can angle downward to 45 degrees.
- Herniation of orbital contents favors development of a narrow orbit.
- Descent of lateral process of frontal due to absent FZ supported by postorbital field.

- Lateral orbital wall is gone so alisphenoid is displaced anteriorly.
- Transverse frontal sinus in orbital roof is extensively pneumatized.

Orbital soft tissues

- Visual impairment: vision loss (37%).
  - Eye muscles: strabismus (37%) secondary to orbital malformation.
  - Refractive errors.
  - Central: post-chiasmal.
  - Lid dysfunction.
- Upper eyelid: occasional coloboma in lateral third, zone 9.
  - Boundary zone r1–r2.
- Lower eyelid coloboma: notch between medial and lateral zones with hypoplastic tarsus and absent eyelashes.
  - This spot lies exactly above the boundary between maxillary process of jugal and the maxilla. Bone hypoplasia here reduces local [BMP4] with epithelial consequences.
  - Lower eyelid lash loss can be complete with a normal lid margin. Neural crest renters the lower lid from both medial and lateral margins.

    The notch represents NC failure to enter the zone or the absence of local trophic factors required for existent neural crest to form hair.
- Lower eyelid punctum can be absent: failure of epithelium to break down appropriately. Faulty signal.
- Lateral canthus is unattached, causing the lid margin (no longer under stretch) to spring medially, giving a "pseudoblepharophimosis."

Nose

- Nasal deformity: "Birdbeak nose," 73% have dorsal hump indicating r1 neural crest accumulation [128].

Maxilla

- Normal shape but overall reduction in size, affecting nasal airway
- Normal complement of teeth

Alisphenoid

- Relates to maxilla as r2 bone field
- Unappreciated **hypoplasia** and **distortion**
  - Reduction of greater wing
  - Loss of lateral plate
  - Blood supply from embryonic StV2 recurrent meningeal through the meningo-orbital foramen supplying zone 9 in lateral orbit, anastomosing with lacrimal artery to supply lacrimal gland

    Explains dysplasia of lacrimal gland

Mandible

- Body of normal length with intact dentition.
  - Inferior border is dysplastic: this was the ancient infradentary zone in *Panderichthys* (the bone fields were later called splenials). Note how modern mandible incorporates the splenials put they remain distinct from the tooth-bearing zone.
- Coronoid and condylar fields develop at the same time: both are dysplastic. TMJ and glenoid fossa the least affected compared to Goldenhar and CFM.

Zygoma complex (jugal + postorbital)

- Spectrum of deficiency runs in four directions:
  - Jugal-maxillary: lateral one-third orbit, ranging from a crease to full thickness (this is the source of confusion with #6 cleft)
  - Jugal-squamosal: not biologically related to the glenoid

    Glenoid belongs to squamosal field
  - Postorbital frontal
  - Postorbital-alisphenoid

Ear

- Microtia
- Hearing loss in 50% of patients
  - Mostly conductive at 50–70 dB [129]

Nerve

- No facial paralysis, i.e., r4–r5 facial motor nuclei spared

Soft tissues

- Characteristic facial cleft involves subcutaneous tissues with intact skin. It extends downward in an arc from the orbital rim at the zygomaticomaxillary boundary (missing or dysplastic zygoma) to the commissure. This problem is vexing for patients and difficult to correct, requiring multiple stages.
- Commissure shows *forme fruste* of macrostomia.
- No lateral cleft.
- Unusual hair pattern with sideburn shift forward into the r2 postorbital deficit.
- High-arched palate (intact): r2 vomer deficiency with vertical defect worse posteriorly, pulling the vomer out of the palatal plane.

In sum: Neuromeric diagnosis is hypaxial r1–r5 with significant involvement of r1. All quadrants of first and second arches represented: r2 is dominant of r3. Epaxial r1 affects eye and coordinates with midline cerebral structures.

- Zygomatic loss affects both jugal and postorbital fields.
- Alisphenoid deformity is *completely* unreported.
- Ophthalmic features prominent.
- Otologic deficits profound.
- Unique form of r2 skin displacement in the alisphenoid zone of anterior temple pulling sideburn forward and downward.
- Can have developmental delay, proportional to intrinsic ocular findings.

Nager Syndrome

Nager syndrome has facial features in common with Treacher-Collins but is distinguished by limb abnormalities on the radial side of the upper extremities and foreshortened lower extremities with small stature. The syndrome is rare. In a 2014 review only 70 cases were found in the literature. Nevertheless, Nager represents an important point on the craniofacial spectrum (Figs. 13.158, 13.159, 13.160, and 13.161).

Orbit

- Normal eyes
- Eyelids can have inverted lashes
- Ptosis
- Coloboma of lower eyelid (a case was reported by [130])
- Flattened nasal bones (indicating r1 neural crest)

Nose

- Midline (since Nager is bilateral) with normal structures
- Narrowing of airway, ?choanal atresia (r1 dysplasia)

Maxilla

- Hypoplastic with a normal complement of deciduous teeth but a reduction in permanent dentition of 2–10 dental units
- Incisors labially tipped (outward flaring)
- Soft palate absence: third arch involvement and Sm6 muscle malformation
- Distal malocclusion: molar deficiency
- Cleft palate defect consisting of severe hypoplasia or complete absence of the soft palate [131]. Tensor veli palatine originates in Sm4 with the first arch while levator veli palatine develops in Sm6 which supplies the third arch. Thus palatal agenesis is an r6–r7 defect.

Zygoma

- Severe hypoplasia, absent arch

Alisphenoid

- Severe hypoplasia, aplasia

Mandible

- Severely hypoplastic, requiring high rate of endotracheal intubation/tracheostomy
- Ramus 1/3 normal height at birth [132]
- TMJ ankylosis
- Distal malocclusion: molar deficiency

Ears

- Low-set ears (if present), severe microtia or anotia
- Ossicular involvement hypoplasia of the temporal bone complex (Sm4–Sm6)
- Conductive hearing loss (50–70 dB) similar to Treacher-Collins

Nerve

- Facial nerve normal (like TCF)

Soft tissues

- Anterior temporal r2 skin deficit with forward displacement of sideburn (like TCF)

Plus (extracranial)

- Larynx and epiglottis hypoplasia fourth arch (r8–r9)
- Preaxial defects
  - Short forearms with radial deficiency or fused forearm bones
  - This differentiates Nager syndrome from the rarer-still **Miller syndrome**, or postaxial acrofacial dysostosis, which is a rare autosomal recessive disorder characterized clinically by severe micrognathia, cleft lip and/or palate, hypoplasia or aplasia of the postaxial elements of the limbs, coloboma of the eyelids, and supernumerary nipple. https://www.omim.org/entry/263750
- Can be normal intelligence

In sum: Neuromeric diagnosis of Nager syndrome is hypaxial r1–r9. Absent palate = third arch (r6–r7), larynx and epiglottis = fourth arch (r8–r9). Involvement of r1 neural crest minimal. Cervical neuromeres c5–c7 involved in radial deficit, multiple other levels.

Parry–Romberg Syndrome

Parry–Romberg disease is the final condition we shall consider because of how it affects the orbit and of equal importance in how it does *not* affect the orbit. This disease is well reviewed so our comments will be focused on a few important features, followed by a neuromeric summary (Figs. 13.162, 13.163, and 13.164).

Parry–Romberg disease (PRS) involves the progressive atrophy of neural crest tissues in the face, usually unilateral. It usually occurs in childhood or adolescence and is more frequent in women (3:2). Facial atrophy is striking as it initially is confined to V1 and V2 regions but later in the disease can affect mandible as well. The final distribution is V2, 80%; V1, 65%; and V3, 50%. Maxilla and zygoma become globally reduced with frontal bone being more resistant. Dermis and fat are thinned out and pigmentation changes can occur (cf. the iris). Electron microscopy of facial soft tissues demonstrates a lymphocytic infiltrate of neurovascular bundles affecting the vascular endothelium and basement membranes. Lipid droplets from the degenerated basement membranes were observed intercalated within the substance of reduplicated basement membranes. The implication is one of chronic vascular damage leading to ineffectual repair and eventual replacement of tissue loss [133].

The bony orbit maintains its shape but thinning of r2 bones leads to enophthalmos zygoma. As the primary contributor to the orbital floor, zygoma is the main target. Resorption of greater wing may occur. Atrophy of r1 orbital fat exacerbates the situation. Pigmentation changes of hair and skin can occur in all zones.

Globe structures (hp2 optic cup, the optic placode, and lens) are spared and visual acuity is preserved. Infiltrating neural crest cells that form anterior structures are affected with changes in pupil size and pigment abnormalities of the iris [134]. Eyelid atrophy is more pronounced in the lower lid up to and including coloboma formation, which would imply a reassignment in neural crest fate at the affected site. Limitations of motion affecting the lateral rectus (Duane syndrome) could reflect fibrosis in the r1 fascia around the muscle leading to compromise of local blood supply to the muscle or abducens itself without invoking an intrinsic myo/neuropathy.

Degeneration of cartilage structures has been in diffuse neuromeres including the nose (r1), ear (r3–r4), and larynx (r8–r9). Microvascular supply to the tongue guided by neural crest fascia (r3) guiding the vascularity can cause atrophy in previously normal muscles [135]. Neurologic manifestations are varied, including trigeminal neuralgia, degeneration of white matter, and, in particular, r1 cerebellar atrophy. These findings are suggestive of concomitant developmental lesions, both intra and extracranial, originating from the same neuromere. MRI and CT studies of Parry–Romberg are essential for complete diagnosis [136–138].

In sum: The neuromeric diagnosis of Parry–Romberg involves structures that bear some relationship with neural crest, either as a source of mesenchyme, or in relation to the sympathetic nervous system, the entire construction of which is (1) neural crest neurons and (2) a neuromere-by-neuromere distribution. Thus, it is not surprising that abnormalities as diverse as the kidney (very neuromeric in its development) or the diaphragm have been described. For our purposes we shall concentrate on r1–r3 hypaxial with the possibility that epaxial r1–r3 populations could affect circulation to the cerebral cortex white matter or branches to the r1 cerebellum (Table 13.8).

CODA

The orbit is truly at crossroads of development. Multiple approaches are required to understand it, including neuroembryology, paleontology, comparative anatomy, and genetics. The orbit of humans and high anthropoids is unique in its osseous composition. Its shape was modified over evolution by the demands of brain growth. Binocularity and precision coordination of the hands were important factors as well. The ancient lineage of the stapedial artery system, dating back to the gnathostome revolution, allows us to trace out the history of orbital tissues and determine their source of origin. As a showcase of pathologies, the orbit is unique. The conditions examined here all shed light on the orbit as a functional milieu. Surgical applications such as hyoid advancement, based on the embryologic role of the pretracheal fascia, illustrate the importance of clinical observations, what works and

Table 13.8 Neuromeric differential diagnosis of laterofacial dysostoses

Syndrome	Neuromeric dx	Pathology	
CFM	r2–r5	Bone > soft tissue	
Goldenhar	r1–r5	soft tissue > bone, dermoid	
TCF	r1–r8	Clefts orbit, coloboma Usually lower lid	Pretracheal fascia airway
Nager	r1–r8 ++	Clefts orbit, bilateral, flat nose, inverted eyelashes	Pretracheal fascia airway worse than TCF
Romberg	r1–r5	Unilateral atrophy, soft tissues and bone	Other anatomic sites can be involved

what does not. These interventions serve as signposts about the consequences of anatomic reconstruction. As such, using a basic science approach, a deeper understanding of the orbit will emerge, thus laying the groundwork for future innovations in treatment.

Commentary: David Matthews

This work is a massive effort spanning at least 20 years that I have known Michael Carstens. We first met in a workshop and ASPS meeting in Orlando. There were only two people in attendance, Leonard Furlow and me. My connection with clefts and craniofacial surgery took off with my meeting with Paul Tessier in 1983. We worked together until his death. After Michael made his presentation, I asked if he would let me send this to Paul Tessier. He did, and I sent it to France. Tessier called me a month or so before his death to say that it was very interesting and he would like to spend more time on it. Unfortunately, he died before he could finish.

At Tessier's funeral and reception, I met Jean-Claude Talmant. I had heard much about him from Paul Tessier and also from his teacher Dr. Jean Delaire. We agreed to stay in contact. Several months later, while on a trip to Paris, Tony Wolfe and I went to meet him in Nantes for an afternoon of talks and dinner. That meeting was an eye-opener for me. I have always wondered why the sidewalls of the nasal pyramid were different in cleft children. The cleft side was flat and the noncleft side rounded. Understanding the amniotic fluid flow during pregnancy (fetal ventilation) gave me the answer. The outcome of the meeting was to stimulate the need to go back and review about 100 of my cleft patients. I looked at their facial growth, amount of orthodontics they received, and their ability to breathe through the nose. What I found was that those who could breathe nasally had much better facial growth. They had no need for reversible headgear or LeFort I osteotomies. I communicated this to Jean-Claude Talmant: that nasal breathing seemed to lead to normal facial growth.

As Tessier always said, "first the bone, then the soft tissue … but always restore function." But as we look at the facial clefts, there is a tremendous variability in their expression and presentation. The common cleft lip and palate, a #2 cleft, will behave differently than the lateral facial cleft numbers 3, 4, and 5. The anatomy is different. We have the eye, orbit, lacrimal system, malar bone, and alveolus involved. The cleft numbers 4 and 5 may well have absence of the medial portion of the facial muscles.

Lateral clefts will usually require a bone graft for overall stability, plus/minus orbital enlargement to prevent exophthalmos, repair of eyelid clefts, possible canthopexy, and lacrimal system. The priority is the functional reconstruction of the orbital floor and rim. This is required to support the globe. Bony defects also involve the malar process. Because of limited bone sources in infants and young children, primary bone grafting is for the orbit. Tessier and I found that these clefts are bone hungry and will have to be repeated as the child grows. Primary grafting is for the orbit leaving the malar processes and the alveolus for later as facial growth is completed. This gives more available bone sources. The functional repair of the soft tissue is modified as growth finishes for the best aesthetic result.

Functional nasal breathing for facial growth is very important in Tessier #2 and 3 clefts. Lateral clefts #4 and 5 do not involve breathing as much. We know that bone grafts will not grow, and we need to wait until growth is finished for a stable reconstruction.

The real challenges come as we try to understand why and how these clefts develop. For years, I thought that nutritional deficiency and environmental toxins—smoking and carbon monoxide—would explain the etiology. I carried out a nutritional screening using SpectraCell looking at the intracellular levels of nutrients inside lymphocytes. We did this on a population of pregnant women carrying a child with a cleft. This was done after a positive ultrasound. We found that there is no consistent nutritional deficiency present. Folic acid was normal. B12 levels varied. Intracellular magnesium was consistently low.

Current research using light forces us into quantum biology. The sunlight frequencies modulate our own vitamin D and vitamin A in neural crest development. There is an ideal balance between vitamin A and vitamin D in our tissue and neural crest development. We regulate that with photoreceptors in skin and eyes. The sunlight stimulates the production in melanocytes, and neuropsin and melanopsin receptors. This modulates neuro- and mucosal health even in the intestine as well as the brain. It will help set the circadian cycle to balance the metabolism. These receptors are found in neuroendocrine as well as neural crest cells. This begs the question as to their role in cleft formation.

Neuro-crest migration research does not talk about balancing vitamin A and vitamin D. Research has shown that elevated levels of vitamin A can block neural crest migration. Interaction between vitamin A and vitamin D may help balance the migration and prevent clefts. Typical lab setups ignore the importance of natural light and circadian rhythms that the sunlight frequencies set. Circadian rhythm helps the transcriptional programming of lipids and amino acids. Normal circadian rhythms cannot be obtained without using natural light sources. It is possible that using this type of research, we can find the answers.

Transcriptional programming of lipid and amino acid metabolism by the skeletal muscle circadian clock—PubMed (nih.gov)

A Note from Dr. Carstens

David C Matthews

Known to but a fortunate few, the lifework of Dr. David Matthews, pediatric plastic and craniofacial surgeon, can be summarized in two words: semper fidelis. A survivor of Vietnam, he transformed the scars of that experience into a lifetime of commitment to children and the underserved. David was the youngest of Paul Tessier's fellows; their emotional bond was strong. Possessed of great technical skills, he never forgot his mission, to provide service to all. He moved to Charlotte, North Carolina. When his practice group there demanded that he change his priorities to augment the bottom line, David went set up a solo practice of reconstructive plastic service for the poorest of the poor. Year after year, he took the hits for the trauma service at Carolinas Medical Center, supporting his dedicated office staff and living on a shoestring in an apartment above his office. The constancy of his service to the community was unmatched. Despite the human toll exacted by these choices, David's commitment never wavered, nor did his creativity. David's insights into cleft care and bone biology are the best in the world. When we met, as Tessier's disciple, he immediately grasped the significance of development field theory. We share common values as well. I no longer felt alone outside the box, and we became friends. With David and his wife Anna, I discovered friendship in another dimension, one that has remained constant through the years. With him, it is semper fi, regardless of the circumstances. When the chips are down, David is there. That is just the way he is. But David Matthews is not done. His thinking has pushed forward into wholistic medicine, acupuncture, nutrition, and electrical properties of cells. He remains ahead of the curve. The rest of us can only try to keep up.

References

1. Abou-Rayyah Y, Rose GE, Chawla SJ, Mosely IF. Clinical, radiological and pathological examination of periocular dermoid cyst: evidence of inflammation from an early age. Eye. 2002;16:507–12.
2. Reissis D, Pfaff MJ, Patel A, Steinbacher DM. Craniofacial dermoid cysts: histologic analysis and inter-site comparison. Yale J Biol Med. 2014;87:349–57.
3. Pratt LW. Midline cyst of the nasal dorsum: embryologic origin and treatment. Laryngoscope. 1965;75:968–88.
4. Tessier P. Plastic surgery of the orbit and eyelids (trans. SA Wolfe). Philadelphia: Mosby; 1981.
5. Ewings EL, Carstens MH. Neuroembryology and functional anatomy of craniofacial cysts. Indian J Plast Surg. 2009;42(Suppl):S19–34.
6. Creuzet S, Vincent C, Couly GF. Neural crest derivatives in ocular and periocular structures. Int J Dev Biol. 2005;49:161–71.
7. Gilbert PW. The origin and development of the human extrinsic ocular muscles. Contrib Embryol Carnegie Inst. 1957;36:59–78. www.embryology/index.php/Book_-_Contributions_to_Embryology
8. Jacob M, Jacob HJ, Wachtler F, Christ B. Ontogeny of avian extraocular muscles. Cell Tissue Res. 1984;237(3):549–57.
9. Whitnall SE. An instance of the retractor bulbi muscle in man. J Anat Physiol. 1911;46(Pt1):36–40.
10. Hutson KA, Glendenning KK, Masterton RB. Abducens nucleus and its relationship to the accessory facial and posterior trigeminal nucleus in the cat. J Comp Neurol. 1979;188(1):1–16.
11. Bain RS. Clinical review 157: pathophysiology of graves ophthalmopathy: the cycle of disease. J Clin Endocrinol Metab. 2003;88(5):1939–46.
12. Caplan A, Correa D. The MSC: an injury drug. Cell Stem Cell. 2011;9(11):11–5.
13. Albert DM, Miller JW, editors. Albert and Jakobiec's principles and practice of ophthalmology. 3rd ed. Philadelphia: WB Saunders; 2008. p. 2913–37.
14. Murthy R. Congenital dystrophic medial rectus muscles. Indian J Ophthalmol. 2017;65(1):62–4.
15. Kakizaki H, Takahasi Y, Nakano T, et al. Müller's muscle: a component of the peribulbar smooth muscle network. Ophthalmology. 2010;117:2229–32.
16. Kakizaki H, Zhao J, Nakano T. The lower eyelid retractor consists of definite double layers. Ophthalmology. 2006;113:2346–50.
17. Kakizaki H, Malhotra R, Madge SN, Selva D. Lower eyelid anatomy: an update. Ann Plast Surg. 2009;63:344–51.
18. Kakizaki H, Malhotra R, Selva D. Upper eyelid anatomy: an update. Ann Plast Surg. 2009;63:336–43.
19. Sires BS, Lemke BN, Dortzbach RK. Characterization of human orbital fat and connective tissue. Ophthal Plast Reconstr Surg. 1998;14:404–14.
20. Wolfram-Gabel R, Kahn JL. Adipose body of the orbit. Clin Anat. 2002;15:186–92.
21. Hwang K, Joon KD, Chung RS. Pretarsal fat in the lower eyelid. Clin Anat. 2001;14:179–83.
22. Fraser GJ, Cerny R, Soukup V, Bronner-Fraser M, Treelman T. The odontode explosion: the origin of tooth-like structures in vertebrates. Bioessays. 2010;32(9):808–17. https://doi.org/10.1002/bies.200900151.
23. Daeshler EB, Shubin NH, Jenkins FA. A Devonion tetrapod-like fish and the evolution of the tetrapod body plan. Nature. 2006;440:757–63. https://doi.org/10.1038/nature04639.
24. Clack JA. Gaining ground: the origin and evolution of tetrapods. Bloomington: Indiana University Press; 2012.
25. Ross C. Muscular and osseous anatomy of the primate anterior temporal fossa and the functions of the postorbital septum. Am J Phys Anthropol. 1995;98:275–306.
26. Bowling B, editor. Kanski's clinical ophthalmology. 8th ed. Philadelphia: WB Saunders; 2015.
27. Butler AB, Hodos W. Comparative vertebrate neuroanatomy. New York: Wiley-Liss; 2005.
28. Hughes A. The topography of vision in animals of contrasting life-styles. In: Crescitelli F, editor. Handbook of sensory physiology: the visual system in vertebrates, vol. 7/5A. Berlin: Springer; 1977. p. 613–756.
29. Nityananda V, Read JCA. Steropsis in animals: evolution, function, and mechanisms. J Exp Biol. 2017;220:2502–12. https://doi.org/10.1242/jeb.143883.
30. Descartes R. L'homme et un traitté de la formation du foetus du mesme autheur. Paris: Christine Angot; 1664.
31. Heesy CP. On the relationship between orbit and binocular visual field overlap in mammals. Anat Rec Part A. 2004;281A:1104–10.

32. Hubel DH, Wiesel TN. Brain and visual perception: the story of a 25-year collaboration. Oxford: Oxford University Press; 2004.
33. Pettirgrew J. Evolution of binocular vision. In: Pettigrew J, Sanderson K, Lewick W, editors. Visual neurosciences. New York: Cambridge University Press; 1986.
34. Pettigrew JD. Neurophysicology of binocular vision. Sci Am. 1972;227:84–95.
35. Asssee A, Qian N. Solving da Vinci stereopsis with depth-edge-selective V2 cells. Vis Res. 2007;47:2585–602.
36. Goodale MA, Milner AD. Separate visual pathways for perception and action. Trends Neurosci. 1992;15(1):20–5.
37. Milner AD. How do the two visual streams interact with each other? Exp Brain Res. 2017;235:1297–308. https://doi.org/10.1007/s00221-017-4917-4.
38. Larsson M. Binocular vision and ipsilateral retinal projections in relation to eye and forelimb coordination. Brain Behav Evol. 2011;77(4):219–30. https://doi.org/10.1159/000329257.
39. Larsson M. The optic chiasm: a turning point in the evolution of eye-hand coordination. Front Zool. 2013;10:41. http://www.frontiersinzoology.com/content/10/1/41
40. Larsson ML. Binocular vision, the optic chiasm, and their associations with vertebrate motor behavior. Front Ecol Evol. 2015;3:89. https://doi.org/10.3389/fevo.201500089.
41. Sussman RW, Rasmussen RT, Raven PH. Rethinking primate origins, again. Am J Paleontol. 2013;75:95–105.
42. Allman AJ, McGuinnes EL. Visual cortex in primates. In: Comparative primate biology, Vol. 4: Neurosciences. New York: Liss; 1988. p. 279–326. http://www.allmanlab.caltech.edu/PDFs/AllmanMcGuinness1988.pdf.
43. Cartmill M. Rethinking primate origins. Science. 1974;184(4135):436–43.
44. Kirk EC. Comparative morphology of the eye in primates. Anat Rec Part A. 2004;281A:1085–102.
45. Vega-Zúniga T, Medina FS, Fredes F, Zúniga C, Severin D, Palacios AG, Karten HJ, Mpodozis J. Does nocturnality drive binocular vision? Octodontine rodents as a case study. PLoS One. 2013;8(12):e84199. https://doi.org/10.1371/journal.pone.0084199.
46. Bloch JI, Boyer DM. Grasping primate origins. Science. 2002;298:1606–9.
47. Napier JR. Studies of the hands of living primates. Proc Zool Soc London. 1961;134:647–56.
48. Tanaka G, Parker AR, Hasegawa Y, et al. Mineralized rods and cones suggest colour vision in a 300Myr-old fossil fish. Nat Commun. 2014;5:5920. https://doi.org/10.1038/ncomms6920.
49. Bowmaker JK. Evolution of color vision in vertebrates. Eye. 1998;12(3b):541–7.
50. Dulai KS, von Dornum M, Mollon JD, Hunt DM. The evolution of trichromatic vision by opsin gene duplication in New World and Old World primates. Genome Res. 1999;9(7):629–38.
51. Carroll R. The rise of amphibians: 365 million years of evolution. Baltimore: Johns Hopkins University Press; 2009. p. 360.
52. Ducasse A, DeLattre JF, Flament JB, Hueau J. The arteries of the lacrimal gland. Anat Clin. 1984;6:287–93.
53. Scott G, Balsiger H, Kluckman M, et al. Patterns of innervation of the lacrimal gland with clinical application. Clin Anat. 2014;27:1174–7.
54. Van der Werf F, Baljet B, Prins M, Otto JA. Innervation of the lacrimal gland in the cynomolgous monkey: a retrograde tracing study. J Anat. 1996;188:591–601.
55. Byun TH, Kim JT, Park HW, Kim WK. Timetable for upper eyelid development in staged human embryos and fetuses. Anat Rec. 2011;294:789–96.
56. de la Cuadra-Blanco C, Peces-Peña MD, Merida-Velasco JR. Morphogenesis of the human lacrimal gland. J Anat. 2003;203:531–6.
57. Posnick JC, Costello BJ. Dermoid sinus cysts, gliomas, and encephalocoeles: evaluation and management. Atlas Oral Maxillofac Surg Clin North Am. 2002;10(1):85–99.
58. Pratt LW. Midline cysts of the nasal dorsum: embryologic origin and treatment. Laryngoscope. 2015;125(3):533.
59. Kardong K. Vertebrates: comparative anatomy, function, evolution. 7th ed. New York: McGraw Hill; 2015. p. 249. Fig 7.10
60. Müller F, O'Rahilly R. Cerebral dysgraphia (future anencephaly) in a human twin embryo, stage 13. Teratology. 1984;30(2):167–77.
61. Müller F, O'Rahilly RO. Development of anencephaly and its variants. Am J Anat. 1991;190:193–218.
62. Asidi K, Gunduz Y, Yaldiz C, Aksoy YE. Intraorbital encephalocele presenting with exophthalmos and orbital dystopia: CT and MRI findings. J Korean Neurosurg Soc. 2015;57(1):58–60. https://doi.org/10.3340/jkns.2015.57.1.58.
63. Lim ZM, Friedland PL, Boeddinghaus R, Thompson A, Rodrigues SJ, Atlas M. Otic meningitis, superior semicirgular canal dehiscence, and encphalocoele: a case series. Otol Neurotol. 2012;33:610–2.
64. Tirumandas M, Sharma A, Gbenimacho I, Shuja M, Tubbs RS, Oakes WJ, Loukas M. Nasal encephalocoeles: a review of etiology, pathophysiology, clinical presentations, diagnosis, and treatment. Childs Nerv Syst. 2013;29:739–44.
65. Shokumbi T, Adeloye A, Olumide A. Occipital encephalocoeles in 57 Nigerian children: a retrospective analysis. Childs Nerv Syst. 1990;6:99–102.
66. Renshaw A, Borsetti M, Nelson RJ, Orlando A. Massive plexiform neurofibroma with associated meningo-encephalocoele and occipital bone defect presents as a cervical mass. Br J Plast Surg. 2003;56(5):514–7.
67. Arévalo-Pérez J, Milán-Juncos JM. Parietal intradiplöic encephalocoele: report of a case and review of the literature. Neuroradiol J. 2015;28(3):264–7. https://doi.org/10.1177/1971400915592554.
68. Mulcahy MM, McMenomey SO, Talbon JM, Delashaw JB. Congenital encephalocoele of the medial skull base. Laryngoscope. 1997;107:910–4.
69. Sanna M, Paolo F, Russoa A, Falcioni M. Management of meningoencephalic herniation of the temporal bone: personal experience and literature review. Laryngoscope. 2009;119:1579–85.
70. Kwon JE, Kim E. Middle fossa approach to a temporosphenoidal encephalocoele. Neurol Med Chir (Tokyo). 2010;50:434–8.
71. Saavalainen T, Jutila L, Mervala E, Kälviäinen R, Vanninen R, Immonen A. Temporal anteroinferior encphalocoele: an underrecognized etiology of temporal lobe epilepsy? Neurology. 2015;85:1467–74.
72. Puvabanditsin S, Malik I, Garrow E, Francois L, Mehta R. Clival encephalocoele and 5q15 deletion: a case report. J Child Neurol. 2015;30(4):505–8.
73. Acherman DS, Bosman DK, van der Horst CMAM. Sphenoethmoidal encephalocele: a case report. Cleft Palate Craniofac J. 2003;40(3):329–33.
74. Salinas-Torres VM. Say-Meyer syndrome: additional manifestation in a new patient and phenotypic assessment. Childs Nerv Syst. 2015;31:1181–7.
75. Say B, Meyer J. Familial trigonocephaly associated with short stature and developmental delay. Am J Child Dis. 1981;135:711–2.
76. Sargent C, Burn J, Baraitser M, Pembrey ME. Trigonocephaly and the Opitz C syndrome. J Med Genet. 1985;22:39–45.
77. Vu HL, Panchal J, Parker EE, Levine NS, Francel P. The timing of physiologic closure of the metopic suture: a review of 159 cases using reconstructed 3D-CT scans of the craniofacial region. J Craniofac Surg. 2001;12(6):527–62.

78. Karabagil P. Pathology in metopic synostosis. Childs Nerv Syst. 2013;29:2165–70.
79. Shimoji T. Mild trigonocephaly with clinical symptoms: analysis of surgical results in 65 patients. Childs Nerv Syst. 2002;18(5):215.
80. Azimi C, Kennedy SJ, Chiayat D, Chakraborty P, Clerke JTR, Forrest C, Teebi AS. Clinical and genetic aspects of trigonocephaly: a study of 25 cases. Am J Med Genet. 2003;117A:127–35.
81. Denis D, Saracco JB, Genitori L, Choux M, Bardot J, Maumenee IH. Ocular findings in trigonocephaly. Graefes Arch Clin Exp Ophthalmol. 1994;232(12):728–33.
82. Anderson FM, Gwinn JL, Todt JC. Trigoncephaly: identity and surgical management. J Neurosurg. 1962;19:723–30.
83. Currarino G, Silverman FN. Orbital hypotelorism, arrhinencephaly, and trigonocephaly. Radiology. 1960;74:206–17.
84. Pellerin P, Vinchon M, Dhellemes P, Wolber A, Guerreschi P. Trigonocephaly: Lille's surgical technique. Childs Nerv Syst. 2013;29:2183–8. https://doi.org/10.1007/s00381-013-2229-y.
85. Marsh JL, Gado MH, Vannier MW, Stevens WG. Osseous anatomy of unilateral coronal synostosis. Cleft Palate Craniofac J. 1986;23(2):87–100.
86. Forte AJ, Steinbacher DM, Persing JA, Brooks ED, Andrew TW, Alonso N. Orbital dysmorphology in untreated children with Crouzon and Apert syndromes. Plast Reconstr Surg. 2015;136(5):1054–62. https://doi.org/10.1097/PRS.0000000000001693.
87. Tokumaru AM, Barkovich J, Cincillo SF, Edwards MSB. Skull base and calvarial deformities: association with intracranial changes in craniofacial syndromes. AJNR. 1996;17:619–30.
88. Yasonov SA, Lopatin AV, Kugushev AY. Craniosynostosis of the sphenoidal suture definition of the main signs of deformity. Ann Maxillofac Surg. 2017;7:222–7.
89. Lo LJ, Marsh JL, Kane AA, Vannier MW. Orbital dysmorphology in unilateral coronal synostosis. Cleft Palate Craniofac J. 1996;33(3):190–7. https://doi.org/10.1597/1545-1569_1996_033_0190.
90. Beckett JS, Persing JA, Steinbacher DM. Bilateral orbital dysmorphlogy in unicoronal synostosis. Plast Reconstr Surg. 2013;131(1):125–30.
91. Stavropoulos D, Bartzela T, Tarno P, Mohlin B. Dental agenesis patterns in Crouon Syndrome. Swed Dent J. 2011;41:195–201.
92. Pickrell BB, Lam SK, Monson KA. Isolated frontosphenoidal craniosynostosis: a rare cause of anterior plagiocephaly. J Craniofac Surg. 2015;26(6):1944–6.
93. Lloyd MS, Rodrigues D, Nishikawa H, White N, Solanki G, Noons P, Evans M, Dover S. Isolated unilateral frontosphenoid suture synostosis in six patients: lessons learned in diagnosis and treatment. J Craniofac Surg. 2016;27(4):871–3.
94. de Ribaupierre S, Czorny A, Pittet B, Jacques B, Billiet B. Frontosphenoidal synostosis: a rare cause of unilateral anterior plagiocephaly. Childs Nerv Syst. 2007;23:1431–8.
95. Sauerhammer TM, Oh AK, Boyajian M, Maggee SN, Myseros JS, Keating RF, Rogers G. Isolated frontosphenoidal synostosis: a rare cause of synostotic frontal plagiocephaly. J Neurosurg Pediatr. 2014;13:533–58.
96. Burdi A, Kusnetz AB, Vernes JL, Gebarski SS. The natural history and pathogenesis of the cranial coronal ring. Cleft Palate Craniofac J. 1986;23(1):28–39.
97. Dundulis BS, Becker DB, Grovier DP, Marsh J, Kane AA. Coronal ring involvement in patients treated for unilateral coronal synostosis. Plast Reconstr Surg. 2004;114(7):1695–703.
98. Ranger A, Chaudhary N, Matic D. Synostosis involving the squamosal temporal sutures: a rare and possibly underreported etiology for cranial vault asymmetry. J Craniofac Surg. 2008;21:1547–50.
99. Bialocerkowski AL, Viadusic SL, Wei Ng C. Prevalence, risk factors, and natural history of positional plagiocephaly: a systematic review. Dev Med Child Neurol. 2008;50(8):577–86.
100. Linz C, Collmann H, Meyer-Marcotty P, et al. Occipital plagiocephaly: unilateral lambdoid synostosis versus positional plagiocephaly. Arch Dis Child. 2015;100:152–7.
101. Menard R, David DJ. Unilateral lambdoid synostosis: morphologic characteristics. J Craniofac Surg. 1998;9:240–6.
102. Mullikin JB, Vander Woude DL, Hansen H, et al. Analysis of posterior plagiocephaly: deformational versus synostostic. Plast Reconstr Surg. 2007;120:993–1003.
103. Rhodes JL, Tye GW, Fearon JA. Craniosynostosis of the lambdoid suture. Semin Plast Surg. 2014;28:138–43.
104. Rhodes JL, Kolar JC, Fearon JA. Mercedes Benz pattern craniosynostosis. Plast Reconstr Surg. 2010;125(1):299–304.
105. Cohen MM Jr. Pfeiffer syndrome: update, clinical subtypes, and guidelines for differential diagnosis. Am J Med Genet. 1993;45:300–7.
106. Schaefer F, Anderson C, Can B, Say B. Novel deletion in the FGFR2 gene at the same codon as the Crouzon syndrome: mutations is a severe Pfeiffer syndrome type 2 case. Am J Med Genet. 1998;75:252–5.
107. Glaser RL, Jiang W, Boyadiev SA, Jabs EW. Paternal origins of FGFR2 mutations in sporadic cases of Crouzon syndrome and Pfeiffer syndrome. Am J Hum Genet. 2000;66(3):768–77. https://doi.org/10.1086/302831.
108. Lewanda AF, Jabs EW. Genetics of craniofacial disorders. Curr Opin Pediatr. 1994;6(6):690–7. https://doi.org/10.1097/00008480-199412000-00014.
109. DeMyer W, Zeman W, Palmar CG. The face predicts the brain: diagnostic significance of median facial anomalies for holoprosencephaly (arrhinencephaly). Pediatrics. 1964;34:256–63.
110. Tessier P. Fentes orbito-faciales verticals et obliques (colobomas) completes et frustes. Ann Chir Plast. 1969;19:301–11.
111. Tessier P. Anatomical classifications of facial, cranio-facial and laterofacial clefts. J Maxillofac Surg. 1976;4:69–92.
112. Carstens MH. The pathologic anatomy of the soft palate, part 2: the soft tissue lever arm, pathology, and surgical correction. J Cleft Lip Palate Craniofac Anomal. 2017;8:84.
113. Gougoutas AJ, Singh DJ, Low DW, Bartlett SP. Hemifacial microsomia: clinical features and pictographic representations of the OMENS classification system. Plast Reconstr Surg. 2007;120(7):112e–20e.
114. Tuin J, Tahira Y, Paliga JT, Taylor JA, Bartlett SP. Distinguishing Goldenhar syndrome for craniofacial microsomia. J Craniofac Surg. 2015;26(6):1887–92.
115. Tamas DE, Mahoney RS, Bowie JD, Woodruff WW III, Kay HH. Prenatal sonographic diagnosis of hemifacial microsomia (Goldenhar-Gorlin syndrome). J Ultrasound Med. 1986;5:461–3.
116. Martinelli P, Maruotti GM, Angli A, Mazzarelli LL, Bifulco G, Paldini D. Prenatal diagnosis of hemifacial microsomia and ipsilateral cerebellar hypoplasia in a fetus with oculoauriculovertebral spectrum. Ultrasound Obstet Gynecol. 2004;24:199–201.
117. Mendes Pegler JR, et al. Clinical presentation of 41 Brazilian patients with oculo-auriculo-vertebral dysplasia. Rev Med Bras. 2016;62(3):202–6.
118. Mitchell RS, Salzman BS, Norton SJ, et al. Hearing loss in children with craniofacial microsomia. Cleft Palate Craniofac J. 2017;54(6):656–63.
119. Carlson BM. Human embryology and developmental biology. 6th ed. New York: Elsevier; 2015.
120. Kokavec R. Goldenhar syndrome with various clinical manifestations. Cleft Palate Craniofac J. 2006;43(5):628–32.
121. Mansour AM, Wang F, Henkind P, Goldberg R, Shprintzen R. Ocular findings in facioauriculovertebral sequence (Goldenhar-Gorlin Syndrome). Am J Ophthalmol. 1985;100:555–9.
122. Hennersdorf F, Friese N, Löwenheim H, et al. Temporal bone changes in patients with Goldenhar syndrome with special emphasis on inner ear abnormalities. Otol Neurol. 2014;35:826–30.

123. Jahrsdorfer RA, Jacobson JT. Treacher Collins syndrome: otologic and auditory management. J Am Acad Audiol. 1995;5:93–102.
124. Scholtz AW, Fish JH, Kammen-Jolly K, et al. Goldnehar's syndrome: congenital hearing deficit of conductive or sensorineural origin? Temporal bone histopathologic study. Otol Neurotol. 2001;22:501–5.
125. Kumaresan R, Srinivasan B, Narayanan M, Cugati N, Karthikeyan P. Craniofacial abnormalities in Goldenhar syndrome: a case report with review of the literature. Plast Aesthet Res. 2014;1(3):108–13.
126. Schorderet DF, Nichi O, Boisset G, et al. Mutation in the human homeobox gene *NKX5-3* causes an oculo-auricular syndrome. Am J Hum Genet. 2008;82:1178–84.
127. Timme-Laragy AR, Karchner SI, Hanh ME. Gene knockdown by morpholino-modified oligonucleotides in the zebrafish model: applications for developmental toxicology. Methods Mol Biol. 2012;889:51–71. https://doi.org/10.1007/978-1-61779-867-2_5.
128. Plomp RG, Mathijssen IM, Moolenburgh SE, van Montfort KA, van der Meulen JJ, Poublon RM. Nasal sequelae of Treacher-Collins syndrome. J Plast Reconstr Aesthet Surg. 2015;68(6):771–81.
129. Marres HA. Hearing loss in the Treacher Collins syndrome. Adv Otorhinolaryngol. 2002;61:209–15.
130. Malik R, Goel S, Aggarwal S. Limbal dermoid in Nager acrofacial dystosis: a rare case report. Indian J Ophthalmol. 2014;62(3):339–41.
131. Jackson IT, Bauer B, Saleh J, Sullivan C, Argenta CL. A significant feature of Nageer's syndrome: palatal agenesis. Plast Reconstr Surg. 1989;84(2):219–26.
132. Halonen K, Hukki J, Arte S, Humerinta K. Craniofacial structure and dental development in three patients with Nager syndrome. J Cranaiofac Surg. 2006;17(6):1180–7.
133. Pensler J, Murphy JF, Mulliken JB. Clinical and ultrastructural studies of Romberg's hemifacial atrophy. Plast Reconstr Surg. 1990;85(5):669–74.
134. Muchnik RS. Ocular manifestation and treatment of hemifacial atrophy. Am J Ophthalmol. 1979;85:889–97.
135. Verma R, Ram H, Gupta M, Vidhate MR. A case of extensive left-sided atrophy of Romberg. Natl J Maxillofac Surg. 2013;4(1):77–80. https://doi.org/10.4103/0975-5950.117881.
136. Moko SB, Mistry Y, Blandain de Charlain TM. Parry-Romberg syndrome: intracranial MRI appearances. J Craniomaxillofac Surg. 2003;31:321–4.
137. Okumura A, Ikuta T, Tsuji T, et al. Parry-Romberg syndrome with a clinically silent white matter lesion. Am J Neuroradiol. 2006;27:1729–31.
138. Wong M, Phillils CD, Hagiwara M, Shatzkes DR. Parry Rombér syndrome: 7 cases and literature review. Am J Neuroradiol. 2015;36(7):1355–61. https://doi.org/10.3174/ajnr.A4297.
139. Noden DM, Evans DJR. Spatial relations between avian craniofacial neural crest and paraxial mesoderm cells. Dev Dyn. 2006;235(5):1310–25.
140. Ziermann JM, Diogo R Noden DM. Neural crest and the patterning of vertebrate craniofacial muscles. Genesis. 2018;56:e23097. https://doi.org/10.1002/dvg.23097.

Suggested Reading

Allam KA, Lim AA, Elsherbiny A, Kawamoto HK. The Tessier number 3 cleft: a report of 10 cases and review of literature. J Plast Reconstr Aesthet Surg. 2014;67(8):1055–62.

Alonso A, da Silva-Freitas R, Azor de Oliveira G, Goldenberg D, Tolazzi ARD. Tessier no. 4 facial cleft: evolution of surgical treatment in a large series of patients. Plast Reconstr Surg. 2008;122(5):1505–13.

Amirjamshidi A, Abbasioun A, Shams Amiri R, Ardalan A, Ramak Hashe SM. Lateral orbitotomy approach for removing hyperostosing en plaque sphenoid wing meningiomas. Description of surgical strategy and analysis of findings in a series of 88 patients with long-term follow up. Surg Neurol Int. 2015;6:79–91.

Andrews JA, Modrell MS, Baker CVH. Developmental evidence for serial homology of the vertebrate jaw and gill arch skeleton. Nat Commun. 2013;4:1436.

Barton RA. Binocularity and brain evolution in primates. PNAS. 2004;101(27):10113–5. https://doi.org/10.1073/pnas.0401955101. www.pnas.org.cgi

Beard C. Müller's superior tarsal muscle: anatomy, physiology, and clinical significance. Ann Plast Surg. 1985;14:324–33.

Benton M. Vertebrate paleontology. 3rd ed. New York: Wiley-Blackwell; 2005.

Bilkay U, Gundogan H, Ozek C, Gurter T, Akin Y. A rare craniofacial cleft: bilateral Tessier no. 5 cleft accompanied by no. 1 and no. 6 clefts. Ann Plast Surg. 2000;45(6):654–7.

Binnert Goetze T, Sleifer P, Machado Rosa F, et al. Hearing characteristics in oculoauriculovertebral spectrum: a prospective study in 10 patients. Am J Med Genet A. 2017;173A:309–14.

Bohnsack BL, Gallina D, Thompson H, et al. Development of extraocular muscles require (sic) early signals from periocular neural crest and the developing eye. Arch Ophthalmol. 2011;129(8):1030–41.

Boyer DM, Yapunich GS, Chester SGB, Bloch JI, Godinot M. Hands of early primates. Am J Phys Anthropol. 2013;37:33–78.

Bradley J, Kawamoto HK. Genioplasty distraction osteogenesis and hyoid advancement for correction of upper airway obstruction in patients with Treacher Collins and Nager syndromes. Plast Reconstr Surg. 2006;117(7):2389–98.

Bradbury JW, Vehrencamp SL. Principles of animal communication. Sunderland, MA: Sinauer Associates; 1998.

Bubanale SC, Kurbet SB, De Piedade Sequeira LMG. A rare case of cleft number nine associated with atypical cleft number two. Indian J Ophthalmol. 2017;65(7):610–2.

Bush EC, Allman JM. Three dimensional structure and evolution of primate primary visual cortex. Anat Rec Part A. 2004;281A:088–1094.

Carstens MH. Mechanisms of cleft palate. In: Bennun R, editor. Cleft lip and palate management: a comprehensive atlas. New York: Wiley-Blackwell; 2014a. p. 3–21.

Carstens MH. Developmental field reassignment in cleft surgery: reassessment and refinements. In: Bennun R, editor. Cleft lip and palate management: a comprehensive atlas. New York: Wiley-Blackwell; 2014b. p. 81–111.

Carstens MH. The pathologic anatomy of soft palate, part 1: embryology, the hard tissue platform, and evolution. J Cleft Lip Palate Craniofac Anomal. 2017;4:37–64.

Cassi D, Magnifico M, Gandolfini M, et al. Early orthopaedic treatment of hemifacial microsomia. Case Rep Dent. 2017;2017:7318715. https://doi.org/10.1155/2017/7318715.

Catalfumo FJ, Golz A, Westerman ST, Gilvert LM, Joachims HZ. The epiglottis and obstructive sleep apnoea syndrome. J Laryngol Otol. 1998;112:940.

Charrier JB, Rouillon I, Roger G, Denoyille F, Josset P, Garabeduab EN. Craniofacial dermoids: an embryologic theory unifying nasal dermoid sinus cysts. Cleft Palate Craniofac J. 2005;42(1):51–7.

Chattopadhay D, Murmu MB, Gupta S, et al. Bilateral Tessier number 5 facial cleft with limb constriction ring: the first case report with an update of literature review. J Oral Maxillofac Surg Med Pathol. 2013;25(1):43–5.

Chauhan DS, Grurprsad Y. Goldenhar syndrome with Tessier's 7 cleft: report of a case. J Maxillofac Oral Surg. 2015;14(Suppl 1):42–6. https://doi.org/10.1007/s12663-011-0279-9.

Cobb ARM, Green B, Gill D, Ayliffe P, Lloyd TW, Bulstrode N, Dunaway DJ. The surgical management of Treacher Collins syndrome. Br J Oral Maxillofac Surg. 2014;52(7):581–9.

Couly GF, Coltey PM, Le Dourain NM. The triple origin of skull in higher vertebrates: a study in quail-chick chimeras. Development. 1993;117:409–29.

Crompton AW, Jenkins FA. Mammals from reptiles: a review of mammalian origins. Annu Rev Earth Planet Sci. 1973;1(1):131–55.

Crompton AW, Rougier G, Musinsky C. Origin of the lateral wall of the mammalian skull: fossils, monotremes and therians revisited. J Mamm Evol. 2017;25:301–13. https://doi.org/10.1007/s10914-017-9388-7.

Creuzet S, Couly G, Le Douarin NM. Patterning the neural crest derivatives during development of the vertebrate head: insights from avian studies. J Anat. 2005;207:447–59.

Creuzet SE, Martinez S, Le Douarin NM. The cephalic neural crest exerts a critical effect on forebrain and midbrain development. PNAS. 2006;103(38):14033–8. https://doi.org/10.1073/pnas.0605899103.

da Silva-Freitas R, Alonso N, Shin J. The Tessier number 5 facial cleft: surgical strategies and outcomes in six patients. Cleft Palate Craniofac J. 2009;46(2):179–86.

Darzi MA, Chodri NA. Oblique facial clefts: a report of Tessier numbers 3, 4, 5, and 9 cleft. Cleft Palate Craniofac J. 1993;30(4):414–5.

David DJ, Moore MH, Cooter RD, Chow S-K. The Tessier number 9 cleft. Plast Reconstr Surg. 1989;83(3):520–5.

DeLeon VB, Smith TD, Rosenberger AL. Ontogeny of the postorbital region in Tarsiers and other primates. Anat Rec. 2016;299:1631–45.

Demetriades AM, Seitzman GD. Isolated lacrimal gland agenesis presenting as filamentary keratopathy in a child. Cornea. 2009;28(1):87–8.

Diamond MK. Homologies of the meningeal-orbital arteries of human: a reappraisal. J Anat. 1991a;178:223–41.

Diamond MK. Homologies of the stapedial artery in humans, with a reconstruction of the primitive stapedial configuration in Euprimates. Am J Phys Anthropol. 1991b;84:433–62.

Depew MJ, Simpson CA, Morasso M, Rubenstein JLR. Reassessing the *Dlx* code: the genetic regulation of branchial arch skeletal pattern and development. J Anat. 2005;207:501–61.

Donaldson IML. Ex libris RCPE: the Treatise of man (De homine) by René Descarte. J R Coll Physicians Edinb. 2009;39:375–6.

Duncan JEM, Dong X-P, Repetski JE, et al. The origin of conodonts and vertebrate mineralized structures. Nature. 2013;502:546–9.

Ezin M, Barembaum M, Bronner ME. Stage-dependent plasticity of the anterior neural folds to form neural crest. Differentiation. 2014;88(2–3):42–50. https://doi.org/10.1016/j.diff.2014.09.003.

Finarelli JA, Goswami A. The evolution of orbit orientation and encephalization in the Carnivora (Mammalia). J Anat. 2009;214(5):671–8. https://doi.org/10.1111/j.1469-7580.2009.01061.

Francel C, Park TS, Marsh JL, Kaufman BA. Frontal plagiocephaly secondary to synostosis of the frontozygomatic suture: case report. J Plast Recnstr Aesthet Surg. 2009;83:733–6.

Fuente del Campo A. Surgical correction of Tessier number 9 cleft. Plast Reconstr Surg. 1989;86(4):658–61.

Gal Z, Yu X, Zhu M. The evolution of the zygomatic bone from agnatha to tetrapoda. Anat Rec. 2017;300:16–29.

Gross JB, Hanken J. Review of fate-mapping studies of osteogenic cranial neural crest in vertebrates. Dev Biol. 2008;317:389–400.

Guercio JR, Martyn LJ. Congenital malformations of the eye and orbit. Otolaryngol Clin North Am. 2007 Feb;40(1):113–40.

Gupta PC, Foster J, Crowe S, Papay FA, Luciano M, Traboulsi EI. Ophthalmologic findings in patients with nonsyndromic plagiocephaly. J Craniofac Surg. 2003;14(4):529–32.

Herrera E, Mason CA. The evolution of crossed and uncrossed retinal pathways in mammals. In: Krubitzer I, Kaas J, editors. Evolution of the nervous system. New York: Elsevier; 2007. p. 307–17.

Hertel RW, Ziylan S, Kataowitz JA. Ophthalmic features and visual prognosis in the Treacher-Collins syndrome. Br J Ophthalmol. 1993;77:642–5.

Heuzé Y, Kawasaki K, Schwartz T, Schonenebeck JJ, Richtsmeier JT. Developmental and evolutionary significance of the zygomatic bone. Anat Rec. 2016;299(12):1616–30.

Hopson JA, Crompton AW. Origin of mammals. New York: Meredith Corporation; 1969. www.researchgate.net/publication/275344897

Hubbard BA, Gorskin JL, Muzaffar AR. Unilateral frontosphoidal craniosynostosis with achrondoplastia: a case report. Cleft Palate Craniofac J. 2011;48(5):631–5.

Hunt JA, Hobar PC. Common craniofacial anomalies: the facial dysostoses. Plast Reconstr Surg. 2002;110(7):1714.

Khanna PC, Thapa MM, Iyer RS, Shashank RS, Prasad SS. The many faces of craniosynostosis. Indian J Radiol Imaging. 2011;21(1):49–56.

Krastinova D, Kelly MB, Mihaylova M, Kronish JW, Gonnering RS, Dortzbach RK, Rankin JH, Reid DL, Phernetton TM. Pathophysiology of the anophthalmic socket. Part I. Analysis of orbital blood flow. Ophthalmic Plast Reconstr Surg. 1990;6(2):77–87.

Krastinova D, Kelly MB, Mihaylova M. Surgical management of the anophthalmic orbit, part 1: congenital. Plast Reconstr Surg. 2001;108(4):817–2.

Kronish JW, Gonnering RS, Dortzbach RK, Rankin JH, Reid DL, Phernetton TM, Pitts WC, Berry GJ. The pathophysiology of the anophthalmic socket. Part II. Analysis of orbital fat. Ophthalmic Plast Reconstr Surg. 1990;6(2):88–95.

Kuratani S, Noritaka Adachi N, Wada N, Oisi Y, Sugahara F. Developmental and evolutionary significance of the mandibular arch and prechordal/premandibular cranium in vertebrates: revising the heterotopy scenario of gnathostome jaw evolution. J Anat. 2013;222:41–55.

Langenberg T, Kahana A, Wszalek JA, Halloran MC. The eye organizes neural crest migration. Dev Dyn. 2008;237(6):1645–52.

Laure B, Picard A, Bonin-Goga B, et al. Tessier number 4 bilateral orbito-facial cleft: a 26-year follow-up. J Craniofac Surg. 2010;38(4):245–7. https://doi.org/10.1016/j.jcms.2009.06.008.

Leong ASY, Shaw CM. The pathology of occipital encephalocoele and a discussion of the pathogenesis. Pathology. 1979;11(2):223–34.

Liem KF, Bemis WE, Walker WF, Grande L. Functional anatomy of the vertebrates: an evolutionary perspective. 3rd ed. Belmont: Brooks-Cole; 2001.

Lin J-L. Nager syndrome: a case report. Pediatr Neonatol. 2012;53:147–50.

Littlewood AH. Congenital nasal dermoid cysts and fistulae. Plast Reconstr Surg. 1961;27:169–79.

Lucas SG, Luo Z. *Adelobasileus* from the upper triassic of west Texas: the oldest mammal. J Vertebr Paleontol. 1993;13(3):309–34.

Machado KFS, Garcia MM. Oftalmopatia tireoidea revisitada. Radiol Bras. 2009;42(4):261–6.

MacIver MA, Schmitz L, Mugan U, Murphey TD, Mobley CD. Massive increase in visual range preceded the origin of terrestrial vertebrates. Proc Natl Acad Sci U S A. 2017;114(12):E2375–84. https://doi.org/10.1073/pnas.1615563114.

Metzler P, Zemann W, Jacobsen C, Grätz KW, Obwegeser JA. Cranial vault growth patterns of plagiocephaly and trigonocephaly patients following fronto-orbital advancement: a long-term anthropometric outcome assessment. J Craniomaxillofac Surg. 2013;41(6):e98–e103. https://doi.org/10.1016/j.jcms.2012.11.035.

Mishina Y, Snider NT. Neural crest cell signaling pathways critical to cranial bone development and pathology. Exp Cell Res. 2014;325(2):138–47.

Moore MH, Lodge ML, David DJ. Basal encephalocoele: imaging and exposing the hernia. Br J Plast Surg. 1991;46:497–502.

O'Rahilly R, Müller F. The meninges in human development. J Neuropathol Exp Neurol. 1986;45(5):588–608.

Onbas O, Aliagaoglu C, Kantarci M, Atasoy M, Alper F. Absence of a sphenoid wing in neurofibromatosis type 1 disease: imaging with multi-detector computed tomography. Korean J Radiol. 2006;7:70–2.

Ortiz Monasterio F, Fuente del Campo A, Dimpoulus A. Nasal clefts. Ann Plast Surg. 1987;18:377–97.

Ozuk C, Gundogan H, Bilkay U, et al. Rare craniofacial anomaly: Tessier no 2 cleft. J Craniofac Surg. 2001;12(4):355–61.

Padget DH. The development of the cranial arteries in the human embryo. Contrib Embryol (Carnegie Institution, Washington). 1948;32:205–61.

Parker AJ. Binocular depth perception and the visual cortex. Nat Rev Neurosci. 2007;8:379–91.

Pereira FJ, Milbratz GH, Cruz AAV, Vasconcelos JJA. Opthalmic considerations in the management of Tessier cleft 5/9. Ophthal Plast Reconstr Surg. 2010;26:450–3.

Porro LB, Rayfield EJ, Clack JA. Descriptive anatomy and three-dimensional reconstruction of the skull of the early tetrapod *Acanthostega gunnari* Jarvik, 1952. PLoS One. 2015;10(3):e0118882. https://doi.org/10.1371/journal.pone.0118882.

Posnick JC, Bortoluzzi P, Armstrong DC, Drake JM. Intracranial nasal dermoid sinus cysts: computed tomographic scan findings and surgical results. Plast Reconstr Surg. 1994a;93(4):745–54.

Posnick JV, Bortoluzzi P, Armstrong DC. Nasal dermoid sinus cyst: unusual presentation, CT scan findings, and surgical results. Ann Plast Surg. 1994b;32(5):519–23.

Presley R, Steel FLD. On the homology of the alisphenoid. J Anat. 1976;121(3):441–59.

René C. Update on orbital anatomy. Eye. 2006;20:1119–29.

Resnick NI, Kawamoto HK. Rare craniofacial clefts: Tessier no. 4 clefts. Plast Reconstr Surg. 1990;85(6):845–9.

Reynolds JM, Tomkinson A, Grigg RG, Perry CF. A Le Fort I osteotomy approach to lateral sphenoid sinus encephalocoeles. J Laryngol Otol. 1998;112:779–81.

Rodríguez Vásqiez JF, Mérida Velasco JR, Jimenez Collado J. Orbital muscle of Müller: observations on human fetuses measuring 35-150 mm. Acta Anat (Basel). 1990;139(4):300–3.

Rootman J, Patel S, Berry K, Nugent R. Pathological and clinical study of Müller's muscle in Graves' ophthalmopathy. Can J Ophthalmol. 1987;22(1):32–6.

Sahinoglu N, Tuncer S, Alparsian N, Peksayer G. Isolated form of congenital bilateral lacrimal gland agenesis. Indian J Ophthalmol. 2011;59(6):522–3.

Samra F, Paliga JT, Tahiri Y, Whitakaer LA, Bartlett SP, Forbes BJ, Taylor JA. The prevalence of strabismus in unilateral coronal synostosis. Childs Nerv Syst. 2015;31:589–96.

Schenck TL, Koban K, Sykes J, Tagosinski S, Earlbacher K, Cotofana S. The functional anatomy of the superficial fat compartments of the face: a detailed imaging study. Plast Reconstr Surg. 2018;141(6):1351–9.

Shete P, Tupkari J, Benjamin T, Singh A. Treacher Collins syndrome. J Oral Maxillofac Pathol. 2011;15(3):348–51. https://doi.org/10.4103/0973-029X.86722. https://openi.nlm.nih.gov/detailedresult.php?img=PMC3227269_JOMFP-15-348-g007&req=4

Shoenwolf GC, Bleyl SB, Brauer PR, Frances-West PH, editors. Larsen's human embryology. 5th ed. Philadelphia: Elsevier; 2015. Fig 3-13

Stelnicky EJ, Lin WY, Lee C, et al. Long-term outcome study of bilateral mandibular distraction: a comparison of Treacher-Collins and Nager syndromes to other types of micrognathia. Plast Reconstr Surg. 2002;109:1819–25.

Surridge AK, Osorio D. Evolution and selection of trichromatic color vision in primates. Trends Ecol Evol. 2003;18:198–205.

Suzuki DG, Fukomoto Y, Yoshimura M, Yamazaki Y, Kosaka J, Kuratani S. Comparative morphology ad development of extraocular muscles in the lamprey and gnathostomes reveal the ancestral state and developmental patterns of the vertebrate head. Zool Lett. 2016;2:10–24.

Talsania SD, Robson CD, Mantagos S. Unilateral congenital lacrimal gland agenesis with contralateral lacrimal gland hypoplasia. J Pediatr Ophthalmol Strabismus. 2015;52:E52–4.

Tan AP, Mankad K. Apert syndrome: magnetic resonance imaging (MRI) of associated intracranial anomalies. Childs Nerv Syst. 2018;34:205–16.

Tucker MS, Sapp N, Collin R. Orbital expansion of the congenitally anophthalmic socket. Br J Ophthalmol. 1995;79(7):667–71.

van der Meulen JC. Metopic synostosis. Childs Nerv Syst. 2012;28:1359–67.

Walls GL. The vertebrate eye and its adaptive radiation. New York: Hafner; 1942. https://doi.org/10.5962/bhl.title.7369.

Williams AE, Mason CA, Herrera E. The optic chiasm as a midline choice point. Curr Opin Neurobiol. 2004;14:51–60.

Zhang Z, Song Y, Zhao X, Zhang X, Fermin C, Chen Y. Rescue of cleft palate in *Msx*-deficient mice by Bmp4 reveals a network of BMP and Shh signaling in the regulation of mammalian palatogenesis. Development. 2002;129:4135–40.

Pathologic Anatomy of the Hard Palate

14

Michael H. Carstens

Part 1. Building Blocks of the Palate: The Lego® Model

The objectives of this section are threefold. We shall begin with a discussion of how the bone and soft tissue structures are assembled, based upon the developmental field model. Because we have covered the anatomy and origins of mesoderm and neural crest (NC) elsewhere, these will be summarized telegraphically but with sufficient detail that you won't get lost. Our next goal is to consider how this normal process is altered when a disruption of the neurovascular pedicle to an individual field results in a deficiency state such that the affected field is unable to fuse with its partner fields. Attention will also be given to the effect that such a deficiency state has on the subsequent development of the partner fields. We conclude with a discussion of the mechanism: how neuroangiosomes can fail and what is potential role the stem cells of the growth cone have the establishment of functional relationships.

The anatomic structures of the head and neck are assembled from tissue units known as *developmental fields*, each of which has a distinct neurovascular pedicle providing sensory and/or autonomic control and blood supply. Fields are often composite structures containing mesenchymal elements such as cartilage, bone, fascia, and muscle. They may have an associated epithelium such as skin or mucosa. Adjacent fields interact. Muscles with a primary attachment to bone or cartilage within one field may have a secondary attachment site in an adjacent field.

Fields develop in a strict spatiotemporal sequence. Congenital conditions that reduce the size or content of a field will affect subsequent growth. In the Pierre Robin sequence, the relative decrease in volume of the mandibular ramus leads to a posterior position of the chin and subsequent relationships of the infrahyoid musculature. The reduction of the frontal process of the premaxilla seen in the typical orofacial cleft causes a distortion of the nasal fossa, malposition of the internal nasal valve, and respiratory dysfunction (Fig. 14.1).

The anatomic defects seen in clefts of the hard and soft palate present as a spectrum involving several fields. Many cases involve deficiency states of the piriform fossa and/or premaxilla and soft tissues of the lip and nose. In other, rarer

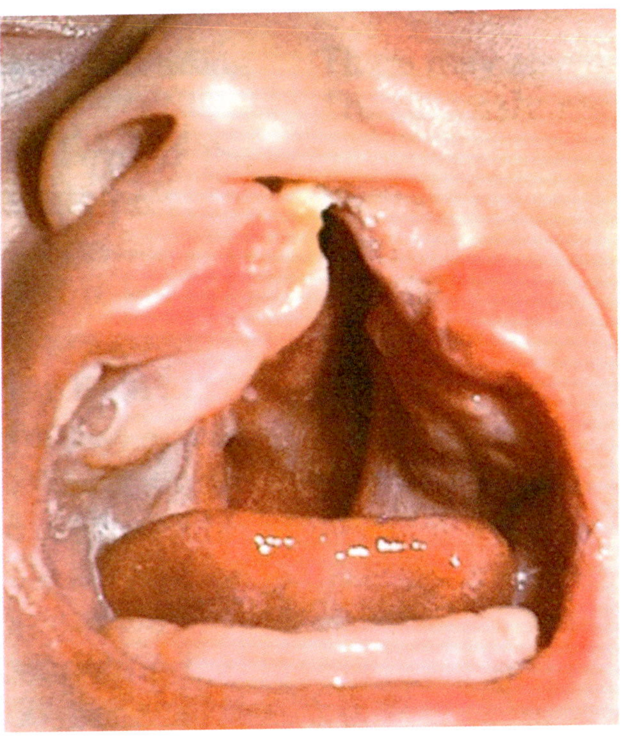

Fig. 14.1 Unilateral cleft lip and palate, both primary and secondary. Note premaxilla with missing frontal process, field and lateral incisor field exceeding critical contact distance (CCD). Premaxilla also displaced and distorted, twisted away from the midline exacerbating the CCD. Ipsilateral vomer deficiency is coupled with deficiency of left palatine fields causing forward displacement of soft palate. [Courtesy of Michael Carstens, MD]

M. H. Carstens (✉)
Wake Forest Institute of Regenerative Medicine, Wake Forest University, Winston-Salem, NC, USA
e-mail: mcarsten@wakehealth.edu

M. H. Carstens (ed.), *The Embryologic Basis of Craniofacial Structure*, https://doi.org/10.1007/978-3-031-15636-6_14

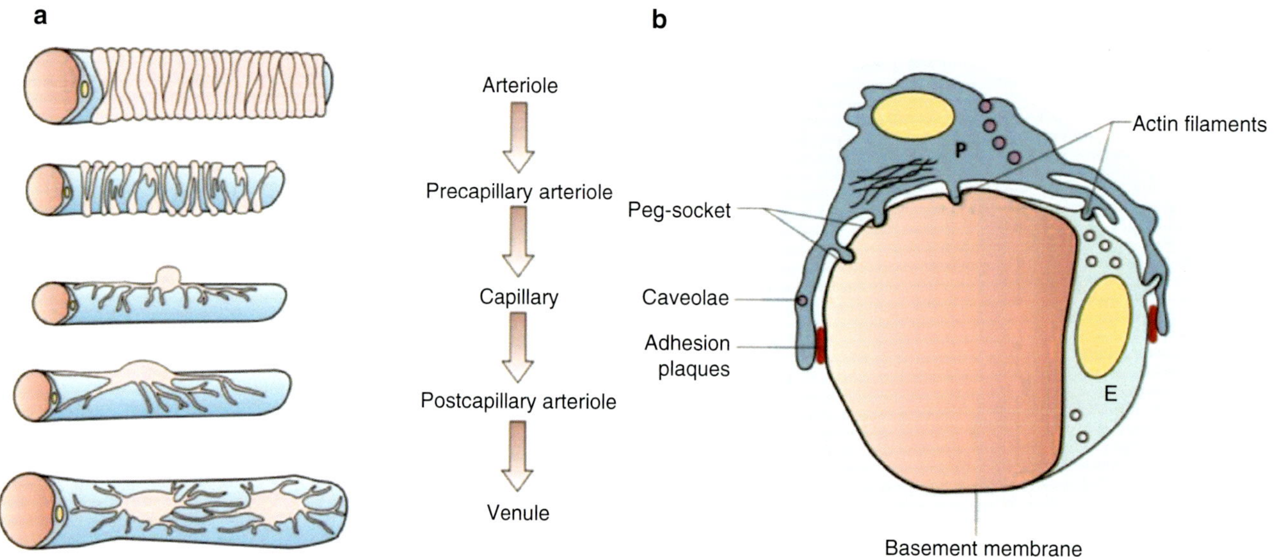

Fig. 14.2 (**a**) Pericytes are ubiquitous throughout the body. They surround all vessels, especially capillaries, providing control of diameter and permeability. Pericytes have contractile fibers. They are interconnected, including between adjacent vessels. Pericytes may have a connection with neural crest, can detach under conditions of inflammation, are the source of white fat, and also give rise to all mesenchymal stem cells of the body. (**b**) Demonstrated are the multiple physiologic functions of pericytes. [Reprinted from Carstens MH. Mechanisms of cleft palate: developmental field analysis. In: Bennun RD (ed). Cleft lip and Palate Management: A Comprehensive Atlas. 2015. with permission from John Wiley & Sons]

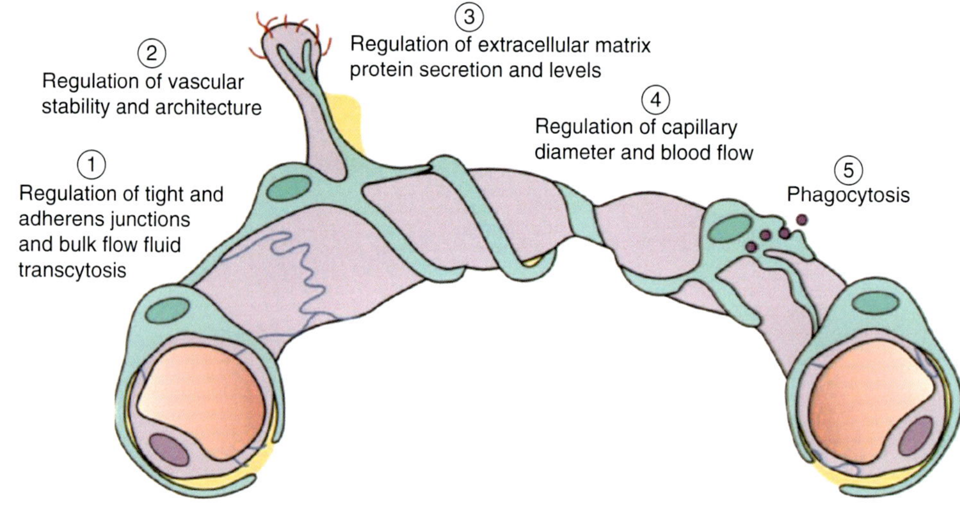

Fig. 14.3 Physiologic functions of pericytes. Of greatest importance is their secretory capability. They are vasoinductive and neuroinductive. [Reprinted from Carstens MH. Mechanisms of cleft palate: developmental field analysis. In: Bennun RD (ed). Cleft lip and Palate Management: A Comprehensive Atlas. 2015. with permission from John Wiley & Sons]

conditions, such as the Tessier 3 cleft, a cleft palate defect coincides with defects in seemingly unrelated anatomic zones, such as the inferior turbinate and medial maxillary wall. For this reason, it is necessary to have a comprehensive picture of the neurovascular anatomy of the oronasopharynx.

Cleft zones, previously discussed, are all supplied by arterial axes running in parallel with the various sensory branches of V1 and V2. Development of the pedicles is a reciprocal process. Neuronal growth cones secrete vascular endothelial growth factor (VEGF) while the arterial growth cone secretes nerve growth factor (NGF). Like all cranial nerves, the trigeminal complex is constructed from neural crest, whereas the histologic composition of the arteries consists of a tubular conduit of endothelial cells made from paraxial mesoderm (PAM) embraced by *pericytes*. These latter cells are contractile and control capillary permeability (Figs. 14.2, 14.3, 14.4 and 14.5).

- Pericytes are ubiquitous throughout the human body.
- They characteristically occupy a perivascular niche.
- Pericytes are the precursor for mesenchymal stem cells and the putative source for white adipose tissue.
- Continuity exists among pericytes; they communicate.
- Pericytes are contractile; they are innervated by the SANS and thus related to neural crest cells.
- Pericytes are biologically related to neural crest cells.
- Pericytes populations are therefore neuromeric.

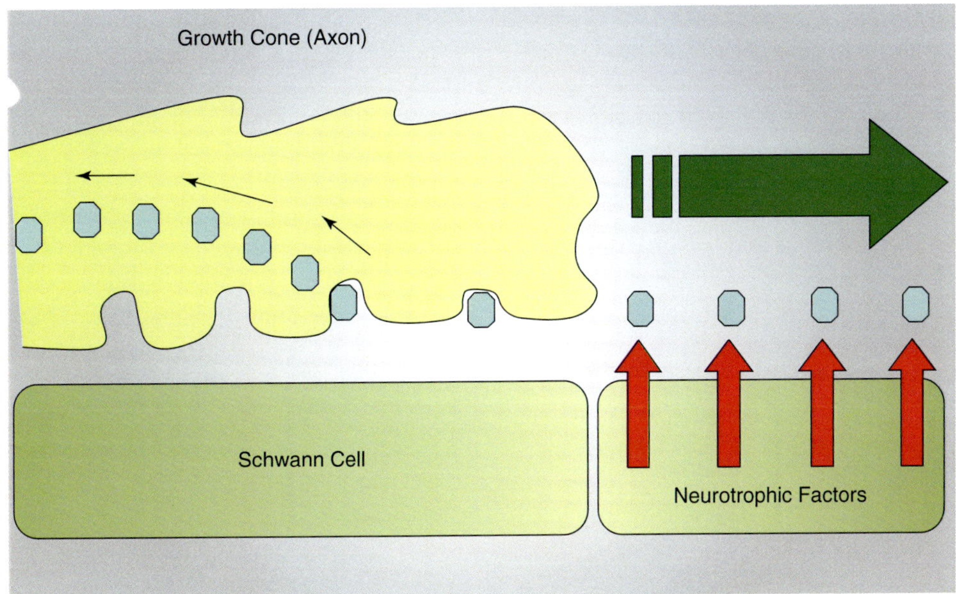

Fig. 14.4 Neuroangiosome failure 1. Schwann cells pick up nutrients and growth factors from the environment and transmit them back into the axon. NGF (nerve growth factor) produced by the pericytes promotes growth of the neural axis in register with the vascular axis. [Reprinted from May F et al. Nerve replacement strategies for cavernous nerves. *Europ Urol* 2005; 48(3):372–378. With permission from Elsevier]

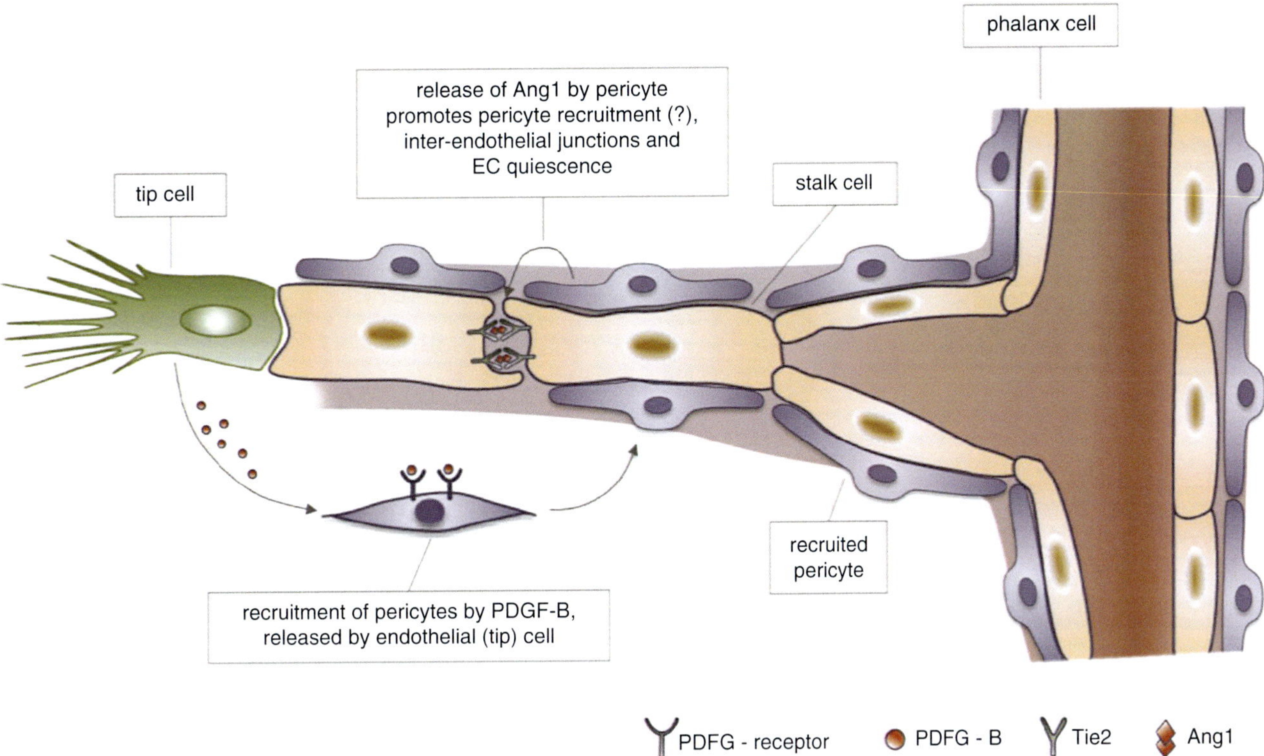

Fig. 14.5 Neuroangiosome failure 2. Vascular endothelial growth factor (VEGF) produced by the nerve cone causes outgrowth of endothelial cells from a vascular axis. These endothelial cells form the core of the new vessel or they add on to an existing vessel to elongate it. But stabilization of the new vessel by pericytes is now required. Tip cells produce PDGF-beta that recruits pericytes to come alongside the endothelial axis and stabilize it. [Reprinted from Quaegebeur A, Lange C, Carmeliet P. The neurovascular link in health and disease: molecular mechanisms and therapeutic implications. *Neuron* 2011; 71(3):406–424. With permission from Elsevier]

- Pericytes develop vascular structures within the context of the pharyngeal arch or frontonasal process in which they are located; they are therefore organized anatomically by local signals, for example, the *Dlx* system.
- Paracrine factors from pericytes at the tip of the growth cone are essential for the survival of the vascular pedicle.
- Pericyte dysfunction at the growth cone will lead to failure of the angiosome to progress.
- Angiosome failure occurs in the context of a defective pericyte population or its response to faulty signals from the local environment.

Thus, we come to a very simple and powerful idea: dysfunction of a vascular growth cone will result in either a *reduction of mesenchymal volume* in the target field or even *outright loss* of the field itself. In the first case, the physical effect of the small field is to constrain subsequent growth of surrounding fields. If a frank tissue defect exists (i.e., a cleft), adjacent fields actually collapse into the site.

In the following section, we shall consider the tissue composition of developmental fields, how they are arranged in the intermediate state as pharyngeal arches, and how, with growth-driven folding of the embryo, these fields become physically repositioned and interactive. The reader will note here terminology that may be unfamiliar: it harkens back to those embryology lectures that we endured ... an endless list of structures that morphed into a final result via mechanisms that were unknown. The molecular revolution transformed the science into developmental biology with a tight connection to genetics. Previously, these two scientific fields had coexisted in virtual isolation from each other. Recommended references are: Benton, Carlson; Gilbert, Kardong, and Standring.

The Manufacture of Pharyngeal Arch Mesenchyme

The embryonic period lasts 8 weeks and is divided into 23 anatomic stages [1] (Fig. 14.6). In the first three stages, the embryo is a rapidly dividing ball of cells. Stages 4 and 5 are all about survival as the embryo implants itself into the uterine wall and begins the process by which blood supply will come from the mother. The stage 4 embryo secretes fluid into its center, becoming a hollow blastocyst with a single layer of cells, the epiblast, becoming segregated to one side of the ball. Thus, there is an inner cell mass (the future organism) and enveloping wall (the trophoblast) that will eventually form the extraembryonic structures, such as the placenta The tightly bound cells of the epiblast then become transiently "loose," allowing some of the epiblast cells to drop down below their previous plane, coalesce, and form a new second layer, the *hypoblast*. By the end of stage 5, the hypoblast has proliferated and formed a lining layer around the inner wall of the trophoblast. The hypoblast now secretes a new layer, extraembryonic mesoderm (EEM), interposed between it and the trophoblast. This geometry allows the EEM to surround the entire embryo and move into the zone of the future placenta. Since blood vessels are formed exclusively from mesoderm, the EEM becomes the source for the entire extraembryonic blood supply (Fig. 14.7).

At stages 6 and 7, intraembryonic tissues are transformed into three layers (ectoderm, mesoderm, and endoderm) via a process called *gastrulation*: "the single most important event in the life of every organism." There are excellent videos of gastrulation available on YouTube. Note that at the completion of gastrulation, the hypoblast is pushed out of the way; it has no role in the formation of the organism per se. In point of fact, the epiblast contributes first to endoderm, then to mesoderm. When the gastrulation process is complete, the

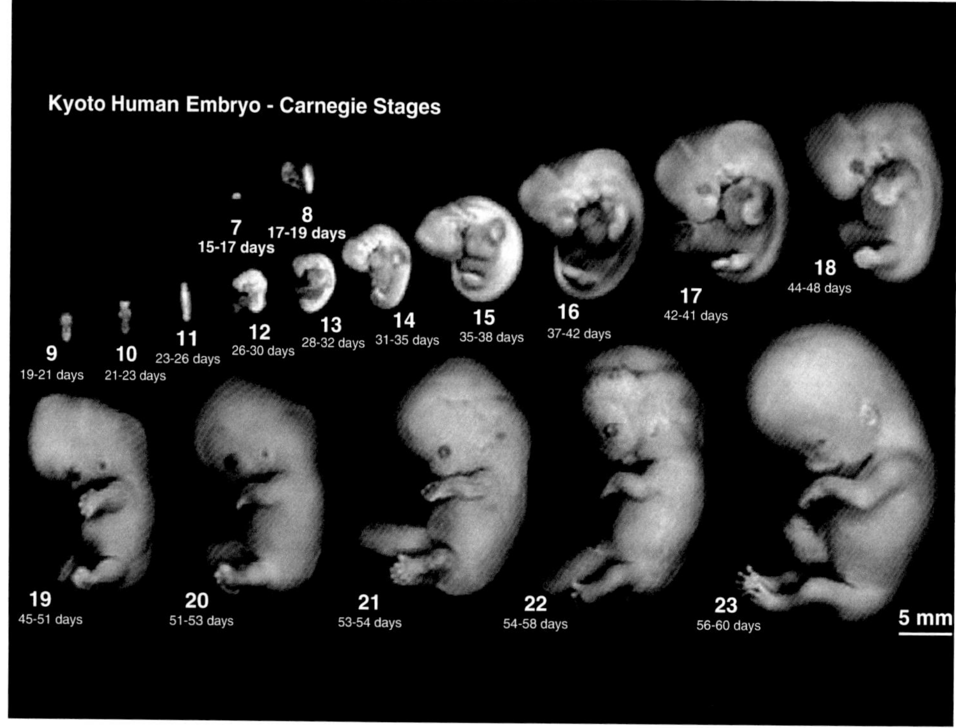

Fig. 14.6 Carnegie staging system. Embryonic period consists of 23 anatomic stages as defined by the Carnegie Institution of Washington, DC. These are well visualized at www.embryology. [Reprinted from The Kyoto Collection, Kyoto University Graduate School of Medicine. Courtesy of Prof. Kohei Shiota and Shigehito Yamada]

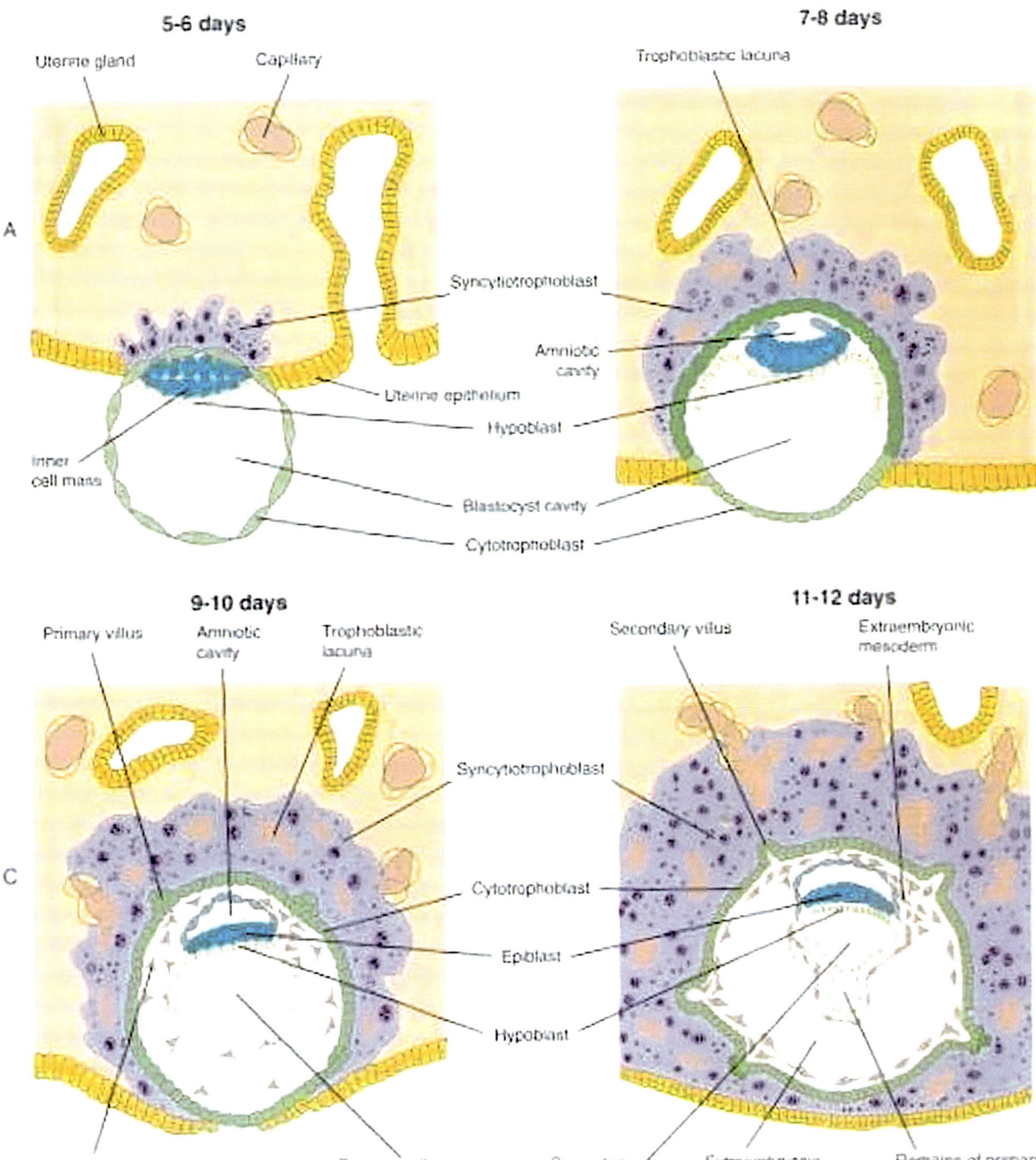

Fig. 14.7 Carnegie state 5: establishment of blood supply. In 5–6 days, hypoplast (yellow) forms second layer of the bilaminar embryo. Cytotrophoblast, CTB (green), forms the walls of blastocele. Syncytiotrophoblast, STB (purple), invades the uterine lining. In 7–8 days, the hypoblast spreads to line the walls of the blastocele. In 9–10 days, hypoblast differentiates into extraembryonic mesoderm, EEM (red), which quickly surrounds the entire conceptus. Yolk sac forms. Primary villi of CTB invade STB but they do not have a mesodermal core. In 11–12 days, yolk sac is ensheathed with EEM, which differentiates into blood islands with hemangioblasts and angioblasts. [Reprinted from Carlson B. Human Embryology and Developmental Biology, 6th Edition. St. Louis, MO: Elsevier; 2019. With permission from Elsevier]

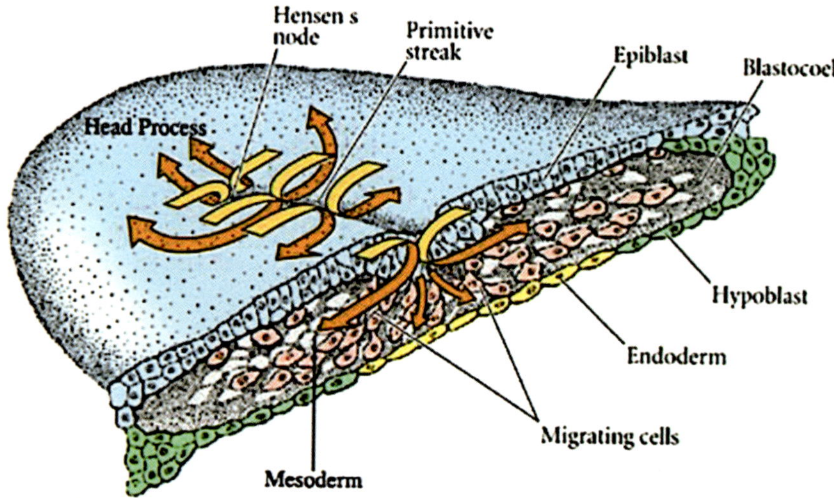

Fig. 14.8 Gastrulation is bidirectional. From initiation (t0) Hensen's node produces from its dorsal lip a column of prechordal mesoderm (PCM), here labeled "head process." PCM will underlie the forebrain. Surround. The following sequence ensues: at t1 Hensen's node sends from its ventral lip, a narrow column of notochord which exactly matches the forebrain. It binds with underlying extraembryonic endoderm. Notochord programs neuromeres of the forebrain, with help from gene products of PCM. Also at t1, gastrulation begins at r0 and proceed caudally. As this happens the node retreats backwards, leaving a continuous midline deposition of notochord in its wake, much as a slug leaves behind a slime trail. [Reprinted from Carstens MH. Mechanisms of cleft palate: developmental field analysis. In: Bennun RD (ed). Cleft lip and Palate Management: A Comprehensive Atlas. 2015. With permission from John Wiley & Sons]

cells remaining behind on the surface are known as the ectoderm proper (Fig. 14.8).

The concept of three germ layers is outmoded and inapplicable to understanding craniofacial development. For simplicity, let's leave the epithelial germ layers (ecto- and endoderm) behind and concentrate on mesoderm. This layer outside the head and neck is responsible for all striated muscles, bone, cartilage, brown fat (white fat is more complex), fascia, and the nonneural internal organs. Furthermore, as mesoderm fans out over the surface of the embryo, its identity becomes determined by the interplay of gene products expressed either from the midline (i.e., the neural tube) such as sonic hedgehog (SHH) and wingless (WNT) or from the peripheral epithelial surfaces of future skin (ectoderm) and mucosa (endoderm) such as BMP4.

Mesoderm Originates by Two Distinct Mechanisms

From the rostral lip of primitive node arise prechordal mesoderm (PCM) and notochord

- *Prechordal mesoderm* (PCM) is the first product of gastrulation. PCM is extruded like toothpaste from the rostral lip of the primitive node; it subsequently passing forward beneath the developing prosencephalon.
- *Notochord* is produced from the same site. It also passes forward in the midline directly beneath the base of the forebrain.
  - The primitive node then retreats from r0, laying down more notochord in its wake, like a slug leaving behind a slime trail.

From the remainder of the primitive node and streak arise three zones* of mesoderm

- *Paraxial mesoderm* (PAM) lies next to the neural tube. It becomes segmented into *somites*, each one of which is developmentally related to its neuromere.
  - Derivatives: axial skeleton, striated muscles, and the major axial arteries (aorta, carotids, etc.).
- *Intermediate mesoderm* (IM) is also neuromerically organized.
  - Derivative: genito-urinary system.
- *Lateral plate mesoderm* (LPM) is not overtly segmented but its zones are in neuromeric register with their respective somites and with the CNS as well.
  - Derivatives: appendicular skeleton, cardiovascular system, smooth muscle, and the viscera.
    - Somatic lateral plate (LPM_S) interacts with the ectoderm.
    - Visceral lateral plate (LPM_V) interacts with the endoderm.

Neural Crest Arises from Three Distinct Sites of the CNS

Left out from this equation is the fourth germ layer, the *neural crest* (NC). The vast majority of all facial soft tissues in the

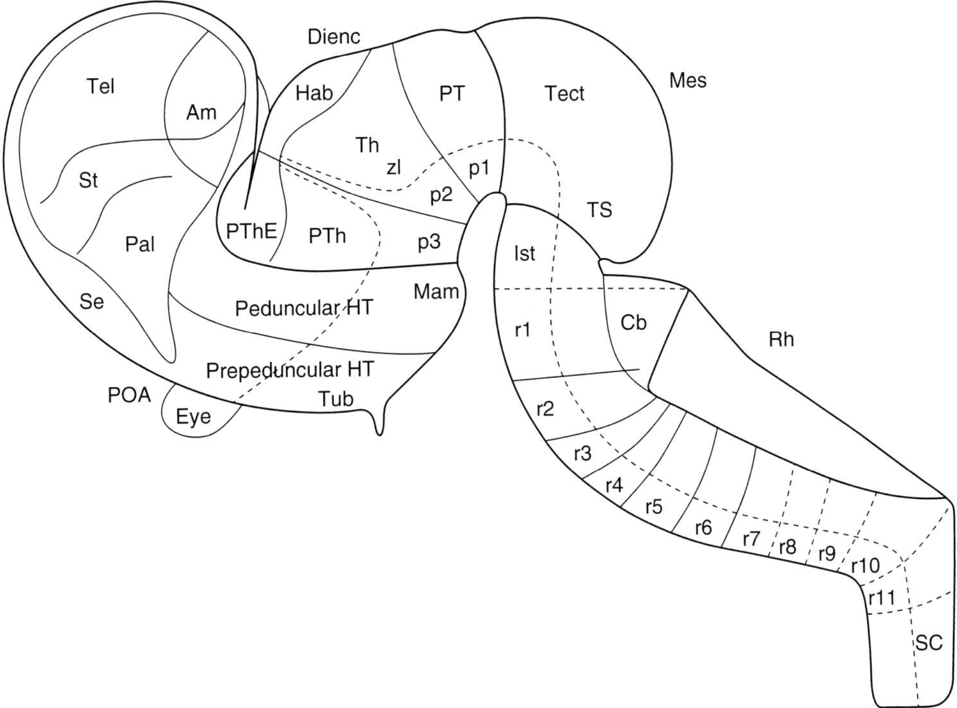

Fig. 14.9 Prosomeric model 2003. The forebrain lies to the left. Note axial bending at cephalic flexure. The longitudinal alar–basal boundary is present throughout the lateral wall of the neural tube, symbolizing all longitudinal components (floor and roof plates not represented); a singularity known as zona limitans (zl) is a transversal spike-like deviation of the general alar–basal boundary. The secondary prosencephalon (Sec.Pros.) is the rostralmost and most complex unit, consisting of telencephalon (Tel), eye, and hypothalamus (HT: divided in two parts). Septum (Se), striatum (St), pallidum (Pal), preoptic area (POA), and amygdala (Am) regions are identified within the telencephalon; the pallium lies under the label Tel. Tuberal (Tub) and mammillary (Mam) subregions of the hypothalamus are marked. The caudal forebrain or diencephalon consists of three prosomeres (p1–p3), whose alar regions include the pretectum (PT), the thalamus and habenula (Th–Hab), and the prethalamus and prethalamic eminence (PTh, PThE); a specific tegmental domain corresponds to each of them (under p1–p3 labels). A simplified view of the large mesencephalic alar plate (Mes) divides it into superior colliculus or tectum (Tect) and inferior colliculus or torus semicircularis (TS); "colliculi" are mammalian terms. The hindbrain or rhombencephalon (Rh) contains 12 neuromeric units, from the isthmus (Ist) and rhombomere 1 (rl) down to rhombomere 11 (r11), which limits with the spinal cord (SC). Note the cerebellum (Cb) forms mainly across isthmus and r1. [Reprinted from Puelles L, Rubenstein JLR. Forebrain gene expression domains and the evolving prosomeric Model. Trends in Neurosciences 2003;26(9):469–476. With permission from Elsevier]

face and all the craniofacial membranous bones arise from neural crest. These cells substitute for mesoderm in the head. They also form the ensheathing Schwann cells of the peripheral nervous system and the entire autonomic nervous system. Neural crest cells are organized according to the neuromeric system (see Chap. 5) [2] (Figs. 14.9, 14.10 and 14.11).

Forebrain Neural Crest

Prosencephalon subdivides into diencephalon and secondary prosencephalon

- Diencephalon has three prosomeres, p1–p3. Its neural folds produce PNC, which provides the dermis for frontonasal skin
- Secondary prosencephalon has two hypothalamic prosomeres, hp1 and hp2
  - hp1 contains caudal hypothalamus and the evaginated telencephalon
  - hp2 contains rostral hypothalamus and the non-evaginated telencephalon
  - *Note*: The neural folds above secondary prosencephalon are *sterile* for neural crest due to blocking signals arising from the notochord below

Midbrain Neural Crest

Mesencephalon is very large

- Mesencephalon has two mesomeres. We shall treat them as one entity
  - m1 contains all oculomotor nuclei
  - m2, isthmic mesomere, is produced from m1 in response to the r0 isthmus

Hindbrain Neural Crest

Neural crest behavior is zone specific. Recall that the original Hox series begins at r3.

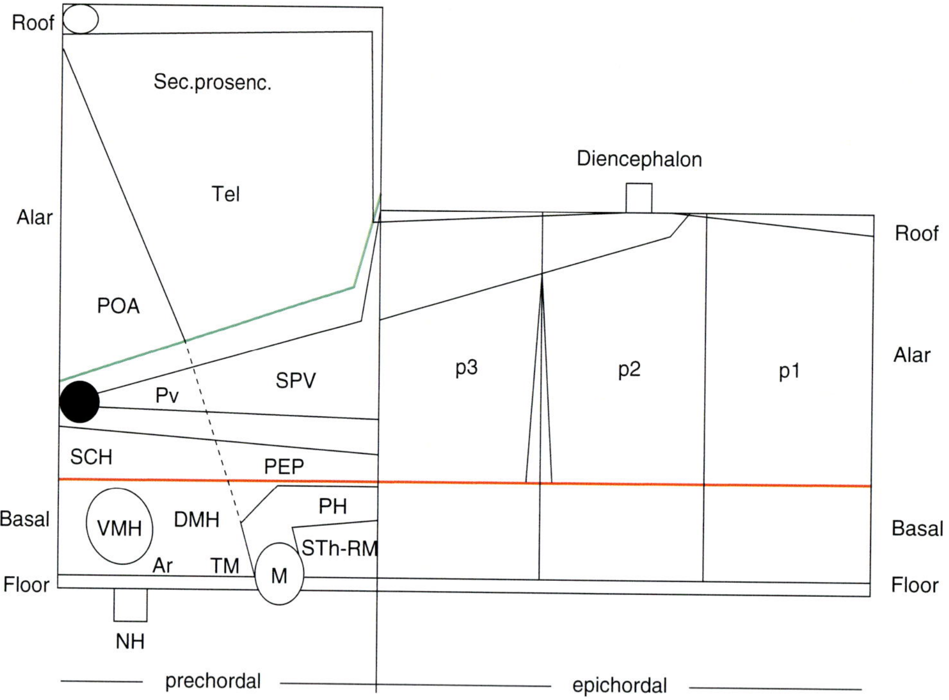

Fig. 14.10 Prosomere 2003 map. The primary anteroposterior (AP) divisions, including epichordal diencephalon and prechordal secondary prosencephalon, are indicated. Primary dorsoventral (DV) zones are marked at the caudal end (right margin). The alar–basal boundary appears as large dashes superposed on a longitudinal thin line. *Note* the roof longitudinal zone ends beyond the choroid plexus (ch, gray shading) and commissural septal areas at the anterior commissure (ac). Telencephalic subdivisions agree with gene patterns and the avian fate map of Cobos, 2001. The pallial–subpallial boundary is drawn as a red line. The septal (se), amygdaloid (amyg), and stria terminalis (ST) regions are histogenetic complexes (small dashes). They get more complex in adults. The optoeminential domain under the telencephalic stalk shows tentative internal subdivisions and has at its caudal end spike-like dorsal expansions into the amygdala. The hypothalamus appears secondarily divided into caudal (CHy) and rostral RHy) marked as medium-large dashes. *Note* the peduncular tract (ped) and fornix tract (fx) are indicated by yellow arrows. Ped connects posterior hypothalamus (PH) with ventral pallium (VP). Fornix connects basal stria terminalis (Bst) to mammillary body (M). Retromammillary area is repositioned where the subthalamic nucleus arises and is redefined as the posterior hypothalamus. Two AP subdivisions of the prethalamus are indicated [i.e., with higher Dlx5 signal caudally and higher Pax6 signal rostrally, forming rostral and caudal parts of the zona incerta (RZI, CZI)]. Rostral prethalamus has the reticular nucleus (Rt) forming dorsal to the rostral zona incerta, whereas the ventral geniculate nucleus (VG) lies in caudal prethalamus, dorsal to the caudal zona incerta. The p3 tegmentum corresponds to the classic Forel fields (FF). Subdivisions shown in the thalamus are limited to its main three histogenetic tiers (dorsal, intermediate, and ventral) and the lateral geniculate (LG) and medial geniculate (MG) primordia. Core and shell domains of the zona limitans intrathalamica are not indicated. The pretectum shows three AP subdivisions, described as precommissural (**p**), juxtacommissural (**j**), and commissural (**c**) domains. Tegmental structures of p1 and p2 include iterated parts of the substantia nigra (SN) and the ventral tegmental area (VTA), a dopaminergic complex that clearly extends beyond the diencephalon–mesencephalon boundary. *ac* anterior commissure, *AEP* anterior entopeduncular area, *AHA* anterior hypothalamus, anterior area, *AHC* anterior hypothalamus, central part, *AHP* anterior hypothalamus, posterior area, *Amyg* amygdala, *Ar* arcuate nucleus, *av.* anteroventral area of thalamus, *Bst* bed nucleus of stria terminalis, *c* commissural pretectum, *ch* choroidal tela, *CHy* caudal hypothalamus, *CZI* caudal zona incerta, *D* nucleus of Darkschewitsch, *d* dorsal tier of thalamus, *DMH* dorsomedial hypothalamic nucleus, *DP* dorsal pallium, *Em* eminentia thalami, *FF* forel fields, *fx* fornix tract, *Hb* habenula (epithalamus), *i* intermediate tier of thalamus, *IC* interstitial nucleus of Cajal, *j* juxtacommissural pretectum, *LG* lateral geniculate nucleus, *LP* lateral pallium, *M* mammillary complex, *MG* medial geniculate nucleus, *MP* medial pallium, *NH* neurohypophysis, *OB* olfactory bulb, *p* precommissural pretectum, *p1–p3* prosomeres 1–3, *Pal* pallidum, *pc* posterior commissure, *ped* telencephalic peduncle, *PEP* posterior entopeduncular area, *PH* posterior hypothalamus, *POA* preoptic area, *PT* pretectum, *PTh* prethalamus (previously known as ventral thalamus), *Pv* anterior periventricular nucleus, *RHy* rostral hypothalamus, *RI* rostral interstitial nucleus, *RM* retromammillary area, *Rt* reticular nucleus, *RZI* rostral zona incerta, *SCH* suprachiasmatic area, *Se* septum, *sm* stria medullaris, *SN* substantia nigra, *SP* secondary prosencephalon, *SPV* supraopto-paraventricular area, *ST* striatum, *STh* subthalamic nucleus, *TH* thalamus (previously known as dorsal thalamus), *TM* tuberomammillary area, *v* ventral tier of thalamus, *VG* ventral geniculate nucleus, *VMH* ventromedial hypothalamic nucleus, *VP* ventral pallium, *VTA* ventral tegmental area, *zli* zona limitans intrathalamica. [Reprinted from Puelles L, Rubenstein JLR. Forebrain gene expression domains and the evolving prosomeric model. Trends in Neurosciences 2003;26(9):469–476. With permission from Elsevier]

- Zone r0–r1 neural crest does not migrate into pharyngeal arches. From a functional standpoint
  - The cerebellum arises from r0 and r1. These two rhombomeres behave as intermediaries between midbrain and the pharyngeal arch system. They act in midbrain fashion to supply tissues to the orbit and the upper one-third of the face
- Zone r2–r11 supplies the five pharyngeal arches (two neuromeres/arch)

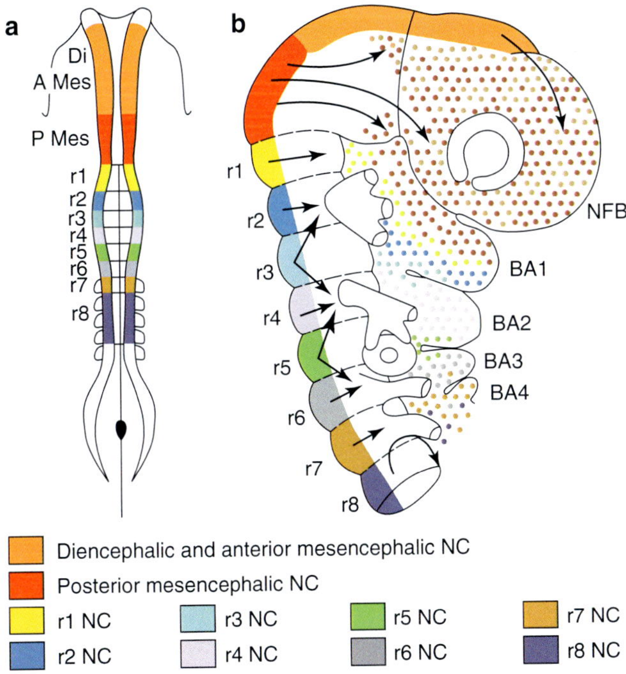

Fig. 14.11 Neural crest. During formation of the neural tube, neural folds running along the entire central nervous system produce neural crest cells. Craniofacial neural folds have four populations. (1) Anterior forebrain nonneural *crest* makes fronto-naso-orbital epidermis. (2) Posterior forebrain (diencephalon) neural crest makes fronto-naso-orbital dermis. (3) Midbrain neural crest makes fronto-naso-orbital and sphenoid mesenchymal structures: bone, cartilage, and dura. (4) Hindbrain neural crest populates the pharyngeal arches. Craniofacial mesoderm from the somitomeres and somites makes only striated muscle. Note that fronto-naso-orbital epidermis is a nonneural crest product of the anterior neural folds over telencephalon. [Reproduced with the permission of UPV/EHU Press from Creuzet S, Couly G, Le Douarin NM. Patterning the neural crest derivatives during development of the vertebrate head: insights from avian studies. J Anat 2005; 207:447–459]

Pharyngeal Arch Formation: The Basics

During stages 7 and 8, *neurulation* takes place. First, a flat neural plate is formed; this then rolls up like a cigar to form the neural tube. Neural crest cells develop at the interface between the outlying ectoderm and the neural plate. These rapidly multiply and are distributed over the entire organism, immediately subjacent to the epithelia (ectoderm and endoderm) and throughout the mesoderm. Neural crest cells are critical for craniofacial development (Figs. 14.12, 14.13, 14.14 and 14.15).

Stages 9–14 are characterized by the formation of five (not six) pharyngeal arches (one per stage). At each stage, a new arch forms and previously formed arches undergo morphogenetic changes. At no time are all five arches distinctly visible. Each arch receives neural crest cells from two rhombomeres with which it is in genetic register. Each arch is divided longitudinally by a genetic axis; the cranial zone is populated by even-numbered neural crest and the caudal zone by odd-numbered neural crest. Within the arches, the neural crest responds to positional genes such as the distal-less (Dlx) system to organize and compartmentalize the arches into mesenchymal subsets that, when the arches are repositioned into the face, will differentiate to assume distinct fates such as bone, fascia, and blood vessels. The biology of neural crest is recounted in monographs by LeDourain, Hall, and Trainor.

Cranial to the First Arch Are Two Types of Mesenchymal Tissue with Unique Composition and Innervation

Mesencephalic neural crest (MNC) constructs the nonneural tissues of the orbit and all frontonasoethmoid structures. These arise from the midbrain proper (mesomeres, m1 and

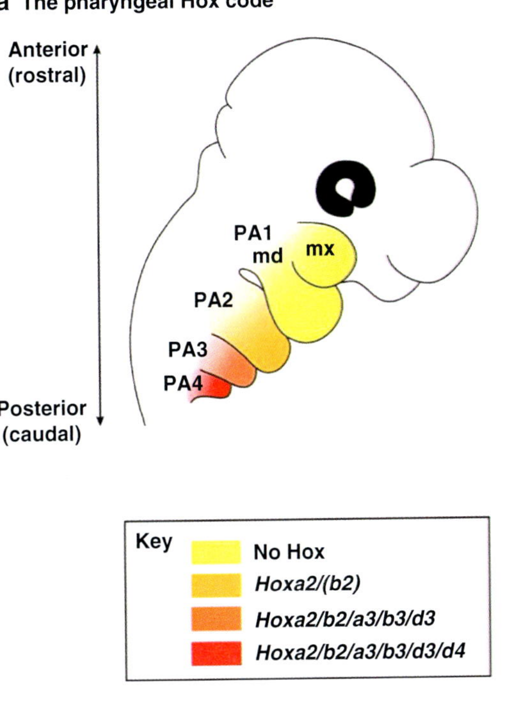

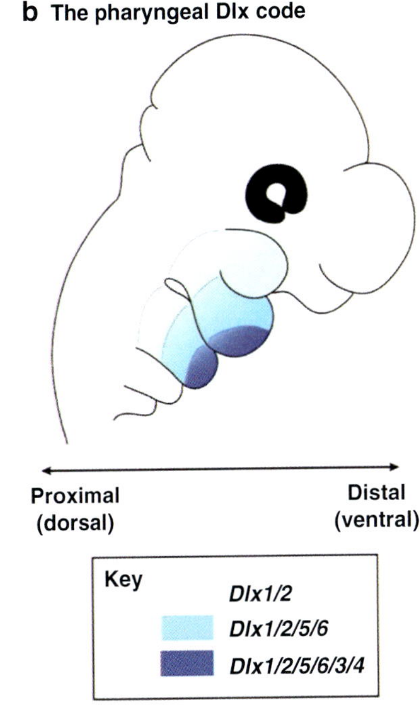

Fig. 14.12 Genetic organization of the pharyngeal arches. Homeotic coding distinguishes each neuromere with two rhombomeres per pharyngeal arch. The Dlx code provides a uniform mapping system *within* each arch. [Reprinted from Minoux M, Rijli JM. Molecular mechanisms of cranial neural crest migration. Development 2010; 137: 2605–2621. With permission from Company of Biologists]

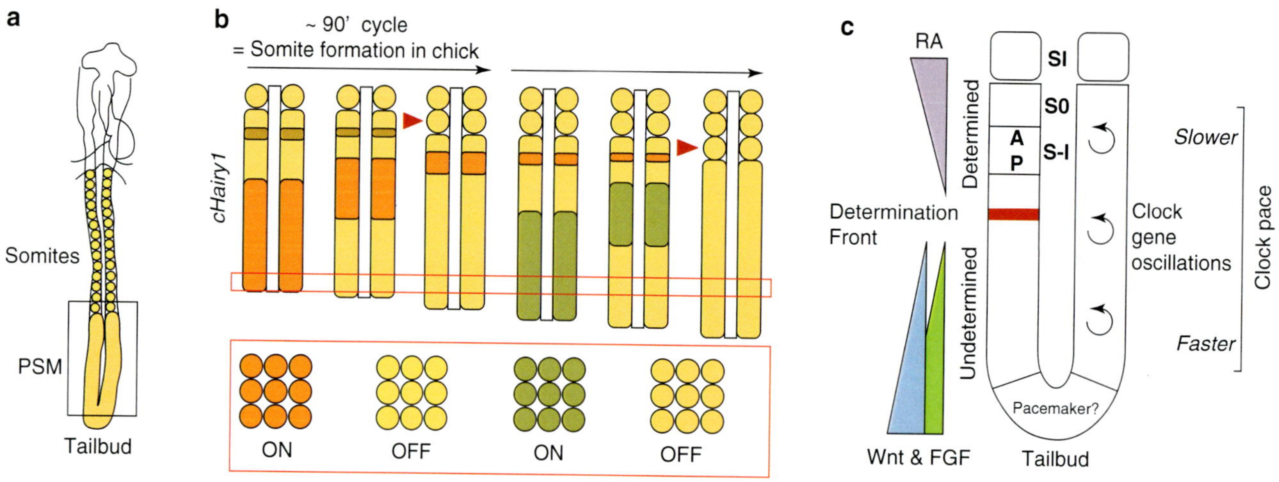

Fig. 14.13 Zones of mesoderm and the function of the vertebrate segmentation clock oscillator. The fate of the paraxial mesoderm with respect to head segmentation. As shown on the left, only two pulses of cyclic gene expression are detected for the whole mesodermal territory anterior to the first somite. The prechordal mesoderm shares with the paraxial mesoderm the ability to give rise to skeletal muscle derivatives. Moreover, the prechordal mesoderm precursors experience the first chevron of cyclic genes in the streak and they subsequently maintain this expression. We propose that the first pulse of cyclic gene expression (W1) marks the production of the prechordal territory, while the second one (W2) marks the production of the whole paraxial head mesoderm. **Note 1**. Segmentation only applies to paraxial mesoderm. Intermediate mesoderm demonstrates overt neuromeric segmentation in the form of *nephrotomes*. Lateral plate mesoderm is neuromerically organized; just like PAM and IM it is in register with the CNS but there is no overt segmentation. **Note 2**. A backward extension of head mesoderm exists which in humans reaches level c6. This mesoderm can been falsely called LPM but it is *branchiomeric* and represents muscles formerly assigned to branchial arches 6–8. (**a**) Schematic showing the dorsal view of a 2-day-old chicken embryo, and the position of somites and the PSM that flank the axial neural tube. As somites bud off the anterior end of the PSM, new cells are recruited into the posterior PSM from the progenitor cells in the tail bud. (**b**) The PSM tissue is magnified in b to illustrate the evidence for an oscillator underlying vertebrate segmentation. Periodic waves of transcriptional expression of the *cHairy1* gene (successive waves shown in different colors) across the PSM share the same periodicity as somite formation, 90 min in chick. The red box is magnified at the bottom of this figure to illustrate what this process means at the level of individual PSM cells. During each oscillation, individual cells within the PSM turn on and off the gene. This dynamic expression at the level of single cells, by virtue of being synchronized across the PSM, results in apparent "waves" of gene expression that "move" across the PSM (top part of panel). The cells themselves suffer very little anterior movement at all. However, as somites bud off, the rostral PSM and new cells enter the caudal PSM; individual cells within the PSM become progressively more anteriorly displaced in the PSM (see the red box in the top part of the panel). (**c**) A schematic diagram integrating the domains of various signaling activities in the PSM: the wavefront of determination on the left hand side, and the clock on the right. The system of opposing gradients of Fgf (green), Wnt (blue), and retinoic acid (RA, purple) signaling in the PSM positions the determination front (red) along the PSM. The determination front marks the position where the next prospective boundary will form, thereby defining somite size. As these cells mature, the anterior (A) and posterior (P) somite compartments become specified. In the most rostral PSM, the definitive morphological boundary of the next prospective somite forms. As indicated on the right side of the diagram, within this same PSM tissue, waves of Notch, Fgf, and Wnt cyclic gene expression controlled by the segmentation clock oscillator traverse the PSM periodically (black spiral symbol). The oscillations slow down as they reach the rostral PSM. Wnt activity appears to act as (part of) the pacemaker mechanism to regulate the periodicity of cyclic gene oscillations. Prospective somites in the PSM are numbered with somite S0 being the forming somite and the somites next to form labeled in negative Roman numerals, S-I, and so on. Segmented somites are numbered in positive Roman numerals, with SI being the most recently formed somite. [Reprinted from Gibb S, Maroto M, Dale JK. The segmentation clock mechanism moves up a notch. Trends Cell Biol 2010; 20(10): 593–600. With permission from Creative Commons License 3.0: https://creativecommons.org/licenses/by/3.0/]

m2), from the isthmus (r0) that connects midbrain with hindbrain, and from r1. Note that, in neuroembryology, rhombomeres r0 and r1 are considered part of the hindbrain but are functionally more closely related to midbrain.

Prosencephalic neural crest (PNC) constructs the frontonasal skin. Caudal PNC (prosomeres pe–p1) makes the dermis. The epidermis comes from the neural folds of hp1 and hp2 but is nonneural ectoderm, not neural crest [3, 4].

Nota Bene for the reader: The identity of the neuromeres that produce the midbrain neural crest (MNC) is poorly appreciated. The midbrain contains two mesomeres marked by external structures: m1 has the superior colliculus associated with vision and m2 has the inferior colliculus associated with hearing. Some authors [2, 5] map the midbrain with a single mesomere. The midbrain longitudinal fasciculus residing in the midbrain coordinates eye movements; therefore, it is not surprising that the mesomeres are physiologically related to the first two rhombomeres, r0 and r1. All four neuromeres contain nuclei for the extraocular muscles. Oculomotor nuclei reside in m1 while rostral r0 (the isthmus) contains the decussation of trochlear nerve, and r1 the trochlear nucleus itself.

The anatomic pathway of V1 sensory nerves and their corresponding V1 stapedial arteries provides a map showing the migration routes of MNC into the orbit and face. The StV1 arteries constitute an "add-on" to the primitive ophthalmic artery. They supply all extraocular structures of the orbit, whereas prOA supplies the globe and optic nerve.

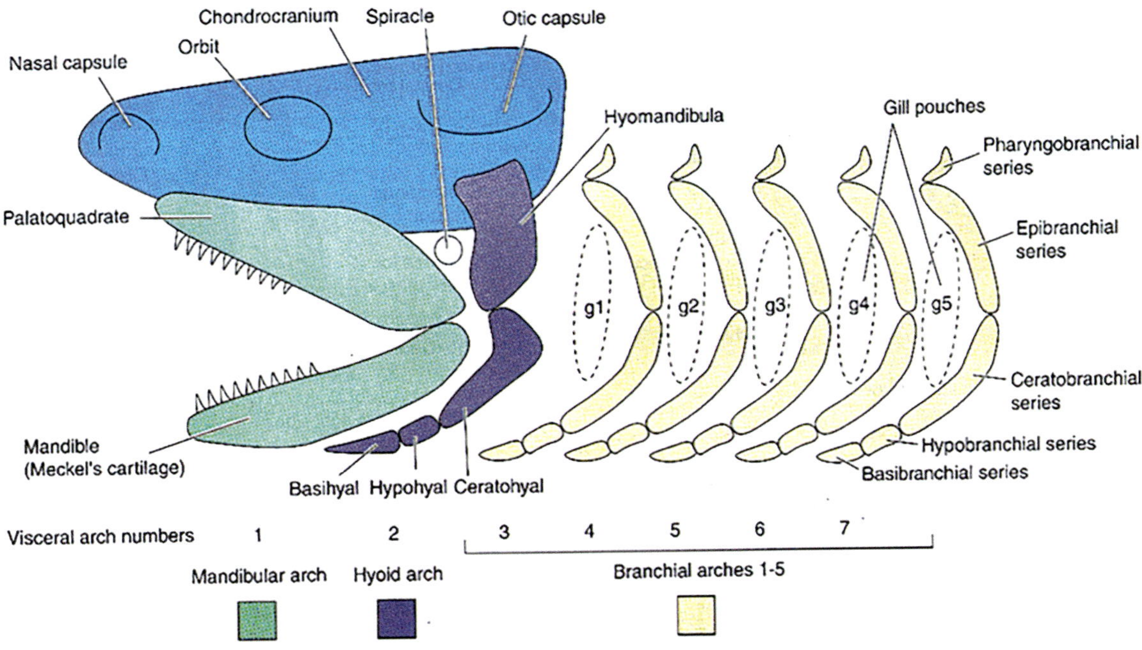

Fig. 14.14 Somitomeres and somites. The gastrulation process starts at r0 and proceeds cranial to caudal. As the cells exit, they acquire a Hox code "tattoo" according to the neuromeric level at which they exit. This applies nicely to endoderm and mesoderm. Paraxial mesoderm located just outside the neural tube is quickly organized into seven partially segmented somitomeres and 36–7 separate somites surrounded by epithelium. Segmentation of paraxial mesoderm takes place at the rate of about 4 units/day. Each somitomere/somite is innervated from the same neuromeric level and this neuroanatomy is shared with all its derivative tissues outside of paraxial mesoderm. [Reprinted from Carstens MH. Pathologic anatomy of the soft palate, part 1: Embryology, the hard tissue platform, and evolution. J Cleft Lip Palate Craniofac Anoml 2017; 4:37–64. With permission from Wolters Kluwer Medknow Publications]

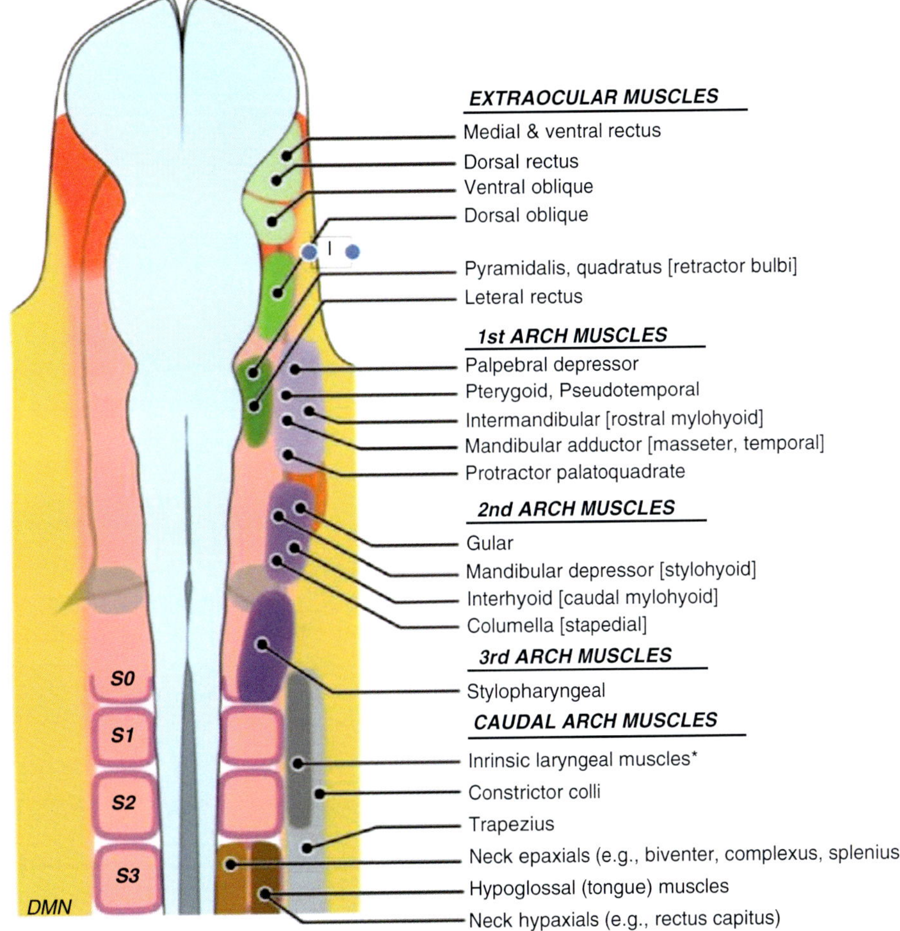

Fig. 14.15 Mesodermal map. Based on the avian model but with some muscles listed with their mammalian counterparts where necessary. [Reprinted from Ziermann J, Noden D, Diogo R. Neural crest and the patterning of vertebrate craniofacial muscles. Genesis 2018; 56(6–7):e23097. With permission from John Wiley & Sons]

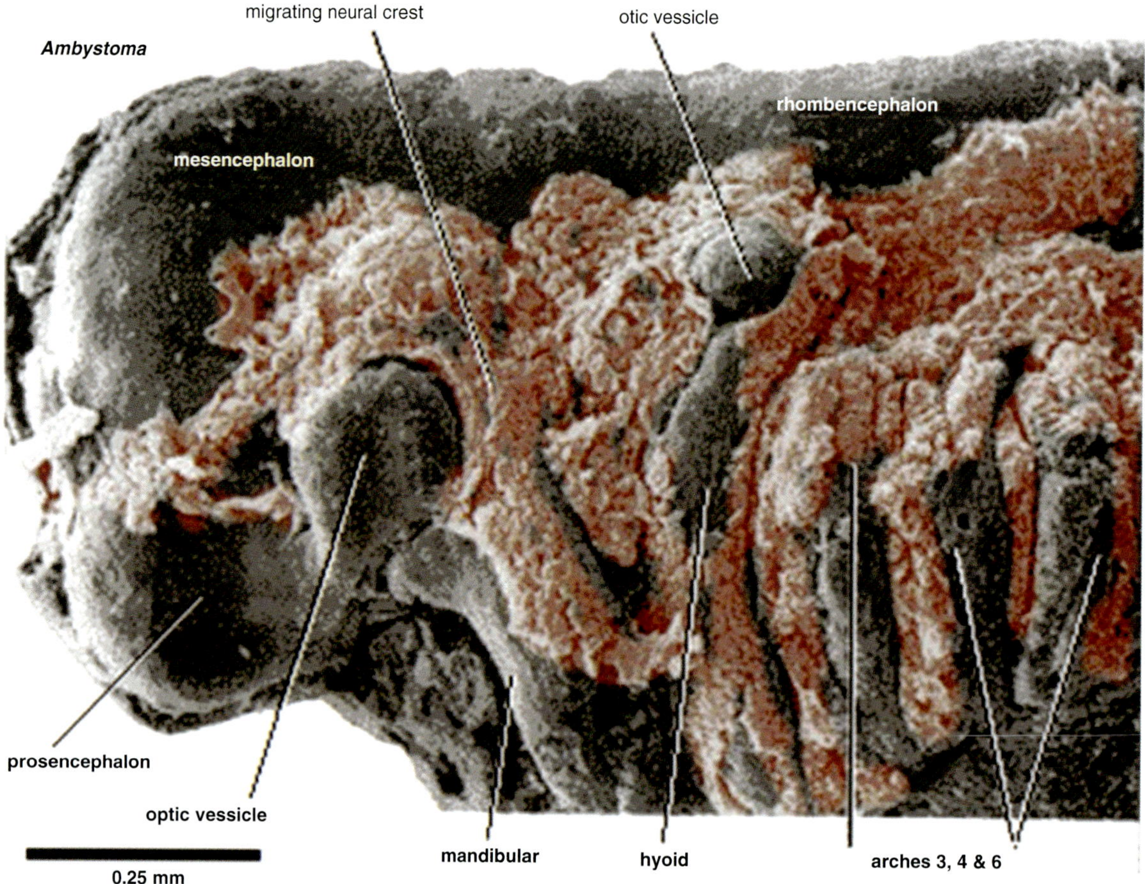

Fig. 14.16 Neural crest migration into the pharyngeal arches. Scanning electron microscopy of neural crest migrating over the surface of the brain and into the pharyngeal arches. Prototypical agnathan *Ambystoma mexicanum* has seven arches: the first two have not been converted into jaws. Nasal cannula cells are stellate-shaped and are seen arching over the optic cup and the nasal placode to reach the midline, achieving forebrain coverage. Additional nasal cannula cells are seen as the top of the fold flowing downward as future primary meninges. Neural crest provides all mesenchyme of pharyngeal arches except the muscles. As a general rule, starting with r2, each arch is in register with two rhombomeres per arch. Each arch has an axis defining cranial tissues versus caudal tissues, with the even-numbered rhombomere supplying the cranial half and the odd-numbered rhombomere supplying the caudal half. In gnathostomes, PA1 and PA2 are confluent. Lateral facial clefts are located along the shared axes of these two arches. [Reprinted from Falck P, Hanken J, Olsson L. Cranial neural crest emergence and migration in the Mexican Axolotl (Amystoma mexicanum). Zoology 2002; 105(3): 195–202. With permission from Elsevier]

- V1-innervated neural crest has a truly massive distribution. It is a major source of dura. It provides the prechordal mesoderm with specific contributions (fascia) to the intrinsic muscles of the eye. It forms the bones of the medial wall and roof of the orbit. All nonmuscular extraocular structures such as fat, fascia, conjunctiva, and sclera are derivatives of r1. The extraorbital distribution of r1 neural crest contributes to the entire fronto-orbital nasal skin envelope, itself a derivative of PNC (a subject covered in Chap. 11 on craniofacial skin and meninges). V1-innervated mucosa of the naso-oropharynx is a derivative of V1 neural crest.

Organization of Mesoderm: Getting Ready to Construct the Face

Stages 8 and 9 are also notable for the organization and segmentation of the zone of paraxial mesoderm flanking the neural tube and forebrain. Prechordal mesoderm and PAM from levels r0 to r7 are referred to as *cephalic (head) mesoderm*. It demonstrates a primitive form of quasi-segmentation into seven somitomeres (Sms), hollow balls of PAM surrounding a fluid-filled cavity. Somitomeres continue to be produced at the rate of 4 per day but at level r8, they begin an additional transformation into somites: 4 occipital, 8 cervical, 12 thoracic, 5 lumbar, 5 sacral, and 3–4 coccygeal. We have discussed this process previously in Chaps. 1, 2, and 9 (Figs. 14.16 and 14.17).

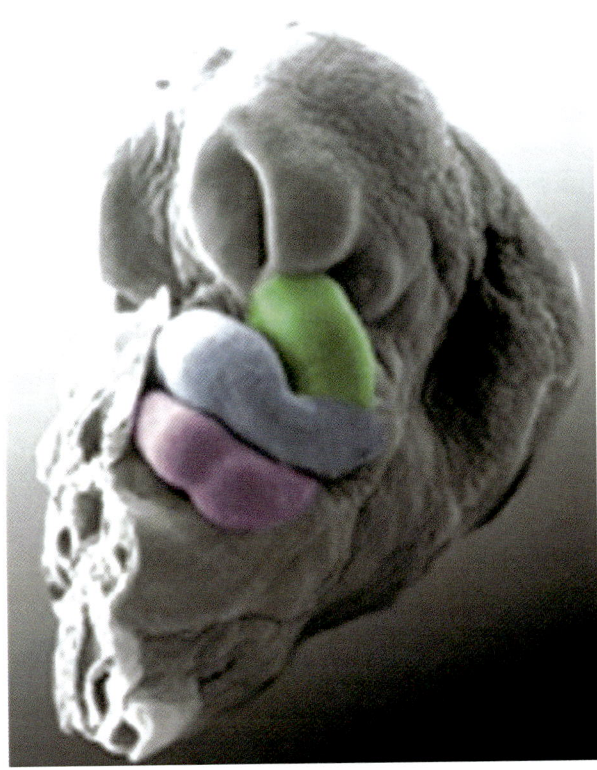

Fig. 14.17 Pharyngeal arches. Note the longitudinal fissure in the first pharyngeal arch separating the maxillary (green) and mandibular (blue) fields. Each pharyngeal arch has a similar axis which is specified by the distal-less (Dlx) genes. [Courtesy of Dr. Kathleen Sulik, University of North Carolina]

- Somites produce muscle, tendon, cartilage, fat, and dermis.
- Somitomeres (Sms) have very limited developmental potential. They produce the striated muscles for pharyngeal arches PA1–PA3. PCM is notable for having Sm1 (SR, MR, IR LPS) and Sm2 (IO) as precursors of extraocular muscles.
- Muscle mapping is based on motor nerve and neuromere (Fig. 14.18) [6, 7].

Muscle Derivatives of Somitomeres and Somites

So what we now have is a critical mass of tissues that will mix together to form a series of five intermediate structures, the *pharyngeal arches*, each of which is innervated by a specific cranial nerve (Table 14.1). These structures first develop at a stage when the embryo is still flat; at stage 9, the embryo begins a complex folding process. The first pharyngeal arch can be seen at this stage. It hangs downward like a sock filled with neural crest and PAM. The neuromuscular anatomy of the arches is the subject of Chap. 9.

Beginning with stage 9, and at each stage thereafter, a new pharyngeal arch makes its formal appearance. Each arch receives neural crest cells from two rhombomeres. The process of making the five pharyngeal arches is thus complete by stage 14. All pharyngeal arches are organized into distinct zones by a series of distal-less (Dlx) genes: proximal/distal, cranial/caudal, and medial/lateral. All arches have the same

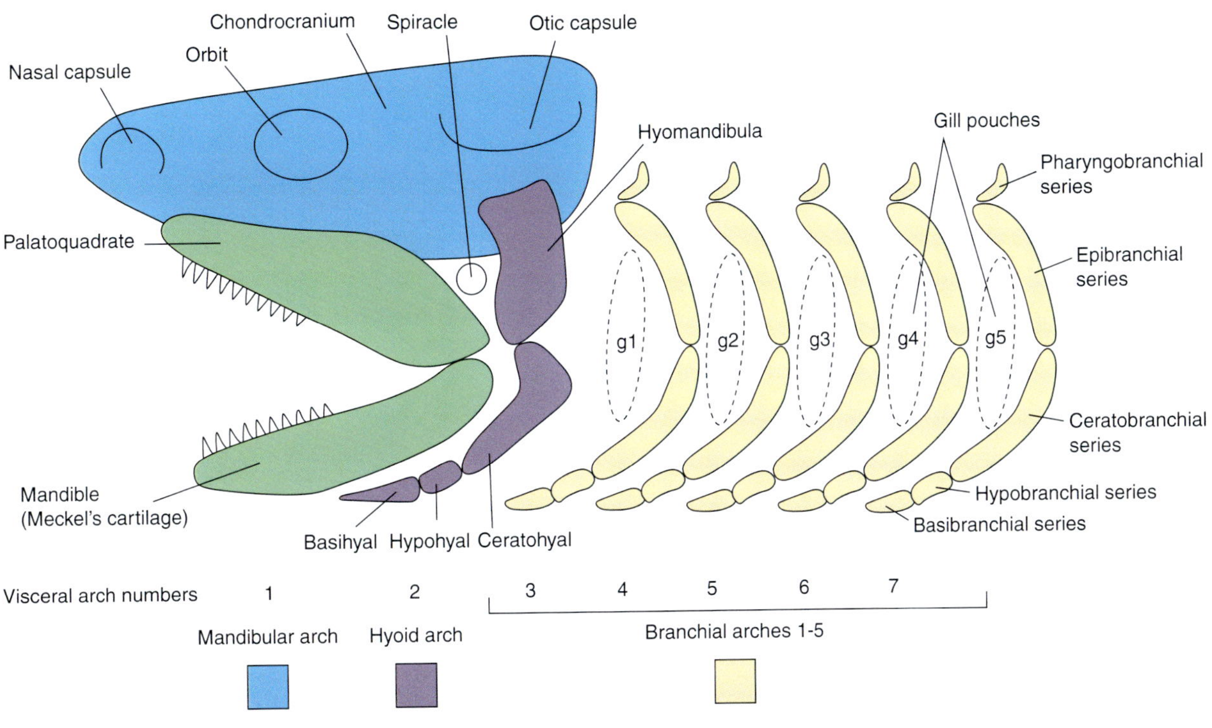

Fig. 14.18 Pharyngeal arches in evolution. Agnathic fishes had seven purely respiratory branchial arches. They had the ability to make bone. Gnathostomes modified the first arch but not the hyoid arch, thus remaining with six respiratory arches (1 + 6). Chondrichthyans (sharks) convert two arches to have five respiratory arches (2 + 5). Osteichthyes (bony fishes) lost a branchial arch, possibly by transformation to the swim bladder. They therefore have four respiratory arches (2 + 4). Tetrapod amniotes have three nonrespiratory arches (2 + 3). [Courtesy of William E. Bemis]

Table 14.1 Muscles mapping based on motor nerve and somitomere (Sm)

Somitomere/somite	Muscle
Sm1	Inferior rectus, medial rectus
Sm2	Inferior oblique, dorsal rectus Levator palpebrae superioris, Müeller's muscle
Sm3	Superior oblique
Sm4	First arch: mastication, tensors
Sm5	Lateral rectus (Sm5 positioned internal to Sm4)
Sm6	Second arch: facial muscles, ear
Sm7	Third arch: palate, superior and middle constrictors
Sm8/S1	Fourth arch: larynx, inferior constrictor
Sm9/S2	Fifth arch: larynx, tongue
Sm10/S3	Tongue, SCM, trapezius
Sm11/S4	Tongue, SCM, trapezius
Sm12/S5	Tongue, SCM trapezius, C1 muscles
Sm13/S6	Tongue, SCM, trapezius, C2 muscles
Sm14/S7	SCM, trapezius, C3 muscles

Dlx organizational pattern [8]. Thus, arch 1 has a rostral maxillary zone innervated by V2 and a caudal mandibular zone supplied by V3; the zones have different *Dlx* coding. Muscles of mastication in the first arch all arise from somitomere 4. Arch 2 fuses with arch 1; it contains the muscles for facial expression in somitomere 6. Again, following the *Dlx* code, the upper division of VII supplies facial muscles from somitomere distributed over the maxillary zone while the lower division of VII innervates muscles from somitomere 6 distributed over the mandible [9, 10] (Figs. 14.12, 14.16, 14.17 and 14.18).

Embryonic folding driven largely by explosive brain growth causes the pharyngeal arches to swing upward into the adult position. By stage 15, although the fourth and fifth arches are internalized with respect to the first three, epithelial seams are present. At stage 16, all of these fuse. Growth of the vascular channels into the arches proceeds concomitant with the penetration of the arches by cranial nerves from the brain. We will now briefly explore how the individual neurovascular pedicles associated with the nasopharynx and oropharynx are organized. This will give us an insight into the manner in which the developmental fields are assembled.

Blood Supply to the Pharyngeal Arches

Key Concept The external carotid system supplies all derivatives of all the pharyngeal arches, except the jaws and structures connecting the jaws with the skull. Table 14.2 gives the system in a nutshell, arch-by-arch.

Table 14.2 ECA = aortic arches + ventral pharyngeal artery

First arch	Lingual, max-mand, second zone (directed to nonneural crest structures)
Second arch	Facial, transverse facial, superficial temporal, post auricular, occipital
Third arch	Ascending pharyngeal
Fourth arch	Superior thyroid
Fifth arch	Superior thyroid (borrowed from AA4)

An Acknowledgement

We are indebted to D. H. Padget at the Carnegie Institution for describing the development of craniofacial vessels. Trained in medical art at Vassar and by Max Bròdel at Johns Hopkins, he never went on to medical school but became a world-class self-taught embryologist. Padget's illustrations based on staged human embryos constitute the groundwork for our understanding of vascular development. The sequential development, involution, and remodeling of the embryonic aortic arches are beautifully demonstrated with scanning electron microscopic studies by Hiruma.

Construction of the External Carotid System

Let's recount how the pharyngeal arches obtain their blood supply. What about blood supply for the arches? In the beginning, as soon as mesoderm is formed, it contains angioblasts and endothelial precursor cells. These are wildly invasive. These cells join together to form the first vesicles and then cords, which remodel as the embryonic blood vessels. This process is first seen at stages 5 and 6 in the extraembryonic mesoderm of the yolk sac and gives rise to the placental circulation. With gastrulation, intraembryonic mesoderm follows the same rules, setting up the intraembryonic circulation. Because these two zones of mesoderm are in physical continuity at the periphery of the embryo, blood supply reaches the conceptus from the mother.

At stage 8, the primitive central intraembryonic arterial system forms by spontaneous aggregation of LPM into paired dorsal aortae running on either side of the neural axis for the entire length of the embryo. Just anterior to the BPM, the LPM is known as cardiogenic mesoderm. Here, dorsal aortae form a U-shaped loop. Fusion of the most anterior part of the loop produces paired primitive heart tubes. The distal tubes will form the atria and the proximal tubes the ventricles and the truncus arteriosus, leading backward to the embryo proper. Brain growth forces this anterior zone to fold downward in a 180° arc such that the heart tubes are tucked underneath the future face. The atria now point backward and the truncus arteriosus points forward.

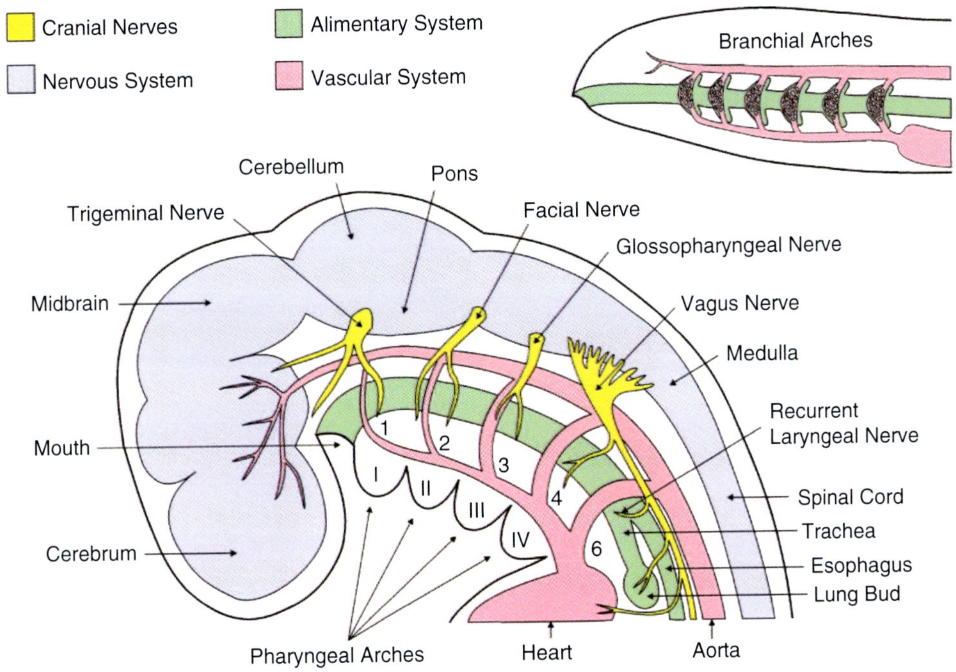

Fig. 14.19 Beginnings of the aortic arches. Stage 7 vascular system established (dorsal aortae). Stage 8 primitive heart anterior to brain. Stage 9 head folding brings heart below the pharynx; connecting segments of dorsal aortae = first aortic arches. Subsequently one arch per stage. Sixth aortic arch artery appears at stage 15; it is exclusively dedicated to the pulmonary circuit. Pharyngeal arches develop around the arterial axis and are recognizable one stage later in time. [Reprinted from Carstens MH. Pathologic anatomy of the soft palate, part 1: Embryology, the hard tissue platform, and evolution. J Cleft Lip Palate Craniofac Anoml 2017; 4:37–64. With permission from Wolters Kluwer Medknow Publications]

The Pharyngeal Arch Phase

It is now time for the pharyngeal arches to make their appearance. We start with the precursor structures, the aortic arch arteries. As they develop, the surrounding mesenchyme clumps around them to form a pharyngeal arch. The walls of these arteries are formed from LPM and neural crest (pericytes). Under the microscope, the aortic arches are observed running up though the center (eccentric position) of each pharyngeal arch [11, 12] (Fig. 14.19).

The first pair of aortic arch arteries (AA1) is unique. Unlike the rest, it does not form de novo. They are part of the original vascular circuit that connects the truncus arteriosus of the heart to the dorsal aortae. With head folding, they are stretched and bent downward. At stage 10, the first pharyngeal arch is clearly seen and consists of a vascular core (AA1) ensheathed by neural crest, ectoderm, and endoderm, all in genetic register with and innervated from rhombomeres r2 and r3 (Fig. 14.20, PA1 @ stage 9).

At each successive stage, a new pair of aortic arch arteries arises, establishing a connection between a more proximal site along the primitive outflow tract and a more proximal site along dorsal aortae. Around each arterial axis, a new pharyngeal arch is organized. Within the pharyngeal arches, the aortic arch arteries constitute primitive vascular cores. These rapidly involute, each one being replaced with a plexus, the confluence of which forms the external carotid system.

In humans AA5 quickly involutes, thus forcing the blood supply for all fifth pharyngeal arch derivatives to become dependent on AA4 and its subsequent iteration as the inferior thyroid artery. The pharyngeal arch period is thus complete by stages 13 and 14. AA6 is dedicated to the pulmonary circulation and has no relationship with the pharyngeal arch system (Fig. 14.21, SEM of aortic arches; Table 14.3).

The physical positioning of the five pharyngeal arches can be likened to that of a Japanese fan. Each arch is tucked inside its predecessor: PA1 and PA2 are melded together, inside of which are PA3 > PA4 > PA5 (Fig. 14.16). The spatial interrelationships of the pharyngeal arches are reflected in their sensory representation and muscle distribution. The skin of the first arch covers over all the remaining arches such that it abuts

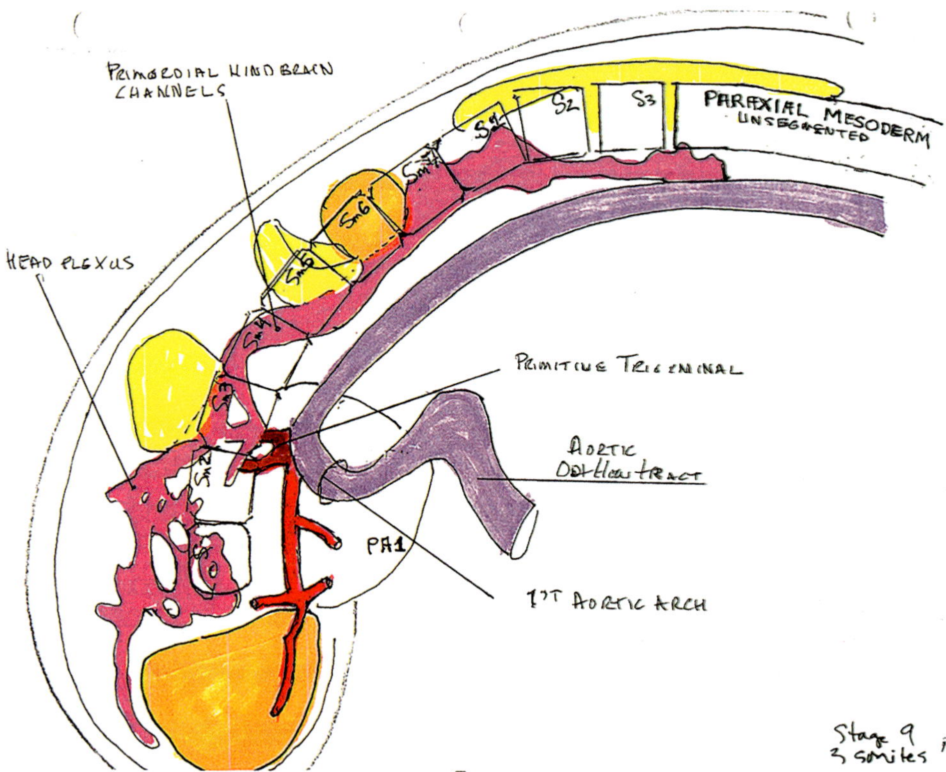

Fig. 14.20 Stage 9 human embryo. Heart is supplied via atria from vitelline circulation (not depicted). When head folds and heart is tucked underneath pharynx the dorsal aortae leading to the head are bent downwards, forming in a passive manner the first aortic arch. Neural crest and lateral plate mesoderm immediately surround AA1 to form the first pharyngeal arch, into which paraxial mesoderm myoblasts migrate. [Courtesy of Michael Carstens, MD]

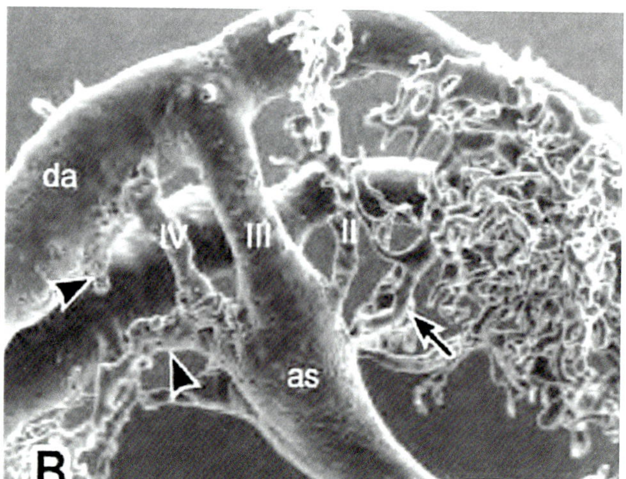

Fig. 14.21 SEM of aortic arches. Scanning electron microscopy shows embryo in reversed position (head to right). At this stage, the pharyngeal arch plexus is very dense. Just caudal to the fourth aortic arch artery, arrowheads indicate the stumps of the involuting fifth aortic arch artery. AA4 will subsequently supply both pharyngeal arches 4 and 5. There is no sixth pharyngeal arch in mammals (in fish, yes). The artery assigned to this mesenchyme, AA6, will be incorporated into the pulmonary circulation. [Reprinted from Hiruma T. Formation of the pharyngeal arch arteries in the chick embryo. Observations of corrosion casts by scanning electron microscopy. *Anat Embryol* 1995; 191(5): 415–424. Reproduced with permission from Springer Nature]

Table 14.3 Pharyngeal arch development by stages

Stage 9	First arch is fully formed with a functional arterial system
Stage 10	Second arch forms
Stage 11	Third arch forms, PA2 has fused with PA1
Stage 12	Fourth arch makes its appearance
Stage 13	Fifth arch, AA5 fails
Stage 14	Sixth (pulmonary) aortic arch artery
Stage 15	PA3 had moved internally to PA2 but remains visible on the surface. Fourth and fifth arches internalized. Epithelial seams are visible separating the arches
Stage 16	*Ventral pharyngeal artery* arrives at V3 where it connects with extracranial stapedial. Proximal to that site, external carotid now develops. Distal to the anastomosis, IMMA, the final component of the ECA arises. Facial development now proceeds rapidly through: (1) *Nonstapedial IMMA*: muscles of mastication/facial expression; and (2) Stapedial StV2/StV3 IMMA: to jaws and supporting structures *External fusion of the arches* takes place, perhaps as a result of the increased perfusion
Stage 17	Definitive external carotid system, AA3 and AA4 still attached to dorsal aorta and thus remain interconnected with each other
Stage 18	Dorsal aortae breaks taking AA4 with it as the beginning of the aortic outflow tract. AA3 remains the sole survivor of the aortic arch system. The ventral base of AA3–AA4 serves as the stem of common carotid

Note: Pharyngeal arch sequence is very compact

against the cervical skin and scalp. Thus, the sensory representation of the midface and lower face is provided by V2 and V3 (the upper face is innervated by V1). The oronasal pharyngeal representation of r2 first arch mucosa makes a boundary with r1 frontonasal mucosa in the nasopharynx and with r6 third arch mucosa in the oropharynx. The first arch mucosa from r3 covers the caudal oropharynx and abuts backward against r7 third arch mucosa of the fauces. *There is no mucosal representation of the second arch in the oropharynx.*

The second arch muscles are sandwiched between the superficial and deep layers of the first arch tissues; they are positioned superficial to the tissues of the third arch.

- The second arch produces *two functional groups* of muscles: a limited number of muscles for mastication and an extensive catalog of muscles for facial expression.
  - The first group shares with the first arch muscles a common deep investing fascia.
  - The second group is enclosed in a separate superficial investing fascia (SIF) that encompasses the entire head, face, and anterior neck (where it is known as the epicranius or galea). This plane of the SIF is of great surgical importance. Paul Tessier recognized it as the superficial musculoaponeurotic system (SMAS).

The migration of "new" facial expression muscles of the second arch is instructive. Anteriorly, they are entirely incorporated within the "envelope" of the first arch because the second arch skin and mucosa disappear. Posteriorly, the second arch muscles lie deep to the cervical skin and external to the skull. The boundary between PA2 and PA3 is seen between (1) buccinator and superior constrictor and (2) stylohyoid and posterior belly of digastric.

Blood Supply to *Non*pharyngeal Arch Tissues of the Face

At Carnegie stage 17, an entirely new system of stapedial arteries develops, the initial stem of which runs upward from internal carotid through the temporal bone. The timing of stapedial development precisely matches the emergence of the cranial nerves, each nerve serving as the template for a respective artery. These dural arteries connect externally to the external carotid system via the trigeminal ganglion. The details of this system have been recounted previously in Chaps. 6, 12, and 13 [13] (Figs. 14.22 and 14.23, stapedial in situ; Table 14.4).

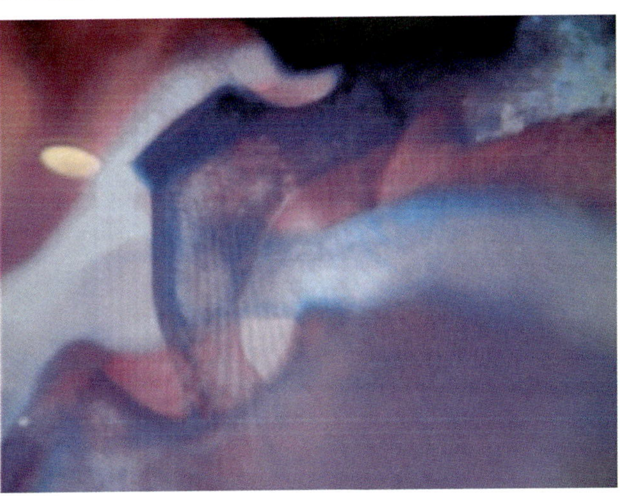

Fig. 14.22 Stapedial stem in the tympanic cavity. The stapedial system develops at stage 17 precisely when the cranial nerves emerge. The stem ascends through the tympanic cavity, goes intracranial, and follows the intracranial sensory nerves throughout the dura. A forward branch to the trigeminal ganglion picks up intracranial V1 and gains access to the orbit where it supplies all extraocular structures. From the trigeminal ganglion, a branch goes extracranial to connect to the maxillary system as the middle meningeal. From the tympanic cavity, a branch tracks extracranial following chorda tympani until it connects with the external carotid system. When the stem involutes, all the branches of stapedial survive on anastomoses with other systems, such as the ophthalmic from the internal carotid, the external carotid distal to facial artery to form the internal maxillary, and multiple connections with other branches of the external carotid system, such as occipital. [Reprinted from Huang C-H, Hu DK. Noggin heterozygous mice: an animal model for congenital conductive hearing loss in humans. *Hum Mol Genet* 2008; 17(6):844–853. Reproduced with permission of Oxford University Press]

- In the orbit, the V1-related stapedial (StV1) plugs into the ophthalmic. The latter supplies the ocular apparatus while the former supplies all the periocular structures (muscle, fascia, bone, etc.). When the parent stem of stapedial involutes, the dural arteries are supplied by the external carotid system. The ophthalmic is a hybrid; the original stem from the internal carotid is dedicated to the eye while StV1 serves the orbital structures exclusively.
- The extension of the stapedial downward through the trigeminal ganglion gives it access to the distal extent of the external carotid system, to which it attaches just beyond the take-off of the facial artery. Those branches associated with StV2 supply the maxillary–zygomatic complexes. Each of these is distributed through the pterygopalatine plexus while StV3 supplies the mandible.

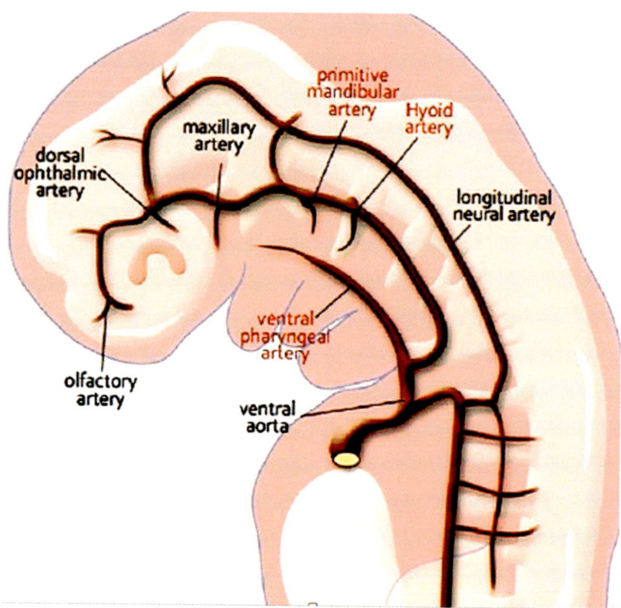

Fig. 14.23 Primitive circulation immediately prior to stapedial system, ca. stages 15 and 16. A portion previously shared between AA3 and AA4 is common carotid. AA3 is directed forward as internal carotid and gives off external carotid. AA4 is now directed backward to the dorsal aortae. AA3 and AA4 are no longer connected by dorsal aortae. Maxillary artery is the first branch of ICA; it is initially directed to the eye but later is reassigned to the hypophysis. Primitive dorsal ophthalmic is replacing maxillary. One stage later, the primitive ventral ophthalmic will develop. ICA divides into anterior and posterior. Anterior terminates as olfactory. The primitive stem of anterior cerebral is seen. Posterior division is seen giving off diencephalic stem and, at the dorsum, the mesencephalic stem. At that juncture, ICA connects with longitudinal neural artery. Above the second arch, hyoid artery can be seen directed downward. Above it is a small upward-directed stem, the stapedial. Breakdown of first and second aortic arches leaves two remnants: primitive mandibular and hyoid, respectively. Hyoid gives off stapedial, then remains as carotico-tympanic artery. [Reprinted from Tanoue S, Kayoso H, Mori H. Maxillary Artery: Functional and Imaging Anatomy for Safe and Effective Transcatheter Treatment. Radiographics 2013; 33(7): 209–229. With permission from The Radiological Society of North America (RSNA)]

Table 14.4 Stapedial development by stages

Stage 13	Hyoid stem moves into position
Stage 14	Primitive dorsal (temporal) ophthalmic distal ICA, stapedial stem
Stage 15	Primitive ventral (nasal) ophthalmic proximal ICA
Stage 16	Ventral pharyngeal artery arrives V3, maxillomandibular div as well
Stage 17	IMMA develops
Stage 18	Tympanic aa appears, middle meningeal annexes supraorbital div.
Stage 19	StV1 makes orbital aa
Stage 20	StV1 annexed by carotid
Stage 21	Involution of stapedial system

Part 2. Developmental Fields of the Palate

Bone Fields

We now come to a crucial concept. Each of the StV2 branches is responsible for supplying one or more fields within the maxillary–zygomatic complex. Thus, a reduction or knockout in any one of these branches will create a tissue-deficiency state (cleft). Each of the *Tessier cleft zones* contains specific bone and cartilagenous structures. Using clinical examination and 3D CT scan or cone beam CT, one can determine which field(s) demonstrates a reduction in size or absence, thus defining the cleft zone. Each angiosome belongs to a Tessier cleft zone. Clefts are due to knockouts, total or partial in a specific angiosome.

***StV1 angiosomes* relevant to cleft palate**

- Internal nasal, medial (**12**): upper septum
- External nasal, medial (**12**): intermediate crus alar cartilage
- Internal nasal, lateral (**11**): superior and middle turbinate, nasal bone
- External nasal, lateral (**11**): lateral crus alar cartilage

***StV2 angiosomes* relevant to cleft palate**

- Medial nasopalatine (**1/2**): lower septum, vomer, premaxilla
- Lateral nasopalatine (**3**): inferior turbinate, medial maxillary wall, nasal mucosa of the maxillary palatal shelf
- Descending palatine (**3**): palatine bone, oral mucosa of the maxillary palatal shelf
- Anterior superior alveolar, medial br. (**4**): frontal process of maxilla, maxillary incisor, canine, maxillary wall medial to foramen
- Anterior superior alveolar, lateral br. (**5**): premolars, maxillary wall lateral to foramen
- Posterior alveolar (**6**): molars, posterior maxillary wall/buttress
- Zygomatico—facial (**7**): jugal bone field (malar eminence and arch)
- Zygomatico—temporal (**8**): postorbital bone field (upper zygoma)

NB Tessier zone **9** is an extremely rare cleft, due to the intersection of multiple angiosomes. On the forehead, this is between lacrimal and Z-T. Transverse facial and anterior deep temporal provide collateral flow. Greater wing is r2 and its angiosome is StV2 anterior branch of meningeal. StV1 meningeal branch penetrates alisphenoid (as described previously) but is usually absorbed.

***StV3 angiosomes* relevant to cleft palate**

- Inferior alveolar supplies the tooth-bearing dentary bone field, which has specific fields for the dental units (incisors canines, bicuspids, molars). These mesial fields get priority. You can have an absent ramus and still have dentition. More proximally are the ramus and condylar fields. These are supplied through the marrow by inferior alveolar,
- Anterior tympanic from the first segment of IMMA contributes to the condyle.

All of the above fields contain one or more membranous bones, *all of which are derived from neural crest.* [NB: In craniofacial development, the only bones that are derived from mesoderm (lateral plate) are the basisphenoid, part of the temporal bone, and the occipital bone complex below the superior nuchal line. All remaining bones arise from neural crest.]

Muscle Fields

What about fields that are composed exclusively of muscle? This would include the muscles of the soft palate and pharynx. How do we categorize their blood supply? What could go wrong in development to produce the pathologies associated with cleft palate?

The physiology of speech depends upon muscles that are both intrinsic to the soft palate and extrinsic to it. The following concepts are essential for understanding the developmental anatomy of this integrated muscle system:

- Craniofacial muscles arise from paraxial mesoderm (PAM).
- PAM is segmentally organized into 7 somitomeres and 35 somites. The process of mesoderm segmentation takes place during stages 7 and 8 and proceeds in a cranio-caudal direction. Each segment will be supplied by a designated motor nerve.
- Craniofacial mesenchyme is predominantly neural crest, not mesoderm. From stages 9 to 14, this mesenchyme becomes itself segmented into five pharyngeal arches.
- Located in the core of each arch is an *aortic arch artery* that spans from the cardiac outflow tract located below the future pharynx to the dorsal aorta lying above the pharynx. The fifth aortic arch artery involutes and the fourth takes over the supply for the structures of both pharyngeal arches 4 and 5.
- Despite textbook dogma, there is *no* sixth pharyngeal arch. The sixth aortic arch artery becomes incorporated into the pulmonary circulation.
- The muscles that develop within the pharyngeal arches originate from paraxial mesoderm that becomes physically incorporated into the arch system. Hypoplasia or aplasia of palate musculature can occur from an intrinsic defect of the mesoderm or from a failure of the arterial axis that supplies it. Such defects can be isolated to a single muscle or can be more global as in hemi-palate.

Craniofacial mesoderm assigned to the orbit and to the first three pharyngeal arches comes from seven somitomeres (Sms). These are incompletely separated balls of mesoderm with a hollow center. The first three somitomeres produce six of the seven extraocular muscles. The remaining somitomeres are assigned as follows:

- Sm4 gives rise to the muscles of mastication in the first pharyngeal arch
- Sm5 and Sm6 produce the lateral rectus and muscles of facial expression in the second pharyngeal arch
- Sm7 contains the muscles of the soft palate and superior constrictor. In contrast to dogma, these are innervated from r6 to r7 nucleus ambiguus via CN X, the vagus, vide infra
- Sm8–Sm11 become somites S1–S4
  - Medial lamina of S1–S4 (somatic motor): tongue muscles
  - Lateral lamina of S1–S4 (branchiomotor): middle and lower pharyngeal constrictors. S3 and S4 provide the muscles of phonation to the fourth and fifth pharyngeal arches

The pharyngeal arch system is transient; so too are the aortic arch arteries that originally supplied each pharyngeal arch. These break down, reorganize, and eventually are connected to the carotid; thus is formed the external carotid artery and all its derivatives serving the pharyngeal arch structures of the face—but not the maxilla. Recall that the V2-associated stapedial artery system forms an anastomosis distal to the facial artery and to the superficial temporal artery; thus is born the system of arteries emanating from the pterygo-palatine fossa that supplies the maxilla and inferior oro-nasopharynx. These are responsible for the bone-bearing fields of the maxilla.

Let us look at the arterial axes that supply the soft palate. The *primary axis of the soft palate* is the ascending pharyngeal artery. It supplies the most primitive muscles, those that develop first, but are last to be affected. The absence of a soft palate is a survivable condition; not so the absence of pharyngeal muscles.

- *Palatal branch* of ascending pharyngeal supplies the palatoglossus.
- *Pharyngeal branch* of the ascending pharyngeal supplies the superior constrictor, middle constrictor, salpingopharyngeus, stylopharyngeus, and palatopharyngeus. These are all important for swallowing; the first two constrict the pharynx while the latter two function as elevators of the pharynx.
- The *secondary axes of the soft palate* are
  - Descending (a.k.a. greater) palatine branch
  - Ascending palatine branch of the facial artery*
  - These axes supply the more recent additions to the soft palate: levator veli palatini, tensor veli palatini, and uvulus. The first two have attachments to the tympanic tube, while uvulus relates to the posterior spine of the palatine bone.

Where do the muscles of the palate and pharynx originate? In trying to answer this question, we run headlong into a *glaring contradiction in anatomy*, one which is resolved by the neuromeric system.

- Tensor veli palatini arises from somitomere 4; it is supplied by the V3. TVP shares a common blastema with the chewing muscle, medial pterygoid.
- Levator veli palatini, palatoglossus, palatopharyngeus, uvulus, and superior constrictor all develop from somitomere 7, which is in neuromeric register with the third arch. LVP shares a common blastema with superior constrictor.
  - Glossopharyngeal nerve IX, the cranial nerve associated with the third arch, is almost exclusively sensory. It supplies only one branchiomotor muscle, stylopharyngeus.
  - The third arch structures of the soft palate are obviously cranial to those of the fourth and fifth arches, that is, the larynx, its muscles, plus the middle and inferior constrictors.
  - Yet we know the motor supply for the palatal series is X.
  - The third arch palate lies cranial to the larynx, plus the middle and superior constrictors.
  - How can the third arch muscles be supplied by a fourth arch nerve?
  - Why do anatomy texts describe the palate as being a fourth arch structure when it makes no sense at all?

The solution for this dilemma lies in recognizing the functional role of *nucleus ambiguus*. NA is a *branchiomotor nucleus*. It runs from r6 to r11; as such it supplies third, fourth, and fifth arches. It is lateral to the *somatic motor nucleus* of hypoglossal, r8–r11. Furthermore, NA is continuous into the cervical spinal cord down to level c4.

Key Fact Vagus extends the length of nucleus ambiguus (Figs. 14.24 and 14.25).

Key Fact

- Nucleus ambiguus supplies the following structures
- r6 and r7 via glossopharyngeal: stylopharyngeus,
- r6 and r7 via vagus: soft palate (except tensor) and superior constrictor,
- r8 and r9 via vagus: middle constrictor,
- r10 and r11 via vagus: laryngeal muscles via inferior constrictor,
- c1–c3 via spinal branch of accessory: sternocleidomastoid,
- c2–c4 via spinal branch of accessory: trapezius.

Neuromeric Diagnosis of Field Defects

Defects of Bone Fields

All defects of the hard palate, either primary (the premaxilla), from the incisive foramen forward, or secondary, from the incisive foramen backward, involve deficiencies in membranous bone. These are readily diagnosed by physical examination and 3D CT scanning. Such bone defects involve one or more arterial axes of the StV2 stapedial system supplying the maxilla. The nucleus of V2 resides within the second rhombomere of the hindbrain (r2). Thus, defects of the maxillary complex represent deficiencies in the population of neural crest cells arising from that segment of the neural fold in genetic register with r2. Strictly by logic, a bone defect occurs when something is intrinsically wrong with the r2 mesenchymal population or there is defective formation of a neurovascular axis supplying a portion of the r2 population. The latter mechanism is more specific; it explains isolated bone defects at later stages in development rather than a global failure of neural crest mesenchyme in the first arch.

Mandibular defects associated with cleft palate, as in the Pierre Robin sequence, can be diagnosed in a similar way. Defects involving any developmental zone of the mandible involve one or more arterial axes of the StV3 stapedial system. The nucleus of V3 resides within the third rhombomere of the hindbrain (r3). Thus, defects of the mandibular complex represent deficiencies in the population of neural crest cells arising from that segment of the neural fold in genetic register with r3. Tensor veli palatini is the sole palate muscle belonging to the first arch; it comes from somitomere Sm4

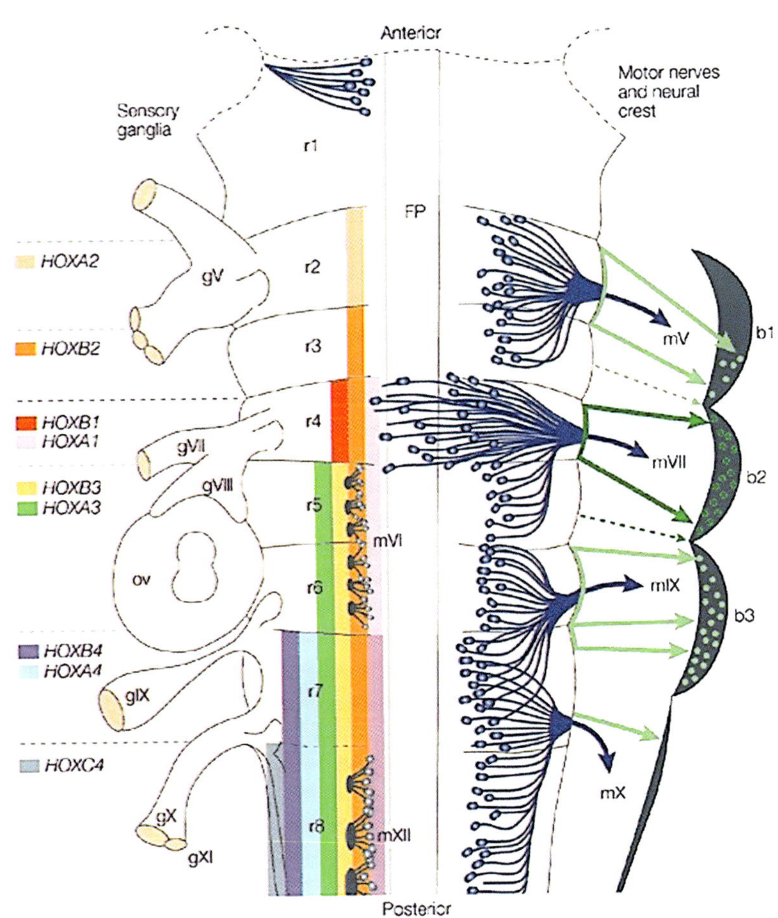

Fig. 14.24 Nucleus ambiguus. Nucleus ambiguus is somatic motor lateral (branchial). Hypoglossal is somatic motor medial to tongue from somites S1 to S4. Nucleus ambiguus has three motor zones: *anterior one-third* (r6–r7) supplies soft palate, superior constrictor. It has nuclei for both IX and X. *Middle one-third* (r8–r9) supplies superior laryngeal nerve to fourth arch and somatic motor to middle constrictor. *Posterior one-third* supplies fifth arch and inferior constrictor. Thus motor supply to soft palate is really based on rhombomeres and can travel via either IX via X. Note: sensory supply for third arch is very accurate and demonstrates the extensive distribution of IX. [Reprinted from Kieker C, Lumsden A. Compartments and their boundaries in vertebrate brain development. *Nature Reviews Neuroscience* 2005; 6:553–564. With permission from Springer Nature]

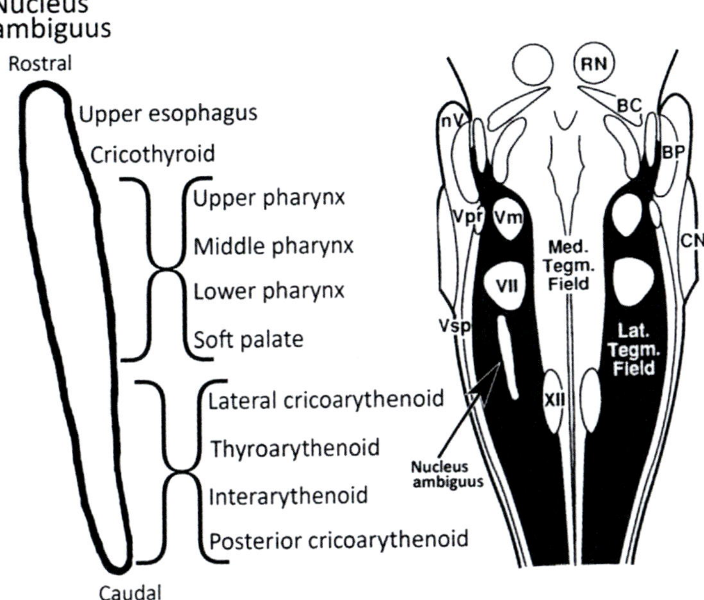

Fig. 14.25 Brainstem nuclei. Nucleus ambiguus (brown) in the brainstem is r6–r11. It supports branchiomeric (pharyngeal arch muscles), therefore is in lateral position corresponding to somatic efferent column. Anterior components of cucullaris arising from somites 1 to 4 is represented by r8–r11. It continues into the spinal cord as the nucleus for accessory nerve; thus, it retains the same function. In the neck, *somatic motor medial* C1–C3 spinal nerves supply strap muscles; *somatic motor lateral* C1–C3 supply cucullaris (sternocleidomastoid and trapezius). Cucullaris is an extracranial muscle, not a branchiomeric muscle. [Reprinted from Holstege G, Subramanian HH. Two different motor systems are needed to generate human speech. J Comp Neurol 2016; 524(8): 1558–1577. With permission from John Wiley & Sons]

(which also bears the muscles of mastication). For this reason, any form of cleft palate involving TVP implies a more proximal "hit" to the StV2–StV3 system or a more global involvement of first arch neural crest.

Defects of Muscle Fields

We shall discuss the development, evolution, and pathologies of the soft palate, in detail in Chap. 16. For our purposes here, suffice it to say that muscle testing using a Peña® muscle stimulator (Integra LifeSciences, Plainfield, NJ) provides accurate assessment. The testing is simple and can be carried out in the outpatient setting in cooperative patients under topical anesthesia.

Neuromeric Summary of Cleft Palate Fields

- r0/r1 = bone fields from stapedial V1: perpendicular plate of the ethmoid, septum, columella, prolabium,
- r2 = bone fields from stapedial V2: vomer, premaxilla, palatal shelf, inferior turbinate,
- r3 = bone fields from stapedial V3: mandible,
- r4–r5 = neural crest to arterial wall of ascending palatine branch of facial > soft palate muscles,
- r6–r7 = neural crest fascia to superior constrictor, neural crest to arterial wall of ascending pharyngeal to soft palate/pharynx muscles,
- Sm4 = mesoderm of tensor veli palatini.
- Sm7 = mesoderm of levator veli palatini, palatoglossus, palatopharyngeus, uvulus.

Part 3. How Fields Fail: Angiosome Disruption

Now that we have an idea of the various neurovascular axes, let us consider mechanisms of failure, concentrating on the *growth cone*. Because growth of the axes proceeds outward, the earlier in time the failure occurs the more structures will be affected. For example, the premaxilla is the terminal field for the medial nasopalatine axis. It has three subfields: the central incisor, the lateral incisor, and the frontal process of the premaxilla (which lies tucked beneath the frontal process of the maxilla). Any perturbation of the nasopalatine axis *will show up first as a deformation in the inferolateral rim of the piriform fossa*. Next the lateral incisor and its bony housing are affected. Finally, a total loss of premaxilla can occur. A simplifying concept involves the *neurovascular growth cone* discussed at the beginning of this chapter. The growth cone of the artery consists of an endothelial tip sprout that produces PDGF-*B* which is chemoattractive for pericytes and positive for the receptor PDGFR*B*. These cells distribute themselves along the abluminal wall of the vessel along which they secrete cytoplasmic processes. Failure in this mechanism will lead to arrest of the vascular growth cone. Recall that the growth cone of the accompanying sensory nerve produces VEGF. Failure in this mechanism can also lead to arrest of the vascular growth. Either or both of these defects will lead to inadequate production of mesenchyme in the field to which the growth cones are assigned.

It is useful to think of field development as the expansion of mesenchyme to fill a genetically designated geometric space. The field is organized around its neurovascular axis. As growth proceeds, mesenchyme laid down on either side of the axis requires blood supply. This prompts the production of **side branches** from the artery. The further away the mesenchyme is from the axis, the longer it takes for its development. Based on this principle, we shall see that both the vomer field and the palatine bones fields are isosceles triangles with arterial axis running along the hypotenuse. The posterior zones develop last. Thus, expansion of these fields appears to follow an anterior-to-posterior gradient. Deficiency of the bone field will always manifest itself in the "**newest**" zone, that is, *posteriorly*.

Palate Closure Is Bidirectional: Why?

The anatomic variations of cleft palate are based on two primary mechanisms, both involving a *reduction in mesenchymal mass*.

- *Critical contact distance.* Fusion of these mesenchymal tissue units (*partner fields*) requires that they be physically positioned relative to each other within a *critical contact distance*. If the tissue volume of one of the fields is reduced, and the critical contact distance is exceeded, fusion will not occur and a cleft will result.
- *Altered fusion potential of the epithelial surface.* The stability of the epithelium is controlled by sonic hedgehog (SHH). As long as this is active, the epithelial surface will be incapable of fusion. SHH is itself inhibited by BMP4 of the underlying mesenchyme. The total amount of BMP4 produced by a field is proportional to its mesenchymal volume. Thus, any reduction in BMP4 production will promote the stability of the epithelial surface and prevent its fusion.

Fusion of the palate is *bidirectional*, proceeding both forward and backward from the incisive foramen. The reason for this has to do with the developmental anatomy of three neuroangiosomes. As a generalization, StV2 internal maxillary divides in the pterygopalatine fossa to two common axes that supply (1) the zygomaticomaxillary complex and (2) the naso-palatine complex. As the latter approaches the sphenopalatine foramen, it divides. One branch, *descending pala-*

tine artery (DPA), dives into the substance of orbital process of palatine running downward through the bone until it reaches the plane of the hard palate. The other branch, *nasopalatine artery* (NPA), runs beneath the mucoperiosteum of the nasal cavity where it divides again into lateral branch (NP_L), and a medial branch (NP_M).

- DPA supplies the *palatine bone complex*. DPA proper descends through orbital plate of palatine bone (P3) and then bifurcates.
  - Greater palatine artery goes forward to the P1–P2 *palatal plate* of palatine bone and to the *lingual wall of alveolus*.
  - Lesser palatine artery goes backward to the P3 *palatal plate* of palatine bone.
- NPA supplies the *walls of the nasal cavity* with two branches:
  - NP_L lateral branch serves inferior turbinate, inferior maxillary wall, and *palatal plate of* maxillary bone.
  - NP_M medial branch supplies vomer and premaxilla.

Let's begin with the *midline components of the hard palate*: vomer and premaxilla. As we follow the pathway of medial nasopalatine artery NP_M, we see that it ascends along the back wall of nasopharynx to reach the midline—at the insertion of the sphenoid process of vomer. From there, it runs obliquely downward and forward in the *nasopalatine groove* along the upper margin of the vomer field. This groove borders the perpendicular plate of ethmoid and, subsequently, the septum. At the nasopalatine foramen, the *vomerine sector* of NP_M terminates; the angiosome continues forward as the *premaxillary sector* (Fig. 14.26).

- The vomer (V) is shaped like an *isosceles triangle* and the axis of NP_M runs along its hypotenuse. It gives off a series of *inferiorly directed side branches* that supply the triangularly shaped vomer mesenchyme. These side branches dead-end at the inferior border of vomer (the palatal plane). They are very long posteriorly and shorten progressively as the artery approaches the incisive foramen. If we imagine the development of vomerine mesenchyme as wax dripping downward from an oblique string, by sheer physics the posterior zones of vomer will develop slower than the anterior zones. Thus, the *biological maturation of the palatal border of vomer is from front to back*. Defects in this process show up first in its posterior zone.
- The premaxilla (PMx) is an L-shaped bookend. It has two fields, central incisor (PMxA) and lateral incisor (PMxB); from the latter sprouts the frontal process of premaxilla

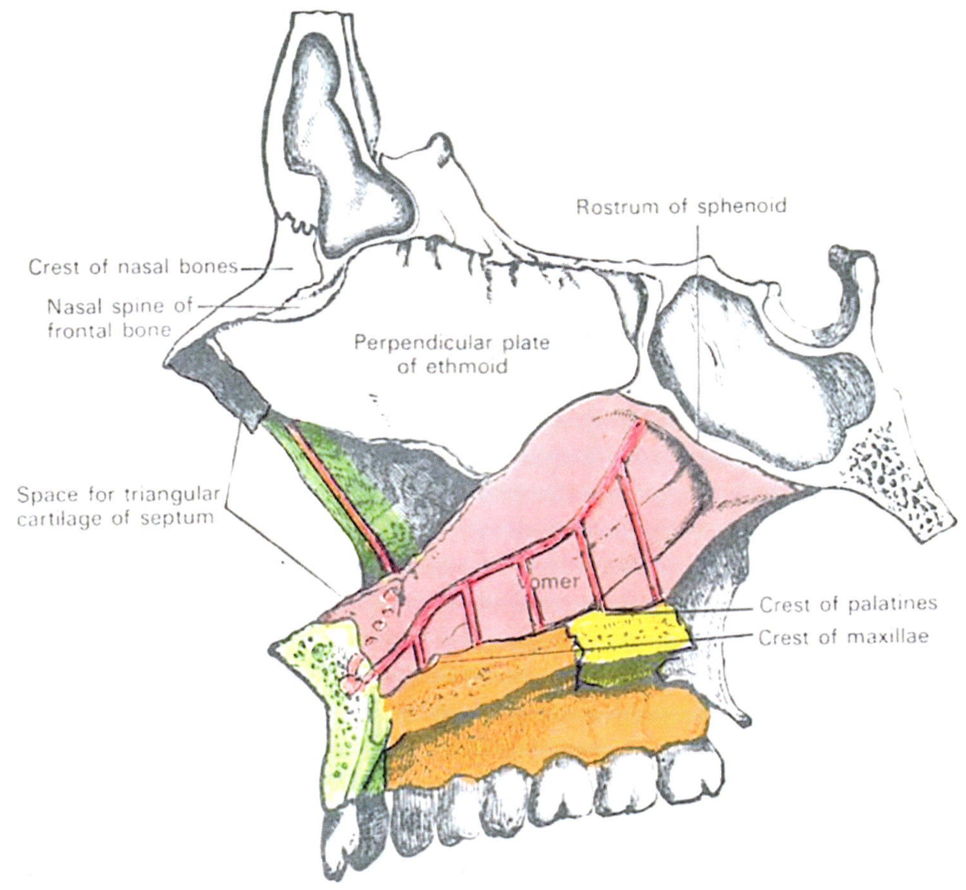

Fig. 14.26 Vomer: medial nasopalatine neuroangiosome. ***Note***: as axis of NPM descends, side branches get progressively shorter. Therefore the anterior mesenchyme develops more quickly. This triangular pattern is replicated by the palatine bone fields, P1–P3. [Reprinted from Lewis, Warren H (ed). Gray's Anatomy of the Human Body, 20th American Edition. Philadelphia, PA: Lea & Febiger, 1918]

(PMxF). Development of PMxF takes place last so defects of the axis affect the piriform fossa first. The arterial supply of the alveolar bone runs dorsal to the roots of the incisors. Alveolar bone formation is medial–lateral, lingual-to-buccal, and dorsal–ventral. The "business end" of premaxilla, where it fuses with maxilla, is the distal margin of PMxB. Thus, the order of pathology is PMxF > PMxB > PMxA. Thus, *the biological maturation of the palatal shelf of premaxilla is from back to front*. In general, we see the following sequence: deficiency of the piriform fossa followed by loss of bone stock around the lateral incisor and a partial alveolar cleft followed by loss of the dental unit and a complete alveolar cleft.

- Although the vascular axis of NPM is continuous from vomer to premaxilla, the latter always has priority. We never observe a missing premaxilla with an intact vomer but the latter situation can exist. Because the arterial axis tracks all the way forward immediately, laying down side branches subsequently, the patient can have an absent vomer and still have a premaxilla, although it is usually rudimentary.

What about the *lateral side of the hard palate*? The horizontal plate of the three palatine bone fields also describes an *isosceles triangle*. Its major sources of blood supply, the greater palatine artery, also follows the hypotenuse, running deep to the oral mucoperiosteum close to the junction between the horizontal plate (hard palate) and the vertical plate (lingual wall of alveolus). The greater palatine artery (GPA) axis sends out medially directed side branches to the medial border of the palate and downward-directed side branches supplying the entire lingual wall of the alveolus. The side branches are long posteriorly and diminish as one proceeds anteriorly. As a consequence, the posterior zones of the palatal shelf take longer to develop, later than the anterior zones. Thus, the *biological maturation of the palatal border of palatine shelf is from front to back*. Defects in this process show up first in its posterior zone.

- Note that the lesser palatine artery serves the horizontal plate of palatine fields P3. This field can manifest varying degrees of failure, occasioning an isolated soft palate cleft, in the presence of a normal P1–P2. Note, however, that growth of P3 is *not independent* of P1–P2. The dimensions of P2 are a template that determines those of P3. Thus, the horizontal plate of palatine bone can only grow as far medial as P2 permits. Furthermore, defects in P3 can be completely independent of the more anterior horizontal plate.

Part 4. Neurology of the Palate

The purpose of this section is to clarify confusion that exists in the innervation of pharyngeal structures and pharyngeal arches, concepts critical to achieving a realistic surgical map of the palate. This information is relevant to Chaps. 15 and 16.

Many biology texts refer to r8 as the terminal rhombomere, lumping together a wide variety of neuroanatomic relationships. Extensive evidence breaks up r8 into four *pseudorhombomeres* (r8–r11) because of individual relationships between the medullary nuclei and the extraneural structures. For purposes of simplification, and because Puelles' work gives a more accurate description, we use this terminology [3].

Apart from the oculomotor nerves (III, IV, and VI) and the hypoglossal nerve to the tongue (XII), all cranial nerves are *mixed*, that is, they contain two or more functional components. In the rhombencephalon, the various nuclei contributing to each cranial nerve are classified into six longitudinal columns, each one with a specific function. The motor columns (efferent output from the CNS) are in the basal (medial) hindbrain. Sensory columns (afferent into the CNS) are lateral. The columns are organized in mirror-image manner according to their function. For those readers wishing a quick review (or for those with nothing better to do with their time), the anatomy of the columns from medial to lateral is as follows (Figs. 14.24 and 14.25).

Motor Columns of the Brainstem (Basal Plate)

The somatic efferent column (general somatic efferent, GSE) is the most medial. It contains the nuclei serving the oculomotor muscles and tongue. It is discontinuous, with m1 having oculomotor (III), r1 containing trochlear (IV), r4–r5 containing the lateral rectus (VI), and r8–r11 having the hypoglossal nucleus of XII.

The nonpharyngeal arch column (general visceral efferent, GVE) is located lateral to GSE. GVE is a misnomer. It is not strictly visceral because it is motor for the glands of the eye and mouth. These nuclei belong to the parasympathetic autonomic nervous system. They supply highly localized ganglia (relay stations) with preganglionic parasympathetic nervous system (PANS) motor fibers for a wide variety of functions, such as pupillary control, salivation, bronchoconstriction, cardioinhibition, and peristaltic activity of the intestine. The midbrain component of this column contains Edinger–Westphal nucleus of III. Superior salivatory nucleus of VII in register with r4–r5 supplies the lacrimal gland and salivary glands. Inferior salivatory nucleus of IX in register with r6–r7 supplies the parotid gland. The caudal continuation of GVE is the dorsal nucleus of X in register with r8–r11. It supplies all organs of the thorax and abdomen. Thus, the motor component of cranial nerve X does not supply striated muscle.

The pharyngeal arch column (special visceral efferent, SVE) is the most lateral motor column. Commonly known as the branchiomotor column, it supplies the muscles of the pharyngeal arches. SVE is also a complete misnomer, as it contains nuclei for structures that are neither branchial nor special nor visceral. However, we use SVE anyway. Why the confusion?

- First, striated craniofacial muscles originate from either somitomeres or the first four occipital somites, not from arches—myoblasts migrate into pharyngeal arches only secondarily.
- Second, only fishes with gills have branchial arches (branch signifying gills). All the rest of us, land-dwelling tetrapods (throw in a few snakes), breathe air; thus, those embryonic structures formerly assigned to form gills now create the pharynx. Hence, the term pharyngeal arches.
- Third, the concept of "branchial" muscles was somehow assumed, by virtue of their location, to relate to visceral functions, such as swallowing. These were considered "ventral" and therefore had to be LPM: this concept is now disproven. We now know that all craniofacial striated muscles originate from PAM. Having cleared up this mess, SVE nuclei supply striated muscles of pharyngeal arches 3–5, down to and including pharynx and larynx. Note that the muscles controlling head-versus-pectoral girdle (sternocleidomastoid and trapezius) are nonarch muscles arising from somites S1 to S7. Their nerve, spinal accessory, is a peripheral nerve incorporated into the head. Motor components of cranial nerves V, VII, IX, and X and the cranial portion of XI are all SVE. The nucleus ambiguus belongs to the SVE column.
- In the upper cervical spinal cord (c1–c4), this situation is much simpler. There are no branchial arch muscles per se and therefore there is but a single somatic motor column.

Sensory Columns of the Brainstem (Alar Plate)

The parasympathetic afferent column (general visceral afferent, GVA) receives parasympathetic information from the gut and viscera. Pain from these organs is often associated with distention.

The taste column (special visceral afferent, SVA), known as the nucleus of *tractus solitarius*, receives taste information via VII (anterior two-thirds of tongue), IX (posterior one-third of tongue), and X palate/pharynx. Representation is somatotopic. Taste fibers from the posterior one-third of the tongue and the soft palate ascend in the nucleus solitarius and synapse in its gustatory zone. Palate fibers most likely synapse at levels r6 and r7, whereas fibers from the tongue target those neuromeric levels that are in genetic register with the mesoderm and neural crest from which they originated, that is, from r8 to r11.

The nocioceptive column (general somatic afferent, GSA) receives obnoxious sensory input (pain, temperature from the head). Nerve fibers travel as glossopharyngeal and vagus from the palate and pharynx and larynx to levels r6–r11.

The balance and hearing column (special somatic afferent, SSA) receives information from the inner ear. The vestibular nucleus is very extensive running from r5 to r11. The cochlear nucleus is localized lateral in register with r6–r7.

Clarifications Regarding Motor Control of the Soft Palate

Palate muscles get their motor supply from rhombomeric zone r6–r7 in nucleus ambiguus. The sensory distribution to the soft palate is associated with *both* cranial nerves IX and X. However, recall that *dorsal motor vagus is not motor to striated muscle*. Nucleus ambiguus "chooses" to send efferent neurons to the palate via the vagal pharyngeal plexus. Thus, although textbooks say the only muscle of the third arch is palatopharyngeus and it is supplied by IX, the reality is that all palate muscles get neurons from the third arch zone of nucleus ambiguus, r6–r7. These simply travel to palate via sensory vagus of the pharyngeal plexus. The gag reflex is sensory to r6–r7 via glossopharyngeal and motor back to palate from r6 to r7 and to pharynx from r8 to r11 via the vagus.

The same rule holds true for the motor supply to the pharyngeal constrictors. Upper constrictor (r6–r7), middle constrictor (r8–r9), inferior constrictor (r10–r11), and the laryngeals (r8–r11) all travel via vagus.

Cranial nerve XI, spinal accessory, is a continuation of nucleus ambiguus that runs down into the cervical spinal cord from c1 to c6. It is a modified peripheral nerve that innervates the ancient branchiomeric cucullaris muscle, the derivatives of which are sternocleidomastoid and trapezius. In fishes, cucullaris is an epaxial muscle that runs from a dermal bone series attached to the back of the skull (the future pectoral girdle) forward to the dorsal aspects of the branchial arches. In tetrapods, the pectoral girdle is detached from the skull and XI appears for the first time. The mammalian cucullaris arises from somites 5 to 10; therefore it has motor nuclei extending all the way from c1 to c6. But its neural crest fascia retains its ancient allegiance to branchial arches BA6–BA7 (and possibly BA8). Sternocleidomastoid is S5–S7 and attaches to c1–c3 clavicle. Trapezius is S6–S10 and attaches to those zones of scapula.

The cranial root of this misunderstood nerve arises from r10 to r11 and joins vagus. It probably controls the intrinsic muscles of the larynx. For this reason, some texts describe the (branchiomeric) soft palate muscles as under the control of XI but this is impossible as XI really belongs to the branchiomeric system of the sixth, seventh, and perhaps the eighth branchial arches (Fig. 14.27).

A Word About Muscles: Epaxial Versus Hypaxial

Neurons exiting the brainstem and spinal cord are either somatic efferent, somatic afferent, or mixed. They grow in two different directions as determined by gene signals, primarily the presence or absence of BMP4.

- Epaxial nerves supply structures dorsal to the neuraxis. These nerves are quite simple. Epaxial nerves may have

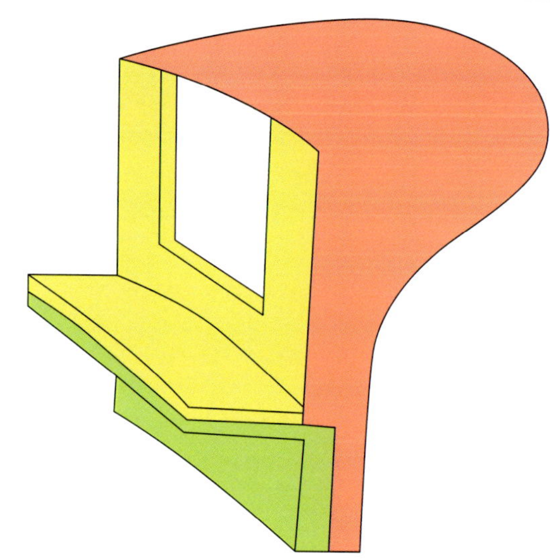

Fig. 14.27 Development of the maxillary–palatine complex. Palatine bone fields (lemon) form the oral shelf of hard palate. The contribution from the maxilla for nasal shelf of hard palate (yellow) is much thinner. The walls of maxilla (pink) proper are initially a lamina but become cup-shaped with growth. [Courtesy of Michael Carstens, MD]

contributions from more than one neuromere, but they travel as individual structures. This is because the migration routes of epaxial muscles are simple; they follow a strict spatiotemporal sequence. Muscles build up on one another, with mononeuromeric muscles before multineuromeric muscles (simple before complex) in craniocaudal, ventro-dorsal, and medio-lateral order. This, in turn, is determined by the fact that such muscles insert into structures that maintain fixed relationships with respect to the body axis.

- Hypaxial nerves supply structures ventral to the neuraxis. In the head and neck, these nerves can be quite complex with contributions from multiple functional columns. Hypaxial nerves can form plexuses, complex "switchyards" giving off branches of which have combinations of cell bodies. This is because the migration routes of myoblasts are complex. Although they follow the same spatiotemporal rules, they insert into appendicular structures, such as the upper limb, which, during development, undergo position changes with respect to the body axis. Structures residing in a fixed position within the body wall, such as intercostal muscles and diaphragm, are innervated by individual hypaxial nerves, not via plexuses.

Note for the Uninterested The term "visceral" is a misnomer when applied to cranial nerves; it really should be read as hypaxial. If we define the viscera as internal structures arising from IM (genitourinary system), ventral LPM (cardiovascular system, gastrointestinal system), and those glands supplying endoderm or ectoderm, then the term viscera is OK. Obviously, the ciliary muscles have nothing to do with the viscera. Moreover, it is difficult to conceive of the larynx as viscera. We just have to keep it in mind that we are using the term viscera rather loosely. All pharyngeal arch structures are hypaxial to the neuraxis as are the cranial nerves that supply them. Pharyngeal arch for tetrapods is equivalent to branchial arch for gill-bearing fishes; both give rise to purely hypaxial structures.

Nerve Supply to the Soft Palate and Pharynx

Having defined our terms, let's review cranial nerves IX, X, XI, and XII in terms of their functional components:

Glossopharyngeal nerve (IX)

- Somatic motor: None, no striated muscles from somites. "Branchial" motor: Striated muscles of the third arch arise from PAM of somitomere 7. Glossopharyngeal nerve per se supplies only one muscle from Sm7, palatopharyngeus. Instead, glossopharyngeal receives motor fibers from XI within the substance of the cranial nucleus ambiguus. (1) stylopharyngeus nerve and (2) communicating nerve to vagus conveys motor fibers from r6 to r7 to the soft palate that are very important for us.
- Visceral motor: Tympanic nerve innervates the tympanic membrane; it carries PANS fibers from inferior salivatory nucleus to lesser petrosal nerve to the otic ganglion just above foramen ovale where they synapse to travel with V3 auriculotemporal nerve to parotid gland.
- Visceral sensory: Carotid sinus nerve carries afferent baroreceptor input.
- General sensory: Tonsillar nerve, lingual nerve, and pharyngeal plexus.
- Special visceral sensory: Taste fibers from posterior one-third of tongue.

Vagus nerve (X)

- Somatic motor: None, no striated muscles from somites.
- Branchiomotor: Striated muscles of the fourth and fifth arches arise from the caudal nucleus ambiguus. Pharyngeal nerve to the constrictors contains contributions from r6 to

r11. The soft palate, superior, and middle constrictors are supplied by r6–r7. Middle constrictor r8–r11: Superior laryngeal nerve from r8 to r9 supplies the PA4 intrinsic muscles of the larynx. Inferior laryngeal nerve from r10 to r11 supplies the PA5 muscles of the larynx.

- Visceral motor: Smooth muscle from the level of c1 downward arises from LPM. All parasympathetic innervation to glands of the pharynx, larynx, neck thorax, and abdomen arise from or are situated in tissues arising from LPM.
- Visceral sensory: Lower two-thirds of esophagus, remainder of the gastrointestinal tract, and all other viscera.
- General sensory: Pharynx from inferior constrictor, larynx, and upper esophagus.
- Special visceral sensory: Taste fibers from epiglottis.

Accessory nerve (XI)*

- Somatic motor: None.
- Branchiomotor: Basal form in sharks is the cucullaris muscle. This muscle functions as levator of the branchial arches extending from pectoral girdle forward to base of the skull, that is, from r10 to r11 to c4/c5. Additional spinal contributions may run as low as c7. Cucullaris becomes sternocleidomastoid and trapezius. In its original form, XI was an exclusively spinal nerve but the backward expansion of the skull brought the first two levels inside the skull.
- Visceral motor: none. Visceral sensory: none.
- General sensory: Somatic sensory from sternocleidomastoid and trapezius.
- Special visceral sensory: none.

*The neuroanatomy of the accessory nerve is of great importance for understanding the transition between the head and the neck.

Part 5. Bones of the Palate

The developmental structure of the hard palate is very different from what you are familiar with. In the midline are the *vomerine bones*, commonly (but incorrectly) described in the singular. The vomers are united on either side with distinct horizontal bone shelves derived from the ancient palatine bone series. Recall from Chap. 8 that the maxilla in sarcopterygian fishes is lined on its lingual side by row of four tooth-bearing palatine bones, P1–P4. Their evolutionary fates are as follows (Fig. 14.28).

- P4, the most posterior field, is called *ectopterygoid* because of its relationship to the pterygoid bone. In evolution, the pterygoid moves forward to basisphenoid with which it fuses, leaving a remnant, the *medial pterygoid plate* (well known for its hamulus). Ectopterygoid remains attached and becomes *lateral pterygoid plate*.

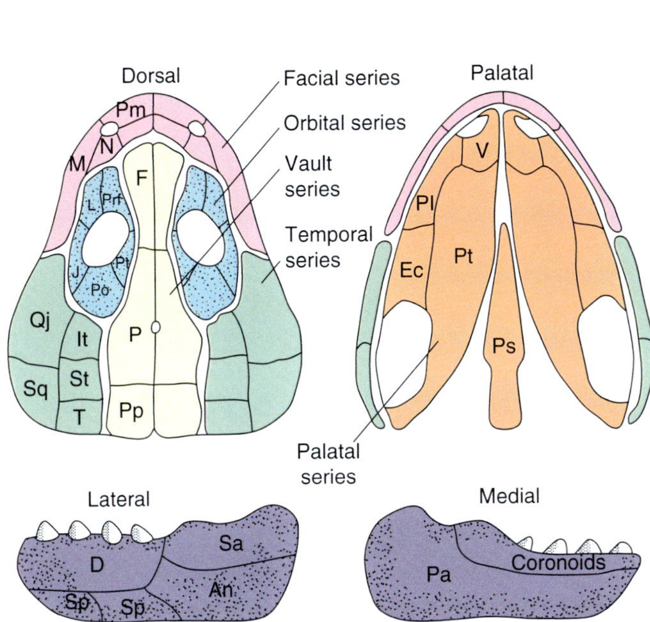

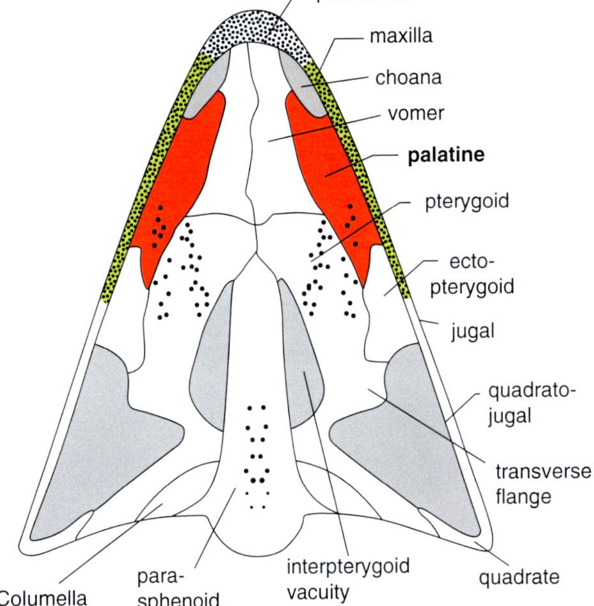

Fig. 14.28 Basal amniote bone fields. Note on palatal view the extent of the pterygoids, connecting vomer with quadrate, the most posterior dermal bone field on the lingual side of palatoquadrate cartilage. P4 ectopterygoid is not in register with the maxilla. It is not tooth-bearing. P1–P3 are aligned with maxilla; they form the horizontal shelf of the hard palate. When pterygoid moves forward to be absorbed by basisphenoid, the transverse flange becomes lateral pterygoid process whereas ectopterygoid, remaining aligned with P3, produces medial plate and/or hamulus. [Reprinted from Kardong K. Vertebrates: Comparative Anatomy, Function, Evolution, 7th ed. New York, NY: McGraw-Hill; 2015. With permission from McGraw-Hill Education]

- P3 becomes the *palatine bone* senso stricto. Its original lamina is horizontal plate but it sends up a perpendicular plate into the orbit. P3 is supplied/programmed by a single neurovascular axis, the lesser palatine artery; hence, it is unilaminar.
- P1–P2 retain their attachment to the maxilla. They form the *lingual lamina of alveolus* and the *horizontal palatine shelf*. Although the terminology is incorrect, we will refer to this as the *palatine process of the maxilla* (MxP) even though it is not.
  - In contrast, MxP is a bilaminar structure. It has a nasal lamina of MxP which is in continuity with the inferior turbinate and inferolateral nasal wall. The oral lamina of MxP is an extension of the lingual wall of the maxillary alveolus.
  - Normally, the two laminae are fused, but on occasion, a sinus can be found and ectopic teeth may be present.

All pharyngeal arch bones are neural crest derivatives. The sensory innervation to the maxillary complex is V2, the nucleus of which resides in the second rhombomere, r2. V2 sends branches to all the developmental fields of the maxillary complex. These are accompanied by vascular pedicles arising from the pterygopalatine fossa—the sensory nerves program the arteries. Defects within this r2 neural crest population—in terms of total cell volume, cellular function, or blood supply—can result in bones that are misshapen, small, or even absent.

Key Concept Each bone field of the palate is supplied by a specific neurovascular pedicle. Deficiency states of a bone field occur in a distal-to-proximal gradient backward along the neurovascular axis.

The most conceptually accurate model of membranous bone development is one in which bone forms in response to a predetermined epithelial program which determines its size and shape. Bone development takes place along its neurovascular axis according to a strict spatial–temporal sequence. Failure of the axis results in a gradient of deficiency that runs from distal to proximal. In the premaxilla, for example, the medial incisor field develops before the lateral incisor field or, later, the frontal process field making up the internal piriform fossa. For example, a minimal defect of the medial nasopalatine axis can leave a smaller or absent frontal process of premaxilla; it can also affect the height of the posterior vomer, causing it to be displaced out of the palatal plane. We now turn our attention to the embryology, in sequence of vomer, maxilla, and the palatine bone.

Vomer

The vomer is a quadrilateral bone that develops in membrane from r2 neural crest. It represents the fusion of paired embryonic vomerine processes. Its shape is programmed from r2 oral ectoderm. Vomer maintains a total of ten articulations with five pairs of bones. Sphenoid and ethmoid are derivatives of r1 midbrain neural crest. Palatine, maxilla, and premaxilla are all derivatives of r2 hindbrain neural crest.

Posteriorly it has a deep groove which receives the (paired and fused) vaginal processes of the r1 sphenoid. Paired alar processes flare outward, forming the letter "T." These articulate posteriorly with the medial pterygoid plates and anteriorly with the sphenoid processes of the r2 palatine bones (Fig. 14.26).

The superior border of vomer receives the fused perpendicular plates of the ethmoid complex. The bones are demarcated by a deep nasopalatine groove, which transports the neurovascular axis of the same name. The inferior border of septum, also of r1 midbrain neural crest derivation, is in direct contact with the anterior-most upper border of the vomer.

The anterior border of vomer articulates with premaxilla and with septum. A collagenous band, the premaxillary-vomer ligament, has been considered by some to be an "engine of growth." In reality, it is just a means to share force vectors as each of the membranous structures grows autonomously.

The inferior border of vomer articulates with the shelves of the maxilla and palatine bones. They form a U-shaped nasal crest into which the vomer is situated.

The vomer contains two ossification centers, one for each lamina. Although they fuse to what is apparently a single bone, proof of previous bilaterality is seen in the bilateral grooves for nasopalatine nerves, the upper midline groove receiving the ethmoid and vomer, and the presence of flared alar processes.

Ossification follows a pattern which traces the trajectory of the nasopalatine axis. Bone is deposited beginning with the penetration of the NP axis through the foramen incisivum. Much like spilling a bucket of paint, mesenchyme for the medial nasopalatine axis "fills out" the premaxilla in three vectors: forward, outward, and downward for the teeth and hence upward for the frontal process. In like fashion, the vomer is synthesized backward and downward. In shape, the vomer can be likened to a scimitar, with a narrow handle hinged at the incisive foramen and a broad blade projecting backward. As the vomer grows, the inferior edge of the scimitar descends, knife-like, into the nasal cavity.

Vomer Growth Pattern and the Medial Nasopalatine Pedicle

Recall that vomer descends obliquely downward as it follows along the inferior border of the ethmoid perpendicular plate and septum. The anterior zone of vomer is very short vertically at the palatal plane whereas its posterior zone is significantly removed from the vascular axis. Therefore, although arterial flow is forward, *the collateral flow takes longer to be established the further posterior you proceed.* For this reason, anterior vomer develops first and its posterior margin is complete later in time, descending into the midline [14–16].

Clinical Correlation: How a Vomer Field Defect Causes Isolated Midline CP

Closure of the hard palate depends on the fusion of the shelves of the palatine and maxillary bones with the vomer. This, in turn, requires that the vomer be physically descended into the palatal plane. The critical contact distance between the two bone fields must be achieved. For this to happen, the vomer must descend into the palatal plane. If the vomer field is insufficient, loss of mesenchyme will be seen at its posterior and inferior borders, it will be out of the palatal plane, the critical contact distance will be exceeded, fusion to the shelves will be impossible, and a cleft will appear.

The fusion process itself proceeds through epithelial dissolution. Sonic hedgehog in the epithelium (which maintains its integrity) is inhibited by BMP4, a soluble morphogen formed as by-product of membranous bone synthesis. Reduction of mesenchymal bone mass means a quantitative reduction in the production of BMP4. The intact epithelial surface will not disintegrate, no fusion is possible, and a "cleft" occurs (Fig. 14.29) [17, 18].

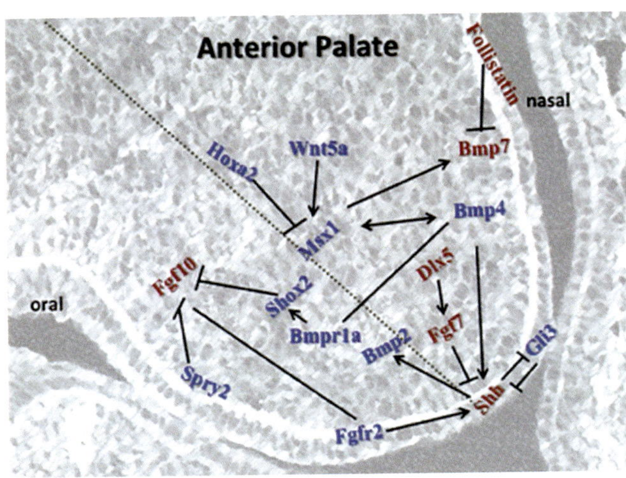

Fig. 14.29 BMP4–Shh model. A model for a genetic pathway integrating *Msx1*, *Bmp4*, *Shh*, and *Bmp2* in the epithelial–mesenchymal interactions that regulate mammalian palatogenesis. In this model, it is proposed that in the anterior palatal shelves, mesenchymally expressed *Msx1*, which can be induced by *Bmp4*, is required for *Bmp4* expression in the palatal mesenchyme. Mesenchymally expressed BMP4 maintains *Shh* expression in the MEE, and Shh in turn induces *Bmp2* expression in the mesenchyme. BMP2 functions to induce cell proliferation in the palatal mesenchyme, which leads to palatal growth. [Reprinted from Smith TM, Lozanoff S, Iyyanar PP and Nazarali AJ (2013) Molecular signaling along the anterior–posterior axis of early palate development. Front. Physiol. 3:488. With permission from Creative Commons License 3.0: https://creativecommons.org/licenses/by/3.0/]

Clinical Correlation: Septovomerine Articulation in Septoplasty

Correction of septal deviation is a common surgical procedure requiring elevation of the mucoperichondrium surrounding the septum. This maneuver presents a technical problem. As one proceeds from above, inferiorly and posteriorly, tears in the mucoperiosteum can occur along the septovomerine junction. This is due to the confluence of periosteum of the two vomerine lamellae, forming a U-shaped "cradle" that receives the septum. Within this bony trough, a fusion takes place between the mucoperichondrium surrounding the septum and the periosteum of the bone. This confluence of tissue, if not dissected properly, can result in a tear.

On the other hand, when a vomer flap is elevated from below, as in a unilateral cleft lip and palate, an incision is made at the boundary between vomer and maxillary shelf. The resultant mucoperiosteal plane and submucoperichondrial plane are confluent. Exposure of the septum from below is fast, easy, and atraumatic. Why the discrepancy? In this scenario, at the site of confluence, the "lip" of vomer ipsilateral on the cleft side is missing. The U-shaped cup of vomer bones that receives the septum is disrupted on the cleft side. Entry into the subperichondrial plane is uncomplicated.

Clinical Correlation: Congenital Absence of Vomer

This entity is quite rare, with proximately 20 reports in the literature. Cone beam CT by Yan demonstrates the complete absence of the vomer with an intact premaxilla, thus demonstrating the developmental primary of the latter. Intact hard palate indicates that vomer is not essential for palatine shelf closure. How then did premaxilla develop? The answer must be that medial nasopalatine followed perpendicular plate and then the inferior margin of septum. Six cases reported by Mohri and Amatsu [19] confirm the intact status of premaxilla. Ozaki presents a case in which the osseous vomer has failed to develop, as evidenced by presence of only a boneless membranous remnant (Figs. 14.30 and 14.31).

In Sum NPM produced premaxilla primarily and vomer secondarily.

Maxillary shelf of the hard palate: palatine fields P1–P2 (Figs. 14.27, 14.32, 14.33, 14.34 and 14.35, maxilla and palatine fields P1–P2; Figs. 14.28, 14.36 and 14.37 palatine proper, P3).

Components of the Maxillary Complex

The best way to understand the structure of the maxilla is to compare it with that of the mandible. Although they appear very different, their functions are identical, as it is their basic design. Both are bilaminar and consist of multiple bone fields assembled into two functional components: A

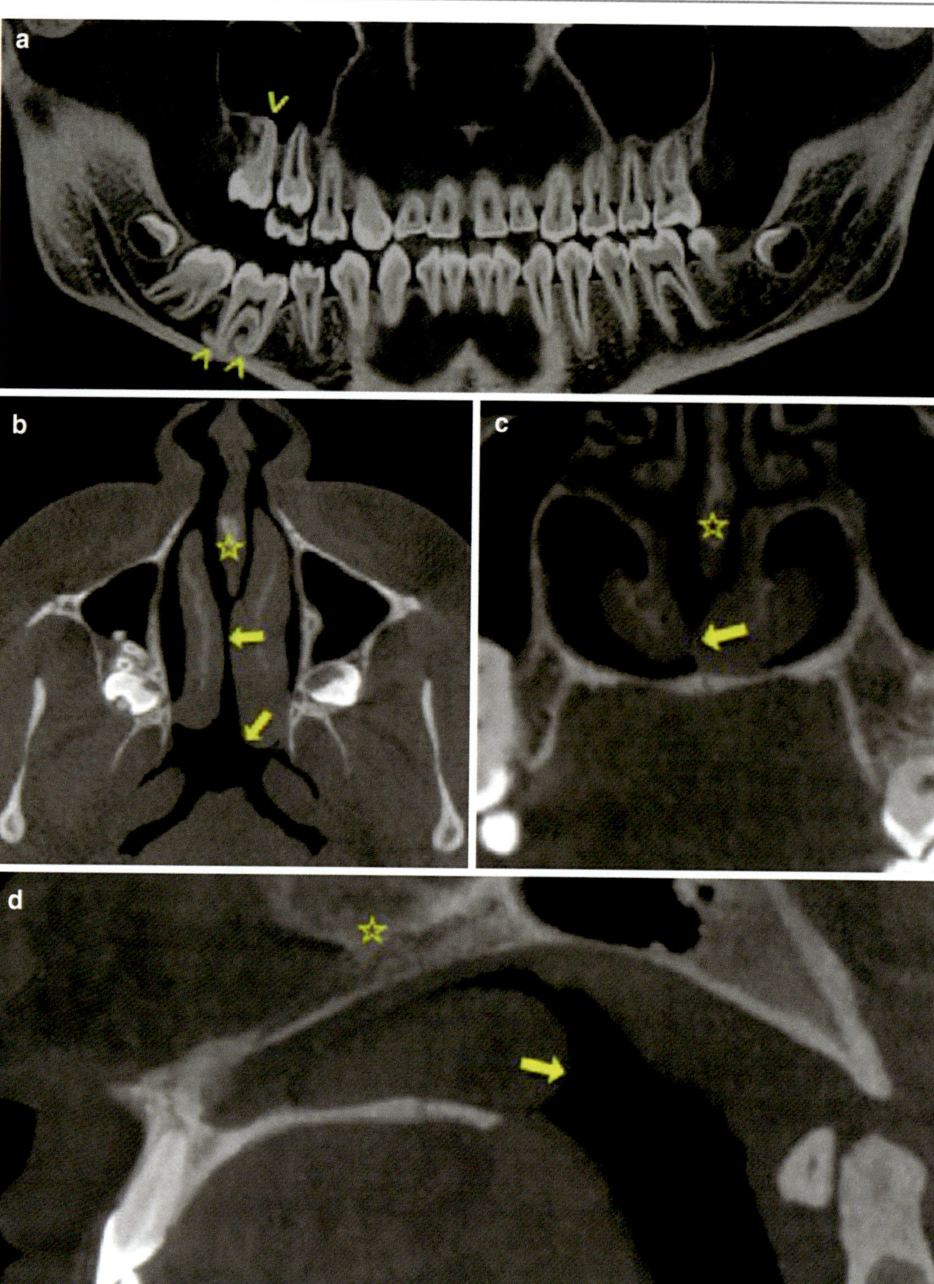

Fig. 14.30 Multiplantar reconstruction using CBCT showing missing vomer 1. Panorex shows the hooked aspect of the root apices of teeth 16 and 46 (arrowheads) with primary failure of eruption. The nasal septum is incomplete (yellow stars) with a large defect in its posterior–inferior part corresponding to complete absence of the vomer. The bony structures of the middle and lower part correspond to complete absence of the vomer. The bony structures of the middle and lower turbinates appear to be preserved. Nevertheless, the left lower turbinate (arrows) is hypertrophied and protrudes through the septal defect. Note contact between the left and right lower turbinate and posterior protrusion of the left lower turbinate into the nasopharynx. [Reprinted from Yan DJ, Lenoir V, Chatelain S, Stefanelli A, Becker M. Congenital vomer agenesis a rare and poorly understood condition revealed by cone beam Ct. Diagnostics 2018, 8:15–19. With permission from Creative Commons License 4.0: http://creativecommons.org/licenses/by/4.0/]

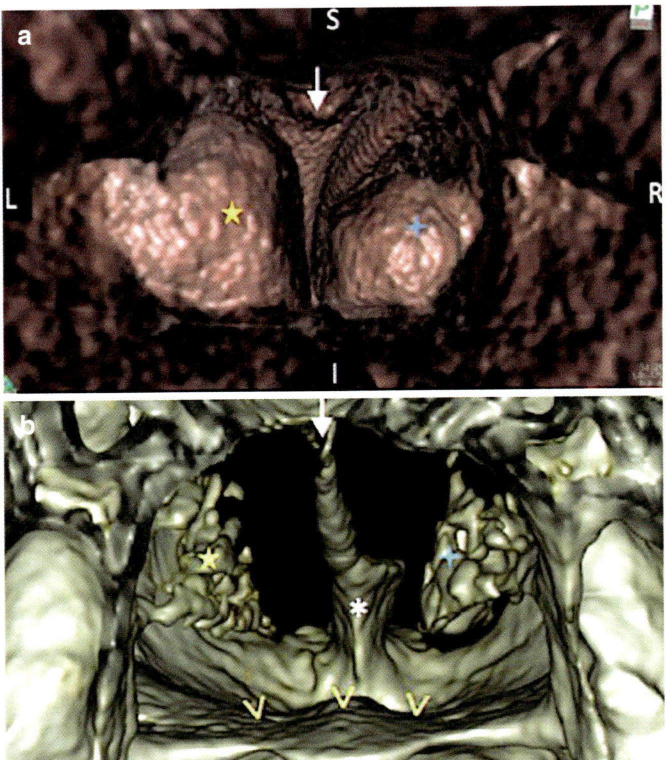

Fig. 14.31 Multiplantar reconstruction using CBCT showing missing vomer 2. 3D virtual endoscopic reconstruction (**a**) and 3D bone reconstruction (**b**) of the CBCT data set. This posterior view of the choanae shows the absence of separation between the left (L) and right (R) nasal fossae in their posterior part. Yellow star: lower left turbinate; blue star: right lower turbinate; white arrow: reversed triangular shape formed by the posterior and inferior edges of perpendicular plate of the ethmoid bone. Yellow arrowheads: floor of the nasal fossae. White asterisk: palatine process of the maxillary bone. [Reprinted from Yan DJ, Lenoir V, Chatelain S, Stefanelli A, Becker M. Congenital vomer agenesis a rare and poorly understood condition revealed by cone beam Ct. Diagnostics 2018, 8:15–19. With permission from Creative Commons License 4.0: http://creativecommons.org/licenses/by/4.0/]

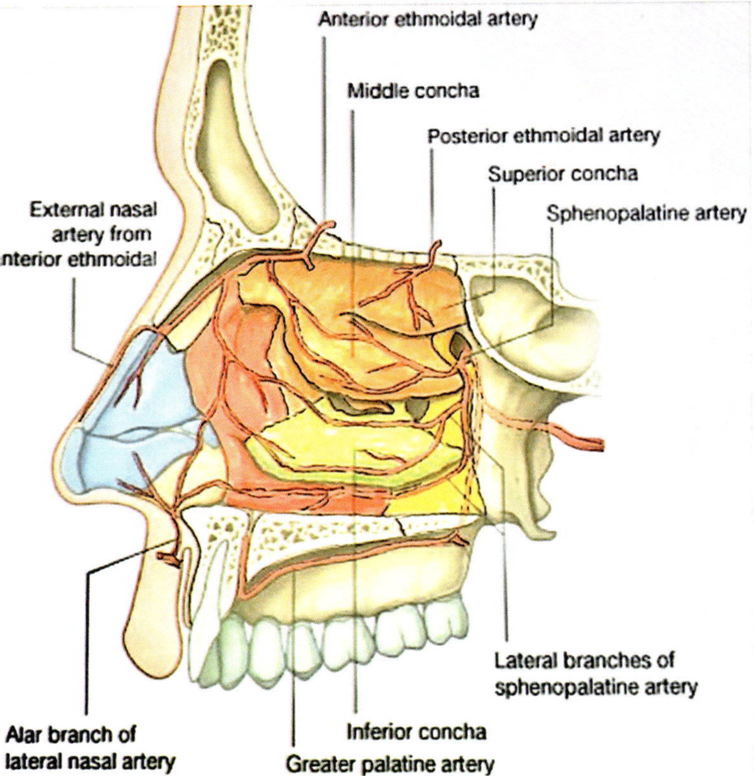

Fig. 14.32 Neuroangiosomes of the hard palate. Palatine bone fields P1–P3 are bilaminar with supply coming from NP_L above and GPA below. In the case of P3, the oral side is LPA. [Reprinted from Carstens MH. Pathologic Anatomy of soft palate, part 1. J Cleft Lip Palate Craniofac Anomal 2017; 4(1): 37–64. With permission from Wolters Kluwer Medknow Publications]

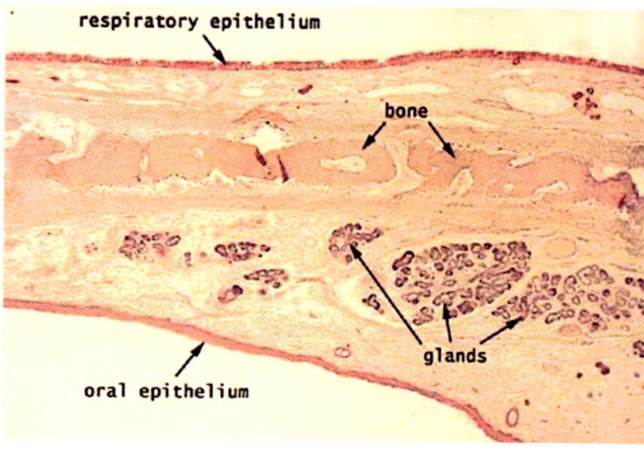

Fig. 14.33 Histology of the hard palate. Nasal mucosa: pseudostratified columnar epithelium (respiratory). Oral mucosa: stratified squamous. [Image by David G. King, used with permission. Retrieved from http://www.siumed.edu/~dking2/crr/CR001b.htm]

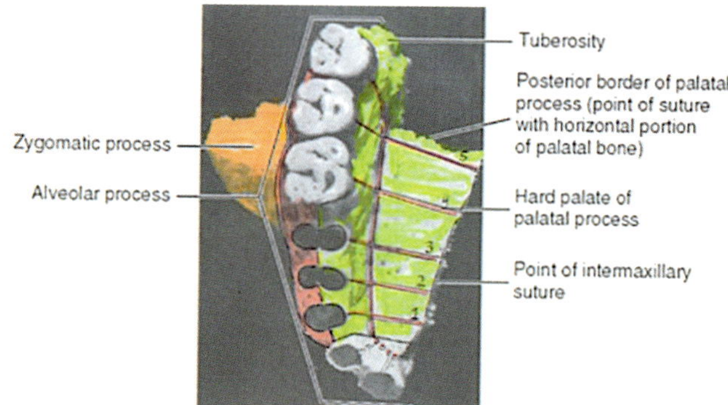

Fig. 14.35 Developmental sequence of the palatine fields, P1–P2. GPA follows a curve forward, giving off side branches in the process. Branch 1 is shortest and this zone of mesenchyme develops first. Thus development proceeds from zone 1 to zone 5. P1 is biologically more mature than P2. Note the alveolus is bilaminar with palatine fields (lemon) and anterior/posterior superior alveolar fields (pink) enclosing the dental units. Zygoma (orange) projects superolaterally. [Courtesy of Michael Carstens, MD]

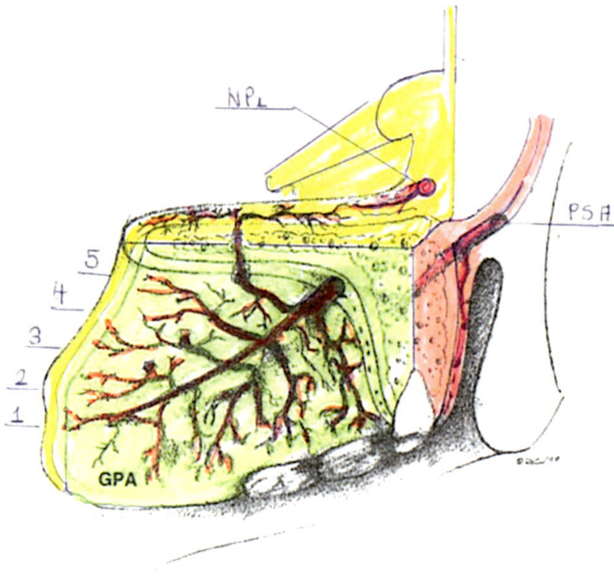

Fig. 14.34 Alveolus and hard palate angiosomes. Palatine fields P1–P2 (green) are supplied by GPA while P3 is supplied by LPA. The axis of NP_L is seen running beneath inferior turbinate all the wall from the nasal side of palatine to the frontal process of maxilla. Posterior superior alveolar artery supplies the lingual mucoperiosteum and molar zone while anterior superior alveolar (not seen) supplied the bicuspids, the canine, and the maxillary lateral incisor. Note the vascular damage inflected on the lingual alveolus by incisions made in the angle between the GPA and the alveolus. All vertical branches are cut. This problem is averted in alveolar extension palatoplasty (AEP), which places the subperiosteal incision at the coronal level of the teeth. See Chap. 15. [Courtesy of Michael Carstens, MD]

housing for dental units and a support structure that articulates with the skull. The upper jaw was not always fixed to the cranial base. In early bony fishes, the tooth-bearing palatoquadrate cartilage (the precursor of the maxilla) had a moveable articulation with the chondrocranium. The mandible of choanate fishes and early tetrapods consists of a tooth-bearing dentary bone supported from below by two splenial bones and a more posterior articular bone that constituted the primitive connection to the skull. These structures have all fused (Herring's law) to make a unitary bone, but both sides of the body and mandible are perfused by distinct arteries supplying the attached muscles, which bear witness to its more complex past. The potential space between the laminae of the dentary bone contains the dental units while that of the supportive lever arm is filled with marrow.

On the opposite side of the bite plane, the original maxillary bone is analogous to the dentary. Its support structure is fixed, but bilaminar, forming a six-sided box, in the center of which is the maxillary sinus. The medial wall of the box (the lateral wall of the nose) is discontinuous. It has a large hole in the center that is partially covered over by the uncinate process of the ethmoid and the inferior turbinate. This is a critical point because the nasal epithelium gains access to the interface between the maxillary laminae and prevents them from fusing. Thus, as growth of the maxillary complex takes place, this potential space expands to form the maxillary sinus. Because the medial lamina of the maxilla is fixed to the cranial base by the ethmoid and palatine bones, expansion of the maxillary sinus must occur laterally, projecting outward beneath the orbit.

SURPRISE: The Hard Palate Does *Not* Belong to Maxilla

When we study the sagittal section of the nasal cavity, we can see that a lamina of bone runs below the inferior turbinate. Above it, under the shelter of the turbinate, we find the aperture into the maxillary sinus. Both are supplied by lateral nasopalatine artery. But this lower lateral nasal field of maxilla is actually L-shaped. It sends out a medially directed lamina based on lateral nasopalatine with its own distinctive respiratory epithelium. This is the nasal lamina of hard palate. At the same time, palatine bone fields P1 and P2 form

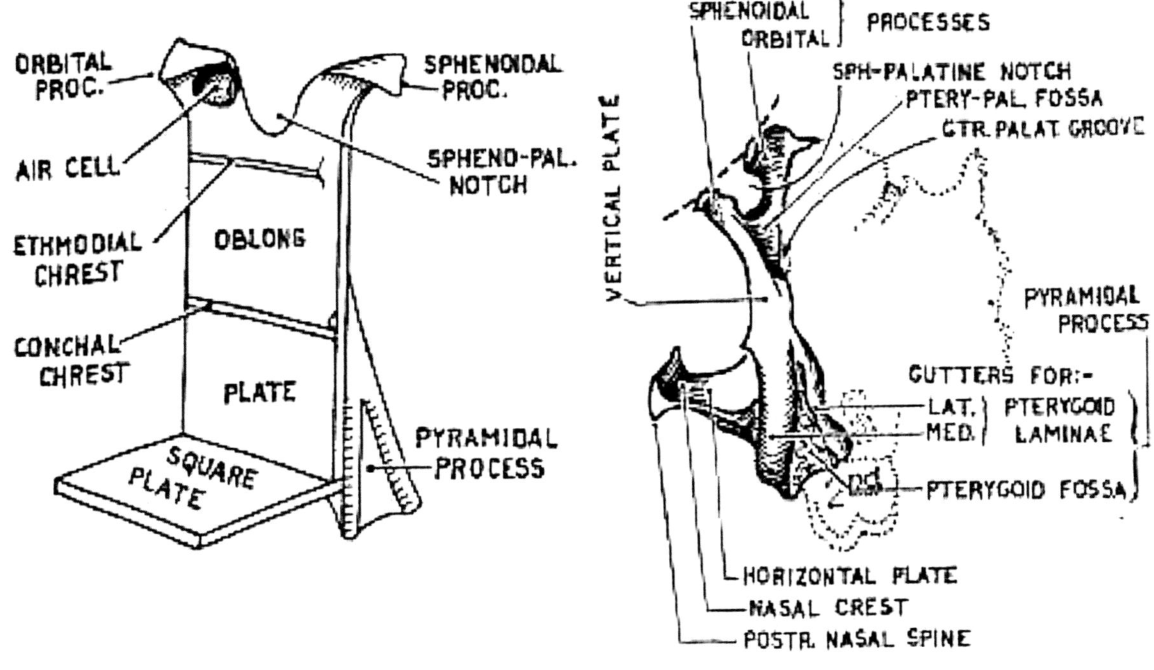

Fig. 14.36 Palatine bone diagram. Medial pterygoid plate represents ectopterygoid. [Reprinted from Frazer JES. Anatomy of the Human Skeleton, 2nd ed. London, UK: J&A Churchill; 1920]

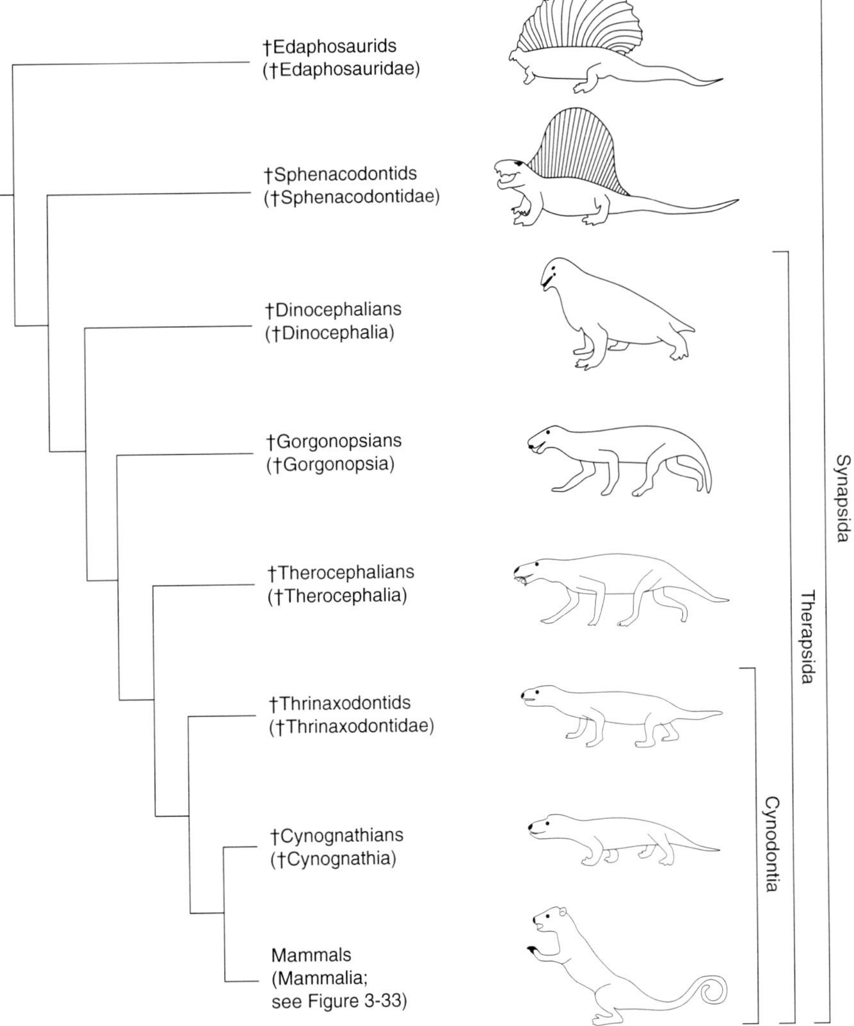

Fig. 14.37 Cladogram showing evolutionary relationships leading up to the tetrapods. Palate evolution began with loss of oral teeth and pharyngeal arches in tetrapods but the bones as we know them began to amalgamate with the cynodont. Soft palate development took place with mammals. [Courtesy of William E. Bemis]

oral lamina of hard palate, supplied by greater palatine artery. This bone field is covered by oral mucoperiosteum. All cleft palate surgeons recognize this difference is tissues as the two epithelial surfaces are different in thickness, color, texture, and histology. Furthermore the oral surface has a well-developed periosteum and salivary glands (Fig. 14.33).

Note The oral mucosa has a unique zone interposed between hard and soft palate, that is, over the anterior palatine aponeurosis "bare area," which contains taste buds supplied by greater superficial petrosal of VII whereas posteriorly they are innervated by glossopharyngeal nerve [20].

There are four lines of evidence for the bilaminar nature of the palatal shelf. First, like other membranous bones such as frontal bone, it has a dual blood supply (and a dual source of programming). Second, the palatal shelf contains marrow (characteristic of flat bones of the skull). Third, formation of a sinus within the palatal shelf has been documented in primates. Fourth, this bone can be tooth bearing. This is not surprising. Primitive vertebrates had multiple rows of teeth [21].

The evolutionary relationship between maxilla and palatine fields is preserved. Naturally each wall of the alveolus shared a common dental unit. These codings extend outward to the palatal shelves.

- Recall that palatine bone fields P1 and P2 are in register with maxilla.
- Recall further that the external alveolar housing of maxilla is supplied by two pedicles: anterior superior and posterior superior alveolar artery.
- This suggests that P1 is in register with maxillary incisor, canine, and the bicuspids while P2 is in register with the molars.

For many readers, the concept of the hard palate shelf as a bilaminar structure may come as a shock, but comparison of the maxillary alveolus with the mandibular alveolus is most reassuring.

- Both structures develop within the first pharyngeal arch in direct opposition to each other (thereby explaining occlusion).
- Both structures are tooth bearing—recall that teeth are always housed in bilaminar bones.
- Both structures, in the evolutionary record, are composed of distinct bone fields.
- Both structures have distinct arterial supplies to their labiobuccal versus lingual walls, and both have distinct arterial supply to the dental units.
  - The posterior superior alveolar artery is the primary supply to the teeth. It is the analog of the inferior alveolar artery of the mandible.
- Thus, the functional design of the maxilla is virtually the same as that of the mandible. The structures have a different shape and different names, but the functional significance remains the same.

Development of P1–P2 Hard Palate: MxP

Based on the histology and the robust blood supply, we determine that the palatine fields supplied by the GPA neuroangiome are the dominant component of the hard palate with the nasal lamina being secondary. Thus, from the developmental standpoint we concentrate strictly on the GPA. Furthermore, in deference to long-standing terminology (and for the sake of brevity) we shall refer to P1–P2 hard palate **MxP**.

Ossification of the maxillary hard palate starts at the incisive foramen and progressively sweeps backward as more mesenchyme is to be added on from proximal to distal (anterior to posterior) and from medial to lateral. The additional mesenchyme demands that new arterioles sprout off from the main axis. For this reason, clefts of the secondary hard palate begin posteriorly and medially. As the mesenchymal deficiency worsens, the cleft extends forward until it terminates at the incisive foramen. In exactly the same way, the neurovascular axis to the lingual lamina of the premaxilla adds mesenchyme from mesial to distal, that is, from proximal to distal. On the buccolabial side, the same pattern is seen. The medial branch of anterior superior alveolar ossifies and sustains eruption before the lateral branch of ASAA.

Clinical Correlation: P1–P2 (MxP) Defects Versus P3 Defects—Which Comes First?

Under almost all circumstances, clefts of the maxillary hard palate occur in the presence of a defect in palatine bone synthesis. This posterior-to-anterior gradient is based on the developmental sequence of the descending palatine artery. DPA supplies the perpendicular plate of the palatine bone. Having reviewed the various processes of the palatine bone, it is clear that a knockout of DPA would affect multiple bone fields of the maxillary complex and the development of the orbit itself. Such a situation would be incompatible with life. However, having synthesized the perpendicular plate, the DPA produces two terminal branches, lesser palatine to the horizontal plate (P3) and greater palatine to the maxillary hard palate (MxP). Although these neurovascular axes are developmentally independent, we know from the ossification sequence that the horizontal plate is constructed on the prior template of the maxillary hard palate. Thus, clefts of P2 result in clefts of P3h, but isolated clefts of P3h do not affect MxP. Isolated defective synthesis of the horizontal plate (maxilla normal) is always accompanied by abnormal soft palate anatomy because LPA supplies both the bone and the tensor aponeurosis.

In the submucous cleft, a palpable notch may extend forward into the posterior maxillary shelf. The presence or absence of soft tissue clefting depends strictly on how closely the bone fields approximate one another in the midline. If the critical contact distance is not exceeded, soft tissue closure will take place.

What about the situation of a normal palatine field constructed on an intact, but foreshortened maxilla? In this case, the soft palate is completely normal but displaced forward. Such a situation can produce noncleft velopharyngeal incompetence (VPI) simply by virtue of an expanded retropalatine space which the soft palate cannot adequately close.

Clinical Correlation: Dental Eruption Sequence

Dental eruption can be rationalized based on a vascular model in which two distinct axes supply neural crest ectomesenchyme with a subsequent remodeling into the final adult configuration. *As a clarification*: infraorbital artery and anterior superior alveolar artery (ASAA) are one and the same. In early development, the medial branch of ASA supplies a potential lateral incisor and the canine, whereas the lateral branch of ASA supplies the primary molars. Dental development spreads out from mesial to distal from the ASA. Thus, medial branch of ASA contributes to an ectopic lateral incisor (9–13 months); it then supplies the primary canine (16–22 months). Lateral branch of ASA supplies successively the first primary molar (13–19 months) and subsequently the second primary molar (25–33 months).

The adult eruption sequence is analogous: lateral incisor—aberrant (8–9 years) followed by canine (11–12 years)/ first premolar (10–11 years) and second premolar (10–12 years). Posterior superior alveolar artery makes its debut at age 6 with the first permanent molar. As bone is added on posteriorly to the alveolus, the remaining adult molars develop, each with its assigned branch of posterior superior alveolar artery (PSAA). The second and third permanent molars appear, respectively, at 11–13 and 17–21 years. In summary, one can easily rationalize the sequence of dental eruption by knowing two principles: (1) Maturation sequence of the three vascular branches (medial/lateral ASAA, PSAA) is mesial to distal and (2) medial ASAA can potentially form an "ectopic" incisor at the same time as the premaxilla forms its own version of the lateral incisor.

The final configuration of the arterial supply to the teeth can be quite different. Within the dental space of the alveolus, the arterial components of ASAA can unite to form a single vascular arcade with PSA. If the vertical contributions of ASAA to the dental units become attenuated, the final anatomy appears as if all the teeth were originally supplied by PSAA. In point of fact, this is just an artifact of vascular remodeling.

Palatine Bone, Proper: P3 (Figs. 14.14 and 14.15)

The palatine bone (P3) develops in membrane from r2 neural crest. The epithelial "program" determining the size and shape of the bone is provided by the r2 nasal and r2 oral mucosa. Descending palatine artery provides its blood supply. This unusual bone plays a key role in connecting the nasal wall, maxilla, and palate with the orbit. It forms the floor of the orbit, inferolateral wall of the nasal cavity, and posterior one-fourth of the hard palate. Its characteristic L-shape results from the intersection of two plates, horizontal and perpendicular, at a slightly acute 80°–85° angle. As we can see, this geometry will have surgical implications for cleft palate repair. It also bears three significant processes (Fig. 14.38).

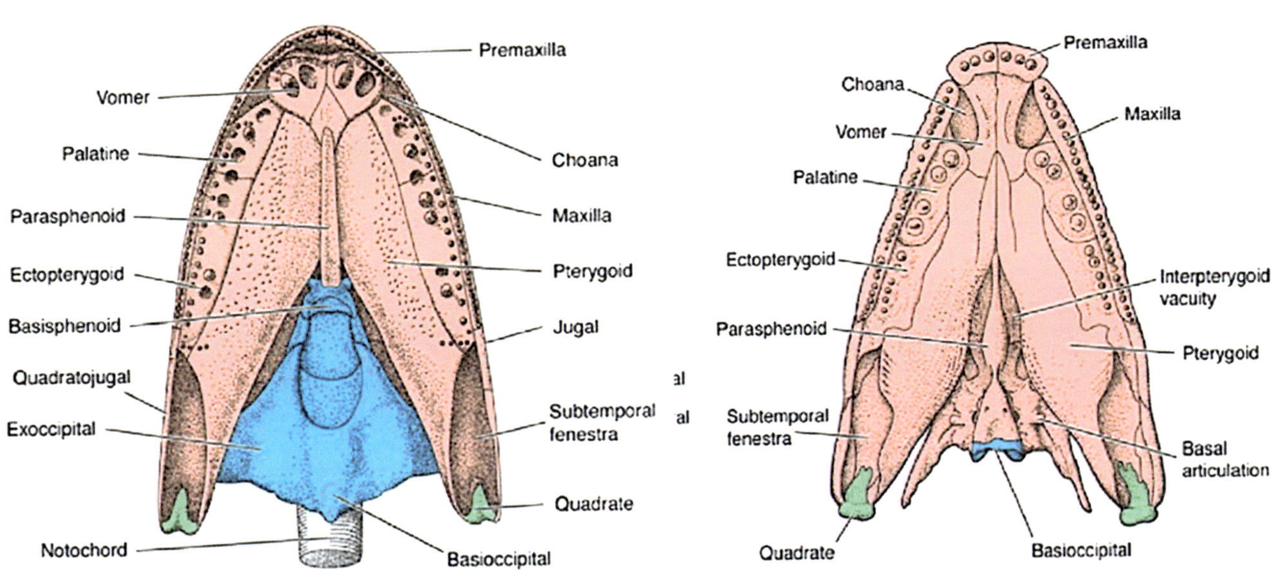

Fig. 14.38 Crown choanate osteichthyan *Eusthenopteron* has the following features. Palatoquadrate cartilage: (1) anterior part covered by dermal bone-maxilla; (2) internally dermal bones ossified to the palate series seen uncovered posteriorly. These give rise to metapterygoid/epipterygoid. [Courtesy of William E. Bemis]

Perpendicular (Orbital) Plate

Perpendicular plate is a thin lamina with medial and lateral sides plus four borders. The medial side is scooped out, forming the posterior part of inferior meatus. Just above that, a transverse ridge, the conchal crest, articulated with the palatine process of inferior turbinate higher up; a second transverse ridge, the ethmoid crest, provides articulation for middle turbinate. Tucked between the two crests is the middle meatus. Still higher up, above ethmoid crest, is a third depression, for superior meatus. The lateral side is rough and irregular; it articulates with maxilla. Along its posterior aspect, a well-chiseled vertical greater palatine groove marks where the maxilla and perpendicular plate enclose the neurovascular bundle. The anterior border is punctuated by a forward projection, the maxillary process, at the same level as the conchal crest. The maxillary process forms the posterior wall of the medial maxillary sinus; it articulates with inferior turbinate. The rough posterior border articulates with the r1 medial pterygoid plate. This expands into the pyramidal process into which medial pterygoid is inserted. The superior border contains a forward-facing orbital process and a backward-directed sphenoidal process. Separating the two processes is the sphenoid notch that, by articulating with sphenoid, forms the sphenopalatine foramen, an all-important landmark because it communicates between the pterygopalatine fossa and the posterior nasal cavity. It transmits the sphenopalatine neurovascular pedicle and the posterior superior nasal nerves into the nasal cavity.

- The *orbital process* contains an air cell. It has two nonarticulatory and three articulatory surfaces. Superior surface forms the posterior orbital floor. Lateral surface contributes to inferior orbital fissure. Anterior surface articulates with maxilla. Posterior surface contains the opening of the air sinus, which can communicate with the sphenoid sinus. Medial surface articulates with ethmoid labyrinth. Sometimes, this can serve as an alternative escape route for the palatine air sinus, which communicates with the posterior ethmoid cells.
- The *sphenoidal process* has three surfaces and three borders. Superior surface connects with medial pterygoid plate and abuts against the sphenoid concha. Medial surface pokes its tiny nose into the nasal cavity. Lateral surface connects with lateral pterygoid plate. Anterior border forms the posterior margin of the sphenopalatine foramen. The medial border of sphenoid process, by virtue of its proximity to the sphenoid bulla, can reach as far as the ipsilateral ala of the vomer where it receives the sphenoid rostrum.

Horizontal Plate

The position of the horizontal plate in space is determined by that of the horizontal plate of the maxilla (MxP). If MxP is reduced in length (its AP dimensions), the horizontal plate of an otherwise normal palatine will likewise be dragged anteriorly. The nasal surface is smooth and concave. The oral surface is rough. It bears near the midline a transverse elevation, the palatine crest, which serves as an attachment for the aponeurosis of tensor veli palatini. The anterior border is irregular, forming a rough articulation with the accessory palatine bone. The posterior border provides attachment for the palatine aponeurosis. The lateral border forms an acute angle (80°–85°) with the perpendicular plate. Two bony prominences project from the fused medial borders. Musculus uvulae inserts into posterior nasal spine. The nasal crest articulates with vomer.

LPA is the principal blood supply to horizontal plate. The mesenchyme to form the horizontal plate is deposited from lateral to medial and from anterior to posterior to form a rectangle. If mesenchymal deficiency occurs, it will reduce the horizontal plate in the opposite directions, creating a triangle base laterally and anteriorly.

The pyramidal process projects backward from the intersection between horizontal and perpendicular plates. Its posterior surface insinuates itself into a crevice between the medial and lateral pterygoid plates. An important landmark is the lateral surface is a roughened zone along its anterior aspect. This will abut against the tuberosity of the alveolus. The lesser palatine foramen is located at the intersection between pyramidal process and horizontal plate. The greater palatine canal is formed by the bilaminar confluence of pyramidal process (lateral) against the perpendicular plate (medial). It transmits the descending palatine artery.

Not to be left out, medial pterygoid plate has a *vaginal process* that articulates with posterior border. The source of medial pterygoid plate is ectopterygoid: palatine bone field P4.

Development of the Palatine Bone

Ossification is observed immediately after the embryonic period at the eighth week of fetal life (stage 23, 20 weeks). The epicenter is between the horizontal plate and the perpendicular plate. It spreads out in three directions: (1) medially directed horizontal plate and (2) posteriorly directed pyramidal process develop first, followed 2 weeks later by (3) superiorly directed perpendicular plate. All three of these axes are independent.

At birth, the horizontal and perpendicular plates are equal in size. Over time, increasing depth of the nasal chamber increases the relative size of the perpendicular plate. As the perpendicular plate extends upward, it encounters the preexisting sphenopalatine neurovascular axis. The osseous mesenchyme splits around the pedicle. The anterior orbital and posterior sphenoid processes result. In point of fact, the complexity of the various processes of palatine bone stems from its relative insertion into previously established bone fields such as sphenoid, ethmoid, and maxilla. The palatine mesenchyme just has to fit in where it can.

As the horizontal plate develops, new mesenchyme is deposited in a medial-to-lateral and anterior-to-posterior direction. Thus, the newest, most vulnerable zone of the horizontal shelf is medial and posterior, that is, the posterior nasal spine.

Clinical Correlations

The mesenchymal masses of perpendicular plate and horizontal plate are developmentally independent. This is reflected in robust blood supply to the perpendicular plate from descending palatine artery. Two or three foramina in the perpendicular plate allow for exit of small neurovascular pedicles to supply the mucosa posterior to it. These anastomose with the blood supply to the tonsil. As descending palatine artery (DPA) nears its junction with the horizontal plate, it splits into two branches. Lesser palatine artery (LPA) is given off as multiple branches before DPA exits as the greater palatine artery (GPA) to the maxillary hard palate. LPA supplies the oral structures of the soft palate and it perforates through two small foramina to supply the nasal surface as well. Thus, deficiency of the vascular axis to horizontal plate can occur independent of the more proximal supply to perpendicular plate or to its other distal terminus, the greater palatine field.

Developmental wipeout of the entire descending palatine axis is structurally incompatible with life. Mesenchymal deficiency can occur in the distal horizontal plate of palatine, in the distal maxillary hard palate, or in both zones. Both neuroangiosomes demonstrate the pattern of tissue loss, according to their developmental gradients. Lesser palatine insufficiency occurs first at its most distal and posterior territory, the posterior nasal spine. Progressive LPA deficiency works its way laterally and anteriorly. Greater palatine insufficiency occurs first at its most distal and posterior territory. The newest branches of GPA, those supplying the medial edge of the posterior palatal shelf, are the most vulnerable. Thus, if failure of bone mesenchyme occurs, it follows a gradient of severity, from medial to lateral and from posterior to anterior, all the way forward to the incisive foramen. This explains the clefting pattern of the secondary hard palate.

Part 6. Evolution of the Palate (Figs. 14.38, 14.39, 14.40, 14.41, 14.42, 14.43, 14.44, 14.45, 14.46, 14.47, 14.48, 14.49 and 14.50)

Fast Forward: From Placoderms to Mammals

Note: Recommended readings for this section can be found in Benton, and Kardong, and Liem (in the reference section).

Elements of the facial skeleton first appeared about 500 million years ago with jawless fishes having seven or more gill arches. Chondrichthyan fishes had the first cartilaginous

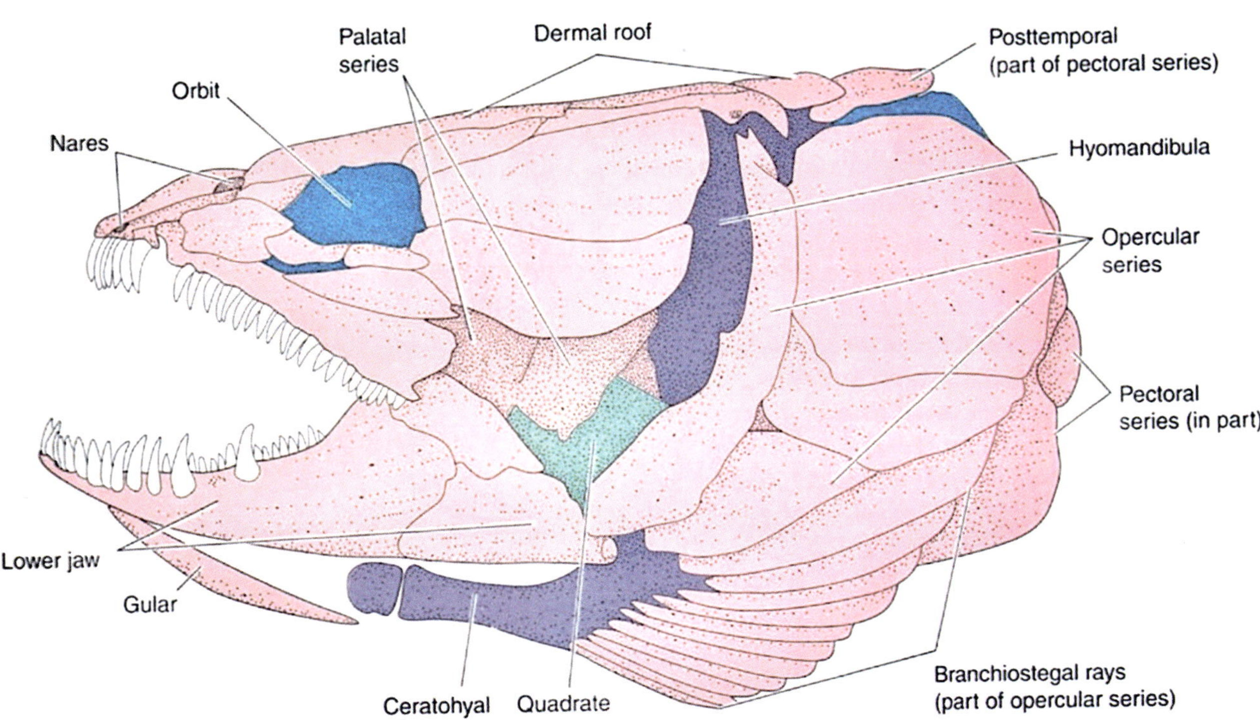

Fig. 14.39 Palate evolution between advanced sarcopterygian fishes and tetrapods. Hard palate of choanate fish *Eusthenopteron* consists of palatine, ectopterygoid, pterygoid, and parasphenoid. (1) Epipterygoid dorsal to pterygoid; (2) epipterygoid suspended from the skull vomer, palatine, and ectopterygoid are tooth bearing. Premaxilla and maxilla are just dermal. Parasphenoid is short, covers presphenoid only. *Paleoherpeton* shows palatal bones in contact with skull base but mobile interpterygoid vacuity: potential space between palate and skull base choanae open into anterior mouth. No separate nasal passage. The vomer is no longer tooth bearing, premaxilla becomes tooth bearing. [Courtesy of William E. Bemis]

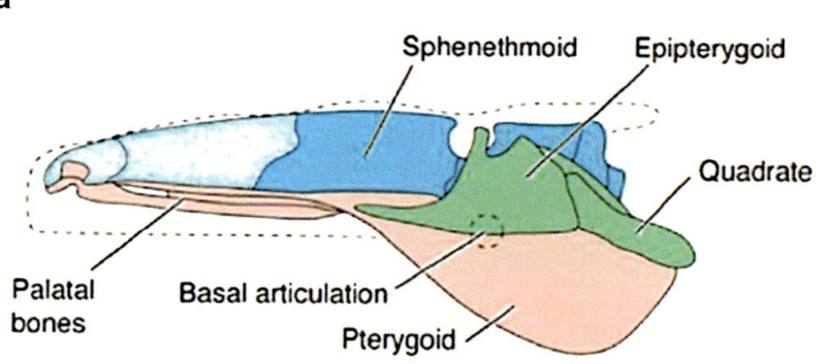

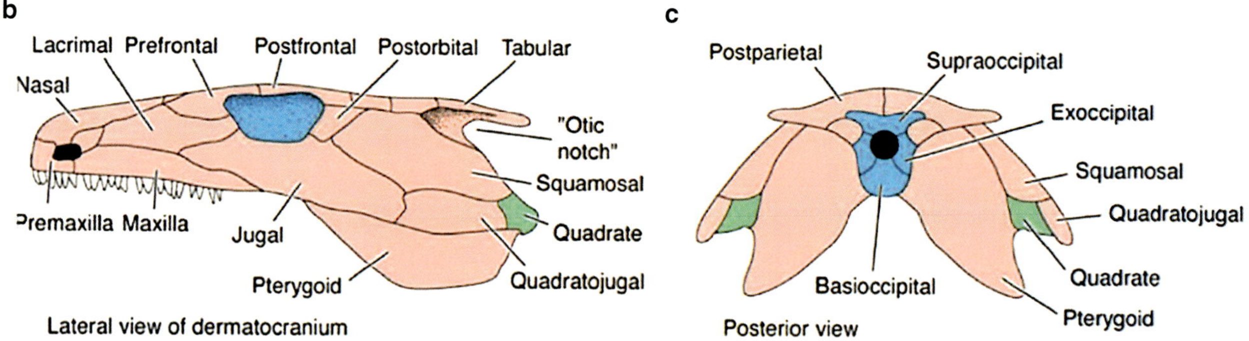

Fig. 14.40 Evolution of epipterygoid (Epi) 1. This key bone (the future alisphenoid) was originally designed as a strut between the palate and the brain case. Lateral side wall of primitive tetrapod shows epipterygoid and quadrate lying deep to external dermal bones (jugal, quadratojugal, postorbital, and squamosal). Biting muscles are deep to roofing parietal/postparietal. With brain expansion the lateral walls undergo changes. Nonmammalian cynodonts have partially ossified sidewalls for the orbital region and braincase. As we have seen, the therian sidewall expands but the adductor mandibulae muscle becomes externalized to the outside of parietal. Here we see the suspension function of epipterygoid (Epi) as it articulates between pterygoid and the roof of the cranium. Epi articulates of the otic capsule behind it. In mammals, Epi will separate from periotic, move forward, and a part will become externalized as the lateral lamina of the temporal fossa. [Courtesy of William E. Bemis]

Fig. 14.41 Evolution of epipterygoid 2. Here we see Epi in a primitive amniote making part of the right lateral sidewall. It is largely isolated, the remainder of the wall being fibrous. In mammals, alisphenoid has an orbital lamina in contact with orbitosphenoid and presphenoid. Its temporal lamina articulates with frontal, parietal, and squamosal. [Courtesy of William E. Bemis]

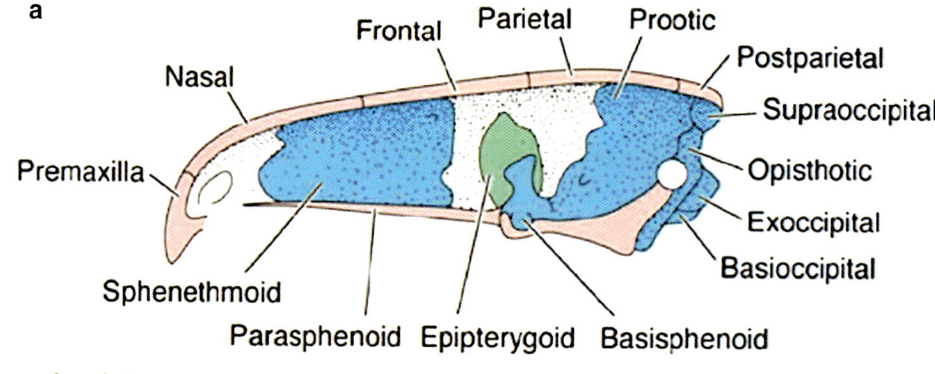

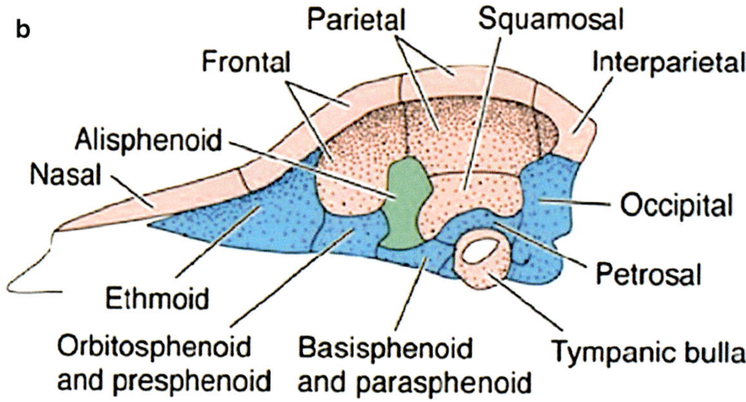

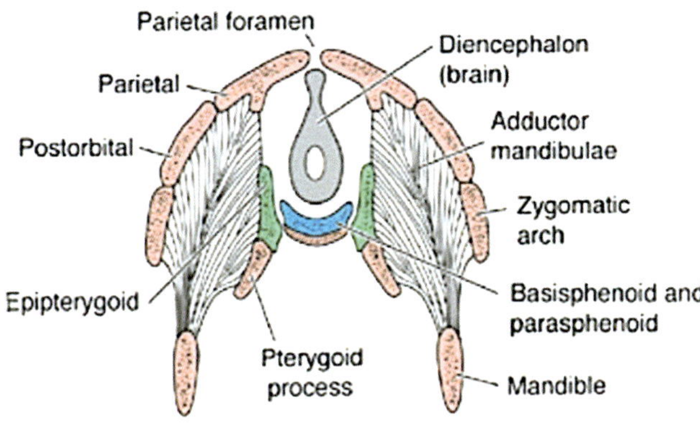

Fig. 14.42 Evolution of the hard palate: anapsid condition seen in reptiliomorphs just before division between reptile/bird line and the mammalian line. Note this coronal section is very posterior. Horizontal shelves not yet produced by maxillae and palatine bones epipterygoid articulates with pterygoid. No temporal fenestra and no zygomatic arch. Parasphenoid (neural crest) articulates with both basisphenoid (mesoderm) posteriorly and presphenoid (neural crest) anteriorly. Defined parasphenoid (neural crest) fuses forward of basisphenoid (paraxial mesoderm) into presphenoid (neural crest). [Courtesy of William E. Bemis]

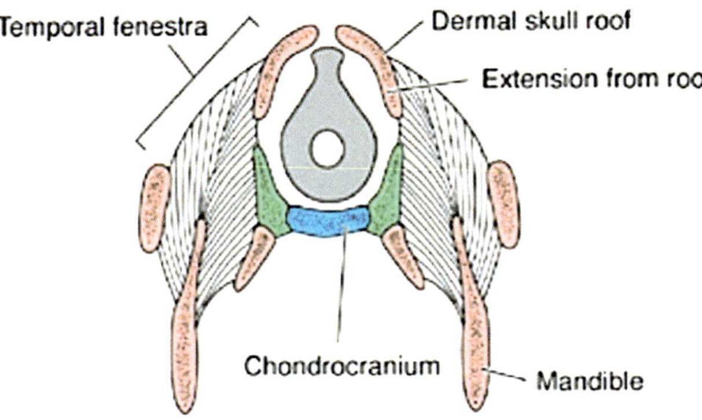

Fig. 14.43 Evolution of the hard palate: synapsid condition (pre-mammal). Hard palate is still kinetic. The postorbital moves forward, temporal fenestra opens up, and zygomatic arch is defined. Parasphenoid (neural crest) shifts forward, is no longer bound to basisphenoid (paraxial mesoderm), and has exclusive contact with presphenoid (neural crest). [Courtesy of William E. Bemis]

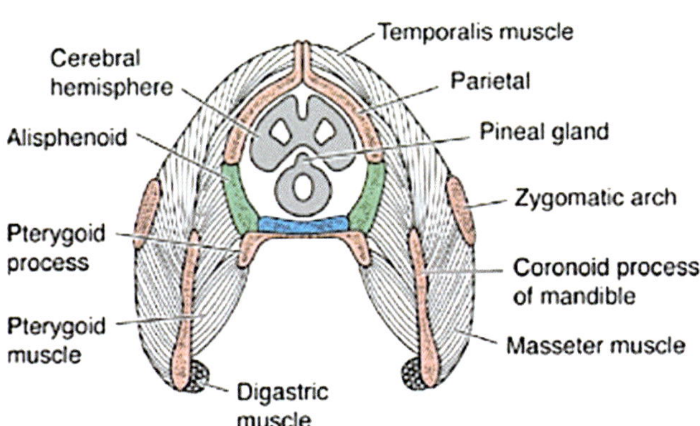

Fig. 14.44 Evolution of the hard palate: mammalian condition. Epipterygoid joins the orbit, becoming alisphenoid. Hard palate is no longer kinetic. Ectopterygoid = pterygoid plate of sphenoid (hamulus). Pterygoid = lateral pterygoid plate of sphenoid. Parasphenoid completely absorbed into presphenoid... possibly the vaginal process that receives the vomer. [Courtesy of William E. Bemis]

skull. They also remodeled the first two gill arches into upper and lower jaws. Shortly thereafter, bony fishes (the Osteichthyes) developed neural crest dermal bones which were used to ensheath the jaws and the brain. Genetic programs embedded in both palatoquadrate and Meckel's cartilages served to program a series of plate-like dermal bones on both the buccal (external) and lingual (internal) sides.

Osteichthyes then diverged into two lines, based on anatomy of the support system for their fins. *Actinopterygians*, the ray-finned fishes, having bony struts controlled by muscles inside the body cavity, became wildly successful. The mammalian line traces back to fossil *sarcopterygian* fishes, named for their fleshy fins with muscles outside the body cavity. These gave rise to the rhipidistians, or lungfishes, a specialization of which, the choanata, developed a mechanism to breathe air into the oral cavity via the nose.

From the choanate fishes, tetrapods emerged on land with critical innovations. Having inherited choanae and lungs from the rhipidistians, tetrapods were air-breathers. The pectoral girdle was completely free from the skull, permitting the evolution of a neck and mobile head of great advantage for capturing prey. The spinal column was strengthened by weight-bearing articulations, zygapophyses. Digit-bearing limbs appeared, with a baseline of six to eight. The next great subdivision arose with the invention of amniotic membranes that permitted birth to take place on land, avoiding the aquatic phase of development. Primitive tetrapods developed into reptiliomorphs that bifurcated into anamniota (amphibians) and amniota (reptiliomorphs: the future reptiles, birds, and mammals). In the latter clade, the digital formula stabilized at five. Changes in predation associated with specialized chewing muscles are manifested to flanges of the sphenoid bone, the modern-day pterygoid plates.

Reptiliomorphs again bifurcated into the line producing true reptiles and birds and the pelycosaurs leading to therapsids and then mammals. The descendants of reptiliomorphs are classified by the fenestration pattern of the temporal fossa. These represent attachments for chewing muscles and give information about control of the mandible. *Diapsids* have both an upper and a lower fenestra behind the postorbital and squamosal bones and are represented by reptiles and birds. All mammals are *synapsids*, having a single temporal fenestra low on the skull and bounded by a bony bar (the future zygomatic arch) running postorbital (the precursor of the upper half of zygoma) and the squamosal bones. Further classification is based on changes in dentition, mastication, and locomotion.

The skull of the basal synapsid dimetrodon shows many changes related to a carnivorous lifestyle with powerful jaw muscles. Many bones changed names. The limbs remained sprawled out, with the femurs being horizontal. Members of the next transition, *Therapsida*, developed large canines. Their limbs were now directly below the body, making them agile runners and effective predators. A subsequent subgroup of therapsids, the cynodonts, is proximal to mammals. The

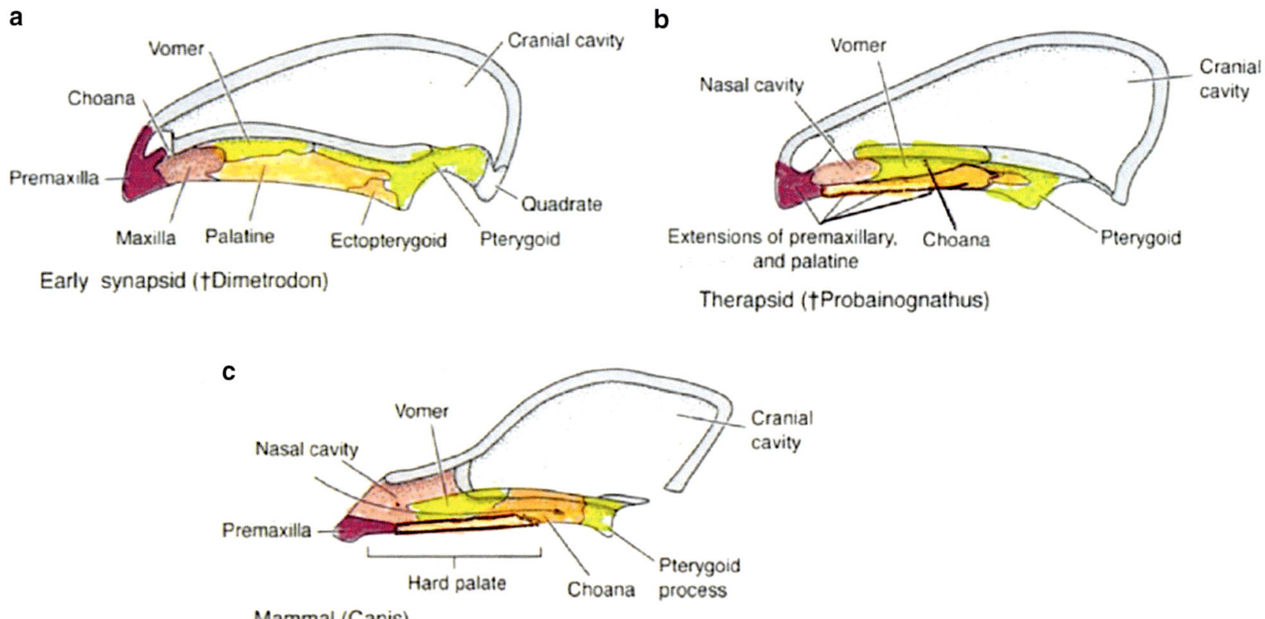

Fig. 14.45 Secondary hard palate. Vomer and palatine bones produce shelves that extend medially. Maxilla produces the nasal lamina with a separate neuroangiosome. Pterygoid/ectopterygoid interposed between pterygoid and maxilla. Secondary hard palate shelves. Vomers come to midline and fuse. Maxilla expands and becomes anterior to palatine. Palatine is reduced and becomes posterior behind the alveolar process. Pterygoid becomes lateral pterygoid plate of sphenoid. Ectopterygoid becomes medial pterygoid plate of sphenoid (hamulus). Perpendicular plate of ethmoid well developed. [Courtesy of William E. Bemis]

dermal bones that ensheath the cynodont mandible meld together for the first time to form a "single" unitary dentary bone. Former bone fields were either suppressed (remaining as genetic entities) or displaced into the temporal region to form the ear bones. Cynodonts developed postcanine teeth with cusps for grinding. Part of temporalis gave rise to temporalis attached to the arch. The presence of a well-developed palate leads one to the conclusion that the cynodonts were endothermic.

Although there is undoubtedly a stem mammilomorph, it has not yet been identified, and perhaps will never be. That being said, from the stem there are three branches. In one direction, we find the extinct rodent-like *multituberculates*. These creatures are interesting because they chewed backward, with a *palinal*, (front-to-back) jaw stroke; so their muscles of mastication were backwards. In the other direction is the mammalian line with a TMJ permitting *propalinal* (back-to-front) and side-to-side jaw strokes for grinding, thus the nonplacentals: *monotremes*, such as the duck-bill platypus. Our ancestral line includes the *monotremats* such as duck-bill platypus and the *theria* (Greek, *therion*, wild beast), the nonplacental *metatherians* (marsupials), and placental *eutherians* (true mammals).

The reason for this detail is that the so-called "first mammal," *Morganucodon*, is also the first to develop a hard palate to full length.

Bones of the Oral Cavity: A Portrait in Broad Brushstrokes

The skull of earliest craniates (jawless fishes) consisted of a hypaxial splanchnocranium, referring to the support structures of the gill arches (up to nine in all) and an epaxial chondrocranium at the brain. Some species were able to ossify their endoskeleton including the skull base. They also produced an exoskeleton in the form of a head shield consisting of dermal bones, the dermatocranium. The original function of splanchnocranium was strictly respiratory, but, with the advent of the chondrichthyans, it became an essential tool for feeding as the first two arches morphed into a dorsal palatoquadrate cartilage and a ventral mandibular cartilage. As we have seen, the first true maxillate gnathostomes, the placoderms developed from their bony oral plates the first three bones of the jaw: premaxilla, maxilla, and mandible. Bone-forming capability was secondarily lost in the cartilaginous fishes but was regained in the osteichthyes.

In the bony fishes, ossification returned with the changes in the skull. The posterior part of both palatoquadrate and mandibular cartilages ossified into two opposing bones, quadrate and articular. These form the jaw joints in all vertebrates except mammals (which have a temporomandibular joint). At the same time, these fishes regained the ability to form a dermatocranium. Multiple dermal bones make up the

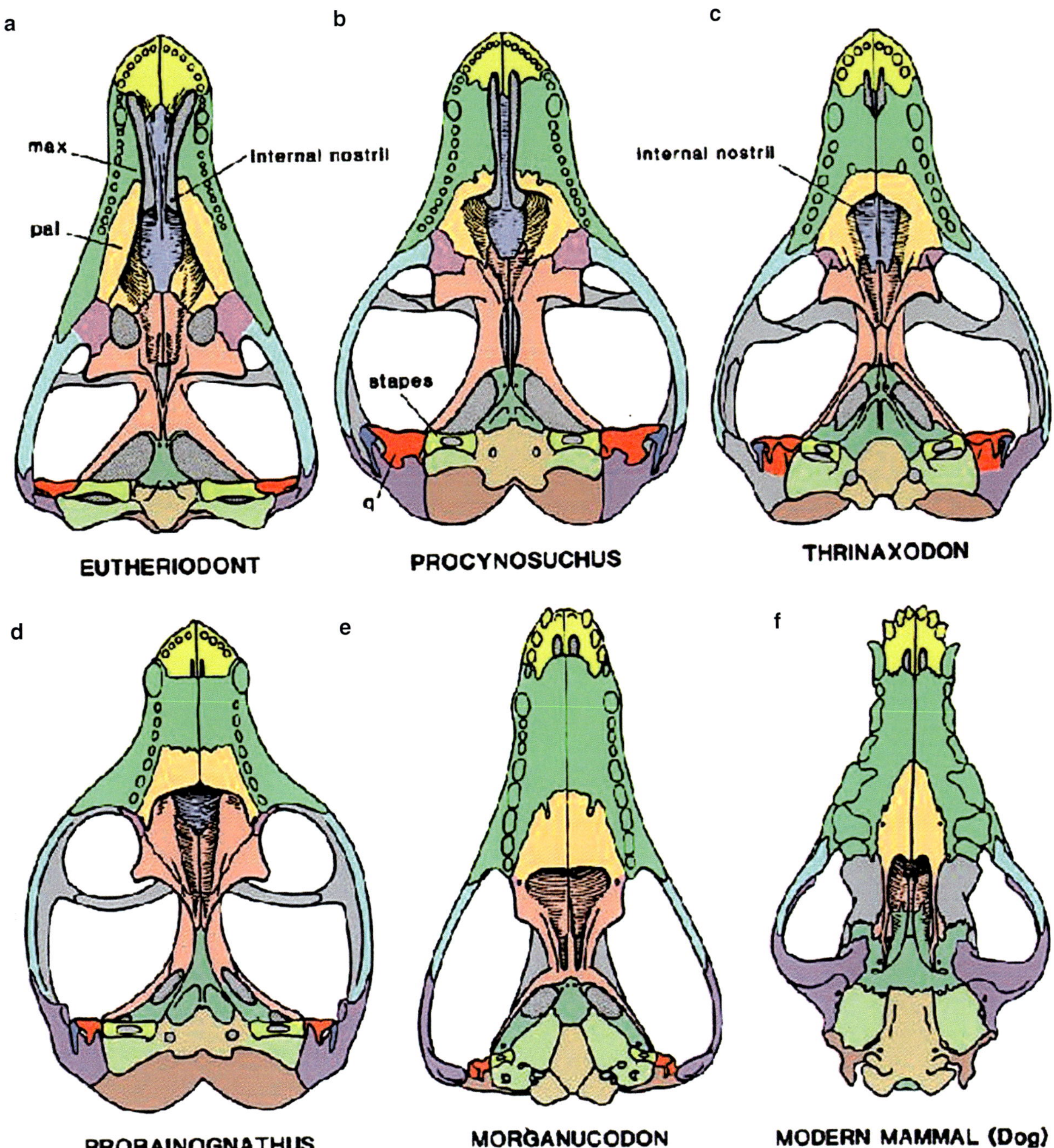

Fig. 14.46 Palatine phylogeny. Color code: premaxilla, yellow; vomer, gray; parasphenoid, lilac; maxilla, green; palatine, tan; pterygoid (medial plate), flesh; ectopterygoid (hamulus), magenta; articular (malleus), blue; quadrate (incus), red; columella (stapes), lemon; prootic/opisthotic (petrosal), purple. Note the extension of pterygoid (brown) in *Eutheriodont* all the way to the occiput. Ectopterygoid (lilac) is just the most posterior of the four palatines. In *Procynosuchus*, the pterygoid moves out of the occiput and insinuates a flange behind ectopterygoid. In *Probainognathus*, pterygoid is incorporating into basisphenoid and ectopterygoid becomes part of future medial pterygoid plate hanging down from sphenoid. Alisphenoid (the future lateral pterygoid plate) does not appear until Mammalia. It is accompanied by another new orbital element, orbitosphenoid. [Courtesy of David Peters]

Fig. 14.47 Similarities of maxilla and mandible. Alveolus of the mandible is a box enclosing marrow, teeth, and a neurovascular bundle. The supporting bone fields are tightly constrained by an extensive muscle envelope. Maxillary-palatine alveolus is the same, with exception of growth of connecting bones. Due to anchorage of the maxilla to the skull base by orbitosphenoid and alisphenoid, expansion of the sinus can only occur laterally and forward. [Courtesy of Michael Carstens, MD]

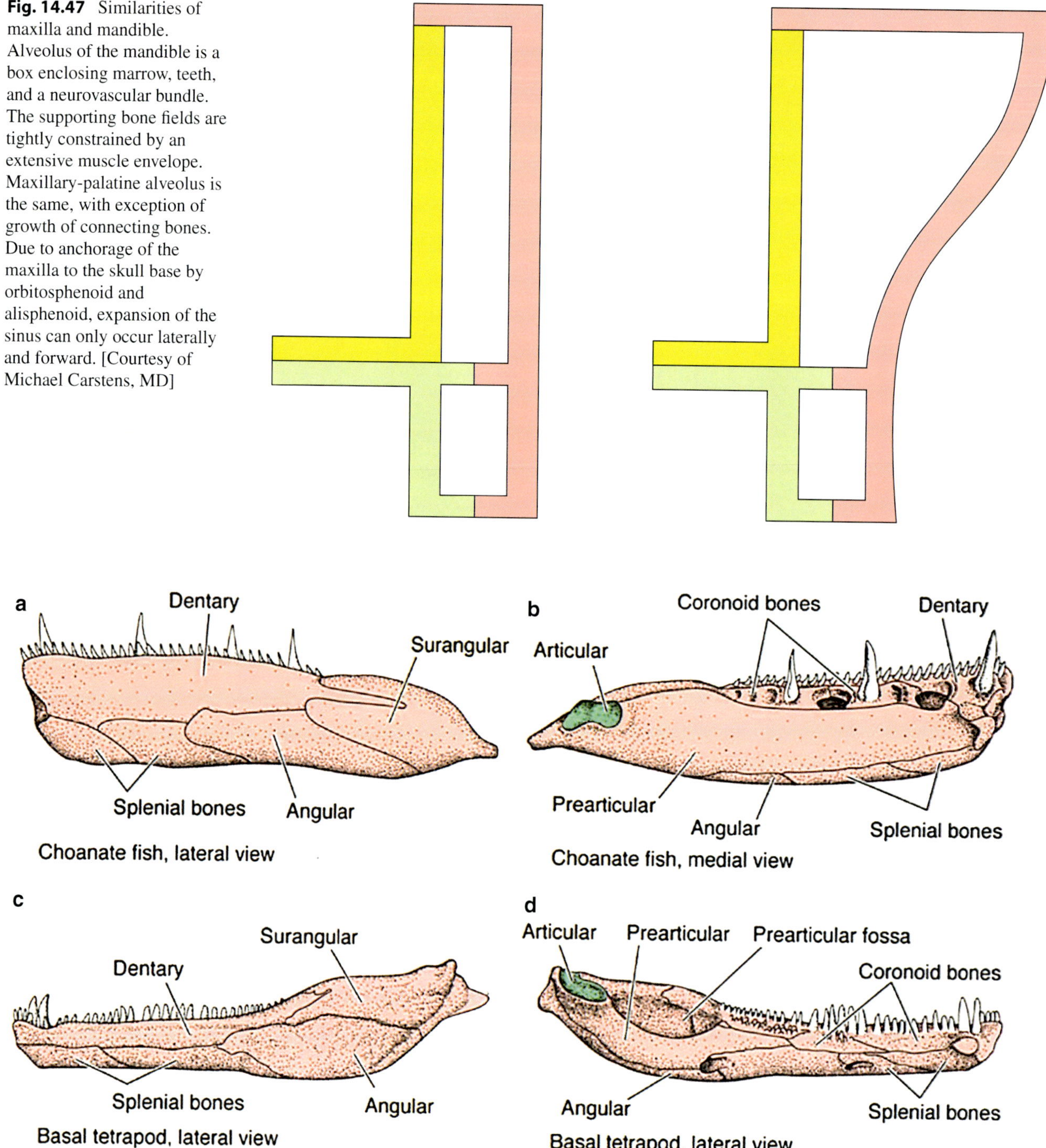

Fig. 14.48 Mandibular bone fields demonstrate bilaminar dermal fields in precursor fish. These consolidate in basal tetrapod. [Courtesy of William E. Bemis]

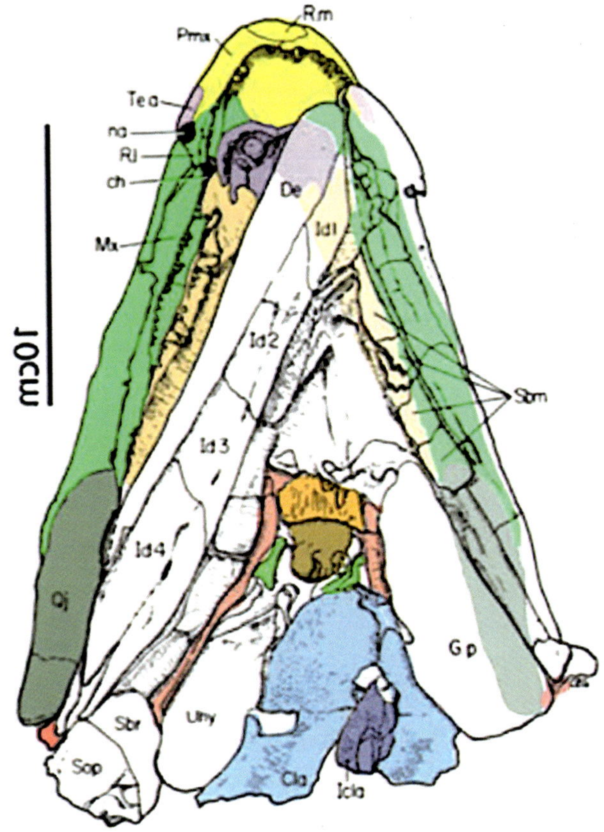

Fig. 14.49 Mandibular bone fields in panderichthys demonstrate the genetic basis of the four dental fields. Sarcopterygian *Panderichnthys* mandible shows four dental fields, Id1–Id4. These zones correspond genetically to those of incisors, canines, premolars, and molars. Note Id1 corresponds in the upper jaw to incisors of premaxilla. Id2–Id4 correspond to palatine bone fields P1–P3, each one of which is supplied by a branch from StV2. Anterior superior alveolar goes to P1 and P2 and posterior superior alveolar serves P3. Descending palatine supplies P4. [Courtesy of David Peters]

Key to Figs. 14.50, 14.51, 14.52 and 14.53

Palate closure: 1–2

Mouse (days post coital) *Human*

10.5–10.75 days pc: Carnegie stage 13

10.75–11.0 days pc: Carnegie stage 14

11.0–11.5 days pc: Carnegie stage 15

11.5 days pc: Carnegie stage 16

1. Medial nasal prominence (MNP)
2. Olfactory pit (becomes primitive anterior naris)
3. Lateral nasal prominence (LNP)
4. Maxillary prominence of first arch (MxP)
5. Mandibular prominence of first arch
6. First pharyngeal groove
7. Second pharyngeal arch
8. Entrance to Rathke's pouch
9. Fusion of LNP + MxP
10. Fusion of MNP + MxP
11. Primitive posterior naris or primary choana. After palate closure this becomes definitive secondary choana
12. First evidence of palatal shelf from maxilla
13. Primitive anterior naris
14. Bulge over primordium of vibrissae
15. Location of primary palate
16. Oblique groove (always seen here)
17. Corneal epithelium
18. Vertically directed medial border of palatal shelf
19. Secondary palate
20. First evidence of rugae
21. The broad ventral surface of the developing vomer
22. Roof primitive nasopharynx
23. Posterior margin palatal shelf
24. Horizontal anterior ½ palatal shelf
25. Vertical posterior 1/2 palatal shelf
26. Short nasal septum, getting narrow
27. Prominent rugae
28. Boundary of maxilla and palatal shelf
29. Bulges overlying primordial whiskers

Key to palate closure: 3–4

1. Bulges over vibrissae (arranged in rows)
2. Location of primary palate
3. Entrance to primary choana (primitive posterior naris). This will subsequently form definitive (secondary) choana after palate formation
4. Horizontal anterior 1/2 palatal shelf
5. Vertical posterior 1/2 palatal shelf
6. Boundary between maxillary body and its palatal shelf
7. Anterior extremity of the narrowing septum
8. Rugae
9. Posterior 1/2 palatal shelf, now horizontal palatal shelf
10. Roof primitive nasopharynx
11. Posterior extremity of palatal shelf
12. Philtrum
13. Initial site of apposition and subsequent fusion of the palatal shelves
14. Posterior (dorsal) extension of apposition and subsequent fusion of the palatal shelves
15. Completion of palate formation (fusion of the posterior halves of palatal shelves across the midline (region of soft palate)
16. Posterior entrance to nasopharynx
17. Naso-palatine canal separating primary palate from secondary palate
18. Longitudinal bulge represents alveolar process containing molars
19. Final site fusion of primary and secondary palates at the incisive foramen
20. Slit-like incisive canal
21. Erupted tip of vibrissa
22. Entrance to nasal canal (external naris)

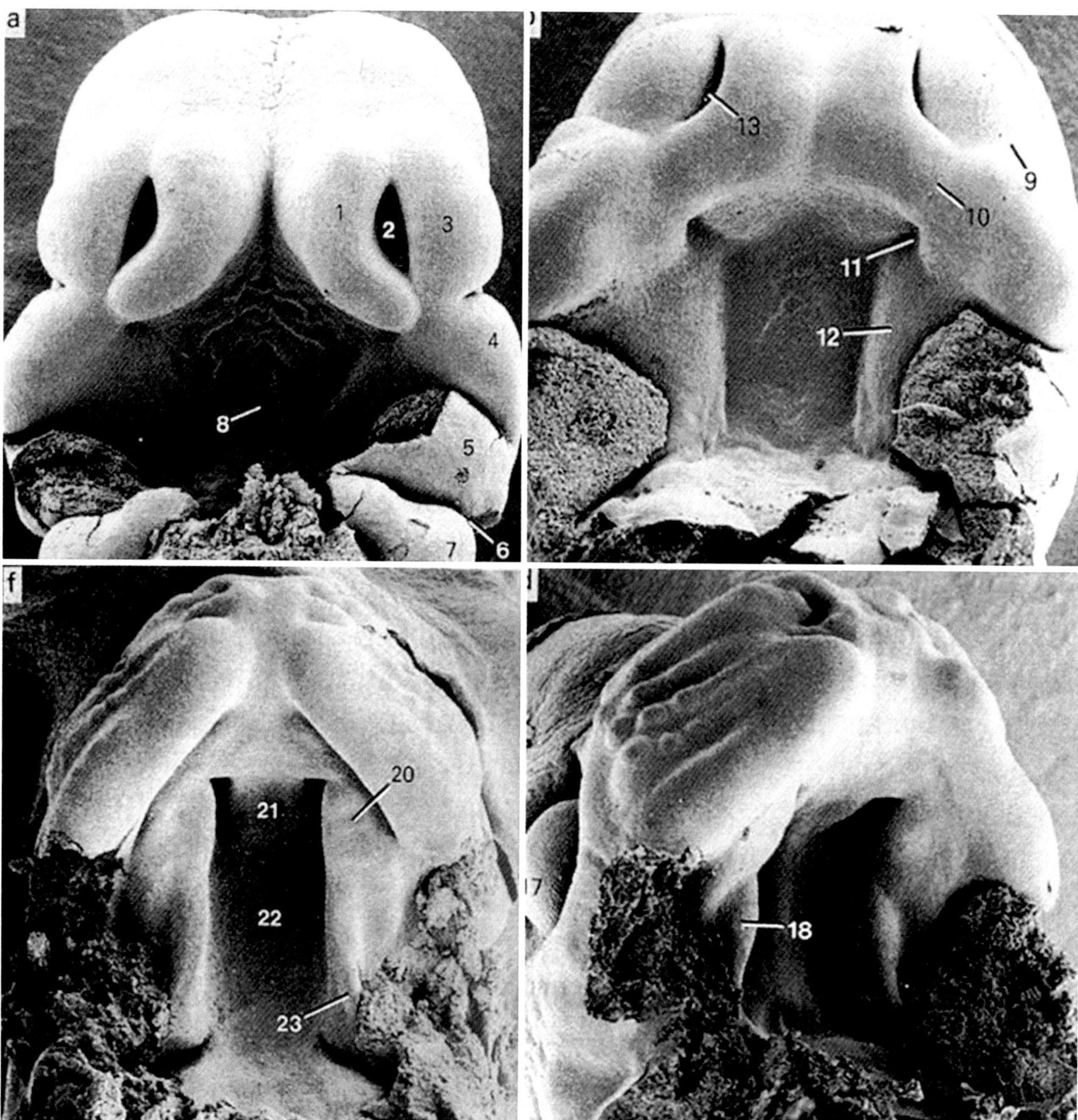

Fig. 14.50 Palate closure 1. [Courtesy of Dr. Kathleen Sulik, University of North Carolina]

roof of the skull and ensheath the jaws. At the time of the choanate fishes such as *Eusthenopteron*, very close to the emergence of tetrapods, these dermal bones cause the face to be elongated for better feeding.

- Multiple bones cover Meckel's cartilage—although these "disappear" in mammals, the genetic "ideas" that programmed them may explain regional differences in differentiation of teeth (Fig. 14.45).
- The same situation obtains to PQ cartilage. Along its outer surface, we find a tooth-bearing maxilla, the back end of which abuts jugal. The inner surface bears a series of four palatine bone fields (P1–P4), pterygoid and parasphenoid bones, all of which are tooth-bearing. P1 and P2 are destined to form the "maxillary" shelf of the hard palate, P3 will become the palatine shelf of the hard palate, and P4 will become ectopterygoid (so named for its relationship to the back corner of pterygoid bone); it is the source of medial pterygoid plate. Paired vomers are in contact with the palatine and pterygoid bones. A vertical intracranial joint at the center of the chondrocranium permitted the mobility of the upper jaw and front of the skull (Fig. 14.23).

We now turn our attention to tetrapods, beginning with the basal *Paleoherpeton*. Choanae appear in the mouth on either side of premaxillae. These are likely preserved today as the nasopalatine canals. Just behind the premaxillae, the vomers no longer have teeth. Lying above the pterygoids (but not attached), epipterygoids represent a persistent mobile articulation with the brain case. They are not seen on the palatal view. They will become alisphenoid. The interpterygoid vacuities between pterygoids and chondrocranium represent the plane of movement. This kinetic skull was not well adapted to predation with heavy bite forces (Fig. 14.46).

Recall from our discussion of alisphenoid and parietal bone evolution that the braincase undergoes a radical change to accommodate the expanding CNS. This is seen best in cross-section (Fig. 14.46).

- The anapsid condition shows the palate suspended from the skull roof by epipterygoids. The lateral skull is made of parietal, postorbital, and zygomatic arch in series. The parietal is *outboard* to the chewing muscles. The lateral skull is open such that the brain is separated from adductor mandibulae by the epipterygoid fascia. Note that cranial base basisphenoid is separated from the epipterygoid. Dermal parasphenoid runs in the midline below chondral basisphenoid.
- The synapsid condition shows the incorporation of parasphenoid into the cranial base, its remnant being the nasohypophyseal duct. Alisphenoid is firmly welded to basisphenoid… this drags AS forward and solidifies its suture with orbitosphenoid. Postorbital is incorporated into the orbit, leaving behind a space between parietal and zygomatic arch: the temporal fenestra. The zygomatic arch becomes bowed outward. Parietal is now isolated *inboard* to the adductor. The brain is kept separate by the new, expanded lateral wall of the braincase, the vertical union between parietal and alisphenoid.
- The mammalian condition demonstrates complete union between parietal and alisphenoid but, whereas AS shifted forward, the petrosal sends an anterior lamina forward to close up the resulting space: this becomes the squamous temporal bone. Pterygoid incorporates into presphenoid producing medial pterygoid plate. Its devoted follower ectopterygoid becomes lateral pterygoid plate.

The lineage of mammals began with the synapsid skull and passed through the sailed reptiles. *Dimetrodon* demonstrates anterior choanae. The primary palate lies dorsal, just beneath the skull base. There are no nasal cavities. Therapsids became more mammal-like with changes in limb positioning and in the cranial skeleton. Perhaps the driving force in the therapsids relates to feeding. The possibility to extract maximum nutrition for high metabolism favored the need to cut up food and chew it before swallowing. As jaw muscles got stronger, the kinetic palate became united to the braincase. This involved the conversion of epipterygoid to alisphenoid (Fig. 14.46).

The advanced therapsid line, the cynodontids (just proximal to true mammals), represented by *Probainognathus*, are the immediate precursors of mammals. They had powerful chewing muscles, requiring a fixed, akinetic skull. It became anchored to the skull, allowing for greater force generation. The epipterygoid (future alisphenoid) unites medially with the remainder of the sphenoid and inferiorly to pterygoid. The latter bone shrinks down to wing-like pterygoid processes projecting downward from sphenoid. Of greatest importance is the development of nasal cavities. These come about from two processes. The choanae shift posteriorly and *the rostrally located primary hard palate shrinks backward* (and will ultimately be eliminated). This process opens up the passage of the nasopharynx. Parasphenoid becomes incorporated upward into basisphenoid. In its place, the vomers move to the midline, fuse, and articulate with perpendicular plate of ethmoid and the nasal septum, which makes its debut at this time.

The floor of the cranial cavity becomes the roof of the nasal cavity. The maxillae and palatine bones now develop medial shelves to form the floor of the nasal cavity. The shelves fuse to the vomer. The nasal choanae are now retro-positioned at the margin of the hard palate. *Fossil evidence for a soft palate appears at this time, with its attachment sites on the pterygoid processes.* The palatine bone ceases to be tooth-bearing.

Let's look at the process of forming the palatal shelves in the progression from synapsid to mammals. First, observe the oral view of the palate in which the bone fields are color-coded (Fig. 14.46).

- Eutheriodont shows non-tooth-bearing palatine bones fields P1–P2 (green) as falsely included with maxilla, while P3 palatine (tan) remains associated with the future molar region and P4 ectopterygoid (magenta) is displaced backward to articulate with pterygoid.
- In basal cynodont *Procynosuchus*, the vomers (gray) are fusing but still seen in the midline, with parasphenoid (lilac) being incorporated. The pterygoid moves out of the occiput and insinuates a flange behind ectopterygoid.
- Fusion of the hard palate occurs in *Thrinaxodon*. The vomers are no longer seen.
- *Probainognathus* shows pterygoid incorporating into basisphenoid and ectopterygoid becomes part of future medial pterygoid plate hanging down from sphenoid.
- *Morganucodon*, the first mammal, shows a reduction in premaxillary teeth. Its hard palate achieves full length… and arguably a fully developed soft palate as well. Alisphenoid (the future lateral pterygoid plate) does not appear until mammalia. It is accompanied by another new orbital element, orbitosphenoid.

We can see the shelves arising on sagittal view in Fig. 14.46. In early synapsid †*Dimetrodon*, the floor of the brain is the roof of the palate. Palatine (P1–P3) and ectopterygoid (P4) lie inside maxilla (not seen for the most part). However, due to expansion of the choana, maxilla projects forward, like a sliding shower door, to form the lateral nasal wall. It is thus seen peeping forward from P1. Just as in *Paleoherpeton*, pterygoid and vomer are internal to the palatine series. The crown cynodont *Probainognathus* shows the development of shelves from the palatine series and premaxilla. What is labeled here as the maxillary shelf is really P1–P2 and palatine shelf is P3 which, as we know, relates to P4 pterygoid process.

Hard Palate Evolution Reveals the Master Plan of Maxilla and Mandible

Our study of the hard palate and its neuroangiosomes has an important take-home message: the *overall plan of maxilla and mandible is the same* (Figs. 14.49 and 14.50).

- Both are constructed around an embryonic cartilage, which performs two functions:
  - The cartilage represents a receptor zone for dentition which receives neural crest in distinct fields to form the dentition.
  - PQ and Meckel's also serve as templates for surrounding dermal bone fields, both lingual and buccal.
- Palatoquadrate and Meckel's cartilages have matching genetic fields which face one another, guaranteeing that like dental units will develop on either side of the midline.
- Both maxilla and mandible are suspended from the skull. In its original iteration, palatoquadrate cartilage had its own hinge, the *amphistylistic condition*. Later on, of course, the Z-M complex becomes fixed to the skull.
- Dental units are supplied by vessels running in a bony trough opposite the roots:
  - In the case of mandible, inferior alveolar artery extends to its entire length.
  - In maxilla, superior alveolar artery also enters posteriorly but is reinforced by two branches from a separate anterior axis.
- Forward growth of the jaws can be blocked by deficiency states in the connecting piece between the tooth-bearing unit and the skull.
  - In the base of mandible, defects in the ramus are responsible for under- or overprojection.
  - Maxilla connects to the skull via the sphenethmoid complex. Deficiency states here result in varying degrees of underprojection: class III status.
- Alveolar walls of both mandible and maxilla are bilaminar.
  - Mandibular alveolus has the splenials laterally and the coronoids medially.
  - Maxillary alveolus has maxillary wall lateral and palatine bone fields (P1–P3) medial.
- Mandible and maxilla have four dental zones defined by neurovascular pedicles: incisors, canines, premolars, and molars. *Note*: Specialization into premolars and molars takes place in mammals only. This dental change may be causative for development of the TMJ as a way to adjust to the *propalinal bite* pattern.
  - Mandible has four zones off a single inferior alveolar axis. These are defined by the original infradentary series (Id1–Id4) seen in sarcopterygians.
  - Maxilla has four zones with four distinct pedicles: NP_M for the incisors, ASA_M for the canine and maxillary incisor, ASA_L for the premolars, and PSA for the molars.

Part 7. Pathologic Anatomy of the Hard Palate

In Chap. 3, we discussed the Carnegie staging system in depth. Our purpose here is to summarize the highlights of primary and secondary palate development, with particular attention to the *critical contact model* of palate closure.

Timeline of Palate Development and Closure

The human primary palate makes its debut at stage 15. Paired frontonasal processes (FNP) (often confusingly described as being singular) descend in the midline. These will ultimately fuse to the combined first/second arch complex to form the alveolus and upper lip. Growth and patterning of the FNP complex is regulated by a BMP-induced frontonasal ectodermal zone (FEZ) characterized by epithelial domains of Shh regulated by BMP4 and FGF8.

At stage 16, the arches are still separated by seams. The third arch can be seen with fourth and fifth tucked inside (cf Fig. 14.17 of pharyngeal arches). *Primary palate closure* takes place in stages 16 and 17 when epithelial seams break down and mesenchymal flow is established. *Secondary palate closure* is much later, in stage 23+. In the late ninth week, the palatal shelves lift above the tongue and in the early tenth week they are horizontal with upper and middle turbinates present. Late in the tenth week, the palatal shelves fuse together with septum and vomer. Inferior turbinate is present indicating the functional presence of lateral nasopalatine angiosome. *Soft palate closure* begins under stimulation of BMP4 of the palatine bone (Pl3) at stage 23, and at stage 23+ (9 weeks) the epithelial seam on the oral side disappears, except at the uvulus. By 10 weeks, that is closed as well.

Scanning EM studies of palate closure in the mouse are shown in Figs. 14.51, 14.52, 14.53 and 14.54.

Critical Contact Distance (CCD) Model

The "business end" of cleft palate involves the inability of adjacent structures to undergo normal fusion. This can result from two circumstances. First, the *mesenchymal volume reduction* of the abnormal structure means that tissues can-

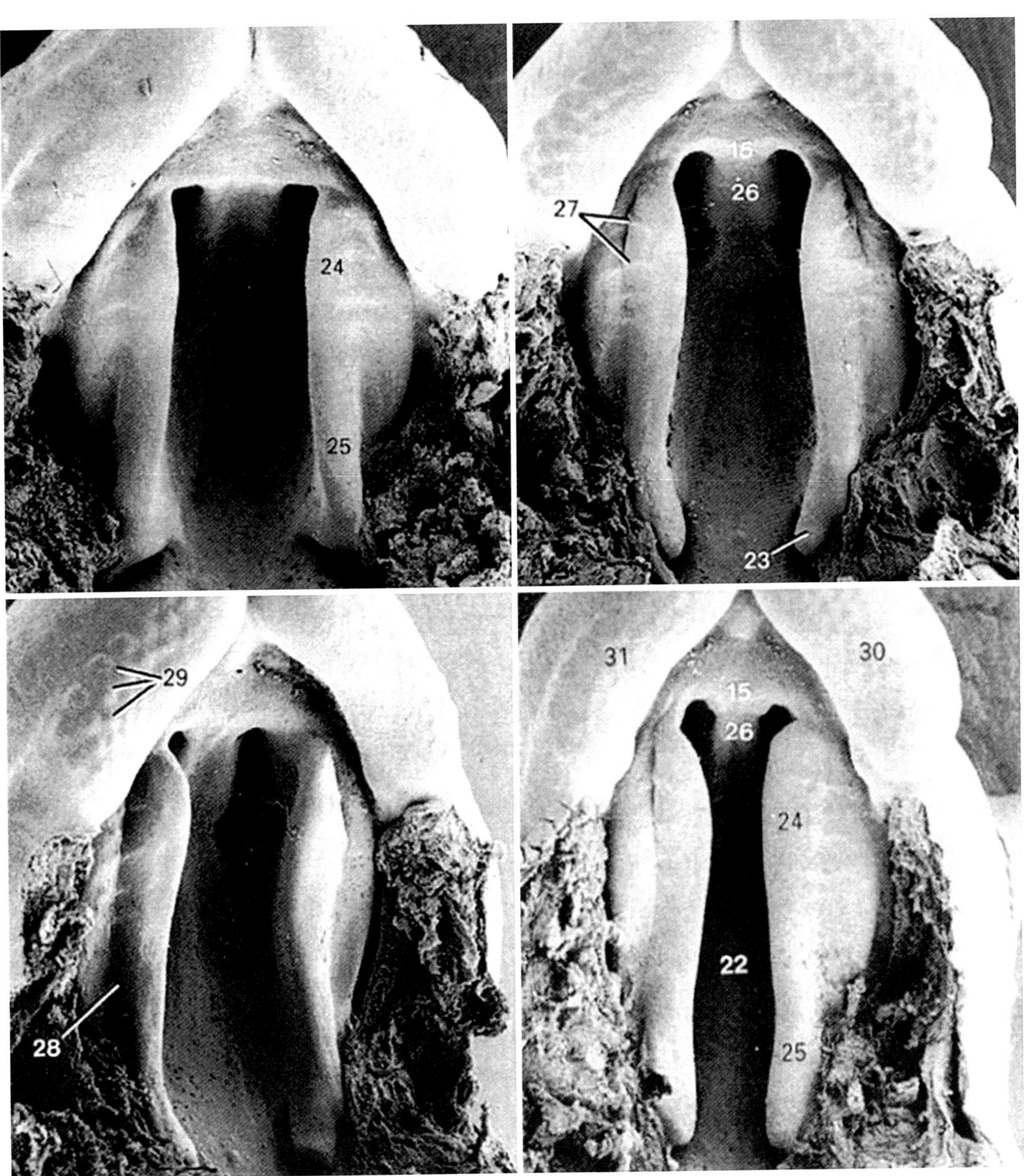

Fig. 14.51 Palate closure 2. [Courtesy of Dr. Kathleen Sulik, University of North Carolina]

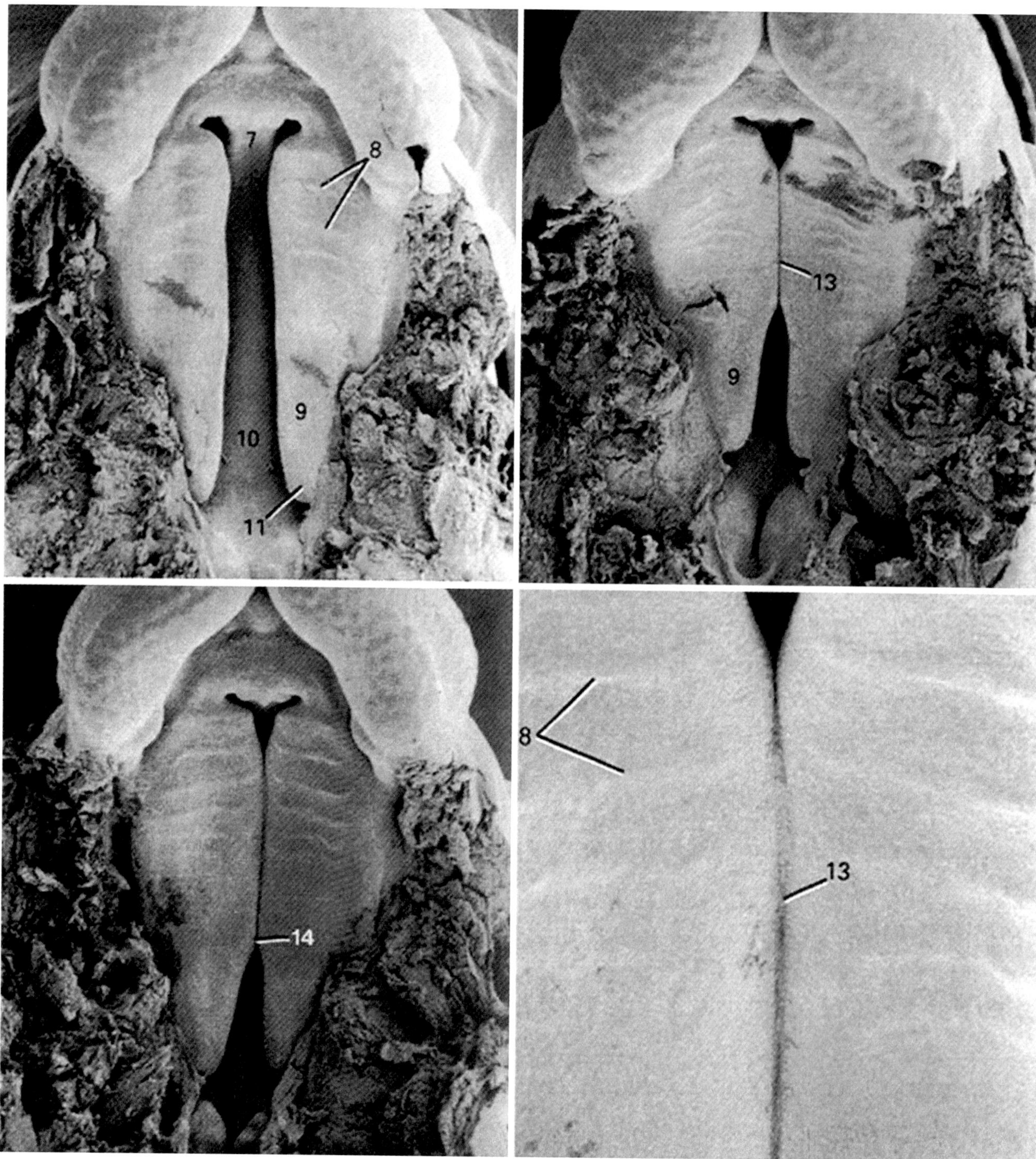

Fig. 14.52 Palate closure 3. [Courtesy of Dr. Kathleen Sulik, University of North Carolina]

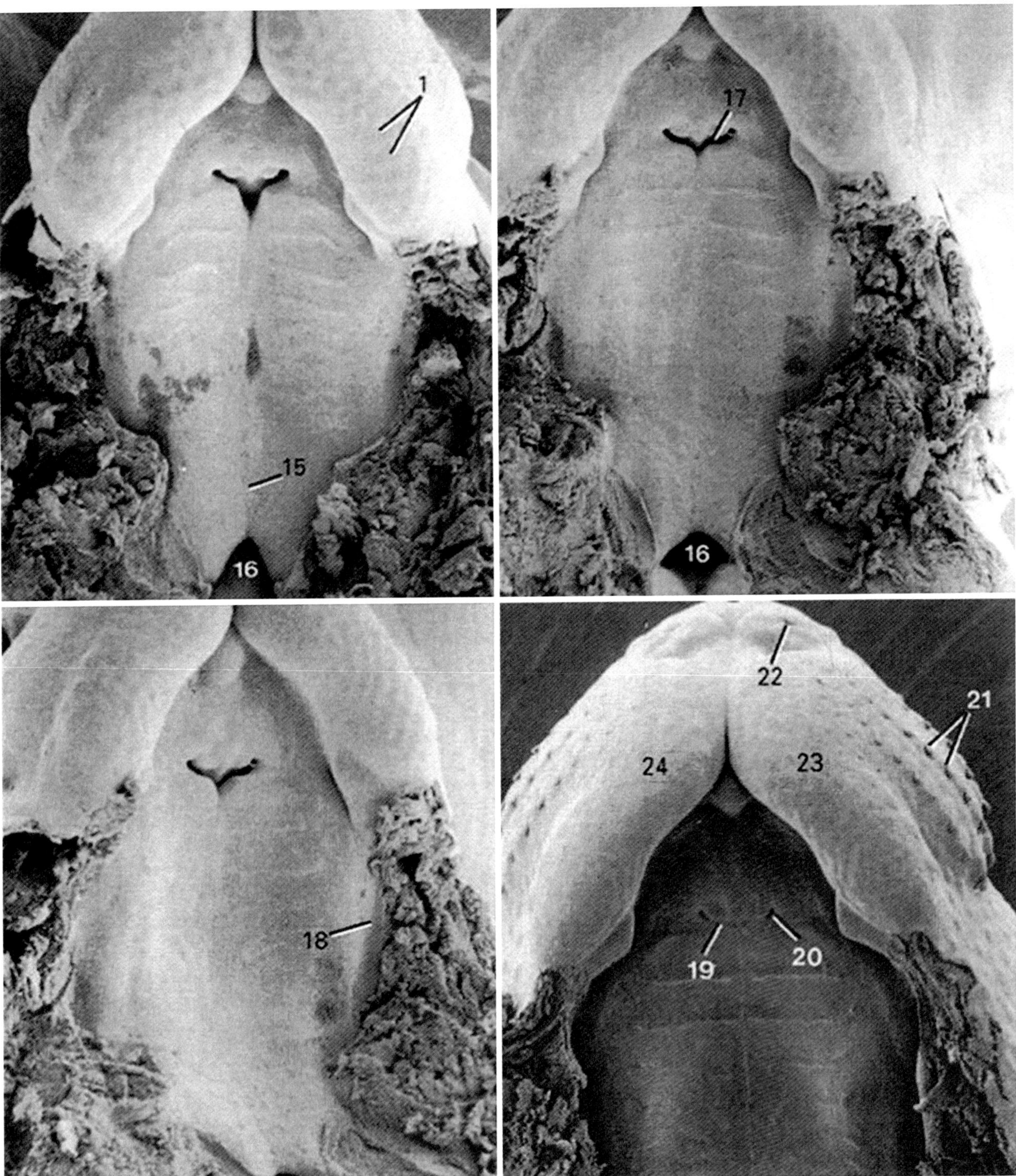

Fig. 14.53 Palate closure 4. [Courtesy of Dr. Kathleen Sulik, University of North Carolina]

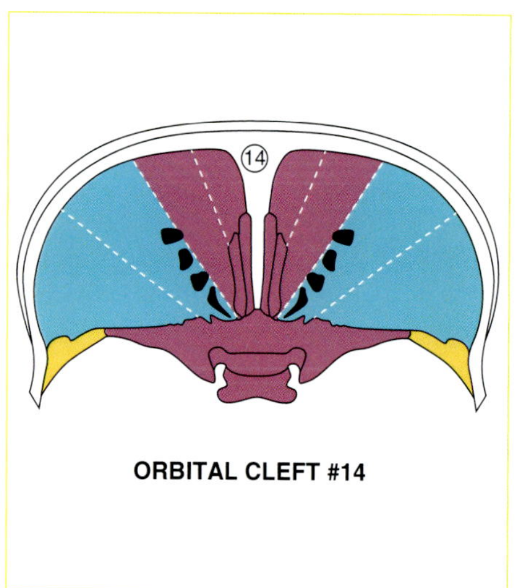

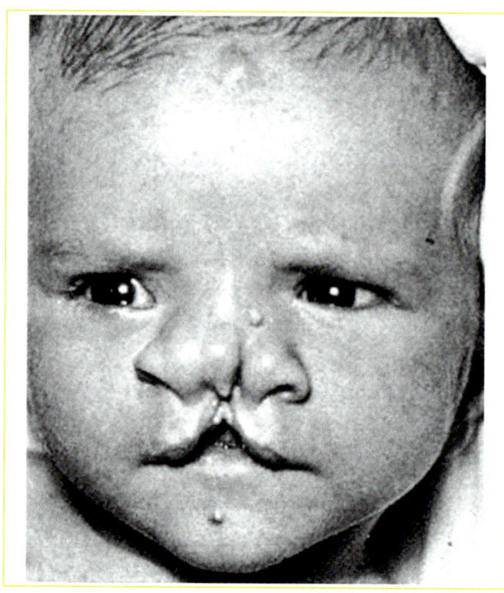

Fig. 14.54 Zone 13 crista galli widening: zone 14 does not exist. Excessive widening of the no. 13 zones (medial purple fields) due to defective sphenethmoid apoptosis (14). Brain is otherwise normal. Note attempt to form columella, probably with rudimental alar cartilage. The fields of perpendicular plate and even of the crista galli are unfused… ***but not "cleft."*** **Zone 13 hypertelorism can also be due to hyperplasia of the sinuses with normal approximation of the midline.** [Courtesy of Michael Carstens, MD]

not make effective contact. Second, there is a *biochemical abnormality* in the tissues that prevents proper fusion from taking place. In reality, these are two sides of the same coin. Using the BMP4/Shh model previously discussed, reduction in mesenchyme of membranous bone causes a reduction in the effective concentration of BMP4 produced during the osteogenic process. This diffusible morphogen cannot reach effective concentration in the epithelium sufficient to inhibit the activity of sonic hedgehog. Shh therefore maintains the stability of the epithelium and fusion cannot take place.

Primary palate CCD

- Primary palate = fusion between premaxillary alveolus (PMx) with maxillary alveolus (Mx). Blood supply is lingual to buccal.
  - Mesial alveolus is bilaminar.
    Both lingual and buccal laminae belong to premaxilla.
  - Distal alveolus is bilaminar.
    Lingual lamina belongs to the palatine bone complex, P1.
    Buccal lamina belongs to zone 4 of maxilla.
- Critical distance between PMx and Mx is required for fusion to take place.
- Developmental fields of normal size are required.
- Fusion process is lingual–buccal and nasal–oral.
- PMx small, CCD is exceeded > **primary palate cleft.**
- Untethered PMx is distracted to normal side, dragging vomer away from midline.
- Vomer ipsilateral to the deficient PMx may itself be deficient.
- If vomer too far away from MxP at incisive foramen > **secondary palate cleft.**

Secondary palate CCD

- Palatal shelf derived the ancient tooth-bearing palatine series.
  - First and second palatine fields (Pl1–Pl2) become what we call "maxillary" palate (MxP).
  - Third palatine field (Pl3) becomes palatine bone proper; it develops an orbital lamina.
  - Fourth palatine field (Pl4) (ectopterygoid) follows the pterygoid bone forward to become lateral pterygoid lamina.
- MxP maturation takes place front-to-back (for vascular explanation, vide infra).
- Vomers fuse from the sphenoid forward.
- Fused V-complex then descends from the incisive foramen backward.
- Normal distance between the two MxPs exceeds their capacity to fuse.
- Interposition of the V-complex *required* to bridge the central gap.
- CCD exists between the V-complex and the palatal plane: inverted-T fusion.
- Failure of the V-complex to reach palatal plane > **secondary palate cleft.**

Part 7. Cleft Zones

Simplistic Preview

Defects of the first rhombomere cause clefts of the palate by two mechanisms

- *Excess states* result from a failure of apoptosis in zone 13. When this occurs in the midline, facial separation occurs.
- *Deficiency states* occur in r1 structures: sphenoid and ethmoid.
  - Suture defects permit escape of CNS contents into the mouth or nose.

- Field defects of zone 13 perpendicular to ethmoid plate affect vomer development.

Defects of the second and third rhombomeres cause cleft palate by *deficiency only*

- r2 deficiencies lead to failure of component fields to achieve fusion.
- Exceeding the critical contact distance due to small size.
- Block of normal epithelial breakdown due to small size.
- r3 deficiencies small mandible.
  - Tongue interposition causes mechanical block of shelves.

Tessier Cleft Zone 14: A Misleading Concept

The 14–0 zone does not exist. This was a *trompe l'oeil* for Tessier. The fontal bone is not singular; it is two halves which fuse in the midline. Each hemi-frontal bone is a bilaminar structure with four developmental fields, all of which are organized around four StV1 neurovascular axes. Zones 13 and 12 are supplied by the supraorbital externally, and internally these zones are supported by the medial meningeal flowing over the ethmoid plate. Zones 11 and 10 are supplied externally by the supraorbital, while internally the lacrimal supplies the lateral orbital (Figs. 14.55, 14.56, 14.57, 14.58, 14.59 and 14.60).

Development of the frontal bone takes place by the membranous ossification of the neural crest which migrates from: rhombomere r0 (zones 13–12) and from rhombomere 1 (zones 11–10).* This ectomesenchyme follows a pathway alongside the midbrain and forebrain, forward over the orbit and downward to form the nasal process. This mesenchyme is the source of the dura and subcutaneous mesenchyme within which the membranous bone fields develop. The neural crest populates and pushes forward a unique ectodermal envelope which is described below.

- Remember that the biologic behavior of m1–m2 and r0–r1 is similar, none of these populations contributes to the first arch—they are lumped together as midbrain neural crest, MNC.
- MNC from r0–r1 forms anterior cranial fossa and frontonasal soft tissues,
- MNC from m1–m2 is distributed to the orbit.

The frontonasal skin is likewise unique. We said repeatedly that the neural folds over secondary prosencephalon do not produce neural crest. Instead, they contain nonneural ectoderm (NNE). This is because these folds are in register with one of two hypothalamic prosomeres: hp1 is caudal and

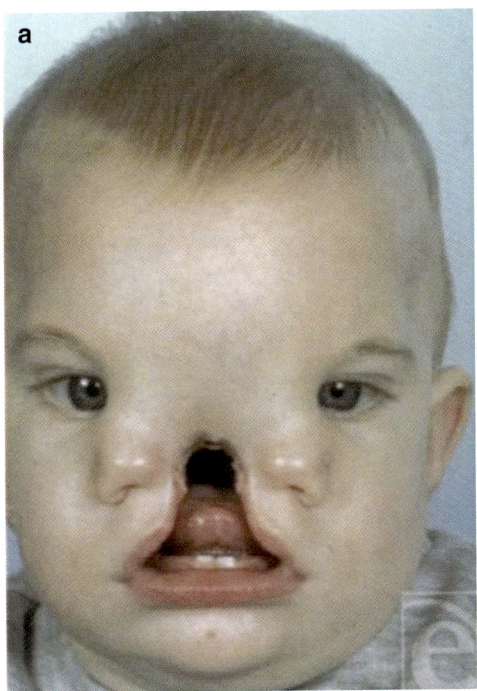

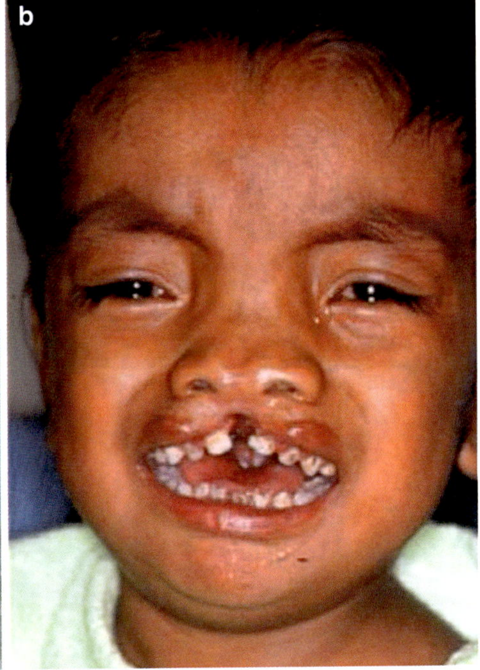

Fig. 14.55 Tessier 0–14 cleft is a misnomer. This really represents failure of apoptosis of sphenethmoid tissue in zone 13. Skull and brain are intact. Zone 1 is always involved with separation of the vomer fields and premaxillary fields. Palate shelves fused to ipsilateral vomers and therefore divided but otherwise normal in dimension. (**a**) Demonstrating a complete separation of 52 mm. (**b**) A more typical presentation. The vomers are fused posteriorly but widened/bifid anteriorly. Premaxillae are split but with normal dental units. The StV2 zones are normal. The issue is a mechanical block caused by excess tissue in the StV1 zone. (**a**) [Courtesy of Anthony S. Wolfe]. (**b**) [Courtesy of F. Gargano]

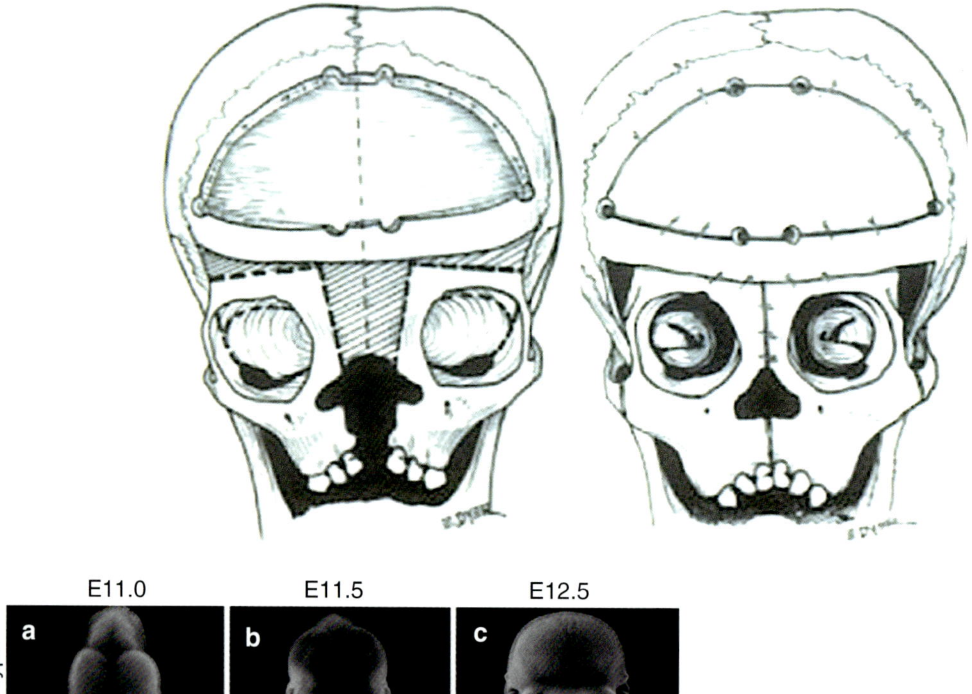

Fig. 14.56 Surgical correction of hypertelorism and excess mesenchyme involves resection of ethmoid excess. [Courtesy of Anthony S. Wolfe]

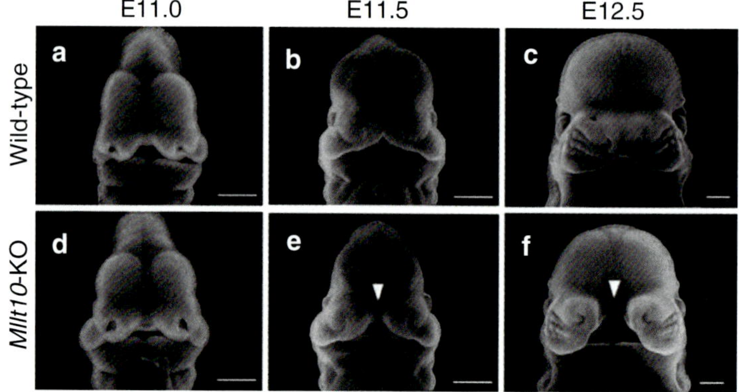

Fig. 14.57 Zone 13 midline "cleft" due to persistence of excess sphenethmoid field, that is, faulty apoptosis. *Mllt10*-KO embryos exhibit severe midline facial "cleft" and hypertelorism. (**a–c**) Normal craniofacial development of wild-type embryos during E11.0–E12.5. (**d–f**) Abnormal craniofacial development of Mllt10-KO embryos, includes midline facial widening (arrowheads). E11.0 the nasal placode fields are in continuity with their respective hemispheres and r1 tissue covers the brain. Forebrain is not interposed. E11.5 the forebrain is encroaching into the sphenethmoid. Note wild-type expansion of forehead. [Reprinted from Ogoh H, Kamagata K, Sakai D, et al. Mllt10 knock-out mouse reveals critical role of AF-10 dependent H3K79 methylation in medfacial development. *Scientific* Reports 2017; 7:11922. With permission from Creative Commons License 4.0: https://creativecommons.org/licenses/by/4.0/]

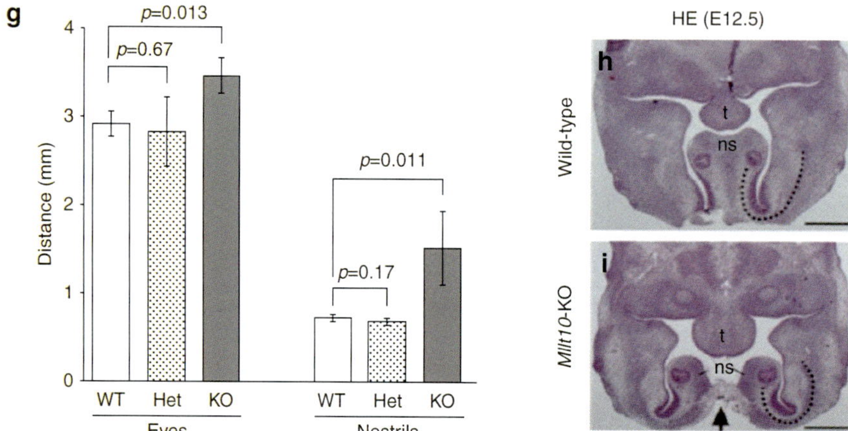

Fig. 14.58 Zone 13 hypertelorism due to apoptosis. Measurements comparing wild type vs. heterozygous vs. *Miit10* knockout demonstrate failure of midline mesenchyme (arrow) to involute as differences in eye–nostril distance. The results in hypertelorism and frank splitting of the nasal septum, **ns**. HE micrographs upside down; note tongue position. [Reprinted from Ogoh H, Kamagata K, Sakai D, et al. Mllt10 knock-out mouse reveals critical role of AF-10 dependent H3K79 methylation in medfacial development. *Scientific* Reports 2017; 7:11922. With permission from Creative Commons License 4.0: https://creativecommons.org/licenses/by/4.0/]

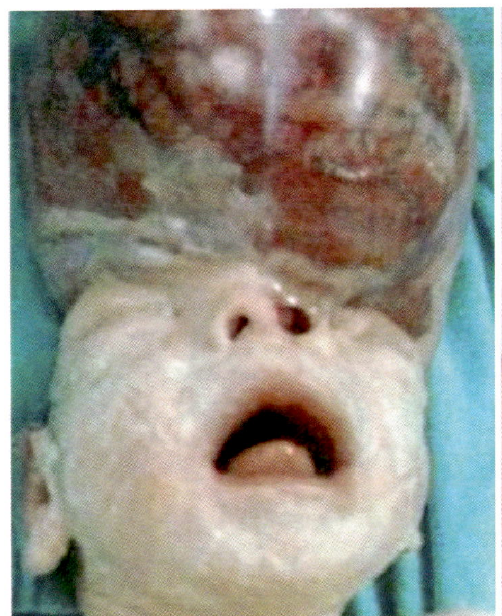

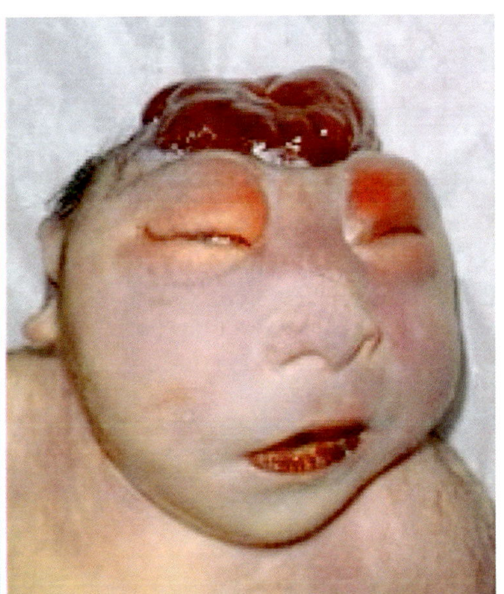

Fig. 14.59 Acrania vs. anencephaly. In acrania the CNS develops normally. Anencephaly implies developmental failure of the brain in addition to calvarial absence. Note that even in these devastating states the ethmoid complex remains united. OPEN (**a**) [Courtesy of ECLAMC—Estudio Cooperativo Latin-American de Malformaciones Congénitas]. (**b**) [Reprinted from Congenital Abnormalities Browser. Retrieved from: http://en.atlaseclamc.org/skull/139-anencephaly-Q00.0#.W29AfC3Mx0. With permission from Instituto de Genética Humana]

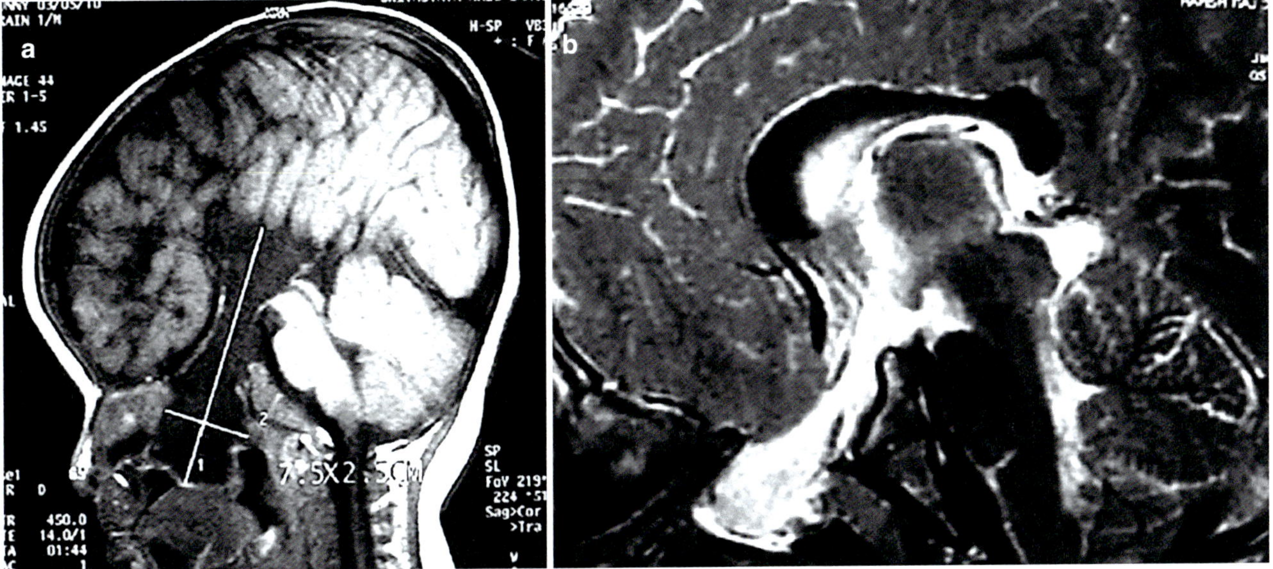

Fig. 14.60 Trans-sphenoidal encephalocoele results from defect in r1 basisphenoid–presphenoid suture. Mass can be seen tracing into the mouth. (**a**) [Reprinted from Sanjari R, Mortazavi SA, Amiri RS, Ardestani SHS, Amirjamshidi A. Intrasphenoidal Meningo-encephalocele: Report of two rare cases and review of literature. Surgical Neurol Int 2013 4:3. DOI:10.4103/2152-7806.106260. With permission from Creative Commons Attribution License.] (**b**) [Reprinted from Sharma RR, Mahapatra AK, Pawar SJ, Thomas C, Al-Ismally M. Trans-sellar, transsphenoidal encephalocoeles: report of two cases. J Clin Neurosci 2002; 9(1):89–92. With permission from Elsevier]

hp2 is rostral. For more details, see Chap. 5 on the neuromeric system. NNE generates *frontonasal epidermis*.

The *frontonasal dermis* comes from diencephalic neural crest of prosomeres p1–p3. It flows forward beneath NNE, completing formation of the skin. At the same time, these populations activate the placodes that are found within the hp1 and hp2 folds. Recall from Chap. 11 that frontonasal skin is unique, with pathologies that occur only there, that is, *frontonasal dysplasia*.

Approximation of the frontal bone zones takes place by a process of *apoptosis*, a controlled breakdown of tissue that allows the frontal fields to move forward into the midline. The driving force for this is the *medialization of the orbits*, a process that is dictated by the growth pattern of the brain and the anterior cranial base. The ethmoid complex is interposed between the orbits. Failure of apoptosis in the ethmoid zones will lead to hypertelorism. Since the ethmoid is a bilateral structure, such hypertelorism can be *unilateral* or *bilateral*. Probably the apoptosis process required of the frontal bone is subject to whatever degree of apoptosis is taking place in the ethmoid complex.

Embryologic proof of the sphenethmoid field complex as the "driver" of facial width comes from cases of *acrania* and *anencephaly*. Note: In acrania, CNS development is normal, but the brain, subsequently exposed to amniotic fluid, deteriorates. Anencephaly has a similar calvarial defect but is a neural tube condition in which normal development of the brain does not occur. In both conditions, even though the calvarium is absent, the orbits are conjoined. Failure of the sphenethmoid fields to fuse, leading to a true "split face" reaching the CNS, is incompatible with development and has not been seen (Fig. 14.60).

Midline pathologies can manifest as a simple excess of tissue or as a fusion failure or cleft (sic). For this reason, it appeared to Tessier as if this were truly an autonomous zone. Recall that the pituitary sits in a cavity between two bones: the anterior neural crest presphenoid and a posterior mesodermal postsphenoid (basisphenoid). The former has a sinus while the latter is solid. Fusion failures extending backward into the sphenoid have a biologic limit to viability. Failure of apoptosis in the midline of zone 13 also leads to a residual excess of mesenchyme in the midline; this is normal orbital approximation and results in hypertelorism.

Medialization can also be altered by the presence of an encephalocoele that can seek out a field failure between any of the zones to achieve an extracranial position and thereby block the closure. The escape routes of encephalocoeles are well documented, be they forward through the frontal bone zones, in the midline through anterior or lateral frontoethmoid sutures, or in paramedian positions (Figs. 14.61 and 14.62).

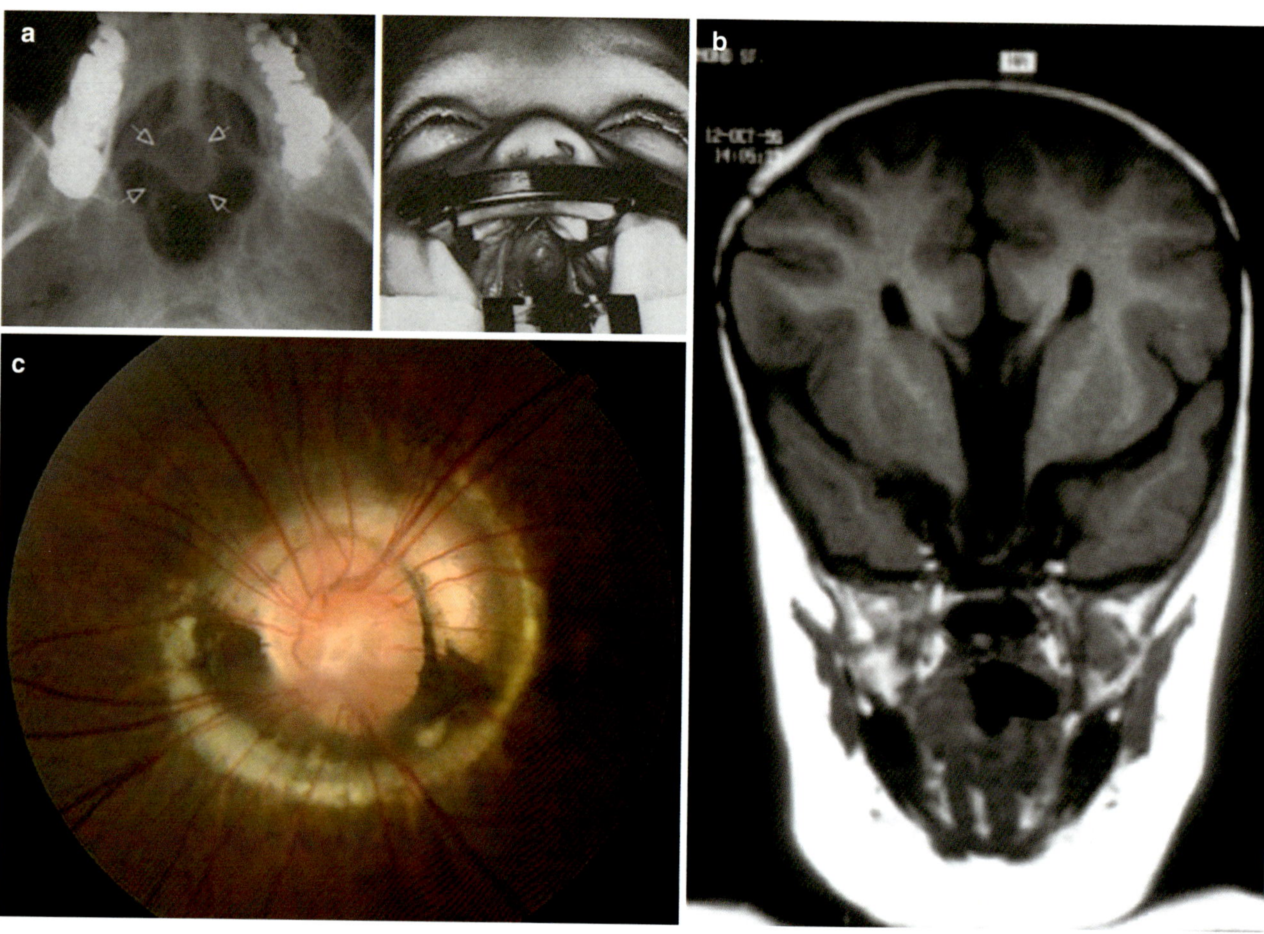

Fig. 14.61 Transethmoidal encephalocoele together with absent corpus callosum, hypertelorism, midline cleft palate, wide nose and midline of the lip can be associated with morning glory optic disc syndrome, which consists of an enlarged excavated disc with fibroglial-appearing tissue at its center, an elevated subretinal peripapillary annulus of chorioretinal pigmentary change, and abnormal vessels radiating outwards. This congenital anomaly was likened to the withering morning glory flower from which it derives its name. Its resemblance to the morning glory flower. It is characterized by an enlarged, funnel-shaped excavation that incorporates the optic disc. The disc itself is enlarged and orange or pink in color within a surrounding area of peripapillary chorioretinal pigmentary changes. The overall diameter of the disc depends on the size of the excavated posterior scleral opening. Occasionally the disc itself may appear elevated. Within the center of the disc is a white glial tuft. Similar to petals on a flower, the blood vessels are increased in number and curve as they emanate radially from the disc, rather than in the usual central branching pattern. (**a**) [Reprinted from Pollock JA, Newton TH, Hott WF. Transsphenoidal and transethmoidal encephalocoeles. Radiology1968; 90(3):442–453. With permission from Radiological Society of North America.] (**b**) [Reprinted from Hodgkins P, Lees OM, Lawson J, Reardon JW, Leitch J, Thorogood P, Winter RM, Taylor DSI. Optic disc anomalies and frontonasal dysplasia. Br J Ophthalmol 1998;82:290–293. With permission from BMJ Publishing, Ltd.] (**c**) [Reprinted from Kindler P. Morning glory syndrome: unusual congenital optic disk anomaly. *Am J Ophthalmol*. 1970; 69(3):376–384. With permission from Elsevier]

Zone 13 medial ethmoid cribriform, dura / forehead
meningeal branch, anterior/posterior ethmoids
supratrochlear

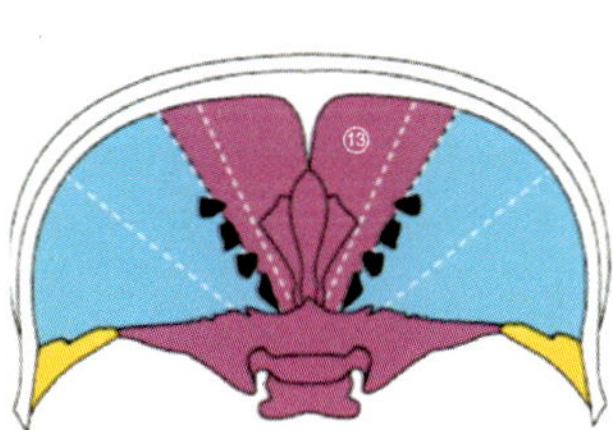

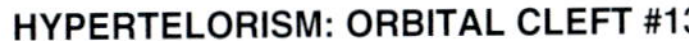

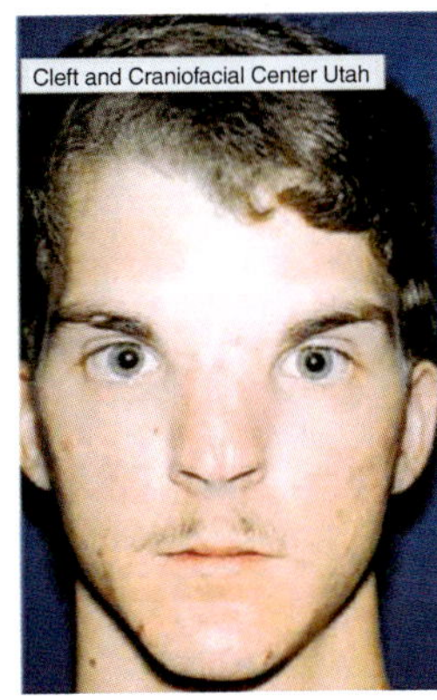

Fig. 14.62 Tessier 13 cribriform XS hypertelorism. XS rl left > right with dysplastic left nasal bone. This isolated cleft, sometimes with absent nasal bone(s) can be involved in the hard palate if accompanied by defective vertical development of the perpendicular ethmoid plate. When vomers are intact and fused no palate cleft occurs. [Courtesy of Michael Carstens, MD]

Neuroembryologic Simplification of Zones 1–13 and Zones 2–12

Tessier originally perceived that a fundamental difference existed between clefts (states of mesenchymal deficiency and/or excess) involving the maxilla and those involving the orbit. At the time, neuroembryology of the area was not recognized. In our correspondence, Tessier was impressed by the correlation between his findings and the regional sensory neuroanatomy. The structures of the nose are unique in that the topology of the fields supplied by the neurovascular axes of StV1 and StV2 is not one of simple planar separation, rather these fields are *interlocking*. This makes the interpretation of congenital field defects difficult.

- The best way to keep things straight is for us to consider these zones by their neuroanatomy, with 1–2 supplied by StV2 axes while those of zones 13 and 12 supplied by STV1 axes.
- This will require violating the traditional way of describing Tessier clefts in this region. Ever the iconoclast, he would doubtlessly agree [22].

Tessier Cleft Zone 13

Zone 13 consists of the midline structures of the cranial base and of the forehead up to, but not including, the eyebrow. It interdigitates with zone 1 because it shares common arterial axes supplying both structures of the upper nasal cavity and the medial external nasal soft tissues of the nose (Figs. 14.63, 14.64, 14.65, 14.66, 14.67, 14.68, 14.69, 14.70 and 14.71).

Let's begin with the internal upper medial structures of the nasal cavity, the perpendicular plates of the ethmoid and septum. These are supplied by the posterior and anterior ethmoid arteries. These two arteries have similar behaviors. They arise within the orbit and enter the air cells (posterior ethmoid gets the posterior sinus and anterior ethmoid supplies middle and anterior sinus). They then (1) track upward through the cribriform plate to access the dura mater as the *meningeal branches* supplying the meninges over the ethmoid complex; and (2) enter the nasal cavity to supply the perpendicular plate of the ethmoid and the superior septum; they also give off lateral nasal branches to the superior and middle turbinates.

- Posterior ethmoid artery is unique in that it passes backward to supply presphenoid.
- Anterior ethmoid artery is also distinct: it passes outward into the midline of the nasal envelope.

Mesenchymal failure in these two angiosomes causes anterior cranial base defects on either side of presphenoid potentially involving two sutures: posteriorly, with basisphenoid, or anteriorly, with the ethmoid.

- *Trans-sphenoidal encephalocoeles* pass into mouth and cause a mechanical disruption of the soft and hard palate.
- *Transethmoidal encephalocoeles* pass into the nose and affect the hard palate. These masses can be associated with alterations in midline neurologic structures such as corpus callosum and have been associated with disruption of visual structures manifested as the *morning glory optic disc syndrome*.

Defects in the ethmoid axis can selectively influence hypertrophy or dystrophy of the medial zone of ethmoid complex, the crista galli and cribriform plate. These in turn can affect the position of the ethmoid sinuses as these belong to zone 12. The result is *ethmoid-based hyper/hypotelorism*. Vertical deficiencies of the perpendicular plate of the ethmoid automatically affect development of the vomer, in most cases, passively by causing a displacement but, in holoprosencephalic state the vomer field can be hypoplastic or absent. Thus zone 13 defects can cause cleft palate states in zone 1.

- Simple vertical deficiency of PPE retracts a normal vomer with successful fusion with the palatal shelves: the *high arched palate*.
- In severe cases, a small ethmoid can retract a normal vomer so far upward out of the palatal plane that fusion is

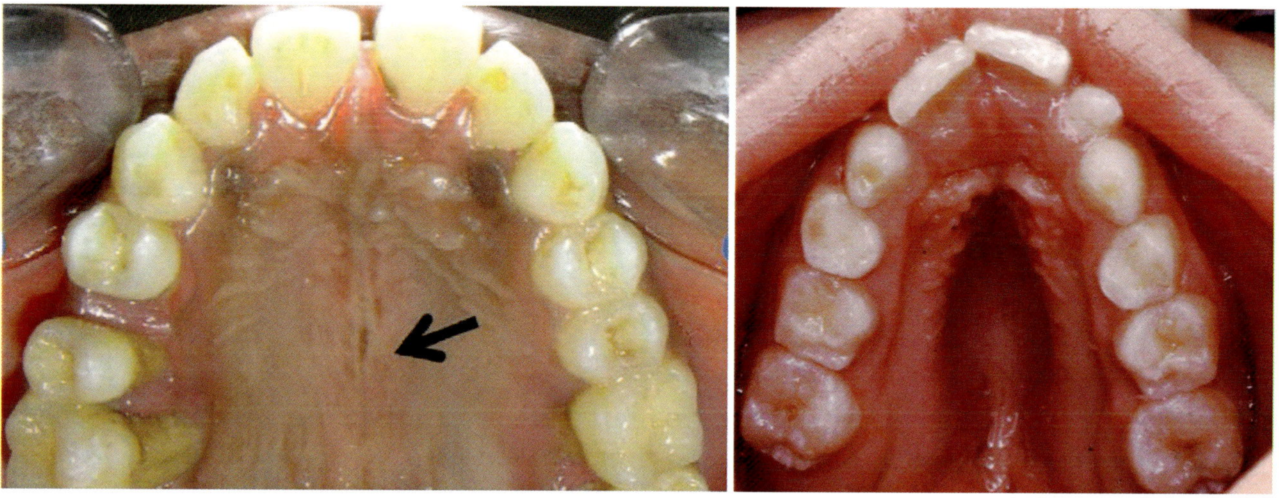

Fig. 14.63 High arched palate with accompanying diastema. More severe case with loss of right lateral incisor. Note deviation of vomer into the left side. Left: [Reprinted from Kasagani SK, Mutthieni RB, Jampani ND, Nutalapati R. Report of a case of Turner's syndrome with aggressive local periodontitis. J Indian Soc Periodontol 2011; 15(2):173–6. With permission from Wolters Kluwer Medknow Publications.]. Right: [Courtesy of Michael Carstens, MD]

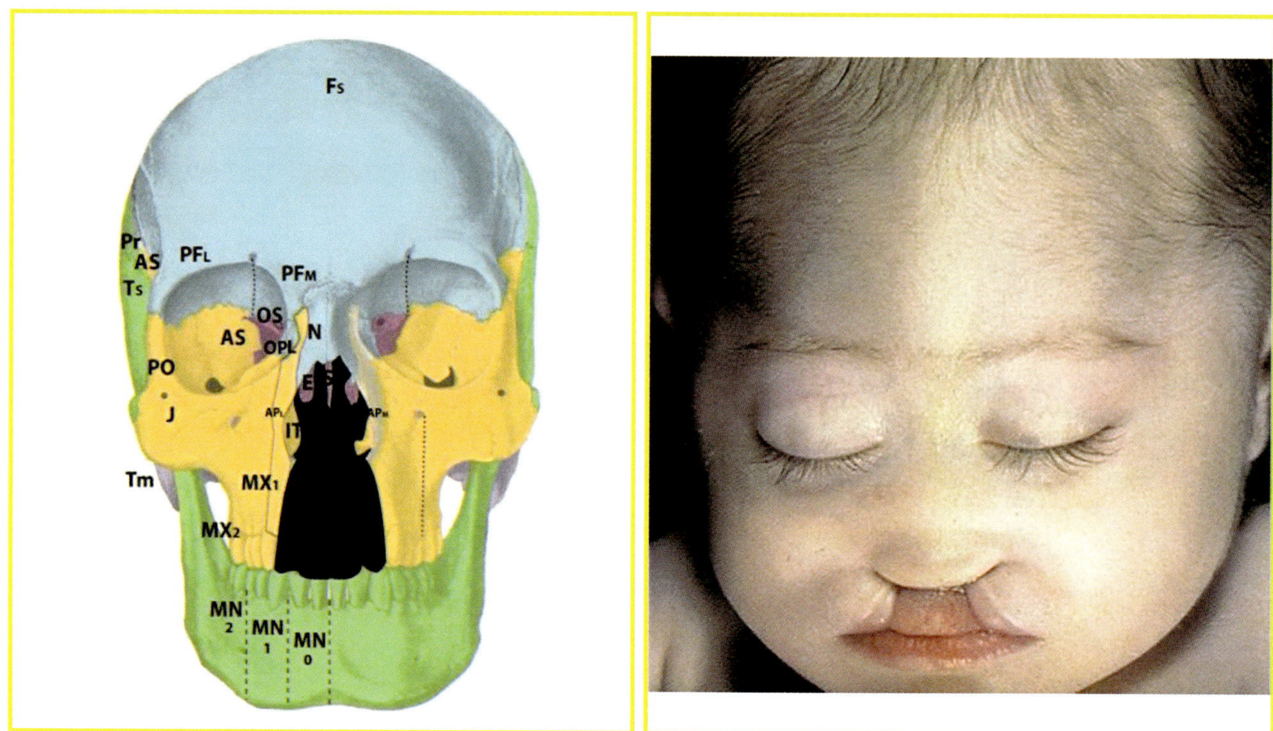

Fig. 14.64 Tessier 13 (holoprosencephaly, HPE). HPE has severe hypoplasia of ethmoid fields leading to failure of formation of premaxillary-vomerine fields. Note flattened nasal bone and dysplastic/absent septum. [Courtesy of Michael Carstens, MD]

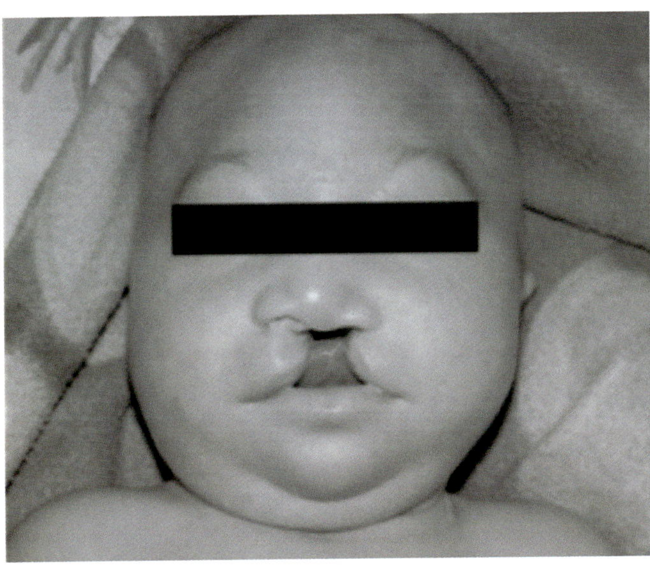

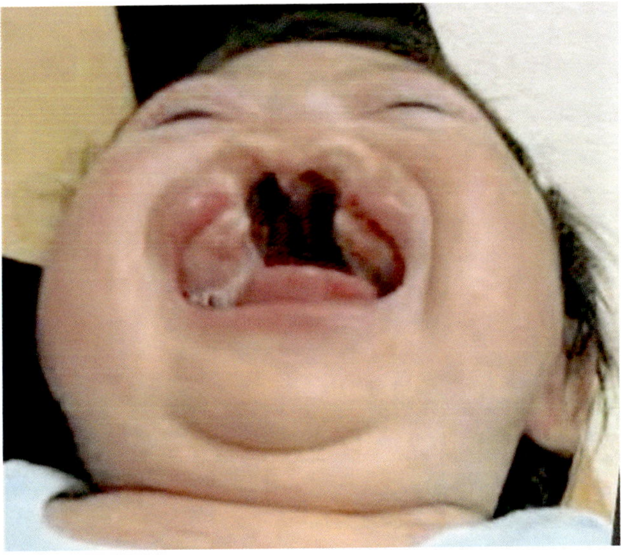

Fig. 14.65 Zone 13 HPE partial. PPE on the left aplastic with hemi-premaxilla. Defect is complete on the right. [Reprinted from ECLAMC Centro Anomalias de Malformaciones Congenital, Professor Ignacio Zarante, Instituto de Genetic Humana, Potifica Universidad Javeriana, Bogotá, Colombia]

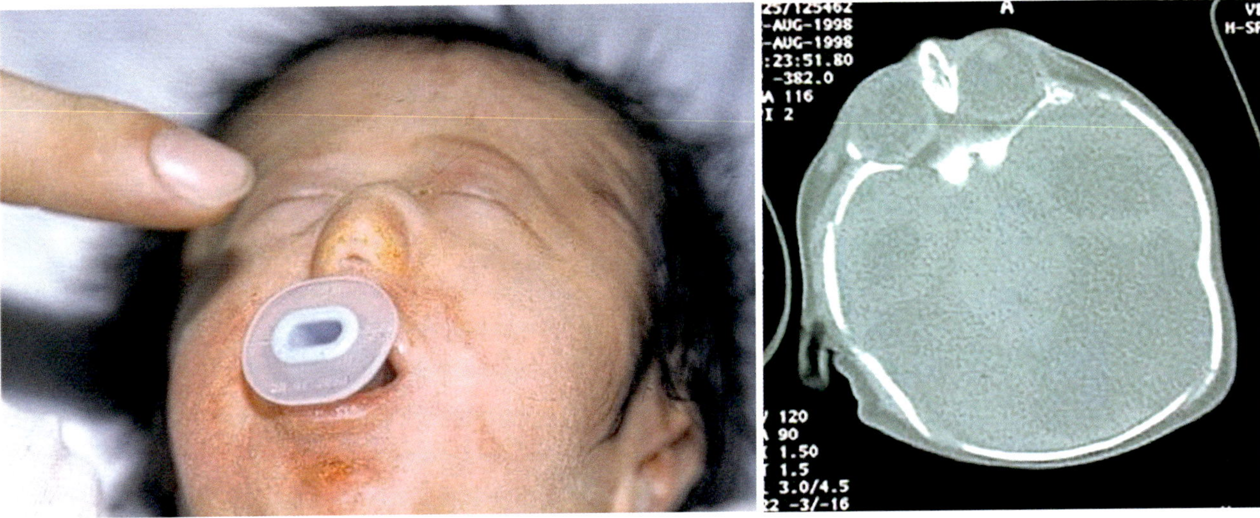

Fig. 14.66 Tessier 13 cebocephaly affecting nasal bones and bilateral absent paranasal sinuses. Ethmoid hypoplasia due to r1 neural crest defect causes absent sinuses and absent nasal bones. Fusion of nasal placodes denotes cebocephaly. Cleft palate results. [Reprinted from ECLAMC Centro Anomalias de Malformaciones Congenital, Professor Ignacio Zarante, Instituto de Genetic Humana, Potifica Universidad Javeriana, Bogotá, Colombia]

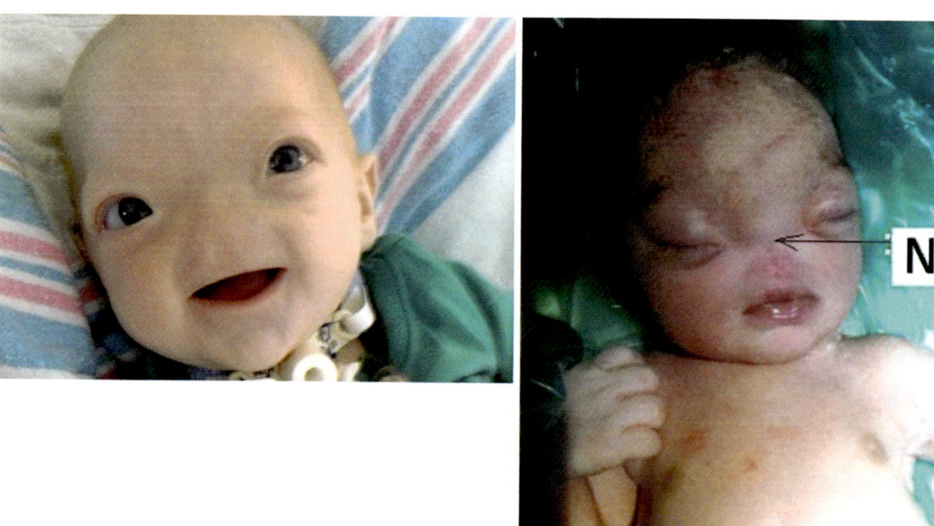

Fig. 14.67 Zone 13 arhinia with absent septum and laryngeal defect requiring tracheostomy: relationship unknown but must involve common homeotic factors between r1 and r8–r11. Ethmoid hyperplasia with adequate facial fields but absent nasal placodes. Severe defect in stillborn with laryngeal stenosis. [Reprinted from ECLAMC Centro Anomalias de Malformaciones Congenital, Professor Ignacio Zarante, Instituto de Genetic Humana, Potifica Universidad Javeriana, Bogotá, Colombia]

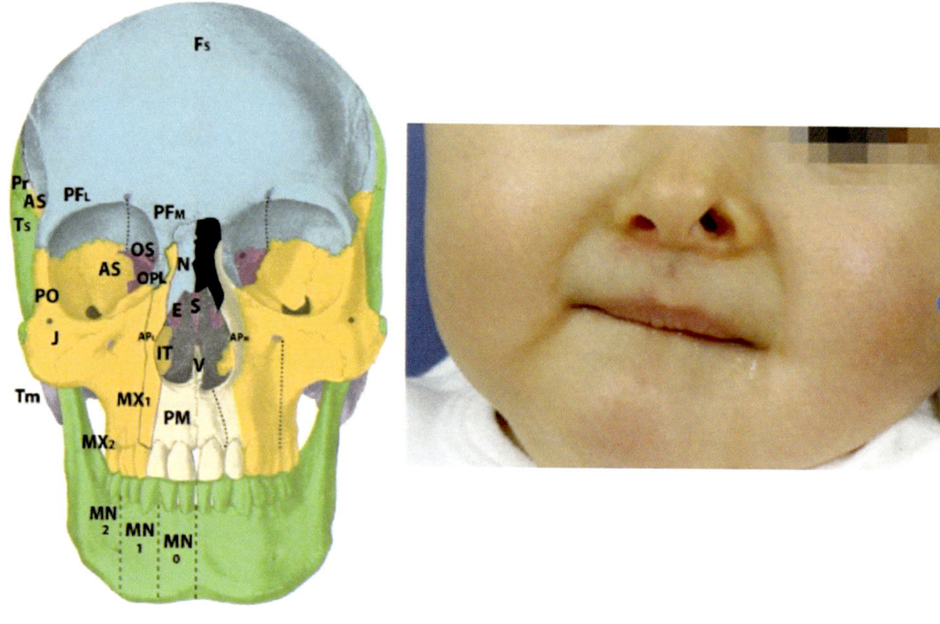

Fig. 14.68 Tessier 13 nasal defect isolated includes hypoplastic, absent nasal bone and notch in nasal dorsum between zones 13 and 12 through the intermediate crus. This can, *if accompanied by deficit in perpendicular plate*, cause arching of the hard palate. [Courtesy of Michael Carstens, MD]

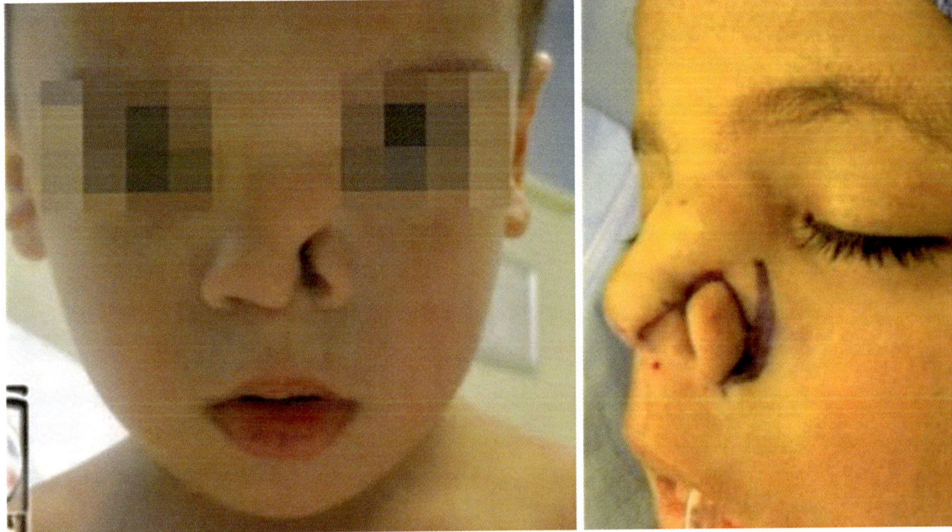

Fig. 14.69 Tessier 13 cleft (zone 1 unaffected). Bony involvement includes the nasal bone at the junction of the nasal process of the maxilla, ethmoid labyrinth, and the olfactory groove of the cribriform plate, which is widened. Soft tissue features are the alar dome notch at the middle third of the nostril rim, and the extension into the nasal dorsum passing medial to the normal canthal structures. The medial portion of the eyebrow can be dystopic without an obvious crease. If zone 1 occurs concomitantly, deep structures are alveolus between the central and lateral incisors, the piriform aperture lateral to the anterior nasal spine. May affect Cupid's bow at the superficial level not sufficient to cause full cleft of the lip. Note: flattening of right nasal bone, the independent alar unit (here being upwardly displaced). [Courtesy of F. Gargano]

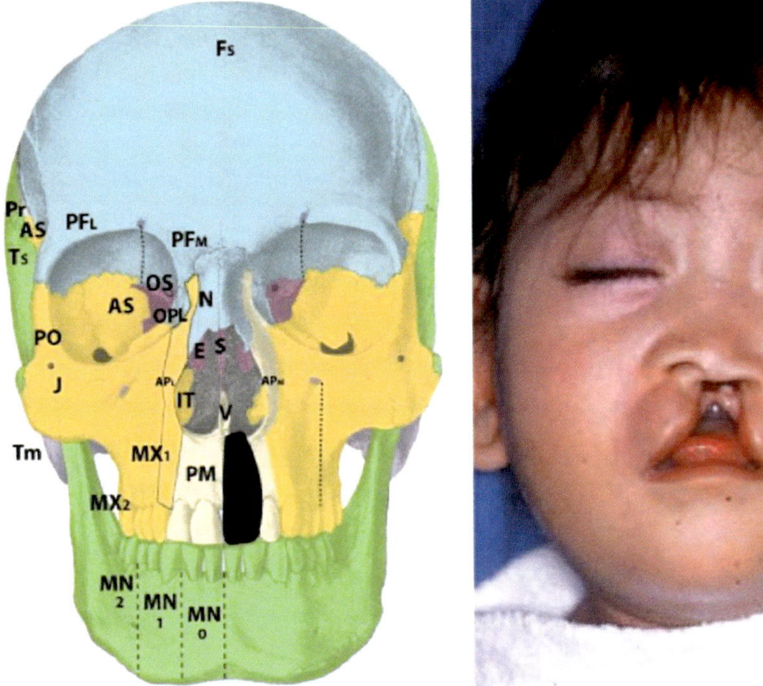

Fig. 14.70 Tessier zone 1 isolated involving premaxilla and lip. Left: Here zone 13 is normal. Faulty development of StV2 NP_M axis leads to hypoplasia/failure of ipsilateral vomer field. Hemivomer is attached to otherwise normal r1 septum with normal width, but deflected. Very rare isolated number 1 with cleft lip. Alveolar notch is paramedian. Vomer is intact. Right: Tessier zone 1 and zone 13. This is much more commonly reported in the literature, because isolated alveolar clefts and vomer deficiencies have not been considered to be part of zone 1. Left: [Courtesy of Michael Carstens, MD]. Right: [Reprinted from ECLAMC Centro Anomalias de Malformaciones Congenital, Professor Ignacio Zarante, Instituto de Genetic Humana, Potifica Universidad Javeriana, Bogotá, Colombia]

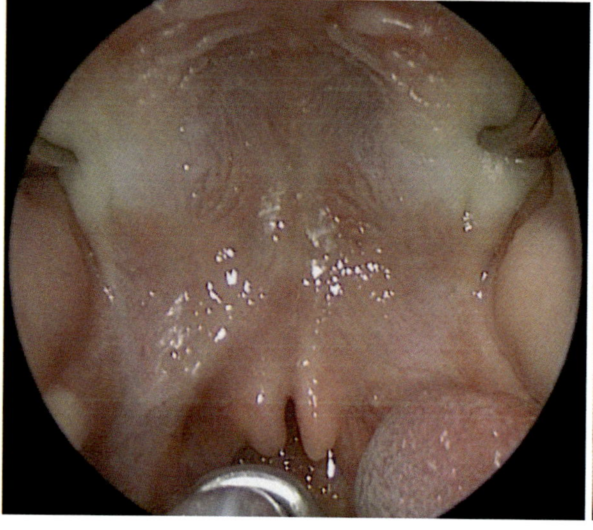
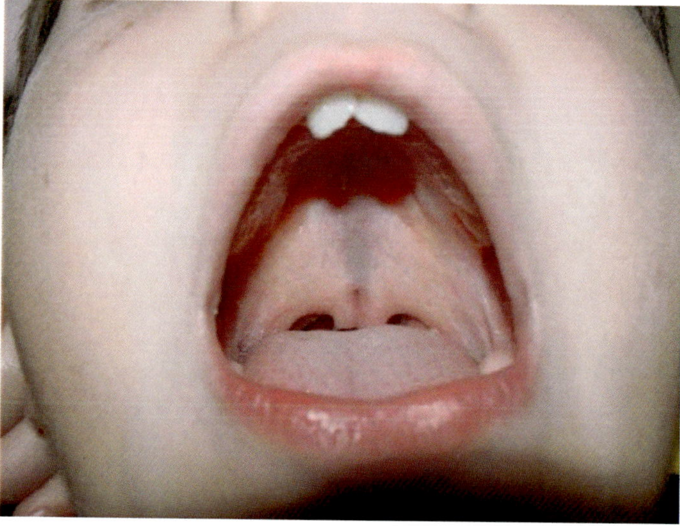

Fig. 14.71 Vomer insufficiency. Palpable notch in P3. Submucous defect is palpable in the midline of hard palate. More advanced submucous cleft with vomer maintaining just enough contact to permit epithelial bridge. Any further deficiency and this would become a midline cleft palate. [Courtesy of Michael Carstens, MD]

impossible, thereby leading to a *midline cleft of the hard palate*. A concomitant r2 defect in the vomer size can be diagnosed depending upon its physical size as discerned either by inspection or using a 3D CT scan.
- In the worst case scenario, as in holoprosencephaly, vomer and premaxilla are cannot develop, leading to characteristic hypoplasia or aplasia of premaxilla This can be unilateral or bilateral (Fig. 14.65).

The StV1 anterior ethmoid artery enters from the orbits to supply the anterior and middle ethmoid air cells and thence proceeds to supply the frontal sinus. It has three destinations (subfields) within zone 1.

- It tracks upward through the cribriform plate to supply the dura underlying zones 13 and 12.
  - These *medial dural branches* to hypertrophy of the anterior cranial fossa between the midline and the medial border of the cribriform plate, that is, the *olfactory groove*. This enlarges the cribriform plate medial to the ethmoid labyrinth. The result is *hypertelorism*.
  - Internal nasal branches descend medially over the *superior septum*. Defects in this axis can explain warping or absence of the septum.
- The anterior ethmoid supplies the *nasal bone* from its internal surface.
  - A defect here can cause absence of the nasal bone.
- The artery emerges between the nasal bone and the upper lateral cartilage to run forward and downward.
  - A defect in this part of the axis causes the nasal skin cleft noted by Tessier in zone 1 as a crease through the intermediate crus of the lower lateral cartilage, that is, between zone 13 medial crus and zone 12 lateral crust.
- The terminus of external anterior ethmoid follows along the upper and caudal border of the septum to supply columella and prolabium.
  - Defects in this part of the axis are responsible for rare cases of *absent columella*, with or without affectation of the prolabium.
  - Note that the prolabium is generally autonomous as it gets a collateral supply from a branch of medial nasopalatine that penetrates through the premaxilla to enter the prolabium from below.

In Sum: Cleft Palate States Involving Zone 13 and Zone 1

Zone 13 defects passively cause zone 1 hard palate clefts through encephalocoeles or perpendicular plate deficiencies. On the other hand, zone 1 defects do not affect zone 13 per se. However… zone 13 and zone 1 defects can and do occur concurrently. The tip-off is the condition of the nasal tip and columella.

Tessier Cleft Zone 1

Caveat The Tessier no. 1 cleft is poorly described in the literature and, as we have pointed out in Chap. 13, the manifestation of the StV1 angiosomes of the nose is conflated with those of the zone 1. What texts describe as zone 1 findings belong strictly to zone 13. These are entirely different mesenchymes with distinct neurovascular supply.

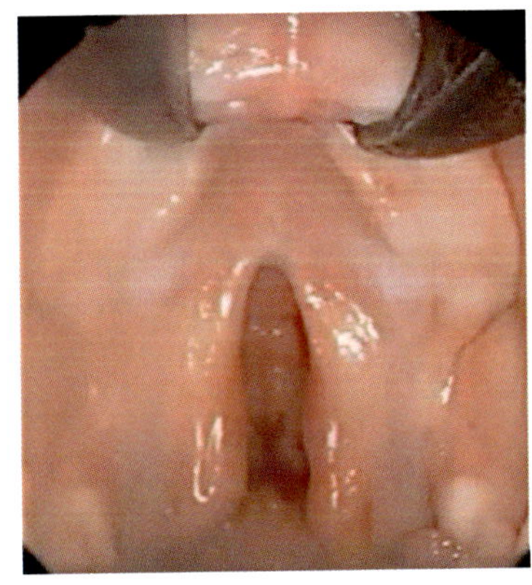
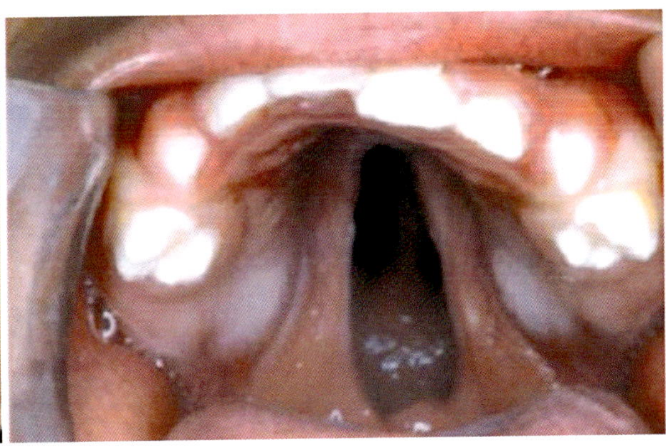

Fig. 14.72 Vomer insufficiency. Partial vertical retraction of the vomer from the palatal plane creates a posterior midline cleft. Complete withdrawal of the vomer from the palatal shelves extends the defect from the back wall of the maxilla to the nasopalatine foramen. [Courtesy of Michael Carstens, MD]

Zone 1 consists of the structures in the midline of the nose and mouth, the vomer and premaxilla. It is supplied by *StV2 medial nasopalatine axis*. Zone 1 interfaces with three other angiosomes (Fig. 14.72).

- Vomer abuts zone 13 perpendicular plate and septum, which are supplied by descending branches from the StV1 posterior and anterior ethmoid arteries.
- Premaxillary frontal process (PMxF) abuts zone 3 frontal process of maxilla (MxF) supplied by medial branch of anterior superior alveolar artery.
- Premaxillary lateral alveolus (PMxB) interfaces with hard palate shelf P1 supplied by greater palatine artery and lateral nasopalatine artery.
  - PMxB is more vulnerable than P1 because it is dependent on a single vascular supply.

Caveat Some confusion exists regarding the boundary between zone 1 of premaxilla, which can produce an isolated alveolar cleft just distal to central incisor, and zone 2 of premaxilla affecting the piriform fossa. We'll discuss this under the lateral incisor section and once again in our treatment of cleft zone 2.

Vomer: An Unrecognized Source of a Common Form of Cleft Palate Paired medial nasopalatine arteries supply paired vomerine bones. Occasionally, these bones may fail to fuse. Alternatively, the intervomerine space may allow for the descent of a tumor or encephalocoele into the oral cavity. Such situations require concomitant pathology of the perpendicular plates of the ethmoid. In the routine case, a deficiency of vomerine mesenchyme will impair the inhibition of sonic hedgehog (SHH) and thus inhibit the normal process of epithelial breakdown required for mesenchymal contact seen as front-to-back fusion of the vomer process with the palatal shelf. If the palatal shelf is normal, the cleft will be narrow. In the minimal state, the palate cleft is very posterior, at the posterior nasal spine, but as the degree of vomer deficiency increases, the palate cleft will extend forward until it reaches the incisive foramen (Figs. 14.73 and 14.74).

When a minimal vomerine deficiency exists in isolation, the width of the palate cleft will be fairly narrow and uniform. The vomer develops in posterior-to-anterior sequence; as it does so, it descends from front to back into the palatal plane like a scimitar. Thus, *the most vulnerable zone of the vomer is posterior and inferior*. Vomerine-based palate clefts can start as posterior notch in which the vomer is literally lifted upward and away from the palatal plane. If the cleft runs all the way to the incisive foramen, compromise of the arterial supply to the ipsilateral premaxilla is likely. The differential diagnosis of these clefts depends on whether the palatal shelf is uninvolved or involved because, in the latter situation, the cleft is wider with the posterior aspect more deficient (wider cleft). Bilateral vomer pathology in the absence of all other field defects results in a bilateral cleft palate that is narrow and can potentially run forward to the incisive foramen. Note that since the vomer is a narrow structure, changes in width of the palatal cleft must be attributed to additional involvement of the palatal shelf based on the StV2 greater palatine artery and/or (rarely) lateral nasopalatine artery.

The Alveolar Cleft in Zone 1: Origins of Lateral Incisor Variation

Tessier described the number 1 cleft as a notch in the primary hard palate between the central and lateral incisors. It is not described as being complete into the secondary hard palate. In such a case, the lateral incisor can be considered to be maxillary rather than premaxillary. *Why is the lateral incisor so variable*? This seemingly trivial point has great significance because it turns on an understanding of dentoalveolar biology between premaxilla and maxilla.

As stated, premaxilla and vomer share a common neuroangiosome, NP_M. During development, the vascular axis always reaches and synthesizes premaxilla and subsequently the vomer. Recall that in evolutions premaxilla is one of the

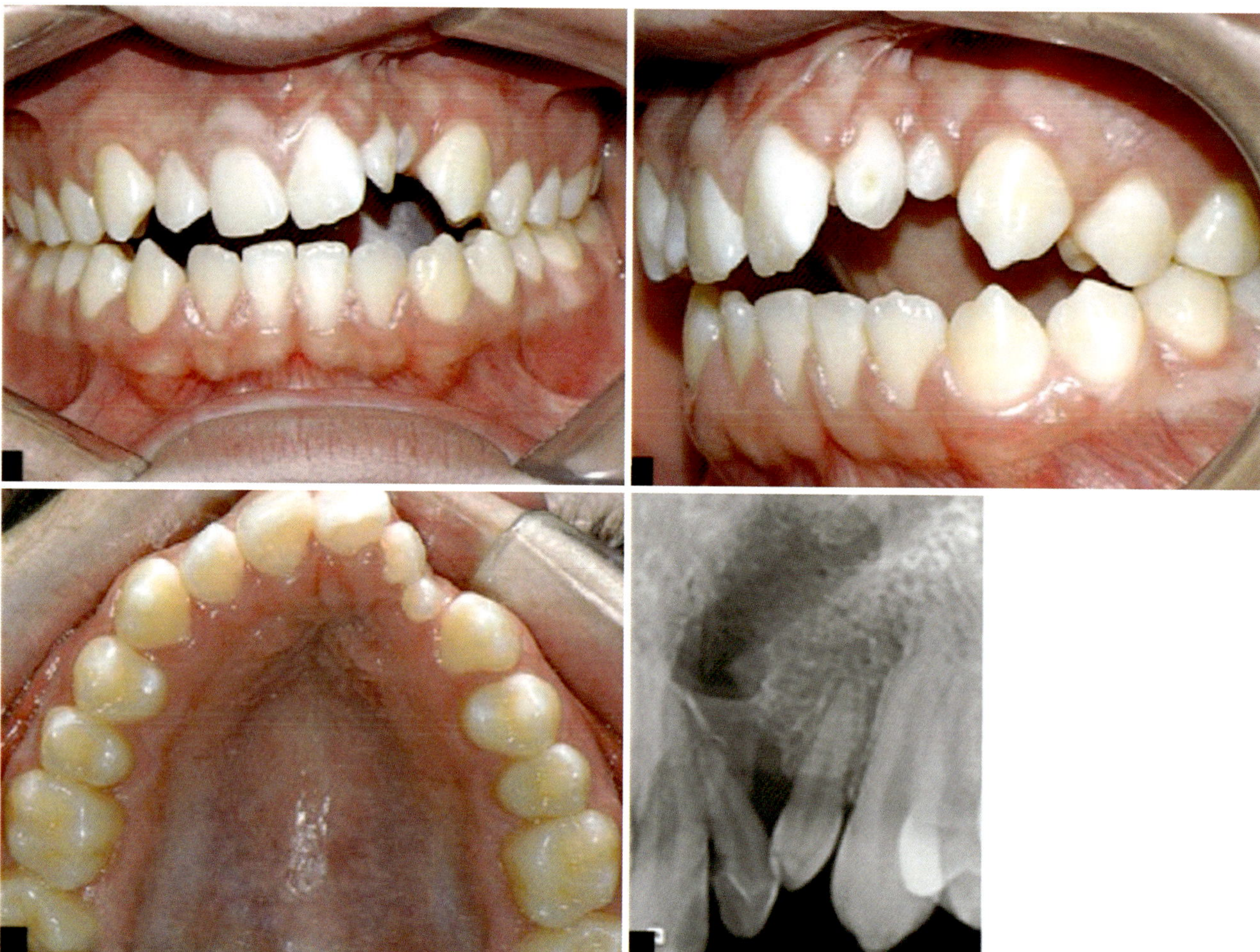

Fig. 14.73 Tessier 1: lateral incisors mesial and distal to cleft. [Reprinted from Garib DC, Rosai JP, Satler R, Ozawa TO. Dual embryonic origins of maxillary lateral incisors: clinical implications in patients with cleft lip and palate. Dental Press J Orthod 2015; 20(5):118–125. With permission from Creative Commons License 4.0: https://creativecommons.org/licenses/by/4.0/deed.en]

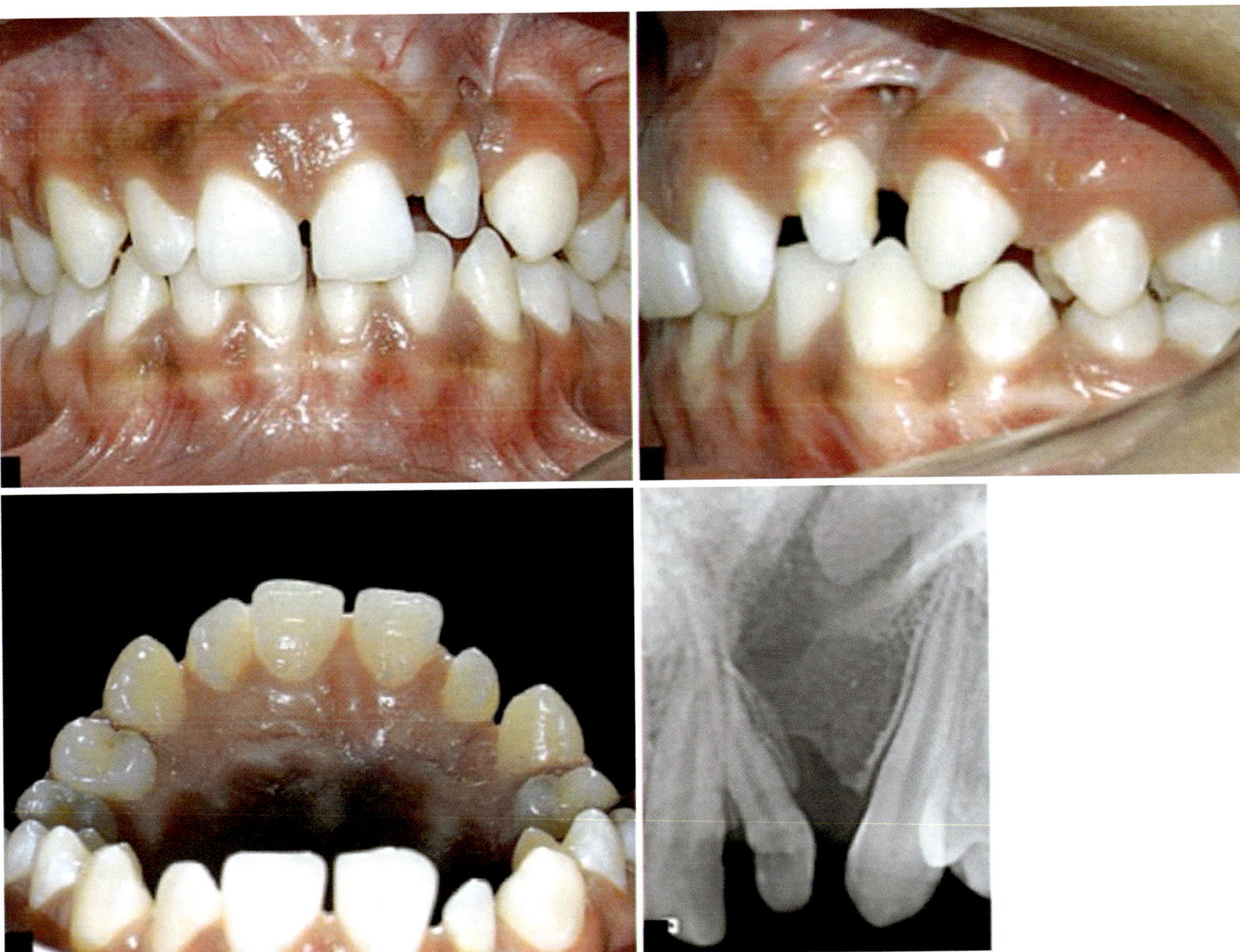

Fig. 14.74 Tessier 1: lateral incisor mesial to cleft. [Reprinted from Garib DC, Rosai JP, Satler R, Ozawa TO Dual embryonic origins of maxillary lateral incisors: clinical implications in patients with cleft lip and palate. Dental Press J Orthod 2015; 20(5):118–125. With permission from Creative Commons License 4.0: https://creativecommons.org/licenses/by/4.0/deed.en]

three jaw bones to appear in placoderms; it predates the appearance of vomer. It is perfectly possible to have a normal premaxilla and a deficient vomer: this is the basis of midline secondary cleft palate (assuming PPE is normal), vide supra. Recall as well that the lateral incisor is the most variable dental unit.

The variations of the alveolar cleft zone are described lucidly by Gamba and Ozawa. Four scenarios are obtained: (1) premaxillary lateral incisor mesial to the cleft, (2) maxillary lateral incisor distal to the cleft, (3) complete absence of incisors, the cleft being bordered by central incisor and canine, and (4) presence of incisors on both sides of the cleft (Figs. 14.75, 14.76, 14.77 and 14.78).

These findings are rationalized by Hovorakova's study which used 3D imaging to follow dental development. As we know, teeth develop by the invasion of an epithelial downgrowth from the alveolar ridge, the dental lamina, into neural crest mesenchyme, which forms the corresponding dental papillae. These interac to form a tooth within the M# zone between the premaxillary alveolus and the maxillary alveolus. This leads to the *dual embryonic origin hypothesis* that the single lateral incisor as typically seen represents a fusion between two sources of mesenchyme, NP_M and NP_L. We are left with an unanswered question: why this should occur exclusively in this zone and nowhere else?

Dental Phylogeny

To answer this question, we must turn backward in time to the phylogeny of teeth. As we have seen, fishes have teeth in many locations through the oral cavity: jaws, branchial arches of the pharynx, and even the tongue. Tetrapods limited the dental units to the alveolus and bones of the palate; over evolution, these palatal teeth disappear in a defined sequence with P4 ectopterygoid being last. Tetrapod bones all the way up to the therapsid ancestors of mammals were rather loosely attached to underlying bone with periodontal ligaments. In the initial iteration, these attachments were *pleurodont*, that is, to the outside edge of the jaws. Pleurodont

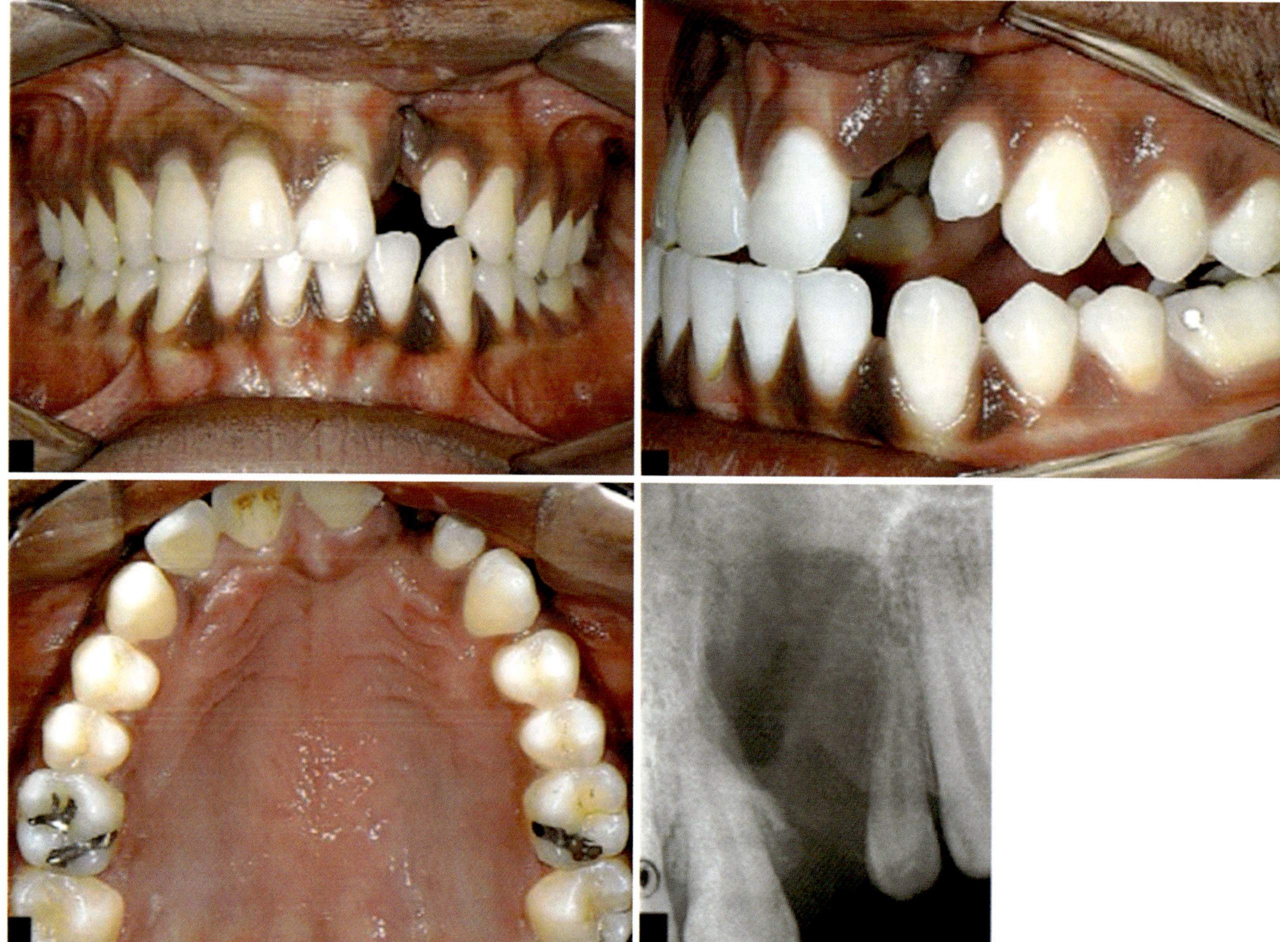

Fig. 14.75 Tessier 1: lateral incisor distal to cleft. This is the *true* Tessier 1 situation. [Reprinted from Garib DC, Rosai JP, Satler R, Ozawa TO Dual embryonic origins of maxillary lateral incisors: clinical implications in patients with cleft lip and palate. Dental Press J Orthod 2015; 20(5):118–125. With permission from Creative Commons License 4.0: https://creativecommons.org/licenses/by/4.0/deed.en]

teeth were rapidly replaced. Later the attachment shifted to the inner edge of the jaws, the *acrodont* condition, which enabled them to withstand more stress with less frequent replacement. Therapsids developed tooth sockets, the *thecodont* condition, which was associated with increased activity and the need for mastication (better nutritional value) requiring stability. As a respectable saber tooth tiger, you just can't spend a lot of time with missing teeth!

Tooth replacement changed over time as well. Up to the mammalian line, most vertebrate are *polyphyodont*, meaning new teeth were formed from retained tooth germs throughout the animal's life. Mammals became *diphyodont* with baby and adult teeth. For the uninterested, toothed whales are *monophyodont*.

Having covered the basics, let's look at the differentiation of teeth. Fishes and most amphibians and reptiles are *homodonts*. Their teeth have a single, monotonous conical shape. Mammals evolved four types of specialized teeth for different functions, the *heterodont* condition. In placental mammals, the basal formula is three incisors (I_1–I_3), one canine (C), four premolars (P_1–P_4), and three molars (M_1). In the universal dental formula, this is written with the number of teeth on each side of the upper jaw as the numerator and those of the lower jaw as the denominator.

- Basal placentals: I 3/3, C 1/1, P 4/4, M 3/3
- Advanced placentals (humans): I 2/2, C 1/1, P 2/3, M 3/3

What evolution tells us is that all humans have the genetic equipment for a third incisor with I_1 being constant and either I_2 or I_3 expressed. If no cleft is present, it will be impossible from inspection to determine if the identity of lateral incisor is I_2 or really I_3, save through angiography or innervation. The presence of a third incisor, I_3 is quite common. These units are not "ectopic" at all, just a different expression pattern. However, the variability we see in size between I_2 and I_3 may tell us that evolution favors the larger dental unit.

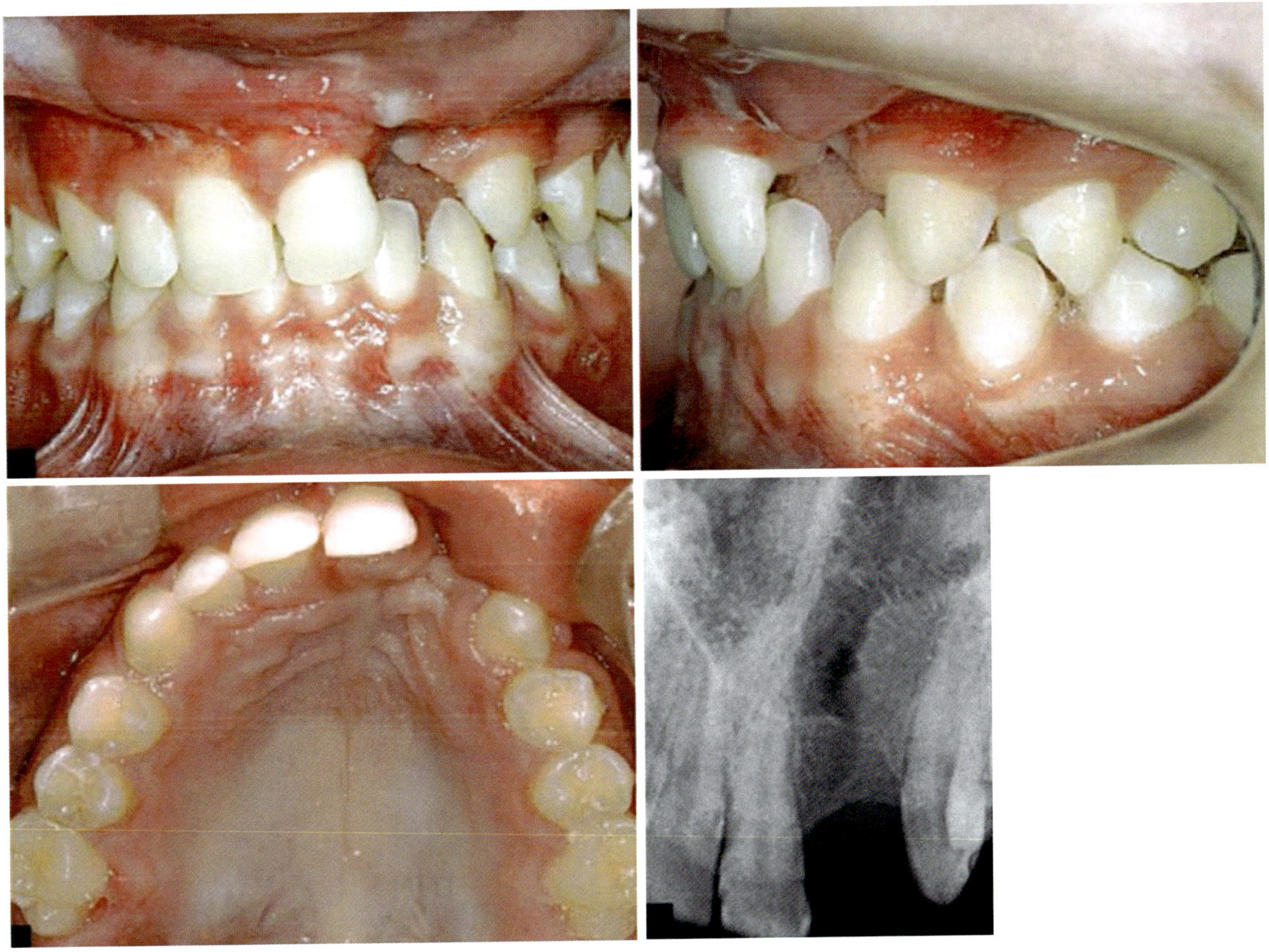

Fig. 14.76 Tessier 1: lateral incisor mesial and distal to cleft. [Reprinted from Garib DC, Rosai JP, Satler R, Ozawa TO Dual embryonic origins of maxillary lateral incisors: clinical implications in patients with cleft lip and palate. Dental Press J Orthod 2015; 20(5):118–125. With permission from Creative Commons License 4.0: https://creativecommons.org/licenses/by/4.0/deed.en]

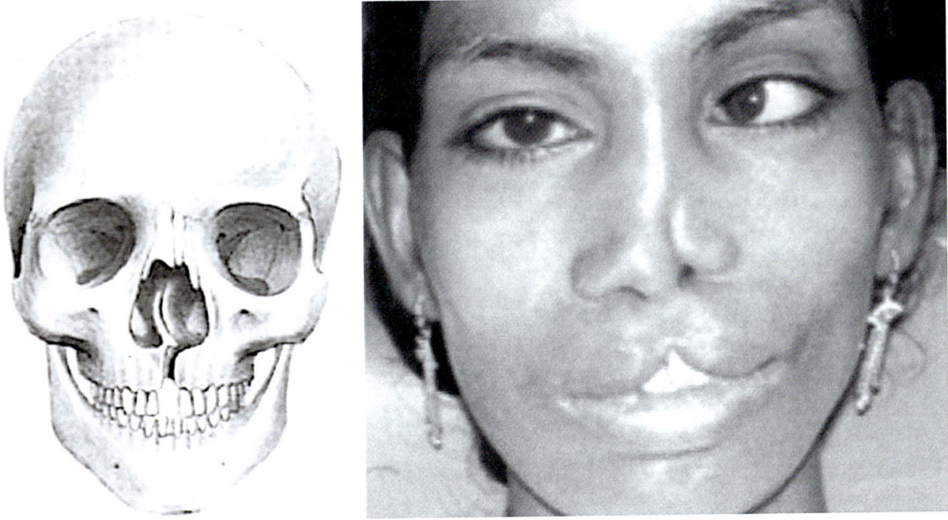

Fig. 14.77 Recap: Tessier zone 1 and zone 13. This is much more commonly reported in the surgical literature because isolated alveolar clefts and vomer deficiencies have not been considered to be part of zone 1. Main feature in zone 1 is the alveolar cleft; zone 13 shows the notch in the midline of the nasal bone. [Reprinted from Jhamb A, Mohanty S. A chronicle of Tessier 0 and 1 facial cleft and its surgical management. J Maxillofac Oral Surg 8(2):178–180. With permission from Springer Nature]

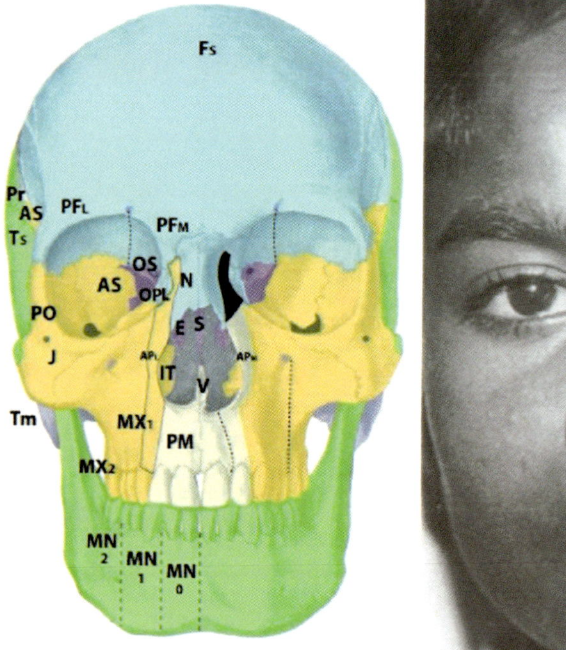

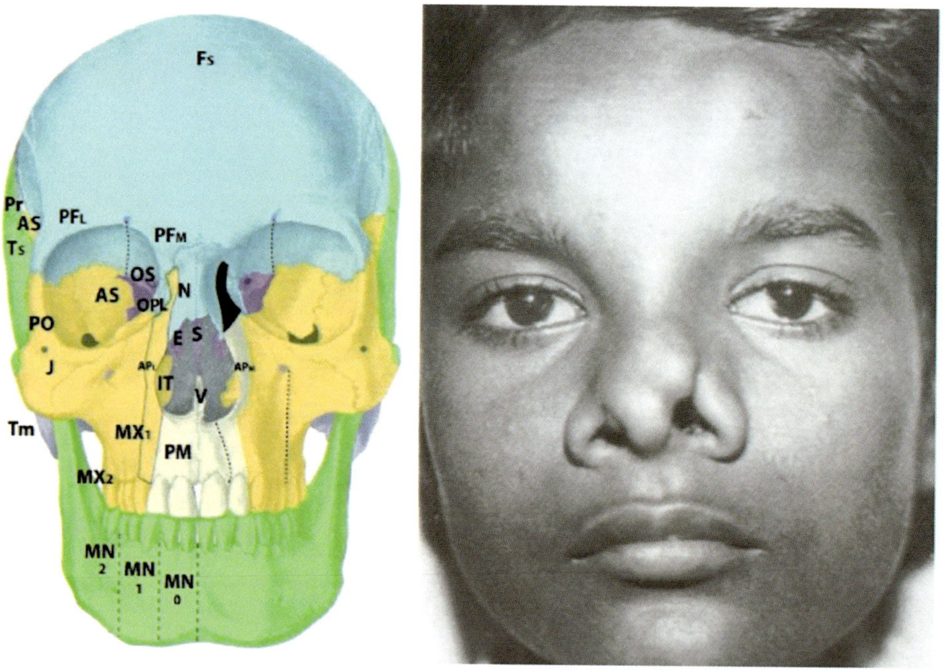

Fig. 14.78 Zone 12, left side. Soft tissue defect through the lateral crus. Bone defect between nasal bone and frontal process of maxilla. Left eye slightly more prominent and left hypertelorism (increased midline to medial canthus distance). Left: [Courtesy of Michael Carstens, MD]. Right: [Reprinted from Tiwari P, Bhatnagar SK, Kalra GS. Tessier number 2 cleft, a variation: case report. J Cranio-maxillo-fac Surg 1991; 19:346–347. With permission from Elsevier]

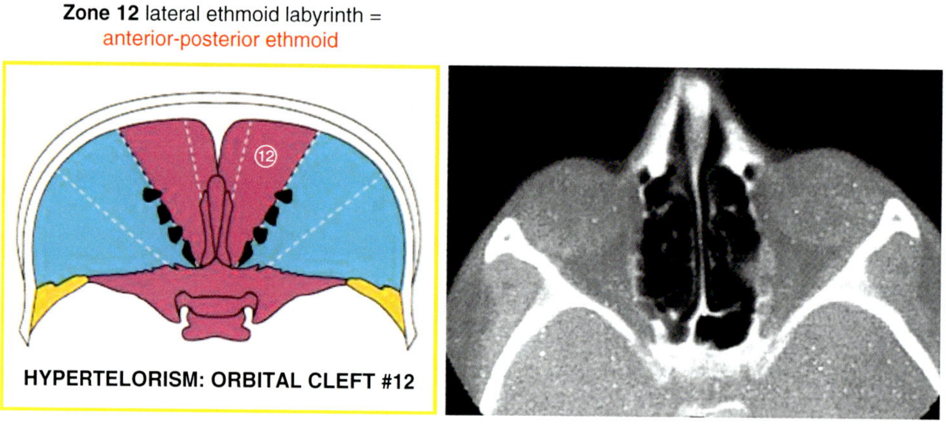

Zone 12 lateral ethmoid labyrinth = anterior-posterior ethmoid

12

HYPERTELORISM: ORBITAL CLEFT #12

Fig. 14.79 Zone 12, left side. CT shows left nasal bone defect. Intraorbital widening of ethmoid sinus, left eye more prominent. [Courtesy of Michael Carstens, MD]

In Sum Alveolar clefts of zone 1 affect the premaxillary alveolus just distal to I_1 causing deficit ranging from a crease to a complete cleft. There is sufficient collateral flow in this situation to maintain the distal segment. The premaxilla does not rotate and the cleft does not affect fusion between vomer and lateral palatine shelf (Fig. 14.79).

Tessier Zone 12

Zone 12 is not causative for cleft palate. We cover it for the sake of completeness. It is analogous to zone 13 (Figs. 14.80 and 14.81).

Zone 12 consists of intracranial and extracranial structures and extranasal structures. The intracranial part of zone 12 corresponds to the internal lamina of the frontal bone overlying the frontal sinus. The anatomy of the dural arterial supply here is uncertain but may involve contributions from the intraorbital part of the ophthalmic artery or of its branches through the anterior ethmoid. Excess tissue here can enlarge the sinus. This in turn can lead to orbital dystopia. These structures are all supplied by the terminal branch of the ophthalmic artery; this gives rise to two branches: the StV1 medial branch of supratrochlear artery supplies the frontal bone and the StV1 infratrochlear artery (dorsal nasal artery) supplies the lateral nasal wall.

The medial supratrochlear supplies the external lamina of the frontal bone and the forehead. Excess in this zone can enlarge the frontal sinus; the glabella can be flat. Deficiencies cause cutaneous defects such as coloboma or notch in the medial one-third of the brow. Inferiorly, the infratrochlear axis supplies the lacrimal sac. There are two descending branches: one runs along the nasal dorsum while the other follows along the lateral nasal wall. This latter is involved in

Cleft zone 2 premaxilla = medial sphenopalatine

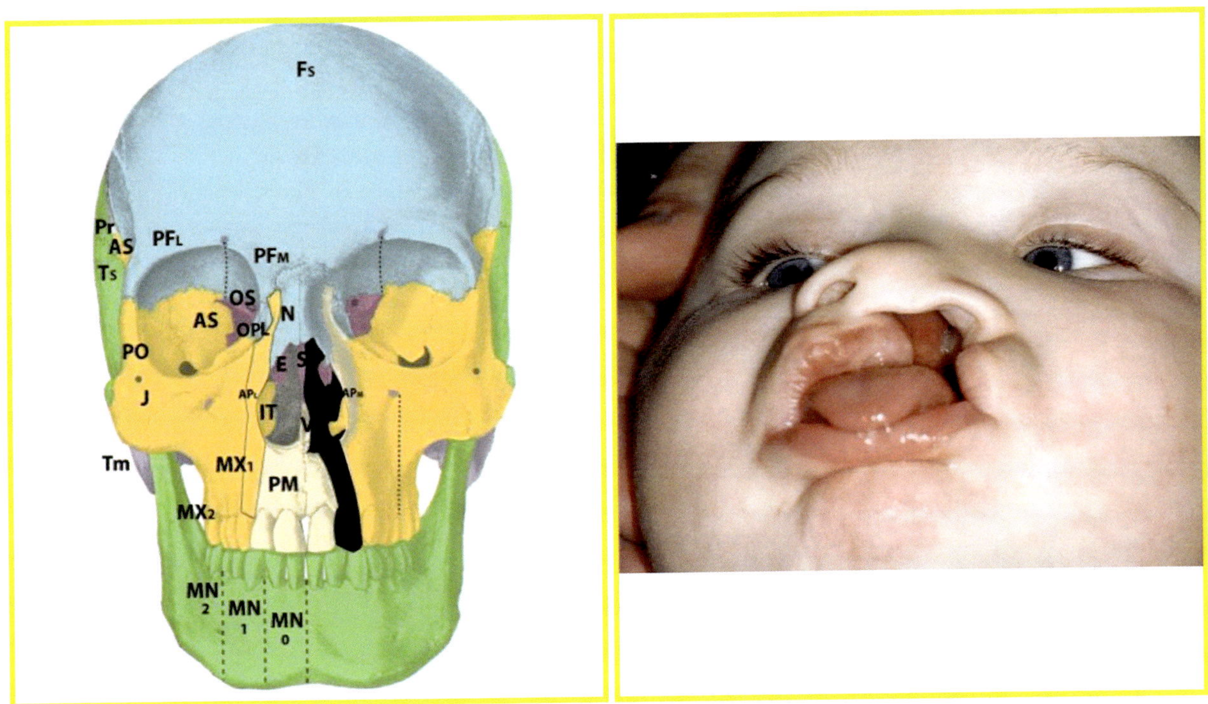

Fig. 14.80 Zone 2 complete unilateral cleft lip/palate. [Courtesy of Michael Carstens, MD]

Zone 3 inferior turbinate = lateral sphenopalatine

massive defects: lateral nasal wall, labio-maxillary cleft, lacrimal

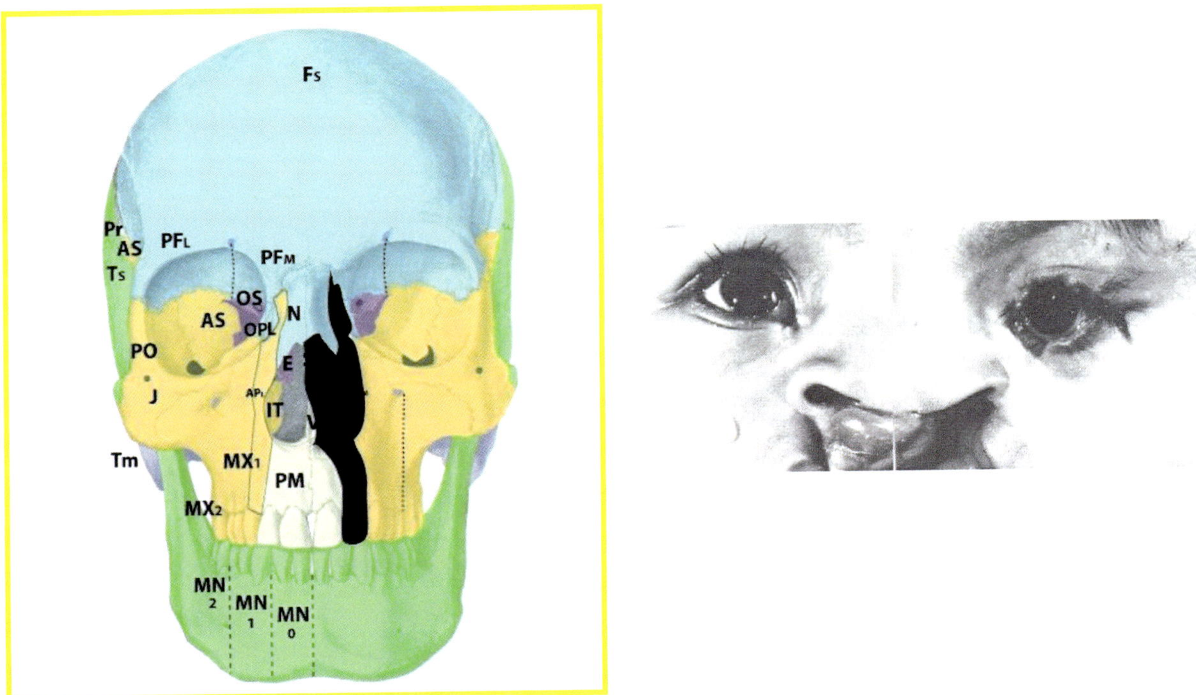

Fig. 14.81 Zone 3–11 cleft is devastating as it wipes out the lateral nasal wall and disrupts the nasolacrimal system. [Courtesy of Michael Carstens, MD]

cleft zone 11. Deficiency of zone 11 shows up most commonly as a mild vertical deficiency of nasal skin. The lacrimal system is unaffected. The formation of the nostril involves induction of soft tissues by the nasal placode. When this process is interrupted, the result is either a proboscis or the outright loss of the structure altogether, a condition called hemi-nose. It is of note that the proboscis, when present, descends from the lacrimal zone. Clefts involving the infratrochlear axis are associated with absence of a recognizable lacrimal apparatus.

Tessier Cleft Zone 2

By definition, cleft zone 2 is the forward extension of the medial nasopalatine axis into the lateral incisor zone of I_2 and I_3 and continuing up the frontal process of premaxilla (Fig. 14.82).

Zone 2 (Figure 1.24) consists of relatively simplistic oral components, the premaxilla and the internal (lingual) lamina of the alveolus housing the central and lateral incisors. These structures are supplied by the StV2 medial nasopalatine artery. The external (labial) alveolar lamina housing the medial and lateral incisors is supplied by the StV2 medial infraorbital. ***Note***: The alveolus of the premaxilla, like that of maxilla and mandible is a bilaminar structure. It also helps support the frontal process of the maxilla. Note that the internal (lingual) lamina of the alveolus housing the lateral incisor territory has an additional collateral supply from the StV2 greater palatine. This explains the great variability in nature seen in the presence or absence of the lateral incisor. On the other hand, the frontal process of the premaxilla appears to be uniquely dependent on the medial sphenopalatine. Because bone is the most distal element of the premaxilla, it is the most vulnerable part of the premaxilla.

The cleft process, when the medial nasopalatine is involved, affects the premaxilla in a very predictable sequence. The frontal process is always involved, even in the most trivial of clefts; this causes a deformity of the piriform fossa. At the lateral incisor zone, an external groove will appear in the alveolus. Further deficiency causes the loss of the lateral incisor. If, for any reason, the most distal aspect of the lateral nasopalatine or that of the greater palatine is affected, the back wall of the alveolus fails to develop and a full-thickness defect of the primary palate ensues. The sequence is always the same. As the defect worsens, the cleft ascends from the incisive surface cephalad and it deepens from labial to lingual. The directionality of pathology is the same for the overlying soft tissues as well.

Tessier Cleft Zone 11

Like zone 12, zone 11 does not directly cause hard palate clefts. We include it because it can occur simultaneously with a zone 3 cleft. This zone consists of the frontal process of the maxilla. It involves the lacrimal groove and the lacrimal segment of the medial lower eyelid. Palate clefts of zone 3 are frequently associated with zone 11 pathology. The arterial axis of zone 11 is the StV1 lateral branch of the infratrochlear artery. For this reason, zone 11 clefts destroy the lacrimal apparatus, which remains intact in other cleft zones. There is marked foreshortening of the lateral nasal skin causing a verticalization of the ipsilateral ala.

Tessier Cleft Zone 3

Zone 3 consists of the inferior turbinate and the corresponding lower lateral nasal wall of maxilla, and the palatine bone

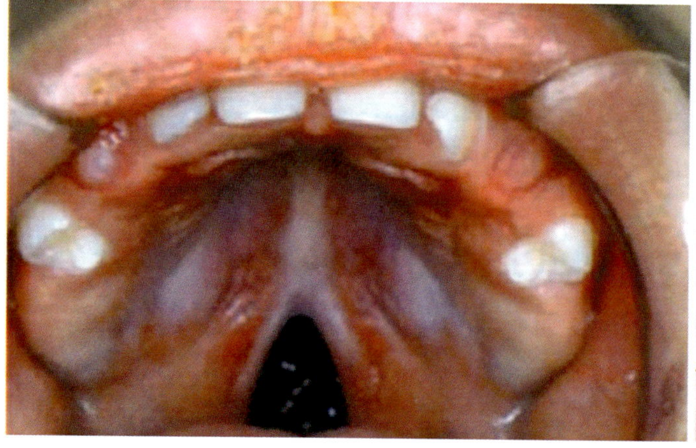

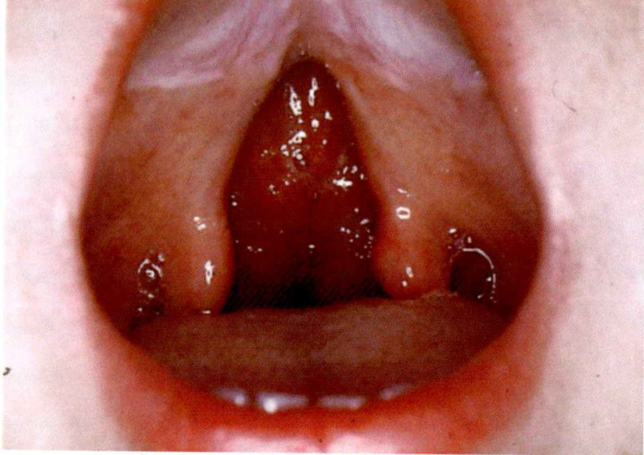

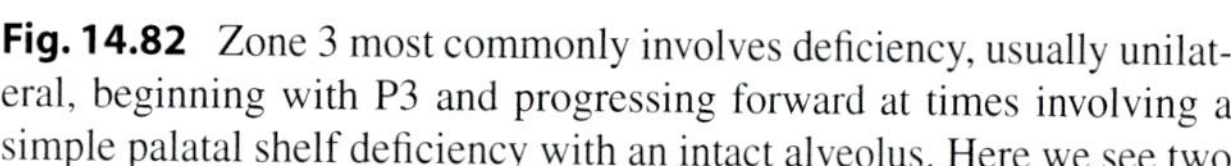

Fig. 14.82 Zone 3 most commonly involves deficiency, usually unilateral, beginning with P3 and progressing forward at times involving a simple palatal shelf deficiency with an intact alveolus. Here we see two examples, the right being more severe, both of which show P3 and P2 deficient with soft tissues drawn forward. Unilateral isolated CP is not common. [Courtesy of Michael Carstens, MD]

fields P1–P3. Blood supply to the lateral nasal wall comes from the StV2 *lateral nasopalatine artery*, including the nasal mucoperiosteum of the palatal shelf, while the StV2 descending palatine artery supplies via the *greater palatine artery* the oral mucoperiosteum of the P1–P2 palatal bones and via the *lesser palatine artery* the P4 palatine bone (Figs. 1.28 and 1.29).

A variety of cleft palate presentations can occur in this zone. The P3 palatine field has a horizontal shelf that develops from lateral to medial and from anterior to posterior. Isolated defects in this bone field range from a simple midline notch to complete elimination of the bone. Because the lesser palatine artery is responsible for the tensor veli palatini, this muscle may be affected as well. But the major blood supply to the muscles in the soft palate is the ascending palatine artery from the cervical division of the facial artery and the ascending pharyngeal artery.

The greater palatine artery sweeps forward to supply the lingual mucoperiosteum of the alveolus as well as the oral mucoperiosteum of the hard palate. The bone field develops from front to back and from lateral to medial. Recall the vascular side branches of vomer and the hard palate shelf conform to an isosceles triangle. Since the closure pattern of the hard palate to the vomer follows the same vectors, an anterior mesenchymal defect will interrupt the "zipper" mechanism from that spot backwards. On the other hand, if the mesenchymal defect occurs further posteriorly in the hard palate field, palate closure will exist anteriorly, while a cleft appears posteriorly.

Zone 3 produces three distinct pathologies. Isolated deficiency of the lesser palatine angiosome will cause a bone defect in the horizontal plate of P3 ranging from mere loss of posterior nasal spine to a frank cleft all the way forward to P2. In Chap. 16, we will explore the pathology of soft palate clefts in greater detail.

Greater palatine deficiency affects hard palate fields P2 and then P1. These always occur in the context of a P3 defect. The lateral wall of the nose and maxillary sinus remain intact. Concomitant lesion of the lateral sphenopalatine artery leads to a devastating cleft that wipes out the piriform fossa, medial wall of maxilla (inferolateral nasal wall), and inferior turbinate.

Mandibular Deficiency States That Cause Cleft Palate

Development of the r3 mandible is similar in plan and execution to that of the maxilla, especially when one considers the anatomy of the alveolar apparatus. In both cases, dental units based upon neural crest blastema occupy a space between two intervening membranous bone units. Running through the membranous bone is either the superior alveolar artery or the inferior alveolar artery. The external (buccolabial) lamina of the maxilla supplied by the medial and lateral branches of infraorbital artery cover, respectively, the incisor/canine zone and the premolar zone while the external branch of superior alveolar artery supplies the molar zone. The internal (lingual) zone is supplied by the medial nasopalatine artery to the incisors while the greater palatine artery supplies all remaining dental zones. The external (buccolabial) lamina of mandibular alveolus is supplied by, posteriorly, the mylohyoid artery and, anteriorly, the submental branch of the facial. The internal (lingual) lamina of mandibular alveolus is supplied by the lingual artery. Thus we see a commonality of bone units supporting the dentition.

The mandible has a third zone, the *ramus*, which is not part of the original embryonic dentary bone. It fact, its evolutionary history is quite distinct. Not surprisingly, the ramus has an entirely different blood supply. Here we find the source of nearly all dentofacial anomalies involving the mandible. Most important, for our purposes, is the situation of cleft palate involving the Pierre Robin anomalad. This entity involves contributions from the first and third arches since the muscle blastema of the palate arises from somitomeres 4 and 7. Associated problems with the fourth arch have also been documented. Retropositioning of the tongue and its potential interference with bony palate shelf development is one possibility. Another possibility is that a similarity in genetic programming exists between the ramus of the mandible and the palatal shelves themselves. Finally in deficiency states of the mandible such as laterofacial microsomias Treacher Collins, Goldenhar, and Nager syndrome, a *deficit in the pretracheal fascia* that links up the ventral aspects of all five pharyngeal arches may be involved in which the syndromes are polyneuromeric.

Soft Palate Defects: Isolated Versus Combined

Diagnosis: The simplest solution is direct testing of individual muscles using a muscle stimulator such as the Peña® device available from Integra Life Sciences (Fig. 1.30).

Any defect affecting the StV2 components of the hard palate (horizontal palatine shelf, palatine bone) is related to the first pharyngeal arch and therefore has the potential to affect the sole muscle arising from somitomere 4: tensor veli palatini (TVP). Because this muscle relates to the cartilagenous lateral wall of the pharyngotympanic tube, it is associated with opening the tube to equalize air pressure, as in swallowing and yawning. First arch pathology affecting the wall of the lateral tube and TVP may explain the relationship seen in unilateral clefts between ipsilateral muscle function and middle ear disease.

Although the developmental fields of the third pharyngeal arch do not contribute membranous bone components to the palate, the cartilagenous medial wall of the pharyngotympanic tube is a third arch derivative. Levator veli palatini arises from somitomere 7 and, when dysfunctional, diminished elevation of the soft palate can be observed. In many cases of microtia, third arch pathology is present. These patients present almost uniformly with a characteristic dampening of physiologic soft palate lift ipsilateral to the microtia.

Failure of midline fusion of the soft palate presents in a posterior-to-anterior gradient. The "cleft" is simply a mesenchymal deficiency—including a reduction in mucosal surface area, with the oral side being more affected than the nasal side. The final muscle pair to form in the soft palate is uvulus. Thus, the first manifestation of the third arch pathology is a midline notch separating these two muscles. As the defect worsens, the levator is affected and the soft palate cleft extends forward toward the posterior nasal spine. Reduction/absence of the helix seen in microtia patients is a manifestation of third arch deficiency. Such patients will frequently demonstrate levator weakness even in the absence of a cleft palate. The combination of third arch with first arch is diagnosed by dysfunction of the tensor veli palatini and/or other osseous deficiency.

Speech problems associated with velopharyngeal insufficiency and/or dysfunction in the absence of cleft palate are not infrequent. Once again, the key to diagnosis is to recognize the existence of individual muscle defects, involving partial or global deficiencies of soft palate muscle. These also include dysfunction of the superior and/or middle constrictor.

Blood supply to the soft palate involves several vessels. The lesser palatine artery from the descending palatine supplies the mucosa. The muscles of the soft palate are supplied by the ascending palatine branch of the cervical division of the facial artery. The ascending pharyngeal artery, proper to the third pharyngeal arch, can also supply the muscles of the palate but its primary target is the superior and inferior constrictors of the pharynx. Isolated defects in the constrictors should be part of the differential diagnosis of noncleft velopharyngeal insufficiency.

Neuromeric Failure

Primary Palate Development: *Critical Contact Model*

1. Primary palate = fusion of lateral premaxilla (PMx) + medial maxilla (Mx)
2. Critical distance between PMx and Mx required for fusion to occur
3. Normal size DFs required for correct positioning of PMx and Mx
4. Fusion sequence: starts at piriform rim, back to front and nasal–oral
5. When PMx small critical contact exceeded = **primary palate cleft**
6. PMx–Mx divergence > *drags vomer away* from midline
7. Vomer ipsilateral to deficient PMx may *also* be *reduced in size*
8. If vomer too far from maxillary palatal shelf (MxP) at incisive foramen, critical contact distance exceeded at incisive foramen > **secondary palate cleft**

Secondary Palate Development: *Critical Contact Model*

1. Palatal shelf (MxP) maturation front to back (20 "older" than 23)
2. MxP elevation follows maturation sequence (front to back)
3. Vomers fuse from sphenoid forward/downward (26 narrower than 21)
4. Fused V complex then descends from incisive foramen backward
5. Normal distance between MxPs *exceeds their capacity to fuse*
6. Interposition of V complex *required* to bridge central gap
7. Critical contact between V complex and MxPs = inverted T fusion
8. Failure of V complex to reach palatal plane > **secondary cleft palate**

Part 8. Neuromeric Classification of Cleft Palate

Anatomic defects

Hp1–hp2 (affects r1) = craniofacial **syndromic CP**

r1 ethmoid (deficient perpendicular plate) = **high arched palate**

r1 ethmoid (absent perpendicular plate) = loss of vomer ± premaxilla = **HPE CL(P)**

r2 vomer reduced height = **midline secondary**

r2 vomer reduced width (palatal shelves normal width) = **submucous CP**

r2 premaxilla defect between central and lateral incisors = **zone 1 alveolar cleft**

r2 premaxilla (partial loss of PMxB) = **primary palate, incomplete**

r2 premaxilla (complete loss PMxB) = **primary palate, complete**

r2 premaxilla + maxillary palatal shelf = **primary and secondary palate**

r2 maxillary palatal shelf (normal vomer) = **isolated secondary palate**

r2 palatine horizontal shelf = **isolated soft palate cleft**
r2 + r3 anterior palatine aponeurosis = **isolated soft CP**
r3 mandible (tongue positioning) = **mechanical "horseshoe" CP palate**
r4–r5 not represented somitomeres 8–11/occipital somites (macroglossia/Downs) = **soft palate**

Functional defects (discussed in Chap. 16, on soft palate pathology)
Somitomere 4 (tensor veli palatini) = **soft palate dysfunction**
Somitomere 7 (LVP, superior constrictor, PP, PG) = **soft palate dysfunction, VPI**

Defects in r1

Ethmoid Perpendicular Plate

Cleft palate due angiosome failure in r1 occurs in the medial nasal process, zone 13. Deficiencies in upper lateral nasal process, zone 12, affect the upper lateral nasal wall and are not relevant for cleft palate.

It presents as spectrum of severity in the medial zone of the ethmoid complex, the lateral zone, and extends distally to the septum. The angiosomes involved are the posterior and anterior ethmoid arteries. In its minimal form, the perpendicular plate (PPE) is vertically foreshortened. Since the mesenchyme of vomer is synthesized along the lower edge of PPE, this pulls the vomer upward, causing a **high arched palate**. The dimensions of the vomer are completely normal and hence the roof of the hard palate remains intact. Although PPE is really a fusion of two laminae, and although the ethmoid complex can be asymmetrically deficient (one set of sinuses being smaller than the other), I have not encountered an asymmetric arched palate.

The holoprosencephaly sequence presents a much more extreme picture. Angiosome failure can wipe out the PPE making descent of the vomer–premaxillary complex impossible. This form of median cleft lip may present with just the soft tissue structures of the columella rolled backwards into the nasal cavity like an elephant tucking its trunk into the mouth. Alternatively, a PPE–vomer–premaxilla can exist on one side only. In this case, there is, of course, a wide-open cleft of alveolus and palate, but the premaxilla is small and contains only two dental units.

Septum

The septum forms concomitantly with perpendicular plate of the ethmoid. In one zone, the mesenchyme forms bone, in the other, cartilage. Posterior ethmoid predominates for perpendicular plate while septum receives both anterior and posterior medial internal nasal arteries. Septal development is a later event, just outside the embryonic time frame. It is present at the time of maxillary palatal shelf formation, just past stage 23 in the beginning of the ninth week. At that time, turbinates are undifferentiated. By the end of 9 weeks, the upper turbinates appear and the palatal shelves begin to elevate. At the early tenth week, the shelves have become horizontal but have not yet fused. Septum is descending. At the end of the tenth week, the shelves have fused with each other and with septum. The inferior turbinates are present. See Chap. 8 for more details about ethmoid embryology.

Isolated congenital absence of the septum is very rare. In one case, complete absence of the septum was documented with adequate development of the ethmoid, including bifidity of the perpendicular plates [23].

On the other hand, a characteristic deformity of the septum accompanies virtually every unilateral cleft affecting alveolar geometry, the distal septum being torqued away to the noncleft side. The situation is more drastic in complete unilateral clefts of the lip and palate because an in utero pressure difference exists between the nasal airway on the noncleft side and that on the cleft side. With fetal breathing of amniotic fluid, pressures are higher on the noncleft side. This causes deviation of the septum and perpendicular plate into the cleft side. At the same time, the premaxilla, its connection to the ipsilateral maxilla rendered asunder, twists away from the cleft side bringing the vomer and septum with it. Thus, the septum in UCLP has an S-shape: away from the cleft anteriorly and into the cleft posteriorly [24, 25].

Defects in r2 (See Figs. 14.82, 14.83, 14.84 and 14.85 for Examples of r2 Defects)

Vomer

The vomer consists of two bone fields supplied by paired medial nasopalatine artery angiosomes. Due to approximation of the pedicles, one side can provide support for the other. Vomerine mesenchyme is laid down anterior–posterior and dorsal–ventral.

Careful study of the vascular anatomy of the vomer explains this developmental sequence. Medial nasopalatine runs obliquely downward giving off perpendicular branches in both directions. The posterior zones of vomer are further away from the axis of the angiosome and take longer to form. At the same time, from nasopalatine foramen, MNPA supplies the premaxilla in the opposite direction, posterior to anterior. Thus, early defects will affect the posterior and ventral zone near just above horizontal palatine shelves. Deficiency of the vomers reduces the width, potentially creating a gap with the hard palate shelves that exceeds the critical contact distance.

The *form fruste* of vomer deficiency is strictly transverse, without affecting the vertical axis. It creates a midline cleft

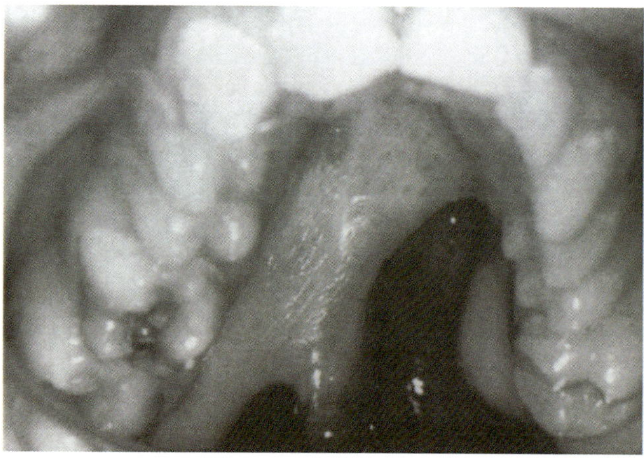

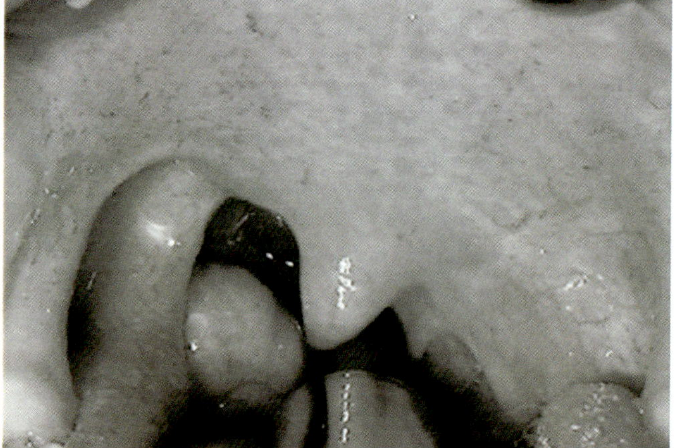

Fig. 14.83 Zone 3 can show unilateral palatal shelf defect. Such cases of "hemipalate" are rare. Left: [Reprinted from Milling MAP, van Straelen P. Asymmetrical cleft palate. Br. J Plast Surg/J Plast Reconstr Aesth Surg 1996; 49:20–23. With permission from Elsevier.] Right: [Reprinted from Tan Y-C, Chen PK-T. Hemipalatal hypoplasia. J Craniofac Surg 2009; 20(4):1150–1153. With permission from Wolters Kluwer Health, Inc.]

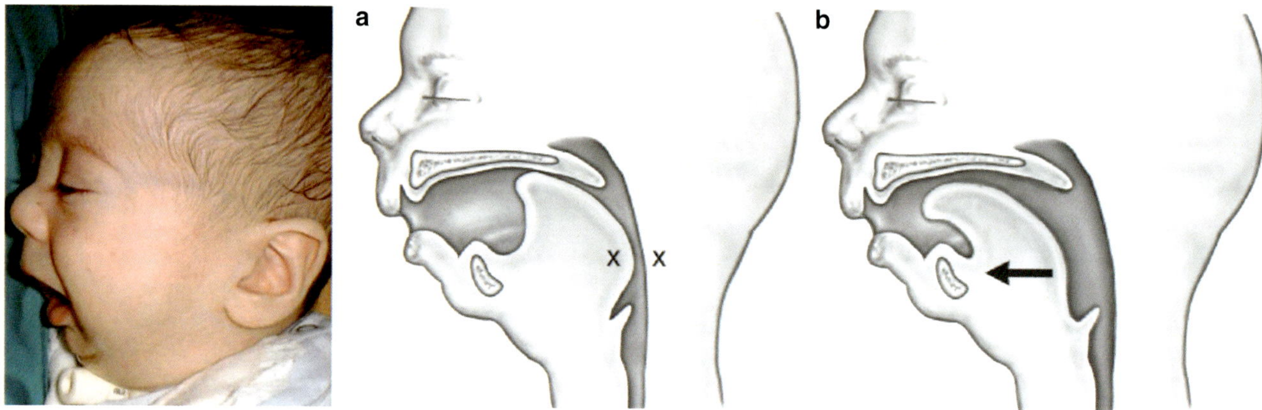

Fig. 14.84 Pierre Robin sequence. Left: Tongue malposition frequently occasions need for tracheostomy. Mandibular hypoplasia has been classically treated with a forward-directed adhesion between the tongue and mandible (see diagram). More recently, this has given way to osteotomies of the mandibular body, followed by distraction. Right: (**a**) Small mandible in Pierre Robin with retropositioned tongue can lead to compromise of the airway. (**b**) Clearance of the airway can be achieved by advancement of the tongue through adhesion to the mandible or advancement of the mandible itself, which carries the entire body of the tongue forward. Left: [Reprinted from Sesenna, E., Magri, A.S., Magnani, C. et al. Mandibular distraction in neonates: indications, technique, results. Ital J Pediatr 38, 7 (2012). With permission from Creative Commons License 2.0: https://creativecommons.org/licenses/by/2.0/]. Right: [Reprinted from Qaqish C, Caccamese JF. The tongue-lip adhesion. Op Tech in Otolaryngol Head Neck Surg 2009; 29(4): 274–277. With permission from Elsevier]

palate in which a thin, almost translucent sheet of ectomesenchyme spans between the hard palate and the vomer, the **submucous cleft of the hard palate**. As vomer is reduced from posterior to anterior, it is pulled upward from the palatal plane and is no longer available for midline fusion. This results in **midline cleft of the secondary palate**. The hard palate cleft thus extends progressively forward from palatine bone shelves to the maxillary bone shelves and finally terminating at the back wall of the premaxilla. This creates a bilateral cleft palate in which the bone gap is small. If the hard palate shelves are likewise reduced, the transverse dimensions of the clefts get wider. As such midline clefts worsen, the vomer shrinks upward progressively.

A mesenchymal deficiency in the vomerine angiosome is independent of the premaxilla. *PMx development precedes that of the vomer*. Phylogenetically, premaxilla came into existence in placoderms along with maxilla and mandible, *prior* to the existence of vomer. In any case, approximately 20 reported cases of congenital absence of the vomer have been reported. In the case cited, the vomer defect does not affect premaxilla, thus confirming the vascular independence of these two fields. Ethmoid complex is also normal [26].

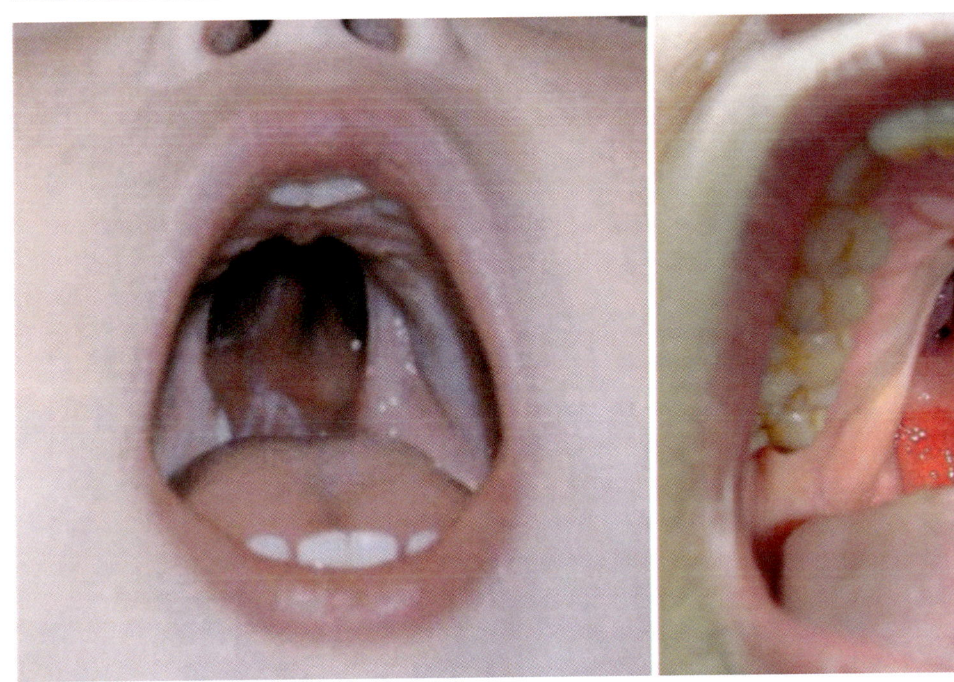
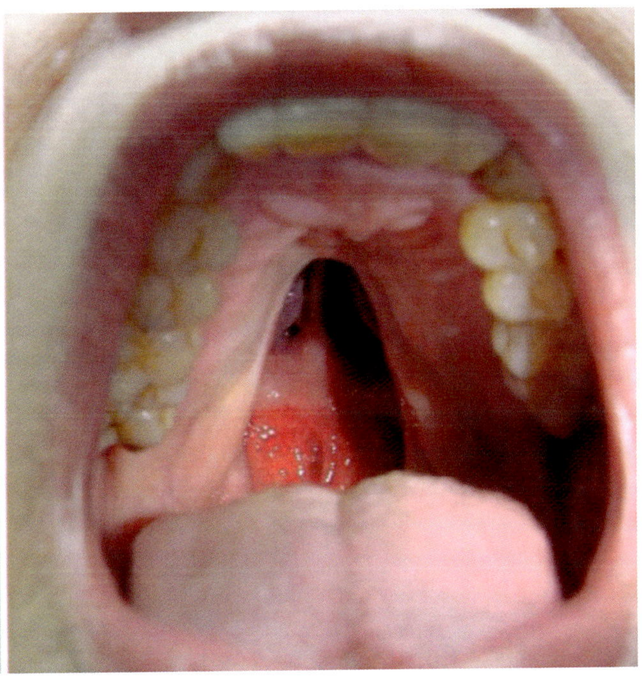

Fig. 14.85 U-shaped cleft palate and V-shaped cleft palate. Typical patterns of cleft palate due to mechanical obstruction in Pierre Robin sequence. Note that in both, the ability of vomer to descend is blocked. [Courtesy of Michael Carstens, MD]

In unilateral holoprosencephaly, cases exist with a hemi-premaxilla. Because the role of PPE is to track the angiosome into position, the ipsilateral vomer must also be missing. Midline cleft palate tends to be narrow. A differential diagnosis can be made on conventional 3D or cone beam CT particularly comparing the axial and coronal views. One can visually determine if the hard palate shelves are of normal width or reduced. If normal, the field defect is isolated to the vomer.

In terms of conventional palate clefts, a perfectly normal vomer can fail to make proper contact with palatal shelf in the presence of a complete alveolar cleft. On the other hand, the fusion process can start normally but if vomer is deficient further posteriorly, in which case it will be elevated, a secondary cleft will develop. If vomer remains in the palatal midline, the defect lies with the palatal shelf.

Premaxilla

The developmental anatomy of the premaxilla has been well described. This ancient bone once housed up to four dental units (side). The human premaxilla represents the distal medial nasopalatine angiosome. It has three fields: central incisor (PMxA), lateral incisor (PMxB), and frontal process (PMxF) which abuts the frontal process from maxilla. In evolution (*vide infra*), premaxilla developed a triangular horizontal *roofing plate* positioned just behind the incisors. The flow of premaxillary mesenchyme is posterior–anterior, medial–lateral, and dorso-ventral. This is enantiomeric with the flow of mesenchyme into the mesial margin of maxilla, zone 4. Recall that the non-tooth-bearing maxillary shelf (descended from the primitive palatine bone field P1) is zone 3. Thus zone 3 of canine and frequently of maxillary incisor lies directly forward of the hard palate shelf.

Deficiency states of the angiosome always begin with PMxF, causing a scooping out of the piriform fossa, and proceed proximally. The bone deficit that results reduces the local [BMP4] with consequent failure to inhibit Shh, a failure that affects the interface between lateral lip element and the prolabium. Soft tissue cleft formation follows a concentration-dependent gradient referenced to the piriform margin. As the BMP4 deficit increases, the cleft ascends up the lip ventral–dorsal envelope and penetrates superficial to deep. The loss of bone stock and the Shh dysfunction mean that PMx cannot fuse with maxilla: the result is an alveolar cleft.

When the cleft is complete, an opening-up of the cleft site becomes possible (but not always). Vomer becomes torqued away from the midline. In this case, when the critical contact distance is exceeded, the "zipper mechanism" of secondary hard palate closure fails… complete cleft results. If alveolar union is achieved, Pl1 (MxP) will contact vomer and the zipper will start. Once the zipper is functional, the hard palate will close until it encounters (1) a vertical defect in the vomer or (2) a transverse defect in the palatal shelf.

Posterior Palatine Bone (Pl3)

The pathologic anatomy of the soft palate is a subject that merits its own chapter. Here we shall summarize the main points. Palatine bone arises from the third bone field of the palatine series. Its nucleation site is in the palatal plane. From there it gives off two fields. *Perpendicular plate* ascends until it encounters the nasopalatine angiosome at which point it divides into a posterior sphenoid process and an anterior orbital process. I have not encountered cases of a deficient perpendicular plate. *Horizontal plate* represents the basal p3 field because it remains in the palatal plane. It is rectangular and develops with mesenchyme being added on from lateral to medial and from anterior to posterior. Posterior nasal spine represents the most recent mesenchyme. Deficits in the horizontal plate occur as a spectrum, in reverse order. Severe palatine deficiency creates a triangle, the base of which is at the maxillary tuberosity and the hypotenuse runs obliquely forward to the back of the maxillary palatal shelf. Defects in the horizontal plate reduce the diffusible BMP4 available for the soft tissue closure, causing submucous cleft palate.

Palatine aponeurosis is attached to the palatine bone. It is an r3 structure which is muscle-free with the exception of uvulus running forward in the midline. Primary defects in the horizontal plate and aponeurosis are usually associated with one another. This pulls the levator complex forward into pathologic position. If the maxillary shelves remain intact, the result is an **isolated soft palate cleft**.

Anterior Palatine Bone (Pl1–Pl2)

The hard palate anterior to palatine bone is *not phylogenetically part of maxilla*. In this text, we have the conventional term, palatal shelf of the maxilla, MxP. However, it would be more accurate to rename this structure as the **anterior palatine bone**. It consists of two fields. The Pl field is in register with the outboard maxilla, anterior superior alveolar axis. Its dental units are lateral incisor, canine, and premolars. Pl2 is in register with the outboard maxilla in the posterior superior alveolar axis. Its dental units are the molars. Blood supply for this bone is dual: on the oral side, *greater palatine* extends all the way forward to premaxilla. On the nasal side, the *lateral nasopalatine* axis supplies nasal side mucoperiosteum and inferior turbinate.

In the development of anterior palatine bone, r2 mesenchyme is deposited anterior–posterior and medial–lateral. Thus, a minimal deficit manifests itself posteriorly as a reduction in the transverse width of the shelf. Horizontal plate of palatine is likewise reduced. Pl3 can only develop as far medially as Pl2 will allow. If GPA is defective in isolation of LPA,

- GPA angiosome failure always leads to soft palate cleft, even if LPA is intact.
- LPA angiosome failure leading to soft palate cleft can occur in isolation from GPA failure. In this instance, the soft palate is cleft and the hard palate remains intact.

Defects in r3

Mandible, Ramus Zone

Recall from our discussion of the phylogeny of the mandible that its original configuration consisted of a core strut, Meckel's cartilage flanked by dermal bones on either side, that develop according to the Meckelian template. Although these are annexed in evolution, the separate binding sites for muscles around the lower jaw reveal their presence. The final amalgam is the tooth-bearing dentary bone. The *postdentary zone* develops into ramus and condyle. This is the site of most cases of retrognathia. A normal complement of tooth structures occupying a standard amount of space is suspended from the skull by a vertical strut that can vary in size, both horizontally and vertically.

Tongue position is determined by the mandible. When vertically short posteriorly, as in the syndromes of Treacher Collins or Nager, mechanical interposition of the tongue can disrupt normal closure of the soft palate and, if more severe, the secondary hard palate as well, causing defects that are characteristically U-shaped or V-shaped (Figs. 14.84 and 14.85).

r3 Fascia of Tensor Veli Palatini

Tensor veli palatini inserts into the lateral margin of anterior one-third of palatine aponeurosis. Its fascia is r3 neural crest and is undoubtedly confluent with the aponeurosis. In lateral facial syndromes such as craniofacial microsomia or Goldenhar's, this zone could be affected.

r8–r11

Occipital somites S1–S4 produce the myoblasts of the tongue. As we shall see further on in Chap. 16, these myoblasts migrate into position as part of the hypobranchial cord. When palatal shelves p1–p3 develop, they are initially vertical on either side of the tongue and then elevate. This requires the tongue to get out of the way. *Macroglossia* can prevent normal shelf development. A *misallocation of mesenchyme* between the sclerotome and myotome can produce a large tongue as seen in *Beckwith–Wiedemann syndrome*. The latter has a triad of macroglossia, >90% growth, and abdominal wall defects such as omphalocele. A minor association exists with enlarged kidneys. Muscle fibers in B–W are perfectly normal, just hypertrophic. Investigation regarding the same process of mesenchymal excess between the tongue and kidney could prove useful.

Final Thoughts

We have completed our review of the developmental anatomy of the oronasopharynx. We have examined the various components that make the mesenchymal structures of the palate. The concept of fields—each with a specific neurovascular axis, each containing neural crest originating from a specific neuromere, each susceptible to failure of formation versus disruption based upon a growth cone dysfunction—has been presented in detail.

From a functional standpoint, the hard palate has several roles. It functions as a support of the masticatory apparatus. It acts as a baffle that separates breathing from the digestive tract. Finally, it constitutes a platform bearing the insertions of the soft palate muscle complex. As such, the position of the soft palate in space is directly dependent upon hard palate platform into which it is inserted. For this reason, efficiency of palate to pharynx contact is not soft tissue dependent but hard tissue dependent. As we shall see in Chap. 16, restoration of length becomes integral for the repair of soft palate cleft.

Developmental fields can contain bone bone-producing factories; these depend on proper stem cell development. The management of cleft palate involves the restoration of osteogenic functional matrix to its proper position on either side of the midline. In cases of unrepaired cleft palate, the growth of facial structures continues in a normal fashion but the *stem cell-containing functional matrices,* being shipwrecked in the wrong position in space, will continue to make bone away from their proper place with respect to the midline. The take-home points are as follows:

- Cleft palate management requires that this displaced development fields be reassigned to their proper anatomic state.
- Developmental fields are defined by their neuroangiosomes.
  - Neuroangiosomes should *never* be subdivided.
- Dissection of developmental fields is subperiosteal.
- Developmental field reassignment (DFR) cleft palate repair is really the restoration of mesenchymal stem cell matrices to their proper relationships.
- Surgical solutions based upon a sound embryologic principles with respect to correct field boundaries offer the best chance for the achievement of harmonious growth and correct function.

Commentary: Ricardo Bennun

Alveolar Extension Palatoplasty: Technical Notes

Ricardo Bennun

Introduction

Successful palate closure became possible in the mid-nineteenth century with the development of the mucoperiosteal flaps by Dieffenbach and von Langenbeck [27].

The fact that simple closure of the palate cleft did not necessarily ensure normal speech was recognized a little later and led to the development of palate-lengthening procedures and various forms of pharyngoplasty.

This chapter serves as a logical extension to the multidisciplinary approach to cleft palate management and reconstruction from genetic factors that precipitate cleft to a thorough discussion of today's approaches.

Surgical Procedure

A complete report of the alveolar extension palatoplasty (AEP) was precisely illustrated in a chapter of our Atlas published in 2015 [28]. The predictive factors of difficulty before reconstruction and the planned strategies were also described by the author [29]. A comparative study between Veau–Wardill–Kilner and the AEP procedure was also published [30].

Technical Recommendations

A short incision of the oral mucosa over the alveolar border must be carried out. Utilizing a Gillies retractor, under direct vision, and with the blade in an oblique direction, the cut must be completed, leaving all the dental germs in their position.

Minimal lateral incisions are planned in simple cases, where the soft palate major gap distance (SPMGD) is inferior to 6 mm, with symmetrical bilateral soft palate length (BSPL) superior to 20 mm. This strategy will also be useful in the presence of easily isolated cleft palates.

When the SPMGD is between 7 and 11 mm, one alveolar extended palatal flap elevation is considered. In that case, the author's choice will be the palatal flap from the sick side. Some surgeons utilizing the AEP procedure will prefer to select the normal-side palatal flap. Outcomes seem to be similar.

In bilateral cleft cases, the suggestion for a complete and cautious hard and soft palate closure will be using both AEP flaps. This recommendation is also valid in patients with SPMGD greater than 12 mm, but having a symmetrical BSPL superior to 20 mm.

In very difficult cases with a SPMGD superior to 12 mm, plus symmetrical/asymmetrical BSPL under 16 mm, or even a short bilateral hard palate length with depressions (BHL&D), our indication would be a nasal plane reconstruction utilizing a vomer flap, complemented with a superior pharyngeal flap [29].

Results

Statistically significant differences were found when comparing the total percentage of complications between both groups (Veau–Wardill–Kilner and AEP). The amount and severity of each complication by groups were also established as significant [30]:

- Bite alterations and transversal collapse with dental malpositioning in group A: 29.84% and group B: 23.155% (*P* value < 0.009)

- Presence of fistulas in group A: 4.11% and group B: 5.0% (*P* value < 0.02)
- Patients with VPI in group A: 6.12% and group B: 0.11% (*P* value < 0.14)

Complications

No additional complications have been reported utilizing this procedure. Outcomes and follow-up during the last 12 years have proved no teeth alteration, in all patients (pictures).

Conclusions

Since 2009, the author has selected the alveolar extension palatoplasty variant, plus the complete muscle dissection and retropositioning, and the posterior pillar's elongation with hemi-uvula rotation/reconstruction, as the technique of choice for primary cleft palate repair [28].

The utilization of the pre-op cited parameters to identify cleft palate diversity and severity seems to be a useful methodology to select the correct surgical strategy [29].

Moving and reducing incisions to protect blood supply, following Carstens' suggestion, allow us to reduce the use of electric coagulation and blood loss. The presence of less raw areas prevents the incidence of retractile scars [31].

Employing regional blocking [32] joint to general and local anesthesia not only decreased intra- and post-op pain and baby neurotoxicity but also allowed us to initiate oral feed and discharge the baby early.

Having bigger palatal flaps in width and length allows us to decrease the incidence of anterior/medial fistulas, velopharyngeal incompetence, and maxillary alterations. Dental malpositioning and misalignment, as medial otitis, were also present in an inferior percentage [33–35].

Note on the Clinical Series

Bilateral cases 4, 5, and 7 and unilateral cases 5, 7, 11, 12, 13, and 14 demonstrate class 1 occlusion.

A Note from Dr. Carstens

Ricardo Bennun

Alveolar extension palatoplasty as a concept is uniquely South American, both in its design and in its surgical verification. AEP, in concept of the palatal version of subperiosteal tissue transfer for cleft lip repair, was drawn out on a paper napkin during an airplane flight to Ecuador. Although the operation was described in 1999, it was subsequently picked up by Dr. Luis Monasterio, a distinguished cleft surgeon in Santiago, Chile. Lucho invited me to do some cases with him at the Fundación Gantz. Since the incisions required to reposition the entire embryonic field of the hard palate were made on the alveolar ridge, the effect of these on dental eruption was an issue of debate. Dr. Monasterio did his own series, following dental development for several years. He determined that eruption was unaffected (which he subsequently reported to the 12th American Cleft Palate Association meeting). He also introduced me to Dr. Ricardo Bennun from the University of Buenos Aires and director of the Fundación Piel. Ricardo subsequently took the operation to the next level, beginning a case series which now extends to over a decade. Ricardo is a true biologic surgeon, reflecting his long-time commitment to burn care; he keeps on asking questions and seeking answers. I had the opportunity to contribute two chapters for his 2015 work, "Cleft Lip and Palate Management: A Comprehensive Atlas"; in the process, his attention to detail and insistence of quality forced me to think about the issues more deeply than before. It is his influence that really pushed me over the edge, daring me to write this book, a task that seemed overwhelming to me at the time. True to form, his atlas of AEP cases faithfully recorded herein will stand the test of time as a surgical proof that developmental biology can win out over dogma for better patient outcomes. For this, all cleft surgeons will be grateful.

References

1. O'Rahilly R, Müller F. Developmental stages in human embryos: revised and new measurements. Cells Tissues Organs. 2010;192:73–84.
2. Puelles L, Rubenstein JL. Forebrain gene expression domains and the evolving prosomeric model. Trends Neurosci. 2003;26:469–76.
3. Puelles L, Harrison M, Paxinos G, Watson C. A developmental ontology for the mammalian brain based on the prosomeric model. Trends Neurosci. 2013;36:570–8.
4. Tomás-Roca L, Corral-San-Miguel R, Aroca P, Puelles L, Marín F. Crypto-rhombomeres of the mouse medulla oblongata, defined by molecular and morphological features. Brain Struct Funct. 2016;221:815–38.
5. Puelles L. Forebrain development: the prosomeric model. 2009. http://www.nslc.wustl.edu/courses/bio3411/woolsey/Readings/Lecture7/Puelles%202009.pdf.
6. Noden DM, Francis-West P. The differentiation and morphogenesis of craniofacial muscles. Dev Dyn. 2006;235:1194–218.
7. Sambasivan R, Kuratani S, Tajbakhsh S. An eye on the head: the development and evolution of craniofacial muscles. Development. 2011;138:2401–15.
8. Depew MJ, Simpson CA, Morasso M, Rubenstein JL. Reassessing the Dlx code: the genetic regulation of branchial arch skeletal pattern and development. J Anat. 2005;207(5):501–61.
9. Hall BK. The neural crest in vertebrate development and evolution. 2nd ed. New York: Springer; 2009.

10. Trainor P. Neural crest cells: evolution, development, and disease. Cambridge: Academic Press; 2014.
11. Barry A. The aortic arch derivatives in human adult. Anat Rec. 1951;111:221–38.
12. Noden DM. Origins and assembly of avian embryonic blood vessels. Ann N Y Acad Sci. 1990;588:236–49.
13. Diamond MK. Homologies of the meningeal-orbital arteries of humans: a reappraisal. J Anat. 1991a;178:223–41.
14. Hansen L, Nolting D, Holm G, Hansen BF, Kjaer I. Abnormal vomer development in human fetuses with isolated cleft palate. Cleft Palate Craniofac J. 2004;41:470–3.
15. Kimes KR, Mooney MP, Siegel MI, Todhunter JS. Growth rate of the vomer in normal and cleft lip and palate human fetal specimens. Cleft Palate Craniofac J. 1992;29:38–42.
16. Sandikcioglu M, Mølsted K, Kjaer I. The prenatal development of the human nasal and vomeral bones. J Craniofac Genet Dev Biol. 1994;14:124–34.
17. Zhang Y, Zhang Z, Zhao X, Yu X, Hu Y, Geronimo B, et al. A new function of BMP4: dual role for BMP4 in regulation of sonic hedgehog expression in the mouse tooth germ. Development. 2000;127:1431–43.
18. Zhang Z, Song Y, Zhao X, Zhang X, Fermin C, Chen Y. Rescue of cleft palate in Msx1-deficient mice by transgenic Bmp4 reveals a network of BMP and Shh signaling in the regulation of mammalian palatogenesis. Development. 2002;129:4135–46.
19. Mohri M, Amatsu M. Congenital defects of the vomer. Ann Otol Rhinol Laryngol. 2000;109:497–9.
20. Miller IJ, Spangler KM. Taste bud distribution and innervation on the palate of the rat. Chem Senses. 1982;7(1):99–108.
21. Benton M. Vertebrate paleontology. 4th ed. Oxford: Wiley-Blackwell; 2015.
22. Ewings E, Carstens MH. Neuroembryology and functional anatomy of craniofacial clefts. Indian J Plast Surg. 2009;42(Suppl):S19–34. https://doi.org/10.4103/0970-0358.57184.
23. Tuncbilek G. Congenital isolated absence of the nasal cartilaginous septum. J Plast Reconstr Aesthet Surg. 2008;61:e1–4.
24. Bocconi L, Boschetto C, Cerani F, Kustermann A. Fetal breathing movements. In: Piontelli A, editor. Development of normal fetal movements. Milan: Springer Italia; 2010.
25. Talmant JC, Talmant JC. Cleft rhinoplasty, from primary to secondary surgery. Ann Chir Plast Esthet. 2014;59(6):555–84. https://doi.org/10.1016/j.anplas.2014.08.004.
26. Yan DJ, Lenoir V, Chatelain S, Stefanelli A, Becker M. Congenital vomer agenesis a rare and poorly understood condition revealed by cone beam Ct. Diagnostics. 2018;8:15–9.
27. Naidu P, Yao CA, Chong DK, Magee WP III. Cleft palate repair: a history of techniques and variations. Plast Reconstr Surg Glob Open. 2022;10(3):e4019. https://doi.org/10.1097/GOX.0000000000004019.
28. Bennun RD, Monasterio AL. Chap. 11: Cleft palate repair. In: Bennun RB, Harfin JF, Sandor GKB, Genecov D, editors. Cleft lip and palate management: a comprehensive atlas. New York: Wiley Blackwell; 2015. p. 163–73.
29. Bennun RD. Cleft palate repair: predictive factors of difficulty and planned strategies to solve it. J Craniofac Surg. 2020;31:1664–7.
30. Astrada S, Bennun RD. Cleft palate repair: a study between two surgical procedures. J Craniofac Surg. 2020;31:2280–4.
31. Carstens MH. Chap. 1: Mechanisms of cleft palate: developmental field analysis. In: Bennun RB, Harfin JF, Sandor GKB, Genecov D, editors. Cleft lip and palate management: a comprehensive atlas. New York: Wiley Blackwell; 2015. p. 3–21.
32. Moggi LE, Ventorutti T, Bennun RD. Cleft palate repair: a new maxillary nerve approach. J Craniofac Surg. 2020;31:1547–50.
33. Rivelli RA, Casadio V, Bennun RD. Audiological alterations in patients with cleft palate. J Craniofac Surg. 2018;29:1486–9.
34. Harfin JF, Bennun RD. Chap. 15: Strengthening surgical/orthodontic interrelationships. In: Bennun RB, Harfin JF, Sandor GKB, Genecov D, editors. Cleft lip and palate management: a comprehensive atlas. New York: Wiley Blackwell; 2015. p. 227–41.
35. Harfin JF. Chap. 16: To what extent dental alveolar osteogenesis can be achieved solely with orthodontic treatment in cleft patient? In: Bennun RB, Harfin JF, Sandor GKB, Genecov D, editors. Cleft lip and palate management: a comprehensive atlas. New York: Wiley Blackwell; 2015. p. 245–52.

Further Reading

Benninger B, McNeil J. Transitional nerve: a new and original classification of a peripheral nerve supported by the nature of the accessory nerve (CN XI). Neurol Res Int. 2010;2010:476018.

Burdi AR. The premaxillary-vomerine junction: an anatomic viewpoint. Cleft Plate J. 1971;8:364–70.

Carlson BM. Human embryology and developmental biology. 5th ed. New York: Elsevier; 2013.

Celebi AA, Uncar FI, Sekerici AE, Caglaroglu M, Tan E. Effects of cleft lip and palate on the development of permanent upper central incisors: a cone-beam computed tomography study. Eur J Orthod. 2015;37:544–9. https://doi.org/10.1093/ejo/cju082.

Diamond MK. Homologies of the meningeal-orbital arteries of humans: a reappraisal. J Anat. 1991;178:223–41.

Diogo R, Kelly RG, Christiaen L, Levine M, Ziermann JM, Molnar JL, Noden DM, Tzahor E. A new heart for a new head in vertebrate craniofacial evolution. Nature. 2015;520(7548):466–73. https://doi.org/10.1038/nature14435.

Dupin E, Creuzet S, Le Douarin NM. The contribution of the neural crest to the vertebrate body. In: Madame Curie bioscience database. Austin: Landes Bioscience; 2000.

Garib DG, Rosae JP, Sahler LR, Ozawa TG. Dual embryonic origin of maxillary lateral incisors: clinical implications in patients with cleft lip and palate. Dental Press J Orthod. 2015;20(5):118–25. https://doi.org/10.1590/2177-6709.20.5.118.128.sar.

Gilbert S, Barresi MJ. Developmental biology. 11th ed. Sunderland: Sinauer Assoc; 2016.

Hiruma T. Formation of the ocular arteries in the chick embryo: observations of corrosion casts by scanning electron microscopy. Anat Embryol (Berl). 1996;193:585–92.

Hiruma T, Hirakow R. Formation of the pharyngeal arch arteries in the chick embryo. Observations of corrosion casts by scanning electron microscopy. Anat Embryol (Berl). 1995;191:415–23.

Hiruma T, Nakajima Y, Nakamura H. Development of pharyngeal arch arteries in early mouse embryo. J Anat. 2002;201:15–29.

Hovoraskova M, Lesot H, Peterkova R, Peterka M. Origin of the deciduous upper lateral incisor and its clinical aspects. J Dent Res. 2006;85(2):167–71.

Kardong K. Vertebrates: comparative anatomy, function, evolution. 7th ed. New York: McGraw Hill; 1997.

LeDourain NM, Kalchaim C. The neural crest. 2nd ed. Oxford: Cambridge University Press; 2009.

Liem K, Bemis W, Walker WF, Grande L. Functional anatomy of the vertebrates. 3rd ed. Boston: Cengage; 2000.

Liem KF, Bemis WE, Walker WF, Grande L. Functional anatomy of the vertebrates. Belmont: Brooks-Cole; 2001. p. 99–101.

Mavelli ME, Ayurek M. Congenital isolated absence of the nasal columella: reconstruction with internal nasal vestibular skin flap and bilateral labial mucosa flaps. Plast Reconstr Surg. 2000;106:393–6.

Noden DM, Trainor PA. Relations and interactions between cranial mesoderm and neural crest populations. J Anat. 2005;207:575–601.

O'Rahilly R, Müller F. Human embryology and teratology. 3rd ed. Somerset: Wiley-Liss; 2001.

Okazaki M, Sarukawa S, Fukuda N. A patient with congenital defect of nasal cartilaginous septum and vomeral bone reconstructed with costal cartilaginous graft. J Craniofac Surg. 2005;16(5):819–22.

Padget DH. The development of the cranial arteries in the human. Contrib Embryol. 1948;32:205–61.

Standring S, editor. Gray's anatomy. 40th ed. New York: Elsevier; 2006.

Tada MN, Kuratani S. Evolutionary and developmental understanding of the spinal accessory nerve. Zool Lett. 2015;1:4.

Talmant JC, Talmant JC, Lumineau JP. Primary treatment of cleft lip and palate, its fundamental principles. Ann Chir Plast Esthet. 2016;61(5):348–59. https://doi.org/10.1016/j.anplas.2016.06.007.

Talmant JC, Talmant JC, Lumineau JP. Secondary treatment of cleft lip and palate. Ann Chir Plast Esthet. 2016;61(5):360–70. https://doi.org/10.1016/j.anplas.2016.06.012.

Valencia MP, Castillo M. Congenital and acquired lesions of the nasal septum: a practical guide for differential diagnosis. Radiographics. 2008;28(1):205–23. https://doi.org/10.1148/rg.281075049.

Wei X, Senders C, Owrn GO, Liu X, Wei Z-N, Dillard-Telm L, McClure HM, Hendricks AG. The origin and development of the upper lateral incisor and premaxilla in normal and cleft lip/palate monkeys induced with cyclophosphamide. Cleft Palate Craniofac J. 2000;37(6):571–83.

Alveolar Extension Palatoplasty: The Role of Developmental Field Reassignment in the Prevention of Sequential Vascular Isolation and Growth Arrest

15

Michael H. Carstens

Bone does not "grow itself"; growth is produced by the soft tissue matrix that encloses each whole bone. The genetic and functional determinants of bone growth reside in the composite of soft tissues that turn on and turn off and speed up and slow down the histogenic actions of the osteogenic connective tissues (periosteum, endosteum, sutures, and periodontal membranes).

—*Donald E. Enlow*

Introductory Remarks

The surgical correction of cleft palate is accompanied by five frequent sequelae necessitating secondary interventions: (1) dental arch collapse and malocclusion, (2) anterior fistula, (3) dentofacial growth disturbance, (4) velopharyngeal incompetence, and (5) posterior fistula. A voluminous literature documents the negative consequences of palatoplasty for development of the arch and its dentition [1]. But how?

Let's categorize these sequelae in terms of what we know, and don't know

- *Dental arch collapse* involving the cleft segment has long been attributed to releasing incisions at palatoplasty.
  - What is the mechanism and how can it be prevented or minimized?
- *Anterior fistula* is due to short flaps and absent repair of the nasal floor.
  - What is the mechanism and how can it be prevented (or corrected secondarily)?
- *Maxillary retrusion* occurs in the context of *staged* procedures. In particular, adult patients with unoperated palate clefts display normal dental arch growth [2, 3].
  - Why is the combination of CL + CP so much more destructive than either procedure alone?
  - What is the explanation for differential effects when palatoplasty is delayed?
- *Velopharyngeal incompetence* relates to soft palate—see Chap. 16.
  - What is the mechanism?
  - Is there a specific field defect?
  - How and when is VPI best prevented or secondarily corrected?
- *Posterior fistula* relates to soft palate—see Chap. 16.
  - What is the mechanism?
  - Is there a specific field defect?
  - How can it be prevented?

Alveolar extension palatoplasty (AEP) was developed in response to these questions. It is based on three premises: (1) The combined effect of conventional cleft lip and cleft palate operations causes damage to neuroangiosomes responsible for supplying the stem cell population of the dentofacial soft tissues leading to palatal fistula and negative patterns of growth. (2) Our model of cleft lip/cleft palate pathology is not based on molecular embryology; it therefore lacks an appreciation for developmental fields. (3) The facial tissues of cleft patients have a normal complement of stem cells distributed incorrectly; when these developmental fields are correctly reassigned, they are capable of normal growth.

The purpose of this chapter is to examine in detail the biologic basis for alveolar extension palatoplasty. In particular we will look at the problems of fistula, arch collapse, and altered growth and see how AEP addresses each

M. H. Carstens (✉)
Wake Forest Institute of Regenerative Medicine, Wake Forest University, Winston-Salem, NC, USA
e-mail: mcarsten@wakehealth.edu

M. H. Carstens (ed.), *The Embryologic Basis of Craniofacial Structure*, https://doi.org/10.1007/978-3-031-15636-6_15

of these. Particular attention will be given to how a new concept, *sequential vascular isolation*, helps us understand pathologic membranous bone growth and how to avoid it.

We shall cover the following topics

- The historical development of alveolar extension palatoplasty
- Surgical technique of alveolar extension palatoplasty
- Arterial anatomy of the alveolus and surrounding fields
- Consequences of cleft lip/cleft palate surgery
- Pathophysiology of insult
- Vascular isolation and dentoalveolar development
- DFR in cleft palate surgery: surgical sequence and clinical performance

Figures for this chapter are arranged as follows

- Comparative palatoplasty techniques
  - Von Langenbeck, Veau, Wardril-Killne, Bardach two-flap, AEP
  - Why are AEP flaps longer?
- Technical aspects of the AEP procedure
  - Total maxillary block
  - Elevating AEP flaps
  - Maximizing extension
- Vascular anatomy of the alveolus
  - Sequential vascular isolation
- Clinical performance of AEP
  - Primary palatoplasty
  - Secondary palatoplasty: anterior fistula reconstruction

The Origin of Alveolar Extension Palatoplasty (Figs. 15.1, 15.2, 15.3, 15.4, 15.5, 15.6, 15.7, 15.8, 15.9, 15.10 and 15.11)

During residency and craniofacial fellowship at the University of Pittsburgh I was influenced by the work of Dr. George Soteranos in the management of secondary alveolar clefts [4]. The procedure we used, a complete subperiosteal mobilization of the anterior alveolus and maxilla carried up to the infraorbital foramen, with a backcut incision up the buttress, uniformly resulted in an advancement over the anterior cleft of two dental units and a tension-free hermetic closure over the bone graft. What's more, it always gave a more symmetrical midface appearance. The lip volume on the cleft side was full and there was better soft-tissue projection in the midline.

From 1990 to 1991 while working for PAHO (Pan American Health Organization) in Nicaragua I saw large numbers of secondary cleft cases, all of which manifested soft-tissue asymmetry of lip volume. Virtually 100% of alveolar clefts were ungrafted. A high percentage of patients had anterior fistulae, were in crossbite, or had class III occlusion with maxillary retrusion. These findings were present despite the large number of well-trained surgeons involved in the primary cases. Why were the lip–nose complexes (including my own) asymmetrical? Why were the palatoplasty flaps so consistently inadequate? I was haunted by the idea that perhaps our principles were not working as intended, that somehow the patients were *growing out of their repairs and into relapse*.

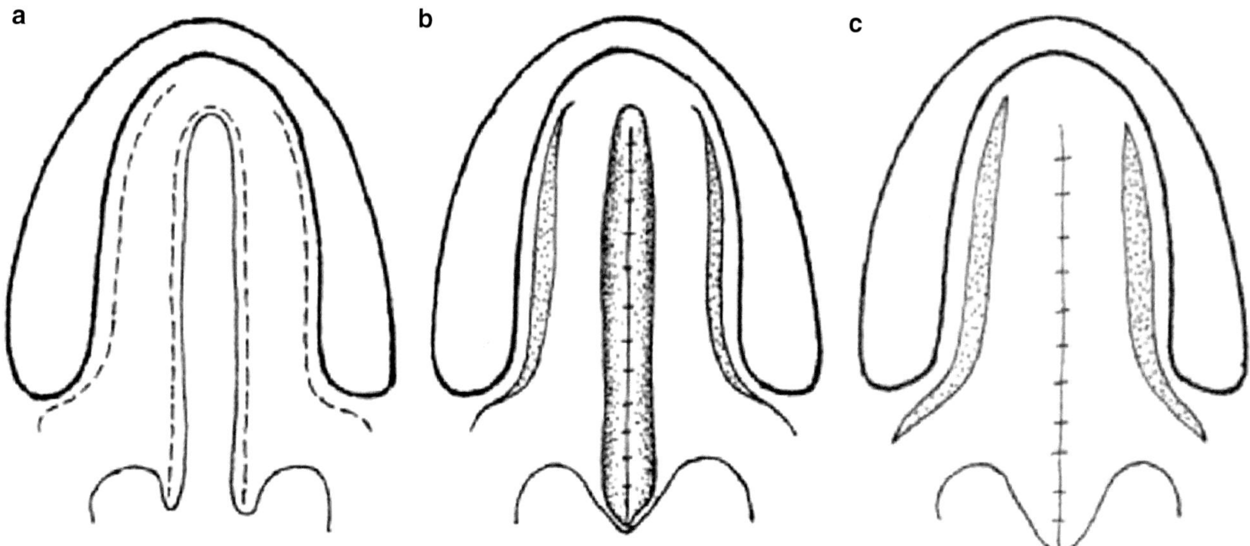

Fig. 15.1 Von Langenbeck palatoplasty. Not designed to treat complete palate cleft. Raw areas of denuded palate heal by scar. (**a**) Incision design, (**b**) subperistoeal mobilization of flaps, (**c**) closure. [Reprinted from Agrawal K. Cleft palate repair and its variations. IJPS 2009 42(Suppl): S102–S109. With permission from Creative Commons License]

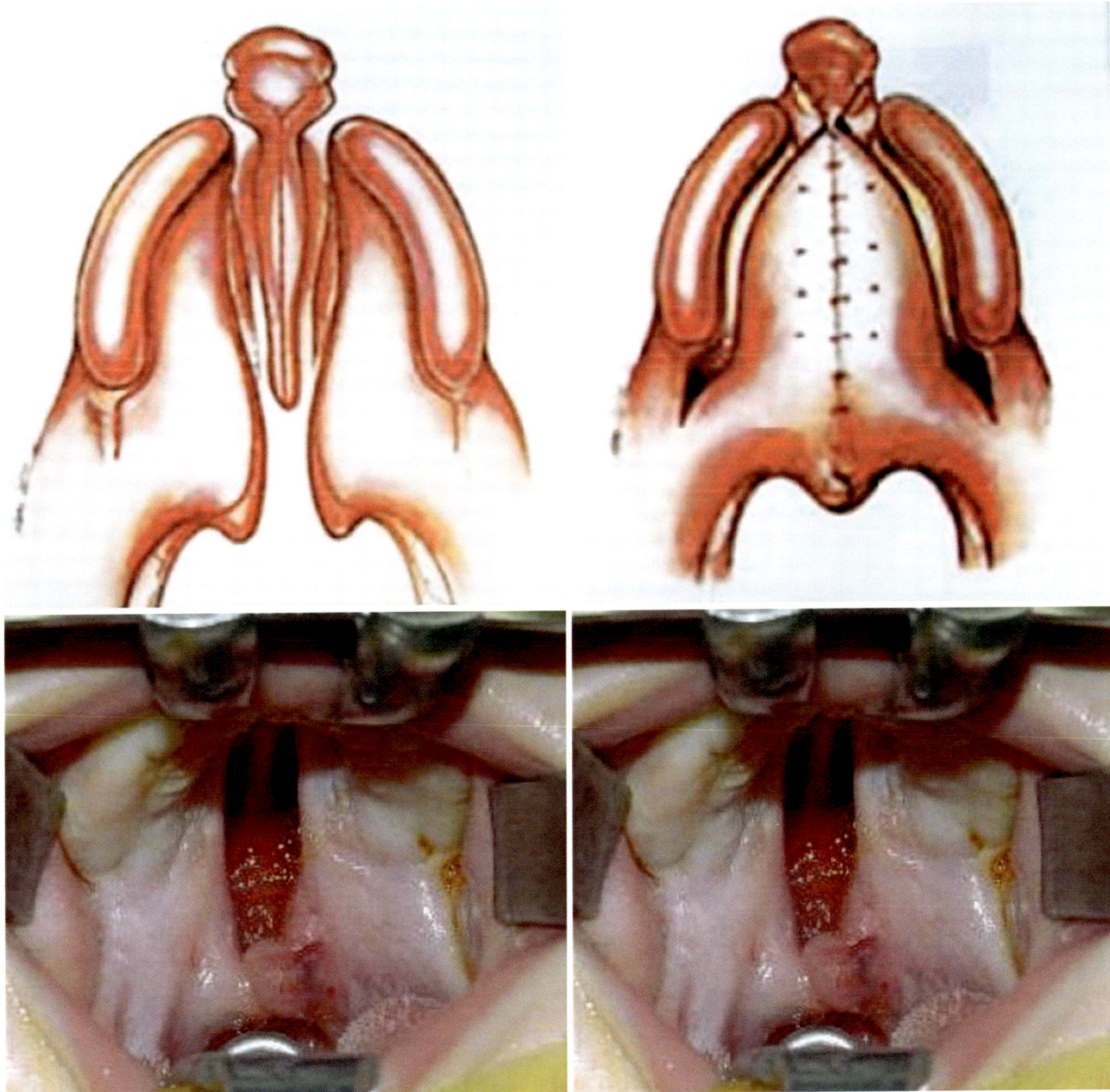

Fig. 15.2 Veau. The incision is made in the junction of the alveolar wall and the palatal shelf limited anterior reach. Wide areas of denuded palate. [Courtesy of Dr. Luis Monasterio, Funación Ganz, Santiago, Chile]

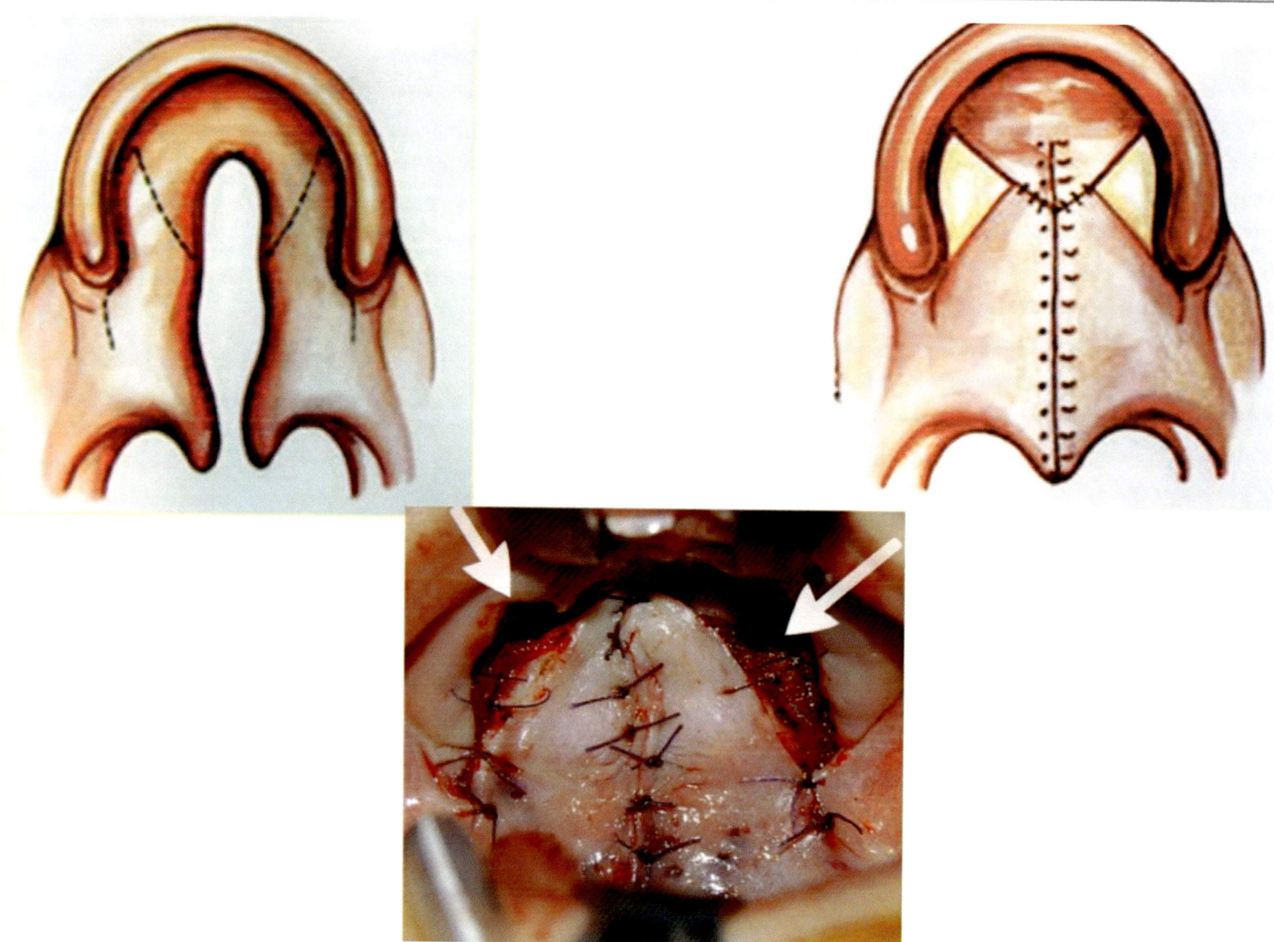

Fig. 15.3 Wardill–Kilner. Not applicable to complete clefts. Completely unanatomic. Anterior flaps depend on backflow from medial nasopalatine. Raw areas seen. [Courtesy of Dr. Luis Monasterio, Funación Ganz, Santiago, Chile]

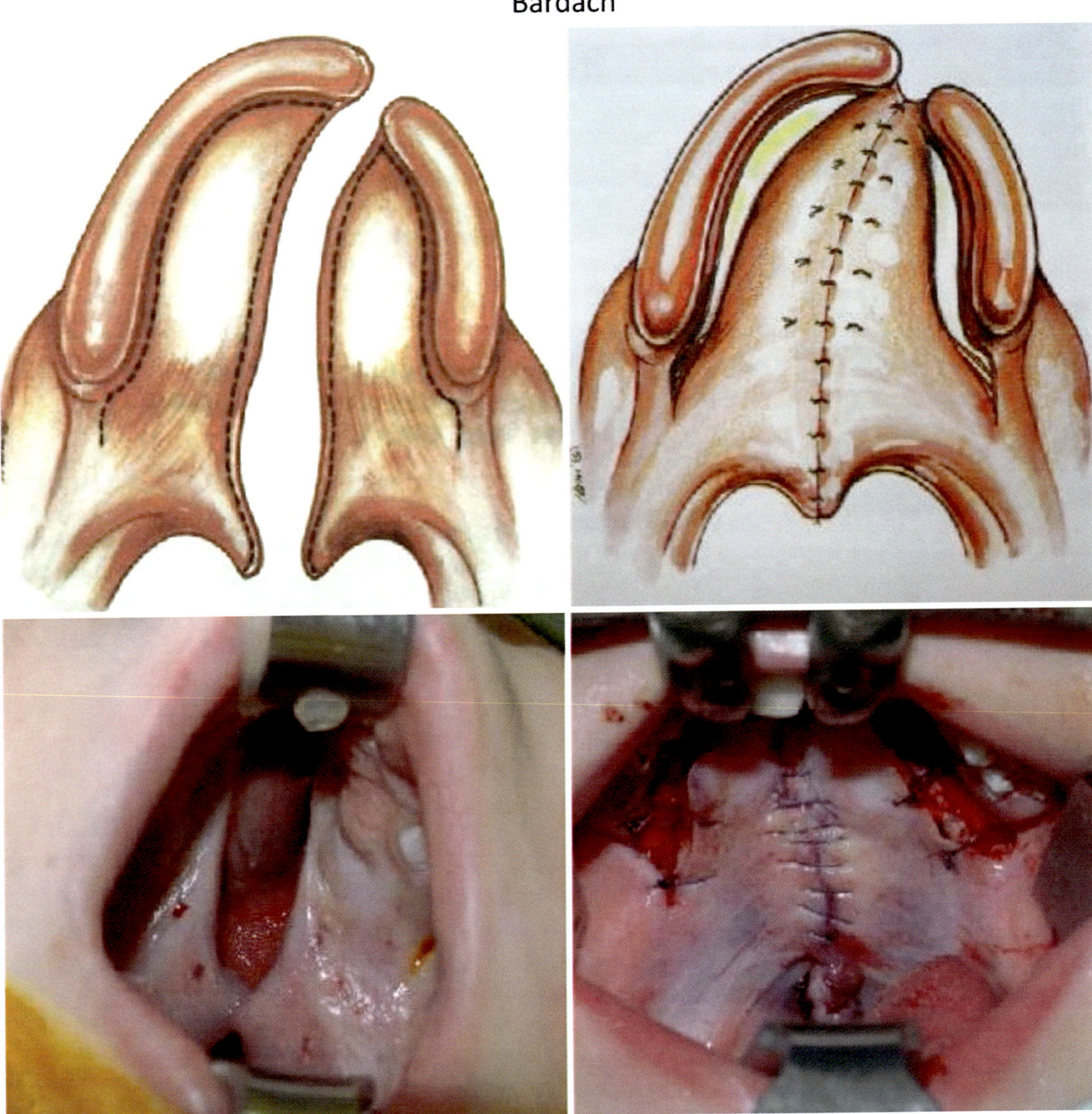

Fig. 15.4 Bardach two-flap palatoplasty. Raw areas, inadequate length. [Courtesy of Dr. Luis Monasterio, Funación Ganz, Santiago, Chile]

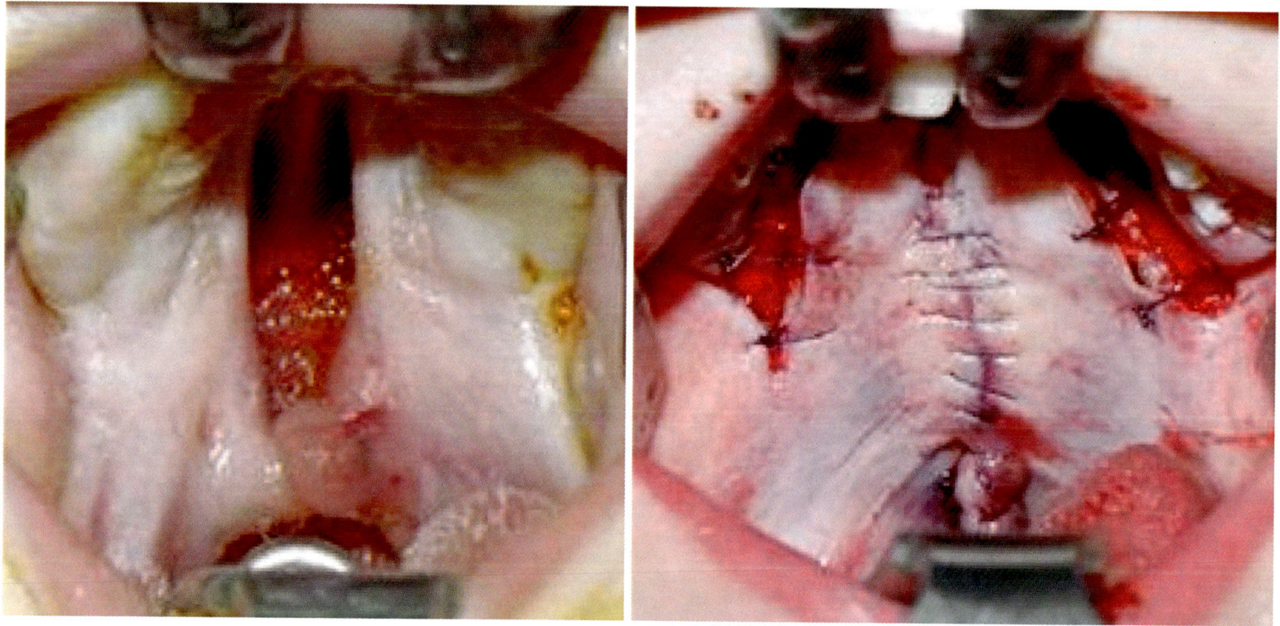

Fig. 15.5 Bardach for bilateral cleft palate. [Courtesy of Dr. Luis Monasterio, Funación Ganz, Santiago, Chile]

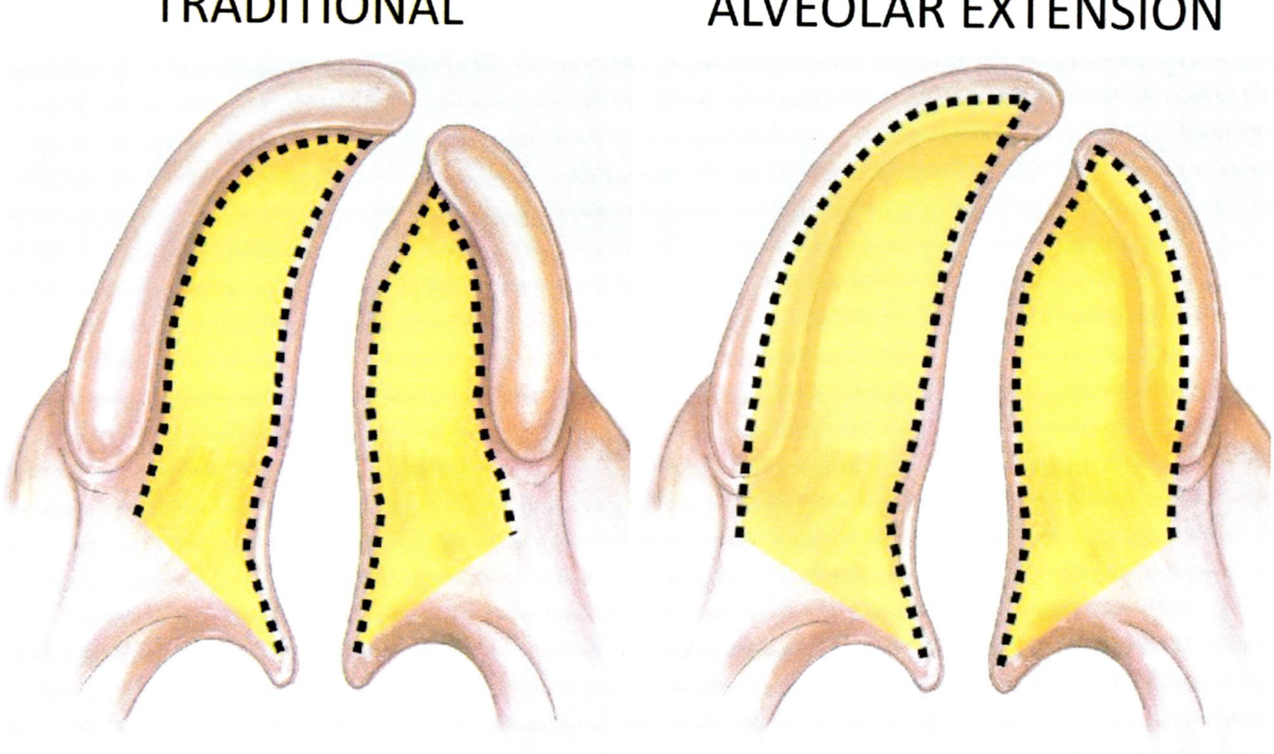

Fig. 15.6 Alveolar extension palatoplasty incorporated alveolar side wall and gingival tissues producing flaps that are longer and wider. The release area is behind the tuberosity when secondary healing will not affect the alveolar shelves. [Courtesy of Michael Carstens, MD]

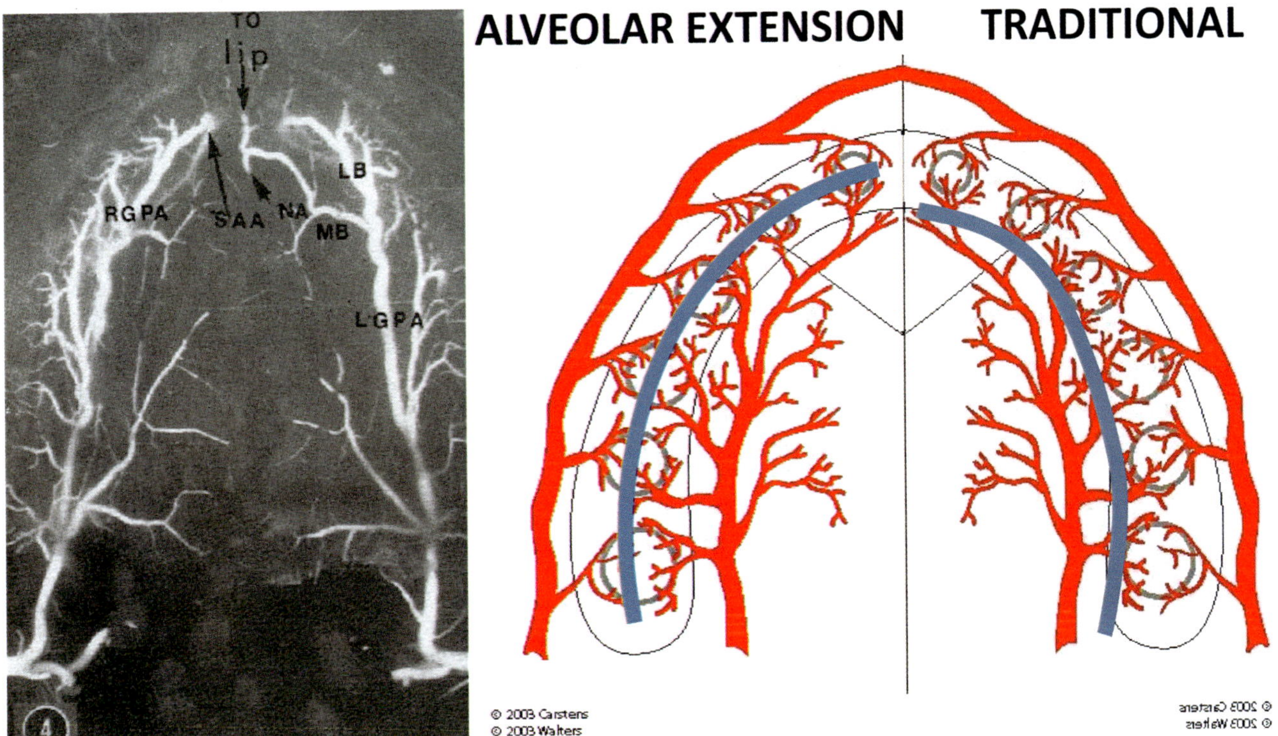

Fig. 15.7 Blood supply is done by the greater palatine artery, multiple branches that are spread laterally until the alveolar ridge (there is a delicate arborization). The entire angiosome is utilized. Left: [Reprinted from Maher WP. Distribution of palatal and other arteries and cleft and noncleft human palates. Cleft Palate J 1977 14(1): 1–12. With permission from Sage Publishing.] Right: [Courtesy of Michael Carstens, MD]

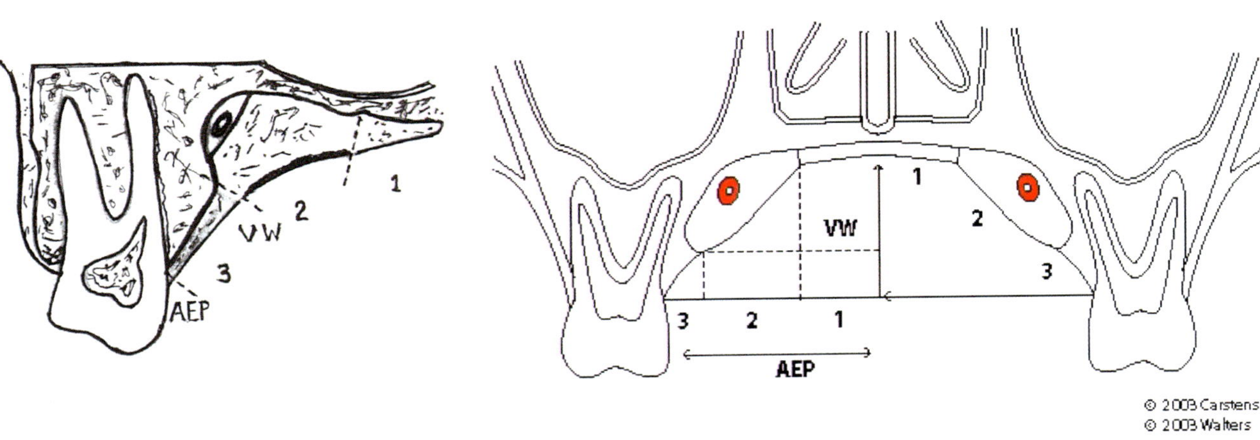

Fig. 15.8 Full angiosomes have extra width for full palatal extension. AEP width is 1–3 mm versus VW width 1–2 mm. When the flaps are lowered they gain additional width which translates to length. [Courtesy of Dr. Luis Monasterio, Funación Ganz, Santiago, Chile]

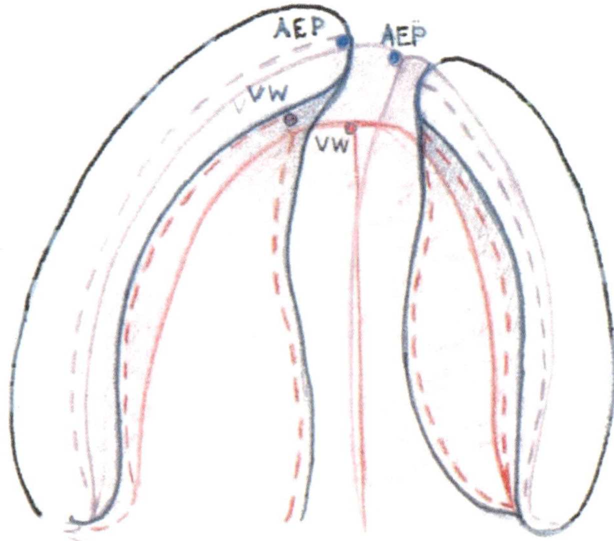

Fig. 15.9 Extension effect. *AEP* = purple; *Veu* = red. [Courtesy of Michael Carstens, MD]

The first clue was the finding of lip tethering. Unilateral cases displayed a thinning of the upper lip on the cleft side from medial to lateral. Dr. Morton Goldstein from Princeton taught me about his *elastic flap* concept and I realized that the soft tissues were somehow tethered laterally. The very act of lip–cheek advancement was causing the thinning. The question was: how to address that?

Back home in Berkeley I recall vividly in 1996 staring out the window at the Golden Gate Bridge across the bay. The window had no curtains; the light was intense. I recall thinking about curtains and realized that if they were pleated and anchored laterally, drawing them together would cause an asymmetry of pleating with the lateral parts being drawn out and thinned out. And if the curtains were released to travel over a frictionless curtain rod, the thinning phenomenon would be eliminated. At that precise moment it was clear that the Sotereanos subperiosteal mobilization with the backcut was responsible for the outcomes at Pittsburgh.'

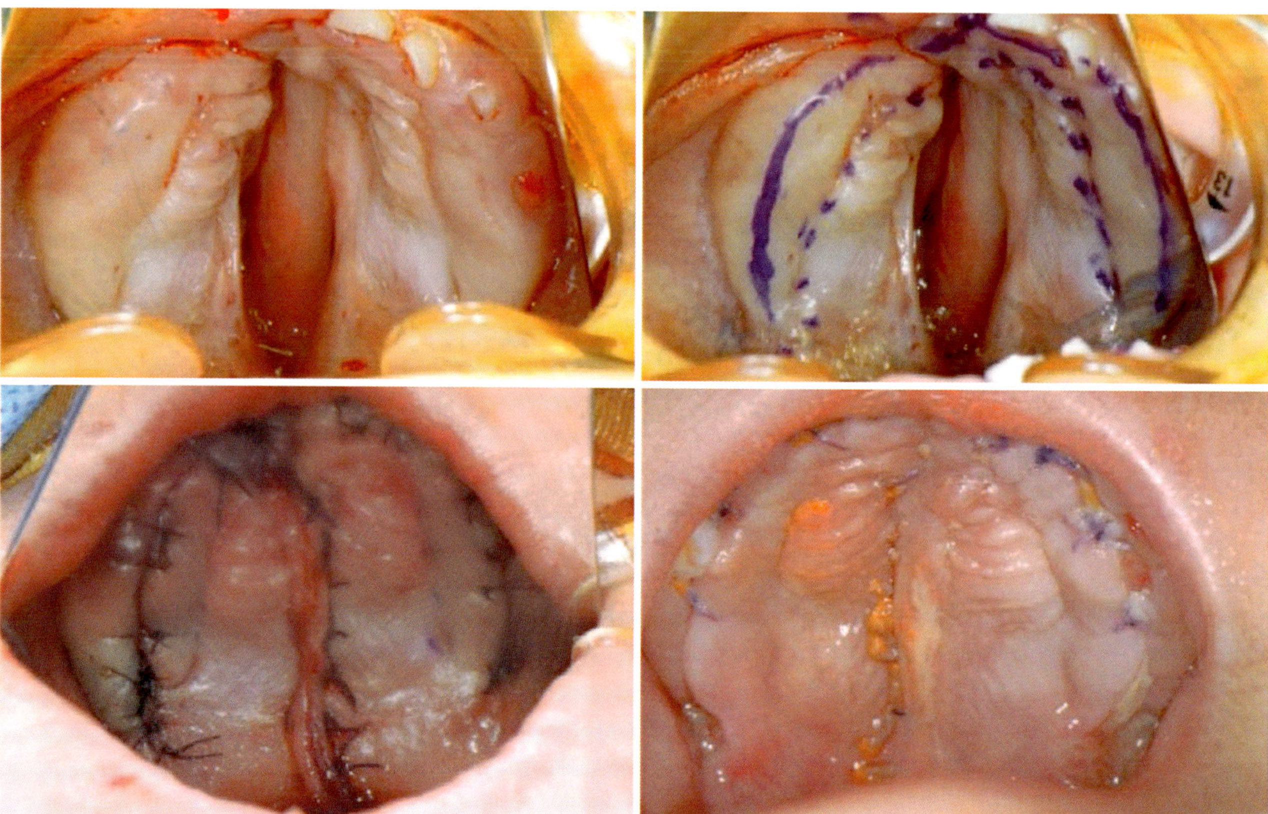

Fig. 15.10 The AEP incision (solid blue line) is made along the lingual aspect of the alveolar ridge or directly over the midline, but quickly skirting laterally to avoid tooth germs. Incision then extends around the tuberosity. *Note the difference in flap size between AEP (solid blue line) and Bardach two-flap (broken blue line).* Subperiosteal dissection of the mucoperiosteal flap and its alveolar extension is carried out on both sides. Advancement of the flaps along the alveolar margin permits to close the anterior cleft margin and the lateral incision; minimal or usually no denuded areas are left. Note the S-shaped closure. [Courtesy of Dr. Luis Monasterio, Funación Ganz, Santiago, Chile]

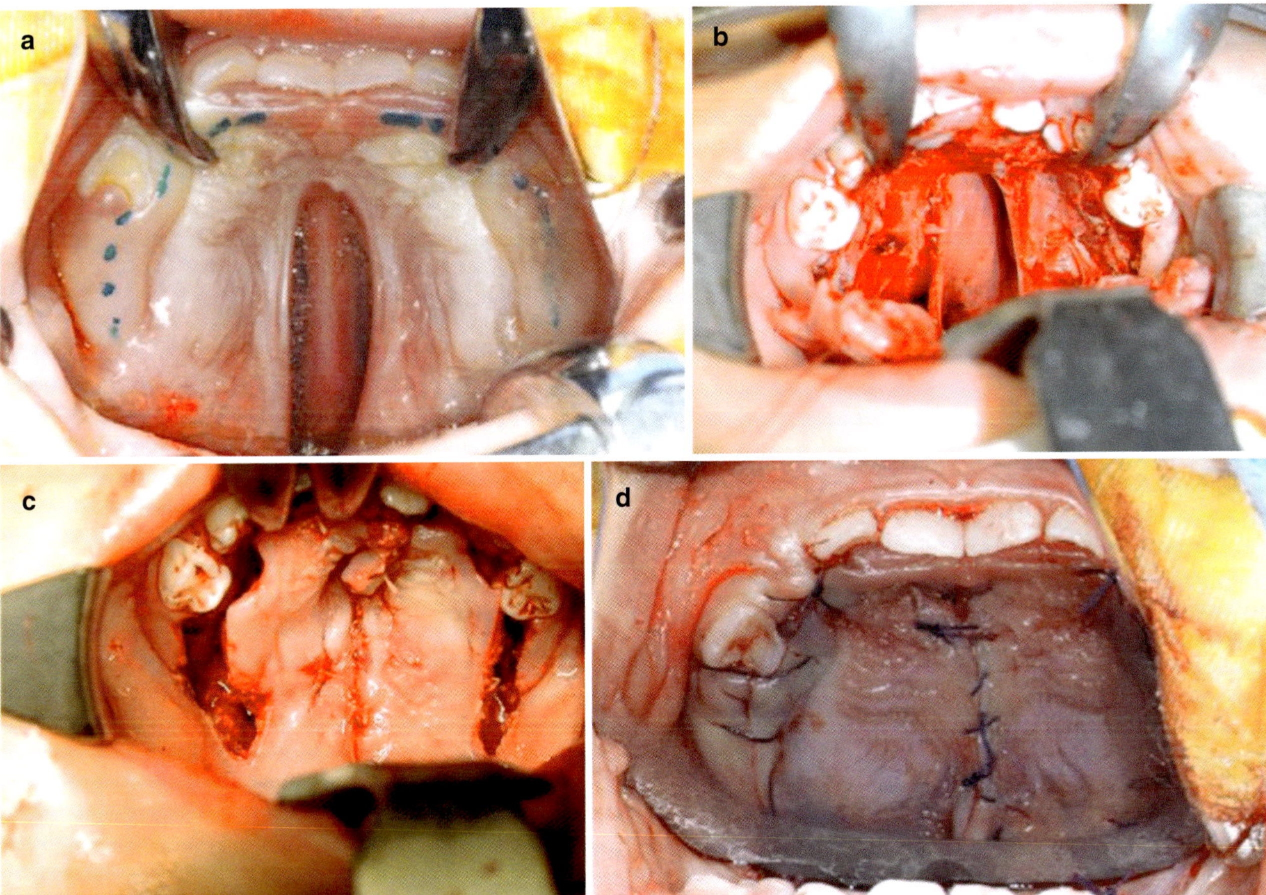

Fig. 15.11 Midline palate cleft with intact alveolus. (**a**) Incisions are pericoronal; (**b**) flaps elevated with good exposure of nasal mucosa; (**c**) AEP flaps close this without raw areas; (**d**) at final closure, alveolus permits elimination of dead space. In comparison, either the Veau or the Wardill–Kilner would devascularize the lingual alveolus and leave raw areas for secondary healing. The flaps are advanced along the alveolar margin. Sutures through the gum or around teeth (if present) permit a secure closure of both the anterior cleft margin and the lateral incision with minimal or no areas of residual denudation. [Courtesy of Michael Carstens, MD]

- The theoretical basis of subperiosteal cleft repair, the *sliding sulcus procedure*, was based on the concept that the lip, sulcus, and gingivomucoperiosteum of the anterior alveolus all constitute a single *functional matrix* (using the concept of Melvin Moss). It was a *neurovascular/aesthetic unit* [5]. Later on, I would realize it to represent a complex of *neuromeric developmental fields*.

From 1997 onward in all cases, I used the Sotereanos technique and observed the same symmetry to result. It appeared to me that the cause of relapse after cleft lip repair was a failure to recognize and reverse unidentified pathologic processes unleashed by the clefting event. With this new perspective fundamental goal of lip repair changed from achieving a geometric "cookie cutter" solution to a biologic repair involving a *reversal* of the processes of deficiency, division, displacement, and distortion. This was a paradigm shift to be sure but it was clear that something more fundamental was at play. I was sure that facial growth would somehow be involved and set out to understand the problem as best I could.

Molecular embryology as we know it today was in its infancy. Lumsden and Krumlauf's landmark study describing homeotic genes in the vertebrate neuraxis had just been published. But solid anatomic and clinical studies were extent. Injection arteriography by Siebert in cadavers had demonstrated that buccal sulcus incisions disrupt the blood supply to the labiobuccal gingiva. Because this tissue was known to contain important osteogenic cells in the cambium layer, I surmised that a reduction in blood supply might be of importance to understand the effect of lip repair on alveolar development. Based on extensive work by Delaire, Markus, Smith, and precious others, subperiosteal dissection had been demonstrated to respect this blood supply. The resultant mobilization of tissues permits a primary two-layer closure of the alveolar cleft, when present.

Maxillary growth is a complex process of simultaneous deposition and resorption of bone along multiple surfaces. The importance of the interface between bone and soft tissues where this process takes place had been emphasized by Enlow. Tissue growth and osteogenesis would be more natural if the soft-tissue envelope that was responsible for membranous bone growth were centralized instead of being left behind. I reported this technique, which included a pericoronal incision and gingival transfer, as the *sliding sulcus procedure*.

Amidst these changes in thinking and technique one question kept bothering me. What might the relationship be between subperiosteal lip repair and palatoplasty? If preservation of periosteal osteogenesis is important on the labiobuccal surface of the alveolus ***should this not be true for the lingual surface as well?*** After all, both the Veau-Wardill and Bardach two-flap operations were done in the subperiosteal plane. What was I missing?

The answer came on an airplane en route to Ecuador. I was considering how the vessels from the infraorbital descended all the way to the occlusal edge of the alveolus and how they entered the dental sacs from the buccal side. The lingual side had to play by the same rules. Drawing this out on a napkin it was obvious that the greater palatine artery would supply each individual dental sac in the same manner as the infraorbital. And it was equally obvious that traditional cleft palate repairs, by virtue of placing an incision lateral to the GPA, at the border between the horizontal palatine shelves and the alveolar wall, would irrevocably cut all connection between the lingual alveolus and the GPA. *Could one make a sliding sulcus type flap on the palate using a pericoronal incision on the lingual side as well?*

Paper napkins are soft and pliable. I drew out an oral view of two right-sided complete palate clefts, putting a Bardach incision on the major segment in one same and a midline incision along the alveolar ridge on the other. Using two napkin furnished courtesy of the airline, I drew out on each one an oral view of two right-sided complete palatal clefts. On one of these I placed a Bardach incision on the major segment. On the other, I drew an incision centered on the midline of the alveolar ridge. I then cut out the flaps and pulled them forward. The second flap was longer and wider and project directly over the alveolar cleft, covering it completely. The first flap fell short of the alveolus, leaving an anterior fistula. Moreover, by virtue of including alveolar tissue, the second flap represented the entirety of the GPA neuroangiosome. Its greater dimensions resulted from extending the flap over the entire hemi-alveolus. The name for this was obvious: **alveolar extension palatoplasty** (AEP).

Fast-Forward to the Present Initially, I admit to succumbing to the siren call of alveolar closure at primary repair, if only to avoid leaving a fistula. Periosteoplasty and the existence of stem cells in the flaps made a lot of sense but it would not become a reality until the addition of recombinant human bone morphogenetic protein-2, a phenomenon we'll consider in Chap. 20. A gingival release at primary surgery was technically challenging and difficult to teach. There seemed to be little to recommend the procedure so I switched the release to a medial incision over the maxillary alveolus, leaving the gingiva intact but with a buttress backcut to allow the mucoperiosteal "curtain" to "slide" forward. As it turns out, the soft-tissue mobilization was just as good. Although the gingiva remained in place the entire nasolabial complex was centralized. Alveolar closure with the pericoronal incision for gingival transfer could then be reserved for the secondary repair of alveolar defects with bone grafts and/or rhBMP-2, ideally (per the Talmant–Lumineau protocol) at age 4.

- As we shall see, the combination of *DFR subperiosteal lip repair* (a technique combining concepts described by Carstens, Matthews, and Talmant) with *DFR alveolar extension palatoplasty* protects the blood supply for maxillary growth, avoids scarring on the hard palate, and provides a water-tight closure of the alveolar cleft.

Alveolar Extension Palatoplasty: Technique

Water-Tight Closure of the Alveolar Cleft: The Five-Sided Solution

As we shall see, AEP, being embryologically based, provides a comprehensive solution to the central problem of clefts involving the primary and secondary palate: achieving a complete closure, preventing fistula, and thus ensuring optimal conditions for bone grafting. The latter procedure, best done at 4 years before eruption of the canine, involves the final closure of the sixth side of the alveolar cleft. The combination of DFR superiosteal cheiloplasty and AEP results in a five-sided box. DFR provides the roof and AEP the remaining four sides. Only the "front door" needs to be closed. This is done with a Sotereanos-type sliding gingivoperiosteoplasty (see Chaps. 18 and 19).

The key ingredient to this concept is the roof. Primary cleft repair allows for the mobilization of nasal mucoperiosteum via the subperiosteal exposure of the piriform fossa. The nasal mucosa of the secondary palate is elevated from beneath the inferior turbinate laterally for about 1 cm. This flap is in continuity with the floor of the nostril sill. Into this slot are inserted tissues of the nasal floor on the medial side which are derived from the nonphiltral prolabium and medial premaxillary mucoperiosteum. Unification of these tissues will re-establish the nasal floor.

Once palate repair is complete, access to the nasal floor of the alveolar cleft is blocked. Failure to achieve closure of the alveolar cleft at primary surgery creates a situation of high risk for *bone graft loss* and *fistula formation* at such time as secondary closure is attempted. So the first order of business for the AEP surgeon is to "fix the roof before fixing the basement."

AEP can be readily carried out after 9–12 months but I recommend 18 months for the purposes of dental definition and subperiosteal dissection of the mucoperiosteal flap from the alveolar ridge. If soft palate adhesion was done at cheiloplasty, the additional time will allow for better approximation of the segments; 18 months give enough time for speech mechanisms to be properly established.

Total Maxillary Block: The "Brazilian Backdoor" (Figs. 15.12, 15.13, 15.14, 15.15 and 15.16)

Blocking of V1, V2, and V3 in the pterygopalatine fossa prior to surgical trauma achieves total pain control from the operative site, thus reducing anesthesia requirements. Moreover, it prevents the release of substance P. This has long-term consequences because once the block has worn off at 6 h* postop, the entire pain cascade is different, frequently eliminating the need for narcotic analgesics. *Dosing via the AEP regimen vide infra makes use of an optional "booster dose" at the conclusion of the procedure done.

The DFR/AEP block uses 0.25% bupivacaine (Marcaine®) with epinephrine. The **toxic dose for bupivacaine** at this concentration *without epinephrine* is 1 cc per kg. The italics indicate an *additional safety margin* by using Marcaine *with* epinephrine. For a 10-kg patient, the toxic dose would be 10 cc. In this case one simply cuts the dose by 50%. Thus 2.5 cc is used per side. For ease of administration the total volume of solution should be 5–10 cc per side to guarantee diffusion. Hence, in the case of the 10-kg infant, one dilutes the 2.5 cc bupivacaine solution with saline to get the desired volume.

The pterygopalatine fossa can be readily accessed via the Brazilian backdoor technique. I learned this from Sergio Vieira, MD, DDS, from Belo Horizonte, Brasil, when we fellows were together at the University of Pittsburgh. It can be done without an assistant prior to placing the mouth gag. The entry site is directly behind the maxillary buttress.

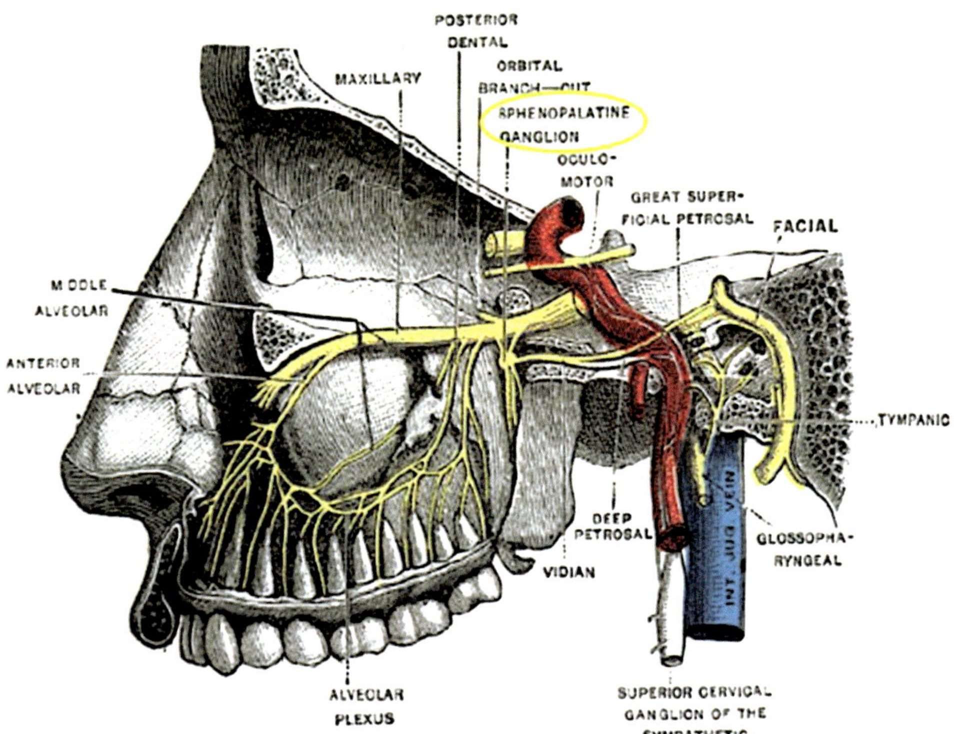

Fig. 15.12 Spenopalatine (pterygopalatine) fossa. The pterygopalatine fossa is a teardrop-shaped space that exists between maxillary tuberosity and lateral pterygoid plaite" the pterygomaxillary fissure. Autonomic connections: PANS from r4 to r5 is greater petrosal nerve of VII (GPN). SANS from the superior cervical ganglia ascends along the carotid, sending fibers as the deep petrosal nerve (DPN). Combination of GPN + DPN connects with SP ganglion. Alveolar plexus is anterior; middle and superior alveolar nerves make the alveolar plexus. Note: anterior and middle superior alveolar neurovascular pedicles are generally given off together just before exit as the infraorbital axis. [Reprinted from Lewis, Warren H (ed). Gray's Anatomy of the Human Body, 20th American Edition. Philadelphia, PA: Lea & Febiger, 1918]

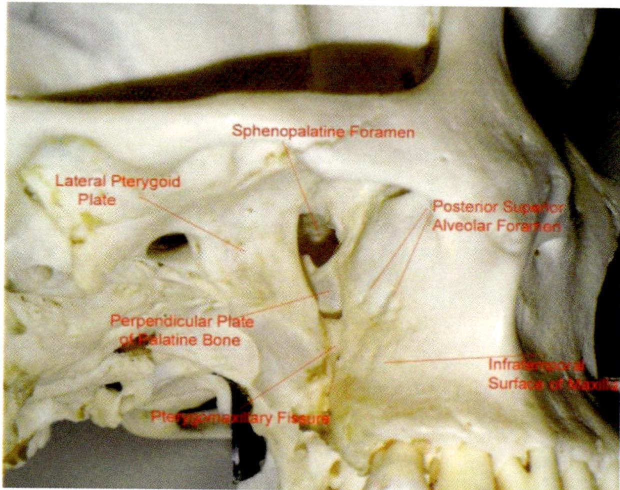

Fig. 15.13 Sphenopalatine foramen is located posterior, medial, and cranial to the pterygomaxillary fissure. [Courtesy of Dr. Ashwin Menon]

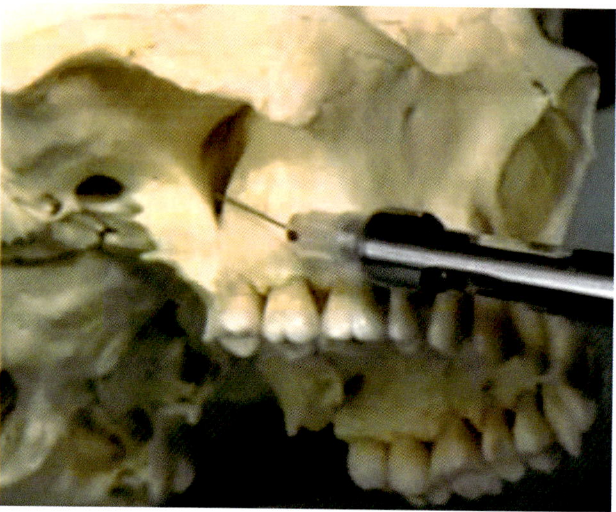

Fig. 15.14 Sphenopalatine block. Needle (bent 45° at the tip) must advance medial backward and upward to enter the foramen. First pass is directed toward the contralateral eye and blocks V2–V3. The needle is withdrawn and re-directed toward the nasofrontal junction. Injection is vigorous to send a cloud of bupivacaine up to V1. [Courtesy of Dr. Ashwin Menon]

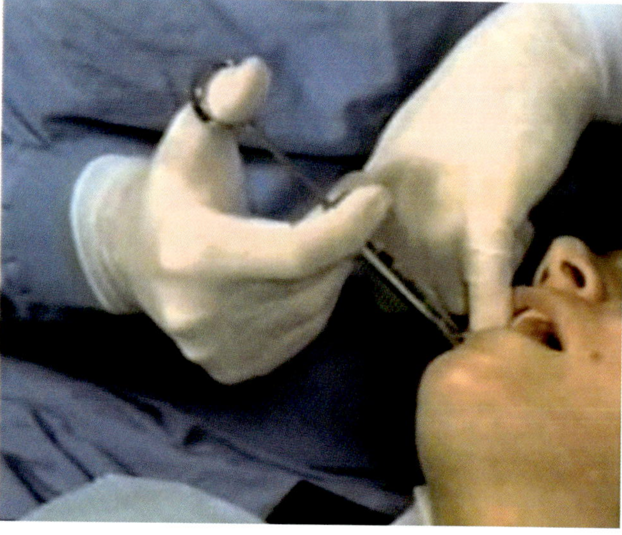

Fig. 15.15 Hand position for the sphenopalatine block: two techniques. Sphenopalatine block can be down from above (as shown) with small finger retracting the cheek. I prefer to administer it from below, looking upward into the mouth. In this position the tip of the index finger faces the pterygomaxillary fissure. As the needle is advanced, the index finger controls the direction of the needle. [Courtesy of Dr. Ashwin Menon]

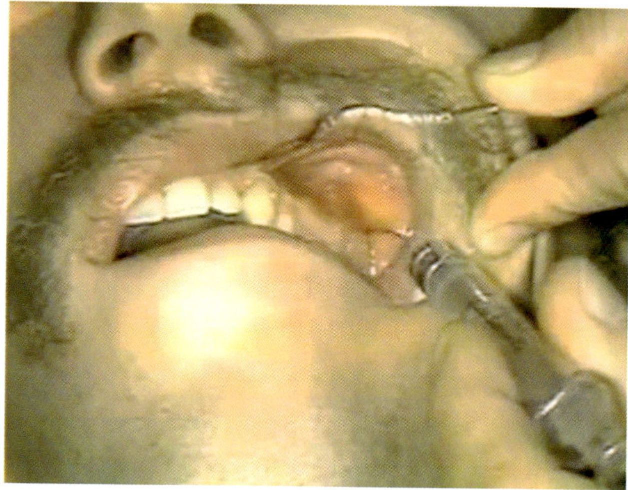

Fig. 15.16 Sphenopalatine block using a cheek retractor. An alternative way to give the block. I don't think it gets V1 very well. Dental needles are flimsy. In order to get to V1, precise control is needed. A #19 or #18 gauge needle works better to accomplish this. Use of the index finger in the mouth ensures palpation of bony anatomy and redirection of the needle tip. [Courtesy of Dr. Ashwin Menon]

- Bend a #18 or #19 gauge needle 45° at its distal 2 cm. This is equivalent to about half the length of the distal phalanx of your index finger. Needles of #20 gauge and higher are too flimsy
- Place the index finger of one hand all the way back to the maxillary buttress and retract the cheek backwards. This places your distal phalanx directly over the access site
- Drop the needle straight backwards in parallel with the teeth or gum at the level of the buccal sulcus, "feeling your way" along the alveolar bone until the needle falls behind the tuberosity
- Using your fingertip, direct the needle into the space
- Advance straight in about 1–2 cm aiming for the contralateral eye, aspirate, and inject half the volume
  - This blocks V2 and V3

- Withdraw and re-advance 102 cm aiming for the nasal frontal junction, aspirate, and inject the remaining half of the volume
  - This blocks V1

To mitigate against the effects of the mouth gag one can optionally block V3 at the lingual nerve as well.

Simultaneous exposure of the palate and alveolar arch is provided by a mouth gag of choice. Infiltration, including the nose and nasopharynx is done with 0.5% lidocaine with epinephrine 1:200,000. Note: the toxic dose of lidocaine at this concentration without epinephrine is 1 cc per kg. One can use the same calculations as with bupivacaine with care to add the amount of saline necessary to (1) elevate the periosteum from the bone and (2) separate soft-tissue planes.

AEP Dissection

Initial Management of the Soft Palate

If soft palate adhesion has not been carried out at cheiloplasty, the mucosal margins of the soft palate are pared. One can manage the uvulus as one wishes. I personally prefer a thicker, more robust uvulus so pare off the nasal mucosa from one side, then the oral mucosa from the other so that the two can be stacked on top of the other. One advances forward splitting the oral and nasal mucosa until one encounters the posteromedial corner of palatine bone (P3). Here, longitudinal muscle fibers from uvulus insert into the bone, often for a distance of 5–10 mm. These are cleaned off from the bone sharply and then with the amalgam packer. The result is complete exposure of the bony "corner" of the cleft. From this point dissection of the nasal mucoperiosteum can be done safely, to maximize a clear entry into the superiosteal plane on the nasal side and avoid inadvertent tears. All remaining soft palate muscle dissection is now deferred until the palate AEP flaps are mobilized, the back wall of the horizontal palatine plate is cleaned off, and the subperiosteal mobilization of the nasal mucoperiosteum is complete.

Elevation of Palatoplasty Flaps (Figs. 15.17, 15.18, 15.19, 15.20, 15.21, 15.22, 15.23, 15.24, 15.25 and 15.26)

AEP flaps are elevated first from the alveolus and then bony palate. I like to start with the cleft side, beginning within the alveolar cleft. Here a vertical incision was made in the mucoperiosteum of the lateral wall, splitting into buccal and lingual halves. It then ascends over the alveolar ridge, just off the midline and proceeds straight backwards. When teeth have not erupted the goal of the alveolar ridge incision is not to injure the dental sacs. One initially cuts vertically through mucosa and submucosa, not all the way down to bone. The incision extends all the way back to the tuber-

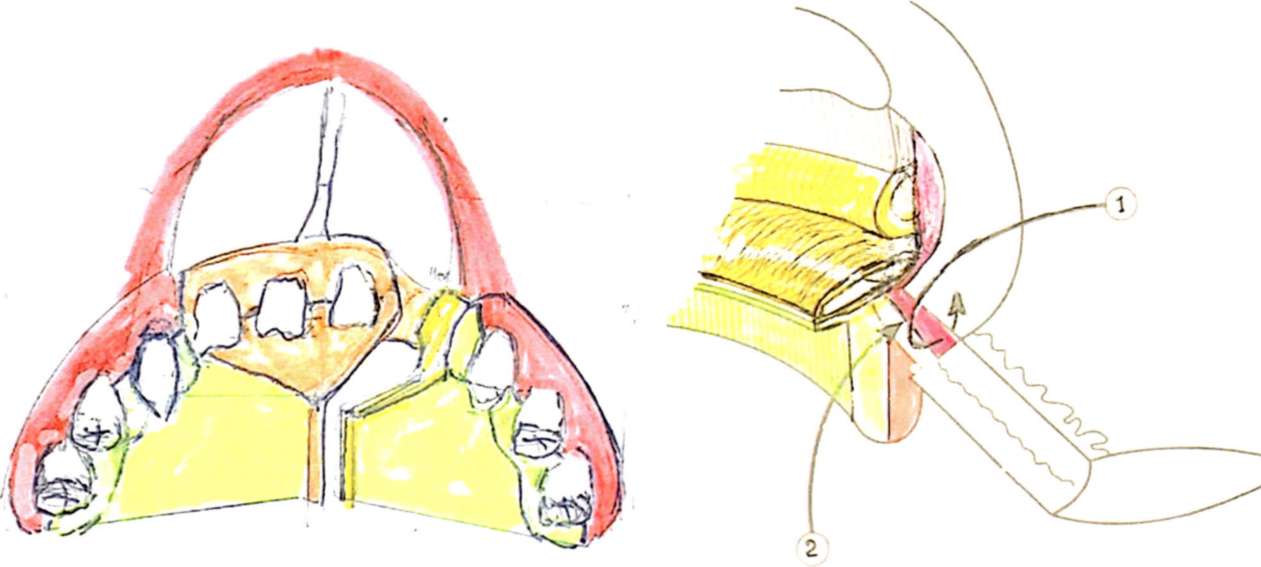

Fig. 15.17 Closure of the anterior nasal floor is an essential step prior to palatoplasty. It is the key step to avoid fistula formation at bone grafting. At primary repair, a flap of non-philtral prolabium and mucoperiosteum is elevated and sutured to a flap of the premaxillary wall is elevated and sutures to flap combining mucoperiosteum from the nasal side of the hard palate with interior of nostril sill. The mesial flap is supplied by medial nasopalatine (orange) and the lateral flap is supplied by lateral nasopalatine (yellow). Note territory of GPA (lemon). The difficult-to-dissect nasal lining of anterior palate can be easily accessed when reflecting the lateral lip element and nose. The incision is anterior just below inferior turbinate. A subperiosteal tunnel can be rapidly developed backward, hugging the lower border of the lateral nasal wall. One then sweeps the elevator medially, all the way to the palatal margin. At cheiloplasty the GPA tissues (lemon) are left in place. If this maneuver is done at palatoplasty the mucoperiosteum of the wall of the alveolar cleft provides an extension of the flap. [Courtesy of Michael Carstens, MD]

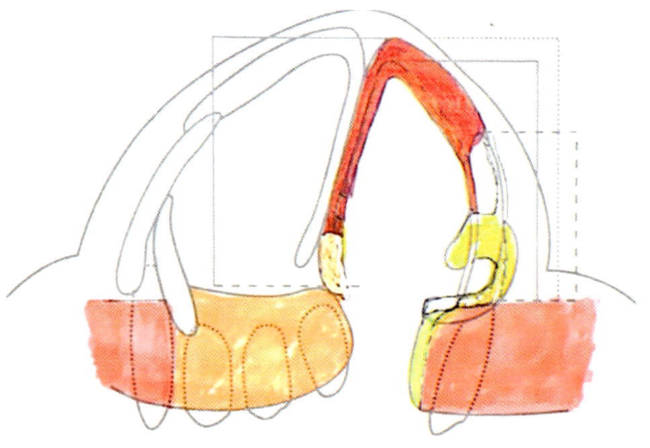

Fig. 15.18 AEP incision is initiated within the alveolar cleft. It is vertical and can either split the angiosomes (lemon versus pink) in the midline or extend to the labial border to get additional length. The incision then proceeds up over the alveolar ridge and backward. Note from this projection that the starting point for the incision of the alveolar ridge is the same as the starting point for separating oral and nasal mucoperiosteum from the medial border of the palatal shelf: GPA (lemon) versus lateral nasopalatine (yellow). Note medial nasopalatine covers premaxilla and vomer (orange) while the STV1 ethmoid supplies the upper two-thirds of both the medial and lateral nasal chamber. [Courtesy of Michael Carstens, MD]

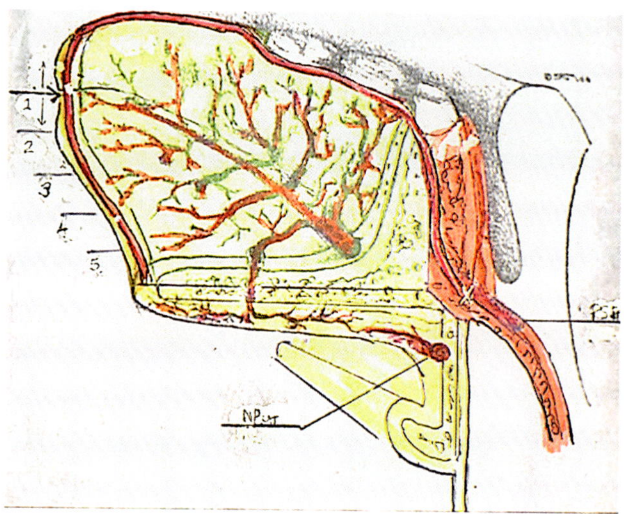

Fig. 15.19 AEP incision #1 is initiated. Incisions (pink) begin at point 1 in the alveolar cleft and proceeds in two directions in three separate steps. Step 1: The incision proceeds backward along the alveolar ridge belonging to palatine fields P1 and P2 supplied by GPA (lemon). If *teeth are present*, the incision is pericoronal and made sharply with a #15 blade straight down to bone. If teeth are not present, the incision is through mucosa and submucosa over the alveolar ridge and then slants lingually. Step 2: Half-way between the midpoint of the alveolus and the lingual border, the incision is abruptly directed downward to bone. This maneuver avoids the dental sacs. Step 3: At the maxillary tuberosity the incision extends downward (from the surgeon's point of view), i.e., cephalad to the plane of the horizontal palatine plate, P3. When the AEP flap is advanced, any dead space will involve the tuberosity as far forward as the third molar. The bone in this zone is thick and unyielding. [Courtesy of Michael Carstens, MD]

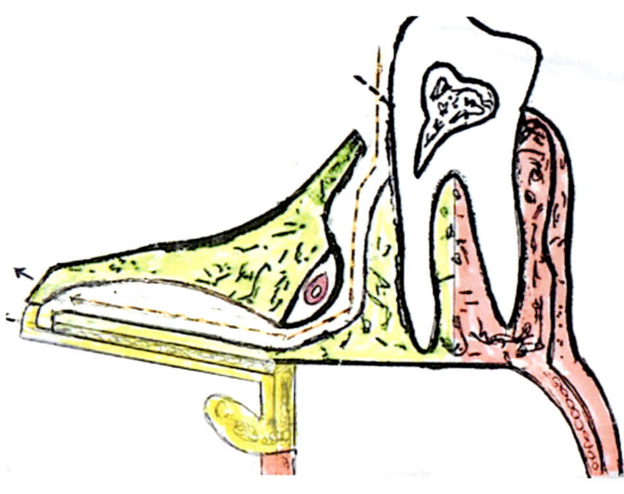

Fig. 15.20 Elevation of the AEP flap and medial border. Molt #9 elevator is passed downward until it hits the hard palate floor. Keep the curve of the spatulated end downward, pushing forward like shoveling snow. Upon reaching the medial border, flip the curve around and lever upward. This will pop the periosteum from the edge (white area). You can the clean the border of the bone with the small head of an amalgam packer. This safety maneuver combines with elevation on the nasal side, beginning behind at the palatine bone where the tissues are easy to access, proceeding forward and then sweeping medially. The incision along the border is the last step. [Courtesy of Michael Carstens, MD]

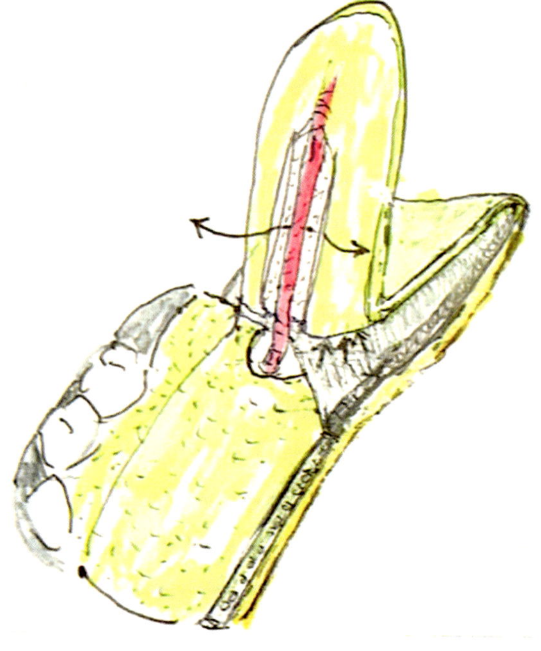

Fig. 15.21 Gaining mobility for the AEP flap. *Step* 1 (right) is to free the muscle from the back of the flap. This requires skeletonizing the lower one-third of the pedicle using parallel incisions through periosteum and then passing a baby right-angle clamp, amalgam packer, or elevator of choice around the artery. One then bluntly reflects downward, separating the muscle away. [Courtesy of Michael Carstens, MD]

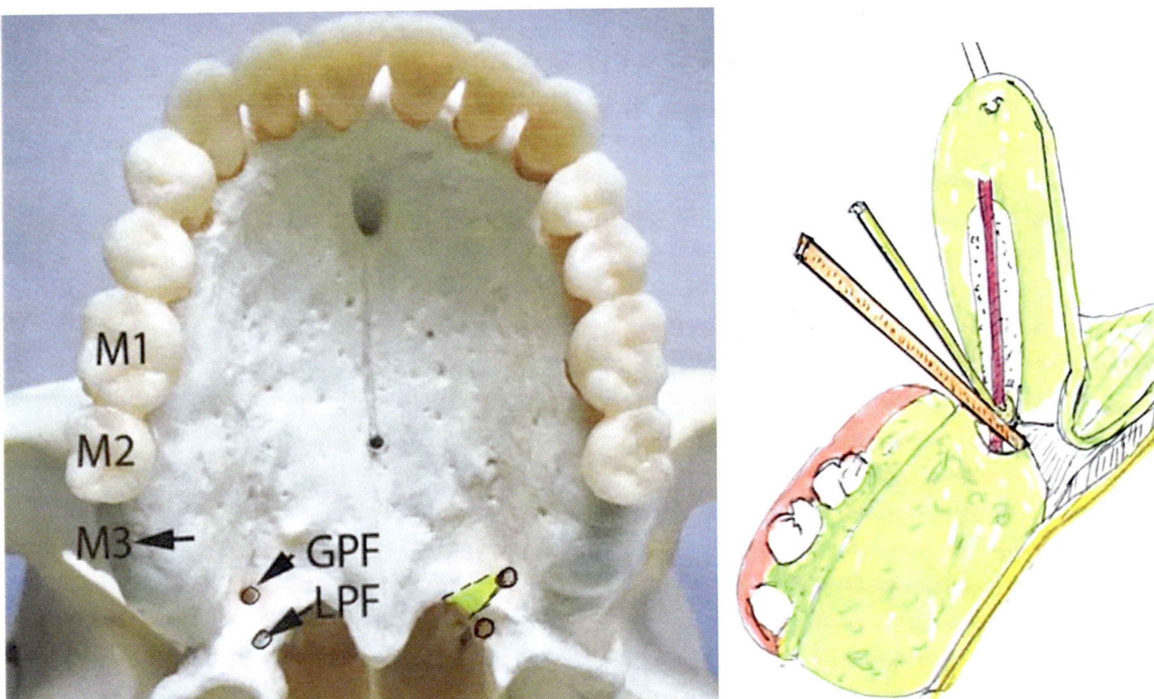

Fig. 15.22 Greater palatine foramen. Labeled cadaver bony palate (right): Often there is a bone spur bordering the pedicle. This can be a problem on medial mobilization, causing kinking or even impalement. A 25% osteotomy of the foramen directed posteriorly and medially avoids the lesser palatine axis. It removes the bone spur. With care the pedicle can be teased out of the foramen altogether. *Diagram (left) Step* 2 is osteotomy of the greater palatine foramen. The flap is held under tension by the assistant who will also tap the osteotome. The pedicle (red) is protected by a curved clamp (yellow). The 2-mm osteotome passes medial to the clamp and is embedded into the posterolateral foramen. When this cut is completed, a second cut is made at the medial corner. The effect is to remove a pie-shaped wedge of bone from the encircling foramen, thus freeing the pedicle and preventing kinking when it is advanced forward and medially into position. Note the extension of uvulus along the extreme medial border all the way to P2–P3 border. This should be released as an initial maneuver in the soft-tissue dissection. [Courtesy of Michael Carstens, MD]

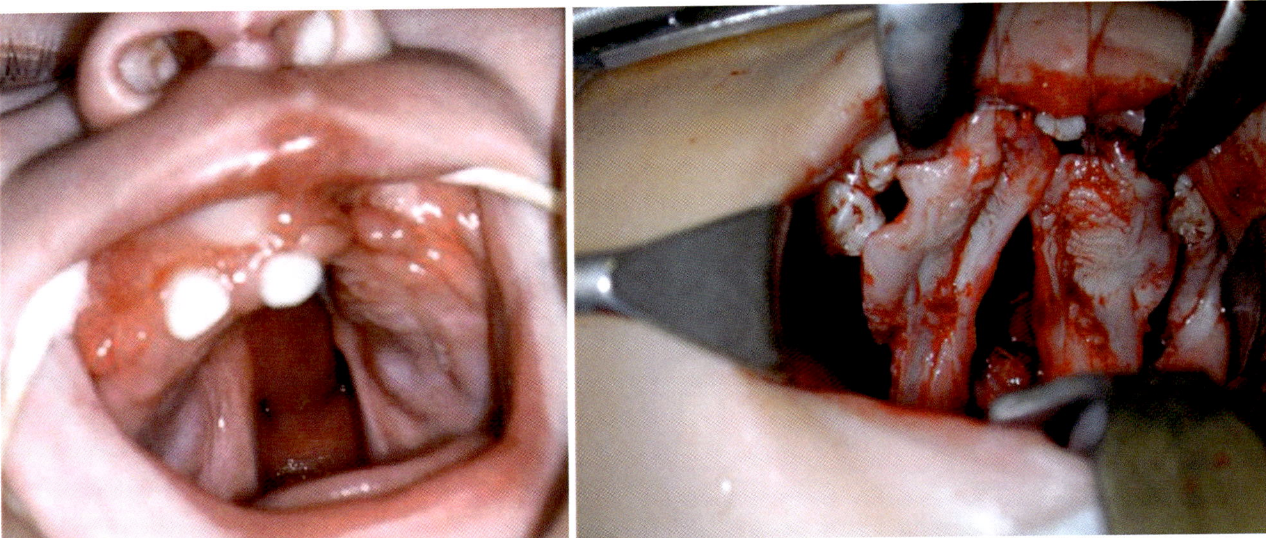

Fig. 15.23 Case 1: UCL/P complete, left. Left: Previously closed complete cleft with an established fistula. Right: AEP flaps mobilized to position forward into the floor of the fistulae. Note extent of the flaps gives closure of the alveolar cleft floor. If combined with roof repair, the alveolar cleft is closed on five sides. [Courtesy of Michael Carstens, MD]

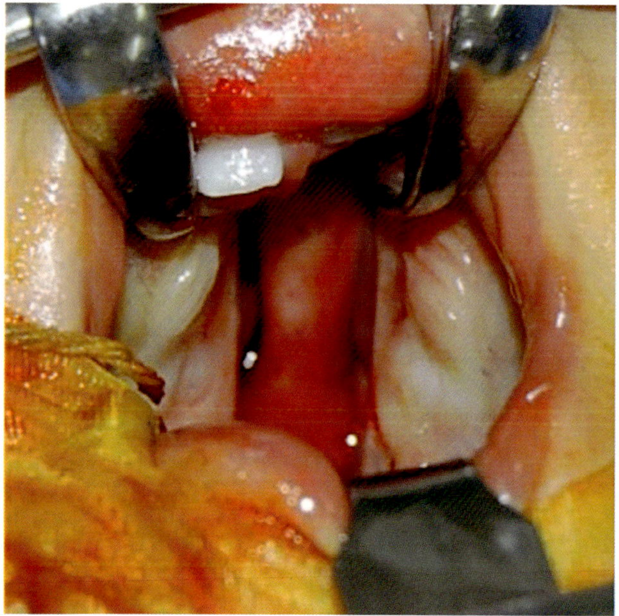

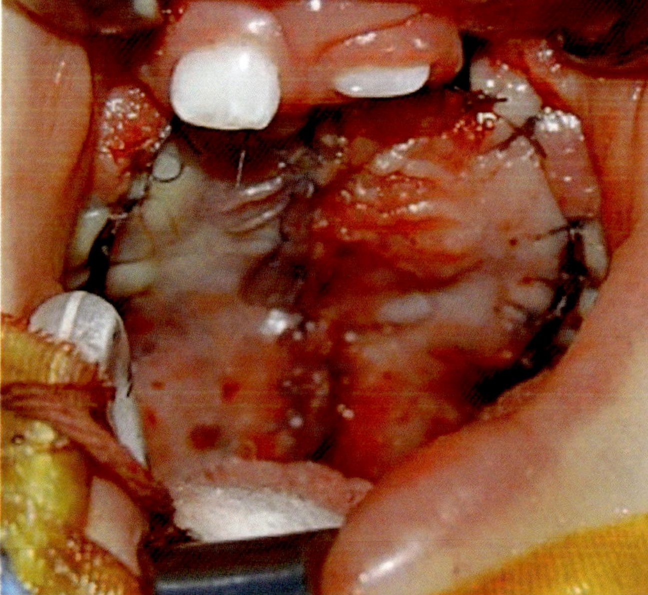

Fig. 15.24 Case 2: UCL/P complete, left, showing flaps advanced. Left: Wide complete unilateral CP with fistula. Right: Complete closure with AEP flaps and no exposed raw areas laterally. Complete closure of alveolar cleft. S-shaped closure. Note advancement sutures around the teeth. [Courtesy of Michael Carstens, MD]

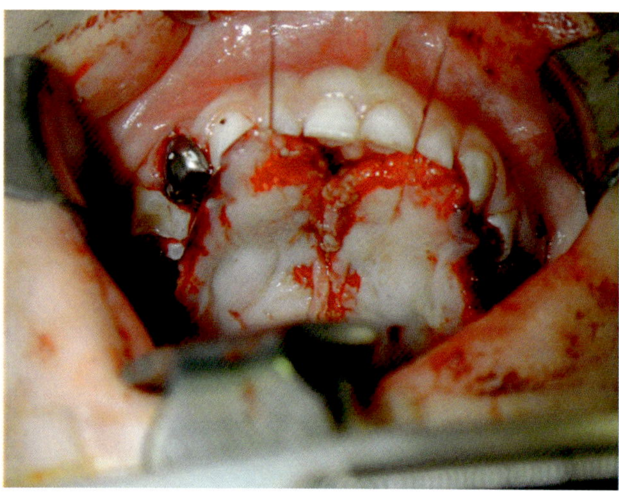

Fig. 15.25 Case 3: U-shaped cleft showing bilateral flaps extending *beyond* the dental arch. [Courtesy of Michael Carstens, MD]

osity. At that point it continues cephalad along the backside of the tuberosity until reaching the plane of the horizontal palatine bone.

The next step is to safely carry the incision down to the bone. Even though one may choose to make the incision directly over the teeth one can easily avoid the dental sacs by angling obliquely lingual and then straight downward, gently breaking the periosteum to enter the subperiosteal plane. An excellent instrument for this is the *amalgam packer*. It is readily available and costs a pittance. It has two heads of different diameters; these have cross-hatched serrations which are capable of separating mucoperiosteum from bone using blunt dissection in the plane (a Tessier concept). This is accurate and atraumatic.

One can now pass over the alveolar ridge, heading straight down to the hard palate. It is rather like falling over a cliff. The curve of a large McIndoe or Molt elevator is directed toward the lingual face of the alveolar bone, hugging it until reaching the plane of the hard palate. One then elevates the palatal mucoperiosteal flap from lateral-to-medial. Upon reaching the border, flip the instrument around 180°, like a turtle on its back, and gently elevate the flap margin in a prying motion. Voila! That hard-to-dissect medial border will balloon outward. One can even get the small head of the amalgam packer into that space, rotate the head so it faces lateral toward the alveolus and, with a gentle motion like a scrub brush, break up the periosteum from the edge.

One now incises the oral margin, elevating the flap from proximal-to-distal until reaching the posterior border of palatine bone medial to the artery. Put it on stretch with a stay suture of choice so the assistant can hold it up in the air. Proceed over the edge of the palatine bone, scrubbing off the periosteum with the packer or elevator until the bony shelf is completely clear from medial to lateral. This will take one safely *behind* the pedicle. From this point one elevates the abnormally positioned muscle backward away from the pala-

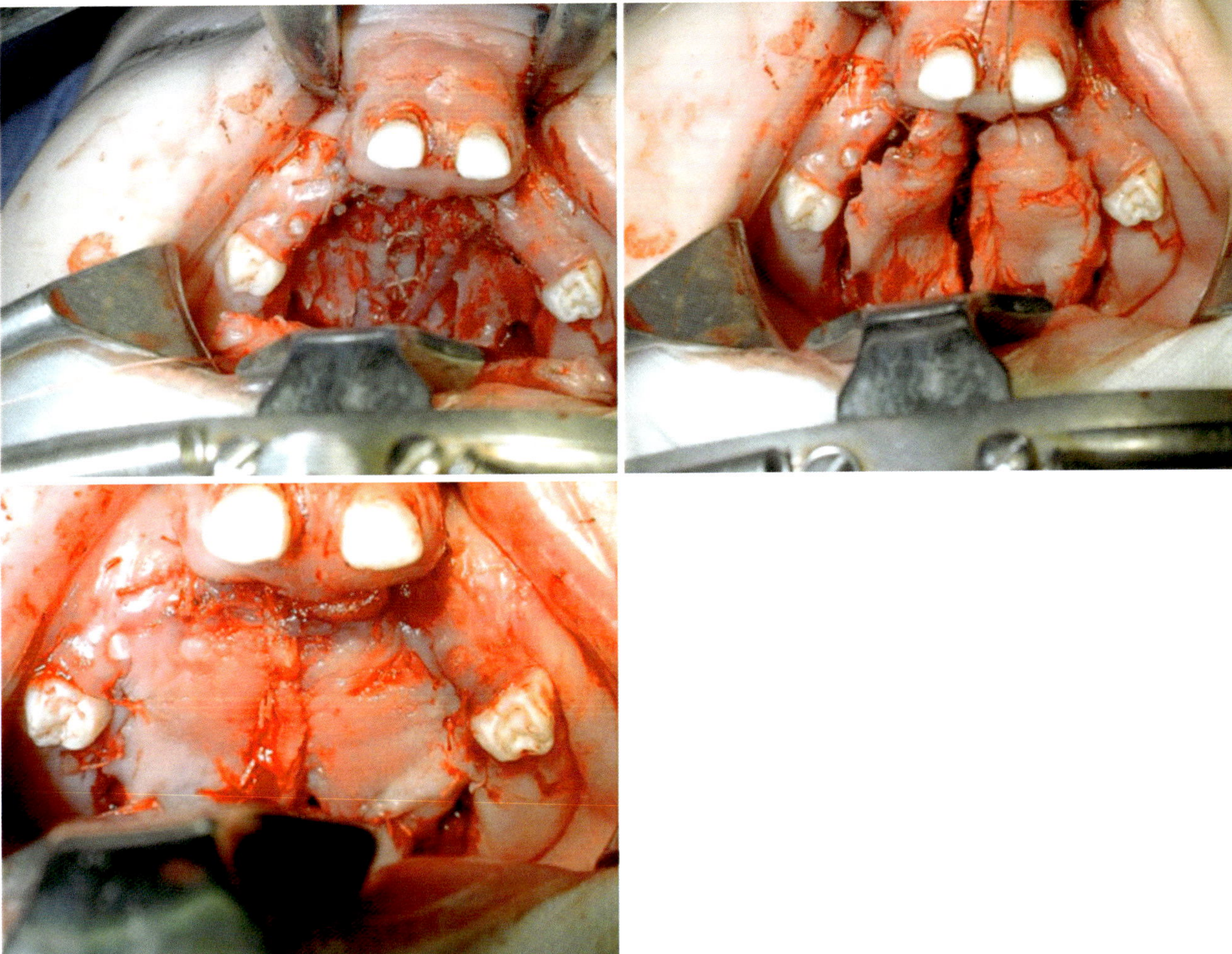

Fig. 15.26 Complete bilateral CP demonstrating reach of the AEP flaps and complete closure without fistulae. Note the release incisions are located at the tuberosities. [Courtesy of Michael Carstens, MD]

tine bone. One is now in a perfect position to complete the muscle dissection of the soft palate.

Up to this point the nasal mucosa has been left in contact with the back side of the palatine bone because it constitutes a point of leverage while one finishes with the soft palate muscles. Now one begins to elevate the nasal mucosa. The small head of an amalgam packer or curved tip of a delicate elevator is pressed upward against the under surface of the bone. The nasal mucosa will come free. One can then proceed forward in the nasal place to complete elevation of the nasal periosteum.

The final step with the nasal mucosa is to free it up so it translates medially. The secret to this is to elevate it from the orbital plate of palatine bone. Push the McIndoe with the curved tip directed upward along the undersurface of the horizontal plate of palatine bone until making contact with orbital plate. Then, maintaining pressure push the elevator cephalad (downward from the surgeon's perspective) toward the roof of the pharynx. Once you reach the limit, gently pry the elevator laterally and the mucosa will strip away, relaxing into the mouth.

Mobilization of the AEP Flap

It is now time to maximize the extension of the AEP flap with respect to two structures: The greater palatine artery and soft palate musculature. With the assistant maintaining traction on the flap with 4–0 silk mattress sutures, the pedicle is under tension. Proceed around the pedicle, cleaning off and defining the origin of the GPA foramen. Parallel incisions are made through the periosteum on the undersurface of the flap on either side of the pedicle. These permit passage of an instrument *behind* the artery. An elevator, packer, or delicate mosquito clamp is passed behind and the pedicle gently spread away from the flap. For that point downward (toward the palatal plane) the final attachments of muscle to the base of the flap are separated. The AEP flap is now in direct continuity with the oral mucosa of the soft palate. The remaining muscle dissection required in the soft palate will be under direct vision.

Follow the palatine bone directly lateral to the pedicle and dissect over its edge. The result is increased mobility of the flap with respect to the tuberosity. Elevate the soft-tissue envelope bluntly upward, backward, and medially. The space of Ernst is *not* entered [6, 7].

The final step is osteotomy of the posterior margin of GPA foramen using a 2-mm straight osteotome. The assistant maintains traction with one hand and will tap the osteotome with the other. The surgeon uses a small instrument, as the amalgam packer or the McIndoe elevator, to clean off the border of the foramen. With the ipsilateral hand the amalgam packer is then placed against the medial wall of the GPA and wedged against the center of the foramen, thus protecting the pedicle. The contralateral hand directs the osteotome straight back to open the center of the foramen, one that directs the osteotome 90° medially. This cut produces a pie-shaped wedge of palatine bone. The foramen is opened up, the pedicle is free to move medially; at times it can actually drop out of the foramen altogether.

The flap is now translated forward and its arc of rotation is checked. This will inform the decision as to how much mobilization will be required from the noncleft side.

Attention is now directed to the noncleft side. Here the maneuvers are the same save for the planning of the medial margin mucosal flap. The alveolar extension increases the dimensions of the mucoperiosteal flap. Thus, in wide clefts, additional medial tissue will be available for nasal closure in the anterior one-quarter to one-third of the hard palate defect. Thus, planning the dimensions of the noncleft nasal flap should be done with the cleft-side flap fully rotated into position.

Final Management of the Soft Palate

As described in Chap. 16, complete clefts of the secondary palate almost invariably involve a mesenchymal defect of the horizontal plate of the palatine bone, zone P3. This defect in the anterior one-third of the aponeurosis is responsible for the creeping forward of the levator complex toward the posterior margin of P3, a situation that is unanatomic. *Nota bene*: the biologic relationship between the musculature and the nasal mucosa is completely normal and should not be disturbed. Thus, freeing the muscles from the nasal lining is *not embryologic*, achieves little (because the deficiency site is in the anterior one-third), and simply results in unnecessary scar.

What one *does* accomplish is to completely separate the musculature from the bone clearing all the way down to what remains of the anterior aponeurosis and gently pushing the muscle backward, re-creating what should be the "muscle-free" zone. One is left staring into the deficiency site. If the anterior–posterior defect is significant, the prognosis for VPI worsens. Buccinator interposition is a logical response but now is not the time for it. Remember: "If it ain't broke, don't fix it!" If one gets good closure, even with difficulty, the patient is not necessarily condemned to VPI. The physiologic response of each patient to palatoplasty is different and cannot be predicted. What appears to be a short soft palate may prove very active and capable of pharyngeal closure. A lively minnow may prove more reliable than a phlegmatic whale. One simply follows along to observe function over time. If needed buccinators should probably be placed at or before age 6. We'll cover this issue a bit more in Chap. 17.

Closure of the cleft is now carried out beginning with the nasal layer. Beginning anteriorly, one uses 4–0 chromic on a small atraumatic P-2 needle. Extreme care is taken at the most anterior aspect of the primary palate to close up the alveolus. It is at this juncture that previous closure of the anterior nasal floor pays off! Vomer flaps are positioned and sutured. Upon reaching the posterior border of the hard palate we switch gears. The two halves of the uvulus are sutured like two pancakes, one atop the other with 5–0 chromic. I like to proceed forward using 5–0 chromic until tension arises. At this point, additional mobilization of nasal mucosa can be done laterally from the orbital plates of P3.

The anterior "shoulder" of the soft palate musculature is closed with a mattress suture of 4–0 Vicryl. The remainder of the muscle layer is closed in continuity with the oral mucosa using interrupted mattress sutures. As the oral mucoperiosteum is sutured, care is taken to approximate it to the nasal layer to eliminate dead space.

Development of the cleft-side segment AEP flap yields one + dental units of additional length. Such flaps routinely project forward to the backwall of the alveolar cleft. The greater segment mucoperiosteal flap yields two + dental units of additional length. Such flaps routinely project beyond the alveolar arch. The tips of the flaps are sustained forward in the midline by stay sutures. They are then closed back-to-front with mattress sutures of 4–0 Vicryl. One of two of these sutures should also pass through the vomer-nasal floor to close dead space.

At the anterior margin, the flap tips are sutured to the gum. Along the lateral margin of the flaps three mattress sutures are placed back-to-front in mattress fashion to the alveolar side walls to achieve a differential advancement and relieve tension anteriorly. The chromic sutures are placed right through the gums with no long-term effects. If teeth are present the flaps are secured to them. These sutures are intended for approximation only but in most cases the AEP flaps and the alveolar walls are in direct contact.

The rationale for mucoperiosteal flap mobilization is the repositioning of the functional matrix to the center of the palatal cleft. Scar location is transferred to the posterior and mesial to the tuberosities. The bone stock here is substantial.

The tuberosity also constitutes the "pivot point" for the maxillary shelf. Forces exerted here are far removed from the cleft epicenter at the incisive foramen.

AEP produces a palatal scar that is very different in its geometry of traditional repairs, being a lazy S. The noncleft-side flap extends uniformly over the alveolar cleft and beyond it for 2–3 mm. It comfortably fills the space without tension. AEP design leaves its tissue defect near the tuberosity, not along the lateral alveolar shelves. The cleft-side flap comes up to the backwall of the alveolar cleft. Together the two flaps provide a secure closure with *no fistula*.

Certain situations may arise in which a complete mucoperiosteal closure of the alveolar cleft is desirable, i.e., an untreated palatoalveolar cleft in an adult. In such case, a sliding sulcus periosteoplasty flap from the cleft side as per Sotereanos will easily close the front wall. A gingival incision is carried back to the buttress and a back-cut placed up the buttress. Wide subperiosteal dissection gives an improved aesthetic symmetry to the face and at the same time creates a tension-free closure of the anterior alveolar wall. If lip revision is required, the SSP can be combined with DFR cleft lip-nasal repair.

Arterial Anatomy of the Alveolus: Sequential Vascular Isolation (Figs. 15.27, 15.28, 15.29, 15.30, 15.31, 15.32, 15.33, 15.34, 15.35, 15.36 and 15.37)

Developmental Anatomy of Alveolus and Maxilla: A Bilaminar Sandwich

The maxilla and the mandible are like two boxes, stacked one on top of the other.

The alveolus up the upper jaw is a *rectangular box* organized around the ancient tooth-bearing *dentary* component of the palatoquadrate cartilage (PQ). The mammalian dentary field contains seven* teeth in three neuromeric zones: maxillary incisor and canine (I3 and C), two premolars (Pm1–Pm2), and three molars (M1–M3). PQ serves as a template for a series of dermal bones that develop like armored plates on either side. The dermal bones are maxilla, quadrate (the eventual incus), the palatine bone series (P1–P4), and epipterygoid. Their function is to suspend the dentary field from the skull.

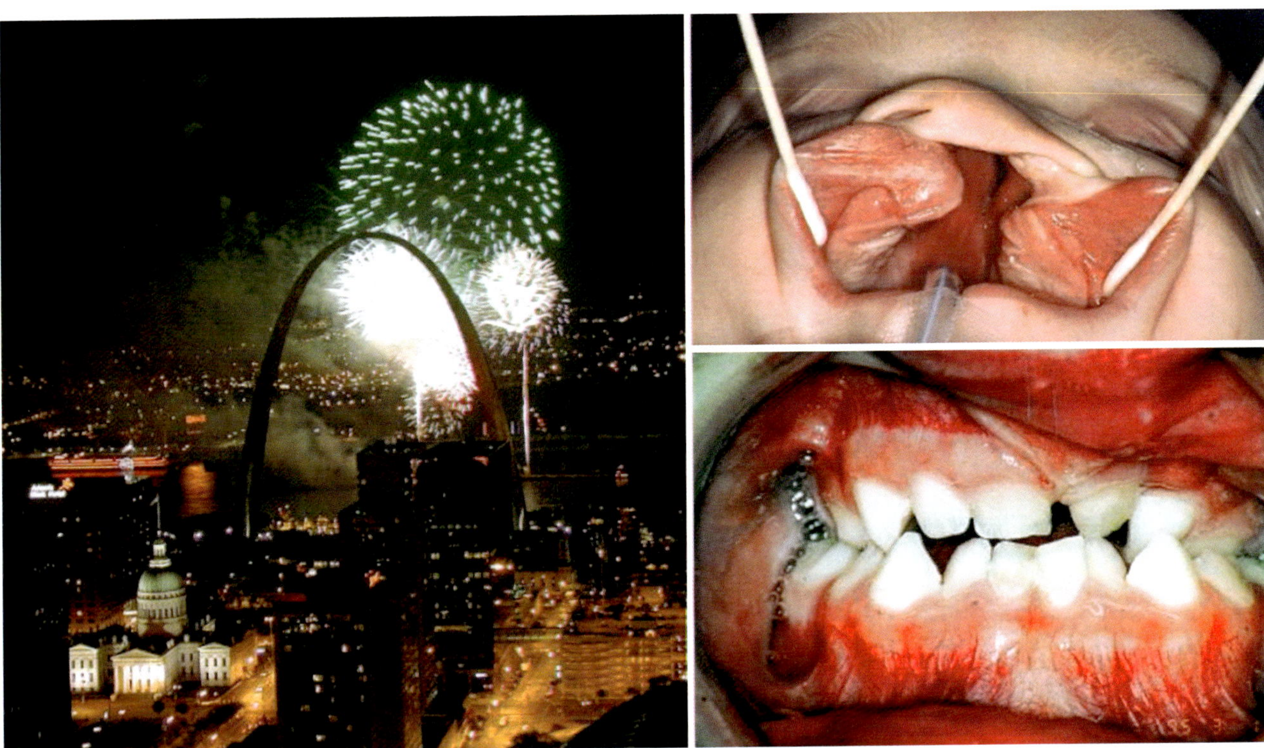

Fig. 15.27 Dead space resulting from inadequate flaps and lateral raw areas causes two problems. The cleft-side arch is intrinsically unstable being unsupported by the vomer. A raw area on the lingual surface (1) acts as a mechanical restraint; (2) restricts local blood supply to lingual alveolus for inadequate growth. [Courtesy of Michael Carstens, MD]

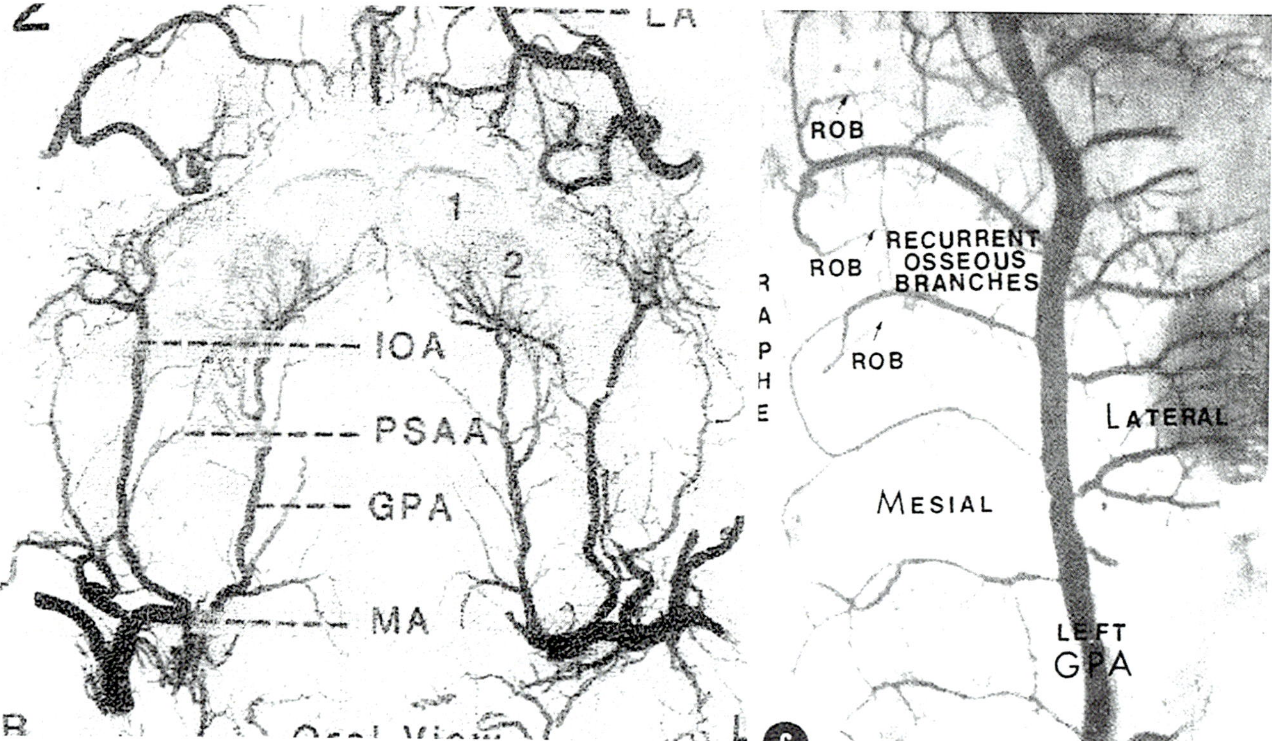

Fig. 15.28 Neuroangiosomes deposit mesenchyme

- Medial sphenopalatine: PMx, vomer
- Lateral sphenopalatine: inferior turbinate + P1–P2
- Descending palatine: >GPA: lingual alveolus + P1–P2 palatine shelf
- Descending palatine: >LPA P3 palatine shelf
- Posterior superior alveolar: posterior one-third of buccal alveolus
- Anterior superior alveolar: anterior two-thirds of buccal alveolus

Left: arteriogram showing distinct blood supply to both sides of the alveolus

- Cheiloplasty interrupts the labiobuccal supply from superior alveolar axis
- Palatoplasty interrupts the lingual supply from greater palatine axis

Right: GPA close-up showing lateral branches entering the dental sac

- Recurrent osseous branches communicate through the palatal shelf with the lateral nasopalatine axis

[Reprinted from Maher WP. Distribution of palatal and other arteries and cleft and noncleft human palates. Cleft Palate J 1977 14(1): 1–12. With permission from Sage Publishing.]

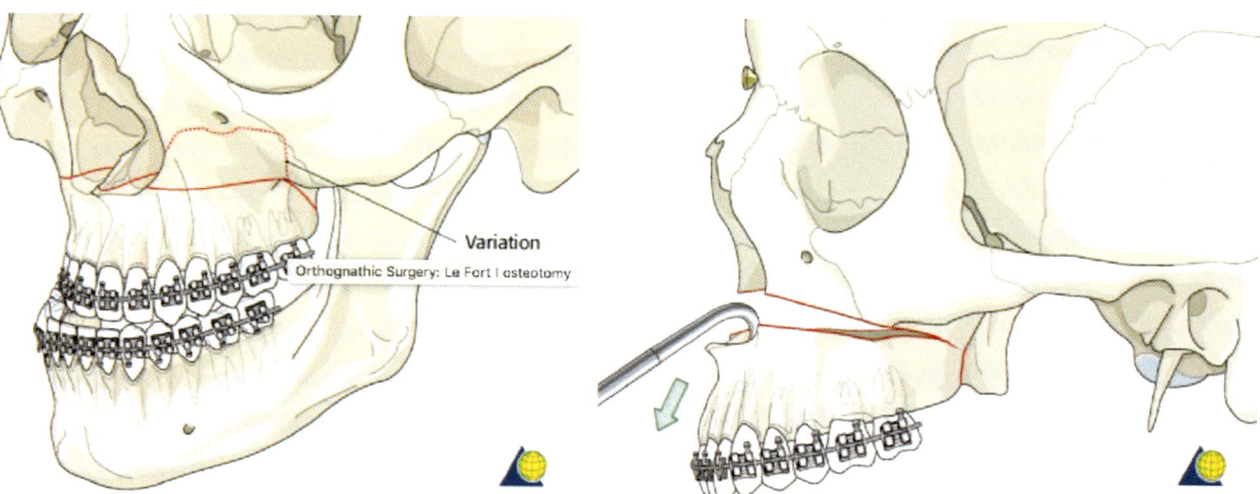

Fig. 15.29 LeFort 1 osteotomy. This procedure mobilizes the maxilla for movement in all three planes in space. [Copyright by AO Foundation, Switzerland]

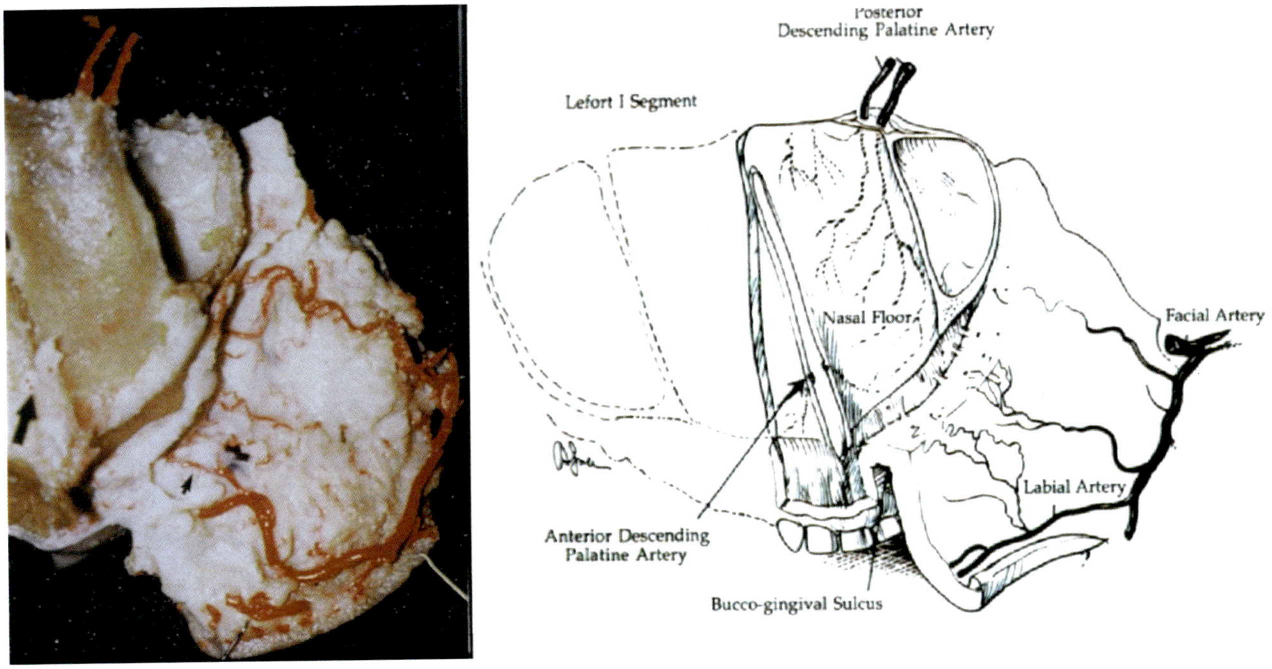

Fig. 15.30 Survival of maxilla after Le Fort I downfracture. What happens to the buccolabial alveolus in *supraperiosteal dissection*? The segment is disconnected from descending palatine, anterior and posterior superior alveolar, and lateral nasopalatine perfusion. Study by Siebert shows maxilla survives based on the soft palate blood supply collateralizing through the palatal flap and thence via the recurrent osseous branches to the nasal side. Modern techniques generally conserve the DPA. However, presence of buccinator interposition flap could be problematic. [Reprinted from Siebert J, Angrigiani C, McCarthy J. Blood Supply of the Le Fort I Maxillary Segment: An Anatomic Study. Plast Reconstr Surg 1997; 100(4): 843–851. With permission from Wolters Kluwer Health, Inc.]

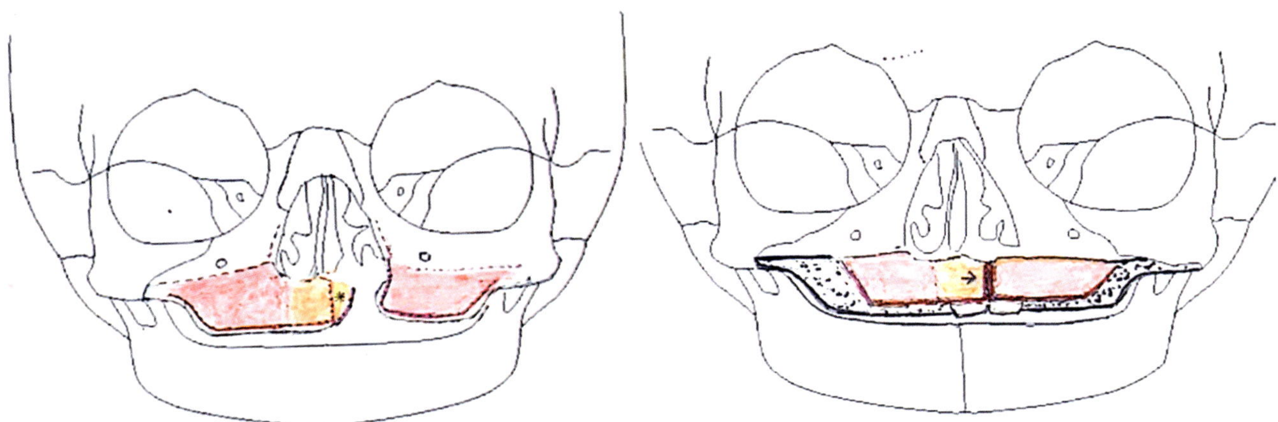

Fig. 15.31 Subperiosteal stem cell transfer: "sliding sulcus" soft-tissue centralization. The original Demas–Sotereanos gingivoperiosteoplasty is a form of stem cell transfer which centralizes osteogenic precursors. It is raised at the level of the alveolar ridge or, if teeth are present, as a pericoronal incision. A back-cut up the buttress between maxilla and zygoma creates a lateral release. The dissection proceeds upward to the infraorbital foramen. Alveolar incision is a solid red line. Parallel to the alveolar incision is a releasing incision *just through periosteum* (dotted red line). Mucoperiosteum is coded as follows: maxilla, zones 4–5 (pink) is supplied by StV2 infraorbital; premaxilla, zones 1–2 (orange) is supplied by medial nasopalatine + crossover from IOA. The flap marked with * is the medial wall of alveolar cleft, i.e., the premaxilla should be elevated at the time of cheiloplasty to close the nasal floor. It can be elevated in an extension of the non-philtral prolabial flap. It can also be included as an anterior extension of the vomer flap. The flap marked with + is the lateral wall of the alveolar cleft. This is supplied by greater palatine artery. It is the anterior extension of the oral mucoperiosteal flap from the hard palate. [Courtesy of Michael Carstens, MD]

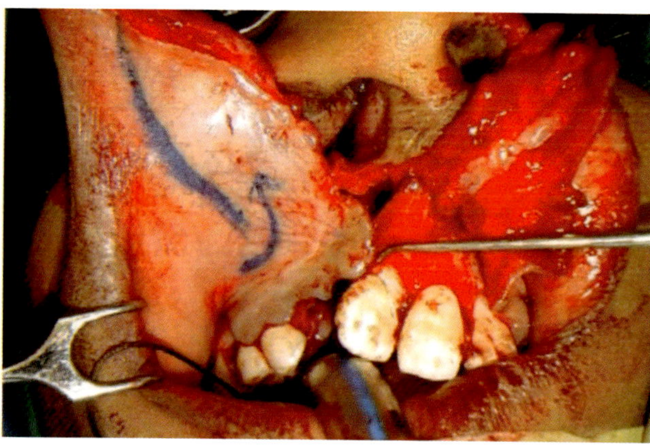

Fig. 15.32 Sotereanos subperiosteal gingivoperiosteoplasty flaps produce a "sliding sulcus" effect that centralizes the soft tissues of the midface. Flap routinely produces *two dental units* of advancement. [Courtesy of Michael Carstens, MD]

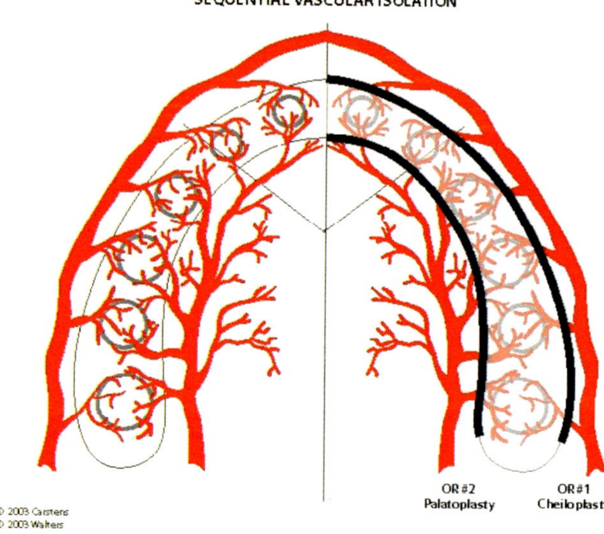

Fig. 15.33 Sequential vascular isolation 1: palatoplasty incisions. Both sequences cheiloplasty > palatoplasty and palatoplasty > cheiloplasty have the same effect. Depends upon the timing between procedures. Maxillary growth with delay of the second stage to somewhere in mixed dentition will be essentially normal. (Left): *Traditional buccal sulcus incision*. This external view shows the sulcus incision interrupting branches from both the StV2 infraorbital and the ECA facial systems which supply the labiobuccal mucoperiosteum. (Right): *DFR incision*(s): One and optionally two vertical incisions are used. The first incision is mesial. It runs along the anterior border of the alveolar cleft. It extends upward to the piriform fossa where it permits subperiosteal access to the lateral nasal wall. Because the medial head of nasalis is falsely inserted into the piriform fossa, this approach enables it to be withdrawn externally—to be re-inserted over the canine (see Chaps. 18 and 19). The mesial incision opens up a broad vertical plane which can be elevated laterally all the way to the maxillary buttress and superiorly to the level of the infraorbital foramen. The second incision is distal and is executed in parallel with the first. It extends up the buttress. The entire mucoperiosteal flap can move forward like a sheet but remains tethered at the gingiva. All vessels are preserved. This attachment can be taken down later as a sliding sulcus flap is needed for anterior alveolar closure. The flap will reattach to the bone in its new position in 72 h. [Courtesy of Michael Carstens, MD]

- *Note*: Upper jaw and lower jaw are *enantiomers*: they follow the same developmental blueprint, albeit with topologic distortions.
- *Note**: This explains why frontal process of maxilla receives a separate branch from medial ASA as the potential supply for the I3, should it develop.

Palatoquadrate Cartilage: the Lateral Program

The maxillary complex develops along the lateral side of PQ. PQ programs maxilla into three developmental fields, organized around its dental developmental zones. These extend upward to contact the skull base. They subsequently morph topologically to form a five-sided box. As described in Chaps. 8 and 14, maxilla is simply an upper structure that opens into the nasal cavity, the dead space being maxillary sinus. Over time the sinus balloons outward while the alveolus remains fixed. The posterior, lateral, and anterior walls of maxilla follow a smooth curve exactly above the alveolus. They have three neuroangiosomes.

- ASAM (anterior superior alveolar, mesial) to the IC zones.
- ASAD (anterior superior alveolar, distal) to the premolar zone.
- PSA (posterior superior alveolar artery) to the M1–M3 zone.
- These arteries supply all dental sacs from the labiobuccal side.
- These three zones laminate over the labiobuccal aspect of PQ to form the **lateral wall of alveolus**.

Posterior to maxilla is the quadrate bone. In all tetrapods except the immediate mammalian line, this functions to suspend the mandible as the *articular-quadrate joint*, the so-called *reptilian jaw joint*. In the cynodonts this gives way to a *dentary-squamosal joint*, which become the mammalian TMJ.

Palatoquadrate Cartilage: The Medal Program

Why is there no sixth side for maxilla? The answer is that the palatine bone complex develops along the medial side of PQ. These three dermal bone fields are L-shaped. P1–P2 are inverted Ls with the vertical limb being the medial wall of alveolus and the horizontal limb the palatal shelf. P3 is an upright L with the horizontal limb being palatine plate and the vertical limb being the orbital plate. Recall that P4 remains as ectopterygoid and eventually forms the hamulus.

- Recall that in evolution palatine bones P1–P4 were originally tooth-bearing and in exact register with maxillary teeth.
- With evolution, teeth decrease in number and get specialized (in mammals).

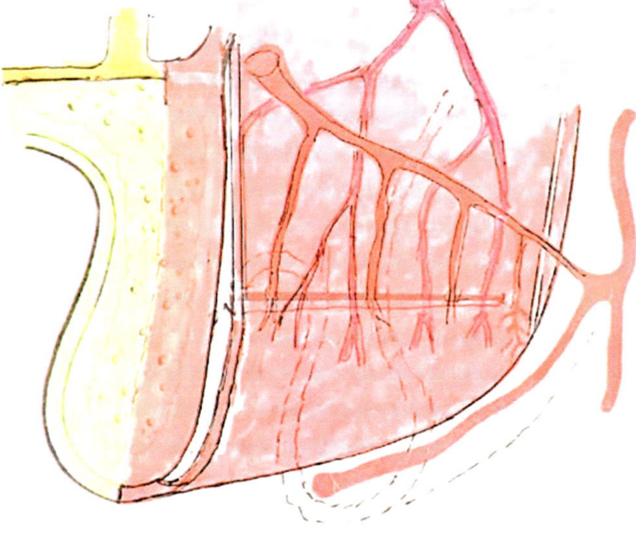

Fig. 15.34 Sequential vascular isolation 2. Comparison of cheiloplasty and palatoplasty incisions is shown: Cheiloplasty uses buccal sulcus (blue) versus functional matrix incision, Carstens (green) with obvious preservation of descending vessels. Differences in palatoplasty flap width between Veau–Wardill (blue) versus AEP (green) are readily appreciated. [Courtesy of Michael Carstens, MD]

PALATOPLASTY INCISIONS

VW + Buccal sulcus incision = Sequential Vascular Isolation
AEP + Functional matrix incision = Vascular Preservation

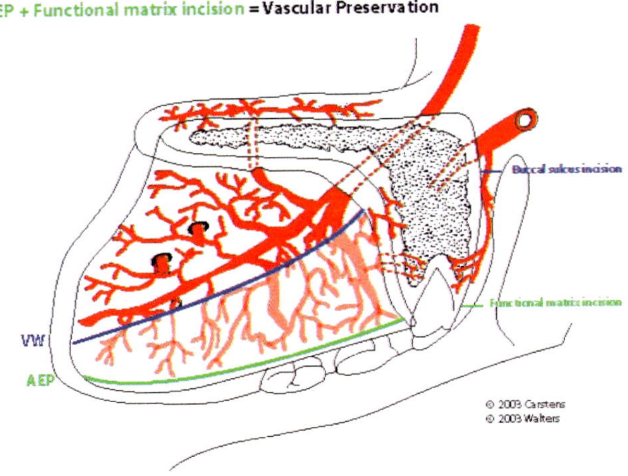

Fig. 15.35 Comparison of cheiloplasty and palatoplasty incisions. Buccal sulcus (blue) versus functional matrix incision, Carstens (green) Veau–Wardill (blue) versus AEP (green). [Courtesy of Michael Carstens, MD]

- In mammals: P1–P2 are in register with the dentition; P3 is pushed out of the way and serves to anchor the palate to the braincase.
- Greater palatine artery supplies all dental sacs from the lingual side.
- The palatine zone laminate over the lingual aspect of PQ to form **medial wall of alveolus**.

Posterior to the palatine series is epiterygoid which joins the pterygoid bone of primitive palate to the braincase. In mammals, this becomes alisphenoid.

How the Angiosomes Gain Access to the Alveolar Bone and Teeth

Access from the Maxillary Complex

Posterior superior alveolar artery (PSAA) is given off from internal maxillary proximal to the axes of zygomatic artery and anterior superior alveolar artery (ASAA). It then bifurcates, with one branch entering the alveolar bone above the tuberosity of Ollier to run in the marrow and supply the molars. The other branch remains with mucoperiosteum, snaking its way laterally around tuberosity beneath the buttress to access the gingiva of the molar zone. Note the lingual periosteum and gingiva are physically not part of the maxilla. These belong to the palatine fields and require a separate blood supply from GPA.

Anterior superior alveolar artery (ASAA) exhibits the same behavior. It gives off medial and lateral branches inside the sinus. These penetrate the bone to supply their respective dental zones. As a footnote, within the sinus, medial branch supplies the frontal process of maxilla, site of the maxillary lateral incisor. The ASAA axis then exits the sinus through infraorbital foramen and divides to supply labiobuccal mucoperiosteum and gingiva. It cannot supply lingual soft tissues.

- **Incisions placed in the buccal sulcus interrupt the blood supply to the gingivomucoperiosteum of the lateral alveolus.**
- That both ASAA and PSAA have distinct branches for the teeth and for gums has surgical implications. Incisions will damage the gums but not the teeth: *alveolar growth is affected but not dental development.*

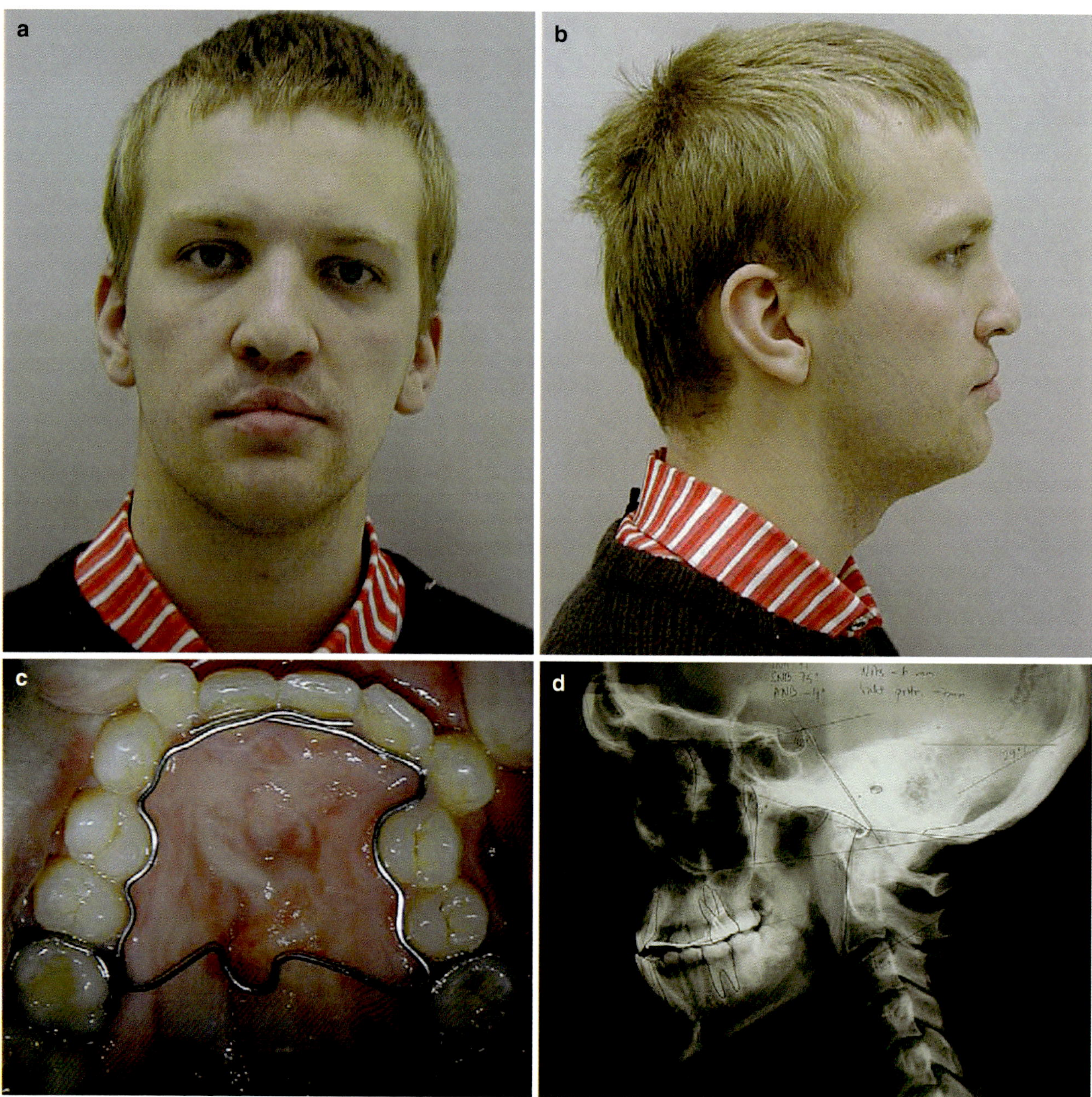

Fig. 15.36 Sequential vascular isolation: consequences. (**a**, **b**) Bilateral cleft lip showing typical sequelae. (**c**) Note class III occlusion and left-sided maxillary collapse. (**d**) Cephalogram showing maxillary retrusion. [Reprinted from Sandor GKB, Ylikontiola LP, Carmichael RP. Midfacial distraction osteogenesis. Atlas Oral Maxillofac Clin North America 2008; 16(2):249–272. With permission from Elsevier]

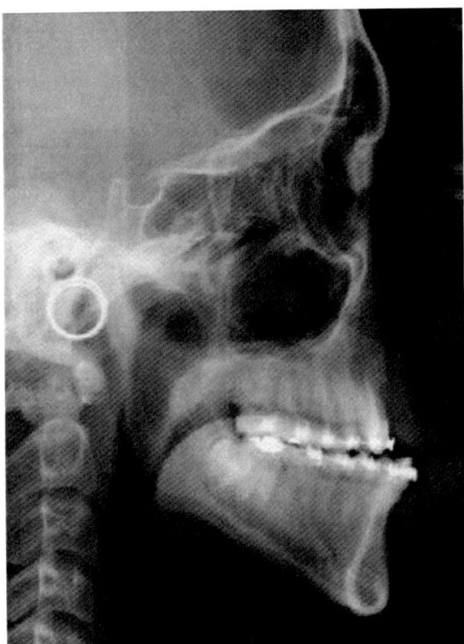
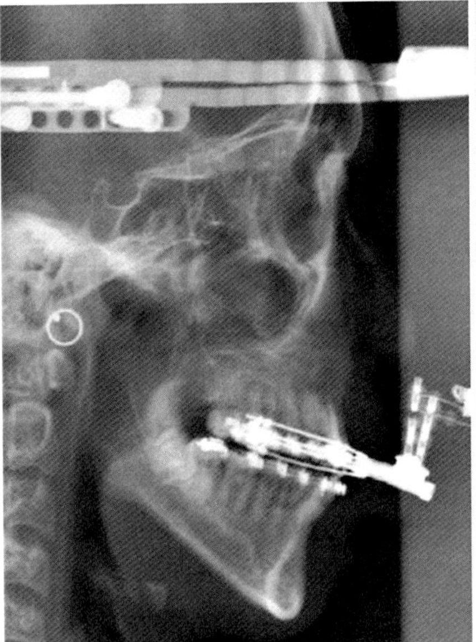
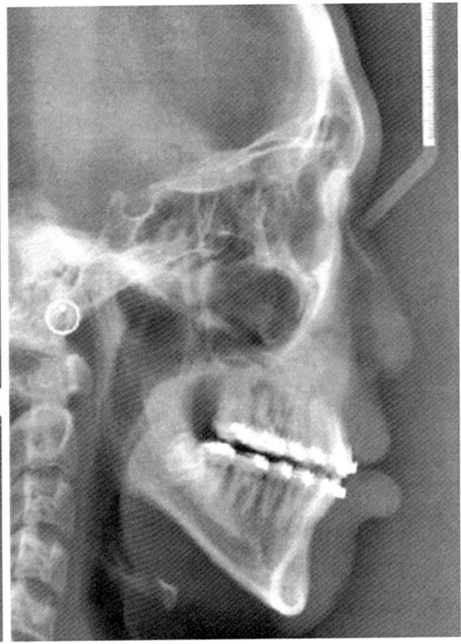

Fig. 15.37 Sequential vascular isolation. Treatment of midface deficiency and class III malocclusion requires Le Fort I osteotomy ± down-fracture. [Reprinted from Yang L, Suzuki EY, Suzuki B. Biomechanical comparison of two intraoperative mobilization techniques for maxillary distraction osteogenesis: Down-fracture versus non-down-fracture. Ann Maxillofac Surg 2014; 4(2):162–170. With permission from Wolters Kluwer Health, Inc.]

Access from the Palatine Complex (Fig. 15.28)

The blood supply to the oral mucoperiosteum of the palate has been well documented angiographically by Maher [8–10]. The pattern of the GPA and nerve is consistent with the concept of neurovascular territories as per Taylor. The angiosome of the GPA comes as StV2 descending palatine and splits into its greater and lesser branches. Studies by Huang in which the overlying mucosa and glandular layers were removed from the palate demonstrate several important findings:

The absence of midline anastomoses is consistent with (1) the embryological separateness of the palatal shelves, and (2) with the presence of paired vomerine fields interposed into the midline. From the axis of GPA, medially directed arterial branches ascend to supply the lingual mucoperiosteum, penetrating into each dental sac. As one "climbs over the gum" the lateral extent of the GPA angiosome is defined sharply along the midline of the alveolar ridge, right here the teeth erupt. These studies have two implications.

- **Incisions placed between the GPA axis and the alveolar wall interrupt the blood supply to the gingivomucoperiosteum of the medial alveolus.**
- Incisions placed exactly over the apex of the alveolar ridge will exactly split the angiosomes of GPA and ASAA/PSAA.

Access from Facial Artery Angiosome: Collaterals

Adjacent to the lateral alveolar envelope are the soft-tissue fields of the lip–cheek complex. These are richly supplied by the facial artery angiosome. Collaterals extend into the buccal sulcus and provide collateral supply for the labiobuccal circulation. This circulation is injured twice in conventional cleft lip repair. First, the buccal sulcus incision severs descending circulation from the infraorbital axis and collateral circulation from the facial system. Second, the blunt mobilization represents additional disruption as it tends to be extensive, in order to eliminate midline tension from the lip repair. Thus, after surgery, the ability of these circulations to revascularize the gingivomucoperiosteum is limited by a wall of scar induced by this unnecessary surgical maneuver. I use the term unnecessary because the subperiosteal mobilization is atraumatic and far more effective. The only reason it is not used is the baseless contention that such dissection will lead to "growth interference." Quite the contrary! Membranous bone growth depends upon a healthy periosteal envelope.

Access from Medial Wall Angiosomes

Recall that the hard palate is a *bilaminar* structure containing marrow. The *nasal surface* of the palatal shelf is covered by a respiratory-type epithelium supplied by the medial nasopalatine artery. When one dissects the nasal surface of the hard palate laterally to beneath the inferior turbinate, no suture is encountered. It looks to all intents and purposes as if the hard palate emanated from maxilla… *which it doesn't.* The inferior wall of maxilla is also L-shaped with a short vertical lamina extending up to the orifice of maxillary sinus just beneath the overhand of inferior turbinate. The horizontal lamina is generally thin and extends medially atop the fields of P1–P2 to which it is fused.

- NP_M does *not* contribute to the alveolus.
- It *does* contribute to maintain the palatal shelf in cases of GPA failure.
- In the Tessier 3 cleft, compromise of both GPA and NP_M occurs, leading to a devastating osseous defect of hard palate and inferior turbinate. Orbital plate of palatine is spared.

Angiosomes and the LeFort I Osteotomy: Survival After Transection of the Maxilla

To understand the alveolar consequences of cleft surgery, the Le Fort I maxillary osteotomy is a useful model. The vascular anatomy involved in this procedure has been extensively investigated by Bell and You. The Le Fort I segment has been shown to be independent from the descending palatine artery. Siebert and colleagues combined latex injections in a cadaver with ink injections in fresh specimens to separate the source of blood supply to osteotomized segments. Despite interruption of the descending palatine artery in all specimens, the *ascending palatine branch of the facial artery* (APFA) entered the soft palate over the levator veli palatini. A second major vessel, the *anterior branch of the ascending pharyngeal artery*, entered more cephalad over the tensor veli palatini (Figs. 15.29 and 15.30). Injection of the APFA in isolation in five cadavers stained the entire hemimaxilla with the overlying mucoperiosteum removed. Thus, elevation of the mucoperiosteum cannot be construed as a vascular insult to the entire maxilla.

In Summation Elevation of the mucoperiosteum from the walls of the alveolus (even if done on both sides simultaneously) will *not* result in bone loss or dental loss.

Consequences of Cleft Surgery

The Consequences of Conventional Cheiloplasty

All forms of cleft lip repair involve three (and frequently four) elements: (1) medial dissection of the prolabium to drop it downward into alignment, (2) lateral dissection of the lip margin so it can attach to the prolabium, (3) a mobilization of the lateral lip element, and (4) an optional dissection of the nose to reposition the cartilages.

For our purposes, the third step is the most critical: In order to achieve midline lip closure with tension, almost all techniques use a *buccal sulcus incision* to achieve medialization of the lip–cheek element. The incision is accompanied by *blunt dissection in the supraperiosteal plane.* These maneuvers are fast, easy, achieve their goal but have unintended and serious consequences.

- The mucoperiosteum covering the external (labiobuccal) lamina of the alveolus contains in its cambium layer a population of mesenchymal stem cells with osteogenic potential critical for growth.
- Two neuroangiosomes supply this envelope: anterior superior alveolar (medial and lateral branches) and posterior superior alveolar. The sulcus incision and mobilization inevitably **INTERRUPT BLOOD SUPPLY**, rendering the alveolar MSC population hypoxic.
- The blunt dissection results in the **INTERPOSITION OF SCAR** between the bone and soft tissues affecting blood supply from the facial artery system to the mucoperiosteum of the anterior maxillary wall.
- Separation of the facial soft tissues from their natural underlying periosteum constitutes an **UNNATURAL MOBILIZATION WITH CONSEQUENT ASYMMETRY** over time.

The Consequences of Conventional Palatoplasty

Von Langenbeck and Veau–Wardill mucoperiosteal flap techniques for closure of the hard palate are well known to every cleft surgeon. Although conceived as representing different designs (bipedicle versus unipedicle), they share common characteristics:

- Blood is supplied directly from greater palatine artery (GPA) and indirectly via the muscles of the soft palate.
- Lateral-releasing incisions are made at the palate–alveolar junction.
- Limited coverage of the anterior-most hard palate results due to **SHORT FLAPS** with **LIMITED ARC OF ROTATION**.
- The releasing incisions create an **OBLIGATORY RAW AREA** that heals with scar.
- The releasing incisions **SECTION THE ANGIOSOME** for the mucoperiosteum of the lingual alveolar ridge.
- If an alveolar cleft is also present, flaps fall short resulting in **ANTERIOR FISTULA**.

Soft palate repairs have their drawbacks as well. All soft palate clefts have some degree of deficiency affecting tissues in the neuroangiosome of the lesser palatine artery. This affects the physical dimensions of three critical structures: horizontal plate of P3 (palatine bone), the anterior one-third of the palatine aponeurosis, and their respective nasal lining. The oral mucosa has a more redundant blood supply. Conventional palatoplasties create resultant tension.

- Mesenchymal deficiency of the P3 zone and anterior one-third of the soft palate results in **POSTERIOR FISTULA.**
- We will cover this topic and its surgical correction in Chaps. 16 and 17.

Physiology of Periosteal Elevation and Transfer (Figs. 15.31 and 15.32)

Uses of Periosteum

Analysis of the consequences of cheiloplasty and palatoplasty must begin with an understanding of the mucoperiosteum. From the time of Veau, the cambium layer of periosteum has been known to contain stem cells capable of bone resorption and deposition. These MSCs are capable of *regeneration and lateral spread.* In bone denuded of periosteum, rapid regeneration of this layer has been observed [11]. In surgically created palatal defects covered with a single-layer mucoperiosteal flap, Bardach and associates demonstrated bone formation occurring from the edges of the osseous defect [12].

Periosteal grafts to the anterior chamber of the eye form bone. This model proves that the grafts contain stem cells which respond to intraocular morphogens by becoming osteoblasts [13]. Tibial periosteum has been used as a free graft to the alveolar cleft [14]. By placing the cambium side directed internally into a prepared alveolar cleft "pocket," Schultz documented bone formation in 15 of 15 children younger than 15 years of age. In two adult patients so treated, free grafting was less successful: one demonstrated inadequate take and the other required reoperation for fistula closure. This is consistent with many studies [15] showing improved results in alveolar bone grafting when performed at a younger age. Probably the best "window" in the face is within 7–8 years of age.

Periosteal flaps have been mobilized at primary repair. In such cases, the cambium layer can be expected to perform in a manner similar to periosteal grafts. Santiago and colleagues reporting an 8-year follow-up of primary alveolar closures using a local mucoperiosteal flap demonstrated that 60% of patients so treated did not require a secondary alveolar bone graft. No adverse effect on maxillary growth was noted. Similar findings have been reported by Massai and by Brusati and Manucci. The likely physiology has to do with blood clot containing BMPs in the local milieu which in turn stimulates the stem cells within the flap to produce bone.

Skoog's "Boneless Bone Graft"

Tord Skoog [16–20] incorporated a superiorly based periosteal flap from the cleft maxilla with broad, supraperiosteal dissection to achieve a two-layer closure of the alveolar cleft. Bone formation was documented by radiographs and on reoperation. The Skoog periosteal flap design presents intrinsic difficulties in its arc of rotation due to its attachment to the infraorbital maxilla. Its vascular basis may vary from random supply to axial depending on the extent of its connection with the overlying soft-tissue envelope. Despite Skoog's excellent results and thoughtprovoking analysis, his technique was not widely adopted, perhaps due to the technical difficulties that subsequent surgeons had with this method. In my opinion the Skoog technique is critically dependent on the extent of separation of the periosteal flap from the soft tissue. In some cases, the transferred periosteum might well have acted as a free graft.

The sliding sulcus mucoperiosteal flap was described by Sitzmann in 1979. The current version of the sliding sulcus flap for secondary alveolar bone grafting is derived from the work of Demas and Sotereanos. It has four components:

- Pericoronal incision from the alveolar cleft back to the buttress
- 45-degree back-cut incision ascending up the buttress from second molar to the zygomatico-maxillary suture
- A broad subperiosteal dissection over the entire face of the maxilla to the infraorbital pedicle
- A horizontal releasing counter-incision (made from below through the periosteum only) running transversely from the floor of the piriform fossa laterally until meeting the buttress release

Using the descending vessels from the anterior superior alveolar axis and the facial-internal maxillary arcade, the Sotereanos flap maintains its relationship to the soft tissues and is transferred mesially, with the cambium layer undisturbed. The periosteal release allows the flap to shift without being tethered to the convexity of the cephalic portion of the maxilla. The average flap transfers *two or more dental units.* The Sotereanos technique creates a raw donor site over the buttresses where the bone is extremely strong. These are healed clinically and are virtually unnoticeable 2–3 weeks postoperatively.

In situ osteogenesis (ISO) using craniofacial periosteum as a stem cell substrate makes use of the exquisite sensitivity of MSC to the morphogen rhBMP-2. Under induction they rapidly form membranous bone identical with that of the local milieu. This has been reported for alveolar clefts [21–

23], mandible [24, 25] distraction [21, 26], and Pierre Robin distraction [27]. We'll explore applications of morphogens as ISO and as distraction-assisted ISO (DISO) in Chap. 20.

Histologic Effects of Periosteal Elevation (Figs. 15.33, 15.34, 15.35, 15.36 and 15.37)

The elevation of sliding sulcus flaps can be predicted by studying the response of palatal mucoperiosteum to surgical elevation. Barro and Latham [28], using histology and fluorescent bone labels, demonstrated in dogs a renewal of osteogenic activity in palatal mucoperiosteum as early as postoperative day 3. A reactive medullary zone characterized by cellular proliferation and bone formation ensues. Although at 14 days bone thickness was greater than the control side, 1–2 weeks later this normalized, possibly due to resorption into the medullary area of the palate.

When bone was resected and the previously elevated palatal mucoperiosteum was placed back directly over the nasal mucoperiosteum, new bone formation was observed at the margins by the third postoperative day. The blood-filled cavity filled with proliferating osteogenic cells. No formation of bone in the periosteum itself was noted at 3 weeks.

The technique of periosteal flap harvest is important. If the integrity of the cambium layer is damaged or disturbed, osteogenic mesenchymal stem cells are compromised. The flaps are most vulnerable along the teeth. Careful elevation with the sharp, curved point of a Molt or elevator of choice followed by use of the amalgam packer is a good way to get the dissection under way.

Here are the important key points about subperiosteal tissue transfer

- The Sotereanos gingivoperiosteoplasty transfers stem cells mesially along the labiobuccal alveolus into a centric position for correct bone synthesis.
- Alveolar extension palatoplasty uses a Sotereanos design to transfer stem cells mesially along the lingual alveolus for correct bone synthesis.
- From the Barro–Latham model: there is no downside to the underlying bone and teeth for temporary elevation of the mucoperiosteal envelope.

In Summation Flaps can be elevated from membranous bone safely. New bone is laid down quickly. The consequences of flap mobilization in the subperiosteal plane are the preservation of cambium layer stem cells and transfer of those stem cells from an abnormal location to a centric milieu where they can function physiologically.

Pathophysiology of Insult

Conventional treatment of complete clefts exerts negative effects on alveolar development in a sequential manner. The first phase of injury occurs at the time of lip repair. The second phase of injury occurs at palatoplasty. The additive nature of injury is the same if the sequence is reversed (palate repair precedes lip repair).

Consequences for Cleft Lip Repair: Facial Soft Tissues

Supraperiosteal cheiloplasty fails to re-establish key anatomic structure required for *symmetrical force distribution* across the cleft. Clefting converts the septopremaxillary ligament spanning from the midline to the maxillary shelves from an inverted T to an inverted L. Subperiosteal repair fixes it. Failure to achieve continuity across the alveolar cleft perpetuates inequality of stimulation to both segments. This is most apparent with the alar base positioning.

An inferior displacement of vestibular lining exists within the nostril in continuity with the lower lateral cartilage. Unless this is released, the lining will remain anchored to a distorted piriform fossa and alar cartilage repositioning, regardless of technique, will not be stable. Subsequent forces of growth continue to act on this distorted envelope in an aberrant fashion.

- The Talmant concept of nasal function should be paramount. In this, restoration of the aberrantly inserted nasalis muscle must be achieved at the first surgery. We'll discuss this at length in Chap. 18 on developmental field reassignment cheiloplasty.

When the periosteal "envelope" into which the bone grows remains displaced bilaterally about the cleft site, it constitutes a dysfunctional matrix. It continues to make bone in the wrong place. This acts ultimately as a restraining "straight jacket," preventing proper forward positioning of the maxilla. Complete subperiosteal release is required (as emphasized by Delaire, Markus, Precious, and Tessier).

Three-dimensional reconstruction of the paranasal (levator) sling and oblique head of the orbicularis is critical for establishment of correct positioning of the alar base, nostril sill, and alar crease. The elements are inserted into periosteum in a lateralized state. They need to be centralized.

Techniques that reposition the muscles but leave their periosteal attachments positioned aberrantly over the maxilla create altered force vectors after repair: The primary insertion changes whereas the secondary insertion (into periosteum) remains the same. This results in a persistent pathological muscle-driven remodeling of the piriform platform which develops within this non-anatomic muscle sling. The unstable lesser segment is molded by these forces postoperatively.

Nasalis muscle is quite the opposite. It should insert above canine to act as a dilator but instead is inserted into the piriform margin and the nostril sill.

- Cleft repairs, be they supraperiosteal or subperiosteal, fail to dissect out nasalis from its recondite state. Instead, they transfer it into the midline and, in so doing, lock it into position as a constrictor.
  - **Key point:** Cleft patients with abnormal nasalis insertion cannot achieve physiologic function of the external valve.

Consequences of Cleft Lip Repair: Hard Palate

Fistula Formation In primary repair, the most critical site for complete continuity of periosteum is to achieve *the nasal floor*, the "roof" of the alveolar cleft, if present.

- Failure to achieve closure significantly increases the risk of anterior fistula and additional risk of failure for future bone grafting.

Stem Cell Damage Supraperiosteal dissection is detrimental to the vascular integrity of the mucoperiosteal envelope supplying the buccolabial surface of the alveolar arch. Because cleftside mobilization is more extensive, the lesser segment is more affected. Sulcus incisions at the time of cleft lip repair interrupt the vertical penetrating branches from the facial-intermaxillary arcade supplying the periosteum. The stem cell population is damaged directly by the dissection and from ischemic insult. Injury to populations on both sides of the alveolus predisposes to disordered membranous growth.

- Cleft lip repair alone cannot cause abnormal dentoalveolar growth.
- The phenomenon of sequential vascular isolation requires the presence of both surgeries, in either order.

Consequences of Cleft Palate Repair

The Three Common Sequelae

The second phase of injury occurs during palate repair. We consider here the four common palatoplasty designs: Von Langenbeck (VL), Veau (V), Veau–Wardill–Kilner (VWK), and Bardach two-flap (B2F) palatoplasty. Of these, VL is not applicable to complete clefts. All four rely on incisions placed at the palatoalveolar junction to elevate mucoperiosteal flaps from the hard palate. These result in three inevitable sequelae:

- **Dead space** with **secondary healing** and **scar** leading to **cleft segment alveolar collapse** (not to be confused with growth failure): all 4 designs.
- **Inadequate length and width** leading, in cases of complete cleft of both primary and secondary palate, to inevitable **anterior fistula.**
- **Devascularization** of the lingual alveolar mucoperiosteum.
  - All lateral branches from greater palatine to the alveolus are cut. Subsequent scar formation does not allow for secondary revascularization to take place.

Sequential vascular isolation requires a *preexisting condition*

- Supraperiosteal lip repair *has not* been done:
  - The end result for dentoalveolar development is scarring of the dead space with collapse of the cleft segment. There is no significant downside for overall facial growth
- Supraperiosteal lip repair *has* been done:
  - The entire mucoperiosteal envelope of the alveolus is now dependent on the bone for its survival

Nota bene: **velopharyngeal insufficiency**, the fourth sequela of palatoplasty is not intrinsic to any of the four designs. It stems rather from the failure of conventional models of cleft palate to incorporate the principles of embryologic developmental fields... and therefore misses the importance of the lesser palatine angiosome deficiency state in the P3 palatine shelf and anterior aponeurosis.

Consequences of Combined CL/P Repairs: Le Fort I Maxillary Advancement

The inevitable result of sequential vascular isolation is a reduced ability of the stem cell envelope to properly respond to the mechanical stresses of compression and distraction. These membranous bone growth mechanisms are well described by Enlow and Proffitt. Recall that a major driver for dental arch growth is the ability of the alveolar envelope to accommodate to the dental units, particularly the premolars and molars. For this reason, relapse into class III occlusion (maxillary dentition with negative overjet) is often not immediately apparent but manifests itself during the phase of mixed dentition when the number of dental units in the arch increases from 20 to 32. In such cases, a combination of preoperative orthodontics, forward repositioning of the maxilla with a LeFort I osteotomy, and another 6–12 months of postoperative orthodontics can achieve a stable relationship between the upper and lower arches.

The design of the LeFort I is well known. It can be accomplished as a one-stage procedure with osteotomy and physical advancement or as a distraction sequence requiring more time and a second procedure to remove the distraction device. Well-known advantages of distraction are its ability to achieve greater degrees of advancement by slowing the conditioning of soft tissues and laying down new bone for stability.

Should Le Fort I maxillary advancement be required, the vascular status of the alveolar envelope may be of considerable importance. The blood supply to the maxilla in such patients may be threatened. You et al. demonstrated in monkeys that ligation of the GPA does not compromise circulation to the bone [29–31]. The actual supply to the maxilla comes from the ascending palatine branch of the facial artery and the anterior branch of the ascending pharyngeal artery. These vessels course within the glandular layer on the oral surface of the soft palate musculature.

This blood supply may be compromised in several circumstances. During palate repair, inadvertent dissection lateral to the hamulus may injure the ascending palatine artery. Entry into the space of Ernst may also injure the same vessel. If carried too far posteriorly, the ascending pharyngeal artery may be injured as well. Both the osteotomy and the soft-tissue dissection required to free the maxilla from the pterygoids at the time of maxillary advancement may also compromise this circulation. Downfracture of the maxilla does not lead ipso facto to avulsion of the GPA. Careful sectioning of the posterior maxillary wall with a small osteotome combined with resection of the ascending portion of the palatine bone preserves the GPA in most instances. Some surgeons skeletonize the palatine bone with a Rongeur to protect the pedicle.

Even if the initial lip repair was performed without buccal sulcus incisions, the approach to the anterior maxilla for the Le Fort I osteotomy commonly uses this exposure. Previous cleft surgeries create a situation in which the alveolar mucosa is converted into a random flap, the blood supply entering the maxilla through the soft palate may be compromised, and the mucosa overlying the nasal aspect of the palate may have been elevated previously. Even with an intact GPA, the blood supply from the oral mucoperiosteum to the alveolar ridge is not direct: it must diffuse from the horizontal maxillary bone across Sharpey's fibers. When a Le Fort I osteotomy is added to these circumstances, a sudden, critical reduction in blood flow to the alveolus may occur. This mechanism may explain reports of maxillary necrosis following Le Fort I advancements in cleft patients [32, 33].

In Summation The sequence of operations for cleft lip and palate constitutes an anatomic insult to the stem cell-containing functional matrix that surrounds the maxilla and palate with consequent negative effects on growth.

Vascular Isolation: Consequences for Dentofacial Development

Unoperated CLP Patients and Isolated CL Patients

Individuals with unilateral cleft lip and alveolus (intact secondary hard palate) do not require a palatoplasty incision. Postoperative consequences of lip repair for the dental arch in these patients are (1) an improvement in overjet resulting from de-rotation of the noncleft segment and (2) worsening of the crossbite [34]. Addition of a complete palatal cleft worsens both lingual crossbite and arch collapse. The clefting mechanics of *facial balance theory* explain these differences well [35]. Comparison of unoperated UCLA and UCLP patients from Mexico and India showed these dentofacial relationships to be similar in both ethnic groups. Bishara summarized the surgical effects of repair in these patients as follows:

- "The tendency for the cleft segment to collapse medially in unoperated cleft individuals is enhanced following palate surgery. On the other hand, lip surgery helps mold the protruding incisors. However, the *combined effects* of surgical repair of the lip and palate cause an *increased incidence* of both anterior and posterior crossbites in operated UCLP individuals."

In the *unrepaired UCLP patient*, the physiological relationship between the soft-tissue envelope and the periosteum remains normal, even though both are malpositioned anatomically. This explains the apparently "normal" growth of unrepaired clefts. Surgical techniques that alter this relationship have negative consequences for growth. Straight-line and triangular repairs do not exact entry into the supraperiosteal plane as the price of closure. Instead, they redirect the soft-tissue envelope mesially with respect to the periosteum. Buccal sulcus incisions achieve a more radical rearrangement of these structures—again with the periosteum left behind. Periosteal bone formation is affected directly by forces of *tension* and *compression*. These forces actually cause the local synthesis of bone morphogenetic protein.

Interposition of an additional layer of scar complicates transmission of these signals. The avulsion of muscle origins from their relation to the periosteum occasioned by supraperiosteal dissection is followed by a reattachment in a different location. Signals transmitted to the periosteum by these facial muscles become reassigned incorrectly, and subjacent bone growth is affected.

Reports from Sri Lanka on unoperated clefts offer a unique opportunity to separate the clinical effects of vascular isolation [36, 37]. Cephalometric studies of 60 male patients older than the age of 13 with complete UCLP were compared with 23 control subjects. The clefts were divided **into** unoperated ($N = 28$), cleft lip repair with unrepaired palate ($N = 18$), and repaired lip and palate. Anteroposterior maxillary dimensions were most normal in unoperated cleft patients; lip repair alone did not affect growth notably. When palate repair was added to UCLP, maxillary growth was affected severely. By more rigid control of the comparison groups, the Sri Lankan data strengthen previous work in unoperated patients [2, 3, 38–42]. The study also supports the contention by Ross that lip repair alone does little to alter anteroposterior maxillary growth.

Growth Inhibition of the Alveolus and Maxilla

Role of the Buccal Sulcus Incision

Although the buccal sulcus incision can be traced back to Veau, Skoog and Millard popularized wide release of the soft-tissue envelope from the maxilla using blunt dissection above the periosteum to mobilize the cheek flap. The technique was easy, fast, and damaging for the periosteal stem cell population. The existence of these cells was, of course, a concept unknown at the time. Unfortunately, this type of repair is now virtually universally accepted. These techniques were critiqued in 1966 by Walker and Colleagues. Writing from Poland in 1967, Bardach advocated avoidance of the buccal sulcus incision, *postulating scar formation* as a mechanism of maxillary growth inhibition. Creation of a false scar plane between the periosteum and the soft tissues is undoubtedly unanatomic. Its subsequent contraction may constitute a distracting force affecting the lip–nose complex. But the real problem is not scar; the real issue is one of blood supply to the periosteal envelope – the biosynthetic engine of osteogenesis and remodeling.

At the heart of the debate is a misunderstanding of what constitutes membranous bone growth. Enlow maintains that there are no specific maxillary growth "centers" against which this scar plane should prove a barrier. Bone deposition and resorption is a multicentric interaction between the maxilla and the soft-tissue envelope into which it must grow. The primary site of membranous bone growth occurs in the cambium layer of periosteum. The metabolic activity of its cellular components is dependent on its blood supply. Maxillary growth aberrations in clefting can be understood in terms of three variables: the physical location of the soft tissue–periosteal complex, the interaction between the soft tissues and periosteum, and the physiological integrity of the periosteum. In the end:

- All three are affected directly by the cleft event
- All three are affected negatively by buccal sulcus incisions
- All three are explained by the osteogenic mesenchymal stem cells

University of Iowa Animal Studies

Lip Pressure Hypothesis

Pioneering laboratory work by Bardach and Eisbach [43] focused on the effects that lip repair alone might have on UCLP configuration. In rabbits with surgically induced clefts, both the Millard (buccal sulcus mobilization) and Bardach (no buccal sulcus incision) techniques were used to repair the clefts. Lip pressures were similarly higher than controls in both types of surgical repairs. Similar findings were later reported using the straight-line technique in beagles. Of note is the absolute width of the lip resection used in these experiments. In nature, the width of the cleft reflects unbalanced and unopposed muscle action on the alveolar segments. Normal clefts have essentially no skin deficit. Bardach's findings regarding lip pressure [43] could be construed as reflecting the intrinsic tension brought about by closing a surgically deficient skin envelope over a skeletal framework that had never "opened up" but is now being forced to rotate back on itself.

Goldstein's elastic flap model can also be applied to explain the lip pressure hypothesis. The alveolar mucosa-sulcus-lip complex is a single anatomic and vascular unit that displaces laterally by the clefting event. It constitutes an abnormal functional matrix. It is maintained in this position by its attachment to the buttress. All repairs that attempt to reposition the lip alone without including the other elements of the anatomic unit create a tension between the centralized lip and the lateralized alveolar mucosa. Lip pressure reflects this tension. Only by a subperiosteal release of the entire complex from the buttress can centralization of the lip be achieved without the elastic effect.

Extent of Supraperiosteal Undermining

Bardach developed an animal model to assess the effects of surgical undermining on midfacial growth. Since no significant variation could be down with palatoplasty, Bardach concentrated on lip repair. In beagles, two groups of surgically created clefts were compared with unoperated controls (group I) and with unrepaired surgical cleft controls (group II). In group III, straight-line closure alone was performed.

Group IV had, in addition to a straight-line closure, wide undermining of maxillary soft tissue to the orbital margin. Although growth aberrations were found in both repaired groups, the group with the most undermining (group IV) was affected most severely [44–47]. This effect was hypothesized to be due to pressure exerted by scar on the developing maxilla. In reality, the issue had to do with stem cells but these were unknown at that time.

Effect of Combined Surgeries: UCL, CP, and UCLP

The simultaneous repair of the experimentally created cleft lip and palate represents a situation in which the vascular consequences of supraperiosteal lip mobilization and conventional palatoplasty incisions are imposed acutely on the patient. The Iowa model was again used to measure craniofacial growth in five groups of beagle skulls to assess the effects of timing of lip and palate closure. Both normal and unrepaired surgically created cleft lip/palate controls (groups I and II) were compared to lip repair (group III), palate repair (group IV), and combined lip and palate repair (group V). Maxillary length was reduced in all repair groups; this seemed to be the primary effect of palate repair alone. Groups III and V demonstrated additional variations in maxillary height, posterior facial width and height, and nasal deflection. Combined lip and palate repair (group V) demonstrated more severe aberrations in all categories than lip repair alone. Maxillary length was shorter in the combined group than in the palate repair group. By creating sequential reductions in blood supply to the alveolar mucoperiosteum and by documenting the resultant impairment in alveolar growth, the *critical role of mucoperiosteal stem cells for maxillary growth was demonstrated* [44, 45].

Lessons from Iowa: What Happens to Periosteum After Supraperiosteal Dissection?

The periosteum is supplied primarily, not from the maxillary bone, but from the overlying soft tissues. When this vascular relationship is destroyed by supraperiosteal dissection, blood flow to the cambium is affected. This does not occur in straight-line or triangular cleft repairs. The buccal sulcus incision leaves the alveolar mucosa undisturbed but makes its blood supply dependent on the palatal angiosome. This factor, above all, "squares the circle" between the experimental findings of Bardach and clinical observations of growth. Patients without buccal sulcus mobilization will likely demonstrate less compromise of postpalatoplasty alveolar osteogenesis. Buccal sulcus supraperiosteal mobilization engenders immediate consequences for the maxilla cephalic to the incision and secondary consequences—after palatoplasty—to the maxilla/alveolus caudal to the incision. Both types of repairs, by failure to recentralize the soft tissue-periosteal envelope, perpetuate the pathological processes unleashed by the clefting event. In the long run these uncorrected bone and soft-tissue relationships will take their toll on the best of repairs, whatever the artistry of the lip incisions might be.

In Summation The Iowa studies demonstrate two key points: (1) the negative effects of supraperiosteal dissection on membranous bone are quantitative—the more extensive the worse the outcome; and (2) the negative effect of lip repair and palate repairs are worse than the effects of the individual procedures.

Sequential Vascular Isolation of the Maxilla: The LeFort I Downfracture

Blood Flow Changes After Buccal Sulcus Incision

Quantitation of these reductions in blood flow was provided by Nelson et al. [48] using macaque monkeys after Le Fort I osteotomies were performed. The effects of buccal incisions and ligation of the GPAs were compared. In animals in which the GPA remained intact, essentially no change in blood flow to either the palatal bone or mucosa was noted after Le Fort I. This is in keeping with the concept of the GPA as a distinct angiosome. Blood flow to the attached gingiva dropped 37% and flow to the alveolar bone dropped 57%. These changes were attributed to transection of the superior alveolar arteries, completely ignoring the facial artery contribution to the mucoperiosteum. Furthermore, the work of You on vessel caliber demonstrates that the blood supply of alveolar bone is primarily intrinsic: very little is contributed by the mucoperiosteum. The primary function of stem cells in the periosteal envelope is in the remodeling of membranous bone via traction (new bone synthesis) and pressure (bone resorption).

Vascular Isolation of the Maxillary Mucoperiosteum: Siebert

Injection studies by Siebert et al. [49] of Le Fort I hemimaxillary segments show the labiobuccal aspect of the attached gingiva to be predominantly supplied by the facial artery. This finding is an *artifact*. Blood supply from infraorbital axis is ignored here because the osteotomy cuts right across it… so the vessels were not injected. The sulcus incision required for the Le Fort I osteotomy isolates the buccolabial mucoperiosteum completely from both the internal branches of ASAA and the external branches of infraorbital, forcing it to depend strictly on the greater palatine artery angiosome (Fig. 15.28).

The next step in vascular isolation involved simultaneous ligation of the GPA: additional effects on the attached gingiva and alveolar bone were observed. As far as the alveolar mucoperiosteum is concerned, deprivation of GPA blood flow due to ligation is tantamount to a Veau–Wardill incision.

It is therefore not surprising that the attached gingival blood flow dropped to 84% of normal, while that of the alveolar bone decreased 89% [48].

Interpretation of alveolar bone flow is complex: severance of the superior alveolar arteries by the osteotomy is not separated from the effects of the pterygoid dissection necessitated by the procedure. However, the straightforward vascular anatomy of the gingiva leaves little doubt that sequential placement of incisions that isolate the mucoperiosteum from its two angiosomes is detrimental to its blood supply.

In Summation Clinical and experimental evidence show that the mucoperiosteal envelope of the maxilla/palate can survive serial ligation of angiosomes as long as flow into the alveolar bone in maintained through the soft palate. In such cases, although mucoperiosteum *can survive by backflow from the alveolar bone*, the overall *blood flow is dramatically reduced.*

Separating Vascular Insults in Time: Cheiloplasty Versus Delayed Palatoplasty

Additional evidence regarding the clinical effects of vascular isolation can be derived from patients treated with veloplasty and delayed hard palate closure: this subject is carefully reviewed by Rohrich and Byrd. It has been long known that patients with palatoplasty performed after 4 years of age demonstrate nearly normal maxillofacial growth. As early as 1921, Gillies and Fry proposed a two-stage technique using obturation of the hard palate. Two long-term studies of this technique demonstrated improvement in arch dimensions [50, 51]. The first age-related comparison study by Rayner suggested that hard palate repair after age 4 caused less arch collapse than when performed at or before 2 years. Hagemann felt that closure should be delayed even further, until secondary eruption was complete. In the United States the impetus for two-stage closure was given by Slaughter (1954) and followed up by Dingman (1971). In reality what these studies indicate is that delaying the palate repair provides a critical period of time for normal maxillary development: the longer, the better.

Analysis of maxillofacial growth by Koberg and Koblin [52] in 1033 patients compared Von Lagenbeck and Veau palatoplasties to two-stage procedures. Hard palate repair after age 12 was found to produce the least alteration in growth. In point of fact, the greatest degree of midface retrusion resulted from palatoplasties performed between ages 8 and 15. Their recommendation was for palatoplasty to be delayed until the completion of secondary eruption by age 15.

These findings were strongly supported by the Schweckendieck's important series which combined soft palate closure followed by lip closure at 7 months with hard palate closure between ages of 12 and 15. Twenty-five-year results (1951–1976) from this series demonstrated a 60–70% approximation of the palatal cleft dimensions with 95% narrowing of the alveolar cleft margins [53]. These minimal consequences for growth following delayed hard palate closure were so impressive that a commission was sent from the United States to Marburg, Germany, to study the results. These were confirmed by Bardach and colleagues in the 1984 Marburg Project.

The vomer flap palatoplasty by Friede and associates showed similar results. In UCLP patients, hard palate closure with a vomer flap at age 7 did not cause midface retrusion and transverse arch collapse when compared with patients treated similarly at 3 months of age.

In Sum The differential effects of staged lip and palate surgeries on dentofacial growth are due to the vascular insults inflicted by these procedures. The best solution is to use the subperiosteal plane and preserve the integrity of all angiosomes.

Clinical Studies of the Vascular Isolation Model

Dental Development in the Alveolus

The principle of vascular isolation is based on principles of dental development. Alveolar growth is rapid in the first 3–4 years of life, reflecting the formation and eruption of primary dentition. From the appearance of the crown until occlusal contact is established, this process is accompanied by the organization of the periodontal fibers, root formation, and alveolar remodeling. This is concomitant with a multicentric pattern of deposition and resorption along the entire maxilla. All primary root development is complete by age 3 [54–56].

As diphyodonts, humans have a second permanent dentition that develops in a lingual position to the primary teeth. By the time resorption begins (4–5 years for the incisors and 6–8 years for the canines), permanent crown formation occurs beneath the primary roots. This shift reflects the overall remodeling of the alveolus, which accompanies dental development. Thus, the roles of the primary dentition are (1) support maxillary bone growth, (2) permit masticatory function in the pediatric jaw, and (3) provide the spatial template for its permanent replacement.

Secondary crown formation of the canines, premolars, and second molars is complete between ages 6 and 8, with eruption occurring between 8 and 13 years. This marks a second, explosive period of maxillary growth during which the maxilla attains its overall size, shape, and relation to the cranial base. As in childhood, the adolescent alveolus reflects the size and position of its new dental units. On the other hand, facial growth of the alveolar bone is requisite for proper eruption to occur. First it creates space to accommodate the secondary dentition. Second, this growth weakens support structures for the deciduous teeth, allowing them to exfoliate normally.

The primary and secondary dental sacs are supplied by penetrating vessels from the alveolar mucoperiosteum. On the labiobuccal side, these are derived from the facial system whereas those on the lingual side come from the GPA. Buccal sulcus and palatoplasty incisions constitute *sequential, permanent interruptions in the blood supply to the growing teeth.* At the same time these incisions reduce blood flow to the osteogenic cambium layer responsible for remodeling the alveolar cortex. When palatoplasty is performed early, these mechanisms disturb the primary dentition and its replacement. When performed in mixed dentition, it should likewise affect the erupting permanent canines, premolars, and second molars. These latter teeth are responsible for much of the additional maxillary volume and, perforce, midfacial projection.

Early Versus Delayed Hard Palate Closure

Vascular isolation following palatoplasty is a very useful model for understanding the conflicting data regarding early versus delayed closure of the hard palate. Preservation of blood flow to the alveolar mucoperiosteum and the developing dentition via the greater palatine angiosome is the basis of the results for the two-stage technique. These studies must be interpreted according to the technique of lip repair (whether a buccal sulcus incision has been made or supraperiosteal undermining has been employed). If the labiobuccal angiosome is undisturbed palate incisions make little or no difference in growth. Sarnat showed in macaques that mucoperiosteal elevation and ligation of the GPA did not cause palatal growth disturbance. Thus, if adequate blood supply is available to the mucoperiosteal cover of the alveolus and to the dental units lying within the subjacent bone then a proper relationship between cambium-mediated surface bone growth at the surface and eruption-mediated bone growth within the alveolus itself can be maintained.

The same principle applies when labiobuccal compromise exists and subsequent palatoplasty is carried out without disturbing the GPA angiosome. This can be accomplished in two ways: (1) paring and advancing the cleft margins, and (2) providing distant tissue via buccal mucosal flaps. Rohrich and Byrd performed direct closure at 15–18 months after first closing the cleft site with a veloplasty at lip repair. Maeda used bilateral, *posteriorly based buccinator flaps* to achieve simultaneous soft palate push-back and coverage of the hard palate. Kaplan's discussion is a valuable critique of the technique. For details on the anatomy of the buccinator flap, see Carstens [57] and Bozola. These protocols vary in three important aspects: (1) orthopedic manipulation of the cleft site, (2) management of midline tension, and (3) vascular design.

In Summation If enough time has elapsed between an initial insult to the buccolabial mucoperiosteum and palate repair such that membranous maxillary bone growth is at a critical stage (somewhere between age 8 and 15), class III relationship is avoided… but speech suffers.

Physical Manipulations of the Soft Palate

Veloplasty at Primary Lip Repair: Approximating the Cleft

The ability of *early veloplasty* to reduce the width of a palatal cleft has been known for some time; this is complementary to the effects of lip repair [58, 59]. Reestablishment of midline soft palate continuity—with or without reorientation of the musculature—corrects the unopposed lateral and cephalic distraction forces exerted on the posterior maxilla. Clinical photographs taken in [53] long-term follow-up depict the change in cleft dimensions that occurs after lip repair and primary veloplasty. This is exactly as predicted by Markus. Velar adhesion veloplasty is crucial to the success of any two-stage closure using marginal flaps alone. Approximation is not always adequate. In this case, Rohrich and Byrd resort to von Langenback mucoperiosteal flaps.

Buccinator Interposition Palatoplasty: Reconstruction of a Deficiency State

We shall discuss this concept and procedure at length in Chaps. 16 and 17. As reported by Maeda the technique is performed as a single-stage procedure at 18 months. It is nonorthopedic and is based primarily on a wound model. Subsequent maxillary growth under these conditions would mimic that of an unoperated cleft with two caveats. *Angiosome failure of lesser palatine will result in a reduction in horizontal shelf surface area and in the anterior zone nonmuscular zone*. The interposition of buccal tissue to "lengthen" the soft palate (and the dissection required to do it) is an accurate means to *replace the tissue deficit present in the anterior one-third of the palatine aponeurosis*. This zone has the same vascular supply as the posterior palatine shelves. A potential nuisance to buccal flap interposition is that it can tether the soft palate motor function. Thus, if the buccal flaps are not separated eventually from the cheek, they will exert a potential "tethering" effect. Patients can potentially bite on the tissue bridge. The release procedure requires 5 min.

Despite the "closing down" of the cleft site after lip repair, a finite lack of mucosa and bone persists. This is a **true deficiency state**, the exact dimensions of which determine whether paring (with possible limited undermining) of the marginal mucosa can be sufficient. The lining of the palatal aspect of the cleft reflects the overall maldistribution of the soft-tissue envelope. If this dysfunctional matrix is simply united, tension will persist at the midline. Furthermore, the pathological relationship between the matrix and the bone is allowed to persist.

Since the anterior one-third of aponeurosis is a muscle-free zone (except for uvulus) the defect is one of mucosa and fascia. Buccal flap interposition certainly obviates central tension; it is a close approximation to replacing "like with like."

Both the Rohrich and Maeda protocols leave the circulation of the GPA angiosome to the alveolus unscathed. By using tissue outside either the facial or GPA angiosomes as a "patch" for the hard palate cleft, the Maeda procedure constitutes a biological "probe" *similar to the unoperated case* but with palatal closure. If this technique was performed using earlier intervelar veloplasty, interesting comparisons of growth and speech results could be made.

DFR in Cleft Palate Surgery

Correction of the Dysfunctional Matrix (Figs. 15.37, 15.38, 15.39, 15.40, 15.41 and 15.42)

There is *no intrinsic growth defect of the cleft maxilla.* The four processes of clefting, deficiency, division, displacement, and distortion, create a *dysfunctional matrix* into which the maxilla must develop. From its inception at Carnegie stage 15–16, the anatomic effects wrought by clefting unleash processes by which normal growth into an abnormal envelope leads to *progressive asymmetry of growth.* The pathogenesis of the cleft deformity is due, in part, to the aberrant orientation of three "rings" of perioral muscles [35]. Abnormal force vectors generated by these muscles in utero literally "pry open" the cleft site. What was a small deficiency of mucosa and bone at *birth* becomes distorted with the passage of time. Although the unrepaired UCLP may grow "normally," its pathological width persists. When Coupe and Subtelny compared the palatal dimensions of unoperated cleft palate subjects with 50 normal subjects, both deficiency and displacement were demonstrated. These were proportional to the severity of the defect. In complete clefts, displacement of the posterior maxilla occurred at the oronasal level [61, 62]. On the other hand, the Schweckendieck trial demonstrated that, by adolescence, lip repair and soft palate closure "narrow down" the entire cleft site, most likely to its original (deficient) dimensions.

CL/P Surgical Sequence

DFR cheiloplasty at 6 months of age combines a (1) subperiosteal centralization of the soft-tissue envelope using a mesial approach *without an alveolar incision*; (2) anatomic muscle repair to achieve a truly functional repair; and (3) correction of the nasal airway as per Talmant. When DFR is sequenced with alveolar extension palatoplasty at 9–18 months of age, maximal preservation of the blood supply to the maxilla is achieved. Early soft palate veloplasty at cheiloplasty may narrow posterior cleft dimensions (making definite repair easier) but has not been shown to optimize speech.

- *Deficiency* of the piriform fossa is recognized
  - Critical correction of the nasal floor defect to prevent fistula

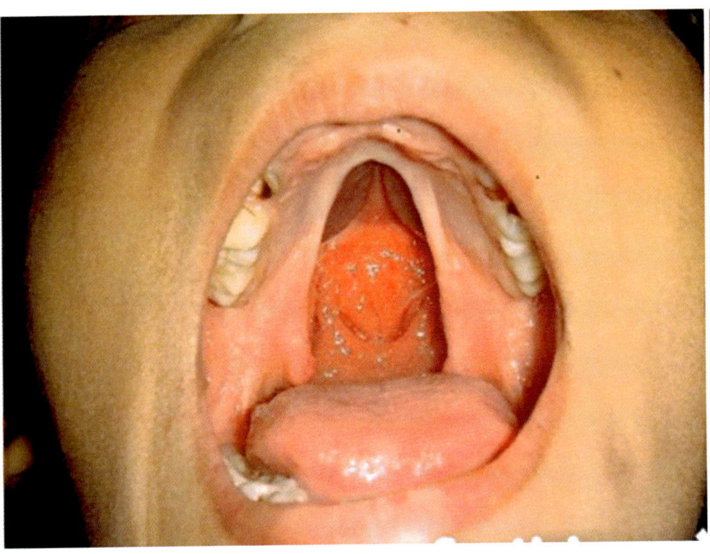

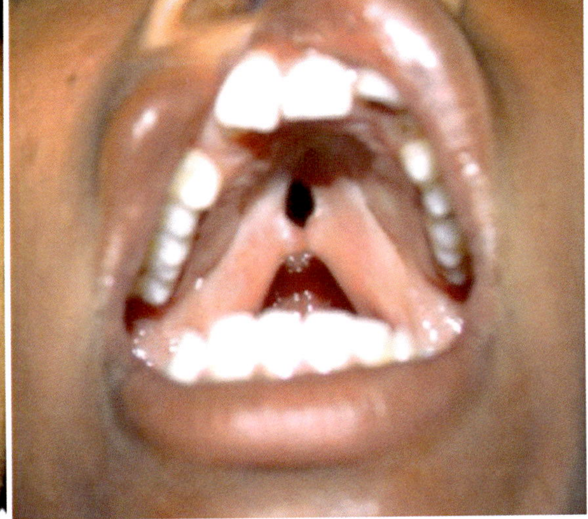

Fig. 15.38 AEP indications: tissue deficiency. Case 1. Midline deficiency. Horseshoe midline cleft palate due to Pierre Robin with small vomer. Lack of midline tissue from vomer presents a challenge. Case 2. Anterior–posterior deficiency, sever in right UCL/P. The use of AEP flaps here can assist in push-back of soft palate but may not be sufficient. Buccinator interposition will add 1.5 cm of length behind the bony platform. [Courtesy of Michael Carstens, MD]

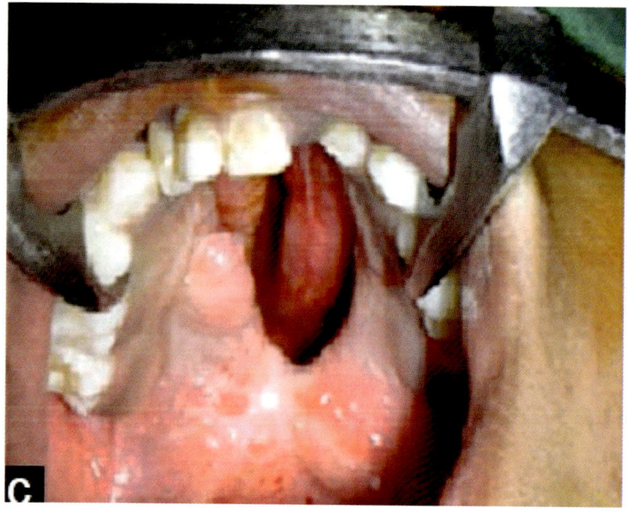

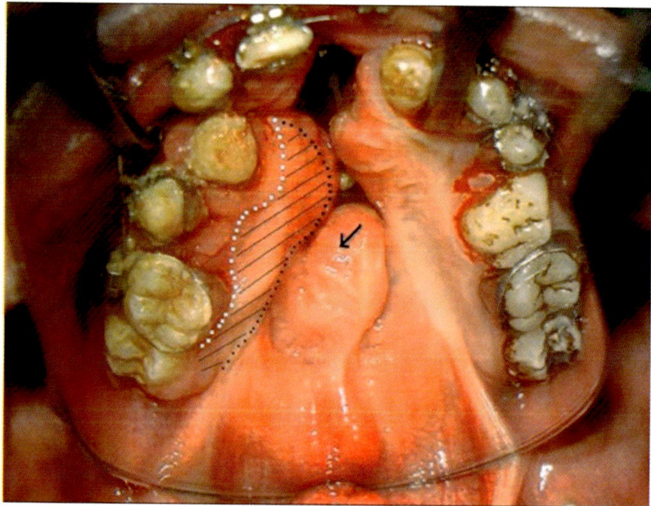

Fig. 15.39 AEP indications: anterior fistula. Case 1 (left). Large anterior fistula with retracted vomer. Note failed right-sided Veau flap. Nasal lining difficult: midline buccinator can be useful to address this problem. Case 2 (right). Bilateral CL/P with torsion of the premaxilla to the right, right-sided collapse, failed Veau-type flap (arrow) and bilateral anterior fistulae. *Note*: The potential tissue available by using AEP flaps is sufficient to fill the space. Centralization of the premaxilla preop would be helpful. The overall goal for reconstruction is to prepare the alveolar clefts for bone grafting and stabilization of the premaxilla. Because nasal floor closure was not performed, a revision DFR cheiloplasty with complete take-down of the previous repair is required in order to locate the angiosomes and achieve anatomic flap transfer. Tongue flap is inadequate and would not enable reconstruction of the alveolar cleft chamber to receive bone graft. Left: [Reprinted from Bonanthaya K, Shetty P, Sharma A, Ahlawat J, Passi D, Singh M. Treatment modalities for surgical management of anterior palatal fistula: Comparison of various techniques, their outcomes, and the factors governing treatment plan: A retrospective study. *National Journal of Maxillofacial Surgery*. 2016;7(2):148–152. With permission from Wolters Kluwer Health, Inc.]. Right: [Reprinted from Sadhu P. Oronasal fistula in cleft palate surgery. Indian J Plast Surg 2009 Oct; 42(Suppl): S123–S128. With permission from Wolters Kluwer Health, Inc.]

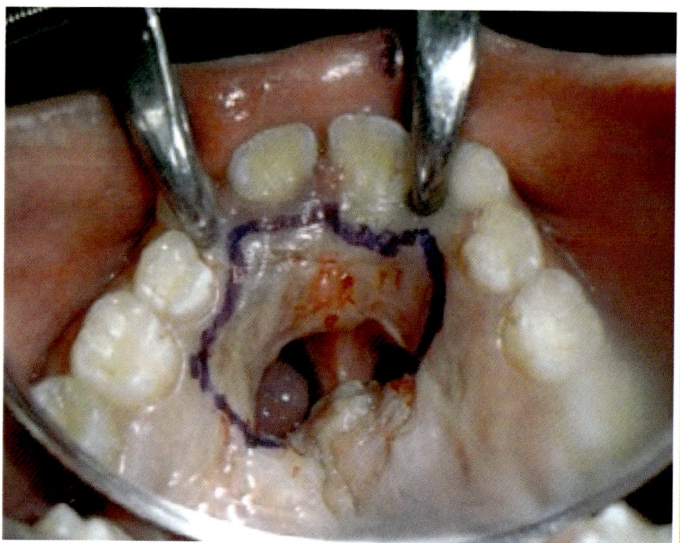

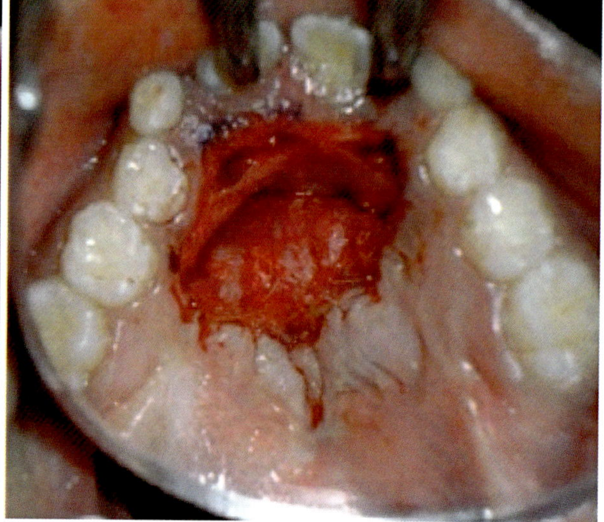

Fig. 15.40 Anterior fistula with turn-over flap. *Left*: Failed Veau flaps seen, retracted. *Right*: Although abundant AEP flaps were available, a turn-over flap was used; this flap is cut-off from the GPA angiosome and provides only a single-layer closure. Double-layer closure likely requires re-entry into the lip. [Reprinted from Sadhu P. Oronasal fistula in cleft palate surgery. Indian J Plast Surg 2009 Oct; 42(Suppl): S123–S128. With permission from Wolters Kluwer Health, Inc.]

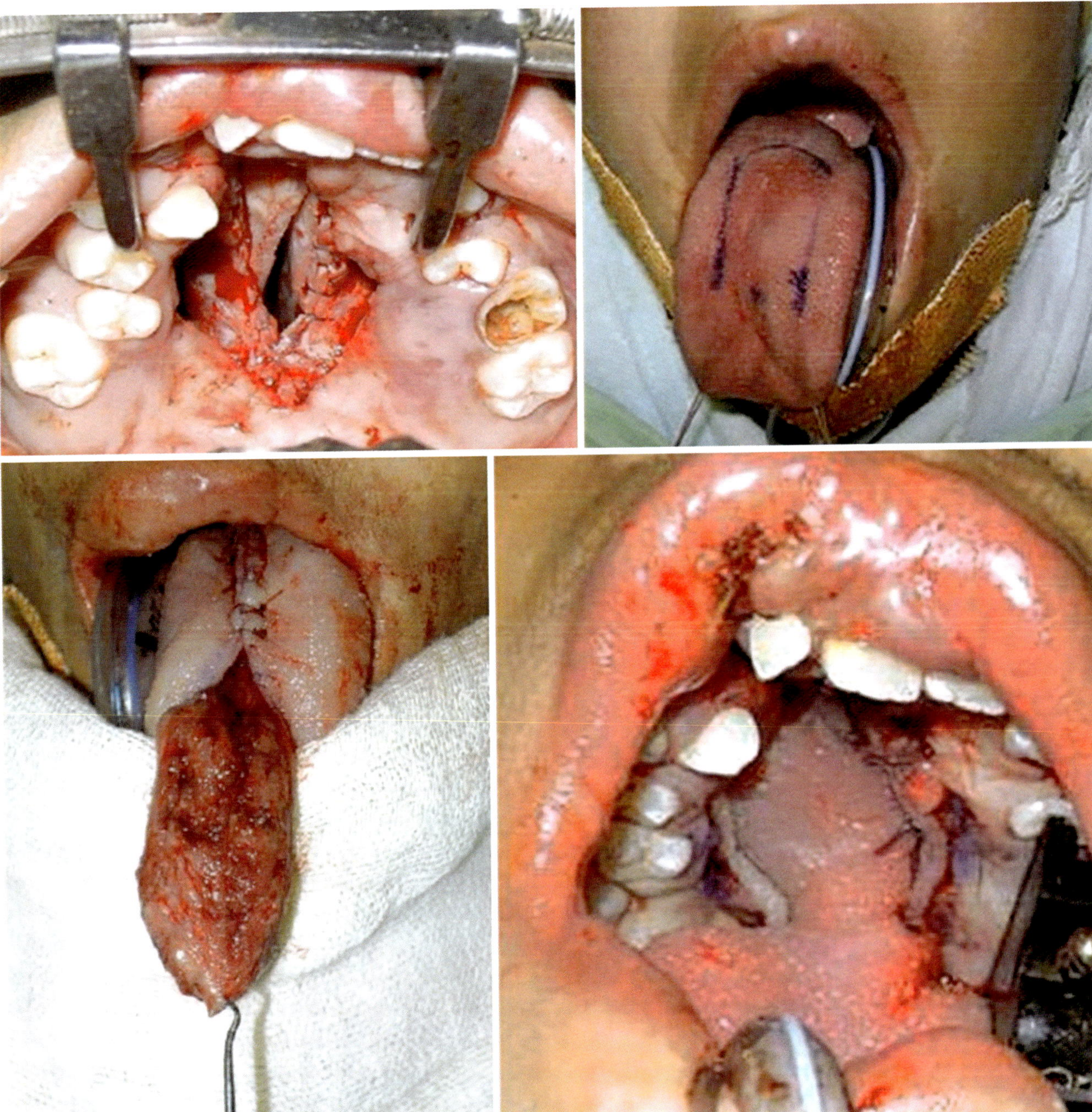

Fig. 15.41 Tongue flap. Fairly reliable two-stage procedure to place soft tissue into the oral defect. *Top left*: Anterior fistula dissected with elevation of mucosal flaps; the oral side reflected nasally. These tissues are random and notoriously unreliable. *Top right*: Design of anteriorly based tongue flap; *Bottom left*: closure of the donor site; *Bottom right*: when the tongue is retracted back into the mouth the flap can be sutured with the raw surface directed nasally. Tongue flaps are hardy. They are sectioned at 21 days and the inset is completed [60]. [Reprinted from Mahajan RK, Chhajlani R, Ghildiyal HC. Role of tongue flap in palatal fistula repair: A series of 41 cases. Indian J Plast Surg 2014; 47(2):210–215. With permission from Thieme Publishers]

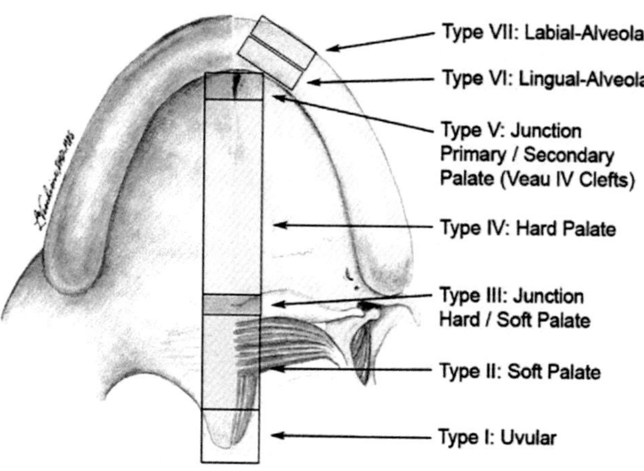

Fig. 15.42 University of Pittsburgh classification of palatal fistulae. [Reprinted from Smith DM, Losee JE, et al. The Pittsburgh fistula classification system: A standardized scheme for the classification of palatal fistulas. Cleft Palate Craniofac J 2007; 44(6): 590–594. With permission from Sage Publishing]

- *Division* of tissues respects the angiosomes of the prolabium and the maxilla
- *Displacements* are corrected
  - Stem cell transfer into centric position
  - Transfer of the lip nose complex into symmetry
  - Nasalis reassignment
  - Nasal vestibular lining
  - Columellar lengthening
- *Distortions* normalized: septal defection anterior (defer posterior for secondary surgery as needed)

DFR palatoplasty (AEP) at 12–18 months of age resolves its mechanisms of its formation

- *Deficiency* at the midline is addressed by the transfer of wide flaps with tension transferred to the tuberosity, out of harm's way.
- *Division* of the functional matrix is done with respect to angiosomes. The flaps are long and wide. Lateral dead space is minimized so that alveolar tissue blood supply is rapidly re-established.
- *Displacement.*
  - Hard palate stem cell populations are moved forward into centric position.
  - Soft palate musculature is corrected by an anatomic intervelar veloplasty, which permits orthopedic realignment of the posterior palate to occur. Tissue deficit of the r2 anterior aponeurosis is replaced as needed. This is usually *deferred*, being performed as a secondary procedure once the function of the soft palate has been assessed.
- *Distortion* of the alveolar segments is addressed with the initial lip repair plus minimizing secondary healing.

Buccinator palatoplasty: speech evaluation and repair age 4–6 after full language acquisition and ability to cooperate

- Videofluoroscopy.
- Muscle stimulation.
- Interposition is deferred as a secondary procedure for VPI.
- ± revision of muscle repair.

Alveolar Extension Palatoplasty: Functional Aspects (Figs. 15.36, 15.37, 15.38, 15.39, 15.40 and 15.41)

Dead Space

As previously discussed, all four conventional palatoplasty incisions produce flaps that are too narrow because they subdivide the GPA angiosome, leaving the entire lingual mucoperiosteum behind, "shipwrecked" on the alveolus. This amounts to approximately 10–15 mm of wasted tissue. The flaps are stiff; they can in no way be stretched all the way from the alveolus to the midline. In the process of flap transfer, an inevitable dead space of 5–15 mm results. This naked bone will heal by secondary intention but remucosalization hides the ugly fact that a wall of scar now exists between the wounded angiosome and its abandoned offspring. Although neovascularization of the scar will take place, blood flow from the GPA flap to the alveolus is dramatically reduced, if not eliminated.

As a consequence, a state of contraction is placed upon the lingual wall of the alveolus. This, on an isolated cleft palate, is sufficient to cause tethering of the cleft segment. If, on the other hand, supraperiosteal cheiloplasty has been performed, sequential vascular isolation will take place as well with more generalized effects on ipsilateral dentoalveolar growth.

AEP flaps are wide enough to cover all the way from alveolus to midline. A release from tuberosity allows for advancement of 1.5–2 cm anteriorly so they can be immediately sutured to the gum in the advanced position. We know from the original work by Barro and Latham that re-establishment of periosteum-to-bone contact elicits rapid angiogenesis and osteogenesis. Bennun reports using a single-flap AEP design for UCL/P based on the cleft segment. Clinical performance of this technique will be discussed in Chap. 16.

Clinical Performance

Dead space is a difficult topic to quantify. Studies looking at palatal dimensions, collapse, arch symmetry, eruption, and occlusion of AEP-operated palates can be compared to historical controls. Monasterio's series reports a Goslon score of 2.2 for AEP versus 2.3 for two-flap palatoplasty, vide infra.

Anterior Fistula

Fistulas, like facts, are stubborn things. If you get the wrong, they don't just go away. Anterior palate fistulae can be particularly symptomatic. Oral–nasal communication leads to chronic sinusitis with consequent serial infections of the respiratory tract. Incompetence affects air flow and intraoral pressure necessary for the production of phonemes such as sibilants. In the presence of alveolar cleft, inadequate dissection at primary repair followed by scarring and collapse creates a communication into the floor of the nose. Secondary intervention to achieve a water-tight closure of the nasal floor becomes difficult. If bone grafting is contemplated, this scenario represents a hostile wound bed predisposed to infection and graft loss.

An alveolar cleft is a six-sided "box" with the lateral walls made up of bone. A well-designed subperiosteal cleft lip repair can achieve coverage over the floor of the nose/roof to set the stage for secondary closure. AEP palatoplasty flaps close the floor and backwall, postponing anterior closure for the time of bone grafting (age 4, Talmant-Lumineau). Unfortunately, traditional palatoplasty flaps are (1) generally too short/small to reach to reach the cleft or (2) under tension and therefore subject to breakdown.

The dimensions of AEP flaps are wider and longer. As demonstrated in the previous figures, inclusion of the lingual alveolar tissues increases length by 15–20 mm and width by approximately 15 mm. In unilateral clefts, the results are as follows: (1) the noncleft segment flap extends about 5 mm *beyond* the alveolar cleft, (2) the cleft segment flap routinely comes to the back wall of the alveolar cleft, and (3) the closure is S-shaped. In bilateral CL/P both flaps come into the back wall but don't project as much.

An obscure but important secret to success in AEP dissection lies in the alveolar cleft. From the lateral wall on the cleft side, mucoperiosteum is elevated backwards in continuity with the remainder of the AEP flap. Not only does this add an additional 5 mm of length, but it also removes tissue from the alveolar wall that would have to be elevated secondarily at the time of bone grafting. Such tissue, if left behind, constitutes a threat to the graft.

Clinical Performance

The literature on palatal fistula is a hodgepodge that lumps together differences in fistula location and preoperative diagnosis. See Hardwicke's review of 44 studies and 9294 patients. In Cohen's series the overall fistula rate was 30%. After initial fistula repair, the recurrence rate was 37%. This indicates the difficulty of reusing local tissue for reconstruction.

Stratification is possible using the 2007 Pittsburgh Fistula Classification System [63]. The most difficult fistulae are type IV (secondary hard palate) and type V (junction of primary and secondary hard palate). In the Pittsburgh series of 255 fistulae, types IV and V constituted 46.7% and 11% of the cases.

Definitive answers can be obtained under two circumstances. Prospective consecutive cases of palatoplasty can be separated into combined primary and secondary CP versus secondary CP alone. The postoperative rate of anterior fistula can be compared between each of these groups with controls from the same center. A second approach centers on the use of AEP as a secondary "rescue" procedure for the treatment of established anterior fistulae.

Monasterio from the Fundación Ganz in Santiago, Chile, reported a prospective comparison of 20 patients operated with AEP versus standard (two-flap) palatoplasty. The groups were similar in time of surgery (13.4 months versus 12.4 months), operating time (110 min versus 128 min), and follow-up (34 months versus 40 months). Raw areas were observed postoperatively. Clinical outcomes (AEP versus two-flap) showed fistulae (0% versus 20%). Dental eruption was unaffected by AEP. The number of teeth at 1 year was 2.75 versus 3.25 and at 4 years was 8.3 versus 9.6. Occlusion by the Goslon yardstick was 2.2 versus 2.3.

Murthy described fistula repair in 194 patients with post-alveolar and secondary hard palate defects constituting two-thirds of all cases: 72% were operated using local flaps; in 28% cases a tongue flap was used due to "tissue shortage." Alveolar extension palatoplasty was used in a total of 48 cases with a success rate of 98% (one case required eventual prosthetic reconstruction).

Retrusive Maxilla

By now you are thoroughly familiar with the principle of sequential vascular isolation. AEP has the most significant impact when used for cases of CL/CP in which damage to both sides of the alveolar stem cell envelope can otherwise occur. Ideally, it should follow subperiosteal cheiloplasty but, if the original cleft lip repair was supraperiosteal, AEP will at least protect the lingual side.

Clinical Performance

Because AEP must coincide with some type of lip repair, we cannot ignore the aesthetic and functional results of cheiloplasty. Subperiosteal lip repair has significant effects on the soft-tissue drape of the lip–nose complex as measured by the Amaratunga Symmetry Index. This overlooked and underappreciated metric is even more useful today given improvements in photography and cone beam CT scanning. Carstens–Matthews–Talmant philosophy of developmental field reassignment includes prolabial conservation, columellar equalization, anterior septal straightening, nasal floor reconstruction, and nasalis relocation with vestibular lining release. Nasal airway reconstruction, using Talmant's techniques and long-term soft silicon stenting, provides for a functional airway in which the dilator function of the nasalis is rescued. Airway measurements, including comparison of airflow, are important prospective parameters which can be compared with those in patients with standard repairs—see Chaps. 18 and 19.

The best test of maxillofacial growth using the combination of DFR lip repair and AEP will be the achievement of stable occlusion with minimization of crossbite. This in turn will depend on the orthodontic protocol used by individual treatment centers. Lumineau's innovative methods of expansion for the alveolar defect at age 4 provide an excellent chance for bone graft placement into an optimized alveolar defect with water-tight soft-tissue coverage. Future studies of DFR/AEP will be based on orthodontic criteria and improved cephalometrics compared with well-known historical controls with crossbite, malocclusion, and class III dentofacial relationships.

Conclusions

Alveolar extension palatoplasty has as its rationale the following principles: (1) hard palate coverage with like tissue, (2) reduction in closure tension, (3) control of the alveolar cleft, (4) prevention of anterior fistulae, (5) minimal exposure of palatal bone, (6) appreciation for the contribution of the mucoperiosteum to alveolar growth, (7) recognition of the duality of the alveolar mucoperiosteal blood supply, (8) recognition that the alveolar ridge represents a watershed between two angiosomes—that of the facial-internal maxillary arcade and that of the GPA, (9) the sequential effects of lip and palate repair on maxillary growth, (10) preservation of alveolar blood supply using the subperiosteal plane, and (11) restoration of the mucoperiosteal functional matrix.

The following clinical observations can be made

- The alveolar ridge is comprised of two angiosomes.
- Vascular compromise of either angiosome can affect alveolar development.
- Palatoplasty repairs preceded by lip repair without buccal sulcus mobilization behave differently than those using such an incision.
- If the labiobuccal angiosome is compromised but surgical isolation of the palatal angiosome is delayed until eruption of the secondary dentition is complete, maxillofacial growth will be impaired minimally, apart from those changes exerted on the alveolus by the lip repair.
- The primary source of molding for the alveolar segments after cleft repair comes from the force vectors of the facial muscles.
- The static effect of the lip repair (lip pressure) is generated by the failure to release the lip-sulcusalveolar mucosa from its lateralized state.
- Maxillary growth into a dysfunctional matrix will continue to be aberrant regardless of the surgical techniques used to correct the cleft.

The fundamental goal of process-oriented cleft lip and palate repair is the conversion of the dysfunctional bone and softtissue matrix to a functional matrix. Long-term clinical studies using arch measurements and cephalometrics are required to quantify its effects on cleft management. In the meantime, it provides a new paradigm from which to consider previously contradictory laboratory and clinical observations.

Commentary: Ricardo Bennun

Alveolar Extension Palatoplasty: Technical Notes

Ricardo Bennun

Introduction

Successful palate closure became possible in the mid-nineteenth century with the development of the mucoperiosteal flaps by Dieffenbach and von Langenbeck [64].

The fact that simple closure of the palate cleft did not necessarily ensure normal speech was recognized a little later and led to the development of palate-lengthening procedures and various forms of pharyngoplasty.

This chapter serves as a logical extension to the multidisciplinary approach to cleft palate management and reconstruction from genetic factors that precipitate cleft to a thorough discussion of today's approaches.

Surgical Procedure

A complete report of the alveolar extension palatoplasty (AEP) was precisely illustrated in a chapter of our Atlas published in 2015 [65]. The predictive factors of difficulty before reconstruction and the planned strategies were also described by the author [66]. A comparative study between Veau–Wardill–Kilner and the AEP procedure was also published [67].

Technical Recommendations

A short incision of the oral mucosa over the alveolar border must be carried out. Utilizing a Gillies retractor, under direct vision, and with the blade in an oblique direction, the cut must be completed, leaving all the dental germs in their position.

Minimal lateral incisions are planned in simple cases, where the soft palate major gap distance (SPMGD) is inferior to 6 mm, with symmetrical bilateral soft palate length (BSPL) superior to 20 mm. This strategy will also be useful in the presence of easily isolated cleft palates.

When the SPMGD is between 7 and 11 mm, one alveolar extended palatal flap elevation is considered. In that case, the author's choice will be the palatal flap from the sick side. Some surgeons utilizing the AEP procedure will prefer to select the normal-side palatal flap. Outcomes seem to be similar.

In bilateral cleft cases, the suggestion for a complete and cautious hard and soft palate closure will be using both AEP flaps. This recommendation is also valid in patients with SPMGD greater than 12 mm, but having a symmetrical BSPL superior to 20 mm.

In very difficult cases with a SPMGD superior to 12 mm, plus symmetrical/asymmetrical BSPL under 16 mm, or even a short bilateral hard palate length with depressions (BHL&D), our indication would be a nasal plane reconstruction utilizing a vomer flap, complemented with a superior pharyngeal flap [66].

Results

Statistically significant differences were found when comparing the total percentage of complications between both groups (Veau–Wardill–Kilner and AEP). The amount and severity of each complication by groups were also established as significant [67]:

- Bite alterations and transversal collapse with dental malpositioning in group A: 29.84% and group B: 23.155% (P value < 0.009)
- Presence of fistulas in group A: 4.11% and group B: 5.0% (P value < 0.02)
- Patients with VPI in group A: 6.12% and group B: 0.11% (P value < 0.14)

Complications

No additional complications have been reported utilizing this procedure. Outcomes and follow-up during the last 12 years have proved no teeth alteration, in all patients (pictures).

Conclusions

Since 2009, the author has selected the alveolar extension palatoplasty variant, plus the complete muscle dissection and retropositioning, and the posterior pillar's elongation with hemi-uvula rotation/reconstruction, as the technique of choice for primary cleft palate repair [65].

The utilization of the pre-op cited parameters to identify cleft palate diversity and severity seems to be a useful methodology to select the correct surgical strategy [66].

Moving and reducing incisions to protect blood supply, following Carstens' suggestion, allow us to reduce the use of electric coagulation and blood loss. The presence of less raw areas prevents the incidence of retractile scars [68].

Employing regional blocking [69] joint to general and local anesthesia not only decreased intra- and post-op pain and baby neurotoxicity but also allowed us to initiate oral feed and discharge the baby early.

Having bigger palatal flaps in width and length allows us to decrease the incidence of anterior/medial fistulas, velopharyngeal incompetence, and maxillary alterations. Dental malpositioning and misalignment, as medial otitis, were also present in an inferior percentage [67].

Note on the Clinical Series

Bilateral cases 4, 5, and 7 and unilateral cases 5, 7, 11, 12, 13, and 14 demonstrate class 1 occlusion.

Alveolar Extension Palatoplasty: Atlas of Cases

Ricardo Bennun and Michael Carstens

Case 1 Bilateral

Primary cleft lip and nose reconstruction—6 months post op [70–75]

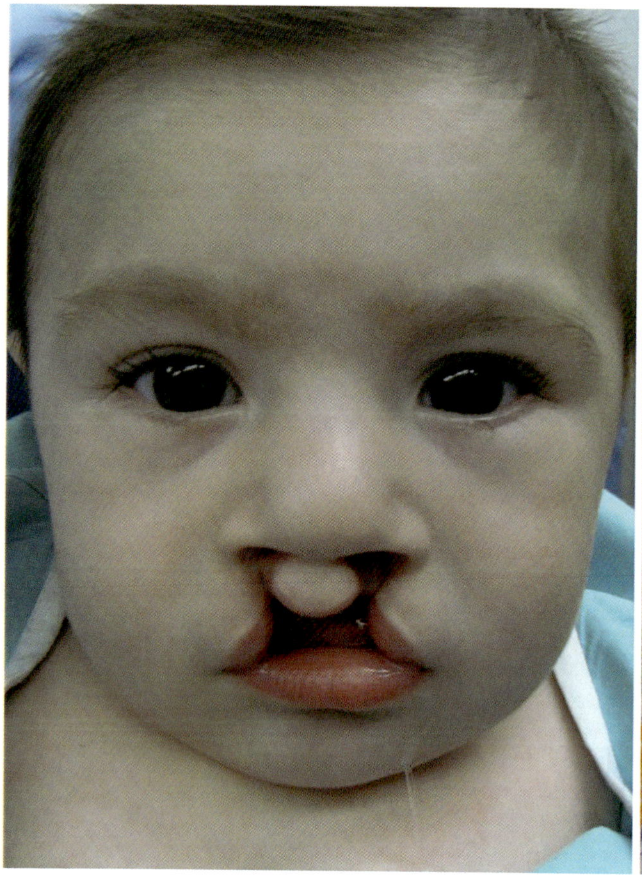

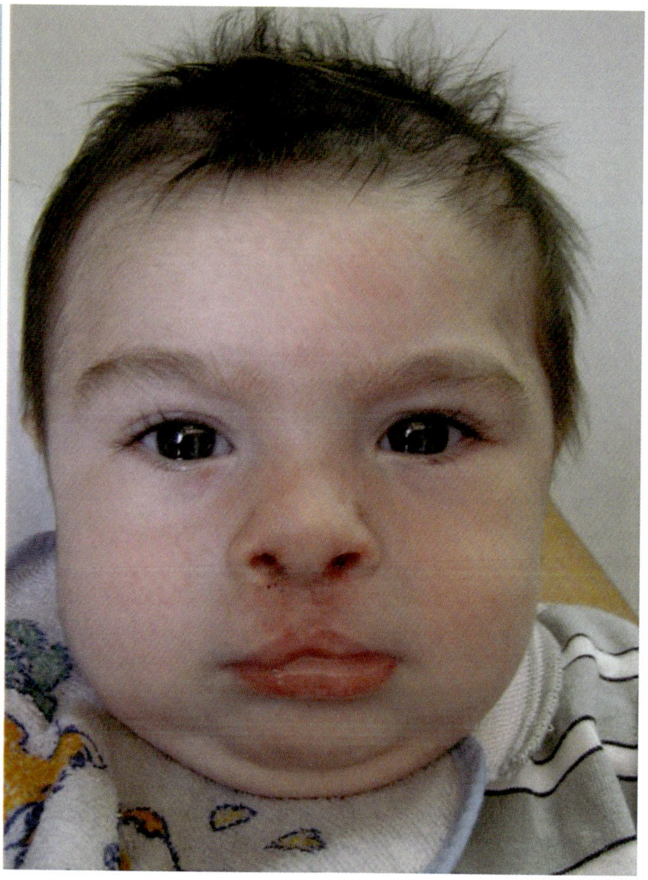

Case 1 Bilateral

Single AEP flap reconstruction (left)

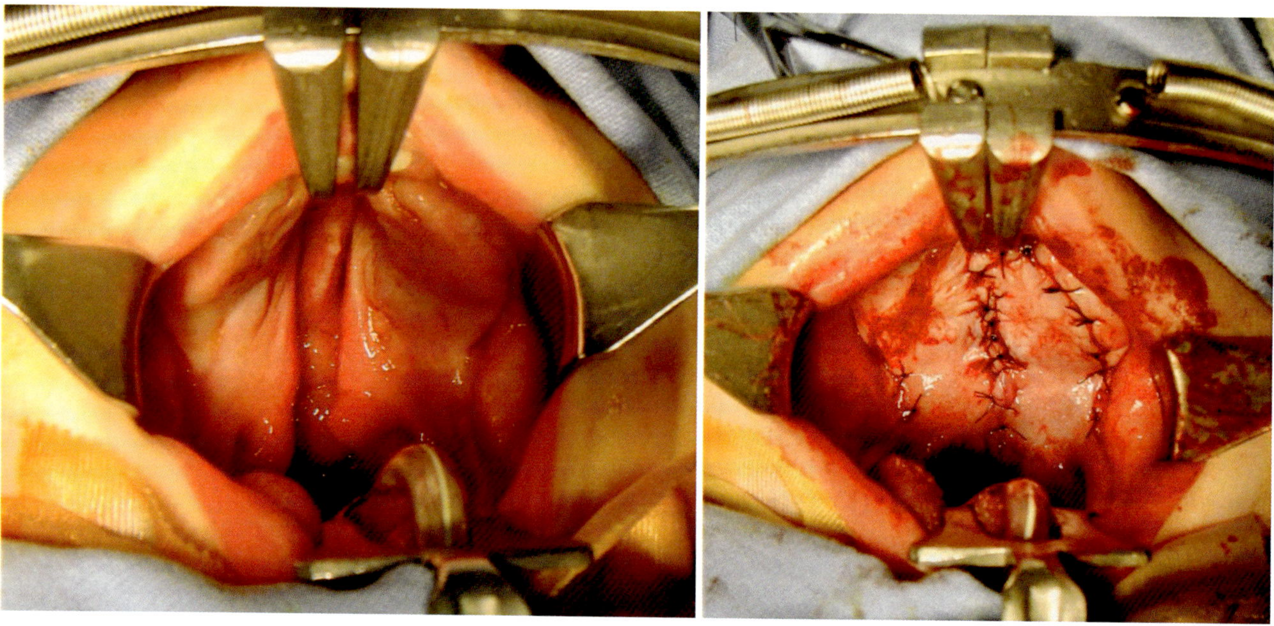

Case 2 Bilateral

Primary cleft lip and nose reconstruction—immediate post op

Double AEP flap reconstruction

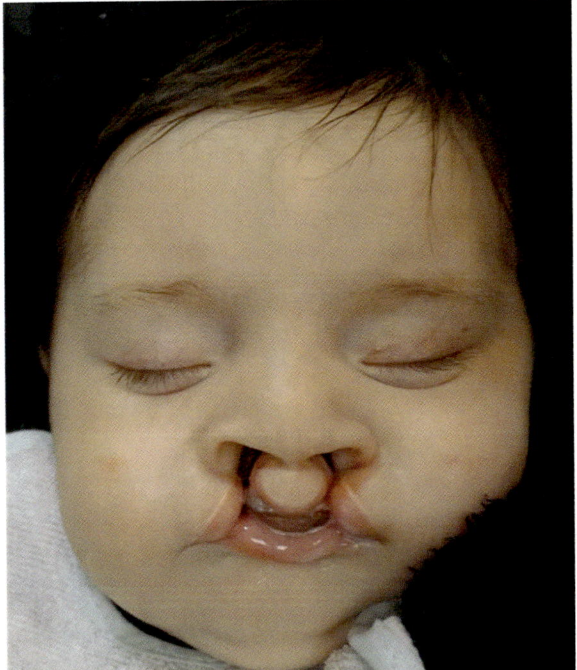

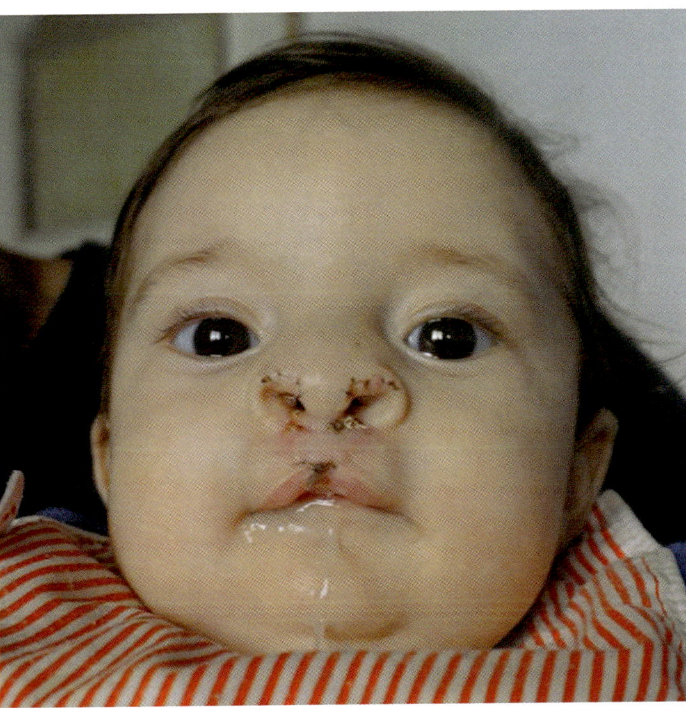

Case 3 Bilateral

Primary cleft lip and nose reconstruction—1 year post op

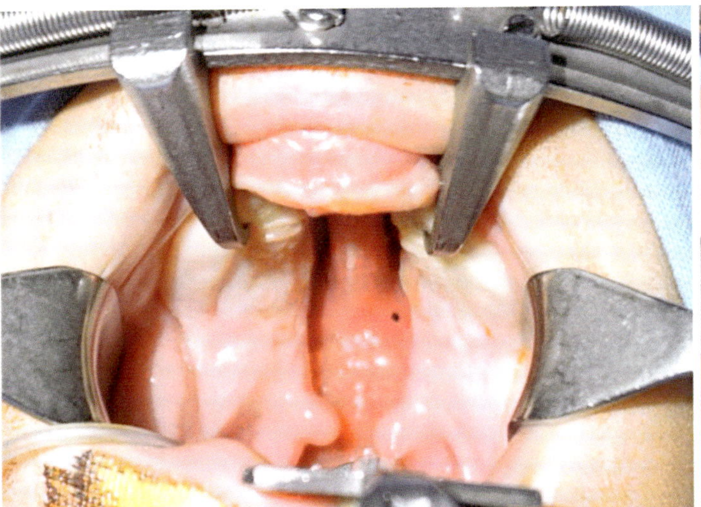

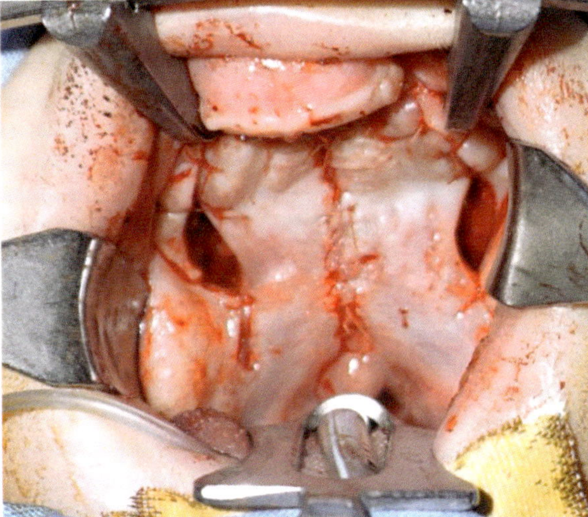

Case 3 Bilateral

Single AEP flap reconstruction (left)

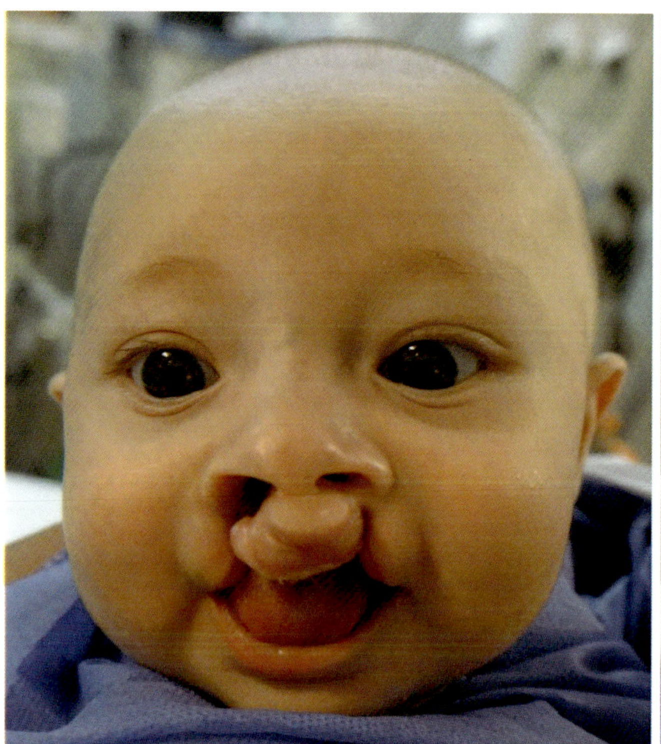

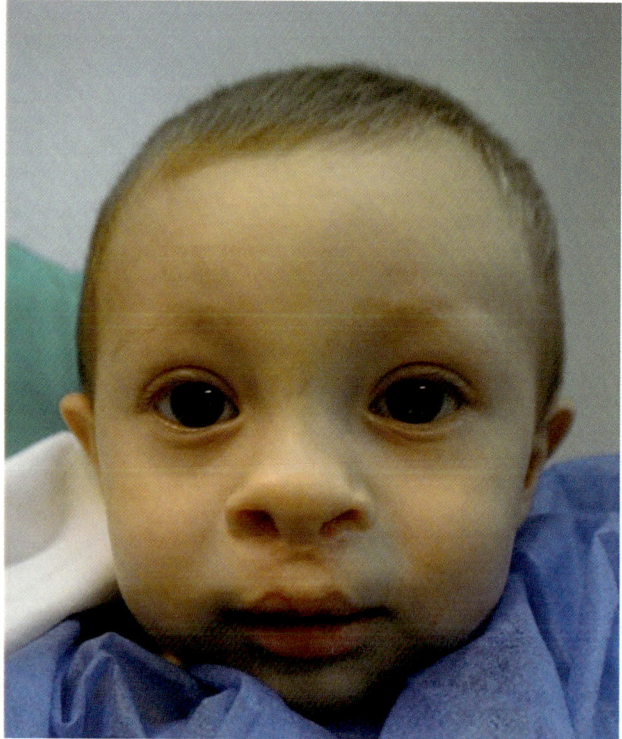

Case 3 Bilateral

3 months post palatoplasty and 2 years frontal view

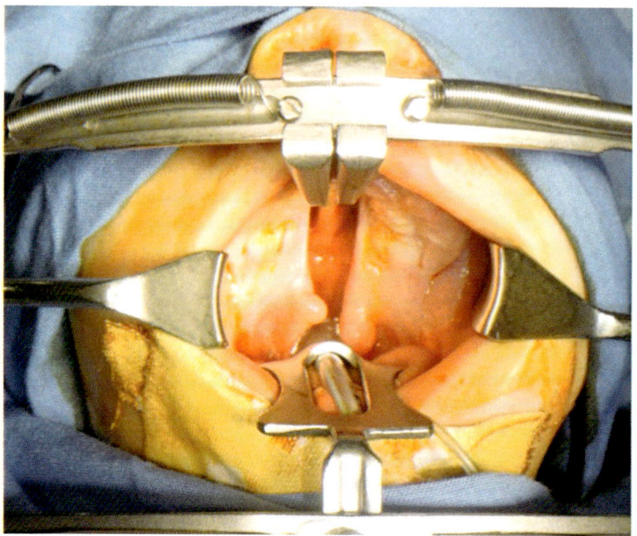

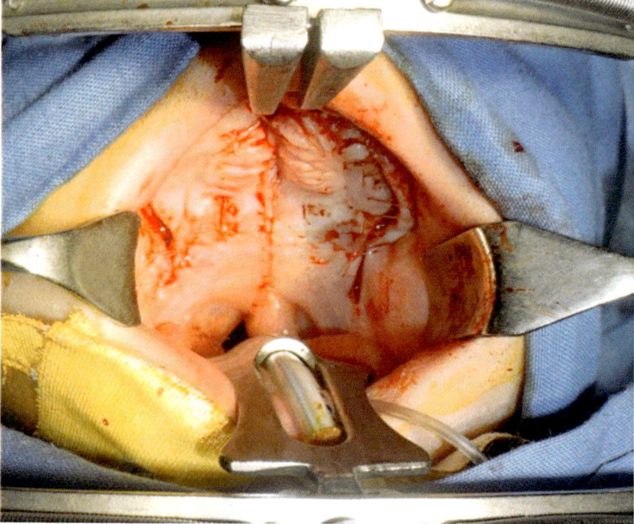

Case 4 Bilateral

Primary cleft lip and nose reconstruction—1 year post op

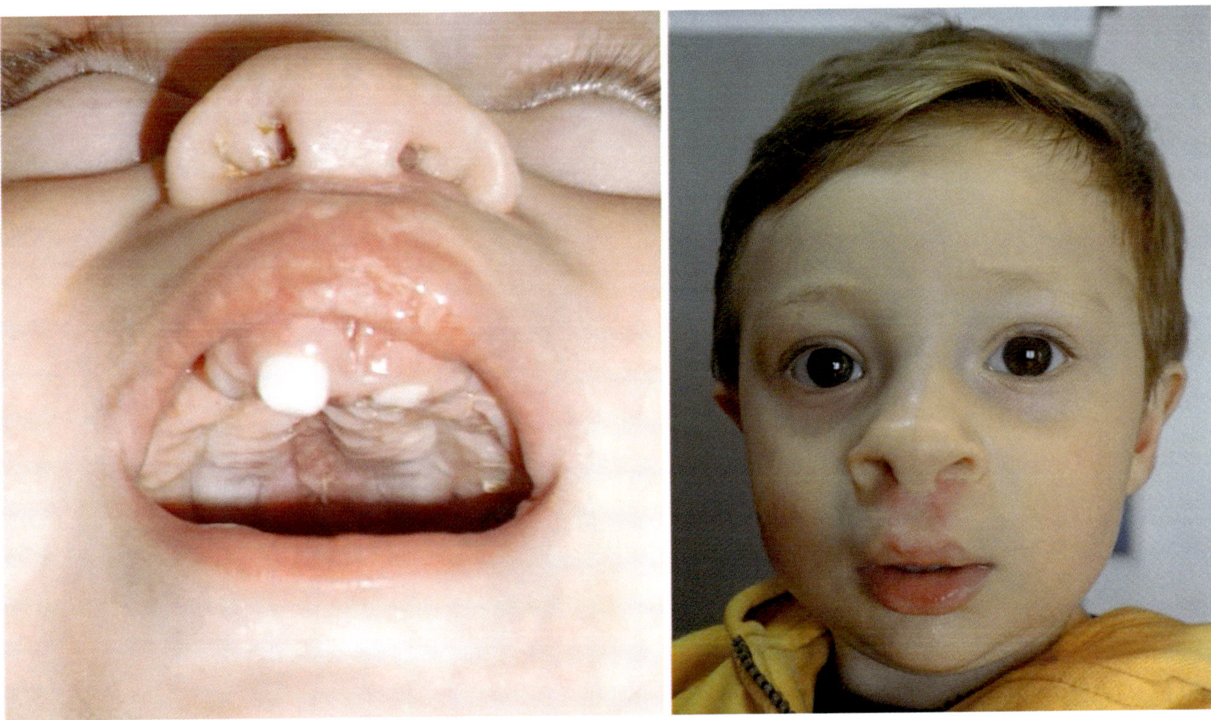

Case 4 Bilateral

Single AEP flap reconstruction (right)

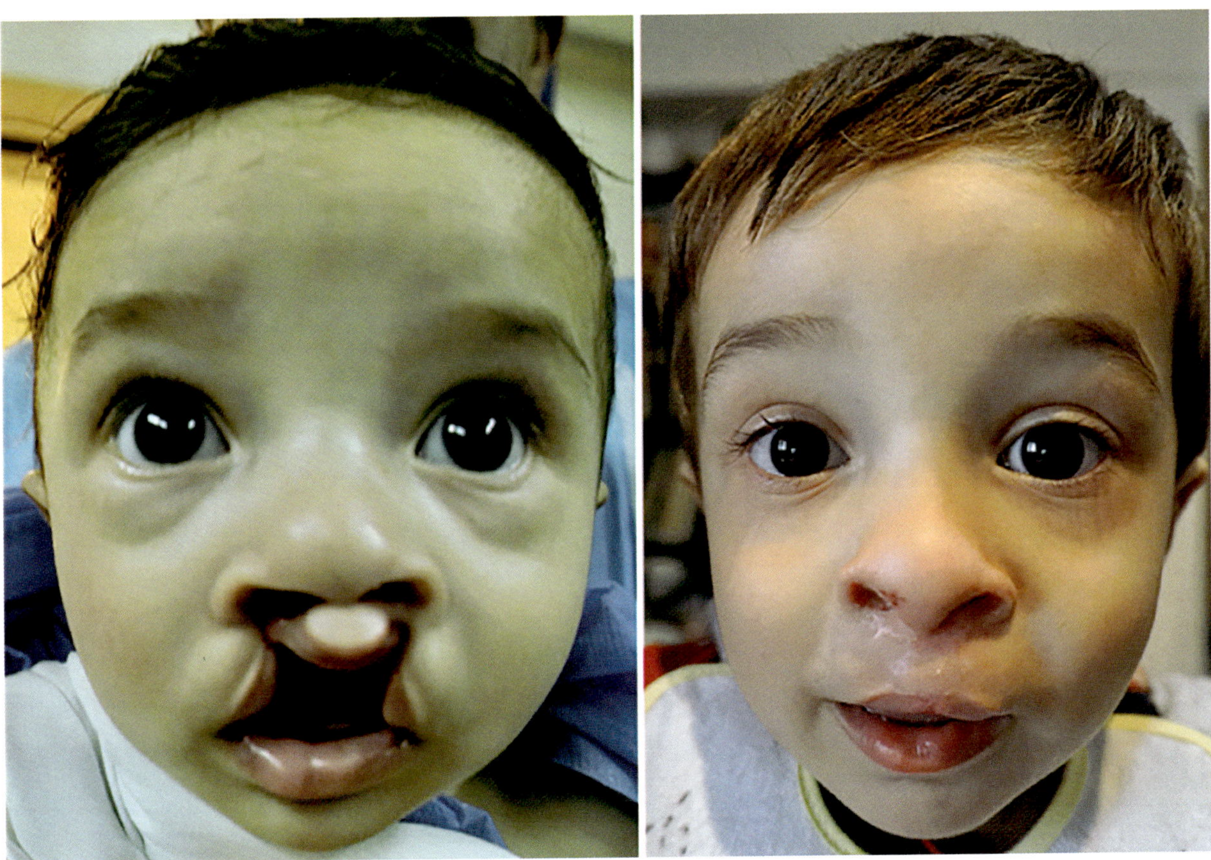

Case 4 Bilateral
Follow up 3 years post op

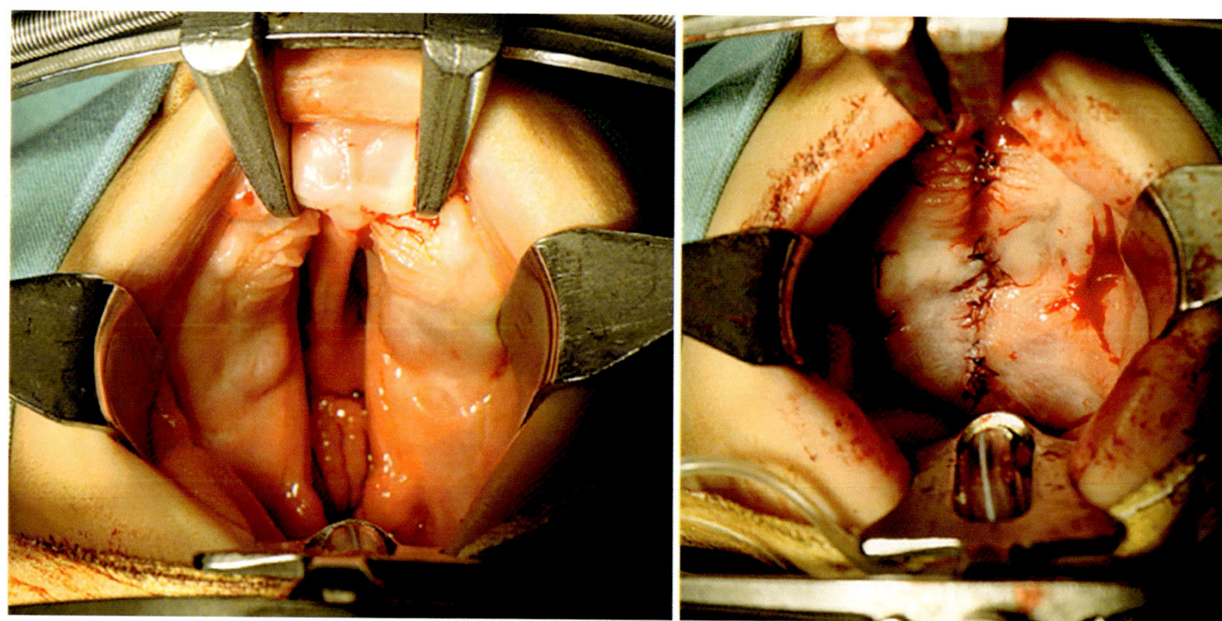

Case 4 Bilateral
3 years post op palatoplasty

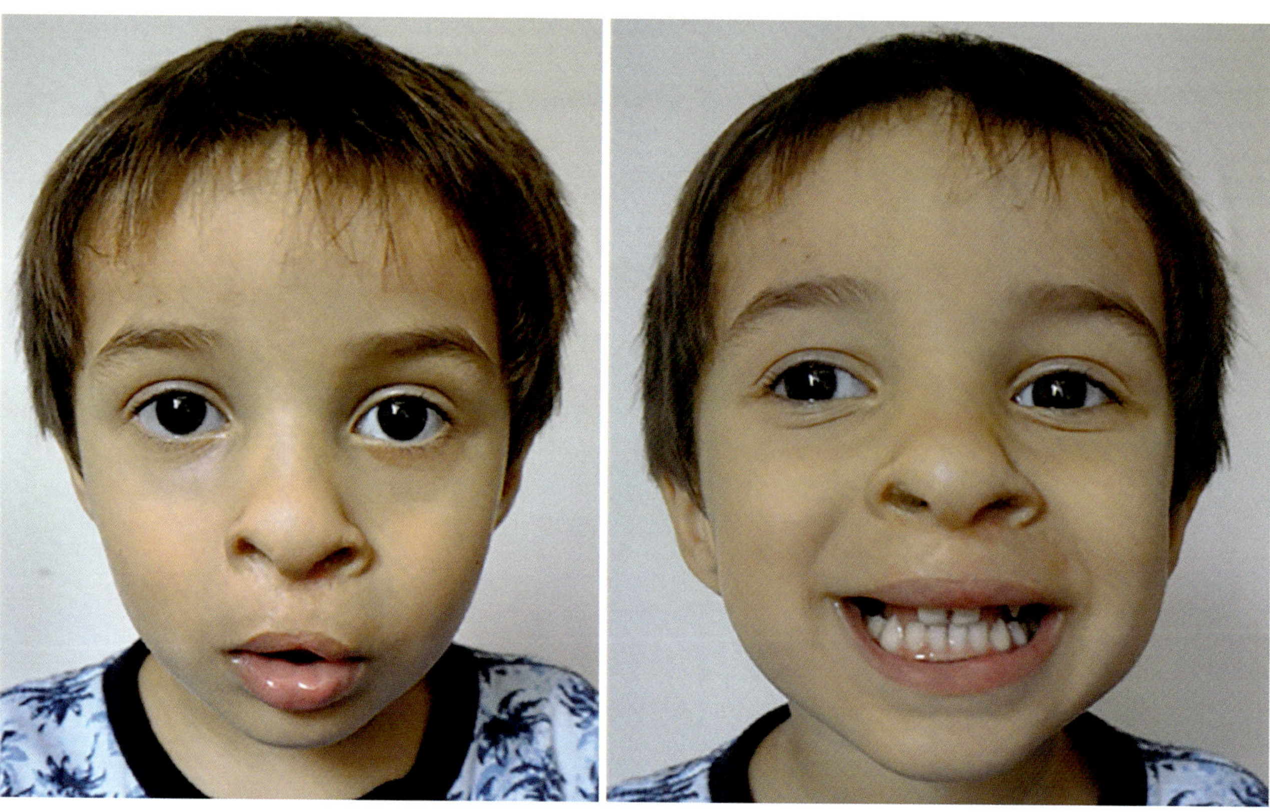

Case 5 Bilateral

Primary cleft lip and nose reconstruction—1 year follow up

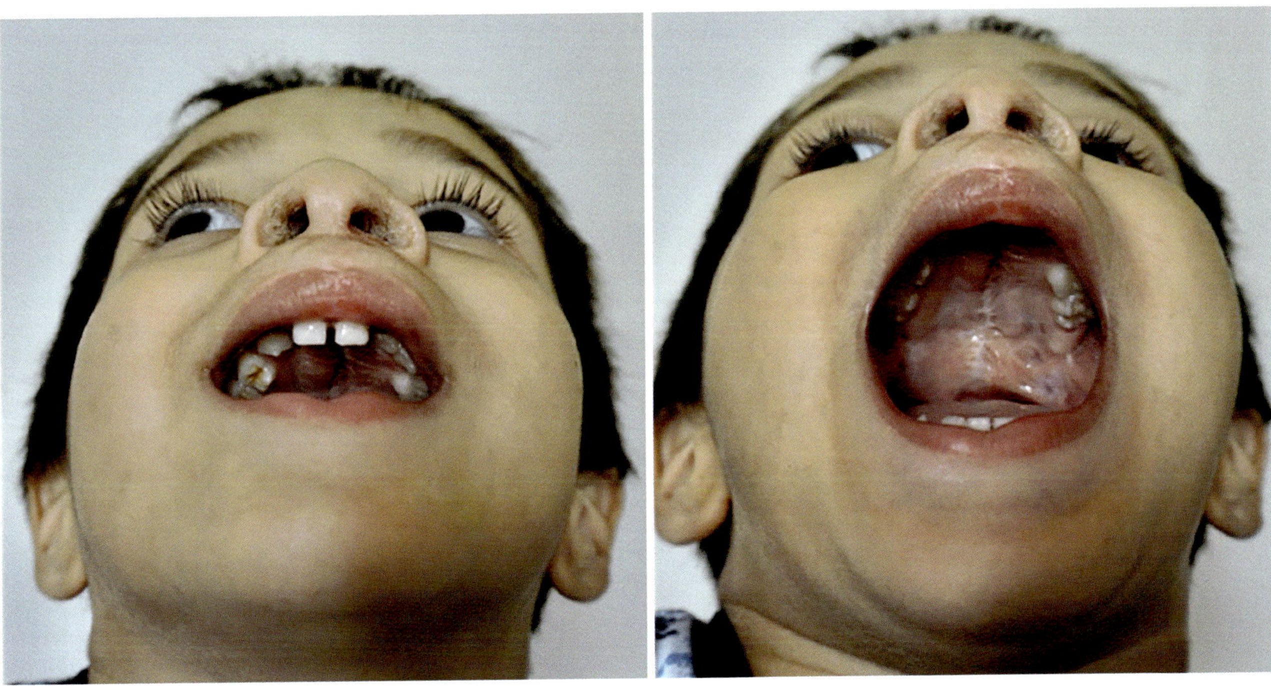

Case 5 Bilateral

Single AEP flap reconstruction (left)

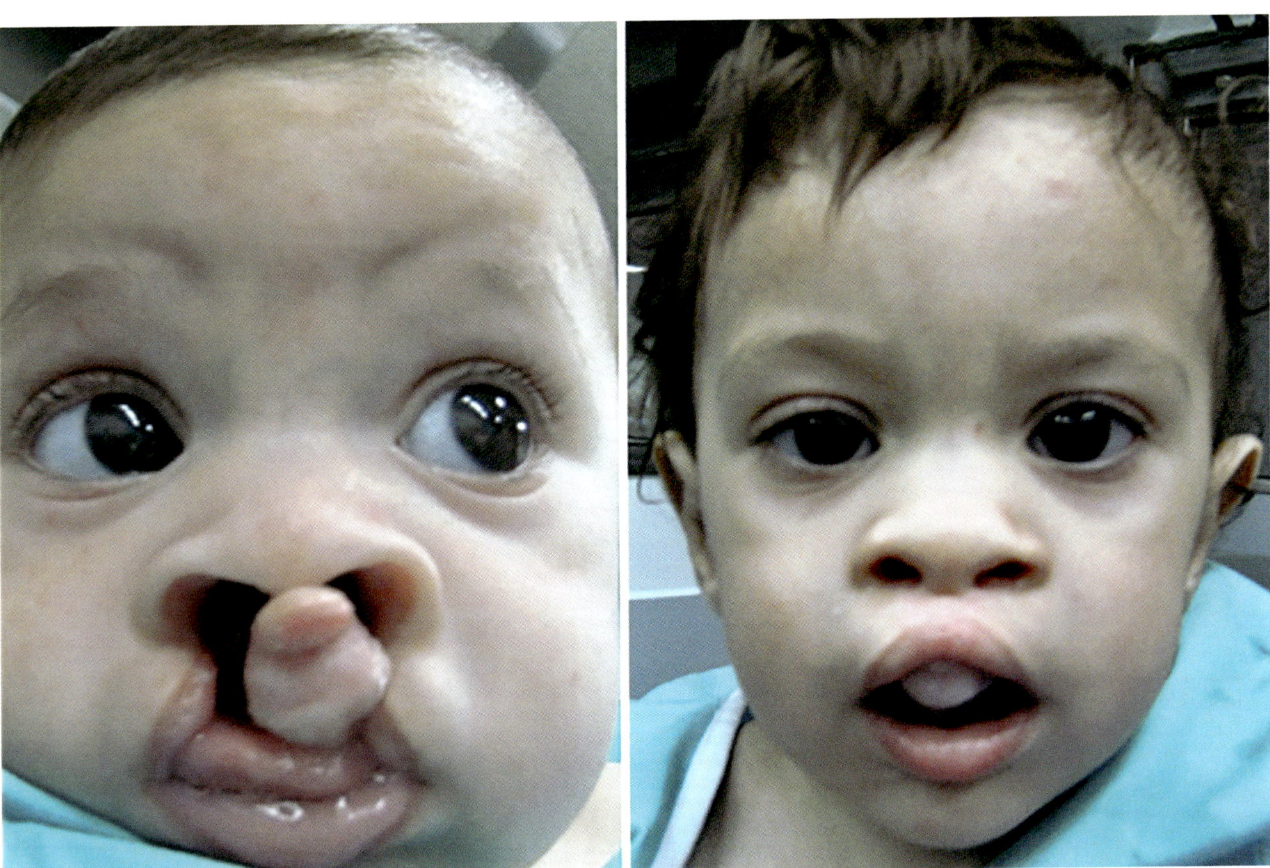

Case 5 Bilateral

6 years post palatoplasty

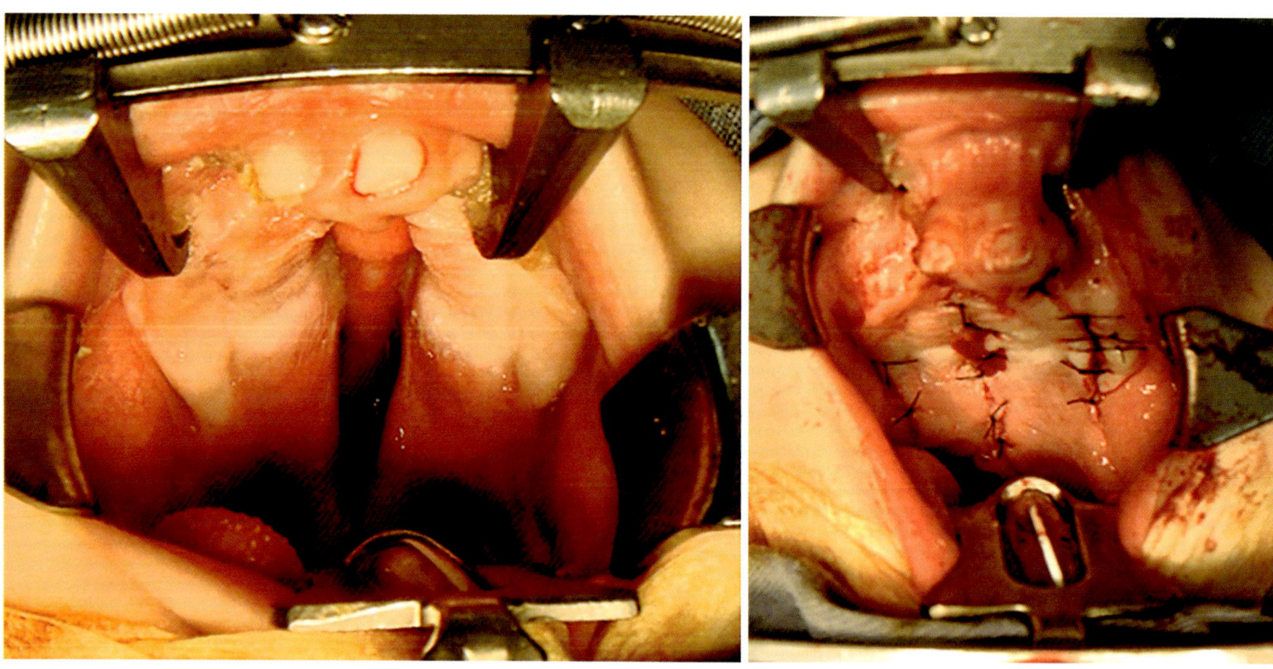

Case 6 Bilateral

Primary cleft lip and nose reconstructions—18 months follow up

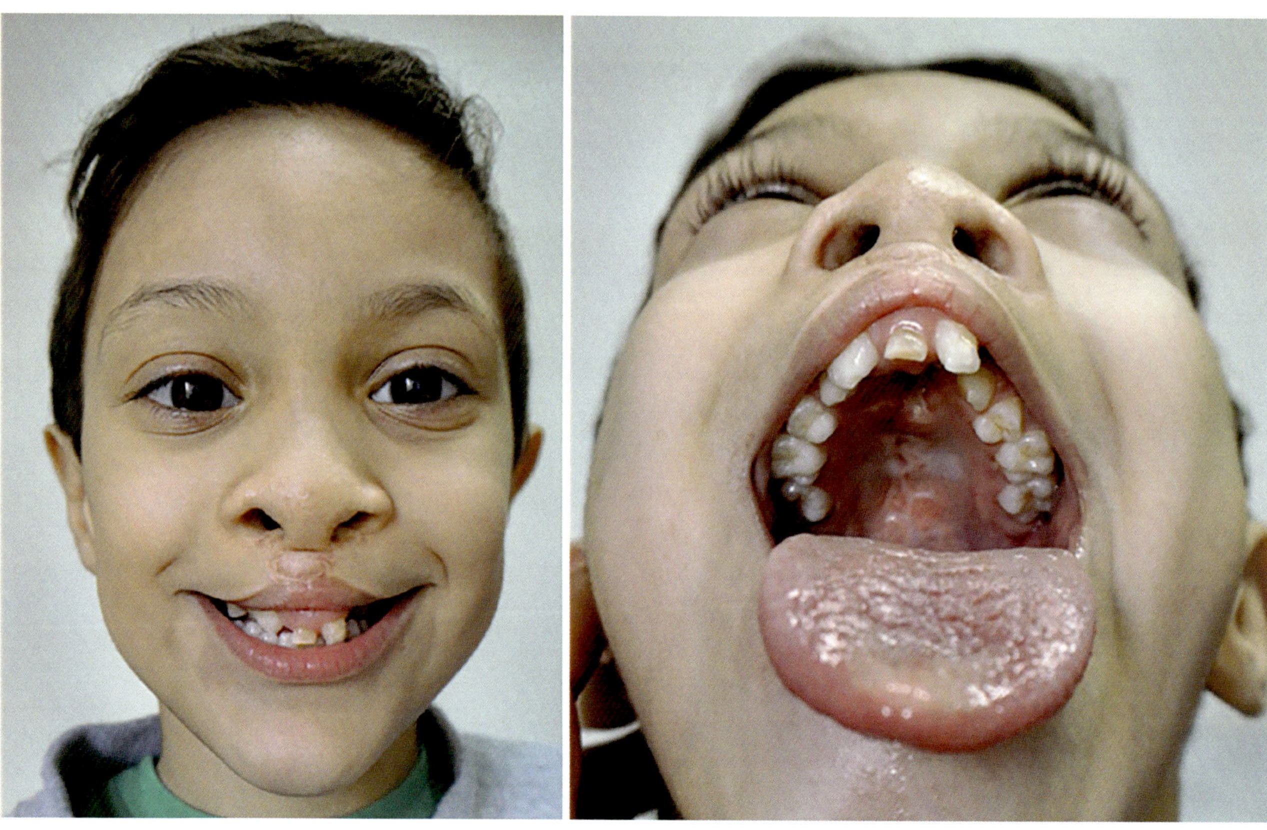

Case 6 Bilateral
Double AEP flap reconstruction

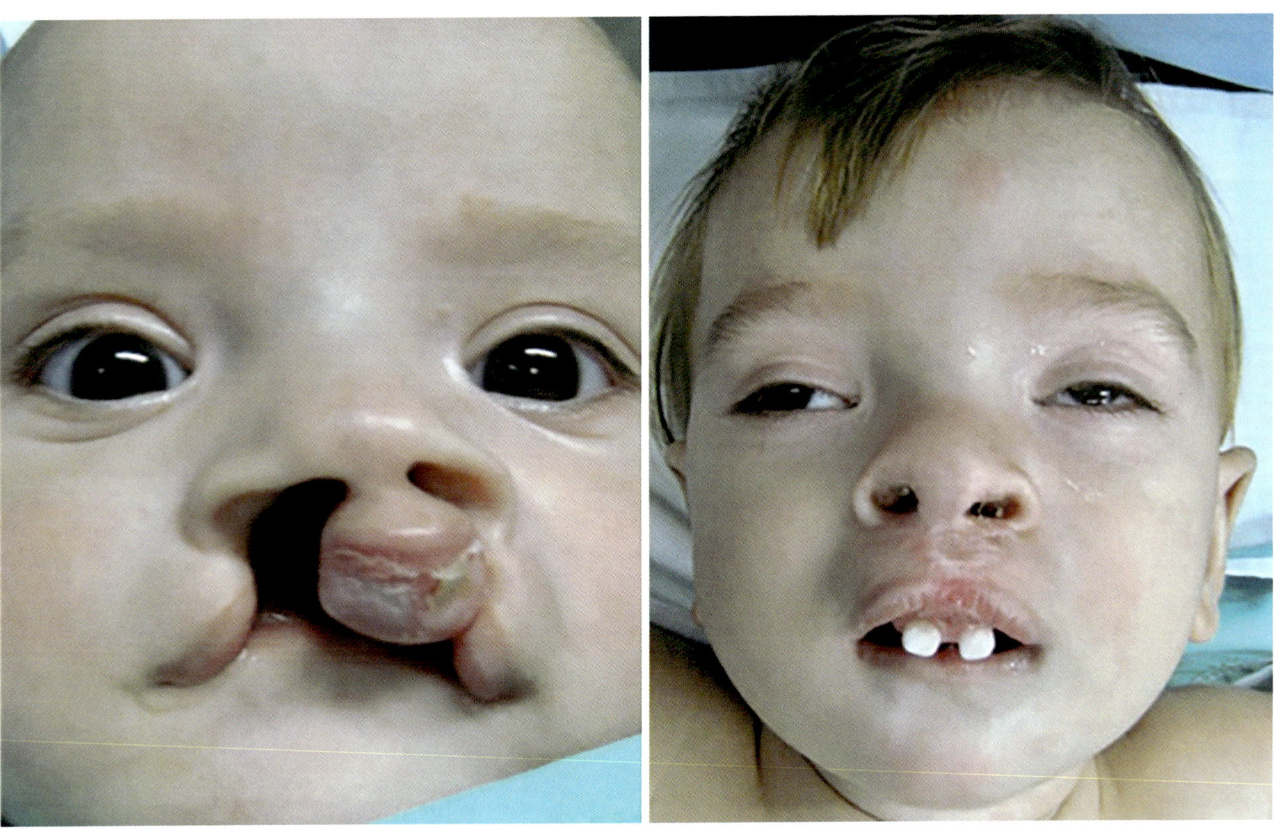

Case 6 Bilateral
7 year post palatoplasty

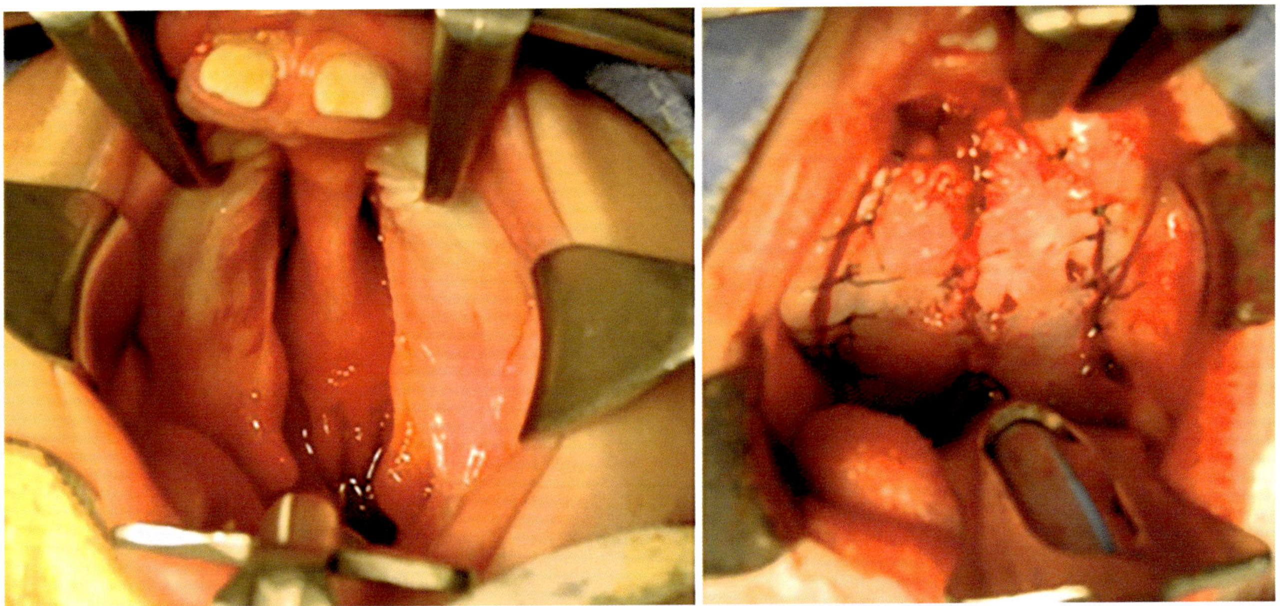

Case 7 Bilateral
Primary cleft lip and nose reconstruction—18 months follow up

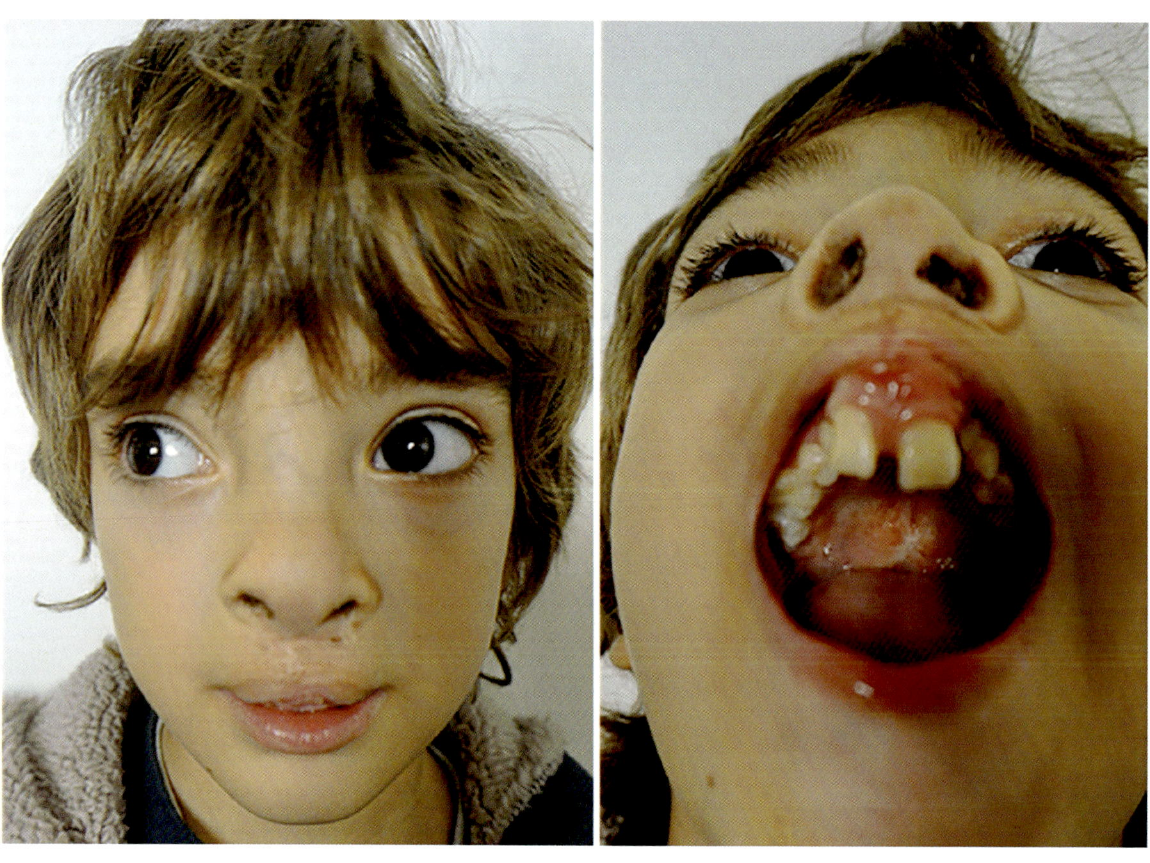

Case 7 Bilateral
Double AEP flap reconstruction

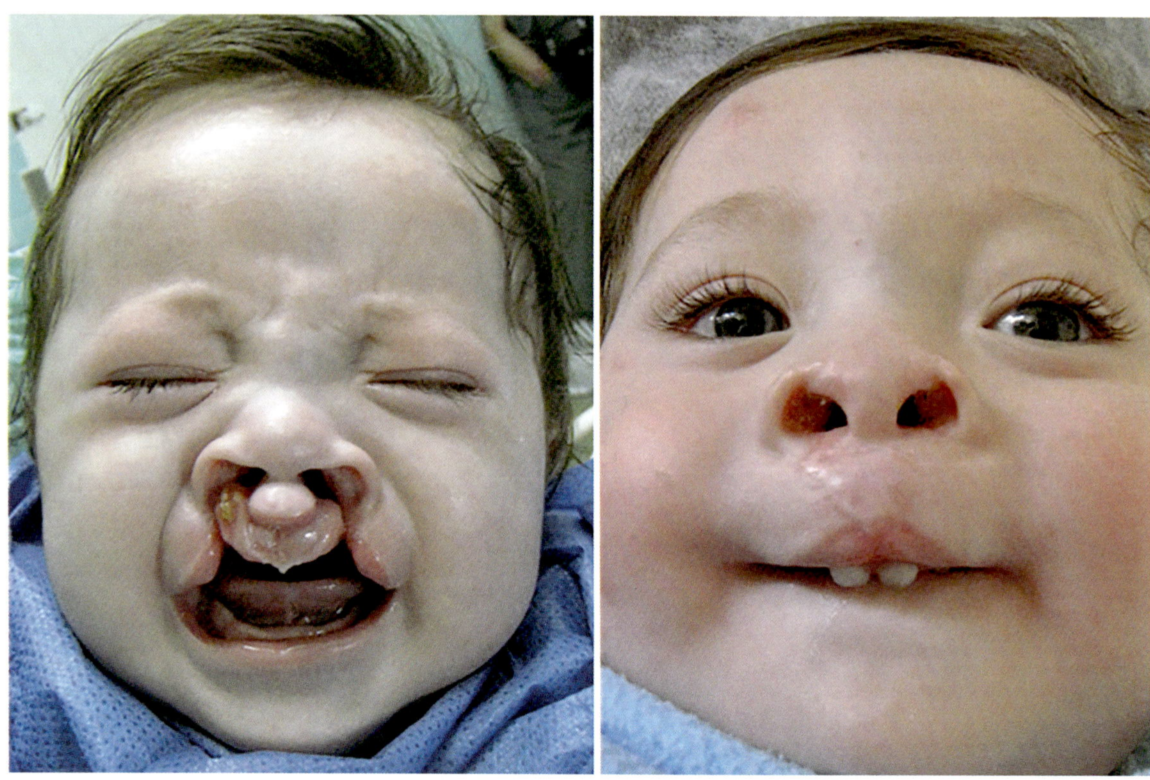

Case 7 Bilateral
8 years post palatoplasty

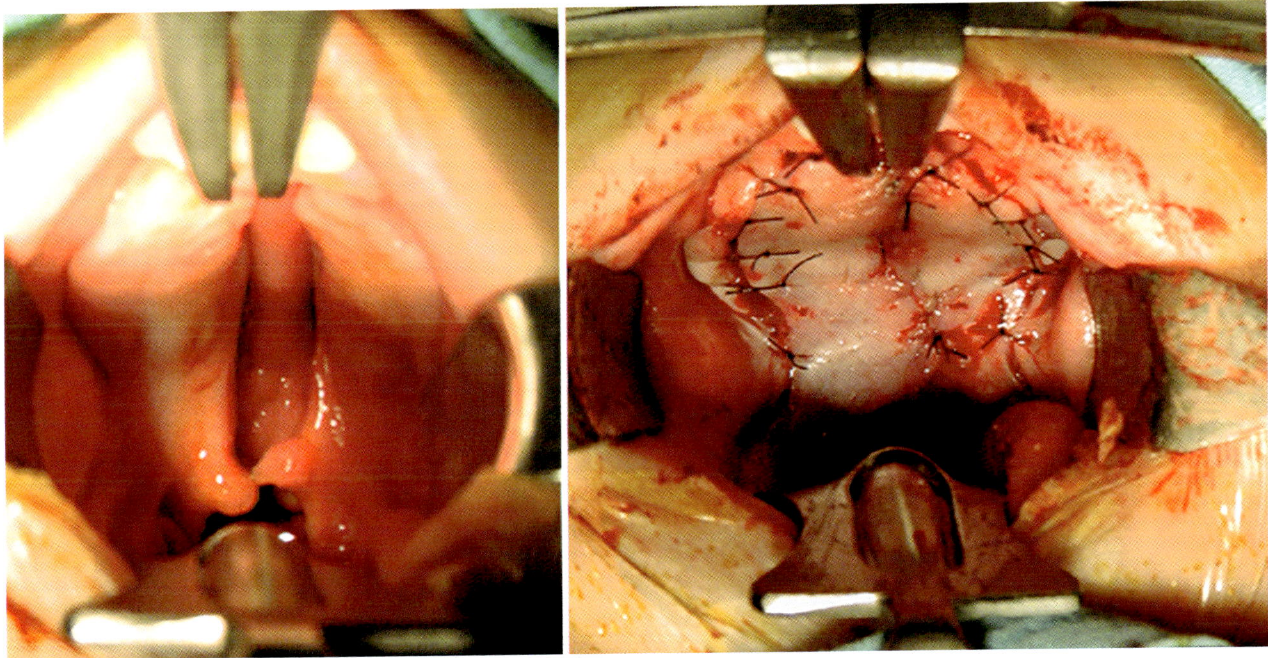

Case 7 Bilateral
8 years post palatoplasty

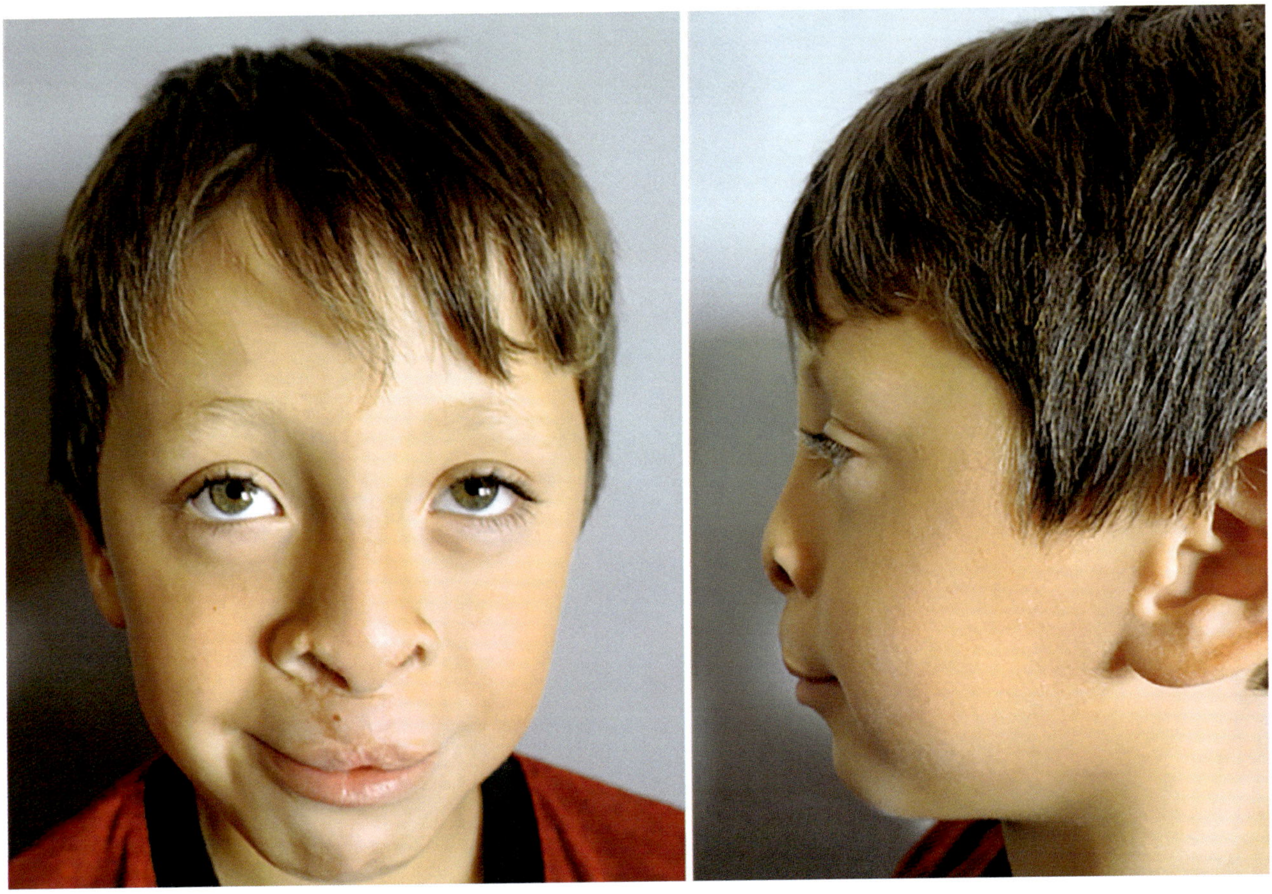

Case 8 Bilateral

Cleft lip and nose reconstruction reconstruction—immediate post op

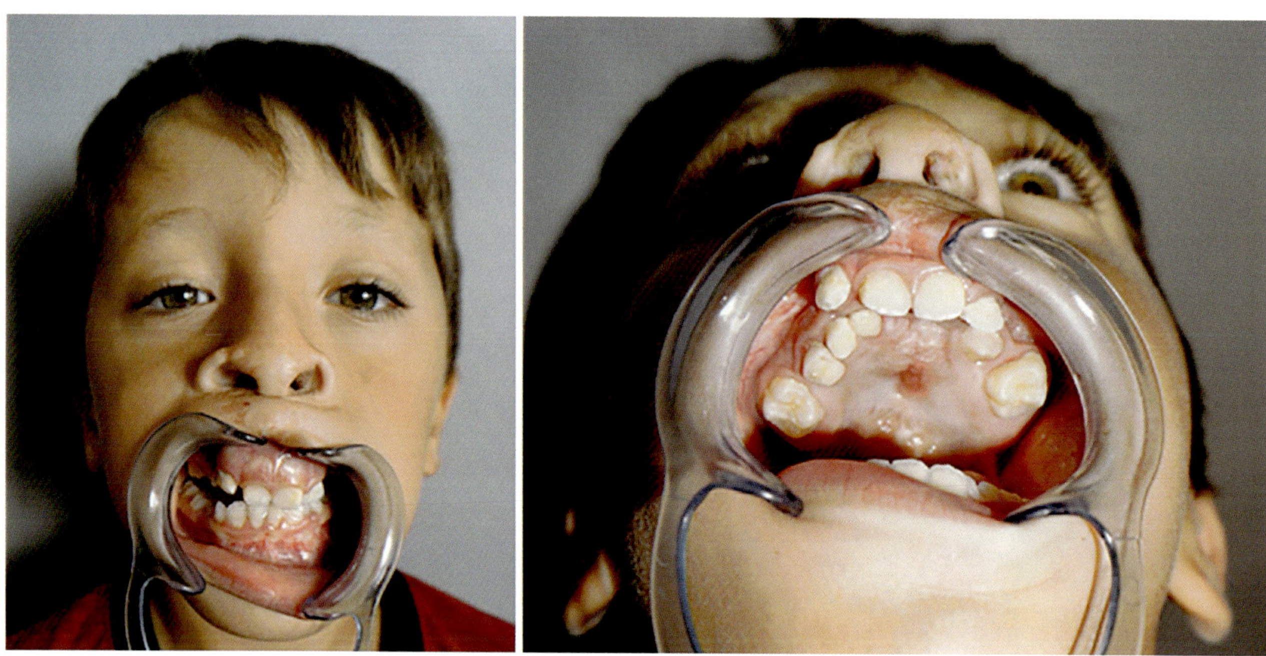

Case 8 Bilateral

Single AEP flap reconstruction (right)

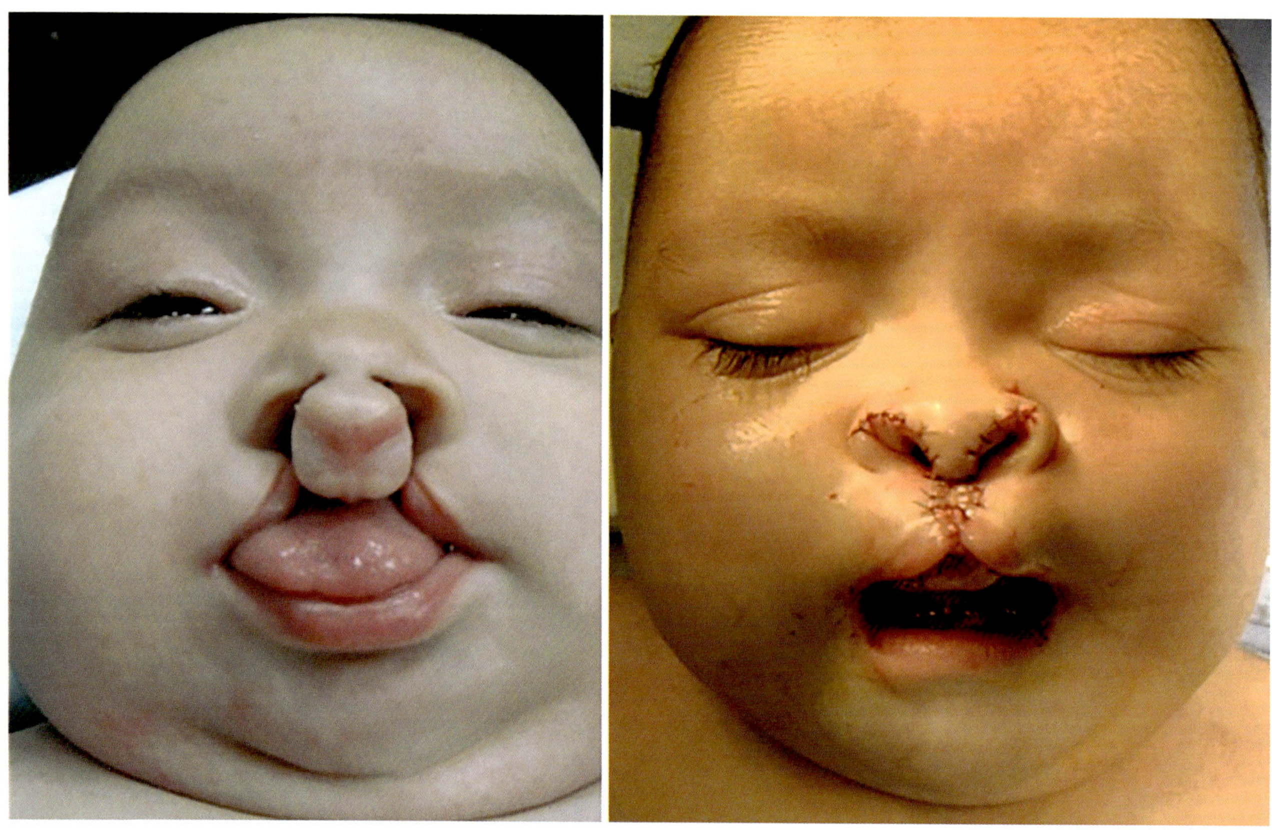

Case 8 Bilateral
9 years post palatoplasty

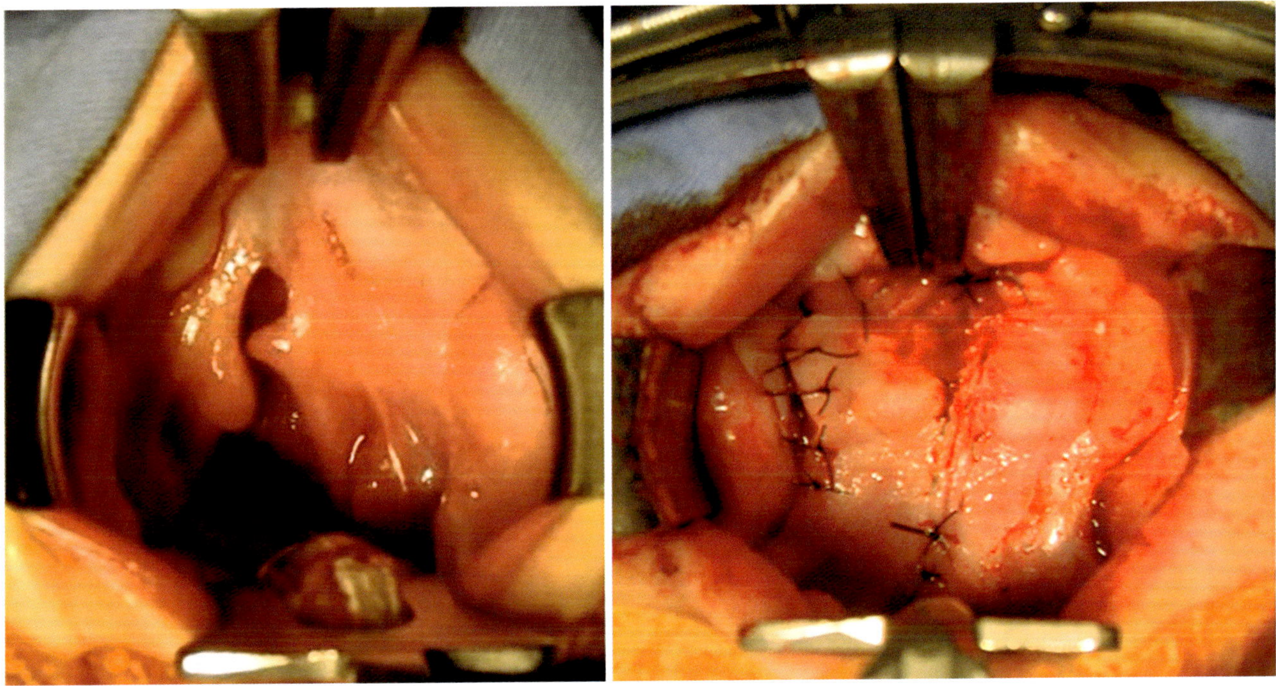

Case 8 Bilateral
9 years post palatoplasty inferior and lateral

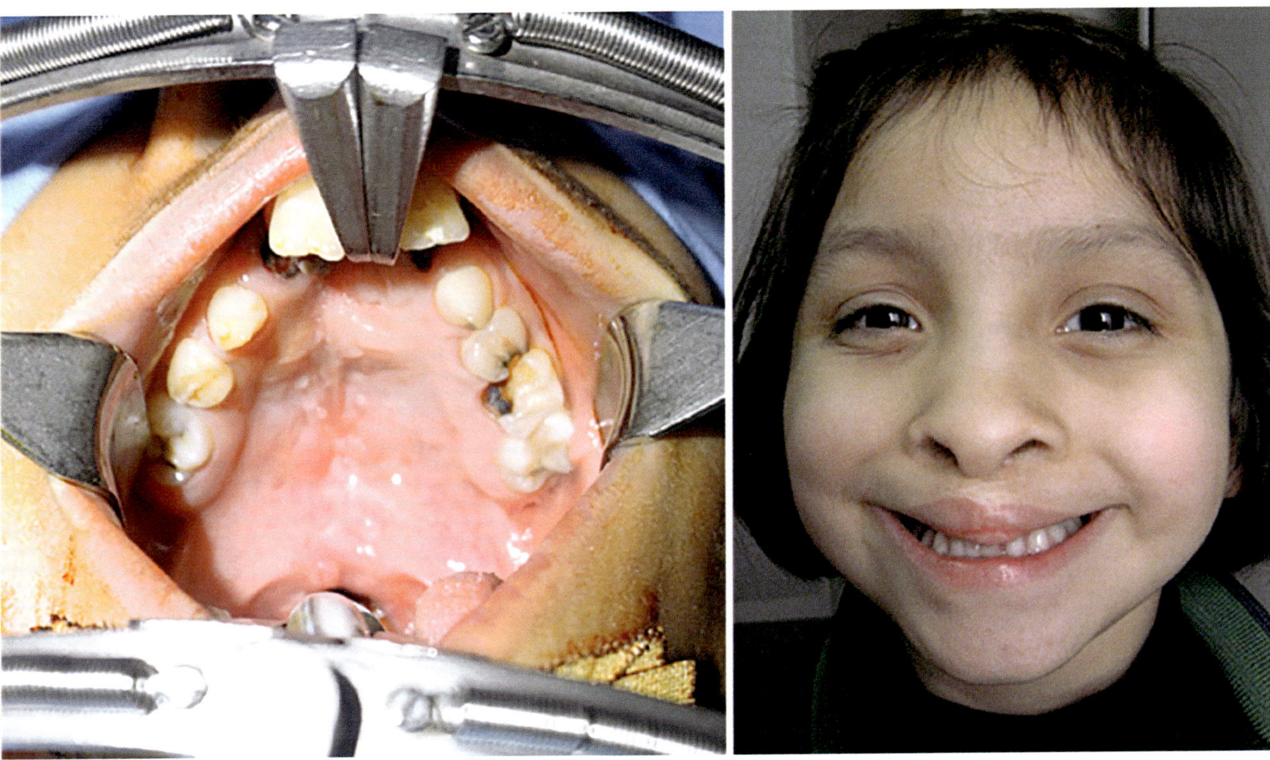

Case 1 Unilateral

Cleft lip and nose reconstruction—1 year follow-up

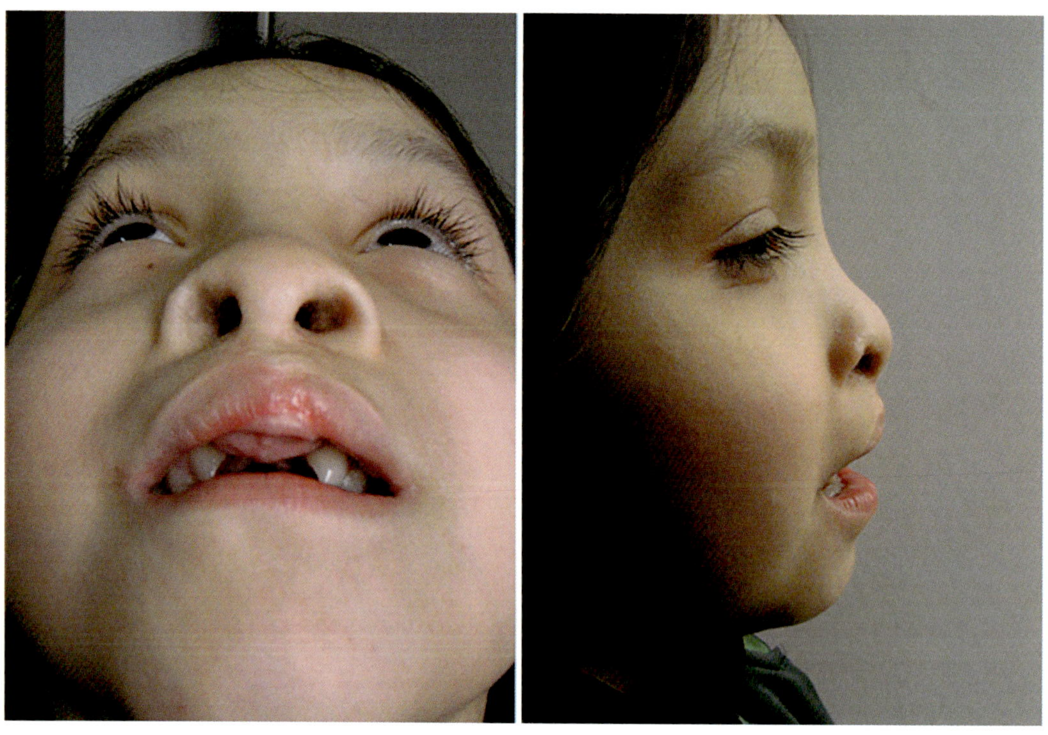

Case 1 Unilateral

Single AEP flap reconstruction (right)

1 month post palatoplasty

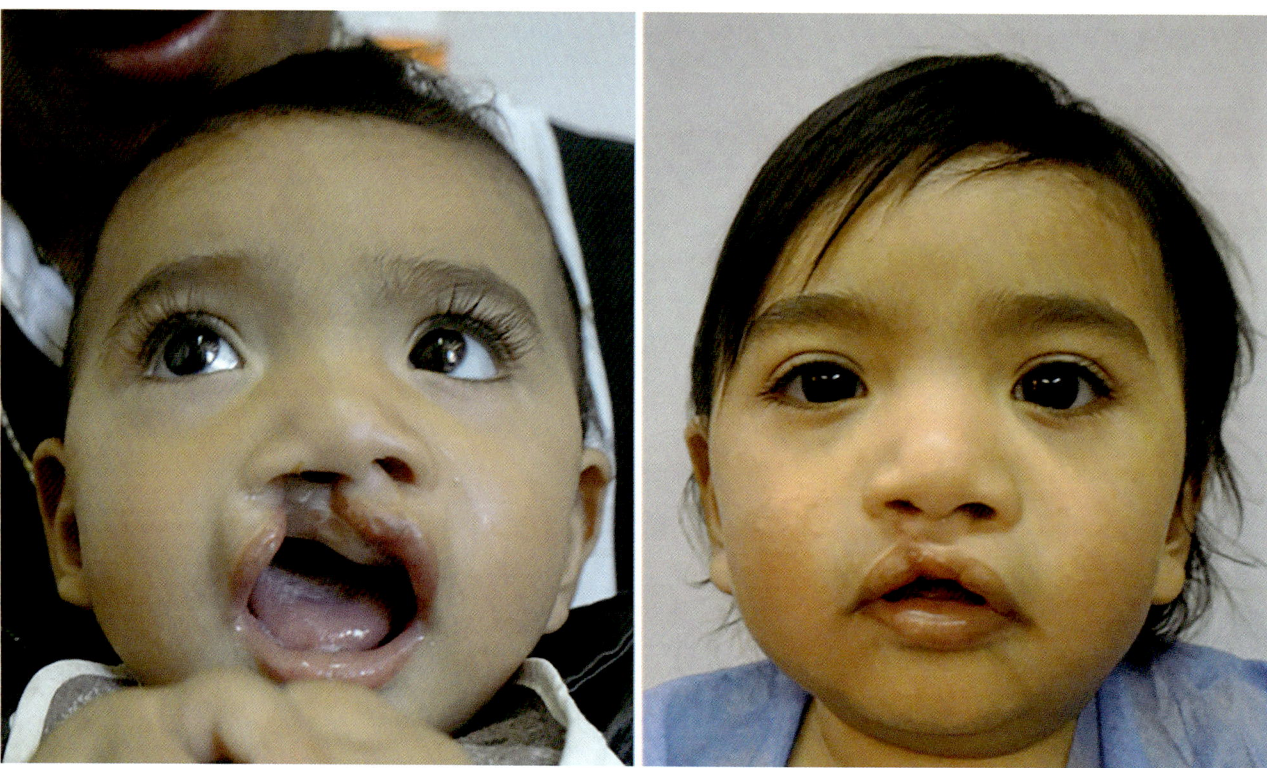

Case 1 Unilateral

Single AEP flap reconstruction (right)

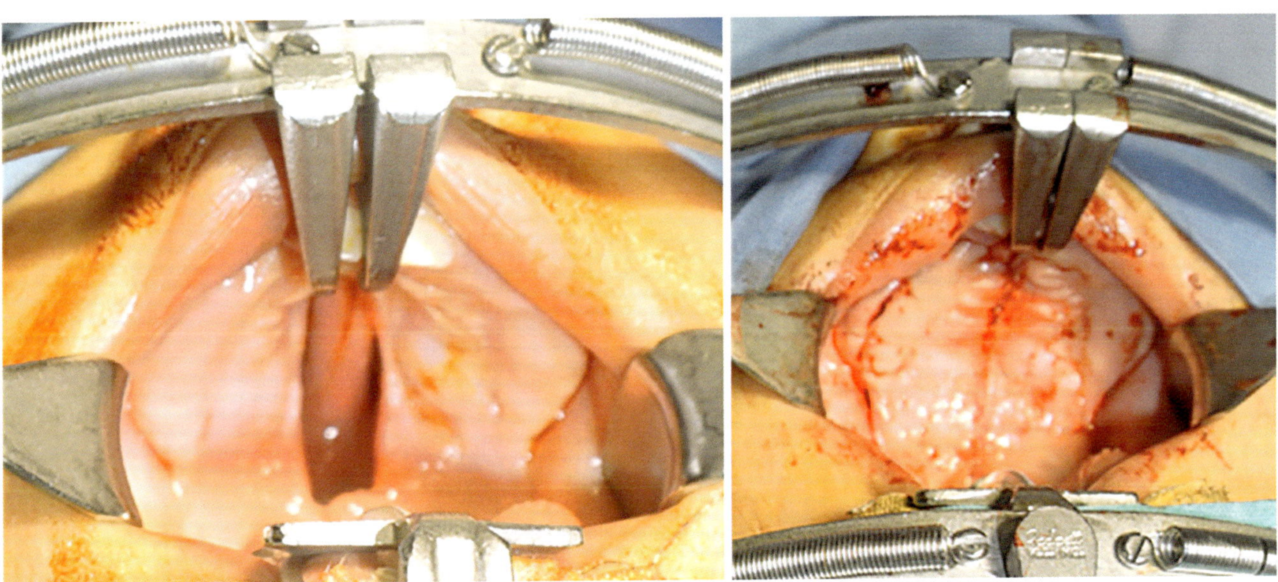

Case 2 Unilateral

Cleft lip and nose reconstruction—1 year follow up

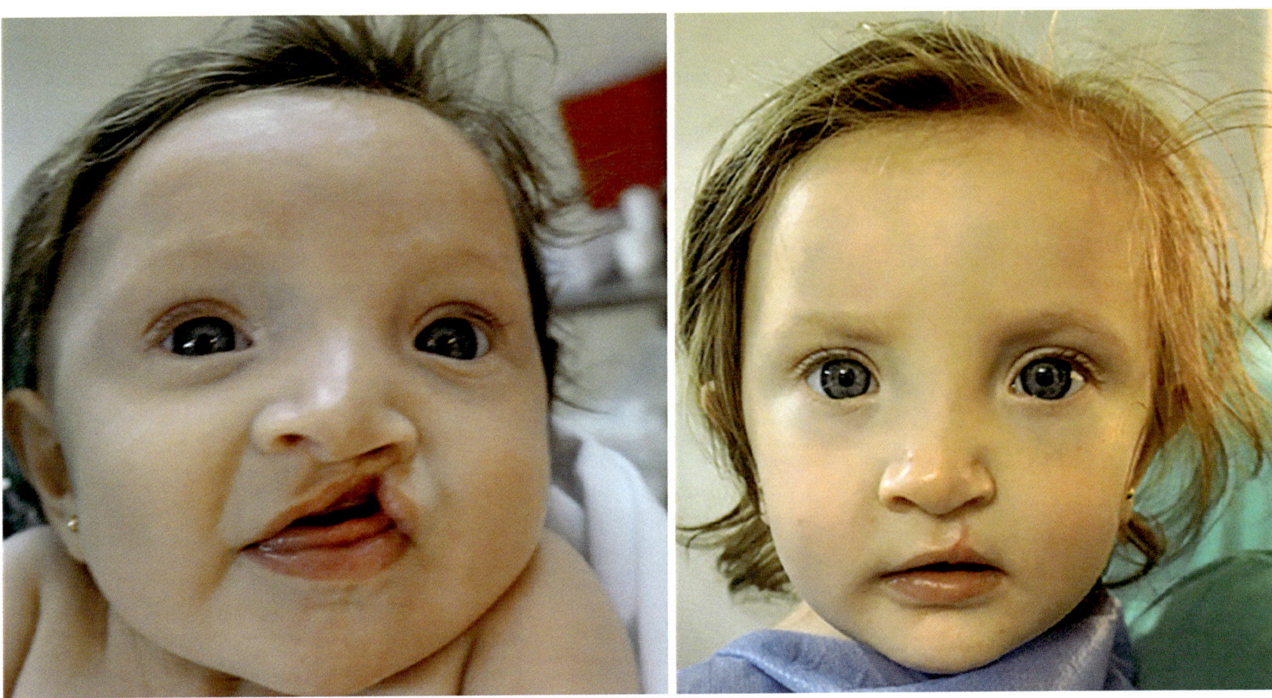

Case 2 Unilateral
Single AEP flap reconstruction (left)

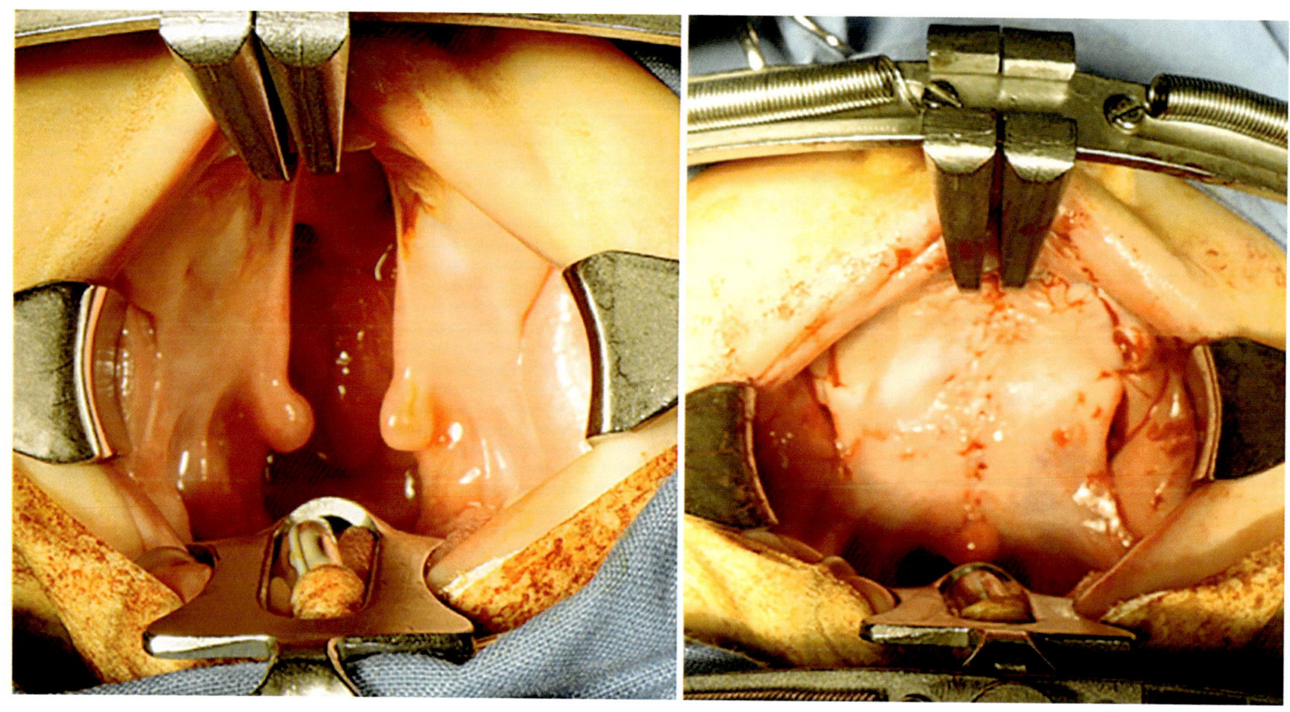

Case 1 Unilateral
3 months post palatoplasty

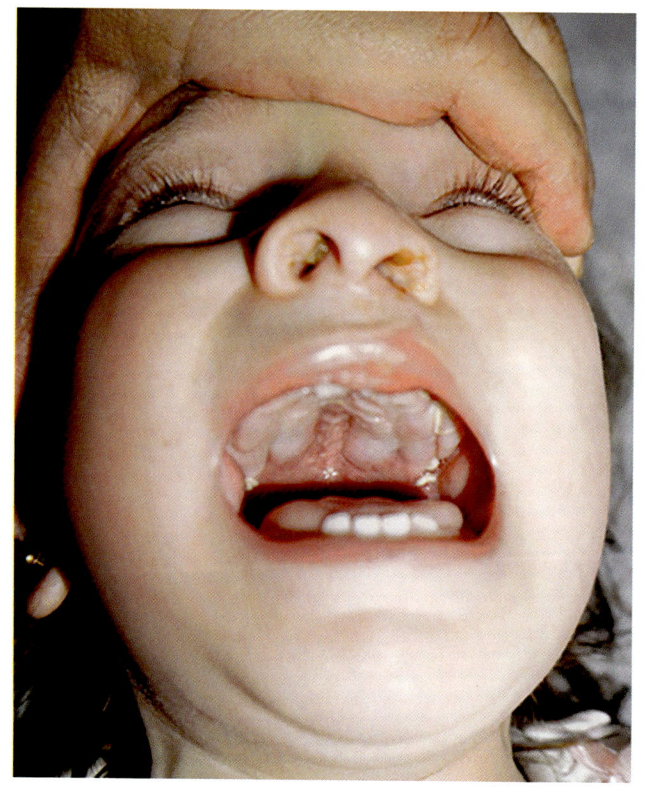

Case 3 Unilateral

Cleft lip and nose reconstruction—1 year follow-up

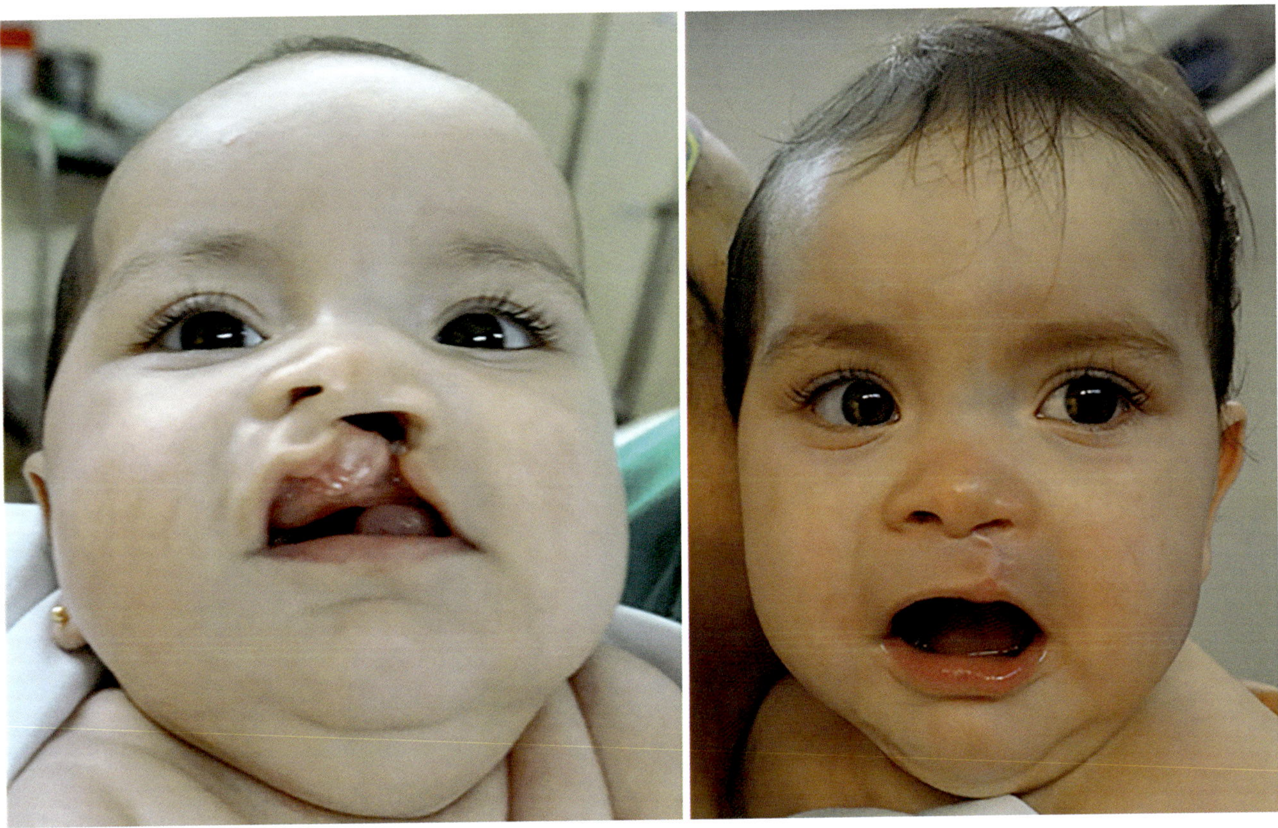

Case 3 Unilateral

Single AEP flap reconstruction (left)

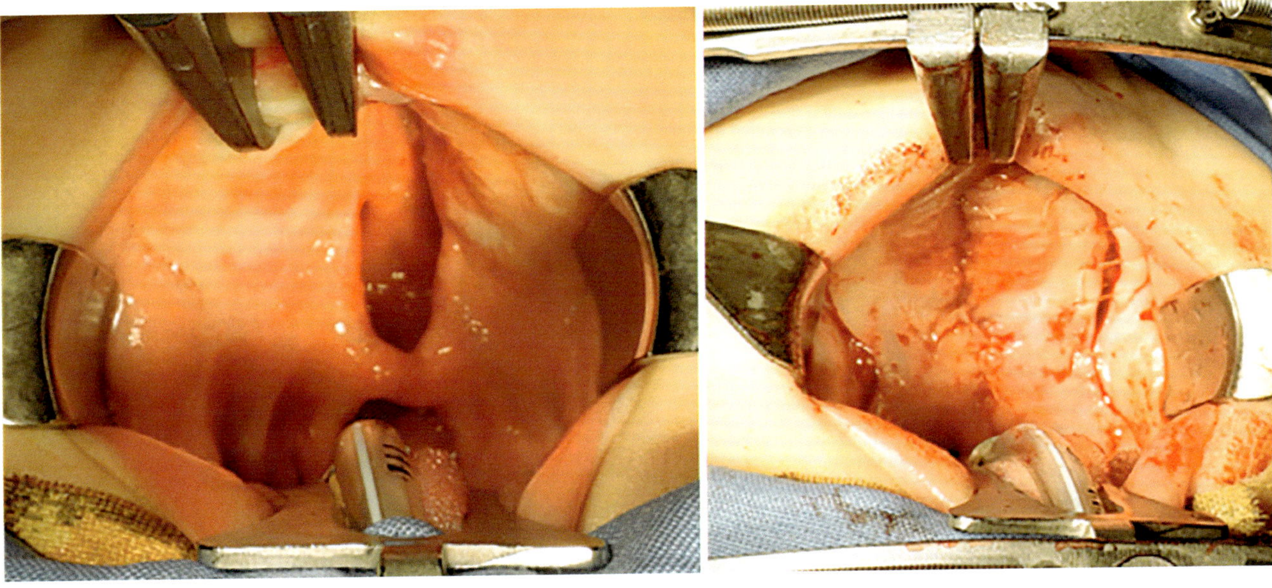

Case 3 Unilateral

2 years post palatoplasty

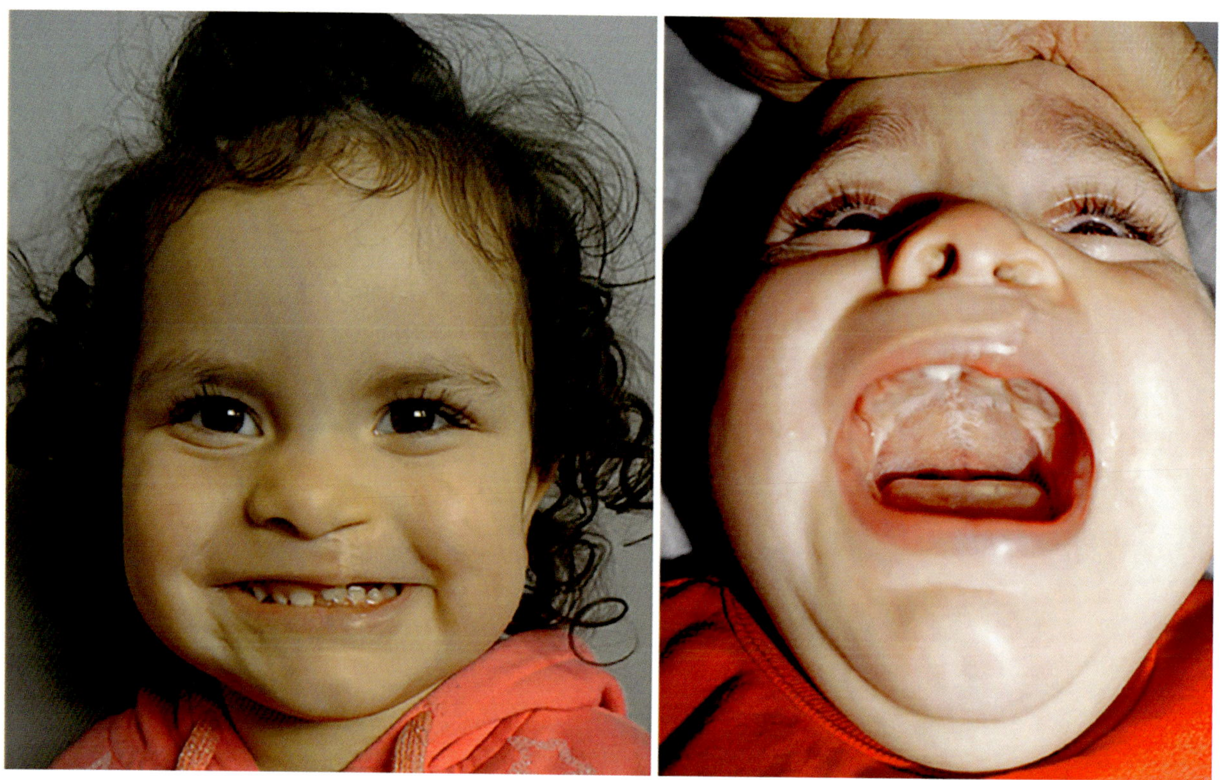

Case 4 Unilateral

Cleft lip and nose reconstruction—3 years post op

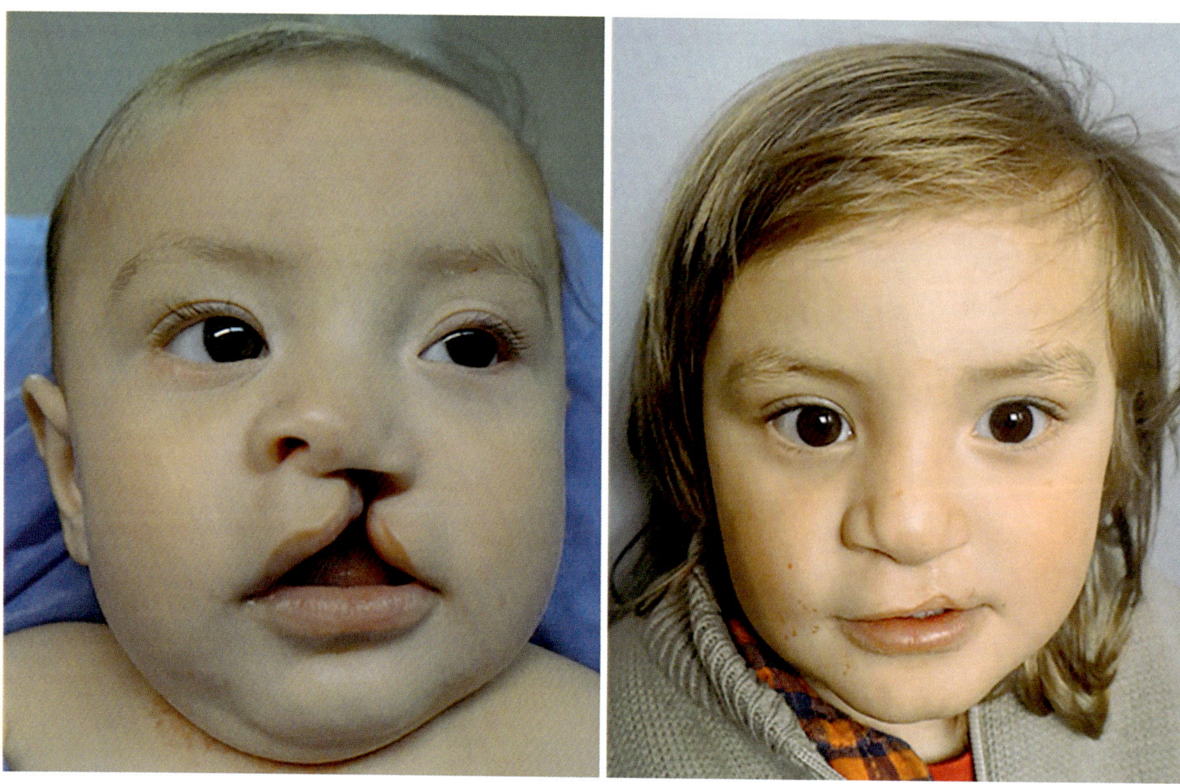

Case 4 Unilateral

Single AEP flap (left) reconstruction

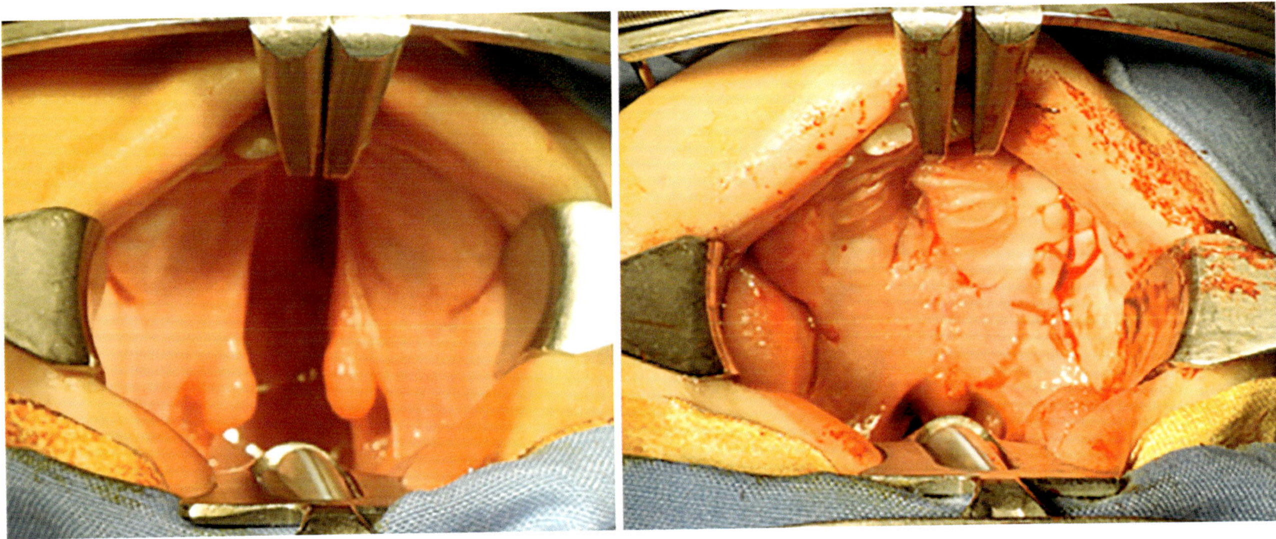

Case 4 Unilateral

3 years post palatoplasty

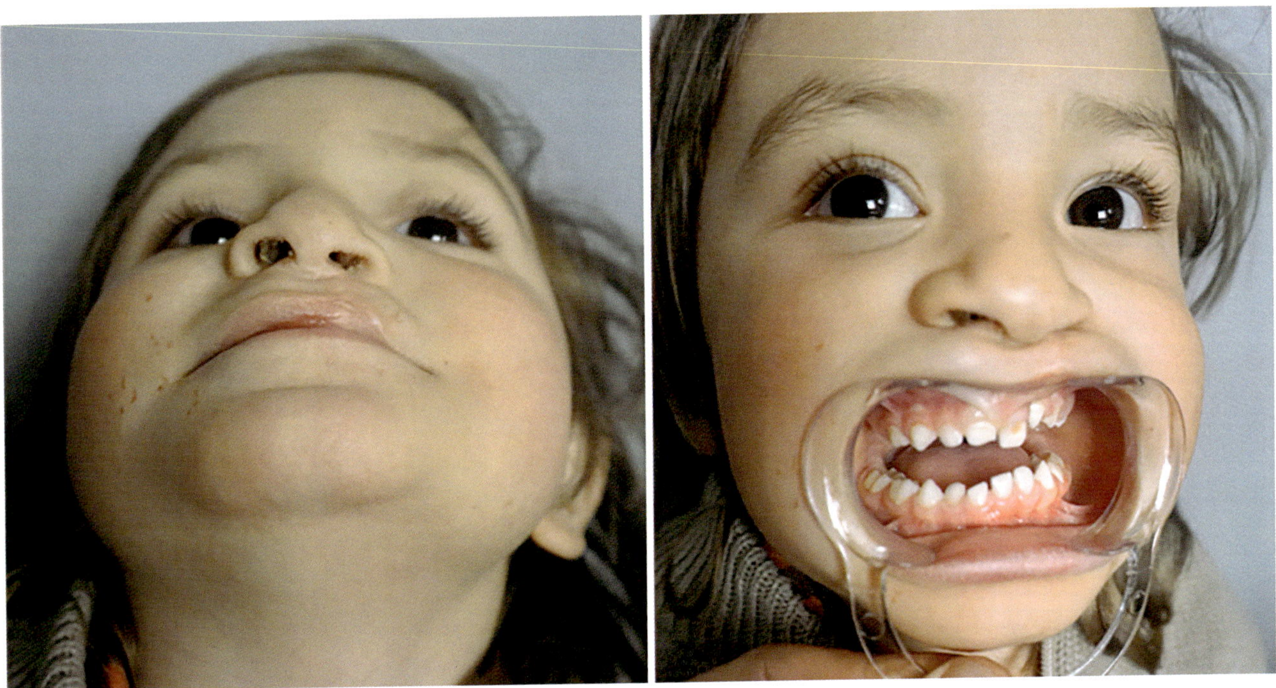

Case 5 Unilateral

Cleft lip and nose reconstruction—1 year post op

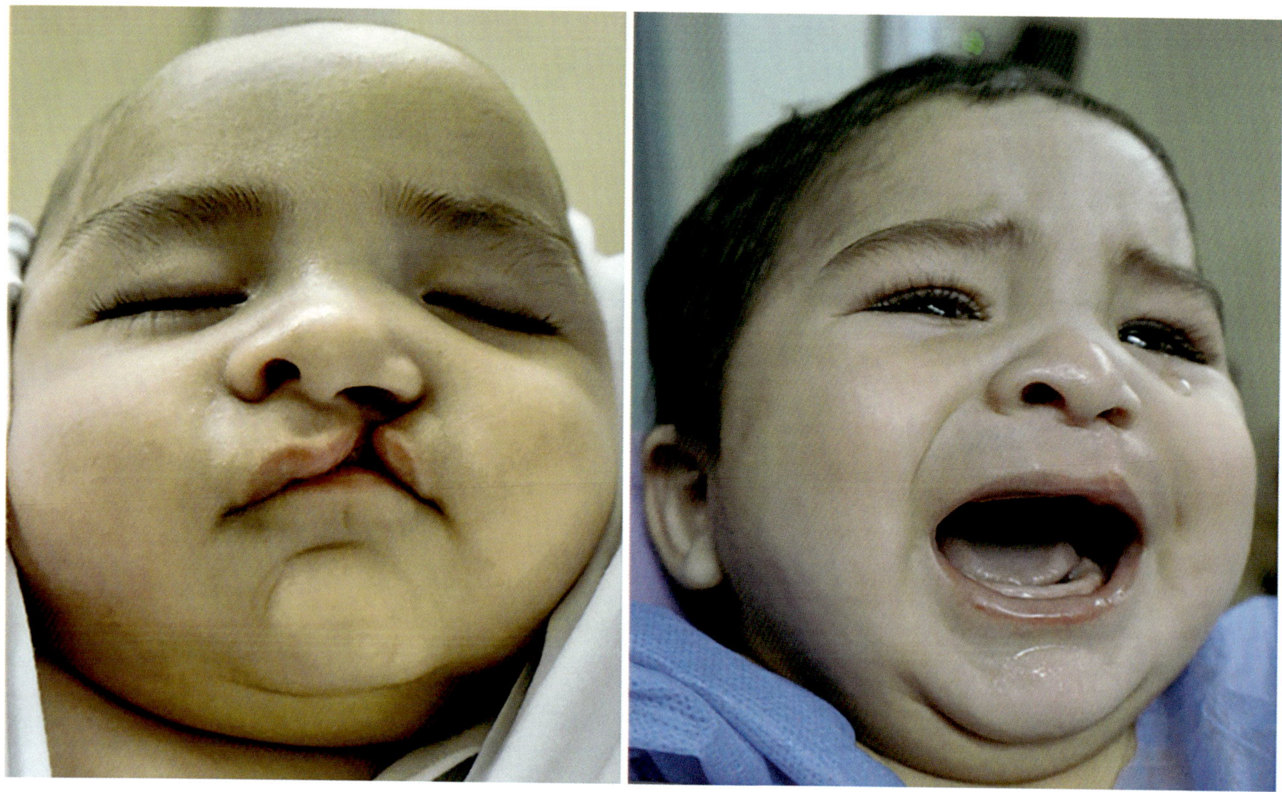

Case 5 Unilateral

Single AEP flap (left) reconstruction

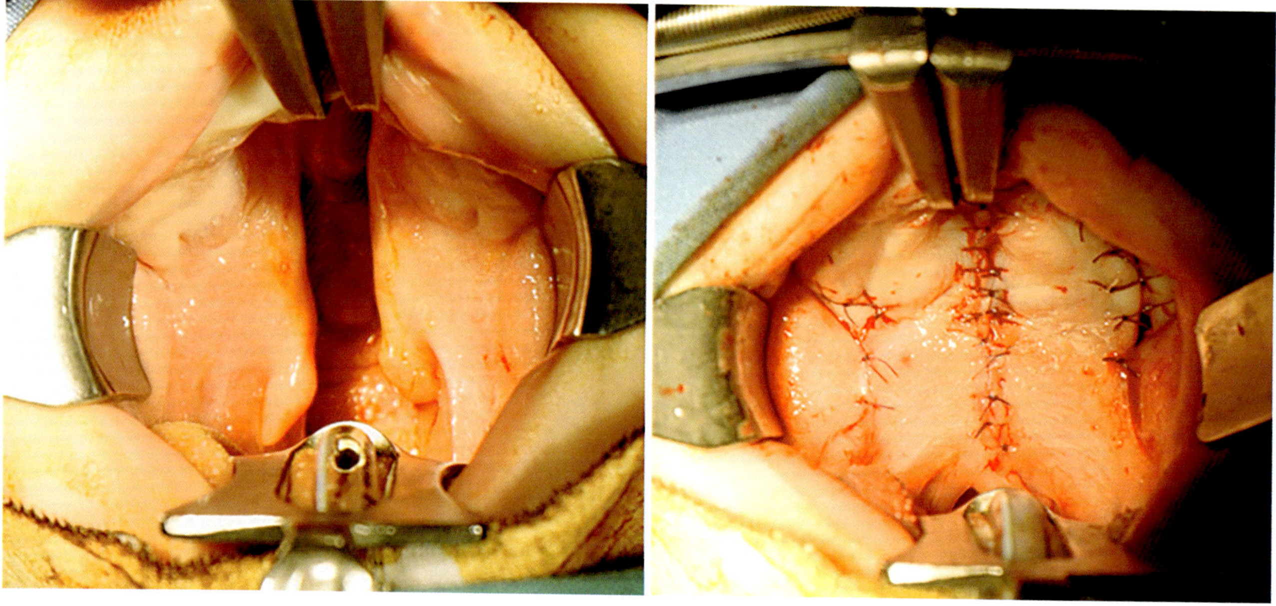

Case 5 Unilateral

4 years post palatoplasty

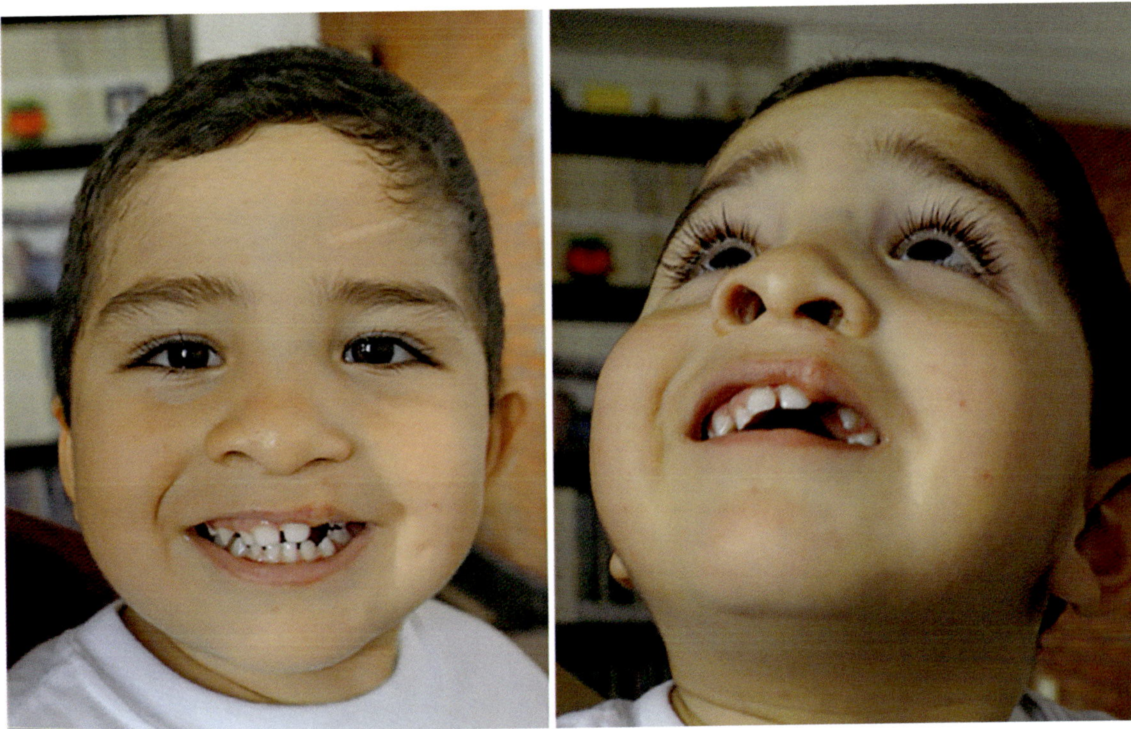

Case 6 Unilateral

Cleft lip and nose reconstruction—5 years follow up

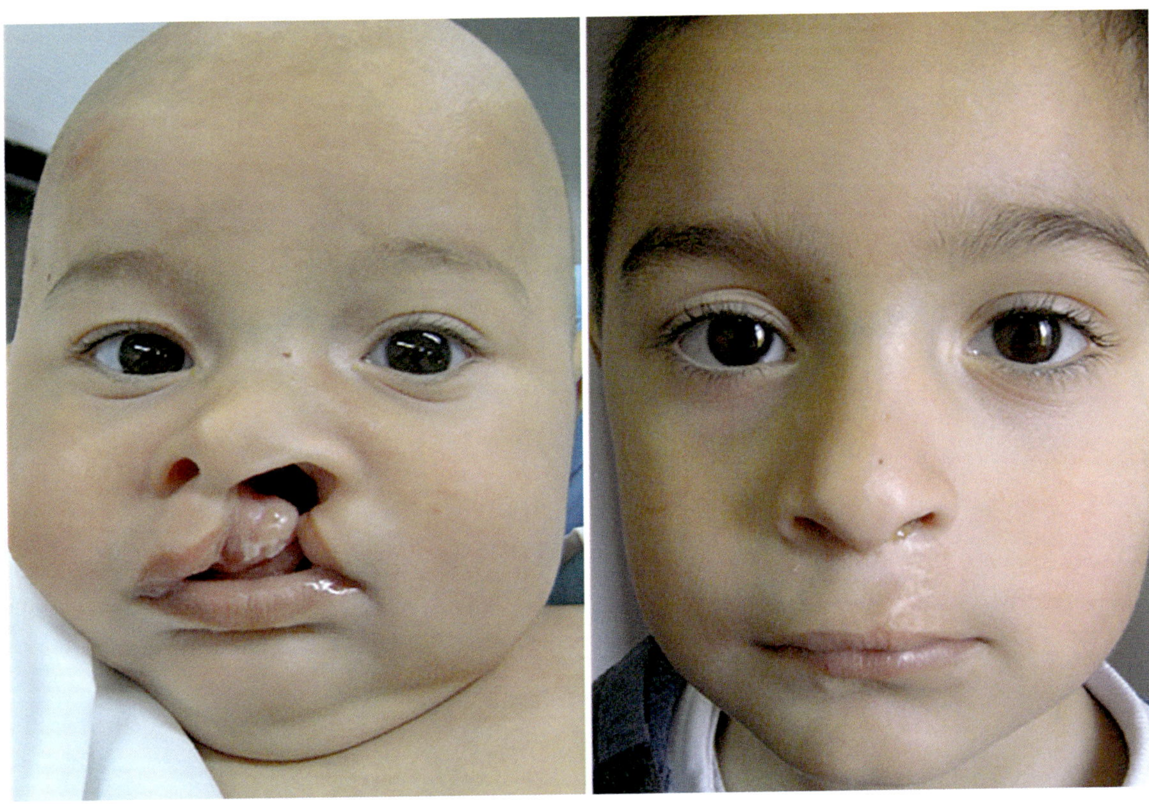

Case 6 Unilateral

Single AEP flap (right) reconstruction

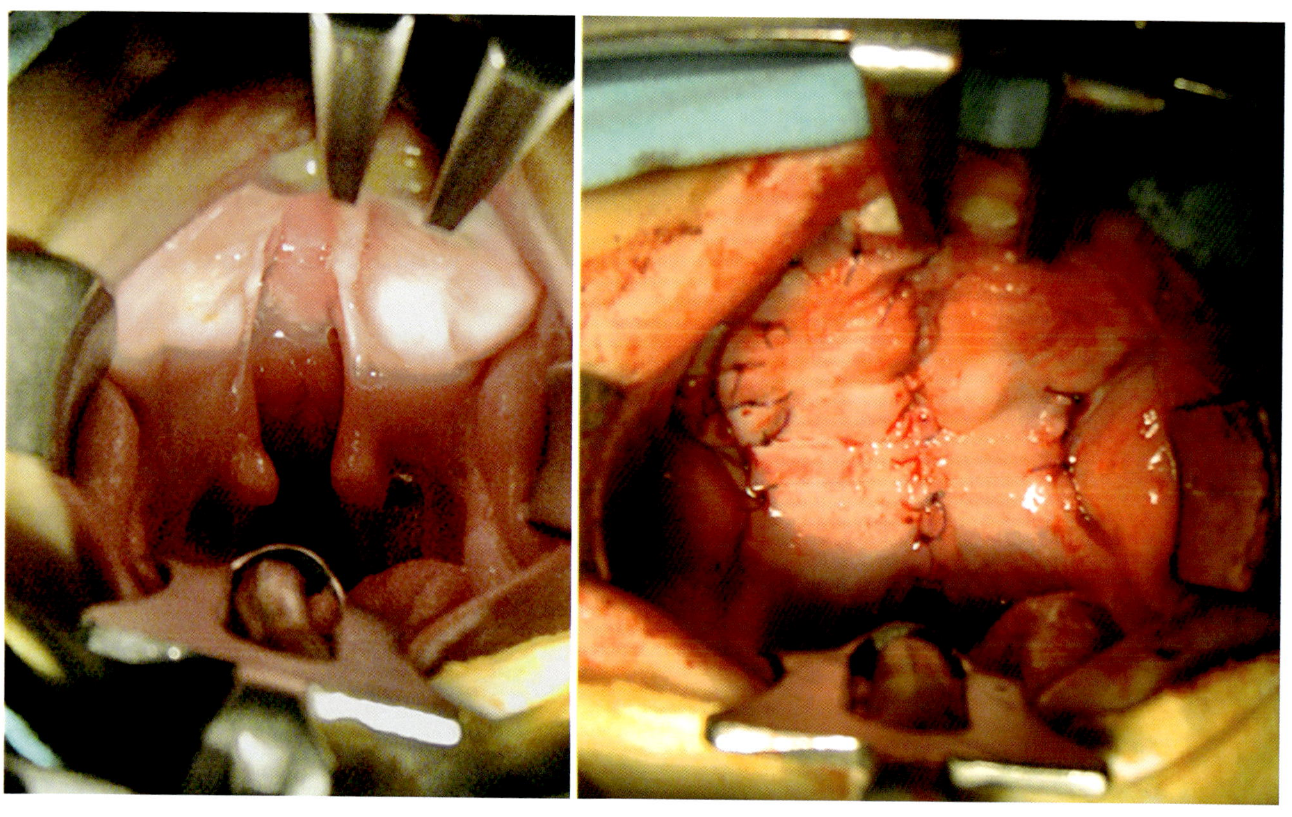

Case 6 Unilateral

5 years post palatoplasty

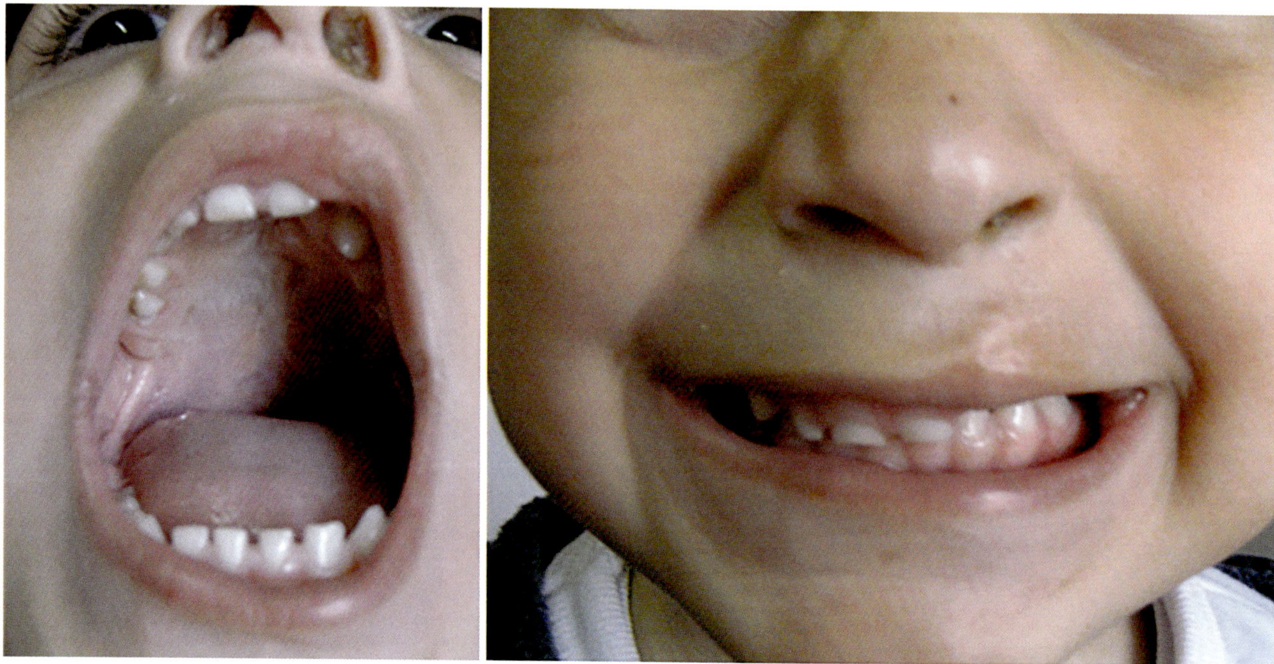

Case 7 Unilateral

Cleft lip and nose reconstruction—1 year post op

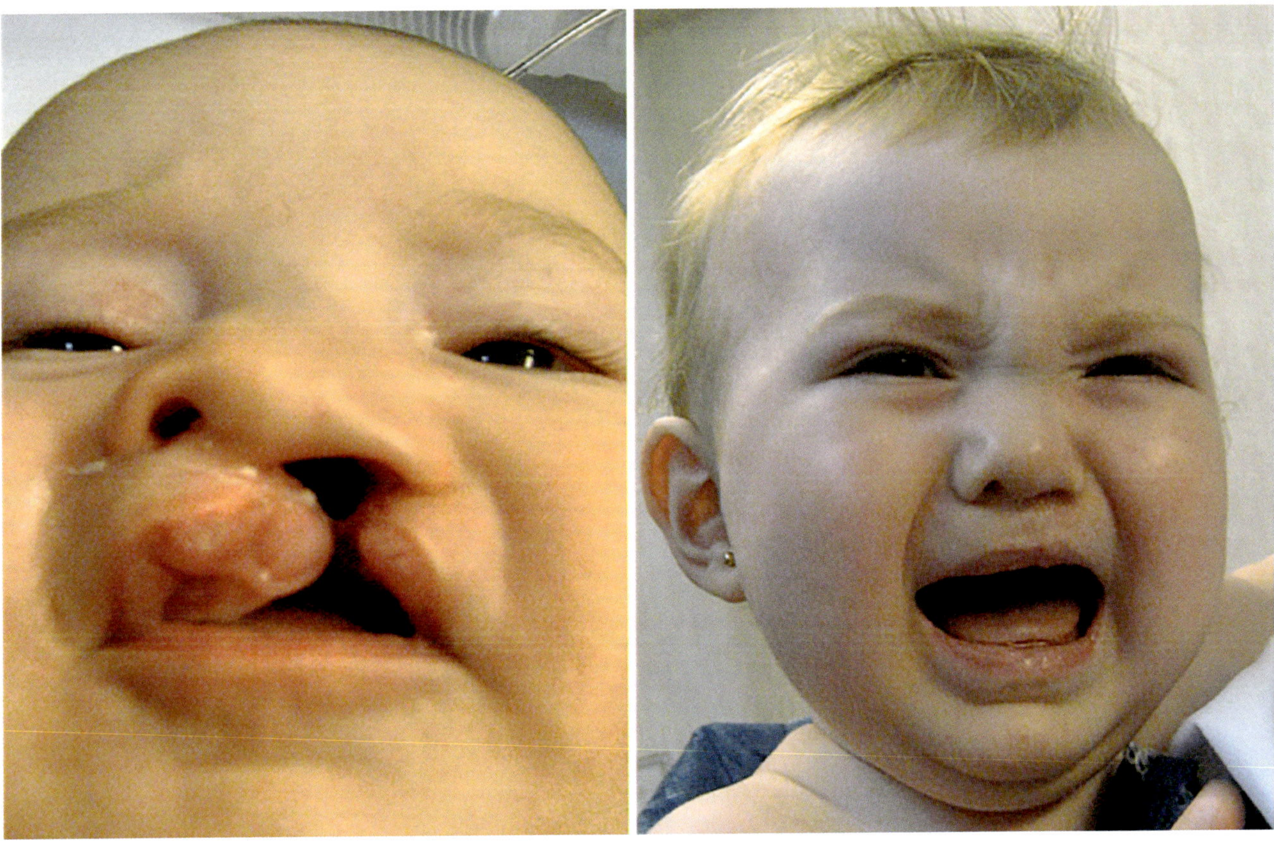

Case 7 Unilateral

Double AEP flap reconstruction

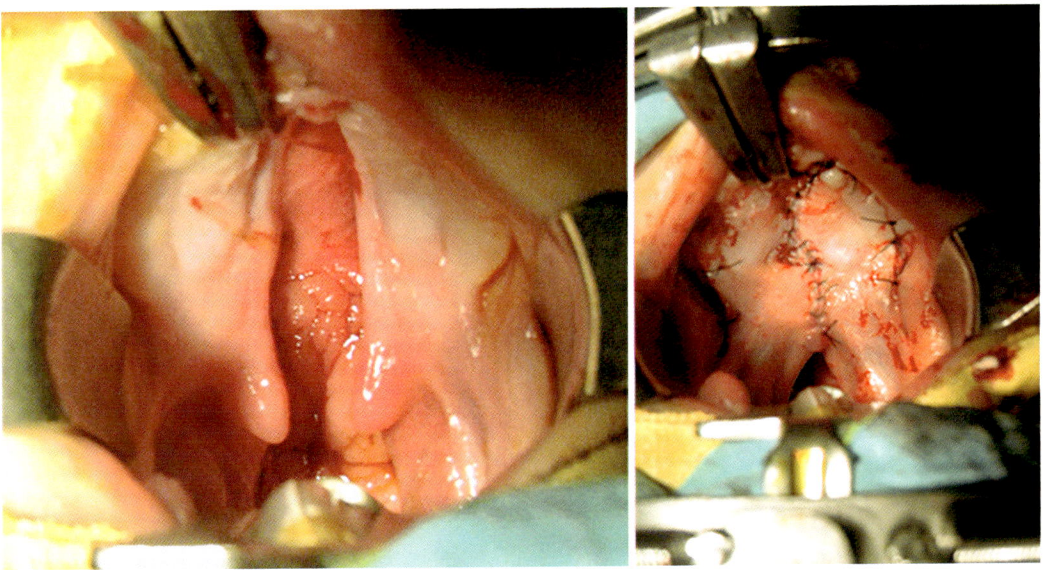

Case 7 Unilateral

6 years post palatoplasty

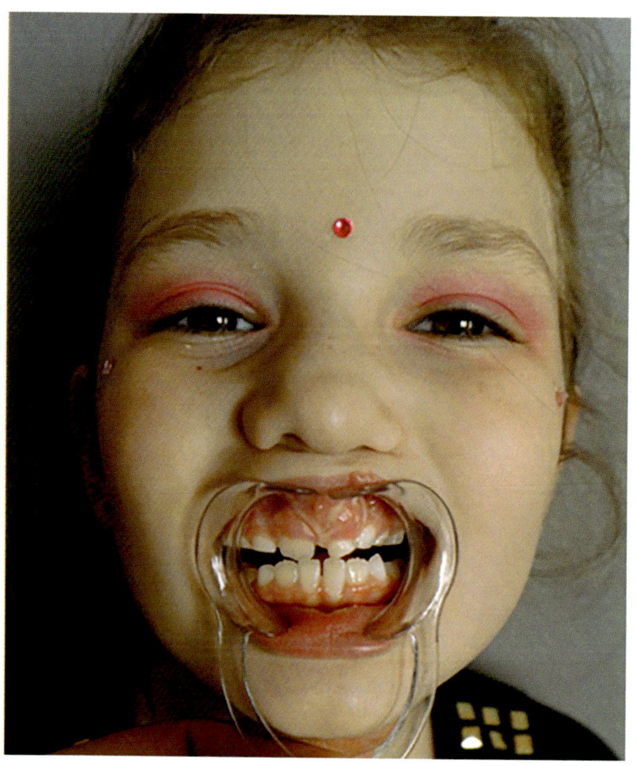

Case 8 Unilateral

Cleft lip and nose reconstruction—7 years post op

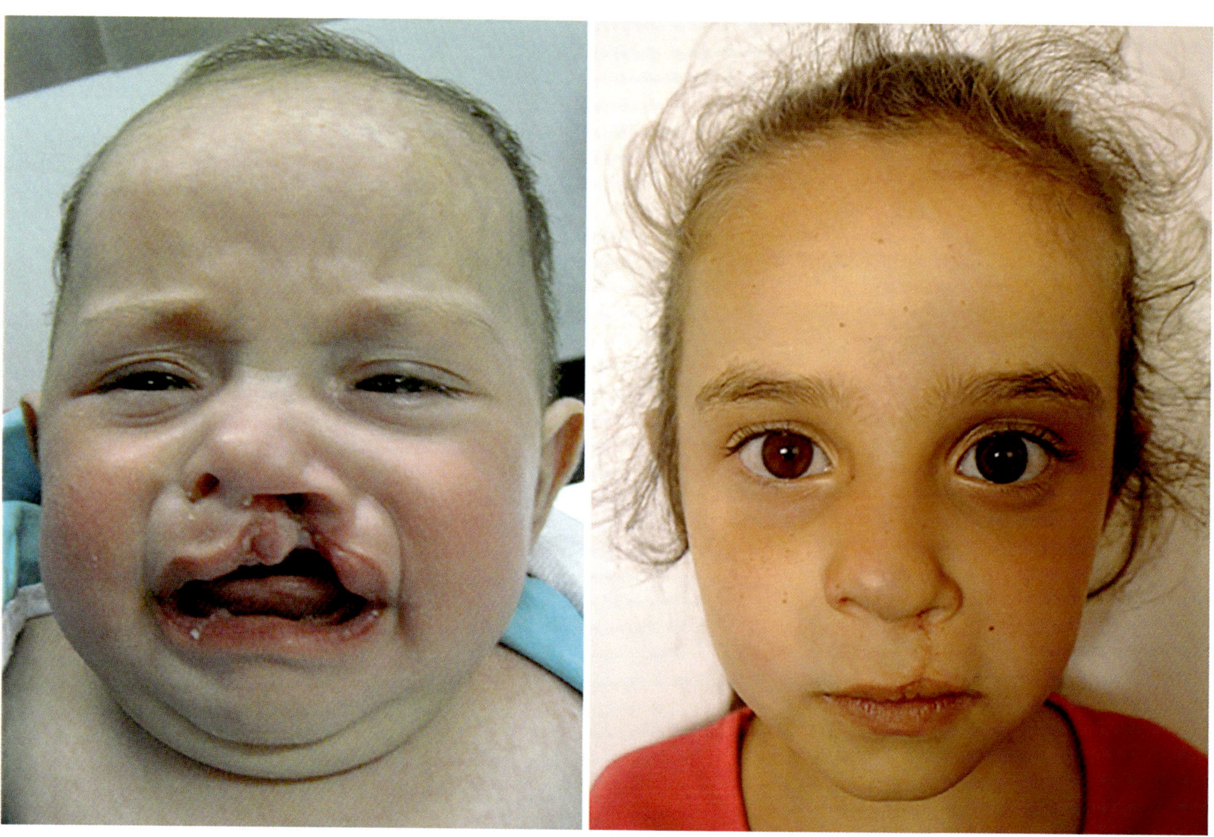

Case 8 Unilateral

Single AEP flap (left)

7 years post palatoplasty

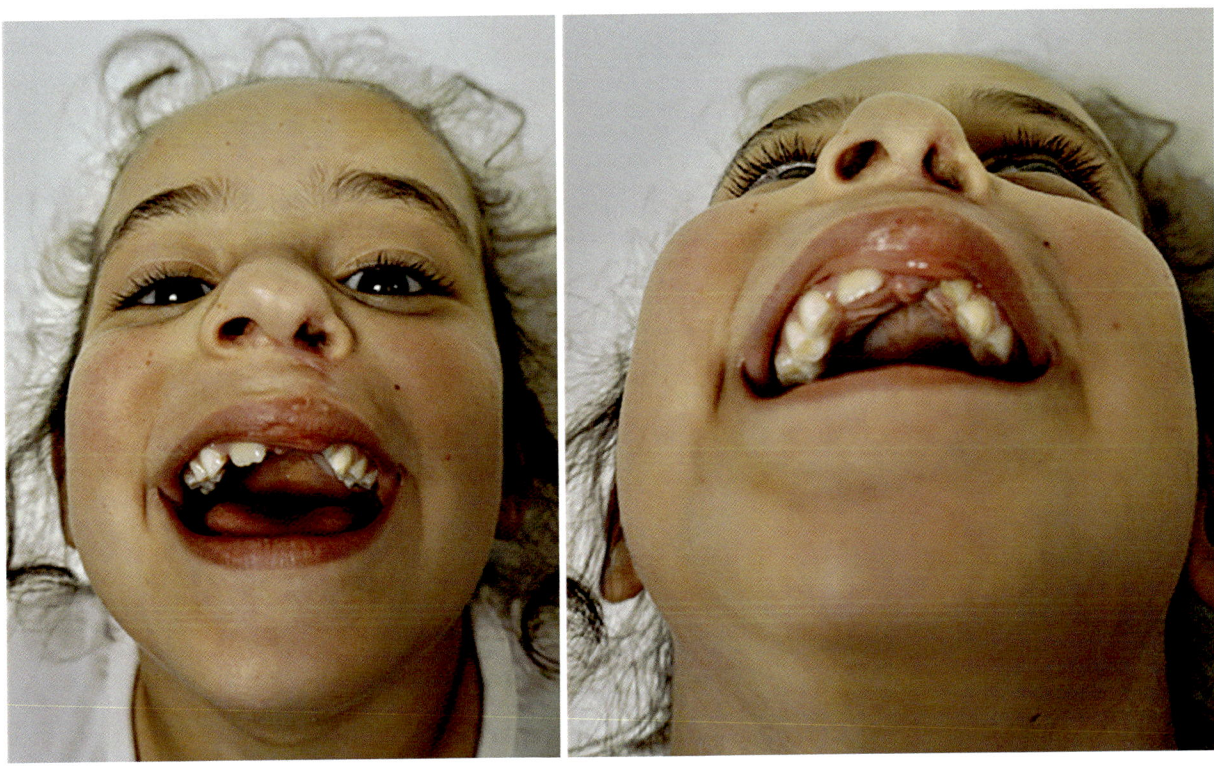

Case 9 Unilateral

Cleft lip and nose reconstruction—1 year follow up

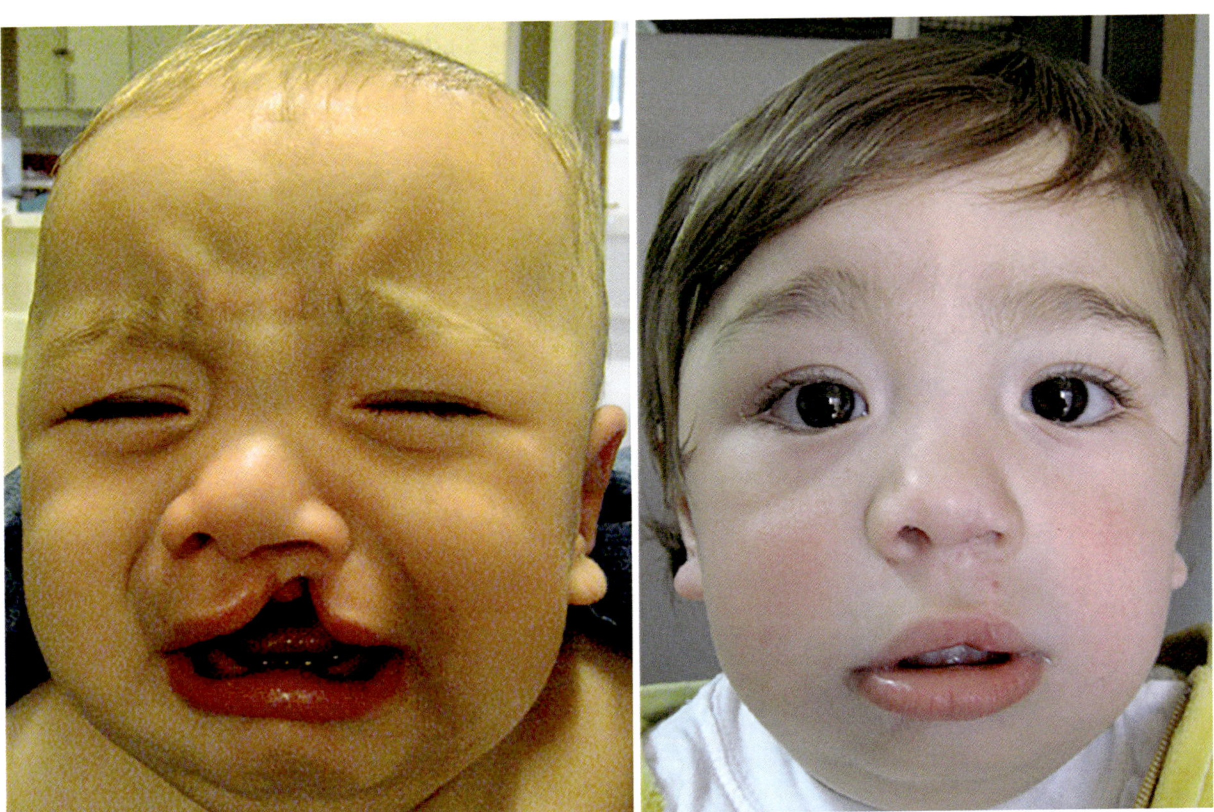

Case 9 Unilateral
Double AEP flap reconstruction

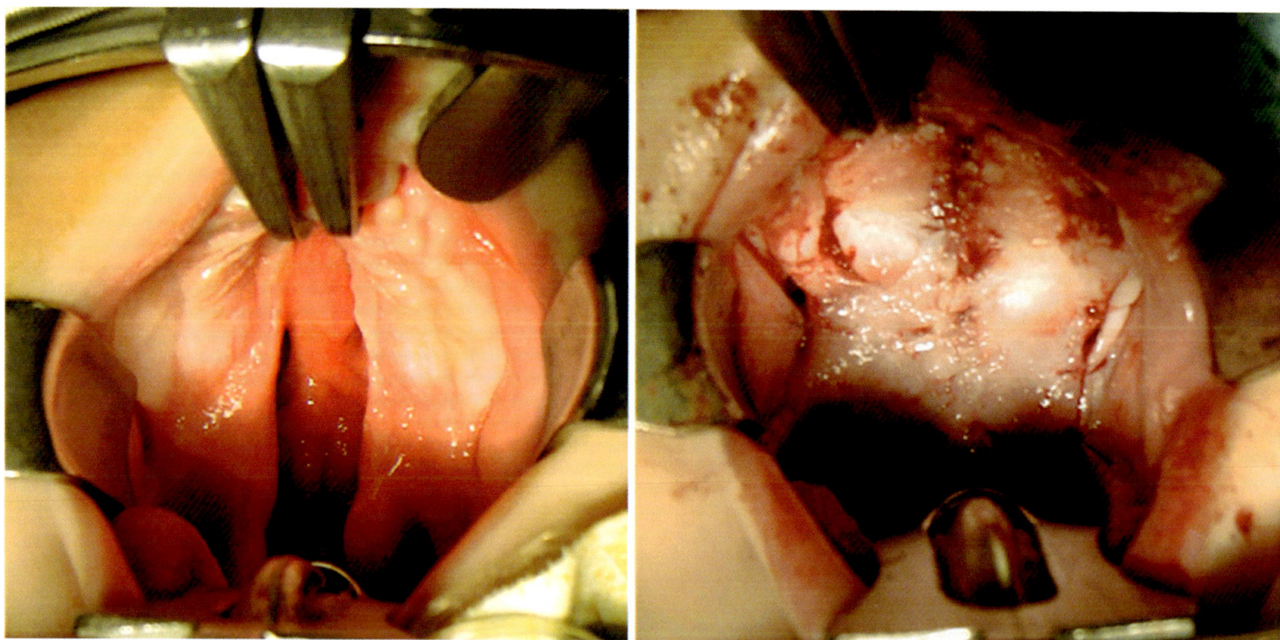

Case 9 Unilateral
7 years post palatoplasty

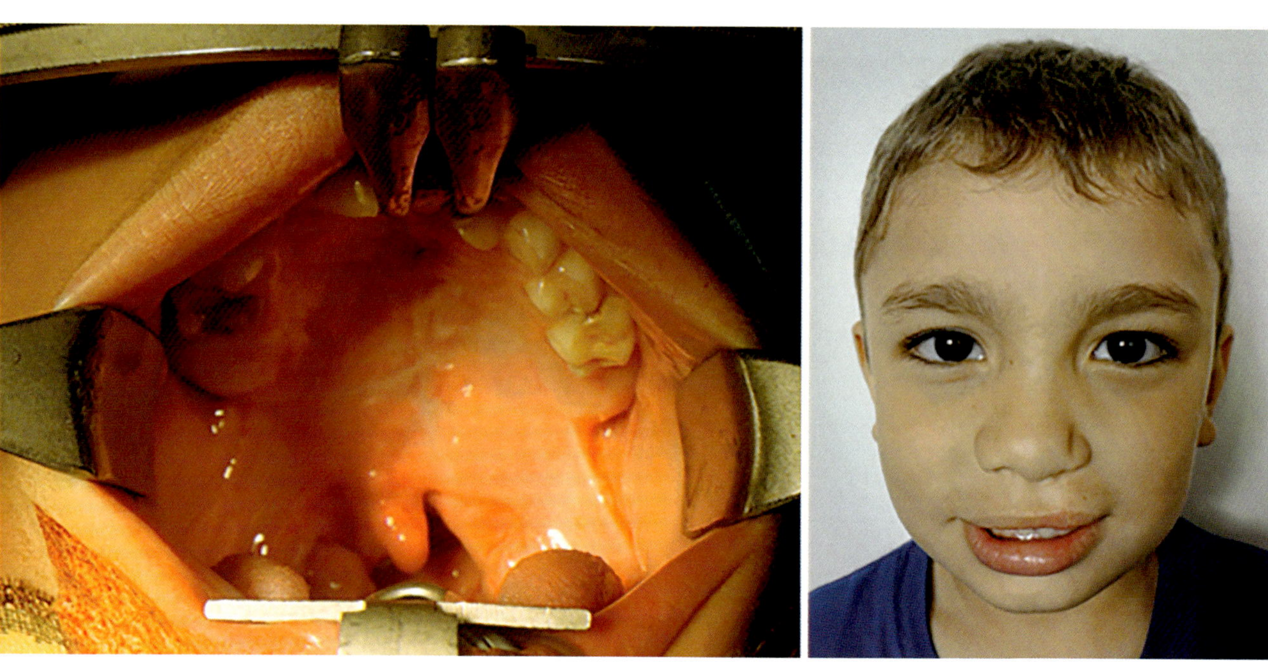

Case 9 Unilateral

7 years post palatoplasty (lateral and inferior)

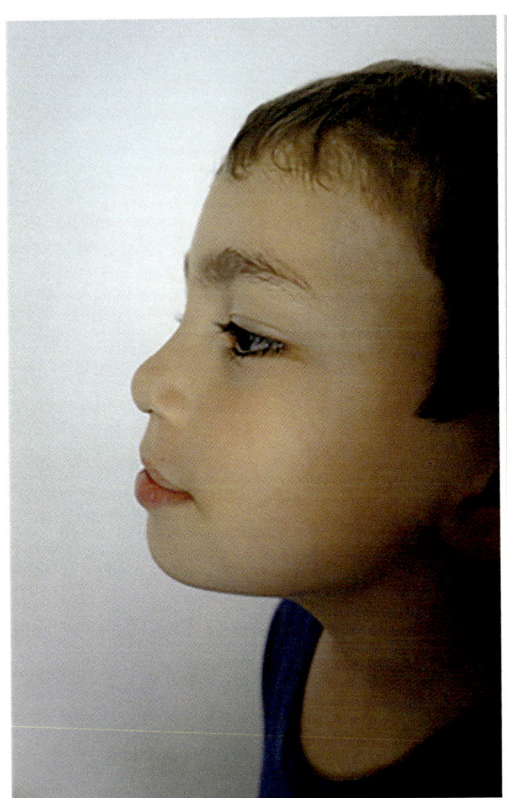
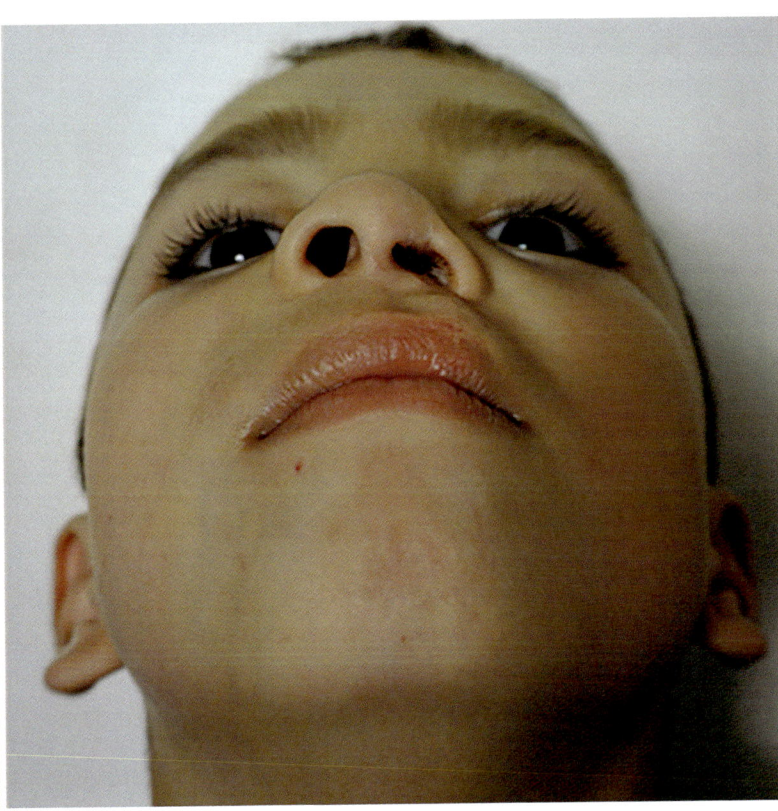

Case 10 Unilateral

Cleft lip and nose reconstruction

Note severe displacement

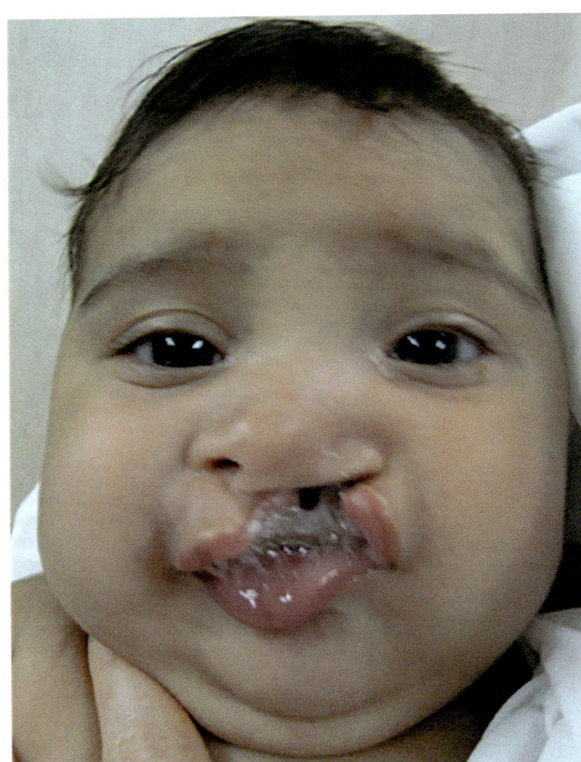
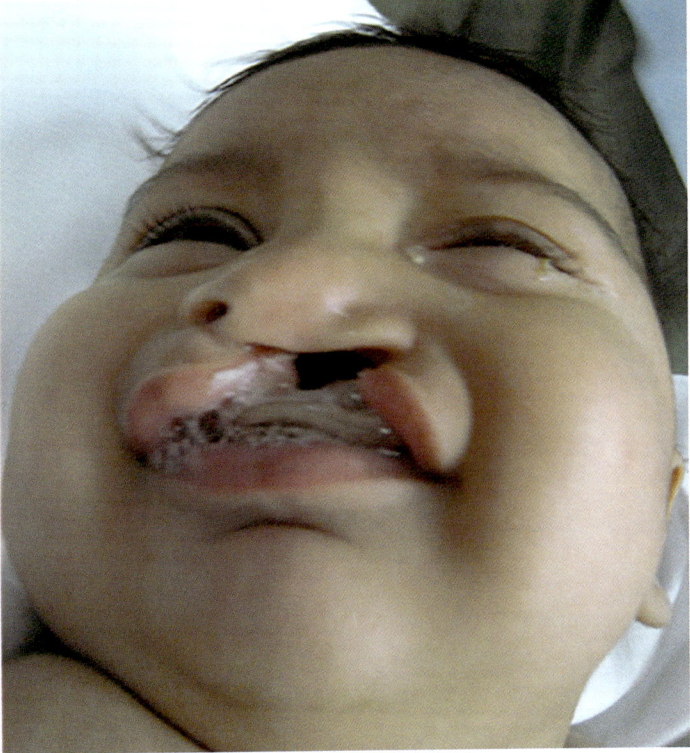

Case 10 Unilateral

Cleft lip and nose reconstruction—1 year follow up

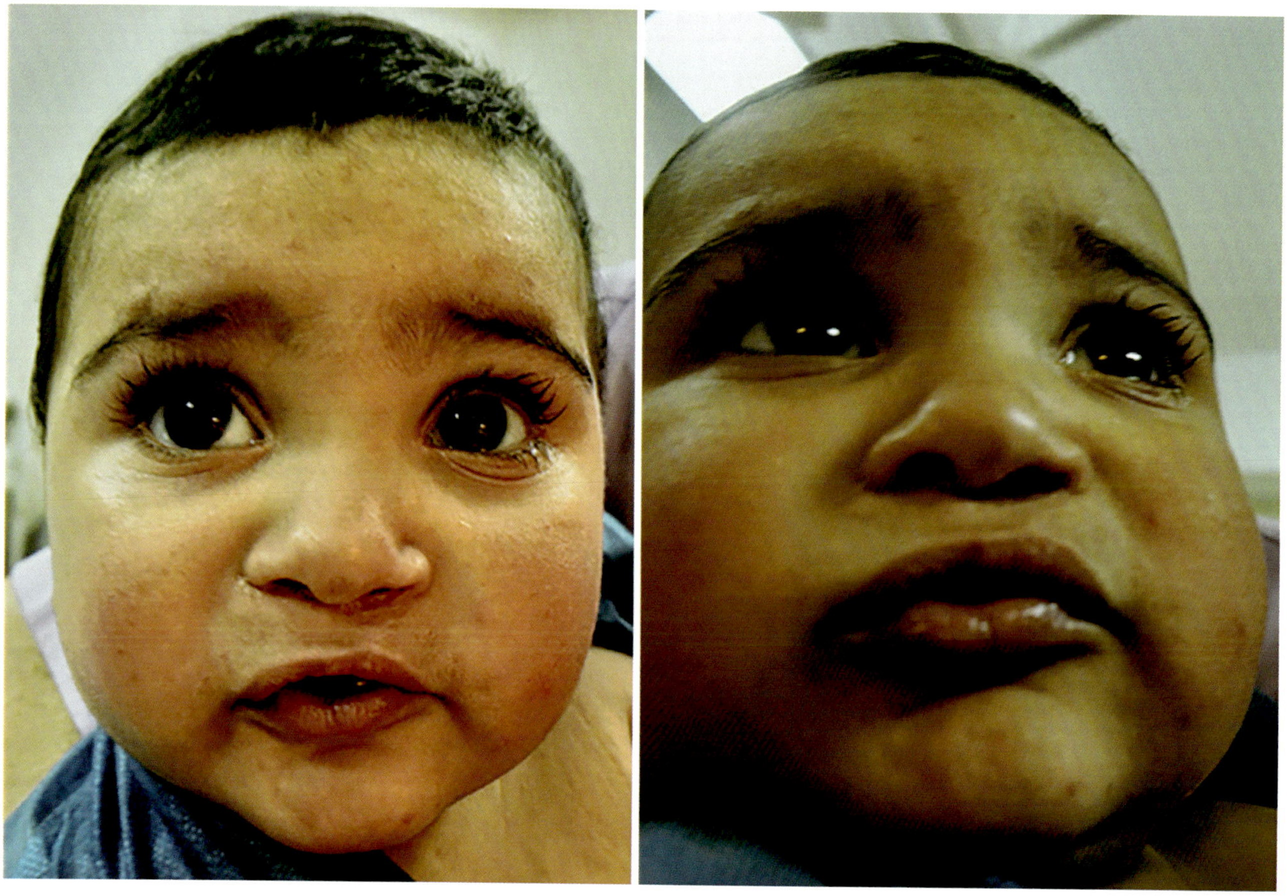

Case 10 Unilateral

Double AEP flap reconstruction

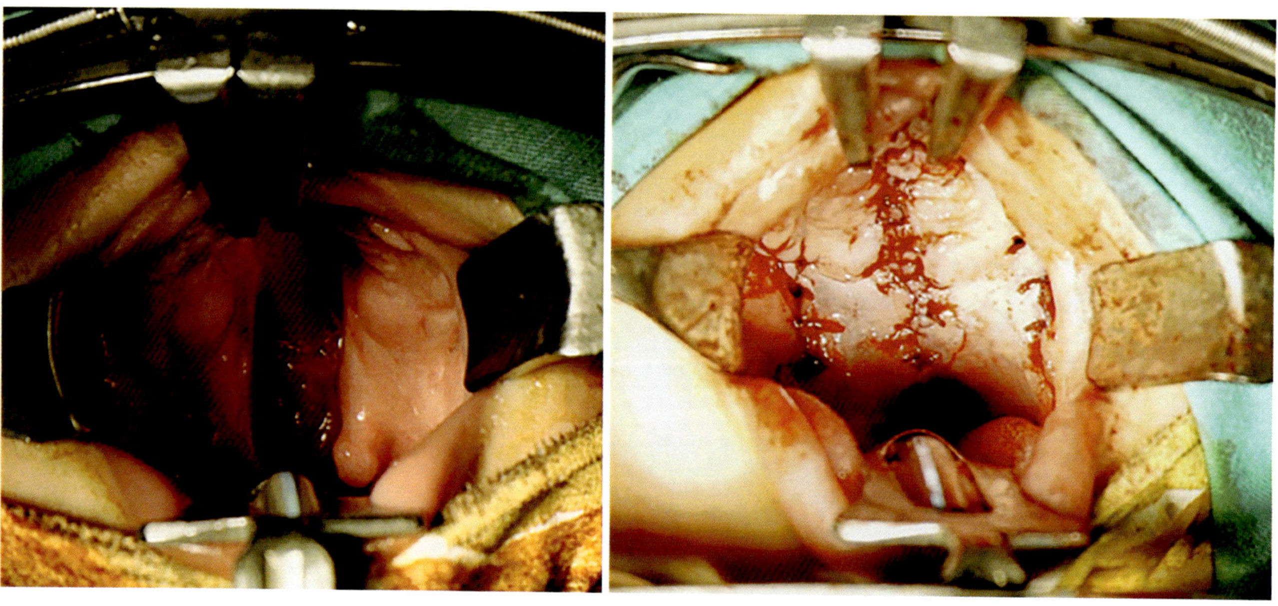

Case 10 Unilateral
7 years post palatoplasty

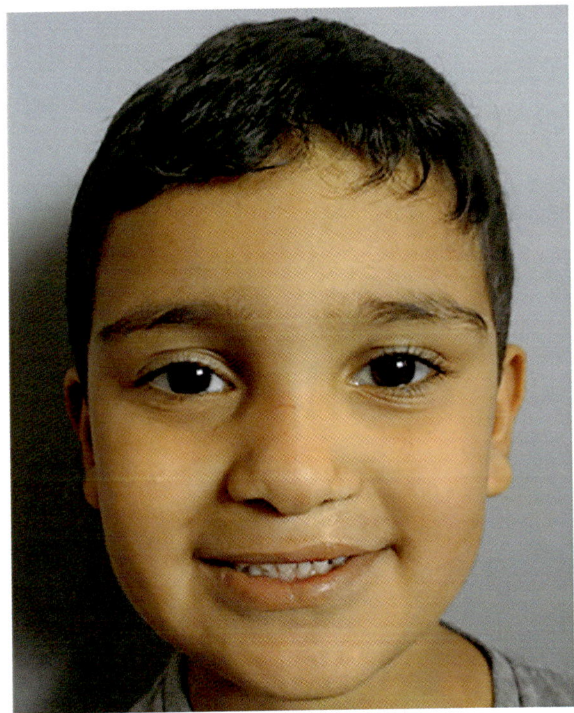

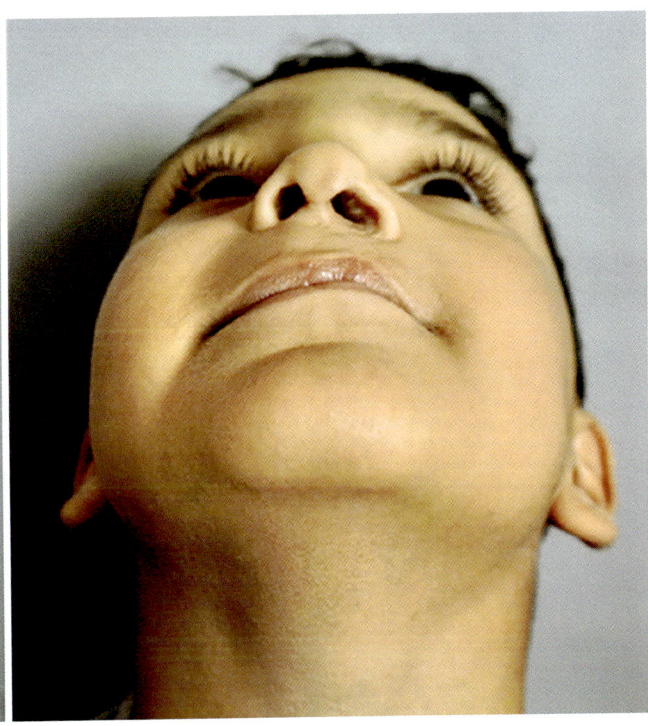

Case 10 Unilateral
7 years post palatoplasty

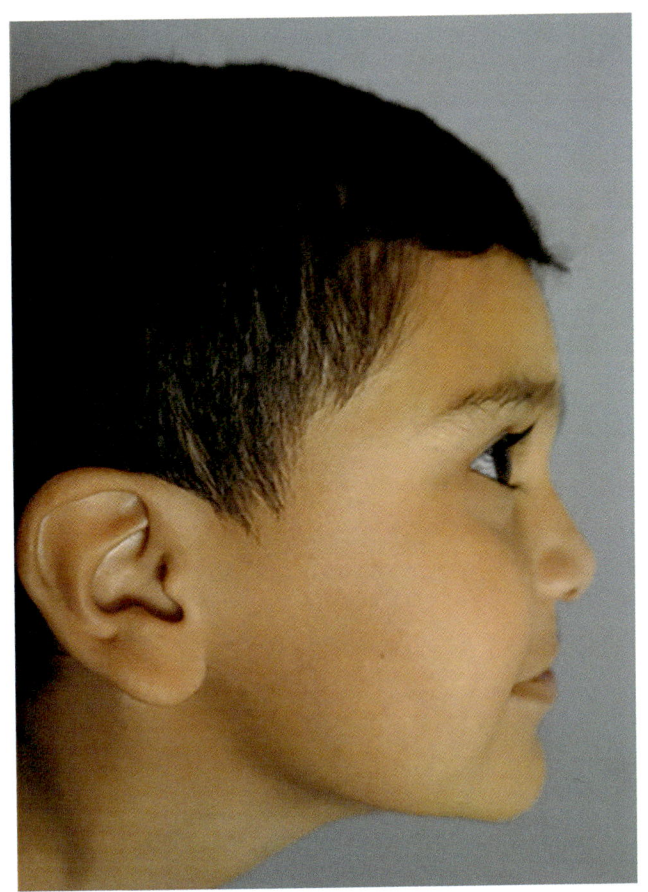

Case 11 Unilateral

Lip and nose reconstruction—7 years follow up

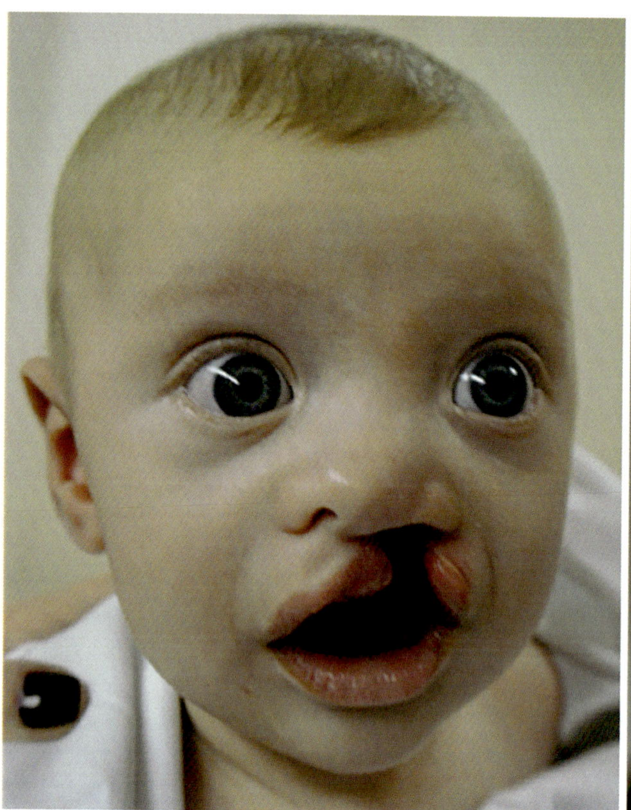

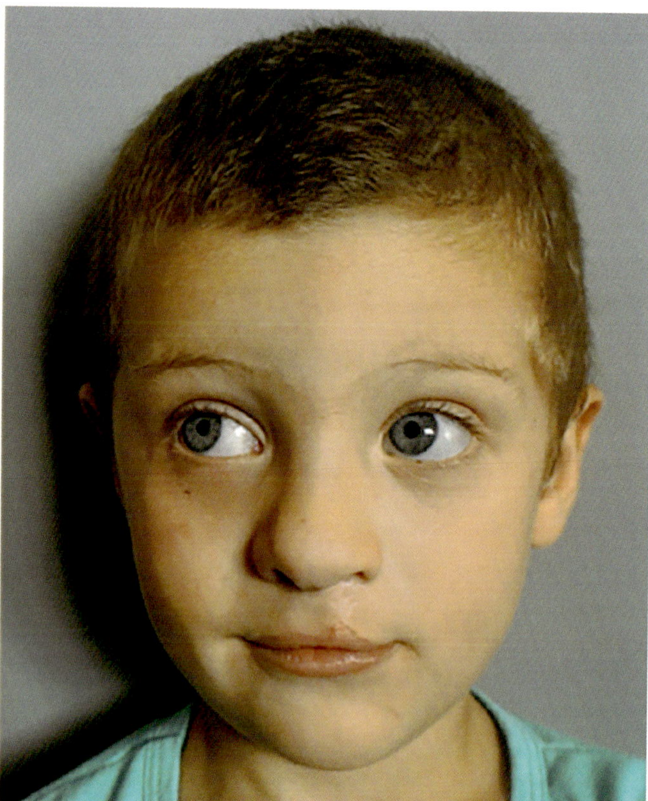

Case 11 Unilateral

Double AEP flap reconstruction

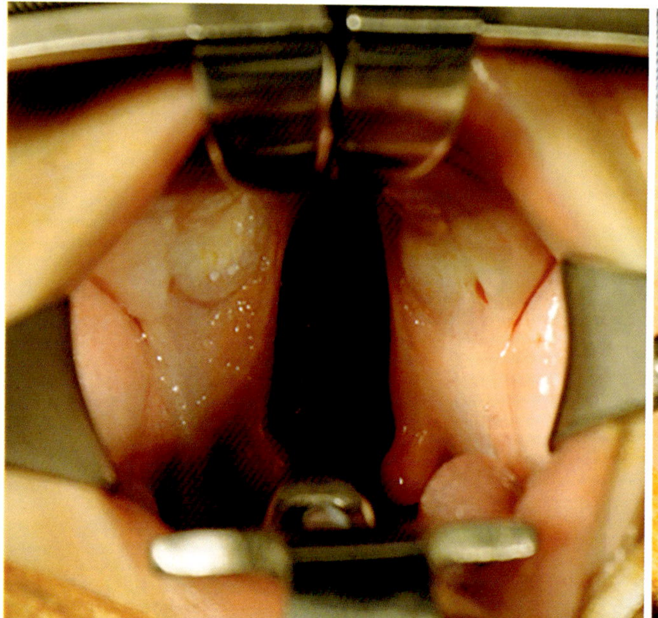

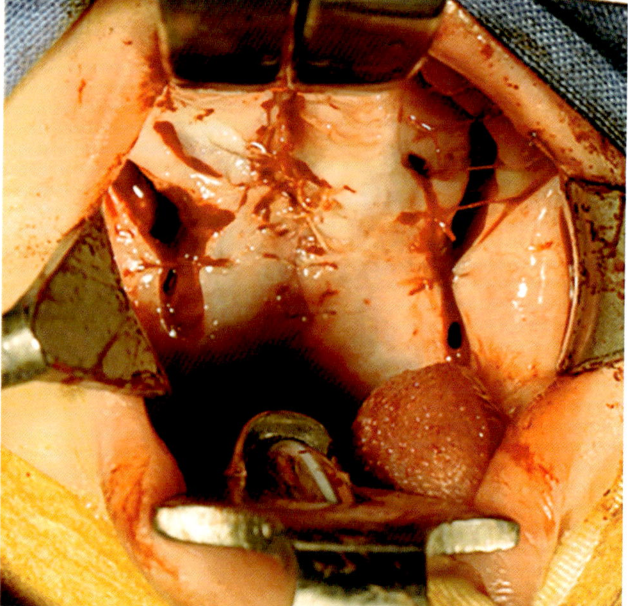

Case 11 Unilateral
7 years post palatoplasty

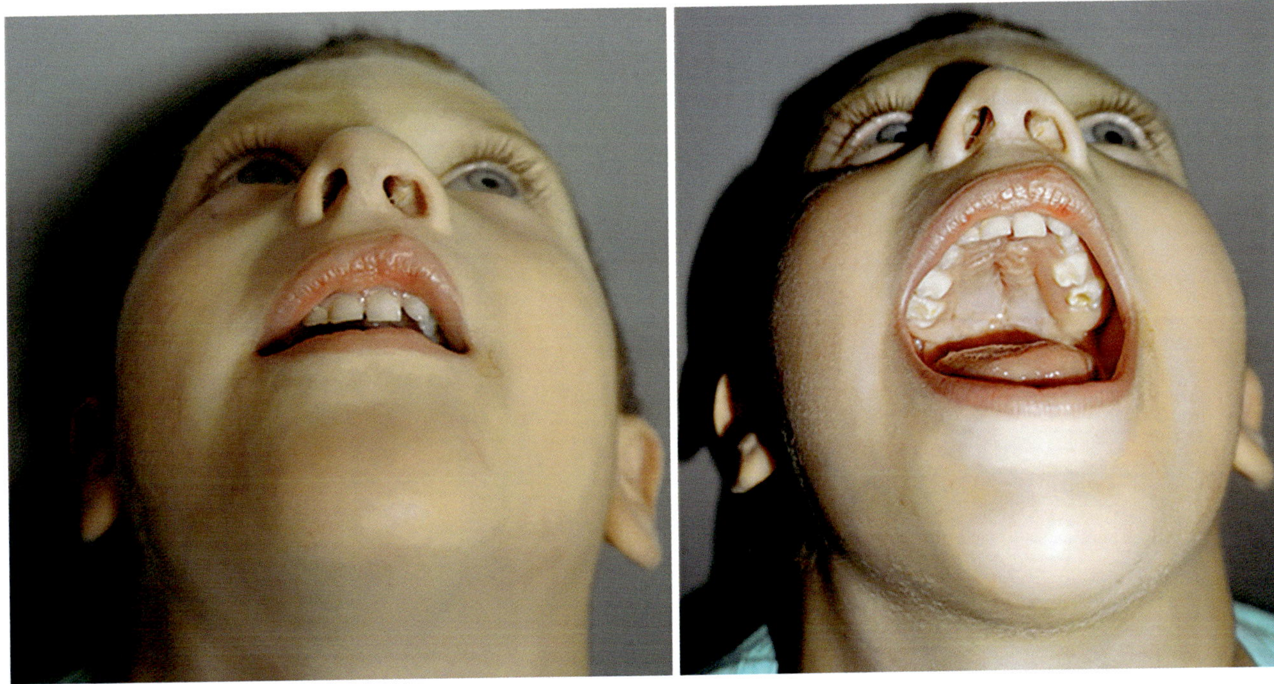

Case 11 Unilateral
7 years post palatoplasty

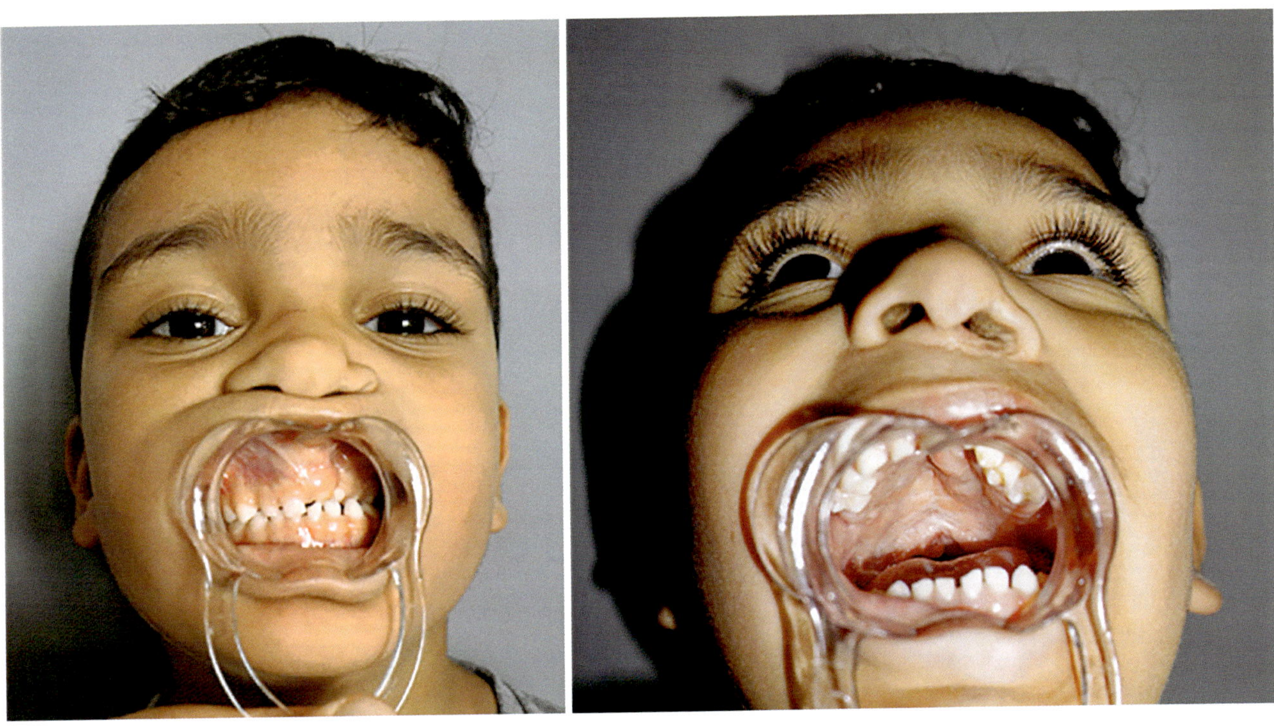

Case 12 Unilateral

Cleft lip and nose reconstruction—1 year post follow up

Follow-up 8 years

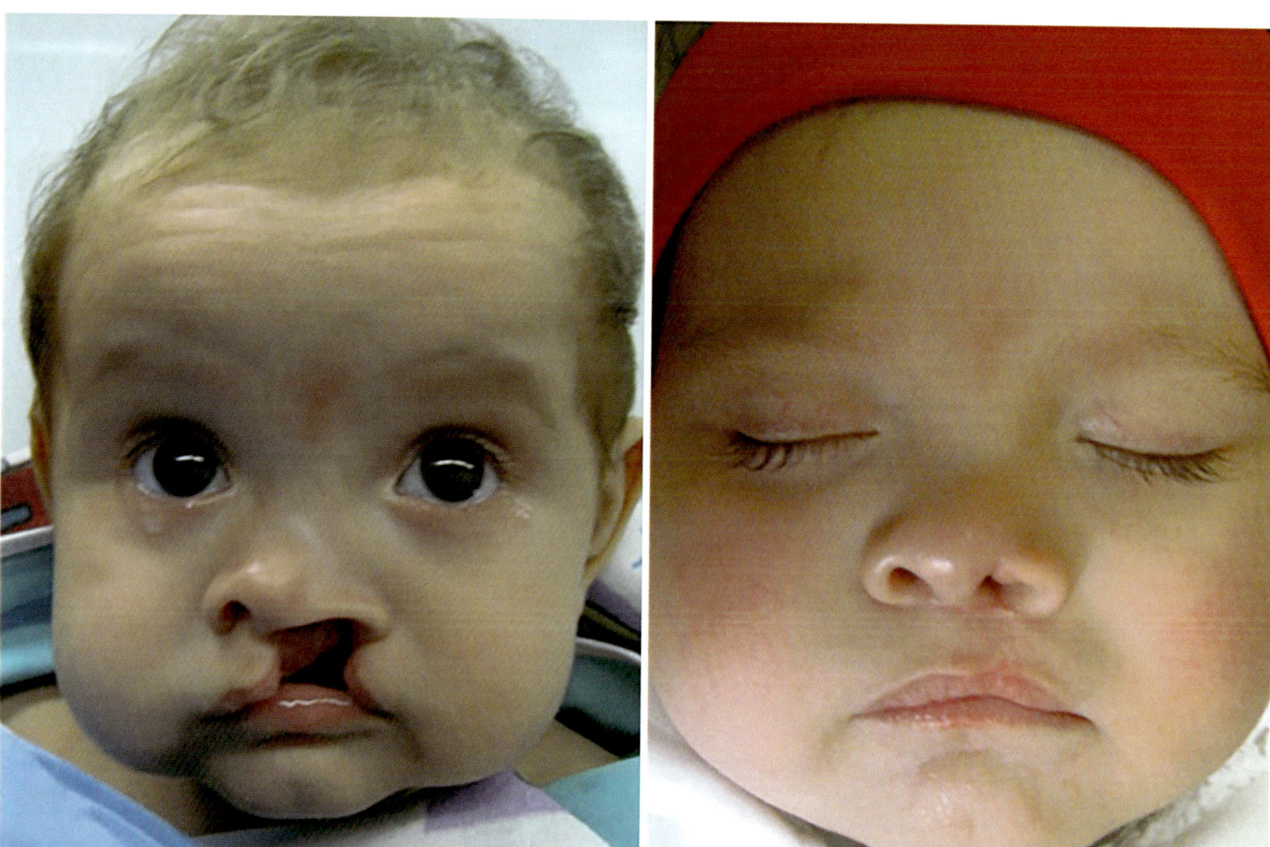

Case 12 Unilateral

Single AEP flap reconstruction (left)

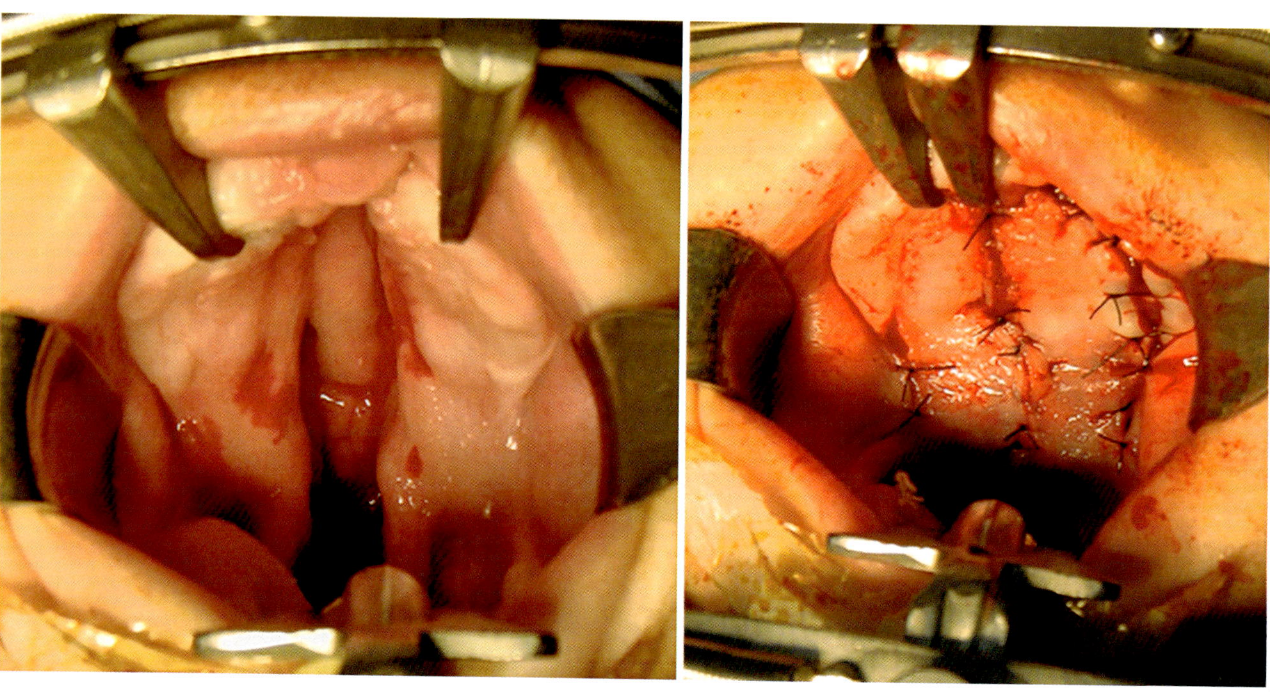

Case 12 Unilateral

8 years post palatoplasty

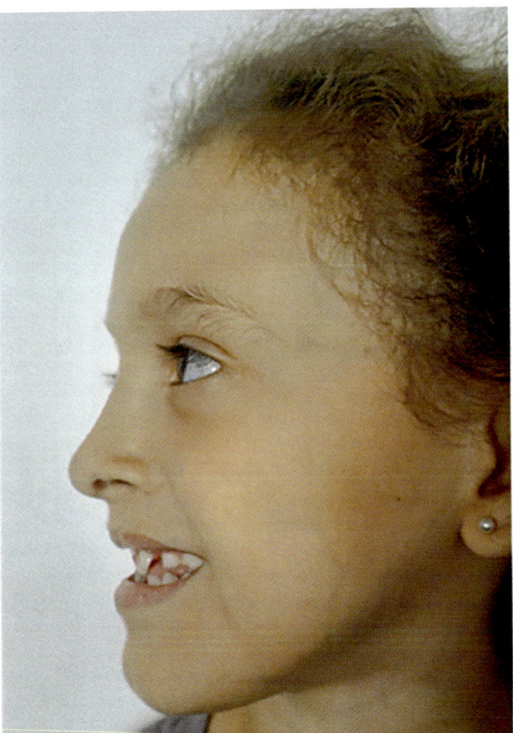

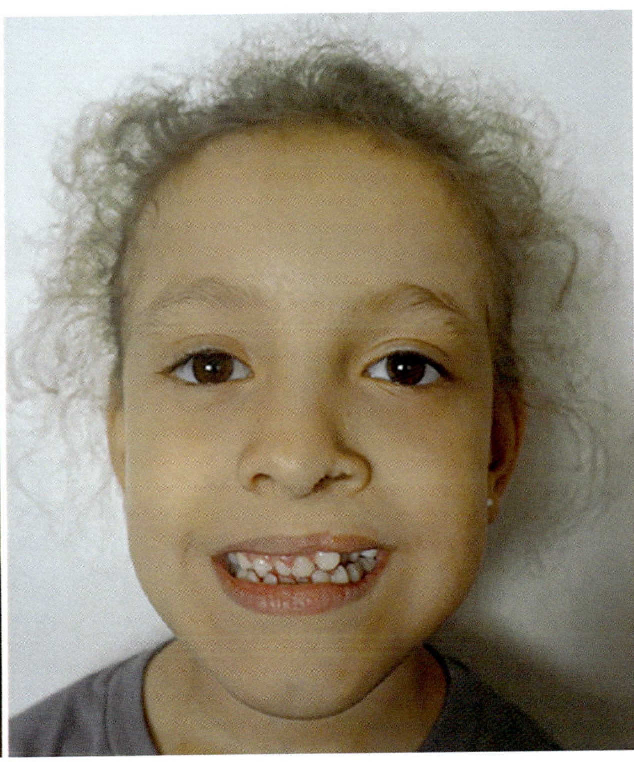

Case 13 Unilateral

Cleft lip and nose reconstruction—1 year post op

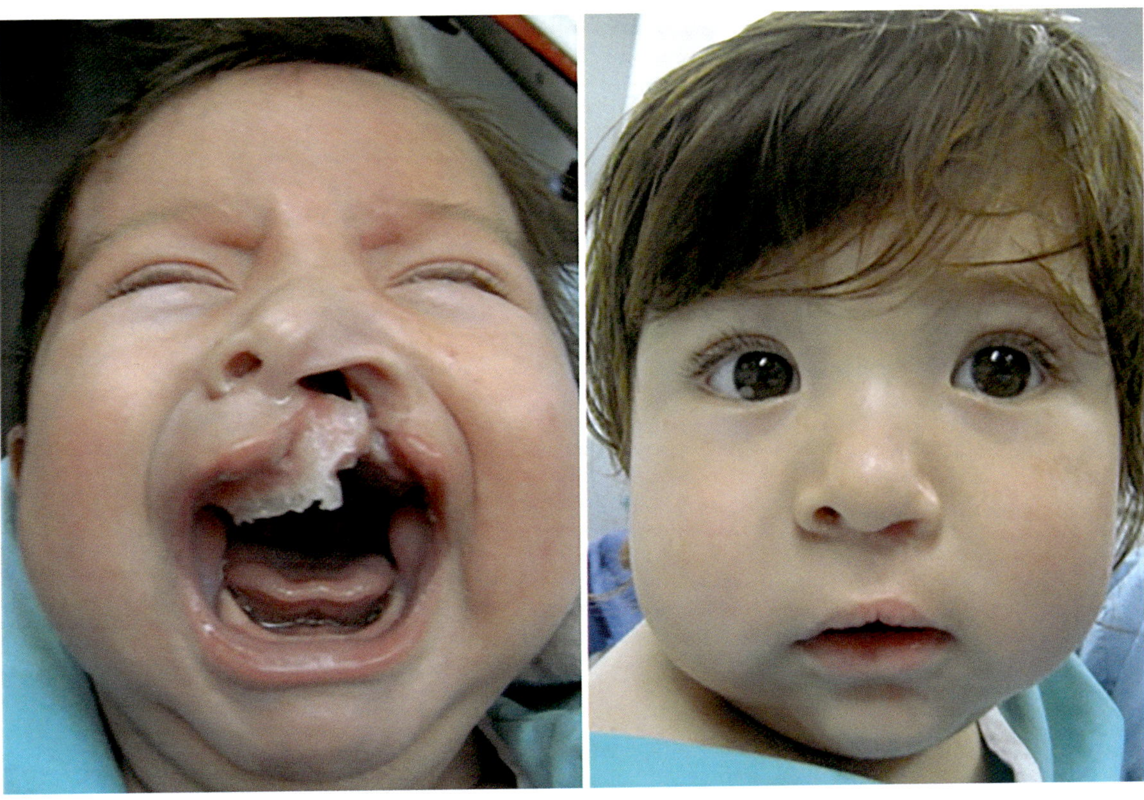

Case 13 Unilateral

Bilateral AEP flap reconstruction (minimal incisions)

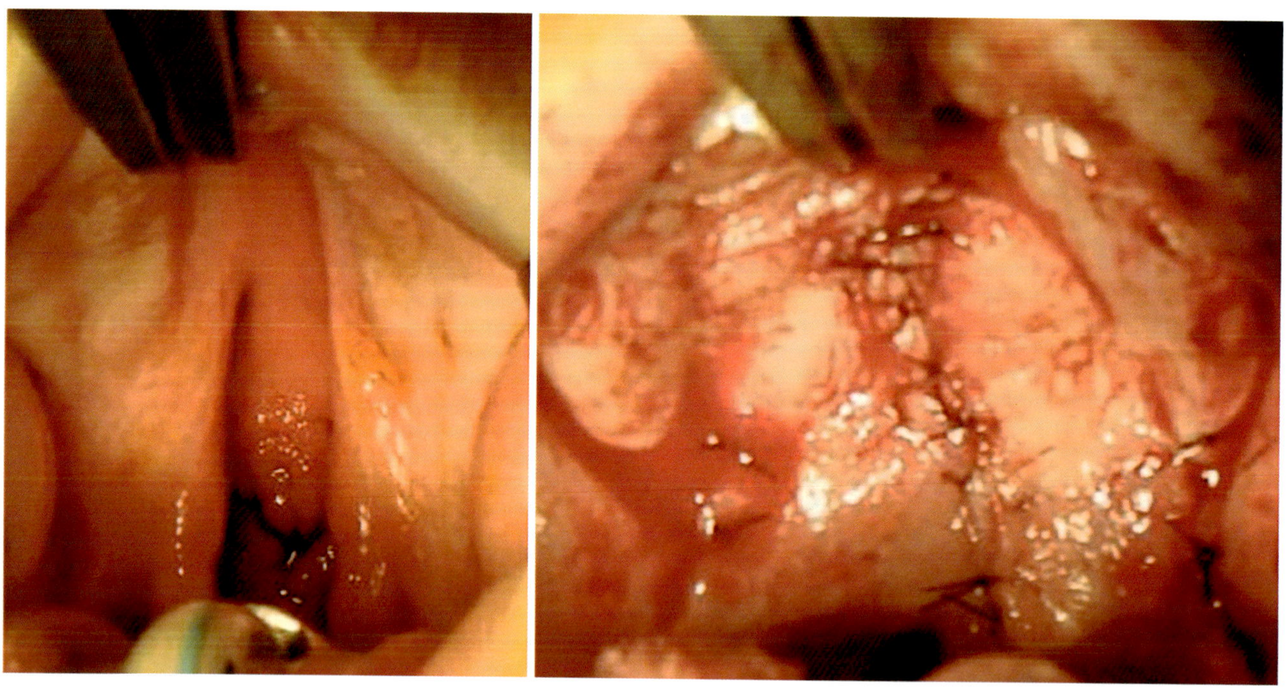

Case 13 Unilateral

9 years post palatoplasty

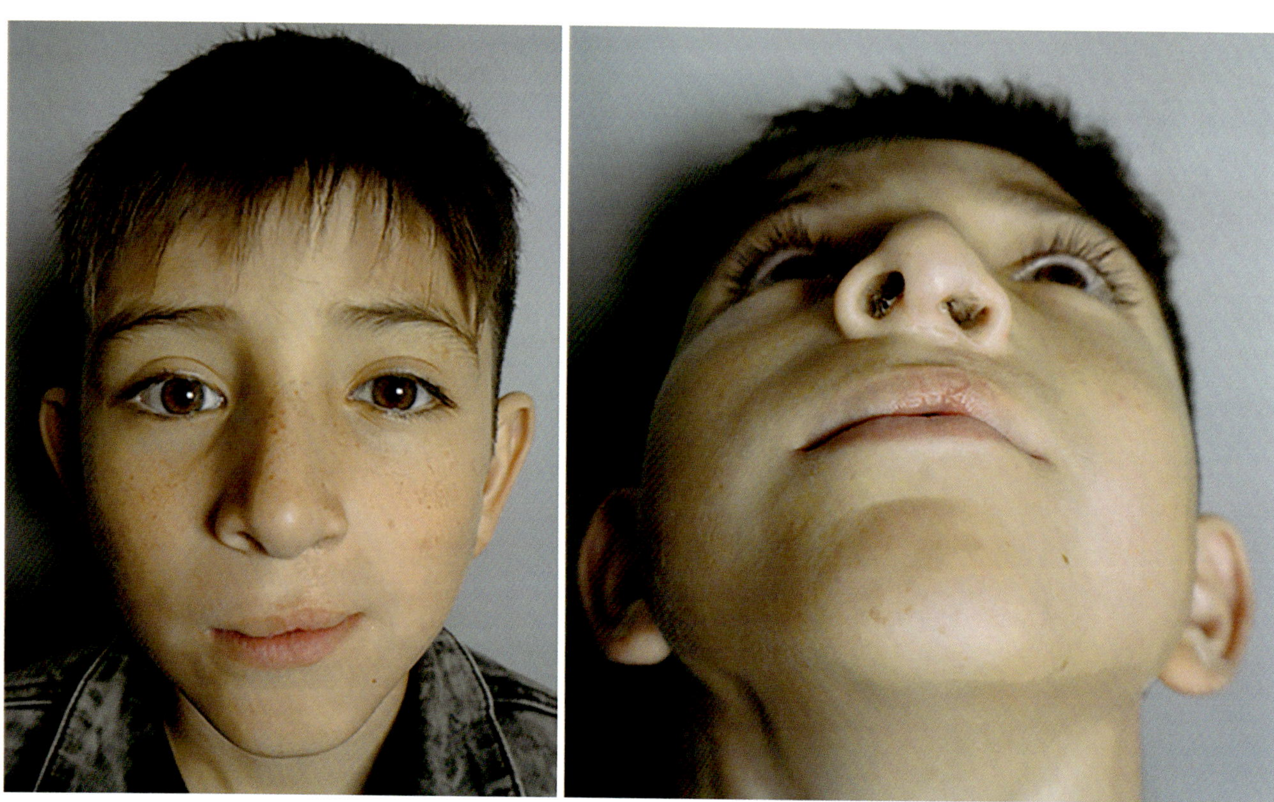

Case 13 Unilateral
9 years post palatoplasty

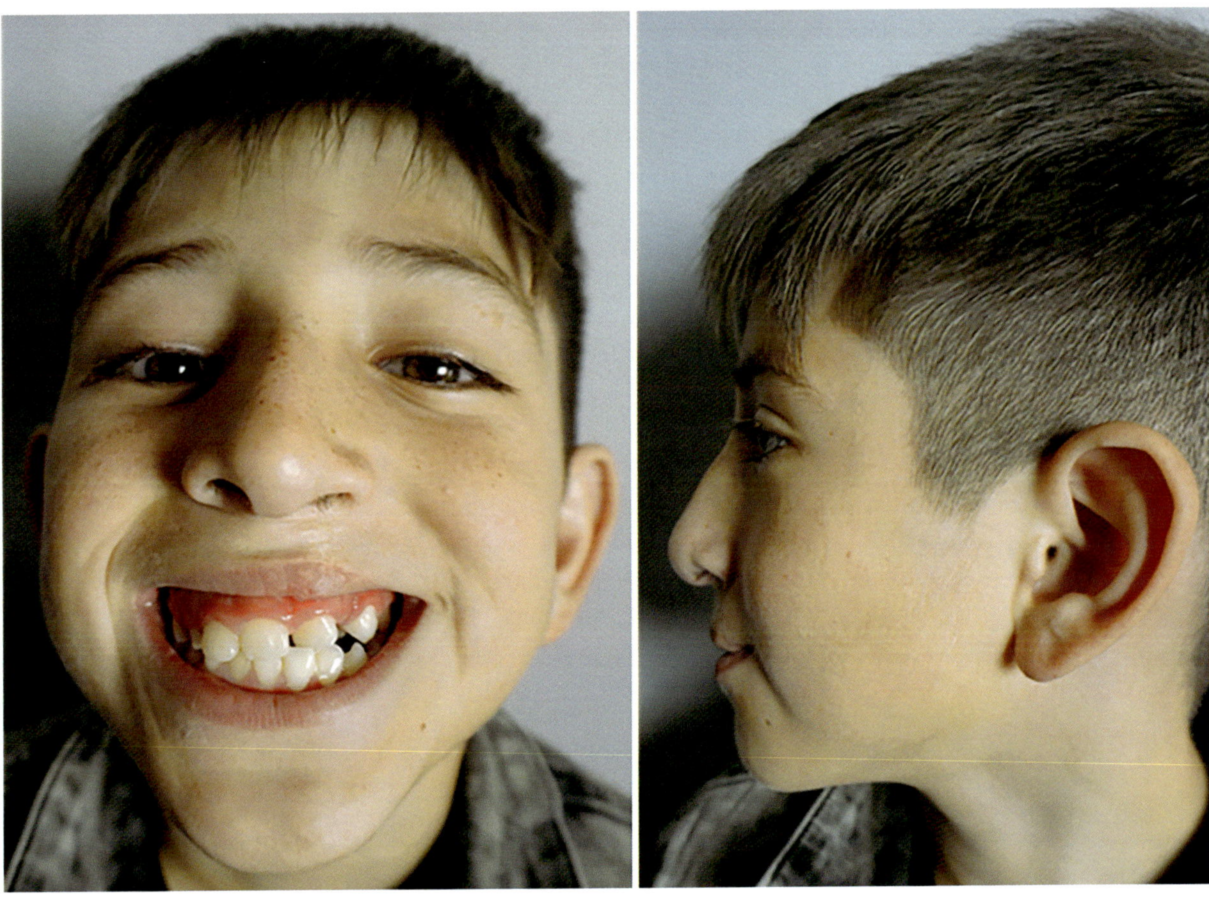

Case 13 Unilateral
9 years post palatoplasty

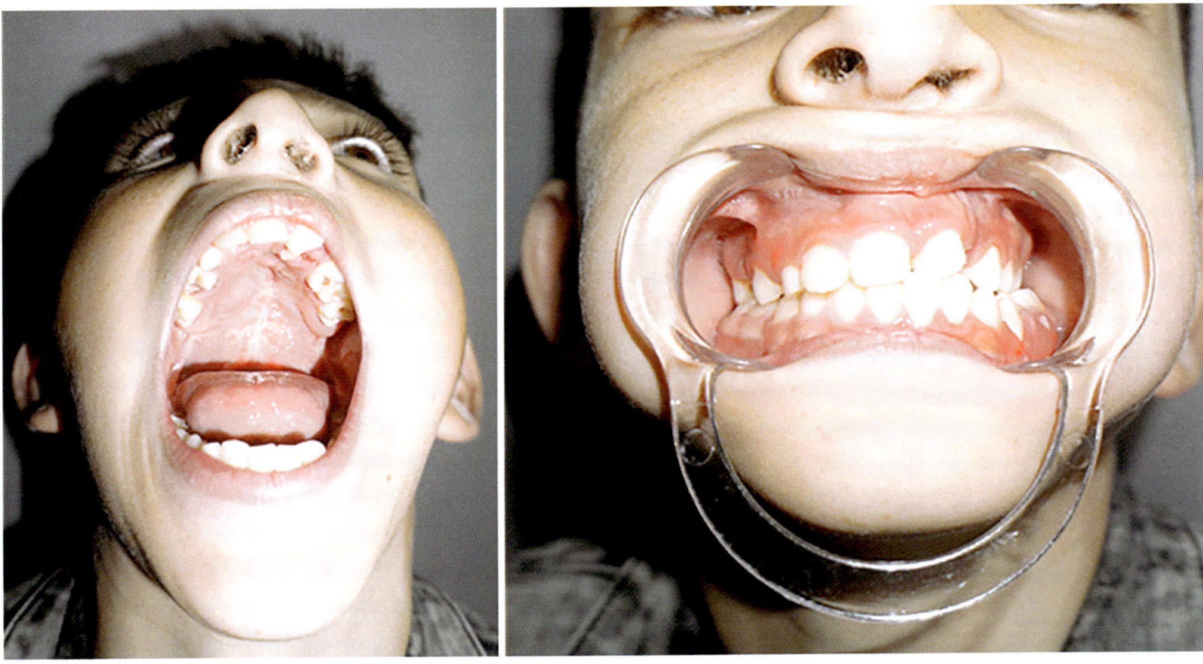

Case 14 Unilateral

Cleft lip and nose reconstruction—12 years post op

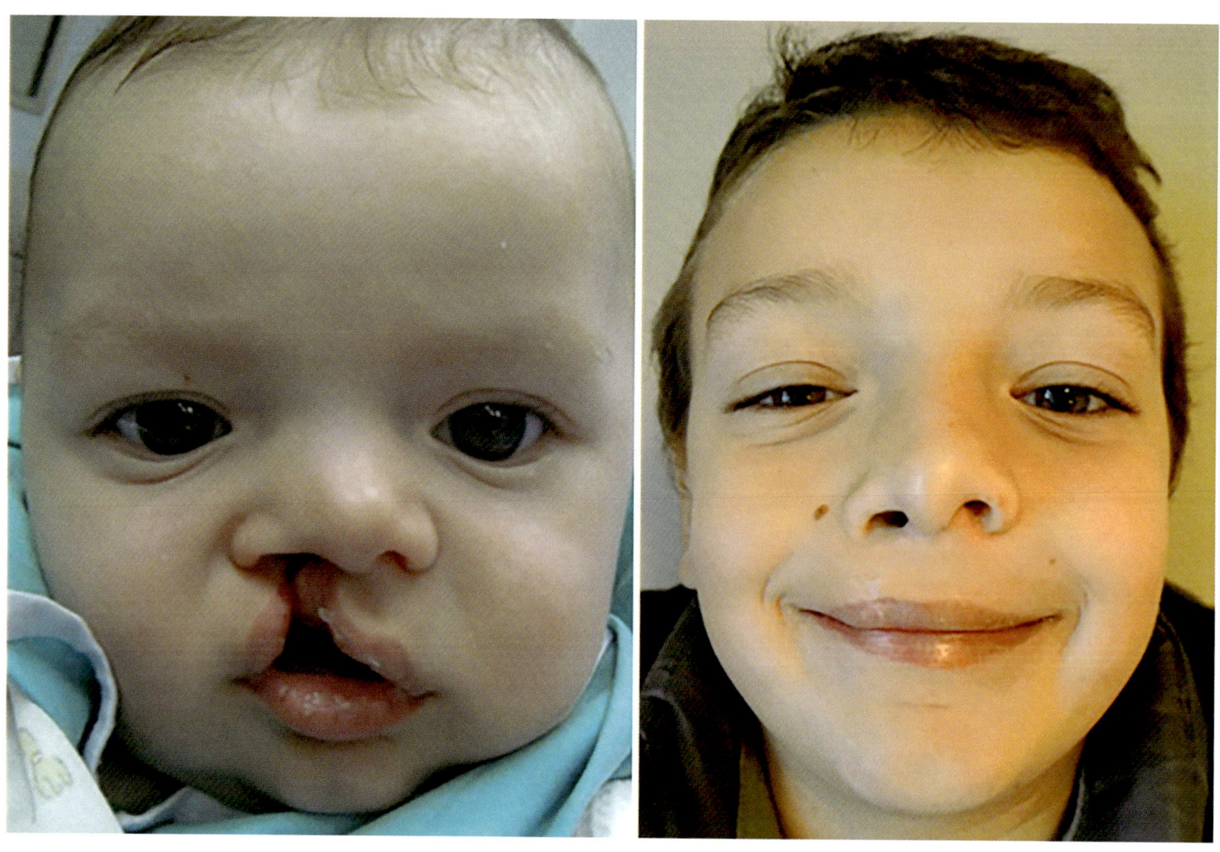

Case 14 Unilateral

Bilateral AEP flap reconstruction (minimal incisions)

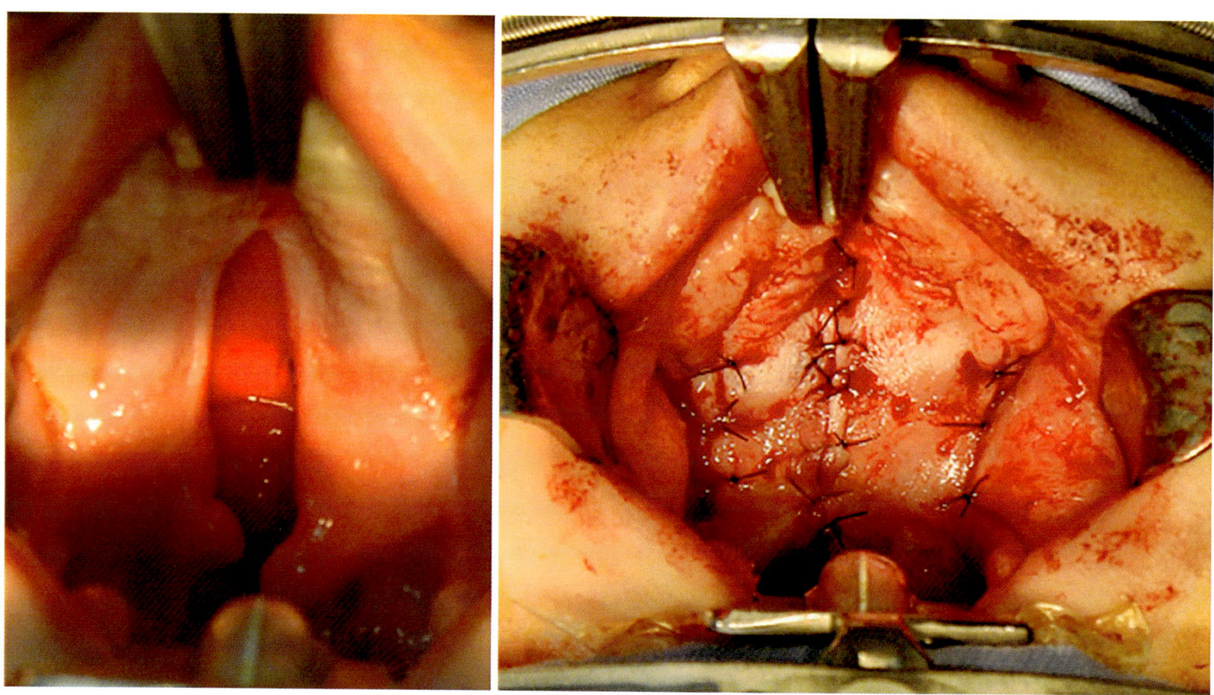

Case 14 Unilateral
12 years post palatoplasty

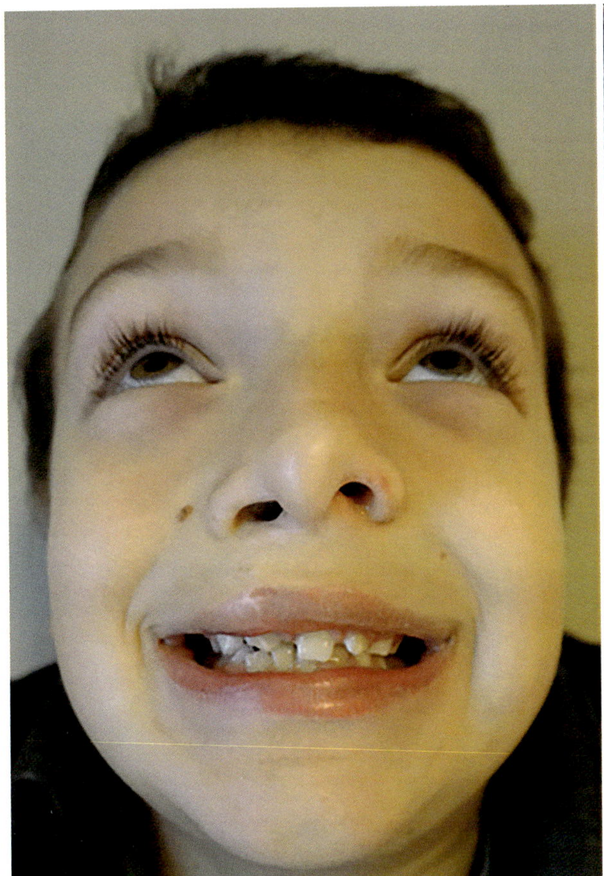

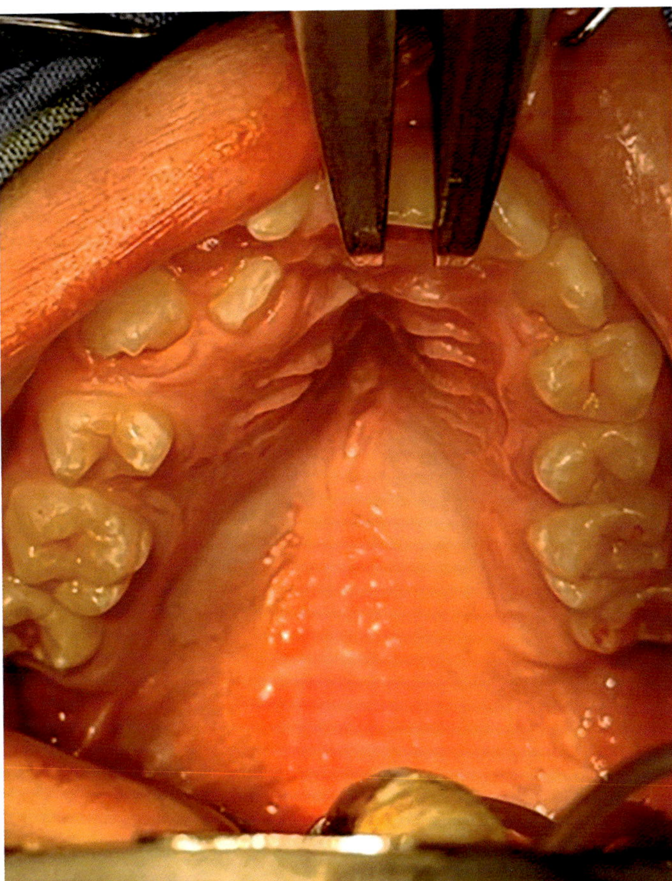

Alveolar Extension Palatoplasty: Conclusions

- Design of AEP is embryologically sound
- Dental eruption is unaffected by incisions
- Avoids transverse retraction of the Veau–Wardill procedure
- Cross-bite 5–10%
- Reduction in class III occlusion
- Reassignment of the mucoperiosteal flap(s) into correct developmental relationships facilitates alveolar remodeling and bone regeneration in the correct anatomic position

Prof. Ricardo D. Bennun, MD, MS, PhD

- Director, Asociación PIEL, Buenos Aires—Argentina
- Director Post Graduate Program in Pediatric Plastic and Craniofacial Surgery—National University of Buenos Aires, School of Medicine—Buenos Aires, Argentina
- Post Graduate Professor, Orthodontic Department, Dental School, Maimonides University—Buenos Aires, Argentina

A Note from Dr. Carstens

Ricardo Bennun

Alveolar extension palatoplasty as a concept is uniquely South American, both in its design and in its surgical verification. AEP, in concept of the palatal version of subperiosteal tissue transfer for cleft lip repair, was drawn out on a paper napkin during an airplane flight to Ecuador. Although the operation was described in 1999, it was subsequently picked up by Dr. Luis Monasterio, a distinguished cleft surgeon in Santiago, Chile. Lucho invited me to do some cases with him at the Fundación Gantz. Since the incisions required to reposition the entire embryonic field of the hard palate were made on the alveolar ridge, the effect of these on dental eruption was an issue of debate. Dr. Monasterio did his own series, following dental development for several years. He determined that eruption was unaffected (which he subsequently reported to the 12th American Cleft Palate Association meeting). He also introduced me to Dr. Ricardo Bennun from the University of Buenos Aires and director of the Fundación Piel. Ricardo subsequently took the operation to the next level, beginning a case series which now extends to over a decade. Ricardo is a true biologic surgeon, reflecting his long-time commitment to burn care; he keeps on asking questions and seeking answers. I had the opportunity to contribute two chapters for his 2015 work, "Cleft Lip and Palate Management: A Comprehensive Atlas"; in the process, his attention to detail and insistence of quality forced me to think about the issues more deeply than before. It is his influence that really pushed me over the edge, daring me to write this book, a task that seemed overwhelming to me at the time. True to form, his atlas of AEP cases faithfully recorded herein will stand the test of time as a surgical proof that developmental biology can win out over dogma for better patient outcomes. For this, all cleft surgeons will be grateful.

References

1. Millard DR. Cleft craft: the evolution of its surgery. Vol. III. Alveolar and palatal deformities. Boston: Little, Brown; 1980.
2. Ortiz-Monasterio F, Serrano A, Barrera G, et al. A study of untreated adult cleft palate patients. Plast Reconstr Surg. 1966;38:36.
3. Ortiz-Monasterio F, Olmedo A, Trigos I, et al. Final results of the delayed treatment of patients with clefts of the lip and palate. Scand J Plast Reconstr Surg. 1974;8:109.
4. Demas PN, Sotereanos GC. Closure of alveolar clefts with cortico-cancellous block grafts and marrow: a retrospective study. J Oral Maxillofac Surg. 1988;46:682.
5. Carstens MH. The sliding sulcus procedure: simultaneous repair of unilateral clefts of the lip and primary palate—a new technique. J Craniofac Surg. 1999;10:415–29.
6. Boorman JF, Friedlander E. Surgical anatomy of the velum and pharynx. Rec Adv Plast Surg. 1992;4:17.
7. Ruding R. Cleft palate: anatomic and surgical considerations. Plast Reconstr Surg. 1964;33:132.
8. Maher WP, Swindle PF. Submucosal blood vessels of the palate. Dent Prog. 1962;2:167.
9. Maher WP. Artery distribution in the prenatal human maxilla. Cleft Palate J. 1981;18:51.
10. Maher WP. Distribution of palatal and other arteries in cleft and noncleft human palates. Cleft Palate J. 1977;14:1.
11. Skoog T. Plastic surgery: new methods and refinements. Philadelphia: WB Saunders; 1974.
12. Bardach J. Cleft palate repair: two-flap palatoplasty. Research, philosophy, technique, and results. In: Bardach J, Morris H, editors. Multidisciplinary management of cleft lip and palate. Philadelphia: WB Saunders; 1990. p. 352–65.
13. Urist MR, McLean FC. Osteogenic potency and new-bone formation by induction in transplants to the anterior chamber of the eye. J Bone Jt Surg Am. 1952;34A:443.
14. Azzolini A, Riberti C, Roselli D, Standini L. Tibial periosteal graft in repair of cleft lip and palate. Ann Plast Surg. 1982;9:105.
15. Denny AD, Talisman R, Bonawitz SC. Secondary alveolar bone grafting using milled cranial bone graft: a retrospective study of a consecutive series of 100 patients. Cleft Palate Craniofac J. 1999;36:144.
16. Skoog T. A design for the repair of the unilateral cleft lip. Am J Surg. 1958;95:223.
17. Skoog T. Modern procedures in uni- and bilateral clefts of the lip, alveolus, and hard palate with respect to primary osteoplasty. In: Schukardt K, editor. Treatment of patients with clefts of the lip, alveolus, and palate. Stuttgart: Thieme; 1966. p. 48–67.
18. Skoog T. Repair of unilateral cleft lip deformity: maxilla, nose and lip. Scand J Plast Reconstr Surg. 1969;2:109.
19. Skoog T. The use of periosteal flaps in the repair of the primary palate. Cleft Palate J. 1965;24:3.
20. Skoog T. The use of periosteum and surgicel for bone formation in congenital clefts of the maxilla. Scand J Plast Reconstr Surg. 1967;1:113.
21. Carstens MH, Chin M, Li XJ. In situ osteogenesis: regeneration of 10-cm mandibular defect in porcine model using recombinant human bone morphogenetic protein-2 (rhBMP-2) and Helistat absorbable collagen sponge. J Craniofac Surg. 2005a;16(6):1033–42.
22. Chin M, Tom WK, Carstens MH. Primary reconstruction of alveolar clefts using recombinant human bone morphogenetic protein-2: clinical and radiologic outcomes. J Craniofac Surg. 2009;20(Suppl 2):1766–7. https://doi.org/10.1097/SCS.0b013e3181b5d0c2.
23. Fallucco MA, Carstens MH. Primary reconstruction of alveolar clefts using recombinant human bone morphogenic protein-2: clinical and radiographic outcomes. J Craniofac Surg. 2009;20(Suppl 2):1759–64. https://doi.org/10.1097/SCS.0b013e3181b5d08e.
24. Carstens MH, Chin M, Ng T, Tom WK. Reconstruction of #7 facial cleft with distraction-assisted in situ osteogenesis (DISO): role of recombinant human bone morphogenetic protein-2 with Helistat-activated collagen implant. J Craniofac Surg. 2005b;16(6):1023–32.
25. Chao M, Donovan T, Sotelo C, Carstens MH. In situ osteogenesis of hemimandible with rhBMP-2 in a 9-year-old boy: osteoinduction via stem cell concentration. J Craniofac Surg. 2006;17(3):405–12.
26. Tom WK, Chin M, Ng T, Bouchoucha S, Carstens M. Distraction of rhBMP-2-generated mandible: how stable is the engineered bone in response to subsequent surgeries? J Oral Maxillofac Surg. 2008;66(7):1499–505. https://doi.org/10.1016/j.joms.2007.12.011.
27. Franco J, Coppage J, Carstens MH. Mandibular distraction using bone morphogenic protein and rapid distraction in neonates with Pierre Robin syndrome. J Craniofac Surg. 2010;21(4):1158–61. https://doi.org/10.1097/SCS.0b013e3181e47d58.
28. Barro WB, Latham RA. Palatal response to surgical trauma. Plast Reconstr Surg. 1981;67:6.
29. You ZH, Zhang ZK, Xia JI. The study of vascular communications between jaw bone casts and the surrounding tissues by SEM off resin casts. West Chin J Stomatol. 1991a;8:235.
30. You ZH, Zhang ZK, Xias JL. Blood supply of jaw bone mucoperiosteum and its role in orthognathic surgery. Chin J Stomatol. 1991b;26:31.
31. You ZH, Zhang ZK, Zhang XE. Le Fort I osteotomy with descending palatal artery intact and ligated: a study of blood flow and quantitative histology. Contemp Stomatol. 1991c;5:71.
32. Lanigan DT, Hey JH, West RA. Aseptic necrosis following maxillary osteotomies. A report of 3 cases. J Oral Maxillofac Surg. 1990;48:142.
33. Lanigan DT. Wound healing after multisegmental Le Fort I osteotomy and transection of the descending palating vessels (discussion). J Oral Maxillofac Surg. 1995;53:1433.
34. Dahl E. Craniofacial morphology in congenital clef t of the lips and palate. Acta Odontol Scand. 1970;28(Suppl):57.
35. Markus AF, Delaire J. Functional primary closure of cleft lip. Br J Oral Maxillofac Surg. 1993;31:281.

36. Mars M, Houston WJB. A preliminary study of facial growth and morphology in unoperated male unilateral cleft lip and palate subjects over 13 years of age. Cleft Palate Craniofac J. 1990;27:7.
37. Mars M, James DR, Lamabadusuriya SP. The Sri Lankan cleft lip and palate project: the unoperated cleft lip and palate. Cleft Palate Craniofac J. 1990;27:3.
38. Bishara SE, Krause JC, Olin WH, et al. Facial and dental relationships of individuals with unoperated clefts of the lip and/ or palate. Cleft Palate J. 1976;213:238.
39. Bishara SE. Cephalometric evaluation of facial parameters in operated and nonoperated individuals with isolated clefts of the palate. Cleft Plate J. 1979;3:239.
40. Bishara SE. Maxillofacial growth and development: the method of analysis. In: Morris HL, editor. The Bratislava project some results of cleft palate surgery. Iowa City: University of Iowa Press; 1978.
41. Mestre JC, De Jesus J, Subtelny JD. Unoperated clefts at maturation. Angle Orthod. 1960;30:78.
42. Ortiz-Monasterio F, Rebel AS, Valderramo M, Cruz R. Cephalometric measurements on adult patients with nonoperated cleft palates. Plast Reconstr Surg. 1959;24:54.
43. Bardach J, Eisbach KJ. The influence of primary unilateral cleft lip repair on facial growth. Part I: lip pressure. Cleft Palate J. 1977;14:88.
44. Bardach J, Kelly KM, Jakobsen JR. Simultaneous cleft lip and palate repair: an experimental study in beagles. Plast Reconstr Surg. 1988;82:31.
45. Bardach J, Kelly KM. The influence of lip repair with and without soft tissue undermining on facial growth in beagles. Plast Reconstr Surg. 1988;82:747.
46. Bardach J, Mooney M, Bardach E. The influence of two-flap palatoplasty on facial growth in beagles. Plast Reconstr Surg. 1982a;69:927.
47. Bardach J, Mooney MP, Giedroc-Juraha ZL. A comparative study of facial growth following cleft lip repair with or with out soft-tissue undermining: an experimental study in rabbits. Plast Reconstr Surg. 1982b;69:745.
48. Nelson RL, Path MH, Ogle RG, et al. Quantitation of blood flow after a Le Fort I osteotomy. J Oral Surg. 1977;35:10.
49. Siebert JW, Agrigiani C, McCarthy JG, Longaker MT. Blood supply of the Lefort I maxillary segment: an anatomic study. Plast Reconstr Surg. 1997;100:843.
50. Poupard B, Coorneart H, Debaere PA. Cleft lip and cleft palate: can the hard palate be left open? A study of 62 cases with a follow up of 6 years or more. Ann Chir Plast. 1983;28:325.
51. Walter JD, Hale V. A study of the long-term results achieved by the Gillies-Fry procedure. Br J Plast Surg. 1987;40:384.
52. Koberg K, Koblin J. Speech development and maxillary growth in relationship to technique and timing of palatoplasty. J Oral Maxillofac Surg. 1973;1:44.
53. Schweckendieck W. Primary veloplasty: long term results without maxillary deformity. A 25 year report. Cleft Palate J. 1978;15:268.
54. Ash MM. Dental anatomy, physiology, and occlusion. 7th ed. Philadelphia: WB Saunders; 1993.
55. Avery JK. Oral development and histology. 2nd ed. New York: Thieme; 1994.
56. Fry WK. The dental aspects of the treatment of congenital cleft palates. Proc R Soc Med. 1921;14:57.
57. Carstens MH. The anteriorly-based buccinator myomucosal flap. Plast Reconstr Surg. 1991;88:39–50.
58. Kaplan EN. Cleft palate repair at 3 months? Ann Plast Surg. 1981;7:179.
59. Markus AF, Smith WP, Delaire J. Primary closure of clef t palate: a functional approach. Br J Oral Maxillofac Surg. 1993;31:71.
60. Mahajan RK, Chhajlani R, Ghildiyal HC. Role of tongue flap in palatal fistula repair: a series of 41 cases. Indian J Plast Surg. 2014;47(2):210–5.
61. Subtelny JD. Studies of the configuration of the nasopharynx and palatal segments in children with clefts as they relate to embryologic studies. In: Pruzansky S, editor. Congenital anomalies of the face and associated structures. Springfield: Charles C. Thomas; 1961.
62. Subtelny JD. Width of the nasopharynx and related structures in normal and unoperated cleft palate children. Am J Orthod. 1955;41:1889.
63. Smith DM, Vecchinone L, Jang S, Ford M, Deleyiannis FW, Haralam MA, Naran S, Worrall C, Dudas JR, Afifi AM, Marazita ML, Losee JE. The Pittsburgh fistula classification system: a standardized scheme for the description of palatal fistulas. Cleft Palate Craniofac J. 2007;44:590–4.
64. Naidu P, Yao CA, Chong DK, Magee WP III. Cleft palate repair: a history of techniques and variations. Plast Reconstr Surg Glob Open. 2022;10(3):e4019. https://doi.org/10.1097/GOX.0000000000004019.
65. Bennun RD, Monasterio AL. Chap. 11: Cleft palate repair. In: Bennun RB, Harfin JF, Sandor GKB, Genecov D, editors. Cleft lip and palate management: a comprehensive atlas. New York: Wiley Blackwell; 2015. p. 163–73.
66. Bennun RD. Cleft palate repair: predictive factors of difficulty and planned strategies to solve it. J Craniofac Surg. 2020;31:1664–7.
67. Astrada S, Bennun RD. Cleft palate repair: a study between two surgical procedures. J Craniofac Surg. 2020;31:2280–4.
68. Carstens MH. Chap. 1: Mechanisms of cleft palate: developmental field analysis. In: Bennun RB, Harfin JF, Sandor GKB, Genecov D, editors. Cleft lip and palate management: a comprehensive atlas. New York: Wiley Blackwell; 2015. p. 3–21.
69. Moggi LE, Ventorutti T, Bennun RD. Cleft palate repair: a new maxillary nerve approach. J Craniofac Surg. 2020;31:1547–50.
70. Bennun RD, Dogliotti PL. Anatomical bases for one stage reconstruction repair in bilateral cleft lip. In: 8th Congress European Association for maxillo-facial surgery, Abstracts book p. 240.
71. Bennun RD. Chap. 3: The nasolabial region: a revision of the vascular anatomy. In: Bennun RB, Harfin JF, Sandor GKB, Genecov D, editors. Cleft lip and palate management: a comprehensive atlas. New York: Wiley Blackwell; 2015. p. 41–6.
72. Carstens MH. Chap. 8: Developmental field reassignment cleft surgery: reassessment and refinements. In: Bennun RB, Harfin JF, Sandor GKB, Genecos D, editors. Cleft lip and palate management: a comprehensive atlas. New York: Wiley Blackwell; 2015. p. 83–112.
73. Rivelli RA, Casadio V, Bennun RD. Audiological alterations in patients with cleft palate. J Craniofac Surg. 2018;29:1486–9.
74. Harfin JF, Bennun RD. Chap. 15: Strengthening surgical/orthodontic interrelationships. In: Bennun RB, Harfin JF, Sandor GKB, Genecov D, editors. Cleft lip and palate management: a comprehensive atlas. New York: Wiley Blackwell; 2015. p. 227–41.
75. Harfin JF. Chap. 16: To what extent dental alveolar osteogenesis can be achieved solely with orthodontic treatment in cleft patient? In: Bennun RB, Harfin JF, Sandor GKB, Genecov D, editors. Cleft lip and palate management: a comprehensive atlas. New York: Wiley Blackwell; 2015. p. 245–52.

Further Reading

Amartunga NA. A comparison of Millard's and Le Mesurier's methods of repair for the complete unilateral cleft lip using a new symmetry index. J Oral Maxillofac Surg. 1988;46(5):353–6.

Bardach J. Rozszczepy Wargi Gornej i Podniebienia. Lekarskich. Warszawa: Panst Zaklad Wydawnictw; 1967.

Bardach J. Facial growth following cleft lip repair: experimental studies in rabbits and beagles. In: Jackson IT, Sommerlad B, editors. Recent advances in plastic surgery. Edinburgh: Churchill Livingstone; 1984a.

Bardach J, Mooney MP. The relationship between lip pressure following lip repair and craniofacial growth: an experimental study in beagles. Plast Reconstr Surg. 1984b;73:544.

Bardach J, Morris HL, Olin WH. Late results of primary periosteoplasty: the Marburg project. Plast Reconstr Surg. 1984c;73:207.

Bell WH. Revascularization and bone healing after anterior maxillary osteotomy: a study using adult rhesus monkeys. J Oral Surg. 1969;27:249.

Bell WH. Revascularization and bone healing after posterior maxillary osteotomy. J Oral Surg. 1971;29:313.

Bell WH, Fonseca RJ, Kennedy JW III, Levet BM. Bone healing and revascularization after total maxillary osteotomy. J Oral Surg. 1975;33:253.

Bell WH, Mannai C, Luhr HG. Art and science of Le Fort I downfracture. Int J Adult Orthod Orthognath Surg. 1988;3:23.

Bell WH, You ZH, Finn RA, Fields R. Wound healing after multisegmental Le Fort I osteotomy and transection of the descending palatine vessels. J Oral Maxillofac Surg. 1995;33:1425.

Bennun RD, Monasterio-Aljaro L. Cleft palate repair. In: Bennun R, Harfin JF, editors. Cleft lip and palate management: a comprehensive atlas. New York: Wiley-Blackwell; 2015.

Bishara SE, Sosa-Martinez R, Vales HP, Jakobsen JR. Dentofacial relationship in persons with unoperated clefts: comparisons between three cleft types. Am J Orthod. 1985;87:481.

Bishara SE, Jakobsen JR, Krause JC, Sos-Martinez R. Cephalometric comparisons of individuals from India and Mexico with unoperated cleft lip and palate. Cleft Palate J. 1986;23:116.

Bonanthaya K, Shetty P, Sharma A, Ahlawat J, Passi D, Singh M. Treatment modalities for surgical management of anterior palatal fistula: comparison of various techniques, their outcomes, and the factors governing treatment plan: a retrospective study. Natl J Maxillofac Surg. 2016;7(2):148–52. https://doi.org/10.4103/0975-5950.201357.

Bozola AR, Gasques JAL, Carriquiry CE, Cardoso de Oliveira M. The buccinator musculomucosal flap: anatomic study and clinical application. Plast Reconstr Surg. 1989;84(2):250–7.

Brusati R, Manucci N. The early gingivoperiosteoplasty. Primary results. Scand J Plast Reconstr Surg. 1992;26:65.

Carstens MH. The buccinator musculomucosal flap: anatomic study and clinical application. Plast Reconstr Surg. 1989;84:250–7.

Carstens MH. Correction of the unilateral cleft lip nasal deformity using the sliding sulcus procedure. J Craniofac Surg. 1999a;10:346–64.

Cohen SR, Kalinowski JK, LaRossa D, Randall P. Cleft palate fistulas: a multi-variate statistical analysis of prevalence, etiology, and surgical management. Plast Reconstr Surg. 1991;87(6):1041–7.

Coupe TB, Subtelny JD. Cleft palate-deficiency or displacement of tissue. Plast Reconstr Surg. 1960;26:600.

Delaire J. La Cheilorhinoplastie primaire pour fente labio maxillaire congentiale unilater ale (essair de schemiatisatione d'une technique). Rev Stomatol Chir Maxillofac. 1975;76:193.

Delaire J. The potential role of facial muscles in monitoring maxillary growth and morphogenesis. In: Carlson DS, McNamara Jr JA, editors. Muscle adaptation in the craniofacial region. Monographs no. 8-cranial growth series. Ann Arbor: University of Michigan; 1978. p. 157–80.

Delaire J. Theoretical principles and technique of functional closure of the lip and nasal aperture. J Maxillofac Surg. 1978;6:109.

Delaire J. L'anatomie et la physiologie des muscles nasolabiaux chez sujet noirmal opere d'une fente labiomaxillaire. Acta Orthod. 1980;8:269.

Delaire J. Considerations sur l'accrosissement du premaxillarie clans les fentes labiomaxillaires. Rev Orthop Dent Fadale. 1991;25:453.

Delaire J, Brunatti S, editors. La Rehabilitation fonctionelle chirugicale et orthopaedique des fentes labio-maxillo palatines congenitales. Premieres resultats de la gingivoperiosteoplastie primaire (avec or sans osteoplastie). Milan: ALPS; 1989. p. 121–31.

Delaire J, Chateau JP. Comment le septum nasale influencet-il la croissance premaxillaire et maxillaire? Rev Stomatol Chir Maxillofac. 1977;78:241.

Delaire J, Precious DS. Influence of the nasal septum on maxillofacial growth in patients with congenital labiomaxillary clefts. Cleft Palate J. 1986;23:270.

Delaire J, Precious DS. Interaction in the development of the nasal septum, the nasal pyramid, and the face. Int J Paediatr Otorhinolaryngol. 1987;12:311.

Delaire J, Talment JC, Billet J. Evolution des techniques de cheilorhinoplastie pour fentes labiomaxillaires. Rev Stomata! Chir Maxillofac. 1977;79:241.

Delaire J, Precious D, Gordeef A. The advantage of wide subperiosteal exposure in primary surgical correction of labial maxillary clefts. Scand J Plast Reconstr Surg. 1988;22:147.

Dingman RO, Argent LO. The correction of cleft palate with primary veloplasty and delayed repair of the hard palate. Clin Plast Surg. 1985;12:677.

Dingman RO, Crabb WC. A rational program for surgical management of bilateral cleft lip and palate. Plast Reconstr Surg. 1971;47:239.

Enlow DH. Facial growth. 3rd ed. Philadelphia: WB Saunders; 1990.

Enlow DA, Bang S. Growth and remodeling of the human maxilla. Am J Orthod. 1965;51:446.

Friede H, Lilja J, Johanson B. Cleft lip and palate treatment with delayed closure of the hard palate. Scand J Plast Reconstr Surg. 1967;21:65.

Gillies HG, Fry WK. A new principle in the surgical treatment of "congenital clef t palate" and its mechanical counterpart. BMJ. 1921;1:335.

Goldstein MH. The elastic flap for lip repair. Plast Reconstr Surg. 1990;85:446.

Hagemann R. Uber Spatfolgen des Operationen Gauments paltenverschlusses. Bruns Beitr Kin Chir. 1941;79:573.

Hardwicke JT, Landini G, Richard BM. Fistula incidence after primary cleft palate repair: a systematic review of the literature. Plast Reconst Surg. 2015;134(4):618e–27e.

Huang MHS, Lee ST, Rajendran K. Anatomic basis of cleft palate and velopharyngeal surgery: implication from a fresh cadaveric study. Plast Reconstr Surg. 1998;101:613.

Huang MS, Lee ST, Rajendran K. Clinical implications of the velopharyngeal blood supply: a fresh cadaveric study. Plast Reconstr Surg. 1998;102:655.

Joos U. Muscle reconstruction in primary cleft lip surgery. J Craniomaxillofac Surg. 1989;37:8.

Kaplan EN. A T-shaped musculomucosal buccal flap for cleft palate surgery (discussion). Plast Reconstr Surg. 1988;79:896.

Kriens OB. An anatomical approach to veloplasty. Plast Reconstr Surg. 1969;42:29.

Lumsden A, Krumlauf R. Patterning the vertebrate neuraxis. Science. 1996;274(5290):1109–15.

Maeda K, Ojimi H, Utsurgi R, Ando S. A T-shaped musculomucosal buccal flap method for cleft palate surgery. Plast Reconstr Surg. 1987;79:888.

Maher WP, Swindle PF. Variation in the network of submucosal arteries in the human fetal palate. J Am Dent Assoc. 1969;69:106.

Markus AF, Delaire J, Smith WP. Facial balance in cleft lip and palate. I. Normal development and cleft palate. Br J Oral Maxillofac Surg. 1992a;30:290.

Markus AF, Delaire J, Smith WP. Facial balance in cleft lip and palate. II. Cleft lip and palate and secondary deformities. Br J Oral Maxillofac Surg. 1992b;30:290.

Massa A. Reconstruction of the cleft maxilla with periosteoplasty. Scand J Reconstr Surg. 1986;20:41.

Millard DR. Cleft craft. The unilateral deformity, vol. I. Boston: Little, Brown; 1976.

Murthy J. Descriptive study of management of palatal fistula in one hundred and ninety-four cleft individuals. Indian J Plast Surg. 2011;44(10):41–6.

Oilier L. Sur l'developpement et la crue des os des animaux. Mem Acad R Sci (Paris). 1742;55:354.

Precious DS, Delaire J. Balanced facial growth: a schematic interpretation. Oral Surg Oral Med Oral Pathol. 1987;63:637.

Precious DS, Delaire J. Surgical consideration in patients with cleft deformities. In: Bell WH, editor. Modern practice in orthognathic and reconstructive surgery, vol. I. Philadelphia: WB Saunders; 1992. p. 390–425.

Rayner HH. The operative treatment of cleft palate: a record of results in 125 consecutive cases. Lancet. 1925;205:816.

Rohrich RJ, Byrd HS. Optimal timing of cleft palate closure: speech, facial growth, and hearing considerations. Clin Plast Surg. 1990;17:27.

Ross BR. Treatment variables affecting facial growth in complete unilateral cleft lip and palate. Cleft Palate J. 1987;24:45.

Sadhu P. Oronasal fistula in cleft palate surgery. Indian J Plast Surg. 2009;42(Suppl):S123–8. https://doi.org/10.4103/0970-0358.57203.

Sandor GKB, Ylikontiola LP, Carmichael RP. Midfacial distraction osteogenesis. Atlas Oral Maxillofac Clin N Am. 2008;16(2):249–72.

Santiago PE, Grayson BH, Cutting CB, et al. Reduced need for alveolar bone grafting by presurgical orthopedics and primary gingivoperiosteoplasty. Cleft Palate Craniofac J. 1997;35:77.

Sarnat BG. Palatal and facial growth in *Macaca rhesus* monkeys with surgically produced palatal clefts. Plast Reconstr Surg. 1958;22:29.

Schultz RC. Free periosteal graft repair of maxillary clefts in adolescents. Plast Reconstr Surg. 1984;73:556.

Schweckendieck H. Zur Frage der Frun-und Spatoperationen der angeborenen Lippen-Keifer-Gaumens palten (mit Demonstrationen). Z Laryngol Rhinol Oto! 1951;30:51.

Schweckendieck H. Die Ergebnisse der Kiefer und die Sprache nach der pimaren veloplastik. Arch Ohr Nas Ke belkopfheilk. 1962;180:541.

Schweckendieck H. Der Zeitplan bei der Behandlw1g der Lippe-Kiefer Gaumenspalten. Laryngo Rhino Otologie (Stuttgart). 1964;43:246.

Sitzmann F. The alveolar flap for the repair of the cleft alveolus related to the development of the upper jaw. J Maxillofac Surg. 1979;7:81.

Skoog T. Skoog's methods of repair of unilateral and bilateral cleft lip. In: Grabb WC, Rosenstein SW, Bzoch KR, editors. Cleft lip and palate. Boston: Little, Brown; 1971. p. 288–304.

Slaughter WB, Pruzansky S. The rationales for velar closure as a primary procedure in the repair of cleft palate defects. Plast Reconstr Surg. 1954;13:341.

Smith WP, Markus AF, Delaire J. Primary closure of the cleft alveolus: a functional approach. Br J Oral Maxillofac Surg. 1993;33:156.

Subtelny JD. A cephalometric study of the growth of the soft palate. Plast Reconstr Surg. 1957;19:49.

Tajima S. The importance of the musculus nasalis and the use of the cleft margin flap in the repair of complete unilateral clef t lip. J Maxillofac Surg. 1983;11:64.

Taylor GI, Palmer JH. The vascular territories (angiosomes) of the body: experimental study and clinical applications. Br J Plast Surg. 1987;46:113.

Taylor GI, Gianoutsos MP, Morris SF. The neurovascular territories of the skin and muscles: anatomic study and clinical implications. Plast Reconstr Surg. 1994;94:1.

Uddstromer L, Ritsila V. Healing of membranous and long bone defects. Scand J Plast Reconstr Surg. 1979;13:281.

Veau V. Division palatine. Paris: Masson; 1931.

Veau V. Bec-de-lievre. Paris: Masson; 1936.

Walker CW, Collito JCMB, Mancusi-Ungaro A, Meijer R. Physiologic considerations in cleft lip closure: the C-W technique. Plast Reconstr Surg. 1966;37:552.

Wisnia P, Monasterio L. Palatoplastīa con extension alveolar (Carstens). Alveolar extension palatoplasty. Palm Springs: American Cleft Palate Association; 2022.

Wood RJ, Grayson BH, Cutting CB. Gingivoperiosteoplasty and midfacial growth. Cleft Palate Craniofac J. 1997;34:17.

Yang L, Suzuki EY, Suzuki B. Biomechanical comparison of two intraoperative mobilization techniques for maxillary distraction osteogenesis: down-fracture versus non-down-fracture. Ann Maxillofac Surg. 2014;4(2):162–70.

16 Pathologic Anatomy of the Soft Palate

Michael H. Carstens

Note to the Reader

This chapter is designed to give you an in-depth appreciation of the soft palate in terms of its component parts and its functional organization as an elaborate *biological pinball machine* consisting of (1) a *bony platform* and (2) a *soft tissue lever arm*. If you have been moving sequentially through this book, you will have digested a great deal of information about mesenchymal mapping and vascularity (Figs. 16.1 and 16.2).

There are surprises in store! Our analysis will reveal (1) that the soft palate is a three-pharyngeal arch structure; (2) that its developmental mechanism is directly related to that of the tongue; (3) that paralysis/paresis of the facial nerve can affect speech; (4) that deficiencies of the horizontal plate of palatine bone, P3, are responsible for virtually all forms of soft palate clefts; (5) that an underappreciated zone, the anterior palatine aponeurosis, is the master key to understanding the abnormal positioning of the soft palate muscles against the hard palate; and (6) that the foreshortened soft palate requires a restoration of length that corresponds to a developmental field defect in the anterior palatine aponeurosis. So buckle up for the ride… Let's take the plunge.

M. H. Carstens (✉)
Wake Forest Institute of Regenerative Medicine, Wake Forest University, Winston-Salem, NC, USA
e-mail: mcarsten@wakehealth.edu

M. H. Carstens (ed.), *The Embryologic Basis of Craniofacial Structure*, https://doi.org/10.1007/978-3-031-15636-6_16

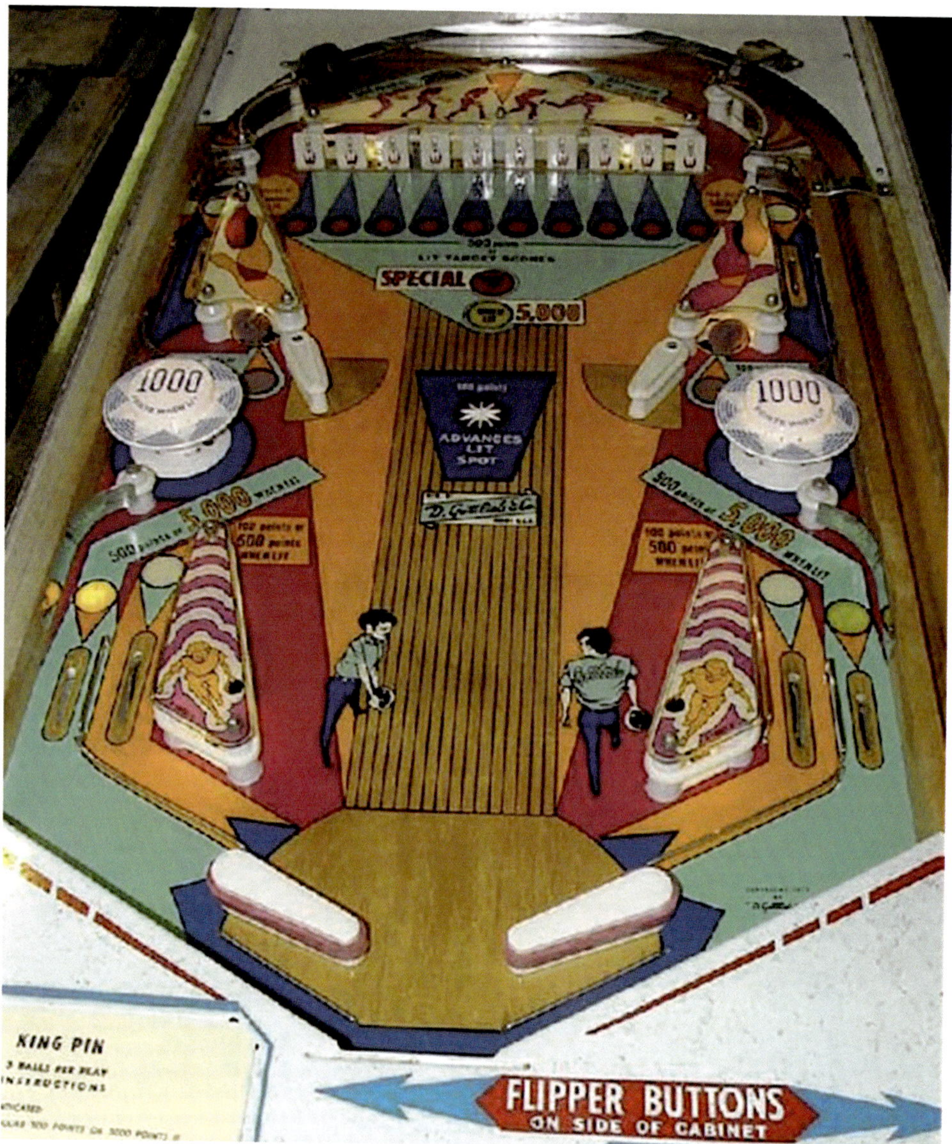

Fig. 16.1 Pinball model. The palate is a functional unit consisting of a bony platform and paired (fused) soft tissue lever arms designed to control speech and swallowing. [Courtesy of Michael Carstens, MD]

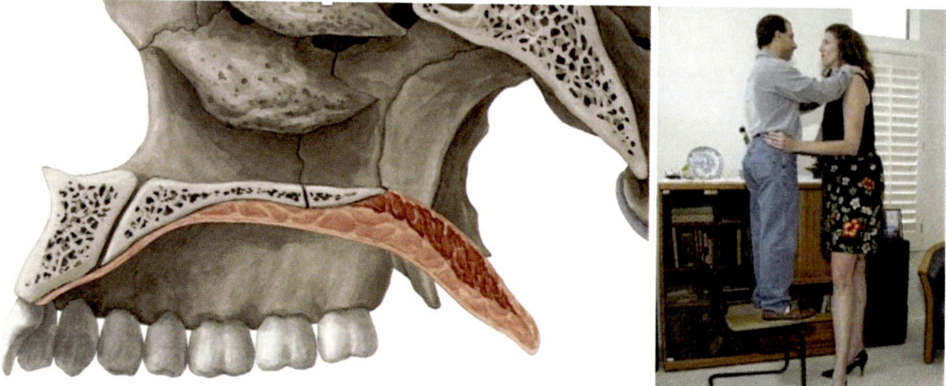

Fig. 16.2 Size discrepancy. The function of the soft palate depends upon its ability to close a critical contact distance with the pharynx. In cleft palate, the platform to which the lever arm is attached, the palatine bone and the anterior palatine aponeurosis are foreshortend. Despite an otherwise technically sound repair, if this tissue loss exceeds the critical contact distance, velopharyngeal incompetence can result. In this circumstance, tissue interposition may be required. [Courtesy of Michael Carstens, MD]

Introduction

Clefts of the soft palate in the presence of an *intact* hard palate exist in a spectrum of anatomic severity, from the *forme fruste* submucous variant to a complete separation of both sides extending forward to the horizontal shelves of the palatine bones and finally into the palatal shelves of the maxilla. What are the underlying mechanisms common to all these presentations? Careful examination of the anatomy of the submucous cleft, as compared with that of the normal soft palate, can yield valuable insights into the embryologic defects responsible for this deformity. The developmental field model provides a rationale for surgical approaches to its repair.

The submucous cleft and indeed all other variations of soft palate clefts share three invariable features: (1) *deficiency of palatine bone P3* causing *universal absence of posterior nasal spine*, (2) concomitant *deficiency of the palatine aponeurosis* causing anterior foreshortening and anterior displacement of soft palate muscles, and (3) a *midline soft tissue defect* ranging from a groove to complete separation. Although in more advanced forms of soft palate cleft these features become more exaggerated, in the submucous form, all three conditions are grossly abnormal.

The Seven Principles of Soft Palate Clefts

Principle 1 ***The physical position of the normal soft palate with respect to the posterior pharynx is absolutely determined by the relative sizes of the bony components to which it is attached.*** Normal palatine bones have horizontal shelves extending medially to fuse with one another. Here they form a small *posterior nasal spine* associated with the insertion of paired uvulus muscles. Furthermore, the physical location of the palatine bones in space is directly related to the anteroposterior length of maxillary palatal shelves. Thus, foreshortening of the maxilla palate and/or the palatine bone will create conditions of velopharyngeal insufficiency *even in the presence of a normal soft palate.*

Surgical repairs of the soft palate in the presence of an abnormally short bony platform typically involve trying to gain length by rearrangements of existing tissue relationships, such as intervelar veloplasty and Z-plasties. Such procedures do not address anteroposterior spatial deficit. This predisposes to postoperative velopharyngeal insufficiency. An alternative is the interposition of tissue (buccal flaps) between the deficient bony platform and the soft palate musculature, thus gaining length while leaving the soft tissue relationships between muscle and aponeurosis undisturbed.

Principle 2 ***The fundamental defect in the soft palate cleft involves the lesser palatine neurovascular axis and its territory of distribution.*** The palatine bone and anterior 25% of the palatine aponeurosis develop from a common embryologic tissue source: neural crest from the second rhombomere. These structures are supplied by a common neuroangiosome: the *lesser palatine artery* and its accompanying branch of V2. Deficiency states of these two structures are shared. Thus, a defective palatine shelf is accompanied by tissue reduction or outright loss of the palatine aponeurosis in its nonmuscular *anterior zone* attached to the posterior margin of the bone. This will bring tensor veli palatini and associated muscles into abnormal close physical contact with the horizontal palatine shelf.

Principle 3 LPA defect causes a ***triangular foreshortening of P3.***

Principle 4 A defect in anterior palatine aponeurosis ***drags the muscles of the levator complex forward*** to approximate the horizontal shelf of P3.

Principle 5 ***Mesenchymal deficiency in the horizontal of the palatine bone unmasks a muscle-binding site at which LVP forms an opportunistic false insertion.*** All soft palate muscles insert into the palatine aponeurosis at programmed binding sites. These are located in the middle half of the soft palate. The program arises from interaction between the r2 neural crest aponeurosis and the r2 endothelium. The anterior 25% of aponeurosis has no binding sites. Uvula muscle lies nasal to the levator and arches over it to insert into the posterior nasal spine. These relationships are always normal. The pathologic insertion of levator into the posterior palatine shelf results from (1) a *foreshortened aponeurosis* that brings the muscles forward into contact with the bone and (2) an abnormal bony shelf with *missing PNS* and *no uvular insertion.*

Principle 6 **Normal function of the soft palate cannot take place** *when the levator sling is falsely inserted into the palatal shelves.*

Principle 7 *The individual* ***soft palate muscles arise from different mesodermal structures*** in the embryo. The functional status of each muscle must be determined prior to surgery.

Principle 8 *Midline union of the soft tissue envelope is controlled by* ***diffusion of morphogens (BMP4)*** *produced by P3.* Bone deficiency results in reduced [BMP4] leading to failure of epithelial breakdown and fusion failure.

Soft Tissues of the Palate: The Lever Arm

In this section, we consider the soft tissue components of the soft palate: epithelium, fascia, muscles, arterial supply, and innervation. These velar tissues constitute a functional "lever arm" for control of speech and swallowing. Fascia and peripheral nerves arise from neural crest originating from rhombomeres 2 to 7. Muscles arise from paraxial mesoderm (PAM) of somitomeres 4, 6, and 7. Lateral plate mesoderm lying outside of PAM provides the building blocks of the circulatory system. Neurovascular analysis discloses the soft palate to have three developmental zones with distinct sources of neurovascular supply. Emphasis is placed upon the anterior third of the palatine aponeurosis; this critical structure determines where the levator complex will insert. The basic field defect of soft palate clefts arises from insufficiency of the lesser palatine neurovascular pedicle affecting the posterior palatine shelf and anterior one-third of the palatine aponeurosis. This leads to forward displacement of the levator complex and pathologic insertion onto the bony margin of the cleft site. Soft tissue disruption will then be presented in terms of simple genetic loop between bone *morphogenetic protein 4* and *Sonic hedgehog*. The migration of soluble factors such as BMP4 from their origin with developing bone to the free border of the epithelium permits fusion of adjacent structures.

The detailed mapping of soft tissue structures has clinical consequences: management of soft palate clefts based on developmental fields model enables proper analysis of the deficiency state, which determines the strategies required to achieve proper reassignment of tissues into normal anatomic relationships. Although nothing can be done to lengthen an otherwise deficient osseous platform, proper attention to angiosome conservation will result in hard palate closure without tension and without anterior fistulae. At the opposite end, proper repositioning of the muscular sling with respect to the pharynx depends upon assessment of the aponeurotic deficiency state. Tissue interposition may or may not be required and, if so, judgment is required as to the timing of such reconstruction.

The following references are useful for an overall orientation to palate anatomy and soft palate development [1–4].

Histology of the Soft Palate

On *sagittal section*, the soft palate has three distinct longitudinal zones: (1) an anterior muscle-free aponeurosis, traversed along the nasal side by uvulus and bearing insertion of tensor veli palatini into the side margins, (2) a central muscle-bearing zone for the levator complex, and (3) a posterior zone which is primarily glandular. *Coronal section* shows differences between the nasal and oral sides. The composition of the mucosal epithelium is different. The oral side has a thick lamina propria, abundant mucus glands, and transverse fibers of the levator complex. Oral side has a thinner lamina propria with smaller volume of mucous glands and longitudinal fibers of uvulus. The two muscle layers are separated by an intervening third layer of glands [5] (Fig. 16.3).

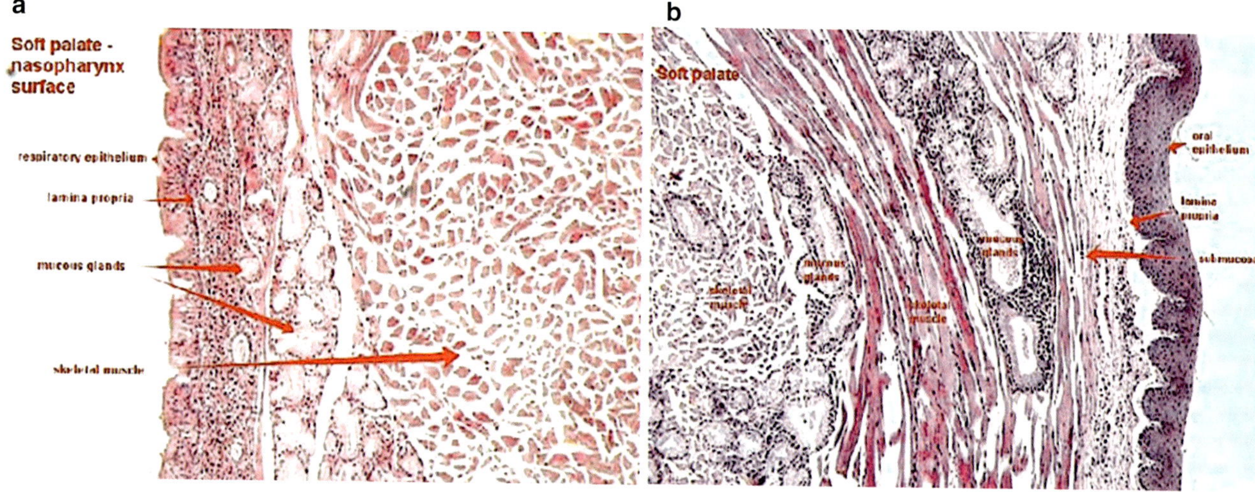

Fig. 16.3 Histology of the soft palate. (**a**) Hard palate *nasal epithelium:* pseudostratified ciliated (respiratory) columnar. *Oral epithelium:* keratinized stratified squamous. (**b**) Soft palate *oral* epithelium: *non*keratinized stratified squamous. *Nasal epithelium, anterior*: pseudostratified ciliated (respiratory) columnar. *Nasal epithelium, posterior*: nonkeratinized stratified squamous epithelium [5]. (**a**) [Courtesy Prof. Brett Kuss University of the Cumberlands. https://inside.ucumberlands.edu/academics/biology/faculty/kuss/courses/Histology/histology02/histo/soft1.htm]. (**b**) [Courtesy Prof. Brett Kuss University of the Cumberlands https://inside.ucumberlands.edu/academics/biology/faculty/kuss/courses/Histology/histology02/histo/soft.htm]

Mucosa and Submucosa

The mucosa of the soft palate is an envelope with two distinct zones denoted on sagittal section by changes in histology and innervation. Let's first contrast it with that of the hard palate. Recall that the mucosa of hard palate on *both* sides is derived from first arch r2 ectoderm; its subepithelial tissues are r2 neural crests. On the other hand, the sources of mucosa and submucosal tissue for the soft differ by anatomic zone.

The key to understanding palate zones is sensory neuroanatomy (Figs. 16.4 and 16.5)

- Posterior palatine: nasal, anterior two-thirds
- Lesser palatine: oral, anterior two-thirds
- Greater petrosal (via posterior and lesser palatine): taste buds 85%
- Glossopharyngeal: mucosa, oral, and nasal, posterior two-thirds, taste buds 15%

Hard Palate *Nasal epithelium:* pseudostratified ciliated (respiratory) columnar. *Oral epithelium:* keratinized stratified squamous

Soft Palate *Oral* epithelium: *non*keratinized stratified squamous for the entire length of soft palate. *Nasal epithelium, anterior*: pseudostratified ciliated (respiratory) columnar. *Nasal epithelium, posterior*: nonkeratinized stratified squamous epithelium

- ***In sum***: The source material for soft palate epithelium changes about half-way back from first arch to third arch. This corresponds to a change in sensory innervation. This explains soft palate pain (V2) and the gag reflex (IX).

The submucosal contents of the soft palate are also different. The neural crest-forming salivary gland structures come from two distinct sources: first arch and third arch.

- *Salivary glands of the soft palate* are unique. Although all intraoral glands are by PANS fibers from superior salivary nucleus, those of the mouse are supplied via V3 lingual nerve whereas the palatine glands are supplied by a direct V2 branch from sphenopalatine ganglion, the *superior palatine nerve* which runs in a bundle along with descending palatine nerve.
- *Taste buds* are also unique. They are located in three distinct zones. Facial nerve VII supplies 85% of taste buds in the rat via greater superficial petrosal nerve whereas glossopharyngeal nerve IX supplies 15% of the glands. There does exist a distinct population of taste buds at the precise boundary zone between the hard and soft palates, the so-called *Geshmacksstreifen*. These can survive bilateral section of the GSP. This zone corresponds to the irrigation of the palatal branch of facial artery.
  - **Take-home point**: Taste bud populations are evidence that second arch neural mesenchyme participates in soft palate development, albeit in *sotto voce*.
  - **Take-home point**: Developmental mechanisms of the soft palate and tongue are similar. We'll explore this point a bit later on.

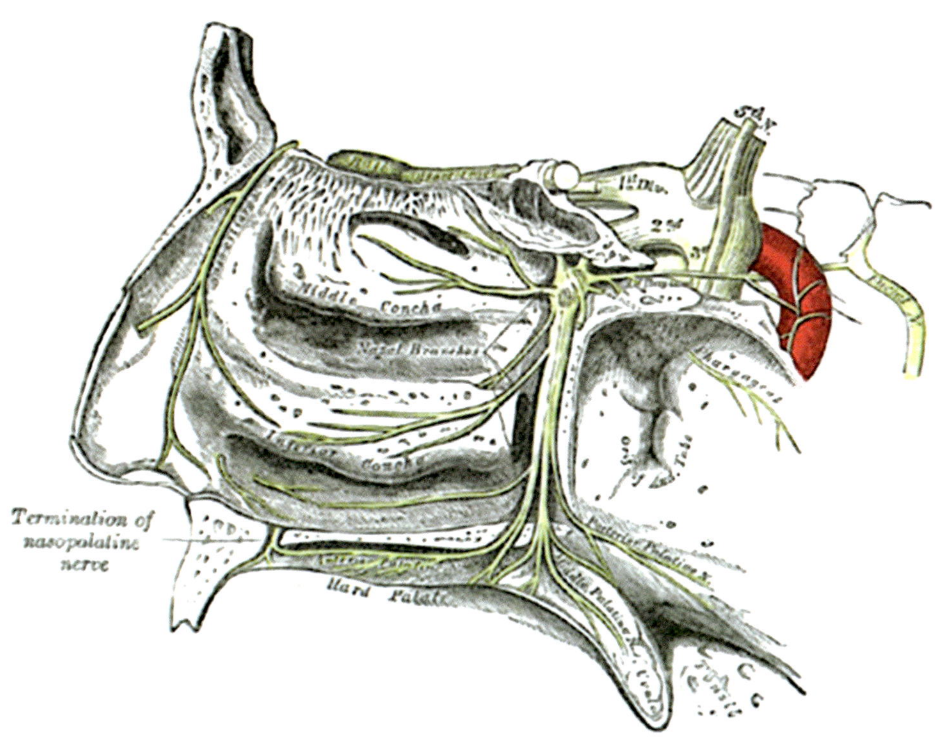

Fig. 16.4 Soft palate innervation. Gray 1918 Fig. 780. Descending palatine gives off lesser palatine branch to the soft palate. Posterior palatine branch supplies collaterals to the nasal surface. Density of innervation on the nasal side is 78% that of the oral side. Glossopharyngeal n. supplies the posterior half. [Reprinted from Lewis, Warren H (ed). Gray's Anatomy of the Human Body, 20th American Edition. Philadelphia, PA: Lea & Febiger, 1918]

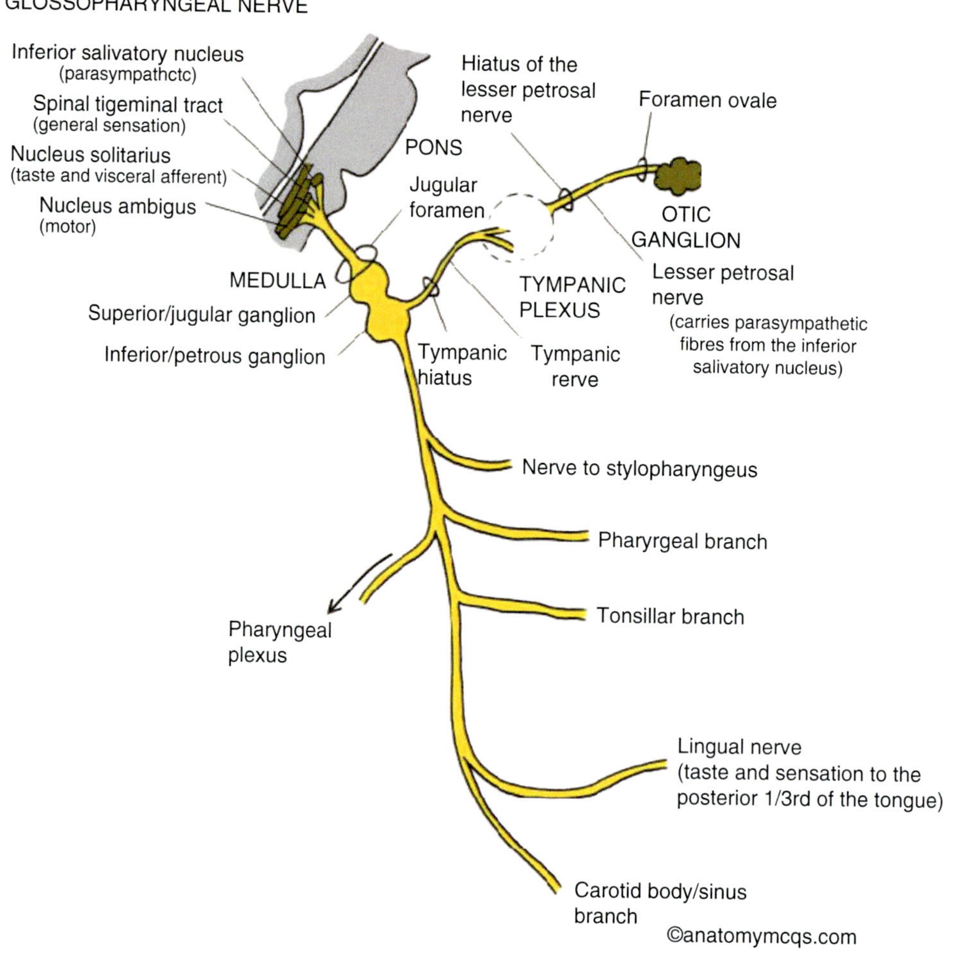

Fig. 16.5 Glossopharyngeal has five components, from proximal-distal: Tympanic: (1) sensory to the eardrum; (2) carries PANS fibers from the inferior salivary nucleus forward via lesser petrosal nerve. Stylopharyngeus pharyngeal: (1) forward to the mucosa and medial wall of the tympanic tube; backward to pharyngeal plexus. Oral: (1) lingual mucosa of posterior one-third tongue and vallate papillae, (2) tonsil, (3) lingual side of epiglottis, and mucosa of aryepiglottic folds carotid body. [Courtesy of Cambridge Questions. © anatomymcqs.com. URL: http://www.cambridgequestions.co.uk/DisplayQuestion.aspx?id=532]

Mesenchymal Structures of the Palate

Palatine Aponeurosis

The aponeurosis is often described as an "extension" of the tendon of TVP. This is a developmental impossibility because the aponeurosis is an r2 neural crest structure, whereas the tendon of TVP is r3 enclosing a mesodermal muscle belly from Sm4 [6]. The aponeurosis results from an epithelial–mesenchymal interaction between the r2 oral epithelium and r2 neural crest submucosa; it covers the *entire surface area* of the soft palate oral epithelium. *The gradient of aponeurotic mesenchyme is medial to lateral and proximal to distal.* Thus, the palatal muscles all move into position following these gradients (Figs. 16.6, 16.7, 16.8 and 16.9).

The blastema of Sm4 (TVP) is first seen lateral to the soft palate processes at stage 20. At this stage, the shelves are vertical and demonstrate a mesenchymal condensation. The blastema of Sm7 (PG, PP, LVP) appears one stage later. By stage 23 (8 weeks) the aponeurosis is present as a separate structure. Insertion of TVP into aponeurosis is complete at 9 weeks. The remaining third arch muscles follow. Uvulus is last to appear.

Neural crest fasciae covering the individual muscles of the soft palate recognize signals produced by the aponeurosis. The muscles insert in a fixed spatial–temporal sequence according to the biologic "maturity" of the target. Those muscles arriving first insert anteriorly; those that come later must accept a posterior binding site. The direction of spread of the muscles (differentiation process) is progressively lateral to medial.

The soft palate is suspended in space by four muscle slings: two are oral (TVP and LVP) and two are nasal (PG and PP). The first nasal sling is TVP. Its insertion is limited, stopping short at the anterolateral border of the aponeurosis. Palatoglossus forms the first nasal sling. This muscle is rather thin; it inserts just behind the "bare area" of aponeurosis. The more substantial palatopharyngeus forms the second nasal sling. It covers two-thirds of the remaining muscular aponeurosis. LVP swoops downward as the second nasal sling of the soft palate. It normally inserts into the middle one-third of the aponeurosis. A second layer of palatopharyngeus overlaps the fibers of LVP. Uvulus runs paired in the midline. It is attached posterior to the raphe; uvulus arcs over levator and attaches anteriorly (again) to the raphe and posterior nasal spine. The presence of uvulus physically pre-

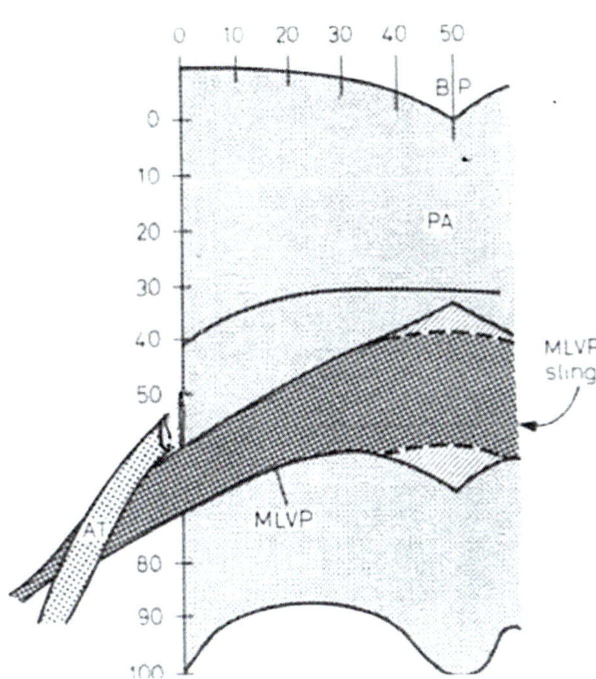

Fig. 16.6 Palatine aponeurosis. Anterior one-third is non-muscle-bearing. Tensor inserts into its lateral border only. Middle one-third bears the insertions of the muscles. Posterior one-third is mostly glands. [Reprinted from Carstens MH. Pathologic anatomy of the soft palate, part 2: The soft tissue lever arm, pathology, and surgical correction. J Cleft Lip Palate Craniofac Anomal 2017;4:83–108. With permission from SAGE Publications]

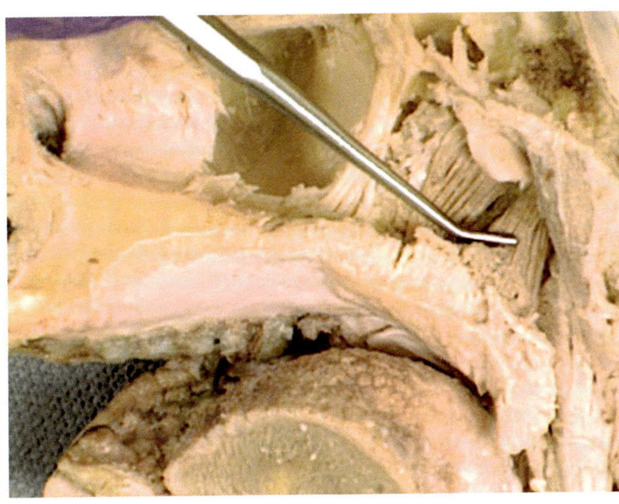

Fig. 16.8 Sagittal section of soft palate shows clamp on the LVP, the TVP is lateral to aponeurosis, and the anterior one-third is muscle-free. [Courtesy of Dr. Samuel Marquez, SUNY Downstate Anatomy]

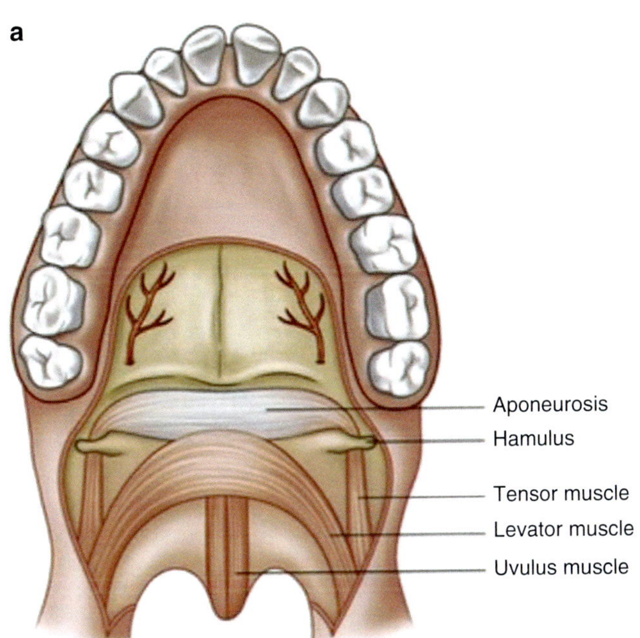

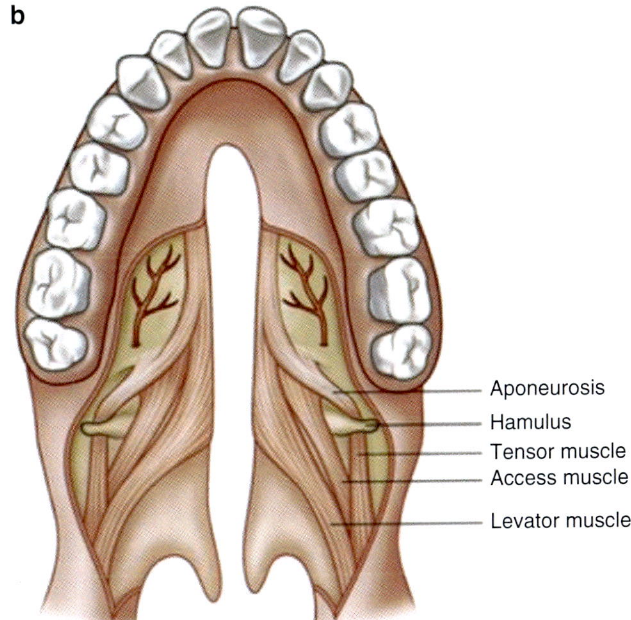

Fig. 16.7 Palatine aponeurosis is an autonomous r2 neural crest structure… not a mere continuation of tensor veli palatini. (**a**) Note that the fascia of associated with the TVP is r3. Note: palatine bone fields P1 and P2 form the horizontal plate of the maxillary bone. Bone field P3 forms the horizontal plate of the palatine bone. (**b**) Soft palate clefts always involve some degree of deficit in the P3 field. Absence of the posterior nasal spine is pathognomonic and the aponeurosis is always deficient. Foreshortening of the aponeurosis brings the levator complex into contact with the hard palate. [Reprinted from Carstens MH. Pathologic anatomy of the soft palate, part 2: The soft tissue lever arm, pathology, and surgical correction. J Cleft Lip Palate Craniofac Anomal 2017;4:83–108. With permission from SAGE Publications]

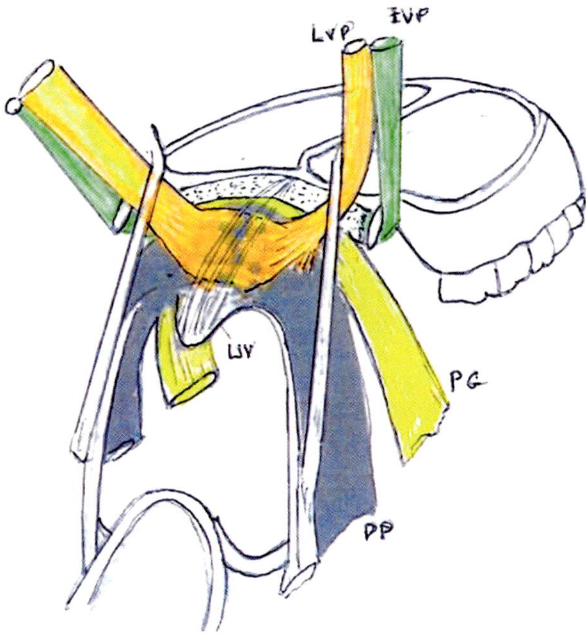

Fig. 16.9 Spatial-temporal order of muscle insertions into aponeurosis. Nasal layer: TVP, palatoglossus, palatopharyngeus (nasal lamina). Oral layer: LVP, palatopharyngeus (oral lamina covers LVP). Note that uvulus is nasal to the levator complex. [Courtesy of Michael Carstens, MD]

cludes any contact between levator and the posterior margin of the palatine bone. In the soft palate, cleft PNS is absent: uvulus cannot insert. Instead it will divide and run along both cleft margins.

Clinical Significance of the Palatine Aponeurosis

As we shall see in our subsequent discussion of the vascular zones of the soft palate, the anterior zone is supplied by the lesser palatine artery associated with first arch. Into its lateral border is inserted tensor veli palatini. The aponeurosis is a distinct structure; it does not represent any transverse extension of TVP.

Soft plate dissections by Vacher comparing normal cadavers and cleft palate demonstrated several important points. (1) In normal specimens, the posterior edge of the hard palate had no muscular insertions save a few tendinous fibers of TVP located at the lateral margin. (2) The orientation of the muscles fibers posterior to the "bare area" of the palatine aponeurosis was transverse. (3) The palatine aponeurosis was not found to be a continuation of the TVP tendon but was a separate anatomic entity. The aponeurosis was related to the oral mucosa but was attached to the posterior palatine margin, from which it could be readily detached. (4) The muscular sling that controls soft palate position is located in the middle one-third of the soft palate, as was previously described by Huang [7, 8].

Cleft palate patients displayed three critical differences from the normal. (1) The overall orientation of muscular fibers was not transverse, but anteroposterior. They were observed sweeping forward to insert into the palatine shelf. (2) *The palatal aponeurosis (the nonmuscular part) was not observed in any of the cleft specimens.* (3) These same findings were repeated in isolated soft palate clefts and in submucous clefts.

Credit Where Credit is Due *The basic defect in soft palate clefts is a deficiency state in the neuroangiosome that supplies the horizontal palatine shelf and the anterior "bare area" of the palatine aponeurosis.* This phenomenon was correctly described in the past. Victor Veau, in his ground-breaking 1931 monograph, attributed the shortening of the soft palate in patients with CP to the absence of the aponeurosis interposed between the palatine bones and the muscles of the soft palate sling. Kriens [9] defined the anterior aponeurosis as a distinct entity, not an extension of TVP. Furthermore, he defined the cleft palate state as (1) the incomplete development or absence of the anterior palatine aponeurosis; (2) forward displacement of anterior components of the musculo-aponeurotic velum; and (3) pathologic fusion with the remaining bony elements [9].

As we shall see, these concepts generate a new algorithm for the management of soft palate clefts with two distinct approaches. Traditional techniques rearrange the existing structures, attempting to gain length with the tissues at hand. The anterior palatine aponeurotic defect is not addressed. If velopharyngeal insufficiency results, pharyngeal tissues are used to reduce the space. Fat grafts to the soft palate or pharynx have been described [10]. Developmental field techniques reconstruct the missing Leg® piece, the aponeurosis with interposition buccal flaps. This approach is a logical response to VPI after an otherwise satisfactory primary repair.

Muscles and Fasciae of the Soft Palate

The fascia of tensor veli palatini comes from r3 neural crest. Proprioception from this muscle reports back to V3. All remaining palate muscles from Sm7 are ensheathed by r6 to r7 neural crest fascia. Proprioception is referred back to the pharyngeal plexus via the glossopharyngeal nerve.

First Pharyngeal Arch: Mastication and Ear Drainage

Tensor Veli Palatini (Figs. 16.10, 16.11 and 16.12)

Tensor veli palatini and medial pterygoid are considered to arise from a common embryonic anlage. Thus, TVP is

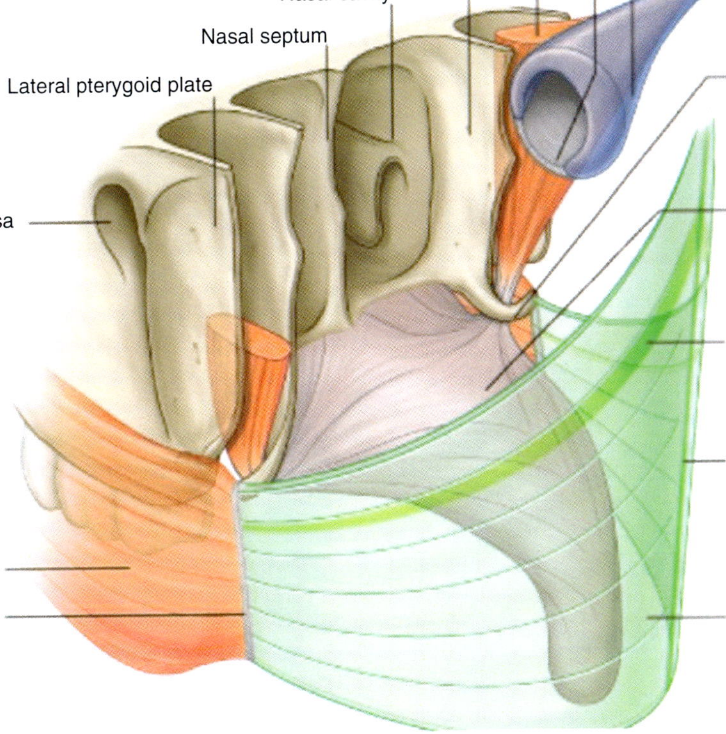

Fig. 16.10 TVP is shown inserting into aponeurosis (lilac). All other muscles are removed. Soft palate muscles. Note "bare area" of anterior one-third of palatine aponeurosis. [Reprinted from Carstens MH. Pathologic anatomy of the soft palate, part 2: The soft tissue lever arm, pathology, and surgical correction. J Cleft Lip Palate Craniofac Anomal 2017;4:83–108. With permission from SAGE Publications]

a muscle of mastication. *By stabilizing the soft palate, it assists in swallowing*. The muscle arises from somitomere 4 and is innervated by V3 palatine branch of the nerve to medial pterygoid. Its fascia is made up of r3 neural crest. TVP has two distinct heads: scaphoid and tubal [11, 12].

The scaphoid TVP gains access to the soft palate by passing over superior constrictor. Its secondary insertion is via a tendon passing around the pterygoid hamulus; it does *not* insert into the hamulus. The tendon of TVP subsequently fans out into the anterolateral palatine aponeurosis. In about 33% of cases, TVP forms an attachment to the maxillary tuberosity [13]. Under normal circumstances, TVP does *not* insert into the palatine shelf.

Although TVP inserts into palatine aponeurosis, it does not contribute to the aponeurosis—these are *developmentally distinct structures* [14]. In patients with cleft palate, abnormalities of TVP insertion have been documented. These include attachments to the horizontal palatine plate and even into the lateral pterygoid process, including hamulus. It is thought that such a slackened or misdirected insertion of TVP may predispose to auditory tube dysfunction.

TVP is the first muscle to develop in the soft palate and is enveloped by r3 neural crest fascia. The palatine aponeurosis is also a neural crest structure into which all the remaining four palatal muscles insert. *Deficiency states of r2 and the lesser palatine neuroangiosome that affect the horizontal palatine plate are concomitantly shared with the anterior palatine aponeurosis*, making it foreshortened. In this case, LVP, which normally inserts into the middle one-third of the aponeurosis, posterior to the fibers from TVP, will be guided forward. *The deficiency of the aponeurosis is readily confirmed when placing interposition buccinator flaps*. The resultant gap is often 1.5–2 cm.

The tubal TVP has its primary insertion along the membranous anterolateral wall of the pharyngotympanic tube including the isthmus, where the bony lateral one-third meets the cartilaginous medial two-thirds. Some studies describe TVP as exerting traction on the tube to open it. This makes sense because the lateral half of the Eustachian tube originates from first arch. Maneuvers such as jaw opening and swallowing to alleviate ear pain from changes in pressure when flying are based upon mechanical traction to open the tube. TVP and tensor tympani do not represent a digastric muscle. They are anatomically distinct [15].

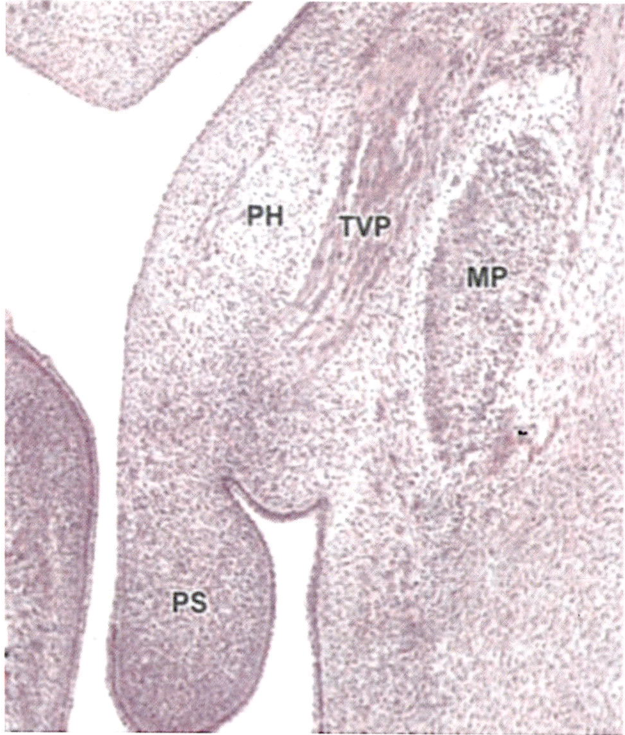

Fig. 16.11 Stage 20: TVP appears with medial pterygoid. Palatine shelf: vertical with tissue condensation = palatine aponeurosis. [Reprinted from Carstens MH. Pathologic anatomy of the soft palate, part 2: The soft tissue lever arm, pathology, and surgical correction. J Cleft Lip Palate Craniofac Anomal 2017;4:83–108. With permission from SAGE Publications]

Tensor Tympani

Tensor tympani also arises from somitomere 4 and is innervated by V3 palatine branch of the nerve to medial pterygoid. Its fascia is made up of r3 neural crest. TT is related in function to muscles of mastication acting on the mandible. This is because the bone fields of the primitive tetrapod mandible (derived from r3 neural crest) become incorporated into the middle ear. Angular bone forms the ectotympanic ring around the eardrum. Prearticular becomes the goniale. Articular forms the actual malleus; it is in physical contact with quadrate, a derivative of the primitive maxilla known as the palatoquadrate cartilage. As one would expect, quadrate becomes the incus, with which malleus articulates. In practical terms, tensor tympani acts to stabilize and protect the eardrum.

Second Pharyngeal Arch: Embryologic Linkage Between the Second and Third Arches

Levator Veli Palatini: Two Different Components? (Fig. 16.13)

Several lines of evidence point to LVP as having components from both the second and third arches. From a clinical standpoint, the effect of a Bell's palsy on palate function was beautifully described by Sanders in 1865 (his account is available online) [16] (Fig. 16.14). Dellon documented this finding in cases of craniofacial microsomia. As such, the

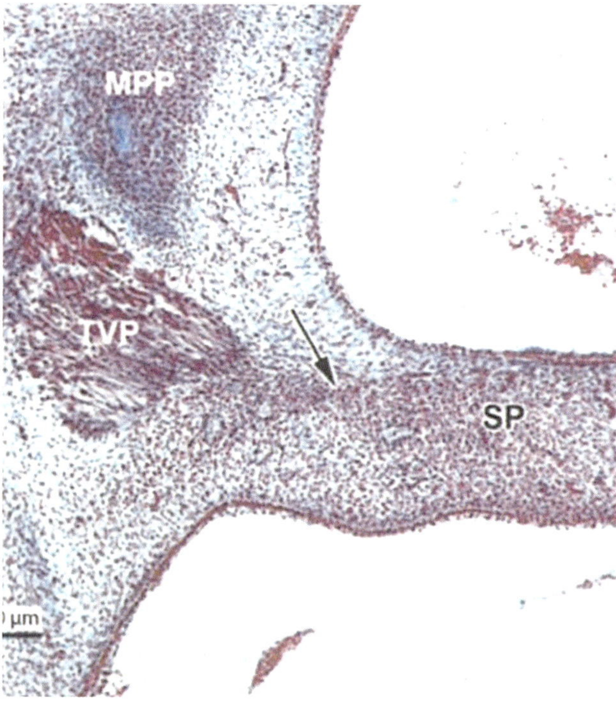

Fig. 16.12 Stage 24. TVP appears with medial pterygoid. Palatine shelf: vertical with tissue condensation = palatine aponeurosis. [Reprinted from Carstens MH. Pathologic anatomy of the soft palate, part 2: The soft tissue lever arm, pathology, and surgical correction. J Cleft Lip Palate Craniofac Anomal 2017;4:83–108. With permission from SAGE Publications]

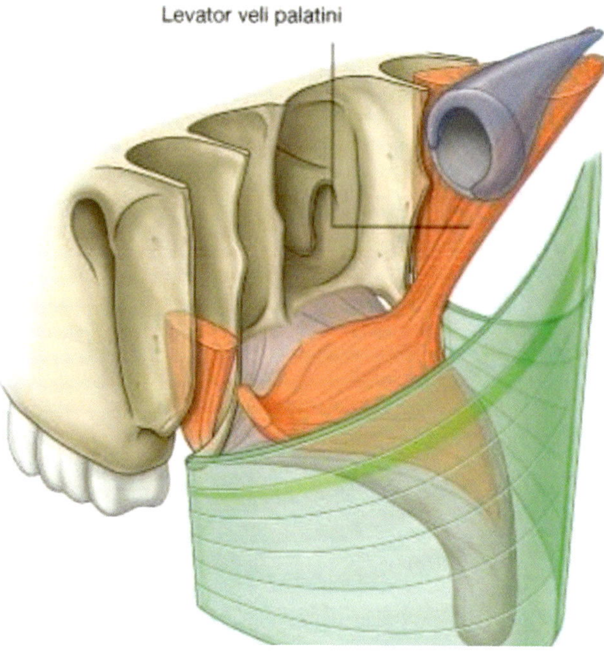

Fig. 16.13 Levator veli palatini: Sm7 innervated by X from rhombomere 6. Inserts middle one-third of palatine aponeurosis; note intervening bare zone of aponeurosis between horizontal plate and LVP. Note: TVP seen, PG, PP, uvulus removed. [Reprinted from Carstens MH. Pathologic anatomy of the soft palate, part 2: The soft tissue lever arm, pathology, and surgical correction. J Cleft Lip Palate Craniofac Anomal 2017;4:83–108. With permission from SAGE Publications]

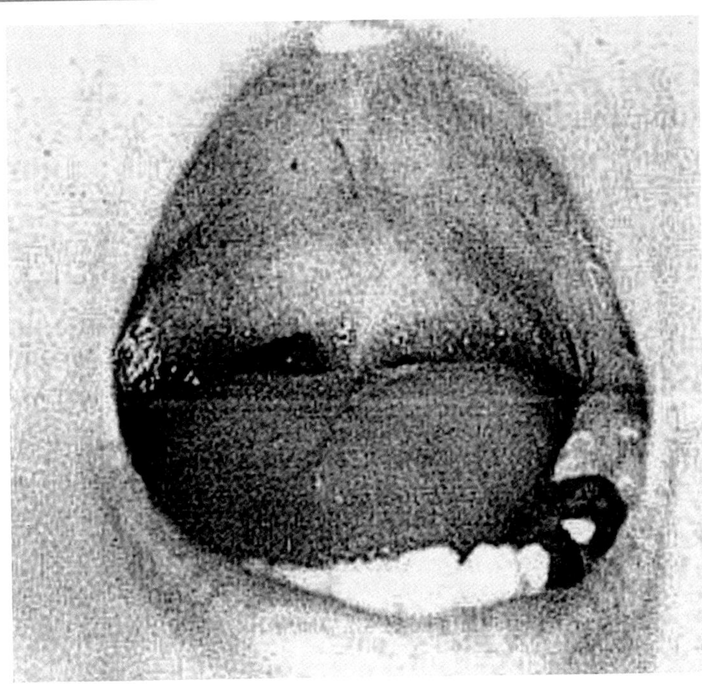

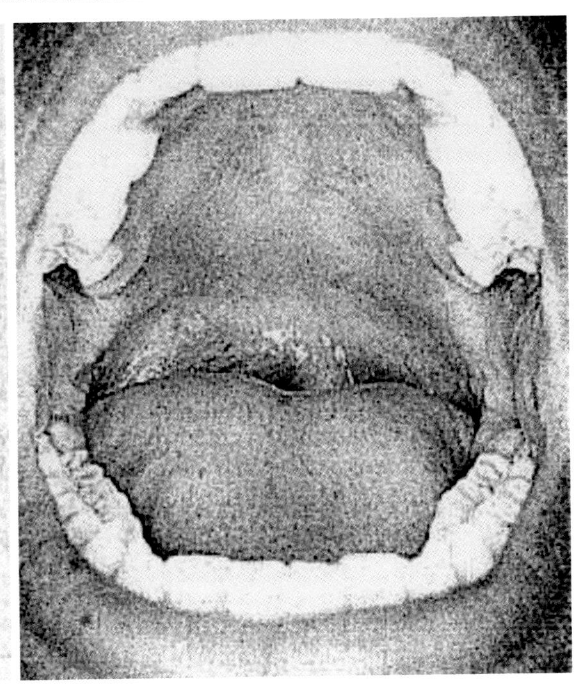

Fig. 16.14 Soft palate manifestations of microtia and VII palsy. Case 1. Right craniofacial microsomia with hemipalatal palsy. Soft palate deviates to the contralateral side. Note "pull" of uvulus to left. Case 2: Right microtia and soft palate dysfunction. Note "dimple" at the insertion of LVP pulling soft palate upward to the left. Left: [Reprinted from Ying-Chien T, Chen, P. Hemipalatal Hypoplasia. Journal of Craniofacial Surgery 209; 20(4): 1150–1153. With permission from Wolters Kluwer Health, Inc.]. Right: [Reprinted from Dellon AD, Claybaugh GJ, Hoopes JE. Hemipalatal palsy and microtia. Ann Plast Surg 1983; 10(6):475–477. With permission from Wolters Kluwer Health, Inc.]

muscle would develop from both somitomeres 6 and 7. LVP forms part of the second arch muscles of deep investing fascia, muscles involved with mastication such as posterior digastric and stylohyoid. It is also functionally related to TVP at the Eustachian tube. Note that directly next to stylohyoid is second arch stylopharyngeus. Thus, LVP arises from myoblasts to form the most posterior sector of Sm6 and the most anterior sector of Sm7. Thus innervation of third arch muscles from Sm6 and Sm7 also receive motor contributions from VII facial nerve. This explains why contralateral deviation of the soft palate can be seen in cases of craniofacial microsomia (Fig. 16.14).

Innervation of LVP by facial nerve has been anatomically described [17]. Stimulation of cranial nerve VII produces TVP movement in the upper nasopharynx while stimulation of IX/X causes upper movement of the velum [18].

Buccinator from second arch and superior constrictor from third arch abut against each other with an intervening raphe to form a common oropharyngeal sphincter. This juxtaposition may imply that myoblasts of distal levator veli palatini are derived from somitomere 6 and retain a motor relationship with facial nerve as the *most posterior component of the buccinator complex*.

This relationship explains the contribution of the ascending palatine branch from facial artery, a derivative of the second aortic arch artery. It also explains the association of levator weakness (not paralysis) with microtia, the latter condition being primarily a second arch defect. Recall that the posterior three cartilaginous hillocks of the auricle are second arch derivatives. When these are absent, the ipsilateral LVP does not elevate equal with that of the opposite side.

Third Pharyngeal Arch: The Workhorse of the Soft Palate

Levator Veli Palatini (Fig. 16.13)

Levator veli palatini is only present in mammals. It has a dual motor supply. Primary motor control is from r6 and r7 via cranial nerves IX and X to the pharyngeal plexus (see below). It is also supplied by lesser palatine nerve, a branch of the V2 complex. But *LPN is not purely sensory*; it receives two types of efferent fibers from the facial nucleus of cranial nerve VII to the soft palate. (1) *Visceral motor* fibers bring parasympathetic control to the palatine salivary glands. (2) *Branchial motor* fibers supply the distal portion of LVP. These fibers originate in facial nucleus and travel to the geniculate ganglion and then are conducted via the superior petrosal nerve to the pterygopalatine ganglion, at which point they gain access to V2 and then travel via LVP to the soft palate. The unappreciated contribution of VII to the soft palate causes *lateral motion in the palatal plane*, a function perhaps related to speech.

Levator veli palatini and superior constrictor are considered to arise as a common embryonic anlage from somitomere 7 with neural crest from r6 to r7, corresponding to third arch. The muscle has been identified by CT scan at stage 21. It becomes readily identifiable by stage 23.

This muscle is only present in mammals. LVP and SC share a common motor supply with three variations: pharyngeal branch of IX, communication branch between pharyngeal branch of IX and X, and the pharyngeal branch of X. In all cases, the cell bodies for these nerves are located in the anterior sector of the nucleus ambiguous corresponding to r6–r7.

The primary insertion of LVP is from the inferior surface of petrous temporal bone just in front of the carotid foramen. It is also inserted into the inferior/medial cartilaginous part of the pharyngotympanic tube. This makes sense because the Eustachian tube, representing the first pharyngeal pouch, is buried on its medial side by the third arch. Contraction of LVP displaces the cartilage superior, medial, and posterior. Its secondary insertion passes through the fibers of palatopharyngeus to join with its contralateral muscle. LVP is above (nasal) to the bulk of palatopharyngeus fibers. In doing so, the levator sling attaches to the palatine aponeurosis in middle 50% of the soft palate (including the uvulus). LVP does *not* normally insert anteriorly into bone. But in cleft palate states, LVP is uniformly pulled forward *because the palatine aponeurosis is foreshortened.* As a consequence, LVP is pathologically inserted into the horizontal lamina of the palatine bone [19–23].

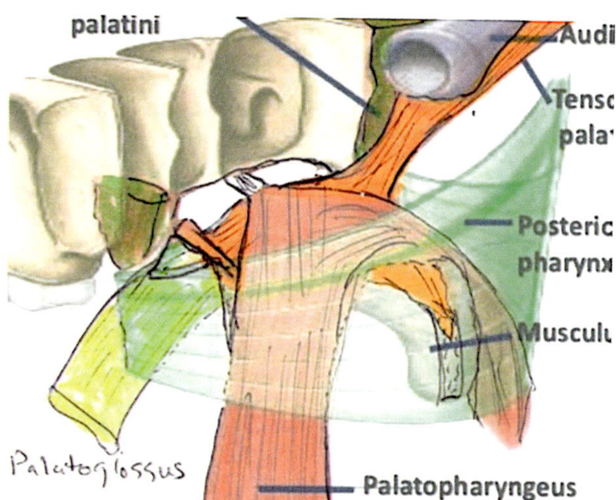

Fig. 16.15 Soft palate muscle sling occupies the middle one-third of the aponeurosis. Palatoglossus, palatopharyngeus, uvulus all originate from somitomere 7. PG and nasal layer of PP form slings posterior to tensor fibers. PG is deep to PP. Oral layer of PP overlaps levator. Only muscle truly inserted into horizontal plate is uvulus. [Reprinted from Carstens MH. Pathologic anatomy of the soft palate, part 2: The soft tissue lever arm, pathology, and surgical correction. J Cleft Lip Palate Craniofac Anomal 2017;4:83–108. With permission from SAGE Publications]

Palatoglossus (Fig. 16.15)

Palatoglossus has its primary insertion in the midsection of the palatine aponeurosis, forming a sling anterior to palatopharyngeus and posterior to the TVP insertion zone into the aponeurosis. It is in direct contact with the oral surface, *deep to palatopharyngeus* and levator veli palatini [22]. Toward the midline, fibers of PLG pass upwards to intertwine with those of PLP and LVP. As it leaves the palate, it passes downward, forming the lateral boundary of the tonsillar fossa, and inserts into the intrinsic transverse muscle fibers of the tongue. The insertion is superficial/cranial to that of styloglossus and of glossopharyngeus. The function of palatoglossus is depression of the soft palate and, to a lesser extent, elevation of the tongue. This assists in the latter stages of swallowing to direct the food bolus to the pharynx [24].

Palatopharyngeus (Fig. 16.15)

Like LVP and PP, palatopharyngeus arises from somitomere 7 and shares a common innervation with LVP. Like LVP, palatopharyngeus forms a common anlage with superior constrictor, with which it forms a common sheet. In the lateral and posterior walls of the pharynx, dense interconnections between the two muscles have been described. A unique feature of this muscle is the two primary insertions into two distinct planes. The split is to embrace LVP. The anterior fasciculus is thicker and runs longitudinally. It is attached to the palatine aponeurosis medial to TVP and below (oral) to levator. The posterior fasciculus is nasal to levator and unites transversely in the posterior midline with the contralateral muscle. At the posterolateral margin of the soft palate, the layers converge and descend, forming the medial boundary of the tonsillar fossa. The fibers are closely associated with stylopharyngeus. The two muscles attach to the posterior thyroid cartilage (a fourth arch derivative). Thus, palatopharyngeus spans the entire posterior border of third arch to insert into the boundary with fourth arch [25].

Uvulus (Figs. 16.16, 16.17 and 16.18)

These thin, paired muscles arise from somitomere 7. They extend backward from the posterior nasal spine along the midline of the palate to gain a secondary insertion into the palatine aponeurosis. The muscles arch over the palatal muscles in the dorsal midline. Immediately deep to the palatal muscles and in parallel the two musculus uvulae are paired longitudinal mesenchymal masses (labeled MM in Fig. 16.18). These form the substance of the fibrous *palatine raphe*. Uvulae run at right angles to levator and are thought to create a "levator prominence" that assists the levator in closing off the nasopharynx [26–28].

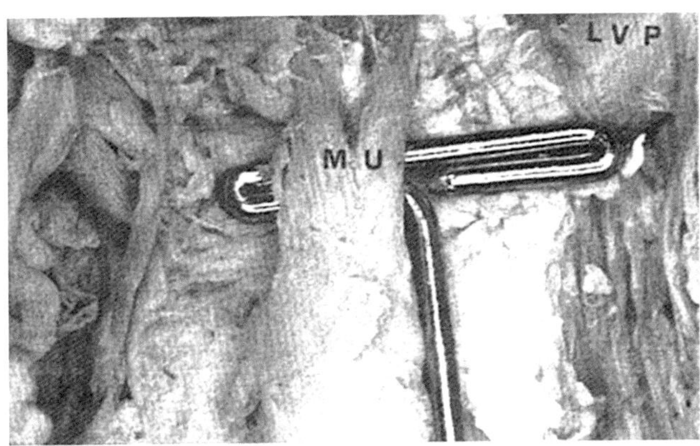

Fig. 16.16 Musculus uvulae consists of paired muscle bellies running forward in the midline, arching over levator and inserting into the posterior nasal spine. Blood supply to this zone is an anastomosis between lesser palatine artery (anterior one-third) and ascending palatine branch of the facial artery with additional supply from ascending pharyngeal. LPA represents first arch, APF represents the second arch, and APh is derived from third arch. Recall that first and second arches are a composite unit. The character of the soft palate changes with APh. [Reprinted from Azzam NA, Kuehn DP. Morphology of musculus uvulae. Cleft Palate J 1977; 14 (1): 78–87. With permission from SAGE Publishing]

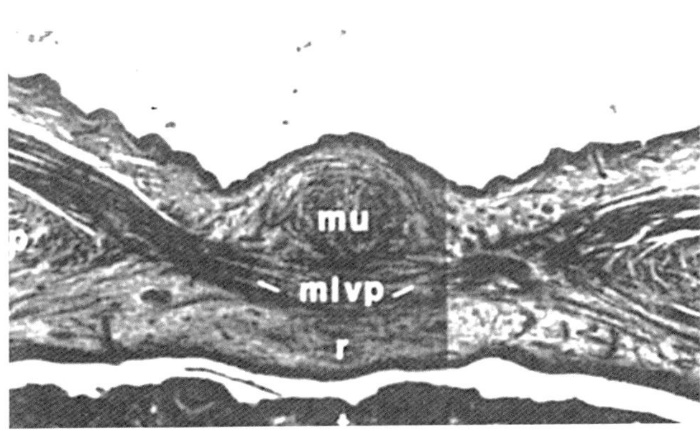

Fig. 16.17 Coronal section of middle one-third shows the midline longitudinal raphe on the oral side, the intervening transverse fibers of levator and paired uvulus muscles on the nasal side. [Reprinted from Langdon HL, Kleuber K. The longitudinal fibromuscular component of the soft palate in the Fifteen-Week Human Fetus: Musculus uvulae and palatine raphe. Cleft Palate J 1978; 15(4): 337–348. With permission from SAGE Publishing]

In cleft situations, where no PNS is present, *wisps of uvulus are always seen running all the way forward along the medial border of the cleft and attaching to bone*. It is important to release them because they tend to tether the levator at the cleft margin. The fact that, in cleft palate, musculus uvulae retains its bulk at the uvulus while becoming attenuated at the bone margin attests to the overall posterior-to-anterior growth pattern of the muscle. *The absence of PNS creates the conditions for a false insertion of levator into the palatine bone*.

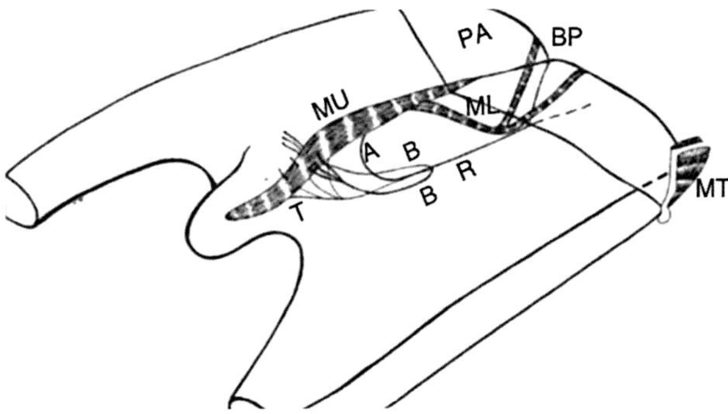

Fig. 16.18 Relationship of uvulus and the palatine raphe. Uvulus arches over the levator complex. [Reprinted from Langdon HL, Kleuber K. The longitudinal fibromuscular component of the soft palate in the Fifteen-Week Human Fetus: musculus uvulae and palatine raphe. Cleft Palate J 1978; 15(4): 337–348. With permission from SAGE Publishing]

Superior Constrictor

The insertion patterns of all three constrictors follow from the biologically "older" zone of pharyngeal arch structures dorsally to the biologically more recent cervical zones. These are neuromerically more distal. The thinnest of the three constrictors, the superior constrictor, arises within somitomere 7. Its motor supply comes primarily from glossopharyngeal or from IX via X. Its primary insertion is bony (to the pterygoid hamulus) and more caudally to the fibrous structures of the pterygomandibular raphe and the mylohyoid line of mandible. Its secondary insertion sweeps backward to the pharyngeal tubercle of the occipital bone and follows the median pharyngeal raphe. Thus, SC spans from between r2 and r3 (buccinator) back to levels c1–c4. Its upper border with the skull base transmits tensor veli palatini and the Eustachian tube. Its lower border with middle constrictor contains stylopharyngeus and the glossopharyngeal nerve... evidence for the third arch relationship of this muscle [23, 29].

Middle Constrictor

Migration of this Sm7 muscle takes place after that of superior constrictor. Its *primary insertions* are therefore more caudal: (1) *superior fibers* track external to SC to seek out hyoid bone, attaching to the superior and inferior cornu; (2) *middle fibers* go out transversely; and (3) the *inferior fibers* insert into the pharynx, passing deep to inferior constrictor. The *secondary insertion* of MC tracks dorsal to SC and extends inferior to it along the median pharyngeal raphe. Thus, MC spans from r5 (lesser cornu) and r6 (greater cornu) distally to levels c3–c6. The space between middle constrictor and inferior constrictor transmits the internal branch of laryngeal nerve and the laryngeal branch of superior thyroid artery (the primary blood supply for the fourth arch). *This space is the boundary between the third arch and the fourth arch*. Nota bene: (1) pharyngeal plexus lies along the lateral surface of middle constrictor. The muscle may also receive fibers from IX via X. (2) At the hyoid cornu, MC lies deep to hyoglossus, a derivative of occipital somite 2. This indicates that as the tongue muscles migrate from somites 2–5, they must skirt around the pharyngeal arches before accessing the floor of the mouth [29].

Fourth Pharyngeal Arch: Separation of the Gut from the Airway

Though not technically part of the soft palate, a few comments are worthwhile to better understand the anatomic boundaries of the third and fourth arches. These are best mapped out from a sensory standpoint. The base of the tongue and the valleculae arise from the third arch. Their mucosa leads downward to the larynx and is supplied by IX, as are the tastebuds of the posterior one-third of the tongue. The epiglottis is a fourth arch structure and has taste fibers supplied by X. The entire laryngeal cartilage is a derivative of r8–r9 neural crest. Vagus from r8–r9 sector of nucleus ambiguus supplies the *external laryngeal nerve* that is motor to both inferior constrictor and cricothyroid.

Inferior Constrictor

The muscle arises from somitomere 8 and relates to the fourth and fifth pharyngeal arches. Its motor supply is exclusively vagus. Inferior constrictor has two parts. (1) *Thyropharyngeus* relates to the fourth arch. It inserts into the oblique line of the thyroid cartilage. Its secondary insertion is into the pharyngeal raphe dorsal and distal to its predecessor. Thus, inferior constrictor spans from r8 to r11 backward to attach at levels c5–c8. (2) *Cricopharyngeus* relates to the fifth arch. Its primary insertion is the attachment between the articular facet of the thyroid cornu and the cricothyroid cartilage. The caudal margin of inferior constrictor is embraced by the thyroid and abuts against esophagus. At this point, roughly at neuromeric level c7, the pharyngeal raphe terminates. So cricopharyngeus inserts into the esophagus.

Two zones of cricopharyngeus have been described. These represent potential *anatomic weak spots*. The upper *pars oblique* continues to insert into the pharyngeal raphe whereas the lower *pars fundiformis* merely forms a circular band around the esophagus, without the reinforcement of the raphe. *Killian's triangle* lies between these two zones. *Laimer's triangle* is located inferior to pars fundiformis and the circular esophageal fibers. At both these sites diverticula can occur. Note that external laryngeal nerve to the larynx descends superficial to IC and penetrates it to supply cricothyroid whereas internal laryngeal nerve supplying the arytenoid muscles lies deep to IC. Thus, IC constitutes a plane separating motor nerves supplying fourth arch muscles from fifth arch muscles.

Blood Supply of the Soft Palate

The soft palate is divided into distinct developmental fields, each of which is defined by a separate neurovascular pedicle. Fields defined in this way are called neuroangiosomes, based on the pioneering work of Taylor [30–32]. Mercer [33] demonstrated that the predominant neuroangiosome of the soft palate has been shown to be the ascending palatine branch of facial pharyngeal artery (APF) and ascending pharyngeal artery (APh).

These findings were confirmed by Maistry's examination of the vascular anatomy of the soft palate in 100 specimens. His study describes three divisions.

- *Proximal soft palate* runs from the posterior border of the hard palate to the point of attachment of superior margin of levator veli palatini and the origin of superior border of palatoglossus. The proximal soft palate thus consists of insertions of TVP into its anterolateral borders and a bare area of fascia without muscle sling, save for midline fibers of uvulus. The predominant blood supply is ascending pharyngeal.
- *Middle soft palate* lies between the superior and inferior margins of LVP and PG. Its blood supply is mixed between ascending branch of facial (second arch) and ascending pharyngeal (third arch).
- *Distal soft palate* lies inferior to the inferior margins of insertion of LVP and PG. Very importantly, the vessels run with the nerves on the *oral surface* of the soft palate. Its predominant supply is ascending pharyngeal (third arch). These findings are consistent with developmental studies by Grimaldi [34] and with the origin of soft palatal musculature in somitomere 7 as mapped by Noden [35].

Lesser Palatine Artery

Recall that all StV2 and StV3 branches arise from a common trunk that is distal to the anastomosis of the extracranial stapedial system to the stump of the maxillary artery from external carotid. IMA serves the first pharyngeal arch structures. Distal to this anastomosis, the hybrid *maxillo-mandibular artery* runs forward to the sphenopalatine fossa where it gives rise to multiple branches, each of which follows a branch of V2. LPA supplies the palatine aponeurosis of tensor veli palatini, along with TVP proper, the sole first arch muscle in the soft palate. Thus, its territory of distribution is specific to proximal soft palate but it shares this with ascending palatine branch of the facial artery. LPA does not supply the levator per se. LPA selectively supplies the oral surface of the soft palate mucosa (Figs. 16.19 and 16.20).

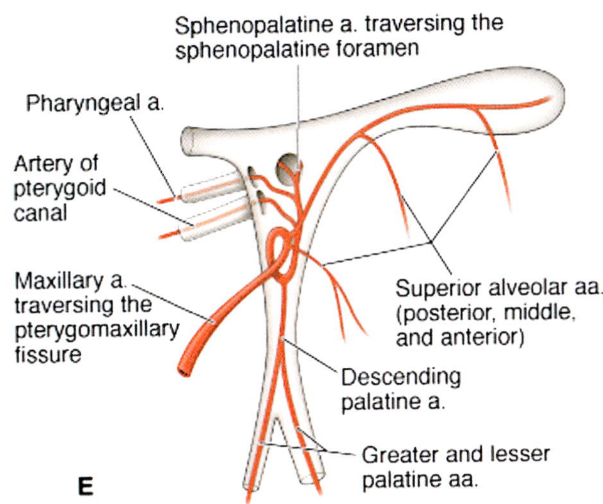

Fig. 16.19 Descending palatine artery gives off greater and lesser palatine arteries. The LPA supplies tissues via multiple foramina directed posteriorly. [Reprinted from Carstens MH. Pathologic anatomy of the soft palate, part 2: The soft tissue lever arm, pathology, and surgical correction. J Cleft Lip Palate Craniofac Anomal 2017;4:83–108. With permission from SAGE Publications]

Anastomoses between descending palatine artery and ascending pharyngeal enhance the perfusion of the latter (Fig. 16.21).

Ascending Palatine Branch of Facial Artery (Figs. 16.22 and 16.23)

Ascending palatine artery (APA) arises from the proximal facial artery. It is *not* a stapedial derivative. Facial artery serves the second pharyngeal arch structures and anastomoses with first arch structures as well. It represents a remodeling of the intermediate part of ventral pharyngeal artery, *itself derived from the original second aortic arch artery* [36, 37].

APA passes upward along stylopharyngeus (a second arch muscle) until it reaches LVP, at which point it divides. One branch is directed to the middle third of the soft palate. The other proceeds upward to supply superior constrictor. This attests to a continuous fascial plane extending backward from buccinator (second arch) to superior constrictor.

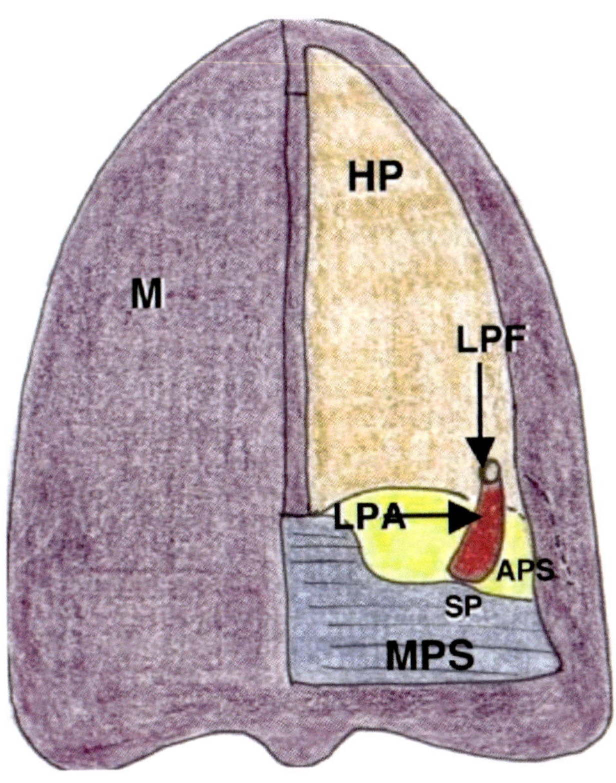

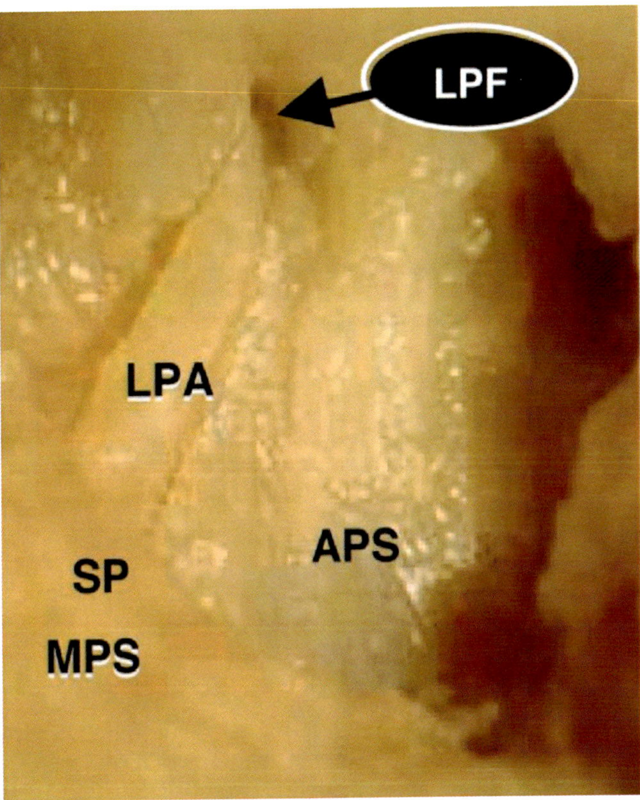

Fig. 16.20 Lesser palatine artery. LPA exits lesser palatine foramen (LPF) and crosses the anterior one-third of the soft palate (APS), to enter the middle one-third of palate, the muscular soft palate (MPS). Note that APS is muscle-free. [Reprinted with permission from Maistry T, Lazarus L, Partab P, Satyapal KS. An anatomical study of the arterial supply to the soft palate. Int. J. Morphol 2012; 30(3):847–857. With permission from the International Journal of Morphology]

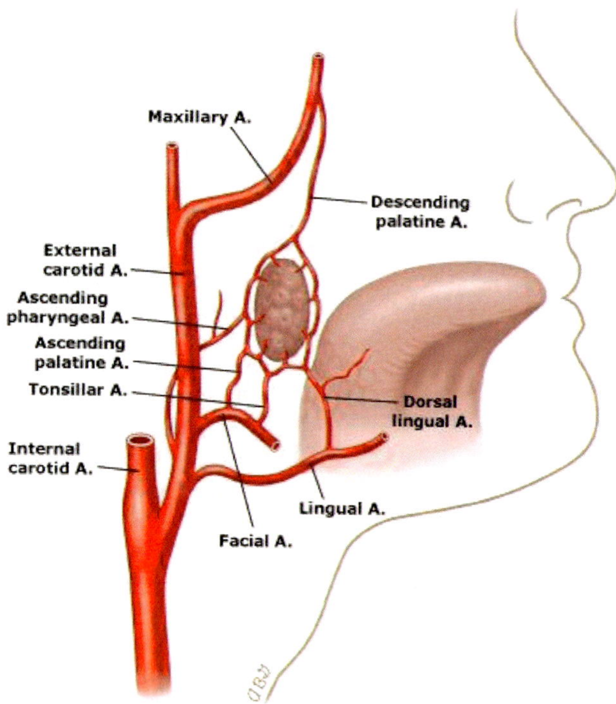

Fig. 16.21 The tonsil is "crossroads of the arches." First arch descending palatine comes from the second sector of internal maxillary and supplies dorsal-anterior tonsil. First arch dorsal lingual supplies ventral anterior zone. Second arch ascending branch of facial embraces the posterior aspect of tonsil and proceeds to the soft palate. It is supported by collateral from third arch ascending pharyngeal. Tonsil is sandwiched between palatopharyngeus and palatoglossus. Mucosa in front of tonsil is first arch (V3), behind tonsil is third arch (IX). Tonsil: endoderm in caudal second pharyngeal pouch between second arch and third arch. Levator veli palatini has dual innervation: VII + IX. [Reprinted with permission from Maistry T, Lazarus L, Partab P, Satyapal KS. An anatomical study of the arterial supply to the soft palate. Int. J. Morphol 2012; 30(3):847–857. With permission from the International Journal of Morphology]

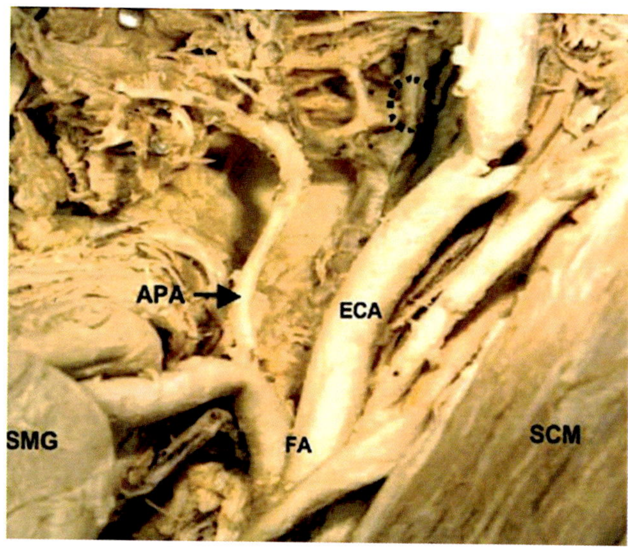

Fig. 16.22 Ascending palatine branch of facial artery (APA). Second arch APA enters middle zone of the palate supplying the lower superior constrictor and, indirectly, levator veli palatini. [Reprinted with permission from Maistry T, Lazarus L, Partab P, Satyapal KS. An anatomical study of the arterial supply to the soft palate. Int. J. Morphol 2012; 30(3):847–857. With permission from the International Journal of Morphology]

The space between superior and middle constrictor admits the passage of IX and therefore constitutes a biologic boundary between the neural associated with r4–r5 (second arch) and that of r6–r7 (third arch). APA is not part of the vascular territory of middle and inferior constrictor—these belong to APh.

With the exception of LVP, the soft palate does not contain second arch muscles, so why should it be irrigated by APA? The answer may lie in the neurologic components of the soft palate. From the geniculate ganglion of the seventh cranial nerve, preganglionic PANS motor fibers for salivary glands and special sensory fibers for taste travel forward in the *greater petrosal nerve*; this eventually targets the pterygopalatine ganglion. From this site, postganglionic PANS fibers and taste fibers flow downward into the palate via the lesser palatine nerve (sensory V2). Thus, the soft palate receives motor supply for the salivary glands and taste receptors all the way down the *oral surface*.

Greater petrosal nerve conveys motor fibers as well. These have been demonstrated in the distal levator veli palatini. LVP thus has a small component contributed by somitomere 6 and represents a posterior connection between the second arch buccinator and the third arch superior constrictor. The motor response of the soft palate to isolated seventh cranial nerve stimulation suggests an additional form of control useful for speech.

Ascending Pharyngeal Artery (Figs. 16.24 and 16.25)

Ascending pharyngeal artery (APhA) is the artery of the third pharyngeal arch. It represents the remodeling of the most proximal part of ventral pharyngeal artery, itself derived from the third aortic arch artery. APhA arises from the medial aspect of external carotid artery. The rationale for APhA in the soft palate is based upon origin of the soft palate muscles (save TVP) which arise from somitomere 7 and (subsequently) populate the third pharyngeal arch. Shortly after its take-off from the external carotid, ascending pharyngeal splits into two main trunks. Careful study of this anatomy provides valuable insights into the derivatives of the third arch.

APhA selectively supplies the nasal surface of the soft palate mucosa. It functions in a gradient with ascending palatine from the facial. APF and APhA both supply the muscular zone of the middle third of the soft palate. The posterior third is selectively supplied by APhA.

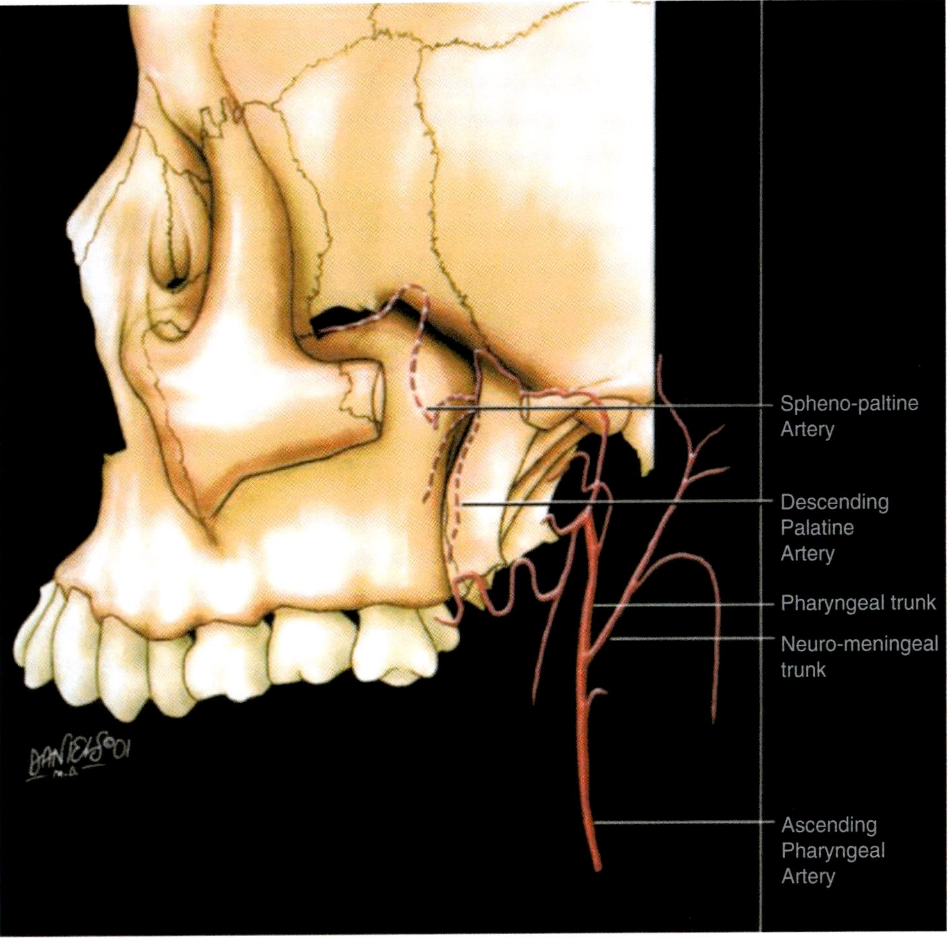

Fig. 16.23 Ascending pharyngeal artery (APhA) divides into pharyngeal and neuromeningeal trunks. The anastomosis with descending palatine (via lesser palatine) is critical to the anterior zone of aponeurosis and the horizontal plate of palatine bone. [Reprinted from Hacein-Bey L, Daniels DL, Ulmer JL, Mark LP, Smith MM, Strottmann JM, Brown D, Meyer GA, Wackym PA. The ascending pharyngeal artery: branches, anastomoses, and clinical significance. Am J Neuroradiol 2002; 23: 1246–1256. With permission from American Society of Neuroradiology]

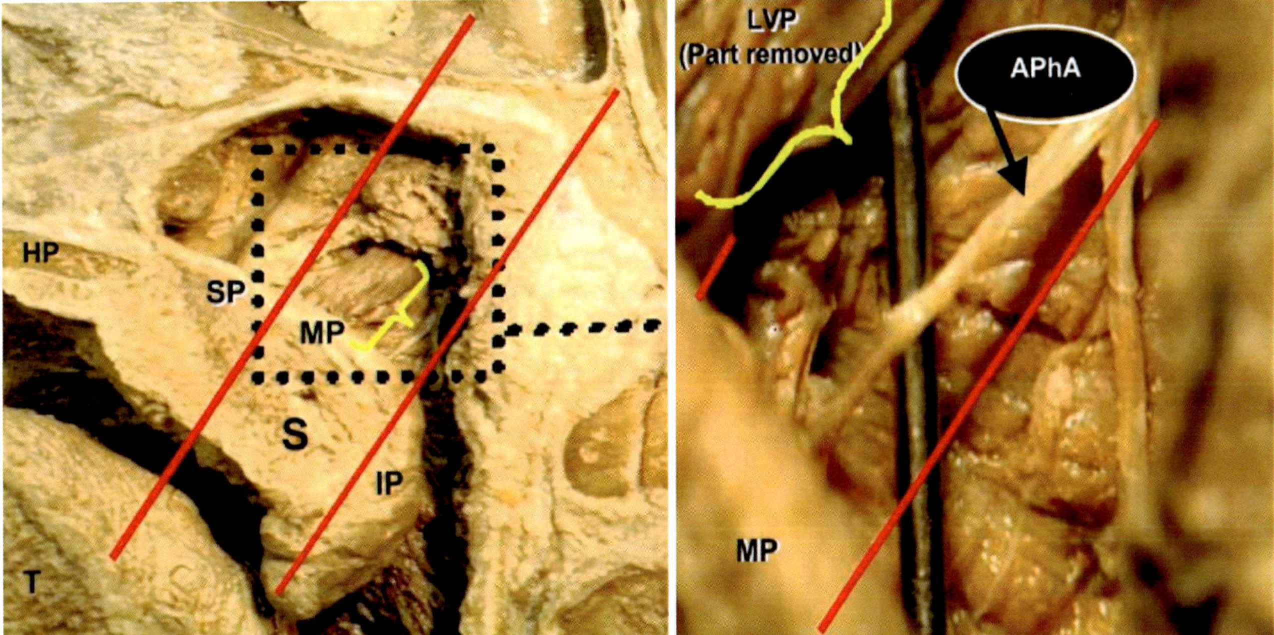

Fig. 16.24 Ascending pharyngeal artery (APhA) vs. palatine branch of the facial (APA). *Second arch APA* enters middle zone of the palate supplying the lower superior constrictor and, indirectly, levator veli palatini. Note that APhA enters palate from above while APA enters palate from below. This belies the overlap between the double origin of superior constrictor (Sm6–Sm7) and LVP (Sm7), i.e., second arch from below and third arch from above. [Reprinted with permission from Maistry T, Lazarus L, Partab P, Satyapal KS. An anatomical study of the arterial supply to the soft palate. Int. J. Morphol 2012; 30(3):847–857. With permission from the International Journal of Morphology]

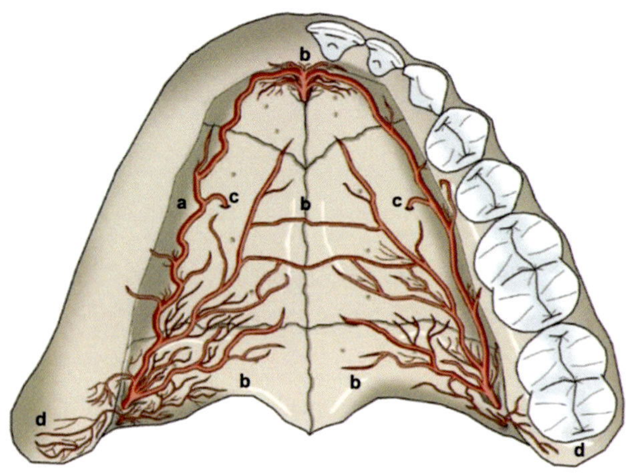

Fig. 16.25 Blood supply to the palate. Descending palatine gives off two branches. LPA (**b**) supplies lingual posterior alveolus and tuberosity, the horizontal shelf of the palatine bone and the soft palate. GPA supplies lingual maxillary alveolus and to horizontal shelf of maxillary bone. Note penetrating branches of GPA (**c**) anastomosing with blood supply to nasal mucoperiosteum (lateral nasopalatine artery). Medial nasopalatine gains oral access (**b**) to supply premaxilla and anastomose with GPA. Ascending palatine br. of facial (**d**) anastomoses with lesser palatine. Ascending pharyngeal (not shown) supplies middle one-third and posterior one-third of the soft palate. [Reprinted from Shahbazi A, Grimm A, Feigl G. et al. Analysis of blood supply in the hard palate and maxillary tuberosity—clinical implications for flap design and soft tissue graft harvesting (a human cadaver study). *Clin Oral Invest* 2019;23:1153–1160. https://doi.org/10.1007/s00784-018-2538-3]

The *pharyngeal trunk* is anterior and extracranial. It gives off three divisions: the *superior, middle, and inferior branches* are distributed into the pharyngeal submucosa space. These supply, respectively, the three corresponding constrictor muscles... inferior constrictor receiving superior thyroid artery as well. The constrictors arise from Sm7, Sm8, and possibly the first somite. Superior or medial pharyngeal artery supplies the medial wall of the Eustachian tube, the site of primary insertion for levator veli palatini. The lateral wall of Eustachian tube is the primary insertion of tensor veli palatini. Thus, the auditory tube is a sandwich, with first arch elements lateral and third arch elements medial. The distal extension of superior pharyngeal, the *pterygo-vaginal artery*, makes a clinically important anastomosis with descending palatine artery. In LeFort osteotomies, trauma to the DPA is compensated by the PVA, providing an important source of blood supply for adequate wound healing.

The *neuromeningeal trunk* is posterior and intracranial, entering the skull via the foramen magnum, where it divides into hypoglossal and jugular branches. *Hypoglossal branch* is distributed to the posterior fossa via the hypoglossal canal where it supplies the meninges and the vasa nervorum of cranial nerve XII. It also sends a descending branch to the odontoid process and richly supplies the dens. It supplies cervical vertebrae C1, C2, and C3, where it anastomoses with the vertebral artery.

Readers familiar with the material in previous work by this author will recall the evolutionary significance of the absorption of proatlas (originally a fifth somite derivative) into the skull base. This explains anatomic features such as the condyles (the transverse processes of proatlas). It is also the rationale for the seeming contradiction of eight cervical somites but only seven cervical vertebrae. The "lost" vertebral body, the proatlas, now incorporated into the mammalian skull base, is faithfully served by this system. Thus, the interface between the APhA and the vertebral system represents the union of derivatives from pharyngeal arch and occipital somites with those derived from cervical somites. *Jugular branch* is also distributed to the posterior fossa via the jugular foramen. It supplies the vasa nervorum of cranial nerves IX, X, and XI. A small branch serves the internal auditory canal. *Inferior tympanic branch* provides an interesting anastomosis between the petrosquamous branch of middle meningeal (a first arch derivative) and the styolomastoid artery (a second arch derivative) supplying the facial nerve.

Neurology of Soft Palate

Nota Bene Innervation of the soft palate is reviewed by Breugem et al. [38] and is well illustrated by Figs. 16.14, 16.26, 16.27, 16.28, 16.29 and 16.30. The classification of pharyngeal muscles based on innervation is discussed by Sakamoto [29].

A Note to the Reader

This section, despite its seemingly arcane detail, is extremely important to understand the structural layout of the palate… and indeed, the entire pharynx. We have all learned that each pharyngeal arch has its own assigned cranial nerve. This system seems to hold true for the first and second arches—although from a functional standpoint, the VII has no somatic sensory role. The third arch marks a critical transition point. We have seen that striated muscle contents of the third arch, originating in Sm7, are much more extensive than appreciated. The entire sensory supply of the third arch mucosa and glands is IX, yet only a single muscle is considered to be innervated by the glossopharyngeal nerve. Instead, we find the motor supply for these muscles to be ascribed to the vagus. Yet the vagus nerve is officially assigned to the fourth and fifth arches. Hence, we have a dilemma. Now that we know where the soft palate muscles come from, how can they be innervated by the fourth arch?

The answer to this resides in inadequacies of conventional neuroanatomy which was conceived before the era of molecular neuroembryology. These contradictions are corrected

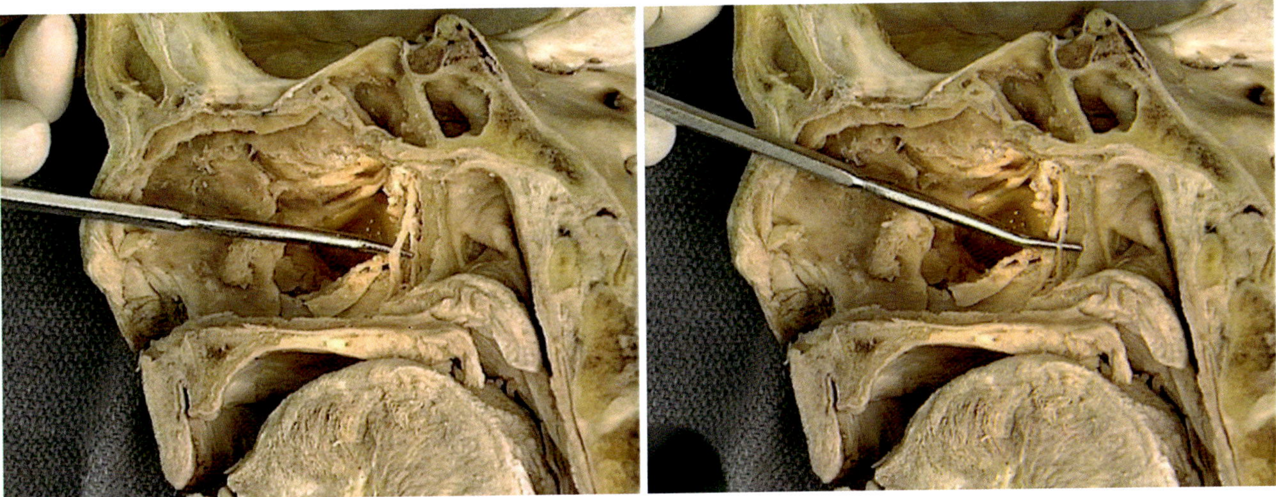

Fig. 16.26 Greater palatine and lesser palatine nerves dissected in situ. [Courtesy of Dr. Samuel Marquez, SUNY Downstate Anatomy]

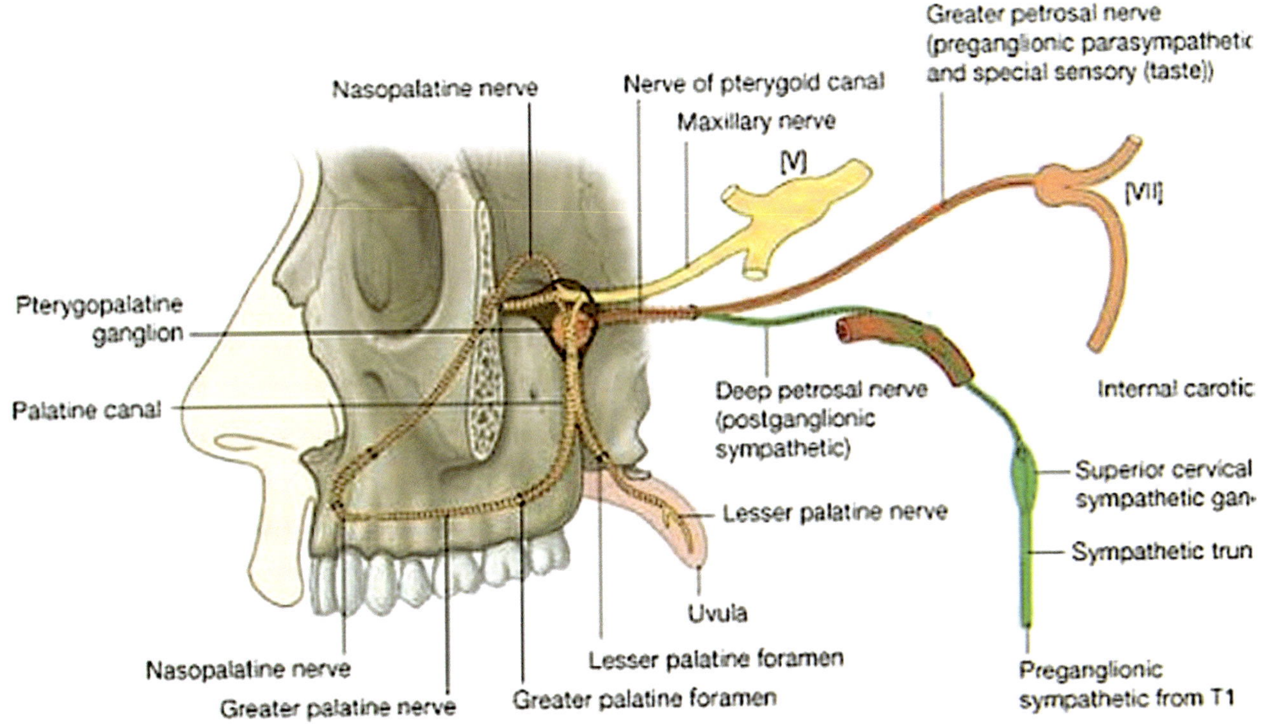

Fig. 16.27 Facial nerve to the soft palate. VII supplies visceral motor (parasympathetic) fibers to glands and somatic motor to the lower (palatal) aspect of levator. [Reprinted from Carstens MH. Pathologic anatomy of the soft palate, part 2: The soft tissue lever arm, pathology, and surgical correction. J Cleft Lip Palate Craniofac Anomal 2017;4:83–108. With permission from SAGE Publications]

and explained by the neuromeric model. They help us to understand the organization of tissues supplied by the medulla r6–r11. To get oriented, recall that neural crest from r6–r7 supplies the third arch, all the muscles of which arise from somitomere 7. Neural crest from r8–r11 supplies the fourth and fifth arches and flows into the occipital somites (S1–S4) as well.

Many Anatomic Changes Take Place at Level r8: The Medulla is a Very Busy Place

1. *Changes in segmentation.* The nonsegmented cephalic mesoderm of somitomeres Sm1–Sm7 gives way to the segmented somite system. This starts with the conversion of the eighth somitomere into the first somite, a process

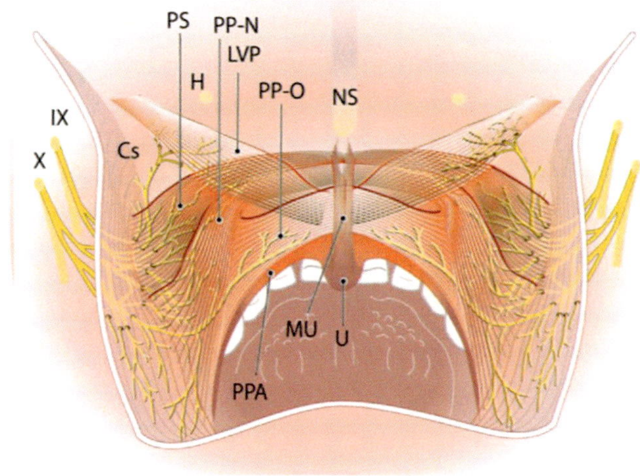

Fig. 16.28 Innervation of the soft palate 1. Distribution of the pharyngeal plexus in the superior-extravelar part of the LVP and nasal and oral parts of the PP. *LVP* levator veli palatini muscle, *PP-N* palatopharyngeus muscle nasal strand, *PP-O* palatopharyngeus muscle oral strand, *PS* palatopharyngeal sphincter, *MU* uvulae muscle, *PPA* palatopharyngeal arch, *Cs* constrictor superior muscle, *H* hamulus, *NS* nasal septum, *U* uvula, *IX* glossopharyngeal nerve, *X* vagus nerve. [Reprinted from Logjes RJH, Bleys RLAW, Breugem CC. The innervation of the soft palate muscles involved in cleft palate: a review of the literature. Clin Oral Invest 2016; 20:895–901. With permission from Creative Commons License 4.0: https://creativecommons.org/licenses/by/4.0/]

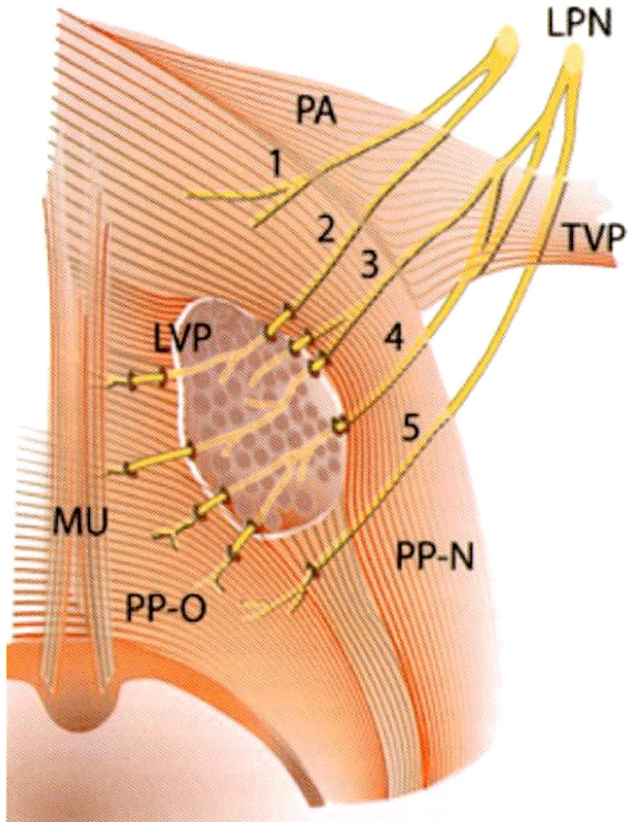

Fig. 16.29 Soft palate nerves 2. A part of the LVP is removed for better view on the five nerve fibers of the lesser palatine nerve, which were found in the human cadaver study by Shimokawa et al. [14]. These nerves run underneath the palatine aponeurosis and the nasal part of the PP and penetrate the inferior-velar part of the LVP on its lateral surface close to the insertion of the LVP in the midline of the velum. *LPN* lesser palatine nerve, *LVP* levator veli palatini muscle, *PA* palatine aponeurosis, *PP-N* palatopharyngeus muscle nasal strand, *PP-O* palatopharyngeus muscle oral strand, *TVP* tensor veli palatini muscle, *MU* uvulae muscle. [Reprinted from Logjes RJH, Bleys RLAW, Breugem CC. The innervation of the soft palate muscles involved in cleft palate: a review of the literature. Clin Oral Invest 2016; 20:895–901. With permission from Creative Commons License 4.0: https://creativecommons.org/licenses/by/4.0/]

that continues all the way backward to the tail of the embryo.

2. *Somites are more anatomically complex*. Somitomeres are simple balls of paraxial mesoderm. Somites are compartmentalized. The occipital somites (S1–S4) have sclerotomes which are assembled into the chondral bones of the posterior cranial fossa and myotomes dedicated to muscles of the tongue. Dermatomes do not appear until the second cervical somite, S6.
3. *Positioned just lateral to medulla are two different mesenchymal structures*: pharyngeal arches 4 and 5 and the first four occipital somites. They have different innervations and different blood supplies.
4. *Neural crest migrates in two ways*. Neural crest emanating from r8–r11 flows into the fourth and fifth pharyngeal arches, using a **2:1 ratio**. Thus, rhombomeres 8 and 9 send their crest to the fourth arch, whereas rhombomeres 10–11 supply the fifth arch. Within the arches, the crest cells organize the mesoderm into distinct muscle masses. Neural crest enters somites on a **1:1 ratio**. From rhombomere 8 backwards, each neuromere supplies its own designated somite. Within the somites, neural crest also organizes myoblasts within the myotome.
5. *The medulla innervates two functional groups of muscles*, innervated by two spatially distinct motor columns. (1) *Branchiomeric muscles* arise when PAM migrates outward at the somitomeric stage, prior to the somite transition. They originate from somitomeres 7–9. Those from Sm8–Sm9 probably develop and migrate *prior to the somitomere-somite transition*. This mesoderm populates arches 3, 4, and 5 to form the muscles of the palate, pharynx, and larynx. Their nuclei are located in the lateral motor column of the brainstem, the *nucleus ambiguus*; it supports cranial nerves IX, X, and XI. (2) A second population of PAM organizes slightly later within somites 2–8. These myoblasts constitute the *hypobranchial cord* that migrates downward and forward in the midline beneath the arches to form the strap muscles and the tongue. These are supplied by the medial motor column of the brainstem, *hypoglossal nucleus*, which runs down into the spinal cord as low as C4. The medial motor column supports cranial nerve XII and the motor branches of the cervical plexus, C1–C4.

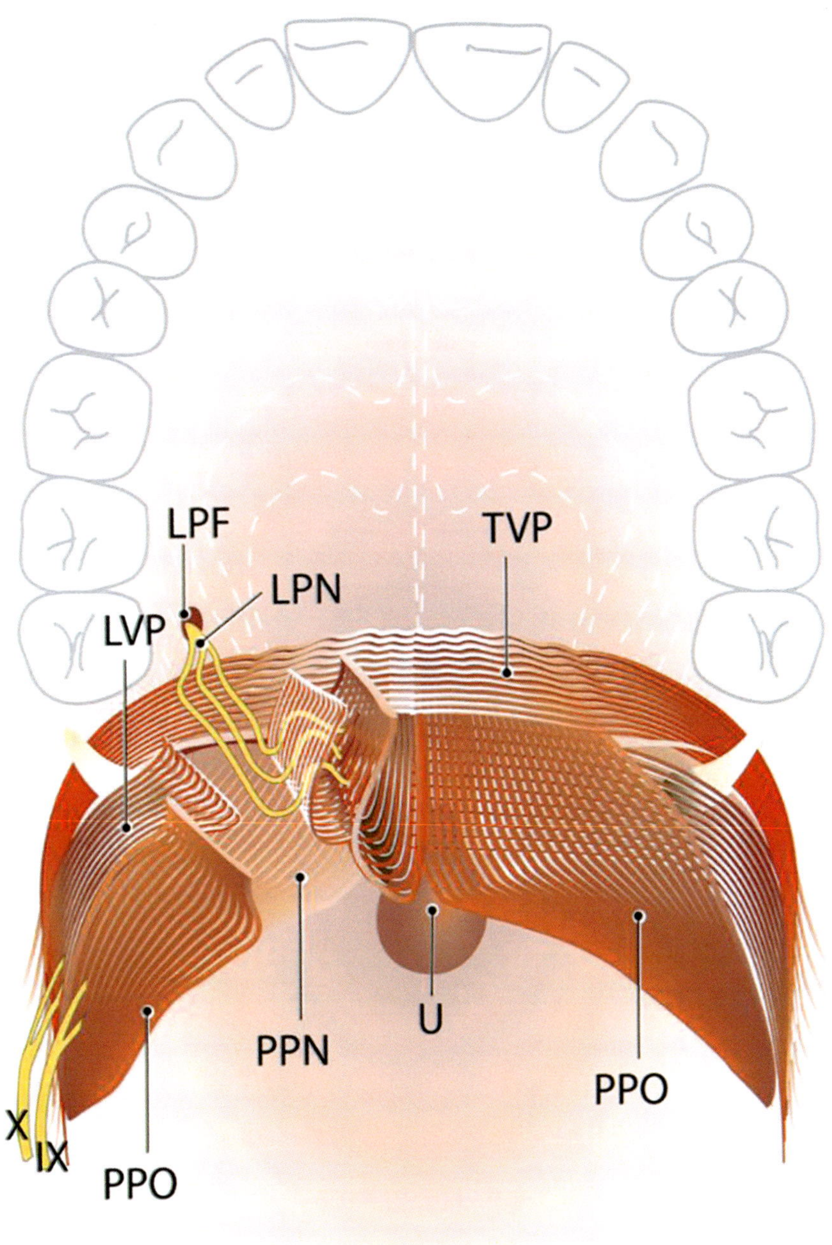

Fig. 16.30 Soft palate nerves 3. The dual innervation of the soft palate by the LPN and the pharyngeal plexus (idea and design by RJH Logjes and CC Breugem after combining the two innervation patterns shown in Figs. 16.1 and 16.3, illustrated by I Janssen). View of the plastic surgeon on the soft palate; the pharyngeal plexus penetrates the superior-extravelar part of the LVP on the lower lateral border. The lesser palatine nerve runs through the lesser palatine foramen and runs over the palatine aponeurosis of the TVP and the nasal part of the PP to enter the inferior-velar part of the LVP on its lateral surface. Here, the LPN innervates the small inferior-velar part of the LVP and the anterior part of the oral part of the PP, together referred to as the anteromedial region of the soft palate muscles. *LPF* lesser palatine foramen, *LPN* lesser palatine nerve, *IX* glossopharyngeal nerve, *X* vagus nerve, *TVP* tensor veli palatine muscle, *LVP* levator veli palatini muscle, *PP-N* palatopharyngeus muscle nasal strand, *PP-O* palatopharyngeus muscle oral strand, *U* uvula. [Reprinted from Logjes RJH, Bleys RLAW, Breugem CC. The innervation of the soft palate muscles involved in cleft palate: a review of the literature. Clin Oral Invest 2016; 20:895–901. With permission from Creative Commons License 4.0: https://creativecommons.org/licenses/by/4.0/]

The hypobranchial muscles arise from seven consecutive somites and consist of two groups which behave in similar fashion. *Tongue muscles* originate in somites S2–S5 and are supplied by r9–r11 and c1. They sweep downward and skirt laterally around and beneath preexisting pharyngeal arches and thence forward in the midline to access the floor of the mouth. The tongue muscles are *Strap muscles* that originate from somites S6–S8 and are supplied by c2–c4. They descend to the interclavicle (which becomes manubrium) and thence forward in the midline to attach to the anterior surface of the third and fourth arches.

Hypobranchial first appeared with the primitive jawed fishes. They represent a new application of hypaxial trunk muscles for jaw control. In sharks, these consist of two groups. *Prehyoid muscles* refer to those that insert anterior to the second arch. Their generic precursor in cartilaginous fishes (sharks) is *coracomandibularis*. In tetrapods, this breaks into the genioglossus group (control of the tongue from the first arch mandible) and the geniohyoid group (control of the tongue from the second and third arch structures, styloid process, and hyoid bone). *Posthyoid muscles* refer to those inserting caudal to the second arch. In sharks, these are represented by *coracohyoideus*. In tetrapods, this complex becomes rectus cervicis and eventually the strap muscles.

Origin and Insertion of Muscles: The Functional Anatomy of Strap Muscles

Questions for the curious: Why do strap muscles exist? How do they "know" where to insert? First, let's clarify some definitions. The *origin* of a muscle corresponds exactly to the *zone or zones of mesoderm in which it develops…* This in turn is corresponded to the neuromeric levels of its motor nerve. The *insertion* of a muscle refers to its attachment into a structure such as bone or cartilage when that structure elaborates a signal, such as BMP4, that acts like the landing lights of an airport and instructs the fascial envelope surrounding the muscle to make an attachment. The *primary insertion* of a muscle is always to a binding site on a structure that is produced from the *same neuromeric level* as the muscle. The *secondary insertion* takes place to the next available binding site in an adjacent structure.

The process of primary and secondary insertions is strictly mathematical. The motor nerves to the strap muscles are C1–C3 so these muscles all arise from the first three cervical somites (S5–S7). The clavicle develops from neural crest that originates at the first three neuromeres immediately after the hindbrain, that is, from levels c1–c3. It is part of the pectoral girdle. The strap muscle myoblasts migrate outward from their somites and make primary insertions into sternum and clavicle *because these structures come from the same neuromeric levels*. They then travel forward beneath the arches to seek out the inferior border of hyoid bone, a third arch structure.

Why does this peculiar arrangement make sense? The system first appears in the primitive bony fishes, such as *Polypterus*, in which jaw opening is accomplished via the primitive hypobranchial strap muscle complex *rectus cervicis* (*coracohyoideus*, and the *coracobranchials*). These muscles connect the pectoral girdle with the hyoid and branchial arches; they function as depressors of the lower jaw and the gills. The piscine pectoral girdle consists of a series of dermal bones fixed proximally to the skull by posttemporal bone and extending outward: posttemporal, supracleithrum, postcleithrum, cleithrum, and finally clavicle. Attached to cleithrum is the chondral scapulocoracoid to which the fin is attached.

Over the course of evolution, the multiple bones of the pectoral girdle disappear. With the emergence of tetrapods on land, the attachment of the pectoral girdle to the skull is lost, creating mobile neck with great advantages for food gathering. At the same time, paired ventral bone units, the *interclavicles*, united in the midline below the neck to stabilize the pectoral girdles. These persist in birds as the furcula (wishbone). In mammals, the interclavicles morphed into the manubrium sterni. The three ossification centers seen in the clavicle (one medial at the manubrial joint and two laterally along the shaft) may represent contributions of three neural crest populations, c1–c3, with c1 being assigned to interclavicle.

In sum, modern strap muscles originate from somites C1–C3 (S5–S7). They are enclosed by neural crest fascia from c1–c3. Their attachments include both manubrium (interclavicle) and the clavicle. The unusual serial arrangement of stenothyroid and thyrohyoid (both sharing an attachment to the oblique line of fourth arch thyroid cartilage with ultimate insertion into third arch hyoid) can be explained by the original branchial arch insertions of rectus cervicis, coracobranchialis.

Cranial Nerves and Their Targets (Figs. 16.31 and 16.32)

Cranial nerves of the midbrain and hindbrain are laid out in individual nuclei and columns. These are arranged from medial to lateral according to their function. Table 16.1 presents a simple model of the motor nuclei and columns. The contents of the nucleus ambiguous are in red.

The first arch receives its muscles contents from somitomere 4. All are innervated by V3, the nucleus of which resides in r3. Only one, tensor veli palatini, is assigned to the palate. TVP is considered a muscle of mastication. It shares a common blastema with medial pterygoid and is supplied by the palatine branch of the nerve to medial pterygoid.

The second arch is supplied with muscle precursors from somitomeres 5 and 6. These are innervated by VII, the nuclei of which are located in r4 and r5. Neural crest originating from the two rhombomeres may be represented as the upper and lower divisions of the facial nerve. The facial muscles are organized in two planes, superficial and deep. Buccinator is a deep plane muscle in continuity with superior constrictor from the third arch. Tensor veli palatini and superior constrictor develop from a common anlage. A small component of levator veli palatini receives innervation from the geniculate ganglion via the greater petrosal nerve which connects to lesser palatine nerve. It is possible that Sm6 contributes to the muscle mass of the palate.

Nota Bene The eye has a number of accessory protective muscles; these arise in the fifth somitomere. These include avian *quadratus nictitans*, the mammalian *retractor bulbi*, and lateral rectus. Innervation for this blastema comes from r4 and r5. This blastema is physically separate from the

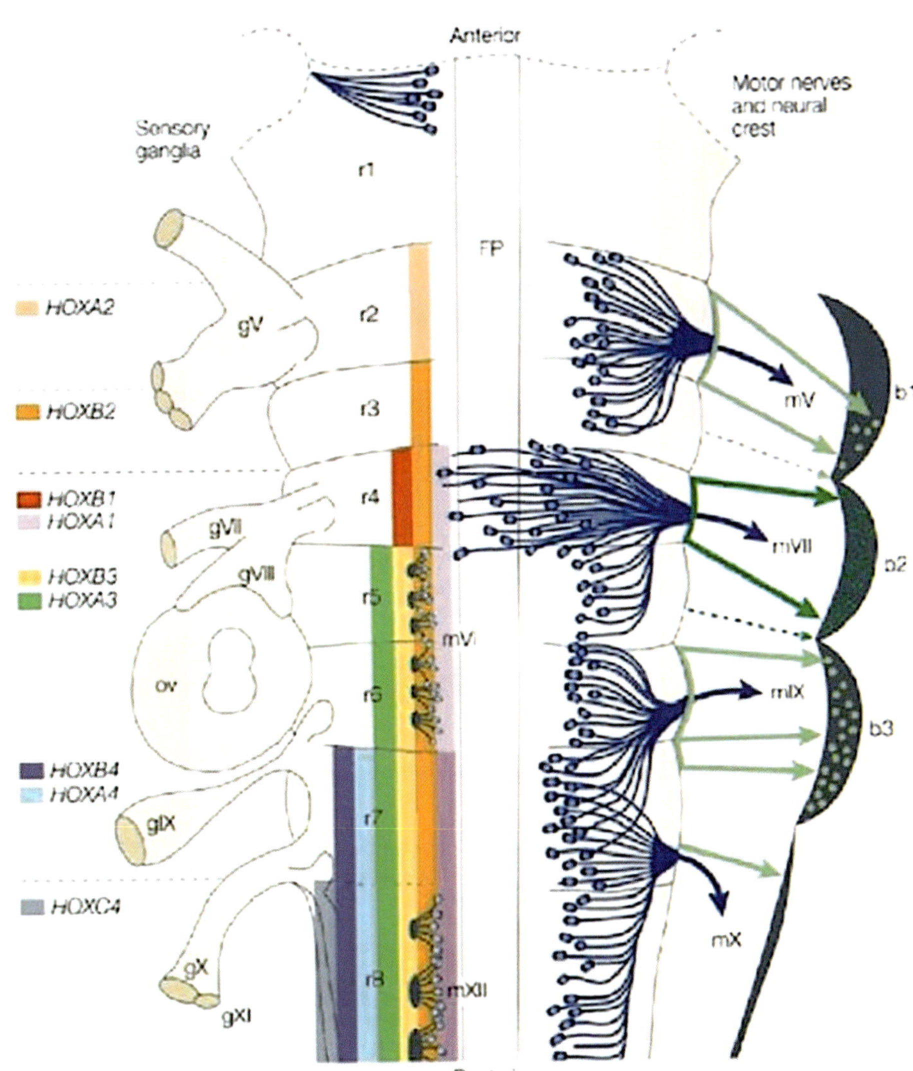

Fig. 16.31 *Eye muscles (yellow)*: *not* pharyngeal arch derivatives; they come from somitomeres Sm1, Sm2, Sm3, and Sm5. *Tongue muscles (yellow)*: *not* pharyngeal arch derivatives; they come from somites S2–S5. *Branchial muscles (brown)*: true pharyngeal arch derivatives—myoblasts arise from

- First arch Sm4 CN V
- Second arch Sm5–Sm6 CN VII
- Third arch Sm6–Sm7 CN IX, X
- Fourth arch Sm8–Sm9 CN X (cranial roots)
- Fifth arch Sm10–Sm11 CN XI (cranial roots via vagus)

[Reprinted from Lumsden A, Keynes R. Segmental patterns of neuronal development in the chick hindbrain Nature 1989; 337: 424–428. With permission from Springer Nature.]

second arch. The nuclei reside in the medial motor column but are not in continuity with the other extraocular muscles because rhombomeres r1–r3 of the pons constitute a "skip area" where medial motor column simply does not exist. Recall that the facial muscles develop in the sixth somitomere (also innervated from r4–r5). This may explain the unusual course of the seventh cranial nerve which has to loop forward around abducens to exit the brain stem.

The muscle contents of third arch are more extensive than generally appreciated. Somitomere 7 produces stylopharyngeus, levator veli palatini, palatoglossus, palatopharyngeus, superior constrictor, and middle constrictor. The neural crest fascia for stylopharyngeus comes from r6; all remaining muscles are supplied from r7.

- The motor supply for the third, fourth, and fifth arches originates from the *nucleus ambiguus* (NA), a nucleus for somatic muscles that extends throughout entire caudal hindbrain from rhombomere 6 to rhomobomere 11.

Nucleus Ambiguus Demystified: Putting an End to Ambiguity

Nucleus ambiguus richly deserves its name because its anatomy is poorly understood. Three nerves emerge from this nucleus, each of which has a contradiction. From the rostral end, glossopharyngeal is said to be the unique cranial nerve for the third arch. Although this is true from a sensory standpoint, IX provides motor control for only one muscle, palatopharyngeus. Vagus nerve exits from the midpoint and appears to supply via the pharyngeal plexus, the soft palate, constrictors, and superior laryngeal muscle to cricothyroid/pharyngeus. But the sensory representation of vagus is exclusive to the fourth and fifth arches. Emerging from caudal portion of nucleus ambiguus are two distinct nerves. Inferior (recurrent) laryngeal nerve belongs to vagus and supplies the intrinsic muscles of phonation. The very same zone gives rise to the cranial roots of accessory nerve. These immediately fuse with the vagus but are considered the possible

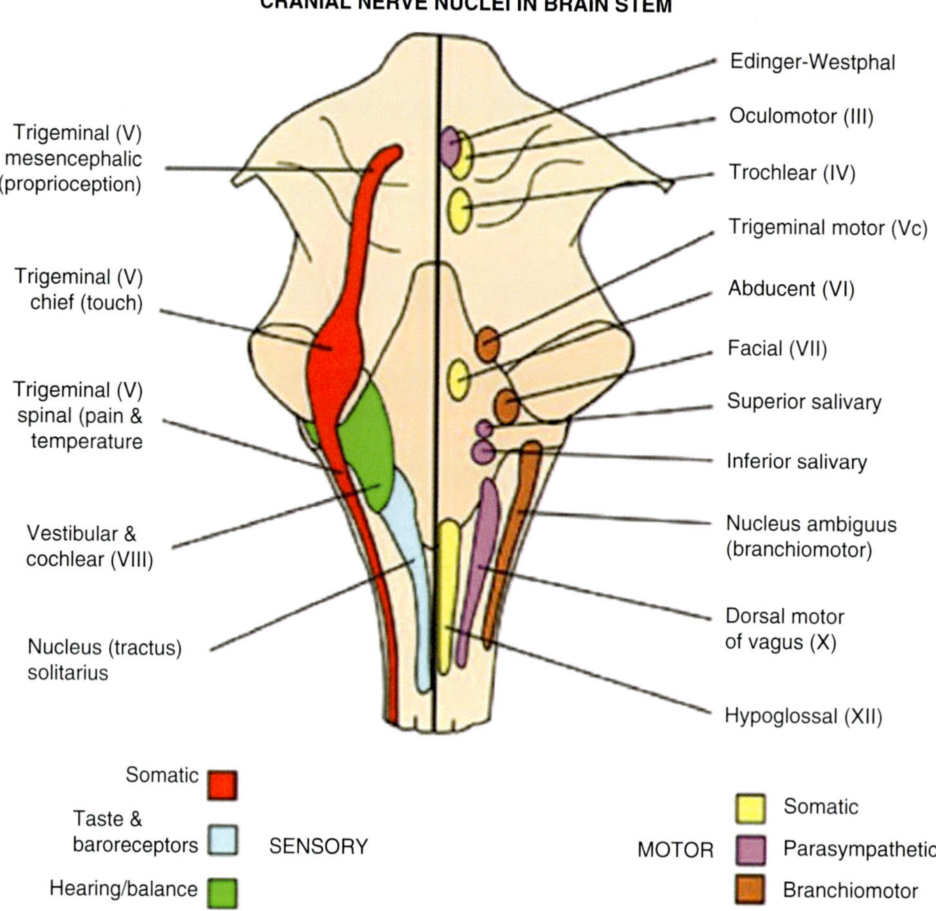

Fig. 16.32 Neuromeric map of cranial nerves IX and X. Soft palate is supplied from nucleus ambiguus which has three motor zones

- Anterior one-third (r6–r7) supplies soft palate, superior constrictor. It has nuclei for both IX and X
- Middle one-third (r8–r9) supplies superior laryngeal nerve to fourth arch and somatic motor to middle constrictor
- Posterior one-third (r10–r11) supplies fifth arch and inferior constrictor. Fibers from r6–r7 are shared between these two nerves.

Conclusion: Thus motor supply to soft palate is really based on rhombomeres and can travel via either IX via X. Therefore, don't code the soft palate as IX or X, just code it as r6–r7. Nuclei of XI are extensive (r10–c6) and supply transitional muscles sternocleidomastoid and trapezius

[Reprinted from Lumsden A, Keynes R. Segmental patterns of neuronal development in the chick hindbrain Nature 1989; 337: 424–428. With permission from Springer Nature.]

Table 16.1 Cranial nerve content of neuromeres

	m2	m1	r1	r2	r3	r4	r5	r6	r7	r8	r9	r10	r11
Midline motor (GSE)	III	IV				VI	VI			XII	XII	XII	XII
Lateral motor (GVE)					V3	VII	VII	IX	IX X	X	X	X XI	X XI
PANS	[a]	[a]				[a]	[a]	[a]	[a]	[a]	[a]	[a]	[a]

Note that the rootlets of the branchiomotor supply to pharyngeal arch 5 muscles have been considered to be the *cranial root of spinal accessory nerve*. In reality, these are physically aggregated with the motor root of vagus as part of the nucleus ambiguus complex. Cranial nerve XI involves neuromeres c1–c6 and is strictly peripheral, representing the backward translation of cucullaris muscle in evolution

GSE general somatic (non-pharyngeal arch) muscles, *GVE* general "visceral" (pharyngeal arch) muscles. The term visceral is a historic misnomer from when arch muscles were thought to be lateral plate mesoderm of the gut. We know all craniofacial muscles are striated and originate from paraxial mesoderm

[a]Denotes motor to parasympathetic system

source of the inferior laryngeal. Nevertheless XI has no sensory distribution in the pharynx so it cannot be considered the motor nerve to the fifth arch. On the other hand, XI supplies two muscles demonstrated to arise from the occipital somites, sternocleidomastoid and trapezius. As these muscles span from the occipital skull to the pectoral girdle, they do not qualify as pharyngeal arch derivatives.

How can these contradictions be resolved? First, nucleus ambiguous is strictly dedicated to the branchiomotor column. It can be conveniently divided into three functional sections containing, at distinct levels, three cranial nerves: glossopharyngeal (r6–r7), vagus (r7–r11), and accessory (r7–r11). The *motor supply for the third arch* is found in the *cranial portion of nucleus ambiguus*, r6–r7. It supplies the muscles of the soft palate, and two constrictors, upper and middle. Located in r6 are the motor fibers glossopharyngeal per se which supply *only one muscle* in Sm7, palatopharyngeus. Located in r7, in addition to IX, vagus is also represented. This is the most anterior nucleus of vagus and it is also dedicated to the third arch. It gains access to all its target muscles via the pharyngeal plexus. Thus, the muscles of the third arch are under control from r6r7 via two nerves, IX and X. The *motor supply for the fourth arch* resides in the *middle portion of nucleus ambiguus*, r8–r9. It supplies the muscles of the Sm8: cricothyroid of the larynx (superior laryngeal nerve) and inferior constrictor. The *motor supply to the muscles of the fifth arch* is located in the *inferior portion of nucleus ambiguus*, r10–r11. These neuromeres contain vagus to the muscles of Sm9: internal laryngeal and arytenoid muscles (inferior laryngeal nerve).

Two structures are critical for understanding the neuroanatomy of the soft palate: (1) the pharyngeal plexus and cranial nerve XI (the accessory nerve). The pharyngeal plexus is responsible for sensory innervation of the oropharynx and the laryngopharynx. It is defined by the sensory distribution of IX and X. The plexus is also responsible for conveying somatic motor fibers from nucleus ambiguus (r6–r11) to target muscles. Gray's 40th edition considers these fibers to have their origin in XI. Specifically, the cranial part of nucleus ambiguus is distributed to the muscles of somitomeres 7 and 8: the soft palate, constrictors, and the cricopharyngeus (superior laryngeal nerve). The caudal part of nucleus ambiguus is distributed to the muscles of somitomere 9: intrinsic laryngeal muscles (inferior laryngeal nerve).

One can ask the question as to why fibers from one nerve connect with those of another, as in the pharyngeal plexus. In the case of the soft palate, motor nuclei that supply three different nerves, IX, X, and XI coexist within the nucleus ambiguus. *When cell bodies for nerves of similar function reside within a common neuromere, the possibility exists for an extracranial communication between those nerves.*

Recall that the sensory "identity" of oral epithelium is derived from the neural crest that forms the submucosa. Neural crest derived from r6 and r7 defines the sensory contribution of the third arch to the oropharynx. The innervation boundary between r6 and r7 corresponds roughly to superior versus middle constrictor. Soft palate submucosa likely originates from r6. Neural crest from r8 and r9 determines the sensory distribution of fourth arch. The fifth arch is in register with r10 and r11 (See Table 16.1).

Given the intimate relationship between cranial nerves IX and X to the third arch it is not surprising that their muscle targets are closely related. Palatopharyngeus (r7, X) passes downward from the soft palate posterior to the tonsil and joins with stylopharyngeus (r6, IX) to insert into posterior thyroid cartilage (arch 4). Also at the boundary between arches 3 and 4 is *Waldeyer's ring* (the insertion of lymphoid tissues) which in turn corresponds to the *site of the buccopharyngeal membrane*. The track of the buccopharyngeal membrane passes through the tonsillar pillars and thence moves backward between superior and middle constrictors. Thus, the tonsillar pillars serve as a boundary between the cranial pharyngeal arches (PA1–PA3) and the caudal arches associated with the pharynx and larynx (PA4–PA5).

To test out these relationships between IX and X, put your finger on the hard palate (sensory supply V3). There is no gag reflex. Now pass it backwards to the oral surface of the soft palate (sensory supply V3 and IX) and you will elicit a gag reflex with a vagus-mediated reflex arc.

Conclusions of This Section (Yes, You Survived)

1. All pharyngeal arches are in register with, receive neural crest from, and are innervated by paired rhombomeres.
2. The functional neuroanatomy of the medulla is better understood on the basis of rhombomeres than by individual nerves.
3. The nucleus ambiguus is no ambiguous at all—in fact, it is the key to the medulla.
4. Muscles supplied by the medial motor column have an unerring mission to move forward in the midline into their insertion sites.
5. The cranial nerve that can be physically identified as providing motor supply to pharyngeal arches 3–5 is vagus via the pharyngeal plexus and separate superior and inferior laryngeal nerves.
6. With the exception of palatopharyngeus, motor fibers from vagus innervate muscles from somitomeres 7–9.
7. The best way to understand the cranial nerves to pharyngeal arches 3–5 is not on the basis of anatomic structures outside the brainstem but on the original neuromeric anatomy of the nucleus ambiguus.
8. True vagus motor neurons may be strictly visceral, i.e., parasympathetic, not somatic.
9. The source of somatic motor neurons traveling via vagus, based upon function, is accessory nerve.

Development of the Soft Palate

The primary and secondary palatal processes begin to develop at stage 17. This is well demonstrated in the monograph by Hinrichsen. The biologic signal for tissues to accumulate and to project outward as the maxillary shelves of the hard palate is located at the interface between two neuroangiosomes. The nasal side is the territory of lateral nasopalatine, while the oral side is the territory of the greater palatine branch of DPA. The biologic signal for the palatine shelves of the hard palate is also an interface between three neurovascular axes. The territory of lesser palatine branch of DPA (first arch) projects the mixed territory of ascending palatine of facial (second arch) and ascending pharyngeal (third arch). It should be emphasized that the representation of second arch in the mid-portion of soft palate is minimal. Intraoral representation of second arch is virtually nonexistent.

The tissues of the soft palate also reflect the apposition of first and third arch derivatives. Representation of second arch is minimal, being relegated to taste buds and glands. This is seen in the juxtaposition of TVP versus LVP. Half-way back in the soft palate the arterial supply to the muscles becomes mixed as well between lesser palatine and ascending pharyngeal. Innervation of the soft palate mucosa is also mixed between somatic sensory from V2 and visceral sensory from IX (Table 16.2).

Bone Growth Drives Soft Tissue Positioning

In conjunction with the medial growth of the horizontal palatine shelf, the mucoperiosteum of the bone drags attached pharyngeal mucosa into the oral cavity. Myoblasts that have accumulated in the lateral pharyngeal wall are spatially positioned according to their final relationships. These are brought along passively. Craniofacial muscle masses develop in conjunction with a fascial envelope or epithelium which provides the program for their boundaries. All their connections to the skull base, Eustachian tube, pharynx, and the third arch base of the tongue are already present. During the process of bone formation, BMP4 signals are elaborated, as previously described. These diffuse outward through the soft palate, permitting epithelial breakdown and fusion to occur in anterior-to-posterior gradient.

The key concept to bear in mind regarding the soft palate muscles is that their spatial relationships are predetermined within their somitomeres of origin. They migrate in a predetermined spatial-temporal order, and assemble themselves at the "jump-off point" in the lateral pharyngeal wall behind the palatine bone. From here they migrate into place over the palatine aponeurosis where they each encounter a binding site and assume their final functional relationships.

Table 16.2 Summary of palatal development by Carnegie stage

Stage	Findings
15	Primary palate (premaxilla) develops
17	Palatal processes begin to fuse
18 (6 weeks)	TVP and medial pterygoid appear as single blastema—palate fusion complete
20	TVP differentiates—soft palate mesenchyme condenses—initial position of palatal shelves is vertical
21	Blastema of LVP, PG, PP
22	V3 approaches blastemal mass of TVP
23 (8 weeks)	LVP differentiates—PG and PL present Palatine aponeurosis develops—pterygoid hamulus ossifies
8 weeks	Bursa between TVP and hamulus, soft palate has an epithelial seam throughout
9 weeks	TVP approaches palatine aponeurosis—soft palate closed except uvulus
10 weeks	TVP attaches to palatine aponeurosis
13 weeks	LVP attaches to palatine aponeurosis
15 weeks	
16 weeks	TVP has 2 distinct parts: scaphoid, tubal

3-D Relationships of Palatal Muscles

As the horizontal palatine shelf begins to proliferate into the oral cavity, five groups of myoblasts are arranged along the lateral pharyngeal wall, awaiting their turn to be positioned into space. They have the same spatial order in the wall and bear the same attachments as in their final state. Uvulus is embedded in the midline. Just lateral, the tensory/levator complex is attached to the primitive pharyngotympanic tube. Palatoglossus extends forward and palatopharyngeus is directed backward. Initially, when the palatine shelf first buds inward, these five myoblast groups are all clustered together. But with rapid embryonic growth, they all are transported away from each other, like an expanding supernova. Thus, the palatal muscles in their final configuration are strung out like guywires suspending the palate in space to the structures that surround it.

All five palate muscles attach to the palatine aponeurosis. Tensor attaches first to the lateral margin. Its fibers of insertion are localized at that site. Although it exerts traction on the aponeurosis, it *does not form a sling* per se. Three muscle slings of the third arch (LVP, PG, PP) all insert into the mid-portion of the palate at about the same time. Palatoglossus lies deep and anterior to palatopharyngeus. The anterior margin of PG respects the "bare area" of the aponeurosis. Palatopharyngeus is much broader and has two insertions which enclose levator. The deep insertion makes direct contact with aponeurosis. The superficial insertion is admixed with levator. Levator swoops downward in the middle one-third of the palate to attach to a diamond-shaped zone, its fibers intermingling with those of palatoglossus and palatopharyngeus. It can be argued that levator has more of a muscular insertion into itself and its companions than into the palatine aponeurosis per se.

Deep to the three muscle slings on the oral side, a midline condensation of fibrous tissue, the *palatine raphe*, runs all the way forward to palatine bone but does not insert into it. Superficial to the slings on the nasal side, paired uvulus muscles run the length of the soft palate to insert into the posterior nasal spine and the anterior and posterior raphe.

Thus, the final order of spatial-temporal development of soft palate muscles is anterior-to-posterior, lateral-to-medial, and oral-to-nasal: tensor, palatoglossus, palatopharyngeus—admixed with levator, uvulus.

Two muscles of the soft palate are directly attached to the palatine shelf. Tensor sends a few fibers very laterally to the maxillary tuberosity. Uvulus inserts into posterior nasal spine. When the posterior nasal spine fails to develop, uvulus cannot attach, leaving an opening for levator to insert opportunistically into the bone.

How Do Mesenchymal Deficiencies of Bone Fields Occur?

Each bone field is supplied by a neuroangiosome. It consists of a genetically paired nerve and artery that run in tandem with each other. Each one has a *growth cone with stem cells* at the tip. The neural growth cone secretes vascular endothelial growth factor while the arterial growth cone secretes nerve growth factor. As tissue development proceeds, "cross-talk" between the nerve and the artery enables them to grow forward into space. As they do so, they continuously nourish the mesenchymal tissues around them. They act as a "spillway" allowing continuous expansion of the tissues they support. In fact, the anatomy of neuroangiosome is just a reflection of how tissues arrive at their destination.

If, for some reasons, growth cone failure occurs from either component of the neuroangiosome, tissue development will be arrested at that very point. Endogenous causes include inadequate stem cell numbers or function so the growth cones stop advancing. Exogenous causes can be due to hypoxia or toxins. Regardless of etiology, growth failure always follows a distal-to-proximal gradient. *The most distal zone of a field is the most vulnerable.* As we have seen, the pattern of hard palate development demonstrates these concepts perfectly. As the artery develops, it progressively adds mesenchyme. The greater palatine axis keeps adding branches from anterior to posterior. Thus, mesenchymal insufficiency of the maxillary shelf shows up posteriorly and medially and progresses forward. The lesser palatine axis supplying the horizontal plate adds bone from lateral to medial and from anterior to posterior. Thus, PNS is the most vulnerable to deficiency.

Phylogeny of Soft Palate Muscles

Mapping the Pharyngo-Palatal and Craniopectoral Muscles

Based on neuroembryology, the neuromeric origin for the muscles of the pharynx and soft palate, as well as the *craniopectoral complex* (sternocleidomastoid and trapezius), can be surmised with great accuracy despite differences with the avian model. In the latter, of course, many (but not all) muscles have been mapped to somitomeres or somites. Furthermore, the soft palate is a completely mammalian invention. Finally, as you will recall Scm and Tpz are descendants of the cucullaris muscle complex, a levator of fish branchial arches BA3–BA7. This would correspond to neuromeric levels r6–r11 and sp1–sp4. These relationships apply neither to birds nor mammals.

- Noden's most recent mapping avian work fails to assign the palatopharyngeal muscles at all and maps Tpz to S4–S5. This has prompted investigators to postulate that sternocleidomastoid and trapezius do not arise within somitic PAM but rather to an undetermined backward extension of unsegmented PAM that is branchiomeric (Fig. 16.33).
- In mammals, Scm and Tpz muscles are transposed backward four neuromeric levels to c1–c6. Recall that neuroanatomically Scm is S5–S7 and Tpz is S6–S10. Their primary insertions are to clavicle (c1–c4) and scapula (c4–c6) with secondary insertions to the mastoid-occiput (r6–r11).

These details are not trivial. Lateral plate mesoderm is functionally distinct from PAM. It produces smooth muscle and striated cardiac muscle but in no instance does it produce striated skeletal muscle. Furthermore, palate–pharyngeal complex and Scm/Tpz are all innervated from a common motor column of the brainstem and spinal cord, nucleus ambiguus. Recall that NA extends posteriorly from medulla into spinal cord down to level c6, extension of nucleus ambiguus into the spinal cord from levels c1–c3 and c2–c6, respectively. So if these muscles don't arise from myotomes, just where do they come from? We can find probable answers based on their spatio-temporal development.

Pharyngo–Palatal Complex

Let's consider first the constrictors. We use the neural crest fascia that surrounds these muscles to help us map them. We know that r4–r5 buccinator is in continuity with r6–r7 superior constrictor. Proceeding backwards is r8–r9 middle constrictor and r10–r11 inferior constrictor. As previously detailed, between each of these muscle pairs we find neurovascular structures passing through the space, such as IX between superior and middle constrictors (Figs. 16.34 and 16.35).

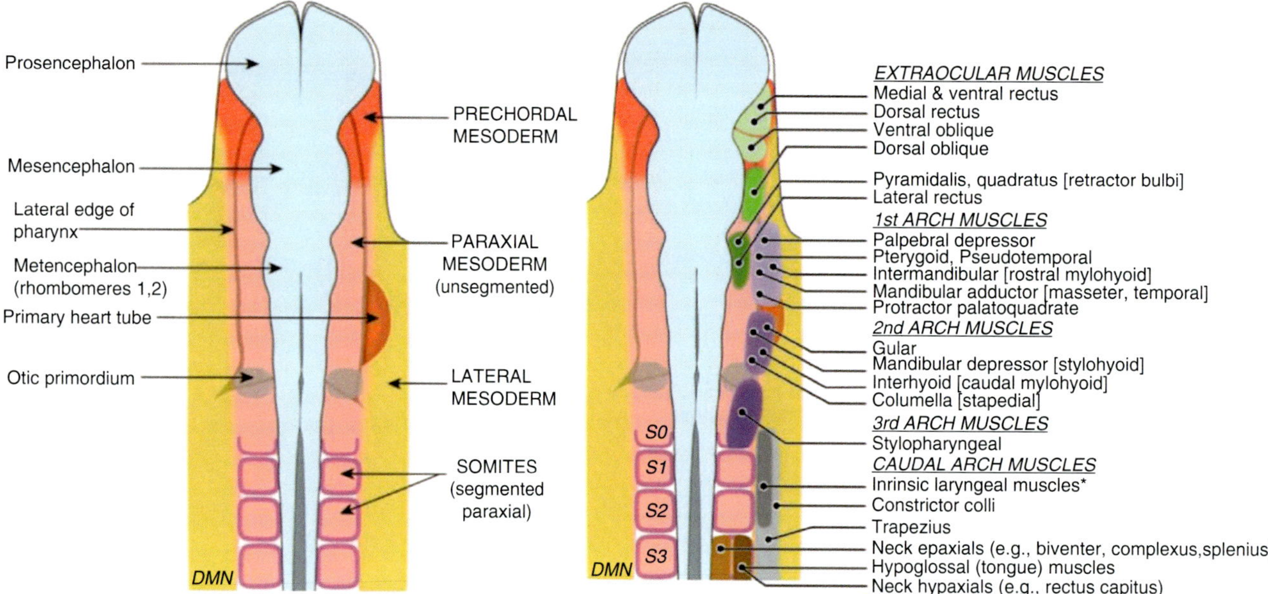

Fig. 16.33 Avian map of craniofacial muscles. Note: The zone of PAM for branchiomeric muscles from levels r8 to c6 is indicated in gray. It is *not* lateral plate mesoderm. Note: Laryngeal muscles are mapped further forward. Human dissections show roots from r10–r11 (the purported cranial root of XI to join with X as a single palato-pharyngo-laryngeal motor nerve of vagus). [Reprinted from Ziermann JM, Diogo R, Noden D. Neural crest and the patterning of vertebrate craniofacial muscles. Genesis. 2018; 56(6–7): e23097. With permission from John Wiley & Sons]

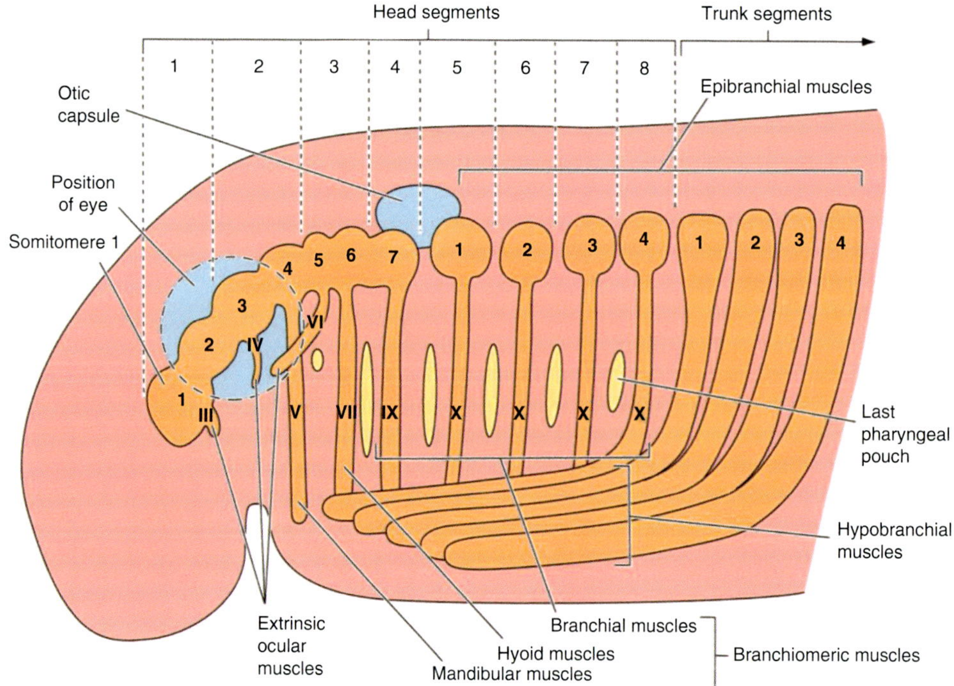

Fig. 16.34 Amniote model of craniofacial muscles. Note: Cranial–caudal pattern of lamination are seen in several muscle groups. The lamination is always medial-to-lateral. Pharyngeal constrictors laminate from superior (Sm6–Sm7) to middle (Sm8–Sm9) to inferior (Sm10–Sm11). Tongue muscle migrates forward from S1–S4. Strap muscles, as derivatives of coracomandibularis, laminate downward to clavicle and scapula and upward to hyoid, maintaining the same relationship in space. [Courtesy of William E. Bemis, PhD]

What somitomeres do these muscles belong to? Let's assume two somitomeres per muscle. We know that superior constrictor is in contact with buccinator so it makes sense that it arises from Sm6 and Sm7. That neatly puts the other two as Sm8–Sm9 and Sm10–Sm11. Let's find confirmation in the following anatomic facts.

- Constrictors make the pharynx before the palate is developed. You can live without a levator but not if you can't swallow or keep your larynx suspended in space.
- The constrictors are in continuity with oral lining buccinators. Facial nerve supplies constrictor and levator develops from a common blastema supplied jointly by VII and IX/X.

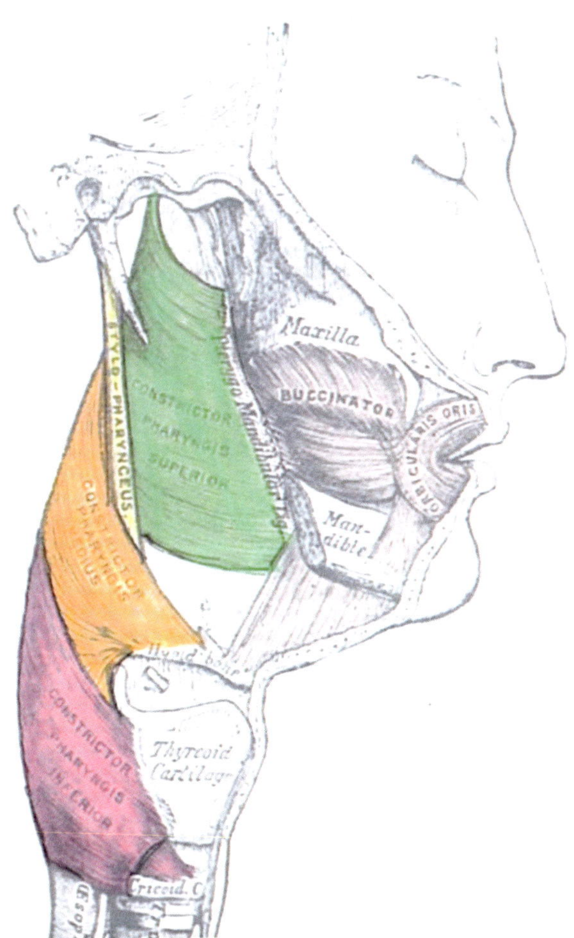

Fig. 16.35 Lamination of constrictors. Superior (Sm6–Sm7) > middle (Sm8–Sm9) > inferior (Sm10–Sm11). Note that key neurovascular structures traverse the space between the constrictors. Superior–middle contains glossopharyngeal nerve IX and stylopharyngeus. Middle-inferior transmits internal branch of superior laryngeal nerve and superior laryngeal branch of superior thyroid artery. [Reprinted from Lewis, Warren H (ed). Gray's Anatomy of the Human Body, 20th American Edition. Philadelphia, PA: Lea & Febiger, 1918]

- Palatal muscles migrate internal to the constrictors.
- All soft palate muscles map to Sm4 or Sm7.
- Tongue muscles from somites S1–S4 develop after the pharynx forms… they are force to migrate around it.
- Somitomeres Sm8–Sm11 predate somites S1–S4.

Let's consider first the constrictors. We use the neural crest fascia that surrounds these muscles to help us map them. We know that r4–r5 buccinator is in continuity with r6–r7 superior constrictor. Proceeding backwards is r8–r9 middle constrictor and r10–r11 inferior constrictor. As previously detailed, between each of these muscle pairs we find neurovascular structures passing through the space, such as IX between superior and middle constrictors.

What somitomeres do these muscles belong to? Let's assume two somitomeres per muscle. We know that superior constrictor is in contact with buccinator so it makes sense that it arises from Sm6 and Sm7. That neatly puts the other two as Sm8–Sm9 and Sm10–Sm11. Let's find confirmation in the following anatomic facts.

- ***In sum***: the constrictors and the soft palate muscles don't map to somites because they are somitomeric… the early bird gets the worm!

Craniopectoral Complex (Figs. 16.36, 16.37 and 16.38)

In the primitive state cucullaris is a longitudinal muscle that spans longitudinally forward in parallel with the epaxial muscles, making serial attachment along the dorsal edges of branchial arches BA3–BA7. The pectoral girdle codes to sp1–sp4 (at least). Thus cucullaris acts as a levator of the arches, bracing them against the scapula of the epaxials.

Recall that sharks elevate their snouts at the cranial joint to widen the gap of their jaws. They do so by using a column of epaxials that runs all the way forward to above the spiracle where they insert into the skull. In tetrapods, this function disappears, as do the branchial arches. Cucullaris is left suspended between levels r6-r7.

But it is noteworthy that Scm/Tpz develops much later than the inboard muscles. For cervical levels c1–c4 the order of formation is intrinsic muscles of the cervical spine and craniovertebral junction > strap muscles > Scm/Tpz. For this reason, one or the other (or both) can be congenitally absent, a condition that is not incompatible with life. Their motor column, the nucleus ambiguus remnant, is spatially, functionally, and temporally distinct from the neuromuscular system controlling the neck.

In order to understand the way in which the pharyngo–palatal complex develops, we have to make a diversion once again into the world of evo-devo. As previously stated, Noden's mapping work debunked the concept that branchiomeric muscles were somehow "visceral." This term is now restricted to the gut wall, viscera, blood vessels, and heart… all of them derivatives of lateral plate mesoderm. With the exception of cardiac muscle, LPM produces only smooth muscle.

We tend to think of somatic muscles as being associated with the extremities and body so the idea that branchiomeric muscles are somatic might seem counterintuitive. But recall that the fascia surrounding these muscles and the connective tissues within them are exclusively neural crest, arising from the surface and migrating into the substance of the pharyngeal arches and, perforce, into the pharyngeal walls.

A schematic drawing of the amniote pattern demonstrates that with the exception of the extraocular muscles of somitomeres 1–3 and 5–6 all branchiomeric muscles are hypaxial (Fig. 16.34). No epaxial muscles are produced in mammals until the level of the sixth somite, c2. Our attention here is on

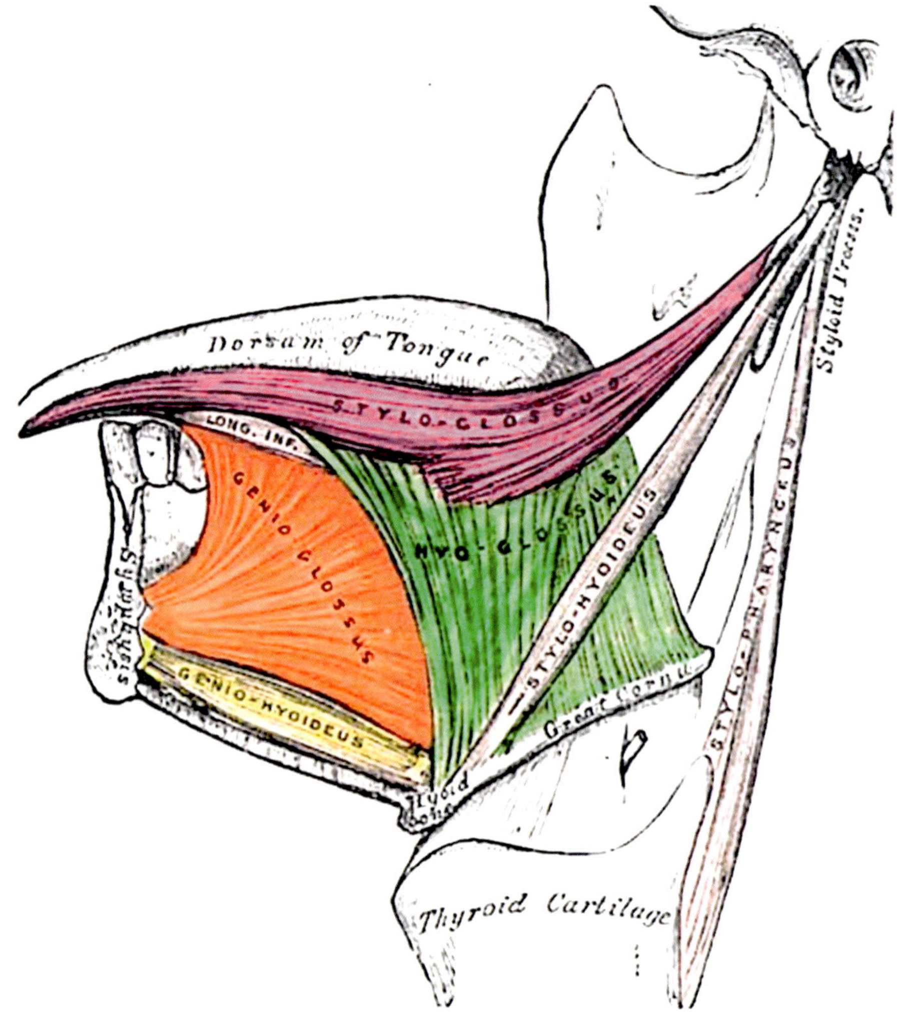

Fig. 16.36 Lamination of the four tongue muscles. Genioglossus (S1) > chondroglossus (S2) > hyoglossus (S3) > styloglossus (S4). Note that styloglossus lies internal to second arch stylohyoid and to third arch stylopharyngeus. [Reprinted from Lewis, Warren H (ed). Gray's Anatomy of the Human Body, 20th American Edition. Philadelphia, PA: Lea & Febiger, 1918]

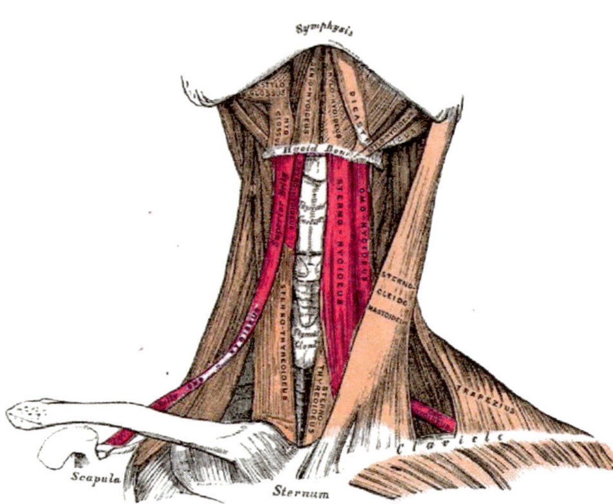

Fig. 16.37 Lamination of the four muscles constituting coracomandibularis. Geniohyoid (C1) > sternothyroid/thyrohyoid (C2–C3) > sternohyoid (C2–C3) > omohyoid (C3–C4). [Reprinted from Lewis, Warren H (ed). Gray's Anatomy of the Human Body, 20th American Edition. Philadelphia, PA: Lea & Febiger, 1918]

somitomeres Sm6–Sm7, the four post-otic somites, and the first four trunk segments (sp1–sp4). Note that Sm6–Sm7 produce superior constrictor and all the third arch palate muscles. These probably begin as a common blastema that subdivides into an internal group and an external group. The critical coordination of swallowing, getting the bolus to advance safely to the level of middle constrictor, requires coordinates with buccinator, the palate, and superior constrictor. The common origin and interrelated neural control of these structures helps in the process. Once the bolus is past the first barrier, it is then a straight shot down to the esophagus. … Assuming no fish bones!

The myotomes of somites S1–S4 produce two sets of myoblasts: Although the sources for the tongue muscles are well-known, no mapping of middle or inferior constrictor exists. However, since these muscles form the pharynx, and since the tongue muscles must migrate around the pharynx we can assume that the constrictors form first; they are really somitomeric, so they develop and *migrate prior to the somitomere–somite transformation.* They are produced in serial fashion each constrictor laminated inside its predecessor.

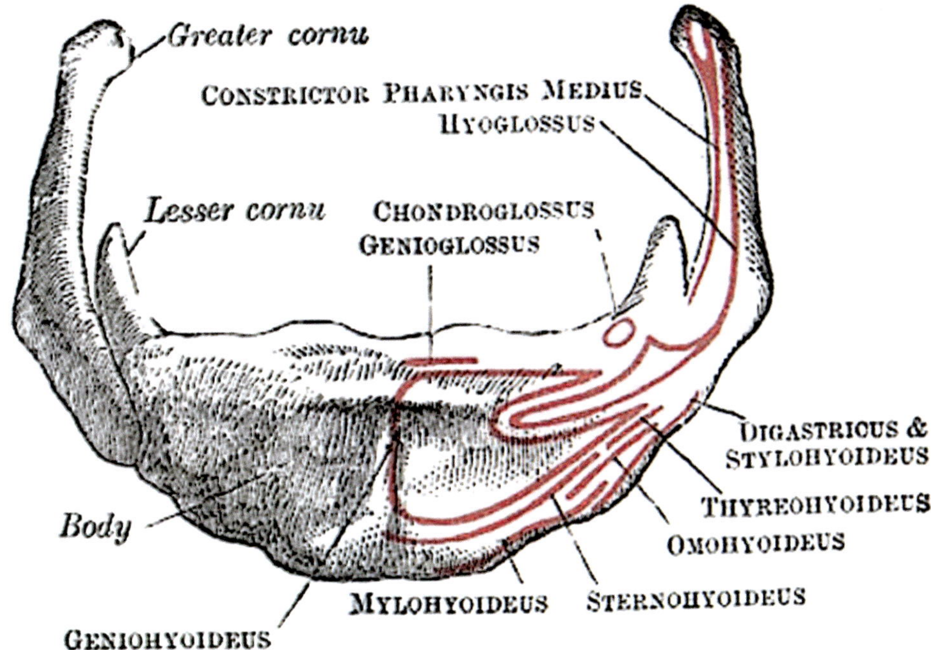

Fig. 16.38 Lamination of the tongue complex and the coracomandibularis complex on the hyoid. The hyoid serves as an intermediary in two muscle chains. The tongue muscles three of four have a primary insertion into the body and greater cornu which places them inside the strap muscles. Note the medial–lateral order: genioglossus, chondroglossus, hyoglossus, which anchors the mobile apparatus of mastication and swallowing to the pectoral girdle. Notes: Geniohyoid is in the midline. Sternothyroid is unlabeled here. It lies medial and deep to sternohyoid. Omohyoid is the most lateral and external of all. [Reprinted from Lewis, Warren H (ed). Gray's Anatomy of the Human Body, 20th American Edition. Philadelphia, PA: Lea & Febiger, 1918]

The *laryngeal muscles* are analogous in position to the palatal muscles. They are internal to the middle and inferior constrictors and migrate medially.

The *tongue muscles of S1–S4* (r8–r11) are termed the *anterior hypobranchial group*. They are non-branchiomeric. They develop later formally within somites. Like the constrictors they also develop from serial sources, with each somite producing a single muscle. Thus S1 produces the midline genioglossus; S2 makes condroglossus; S3 produces chondroglossus, and S4 brings up the rear with styloglossus. This neuromeric lamination is well seen on sagittal section and on the frontal view of the hyoid bone (Figs. 16.37 and 16.38).

The *strap muscles of S5–S8* (sp1–sp4) display a similar behavior and are referred to as the *posterior hypobranchial group*. They are non-branchiomeric. The migration laminates medial-to-lateral: sternothyroid, sternohyoid, omohyoid.

Cranial Muscles Homologies: *Squalus* ≥ *Necturus* ≥ Mammal

V, Sm4

- Lesser palatoquadrate, spiracularis, adductor mandibulae > adductor mandibulae > masseter, temporalis, pterygoids, tensors
- Intermandibularis > Intermandibularis > mylohyoid and anterior digastric

VII, Sm6

- Levator hymandibulae > depressor mandibulae, branchiohyoideus > stapedius, upper facial muscles and platysma, partial superior constrictor
- Interhyoideus > interhyoideus and constrictor colli > lower facial muscles and platysma, posterior digastric, stylohyoid

IX–X–XI, Sm7–Sm11

- Superficial constrictors and interbranchials > dilator laryngeus, subarcuals, transversi ventrales, depressors arcuum > constrictors, palato-laryngeal mm (larynx innervated by r10–r11 root of XI)
- Cucullaris > cucullaris, levatores arcuum > sternocleidomastoid/trapezius (Scm–Tpz complex shifts caudal has c1–c6 spinal root XI)

XII, C1–C3, Somites S1–S4 and S5–S10

- Coracomandibularis > genioglossus, geniohyoid (C1) > all tongue muscles, geniohyoid (C1 via XII)
- Coracoarcuals, coracohyoid > rectus cervicis > strap muscles (sternohyoid, sternothryoid, omohyoid) all C2–C4, ansa cervicalis

Evolution of the Branchial and Hypobranchial Muscles

Now that we know where we are going, in terms of mammalian muscle groups, let's see where we started from. We'll begin with a transverse section through the head of the humble dogfish shark, *Squalus*. Its first two branchial arches amalgamate to form the jaws but the last five are repetition of the same anatomy. BA3–BA7 are separated by four *transverse septae*. Each septum is stiffened by cartilaginous *branchial rays*. The gills are on either side of the septae. Where the septae reach the surface the interbranchial muscles form a ring, the *superficial constrictors*, dorsal and ventral. Each branchial arch forms an arc or ring around the pharynx. The segmented cartilages of the arches can be contracted by *interarcual* or *adductor* muscles. Levator muscles (seen in the primitive fish model) expand the pharynx. Note that in dogfish these have morphed into cucullaris. Note the presence of epibranchial and hypobranchial somitic muscles (Figs. 16.39, 16.40, 16.41, 16.42 and 16.43).

We turn now to a basal primitive fish model, seen from the lateral aspect. Recall that the gill arch is an articulated system of five branchial cartilages: pharyngobranchial, epibranchial, ceratobranchial, hypobranchial, and basibranchial. The ancestral condition of fish gills has two sets of opposing sets of muscles.

- *Levators* open up the hinge and expand the gills.
- *Constrictors* squeeze water through pharynx.
- Adductors bend the arches, causing them to open up, allowing water to escape.

In higher fishes the system simplifies. Chondrichthyans maintain the five-cartilage arches but *coracobranchialis* develops as a fusion of the ventral deep constrictors. Advanced osteichthyans have a fusion of the upper two cartilages with four sets of muscles operating in opposition: constrictors versus adductors and dorsal versus ventral branchials.

Of great importance for us is cucullaris, which relates to the dorsal levators. Cucullaris spans from the dorsal body axis downward to the last branchial arch and to the scapula. The fate of these muscles changes drastically in tetrapods but their genetic assignments persist. The external cucullaris, for example, morphs into sternocleidomastoid and trapezius with similar vectors of action (Figs. 16.37 and 16.42).

Further forward, over the first and second arches, we observe in sharks a complex series of muscles designed to control moveable upper jaw. Sarcopterygians such as the lungfish, *Neoceratodus*, show a fusion of maxilla to the cranial base. The jaw muscles consolidate to adductor mandibulae. As we have seen previously adductor shifts around in evolution to provide progressively greater degrees of biting power.

These relationships help us understand the derivation of the palate-pharyngeal-laryngeal muscles. These muscles have a linear contraction vector. They are not compressors; they lift and tighten. Because the soft palate and larynx are

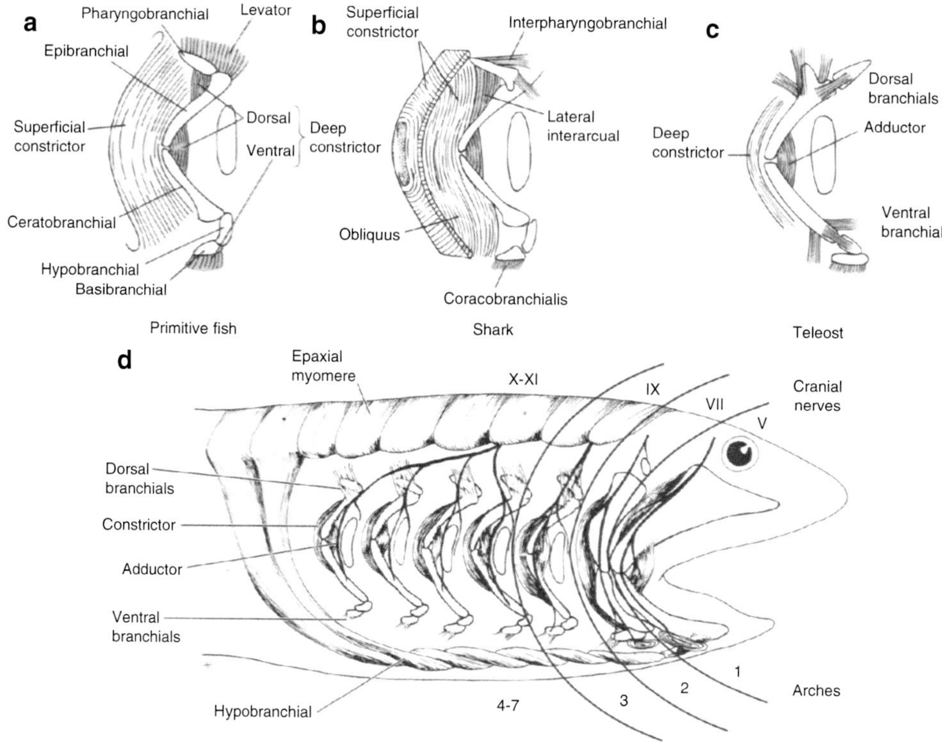

Fig. 16.39 Evolution of branchial arch muscles (pre-tetrapods). (**a**) Basic anatomy of the five branchial arch cartilages: pharyngobranchial, epibranchial, ceratobranchial, hyobranchial, and basibranchial. Note that levator, in the primitive state, is a single muscle. (**b**) In the shark, two sets of fibers appear in superficial constrictor indicating a more complex mechanism of action. (**c**) Advanced bony fishes (teleosts) have complex dorsal and ventral branchial. (**d**) Teleost showing the relationships between muscles and the seven branchial arches. [Reprinted from Kardong KV. Vertebrates: Comparative Anatomy, Function, Evolution, seventh ed. New York, NY: McGraw-Hill. With permission from McGraw-Hill Education]

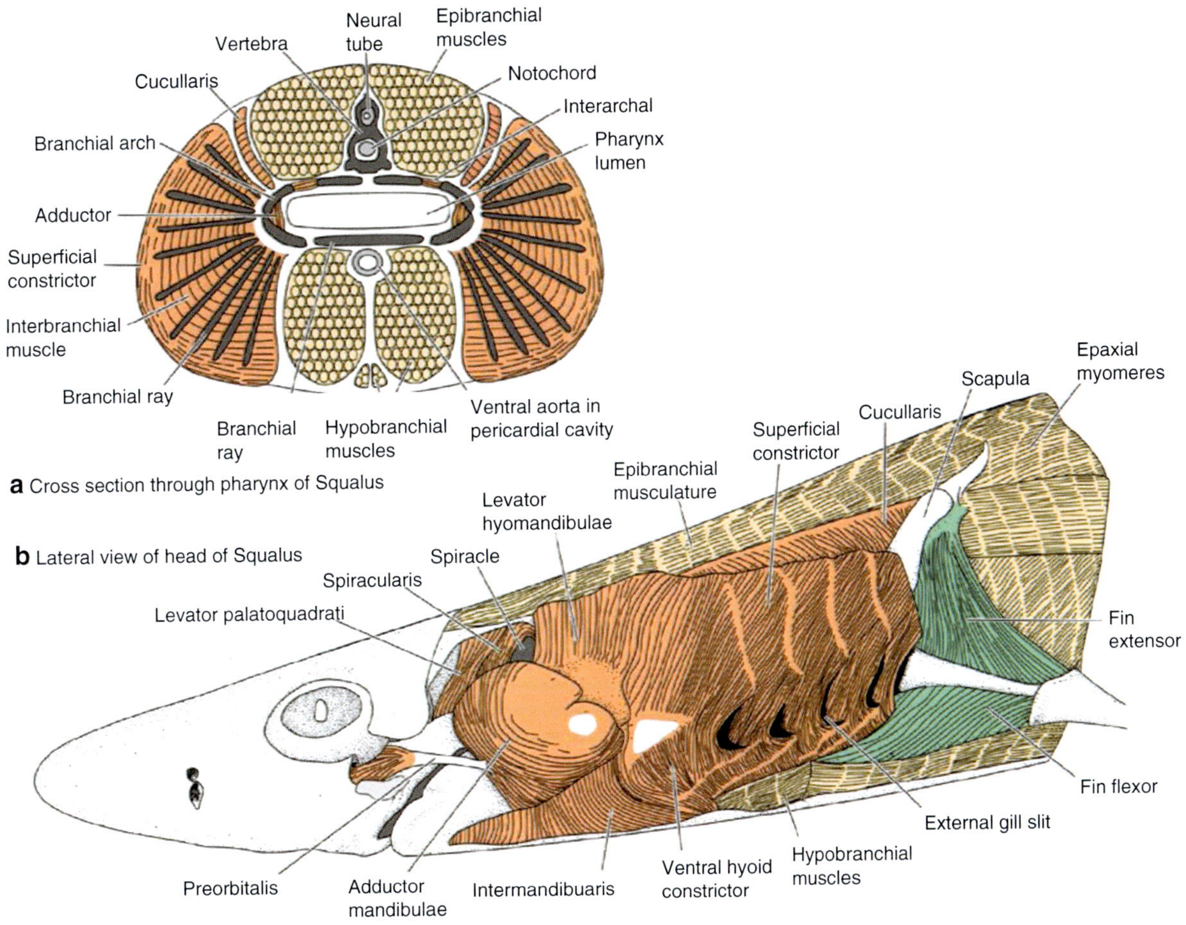

Neural tube
Epibranchial muscles
Vertebra
Cucullaris
Notochord
Interarchal
Branchial arch
Pharynx lumen
Adductor
Superficial constrictor
Interbranchial muscle
Branchial ray
Branchial ray
Hypobranchial muscles
Ventral aorta in pericardial cavity

a Cross section through pharynx of Squalus

b Lateral view of head of Squalus

Epaxial myomeres
Scapula
Cucullaris
Superficial constrictor
Epibranchial musculature
Levator hyomandibulae
Spiracle
Spiracularis
Levator palatoquadrati
Fin extensor
Fin flexor
External gill slit
Hypobranchial muscles
Ventral hyoid constrictor
Preorbitalis
Adductor mandibulae
Intermandibuaris

Fig. 16.40 Cranial muscles of the dogfish shark, *Squalus*. (**a**) Note pharyngeal arch cartilages (gray) encircling the larynx. Interarcual muscles draw them together and adductors (medial to the arch cartilages) cause them to bend and open up. Cucullaris is an example of a levator that straightens out the arch cartilages causing them to close up, thereby squeezing water through the pharynx. (**b**) On lateral view, preorbitalis and levator palatoquadrate retract the head as does the long dorsal column of epibranchials. Intermandibularis is a sling that spans below the mouth. Note the hypobranchial column connecting the coracoid of the pectoral apparatus with the mandibulae. These muscles, and amalgamation derived from a series of ventral deep constrictors, function to open the lower jaw, albeit weakly. The main source for the gaping maw of sharks is the dorsal retractor series. [Courtesy of William E. Bemis, PhD]

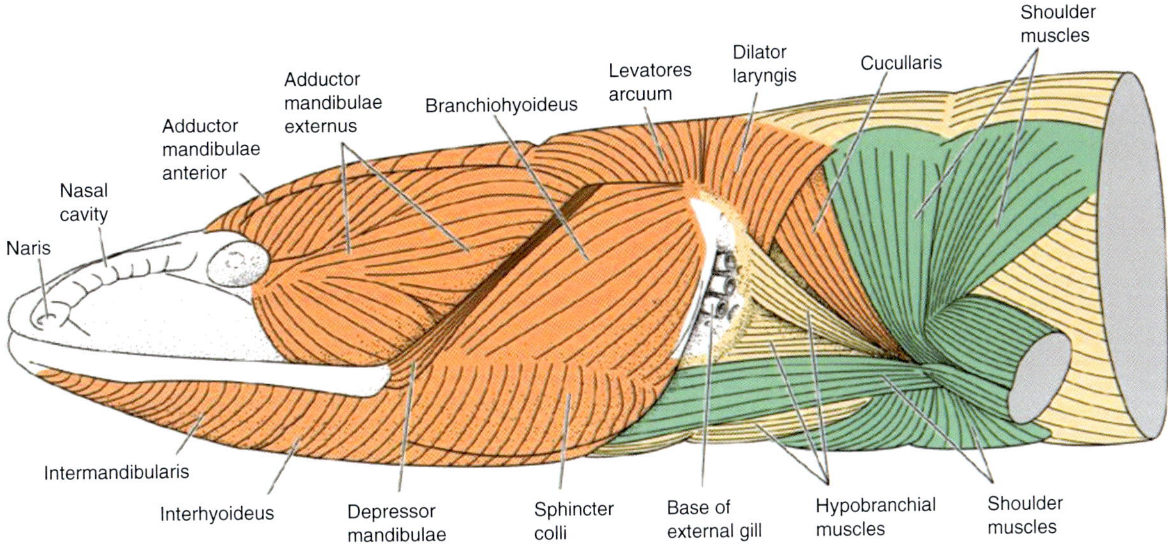

Fig. 16.41 Primitive tetrapod muscles, as seen in *Necturus*. Assuming dilator laryngeus has moved to r10–r11, cucullaris is now inserted into the back of the skull at those levels and into scapula. First arch intermandibularis will become mylohyoid and second arch sphincter colli will become platysma. [Courtesy of William E. Bemis, PhD]

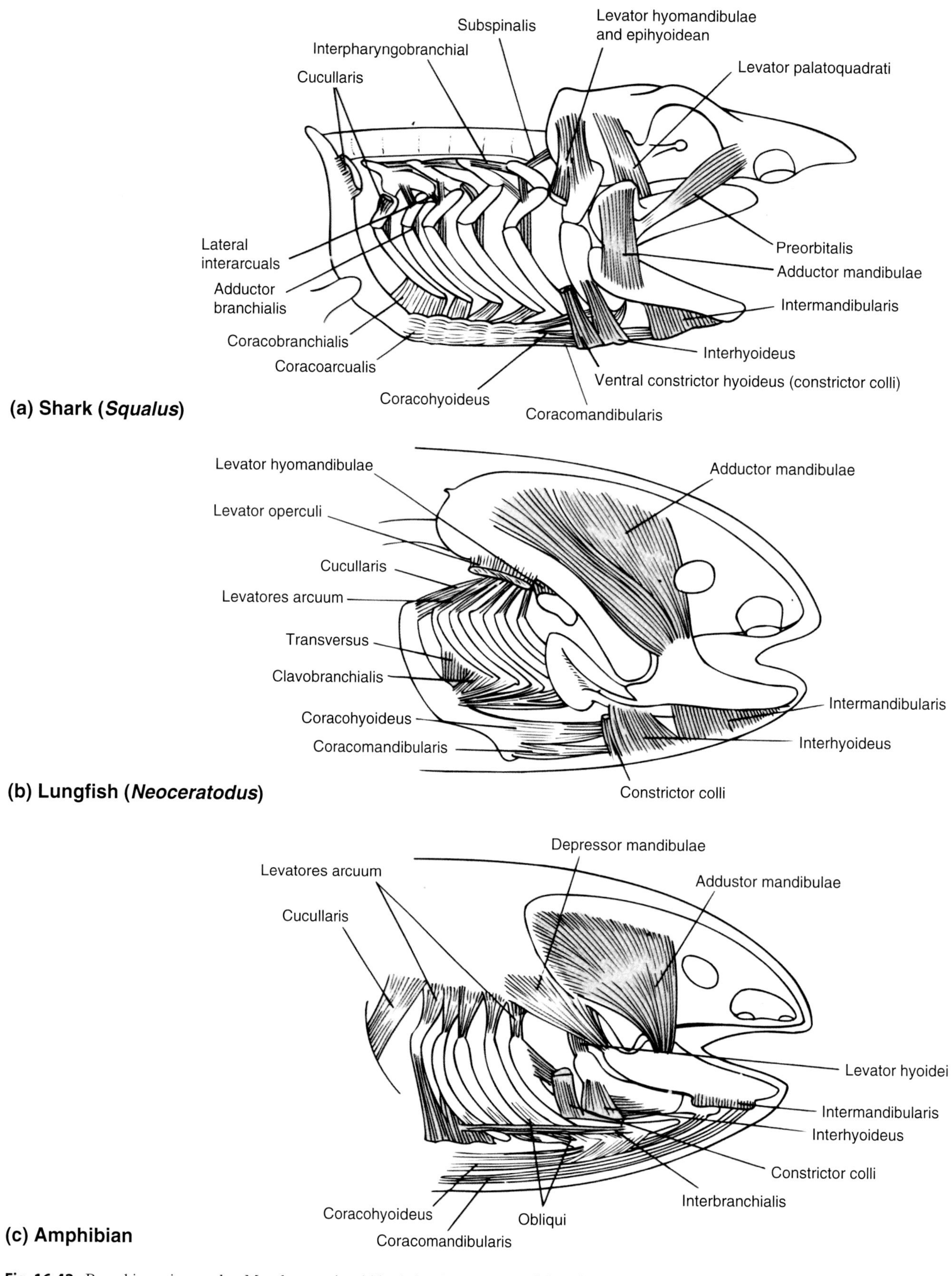

(a) Shark (*Squalus*)

(b) Lungfish (*Neoceratodus*)

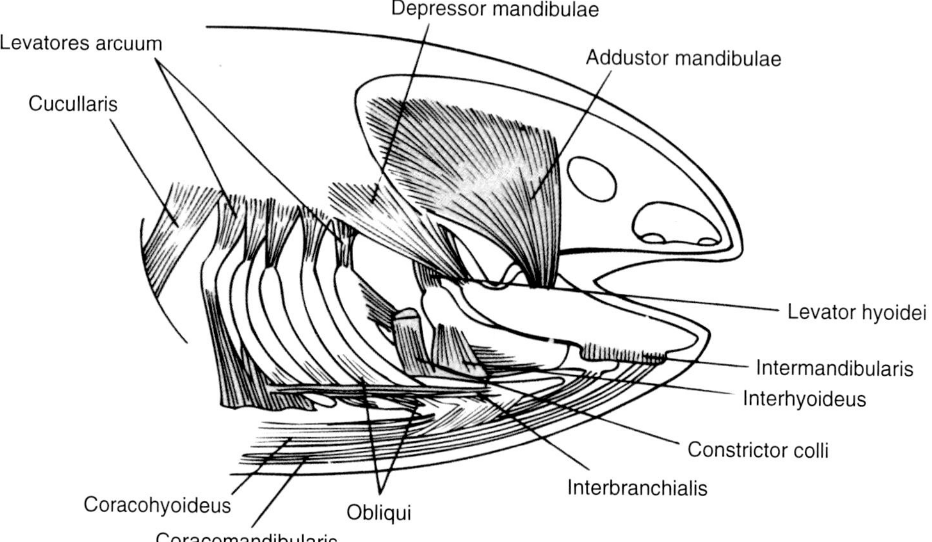

(c) Amphibian

Fig. 16.42 Branchiomeric muscles. Muscles remain within their original arch/somitomeres of origin. CN V, VII, IX supply arches 1–3, respectively. CN X–XI supply arches 4–7; recall that arches 6–7 constitute levels c1–c4. (**a**) Chondrichthyan (shark), (**b**) transitional fish-tetrapod (lungfish), (**c**) tetrapod (amphibian). [Reprinted from Kardong KV. Vertebrates: Comparative Anatomy, Function, Evolution, 7th ed. New York, NY: McGraw-Hill. With permission from McGraw-Hill Education]

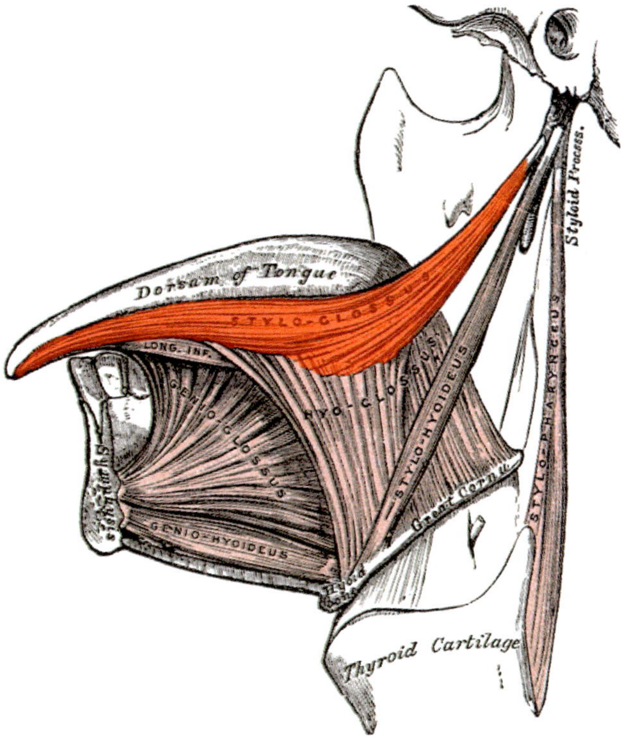

Fig. 16.43 Styloglossus is superficial to genioglossus. The latter, being the most external tongue muscle *must* pass external to pharyngeal arch derivatives. The internal muscles of palate and larynx may come from the ancient adductors but seem to bud off from the medial side of the constrictor series. Levator veli palatini and palatopharyngeus are known derivatives of superior constrictor. Palatoglossus lies within the territory of superior constrictor and is likely derived from it as well. [Reprinted from Lewis, Warren H (ed). Gray's Anatomy of the Human Body, 20th American Edition. Philadelphia, PA: Lea & Febiger, 1918]

internal to the constrictors, we can speculate that they descend from the primitive *branchial arch adductor muscles*.

- Superior constrictor (Sm6–Sm7) is known to produce internal derivatives, the levator and palatopharyngeus, palatoglossus.
- Middle constrictor (Sm8–Sm9) may contribute to palatopharyngeus, due to its lengthy inferior extension beyond the territory of superior constrictor.
- Inferior constrictor (Sm10–Sm11) is the likely source of the laryngeal muscles.

Evolution of the Soft Palate

Basal synapsids gained speed as carnivores by putting the legs under the body. They had *choanal folds* which partially separated the air passage from the mouth. Further along the line we see the first manifestation of a hard palate in therocephalians, fierce carnivores with large skulls and dagger-like teeth. Stiff choanal folds were vital for sustaining respiration while holding down struggling prey. In their descendants, the pre-mammal cynodonts, such as *Procynosuchus*, the folds morphed into palatine shelves. In fact, the completion of the secondary hard palate was essential for the emergence of herbivores. The dental innovations required for this type of mastication proved to be the stimulus for the innovation of the TMJ. In addition to complex molars for grinding, the post-canine dentition of crown cynodonts, the premolars, is provided for shearing of tissue.

Muscles of mastication changed in this transition as well. Adductor mandibulae subdivides in pre-mammal cynodonts into temporalis and masseter. The invention by mammals of the TMJ supports differentiation of masseter into the pterygoids. At the same time *unossified remnants of the choanal folds were invaded* by muscles of different origins.

- Medial pterygoid blastema is the source for *tensor veli palatini*.
- Superior constrictor blastema spatially shelters the future soft palate. It has two dorsal derivatives, *levator veli palatini* and *palatopharyngeus*, and two ventral derivatives, *palatopharyngeus* and *palatoglossus*. All five muscles share a common innervation from the r6–r7 nucleus ambiguus.
- Medial and inferior constrictors are well-positioned to give off the laryngeal muscles. These have a common innervation from r8–r11 nucleus ambiguus.

Pathologies of Soft Palate Cleft

Key Concept Descending palatine artery supplies the hard palate via greater palatine artery and the soft palate via lesser palatine artery. Secondary hard palate clefts represent pathology of the GPA field while soft palate clefts represent pathology of the LPA field (Figs. 16.44 and 16.45).

Recall that the neural crest source of the soft palate is (like the tongue) from the first arch and third arch with some second arch represented as well. It is not surprising that the blood supplies of the three zones of the soft palate reflect this gradient: anterior zone (first + second), middle zone (second + third), and posterior zone (third).

Abnormal Platform/Normal Muscles

Lesser Palatine Neuroangiosome

Isolated cleft palate is characterized by a reduction in the horizontal plate of palatine bone, an r2 neural crest derivative. Its neurovascular axis is the lesser palatine artery. LPA supplies (1) horizontal plate of the palatine bone and (2)

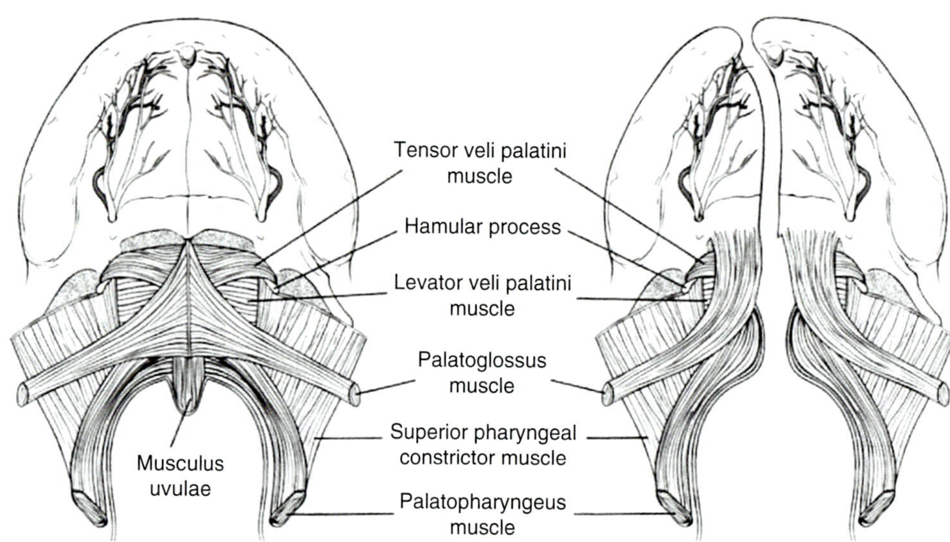

Fig. 16.44 CP muscle insertions. Aponeurosis absent. TVP does not have transverse fibers. Uvulus at the margin, flanked by LVP. [Reprinted from Carstens MH. Pathologic anatomy of the soft palate, part 2: The soft tissue lever arm, pathology, and surgical correction. J Cleft Lip Palate Craniofac Anomal 2017;4:83–108. With permission from SAGE Publications]

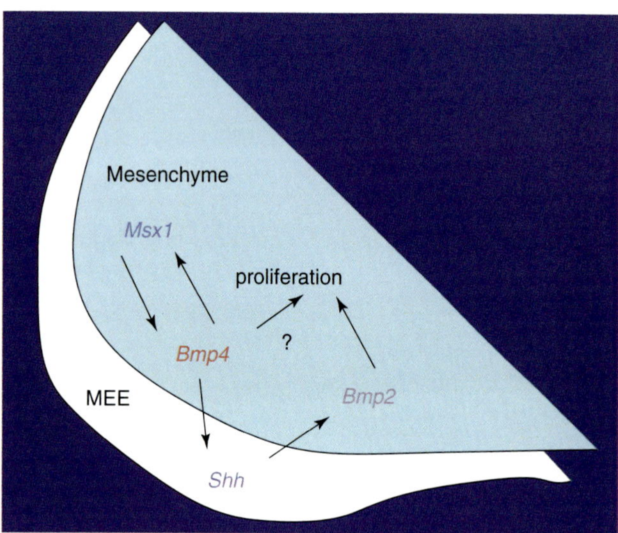

Fig. 16.45 Soft tissue fusion failure of the soft palate. The circuit between bone morphogenetic protein 4 and sonic hedgehog interrelates mesenchymal bone production with stability of the epithelium of the overlying soft tissue envelope. Membranous palatine bone synthesis produced BMP4 which diffuses for the bone edge backward and from nasal to oral to reach the epithelial margins. Here, under normal circumstances, it inhibits the stabilizing effect of sonic hedgehog on the epithelium. Fusion is then permissible. Quantitative reduction in bone mass = reduction in [BMP4] which causes the epithelium to remain intact and not fuse. This process is the reverse of BMP4 diffusion: it thus proceeds from distal-to-proximal. [Reprinted from Zhang Z, Song Y, Zhao X, et al. Rescue of cleft palate in Msx-1-deficient mice by transgenic BMP4 reveals a network of BMP and Shh signalling in the regulation of mammalian palatogenesis. Development 2002; 129 (17): 4135–4146. With permission from the Company of Biologists]

anterior one-third of the palatine aponeurosis. As previously discussed, development of the horizontal plate involved the progressive deposition of mesenchyme by the LPA neuroangiosome in a medial-to-lateral and anterior-to-posterior sequence. The most medial and distal structure of the horizontal plate to form is the posterior nasal spine (PNS).

When a deficiency state occurs in the LPA neuroangiosome, *the pattern of tissue loss is exactly the reverse of its developmental gradient*. Thus, the PNS is the very first structure to be knocked out. *Absence of the PNS is pathognomonic for isolated CP*. Further mesenchymal loss reduces the horizontal plate from its initial rectangular shape to that of a triangle. The palatine aponeurosis, being of the same angiosome, is reduced or absent. This causes forward displacement of LVP, bringing it into contact with the posterior edge of palatine and facilitating its insertion into the bone. Because aponeurosis is an r2 structure, it may itself be *intrinsically reduced*, bringing the entire palatal muscle complex further forward into space (Fig. 16.45).

The soft palate is a form of "muscle sandwich." The spatial order of the soft palate muscles reflects the timing with which these muscles come into position. In every case TVP arrives first and is associated with the oral mucosa. LVP follows next and assumes an intermediate position. Palatoglossus and palatopharyngeus are last. These muscles lie oral/cranial to LVP; they are associated with the nasal mucosa. Uvulus is anomalous but is associated with a secondary insertion into the posterior nasal spine.

Insertion is a process by which the fascial envelope of a muscle "recognizes" a biologic signal, such as BMP4, elaborated by bone or fascia. Under normal circumstances bone or aponeurosis may express BMP4 signals in a spatio-temporal order as its various sectors achieve developmental maturity. Much as landing lights signal the presence of an airport runway, so does BMP4 guide a fascia-muscle unit into its appropriate insertion site. Muscle units arrive in a fixed sequence. Each muscle will respond opportunistically, choosing an insertion site according to its spatial position and the timing with which the "landing lights" are turned on.

Under normal circumstances, uvulus leaps over the levator complex and the "bare area" of aponeurosis to insert into the posterior nasal spine. Recall that each bone has a spatiotemporal maturation sequence. As soon as one zone is mature, it stops producing BMP4 [39]. *Muscle insertion will then take place at the next available maturation site*. In the case of the palatine bone the vertical plate is the "oldest," and its ossification sequence leads to the synthesis of the horizontal plate, from lateral-to-medial and from anterior-to-posterior. The PNS is the "newest" zone and, consequently, the most vulnerable. Under normal circumstances, *LVP assumes its insertion site into at a time when development of the horizontal lamina is complete*, leaving only the PNS as a potential insertion site for uvulus. When the developmental sequence of palatine bone is altered, the PNS is gone, the intervening anterior aponeurosis is absent, LVP is positioned forward, and the horizontal lamina expresses BMP4. Uvulus makes an opportunistic insertion into the edge of the bony cleft, followed by LVP. *Nota bene*: These two muscles can be distinguished. Uvulus is seen as distinct, wispy longitudinal fibers. Levator is just medial and is distinctly more bulky.

This concept explains why, in the submucous cleft, levator is malpositioned, even when the apparent volume of the palatine bone can appear normal. The absence of PNS is the "give-away." It tells us that the developmental sequence of palatine bone is aberrant. And the defect in aponeurosis always precedes that of the bone. Thus, even though only PNS is absent, soft palate muscles are forward-positioned. As the deficit in the palatine bone becomes more pronounced, the normally rectangular horizontal lamina becomes triangular. Ultimately, a pronounced notch will be present. As the palatine pathology worsens, the levator insertion streams further and further forward. If horizontal plate is sufficiently attenuated medially, as in the case of a maxillary cleft, levator fibers will actually pass forward to insert into the posterior-medial margin of the maxilla.

Mechanism of the Soft Tissue Cleft

OK, you might say, I understand how a bone deficiency can pull the soft palate forward. But how does that affect the actual midline fusion of the soft palate envelope? How can a bone defect cause a soft tissue defect distal to it? The answer has to do with the *epithelial stability* of the mucosal envelope. As previously discussed, muscles assigned to the soft palate fill out a potential envelope of mucosa located at the interface of two neuroangiosomes just behind the maxilla. The boundary between third arch r6 mucosa innervated by IX and first arch r2 mucosa supplied by lesser palatine nerve constitutes the biologic "exit sign" for the myoblasts to move into position. The two edges of the mucosal envelope break down and fusion takes place.

Zhang et al. describe an elegant model for soft tissue fusion that depends upon the production of soluble protein products by adjacent developing bone fields [40]. Epithelial integrity of the mucosal envelope depends upon the local production of sonic hedgehog (SHH) gene products. BMP4 inhibits SHH and permits fusion to take place. BMP4 is a byproduct of membranous bone production. Thus, the horizontal palatine shelf produces a fixed amount of BMP that diffuses backward through the tissue of the soft palate. Any reduction in bone mass of the horizontal plate will reduce the amount of BMP4 available for SHH inhibition. When the deficiency of BMP4 reaches a critical point, persistent SHH in the epithelium will prevent the fusion mechanism from taking place. Because the source of BMP4 is anterior, the most vulnerable zone of the soft palate is distal, at the uvulus. Worsening BMP4 deficits simply cause the soft tissue cleft to advance forward until it finally reaches the bony shelf (Fig. 16.45).

Greater Palatine Neuroangiosome

If the biologic problem involves the greater palatine artery neuroangiosome as well, a forward extension of this very mechanism will be seen. If the maxillary hard palate gap is narrow, mucosal fusion can still take place, leaving a palpable defect in the bone, but no visible cleft. But if the critical contact distance is exceeded, if the maxillary shelves are too much reduced, soft tissue closure becomes impossible to achieve. A final variation can occur when both shelves are perfectly normal but are related to a vomer (actually, paired vomerine fields) that is deficient. A very narrow midline cleft can occur. This is perhaps the only situation in which an intact palatine hard palate can be found in the presence of such a midline maxillary hard palate cleft.

Abnormal Bony Platform/Abnormal Muscles

This situation obtains when one or more muscles are dysfunctional. The cause can be mesenchymal, i.e., pathology affecting the mass or quality of myoblasts within their somitomere of origin prior to migration. It can also be neurologic, with an otherwise normal muscle but poor to absent motor innervation.

Tensor veli palatini develops from somitomere 4; its fascia originates from r3 neural crest. Problems with TVP may be a manifestation of a more global neural crest problem affecting both r2 and r3. TVP weakness manifests itself as inadequate opening of the pharyngotympanic tube and predisposition to otitis media. In such cases, cleft palate repair cannot be expected to correct a dysfunctional TVP.

Palatopharyngeal muscles develop from somitomere 7 (or non-LMP head mesoderm; lying just outside Sm7). Their fasciae originate from r6–r7 neural crest. These include superior and (?) middle constrictors. Dysfunction of these muscles may lead to inadequate palate closure. Hypoplasia or paresis of LVP and the remaining muscles of the third arch can be seen in Treacher-Collins and Goldenhar syndrome.

Table 16.3 Palatal muscle action on electrostimulation

	Findings on unilateral stimulation
Tensor VP	Lateral retraction of the soft palate
Levator VP	Elevation and backward extension of posterior soft palate
Palatoglossus	Elevate root of tongue, approximate PG arches to midline
Palatopharyngeus	Pull pharynx up/forward, approximate PP arches to midline
Uvulus	Thicken middle one-third of soft palate (hard to see)
Superior constrictor	Constriction of upper pharynx

Electrostimulation of palatal muscles can diagnose the following functional deficits (Table 16.3).

Note that although stylopharyngeus is also a derivative of Sm7 its fascia is r6. This is consonant with its glossopharyngeal innervation form cranial nerve IX with its nucleus in in the r6 component of the nucleus ambiguus.

Normal Bony Platform/Abnormal Muscles

This is the situation seen in most cases of non-cleft VPI. Presence or absence of the posterior nasal spine determines if the palatine bone is considered borderline deficient. If the ANS to PNS distance is normal but VPI is present, the diagnosis will rest upon which muscles are involved. Abnormalities of Sm4 and Sm7 exist independently but in the latter case, all of its derivatives are usually involved.

A curious observation is seen in cases of microtia when the hard palate is otherwise intact. These demonstrate a near universal weakness of ipsilateral levator function (but not the remaining Sm7 muscles). Recall, from our discussion on pharyngeal arch syndromes and ear development (Chaps. 9 and 11), that the Eustachian tube is trilaminar. The lateral tube, being a derivative of the first arch, provides attachment for Sm4 tensor. The medial tube derived from the third arch, provide attachment for Sm7 levator. The tube itself is the buried remnant of the first pharyngeal pouch. Recall that virtually all representation of second arch within the oral cavity is *obliterated* by a combination of involution of second arch mucosa and overgrowth of mucosa from the first and third arches.

Summary of Pathology by Neuroangiosome

GPA Affected/LPA Unaffected

It is theoretically possible to have a minor GPA deficiency state with an intact soft palate. This presupposes that the maxillary palatal shelves cannot achieve the critical contact distance required to fuse with each other and/or with the vomer. Such rare defects can occur at any place along the length of the maxillary hard palate with one caveat. As long as the width of the maxillary shelves is sufficient posteriorly so that the horizontal palatine shelves can track along them normally and achieve fusion, the soft palate will remain intact. Posterior nasal spine will be present.

If GPA is significantly affected, it will entrap the mesenchyme of palatine bone even if the LPA pedicle is biologically intact. Posterior nasal spine will be absent. If the mesenchymal masses are physically separated beyond the critical contact distance a soft palate cleft will result.

LPA Affected/GPA Unaffected

This is the typical scenario for isolated soft palate clefts. In such cases the defect in the palatine shelves can run forward as far as the posterior border of the maxillary shelves, but no farther.

LPA Affected/GPA Affected

This describes the situation of palate clefts involving a complete separation of the palatine shelves with the defect extending forward into the maxillary hard palate. Since the deposition of bone mesenchyme by the GPA neuroangiosome is anterior to posterior, the most recent zone and the one affected first by any GPA deficiency state will be the posterior margin. Thus, combined clefts of the palatine and maxillary hard palate extend forward toward the incisive foramen with greater degrees of mesenchymal insufficiency.

Correction of Soft Palate Cleft: Developmental Field Algorithm

Is VPI Predetermined?

It should be apparent at this point that in most cases of soft palate clefts the muscles are normal in size and function. Proper spatial repositioning at the appropriate age should lead to a good outcome. An exception to this is velocardiofacial syndrome in which third arch pathology affects the palatal musculature itself. Assuming normal neuromuscular development, why does velopharyngeal insufficiency (VPI) occur in the face of an otherwise uncomplicated repair? Excluding issues of surgical trauma and inadequate dissection, several questions remain: Is the physical position of the soft palate in space inadequate from the very start? What is the developmental sequence of the bony palate?

Surgical Management: A Developmental Field-Based Algorithm

Nota Bene The literature of cleft palate is vast. Anatomic studies by Kriens remain helpful [9, 41]. Management of the hard palate cleft using *alveolar extension palatoplasty* technique (AEP) uses embryologic dissection of the entire greater palatine neuroangiosome to generate longer flaps and prevent fistulae [42]. Unlike the two-flap technique of Bardach, the AEP incision is not placed at the junction of the alveolus with the palatal shelf. The flaps are incised just off the midline of the alveolar ridge, thus preserving all lateral vessels ascending from the GPA to supply the stem cells within the biosynthetic mucoperiosteum responsible for bone growth of the lingual alveolus. The technique is nicely illustrated by Bennun and Monasterio (Figs. 16.46, 16.47, 16.48, 16.49, 16.50, 16.51, 16.52, 16.53 and 16.54).

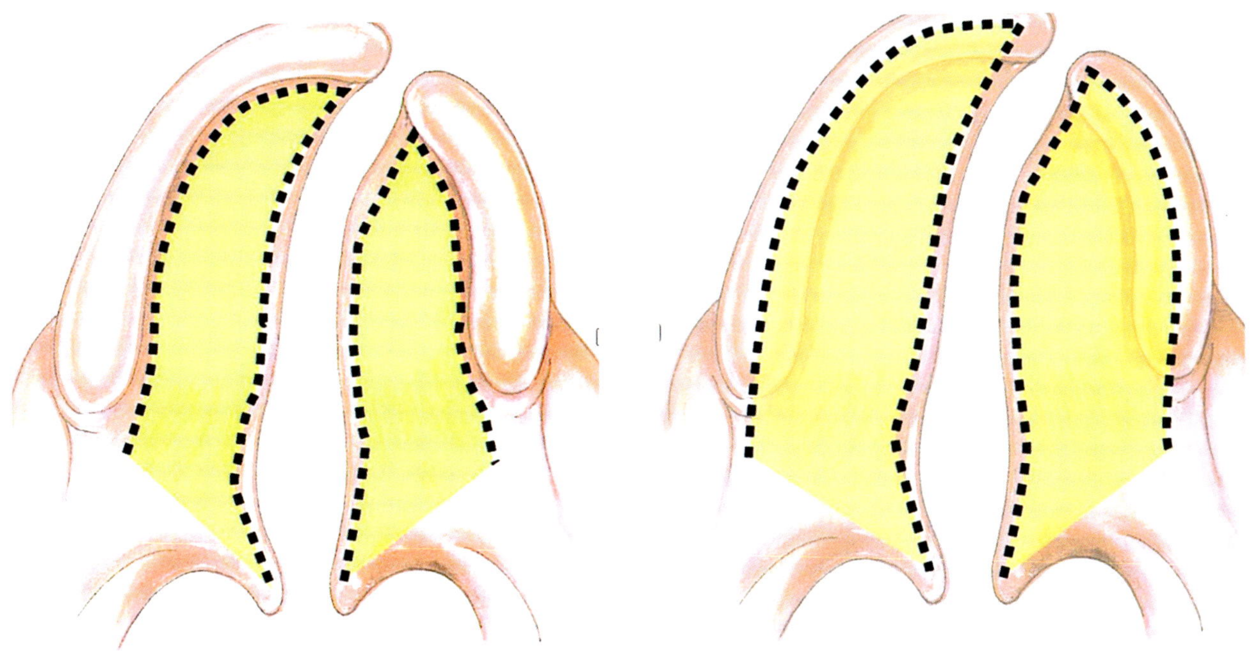

Fig. 16.46 Two-flap palatoplasty (Bardach): lateral incision at border of hard palate with alveolus. Alveolar extension palatoplasty (AEP, Carstens): incision just lingual to the midline of the alveolus. [Courtesy of Michael Carstens, MD]

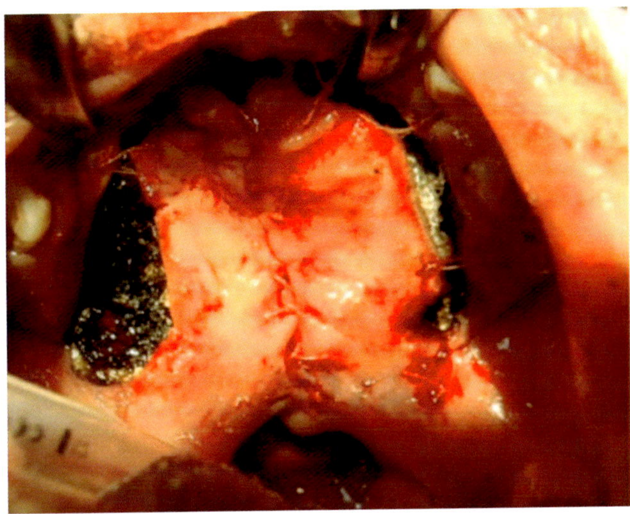

Fig. 16.47 Two-flap palatoplasty flaps intrinsically short and narrow. Cannot reach anterior palate. GPA supply to lingual mucoperiosteum interrupted—alveolar growth. Denuded areas fill with scar. [Courtesy of Michael Carstens, MD]

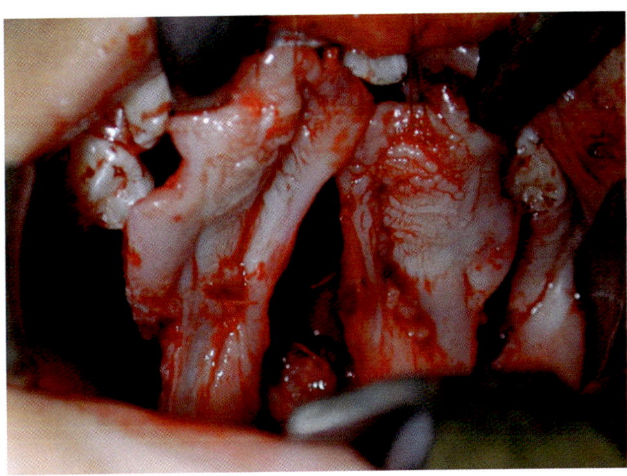

Fig. 16.48 AEP flaps demonstrating gain in length from using the entire neuroangiosome. Non-cleft flap extends 0.5 cm *beyond* the alveolar cleft. Cleft-side flap reaches anteriorly immediately behind the alveolar cleft. Midline flap closure is S-shaped. [Courtesy of Michael Carstens, MD]

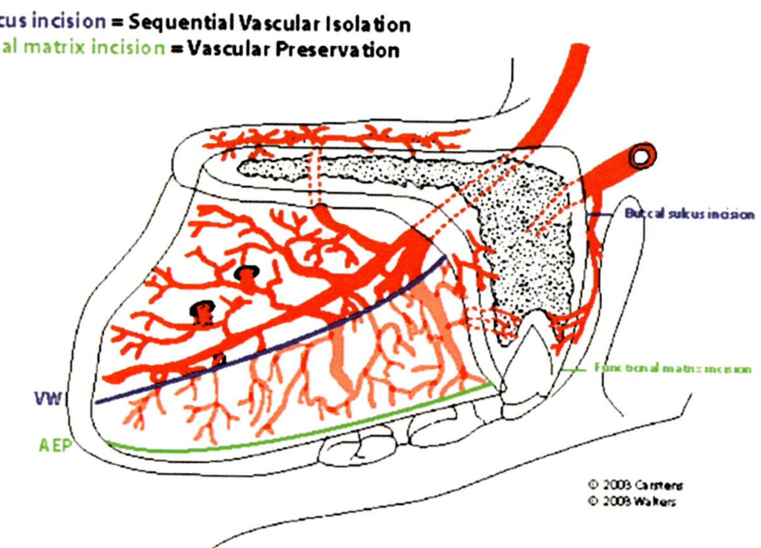

Fig. 16.49 Subdivision of angiosomes damages the osteogenic stem cell envelope. Buccal sulcus incision (blue) interrupts blood supply to the buccal alveolus. Veau–Wardill incision knocks off the blood supply to the lingual alveolus. DFR incision in primary lip repair is vertical from the piriform margin down to the alveolar margin. It does not become pericoronal. In secondary repair of alveolar fistula, at ages 4–6, the mucoperiosteal envelope can be elevated along the inferior margin of the alveolus and transferred mesially to two dental units. [Courtesy of Michael Carstens, MD]

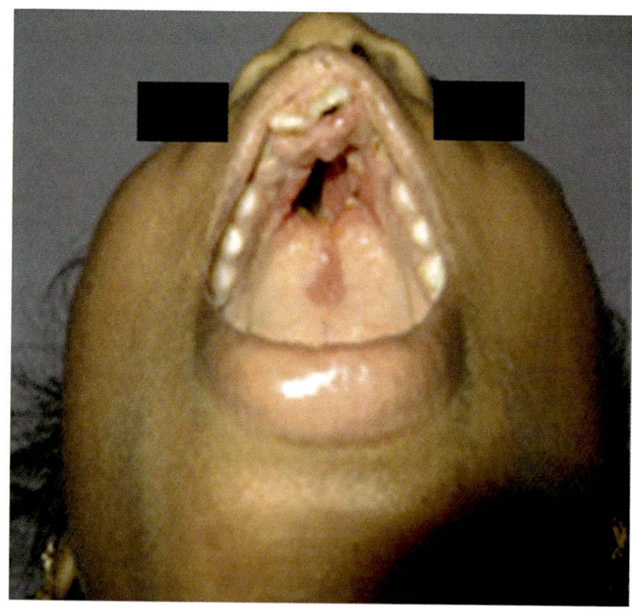

Fig. 16.50 Veau–Wadrill flaps embryologically incorrect; cause iatrogenic damage. Anterior fistula: results from subdividing the GPA angiosome. GPA injury: incision at junction of hard palate and alveolus kills the P1–P2 palatal shelves but spares the alveolus. Vascular anatomy of GPA is sharply demarcated. [Courtesy of Michael Carstens, MD]

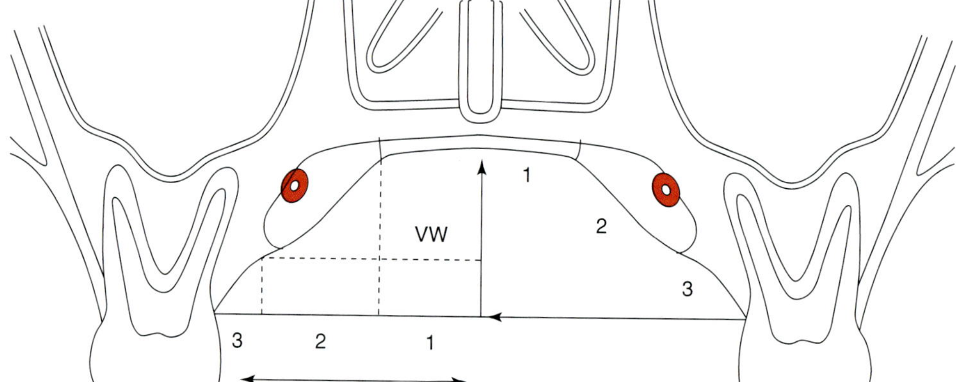

Fig. 16.51 Geometric expansion of AEP flaps vs. Veau–Wardill flaps. Augmentation of both width and height translates to (1) significantly larger anterior reach; and (2) ability to cover lateral releasing incisions without a raw area. Only open site is at the maxillary tuberosity. [Courtesy of Michael Carstens, MD]

Fig. 16.52 AEP "extension effect." Veau–Wardill (red): short of incisive foramen AEP (purple): +30% gain… greater segment exceeds nasal floor. [Courtesy of Michael Carstens, MD]

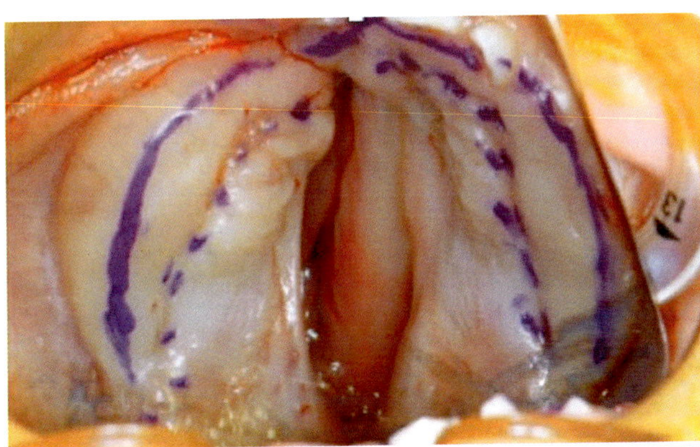

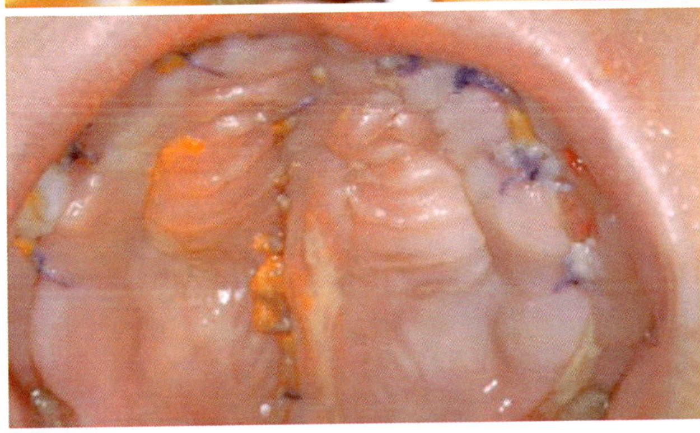

Fig. 16.53 Right-side UCLP closure at 1 year. No raw areas at completion of procedure. [Courtesy of Michael Carstens, MD]

We have covered a great deal of information in this chapter. All of it boils down to a very simple model. The palate is a functional structure designed to assist in the management of food and to control airflow for proper speech. It consists

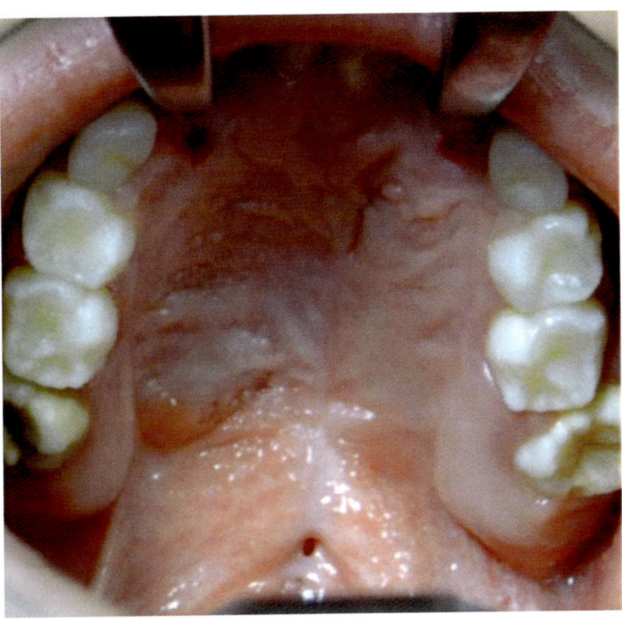

Fig. 16.54 Same patient with primary AEP now at age 5 showing normal eruption. Paramedian incisions have no sequelae, provided that the dissection, once penetrating the mucosal layer, shifts medially to avoid the dental laminae, becoming in the way immediately subperiosteal. [Courtesy of Michael Carstens, MD]

of a static bony platform to which is attached a dynamic muscle effector arm. To function properly, the muscular soft palate must be physically capable of making effective contact with the pharynx. To do so, it must be (1) positioned properly in space and (2) functionally capable of lifting and extending as required.

The management of isolated soft palate clefts depends upon the answers to two fundamental questions

- First, are the dimensions of the hard palate of proper proportions such that the soft palate is correctly positioned in space?
- Second, are the muscles of the soft palate biologically capable of fulfilling their roles?

The answers to these two questions will determine surgical strategy.

Principle 1 *The physical position of the soft palate with respect to the posterior pharynx is absolutely determined by the relative sizes of the bony components to which it is attached.* It is imperative to know if the platform is normal or short. Can we deduce an insufficiency anteroposterior state prior to surgery? If the palatal platform is foreshortened, what can be done about it and when?

The length of the bony palate can be quantitated in several ways. *Physical examination* reveals the geometry of the horizontal palatine plate. A notch or defect can be assessed in terms of its dimensions. *Caliper measurement* at surgery can give the answer as well. The dimensions and shape of the palatine bone can be readily assessed by palpation. When the ANS–PNS distance falls short of normal by 1.5 cm, consideration can be given to immediate versus delayed interposition soft tissue reconstruction.

Imagining studies such as 3D CT scanning in older patients provide an accurate way to measure dimensions of bone and soft tissue. In an ideal world, a study of normal subjects would determine age-related norms of distances in the palatal plane for ANA–PNS, and ANS to posterior pharyngeal wall (PPW). Similar studies of cleft palate subjects would demonstrate the discrepancies between the two groups. In isolated cleft palate subjects, PNS–PPW would be either normal or excessive. The latter situation could be due to foreshortening of the palatine shelf, the maxillary shelf, or both. The diagnosis would rest in assessing the shape of the palatine bone. Thus, as foreshortening increases, so does the risk of VPI, even in the face of technically perfect surgery. Unfortunately, such technologies have limited usefulness for diagnosis in very young patients due to requirements for sedation and radiation exposure. Further advances in imaging may ameliorate this situation.

Principle 2 *The dimensions of the anterior palatine aponeurosis determine the physical position of the levator complex with respect to the bony platform.* Both the palatine bone and the palatine aponeurosis develop from a common embryologic tissue source: neural crest from the second rhombomere. An r2-based deficiency state affects both bone and aponeurosis. All soft palate muscles insert into the aponeurosis at programmed binding sites. The anterior one-third of the aponeurosis is muscle-free (except for uvulus). In the presence of an r2 osseous deficiency, the anterior aponeurosis becomes foreshortened or absent, dragging the otherwise normal soft palate muscles forward into space. This brings TVP into abnormally close contact with the horizontal plate. This is true, even in cases when the length of the hard palate is normal.

In cases of a foreshortened palate, *there will always be deficiency or absence of the anterior palatine aponeurosis.* This situation may demand reconstruction, either at primary surgery to prevent VPI or as the first option in cases of established VPI. If the gap is modest, reconstruction with a single buccinator flap is rational. If the gap is significant buccinator flaps can be overlapped in the midline to allow for retroposition of the muscle envelope without resorting to extensive separation of muscle components from their normal relationships to aponeurosis and mucosa. If VPI supervenes, revisionary surgery consisting of intervelar veloplasty can be carried out. This sequence can also be reversed, the drawback being that intervelar scarring has been created by the first procedure.

Principle 3 *Mesenchymal deficiency in the horizontal plate of the palatine bone unmasks a muscle-binding site into which LVP forms an opportunistic false insertion.* Tethering of LVP prevents normal palatal function. Because the fibers of palatopharyngeus are intermingled with levator, its function is also impaired. *Complete disinsertion from the bone is required.* But LVP has an otherwise normal relationship to aponeurosis. Thus, if the anterior aponeurotic deficit is not addressed, separation of the levator from the palatine aponeurosis (intervelar veloplasty) may simply introduce scarring with little increase in length. *Repairs using z-plasty flaps obliterate otherwise normal embryologic relationships.*

Principle 4 *The muscles of the soft palate arise from different mesodermal structures in the embryo. The functional status of each muscle must be determined prior to surgery.* A perfect anatomic repair in the face of a dysplastic levator sling or an unresponsive superior constrictor will result in failure. Muscle stimulation can be carried out in a cooperative patient with anesthetic spray to the palate mucosa. Alternatively, it can be performed under anesthesia. This information is of prognostic and therapeutic significance. Thus, patients undergoing primary repair can be assessed for their potential to require an additional procedure for VPI. Closure patterns can be reliably reproduced in a non-effort-dependent manner to determine the best procedure for management of VPI, should it arise.

Principle 5 Given a good anatomic repair of the levator sling and normal musculature, *the functional status of the soft palate musculature can only be ascertained after surgery.* The patient will demonstrate the extent of brain–palate coordination. Speech therapy is vital to avoid or correct errors. Sometimes even an apparently short palate can achieve closure by coordination with surrounding pharyngeal elements.

Principle 6 When postoperative VPI is present, buccinator myomucosal flap interposition, either single or bilateral, achieves palatal lengthening by addressing the primary problem: anterior palatine aponeurosis deficiency. The procedure reliably achieves 1.5–2.5 cm of retropositioning with minimal donor site morbidity. Reoperation of an otherwise good palate repair is avoided. It is the logical first line of defense for VPI correction.

Surgical Algorithm

1. Check soft palate length: does P3 defect exceed critical contact distance?

2. Check soft palate function: muscle stimulator.
3. Goals for primary surgery:
   - anatomically sound muscle repair
   - observe postoperative function
4. Goals for secondary surgery:
   - gain additional length.
   - prevent unnecessary operation and fibrosis of the muscular soft palate.

Restoration of Soft Palate Length: Buccinator Palatoplasty

When a length defect exists, buccinator interposition palatoplasty is a high-impact procedure which can be carried out under primary or secondary circumstances. Here is a quick review of the technique. Key references for anterior and posterior flap design and applications are [43–47] (Figs. 16.55, 16.56, 16.57, 16.58, 16.59 and 16.60).

Trigeminal block, either extraoral or intraoral, with 0.25% bupivacaine with max dose of 1 cc/kg is administered. This is without epinephrine. With the addition of epinephrine, the maximum dose is higher so keeping within the range guarantees safety. A transverse incision is made down to the level of the nasal mucosa. If, in one's assessment of the prior repair, further muscle retropositioning and alignment is needed (cases performed elsewhere) this can be done, exposing an

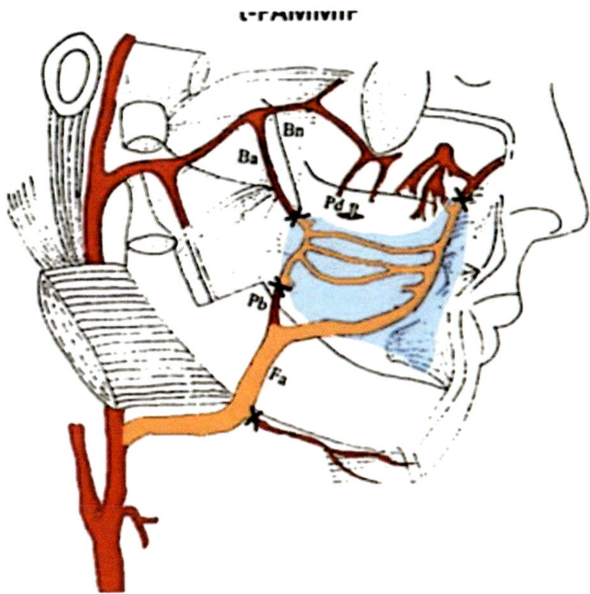

Fig. 16.55 Buccinator anatomy: type 3 flap with extensive anastomoses, anterior—facial artery. Posterior—buccal artery: much more extensive. Can be elevated as an island or with posterior muscle pedicle. [Courtesy of Johannes Fagan, MD]

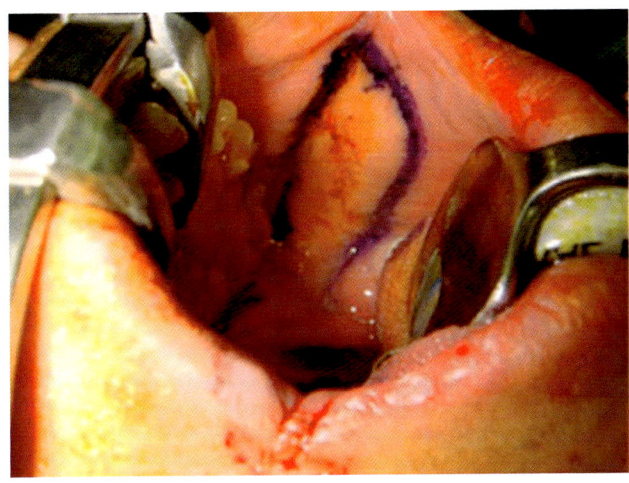

Fig. 16.56 Buccinator design—left cheek. Important to complete mucosal incision posteriorly. Muscle remains intact as posterior pedicle. [Courtesy of Johannes Fagan, MD]

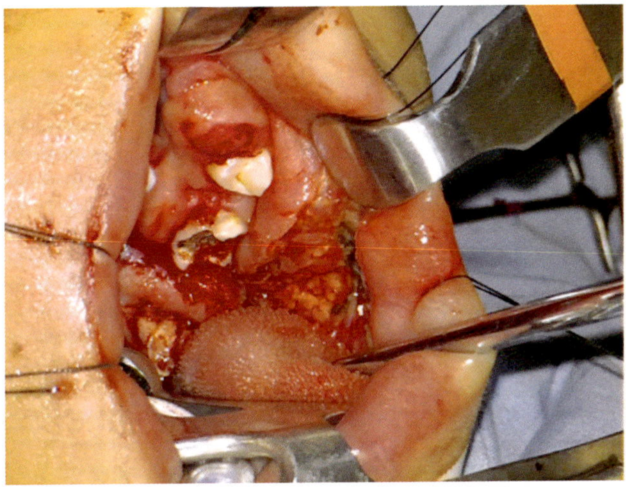

Fig. 16.57 Buccinator (posteriorly based) turned down behind molar. Transverse coverage = reconstruction of palatine aponeurosis. Mucosa sectioned circumferentially—the posterior muscle is left intact. [Courtesy of Johannes Fagan, MD]

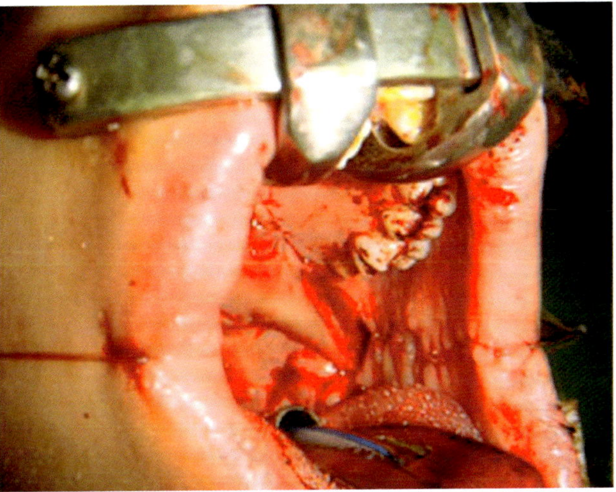

Fig. 16.58 Buccinator donor site closes easily with 4–0 chromic suture. [Courtesy of Johannes Fagan, MD]

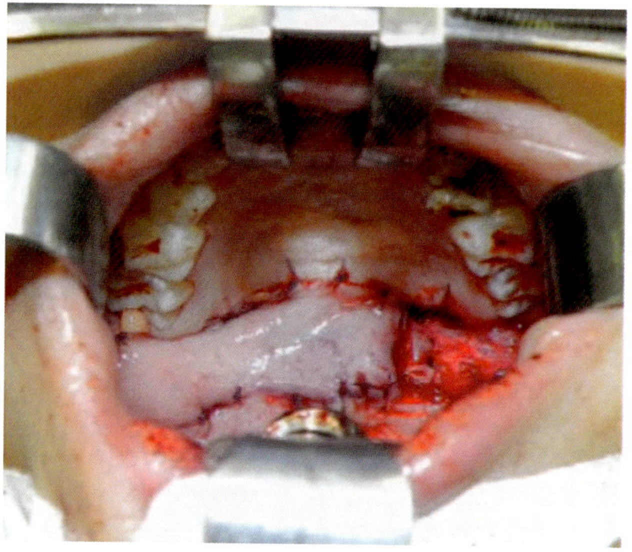

Fig. 16.59 Double opposing transverse buccinators. Designed for both nasal and oral coverage in a case of velopharyngeal insufficiency due to missing palatine aponeurosis. [Reprinted from Raposo-do-Amaral CA. Bilateral buccinator myomucosal flap for velopharyngeal insufficiency: preliminary results. Rev. Bras Cir Plást. 2013;28(3):455–61. With permission from Creative Commons License 4.0: https://creativecommons.org/licenses/by/4.0/]

intact zone of nasal mucosa. Buccinator interposition is then carried out. If the repair is deemed adequate but short, the muscle complex is freed from the nasal mucosa for a distance of 3–4 mm. The nasal mucosa is then divided transversely; the proximal cuff of tissue will serve for proper inset.

The donor site is suspended with two sutures to give it planarity and injected with lidocaine and epinephrine solution for hemostasis. An elliptical design is drawn and the mucosa incised with a knife or cutting cautery. The underlying muscle is incised from proximal to distal and the fat plane superficial to the muscle is entered. The muscle has two pedicles. Anteriorly, the facial artery gives off two to three branches running longitudinally into the muscle from the external (facial) aspect. These are cauterized without compromising the facial artery itself.

The dissection moves distally, gently spreading the fat and cautery as required. No issues with facial nerve are present. The flap involving the middle 50% of the muscle (or more) will be deinnervated. In the posterior one-third of the flap, a series of vessels coming from the buccal artery are seen. These are preserved. Nota bene: In an alternative technique, Robert Mann has shown that simple preservation of flow through the muscle is sufficient to preserve the flap [48].

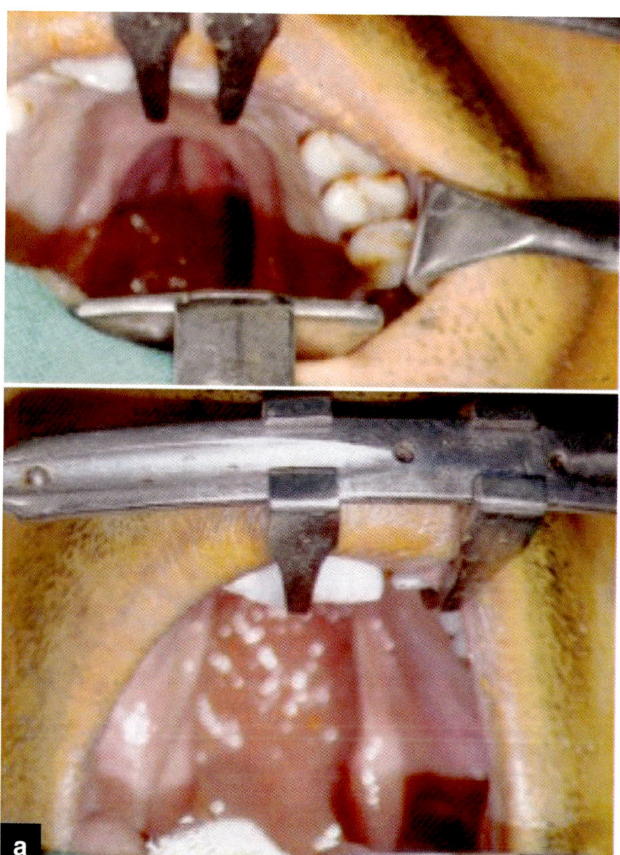

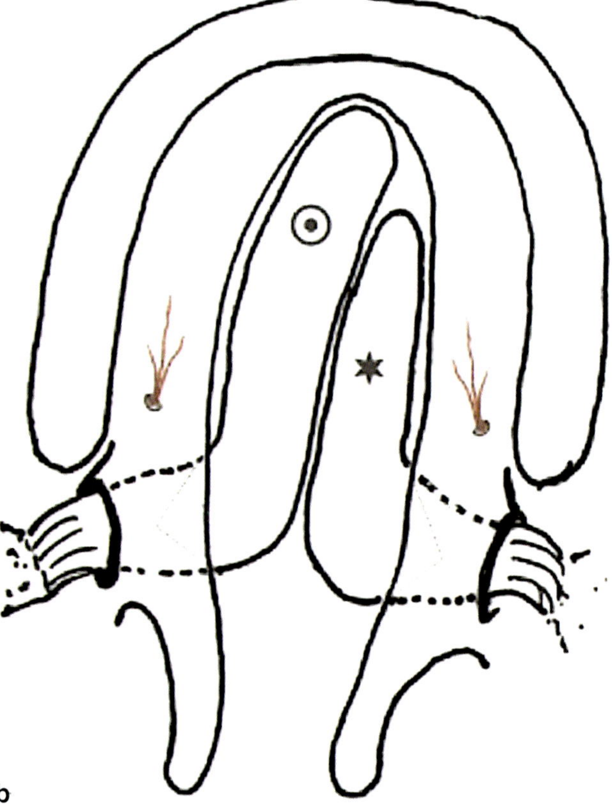

Fig. 16.60 (**a**, **b**) Indications for buccinator 2: midline deficiency. Midline secondary cleft palate with wide horseshoe defect can result from abnormal tongue (Pierre Robin) or very narrow shelves of P1–P3. The vomer is small with inadequate flaps. Double buccinators, placed side-to-side close the defect. The nasal side of the buccinator flaps is raw but will mucosalize in 2–3 weeks. [Reprinted from Bhanyani B. Buccinator myomucosal flap in cleft palate repair: revisited. J Cleft Lip Palate Craniofac Anomal 2014; 1(1):11–16. With permission from Wolters Kluwer Health, Inc.]

Dissection can proceeds posteriorly, all the way to the tail of the flap. This is liberated completely, taking care to sever any attachments to the periosteum of the maxillary tuberosity. Using two forceps, the flap is rotated into position, usually with the proximal tip being brought anteriorly and into the midline—this rotation will move the tail of the flap distal to the vascular pedicle. Suture should be placed below the bite plane of the third molar to achieve recession of the flap. Rotation is accompanied by spreading in the plane of the vessels to achieve tension-free inset. In the Mann technique, the posterior muscle remains intact and constitutes a bridge behind molars. Often this remains insignificant but at times it may require surgical division as a secondary procedure.

The flap or flaps are first inset distally, into the soft palate. The proximal margin thus remains well visualized. The posterior edges are inset. The tail is sutured last. Donor site closure is carried out from proximal to distal using 4–0 chronic or Vicryl. After the mouth gag is removed, with the cheek on stretch, it is usually possible to achieve primary closure of the donor site incision. If a raw area remains, it will mucosalize. Over time, with an active soft palate, the buccinator flaps can stretch out substantially.

Applications and a Preview

Applications of the buccinator include primary palatoplasty, secondary palatoplasty for velopharyngeal insufficiency, and reconstruction of fistulae. Depending upon the circumstances, bilateral flaps can be elevated. These can be inset side-to-side or as double opposing flaps to achieve a mucosal lining on both oral and nasal sides of the defect. Alternatively, the flaps can be placed longitudinally into the midline of the palate.

- Chapter 17 will cover the anatomic basis of the buccinator flap, details of its dissection, and clinical experience.

Conclusions

1. The biologic defect in cleft palate involves neural crest produced by the neural folds at the level of the second rhombomere.
2. The migration pattern of r2 neural crest follows the individual branches of V2 to create neuroangiosomes. These nourish the various developmental fields of the nasal cavity and maxilla.
3. Deficiency of r2 neural crest affects synthesis of the maxillary and palatine shelves. The gradient is always anterior to posterior and lateral to medial.
4. Palatine bone insufficiency is always accompanied by a reduction in the palatine aponeurosis, eliminating the anterior 25% nonmuscle-binding zone, and bringing the muscle-bearing mid-portion into contact with the palatine shelves.
5. Loss of the posterior nasal spine exposes a binding site for muscles along the horizontal palatine shelf.
6. Loss of mesenchymal mass of the horizontal palatine shelf results in a decreased production of BMP4.
7. BMP4 normally diffuses through the soft palate tissue to reach the epithelial border. There it inhibits sonic hedgehog (SHH), a gene product responsible for maintaining epithelial integrity—and prevents fusion.
8. Reduced concentration of BMP4 at the epithelial edges fails to repress SHH resulting in failure of fusion and a soft tissue cleft.
9. Muscle defects from somitomere 4 (TVP) or somitomere 7 (LVP) can affect soft palate function in the presence of a normal bony platform.

Commentary: Robert Mann

There are only a few books in each generation that effect a field so dramatically that they truly change the trajectory of their subject. Michael Carstens' opus fully clicks these boxes. *The Embryologic Basis of Craniofacial Structure: Developmental Anatomy, Evolutionary Design, and Clinical Applications* has been a long time coming but has been well worth the wait.

Dr. Carstens' task was to take the very complex subjects of embryology and anatomy, then meld these two distinct fields into a one-of-a-kind surgical textbook. While every book is the culmination of years of hard work and dedication, this book required Dr. Carstens to fully examine how and when birth defects develop and to then outline the resulting anatomic changes. To this, he added in-depth discussions of specific surgical treatments and how these procedures would impact both the birth defect area and the development of the complete face, over time.

This text contains a treasure trove of information. The topics are complex and interwoven. As you move through, gem after gem emerges explaining how multiple types of birth defects developed, how we have approached their care as we have, and how we might wish to go forward.

One of many examples is his examination of the surgical treatment of the cleft lip and palate. Historically, the understanding was that the defect was a split or gap in the tissue structures that could be simply pulled together for a complete reconstruction. This thought reigned for over 150 years and became the base beneath our reconstructive philosophy. But embryologic research has continued to advance over the decades; this book bridges the disconnect, bringing modern embryology and anatomy into focus with contemporary clinical cleft care.

Years before meeting Michael, I had begun adding tissue during the primary repair of more severe clefts to reduce wound tension and scarring. Over time, I found tissue addition caused the same good results for these clefts as what I had historically seen only with more minimal clefts. This was a significant leveling of the field in cleft repair. A new era of treatment equity was dawning, where all patients, regardless of cleft width or classification at birth, could achieve the same excellent speech and physical growth results at maturity.

It was in coming to a deeper understanding of the inherent tissue deficiencies in a cleft that Michael's insights were so helpful. I began to analyze and visualize each patient's unique cleft defect more completely, designing a patient-specific treatment strategy that led to the consistently better results I was seeing, regardless of cleft severity. How is a birth defect different from normal? The key is embryology. Only by understanding how the malformation affects the anatomy can the reconstructive surgeon appreciate the full extent of the tissue deficiency. If embryology is the key to understanding the cleft defect, then tissue replacement is the gateway to the restoration of the malformation.

This is just one example of the value of the powerful information packed in this amazing text. I congratulate and thank Dr. Carstens for all the hours, months, and years spent researching these very important subjects. Countless people will be the beneficiaries of this work as clinicians and researchers embrace and act on its implications.

A Note from Dr. Carstens

Robert Mann

For cleft surgeons worldwide, the work of Dr. Robert Mann needs no introduction. For over three decades, he has relentlessly explored and refined the concepts of interposition palatoplasty using vascularized buccal tissue, all the while patiently and thoroughly documenting his results. Our collaboration represents a convergence of thought and technique, a phenomenon not uncommon in science. In this case, Robert's work dramatically illustrates the impact of patients when embryologic principles are applied to a clinical problem. It is also a testimony to his unflagging persistence in the face of controversy and to his constant willingness to give of himself to share these new ideas with the coming generation of surgeons. His thinking and the cases herein presented stake out a definitive new standard in the treatment of congenital palate defects.

References

1. Bush JO, Jiang R. Palatogenesis: morphogenetic and molecular mechanisms of secondary palate development. Development. 2012;139:231–43.
2. Carstens MH. Mechanisms of cleft palate: developmental field analysis. In: Bennun RD, Harfin J, Sandor GKB, editors. Cleft lip and palate management: a comprehensive atlas. New York: Wiley-Blackwell; 2016.
3. Doménech-Ratto G. Development and peripheral innervation of the palatal muscles. Acta Anat. 1977;97:4–14.
4. Hilliard S, Yu L, Gu S, et al. Regional regulation of palate growth and patterning along the anterior–posterior axis in mice. J Anat. 2005;207:655–67.
5. Kuehn DP, Kahane JC. Histologic study of the normal human adult soft palate. Cleft Palate J. 1990;27(1):26–34.
6. Oka K, Honda MJ, Tsuruga E, et al. Roles of collagen and periostin expression by cranial neural crest cells during soft palate development. J Histochem Cytochem. 2012;60:57–68.
7. Huang MH, Lee ST, Rajendran K. Anatomic basis of cleft palate and velopharyngeal surgery: implications from a fresh cadaver study. Plast Reconstr Surg. 1998;101:613–27.
8. Vacher C, Pavy B, Ascherman J. Musculature of the soft palate: clinic-anatomic correlations and therapeutic implications in the treatment of cleft palate. Cleft Palate Craniofac J. 1997;34(3):189–94.
9. Kriens OB. Anatomy of the velopharyngeal area in cleft palate. Clin Plast Surg. 1975;2:261–83.
10. Bishop A, Hong P, Bezuhly M. Autologous fat grafting for the treatment of velopharyngeal insufficiency. J Plast Reconstr Aesthet Surg. 2014;67(1):1–8.
11. Barsoumian R, Kuehn DP, Moon JB, Canady JW. An anatomic study of the tensor veli palatine and dilator tubae muscles in relation to Eustachian tube and velar function in relation to Eustachian tube and velar function. Cleft Palate Craniofac J. 1997;34(3):189–94.
12. Ross MA. Functional anatomy of tensor veli palatini. Arch Otolaryngol. 1971;93:1–3.
13. Abe M, Mukukami G, Nogouchi M, et al. Variations in the tensor veli palatini muscle with special reference to its origin and insertion. Cleft Palate Craniofac J. 2004;41(5):474–84.
14. De la Cuadra Blanco C, Peña MP, Rodríguez-Vásquez JF, et al. Development of the human tensor veli palatine. Cells Tissues Organs. 2012;195:392–9.
15. Self S, Dellon AL. Anatomic relationships between the human tensor and levator veli palatine and the Eustachian tube. Cleft Palate J. 1978;15:329–36.
16. Sanders WR. On paralysis of the palate in facial palsy. Edinburgh Med J. 1865;11(2):141–54.
17. Ibuki K, Matsuya T, Nishio J, et al. The course of the facial nerve innervation for the levator veli palatine muscle. Cleft Palate J. 1978;15(3):209–14.
18. Nishio J, Matsuy T, Ibuki K, Miyazaki T. Roles of the facial, glossopharyngeal, and vagus nerves in velopharyngeal movement. Cleft Palate J. 1976;31:201–14.
19. Boorman JG, Sommerlad BC. Levator palatini and palatal dimples: their anatomy, relationship and clinical relevance. Br J Plast Surg. 1985;38:326.
20. Katori Y, Rodríguez-Vásquez JF, Verdugo-López S, et al. Initial stage of fetal development of the pharyngotympanic tube cartilage with special reference to muscle attachments to the tube. Anat Cell Biol. 2012;45(3):185–92. https://doi.org/10.5115/abc2012.45.3.185.
21. Kishimoto H, Yamada S, Kanahashi T, et al. Three-dimensional imaging of palatal muscles in the human embryo and fetus: development of levator veli palatini and clinical importance of the lesser palatine nerve. Dev Dyn. 2016;245:123–31.
22. Klueber K, Langdon HL. Anatomy of musculus levator veli palatine in the 15-week human fetus. Acta Anat. 1979;105:94–105.
23. Shimokawa T, Yi S-Q, Izumi A, et al. An anatomical study of the levator veli palatini and superior constrictor with special reference to the nerve supply. Surg Radiol Anat. 2004;26:100–5.

24. Kuehn DP, Azzam NA. Anatomical characteristics of palatoglossus and the anterior faucial pillar. Cleft Palate J. 1978;15(4):349–59.
25. Sumida K, Yamashita K, Kitamura S. Gross anatomical study of the human palatopharyngeus muscle throughout its entire course from origin to insertion. Clin Anat. 2012;25:314–23.
26. Azzam NA, Kuehn DP. The morphology of musculus uvulae. Cleft Palate Craniofac J. 1977;14:78–87.
27. Langdon HL, Kleuber K. The longitudinal fibromuscular component of the soft palate in the 15-week human fetus: musculus uvulae and palatine raphe. Cleft Palate J. 1978;15(4):337–48.
28. Sumida K, Kashiwaya G, Seki S, et al. Anatomical status of the human musculus uvulae and its functional implications. Clin Anat. 2014;27:1009–15.
29. Sakamoto Y. Classification of pharyngeal muscles based on innervations from glossopharyngeal and vagus nerves in humans. Surg Radiol Anat. 2009;31:755–61.
30. Taylor GI, Palmer JH. Angiosome theory. Br J Plast Surg. 1992;45:327–8.
31. Taylor GI, Pan W-R. The angiosome concept and tissue transfer. St. Louis: QMP Medical; 2014.
32. Taylor GI, Corlett RJ, Ashton MW. The functional angiosome: clinical implications of the anatomic concept. Plast Reconstr Surg. 2017;140(4):721–33.
33. Mercer NS, McCarthy P. The arterial supply of the palate: implications for closure of cleft palates. Plast Reconstr Surg. 1995;96(5):1038–44.
34. Grimaldi A, Parada C, Chai Y. A comprehensive study of soft palate development in mice. PLoS One. 2015;10(12):e0145018. https://doi.org/10.1371/journal.pone.0145018.
35. Noden D, Francis-West P. The differentiation and morphogenesis of craniofacial muscles. Dev Dyn. 2006;235:1194–218.
36. Hacein-Bey L, Daniels D, Ulmer JL, et al. The ascending pharyngeal artery: branches, anastomoses, and clinical significance. Am J Neuroradiol. 2002;23:1246–56.
37. Lasjaunias P, Moret J. Ascending pharyngeal artery and the blood supply of the lower cranial nerves. J Neuroradiol. 1978;5:287–301.
38. Logjes RJH, Bleys RLAW, Bruegem CC. The innervation of the soft palate muscles involved in cleft palate: a review of the literature. Clin Oral Invest. 2016;20:895–901.
39. Baek JA, Lan Y, Liu H, Malthy KM, Mishina Y, Jiang R. Bmp 1a signaling plays critical roles in palatal shelf elevation and palatal bone formation. Dev Biol. 2011;350:520–31.
40. Zhang Z, Song Y, Zhao X, et al. Rescue of cleft palate in Msx-1 deficient mice by transgenic Bmp4 reveals a network of BMP4 and Shh signaling in the regulation of mammalian palatogenesis. Development. 2002;129:4135–46.
41. Kriens OB. An anatomical approach to veloplasty. Plast Reconstr Surg. 1969;43:29–41.
42. Carstens MH. Sequential cleft management with (subperiosteal) sliding sulcus technique and alveolar extension palatoplasty. J Craniofac Surg. 1999;10(6):503–18.
43. Abdaly H, Omranyfard M, Ardekany MR, Babaei K. Buccinator flap as a method for palatal fistula and VPI management. Adv Biomed Res. 2015;27(4):125–37. https://doi.org/10.4103/2277-9175.161529.
44. Bozola AR, Gasques JAL, Cariquirry CE, de Oliveira MC. The buccinators myomucosal flap: anatomic study and clinical applications. Plast Reconstr Surg. 1989;84(2):250–7.
45. Carstens MH, Soteranos GM, et al. The buccinator myomucosal island flap: anatomic study and case report. Plast Reconstr Surg. 1991;88(1):39–50.
46. Jackson IT, Moreira-Gonzalez AA, Rogers A, Beal B. The buccal flap—a useful technique is cleft palate repair? Cleft Palate Craniofac J. 2004;41(2):144–51.
47. Logjes RJ, van den Aardweg MT, Blezer MM, van der Heul AM, Breugem CC. Velopharyngeal insufficiency treated with levator muscle repositioning and unilateral buccinator myomucosal flap. J Craniomaxillofac Surg. 2017;45(1):1–7. https://doi.org/10.1016/j.jcms.2016.10.012.
48. Mann RJ, Martin MD, Eichhorn MG, Neaman KC, Sierzant CG, Polley JW, Girotto JA. The double opposing z-plasty plus or minus buccal flap approach for repair of cleft palate: a review of 505 consecutive cases. Plast Reconstr Surg. 2017;139(3):735e–44e. https://doi.org/10.1097/PRS.0000000000003127.

Buccinator Interposition Palatoplasty: The Role of Developmental Field Reassignment in the Management of Velopharyngeal Insufficiency

17

Michael H. Carstens

We shall cover the following topics

- Historical development of the buccinator myomucosal flap
- Surgical anatomy and function of the buccinator muscle
- Phylogeny and development
- Dissection technique of buccinator flap
  - Anterior-based flap
  - Posterior-based flap
- Clinical applications
- Buccinator interposition palatoplasty

The Origin of Buccinator Palatoplasty

Dans les champs de l'observation le hasard ne favorise que les ésprits preparés.

In a lecture given at the University of Lille, December 7, 1854, Louis Pasteur characterized the experience of discovery as one in which "Chance favors the prepared mind." But this well-known quotation misses in translation the deeper meaning embedded in French which can be read as follows: "In the fields or realm of observation chance will not favor anyone but those with prepared souls." Recognition of the unknown, perceiving an alternative reality, has its basis in an attitude of curiosity, to question what lies before us, unfettered by the dogma, and poses the simple question, why is this here and what does it mean?

Preparing this chapter on the buccinator myomucosal flap has given me the opportunity to remember and relive the experience of discovery, of finding something new… a small part of the human visage which would prove to be of great value for patients and surgeons alike. How would I have known that this small obscure muscle, tucked away in the depths of the cheek, known by most of us for its role in playing a trumpet or in inflating a balloon, should became a key element in the reconstruction of soft palate clefts? In point of fact, the clinical use of the buccinator flap in cleft palate cases would prove for me the key to understanding the nature and location of the embryologic defect underlying the cleft itself.

How the anterior buccinator came to light is not a straightforward story; it was an accident, a moment in time resulting from various threads. Here's how they came together. It was 1988 at the University of Pittsburgh plastic surgery service, and Dr. Bill Futrell was the chief, a mentor who was part football coach (he played for Duke), part gadfly, a philosopher always with a book in his pocket, a teacher who would take us to the most difficult part of a case and make us Figures out the rest, an iconoclast, and an inspiration for a generation. That year he invited David Tolhurst from the University of Leiden as a visiting professor and I was assigned to him to carry out anatomic research on the epicranius, the fascial layers surrounding the entire head. This work, which described the vascularity of the subgaleal-galeal fascia (SGF) gave rise to a number of new clinical applications [1, 2].

Seeing what the SGF could do was fun and exciting but the act of investigation affected me much more. In dissecting the fascial layers one-by-one in a global way, rather than just in a specific region, I saw the anatomy as a whole and I found myself asking the question of *why*… what might be the reason behind this layering? Even then, long before my hegira through developmental anatomy began, it was obvious that structures were laid down in a fixed order with different sources of blood supply. And all of this had to take place according to a master blueprint in four dimensions.

I recall seeing the supply to the very same layers coming from seemingly different sources. Why should frontal scalp be irrigated from the internal carotid artery, associated with the brain while all remaining scalp was supplied by the external carotid?

And why should the frontonasal skin be capable of unique forms of dysplasia associated with the orbital cleft system of Tessier? Years later, when I found Dorcas Padget's work on

M. H. Carstens (✉)
Wake Forest Institute of Regenerative Medicine, Wake Forest University, Winston-Salem, NC, USA
e-mail: mcarsten@wakehealth.edu

M. H. Carstens (ed.), *The Embryologic Basis of Craniofacial Structure*, https://doi.org/10.1007/978-3-031-15636-6_17

the stapedial system, these long-standing questions became important. Curiosity sometimes lives in suspended animation. Years may pass until suddenly, with new piece of information, or with a change in thinking, answers appear as if out of the blue.

Bill Futrell had a habit which was at times, funny, annoying, or provocative… but always with a point in mind. He would bombard us with articles bearing handwritten questions talked about the next day: new ideas, why don't you look into this, food for thought. And so it was, that an obscure paper by a Japanese surgeon regarding the blood supply to the oropharyngeal mucosa led me to the doorway of the pterygopalatine fossa, dog heads, the dissecting laboratory with our SGF specimens, and the buccinator. I can still see Bill's handwriting: "We don't know much about the blood supply here. Why don't you look into this?" Like Alice in Wonderland I was soon down the rabbit hole and into strange territory.

As a consequence of the SGF dissections, we had leftover latex and colored dyes for injection. I commandeered some dog heads from the comparative anatomy lab and brought them home where they resided in an extra fridge in the basement. There, in the comfort and safety of my home, I tied off arterial branches, injected the internal maxillary, took the skull apart to see where the contrast went, and took photographs of different areas. In particular, I was struck by the distribution of the lateral and medial nasopalatine arteries. I also noted the paucity of supply to the buccal mucosa when tying off distal to the second part of internal maxillary, thus isolating the buccal artery. Later on, with ligation of the lingual and injection of the IMA just proximal to the second part, cheek under Stenson's duct lit up. "Oh, that makes sense," I said to myself, "must be the buccal artery. First arch mucosa, first arch artery." But the relationship between second arch facial artery and the buccal mucosa was not on my radar screen. What if I had tied off everything but facial artery, and injected with methylene blue? Would have gotten through the buccinator to the cheek? These questions will never be answered because I soon found myself back in the anatomy lab looking at the fascial planes between the SGF and the parotideomasseteric fascia.

Taking a fresh specimen, I proceeded to remove skin and subcutaneous tissue from the midline, it was the right side, as I recall. This took me over the upper lip, the nasolabial fold, and cheek. Proceeding down through the SIF/SMAS I came to the plane of the DIF and the buccal fat pad. The specimen had been injected with latex. There was facial artery coursing upward, with three distinct branches plunging into the buccinator muscle below. I remembered the dog heads. Suddenly it dawned on me that I was looking at a myomucosal flap. If flow came through the buccal artery into the posterior buccinator and thence into the mucosa, would it not also proceed through the facial in like manner? The double pedicle, facial versus internal maxillary (facial being dominant), would make this a type II flap using the Mathes and Nahai classification. I went backward along the muscle and found the buccal artery which was smaller in size. It dawned on me that if the posterior pedicle were ligated, it would be possible to take the muscle and mucosa right out of the cheek, based on the lengthy axis of the facial artery and vein, and move it about freely. Alternatively, one could drop it back into the mouth for oral coverage. Flush with enthusiasm, I trekked over to Scaife Hall to report my findings to Dr. Futrell.

I shall never forget that afternoon. He was in his office, the afternoon sunlight flooding from behind. I blurted out, "Dr. Futrell, I think I have found a new flap." After hearing me out he replied, "It will never work. But you should do it anyway!" So it was, that we organized dissections with Microfil® and during my craniomaxillofacial fellowship year 1989–1990, under the direction of Dr. George C. Sotereanos, Guy Stofman and I successfully used 14 flaps in 12 patients; our results were reported in 1991. By that time, I was aware of the studies by Reychler [3] and Bozola [4] describing the posteriorly based buccinator flap; both authors having used it for soft palate reconstruction. Guy's presentation of the buccinator flap to the Ivy Society in Hershey, PA won him the first prize for clinical research. Unfortunately, I was on call that weekend in Pittsburgh and so missed his award, which was richly deserved.

The dissection of the anterior flap was challenging at first but became greatly simplified. Guy and I were impressed with the large number of areas that could be covered either inferiorly or superiorly. It seemed as if the facial artery was dominant. This conclusion was quite wrong. As I got into clinical practice, I found the posterior buccinator to be extremely reliable, reclassifying the flap as class III (Figs. 17.1, 17.2, 17.3 and 17.4).

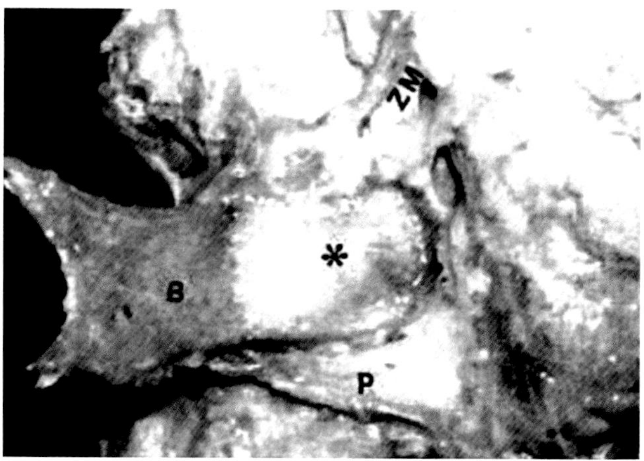

Fig. 17.1 Dissection 1. In this cadaver dissection, buccinator muscle (B) has been reflected away (posterior-to-anterior) from oral mucosa*. Recall that the mucosa is r2–r3 ectoderm (it sits in front of the buccopharyngeal membrane). Neurovascular pedicle (P) is seen ascending from the substance of the submandibular gland. [Reprinted from Carstens MH, Stofman GM, Hurwitz DJ, Futrell JW, Patterson GT, Sotereanos GC. The buccinator myomucosal island pedicle flap: anatomic study and case report. Plast Reconstr Surg 1991 88(1):39–50. With permission from Wolters Kluwer Health, Inc.]

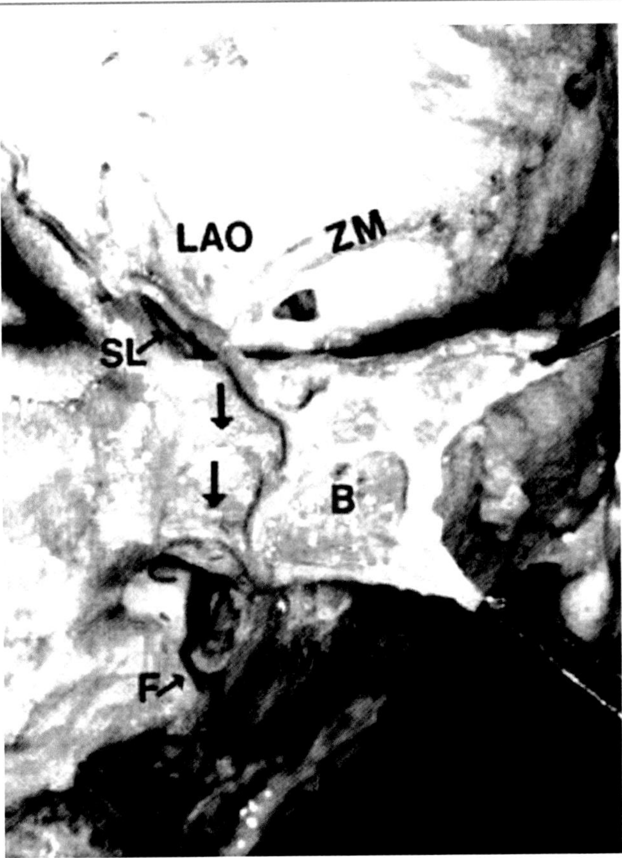

Fig. 17.2 Dissection 2. Cadaver dissection of left buccinator, showing the flap (B) elevated from its fossa in posterior-to-anterior fashion. Forceps hold the upper and lower posterior corners. Facial artery (F) is seen coursing over the external surface of the muscle giving off two branches (arrows) that penetrate the substance of the muscle. Facial artery gives superior labial artery (SL), at which point it continues on between levator labii superioris (here dissected away and not visible) and levator anguli oris (LAO) as the newly renamed angular artery. A portion of zygomaticus major (ZM) is seen inserting into levator anguli oris. Masseter muscle (M) is posterior and superficial to buccinator. [Reprinted from Carstens MH, Stofman GM, Hurwitz DJ, Futrell JW, Patterson GT, Sotereanos GC. The buccinator myomucosal island pedicle flap: anatomic study and case report. Plast Reconstr Surg 1991 88(1):39–50. With permission from Wolters Kluwer Health, Inc.]

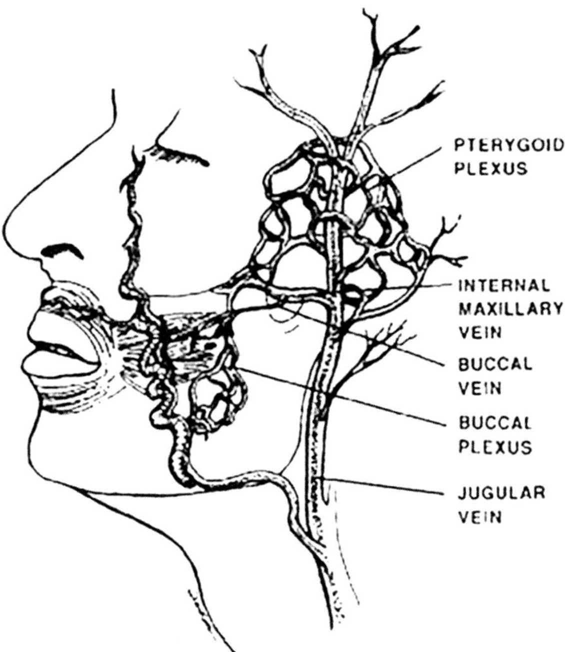

Fig. 17.3 Venous anatomy of the buccinator. Facial vein runs in proximity to the facial artery but just behind it. Buccal vein and facial vein communicate via a plexus. Posterior buccinator flaps, when severed from the anterior supply, do not suffer from venous congestion because of drainage both superiorly and inferiorly, anterior flaps, when disconnected from the buccal plexus, can be traced backward down to the jugular vein and external carotid artery, generating significant length as an island pedicle flap (see next Fig.). [Reprinted from Carstens MH, Stofman GM, Hurwitz DJ, Futrell JW, Patterson GT, Sotereanos GC. The buccinator myomucosal island pedicle flap: anatomic study and case report. Plast Reconstr Surg 1991 88(1):39–50. With permission from Wolters Kluwer Health, Inc.]

From 1990 to the end of 1991 I worked in Nicaragua, Central America with PAHO. During that type, Guy continued to publish about the buccinator along with Dennis Hurwitz. By the time I returned home in 1992, Pribaz had described the same flap giving it a different name, the facial artery musculomucosal flap. The acronym FAMM stuck for a while in the literature, possibly because it had an echo with the well-known TRAM flap (transverse rectus abdominus) pioneered by Carl Hartrampf. I had always thought it best for use to name muscle flaps based on the muscle itself, rather than the artery. TRAM was a good example. For that reason, anterior buccinator seemed more appropriate. With the passage of time, and with multiple authors from different countries reporting their experiences, the traditional anatomic name for this flap has persisted, as acronyms in English do not translate well.

Over the years I have found the buccinator flap, in both its anterior and posterior forms to be reliable and of clinical value. But its greatest contribution to humanity is that pioneered by Bozola and later by Jackson and Mann for the reconstruction of cleft palate. As we shall see, buccinator interposition palatoplasty combined with AEP represent ideal applications of developmental field reassignment to clefts of the soft and hard palate respectively. Not only do these procedures resolve clinical problems but they also offer ongoing scientific confirmation about the pathology and how it can best be managed.

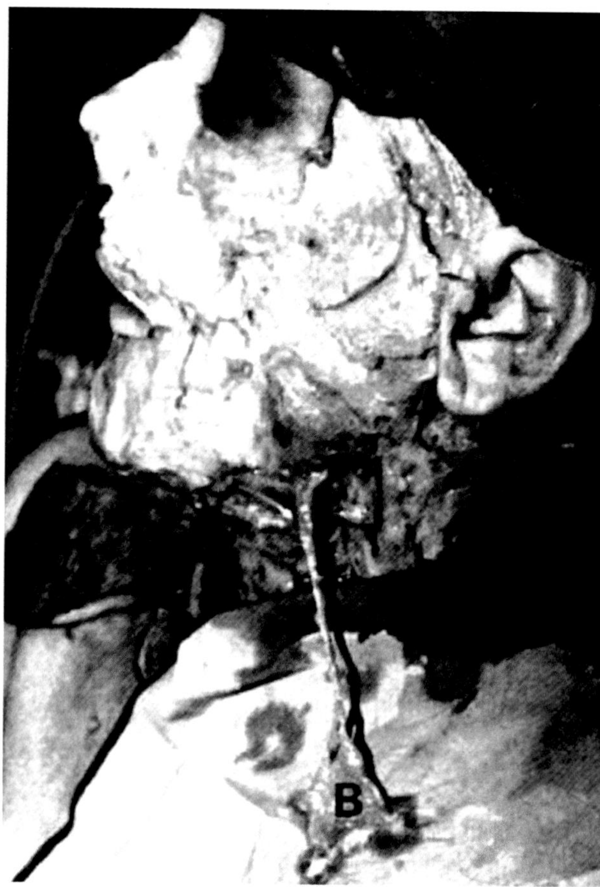

Fig. 17.4 Buccinator as an island pedicle flap. Taking the facial artery and vein down through the submandibular gland generates significant length. The buccinator flap (B) is shown on the chest of the cadaver. [Reprinted from Carstens MH, Stofman GM, Hurwitz DJ, Futrell JW, Patterson GT, Sotereanos GC. The buccinator myomucosal island pedicle flap: anatomic study and case report. Plast Reconstr Surg 1991 88(1):39–50. With permission from Wolters Kluwer Health, Inc.]

Surgical Anatomy and Function

The anatomic details described below are readily appreciated on extraoral dissection, an approach not commonly required for buccinator flap reconstructions. Intraoral dissection, by far and away the more commonly used, is essentially a blind procedure. Thus, we will approach the relevant structures from the outside in.

Arterial Supply (Figs. 17.5, 17.6, 17.7, 17.8 and 17.9)

Vascular anatomy of the buccinator has been documented in fresh cadavers in 8 dissections using Mircrofil® [5] and 20 dissections using red latex for the buccal and facial arteries and blue latex for the veins [6]. The pilot study established the constancy of anterior buccinator branches of facial to the anterior muscle, i.e., the existence of an anteriorly based buccinator flap. Zhao's subsequent work demonstrated additional buccinator branches of facial to the posterior muscle, and documented the relationship between the posterior buccinator branch of facial system and buccinator branch of the internal maxillary artery.

Facial Artery

When approached from the nasolabial incision, the buccinator muscle is immediately found deep to and lateral to orbicularis oris. The fascial planes of DIF and SIF planes are well defined. The muscle develops in contact with the r2/r3 oral mucosa ectoderm. It lies in a plane deep to buccal fat pad which extends forward in the same plane as the anterior border of masseter. Superficial to the fat pad is the buccomasseteric fascia and above that the SMAS, which encloses zygomaticus major and depressor anguli oris. From an external approach it is possible to free the muscle away from the mucosa. It is easier still to dissect the muscle intraorally including an appropriately designed mucosal flap. Once the mucosal incision is complete, one can literally wipe through the muscle with a blade of scissors without disturbing the fat pad. Recall that the facial nerve runs separately beneath and in conjunction with the SIF. Thus, buccinator dissection, being confined to the DIF, spares the seventh nerve.

As the facial artery ascends, it gives off two sets of branches. The *cervical branches* are ascending palatine, tonsillar, glandular (to submandibular gland), and submental. The *facial branches* are inferior labial, buccinator, superior labial, lateral nasal, and angular.

Let us follow the relations of facial artery as it relates to buccinator, beginning at submandibular gland where one finds 3–4 *glandular branches*. Immediately upon exit from the gland, the facial axis gives off the anteriorly directed *submental branch*, the largest of the cervical branches. Submental runs along the surface of mylohyoid just below the mandibular border and deep to the first arch anterior belly of digastric.

The axis of facial artery now winds around the mandible; in so doing it contributes several branches to masseter. At this juncture facial artery ascends obliquely across the buccinator from the midpoint of its inferior border to its anterior superior corner. In so doing, facial supplies the muscle with three sets of buccinator branches.

- *Posterior buccinator branch* measures 0.72–1.26 mm in diameter. It was found in all 20 of Zhao's dissections. FBp ascends from the midline of the inferior border in an oblique and posterior fashion to anastomose with the *buccinator branch of internal maxillary*. Together two vessels supply the posterior half of the buccinator muscle.

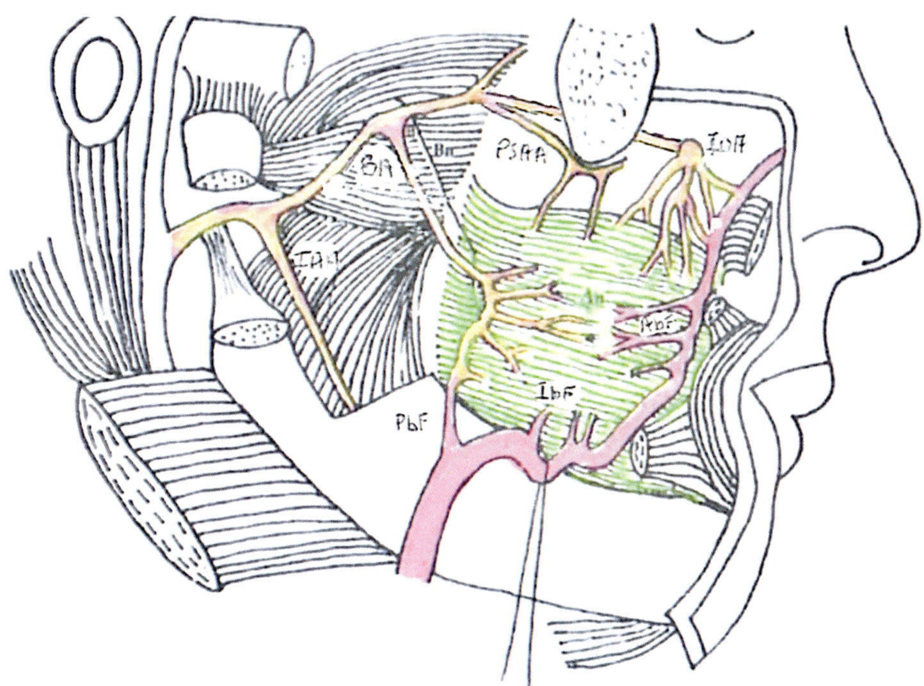

Fig. 17.5 Facial artery. Arterial branches to buccinator based on 20 dissections. Key: Posterior buccal branch of facial (PbF), inferior buccal branch of facial (IbF), and anterior buccal branch of facial (IbF) are found on the surface of buccinator. Buccal branch of IMMA (Ba) is larger in diameter than PbF. Buccal nerve (Bn) from V3 runs parallel to Ba, as expected. Note that Ba is supplemented by a StV3 branch of inferior alveolar artery. Anastomotic network runs (safely) beneath the parotid duct. Note contributions from StV2: posterior superior alveolar artery (Psa) and infraorbital (Ioa). [Reprinted from Zhao Z, Li S, Yan Y, et al. New buccinator myomucosal island flap: anatomic study and clinical application. Plast Reconstr surg 1999; 104(1): 55–64. With permission from Wolters Kluwer Health, Inc.]

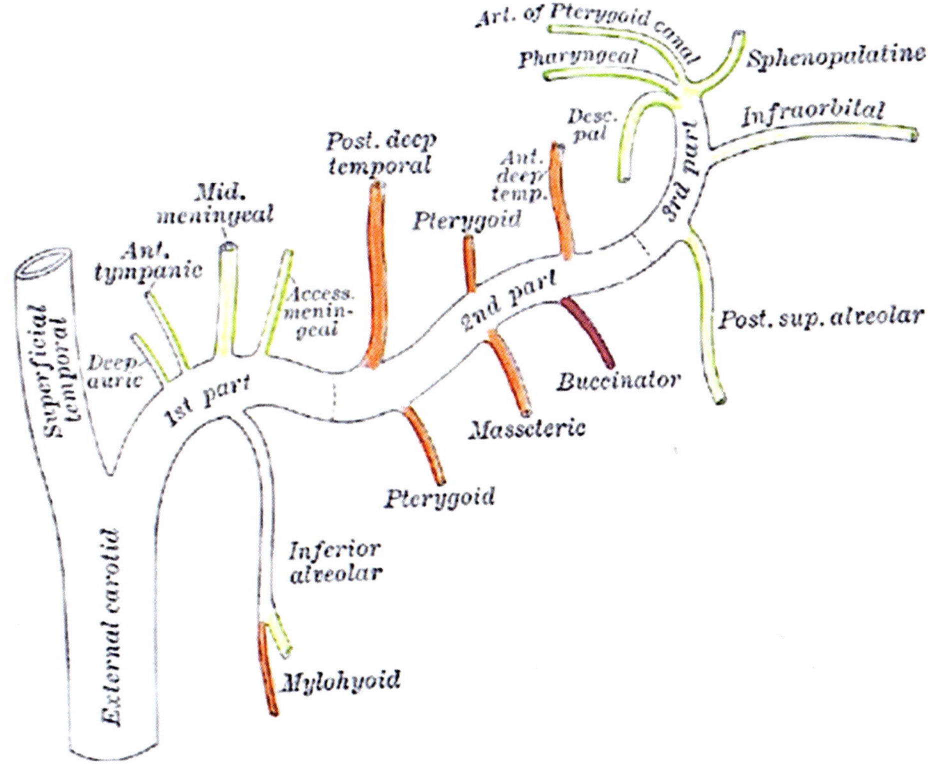

Fig. 17.6 Internal maxilla-mandibular artery (IMMA). IMMA has three zones. Zone 1 dorsal branches are dedicated to the tympanic cavity and the dura. The ventral branch supplies mandible and mylohyoid. Zone 2 branches the muscles of mastication, the most distal of which is buccinator (which belongs to the pharyngeal constrictor system as well). Zone 3 branches of pterygopalatine fossa supply the zygomatic-maxillary complex and nasal cavity. Buccinator is supplied from zone 2 (buccal artery) and zone 1 (posterior superior alveolar artery). [Reprinted from Lewis, Warren H (ed). Gray's Anatomy of the Human Body, 20th American Edition. Philadelphia, PA: Lea & Febiger, 1918.]

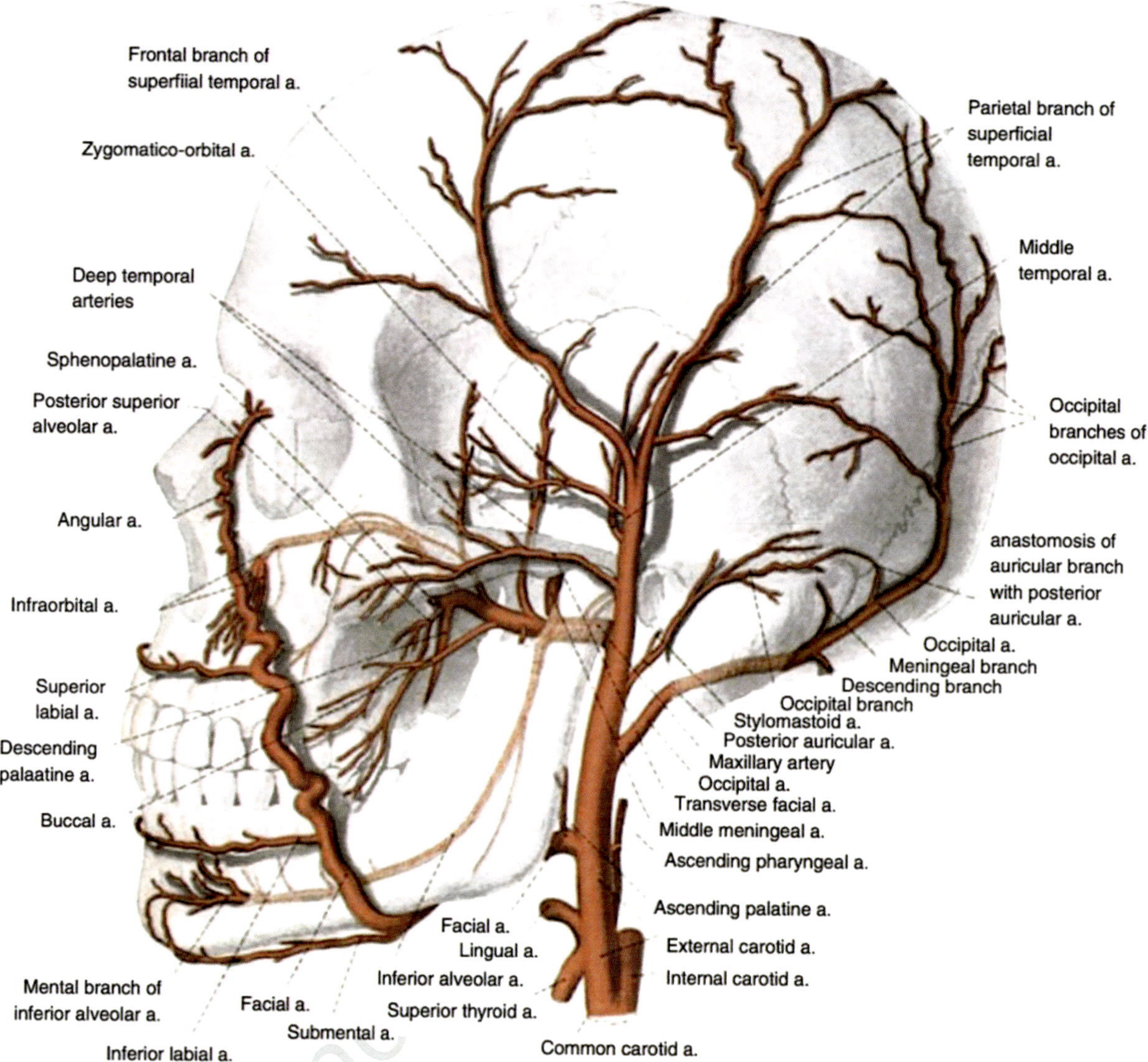

Fig. 17.7 Buccal artery splits into maxillary and mandibular branches. These cover the two leaves of the muscle. [Reprinted from Wikipedia. Retrieved from: https://commons.wikimedia.org/wiki/File:External_carotid_artery_with_branches.jpg.]

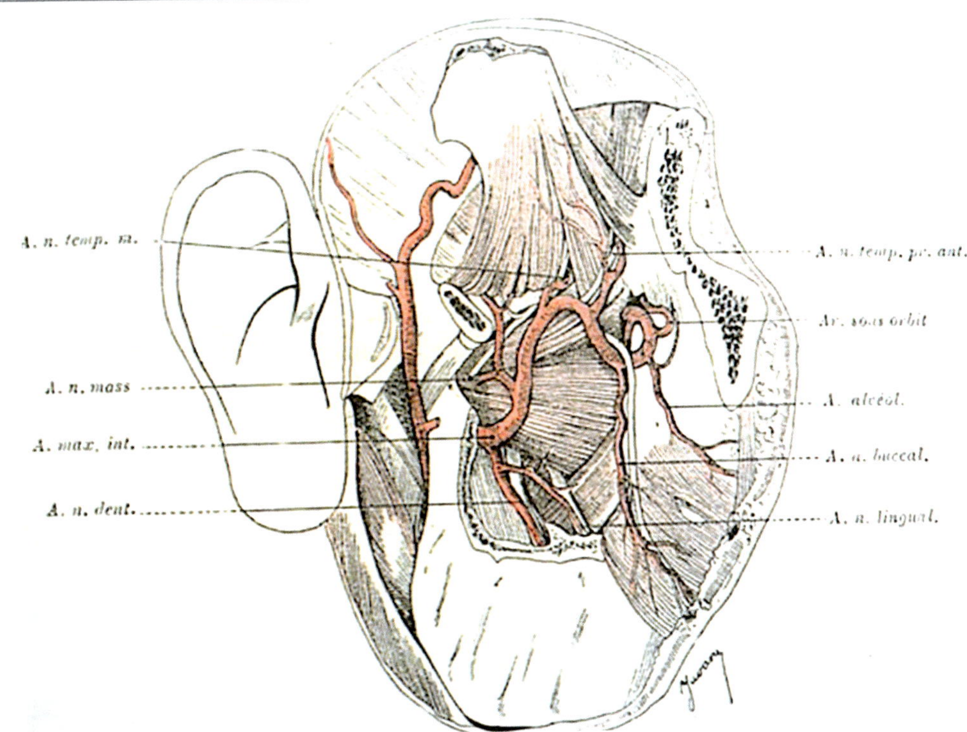

Fig. 17.8 Posterior superior alveolar artery. Original work by Juvara demonstrating contribution of posterior superior alveolar artery to the r2 buccinator contrasted with buccal artery from IMMA supplying the r3 buccinator. Ernest Juvara is credited with the first description of an anatomic space. His doctoral thesis at the Ecole de Médicine in Paris investigated the anatomy of the pterygomaxillary region. Juvara established a standard that future studies would follow when he meticulously defined the structures contained within a space. [Courtesy of the Francis A. Countway Medical Library (Boston, MA). From: Juvara E. Anatomie de la région ptérygomaxillaire. Thèse no. 186. Baittaille and Cie, Paris, 1895.]

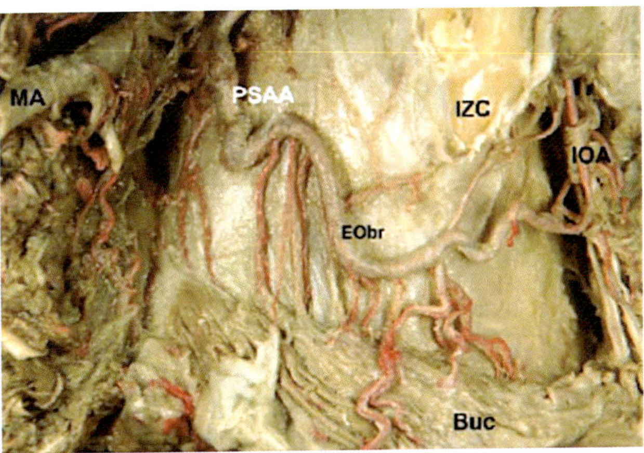

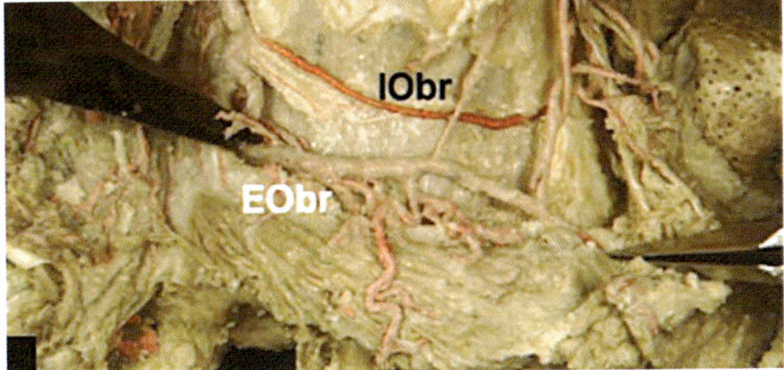

Fig. 17.9 Posterior superior alveolar artery. Decalcified specimens showing the anatomy of posterior superior alveolar artery (PSAA). This is the first branch immediately before the axis of IMMA that enters the pterygopalatine fossa. At the posterior superior alveolar foramen it divides into intraosseous branch IOBr and an extraosseous branch EOBr. IOBr supplies the molars and joins with the anterior superior alveolar artery (ASAA). (Top): IOBr and EOBr run in parallel but their vertical branches can be seen with EOBr clearly directed to the r2 buccinator. (Bottom): Bifurcation of EOBr and IOBr is more clearly seen. [Reprinted from Hur M-S, Kim J-K, Bae HEK, Park H-S, Kim H-J. Clinical implications of the topography and distribution of the posterior superior alveolar artery. J Craniofac Surg 2009; 20(2):551–554. with permission from Wolters Kluwer Health, Inc.]

- *Inferior buccinator branches* of facial are 1–3 in number and 0.3–1.0 mm in diameter. FBi branches are found midway along the anterior half of buccinator.
- *Anterior buccinator branches* of facial are 3–5 in number and 0.4–0.6 mm in diameter. FBa branches were the ones reported in our initial communication.

The remaining branches of facial artery can be defined in reference to the anatomy of the buccinator muscle

- *Inferior labial* is given off directly opposite to the anterior inferior corner of the muscle. It runs deep to depressor anguli oris.
- *Superior labial* is given off opposite to the anterior buccinator branches. It follows the same pattern lying deep zygomaticus major.
- *Lateral* nasal is located just above the superior border of buccinator. It anastomoses with superior labial, dorsal nasal branch of the StV1 system and infraorbital of the StV2 system.
- *Angular* follows an intramuscular course within the fibers of smallest muscle with the longest name, levator labii superiors et alaeque nasi all the way to dorsal nasal branch.

Internal Maxillary Artery

Recall that internal maxillomandibular artery (IMMA) represents the confluence of the first arch branch from ECA plus the extracranial stapedial at the level of V3 just below foramen ovale skull. IMMA has three zones (see Fig. 17.6).

Zone 1 consists of 4 dorsal branches, all of which are designated for neural crest structures of the ear and dura. A single ventral branch supplies neural crest mandible and a Sm4 muscle of mastication, mylohyoid.

- Inferior alveolar artery can supply posterior inferior buccinator (posterior buccinator branch of facial can be absent).

Zone 2 gives rise to six arteries dedicated exclusively to muscles of mastication. Three are dorsal and three are ventral; the most distal is *buccinator artery*.

- Buccinator branch of IMMA descends to the second part of the IMMA axis to the midpoint of the posterior one-third of buccinator muscle. This corresponds to a distance 1–1.5 cm from the pterygomandibular raphe. Here is anastomosis with posterior buccinator branch of facial. This artery measures approximately the same (or greater) compared with the posterior buccinator branch of facial.

Zone 3 supplies the neural crest bone fields of the lower nasal cavity and the maxillary–zygomatic complex.

- Superior alveolar artery, anterior enters the anterior superior border of buccinator.
- Superior alveolar artery, posterior enters the posterior superior border of buccinator.

In Sum Buccinator muscle is constructed from two angiosomes, approximately equal in size. The anterior zone is perfused exclusively by facial artery and the posterior zone is supplied by internal maxillomandibular. This permits dissection of four potential flaps: posterior-based island flap, anterior-based with facial intact, anterior superior island, and anterior inferior island.

Venous Drainage

Veins arising from the buccinator drain into three sites. This situation permits dissection of the four flap variants without congestion. In particular, for the dissection of anterior flaps, ligation of buccal vein posteriorly–superiorly can been done with impunity.

- Posterior drainage.
  - Posterior buccinator drains into pterygoid plexus which is positioned posterior and superior to the muscle. It receives the buccal vein from the posterior muscle.
  - Internal maxillary vein can receive the buccal vein as well.
- Anterior drainage.
  - The anterior muscle drains into the facial vein. Facial vein runs obliquely downward across buccinator, posterior to facial artery. It receives tributaries from buccinator and from both superior and inferior labial arteries.
  - Deep facial vein lies in the buccal fat pad. It also receives branches from the muscle. Deep facial vein drains into the pterygoid plexus anterior to the master muscle.
  - Thus, facial artery and facial vein are isolated with respect to the rest of the flap. This permits the flap to be delivered from the wound.

    Vessels can be ligated superiorly and the flap dropped down onto the neck.

    Vessels can be ligated inferiorly and the flap translated upward to the orbit.

Innervation

Facial Nerve: Motor (Figs. 17.10, 17.11, 17.12 and 17.13)

The predominant motor supply for the buccinator is facial nerve from the cervical division, mandibular branch. This nerve has two branches that divide to supply the upper r4 muscle and the lower r5 muscle. Buccinator branch of VII runs deep to platysma and depressor anguli oris, the two

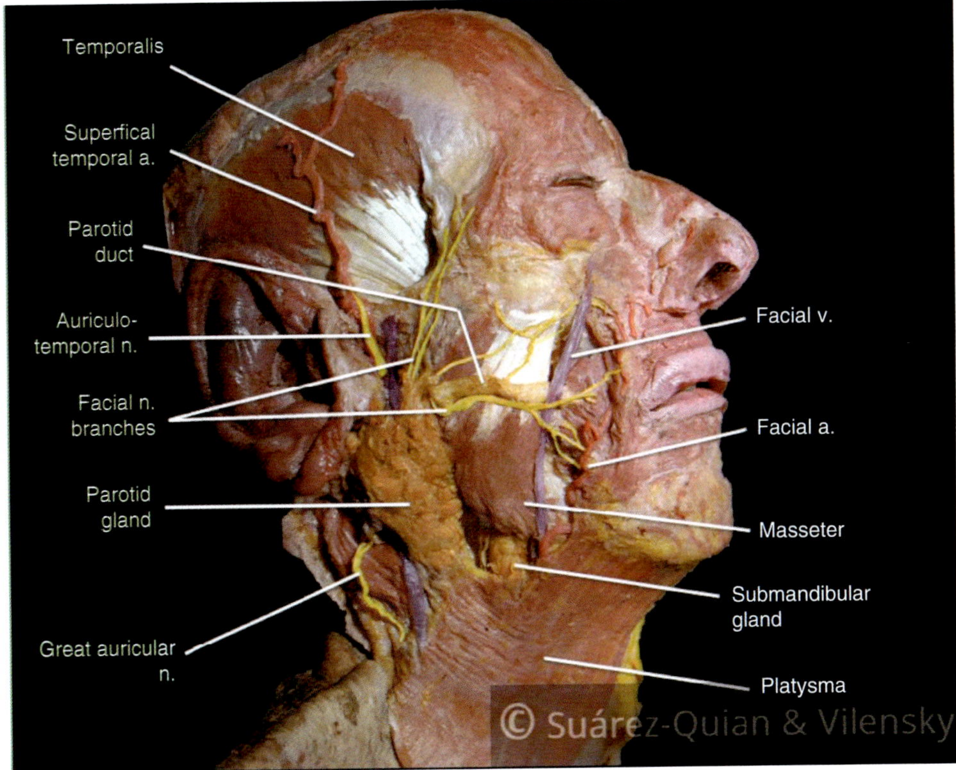

Fig. 17.10 Buccinator branch of facial nerve divides into upper and lower branches. The upper buccinator also supplies zygomaticus and superior orbicularis oris. Lower buccinator supplies inferior orbicularis orgis. Thus, buccinator has components derived from both r4 (upper) and r5 (lower). [Courtesy of Carlos A. Suárez-Quian, PhD]

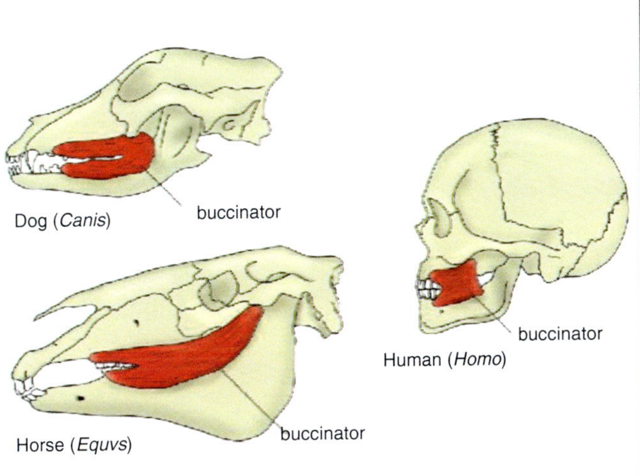

Fig. 17.11 Gape angle Is larger in carnivores due to design of buccinator. Humans 60°–85°, dog 90°–105°. Facial nerve bifurcates: r4 innervates upper portion corresponding to r2 maxilla; r5 innervates lower portion corresponding to r3 mandible. Left: [Reprinted from Herring SW, Herring SE. The superficial masseter and gape in mammals. American Naturalist 1974; 561–576. With permission from University of Chicago Press.] Right: [Courtesy of Viva! Health]

muscles in continuity. As VII is deep to SMAS it reaches the *external* surface of buccinator quite posteriorly, just in front of masseter, thereby avoiding the buccal fat pad. For this reason, one can dissect in the buccal fat pad with impunity, without risking damage to VII.

Buccal Nerve: Sensory and Motor

Mandibular nerve, the third division of trigeminal, has two divisions. *Posterior division* is primarily sensory. *Anterior division* is primarily motor. It supplies all muscles of mastication, save medial pterygoid and mylohyoid, a distinct branch of anterior division, *buccal nerve*, emerges from lateral pterygoid superficial to internal maxillary just before the latter re-emerges from between the two heads of the pterygoids. It then divides into a temporal branch (often not depicted) and a buccinator branch (Fig. 17.10). The lower branch crosses forward obliquely and serves as the program for StV3 buccinator artery.

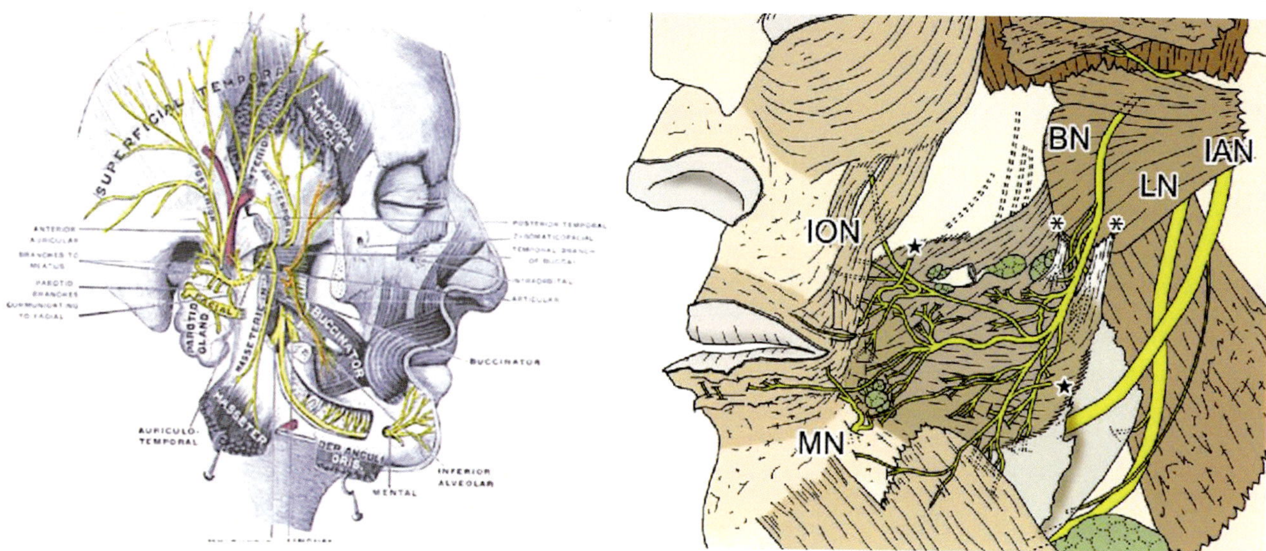

Fig. 17.12 Buccal nerve. (Left): Long buccal nerve (orange) is part of V3 (all remaining branches in yellow). It is (falsely) conceived as being purely sensory. It is motor to lateral pterygoid and to buccinator as well, that muscle functioning as a muscle of mastication. Buccinator may have a dual origin with masticatory mesenchyme originating from Sm4 and expression with Sm6. This is manifested in two ways. First, the buccal nerve splits in two for both the maxillary (r2) and the mandibular (r3) components of buccinator and buccal mucosa. (Right): Buccal nerve splits in two for r3 and r4 components of mucosa and muscle. Mucosa below the parotid duct is supplied by buccal nerve whereas the mucosa above the parotid duct is supplied by superior alveolar nerves. Temporalis tendon (*), inferior alveolar nerve (IAN), lingual nerve (LN), infraorbital nerve (ION). Note IAN entering mandible (star). Left: [Reprinted from Lewis, Warren H (ed). Gray's Anatomy of the Human Body, 20th American Edition. Philadelphia, PA: Lea & Febiger, 1918.] Right: [Reprinted from Takezawa K. The course and distribution of the buccal nerve: clinical relevance in dentistry. Australian Dental 2018; 63(1):66–71. With permission from John Wiley & Sons.]

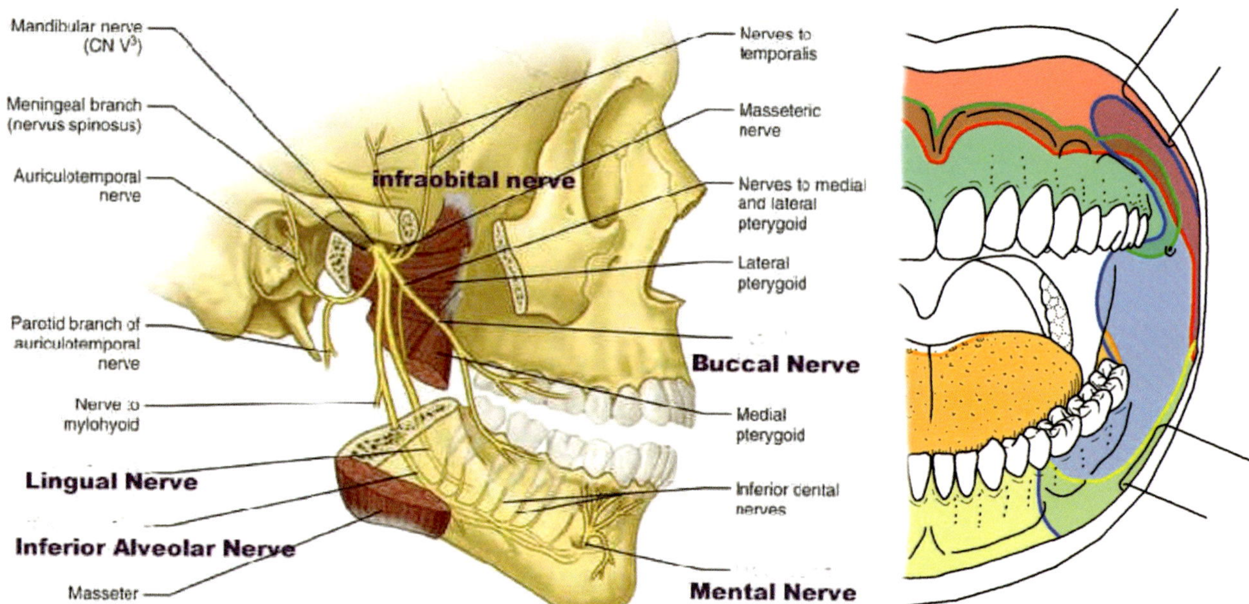

Fig. 17.13 Sensory distribution of buccal nerve over the buccinator muscle is primarily below the parotid duct. Buccal nerve (blue), lingual nerve (orange), mental nerve (yellow), and superior alveolar nerve—proximal (green) and superior alveolar nerve—distal/infraorbital (red). Note that V2 distribution stops at the upper border of parotid duct, placing it between r2 and r3. Buccal nerve distribution and buccinator extension into upper orbicularis suggest possible first arch as an additional source for the buccinator. [Reprinted from Takezawa K. The course and distribution of the buccal nerve: clinical relevance in dentistry. Australian Dental 2018; 63(1):66–71. With permission from John Wiley & Sons.]

- Buccinator nerve of V3 is considered strictly sensory but this is not true: it is *motor to lateral pterygoid* and possibly *motor to buccinator* as well. The fibers of V3 buccinator enter the muscle in tandem with those of VII. The physical separation of buccinator from the pterygomandibular raphe, if done bluntly, spares both nerves.
- Temporal branch of buccal nerve enters the muscle. It is to be considered as a source of referred pain. The association between buccinator, temporalis, and lateral pterygoid suggests a developmental contribution from Sm4 to the substance of buccinator muscle.
- Thus, sensory innervation of buccinator muscle and its fascia is by means of V3, whereas motor innervation is by VII predominates.

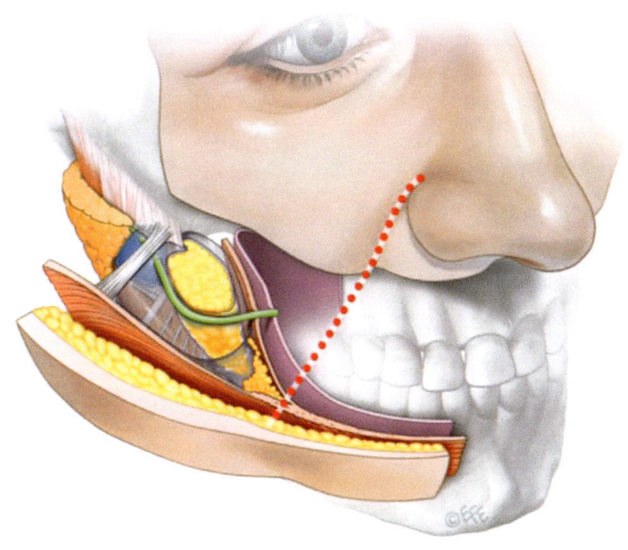

Fig. 17.14 Planes of facial fascia planes (Mendelson). The layers of the mid cheek demonstrate clear separation of the more primitive DIF and the animation layer of SIF (SMAS): 1, skin; 2, subcutaneous, including the retinaculum cutis/superficial fascia; 3, SIF (SMAS) containing second arch muscles of animation; 4, adipose layer/sub-SMAS space; 5, DIF containing muscles of mastication, including buccinator. Note that from blood supply and sensory innervation, buccinator may represent a composite muscle from Sm4 and Sm6. [Reprinted from Mendelson B, Wong C-H. Commentary on: SMAS Fusion Zones Determine the Subfacial and Subcutaneous Anatomy of the Human Face: Fascial Spaces, Fat Compartments, and Models of Facial Aging. Aesthetic Surgery Journal 2016; 36(5): with permission from Oxford University Press.]

Musculofascial Planes

The planes of the cheek faithfully reflect how the first and second arches melded together in the gnathostome revolution (Figs. 17.14, 17.15, 17.16, 17.17, 17.18 and 17.19).

Superficial Fascia (SF) Immediately beneath the dermis lies a layer of subcutaneous fibroadipose tissue. SF is continuous across two transition zones of anatomic importance. Nasolabial fold is not interrupted, but it continues from cheek to lip. At zygomatic arch it provides cover for the facial nerve. Fat is aggregated into discrete anatomic sites

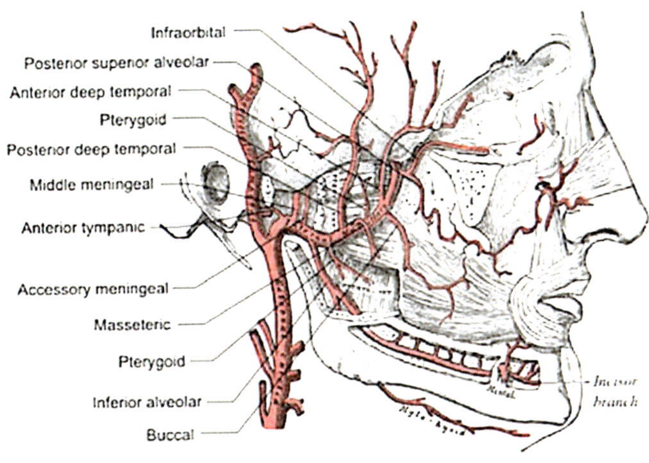

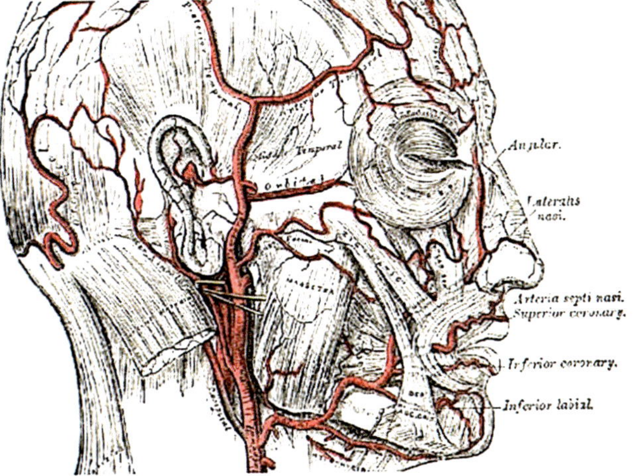

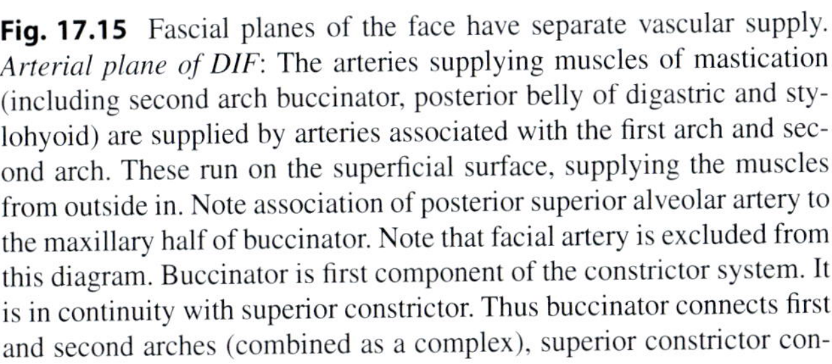

Fig. 17.15 Fascial planes of the face have separate vascular supply. *Arterial plane of DIF*: The arteries supplying muscles of mastication (including second arch buccinator, posterior belly of digastric and stylohyoid) are supplied by arteries associated with the first arch and second arch. These run on the superficial surface, supplying the muscles from outside in. Note association of posterior superior alveolar artery to the maxillary half of buccinator. Note that facial artery is excluded from this diagram. Buccinator is first component of the constrictor system. It is in continuity with superior constrictor. Thus buccinator connects first and second arches (combined as a complex), superior constrictor connects 2rd with third, middle constrictor connects third with fourth, and inferior constrictor connects fourth with fifth arch. Note buccinator being part of the masticatory system lies deep to the facial musculature. It is *not* a muscle of facial expression. *Arterial plane of SIF*: The arteries supplying the muscles of facial expression are supplied by ECA branches associated with the second arch. These run on the deep surface. [Reprinted from Lewis, Warren H (ed). Gray's Anatomy of the Human Body, 20th American Edition. Philadelphia, PA: Lea & Febiger, 1918.]

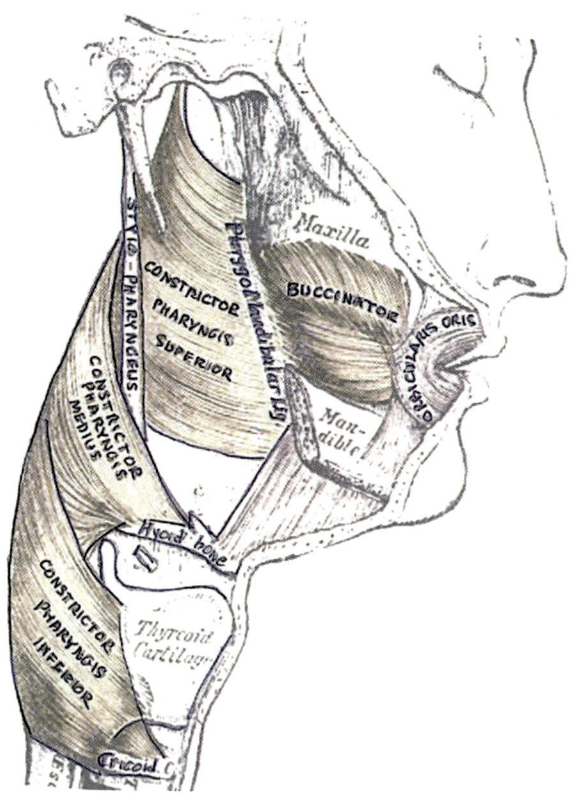

Fig. 17.16 Buccinator, the constrictor series and buccopharyngeal fascia. Buccinator muscle is a part of the constrictor series. It is *not* a muscle of facial animation (at least in part). Superior constrictor (SC) and buccinator (BC) are separated by the pterygomandular raphe. This marks separation between second arch and third arch. Tonsillar branch of facial artery (1) comes in below SC. Space between superior and middle constrictor (MC) transmits glossopharyngeal nerve (2). This space represents the division between r6–r7 and r8–r9, i.e., between third arch and fourth arch. Stylohyoid ligament (3) connects second arch styloid process with second arch upper hyoid bone. Buccinator, superior/middle/inferior constrictor extend through all the arches, terminating with cricoid cartilage at the pharyngo-esophageal junction. They are covered by a continuous sheet of buccopharyngeal fascia. Innervation follows from nucleus ambiguus: SC, r6–r7; MC, r8–r9; and IC, r10–r11. Note: that in the floor of mouth, BPF relates to mylohyoid while PTF relates to hyoglossus (shown here, but not covered with PTF). Note: Styloglossus the first tongue muscle (Sm1) and stylopharygeus (third arch) lie outside the BPF. The *buccopharyngeal fascia* (BPF) fuses with the periosteum on the posterior part of the alveolar process of the maxilla and with the periosteum of the inner plate of the pterygoid process. The fascia extends backwards from this point, over the superior pharyngeal constrictor of the pharynx and then continues into the tunica adventitia of the pharynx and the esophagus. The buccopharyngeal fascia forms the *pterygomandibular raphe*, between the hamulus of the medial pterygoid plate and the posterior end of the mylohyoid line of the mandible, providing attachment for the superior pharyngeal constrictor muscle. BPF is continuous with *pretracheal fascia* (PTF). Not seen: (1) lateral pterygoid inserted into the TMJ and (2) medial pterygoid inserted into the mandibular angle. The pterygoids, as muscles of mastication, are in the plane of the DIF. The plane of BPF (buccinator and superior constrictor) lies deep to that of DIF (masseter). Note: that in the floor of mouth, BPF relates to mylohyoid while PTF relates to hyoglossus (shown here, but not covered with PTF). Note: Styloglossus the first tongue muscle (Sm1) and stylopharygeus (third arch) lie outside the BPF. [Adapted from Wikimedia. Retrieved from: https://commons.wikimedia.org/wiki/File:Musculusconstrictorpharyngissuperior.png]

which are just beginning to be documented. Neuromeric relationships have not been established [7].

Superficial Investing Fascia (SIF) This layer, which contains the muscles of facial expression, is known as superficial musculoaponeurotic system (SMAS). Nowhere it is attached to bone. All Sm6 muscles use the SMAS as their primary insertion; their secondary insertions are to bone or dermis. One fingerbreadth below the zygomatic arch, SMAS becomes indistinct but reasserts itself as galea. SMAS changes its configuration, but not its position, over the StV1 frontonasal zone and the StV2 midface, where it encloses frontalis, orbicularis, and periorbital muscles and zygomaticus major.

Sub-SMAS Plane This is the second arch layer, which is found throughout the face saved over the zygomatic arch. In the scalp it becomes the subgaleal fascia [2]. It functions as a gliding layer. In the neck, sub-SMAS tissue lies between second arch platysma and the underlying first arch DIF (deep cervical fascia). Above parotid sub-SMAS is thoroughly fused with parotid fascia.

Parotideomasseteric Fascia This layer marks the transition to muscles of mastication, of both first and second arch. As such it overlies the facial nerve and the parotid duct as these structures course forward over the masseter muscle. It continues forward to overlie the buccal fat pad. It blends with epimysium of buccinator, another indication of the dual first arch/second arch identity of that muscle. Below the mandible it becomes continuous with DIF deep cervical fascia.

Buccal Fat Pad The BFP consists of three lobes (anterior, intermediate, and posterior, each of which has its own capsule, is suspended by distinct ligaments, and has a dedicated arterial supply). This structure is of great importance; we shall discuss it in a separate section.

Buccopharyngeal Fascia (BPF) This thin layer is branchiomeric; it encloses the buccinator muscle. BPF unites the various components of the ancient gill constrictor system. Thus, buccinator is continuous with the superior constrictor.

Deep Investing Fascia (DIF) The buccomasseteric fascia completely encircles the mandible and encloses masseter almost entirely except at the zygomatic arch to which it is attached. At that site, the deep aspect of the masseteric compartment communicates with temporalis. This makes sense because the two muscles have a common phylogenetic origin from adductor mandibulae.

Buccinator Muscle This flat quadrilateral muscle occupies the deep space of the cheek and is in contact with the oral mucosa. In the adult, buccinator measures 5 × 7 cm. It has

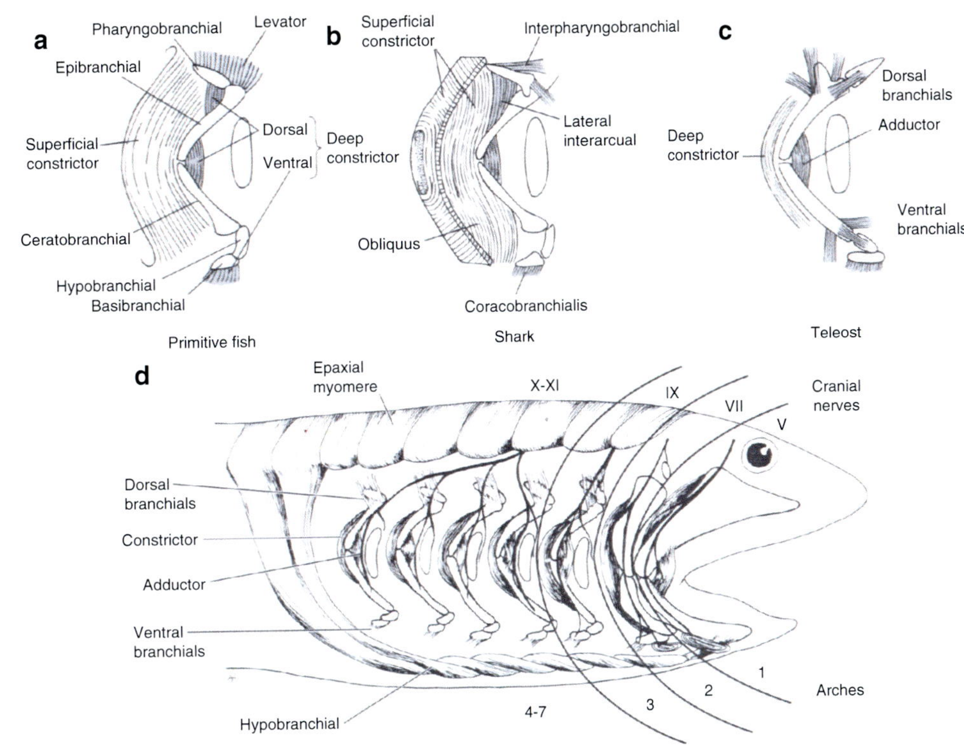

Fig. 17.17 Evolution of the pharyngeal constrictors. Primitive fish muscles get simplified in crown bony fish—and presumably in the sarcopterygians as well. Constrictors straighten the arch causing it to squeeze water through the pharynx. Adductors bend the arch to open it up for water to escape. Dorsal and ventral branchials control moving parts within the arches themselves. Primitive constrictors morph in mammals into pharyngeal constrictors (including buccinator) while the adductors are the likely source for the muscles of the palate and larynx. [Reprinted from Kardong KV (ed). Vertebrates: comparative anatomy, function, evolution. McGraw-Hill Education; 2015. With permission from McGraw-Hill.]

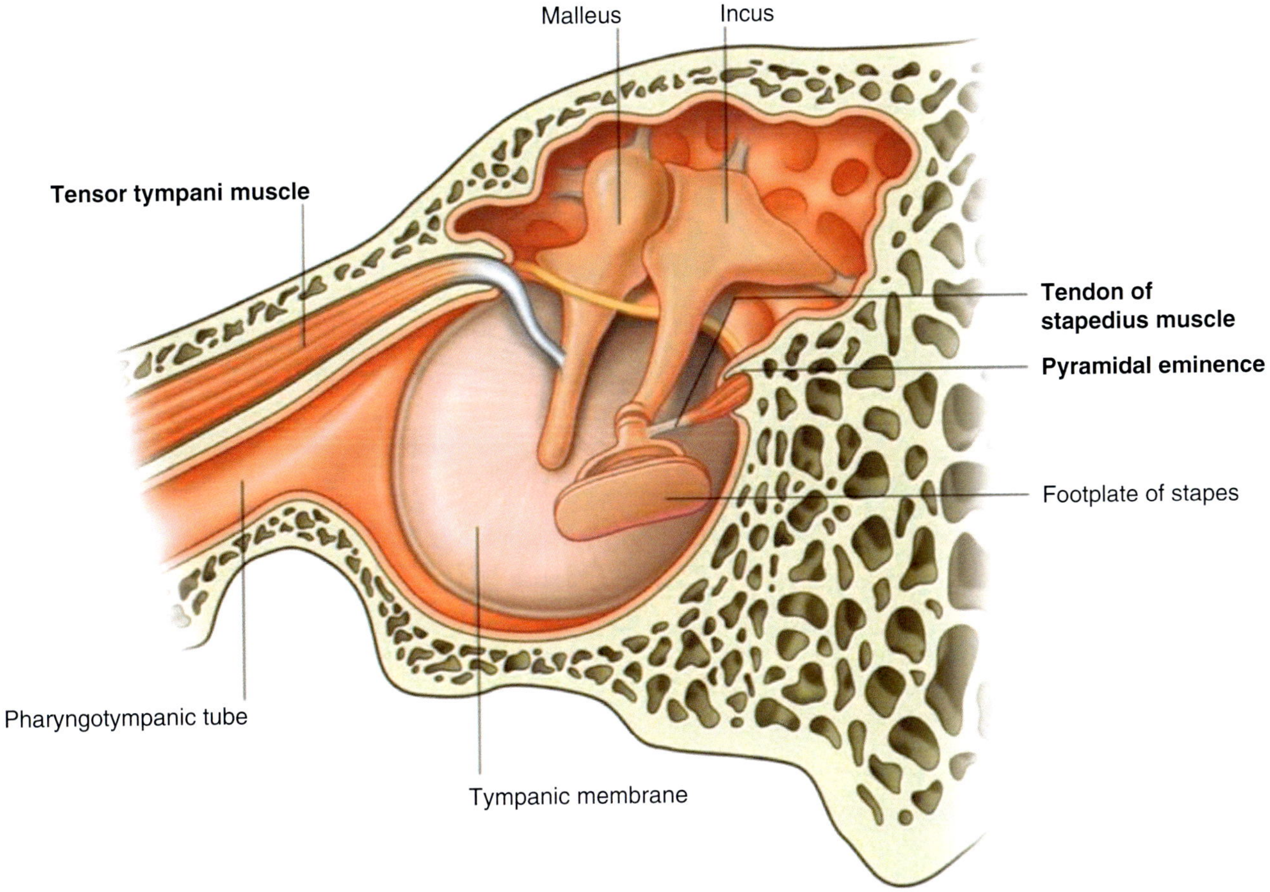

Fig. 17.18 Stapedius muscle: transformation from a muscle of mastication. Stapedius is attached to backwall of the tympanic cavity derived from prootic bone: represents the connection of r5 muscle to r6 bone derivative. [Reprinted from Drake RL, Vogl AW, Mitchell AWM. Gray's Anatomy for Students, third ed. Philadelphia, PA: Churchill Livingstone, 2015. With permission from Elsevier.]

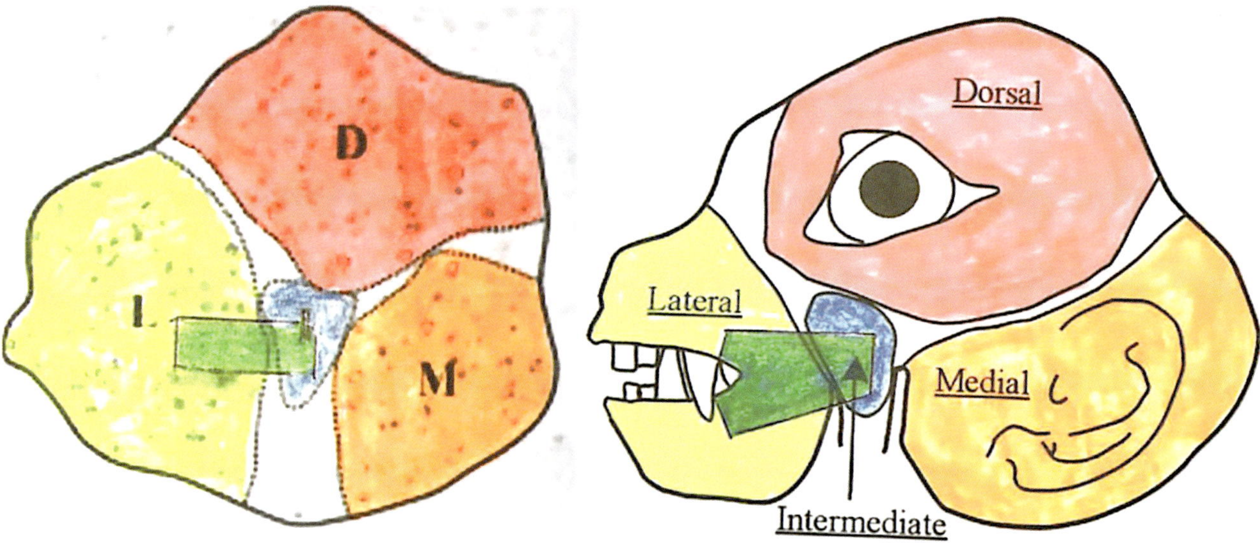

Fig. 17.19 Functional map of facial nerve muscles in the primate facial nucleus. Left: sagittal section, oriented anterior–posterior, of facial nerve motor subnuclei. Buccinator motor nuclei colocalize with those of platysma in the facial map. Right: spatial position of the nuclei with respect to the face. Note inordinately large area dedicated to protection of the eyes and animation of the lips. Dorsal, orbicularis oculi (pink); intermediate, zygomaticus (blue), lateral, orbiculari. Key: oris (orange); intermedio-lateral, buccinator, and platysma (green). [Adapted from Horta-Júnior JA, Tanega CJ, Cruz-Rizzolo RJ. Cytoarchitecture and musculotopic organization of the facial motor nucleus in *Cebus apella* monkey. J Anat 2004; 204(Pt 3):175–190. With permission from John Wiley & Sons.]

two sites of secondary insertions to both the maxillary and mandibular alveolus corresponding to the molar zone, M1–M3. Posteriorly, the buccinator has a primary insertion into the anterior border of the pterygomandibular raphe. Behind the raphe it is in continuity with superior constrictor. Anteriorly, at the oral commissure, buccinator fibers cross one another. Those of inferior half are confluent with superior orbicularis oris, while those of the upper half are confluent with the lower orbicularis oris.

- Recall that posterior buccinator shares a common plane with medial pterygoid plate and with superior pharyngeal constrictor.
- Fibers from tendinous band between maxilla and pterygoid hamulus, i.e., between tuberosity and the uppermost border of pterygomandibular raphe.
  - Behind this band is an aperture for the tendon of tensor veli palatini that permits it to pierce the pharyngeal wall and access the palate.
  - Recall that TVP was originally a muscle of mastication sharing a common blastema with medial pterygoid.

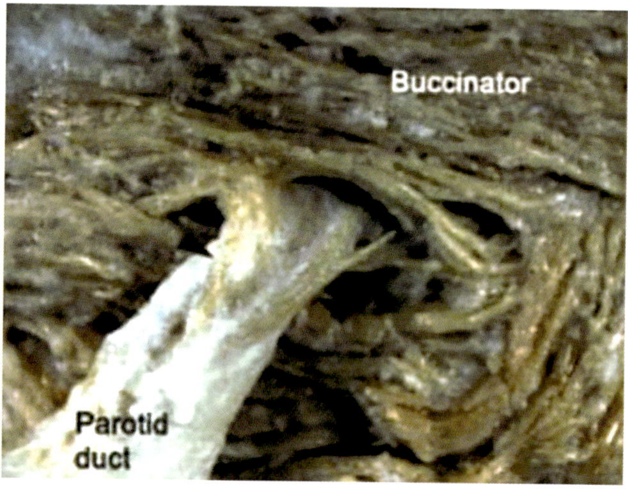

Fig. 17.20 Parotid duct. Previously considered a fibrous structure, the duct has fibers of buccinator that are position to open it up; these are non-sphincteric. [Reprinted from Kang H-C, Kwak H-H, Hu K-S, Youn K-H, Jin G-C, Fontaine C, Kim H-J. An anatomical study of the buccinator muscle fibers that extend to the terminal portion of the parotid duct, and their functional roles in salivary secretion. J Anat 2006; 208: 601–607. With permission from John Wiley & Sons.]

Parotid duct enters buccinator opposite the third maxillary molar and runs obliquely forward through the muscle, with buccinator fibers inserted into it. The duct exits through the mucosa opposite the second maxillary molar (Fig. 17.20).

Oral Mucosa Recall that the epithelium of the oral mucosa is ectodermal. Endoderm is first encountered behind the anatomic boundaries of the ancient buccopharyngeal membrane (BPM), which are in line with the tonsillar fossa and Waldeyer's ring. The oral mucosa has two neuromeric zones. The *maxillary mucosa* per se is quite limited. It originates from r2 and covers the undersurface of the upper lip and the maxillary sulcus. It is supplied by the distal (infraorbital) branches of anterior superior alveolar nerve, whereas the gingiva is supplied by the proximal (alveolar) branches of superior alveolar nerve. The *mandibular mucosa* r3 is much more extensive. It covers the floor of mouth and the tongue and extends upward into the upper buccal sulcus as far forward as the first molar where it comingles with r2. This mucosa is innervated by V3 buccal nerve and V3 lingual nerve. Note that buccal nerve splits into two parts with a superior branch running into the sulcus of the upper cheek and lip.

Identifiable Structures on Intraoral Dissection

In the technical section below, it will be apparent that buccinator flap dissection is simple and safe. Both relevant vascular structures enter the muscle through the buccal fat pad and are readily cauterized or ligated from inside the mouth simply by dividing the muscle fibers, locating the vessels (they are easy to find), and controlling them. The key anatomic point from below is to recognize that the vascular leash supporting the posterior flap parachutes into the center of the muscle in its posterior one-third of the muscle about 2 cm forward from the pterygomandibular ligament. Once one has sectioned the muscle and created an island flap, gentle spreading of the plane of the buccal pad is easy and safe; the vessels are elastic and unwind, yielding good mobility for rotation and inset of the flap.

Phylogeny of the Buccinator and Its Development in Man

Buccinator muscle is classically described as a second arch muscle, meaning that it originates from the sixth somitomere. But how can we explain the duality of its blood supply and (quite possibly) a duality of its innervation. What is its relationship to the constrictor system? What lessons does buccinator have for us regarding how the first and second pharyngeal arches are assembled?

Hypotheses

- Buccinator is neither a muscle of mastication nor it is one of facial expression; it belongs to the constrictor system of the pharynx.
- Buccopharyngeal fascia is deep to, and distinct from, the deep investing fascia of muscles of mastication.
- In ganthostome evolution, the amalgamation of the first and second branchial arches causes a rearrangement of the initial pattern of *constrictors branchiales.*
- Phylogenetic history explains development of the mammalian buccinator.

Phylogeny

First off, let's look at the constrictor muscles as organized in agnathic fishes. Posterior and deep to each gill slit is a constrictor muscle. Collectively they *compress the pharynx* to facilitate the passage of food. They represent an interconnection between the branchial arches.

The advent of jaws involved a radical restructuring of the first two branchial arches in which the second arch tissues are repositioned in the first arch envelope like a sandwich. The constrictors of BA1 and BA2 were reorganized whereas those of BA3–BA7 were not.

- Ectoderm from BA1 acted as the recipient sac for the aqueous environment providing internal epithelium for the mouth and external epithelium for the skin.
  - Ectoderm from BA2 disappears.
- Skeletal elements of the branchial baskets are reassembled into the jaws.
- The muscles originally attached to these cartilages have to look for new work.
  - Gill arch muscles became more specialized with new names.
  - A new class of muscles for mastication arises from both first and second arches and these muscles are all coplanar.
- Branchial muscles attached to the third to seventh arches retain their gill function.

Recall that in pre-mammals chewing muscles arise from both the first and second arches. In mammals this situation changes. Not only does first arch adductor mandibulae subdivide into more sophisticated derivatives capable of driving the complex movements of the TMJ, but also the second arch musculature assumes four distinct fates.

- The original *muscles of mastication* are represented by posterior belly of digastric and stylohyoid. These are ensheathed, along with their first arch partners, in a common deep investing fascia (DIF).
- *Stapedius* acquires a new function, reflecting the transformation of hyomandibula, a supporting bone for the lower jaw in fishes, into stapes which is incorporated into the middle ear. The dorsal muscle connecting the dorsal hyomandibula to the skull becomes the anchorage of r5 stapedius muscle to r6 prootic bone (Fig. 17.18).
- All the muscle blastemas of Sm6 assume an entirely new identity (with one glaring exception). These *muscles of facial animation* develop in an entirely new plane, the superficial investing fascia (SIF) within which they spread widely over the entire head to form the superficial musculoaponeurotic system: the SMAS.

- **Note**: Buccinator is *pharyngeal constrictor* united with its partners in a distinct buccopharyngeal fascia. It is *not* a muscle of facial expression.

Anatomic Manifestations of the Singularity of Buccinator

First, buccinator has its own *distinct fascia*. Vascular planes are established with respect to the muscle layers. DIF muscles are supplied from *superficial to deep* by branches of the ECA specific to first arch and selected branches from second arch. SIF muscles supply the muscles of facial expression from *deep to superficial* by ECA branches specific to second arch. This implies that when the two layers are superimposed their blood supply runs in a common plane of fibro-adipose tissue.

What we observe with buccinator and buccopharyngeal is that lies deep to the DIF. The constrictor muscles of the pharynx are programmed by the epithelium of the mucosa. Thus the r2–r3 oral ectoderm determines the program of the underlying mesoderm, i.e., buccinator. There is no r4–r5 oral mucosa. Thus, the next available mucosa program is r6–r7; it determines superior constrictor. Thus, the oral program for buccopharyngeal fascia is discontinuous: r2–r3, a hiatus for r4–r5, and then r6–r11, all the way to the border of cricoid and esophagus.

Second, studies of *morphology and enzyme histochemistry* demonstrate that buccinator is biochemically unrelated to any other second arch muscle. These second arch muscles have type 1, type IIa, and type IIc fibers but the type 1 fibers of non-masticatory muscles have a distinct myofibrillar ATPase. Furthermore, facial animation muscles, such orbicularis oris, have a 71% predominance of fast-twitch type II fibers whereas buccinator is 55% type I, indicating the capacity for continuous work as low levels of force [8].

Third, neuroanatomy of the *buccinator motor nucleus* within the brainstem is distinct from other facial nerve muscles. The motor nucleus of facial nerve has a unique musculotopic organization in primates. Injection of horseradish peroxidase and fluorescent neuroanatomic tracers into the facial muscles of *Cebus apella* demonstrates that motor fibers of buccinator colocalize with those of platysma [9]. Four longitudinally oriented subnuclei were defined: medial, intermediate, dorsal, and lateral. Columns of orbicularis oculi, zygomaticus, orbicularis oris, and auricularis superior were located in the dorsal, intermediate, lateral, and medial subnuclei. Buccinator and platysma were in a combination of lateral and intermediate subnuclei. It should be noted that these muscles are histochemically, functionally, and anatomically distinct. Column sizes demonstrate a relative hierarchy of size depending on functional importance of the muscle in the primate context. Of interest, two Sm6 muscles that home to frontonasal mesenchyme, frontalis, and levator labii superioris aleqae nasi were not associated with a specific subnucleus (Fig. 17.19).

What is the Origin of Buccinator Muscle Myoblasts?

The Constrictor System: Does the Buccinator Fit the Pattern?

If we look at the constrictor system as exemplified in fishes and extend it to mammals, a striking pattern emerges. The five branchial arches post-jaws (BA3–BA7) are interconnected by four constrictor muscles, each one based on the posterior aspect of the arch. Sm14–Sm15 (sp3–sp4) connects seventh arch to the sixth, Sm11–Sm13 (sp1–sp2) connects sixth arch to the fifth, and so on. In tetrapods the last two arches are reassigned, as we have previously discussed. Thus we have the following situation (Table 17.1).

Where does facial nerve fit into this situation? Time and again we have described its origins from rhombomeres r4 and r5. The Kreisler mouse mutant indicates that the branchiomeric motor neurons are localized primarily in r4 whereas the PANS motor neurons and the abducens nucleus localize to r5 [10–12]. The buccinator muscle, being a mammalian invention, has not been mapped to a specific somitomere. The more important question is, does buccinator arise from a pair of somitomeres? And, if so, does it also represent some form of amalgamation between the arches?

Buccinator is a muscle born out of wedlock, the product of a shotgun marriage between BA1 and BA2, arising from the evolutionary imperative jaw production. The entire environment of this "household" is ad hoc. Take, for example, the relationships between buccinator, the buccal artery, and the oral mucosa. Recall that second arch epithelium disappears, subsumed by that of first arch. That blood supply to the check comes from IMMA makes perfect sense: first arch supplies first arch. Now throw buccinator into the mix. The program for this Sm6 muscle is r2–r3 ectoderm and submu-

Table 17.1 Neuromeric model of the constrictor series

Arch	Neuromere	Somitomere	Muscle	Nerve
5th–4th	r10–r11	Sm10–Smr11	Inferior constrictor	X
4th–3rd	r8–r9	Sm8–Smr9	Middle constrictor	IX/X
3rd–2nd	r6–r7	Sm6–Sm7	Superior constrictor	IX
2nd–1st	r4–r5	Sm5–Sm6	Buccinator	VII

cosa. And buccal artery supplies its posterior half. Is that purely random, a mechanical accident? Or does it mean that some part of the buccinator comes from the first arch itself, an evolutionary throwback to the original BA1 mouth constrictor before the invention of jaws?

A final intriguing issue of buccinator is whether this muscle contains elements of the first arch admixed with those of the second arch. Or it perhaps segregated with r2–r3 elements posterior and r4–r5 elements anterior. Homeotic mapping can resolve these questions in the future. Contrast studies by Han using decalcified cadavers demonstrate that the posterior aspect of buccinator is directly supplied by the extraosseous branch of posterior superior alveolar artery. Zhang demonstrated that PSAA is first directed to the posterior superior lobe of buccal fat pad. Furthermore, facial artery has two distinct sets of branches to the anterior muscle: posterior and anterior. These findings, taken together, suggest that buccinator has **four neuromeric zones**, each with a distinct artery.

- Posterior inferior zone (r3): buccal artery from second part of IMMA
- Posterior superior zone (r2): posterior superior alveolar artery from the third part of IMMA
- Anterior inferior zone (r5): facial artery, posterior buccinator branches of facial
- Anterior superior zone (r4): facial artery, anterior buccinator branches of facial

Development of Buccinator

Following the work of Raymond Gasser, buccinator follows the temporal pattern of the other facial muscles: deep-to-superficial and caudal-to-cranial. It first appears at 26 mm but is difficult to outline, especially the dorsal boundary. At 37 mm. the caudal portion (r3) is large and the cranial half (r2) is small. They are divided by parotid duct [13]. Between 37 and 43 mm the caudal portion extends backward to merge with superior constrictor and the muscle becomes completely innervated by buccinator branch of VII [14].

If the vascular map suggested by above is correct the muscle would be synthesized by fibers from Sm4 that would be laid down in the posterior zone, over the molars. This makes sense because IMMA develops prior to facial. Sm4 muscle attaches first and Sm6 muscles fills in the rest of the space.

Functional Aspects of the Buccinator

Mastication

EMG studies demonstrate that buccinator acts in concert with orbicularis oris for food propulsion through the fauces (boundary of the buccopharyngeal membrane and Waldeyer's ring), and into the retro-faucial pharynx [15]. These muscles close the mouth firmly to keep the food in place, flatten the cheek to press inward, and maintain contact between the food and teeth. As previously mentioned, its fibers are designed for slow, sustained contraction at low velocity. Buccinator physiology during mastication is reviewed by Perlman, Dutra.

Parotid Secretion

Cadaver studies by Kang show muscle fibers from buccinator inserting into the terminal segment of the duct, a structure that had previously been considered a strictly fibrous tube (Fig. 17.20). The fibers insert into two tubal structures at the point of penetration where they act as dilators such that they could control salivary flow. Luminal structures that relate to longitudinal smooth muscle are likely peristaltic with valvular structure preventing reflux. The terminal ampulla, associated with the aggregation of muscle fibers may play a role in allowing saliva to collect, thereby localizing sialolithiasis. No sphincteric role has been demonstrated for buccinator muscle fibers.

Vth Nerve Paralysis

Peripheral facial paralysis (PFP), as in Bell's palsy, offers a good insight into buccinator function. In PFP, these muscles cannot prevent the accumulation of food in the sulci. This causes food, especially liquids, to drool from the paretic side. Alternatively, retained liquid can enter the pharynx later, causing "second swallowing." Bell's palsy patients are frequently symptomatic. During the initial phase of PFP, 79% of patients have difficulty *controlling the oral bolus*. A not-unexpected decrease in *taste* is seen in 39%. Another 43% of patients demonstrate an interesting decrease in *tactile sensation* over the mucosa of the ipsilateral tongue and cheek. The *dysphagia limit* in acute phase PFP, defined by difficulty swallowing 20 cc or less water, is abnormal in 55% of patients, extending to the midline in 43% and to the contralateral side in 28%.

- Recall that the motor supply of superior constrictor is a combination of VII and IX. Although this muscle is controlled from r6 to r7, it receives motor fibers from r5 as well. Thus, the effects of PFP likely spread backward causing discoordination between buccinator with superior constrictor. Middle constrictor, being supplied from r8 to r9 remains normal.
- PANS drive to the submandibular and parotid glands remains unaffected on the paretic side.

The Buccal Fat Pad (Figs. 17.21, 17.22, 17.23 and 17.24)

Function and Structure

Recall that adipose tissue is a neural crest derivative and that it accompanies all neurovascular pedicles. Thus, the buccal

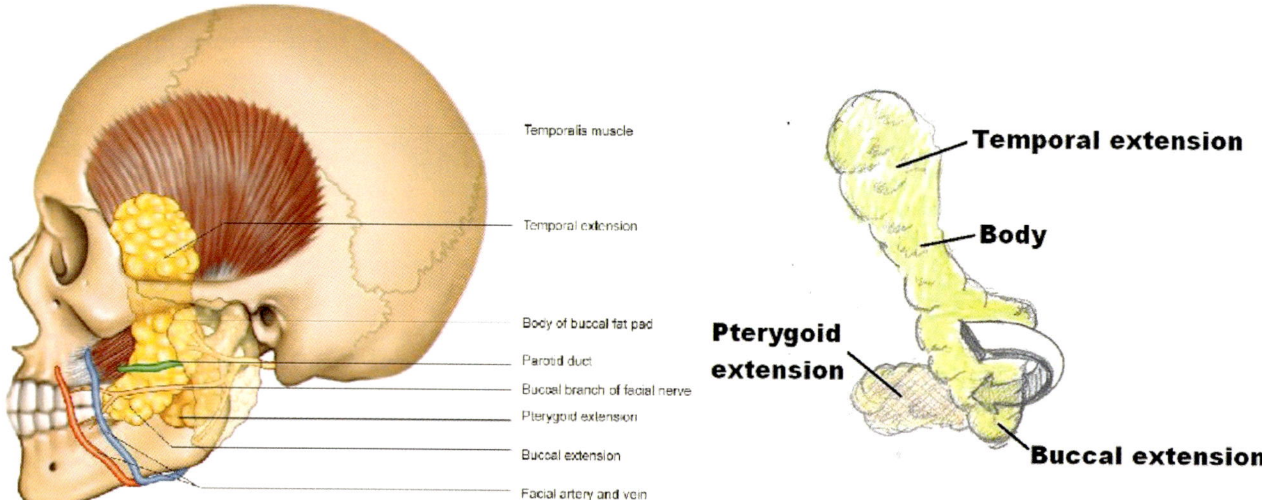

Fig. 17.21 Buccal fat pad. It is a deep fat pad located on either side of the face between the buccinator and several more superficial muscles (including the DIF masseter and the SIF zygomaticus major and zygomaticus minor). The inferior portion of the buccal fat pad is contained within the buccal space. The buccal fat pad has been described as being in various ways. It is divided into three lobes: the anterior, intermediate, and posterior. "According to the structure of the lobar envelopes, the formation of ligaments, and the source of the nutritional vessels".[1] Also, there are four extensions from the body of the buccal fat pad: the sublevator, the melolabial, the buccal, and the pterygoid. The nomenclature of these extensions derives from their location and proximal muscles.[4] The anterior lobe of the buccal fat surrounds the parotid duct, which conveys saliva from the parotid gland to the mouth. It is a triangular mass with one vertex at the buccinators, one at the levator labii superioris alaeque nasi, and one at the orbicularis oris. The intermediate lobe lies between the anterior and posterior lobes over the maxilla. [1,4] The intermediate lobe seems to lose a significant amount of volume between childhood and adulthood. The posterior lobe of the buccal fat pad runs from the infraorbital fissure and temporal muscle to the upper rim of the mandible and back to the mandibular ramus. Left: [Reprinted from Zhang H-M, Yan Y-P, Qi K-M. Anatomical Structure of the Buccal Fat Pad and Its Clinical Adaptations. Plast Reconstr Surg 2002; 109(7): 2509–18; discussion 2519–20. With permission from Wolters Kluwer Health, Inc.] Right: [Reprinted from Wikipedia. File:Buccal Fat Diagram.jpg https://en.wikipedia.org/wiki/Buccal_fat_pad. With permission from Creative Commons License 3.0: https://creativecommons.org/licenses/by-sa/3.0/deed.en.]

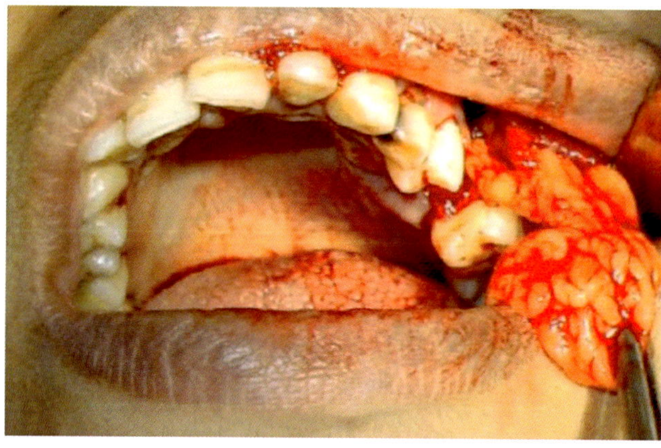

Fig. 17.22 Buccal fat pad delivered from incision. BFP has a temporal extension beneath the zygomatic arch. This continuity allows SGF flaps supplied by superficial temporal artery to be passed into the mouth. This flap is elastic and can cover virtually any site in the oral cavity. [Reprinted from Candamourty R, Jain MK, Sankar K, Ramesh Babu M R. Double-layered closure of oroantral fistula using buccal fat pad and buccal advancement flap. J Nat Sci Biol Med 2012;3:203–5. With permission from Creative Commons License 3.0: https://creativecommons.org/licenses/by-nc-sa/3.0/.]

neuroangiosome brings its own r3 fat into the space above buccinator. Additional fat is synthesized from r4/r5 fat into the same plane. As we shall see, these complex embryonic relationships explain the anatomic features of the buccal fat pad (BFP). Two cadaver studies are particularly useful [16, 17]: the former, being quite clear in defining the space occupied by the BFP but the latter relating the parts of the BFP to blood supply. Braun provides a review of buccomasseteric anatomy via CT imaging [18].

Mucosa and skin are each supplied using two different systems. Buccal mucosa can be supplied by branches traversing the buccinator or by submucosal arteries such as lingual. Skin in many parts of the face, although originating from first arch, is supported by underlying second arch muscles. Nonetheless, penetrating branches such as StV1 supraorbital artery or StV2 zygomaticotemporal artery reach the skin independent of an intervening muscle.

The overall function of buccal fat pad is as a support tissue for neurovascular structures and as a lining tissue for the muscles of mastication, but the portion in contact with buccinator is below the duct. Buccal fat pad is described as hav-

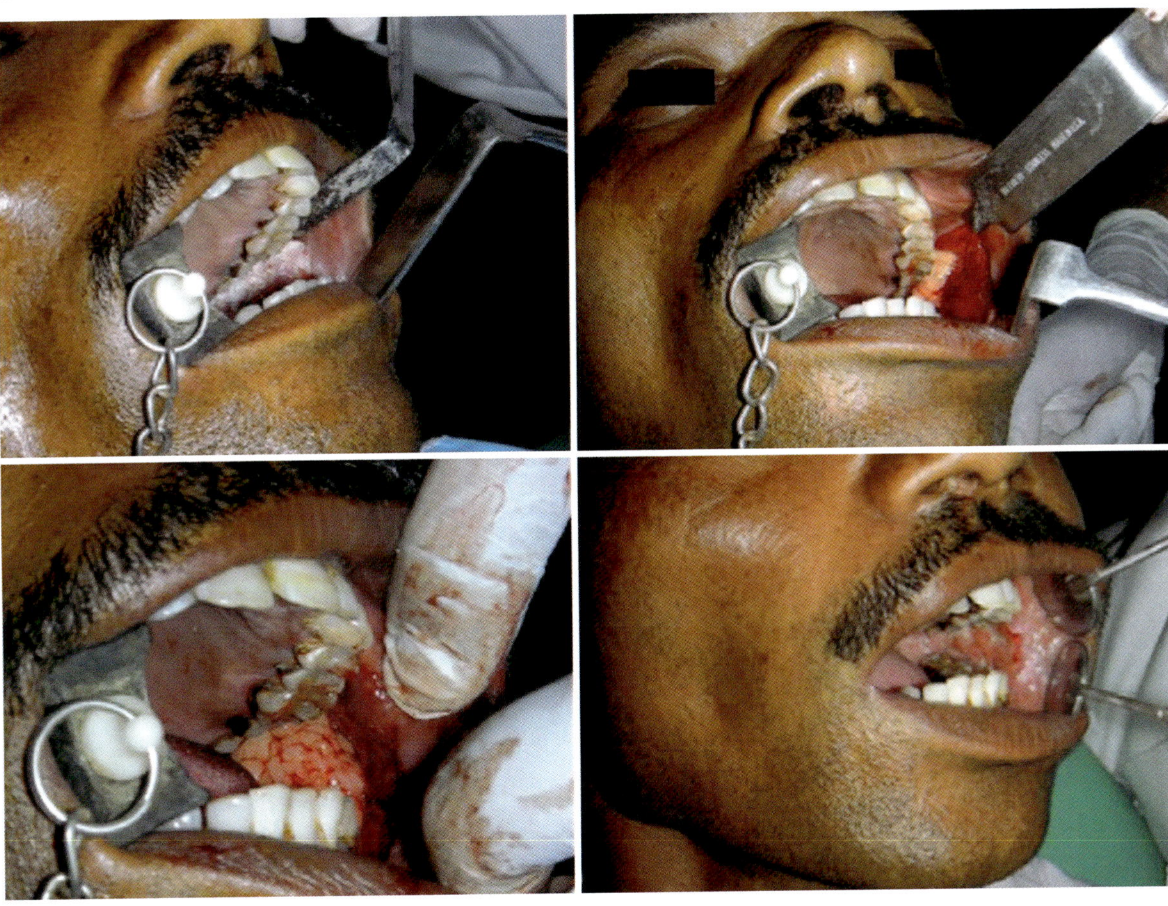

Fig. 17.23 Buccal fat pad as a free graft does not need epithelial coverage. Verrucous leukoplakia is excised, leaving a defect over buccinator muscle. Free BFP graft is sutured into place. At 2 weeks epithelialization is fully complete. [Reprinted from Mohan S, Kankariya H, Harjani B. The use of the buccal fat pate for reconstruction of oral defects: A review of the literature and report of cases. J. Maxillofac. Oral Surg 2012; 11(2):128–131. With permission from Springer Nature.]

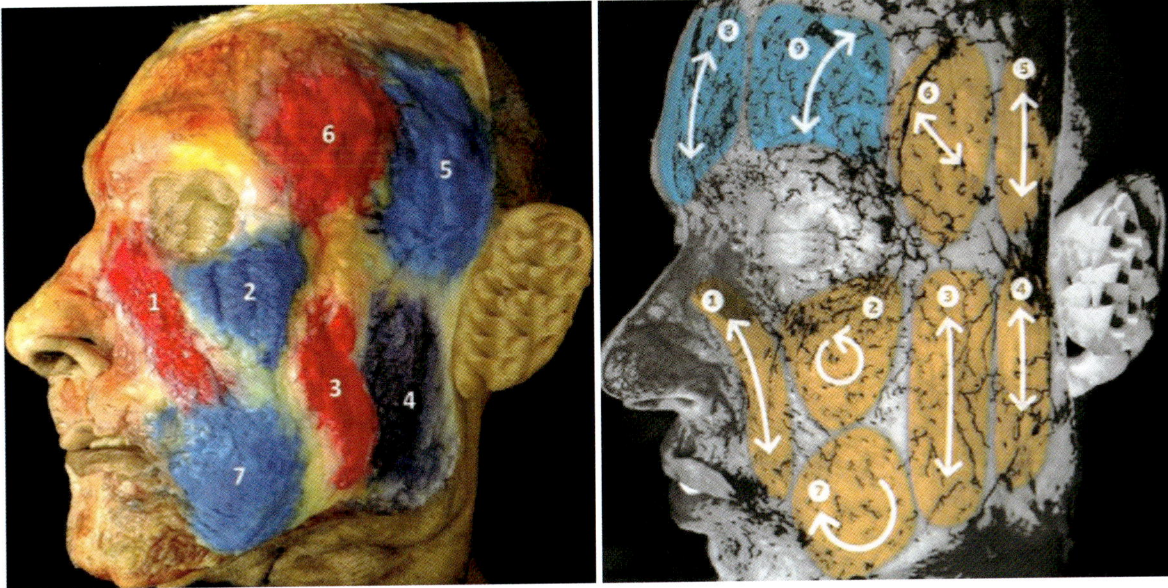

Fig. 17.24 Superficial facial fat maps to individual branches of second arch external carotid and StV1 orbital. Note that branches of superficial temporal artery are excluded from zones 10–11 and 12–13. This fat forms in association with neural crest from two sources: r4–r5 second arch, and r1 frontonasal. [Reprinted Schenck TL, Koban KC, Schlattau A, Frank K, Sykes JM, Targonsinski S, Erlbacher K, Cotofana The functional anatomy of the superficial fat compartments of the face: A detailed imaging study. Plast Reconstr Surg 2018; 141(6):1351–1359. With permission from Wolters Kluwer Health, Inc.]

ing centrally positioned body from which radiate four extensions, three of which are anatomically legitimate.

- The *body* lies along anterior border of masseter. Its greater part lies above Stenson's duct. It extends from the second molar zone of maxilla backward around the tuberosity and into the pterygomaxillary fissure where it serves as a support tissue for the StV2 branches of internal maxillary artery to the zygomaticomaxillary nasal complex.
- A *buccal extension* lies below Stenson's duct and covers buccinator. Its anterior border is defined by the obliquely ascending facial artery.
- The *pterygoid extension* lies posterior to mandibular ramus. It follows V3 neurovascular pedicles of inferior alveolar and lingual nerves.
- The *deep temporal extension* intervenes between zygomatic arch and the temporalis muscle. It extends forward in this space to reach the back wall of alisphenoid. It conducts deep temporal arteries from the second part of IMMA into the muscle. This tissue is clinically useful because it provides a passageway for galeal–subgaleal fascia flaps (SGF) into the oral cavity.
- A *superficial temporal extension* (sic), better known as the superficial temporal fat pad, is a misnomer. It is described as an extension of the buccal fat pad but it is from a different embryonic source. It is supplied by second arch superficial temporal artery, middle branch. Its structure, sandwiched between two leaves of fascia extending upward from zygomatic arch, is explained by results from the splitting and unification of second arch SMAS as it flows around the arch. The fat pad is surgically relevant because it can be entered from above to gain access to the superior border of zygomatic arch, along which subperiosteal dissection can proceed safely into the face, by passing deep to the facial nerve that crosses the zygomatic arch just above the periosteum.

The Embryology of the Buccal Fat Pad

Recall that the pericytes (with a primordial relation to neural crest) have three important roles. First, they are part of the external coat of all blood vessels and lymphatics. Second, they are the source of mesenchymal stem cells throughout the body. Third, they are also the putative source of white adipose tissue (paraxial mesoderm serves that purpose for brown adipose tissue). The BFP intervenes between the differing planes of the DIF and plane separating the DIF from the SIF. Pioneering work mapping the facial fat into anatomic zones is described by Schenck. Differing zonal patterns of facial fat involution are seen in diseases such as AIDS.

Buccal fat pad relates parotideomasseteric fascia (PMF). Recall that masseter is surrounded by its own envelope of DIF, a r3 neural crest product. Parotid gland, being situated between DIF and SIF, is a mixture of mesenchyme from first and second arches and has its own capsule which is intimately fused with that of masseter: thus, PMF is considered as a single unit, which has defied surgical attempts at separation. It provides overlying cover for facial nerve as it ascends through the gland such that it is not in direct contact with SMAS over the gland. At the forward limit of the parotid gland, zygomatic and buccal branches of facial nerve ascend upward through BFP to enter their muscle targets. These branches are often in close association with the parotid duct. A forward extension of PMF underlies the buccal fat pad and separates it from buccinator. For this reason, although the PMF is quite flimsy, buccinator flaps can be developed for intraoral transfer without herniation of the buccal fat pad into the mouth. Thus, buccal fat pad is enclosed within an envelope of parotideomasseteric fascia.

With these anatomic landmarks in mind, one can understand how an infraductal buccinator flap can be elevated away from the overlying buccal fat, teasing the pedicle away, without risk to the facial nerve and with a low probability of BFP prolapse, a situation that is readily addressed with a secure closure of the donor site.

Clinical Applications of the Buccal Fat Pad

We have previously mentioned how the temporal extension of BFP provides an access route for galeal flaps into the mouth. With reference to the buccal fat pad itself, an intraoral approach is capable of delivering 50% of the pad as a flap or as a graft. BFP flaps have been used to support soft tissue support for bone graft places around dental implants in the posterior maxilla [19]. Not only are they biologically superior to allogenic membranes used for the same purpose, but they also contain mesenchymal stem cells which may contribute to the healing process. Another use for BFP flaps is to provide coverage for oral-antral [20] and oronasal fistulae [21]. Flaps and grafts of BFP can be left exposed in the oral cavity as they will mucosalize in 2 weeks. Perhaps the most common use for the BFP has been for aesthetic purposes, either by resection to reduce an excessively full midface or as a graft to improve facial contour [22, 23]. General reviews regarding BFP are extent [24–26].

Dissection Technique: Buccinator Flaps (Figs. 17.25, 17.26, 17.27, 17.28, 17.29, 17.30, 17.31, 17.32, 17.33, 17.34, 17.35, 17.36, 17.37, 17.38, 17.39, 17.40, 17.41, 17.42, 17.43, 17.44, 17.45, 17.46, 17.47, 17.48, 17.49, 17.50, 17.51, 17.52, 17.53, 17.54, 17.55, 17.56 and 17.57)

Local Anesthesia

All buccinator procedures benefit from a preoperative block of V2 and V3 with 0.25% bupivacaine with epinephrine at a recommended dose of 0.5 cc per kg (the toxic dose being

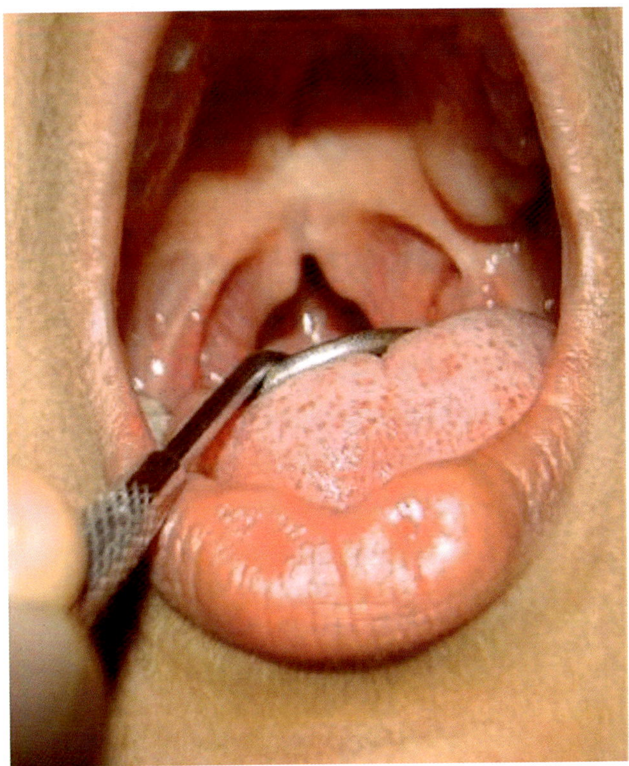

Fig. 17.25 Indications for buccinator 1: foreshortened platform. Transverse deficiency at P3 creates a triangle out of the rectangular horizontal plate, This drags the aponeurosis along with it. What's more, since aponeurosis is part of same neuroangiosome, it is reduced in its own right. The effect is to drag the levator complex forward: the soft palate is foreshortened. [Courtesy of Michael Carstens, MD]

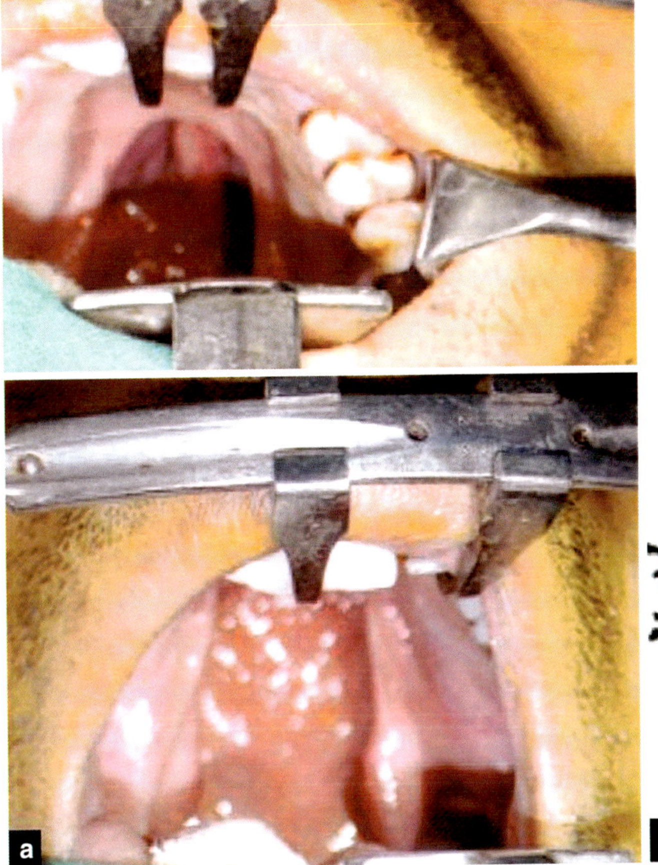

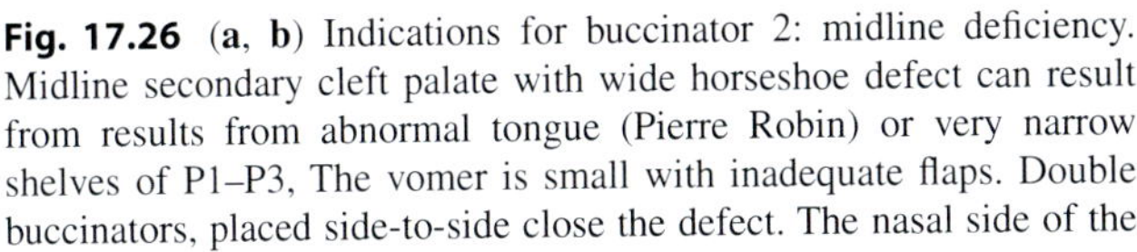

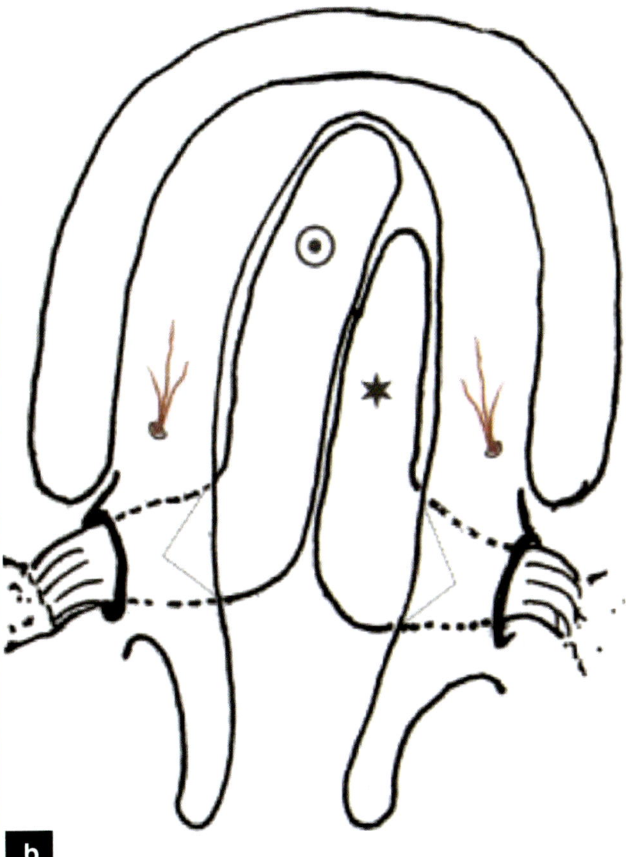

Fig. 17.26 (**a**, **b**) Indications for buccinator 2: midline deficiency. Midline secondary cleft palate with wide horseshoe defect can result from results from abnormal tongue (Pierre Robin) or very narrow shelves of P1–P3, The vomer is small with inadequate flaps. Double buccinators, placed side-to-side close the defect. The nasal side of the buccinators flaps is raw but will mucosalize in 2–3 weeks. [Reprinted from Bhanyani B. Buccinator myomucosal flap in cleft palate repair: revisited. J Cleft Lip Palate Craniofac Anomal 2014; 1(1):11–16. With permission from Wolters Kluwer Health, Inc.]

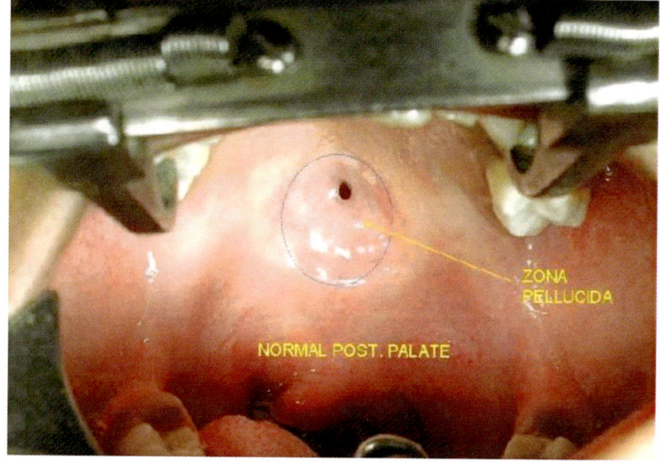

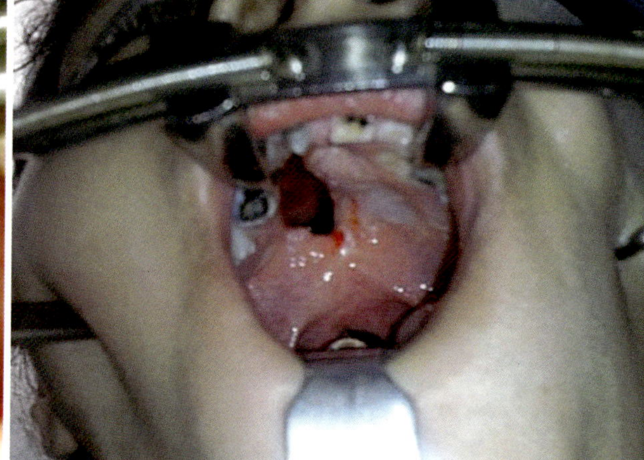

CONGENITAL PALATAL FISTULA WITH NORMAL POSTERIOR PALATE

Fig. 17.27 Indications for buccinator 3: fistula. Left: For "pinhole" anterior fistula, buccinator can reach as far forward as the nasopalatine foramen. Right: massive fistula due to flap loss. Traditional closure with 2-stage tongue flap. Buccinator works well here. [Reprinted from Alsalman AK, Algadiem EA, Alwabari MS Almugarrab FJ. Single-layer Closure with tongue flap for palatal fistula in cleft palate patients. Plast Reconstr Surg glob Open 2016; 4(8):e852. With permission from Wolters Kluwer Health, Inc.]

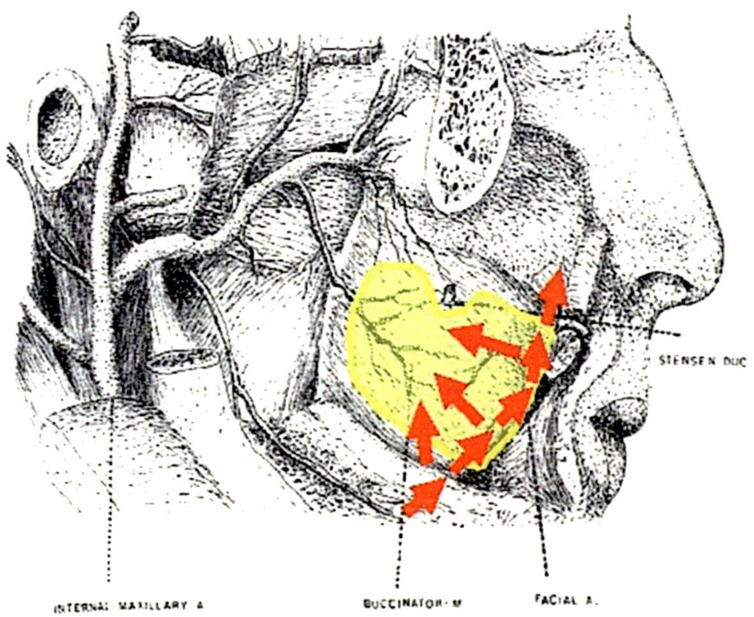

Fig. 17.28 Anterior buccinator flap. The vascular axis flows upward but can be ligated at either end with adequate venous drainage. Alternatively, anterior buccinator flap can be dissected out leaving the entire axis intact and simply flipped medially. Elasticity of the vascular leash lends to great flexibility, even in the "turnover" configuration. [Courtesy of Professor Johannes Fagan]

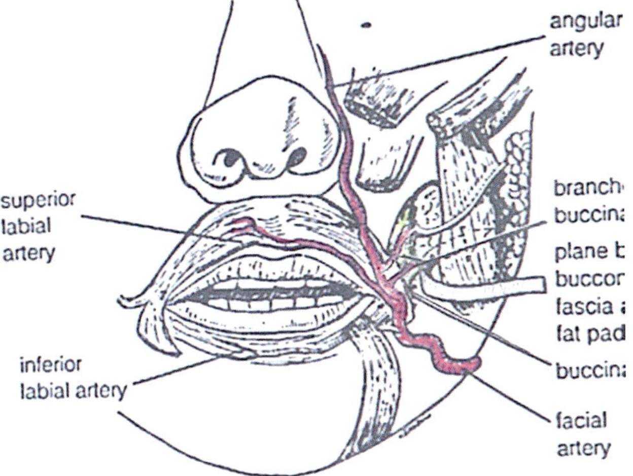

Fig. 17.29 Buccinator anterior incision in nasolabial fold. Facial artery reliably gives off 2–3 posteriorly directed branches that enter the muscle. Dissection in the fat plane above the muscle is easy. [Adapted from Carstens MH, Stofman GM, Soteranos GC, Hurwitz DJ. A new approach for repair of oral-antral-nasal fistulae: the anteriorly based buccinator myomucosal island flap. J Cranio-Maxillo-Fac Surg 1991; 19(2): 64–70. With permission from Elsevier.]

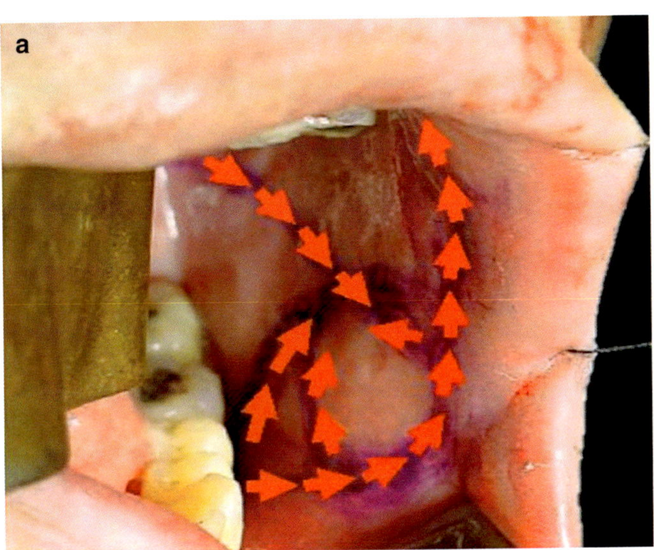

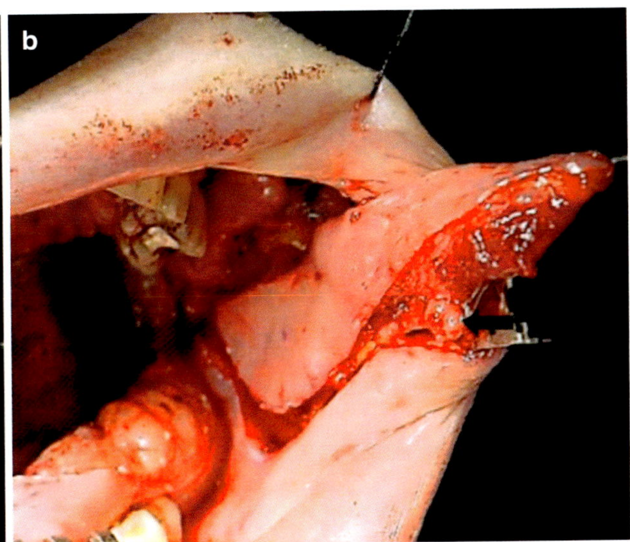

Fig. 17.30 (**a**, **b**) Anterior buccinator can be harvested intraorally with ease. Doppler shows the location of facial artery but it can be easily identified by dissection alone. Elasticity of the flap is demonstrated. Note facial artery (black arrow) inferior labial incision. The proximal end of the flap can be split to create a Y-shaped design as needed. (**a**) [Reprinted from Zhao Z, Li Y, Xiao S, et al. Innervated buccal musculomucosal flap for wider vermilion and orbicularis oris muscle reconstruction. Plast Reconstr Surg 2005; 116(3): 846–852. With permission from Wolters Kluwer Health, Inc.] (**b**) [Courtesy of Professor Johannes Fagan]

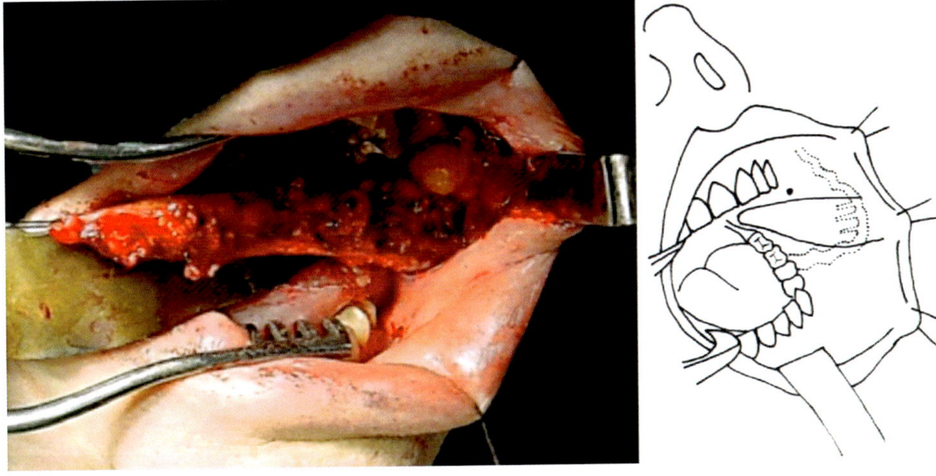

Fig. 17.31 Anterior buccinator can be harvested intraoperative Doppler shows the location of facial artery but it can be easily identified by dissection alone. Elasticity of the fap is demonstrated. Flap can be rotated into position. [Courtesy of Professor Johannes Fagan]

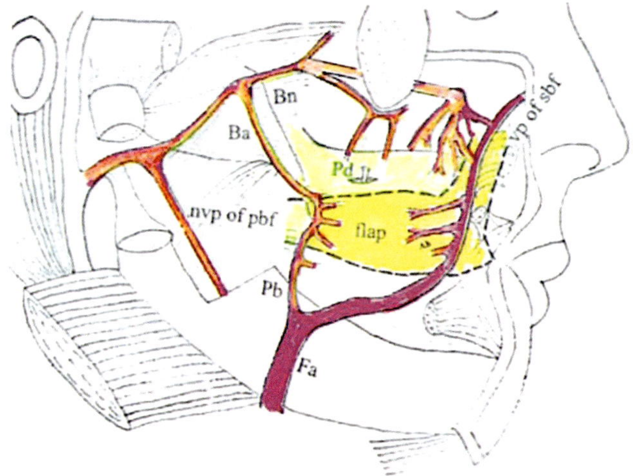

Fig. 17.32 Superiorly and inferiorly based buccinator flaps as per Zhao. Distal facial artery supplies the anterior zone of buccinator through 2–3 distinct branches. Superior design involves section of the facial artery at the level of the sulcus at canine. The inferior-based design requires a section at the opposite side. Note that these flaps are oriented below the parotid duct. [Adapted from Zhao Z, Yan Y, Li Y, et al. New buccinator myomucosal island flap: anatomic study and clinical application. Plast Reconstr Surg 1999; 104(1):55–64. With permission from Wolters Kluwer Health, Inc.]

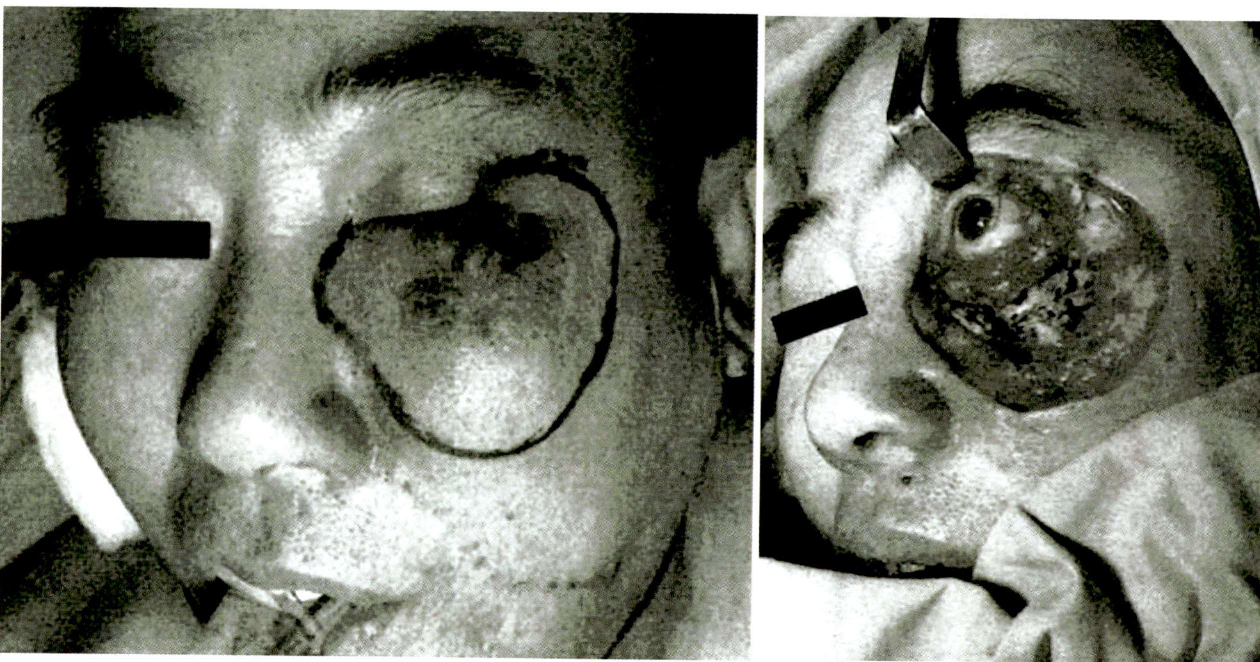

Fig. 17.33 Anterior superior buccinator to orbit 1. Invasive basal cell tumor resected to create periorbital defect. [Reprinted from Zhao Z, Yan Y, Li Y, et al. New buccinator myomucosal island flap: anatomic study and clinical application. Plast Reconstr Surg 1999; 104(1):55–64. With permission from Wolters Kluwer Health, Inc.]

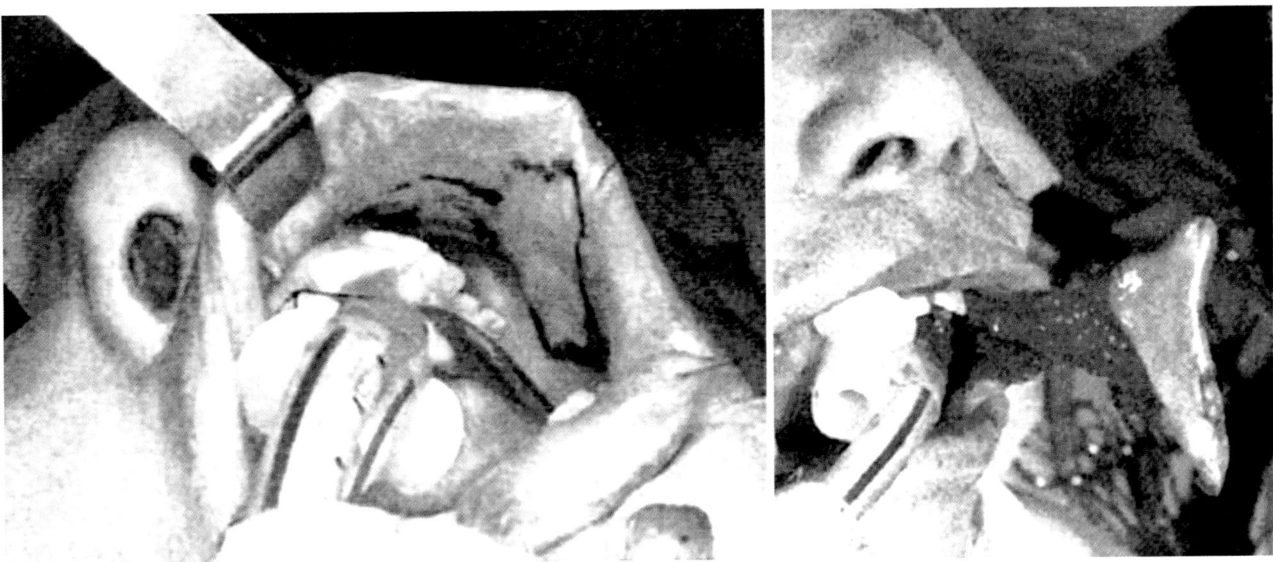

Fig. 17.34 Anterior Superior buccinator to orbit 2. Intraoral flap design and elevation. [Reprinted from Zhao Z, Yan Y, Li Y, et al. New buccinator myomucosal island flap: anatomic study and clinical application. Plast Reconstr Surg 1999; 104(1):55–64. With permission from Wolters Kluwer Health, Inc.]

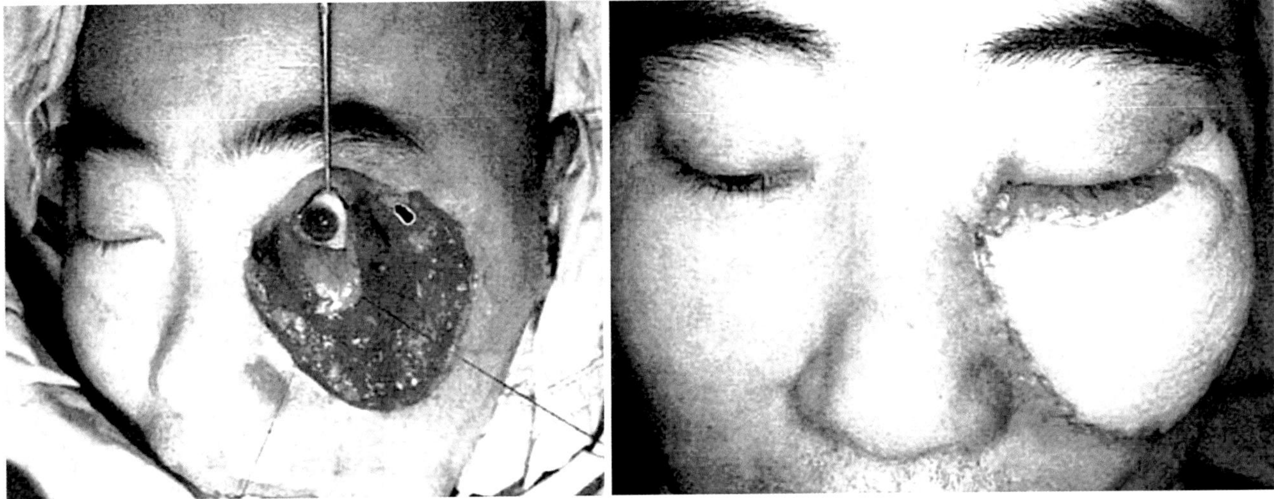

Fig. 17.35 Anterior Superior buccinator to orbit 3. Flap in place with skin graft coverage. [Reprinted from Zhao Z, Yan Y, Li Y, et al. New buccinator myomucosal island flap: anatomic study and clinical application. Plast Reconstr Surg 1999; 104(1):55–64. With permission from Wolters Kluwer Health, Inc.]

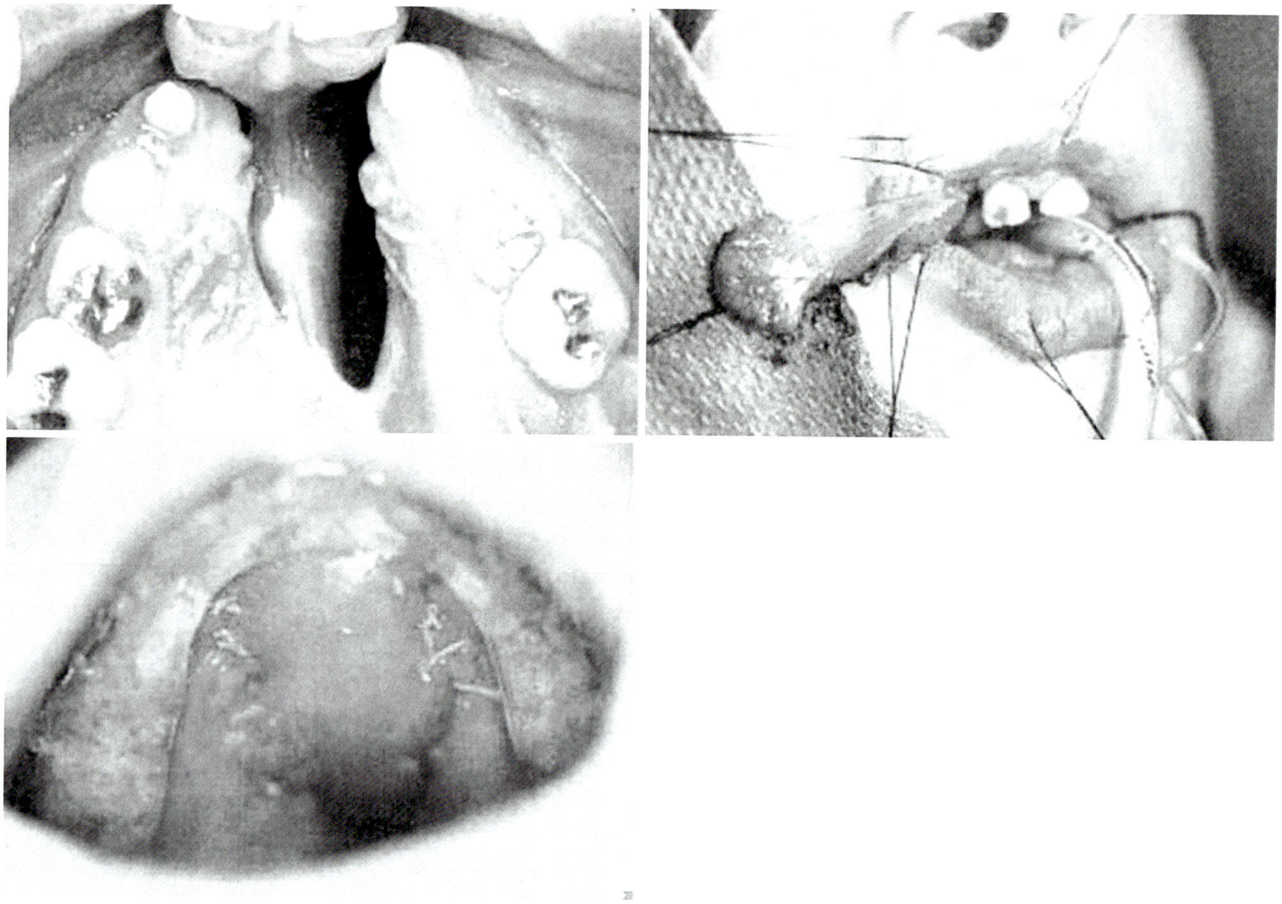

Fig. 17.36 Buccinator anterior superior for palate fistula. Here the flap is passed *backward* through the alveolar cleft. [Reprinted from Pribaz J. A new intraoral flap: facial artery musculomucosal (FAMM) flap Plast Reconstr Surg 1992; 90(3): 421–429. With permission from Wolters Kluwer Health, Inc.]

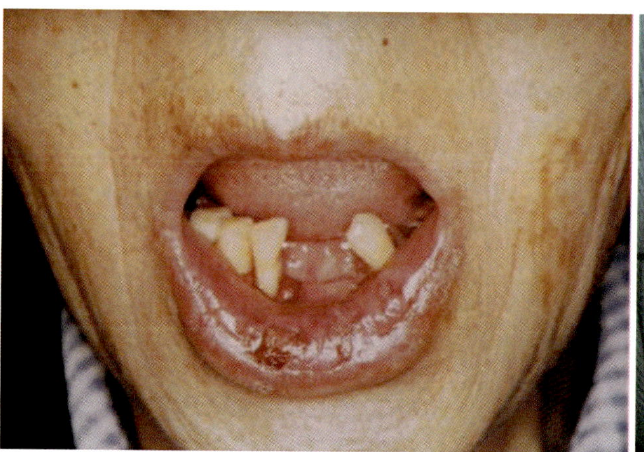

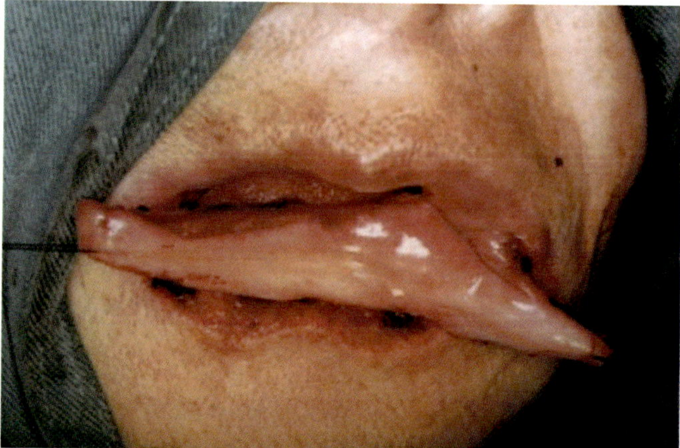

Fig. 17.37 Anterior inferior buccinator flap reconstruction of lower lip. Note elasticity of the flap allows it to cover the entire lower lip. Nice aesthetic result. Left: Squamous cell carcinoma of the lower lip mucosa. Right: Resection defect with inferiorly based buccinator flap elevated. Anterior (distal) margin of flap directed toward right commissure. [Reprinted from Zhao Z, Ying L, Xiao S, et al. Innervated Buccal musculomucosal flap for wider vermilion and orbicularis oris muscle reconstruction. Plast Reconstr Surg 2005; 116(3):846–852. With permission from Wolters Kluwer Health, Inc.]

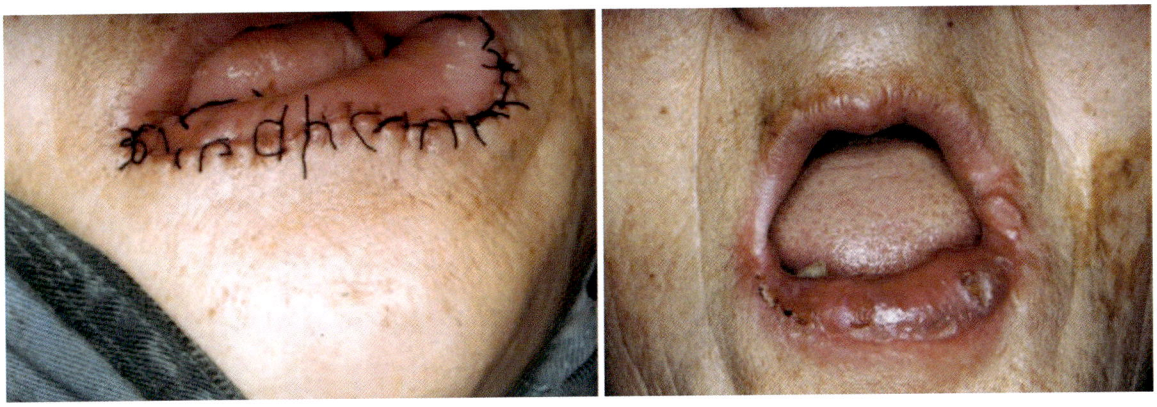

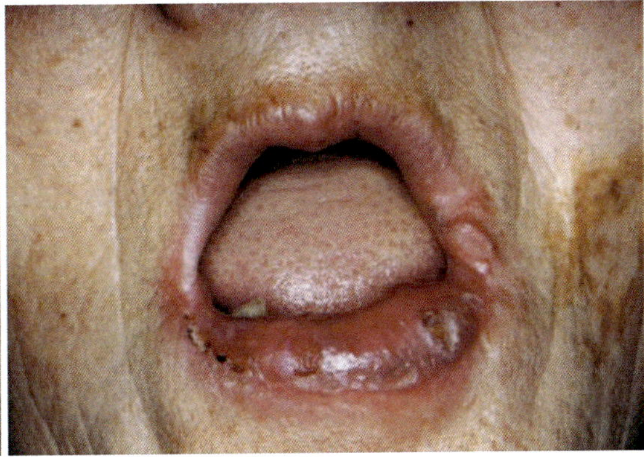

Fig. 17.38 Anterior inferior buccinator flap reconstruction of lower lip. Note elasticity of the flap allows it to cover the entire lower lip. Nice aesthetic result. Left: Inset of flap; distal margin viable. Right: Nerve to buccal mucosa preserves sensation. [Reprinted from Zhao Z, Ying L, Xiao S, et al. Innervated Buccal musculomucosal flap for wider vermilion and orbicularis oris muscle reconstruction. Plast Reconstr Surg 2005; 116(3):846–852. With permission from Wolters Kluwer Health, Inc.]

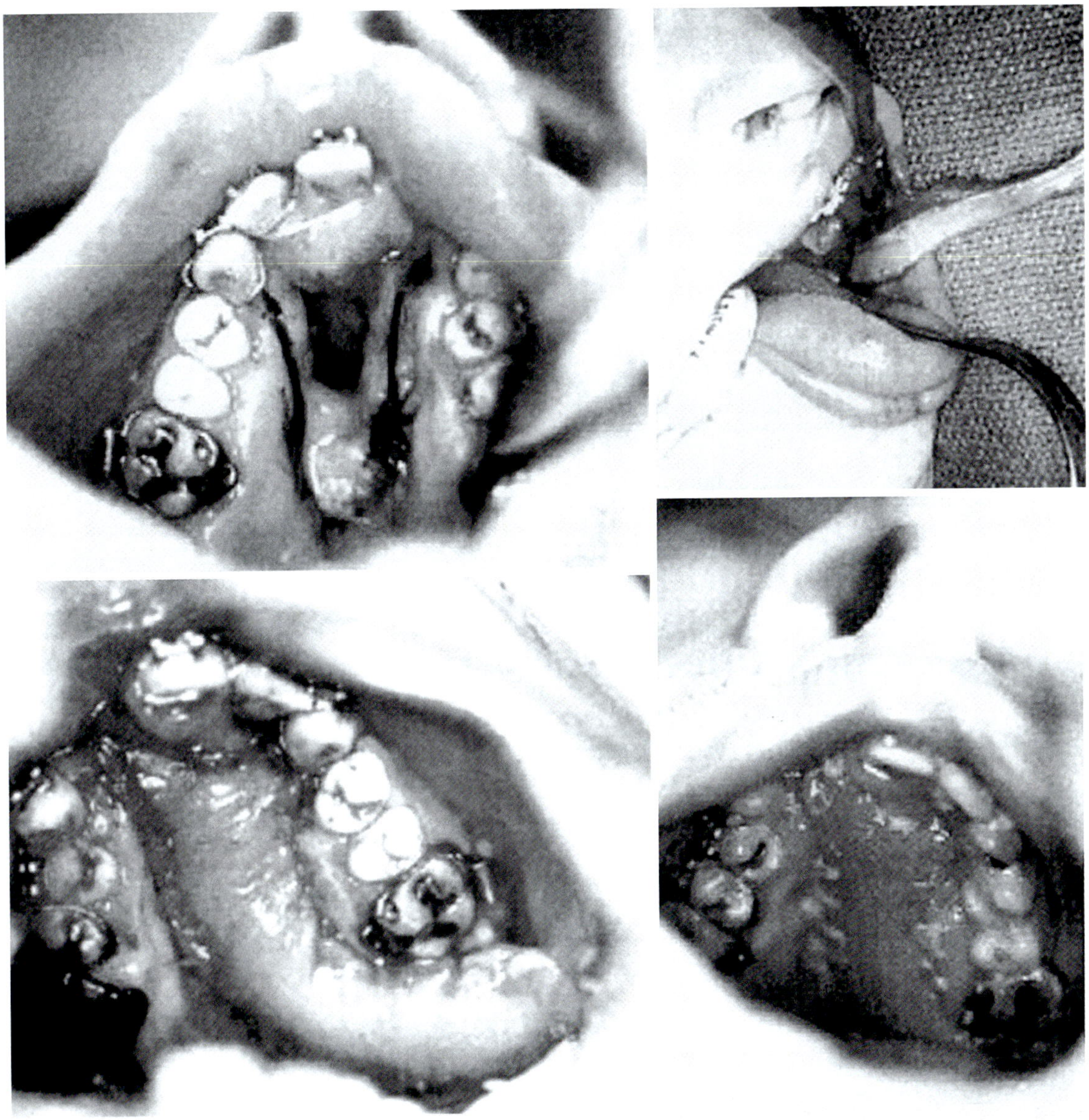

Fig. 17.39 Extreme flexibility of the pedicle allows the entire flap to be transposed 180° *backwards* around the tuberosity. It looks almost like a posteriorly based flap. Facial artery axis is more elastic than the buccal artery axis. [Reprinted from Pribaz J. A new intraoral flap: facial artery musculomucosal (FAMM) flap Plast Reconstr Surg 1992; 90(3): 421–429. With permission from Wolters Kluwer Health, Inc.]

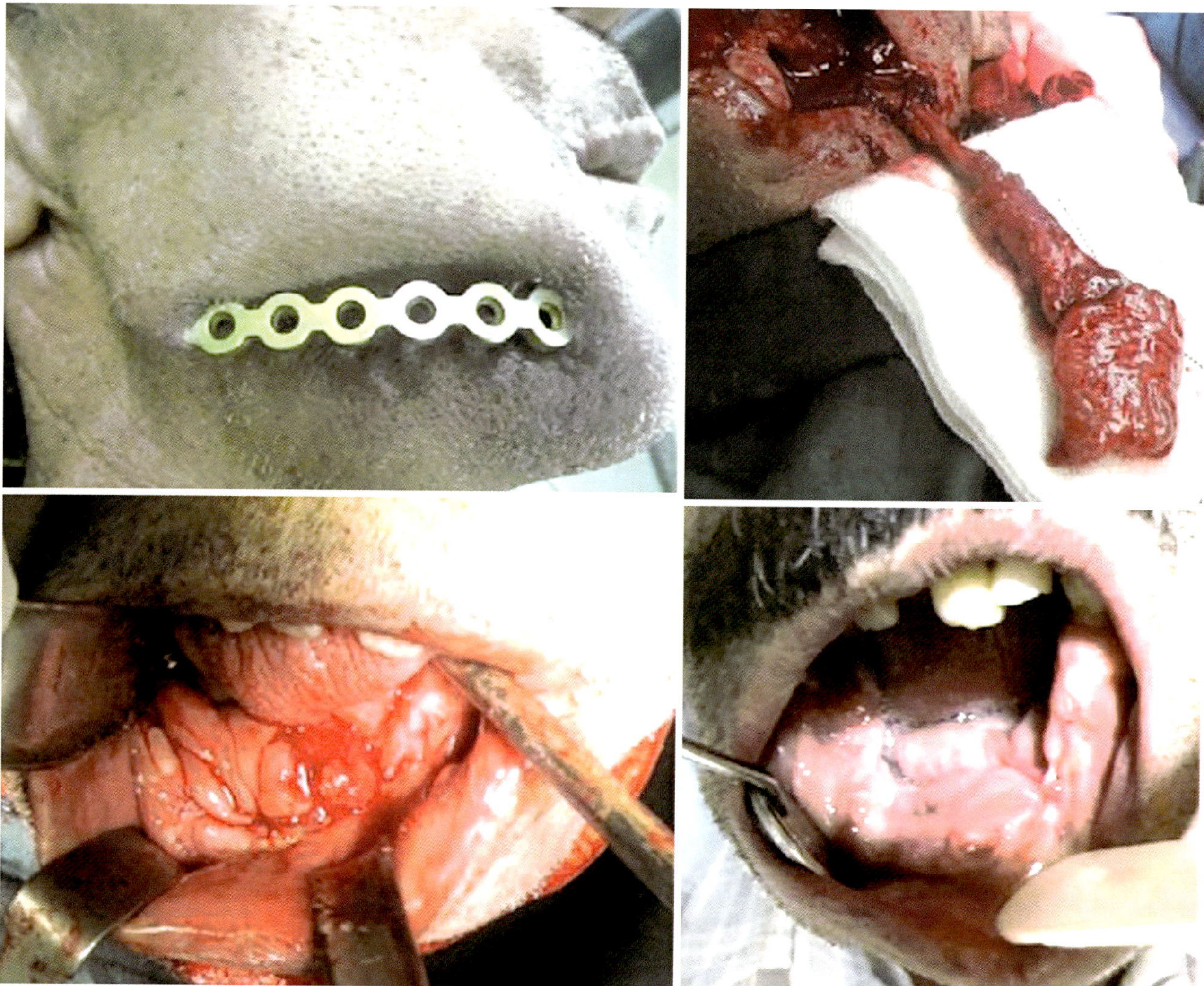

Fig. 17.40 Anterior inferior island flap. (Top left) Exposed right sided reconstruction plate. (Top right) Left sided anterior buccinator island flap, inferiorly based. Long pedicle permits it to be transferred across to the other side. Note left submandibular incision for access to the flap (Carstens, Zhao technique) (Bottom left) Low profile plate replaced, intraoral closure achieved, followed by external right cheek closure. (Bottom right) Follow-up at 3 months with edema down; flap is well-seen. [Reprinted from Rahpeyma A, Khajehajmadi S. Inferiorly based buccinator myomucosal island flap in oral and pharyngeal reconstruction. Four techniques to increase its application. Int J Surg Case Rep. 2015; 14: 58–62. With permission from Elsevier.]

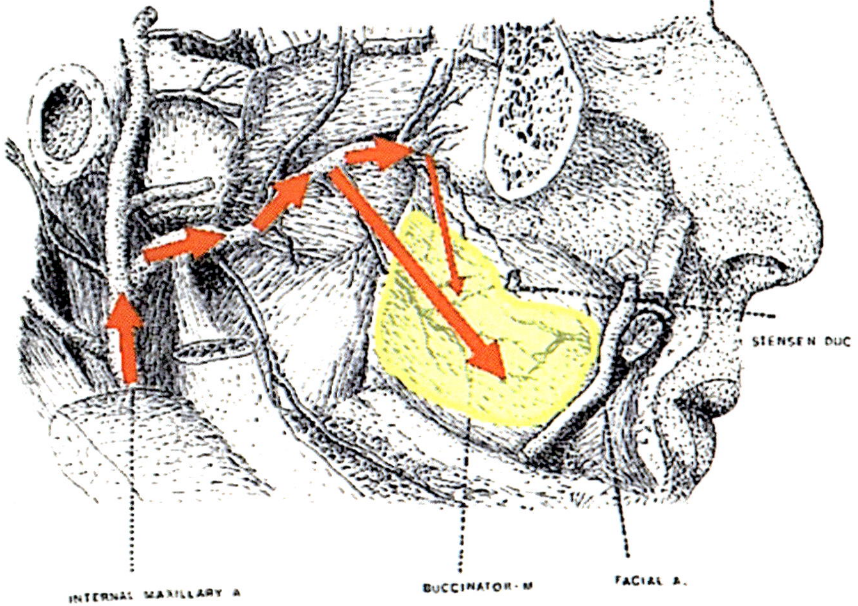

Fig. 17.41 Posterior buccinator flap. Note additional contribution from posterior superior alveolar artery. [Courtesy of Professor Johannes Fagan]

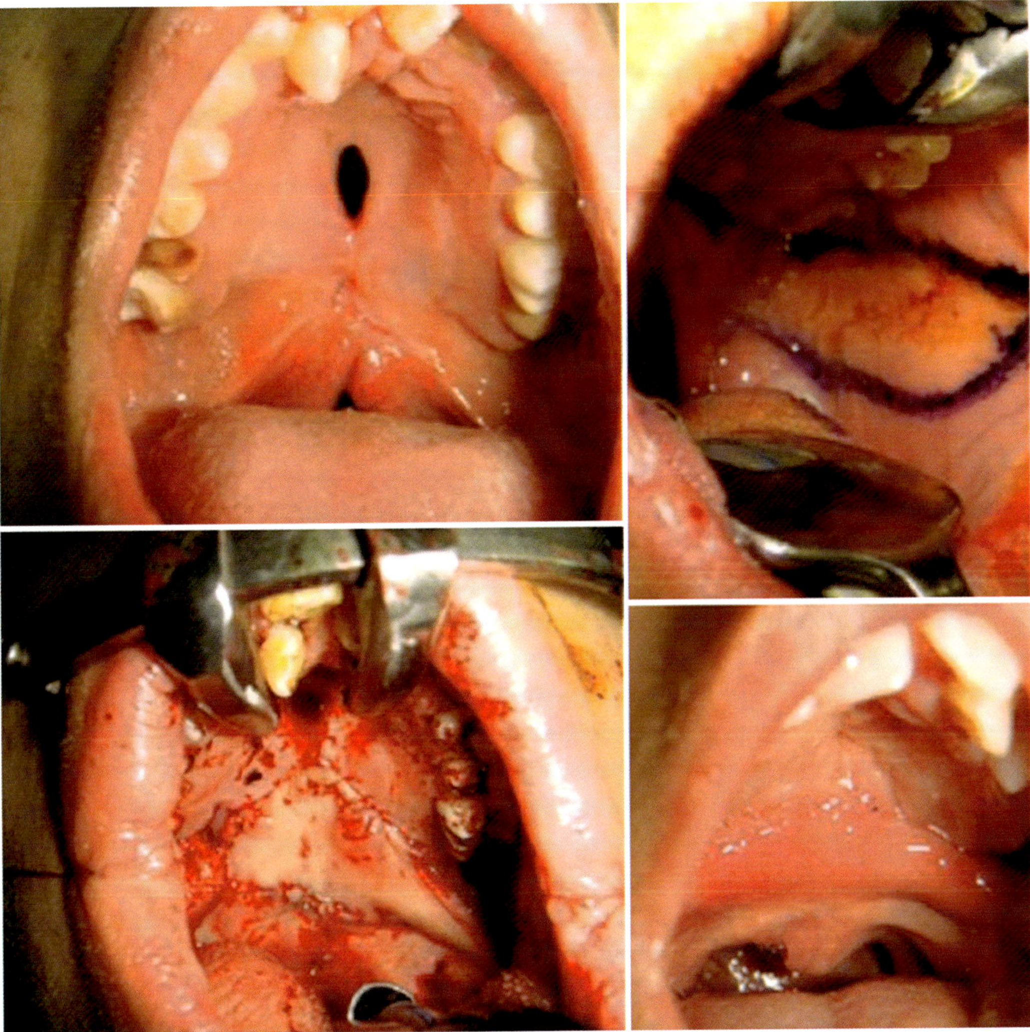

Fig. 17.42 Fistula of hard palate. (*Top left*) Hard palate fistula in the presence of a short velum. (*Top right*) Design of left-sided posterior buccinator flap. (*Bottom* left) Fistula is closed with mucoperiosteal flaps, and buccinator flap inset. (*Bottom right*) Follow-up with retrodisplacement of soft palate, particularly in the midline. [Reprinted from Abdaly H, Omranyfard M, Ardekany R, Babaei K. Buccinator flap as a method for palatal fistula and VPI management. Adv Biomed Res 2015; 4:135. With permission from Wolters Kluwer Health, Inc.]

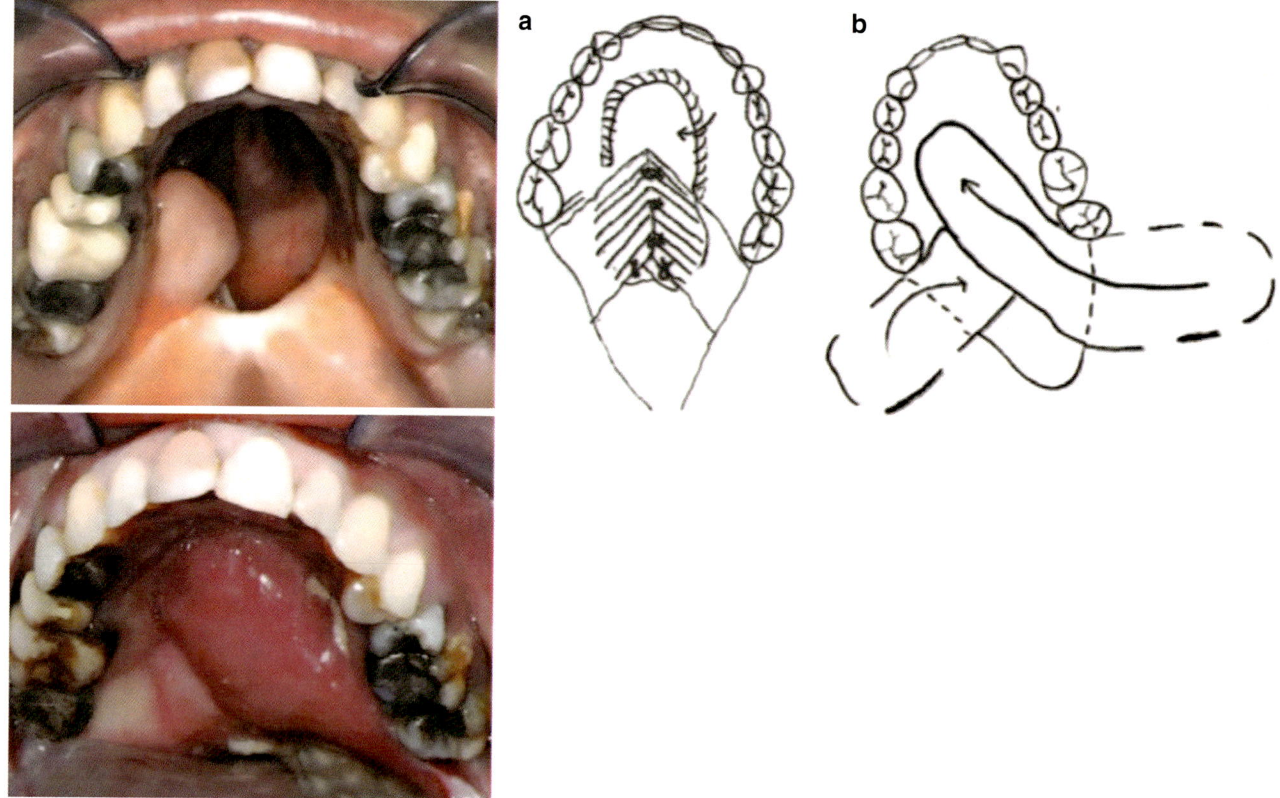

Fig. 17.43 Large anterior palatal fistula. Top left: Three separate turn-over flaps were used for large defect. Top right: (**a**) Nasal side closure of the posterior fistula was achieved with inverted mucosal flap from the soft palate. (**b**) Posteriorly based buccinator myomucosal flaps (long and short). Bottom: Left flap length is long and is used for oral side coverage. The right flap is shorter and covers the raw soft palate defect. Bottom center: postoperative closure of fistula. [Reprinted from Rahpeyma A, Khajehahmadi S. Closure of huge palatal fistula in an adult patient with isolated cleft palate: A technical note. Plast Reconstr Surg Glob Open 2015; 3:e306. With permission from Wolters Kluwer Health, Inc.]

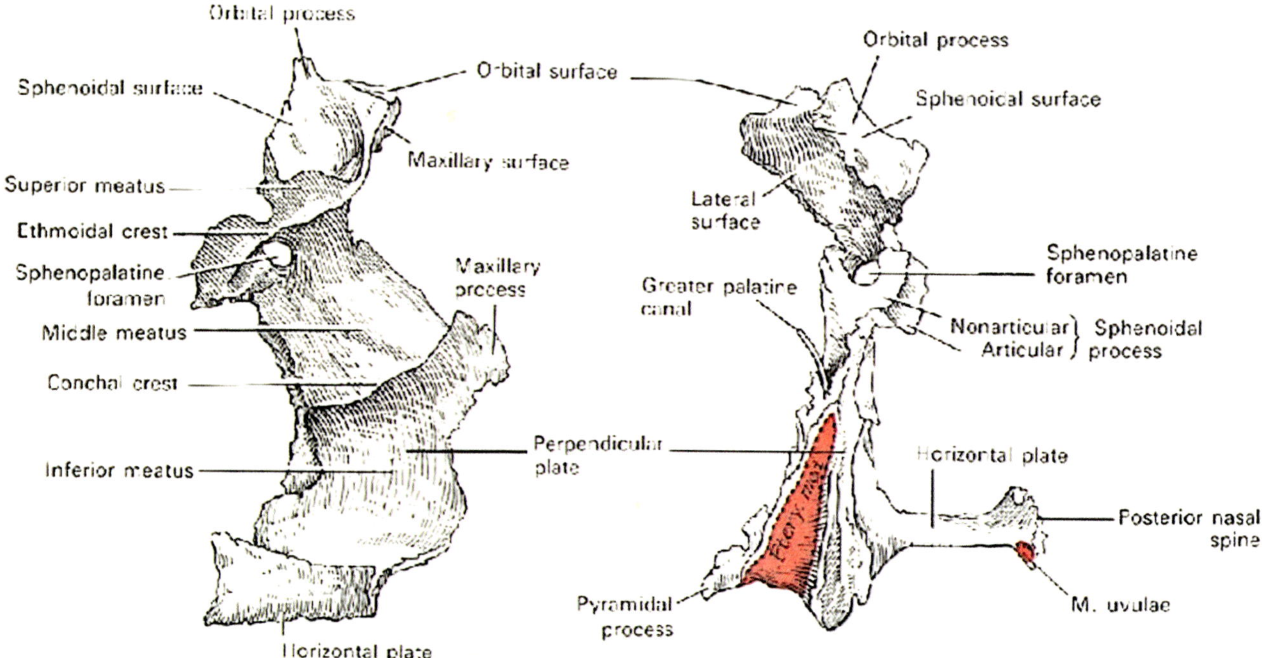

Fig. 17.44 Palatine bone. Horizontal plate develops medial-lateral and anterior-posterior: PNS is final field. Attachments: aponeurosis, (all) uvulus (medial), palatopharyngeus (lateral). [Reprinted from Lewis, Warren H (ed). Gray's Anatomy of the Human Body, 20th American Edition. Philadelphia, PA: Lea & Febiger, 1918.]

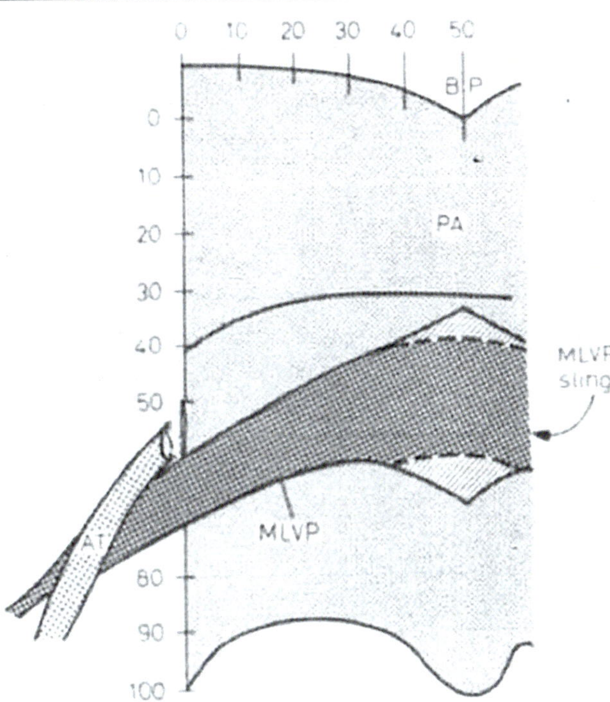

Fig. 17.45 Palatine aponeurosis position of the muscular sling. Palatoglossus deep, palatopharyngeus splits to enclose levator, uvulus is nasal. ANTERIOR: Non-muscular "bare area"

- Midline palatine raphe
- Uvulus arches over

MIDDLE: muscular

- Palatoglossus: oral
- Palatopharyngeus: oral and nasal
- Levator: nasal

POSTERIOR: glandular

- Raphé: oral
- Uvulus: nasal

[Courtesy of Michael Carstens, MD]

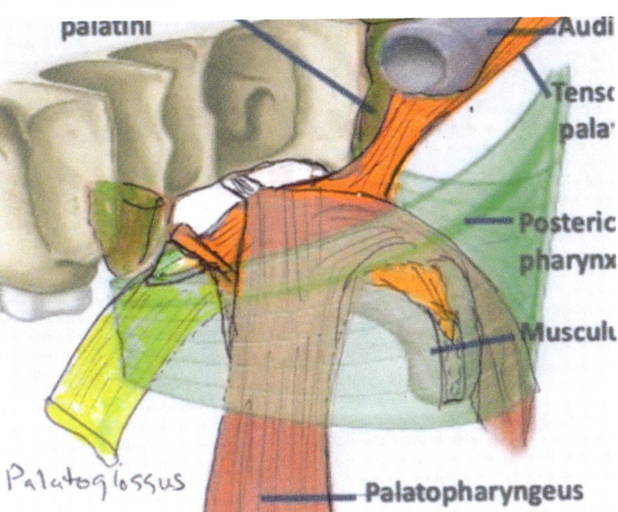

Fig. 17.46 Soft palate muscle sling occupies middle one-third of aponeurosis. Palatoglossus, palatopharyngeus, uvulus all arise from somitomere 7. PG and nasal layer of PP form slings posterior to tensor fibers. PG is deep to PP The oral layer of PP overlaps levator. The only muscle truly inserted into horizontal plate is uvulus at the PNS. Note superior constrictor is continuous with LVP and with buccinator (not shown). Key: TVP, green; LVP, orange; PG, yellow; PP, pink; SC, lemon green. [Courtesy of Michael Carstens, MD]

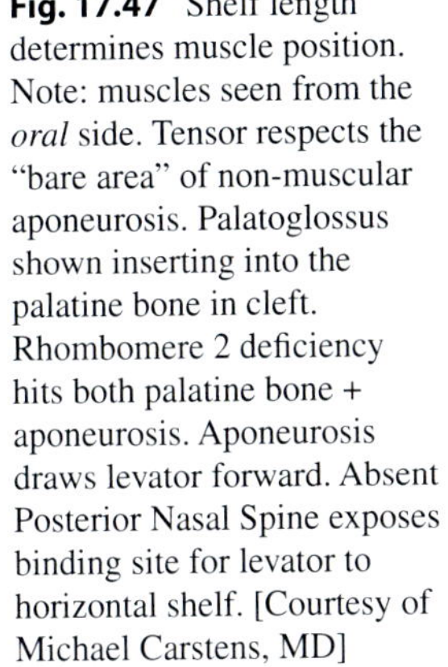

Fig. 17.47 Shelf length determines muscle position. Note: muscles seen from the *oral* side. Tensor respects the "bare area" of non-muscular aponeurosis. Palatoglossus shown inserting into the palatine bone in cleft. Rhombomere 2 deficiency hits both palatine bone + aponeurosis. Aponeurosis draws levator forward. Absent Posterior Nasal Spine exposes binding site for levator to horizontal shelf. [Courtesy of Michael Carstens, MD]

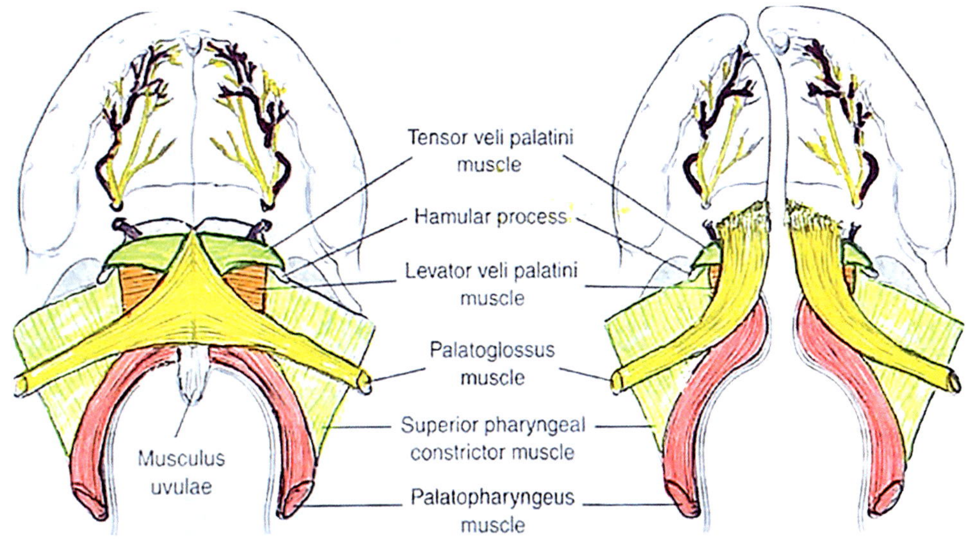

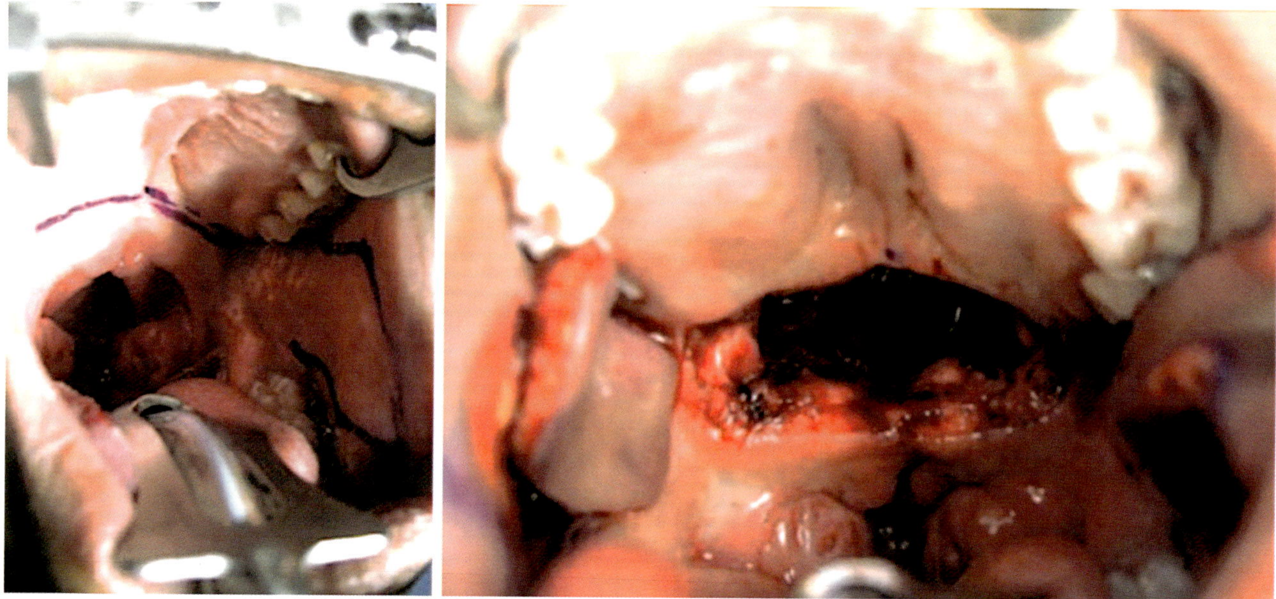

Fig. 17.48 Buccintaor interposition 1. (*Left*) Incision is marked in the hard-palate/soft-palate junction. Continuity with the left buccinator flap is seen. (*Right*) Release allows the soft palate to move posteriorly. Buccinator flaps designed on both sides with a 17-mm base at the retromolar trigone. [Reprinted from Mann RJ, Keating KC, Armstrong S, Ebner B, Bajnrauh R, Naum S. The double-opposing buccal flap procedure for palatal lengthening. Plast Reconstr Surg 2011; 127(6):2413–2418. With permission from Wolters Kluwer Health, Inc.]

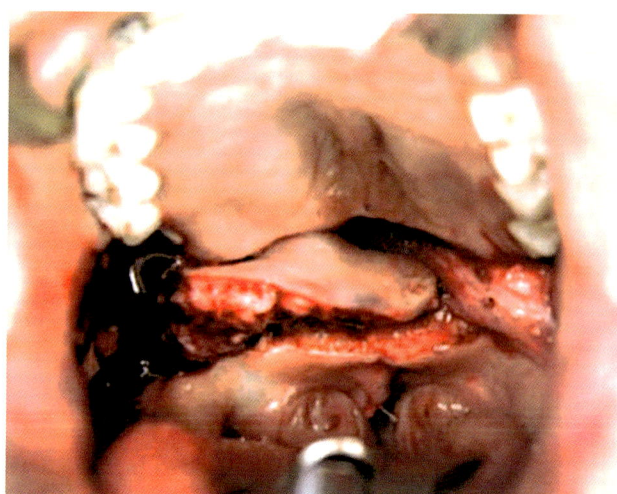

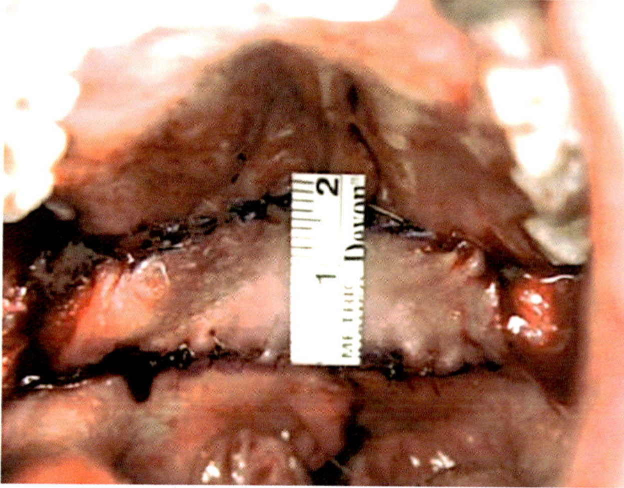

Fig. 17.49 Buccinator interposition 2. (*Left*) Buccinator flaps are elevated, and rotated into position, with one side forming the nasal closure and the other the oral closure. (*Right*) The palate is typically augmented 15–18 mm. [Reprinted from Mann RJ, Keating KC, Armstrong S, Ebner B, Bajnrauh R, Naum S. The double-opposing buccal flap procedure for palatal lengthening. Plast Reconstr Surg 2011; 127(6):2413–2418. With permission from Wolters Kluwer Health, Inc.]

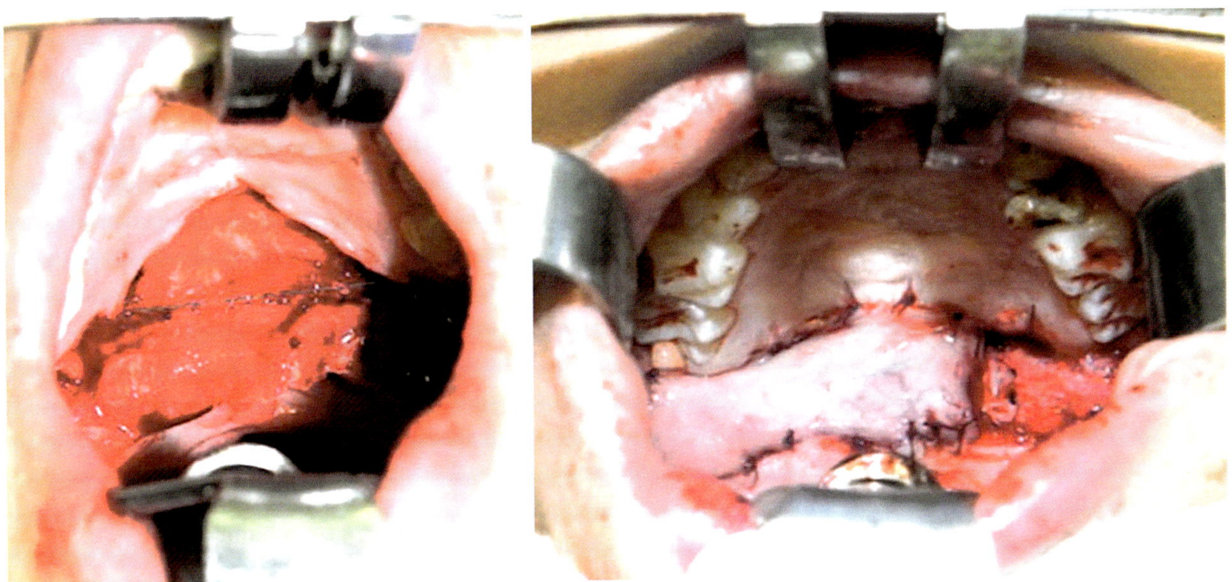

Fig. 17.50 Secondary soft palate setback with double opposing buccinators. [Reprinted from Raposo-do-Amaral CA. Bilateral buccinator myomucosal flap for the treatment of velopharyngeal insufficiency: preliminary results. Rev. Bras Cir Plást. 2013; 28(3):455–61. With permission from Creative Commons License 4.0: https://creativecommons.org/licenses/by/4.0/.]

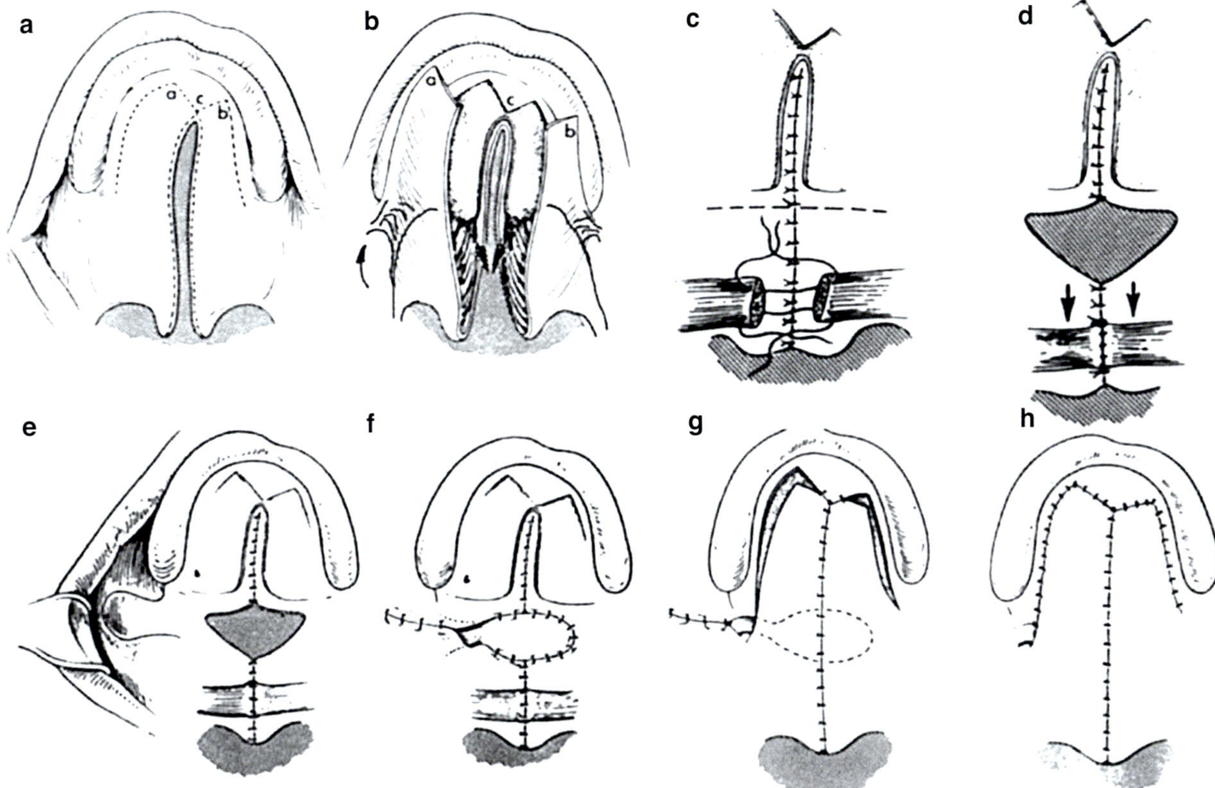

Fig. 17.51 Primary buccinator interposition (IT Jackson). Repair of a complete midline cleft of secondary palate is show. (**a**) Design of repair outlined, without retrotuberosity dissection. (**b**) Veau flaps elevated together with nasal lining. Note the benefit of AEP design. Soft palate muscles being dissected out. Division of the palatopharyngeus tendon is pointed laterally. (**c**) Closure of nasal layer of secondary hard palate completed and palatal musculature is approximated. Dotted line indicates separation of soft palate from hard palate. Incision is full thickness down to nasal mucosa. This an excellent time to assess tension. A cuff of nasal layer is retained transversely for closure. (**d**) Nasal layer is divided and elongation of the soft palate accomplished. (**e**) Design for elevation of buccal flap. (**f**) Buccal flap used to repair the nasal layer defect. This is introduced behind the great palatine vessels. Donor site closed directly. (**g**) Oral layer closed in the midline. (**h**) Closure of the lateral defects. [Reprinted from Jackson IT. The buccal flap—a useful technique in palate repair? Cleft Palate Craniofacial Journal 2004; 41(2): 144–151. With permission from SAGE Publications]

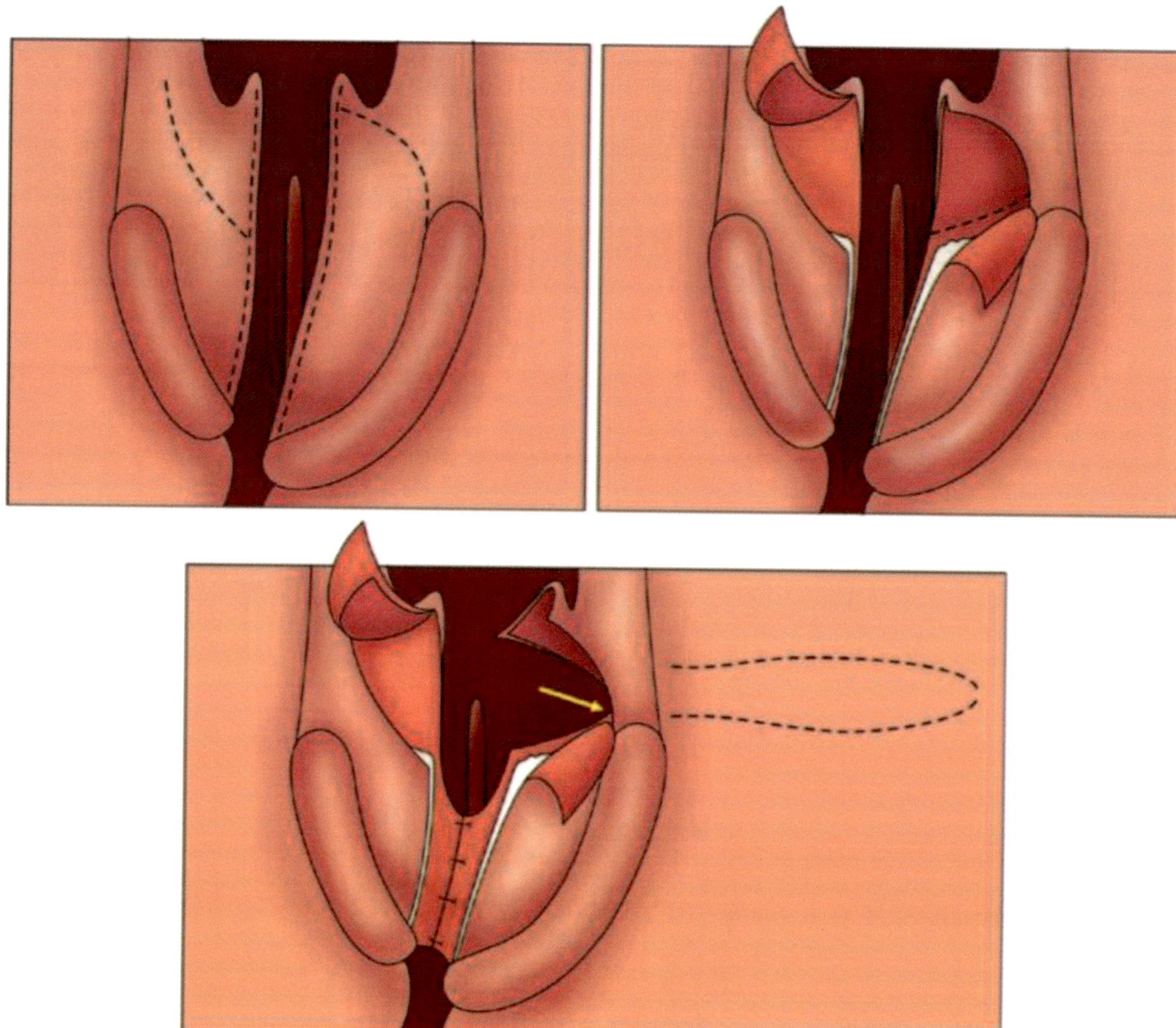

Fig. 17.52 Primary buccinator + double opposing Z-plasty (DOZ). (*Top, left*) Vomer flaps developed, Furlow-type z-plasty but oral mucosa-only is rectangular, rather than triangular. (*Top, right*): oral flaps developed. (*Bottom*) Nasal myomucosal flap (pat left side) is back-cut. Nasal mucosa-only flap (patient right side) is elevated and transposed to Eustachian tube (yellow arrow). If tension present, buccinator flap is developed. [Reprinted from Mann RJ, Martin MD, Eichhorn KC, Sierzant CG, Polley JW, Girotto JA. The double opposing Z-plasty plus or minus buccal flap approach for repair of cleft palate: a review of 505 consecutive cases. Plast Reconstr Surg 2017; 139(3):735e-744e. With permission from Wolters Kluwer Health, Inc.]

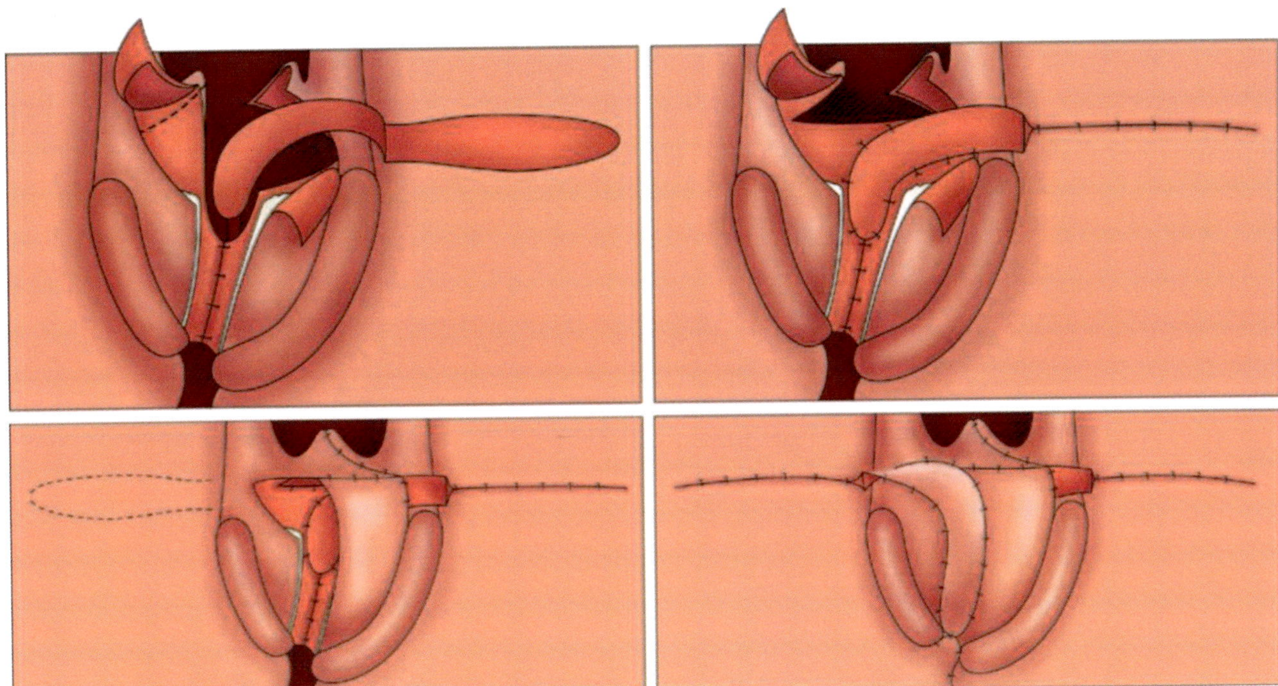

Fig. 17.53 Primary buccinator + double opposing Z-plasty (DOZ). (*Top left*) Left-sided buccinator turned into the defect with the mucosa facing the nasal surface. Note it is sutured to the nasal mucosal margin of P3. (*Top right*) Nasal mucosa-only flap advanced under tension along the backside of the buccinator flap. (*Bottom left*) Muscle sling is repaired. Oral mucosa-only z flap sutured to oral myomucosal z flap. (*Bottom right*) If tension is present on the repair for remainder of palate, a second buccinator flap is elevated and inset. Note: AEP mucoperiosteal flaps are wider and longer. They eliminate midline insufficiency and close the alveolar cleft in a 5-sided box (provided lip repair has properly closed the nasal floor.) [Reprinted from Mann RJ, Martin MD, Eichhorn KC, Sierzant CG, Polley JW, Girotto JA. The double opposing Z-plasty plus or minus buccal flap approach for repair of cleft palate: a review of 505 consecutive cases. Plast Reconstr Surg 2017; 139(3):735e-744e. With permission from Wolters Kluwer Health, Inc.]

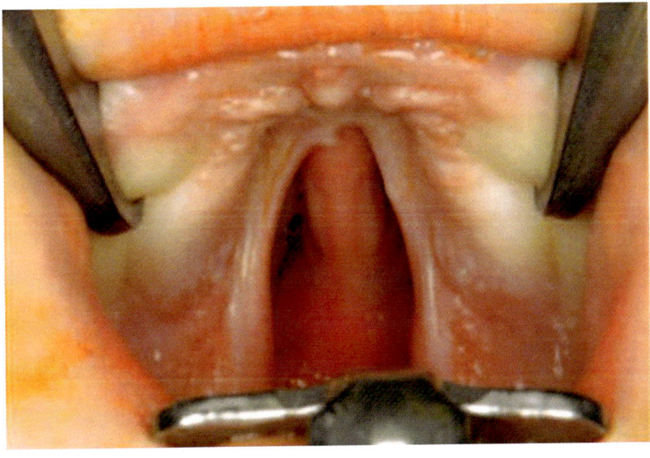

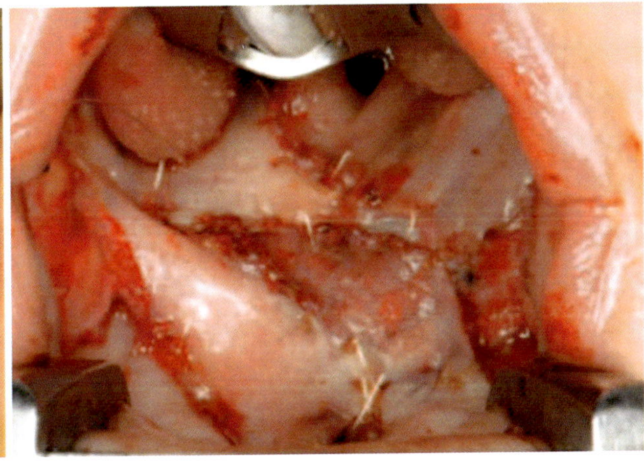

Fig. 17.54 DOZ + buccinator (Mann) 1. (*Left*) Veau class II cleft, wide. (*Right*) Immediately postop. [Reprinted from Mann RJ, Martin MD, Eichhorn KC, Sierzant CG, Polley JW, Girotto JA. The double opposing Z-plasty plus or minus buccal flap approach for repair of cleft palate: a review of 505 consecutive cases. Plast Reconstr Surg 2017; 139(3):735e-744e. With permission from Wolters Kluwer Health, Inc.]

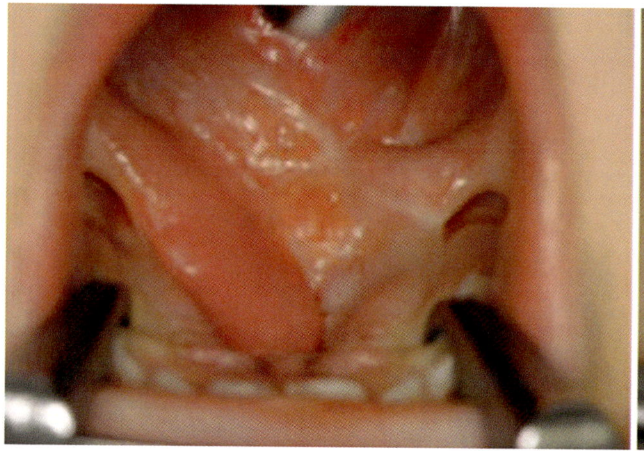

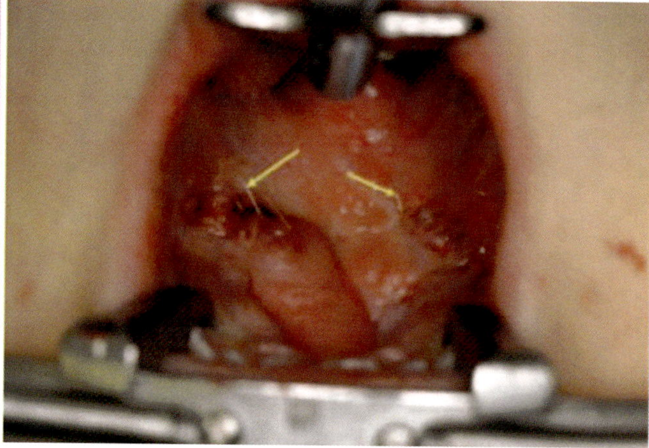

Fig. 17.55 DOZ + buccinator (Mann) 2. (*Left*) Pedicles ready for division as a secondary procedure. (*Right*) Pedicles divided at arrows. The final arrangement of soft tissues is not anatomic. [Reprinted from Mann RJ, Martin MD, Eichhorn KC, Sierzant CG, Polley JW, Girotto JA. The double opposing Z-plasty plus or minus buccal flap approach for repair of cleft palate: a review of 505 consecutive cases. Plast Reconstr Surg2017; 139(3):735e-744e. With permission from Wolters Kluwer Health, Inc.]

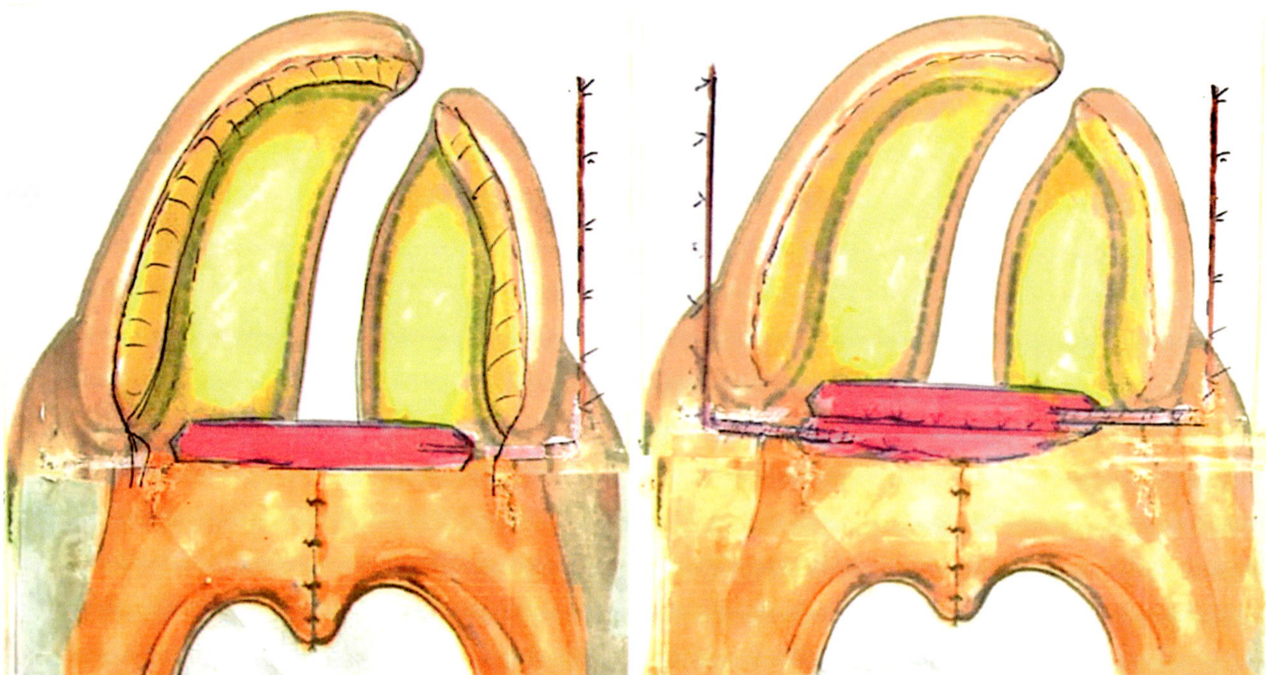

Fig. 17.56 Buccinator interposition as primary palatoplasty. Single flap = 1.0–1.5 cm retroposition. Double flap (side-to-side) = 1.5–1.8 cm retroposition. (Courtesy of Michael Carstens, MD)

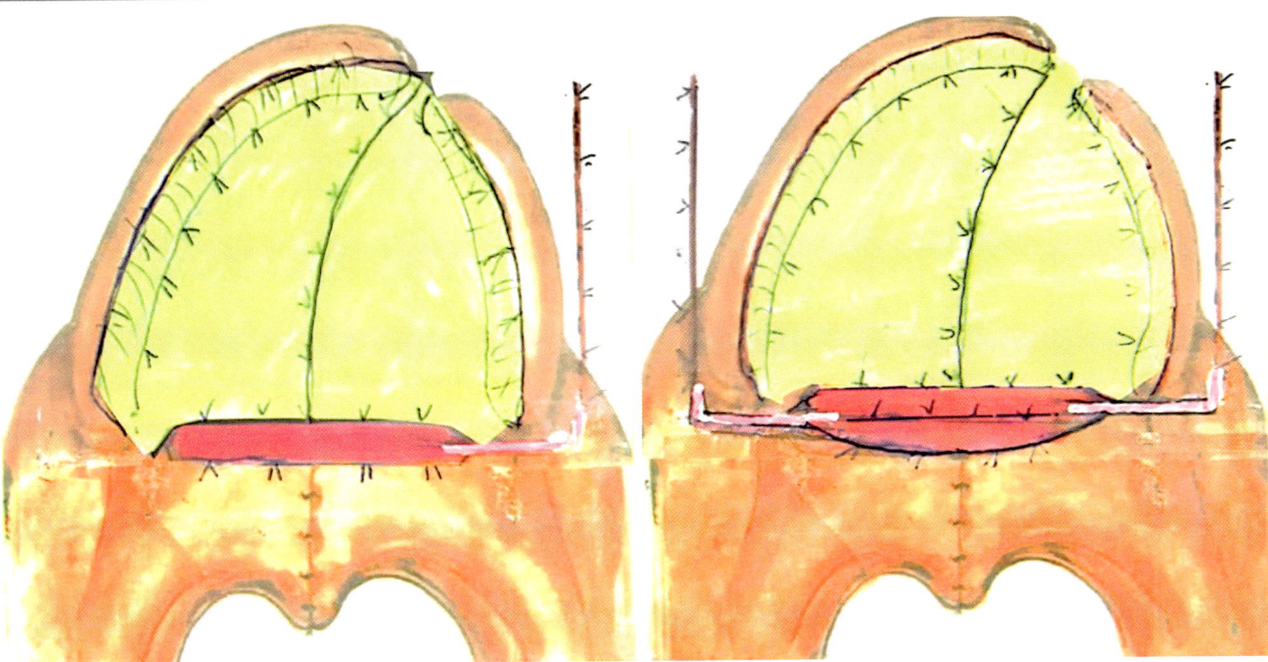

Fig. 17.57 Buccinator interposition + AEP as primary palatoplasty. (Courtesy of Michael Carstens, MD)

1 cc/kg without epinephrine). The block can be administered intraorally as described in Chap. 16. An extraoral approach can also be used. It is fast, simple, and safe. Use a short spinal needle, #19 or #20 gauge. With the distal phalanx of the index finger flexed, palpate the coronoid notch. With the needle perpendicular to the skin surface, place the needle through the skin into the subcutaneous tissue. Now drop the syringe 45° caudal and 45° posterior. Advance straightforward until you hit the back wall of the maxilla. Now, with the contralateral hand controlling the needle at the level of the skin and the ipsilateral hand controlling the syringe, "walk" the tip medially until you feel it slip around the corner of the maxilla into the pterygomaxillary fissure. The "walking" process consists of feeling the periosteum, withdrawing the needle, and advancing further along the wall until the definitive "slip" takes place. Aspirate and then administer the solution into the space.

Extraoral Approach: Anterior Buccinator (Superior or Inferior) (Fig. 17.4)

The indications for an extraoral approach are relatively few, when extensive mobilization is needed. But the details of the dissection are worth thinking about because they enable one to better understand intraoral flap harvest which might otherwise seem to be a rather blind procedure.

External exposure through the nasolabial fold generates an anterior flap that can mobilized either superiorly (for the orbit) or inferiorly (for the neck), depending on where one ligates the facial artery axis. More distal inferior defects can be addressed by preserving the facial artery to level 1b of the neck. Most intraoral defects amenable to anterior buccinator reconstruction involving the lips, alveolus, and hard palate can be readily addressed with the technically simpler intraoral approach (*vide infra*).

- Place 2–0 silk sutures about 2 cm lateral to each commissure. When placed on stretch by the assistant the lip–cheek unit is tensed, facilitating dissection of the surgical planes.
- Design the nasolabial incision, marking out the facial artery with a pencil Doppler.
- Intraoral flap design is then marked with the lip on full stretch. The parotid duct is identified, as it marks the upper border of the flap and is thus to avoid. The surgeon may wish to cannulate the duct with silastic tubing, if desired. With experience, this is seldom needed. In adults flap size can be as large as 7 × 5 cm. If more mucosa is needed, the duct should be cannulated and replanted. After marking, inject the tissues around the incision, but not behind the buccinator, using lidocaine with epinephrine 1:100,000. Conclude by scoring the mucosa with a Colorado® tip needle cautery on cutting, the setting being blend 1.
- The nasolabial incision is made to the level of subdermal fat, using loupe magnification. The facial artery is readily located with blunt dissection. Isolate it and proceed proximally. The artery will lie along the anterior border of the buccinator muscle. Always stay superficial and anterior to the artery. Leave the branches to buccinator, which are located on the posterior side of the artery undissected.

- Small hemoclips are placed on the inferior labial, superior labial, and angular branches of the facial artery, thereby directing blood flow to the buccinator.
- When facial artery dissection is complete, dissect the fatty areolar tissue separating muscle from the overlying buccomasseteric fascia, part of the DIF system. The dissection is fast and easy, terminating at the anterior border of masseter muscle. Bear in mind that you are working on plane below the SIF system so that the facial nerve remains safe. At the conclusion of this step, the muscle is freed along the critical anterior border of buccinator and from relationship to the lateral tissues.
- Intraoral dissection is now resumed. The previous mucosal incision is deepened through the muscle and into the buccal fat pad. Blunt dissection spares the parotid duct. Along the posterior border the buccal branch from IMMA and corresponding vein are cauterized or ligated. In our experience, where the flap is tethered at this point is by the vein along its inferior border. Preservation of the facial vein is the goal at this point. As one lifts out the flap from the bed, keep the vein in association with it. When the dissection is complete, the flap will fall into the mouth, based on the inferior facial axis.
- The flap is then passed outward through the nasolabial incision if required.
- The flap is ready for inset. The donor site is closed with interrupted 4-chromic.
- An alternative strategy for a superiorly based flap simply reverses the ligation of the facial artery and vein.
- Closure of the intraoral donor site up to 2.5 cm width. Wider donor site closures can be achieved by mobilizing the buccal fat pad and letting it epithelialize.

Intraoral Approach: Anterior Buccinator (Superiorly or Inferiorly Based) (Figs. 17.25, 17.26, 17.27, 17.28, 17.29, 17.30, 17.31, 17.32, 17.33, 17.34, 17.35, 17.36, 17.37, 17.38, 17.39, 17.40 and 17.41)

Anterior buccinator flaps can be harvested with either a superior or inferior pedicle of the facial axis. The principles are exactly the same. The principle reconstructive sites are similar to those of the extraoral anterior buccinator, the main difference being a more limited range of access but a simpler procedure. *Flap length is adequate to reach the contralateral canine of either maxilla or mandible.*

- Intraoral flap design is marked out with methylene blue. The location of the facial artery is determined with a Doppler and marked.
- With the lip on stretch the mucosa is incised circumferentially.
- The rectangle of muscle is incised along three of its four sides, the anterior muscle being spared for the moment. Gentle spreading along the superior, inferior, and posterior borders with a baby right angle clamp places one in the correct plane of the buccal fat pad borders.
- The flap is elevated from posterior to anterior.
- Staying anterior to the Doppler markings the anterior border of this wispy muscle is parted by gentle spreading until the facial artery and vein are visualized. Perforating vessels of the facial to buccinator should be left undisturbed.

Intraoral Approach: Posterior Buccinator

This flap is ideally suited to reconstruct the soft palate (our emphasis here), base of the tongue, lateral tongue, and tonsillar fossa

- Mark the buccal with Doppler.
- Incise the buccal mucosa and buccinator. One can transect the posterior border or not initially. As the buccal artery enters the muscle about one-third of the distance to the commissure, posterior dissection safe done bluntly.
- Elevate the flap in the areolar plane from anterior to posterior.
- The posterior border is freed and the flap is rotated. Additional mobility can be gained by spreading as the buccal artery, like the facial, is tortuous and will lengthen.
- When passing behind the molar, create a trough for the pedicle. I don't like to make tunnels. Leaving the site open avoids problems; it will close rapidly.

Clinical Applications of Buccinator Myomucosal Flaps, Anterior and Posterior

Terminology (Figs. 17.42, 17.43, 17.44, 17.45, 17.46, 17.47, 17.48, 17.49, 17.50, 17.51, 17.52, 17.53, 17.54, 17.55, 17.56 and 17.57)

Over the years, a plethora of terms has arisen to describe the buccinator flap. Rahpeyma proposed a simplification based on pedicle location. The diagrams and cases of Fagan are very clear. Using vascular supply, the classification of buccinator flaps simplified down to a simple vascular algorithms:

- Anterior [5]
  - Facial artery inferior: pedicle or island
  - Facial artery superior: pedicle or island
  - Facial artery bipedicle
- Posterior [4]
  - Buccal artery: pedicle or island

Anterior Buccinator Flap

Despite the multiplicity of skin flaps based on the facial artery, its relationship to the underlying buccinator has been rather obscure. Branches perfusing the muscle were first described by Bourgery in 1851 but their ability to support an independent composite flap of buccinator and mucosa was not reported until 1991 [5].

Anterior buccinator flaps are best placed for defects of the anterior maxilla/mandible and lips. Intraoral anterior flaps cover as far as the contralateral canine. The only reason to consider an external approach for a single buccinator flap is if coverage distal to the contralateral canine is needed, when the flap needs extraordinary mobility, i.e., for the floor of the mouth of the flap is to be inset outside the mouth, e.g., the orbit [27–29].

Numerous applications of anteriorly based intraoral flaps that have been described with Zhao are credited with working out the superior facial pedicle flap. Vermilion resurfacing and lip reconstructions using anterior facial artery flaps from either direction are described by Pribaz. Bilateral anterior flaps provide successful coverage for vascularized free fibular bone graft to the maxilla [30]. Inferior-based flap with Zhao design has a long pedicle and can achieve coverage of lateral pharyngeal wall and base of the tongue [31].

Posterior Buccinator Flap

The idea of using a posteriorly based buccal mucosal flap is not new. Culf published use of this flap in 1974 for velopharyngeal incompetence. Kaplan used posteriorly based mucosa as nasal lining for soft palate repair in 1975. But the deliberate harvest of a posteriorly based mucosal flap based on buccinator muscle was first reported by Bozola in 1984 and subsequently with cadaver dissections in 1989. Initial uses were for palatal lengthening in primary cleft surgery, palatal fistula, and reconstruction post tumor resection. Since that time, posterior buccinator flaps, either singular or paired, have been used for a wide variety of defects of the palate (hard and soft), floor of mouth, and tonsillar fossa [32, 33].

The greatest worldwide impact of the posterior buccinator flap lies in its application to congenital palate defects

- Posterior buccinator has the ability to reach forward almost, but a very important not quite, to the incisive foramen.
- It can provide primary midline closure of the secondary hard palate.
- Fistulae of the hard and soft palate are accessible to this flap.
- As a transverse interposition, posterior buccinator adds length to the soft palate and constitutes a developmental field reassignment solution for velopharyngeal insufficiency.

Buccinator Flaps for Cleft Palate Surgery

Padgett performed what was in essence the first buccinator flap to the palate in 1936 but the vascular anatomy was unknown [34]. Maeda was unaware of Bozola's 1984 work in Brazil. Citing previous work on mucosal flaps by Cronin et al. [35], he recognized the potential for a posterior pedicle flap to provide additional nasal mucosa, a modification of Dorrance's hard palate "pushback" applied to the soft palate [36]. What's more, publications by Millard [37], and Kaplan [38] convinced him that a musculomucosal buccal flap could resolve two dilemmas: abnormal insertion of the levator sling into the nasal mucosa at the posterior border of palatine bone; and the problem of postoperative fistula at the junction of the hard and soft palate. In 1987, Maeda reported his T-shaped buccal flap procedure in 30 patients. Maeda noted that the flap could not resolve an alveolar cleft. The variations of his cases are of interest but hard palate reconstruction using AEP, either primarily or secondarily, obviates the need for more convoluted designs.

Hard Palate Procedures

Posterior buccinator has two main uses for reconstruction of the secondary hard palate: fistulas and wide midline defects. Midline breakdown is a rather simple situation requiring one or two flaps as the situation warrants. The reach of a single flap is well illustrated by Abdaly. Double-barrel flaps have been able to cover large eccentric fistulae, incorporating additional mucosal flaps as necessary [39].

In primary repair of wide isolated secondary cleft palate, use of the buccinator may either make a lot of sense or constitute a misapplication depending on one's point of view. Based on the old Von Langenbeck design, simple incision along the medial cleft border minimizes mucoperiosteal flap elevation. One simply elevates the edges and drops in the buccinator. But the feared alveolar growth consequences can be dealt with using AEP that brings the hard palate flaps together in the midline. The drawbacks to this simplistic scheme are not replacing like with like. On the other hand, a turnover buccinator sutured to the nasal lining may resolve a nasty problem of nasal mucosal deficiency. Although the mucosae are dissimilar (respiratory versus oral), they are a better match outcome.

- Buccinators are best for hard palate fistula rescue procedures.
- Primary midline repair should be reserved for selective circumstance.

Soft Palate Procedures

Posterior buccinator is useful for two types of secondary situations: posterior fistula and velopharyngeal insufficiency.

Fistulae can be quite nasty. They occur because one underestimates the severity of r2 defect in two anatomic sites: the horizontal plate of the palatine bone and the amount of tissue in the anterior one-third of the palatine aponeurosis. The nasal side is always tighter than the oral side. Fistula repair is an ideal indication for the posterior buccinator flap.

- The choice of single versus double buccinator depends on the size of the defect and whether or not one wishes to "sandwich" the two flaps to minimize raw surface and consequent contractions.
- In point of fact, buccinator flaps epithelialize well in 2 weeks: to date, the contraction issue has not been proven to be significant.
- Contraction does not seem problematic for exposed buccal fat pad grafts.

Veloharyngeal insufficiency (VPI) is present in 20–30% of cleft palate surgeries. Buccinator interposition palatoplasty is very effective at retropositioning the soft palate. Single flaps in the midline reliably lengthen the midline 1.5 cm. When double flaps are transposed, one in front of the other (side-to-side), they overlap in the midline, creating a diamond-shaped reconstruction with expansion of up to 2 cm. As we shall see, the lessons learned using buccinator interposition palatoplasty for VPI have profound implications for how cleft palate will be managed in the future. Let's look at the clinical results in some landmark papers, considered in chronological order. These are chosen for relevance to illustrate various aspects of the repair. Additional studies that appear over time will help our understanding of its consequences more nuanced.

Buccinator interposition also represents a conceptual breakthrough when compared with operations based on pharyngeal manipulation

- Both the pharyngeal flap and sphincter pharyngoplasty are non-embryologic and non-anatomic as they disrupt otherwise normal anatomic structures.
- They have complications as well, including sleep apnea, mouth breathing (with consequent effects on dentition), and hyponasal speech. Sleep studies done at one-year post VPI treatment with pharyngeal surgery demonstrate obstructive sleep apnea in over 80% of patients [32].
- Pharyngeal manipulation procedures do not address the real cause of VPI, which is the deficiency of r2 mesenchymal structures: P3 palatine bone and anterior one-third of the palatine aponeurosis.
- Instead of addressing the developmental field defect and of replacing the deficit with like tissue, the pharyngeal flap and sphincter pharyngoplasty operations create distorted anatomy and abnormal physiology.

Buccinator Interposition Palatoplasty for VPI

Bozola [4] This paper is valuable for its description of technique. The series consists of 24 primary and 12 secondary palate repairs. No long-term follow-up for VPI or comparison of VPI rates using other techniques is given.

Hill [40] followed 16 patients with secondary VPI prospectively over 3 years. The first flap was sutured with mucosa facing the nasal side and the second flap was reversed. Of 16 patients, 15 (93%) had significant improvement in nasality, with 14 of 15 achieving complete normality. The two less-than-optimal cases had postoperative complications. The authors cite three advantages of Buccinator Interposition Palatoplasty (BIP) over posterior pharyngeal flap and sphincter pharyngoplasty procedures.

- The anatomic response to the actual gap is directly addressed.
- Soft palate muscle anatomy is preserved (versus inset of a pharyngeal flap).
- The pharynx is left unviolated.
- Technique is easy compared with pharyngeal access.

Mann [41] performed a retrospective review of bilateral opposing buccinator flaps placed for post-palatoplasty VPI. The technique used was that of Hill. Sandwich flaps do not augment as much as side-to-side flaps. The average gain in length was 1.5–1.8 cm. Pre-buccinator intelligibility improved from 65.4 to 95.5% post-procedure ($p < 0.0001$). Resonance (nasality) improved as well ($p < 0.05$) (Figs. 17.52 and 17.53).

Hens [42] used a sandwich technique in 32 cases of post-palatoplasty VPI. A total of 81% were considered definitive, requiring no further surgery. Nasality improved ($p < 0.005$). Lateral videofluoroscopy showed physical elimination of the velopharyngeal gap in 77% of patients. Nasality improved interestingly, the mean setback using the sandwich technique was only 7.5 mm, indicating considerable differences in the extent of the flap harvest. We know now that flap widths of 2.5 cm can obtained with the cheek donor site closed primarily.

Dias, reporting from Ceylon [43], evaluated speech results after buccinator repair for VPI in 34 patients divided by age, those less than or greater than 8 years. Younger patients achieved significant reduction in hypernasality ($p < 0.05$). Reduction in hypernasality was only 60% in the older group. All patients achieved improvement in consonant production. Of note, at 1-year follow-up, the younger group showed additional palate lengthening.

Wan and Zhang [17] described a useful application of a single posterior buccinator island flap in secondary palate cases involving children aged 5–9, in which the goal was to achieve maximal retroposition of the levator sling from the nasal mucosa, while leaving the latter structure intact. Levator was elevated off the nasal mucosa using microscopic technique as described by Sommerlad. The oral mucosa was thus retropositioned but not that of the nasal side. Embryologically, the true defect in the palatine aponeurosis involves the nasal mucosa. Follow-up here is short, only 1 year, and speech patterns are already substantially engrained. It would be of interest to see if VPI reappeared in adolescence.

Virtually all previous of posterior buccinator reconstruction have relied on bilateral flaps under the assumption that lining should be provides on both the oral and nasal sides. Robertson [44] was the first to evaluate the use of singular buccinator flaps for VPI in 20 cleft patients. The flaps used were 1.5–2 cm wide by 2.5 cm long. They were inset with the mucosa directed to the nasal surface. A total of 50% of the patients had an additional procedure, which makes the conclusions less straightforward. Nonetheless, no repeat fistulas occurred; of 20 patients, speech improved in 17 (77%), hyperresonance was corrected in 18 (90%), and VPI was eliminated in 18 (90%). All the parameters were significant at $p < 0.00$. Over time the flap dimensions were maintained, as previously with endoscopy by Freedlander and Jackson, in which flaps were readily identified visually on the palatal surface in only 3 of 50 patients.

Logjes and Breugem [45] retrospectively evaluated 42 consecutive patients treated for VPI with a single buccinator interposition palatoplasty. Nasality, intelligibility, and adequacy of velopharyngeal closure were evaluated at anywhere between 1.2 and 4.9 years. Optimal speech was achieved in 35 (83%) of 42 cases, with 7 patients requiring reoperation. In 29 patients evaluated by nasopharyngoscopy, 24 (83%) showed improved closure. These results were similar to those reported previously by Hens (with a success rate of 81%).

Buccinator Interposition Palatoplasty in Primary Cleft Palate Repair

For historical purposes, the use of a posteriorly based cheek flap for primary cleft palate repair was first used in India by Mukherji in [46]. We will here examine two uses of single buccinator interposition as applied to initial cleft palate management.

- Embryologic soft palate dissection, simple setback, and buccinator interposition as described by Ian Jackson
  - Setback extent is determined by mechanical release: "cut-as-you-go"
- Four-flap double-opposing Z-plasty (DOZ) of the soft palate and buccinator interposition as described by Robert Mann
  - Setback extent is determined by Z-pasty flaps

Jackson [47] reports a series of 156 non-syndromic cases operated between 1989 and 2002. In all cases where primary palate cleft was present, it was closed at the first surgery (cheiloplasty). All flaps were *single* and the inset with the mucosa placed into the nasal cavity. The median age at surgery was 6.2 months and the median follow-up time was 5.8 years. With experience, both the fistula rate and the VPI rate dropped, the former from 3.6 to 0.7% and the latter from 8.9 to 3.5%. Speech outcomes were striking. Velopharyngeal closure adequacy was 91%, resonance was normal in 91%, articulation was normal in 97.9%, and speech quality was judged good in 89%. There were no cases of hyper- or hyponasality.

- Jackson's diagram should be studied with care. Note how tension-free soft palate repair is accomplished by releasing nasal mucosa thereby re-creating the defect in the anterior palatine aponeurosis.
- The fundamental relationship between the levator sling and its mucosa is normal. The deficiency site lies ahead in the muscle-free anterior one-third of aponeurosis. Thus, radical elevation of muscle from its natural mucosal insertion is irrational. All the problems are anterior.
- No double-opposing Z-plasty is needed. You don't try to get blood out of a stone, by making backcuts through normal muscle. Length is achieved by replacing the defect in the anterior one-third.
- Design provides "cut as you go" in which the soft palate defect is assessed before harvesting the buccinator. Double-barrel side-to-side buccinators can be placed as needed.
- The AEP/buccinator design reserves one buccinator for future use, should it be needed for either additional setback in the treatment of a fistula.

The issue of how buccinators can be used with the double opposing Z-plasty is addressed definitively by Mann's report of his 27-year experience, with a total of 505 consecutive cases using the Furlow repair, of which 319 (65%) required a buccinator flap [48]. The primary goal of this sequence was to achieve length and improve outcomes for speech. Careful attention should be given to this important study which we shall contrast with Jackson's results.

A little editorial license is required here. For those of us who are plastic surgeons, the Z-plasty comes a tried and true method to gain length at the expense of width. It works great with skin, where different angles between the limbs of the Z's generate different lengths in the central axis. The 60° angles generate a 1.75-fold increase. Cleft surgeons from the get-go had to wrestle with the issue of the short palate. All aesthetics aside, the most devastating consequence of clefts is unintelligible speech, a source of social (and economic) isolation.

Palate function depends on structure. Recall that the speech mechanism is that of a pinball machine in which the hard tissue platform determines to position of lever arms attached to its posterior shelf. The "flippers" move backwards to make velopharyngeal closure, but the ability of the flippers to accomplish this goal is utterly dependent on their attachment to the back of the "box," i.e., the P3 palatine bone field. VPI results from either a short box or defective flippers (usually the former). Chapters 14 and 16 are intended to explore the developmental reasons behind this phenomenon.

AEP flaps have a good developmental rationale and are highly versatile but they are not a solution for a foreshortened P3 bone field and anterior palatine aponeurosis. Without an understanding of that embryonic deficiency zone, the sole means to achieve length with the available velum is to carve it up into chunks. No question that the Z-plasty works, but at the price of disrupting normal anatomy, reducing blood supply, and laying down scar into what was a previously supple structure. And like any apparently simple solution to a complex problem, there is an inevitable "catch." The wider the cleft the more one struggles to get length and still achieve closure without creating a fistula. Failure of a Furlow repair in a wide cleft results in a reconstructive nightmare [48]. There is no "Plan B."

Mann's series included 16 submucous clefts, which were excluded. A total of 170 standalone DOZ procedures were performed. He added buccal flaps to his double-opposing Z-plasty in 319 patients whenever he determined intraoperatively that the standard Furlow repair could not be completed with a tension-free closure, *which was 65% of the time* [48]. Follow-up was lengthy, 7.7 years.

The overall outcomes for both types of repair regarding speech were similar: presence of fistulae, nasal resonance, intelligibility, and the need for secondary surgery for persisting VPI. Speech evaluations were carried out at 5 years by cleft-trained therapists. Normal speech was reported in 90%. Nasal resonance scores were equal as was the need for secondary surgery: 7.1% for the combination versus 5.4% for the Furlow alone [48].

Fistulas were developed in 30 (6.13%) of 489. The postoperative fistula was associated with increased cleft width. Fistulas were more common in the DOZ + buccinator group (8.78%) versus the DOZ group alone (1.18%) with strong significance ($p < 0.0009$). The reason is simple because the buccinators were required for excessive width in the first place. When clefts were graded by width (small, <5 mm), medium (5–10 mm), and large (>10 mm), there were significantly more wide clefts in the combination procedure (55%) than in the DOZ alone (8.2%; $p < 0.001$).

Let's take a close look at the technical execution of the Furlow double opposing Z-plasty + buccinator procedure

- Mann's seven-step diagrams should be studied with care; see Figs. 17.52, 17.53, 17.54 and 17.55.
  - Note that AEP flaps are not included in the design, thus transverse closure of the hard palate oral layer is compromised.
- Oral side Z-flaps are 2.5 cm in length.
  - Oral side mucosa-only Z-flap is designed in continuity with the non-cleft hard palate mucoperiosteal flap. It is more rectangular. It is cut first.
- Nasal side Z-flaps are 1.5 cm in length.
  - Nasal side mucosa + muscle Z-flap is cut first from the non-cleft side.
  - A cuff of mucoperiosteum is left along the posterior border of P3.
- Step 4 is critical.
  - If nasal side mucosa-only Z-flap (cleft side) stretches across the defect to the level of the Eustachian tube without tension, DOZ alone can be done safely.
  - If tension noted, buccinator flap must be added.
- Buccinator from non-cleft side is inset with mucosa facing nasally.
  - If inset is *not* symmetric, the distal margin is inserted into the midline where it incorporates the vomer flap. Note the L-shaped inset.
- Soft palate Z-plasties closed, nasal layer first, then oral layer.
  - If tension noted on the oral side, a second buccinator flap is added. Note the L-shaped inset.
- Alveolar closure with mucoperiosteal flaps is *not* part in Mann protocol.

What are the biologic lessons of the Mann protocol? First, it illustrates the importance of the developmental field defect of post-P3 zone. DOZ alone does not reliably address this problem. It fails 65% of the time. Second, defects of the palatine aponeurosis are corrected with buccinator. Third, speech results in either scenario are the same. Addition of the buccinator does not enhance the setback, it just fills the hole. Fourth, the fact that it resorts to a second buccinator flap to achieve midline hard palate closure demonstrates the limitations of conventional hard palate flap design in generating enough width to close the cleft.

- **In sum**: the extensive and well-documented 27-year experience of DOZ alone versus DOZ + buccinator demonstrates the importance of recognizing and addressing the developmental field defects at the primary surgery, not only in the soft palate but also in the maldistribution of mucoperiosteum with consequences of primary palate closure and long-term growth.

Comparison of Buccinator Interposition Protocols: DFR Versus DOZ

Study Characteristics

Length of study
- DFR: 12 years
- DOZ: 27 years

Number of patients (non-syndromic)
- DFR: 156
- DOZ: 505, with 303 followed up long term for speech/ secondary surgery

Age at time of repair
- DFR: 6 months
- DOZ: 8–12 months

Mean follow-up
- DFR: 5.8 years
- DOZ: long-term >5 years in 303

Outcomes

Velopharyngeal status (adequate closure)
- DFR: 91.1% adequate, 6.2% marginal, 2.7% incompetent
- DOZ: N/A (can be inferred from below)

Nasality
- DFR: 91.1% of all patients with hypernasality had an inadequate VP port.
- DOZ: 93.4%.

Articulation
- DFR: 97.7% early age of repair.
- DOZ: N/A.

Speech quality
- DFR: 89%.
- DOZ: 90%.

Postoperative VPI (age >3 years with videoendoscopy)
- DFR: started at 8.8%; finished at 0.7%; DFR = less scarring.
- DOZ: 6.6%.

Fistula
- DFR (all sizes): overall 3.6%, later years 0.7%.
- DOZ: 6.1% overall; 8.7% in buccinator group (wide clefts), 1.2% (narrow clefts).

Need for secondary VPI surgery (compared with 10–36% incidence reported)
- DFR: 7% (8/113).
- DOZ: 6.6% (20/303).

Biologic Parameters

Soft palate dissection characteristics
- DFR: non-invasive, maintain embryologic relationship of muscle sling to the aponeurotic, preservation of blood supply and innervation, minimal scarring, reserves option for a simple Z-plasty as a secondary procedure
- DOZ: invasive, bridges are burned

Soft palate retropositioning
- DFR: Setback is simple and transverse, extent determined by mechanical release: "cut-as-you-go".
- DOZ: Setback is complex, determined by size of Z-pasty flaps.

Effect of scarring
- DFR: scarring minimal, soft palate tissue expansion, less residual VPI.
- DOZ: extensive scarring, length of the soft palate becomes "locked in".

Buccinator flap geometry
- DFR: straightforward, reconstructs the entire aponeurotic defect with a single-type of tissue, placed transversely.
- DOZ: buccinator is L-shaped, reconstructs aponeurotic defect partially.

Use of a second buccinator flap
- DFR: is not required, sandwich is not required (epithelization is rapid). If double flaps are placed side to side, additional setback is achieved. Hard palate closure relocates stem cells and closes the primary palate. Available for secondary procedures, recurrent VPI, Le Fort I advancement.
- DOZ: second flap is required due to design issues and in hard palate closure is inadequate. Bridges are burned.

Compatibility with secondary pharyngeal procedures
- DFR pharyngeal flap YES, sphincter YES,
- DOZ: pharyngeal flap NO, sphincter YES.

Hard palate repair: compatible with delay
- DFR: YES
- DOZ: NO

Hard palate repair: alveolar cleft closure with stem cell transfer

- DFR: YES
- DOZ: NO

DFR and DOZ: How Do They Stack Up?

Both repairs address the embryologic defect of the anterior aponeurotic deficiency site. DFR does so in a deliberate fashion. In DOZ, it happens by default. Both repairs achieve a similar retropositioning of the muscle sling, but DFR is more flexible and its lack of scarring promotes tissue expansion of the soft palate. Critical speech outcomes, including the need for secondary surgery, are similar. Postoperative sequelae, such as fistula and residual VPI were more common in DOZ due to the complexity of the repair. Finally, and of critical importance, DOZ is a more restrictive procedure which limits options for secondary surgery. DFR is anatomic, simple, and flexible.

- DOZ offers no selective advantage over DFR and has numerous technical drawbacks, especially if revision is needed.

Protocol for Buccinator Interposition Palatoplasty

Personal Perspective

In the confines of these four Chaps. (14, 15, 16 and 17), we have explored the biology of the hard and soft palates. The concept of developmental field reassignment (DFR) surgery applies to oral structures as well as to the face. The four D's of embryonic pathology, deficiency, division, displacement, and distortion, must be identified and addressed using reconstructive principles that conserve neuroangiosomes and achieve transfer of biologically active stem cell populations into their proper anatomic relationships. DFR for the hard palate corrects critical deficiencies in r2 bone fields such as the alveolus, palatine bone complex, and vomer. DFR for the soft palate involves the restoration of length and lining by reconstruction of the anterior aponeurosis and the proper repositioning of the muscle sling.

Soft Palate and Hard Palate Have Differing Biologic Imperatives

The benefits of early soft palate repair are very clear in the literature. Speech development is optimized, especially with regard to articulation. Moreover, soft palate repair has an orthopedic role in that it helps to reverse the cleft-initiation displacement of the posterior alveolar elements, i.e., the hard palate cleft narrows over time. Thus, some form of soft palate repair at 6 months makes eminent sense. This procedure can be carried out in the immediate period before or after cleft lip–nose repair, or as a combined procedure. It is best done in isolation from hard palate repair, which is performed later.

Hard palate repair must take into account the needs for proper dental development and facial growth. Alveolar extension palatoplasty optimally repositions the mucoperiosteum, allowing for primary control of the alveolar cleft, if present. It can be accomplished at 18 months but, depending on one's philosophy, can be delayed until age 4 when combined with bone grafting (see Talmant and Lumineau reference from Chap. 18).

DFR Soft Palate Repair: Buccinator Interposition Palatoplasty

The bottom line for soft palate repair is good speech. We know, based on recent historical controls, that soft palate repairs have a failure rate (as defined by postoperative velopharyngeal incompetence) of 10–36%. The reason for this lies in use of techniques that fail to reconstruct the speech mechanism. Deficiencies in the P3 bone fields and anterior aponeurosis are always present. The flippers of our pinball machine will not work if their hinge points remained attached to a foreshortened platform. How can we make it longer? There are two points of attack.

- Distraction of the palatine bone backward is theoretically clean but does not address the aponeurosis defect. It is also completely impractical.
- Soft tissue interposition can compensate for both the missing bone stock as well as aponeurosis. It is simple to execute.

A simplistic solution would be to do all primary repairs with a buccinator flap.

But our desire to avoid VPI misses those remaining cases (the majority), in which standard repairs achieve adequate velopharyngeal closure. Can we rationalize using DFR soft palate repair as a universal primary procedure? I think the answer is a qualified yes, if the soft palate repair is either: (1) short or (2) under tension.

Assessment of velopharyngeal length at primary repair is a clinical guess. Mukherji [46] reported radiographic studies in 35 children 1.5–2.5 years of age, to determine the distance between the soft palate and the posterior pharyngeal wall. When this exceeded 5 mm (57% of the series), the child was considered at risk for speech problems. There is no good standard to determine the age-dependent length of the hard palate platform. If there were, we could measure the distance from the incisive foramen to the back of the palatine shelves and make the decision. In the absence of metrics (and a flexible wire) our best guess lies in the assessment of the horizontal plate of P3: how severely is it deficient? Another test comes as we dissect out the nasal lining. Simple traction will determine whether it is under tension. So, when in doubt, follow the DFR principle: replace the deficit. There is virtually no downside to doing so.

- Donor site morbidity is minimal. The remaining buccinator muscle continues to function.
- The buccinator flap is forgiving and flexible. With time, if the circumstances warrant, muscle to achieve velopharyngeal closure will drive a process of tissue expansion.
- Hyponasality after buccinator interposition is virtually non-existent (if present, it will improve with additional growth).

DFR Hard Palate Repair: Alveolar Extension Palatoplasty

Our final decision point for the palate regards the timing of AEP. At 18 months the bone stock is good and subperiosteal dissection is simple. A five-sided box is created for the alveolar cleft. There is great security in this construct because the walls of the alveolar "box" are stable. You just wait 3 years for the bone graft at which time the sixth side is completed. The problem is alveolar cleft collapse and the soft tissue resistance.

- DFR lip repair at 6 months will ensure closure of the roof of alveolar cleft.

By 18 months the orthopedic effect of veloplasty will narrow down the hard palate, minimizing tension. An 18 months wait for AEP is well tolerated by families.

- Speech results are unaffected.

Lumineau and Talmant present an alternative, and very compelling, protocol based on alveolar grafting at age 4. It is completely compatible with AEP. The rationale for the delay is dental arch control. Under the influence of the lip repair the arches will have undergone some degree of collapse. Lumineau's simple intraoral wire device is applied. In a matter of weeks, it acts like a spring to open up the alveolar space to full length. Grafting and hard palate repair are completed concomitantly. The results, in terms of dental arch development, are outstanding; the best I have ever seen.

- The protocol is very technique-sensitive. It should be widely taught.

Conclusion

In 1956, at the end of the bus boycott in Montgomery, Alabama, the Rev. Dr. Martin Luther King Jr. paraphrased the words of an 1853 sermon by abolitionist preacher Theodore Parker: "The arc of the moral universe is long, but it bends toward justice." In a similar way, we see over time the slow integration of biologic truth into clinical practice. Care for children with clefts is a good example of this process. Generations of clinicians have sought answers, fought with complications and patiently documented their outcomes. The result has been the accumulation of many disparate protocols and seemingly contradictory conclusions. Ralph Millard's magnum opus Cleft Craft, Vol III bears testimony to their efforts.

When one approaches this subject and plows through the evidence, one can come away feeling overwhelmed by facts and underwhelmed by biologic concepts. Fortunately, we have at hand with molecular embryology a new means to understand the intimate connection between early development and the mechanisms by which craniofacial structures are assembled. As a consequence, we now have a way to understand cleft pathology based on recognizable defects in neuroangiosomes, with specific reference (as we shall see) to stem cell populations.

We have slowly worked our way through the bone fields of the palate, the pathologies that can befall them, to develop a rationale for stem cell population reassignment and better growth. In our study of the soft palate, we have used the same process to construct a new neuromeric model for the velum that better enables us to understand its components, what happens to them under conditions of clefting, and how to achieve repairs that will have reliable functional outcomes with a minimum of complications.

There are important lessons in this process. We should appreciate the work done by those who have gone before us, seek answers relentlessly, and expand beyond the confines of our various disciplines to synthesize knowledge in new ways that will take diagnosis and therapy to a higher level. So, for all patients afflicted with cleft palate the moral arc of therapy will be shaped by our understanding of craniofacial embryology, the why of things, for that is what will bring justice and order to tissues in chaos, through surgical repairs that are no longer based on geometry, or fads, but on the bedrock of development and evolution.

References

1. Carstens MH, Greco RJ, Hurwitz DJ, Tolhurst DE. Clinical applications of the subgaleal fascia. Plast Reconstr Surg. 1991a;87(4):615–26.
2. Tolhurst DE, Carstens MH, Greco RJ, Hurwitz DJ. The surgical anatomy of the scalp. Plast Reconstr Surg. 1991;87(4):604–12.

3. Reychler H. Le lambeau en ilot du buccinateur. Acta Stomatol Belg. 1988;85:63.
4. Bozola AR, Gasques JAL, Carriquiry CE, Cardoso de Oliveira M. The buccinator musculomucosal flap: anatomic study and clinical application. Plast Reconstr Surg. 1989;84(2):250–7.
5. Carstens MH, Stofman GM, Hurwitz DJ, Sotereanos GC, Futrell JW. The buccinator myomucosal island pedicle flap. Plast Reconstr Surg. 1991b;88(1):39–50.
6. Zhao Z, Yan Y, Li Y, et al. New buccinator myomucosal island flap: anatomic study and clinical application. Plast Reconstr Surg. 1999;104(1):55–64.
7. Schenck TI, Koban KC, Schlattau A, et al. Functional anatomy of the superficial fat compartments of the face: a detailed imaging study. Plast Reconstr Surg. 2018;141:1351–9.
8. Stal P, Eriksson PO, Eriksson A, Thronell LE. Enzyme-histochemical and morphological characteristics of muscle fibre types in the human buccinator and orbicularis oris. Arch Oral Biol. 1990;35(6):449–58.
9. Horta-Júnior JA, Tanega CJ, Cruz-Rizzolo RJ. Cytoarchitecture and musculotopic organization of the facial motor nucleus in *Cebus apella* monkey. J Anat. 2004;204(Pt 3):175–90.
10. McKay IJ, Muchamore I, Krumlauf R, Maden M, Lumsden A, Lewis J. The Kreisler mouse: a hindbrain segmentation mutant that lacks two rhombomeres. Development. 1994;120(8):2199–211.
11. McKay IJ, Lewis J, Lumsden A. Organization and development of facial motor neurons in the Kreisler mutant mouse. Eur J Neurosci. 1997;9(7):1499–506.
12. Watson C. Motor nuclei of the cranial nerves. In: Watson C, Paxinos G, Puelles L, editors. The mouse nervous system. Amsterdam: Elsevier; 2012. p. 490–8. https://doi.org/10.1016/B978-0-12-369497-3.10016-0.
13. Gasser RF. The development of the facial muscles in man. Am J Anat. 1967;120:357–76.
14. Gasser RF. Surgical anatomy of the parotid duct with emphasis on the major tributaries forming the duct and the relationship of the facial nerve to the duct. Clin Anat. 2005;18(1):79.
15. Perkins RE, Blanton PL, Biggs NL. Electromyographic analysis of the 'buccinator mechanism' in human beings. J Dent Res. 1977;56:783–94.
16. Stuzin JM, Wagstrom L, Kawamoto HK, Baker TJ, Wolfe SA. The anatomy of clinical applications of the buccal fat pad. Plast Reconstr Surg. 1990;85(1):29–37.
17. Wan YJ, Zhang HC, Zhang Y, Cheng YS, Zhang Y, Wang C. Application of levator eli palatine retropositioning combined with buccinator myomucosal island flap for congenital cleft palate. Exp Ther Med. 2016;12(4):2544–6.
18. Braun IF, Hoffman JC. Computed tomography of the buccomasseteric region. AJNR Am J Neuroradiol. 1984;5:605–10.
19. Peñarrocha-Diago MA, Alonso-González R, Aloy-Prósper A, Peñarrocha-Oltra D, Camacho F, Peñarrocha-Diago M. Use of buccal fat pad to repair post-extraction peri-implant bone defects in the posterior maxilla. A preliminary prospective study. Med Oral Patol Oral Cir Bucal. 2015;20(6):e699–706. http://www.medicinaoral.com/medoralfree01/v20i6/medoralv20i6p699.pdf
20. Manuel S, Kumar S, Nair PR. The versatility of the use of the buccal fat pad in the closure of oral-antral fistulas. J Maxillofac Oral Surg. 2015;14(2):374–7.
21. Alonso-Gonzalez D, Peñarocha-Diago M, Peñarocha-Oltra D, et al. Closure of oroantral communications with Bichat's buccal fat pad. Level of patient satisfaction. J Clin Exp Dent. 2015;7(1):e28–33. https://doi.org/10.4317/jced.51730.
22. Cohen SR, Fireman E, Hewett S, Saad A. Buccal fat pad augmentation for facial rejuvenation. Plast Reconstr Surg. 2017;139(6):1273e–6e. https://doi.org/10.1097/PRS.0000000000003384.
23. Kim JT, Ho SY, Hwang JH, Sung KY. Efficacy of the buccal fat pad graft in facial reconstruction and aesthetic augmentation. Plast Reconstr Surg. 2014;133(1):83e–5e. https://doi.org/10.1097/01.prs.0000436800.27670.dd.
24. Singh J, Prasad K, Lalitha RM, Ranganath K. Buccal pad of fat and its applications in oral and maxillofacial surgery: a review of published literature (February) 2004 to (July) 2009. Oral Surg Oral Med Oral Pathol Oral Radiol Endod. 2010;110:698–705.
25. Tostevin PH, Ellis H. The buccal pad of fat: a review. Clin Anat. 1995;8(6):403–6.
26. Yousuf S, Tubbs RS, Wartmann CT, Kapos T, Cohen-Gadol AA, Loukas M. A review of the gross anatomy, functions, pathology, and clinical uses of the buccal fat pad. Surg Radiol Anat. 2010;32(5):427–36. https://doi.org/10.1007/s00276-009-0596-6.
27. Carstens MH, Stofman GM, Sotereanos GC, Hurwitz DJ. A new approach for repair of oro-antral-nasal fistula. J Cranio Max Fac Surg. 1991c;19:64–70.
28. Stofman GM, Carstens MH, Berman PD, Arena S, Sotereanos GC. Reconstruction of the floor of the mouth by means of an anteriorly based buccinator myomucosal island flap. Laryngoscope. 1995;105(1):90–6.
29. Rahpeyma A, Khajehajmadi S. Inferiorly-based buccinator myomucosal island flap in oral and pharyngeal reconstruction. Four techniques to increase its application. Int J Surg Case Rep. 2015b;14:58–62.
30. Gonzalez-Garcia R, Ruiz-Laza L, Manzano D, et al. Buccinator myomucosal flap as soft tissue covering for vascularized free fibular flap in anterior maxillary bone defects. J Oral Maxillofac Surg. 2010;68:927–30.
31. Khan K, Hinkley V, Cassell O, et al. A novel use of the facial artery-based buccinator musculo-mucosal island flap for reconstruction of the oropharynx. J Plast Reconstr Aesthet Surg. 2013;66:1365–8.
32. Cuesta-Gil M, Romanya P, Navarro-Cuellar C, et al. Reconstrucción de defectos palatinos con el colgajodel músculo buccinador. Rev Esp Cir Oral y Maxilofac. 2005;27(4):206–15.
33. Ferrari S, Ferri A, Bianchi B, Copelli C, Sessena E. Reconstructing large palatal defects: the double buccinator myomucosal flap. J Oral Maxillofac Surg. 2010;68:924–6.
34. Padgett EC. The repair of cleft palates primarily unsuccessfully operated upon. Surg Gynecol Obstet. 1936;63:483–96.
35. Cronin TD, Bruer RO, Alexander JT, et al. Push-back repair using nasal mucosal flaps: results. Cleft Palate J. 1964;1:269.
36. Dorrance GM, Branfield JW. The pushback operation for repair of cleft palate. Plast Reconstr Surg. 1946;1:145.
37. Millard DR Jr. Wide and/or short cleft palate. Plast Reconstr Surg. 1962;29:40.
38. Kaplan EN. Soft palate repair by levator muscle reconstruction and a buccal mucosal flap. Plast Reconstr Surg. 1975;56(2):129.
39. Rahpeyma A, Khajehahmadi S. Closure of huge palatal fistula in an adult patient with isolated cleft palate: a technical note. Plast Reconstr Surg Glob Open. 2015a;3:e306. https://doi.org/10.1097/GOX.0000000000000279.
40. Hill C, Hayden C, Riaz M, Leonard AG. Buccinator sandwich pushback: a new technique for treatment of secondary velopharyngeal incompetence. Cleft Palate Craniofac J. 2004;41(3):230–7.
41. Mann RJ, Keating KC, Armstrong S, Ebner B, Bajnrauh R, Naum S. The double-opposing buccal flap procedure for palatal lengthening. Plast Reconstr Surg. 2011;127(6):2413–8.
42. Hens G, Sell D, Pinkstone M, Birch MJ, Hay N, Sommerlad BC. Palatal lengthening by buccinator myomucosal flaps for velopharyngeal insufficiency. Cleft Palate Craniofac J. 2013;50(5):e84–9.

43. Dias DK, Fernando PDC, Dissanayake RDA. Improvement in quality of speech in patients with velo-pharyngeal insufficiency corrected using a buccinator myomucosal flap. Ceylon Med J. 2016;61(3):130–4.
44. Robertson AGN, McKeowen DJ, Jackson IT, et al. Use of buccal myomucosal flap in secondary cleft palate repair. Plast Reconstr Surg. 2008;122(3):910–7.
45. Logjes RJH, Bleys RLAW, Breugem CC. The innervation of the soft palate muscles involved in cleft palate. Clin Oral Invest. 2016;20:895–901. https://doi.org/10.1007/s00784-016-1791-6.
46. Mukherji MM. Cheek flap for short palates. Cleft Palate J. 1969;6:415–20.
47. Jackson IT, Moreira Gonzalez AA, Rogers A, Beal BJ. The buccal flap: a useful technique in cleft palate repair? Cleft Palate Craniofac J. 2004;41(2):144–54.
48. Furlow LT Jr. Cleft palate repair by double-opposing z-plasty. Plast Reconstr Surg. 1986;78:724–38.

Further Reading

Abdaly H, Omranyfard M. Buccinator flap as a method for palatal fistula and VPI management. Adv Biomed Res. 2015;2015:161529. https://doi.org/10.4103/2277-9175.161529.

Bither SM, Halli R, Kini Y. Buccal fat pad in intraoral defect reconstruction. J Maxillofac Oral Surg. 2013;12(4):451–5. https://doi.org/10.1007/s12663-010-0166-9.

Bozola AR, Cardoso de Oliveira M, Miura O, Gasques J, Marchi MA. The use of a myocutaneous flap of the buccinator muscle for the correction of cleft palates. Rev Soc Bras Cir Plast. 1985;84:1.

Carstens MH, Stofman GM, Sotereanos GC, Hurwitz DJ. A new approach for repair of oro-antral-nasal fistula. J Cranio-Maxillofac Surg. 1991b;19:64–70.

Dutra EH, Caria PHF, Rafferty KL, Herring SW. The buccinator during mastication: a functional and anatomic evaluation in minipigs. Arch Oral Biol. 2010;55:627–38.

Fagan J. Buccinator myomucosal flap. In: Fagan J, editor. The open access atlas of otolaryngology, head and neck operative surgery. Cape Town: University of Cape Town; 2014. www.entdev.uct.ac.za.

Freelander F, Jckson IT. The fate of buccal mucosal flaps in primary palate repair. Cleft Palate J. 1989;26:110.

Han M-S, Kim J-K, Bae HEK, Park H-S, Kim H-J. Clinical implications of the topography and distribution of the posterior superior alveolar artery. J Craniofac Surg. 2009;20(2):551–4.

Hassani A, Sadaat S, Moshiri R, Shahmirzadi S. Hemangioma of the buccal fat pad. Contemp Clin Dent. 2014;5(2):243–6. https://doi.org/10.4103/0976-237X.132368.

Hernanedez-Alfaro F, Vals-Otañon A, Blasco-Palacio JC, Guijarro-Martinez R. Malar augmentation with pedicled buccal fat pad in orthognathic surgery: three dimensional evaluation. Plast Reconstr Surg. 2015;136(5):1063–7.

Jung BK, Song SY, Kim SH, et al. Lateral oropharyngeal wall coverage with buccinator myomucosal and buccal fat pad flaps. Arch Plast Surg. 2015;42(4):453–60.

Juvara E. Anatomie de la région ptérygomaxillaire. Thèse no. 186. Baittaille and Cie, Paris. Boston: Courtesy of Francis A. Countway Medical Library; 1895.

Kang H-C, Kwak H-H, Hu K-S, Youn K-H, Jin G-C, Fontaine C, Kim H-J. An anatomical study of the buccinator muscle fibers that extend to the terminal portion of the parotid duct, and their functional roles in salivary secretion. J Anat. 2006;208:601–7.

Mann RJ, Martin MD, Eichhorn KC, Sierzant CG, Polley JW, Girotto JA. The double opposing Z-plasty plus or minus buccal flap approach for repair of cleft palate: a review of 505 consecutive cases. Plast Reconstr Surg. 2017;139(3):735e–44e.

Mendelson B, Wong C-H. Commentary on: SMAS fusion zones determine the subfacial and subcutaneous anatomy of the human face: fascial spaces, fat compartments and models of facial aging. Aesthet Surg J. 2016;36(5):529–32.

Millar DR. Cleft craft: the evolution of its surgery, vol 3. Alveolar and palatal deformities. Boston: Little Brown; 1980.

Mohan S, Kankariya H, Harjani B. The use of the buccal fat pate for reconstruction of oral defects: a review of the literature and report of cases. J Maxillofac Oral Surg. 2012;11(2):128–31. https://doi.org/10.1007/s12663-011-0217-x.

Perlman AL, Christianson J. Topography and functional anatomy of the swallowing structures. In: Perlman AL, Schultze-Delrieu KS, editors. Deglutition and its disorders. San Diego: Singular Medical Publishing Group; 1997. p. 15–42.

Pessa JE. SMAS fusion zones determine the subfacial and subcutaneous anatomy of the human face: fascial spaces, fat compartments and models of facial aging. Aesthet Surg J. 2016;36(5):515–26.

Pribaz JJ, Stephens W, Crespo L, Gifford G. A new intraoral flap: facial artery musculomucosal (FAMM) flap. Plast Reconstr Surg. 1992;90:421.

Pribaz JJ, Meara JG, Wright S, Smith JD, Stephens W, Breuing KH. Lip and vermilion reconstruction with the facial artery musculomucosal flap. Plast Reconstr Surg. 2000;105(3):864–72.

Rahpeyma A, Khajehahmadi S. Buccinator-based myomucosal flaps in intraoral reconstruction: a review and new classification. Natl J Maxillofac Surg. 2013;4(1):25–32.

Rahpeyma A, Khajehajmadi S. Inferiorly-based buccinator myomucosal island flap in oral and pharyngeal reconstruction. Four techniques to increase its application. Int J Surg Case Rep. 2015b;14:58–62.

Seçil Y, Aydogdu I, Ertekin C. Peripheral facial palsy and dysfunction of the oropoharynx. J Neurol Neurosurg Psychiatry. 2002;72:391–3.

Terada S, Sato T. Nerve supply of the medial and lateral pterygoid muscles and its morphological significance. Okajimas Folia Anat Jpn. 1982;59(4):251–64. https://www.jstage.jst.go.jp/article/ofaj1936/59/4/59_251/_pdf

Van Lierop A, Fagan JJ. Buccinator myomucosal flap: clinical results and review of anatomy, surgical technique and applications. J Laryngol Otol. 2008;122:181–7.

Ziermann JM, Miyashita T, Diogo R. Cephalic muscles of cyclostomes (hagfishes and lampreys) and Chondrichthyes (sharks, rays and holocephalans): comparative anatomy and early evolution of the vertebrate head muscles. Zool J Linn Soc. 2014;172(4):771–802. https://doi.org/10.1111/zoj.12186.

Zhang H-M, Yan Y-P, Qi K-M, et al. Anatomical structure of the buccal fat pad and its clinical applications. Plast Reconstr Surg. 2002;109(7):2509–18.

Zhao Z, Ying L, Xiao S, et al. Innervated Buccal musculomucosal flap for wider vermilion and orbicularis oris muscle reconstruction. Plast Reconstr Surg. 2005;116(3):846–52.

Pathologic Anatomy of Nasolabial Clefts: Spectrum of the Microform Deformity and the Neuromeric Basis of Cleft Surgery

18

Michael H. Carstens

Introduction

The locus of the cleft defect is in the floor of the nose, the upper lip, and the alveolus, rather than the free border of the lip.

—Bard Cosman* [1]

The Microform Deformity and Its Variants: The Rosetta Stone of Cleft Pathology

During Napoleon Bonaparte's Egyptian campaign in 1799, a French soldier discovered a black basalt slab inscribed with ancient writing near the town of Rosetta, about 56 km north of Alexandria. The irregularly shaped stone, which would become known as the Rosetta Stone because of where it was found, contained fragments of passages written in three different scripts: Greek, Egyptian hieroglyphics and Egyptian demotic. The ancient Greek on the Rosetta Stone told archaeologists that it was inscribed by priests honoring the king of Egypt, Ptolemy V, in the second century BC (Fig. 18.1).

Fig. 18.1 The Rosetta stone. The discovery of the stela, written in two forms of ancient Egyptian as well as Greek, enabled scholars to decipher the hieroglyphic system. It currently resides in the British Museum (see text). [Courtesy of Mr. Blake Linder, Journalist for Caxton West Branch. Reprinted from the Roodepoort Record: https://roodepoort-northsider.co.za/278371/today-in-history-french-soldier-discovers-the-rosetta-stone-web/]

Perhaps even more surprising, the Greek passage announced that the three scripts were all of identical meaning. The artifact thus held the key to solving the riddle of hieroglyphics, a written language that had been 'dead' for nearly 2000 years.

When Napoleon, an emperor known for his enlightened view on education, art and culture, invaded Egypt in 1798, he took along a group of scholars and told them to seize all important cultural artifacts for France. Pierre Bouchard, one of Napoleon's soldiers, was aware of this order when he found the basalt stone, which was almost 1.2 m long and 0.8 m wide, at a fort near Rosetta.

When the British defeated Napoleon in 1801, they took possession of the Rosetta Stone. Several scholars, including Englishman Thomas Young, made progress with the initial hieroglyphics analysis of the Rosetta Stone. French Egyptologist, Jean-Francois Champollion, who had taught himself ancient languages, ultimately cracked the code and deciphered the hieroglyphics using his knowledge of Greek as a guide.

Hieroglyphics used pictures to represent objects, sounds and groups of sounds. Once the Rosetta Stone inscriptions were translated, the language and culture of ancient Egypt were suddenly opened to scientists as never before.

The Rosetta Stone has been housed at the British Museum in London since 1802, except for a brief period during World War I. At that time, museum officials moved it to a separate underground location, along with other irreplaceable items from the museum's collection, to protect it from the threat of bombs.

—Blake Linder, Roodepoort *Northsider*

M. H. Carstens (✉)
Wake Forest Institute of Regenerative Medicine, Wake Forest University, Winston-Salem, NC, USA
e-mail: mcarsten@wakehealth.edu

M. H. Carstens (ed.), *The Embryologic Basis of Craniofacial Structure*, https://doi.org/10.1007/978-3-031-15636-6_18

In the study of congenital deformity, whenever a condition exists as a spectrum of presentations, the anatomy of its form fruste will point an accusatory finger at the embryologic defect. In 1938, Victor Veau conceptualized the cleft process as an "uninterrupted chain" of presentations from the subtle to the overt. Thus, the microform variant, the so-called "cleft lip nose" with a normal-appearing lip, constitutes the Rosetta Stone of clefts. Disciplined study of its anatomic features will reveal the following: (1) the exact location of the defect and its neuromere(s) of origin; (2) the effects exerted by the defect on the osseous and soft tissue surround; and (3) the developmental mechanisms by which these pathologies are produced.

The microform cleft established for me once and for all the locus of the deficiency site, the lateral nasal floor and piriform fossa. At the time I had stumbled on the stapedial system, or the premaxillary field as the primary defect. Instead I was conceptualizing cleft pathophysiology as a sequence of four *processes*: *deficiency*, *division*, *displacement* and *distortion*; these were the consequences of some unidentified "event."

"Biologic events at the deficiency site transform the surrounding functional matrix creating asymmetry of force vectors which, acting over time, cause distortions of otherwise normal structures in developmental fields adjacent to the cleft site. The microform cleft deficiency site, an abnormal developmental field, represents the cleft 'lesion' in its purest form. As such it provides insights into all other forms of cleft spatial and temporal aspects of perioral and nasal development."

The objective of this chapter is to explore spectrum of incomplete cleft using the microform variant as focus. Although in the original paper there were many concepts I did not understand, it proved a crucial step forward [2, 3]. Analysis of the *form fruste* variant revealed for the first time the location of the cleft "lesion." Dentoalveolar abnormalities in the microform convinced me that a deficiency in bone stock in the nasal floor was universal in all clefts and that it was somehow related, not only to the nasal deformity, but the lip cleft as well. What's more the deficit was somehow quantitative: the worse the bone deficit, the more pronounced were the soft tissue finings. Using the Carnegie system and SEMs by Hinrichsen I was able to reconstruct the structural events of lip closure [4]. What we can now bring to this story is the model of neuroangiosomes based on the stapedial system, the BMP4/SHH model of epithelial fusion and the insights of Talmant regarding the nasalis muscle complex and its misinsertion in the cleft state.

In sum, we can now explain the entire spectrum of cleft deformity and apply the concepts of DFR surgery to achieve an embryologically accurate correction.

We shall address the following topics

- Pathologic anatomy of the microform cleft
- Developmental sequence of the microform cleft
  - Developmental fields
  - Nasolabial fusion sequence
  - Lateral nasal process as a dysfunctional matrix
  - Biochemical evidence: the carbon monoxide hypothesis
- Muscle in cleft lip and cleft palate: histology and histochemistry
  - Diffusible morphogen model
- Clinical studies of microform cleft
  - Classification: Boo-Chai and Mulliken
- Neurovascular anatomy of the prolabium and premaxilla
- Developmental field reassignment and the incomplete cleft
- The pathology of nasal-labial-maxillary cleft lip and palate

Anatomic Features of the Microform Cleft (Figs. 18.2, 18.3, 18.4, 18.5, 18.6, 18.7, 18.8 and 18.9)

Previous reviews have described the characteristics of the minimal deformity [5–7]. The emphasis was placed upon the nose and soft tissue abnormalities of the lip. Comments about them were based on neuromeric model.

Alveolar abnormalities are basic to the pathological lesion of the minimal cleft. The deficiency of bone stock begins in the frontal process of premaxilla (PMxF) and extends downward to the lateral incisor zone (PMxB), causing a caudal lowering of the floor of the piriform fossa (and its vestibular lining). If something is wrong with the neural crest dental lamina, the teeth on the cleft will will erupt later. An occult cleft of the primary palate is sometimes present. As one move from the microform to the incomplete form, the alveolus may be frankly cleft. Frank deformation of the lateral piriform margin is universal in microform cleft; this represents the defect in PMxF.

Dental disturbances include the following: (1) missing or supernumerary teeth, (2) alterations in tooth size or shape, and (3) abnormal eruption pattern, including overcompensation by the mandibular dentition [8, 9]. The first two conditions arise from the *dental germ line* in situ. The third and fourth conditions reflect the *osseous environment* within which they form. Because the bone arises within a soft-tissue matrix, any deficiency or deformation of that matrix will alter the form and spatial positioning of the bone stock. The physical forces exerted on the alveolar bone (and ultimately on the dentition) are likewise affected.

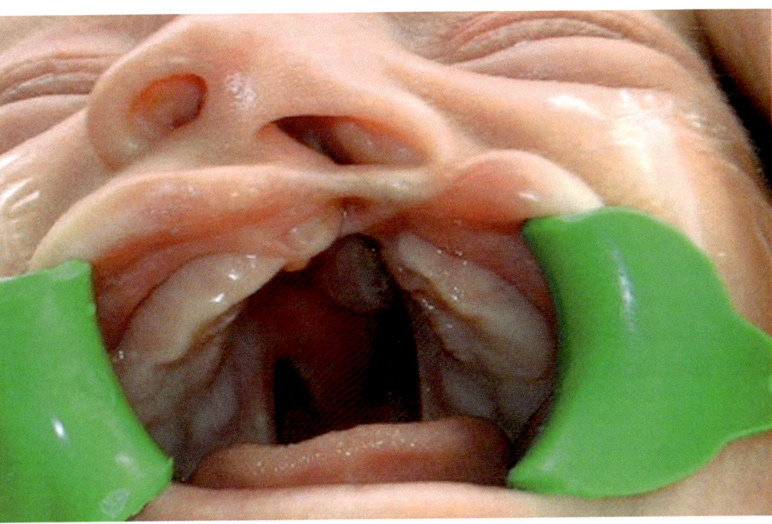

Fig. 18.2 Microform cleft. (Left): Note epithelial "scar does not approximate the left alar footplate. It tracks into the nasal floor. Note how reassignment of non-philtral prolabium "drapes" itself into the nasal floor. Nostril sill is "in-turned" and not readily seen. (Right): **Simonart's band** conducts mesenchyme from lateral-to-medial. It can also create an adhesion between maxillary alveolus and premaxillary alveolus. In this case, there does not seem to be bone between the two alveolar processes, making this an incomplete cleft lip with near complete primary palate cleft and a complete secondary palate cleft. Left: [Reprinted from Kim EK, Khang SK, Lee TL, Kim TG. Clinical features of the microform cleft lip and the ultrastructural characteristics of the orbicularis oris muscle. Cleft Palate–Craniofac J 2010; 47(3):297–302. With permission from SAGE Publishing]. Right: [Reprinted from Biljic F, Sozer OA. Diagnosis and presurgical orthopedics in infants with cleft lip and palate. Eur J General Dentistry 2015; 4(2): 41–47. With permission from Wolters Kluwer Health, Inc.]

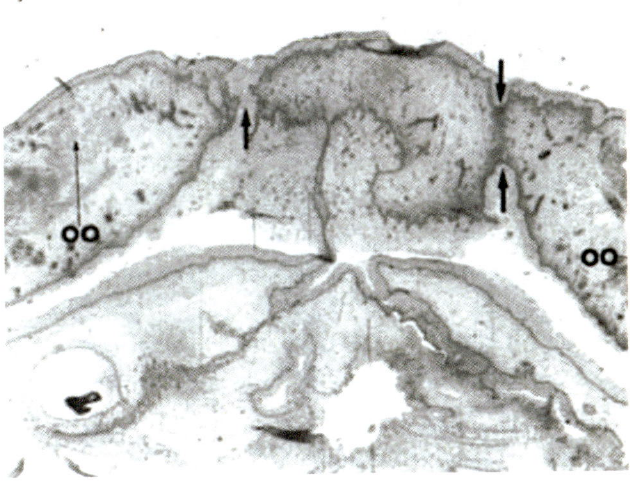

Fig. 18.3 Subepithelial cleft. A 18-week specimen with interruption of orbicularis oris (**00**) bilaterally. Note the complete connective tissue interruption on the left (**arrow**) but only partial interruption on the right (double arrow). The striking feature is the *fibrotic tissue* on the incomplete side: muscle cannot form. [Reprinted from Martin RA, Jones KL, Benirschke K. Extension of the cleft lip phenotype: The subepithelial cleft. Am J of Med Genet 1993; 47:744–747. With permission from John Wiley & Sons]

Nasal deformity is the *sine qua non* of the microform cleft

- Deficiency of the functional matrix (i.e., loss of PMxF and deficiency of PMxB) within the floor of the nose pulls the entire vestibular lining into the site. This sets up traction forces that deviate the nasal tip and the cephalic portion of the columella toward the cleft. The base of the columella is affected as well. Asymmetric muscle insertion of oblique fibers of Delaire (SOO) traction is to the non-cleft side.
- On the cleft side, the nostril floor is pulled downward into the bone deficit exerting *traction on the medial footplate*, drawing it downward and backward. The insertion of the paranasal muscles into the alar base is asymmetrical, causing it to drift laterally away from the midline. The *lateral alar crease* becomes indistinct, blending into the cheek.
- *Nasalis* fails to insert correctly over the canine and lateral incisor fossae. Instead it inserts falsely into the nasal floor, causing downward traction of the lateral crus.
  - Pars transversus inserts over the canine.

    In microform it inserts into the piriform fossa.
  - Pars alaris inserts over the lateral incisor.

    In microform it inserts into skin of the nostril sill.
- The *lower lateral cartilage* develops at 4 months' gestation within a deformed soft-tissue envelope. Thus, it is not surprising that it occupies a lateralized position. Boo-Chai and Tange reported that the alar cartilage separated from its Opposite member by fibrofatty tissue [10]. This is reminiscent of similar findings by Trott in bilateral cases [11]. Although there is no intrinsic change in the size of the alar cartilage, its distortion blunts the tip's definition and flattens out the alar rim.

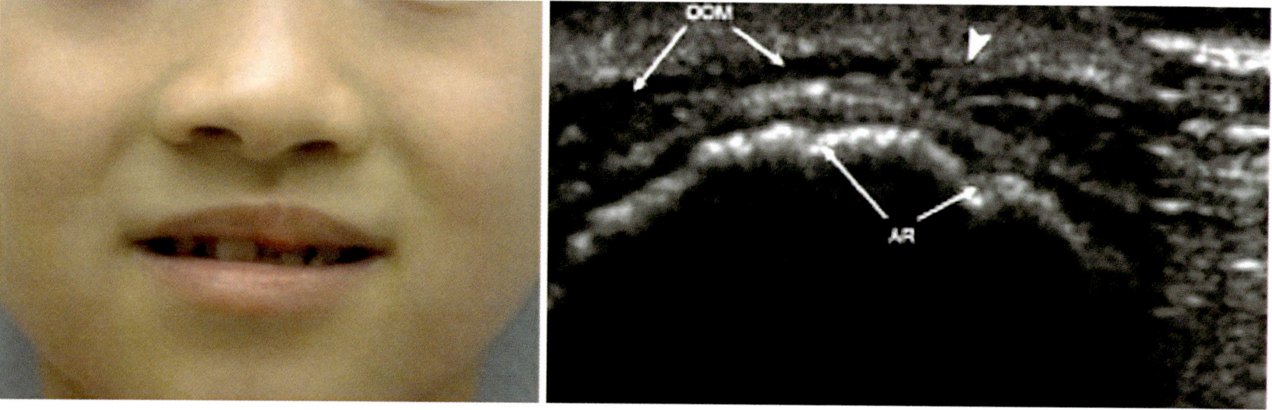

Fig. 18.4 Subepithelial and microform defects. Ultrasound images of the Proband S91C, which has an orbicularis oris muscle defect. Ultrasound images are taken in the transverse plane; the top of the image corresponds to the anterior structures; *OOM* orbicularis oris muscle, *AR* alveolar ridge. The arrowhead in (A) points to the discontinuity in the OOM of the proband. The subject's left is on the right. [Reprinted from Marzita ML. Subclinical features in non-syndromic cleft lip with or without cleft palate (CL/P): review of the evidence that subepithelial orbicularis oris muscle defects are part of an expanded phenotype for CL/P. Orthod Craniofacial Res 2007;10: 82–87. With permission from John Wiley & Sons]

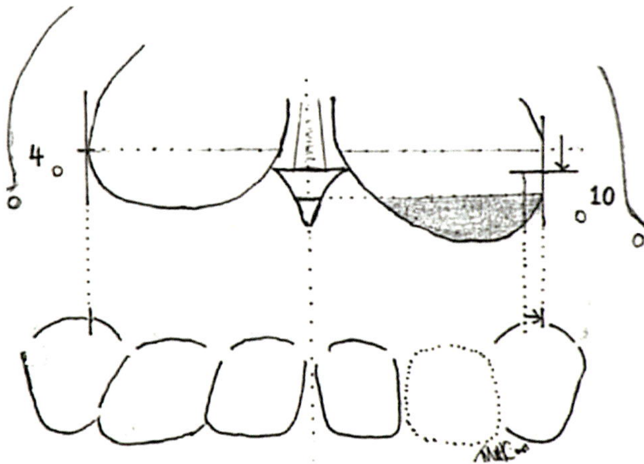

Fig. 18.5 Piriform fossa deficiency is universally present in microform cleft. Although the alveolus remains intact, the alveolus always has: (1) a "scooping out" of the lateral wall due to deficiency or absence of frontal process zone of premaxilla, PMxF; and (2) a depression in the nasal floor due to deficiency in lateral incisor zone of premaxilla, PMxB. Note that PMxF is a vertical outgrowth of PMxb. It marks the terminal perfusion of the medial nasopalatine axis. PMxF abuts against its counterpart, frontal process of maxilla, MxF, which is supplied by a small branch from the medial division of anterior superior alveolar artery before its exit from the sinus cavity. MxF provides housing for the third incisor. Volume deficit clearly seen. The cleft airway is compromised by tightening of the vestibule, abnormal nasalis and posterior septal reflection. Septum is forced over by increased hydraulic pressure in the normal airway during fetal breathing of amniotic fluid. [Courtesy of Michael Carstens, MD]

- The nostril sill is "hidden," being retracted backward into the deficiency site of the lateral vestibular wall.

Orbicularis oris has two layers which behave differently in the microform cleft

- The deep (sphincteric) layer of orbicularis (DOO) is unaffected [12, 13].
- The superficial (dilator) layer of orbicularis (SOO) does not "fill" the philtral columns. Even though the peripheral fibers of the deep layer fill out the vermilion, the superficial deficiency may draw up the mucocutaneous border into a notch.
- The oblique fibers of Delaire may not be inserted, causing a flattening of the upper lip just inferolateral to the columella.

Subepithelial manifestations in microform range from stria to fibrosis with a dimple on animation corresponding to SOO misinsertion. These subtleties are characterized by histologic changes consistent with fibrosis [14]. Ultrasound now demonstrates this fibrosis as a hypoechoic interruption in an otherwise hyporechoic (black) band corresponding to orbicularis [15, 16].

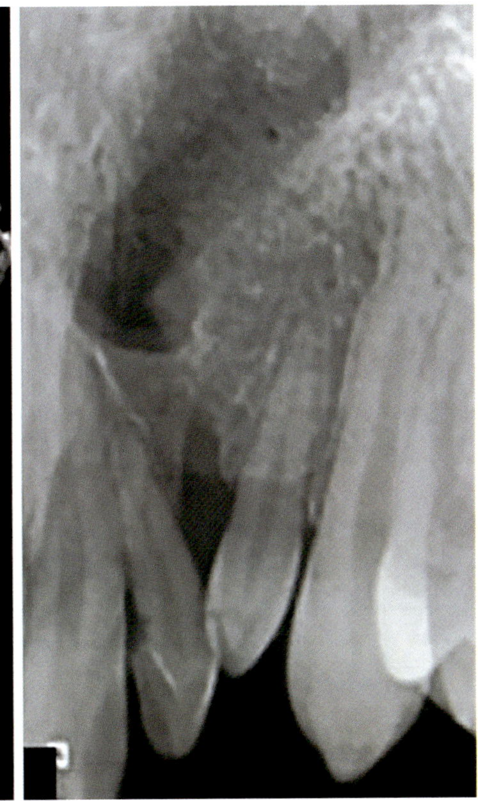

Fig. 18.6 Cone beam CT (CBCT) of primary palate cleft demonstrates typical piriform depression. Panorex of the same patient showing depression of lateral piriform wall and thinning of alveolar bone around the unerupted canine. Note reduction if bone volume in PMxA (central incisor) on the cleft side. [Reprinted from Machado G. CBCT imaging—A boon to orthodontics. Saudi Dental Journal 2015; 27(1): 12–21. With permission from Elsevier]

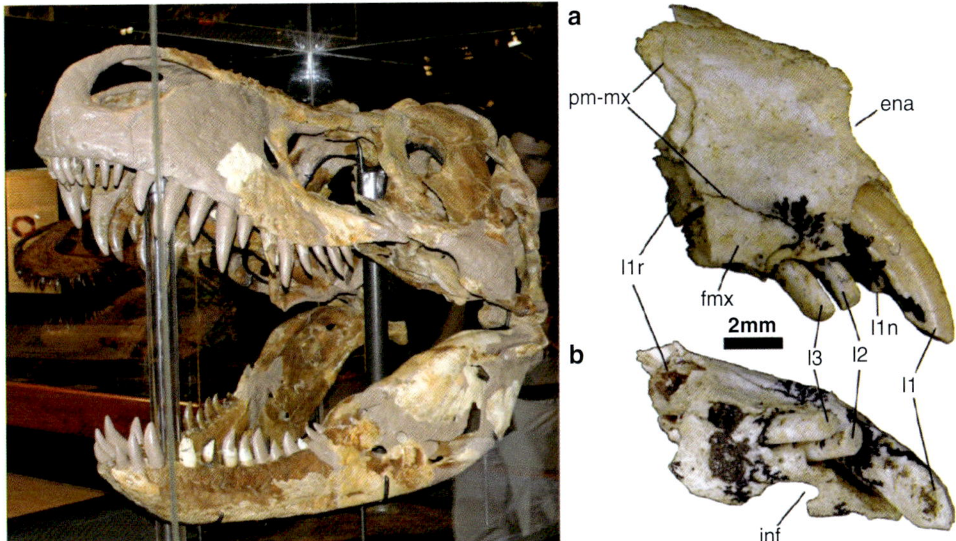

Fig. 18.7 Dental content of the premaxilla changes with evolution. (Left): *Tyrannosaurus rex* premaxilla had four teeth per side but these were not differentiated into incisors. The basal formula for placental mammals has three incisors in the premaxilla. (Right): (**a**, **b**) Prehistoric marsupial *Yalkaperidon*, shows basal dental formula with three incisors in its premaxilla, *Ena* external nasal aperture, *fmx* facial process of maxilla, *L1n* notch on distal surface of first premaxillary upper incisor for contact with first mandibular incisor, *I1r* open root of first upper incisor, *inf* incisive foramen, *pm-mx* premaxillary-maxillary suture. Left: [Reprinted from Wikimedia. Retrieved from: https://commons.wikimedia.org/wiki/File:T._rex_MOR_008.jpg. With permission from Creative Commons License4.0: https://creativecommons.org/licenses/by-sa/4.0/deed.en.] Right: [Reprinted from Beck RMD, Travouillon KJ, Aplin KPA, Godthelp H, Archer M. The osteology and systematics of the enigmatic Australian Oligo-Miocende metatherian, *Yalkaperidon* (Yalkaparidontidae; Yalkaparidontia? Australidelphia; Marsupialia). J Mammalian Evolution 2014; 21:127–172. doi:10.1007/s10914-013-9236-3. With permission from Springer Nature]

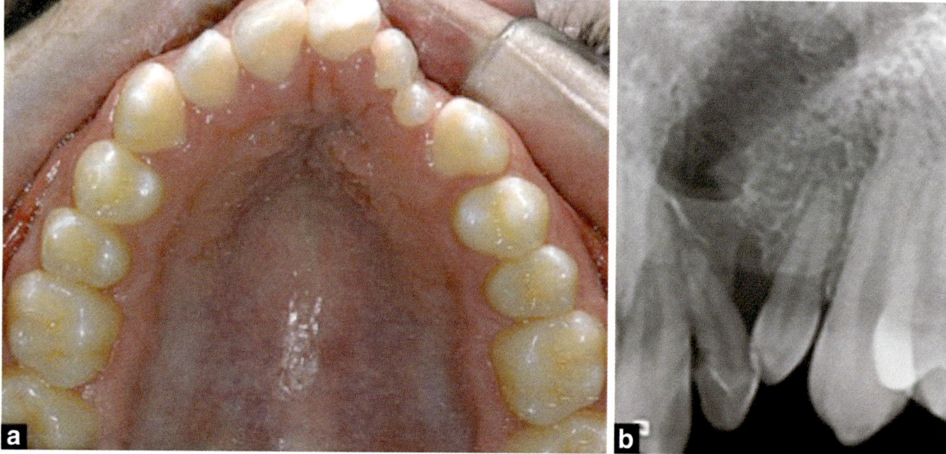

Fig. 18.8 Incisors are located on *either* side of alveolar cleft. (**a**) Oral view palate shows three incisors are present in the case of form fruste alveolar cleft. Permanent central incisor is forced to erupt over the mesial incisor. (**b**) Distal incisor in Mx1 is seen on the other side of the cleft. [Courtesy of Michael Carstens, MD]

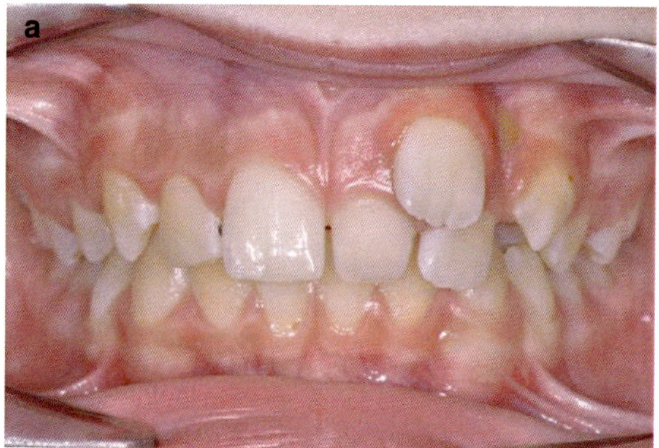

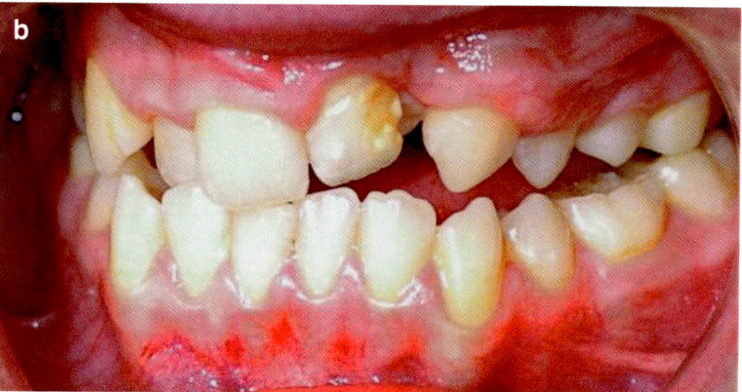

Fig. 18.9 Abnormal eruption. (**a**) Three incisors are present in the case of form fruste alveolar cleft. Permanent central incisor is forced to erupt over the mesial incisor. Distal incisor in Mx1 is seen on the other side of the cleft. Cleft-associated enamel hypoplasia. (**b**) Lateral incisor mesial to cleft shows severe enamel defect; the canine is affected to a lesser degree. (**a**) [Reproduced from Orthodontic Update (ISSN 1756-6401), by permission of George Warman Publications (UK) Ltd]. (**b**) [Reprinted from Pegelow M, Algardi N, Karsten AL-A. The prevalence of various dental characteristics in the primary and mixed dentition in patients born with non-syndromic unilateral cleft lip with or without cleft palate. Eur J Orthod 2012; 34(5):561–570. With permission from Oxford University Press]

Consequences of Muscle Imbalance in the Cleft State

Origin and Migration of Facial Muscles

The muscles surrounding and animating the nose and lip originate when the paraxial mesoderm of the sixth somitomere melds with neural crest from r4–r5 to produce the second pharyngeal arch which appears at Carnegie stage 10. Within the arch, myoblasts are segregated into three functional layers, each defined by its own fascia. *Buccopharyngeal fascia* contains pharyngeal constrictor buccinator and a partial contribution to superior constrictor. *Deep investing fascia* is dedicated to muscles of mastication: anterior digastric, stylohyoideus and evolutionary transformation in mammals, stapedius. *Superficial investing fascia* (SMAS) carries within it the muscles of facial expression and spreads below the subcutaneous fascia to envelop the entire face and head, and the anterior neck.

- It is of evolutionary interest that deep orbicularis, although programmed, just like buccinator, by the oral mucosa, does not share buccopharyngeal fascia with buccinator. The BPF stops at the modiolus.

By stage 12, the second arch has melded into the first arch to produce the maxillary process, MxP. This term, encrusted forever in our anatomic lexicon, is a misnomer because MxP subdivides along its axis. The upper zone containing r2/r4 neural crest becomes the maxilla and its supporting bone field. The lower zone with r3/r5 neural crestbecomes the mandible. The "fault line" between the zones extends from the oral commissures back to the external auditory canal. The embryonic separation can be observed in lateral facial clefts.

As we shall see, during stages 13 and 14, the first arch invades the floor of the frontonasal process and by stage 15 MxP has arrived in position. From here facial myoblasts are carried within the flow of neural crest fascia in two arcs surrounding the orbit. The *supraorbital migration* brings Sm6 muscles to the forehead and glabella. The *infraorbital migration* provides all the rest, with the individual muscles developing in deep-to-superficial and lateral-to-medial sequences [17]. **Note**: *the terminal muscle in each migration inserts last and is the* ***most vulnerable*** *to disturbances at its insertion site.*

- Procerus is the terminal muscle of the supraorbital migration.
- Nasalis is the terminal muscle of the infraorbital migration.

Insertions of Facial Muscles: Two Mechanisms

Proximal-distal They can migrate into position within a structure, such as a fascia, which constitutes the primary insertion. Subsequently they seek out a secondary insertion into bone or dermis, when the latter manifests a BMP4 signal which is detected by the fascial envelope which follows the muscle into contact, sometimes becoming a tendon. Here are two examples. *Frontalis* begins in the SMAS and has its secondary insertion upward into the aponeurotic fascia of epicranium. *Orbicularis* DOO begins with a fascial condensation of SMAS at modiolus and then extends mesially to insert into contralateral DOO.

Distal-proximal In this mechanism, the muscle migrates all the way to the distal site and binds to it. In so doing, it retains its fascial envelope which constitutes a sort of "trail of bread crumbs," which marks the migration route. Subsequently the muscle extends backward along the trail until it encounters the next available insertion site. *Biceps brachii* forms a single muscle mass with primary insertion on the humerus. Secondary insertions are backwards to scapula, the long head to the glenoid and the short head to coracoid process. Nasalis migrates over the nasal dorsum to gain primary insertion to the contralateral muscle. Secondary insertions of nasalis are into two distinct sites of premaxilla: canine fossa and lateral incisor fossa.

Muscle Anatomy in Labionasal Clefts: The Muscle Ring Theory of Delaire

Faulty signals from underlying bone fields affect soft tissue fusion. Naturally the superficial layer is affected first. For this reason, in microform cleft, SOO becomes dysplastic first. Disruption of DOO follows the mucosal defect. The effect of these changes in microform cleft is subtle, seen primarily in the columellar base. But as the deformity increases in severity, lateral drift of the alar base is seen. Based on the anatomy of the complete cleft, alar drift has traditionally been ascribed to lateral traction exerted by periorbital muscles, but this construct cannot occur in the microform cleft because muscle continuity across the midline is preserved. The best explanation for alar drift in the microform is the false insertion of nasalis into the nostril sill.

Delaire and Markus described a system of three rings of facial muscle all converging on the lip. In this model, the pathologic anatomy of complete cleft lip results from the breakup of these rings, generating abnormal force vectors which drag the soft tissues of the face into asymmetry. Proper surgical treatment would consist in re-establishing facial balance. The moulding effect of facial muscles on the underlying maxilla was of concern for Bardach. These studies were important signposts refuting the idea that maxillary "hypoplasia" is intrinsic to the cleft condition. Retrusion of isolated envelope as can be seen in bilateral clefts [18, 19] is the result of unequal force vectors to the midline, leaving the cleft-side maxilla disconnected (Figs. 18.10, 18.11, 18.12, 18.13, 18.14, 18.15 and 18.16).

Nasalis is the *Forme Fruste* Muscle Involved in Nasolabial Clefts

The region above the primate premaxilla gives rise to small muscles with secondary insertion into the fossae above the incisors. As previously discussed, humans have a central and a lateral incisor but a third incisor is frequently present [20]. This is entirely unremarkable. Recall that dental specialization appears with placental mammals due to the complexities

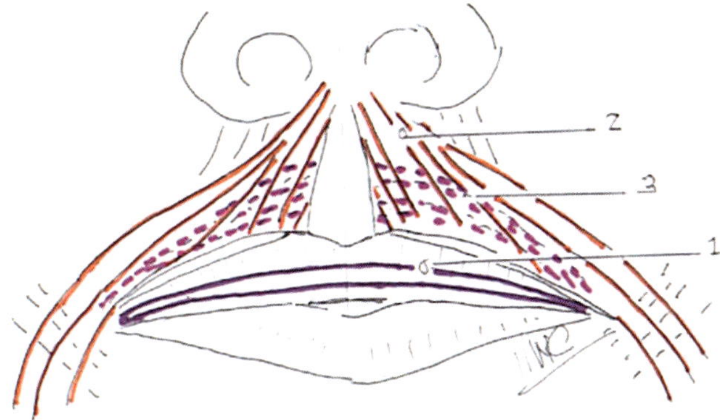

Fig. 18.10 Orbicularis oris fibers. Deep (sphincteric) layer of orbicularis oris (DOO) is almost completely hidden by superficial (dilator) layer of orbicularis oris (SOO) except at the vermilion were it presents as *pars marginalis*. Note that DOO is continuous caudal to Cupid's bow (philtral prolabium). SOO has two parts: *Pars transversalis* fibers (2) do not extend medial to the philtral column. They are blocked by r1 mesenchyme; *Pars obliquus* fibers (2) extend upward to blend with the columellar base. Delaire pointed out the importance of these fibers in establishing the "aesthetic drape" of the lip. Note the potential for confusion between these fibers and those of depressor septi nasi. Orbicularis is superficial to DSN and to the nasalis muscle bellies. DSN actually takes origin from the fossae above the central incisors. [Courtesy of Michael Carstens, MD]

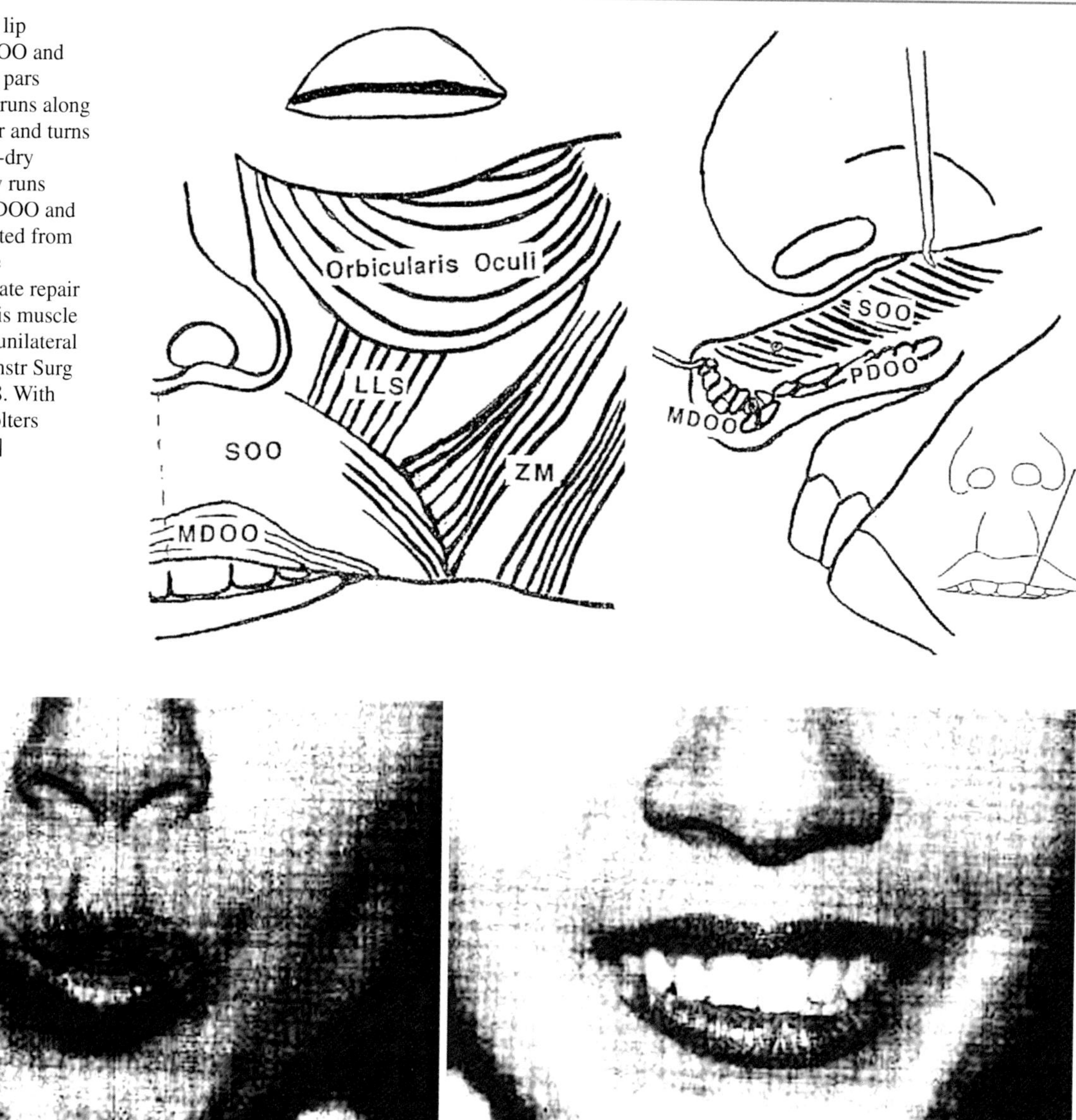

Fig. 18.11 Lateral lip element showing DOO and SOO. Note how the pars marginalis of DOO runs along the vermilion border and turns up to define the wet-dry mucosal line. Artery runs between the "J" of DOO and distal SOO. [Reprinted from Park CG, Ha B. The importance of accurate repair of the orbicularis oris muscle in the correction of unilateral cleft lip. Plast Reconstr Surg 1995;96(4):780–788. With permission from Wolters Kluwer Health, Inc.]

Fig. 18.12 Function of DOO versus SOO on the philtral columns. Note that DOO inserts into modiolus in the same plane as buccinator but does not have SMAS. [Reprinted from Park CG, Ha B. The importance of accurate repair of the orbicularis oris muscle in the correction of unilateral cleft lip. Plast Reconstr Surg 1995;96(4):780–788. With permission from Wolters Kluwer Health, Inc.]

of their chewing apparatus, the baseline formula being I3, C1, P4, M3. There is likely some type of suppression present in the normal state such that, in the presence of PMxB, the dental lamina of the third incisor that would correspond to PMxF does not develop. In development the bone above each of the 3 tooth roots, in response to morphogen BMP4, opens up (within a limited time frame), a secondary binding site for a muscle. These are:

- Central incisor: (PMxA) depressor septi nasi (DSN)
- Lateral incisor 1 (PMxB): medial head of nasalis
- Lateral incisor 2 (PMxF): lateral head of nasalis

Anatomic studies of these muscles have, by virtue of the overlying orbicularis and outdated terminology, resulted in confusion. The embryonic layering of the orbicularis complex is also not appreciated. A good example of this is the so-called caninus muscle, *levator anguli oris*. LAO lies deep to the 4 heads of *quadratus labii superioris* (also called levator labii oris). The primary insertion of LAO is at modiolus and its secondary insertion into the "canine fossa," located high up on the face of the maxilla, just below the infraorbital foramen. This fossa is a misnomer, having nothing to do with canine tooth. It is separated from incisor fossa by the *canine ridge*, an eminence properly due to the root of the canine.

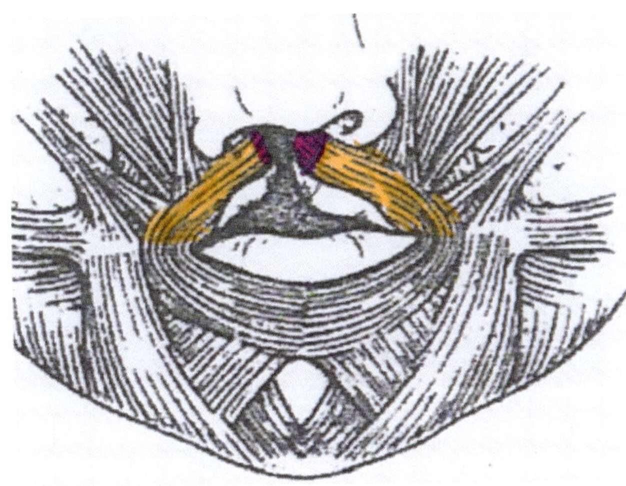

Fig. 18.13 Unilateral cleft lip showing interruption of muscles. Note the unbalanced muscle pull on the columella. [Reprinted from Slaughter WB, Henry JW, Berger JC. Changes in blood vessel patterns in bilateral cleft lip. Plast Reconstr Surg Transplant Bull. 1960 Aug;26:166–79. With permission from Wolters Kluwer Health, Inc.]

- LAO lies lateral to canine ridge
- Lateral head of nasalis inserts medial to canine ridge

Depressor septi nasi and nasalis have been confused with one another due to technical execution of the dissection. De Souza Pinto, began from under the lip, proceeding upward through orbicularis, described depressor septi nasi as having *three* heads, with the most medial being directed to the septum and the lateral two being inserted into the accessory cartilages of the nose. In reality, he actually described the two heads of nasalis, seen in situ on the premaxilla. Because he did not complete the dissection above the cartilages, he missed the functional relationship with the remainder of nasalis. His study correctly recognized DSN to be of significance for the aesthetic management of the nasal lip, but this was not related to the functional significance of the latter two heads (Figs. 18.17, 18.18, 18.19, 18.20, 18.21 and 18.22).

Rohrich provides a good review of DSN studies. Barbosa correctly showed DSN to relate strictly to septum. He noted

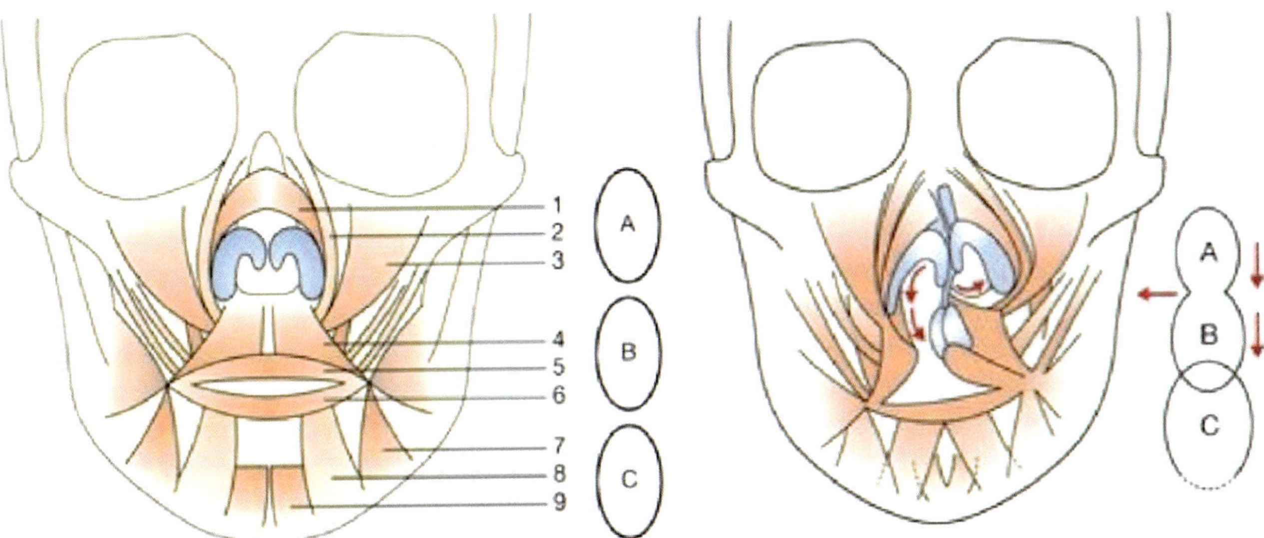

Fig. 18.14 Muscle ring theory of Delaire. A, nasolabial muscles (1–3), B, bilabial muscles (4–6), C, labiomental muscles (7–9). Nasalis pars transversalis (1), levator labii superioris et alaeque nasi (2), levator labii superioris (3), orbicularis obliquus (4), orbicularis transversus (5), orbicularis lower lip (7), depressor labii inferioris (8), mentalis (9). [Reprinted from Markus AF, Delaire J. Primary closure of cleft lip. Br J Oral Maxillofac Surg 1993; 31(5):261–291. With permission from Elsevier]

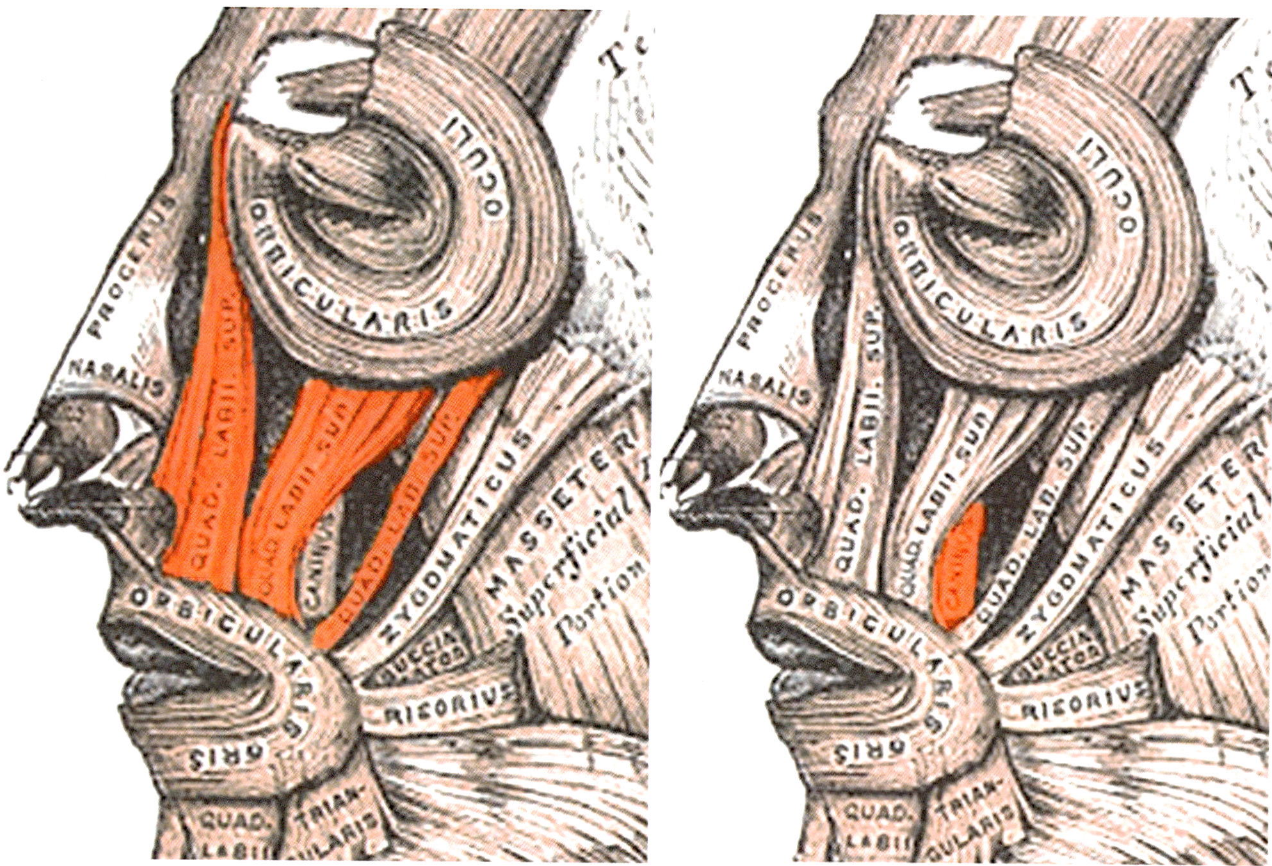

Fig. 18.15 Muscles inserting into upper lip and modiolus. *Superficial layer* Levator labii superioris (red), also known as quadratus, due to its 4 heads. The medial 3 have primary insertion into the SMAS of upper fibers of orbicularis. The lateral head has primary insertion into modiolus and is coplanar with zygomaticus major and risorius. *Deep layer* includes buccinator and caninus, also known as levator anguli oris. LAO has primary insertion at modiolus and secondary insertion into the misnamed canine fossa. Canine root. [Reprinted from Lewis, Warren H (ed). Gray's Anatomy of the Human Body, 20th American Edition. Philadelphia, PA: Lea & Febiger, 1918]

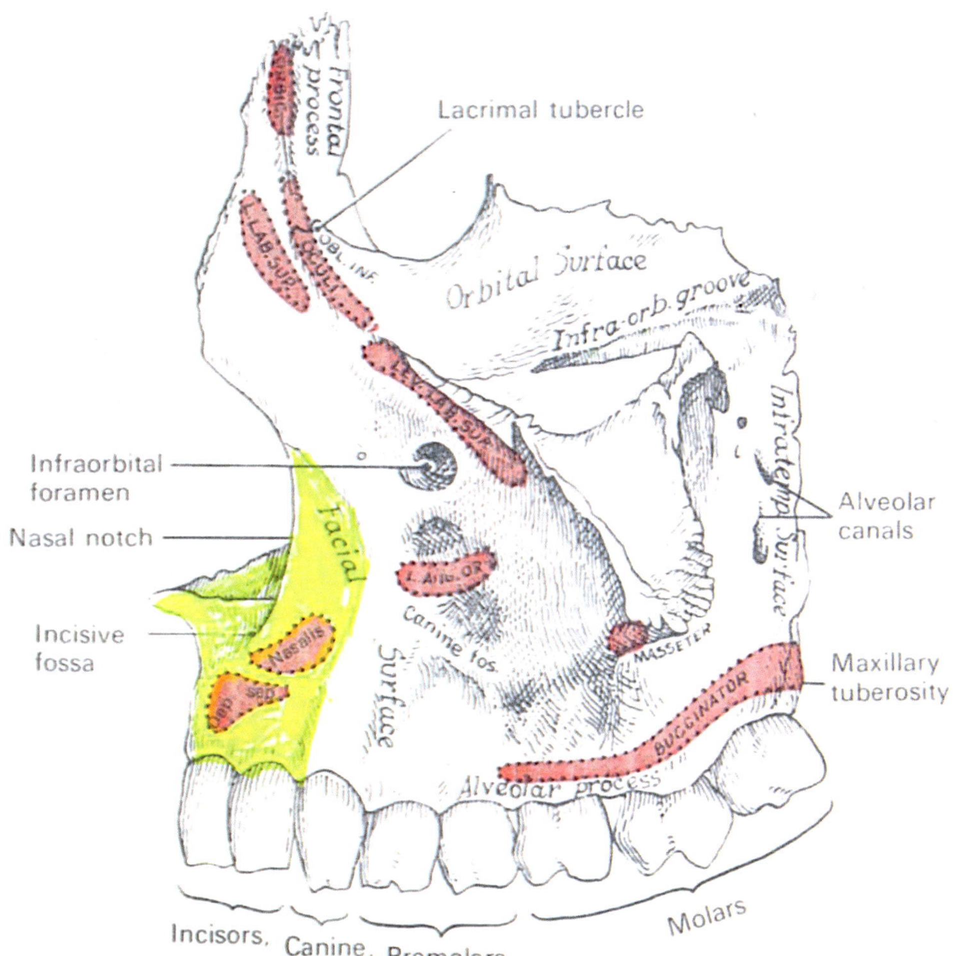

Fig. 18.16 Levator anguli oris and canine fossa. Note: (1) Triangular area over central incisor is insertion for depressor septi nasi. (2) Broad area over lateral incisor extending to canine ridge bears insertion of two heads of nasalis. (3) The canine ridge extends upward in curvilinear path as lateral piriform wall. Medially it shelters lateral head of nasalis. Laterally, it defines the canine fossa, so-labeled, beneath infraorbital foramen as the secondary insertion of caninus, i.e., levator anguli oris. Premaxilla (yellow) has frontal process, PMxF, ascending from lateral incisor process, PMxB, This zone is overlapped by frontal process of maxilla, MxF. This zone, between incisor 2 and canine is biologically capable of have a third incisor mesial to canine. [Reprinted from Lewis, Warren H (ed). Gray's Anatomy of the Human Body, 20th American Edition. Philadelphia, PA: Lea & Febiger, 1918]

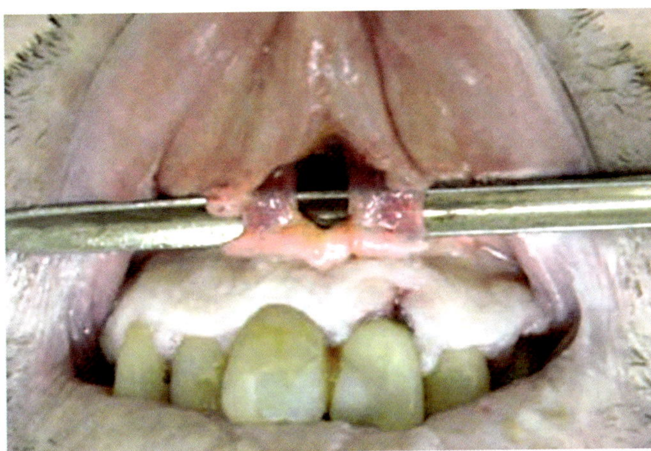

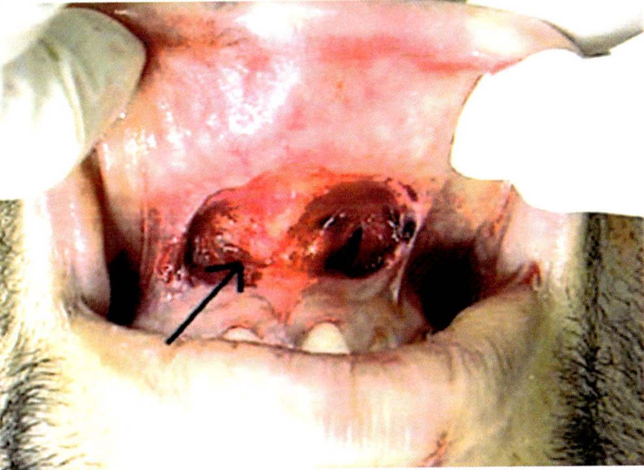

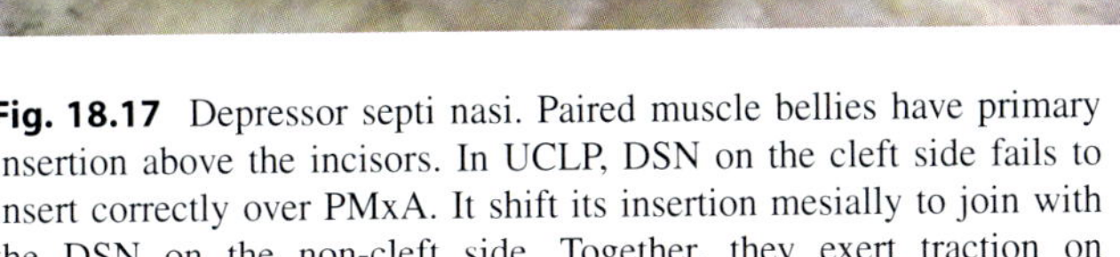
Fig. 18.17 Depressor septi nasi. Paired muscle bellies have primary insertion above the incisors. In UCLP, DSN on the cleft side fails to insert correctly over PMxA. It shift its insertion mesially to join with the DSN on the non-cleft side. Together, they exert traction on Pitanguy's ligament to traction the columellar base toward the non-cleft side. **Unilateral DSN**. [Reprinted from Barbosa MV, Nahas FX, Fereria LM. Anatomy of the depressor septi nasi. J Plast Surg Hand Surg. 2013 Apr;47(2):102–5. With permission from Taylor & Francis]

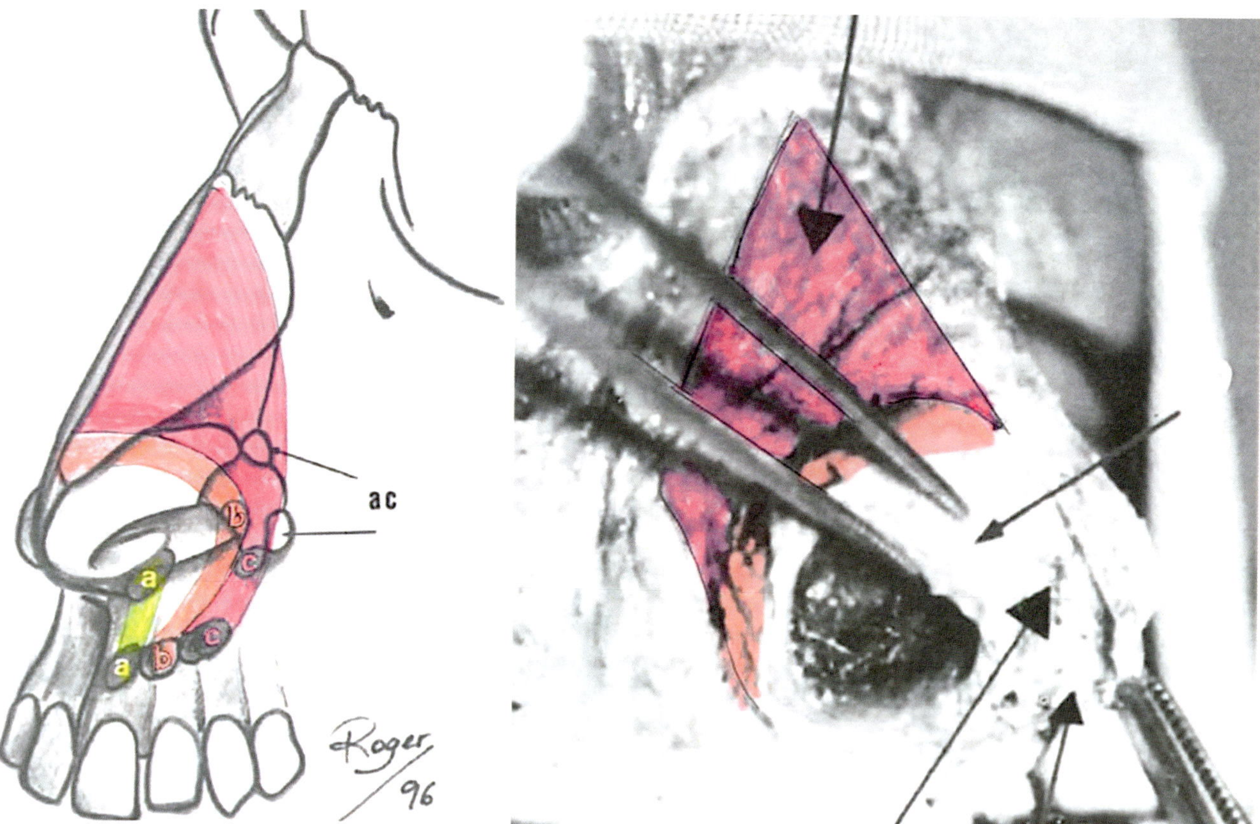

Fig. 18.18 DSN and nasalis. This is a very important concept. De Souza and others confuse lateral muscles as DSN with multiple heads. Nasalis, the rightful occupant of lateral incisor and canine insertion sites, was missed because of the direction of dissection was from below-upward

- DSN above the central incisor (yellow) acts on the mobile septum through the dermocartilagenous ligament (Pitanguy)
- Medial head of nasalis (tangerine) inserts above lateral incisor and extends over the alar cartilage as *pars alaris*
- Lateral head of nasalis (pink) inserts above canine and extends over the dorsum as *pars transversalis*

On the right we have a median section showing DSN (yellow) inserting into orbicularis (orange) and upward into the ligament of Pitanguy (P), i.e., the dermocartilagenous ligament (brown) sweeping in front of the septum (it is easily de identified) and over the midline of the dorsum. Traction on P pulls the nasal tip down with facial animation, especially smiling

[Modified from De Souza Pinto EB, Porto Da Rocha R, Queiroz Filho W, et al. Anatomy of the median part of the septum depressor muscle in aesthetic surgery. Aesth. Plast. Surg. 1998;22:111–115. With permission from Springer Nature]

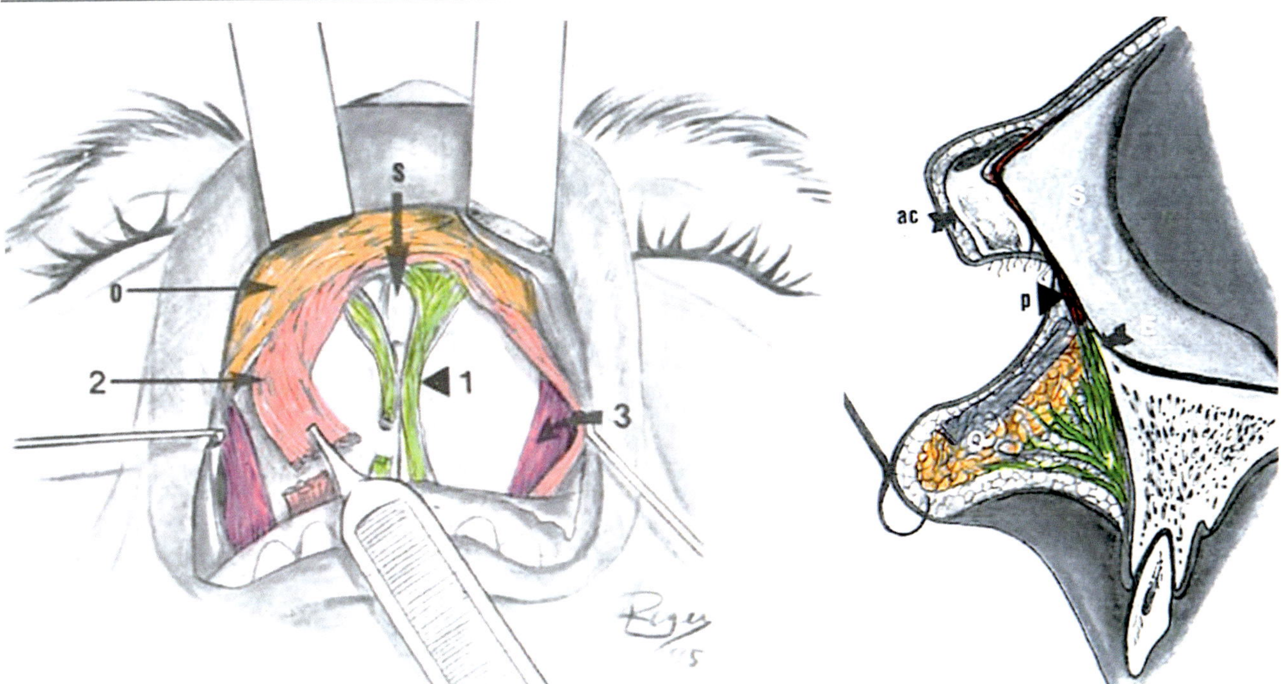

Fig. 18.19 DSN and nasalis. This is a very important concept. De Souza and others confuse lateral muscles as DSN with multiple heads. Nasalis, the rightful occupant of lateral incisor and canine insertion sites, was missed because of the direction of dissection was from below-upward. DSN above the central incisor (yellow) acts on the mobile septum through the dermocartilagenous ligament (Pitanguy). Medial head of nasalis (tangerine) inserts above lateral incisor and extends over the alar cartilage as pars alaris. Lateral head of nasalis (pink) inserts above canine and extends over the dorsum as pars transversalis. On the right we have a median section showing (1) DSN (yellow) inserting into orbicularis (orange) and upward into the ligament of Pitanguy (P), i.e., the dermocartilagenous ligament (brown) sweeping in front of the septum (it is easily de identified) and over the midline of the dorsum. Traction on P pulls the nasal tip down with facial animation, especially smiling. [Modified from De Souza Pinto EB, Porto Da Rocha R, Queiroz Filho W, et al. Anatomy of the median part of the septum depressor muscle in aesthetic surgery. Aesth. Plast. Surg. 1998;22:111–115. With permission from Springer Nature]

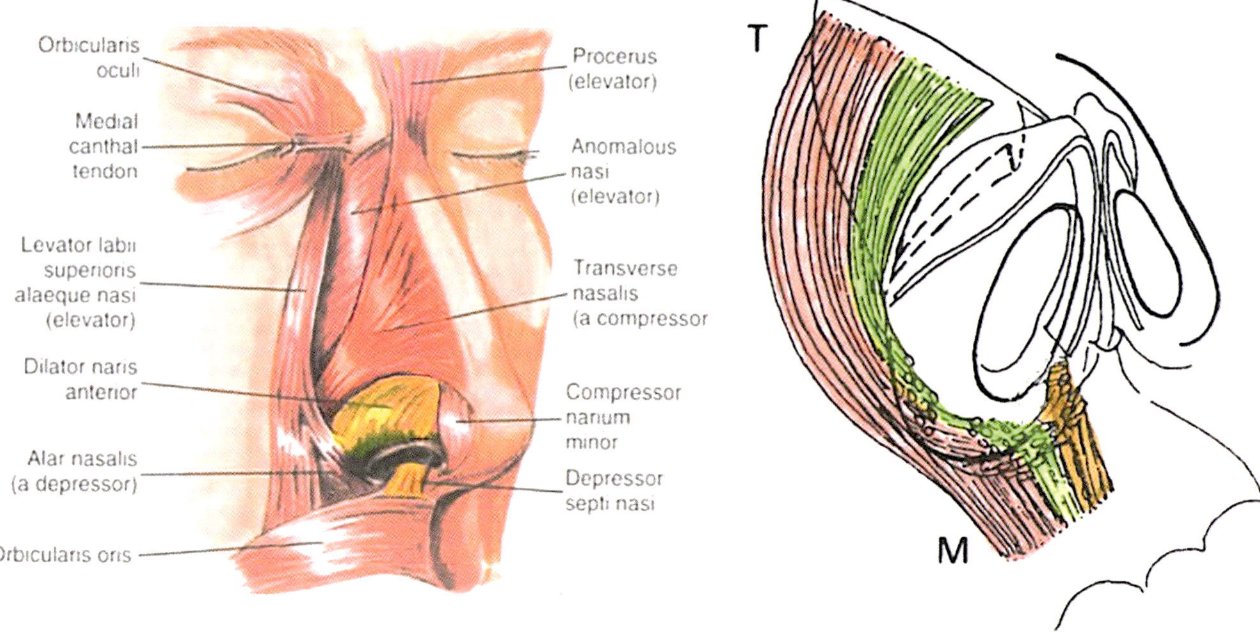

Fig. 18.20 Nasal muscles. Recall that the dental formula for placental mammals is I3, C1, P4, M3. Thus, the premaxilla has three potential binding sites. The site over canine is at the junction of premaxilla and maxilla. The most mesial zone of maxilla frequently houses a third incisor. *Pars transversus* (pink) inserts into the canine fossa. Pars alaris (lemon) inserts above lateral incisor. Depressor septi nasi inserts above the central incisor. The first two muscles act as dilators of the nasal airway. DSN has a midline insertion and tractions on the ligament of Pitanguy. Left: [Reprinted from Markus AF, Delaire J. Functional primary closure of cleft lip. Br J Oral Maxillofac Surg 1993; 31(5):281–291. With permission from Elsevier.] Right: [Courtesy of Michael Carstens, MD]

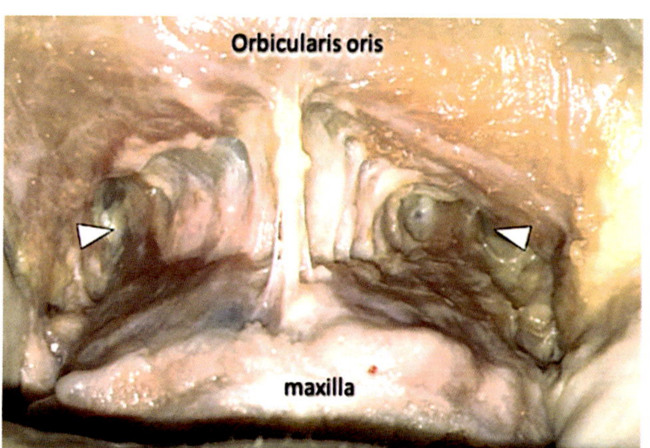

Fig. 18.21 Nasalis and depressor septi nasi in situ. DSN fibers are yellowish and ascend in the midline through orbicularis to seek out the ligament of Pitanguy, not seen because it is on the other side of the orbicularis. Nasalis fibers are a darker red color. The space between the muscles connect the floor of the piriform fossa. [Reprinted from Iwanga J, Watanabe K, Schmidt CK, Voin V, Alonso F, Oskouian J, Tubbs RS. Anatomical study and comprehensive review of the incisivus labii superioris (nasalis) muscle: application to lip and cosmetic surgery. Cureus 2017; 9(9): e1689. DOI 10.7759/cureus.1689. with permission from Creative Commons License 3.0: https://creativecommons.org/licenses/by/3.0/us/]

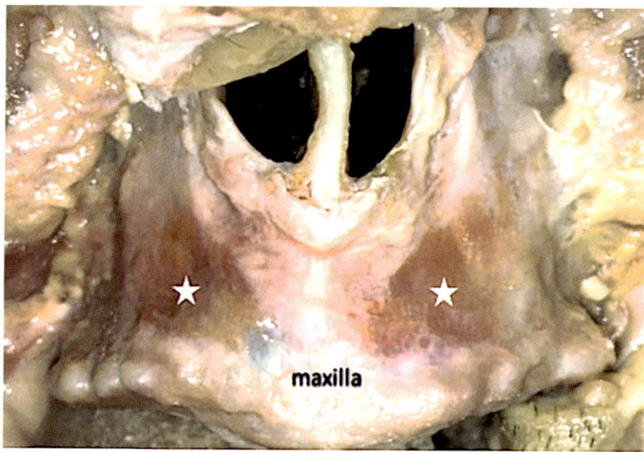

Fig. 18.22 Nasalis in situ over lateral incisor and canine. Fibers of depressor septi nasi (yellow color) over central incisor have been transected. They insert into the ligament of Pitanguy which lies directly in front of the septum. [Reprinted from Iwanga J, Watanabe K, Schmidt CK, Voin V, Alonso F, Oskouian J, Tubbs RS. Anatomical study and comprehensive review of the incisivus labii superioris (nasalis) muscle: application to lip and cosmetic surgery. Cureus 2017; 9(9): e1689. DOI 10.7759/cureus.1689. With permission from Creative Commons License 3.0: https://creativecommons.org/licenses/by/3.0/us/]

the absence of DSN in complete UCL. He also identified a case of rudimentary right-sided DSN, causing deflection of the columellar base to the left. This is of potential value in microform cleft, as it could contribute to deflection of the columellar base. The roles of DSN in clefts are as follows:

- DSN is a deep layer muscle. It forms prior to orbicularis. If a defect in the PMxA binding site exists, DSN from the cleft side could shift over to compete with and bind to the PMxA site on the non-cleft side
- Complete CL, only one DSN is present; complete BCL, it is absent
- The fate of DSN myoblasts in the lateral lip in cleft is unknown

Nasalis has suffered by virtue of terminology as well. Iwanaga and Tubbs provide a comprehensive study of nasalis, which they termed *incisivus labii superioris* (ILS). In their model, they considered ILS to have an inferior part which inserted into premaxilla (the sites are correctly identified) and then inserted into orbicularis. They then describe a superior part of ILS that continues from orbicularis upward to blend into nasalis. Of greatest importance is their demonstration of nasalis function with contraction causing opening of the nasal vestibule at the level of external valve (Figs. 18.23 and 18.24). The rotation-advancement operation misinserts the nasalis with a subsequent compromises in nasal function on the cleft side (Figs. 18.25 and 18.26).

Pathophysiology of Nasalis in the Cleft Condition

In the cleft state, nasalis is always correctly inserted into the nose, with lateral head, *pars transversalis*, draped over the dorsum and medial head, *pars alaris*, sweeping around the margin of lateral crus. The second step is where the error occurs. The binding sites at canine ridge and lateral incisor are not available. Lateral head inserts into the piriform fossa and lateral head into the nostril sill; nasalis now becomes a constrictor.

This condition is readily diagnosed. One observes the patient's breathing while seated with the head tilted backward, from the submental vertex position looking into the nostril and from the vertex submental position, standing behind the patient, looking downward over the dorsum. Observe degree of alar flare on forced breathing. Now, with the patient occluding the aperture of each airway with a fingertip placed gently at the introitus, not compressing the nostril, observe the nostrils and alar motion on forced respiration. Differences between the two sides are readily appreciated and can be photographed.

Standard cleft repairs from Millard's rotation-advancement to Delaire's functional muscle repair have emphasized the correct suture of the lateral lip muscles to the midline, thus creating the "aesthetic drape" of the lip. "Functional closure can now be started, commencing with the closure of the nasal floor from behind forwards… The periosteum, transverse nasalis and myrtiformis muscles are identified and sutured to the midline" [21]. This concept is widely employed by surgeons working in the subperiosteal plane. Those dissecting above the periosteum [22] still anchor the lateral lip element as a unit, often not separating DOO from SOO. In any case, the effect is the same: locking in nasalis as a constrictor, contrary to its function.

Using the outdated concept of nasalis as curving around the ala to blend with orbicularis, I illustrated this incorrectly as well in the 2004 iteration of functional matrix cleft repair before the discovery of the Padget and the angiosome map. The lateral nasal wall was released to free up the nostril sill flap and replace it with a turbinate flap (the latter tissue is still a valuable technique) (Figs. 18.27, 18.28 and 18.29). Despite this, the nasal airway did not improve. In retrospect, I was guilty of reassigning nasalis to the wrong place. Bagatain and Larrabee demonstrate the rescue of nasalis muscle through an incision below the nostril sill, only to anchor them medially. The consequences should be quite clear (Fig. 18.30).

Righting the Wrong

In the surgical technique section of this chapter, we will look at incision design as applied to the microform and incomplete variants and how nasalis can be accessed. The details of developmental field reassignment surgery are left, in large measure to Chap. 19. Suffice it to say that, in revision surgery, it is *always* possible to retrieve nasalis and anchor it correctly to the periosteum and buccal sulcus above the canine.

The decision whether nasalis reconstruction is worthwhile must be based on the perception of the patient of function and aesthetics. If elevating the dome and lateral crus with a cotton tip applicator opens the airway, the diagnosis is made. On occlusion of the non-cleft side the patient should experience a perceptible improvement in breathing. The test is equally valid for bilateral cases, except there is no control… just the sense of better airflow. In cases where the alar cartilage deformity is significant, nasalis reconstruction will also improve the long-term outcome. In most cases, the objective sense of relief experienced by the patient to the first test far outweighs aesthetic concerns.

In Sum

- Microforms not requiring takedown of the lip: access via nostril sill incision and intraoral dissection.
- Microforms and incomplete forms requiring lip takedown: access nasalis in the standard manner via the lateral lip incision and intraoral subperiosteal exposure.

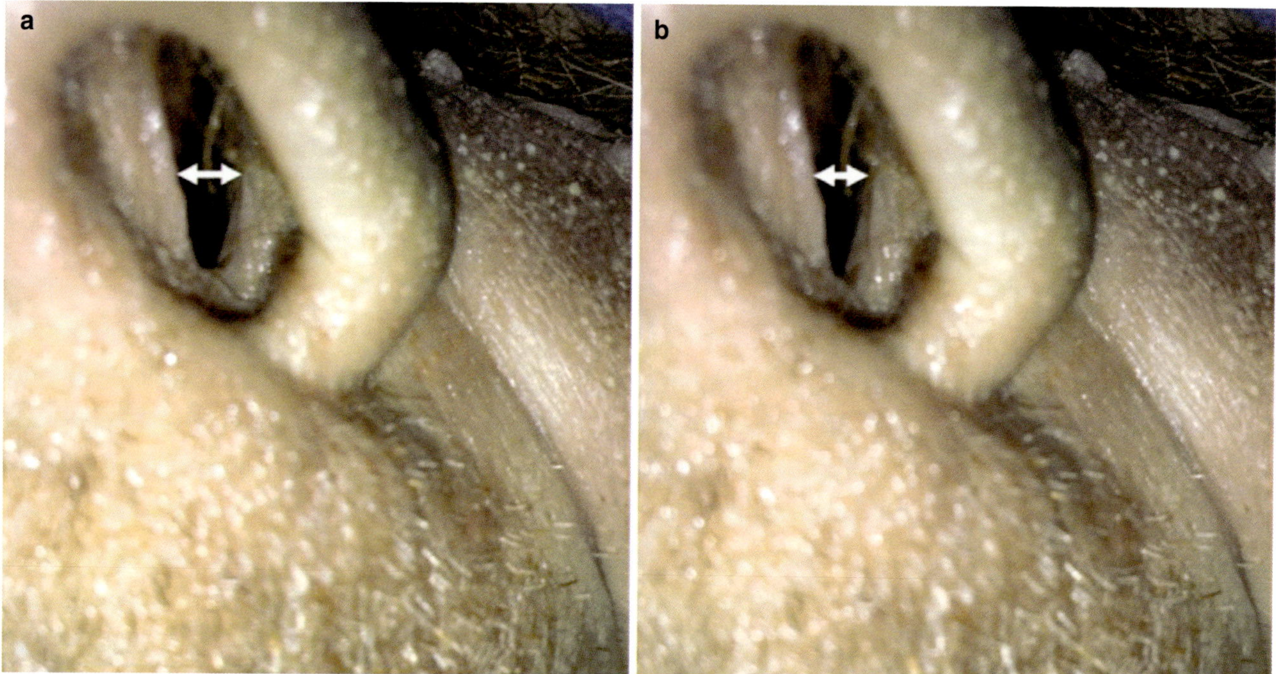

Fig. 18.23 External nasal valve and the function of nasalis. Nasal valve opens with muscle retraction (**a**) and closes in the relaxed state (**b**). [Reprinted from Iwanga J, Watanabe K, Schmidt CK, Voin V, Alonso F, Oskouian J, Tubbs RS. Anatomical study and comprehensive review of the incisivus labbii superioris (nasalis) muscle: application to lip and cosmetic surgery. Cureus 2017; 9(9): e1689. DOI 10.7759/cureus.1689. With permission from Creative Commons License 3.0: https://creativecommons.org/licenses/by/3.0/us/]

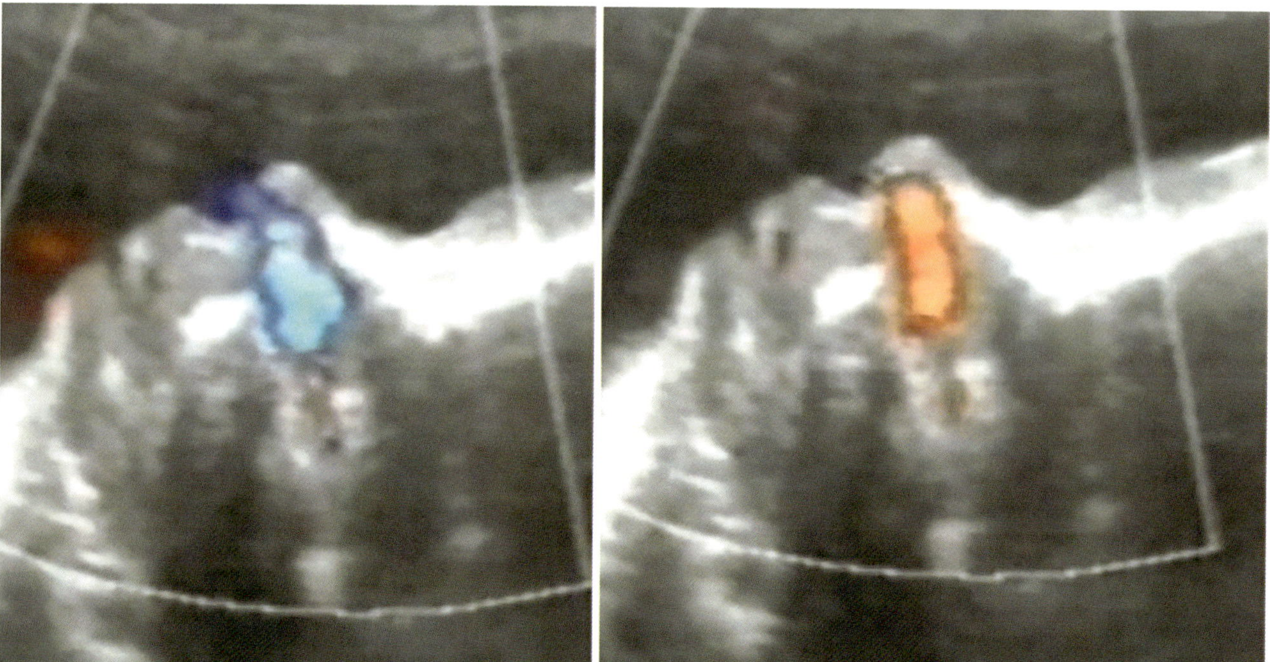

Fig. 18.24 Fetal breathing. Fetal breathing was first reported by Boddy and Dawes in 1975 and reviews by Fox in 1976. It begins between 40 and 60 days. Amniotic fluid exerts a hydraulic pressure effect within the nasal chambers that shape the septum. For a video of this process go to the link below. Knowledge of this process profoundly affected Talmant's concept of airway reconstruction. The link below demonstrates nasal breathing https://doi.org/10.1093/oxfordjournals.bmb.a071237 (nasal only). This link shows both nasal and oral breathing. You can actually see the **nasal chambers expand and contract**. The **heart** can be seen as well. https://www.youtube.com/watch?v=g2voFRimLXw (my favorite). [Courtesy of Jean-Claude Talmant]

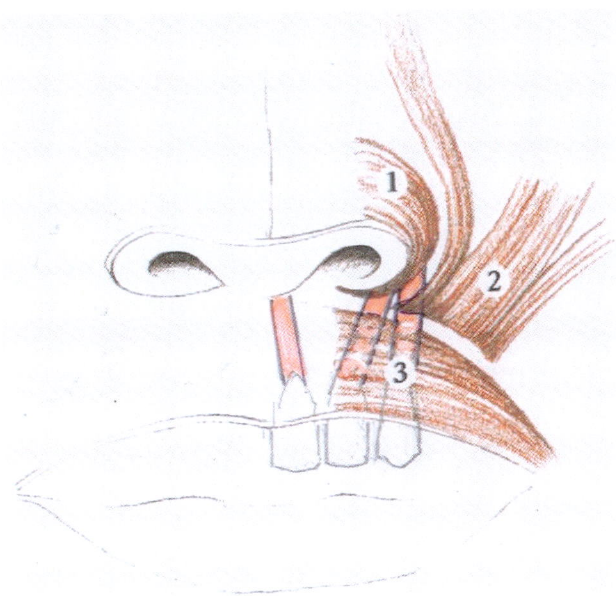

Fig. 18.25 Nasalis misconception. This is a standard model of nasalis, with it curling around the ala and terminating in the sill. The alternative shows depressor septi nasi muscle over the central incisor and the two heads of nasalis over lateral incisor and canine. Note theses are higher that DOO the upper border of which terminates in the sulcus. Key: (1) nasalis; (2) levator labii superioris and zygomaticus major; (3) orbicularis. [Courtesy of Michael Carstens, MD]

Neurovascular Map of the Prolabium: The Philtral Prolabium Versus the Non-philtral Prolabium

In this section, we consider the angiosomes that supply the facial midline, specifically the distribution of r1 mesenchyme originating from the midbrain neural crest in association with the frontonasal process versus r2 mesenchyme originating from the hindbrain in association with the first pharyngeal arch and with the muscles of the second arch. We shall see that the map of these tissues are perfused by the StV1 anterior ethmoid axis and the StV2 medial nasopalatine axis.

Injection Studies

Early arteriography of the entire carotid system by Miroslav Fará demonstrates relevant findings (Figs. 18.31, 18.32, 18.33 and 18.34).

Experimental data are provided here from fetal injection studies reported in 2002. These first disclosed the existence of distinct embryonic components in the columella and prolabium, thereby showing (1) rotation-advancement to be incompatible with developmental fields, and (2) the exis-

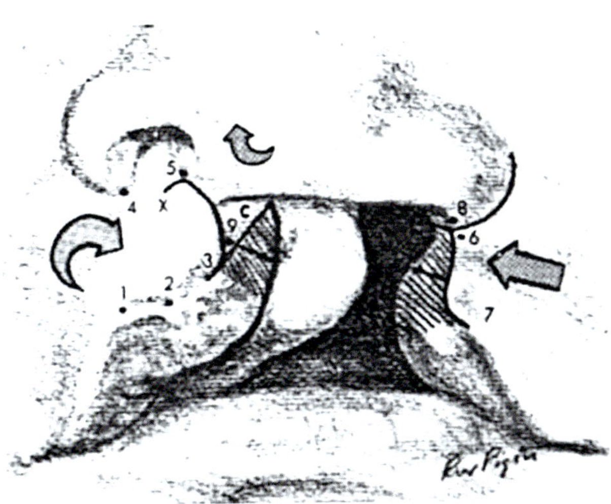

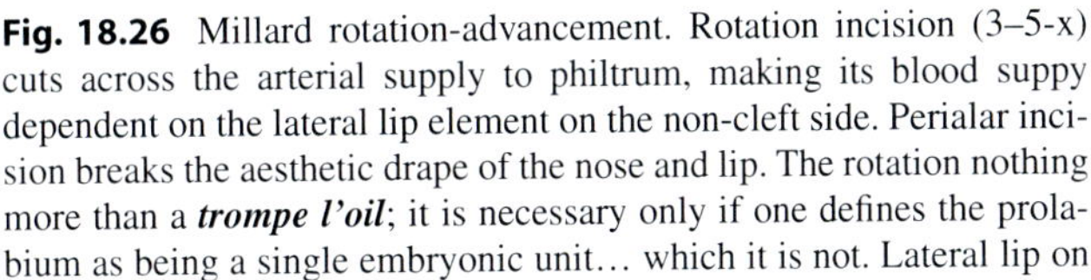

Fig. 18.26 Millard rotation-advancement. Rotation incision (3–5-x) cuts across the arterial supply to philtrum, making its blood suppy dependent on the lateral lip element on the non-cleft side. Perialar incision breaks the aesthetic drape of the nose and lip. The rotation nothing more than a ***trompe l'oil***; it is necessary only if one defines the prolabium as being a single embryonic unit… which it is not. Lateral lip on the cleft side is opened (7–8) and the muscle sutured en-bloc to the base of the columella. ***Note*** DFR measures the philtral width equal to the columella with the cleft side peak very close to Millard point 2. [Reprinted from Fisher DM, Sommerlad BC. Cleft lip, cleft palate, and velopharyngeal insufficiency. Plast Reconstr Surg 2011; 128 (4): 342e–360e. With permission from Wolters Kluwer Health, Inc.]

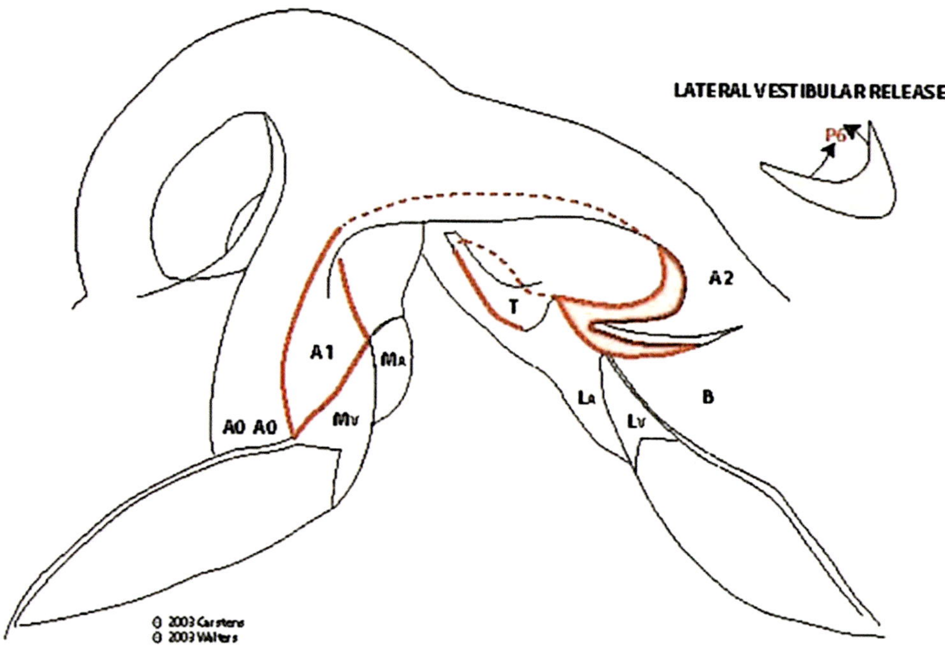

Fig. 18.27 Original design for functional matrix repair recognized the width of the true philtrum as being that of the width of the columella. Remaining non-philtral prolabium (based on medial nasopalatine) was kept in continuity with medial crus based on anterior ethmoid. The combination flap advanced the medial crus and provided tissue based on the columella for the nasal floor. The flap is designated **NPP-LCC**: the non-philtral prolabium lateral columellar chondrocutaneous flap. Releasing incision behind the nostril sill received a turbinate flap… still usefuls (Fisher 2005) but the true advancement of the vestibular lining would not be achieved in DFR until I learned of the Talmant maneuver which would separate it bluntly from the level of piriform fossa all the way up to the undersurface of the nasal bones (see Chap. 19 for details). [Courtesy of Michael Carstens, MD]

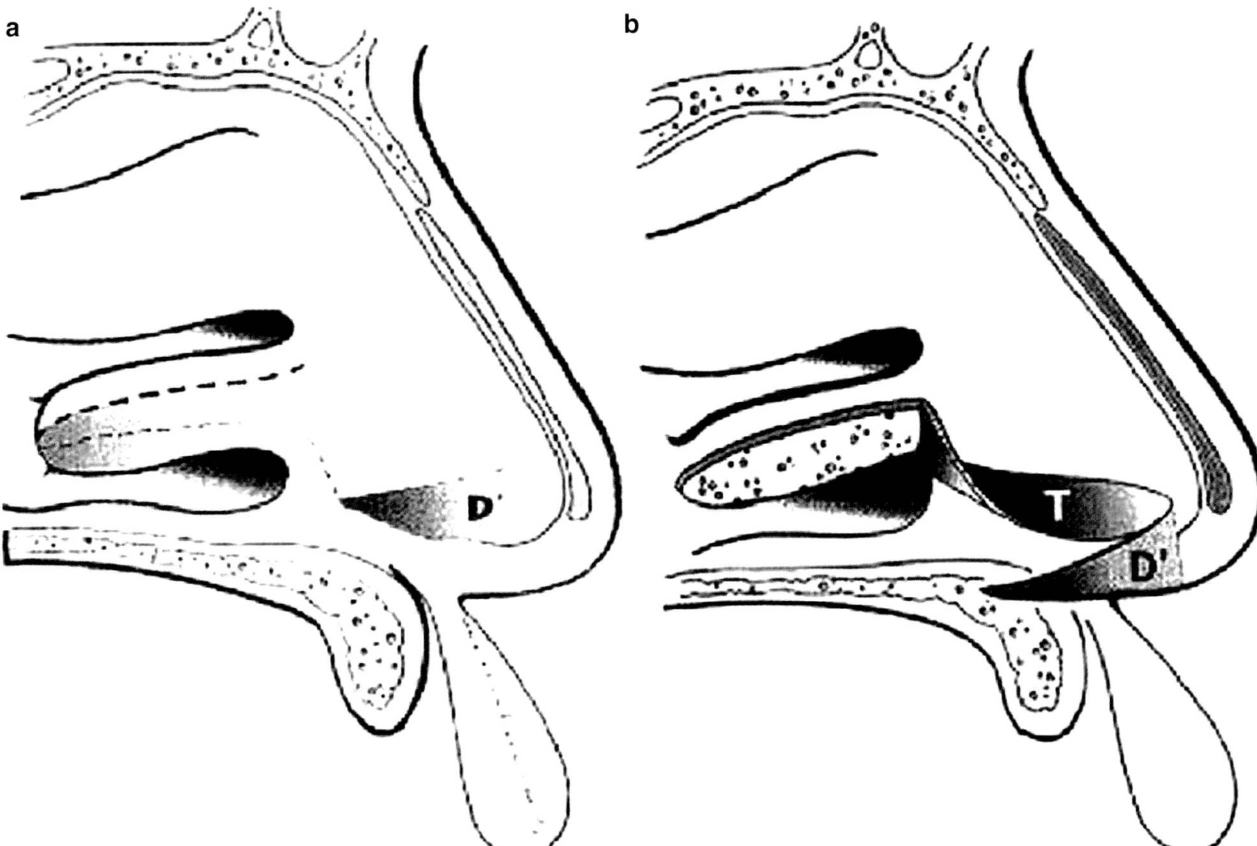

Fig. 18.28 Turbinate flap. (**a**) I started using this flap to replace the deficit left behind when the nostril sill flap D was rotated outward from the lateral nasal wall. The turbinate flap is simple to harvest. The donor side can be cauterized and mucosalizes over in 2–3 weeks. Parallel incisions create a "bucket-handle" flap with an anterior base. (**b**) The NPP-LCC reaches only the medial half of the nasal floor. Suturing it all the way over into the D′ defect will create a contraction band along the nasal floor and compromise the width of the airway. This flap remains quite useful. [Courtesy of Michael Carstens, MD]

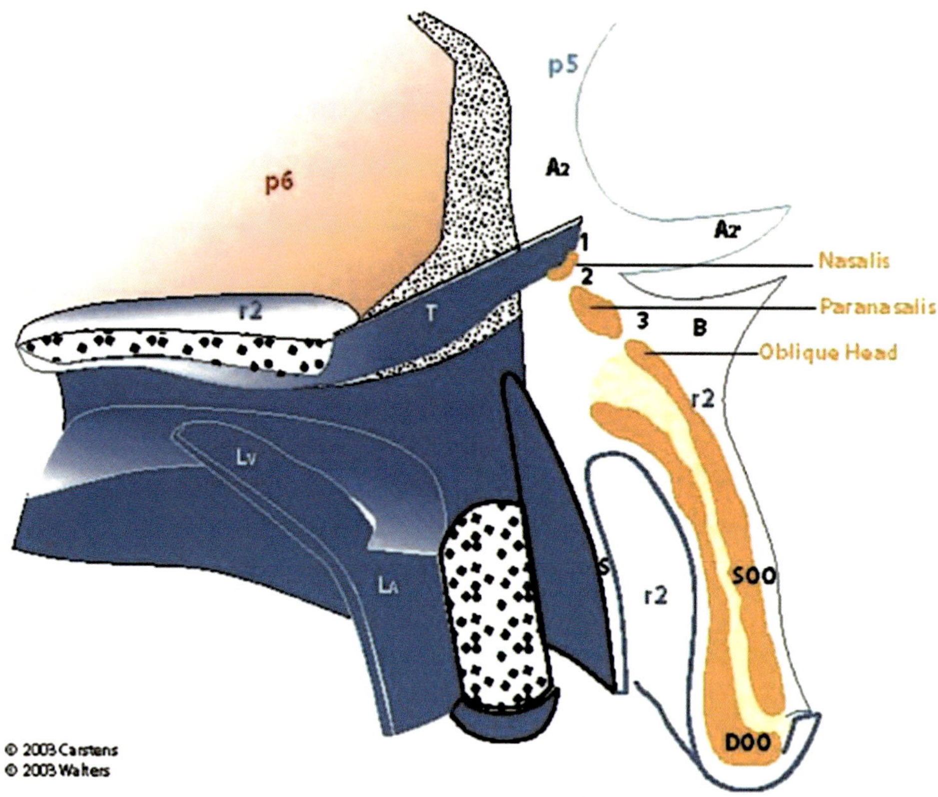

Fig. 18.29 Lateral wall dissection. The design of this dissection shows three mistakes I made along the way to DFR repair

- First, I thought "tightness" of the vestibular lining (p6) was due to a local defect in the lateral piriform fossa; and that this could be released and patched with a turbinate flap (T) or composite graft. This move, combined with alar cartilage repositioning, would result in an improved airway. This was flat-out wrong. Release of p6 takes place using wide mobilization between the layers (Talman maneuver)
- Second, the release of the sliding sulcus flap (S) by a gingival incision and the reflection of mucoperiosteal flap (L_A) from the alveolar cleft margin were used to create a primary gingivoperiosteoplasty (GPP). This is not necessary. The best algorithm is closure of the nasal floor at lip repair followed by AEP closure of all remaining walls except the anterior. These steps eliminate fistula. Grafting at age 4 is completed using an S flap
- Third, nasalis is not small, Its lateral head. Its lateral head (1) must be retrieved from the piriform fossa and the medial head (2) from the nostril sill. The following concept, in my own words (2004) remains in use in most UCL repairs and is embryologically *wrong*
- "Nasalis complex sutured to the ipsilateral membranous septum achieves vertical and anteroposterior alar base positions."

This drawing does correctly show DOO and SOO. It also recognizes the oblique fibers of Delaire at the apex of SOO. Suture of these to base of columella achieves final aesthetic "drape" of the lip

[Courtesy of Michael Carstens, MD]

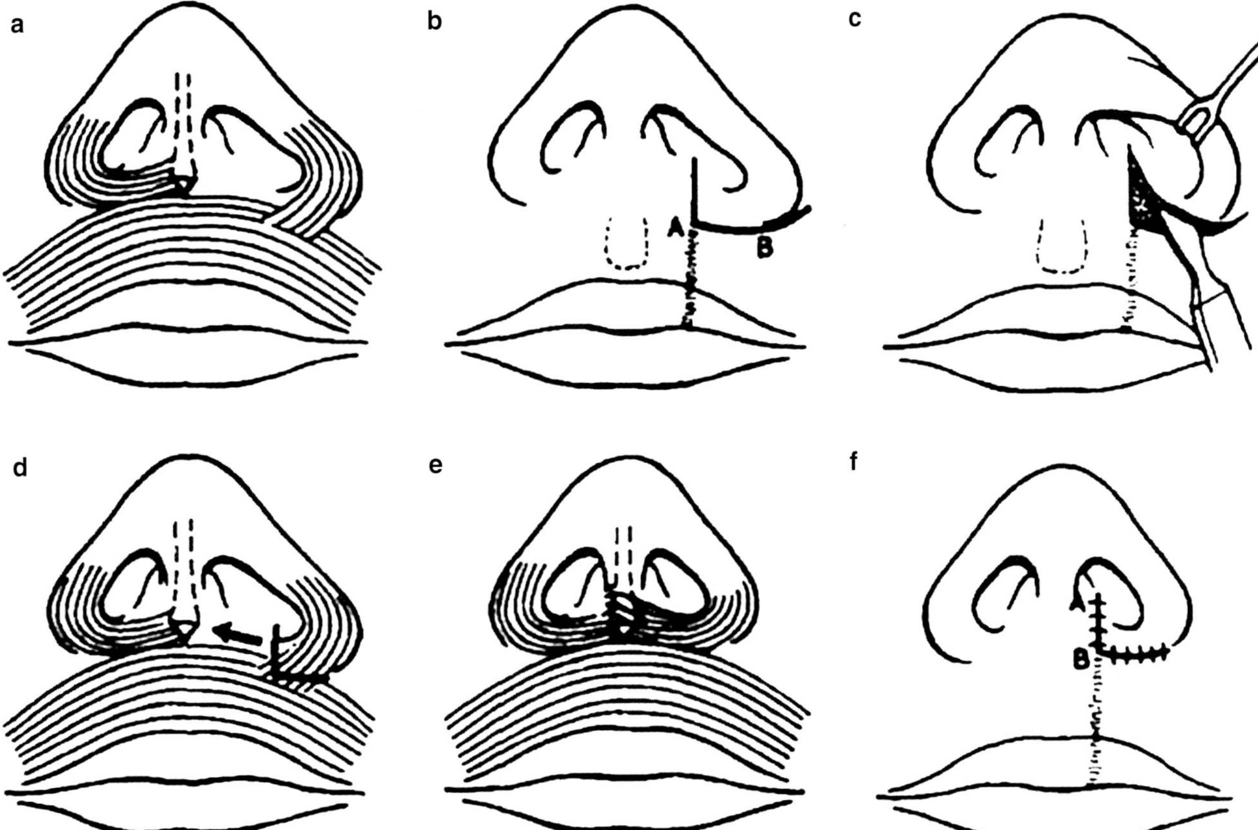

Fig. 18.30 Secondary repair of the nasalis leads to nasal obstruction. (**a**) Nasalis remains mal-positioned after virtually **all** cleft repairs. (**b**, **c**) Nasalis dissection does indeed involve "fishing the muscle out" with an anterior peri-alar incision. The latter should *not* extend laterally around the side of the alar base. (**d**) The three primary mistakes with nasalis dissection are:

- failure to separate it from orbicularis
- failure to free it from the piriform fossa
- suture fixation of the muscle to the midline

(**e**) Using the same incision design, correct repositioning of nasalis to the mucoperiosteum and sulcus at the canine can be achieved [Reprinted from Bagatain M, Khosh M, Nishoka G, Larrabbe WF Jr. Isolated nasalis reconstruction in secondary unilateral cleft lip nasal reconstruction. Laryngoscope 1999; 109 (2 Pt 1):320–323. With permission from John Wiley & Sons]

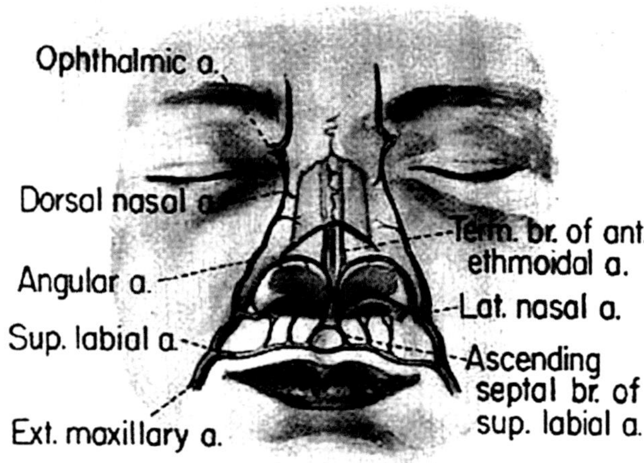

Fig. 18.31 Arterial supply to normal upper lip in normal. Millard popularized Slaughter's drawings. The paired StV1 anterior ethmoids are shown emerging from beneath the nasal bones and travelling in a midline depression just above the septum. StV1 dorsal nasals connect with ECA angular at the level of the alae. ECA and ethmoids anastomose over the alar cartilages, along the nasal floor and descend into the philtrum. Slaughter could have readily guessed about the separate circulations if selective dye injections had been used. Connection between greater palatine and the medial nasopalatine axis occurs via the incisive canal. [Reprinted from Slaughter WB, Henry JW, Berger JC. Changes in blood vessel patterns in bilateral cleft lip. Plast Reconstr Surg 1960; 26(2):161–179. With permission from Wolters Kluwer Health, Inc.]

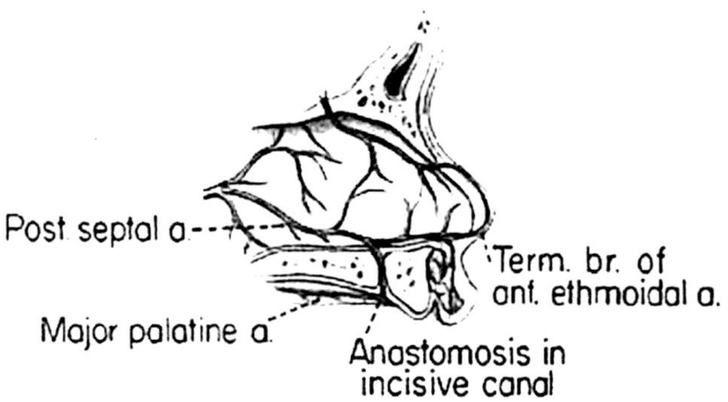

Fig. 18.32 Original drawing by Slaughter showing that the **terminal branch of anterior ethmoid** supplied the midline of the philtru. He but did not recognize the separate circulations of StV1 and StV2. [Reprinted from Slaughter WB, Henry JW, Berger JC. Changes in blood vessel patterns in bilateral cleft lip. Plast Reconstr Surg 1960; 26(2):161–179. With permission from Wolters Kluwer Health, Inc.]

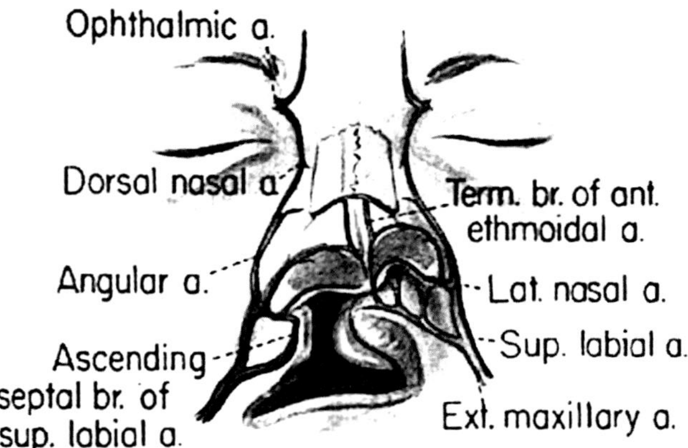

Fig. 18.33 Incomplete right-sided cleft with soft tissue "bridge" between alar base (LNP) and columella/prolabium (MNP). Note the presence of arterial supply in the bridge. [Reprinted from Slaughter WB, Henry JW, Berger JC. Changes in blood vessel patterns in bilateral cleft lip. Plast Reconstr Surg 1960; 26(2):161–179. With permission from Wolters Kluwer Health, Inc.]

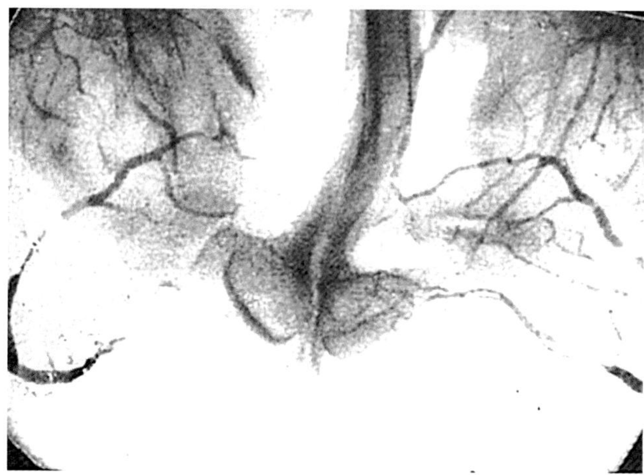

Fig. 18.34 Prolabial blood supply in bilateral cleft lip. Arteriography by Fará but this finding went unrecognized. Here paired anterior ethmoid vessels are seen descending. "The underdevelopment of the muscles and the poorer blood supply in *the half of the philtrum facing the cleft* (**non-philtral prolabium**) suggests that the ability of the orbicularis to grow across the midline is limited. It is as if the muscles of one half of the lip are incapable of supplying with muscle the part that actually belongs to the other side." "in the prolabium there was always a rich vascular network starting in the septal (aa. nasales posteriores septi) and the columellar (aa. ethmoidalis anterior)." [Reprinted from Fará M. Anatomy and arteriography of cleft lips in stillborn children. Plast Reconstr Surg 1968; 42(1):29–36. With permission from Wolters Kluwer Health, Inc.]

tence of an embryonic separation plane using a modified straight-line incision based on mapping the prolabium according to its neurovascular fields, discarding the "obvious" point 3 as assumed for the previous 50 years. The series included several cases of holoprosencephaly, cebocephaly, etc. which I previously described, so we shall concentrate on normal specimens and a fetal specimen with right-sided complete unilateral cleft lip.

Methodology

Fetuses under the age of 22 weeks were presented for necropsy at the Neonatal Pathology Section, Department of Pathology, Kaiser Hospital Oakland, under the direction of Dr. Geoffrey Machin. The common carotid arteries were cannulated with a #25 catheter. Ligation of the external carotid, subclavian and vertebral arteries was performed. Blue dye was injected; 6–10 cc of dye were sufficient per case. Photographs were taken using a copy stand with a Nikon 105 mm lens at f22 and f32 and Kodak ASA 100 Ektachrome film.

After examination of the face, cerebrectomy was performed and the skull base examined. In selected cases all midline structures were resected en-bloc. Each of these specimens consisted of the following: cribriform plate, crista galli and ethmoid sinuses, the lateral wall of the nose, the entire palate, nose including nasal bones, columella, philtrum, septum, vomer and the perpendicular plate of the ethmoid. The specimen was separated posteriorly from the skull base at the spheno-ethmoid junction. Step-wise dissection of the lateral wall of the nose and the palate was carried out to determine the extent of dye perfusion. The final specimen consisted of all midline nasal structures and ethmoid structures, the premaxilla and the lateral wall of the nose to the level of the inferior turbinate (Figs. 18.35, 18.36, 18.37, 18.38, 18.39, 18.40, 18.41, 18.42 and 18.43).

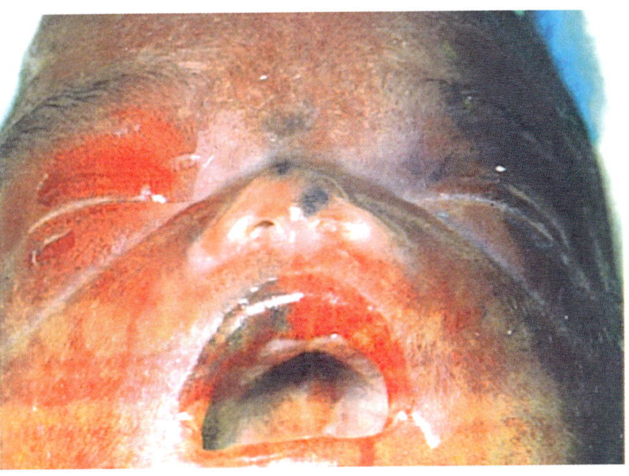

Fig. 18.35 Specimen 1 injection of ICA with ligation of the ECA. Any branches of ECA seen can only come via backflow through anastomoses. Skin is intact. Note the blush of injectate in the nasal tip from anterior ethmoids with backfill along the rim of left nostril and into the right labial artery beneath the wet-dry mucosal junction. [Courtesy of Michael Carstens, MD]

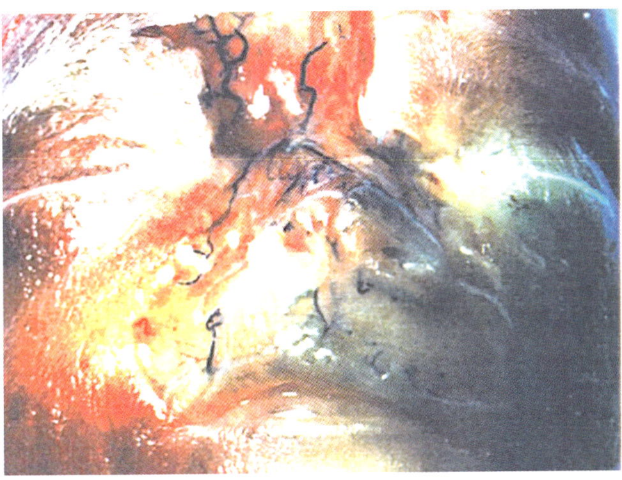

Fig. 18.36 Specimen 2 having the same injection protocol. StV1 arteries are seen after removal of nasal skin. Anterior ethmoids and lateral nasals from infratrochlear artery. Philtral artery is seen in intact skin. Backflow into ECA is seen below the left nostril sill and vertical ascending branch from the labial. [Courtesy of Michael Carstens, MD]

Case 1 Normal Fetus (Fig. 18.35)

Injection of the right common carotid resulted in a blush seen first at the right upper eyelid but then spreading to the supratrochlear skin. As more injectate was administered, pinpoint stains appeared on the scalp (emissary veins), indicating full perfusion of the brain. Similar findings (but to a lesser degree) appeared on the left side indicating midline crossover of the dye. The next finding was a faint blush in the midline of the upper lip at the vermilion of the tubercle followed by lateral spread along the lip. No dye was noted in the skin of the nasal dorsum, cheek or upper lip (see plates).

The nasal skin was split down the midline, revealing paired anterior ethmoid arteries at the upper border of the upper lateral cartilages. These vessels ran in parallel along the upper border of the septum, then turning caudally to pass through the philtrum and thence to the vermilion border. Collateral flow into the external carotid lateral nasal and upper labial branches was noted.

Reflection of the anterior scalp flap revealed an expected perfusion of the supraorbital and supratrochlear vessels. Although the staining pattern was bilateral, the intensity of staining was greater on the right side. At cerebrectomy two normal olfactory nerves were noted. Dye was seen in isolated vessels running in parallel with the supraorbital artery along the interior surface of the frontal bone. Intense staining of the cribriform plate terminated abruptly at the frontoethmoid margins.

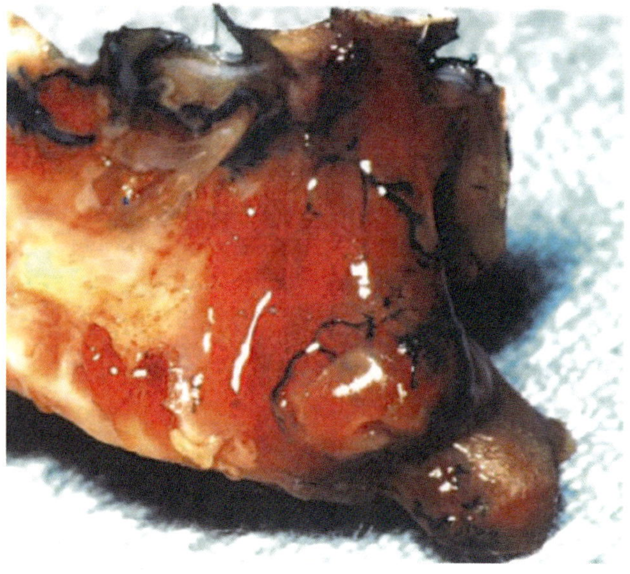

Fig. 18.37 Specimen 3 oblique. Oblique view of en-bloc dissection shows ethmoids coursing forward and then emerging beneath right nasal bone. [Courtesy of Michael Carstens, MD]

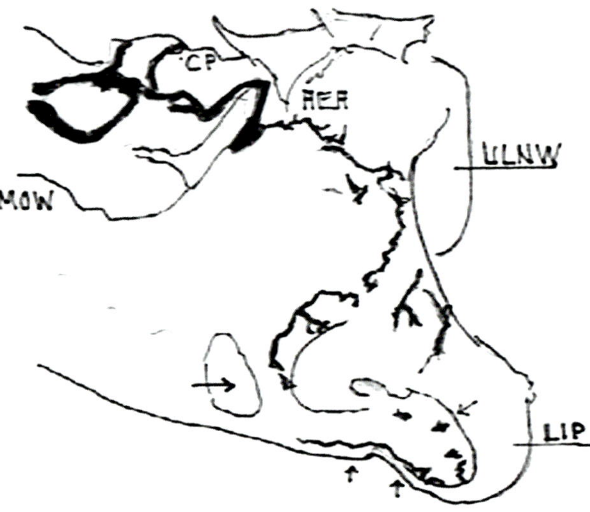

Fig. 18.38 Specimen 3 oblique diagram. [Courtesy of Michael Carstens, MD]

En-bloc resection of the midline resulted in a rectangular box of tissue with the cribriform plate and crista galli constituted its superior margin. The inferior (oral) margin of the specimen disclosed dye running to the anterior nasal spine. The entire hard palate and alveolar ridges (excluding the premaxilla) disclosed no evidence whatsoever of perfusion. When the palate was dissected away from the premaxilla and vomer, no dye was noted at the cut edges of the specimen. Midline section of the lip leaving the nose intact revealed continuity of dye between the philtrum and columella, between the philtrum and premaxilla and between the premaxilla and the base of the nose.

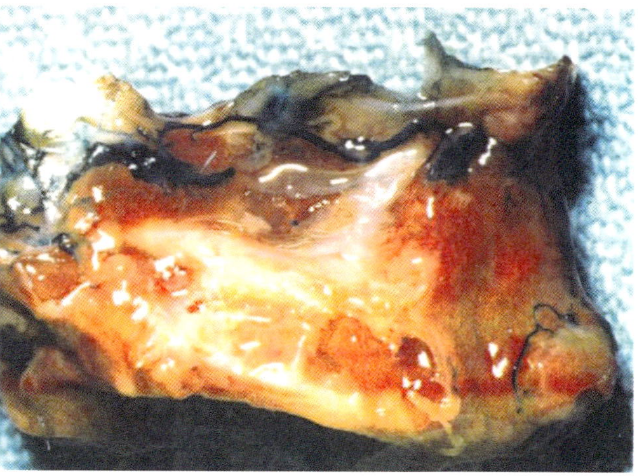

Fig. 18.39 Specimen 3 lateral. [Courtesy of Michael Carstens, MD]

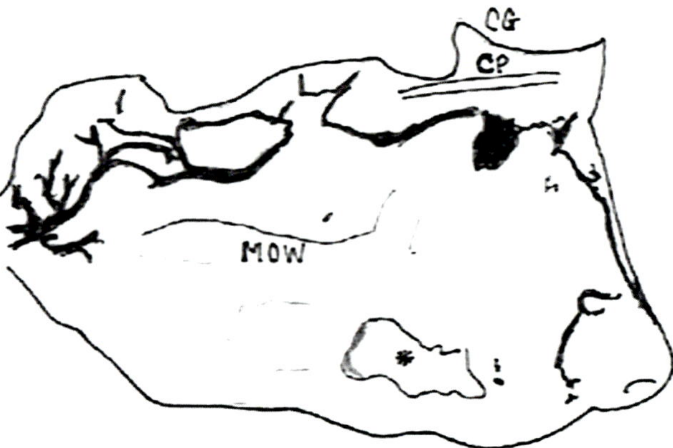

Fig. 18.40 Specimen 3 lateral diagram. [Courtesy of Michael Carstens, MD]

- StV1 in the philtrum has anastomosis with StV2 to premaxilla and anterior nasal spine

Staining of the lateral wall of the nose was limited to its *upper half*, terminating at the level of the orbital floor. The lower half of the lateral margin (consisting of the medial wall of the maxilla) and the palate was resected. No staining was noted along the superior border of the specimen, i.e., inferior turbinate. Thus, the border of StV1 anterior and posterior ethmoids was to the level of middle turbinate.

Along the lateral wall of the nose, dye was noted in the posterior and anterior ethmoid arteries. The latter was seen penetrating into the anterior ethmoid air cells, subsequently to re-emerge from under the nasal bones, running its course astride the septum to penetrate the columella and philtrum as previously described. The lateral wall dissection permitted examination of the perpendicular plate and septum.

Dye was seen in continuity with the nasal side of the cribriform plate, running down the upper border of the septum. It did not perfuse the vomer.

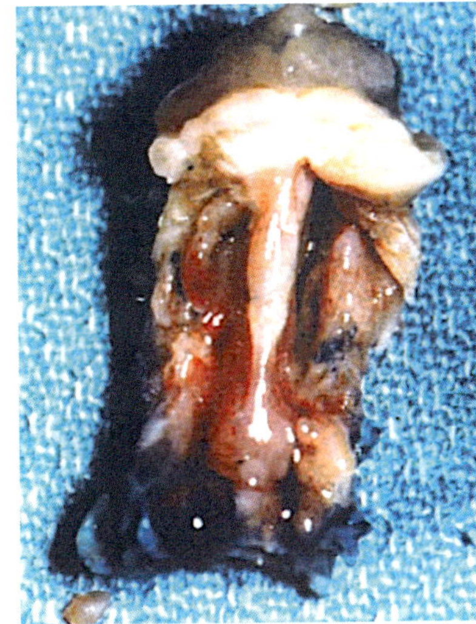

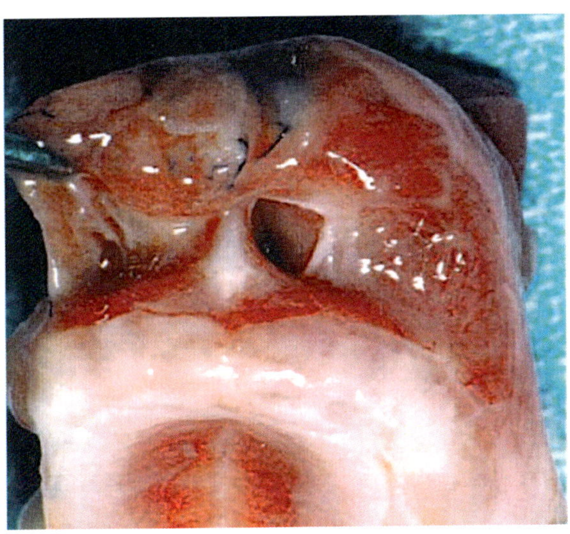

Fig. 18.41 Specimens 1 and 2, palatal view. (Left) Specimen X has a midline cleft palate. The nasal cavity has been dissected free en-bloc with StV1 fields of ethmoid and septum and the lateral nasal walls to the level of inferior turbinate. Hard palate has been removed. (Right) Specimen Y has an intact palate. Mucoperiosteal cover over the premaxilla has been elevated to show the [Courtesy of Michael Carstens, MD]

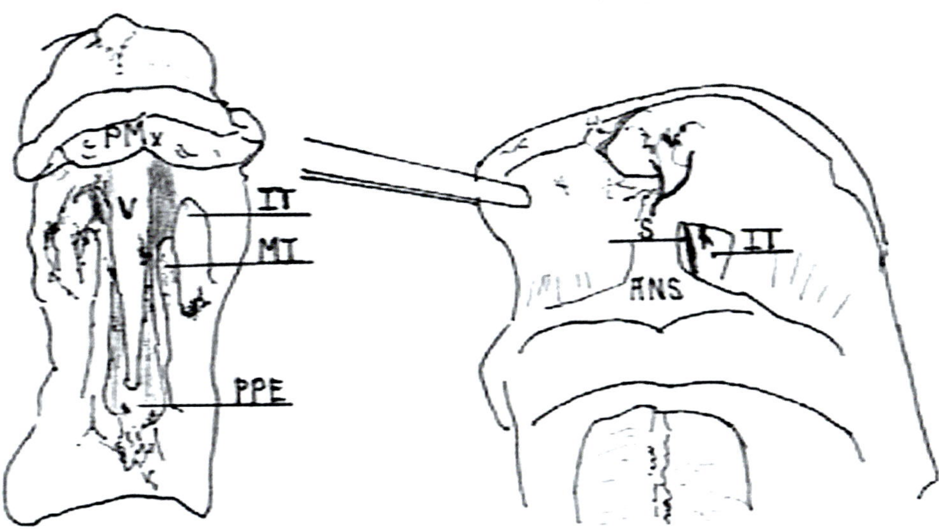

Fig. 18.42 Diagram of 1.1 and 2.1. [Courtesy of Michael Carstens, MD]

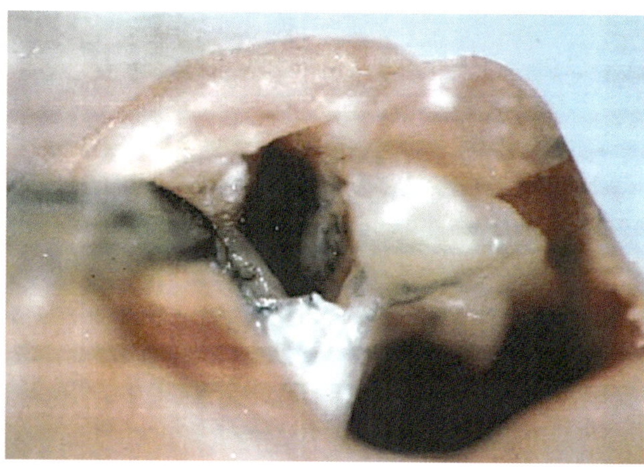

Fig. 18.43 Specimen 4 unilateral cleft lip and alveolus. Secondary palate is intact here. [Courtesy of Michael Carstens, MD]

Case #2 Normal Fetus (Fig. 18.36)

Ten cc of dye divided between both common carotids produced the same sequence of blush production but in equal intensity on both sides. Dye was seen transcutaneously (perhaps due to the greater volume of injectate or symmetry of its application) along the nasal dorsum. *Two vessels were seen running vertically through the columella.* A blush appeared in the center of the vermilion, spreading laterally on both sides to the commissures. This represented backflow from the philtral arteries into the labial arteries. Although the paired vessels were not seen transcutaneously in the philtrum, removal of the skin from the tip of the nose, columella, philtrum and lip demonstrated the paired vessels running from the columella to the vermilion border. Backflow of dye from the medial canthus into the angular artery and from the nasal dorsum into the alae via anastomoses with the external carotid system was consistent with standard anatomic description. Vertically oriented arteries filled from the labial arcade were seen in the late injection phase (as expected).

En-bloc resection of the StV1 fields was performed. Once again, both the lower lateral nasal walls and palate were not perfused. Upon resection of these structures dye was noted in the turbinates and along the upper course of vomer. A diffuse blush involved the septum; along its caudal border the *antero-posterior course of the medial sphenopalatine* artery was noted transmucosally. This occurred from back diffusion.

Case #3 Right Unilateral Cleft Lip and Primary Palate (Secondary Palate Intact) (Figs. 18.37, 18.38, 18.39 and 18.40)

Having the two StV1 fields intact, in this specimen two vessels ran from the columella into the vermilion of the lip on the non-cleft lip side. The vessels defined a perfusion zone as the true philtrum or **philtral prolabium (PP)**. The vessels were 2–4 mm apart. The remainder of the prolabium was *lateral* to these vessels. It could be defined as the **non-philtral prolabium (NPP)**. Continuity of dye between the septum and the premaxilla (via backflow) was present. The brain was normal with a well-defined and perfused cribriform plate; two olfactory bulbs of equal size were present.

Clinical Applications of Blood Supply: Dissection of the Prolabium

Where Does the Non-philtral Prolabium Come From?

As a consequence of the above studies I asked myself the question: If we don't see the non-philtral prolabium in the normal state, why does it appear on the cleft side of the philtrum in the unilateral and on both sides of the philtrum in the bilateral? Where does this tissue really belong? The immediate answer to this question was the realization that in complete clefts, if the alveolar housing of the lateral incisor were gone, so too would been soft tissue overlying that zone, a patch of tissue that resided in the nasal floor. Although the alveolar defect represented a defect of intraosseous perfusion from medial nasopalatine, perhaps the blood supply to that soft tissue remained.

Conclusions

- The blood supply to the non-philtral prolabium is medial nasopalatine artery.
- The base of the NPP flap is located directly below the footplate of the medial crus. This constitutes a field boundary between the StV1 axis of the columella and membranous septum and the StV2 axis of the vomer.
- NPP can be dissection away from the philtrum and transposed into the floor of the nose… where it was supposed to be.

How to Define the Philtral Prolabium (Fig. 18.44)

Although our focus here is the microform cleft, this is a good time to reiterate the measurements that define the philtral and non-philtral zones of the prolabium. In all but the most subtle of microforms, an NPP zone can always be found. It coincides with the skin "scar" but as the microform worsens, NPP gets wider and encompasses more tissue.

For right now, let's take the simple case of a complete UCL. The perfusion territory of the anterior ethmoids runs downward from the columella as philtral prolabium. Its width is equal to the distance x between the footplates of the columella. When marking the prolabium, completely ignore what you always considered point 1 of the Millard system, the so-called "center" of the prolabium. This is based on a geometric *trompe de l'oil*. Do not be fooled by cheap imitations! Instead, mark the philtral column on the non-cleft side… point 2 of the Millard system. Now mark distance x

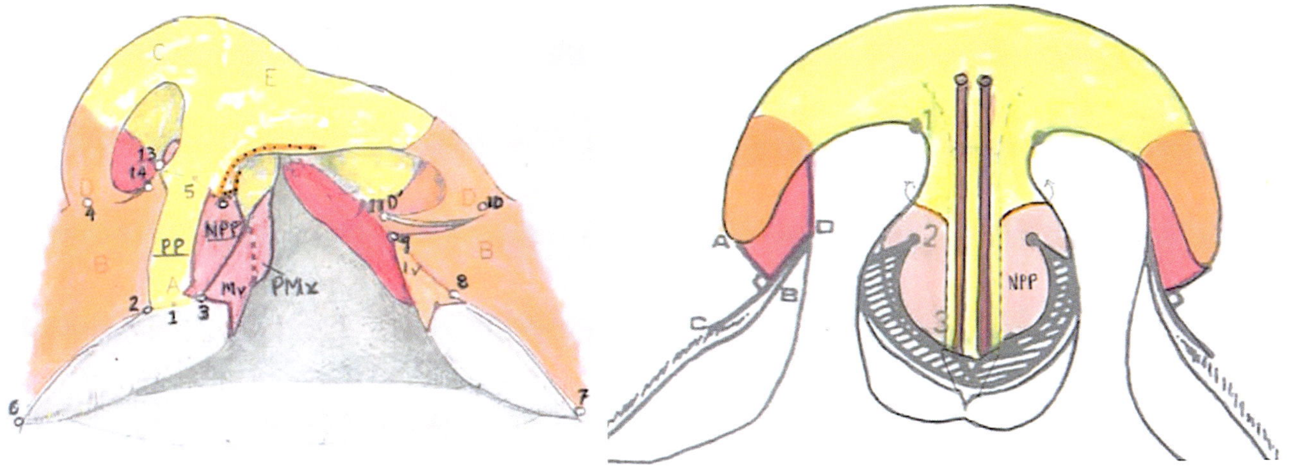

Fig. 18.44 Mesenchymal map of the prolabium. Philtral prolabium (yellow) and non-philtral prolabium (pink) can be elevated en-bloc and then separated. Entire NPP goes in nasal floor:

- Frontonasal process (yellow): r1 neural crest, StV1 neurovascular supply, anterior ethmoids
- Lateral nasal process invaded by PMx (orange): r2 + r4 neural crest, StV2 infraorbital and ECA facial
- Medial nasal process invaded by first arch (pink): r2 neural crest, StV2 medial nasopalatine

Medial crus is elevated as a lateral columellar chondrocutaneous flap to match the normal side and NPP is translated into nasal floor. Technical details are given in Chap. 19
Left: [Courtesy of Michael Carstens, MD]
Right: [Modified from Song R, Lin C, Zhao Yu. A new method for unilateral complete cleft lip repair. Plast Reconstr Surg 1998; 102(6):1848–1852. With permission from Wolters Kluwer Health, Inc.]

from point 2. This is point 3 the true height of the philtral column on the cleft side. You will find that it is often remarkably close to the original point 1! Voila! The true philtral prolabium, is visually about 1–2 mm wider than the columella. This permits a small back-cut right above the white roll for height adjustment with a single z-plasty flap from the lateral side. Everything else is pared away.

NPP is separated from PP with a full thickness straight-line incision extending up the footplate of medial crus. Here the incision *made only through skin* curves straight backward beneath crus. You can slip a scissors into the incision and, hugging the inner surface of medial crus, get right into the nasal tip. On the other side, the NPP flap can be freed up by gentle spreading staying above it at all times. It will rotate without tension into the nasal floor behind the nostril sill or into a releasing incision for the medial crus as the situation warrants.

- In microforms, NPP flap can be just skin and subcutaneous tissue but it can still be rotated. This was originally described by Vissiaronov as a "scar flap."

Arteriography by Fará was carried out in 15 fetuses with various forms of cleft and 1 normal specimen (Figs. 18.31, 18.32, 18.33 and 18.34). The common carotids were injected, thus perfusing arteries of all three systems supplying the face: StV1 from ophthalmic/ICA, StV2 from internal maxillary/ECA and facial from the ECA. In the bilateral cleft, he correctly identified paired anterior ethmoid vessels are seen descending into prolabium: "…in the prolabium there was always a rich vascular network starting in the septal (aa. nasales posteriores septi) and the columellar (aa. ethmoidalis anterior)." He thus described a dual contribution from StV1 into the superficial layer of prolabium and StV2 into its deep layer which, under normal circumstances, is occupied by the DOO.

"The underdevelopment of the muscles and the poorer blood supply in *the half of the philtrum facing the cleft* (***non-philtral prolabium***) suggests that the ability of the orbicularis to grow across the midline is limited. It is as if the muscles of one half of the lip were incapable of supplying with muscle the part that actually belongs to the other side." What he observed was that the DOO of non-philtral prolabium is not normal, as we shall see shortly in our discussion of cleft margin histochemistry.

Developmental Sequence of Microform Cleft

The Outdated (But Useful) Concept of Facial "Processes"

The level of discussion of facial development available in standard texts is inadequate to explain the variations of cleft. Clinicians treating clefts are generally familiar with two schools of thought: the nineteenth century theory of *aberrant ectodermal process fusion* and Stark's twentieth century concept of *defective mesodermal migration* [23]. Although either one can be applied to the emergence of the lateral nasal process and the advancement of the maxilla, they are vague when it comes to the formation of midline structures. How precisely does the premaxilla form and in what temporal relation to the nose? When, where and how does the septum begin? Does the philtrum belong to the nose or to the lip? What makes a Cupid's bow?

Nature continually provides *materia prima* for the surgeon in the form of clefts in their various guises which, if carefully examined, demand an alternative explanation. Consider the supposed emergence of the lip and philtrum from the medial nasal process. The German embryologists described a ventromedial projection associated with the MNP as the ***globular process***; this structure was observed to descend in the midline to form the central lip [24, 25]. However, the blood supply to the columella and philtrum is derived from the ICA, whereas that of the MNP comes from the facial artery [26, 27]. These two structures can be separated surgically [28]. Patients with arrhinia have no nostrils; yet they have a fully developed Cupid's bow [29–32]. Holoprosencephaly presents us with varying degrees of frontonasal dysplasia or aplasia.

It should be noted that controversy exists as to the propriety of the term "**process**." Firmly embedded in the literature from the nineteenth century onward, the term implies an unwarranted degree of separateness. Although externally these embryonic entities appear distinct, they share a common mesenchyme more internally. The term "**prominence**" is much more suited to the situation. It recognizes that the surfaces of embryonic subunits are thrown into relief by underlying concentrations of mesenchyme. The biochemical interactions between the epithelium and the mesenchyme determine the clinical behavior of that process in the developmental process. What's more, prominences are not homogeneous blocks of tissue: they are accumulations of developmental fields defined by neuroangiosomes.

I found the concepts of processes and prominences vague and unhelpful. Moreover, they did not seem to square with neurovascular anatomy. As it turns out, a developmental field refers to tissues supplied by a single neuroangiosome or collection of neuroangiosomes, which behaves in an autonomous manner when separated from partner fields. Fields have identifiable sources of neurovascular support; in the face corresponding to branches of the trigeminal and stapedial systems, supplemented by the external carotid circulation.

As you can see below, the 2000 field concept still vague but it was about to become much more anatomic. In December, 1999 at John Rubenstein's neuroembryology lab I became aware of the prosomeric system. By 2002 this morphed into a neuromeric model for the Tessier cleft classification. Injection studies had separated out the blood supply to the midline between internal carotid and external carotid systems. But the precise definition of neuromeric fields based on specific branches of the stapedial system would elude me until much later upon the discover of Dorcas Padget's work.

"Observations such as these are most compatible with a model of separate pairs of embryonic tissue blocs termed "fields" that interact with each other to produce the nose and mouth. Fields do not display discrete epithelial separations; they are mesodermal prominences with a separate blood supply. They are not static but possess mass, directionality and timing. Partner fields are those required to interact to form a structure. For example, contact between the lateral lip element (B field) draws down the ipsilateral philtral column (A field). Clefts prevent such normal interactions from taking place. Fields are consequently forced to occupy stereotypical aberrant locations; the resulting arrangement defines the appearance of the cleft lip-nasal complex. This is known as **field mismatch**. A fundamental goal of cleft repair is the surgical reassignment of fields into their proper relationships."

Embryonic Contents of Facial Processes After Placode Invagination

Note: Reference the cranial placode map (Fig. 18.45).

MxP: Maxillary process

- first arch (r2): skin/mucosa, muscles of mastication (Sm4), DIF zygomaticomaxillary complex, internal maxillary, V2 sensory
- second arch (r4): no skin/no mucosa, muscles of facial expression (Sm6), SIF/SMAS (r4 neural crest), ECA facial artery, VII motor only

MNP: Medial nasal process

- Skin cover, FNO; mesenchyme (r1), StV1 anterior ethmoid artery, terminal nerve

LNP: Lateral nasal process

- Skin cover, FNO + r2; mesenchyme (r1 + r2), neuroepithelium (hp2), mesenchyme (r1 and r2), StV1 lateral nasal and facial

FNP: Frontonasal process

- Skin cover, FNO; mesenchyme r1, muscles of facial expression (Sm6), SIF/SMAS

Globular process (r1) This r1 mass "extends" from MNP, forms prolabium

Premaxillary process (r2) extends beneath prolabium

DOO: Deep layer of orbicularis oris

- Sub-mucosal *pars peripheralis*
- Sub-vermilion *pars peripheralis*

SOO: Superficial layer of orbicularis oris

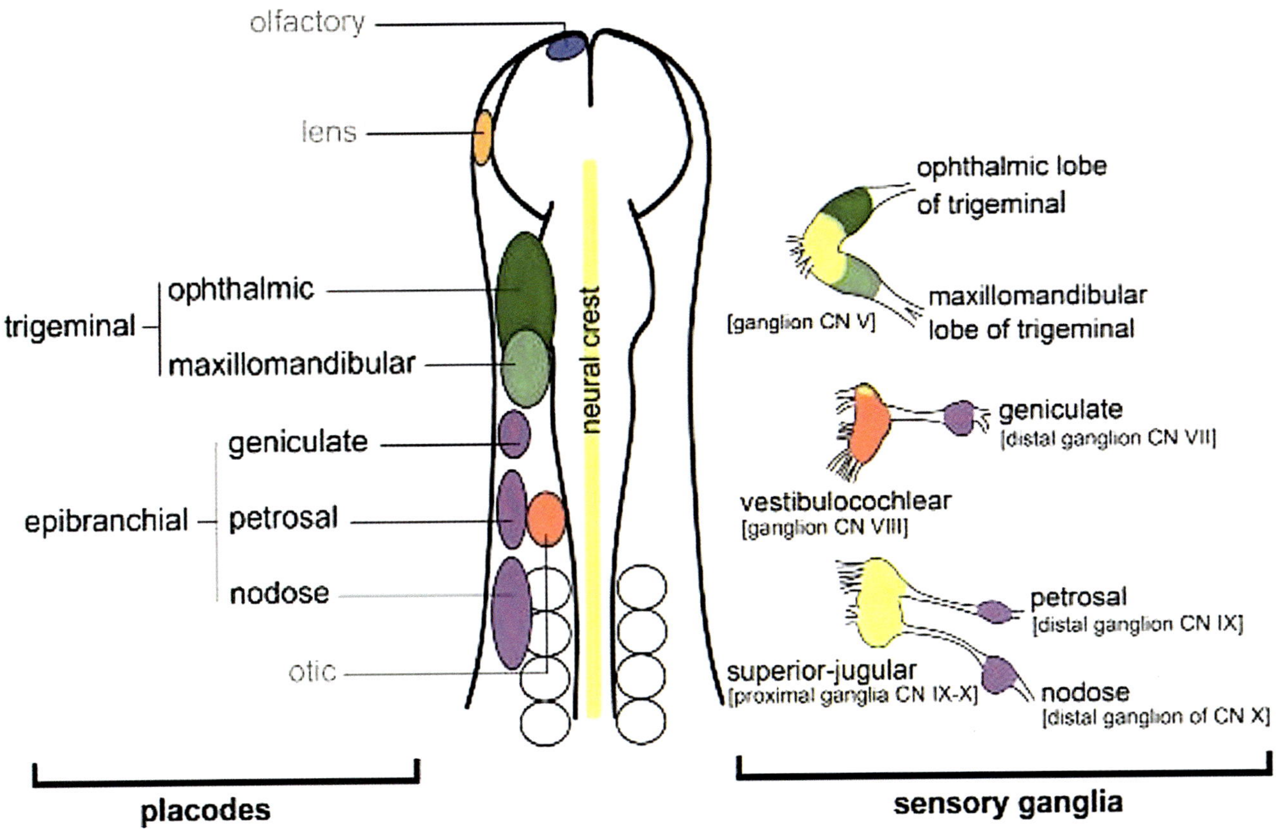

Fig. 18.45 Placodes in chick embryo (The anatomy is analogous in mammals). Nasal placode is the basis for the medial and lateral nasal processes. Trigeminal and epibranchial placodes contribution to sensory neurons. Sensory ganglia and the placodes from which they arise are color-coded. The position of the trigeminal and epibranchial placodes is shown on the left side. Trigeminal at rhombomeres r1–r3. Corresponding sensory ganglia to which they contribute are depicted on the right. The neural crest (yellow) contributes neurons to the proximal aspect of the trigeminal ganglion complex of cranial nerve V, and the distal aspect of cranial nerves VII, IX, and X. Optic appears first at Carnegie stage 11. Forebrain matures later with optic placode forming first then nasal placode. [Reprinted from Baker CV, Bronner-Fraser M. (2001). Vertebrate cranial placodes I: Embryonic induction. Dev Biol 2001; 232: 1–61. With permission from Elsevier]

Frontonasal Process, Prolabium and Premaxilla

This section references Chap. 4 on the formation of the face but goes into greater detail. It is essential material to understand the formation of the lip.

The original concept of FNP was proposed in 1948 by Streeter, O'Rahilly in which paired premaxillary processes were thought to fuse with a frontal field to create a single frontonasal process [33, 34]. This was combined with the notion of a midline septal cartilage originating from the chondrocranium from which the architecture of the oronasal cavity is somehow derived [35–38]. This construct lumps together tissues from different neuromeric sources and blood supplies.

- Tissues originating from the frontonasal process (FNP) remain as the forehead, superior-medial orbit, ethmoid complex, nasal chambers, nasal envelope, septum, columella and philtrum. Its mesenchyme is r1 and its neuroangiosomes arise as StV1 branches from the ophthalmic stem.

Neuromeric Model of Fronto-Naso-Orbital Development

Carnegie Stages to the Time of Lip Formation

- *Stage 8* Recall that FNP projects forward from r0 notochord and Rathke's pouch like a proboscis. FNP results from the rolling-up of the neural tube corresponding to forebrain. The skin cover of FNP is hp2 epithelium and p1–p3 neural crest dermis. Encased within the proboscis is the developing prosencephalon. The space between the skin and brain is occupied by angioblasts, probably derived from the prechordal mesoderm. The make endothelial component of the primitive head plexus. Each side of the FNP has 3 placodes: adenohypophyseal, olfactory and optic
- *Stage 9* first arch forms and neural crest migration begins. RNC from r1–r3 admixes with the head plexus, forms meninges and lays down r1 mesenchyme for the sphenethmoid complex
- *Stage 10* second arch forms

- Invasion of ventral FNP by first arch
- *Stage 11* third arch forms; first and second arches meld together to form B fields
  - Invasion of LNP by first arch
- *Stage 12* fourth arch forms, B fields now in physical contact with FNP which allows r2 mesenchyme to migrate into position under the surface of FNP in the future mouth
  - Invasion of LNP by second arch
- *Stage 13* fifth arch forms
  - The primärer nasenböden (primary nasal floor) falls apart
  - Behind MNP is medial r2 mesenchyme
  - Behind LNP is lateral r2 mesenchyme
  - Continuity of r2 mesenchyme is posterior (the future sphenopalatine notch)
- *Stage 14* lip formation sequence begins

Recall that the nasal placodes sink into the r1 mesenchyme of FNP to set up two nasal chambers. These are widely separated by a central block of r1 which subsequently undergoes apoptosis. There is no mouth; the entire FNP is r1 in front of the brain.

Embryonic folding begins at stage 9. By stage 10 MxP becomes confluent with FNP, first at its posterolateral corner and then fusing progressively further forward. Confluence of MxP with FNP and later with LNP allows for the sharing of mesenchyme. Let's walk through this three-step process.

Recall that nasal placodes are physically connected with the underlying forebrain. As the surrounding FNP mesenchyme expands forward, the placodes stay stable; they therefore appear to be mechanically "retracted" backwards into the r1 mesenchyme of FNP. This creates a nasal chamber lined by specialized hp2 placodal epithelium. The lateral wall contains neurons of the olfactory system (OS) and the medial wall contain neurons of the accessory olfactory system and the terminal nerve (cranial nerve 0). At this time the back wall of the chamber is closed. The nasal cavity is a blind sac exclusively within the FNP. Later on, when the primary nasal floor of the sac (*primärer nasenböden*) breaks down, the two nasal cavities are now in continuity with the developing oral cavity.

The **first interaction** between MxP and FNP occurs at stage 10 and takes place (logically) at site of origin of first arch at the posterior border of the frontonasal process. Recall that the placodes are widely displaced. They come together in the midline as the result of apoptosis within the central block of r1 sphenethmoid mesenchyme. Recall further that the hp2 epithelial "skin" of the FNP comes from the neural folds of the forebrain. As the two side of the FNP approximate each other ventrally, above what will be the mouth, first arch ectoderm and mesenchyme "chase" the retreating hp2 skin until reaching the undersurface of what is the medial wall of the nasal chamber.

Neuroangiosomes invade from StV2. They bifurcate within the floor of the primary nasal sacs. The nasal walls are now two-tiered.

- Medial wall has the StV1 fields of the ethmoid and septum above, and StV2 fields of vomer/premaxilla.
- Lateral wall has StV1 upper and middle turbinates above and StV2 inferior turbinate + palatine bone complex (P1–P3) below.

When the bottom of the primary nasal floor opens, the two branches of the nasopalatine axis are physical separated as medial nasopalatine to vomer and premaxilla and lateral nasopalatine to lateral nasal wall but they remain connected posteriorly at the back wall of the nose.

The **second interaction** between MxP and FNP takes place anteriorly along the surface of the developing face. It involves the invasion of the lateral nasal process first by first arch skin and mesenchyme and later by second arch mesenchyme. Here's what happens.

- Simultaneous with the "invagination" of the placodes, the r1 mesenchyme that was surrounding them expands, throwing the surface of the FNP on either side of the nasal chamber into two prominences, the media nasal process (MNP) and the lateral nasal process (LNP),
- Blood supply of MNP remains StV1 but the LNP is quickly invaded by MxP such that the mesenchymal composition of caudal half switches from r1 MNC to r2 and r4 RNC from the first arch/second arch complex.

In the roof of the future mouth (originally covered by PNC skin) first arch mesenchyme "chases" the retreating FNP to become the floor of the primitive nose. When the primary nasal floors break down, we now have paired nasal chambers and a mouth.

- Laterally, r2 mesenchyme makes up the lower lateral wall of the nose: the **inferior turbinate** and **palatine bone complex.**
- Medially, r2 mesenchyme makes up the lower medial wall of the nose as the **vomer** fields.
- MNC r1 neural crest mesenchyme lying adjacent to MNPs expands, supplied by the anterior ethmoid angiosome running down columella, to form the **philtral prolabial processes**.
- RNC r2 neural crest deep to MNP proliferates, supplied by the medial nasopalatine angiosome. These become the **premaxillary processes**.
- RNC r2 neural crest produces terminal soft tissue buds from the vomer-premaxillary junction. These become the **non-philtral prolabial processes**.

Developmental Fields of the Vomer and Premaxilla

The following sequence is demonstrated in the SEMs of Hinrichsen and the staged drawings by Krause

- *33 days*, r2 mesenchyme in the roof of the future oral cavity presents itself as V-shaped swellings just beneath the medial wall of the nasal chamber. As development proceeds, the V narrows and fuses in a posteroanterior manner. This represents the fusion of the vomer fields. Above the compaction of the medial nasal walls forms a sandwich within which perpendicular ethmoid plate and septum will develop.
- *41 days*, two anterior swellings appear.
  - Premaxilla is a continuation of the vomer field.
  - Prolabium buds from the medioventral corner of MNP.
- *48–50 days*, vomer field fusion has reached the posterior border of the premaxillae (the incisive foramen). At this point, palatal shelf elevation can be observed.
- 56 days, fusion of the premaxillae is complete.
- **In sum**: At 56 days, completed facial midline consists of paired field complexes of stacked-like boxes.
  - Upper boxes (r1): ethmoid complex, nasal envelope, columella and philtral prolabium.
  - Lower boxes (r2): vomer, premaxilla and non-philtral.

Lip–Nose Fusion Sequence

The actual events surrounding the closure of the lip are not well documented in the literature. Nevertheless, SEM work by Hinrichsen and drawings of staged embryos by Krause provided a wonderful opportunity to depict this process in diagrammatic form. What follows, referenced to Figs. 18.46, 18.47, 18.48 and 18.49, is my best guess as to the sequence of individual events and their timing by Carnegie stage. Note carefully the events inside the oral cavity.

Development of the upper lip takes place between Carnegie stages 14 and 18. Let's look at the following visual model. For the sake of brevity, let A = FNP; B = MxP (combining first and second arch tissues); C = MNP; and D = LMP. Invasion of the FNP floor and of the ventral half of LNP by first arch mesenchyme has already taken place. The *primärer nasenböden* has broken down and r2 tissue can be seen inside the nasal cavity. Note: yellow = r1 and pink = r2.

Step 1 D fuses to C. This takes place by means of "Simonart's band." The gap separating the ventral edges of LNP and MNP is bridged by an epithelial strand of r2 tissue produced by D field. This is an energy-dependent process that is oxygen dependent in the mouse. Simonart's band is the form fruste of the nasal sill. It provides a conduit for mesenchymal flow from B medially. It lies internal to the eventual fusion point of D with C [39]. Simonart's band will not form under two conditions:

- A complete cleft of the primary palate positions the LNP too far laterally from the midline: the critical contact distance is exceeded.
- Metabolic failure due to environmental agents blocks synthesis of the band.

Failure: **complete cleft lip**. No further B field tissue migration

Step 2 B fuses to D/C complex. This is a sequential step with MxP fusing first to LNP and then translating to MNP.

Failure: **complete cleft lip + Simonart's band**

Step 3 B fuses to A. Carnegie stage 17. The expansion of the medial and caudal aspect of MNP, previously called "the globular process," creates the philtral prolabium (PP). In the meantime, tucked inside, r2 non-philtral prolabium (NPP) adds to the medial nasal floor. In the presence of a cleft between A (philtral prolabium) and B (lateral lip element), the NPP becomes visible and fuses with PP to create the prolabium we are accustomed to see in bilateral cleft lip.

- When B joins A, the PPP descends.
- The filling up of the lip is deep-to-superficial and dorsal-to-ventral
- Premaxilla differentiates later

Failure: **incomplete cleft lip**

Step 4 Development and insertion of oronasal musculature: The deep layer of the orbicularis is unified well before the superficial layer. The oblique head of the orbicularis levator labii et alaeque nasi and pars alaris of nasalis insert last.

Failure: **minimal cleft lip**

Scientific Importance of the Minimal Cleft Sequence

To the reader: *Often in the process of investigation, we become misled by a misunderstanding of a particular term. At the time of the original paper, I had not yet stumbled upon the neuromeric system. Therefore, I kept referring to a cleft "event" of unknown nature and was seeking to understand the pathology of cleft as a series of "processes" unleashed by the event. As it turned out, processes are useful but I got*

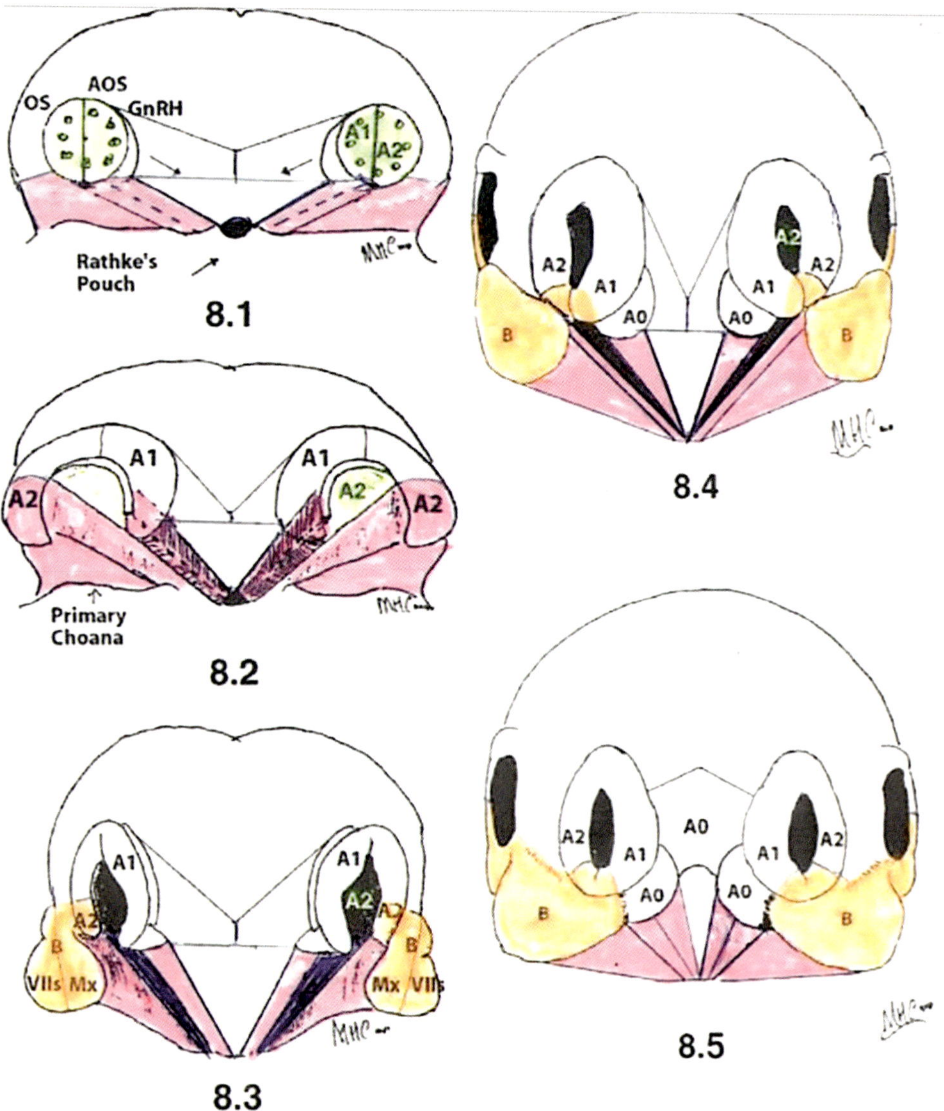

Fig. 18.46 Midline facial development. The developmental pattern nasal cavity is seen here in 3-dimensional representation, begins with two columns of r1 neural crest running forward in divergent fashion from Rathke's pouch to the nasal placodes. This tissue provides the scaffolding by which olfactory neurons from the lateral placode and GnRh neurons from the medial placode will find their way to the brain. Fusion of the StV1 ethmoid fields and the StV2 vomer-premaxillary fields occurs in posteroanterior fashion. Apoptosis of r1 sphenethmoid mesenchyme is requisite for proper medial positioning of the orbits. The following are mouse versus Carnegie (human) stages. Key: r1 mesenchyme, white; r2 mesenchyme from first arch, pink; r2 + r4 mesenchyme from PMx (the amalgams of first and second arches)

8.1 = **Stage 13** (28 days), floor of frontal nasal process, composes of r1 mesenchyme is invaded by r2
8.2 = **Stage 14** (32 days): primary nasal floor is riven, r2 from first arch mesenchyme and StV2 to LNP, first and second arches fused into PMx
8.3 = **Stage 15, early** (33 days): PMx fuses to side of LNP. Not color change to orange
8.4 = **Stage 15, late** (35 days): LNP fuses to MNP. Simonart's band was the intitiating step. MxP advances forward reach caudal tip of LNP. Globular process (A0) emerges caudal tip of MNP
8.5 = **Stage 16, early** (36 days): MxP has moved to caudal tip of MNP and is fused with globular processes (A0) which are much larger. Columns of r2 mesenchyme for vomer and premaxilla are prominent
[Courtesy of Michael Carstens, MD]

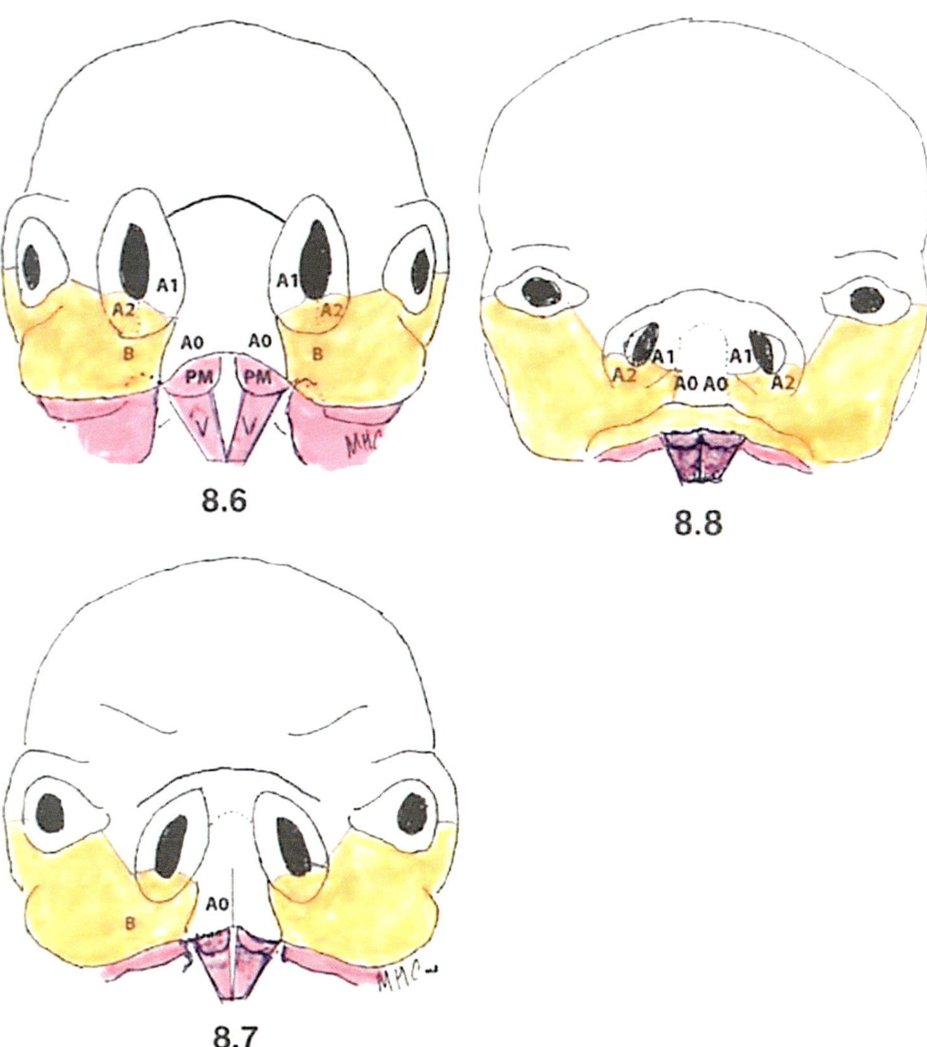

Fig. 18.47 Midline facial development. The developmental pattern nasal cavity is seen here in 3-dimensional representation, begins with two columns of r1 neural crest running forward in divergent fashion from Rathke's pouch to the nasal placodes. This tissue provides the scaffolding by which olfactory neurons from the lateral placode and GnRh neurons from the medial placode will find their way to the brain. Fusion of the StV1 ethmoid fields and the StV2 vomer-premaxillary fields occurs in posteroanterior fashion. Apoptosis of r1 sphenethmoid mesenchyme is requisite for proper medial positioning of the orbits. The following are mouse versus Carnegie (human) stages. Key: r1 mesenchyme, white; r2 mesenchyme from first arch, pink; r2 + r4 mesenchyme from PMx (the amalgams of first and second arches)

8.6 = Stage 16, late (37 days): premaxillae and alveolar processes prominent. Palatal shelves vertical

8.7 = Stage 17 (41 days), vomer-premaxillary columns approximating

8.8 = Stage 18 (44–46 days): eyes frontalized, nasal tip definition

[Courtesy of Michael Carstens, MD]

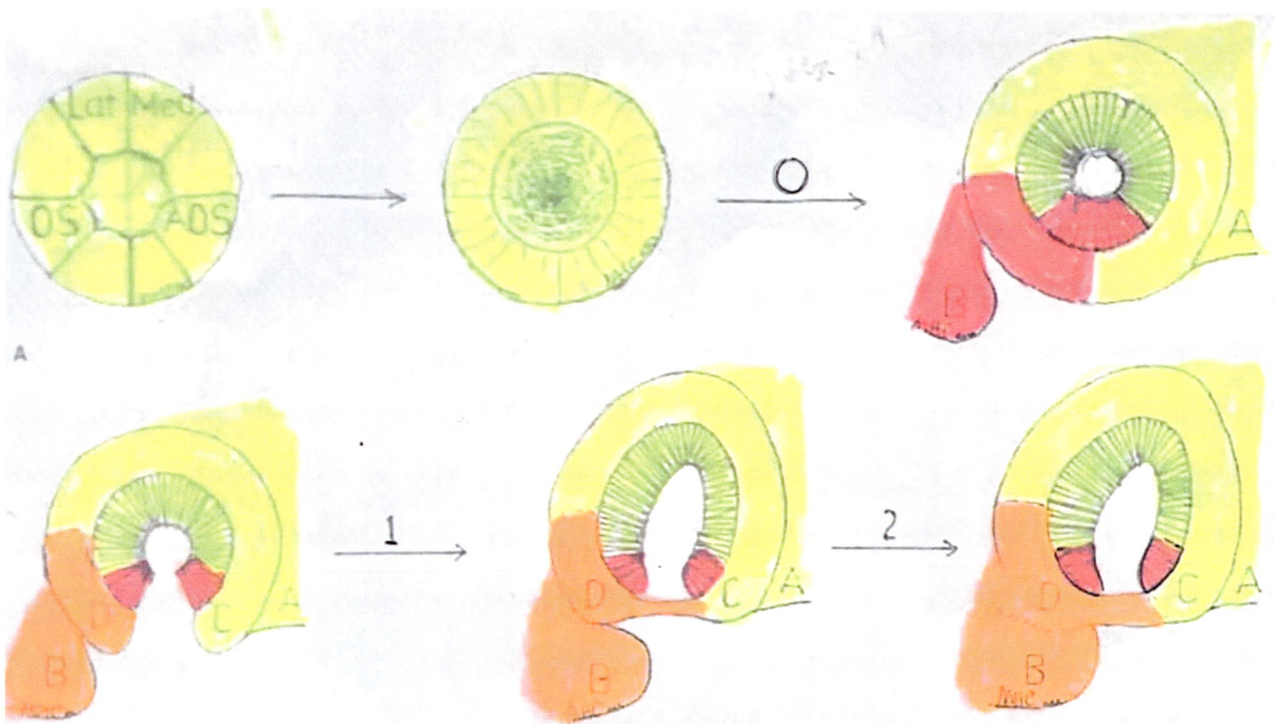

Fig. 18.48 Nasolabial closure sequence 1–2. *Key*: A = prolabium. B = lateral lip element (MxP) containing first and second arch neural crest mesenchyme and facial muscle, C = medial nasal process (MNP), D = lateral nasal process (LNP)
Step 0: preparation. Stages 11–13: Nasal floor, lateral nasal wall, and caudal LNP invaded by first arch (red). Stage 14: dissolution of nasal floor and approximation of MxP. Stage 15 MxP fuses to LNP
Step 1: Simonart's band. LNP, now transformed and activated by blood supply from StV2 and ECA
- Arrest at this point = **complete cleft lip**

Step 2: Migration of lateral lip element PMx to reach MNP
Nota bene: normal lip development can proceed both sides but non-syndromic apoptotic failure will lead to midline widening
[Courtesy of Michael Carstens, MD]

them initially in the wrong order (division, deficiency, displacement and distortion) when in point of fact it was the deficiency in the r1 premaxillary field that led via *a critical lack of BMP4 synthesis to the epithelial fusion failure…and therefore to division. So, to repeat for the umpteenth time, the term "cleft" is itself a misnomer, an intellectual trap. It is simply a failure of normal tissue behavior due to abnormal biochemical signals arising from a site of mesenchymal deficiency. To illustrate this point, I include here the original language. To be sure, microform cleft is important but I kept missing the bigger picture.*

"The sequential nature of the clefting process is best seen in the minimal deformity. The clef ting 'event' happens in the lateral margin of what will be the floor of the nose in step 1. If the event is severe, a full-blown complete cleft lip will result. Lesser degrees of abnormality imply progressively milder forms of expression in the lip. *This is wrong. The problem is not in the lip by in the nasal floor*. The minimal deformity defines the outer border of this envelope, which is the definition of the philtrum at step 4. Median clefts do not show a piriform deformity because the pathologic event occurs in the midline A fields (the r1 neural crest mesenchyme of the FNP)."

It is precisely at this paragraph that I had a change of concept. Except that a fine deficiency of neural crest mesenchyme is the static outcome of angiosome failure, not an "event" in itself. "The fundamental event of cleft formation is the creation of a deficiency state in the floor of the nose. All the remaining processes of cleft (division, displacement and distortion) follow from this single event. When the anatomy of the nasal floor in clefts is compared with normal development, the location of this deficiency site is defined. The known embryonic events that create this site can be analyzed to further define how the deficiency state is created. *This led subsequently to differential dye injections and the discovery of the neurovascular fields in the facial midline*.

- The lessons of the microform cleft were the first step toward the development of the neuromeric model.

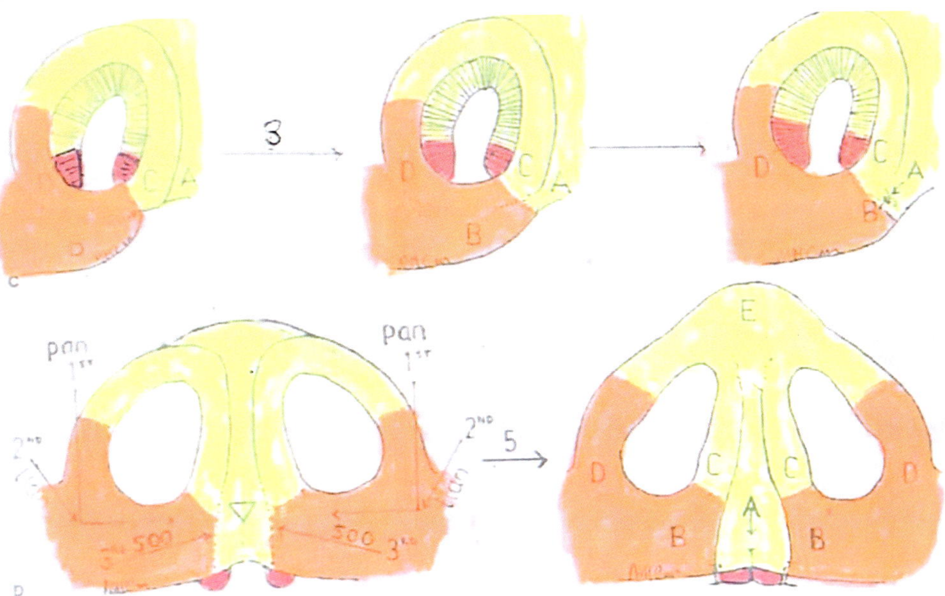

Fig. 18.49 Nasolabial closure sequence 1–2. *Key*: A = prolabium. B = lateral lip element (MxP) containing first and second arch neural crest mesenchyme and facial muscle, C = medial nasal process (MNP), D = lateral nasal process (LNP)

Step 3: Mesenchymal transfer from lateral lip element to prolabium. Transfer begins when B–A reaches 1/3 lip height. Probably reflect the *timing* such that myoblasts are present at the margin of lateral lip element

- Arrest at this point = **incomplete cleft lip**

Step 4: Development of alveolar processes of both maxilla and premaxillae

Step 5: Fusion of alveolar processes, midline fusion and development of the nasal tip

Nota bene: normal lip development can proceed both sides but non-syndromic apoptotic failure will lead to midline widening

[Courtesy of Michael Carstens, MD]

Functional Matrix: Dysfunction in the Lateral Nasal Process

Membranous bone is a Johnny-come-lately in the formation of the face. The evidence that facial bone formation responds to its soft-tissue template has been exhaustively reviewed by Enlow [40]. Distortions in that template are faithfully mirrored by skeletal architecture. *Thus, the morphology of the piriform aperture in the floor of the cleft-lip nose is an "impression" of a dysfunctional matrix.*

- The functional matrix concept of ML Moss describes composite blocks of tissue having independent developmental, origin, vascularity and function [41, 42].
- A narrow zone of deficiency thus exists at the anterolateral border of the piriform; this extends interiorly to the beginning of the inferior turbinate. When the alar complex is sufficiently freed from the piriform so as to assume a normal configuration, this zone will require coverage with mucosa. *This is the first description of the defect in the frontal process of the premaxilla* (Fig. 18.7).

Deformities of the nasal capsule have been documented in studies comparing cleft patients with normal controls [43]. The piriform aperture appears "expanded" and the nasal floor is displaced caudally. These changes occur because the bone has formed within a distorted matrix. Despite this, the cross-sectional area of the airway in adult cleft patients was found by Hairfield and Warren to be reduced 59% [44]. This is caused by alterations in the size and shape of the septum and vomer, enlarged turbinates and vomerine spurs. Kimes demonstrated altered growth rates for these structures in cleft patients.

- *These airway findings anticipate the work of Talmant in the dysfunctional airway but the unrecognized problem he addressed is the pathologic misinsertion of the nasalis muscle.*

The quantitative absence of soft tissue at this juncture is the cleft site accompanied by a proportional deficit of bone. All further manifestations of the cleft follow from this. The periosteal attachment to this deficient and distorted site perpetuates the deformity. As growth proceeds, asymmetries of force vectors continue to open up the cleft site. Surgical reversal of this ongoing process in infancy, while facial growth is explosive, is a fundamental principle of DFR cleft repair.

This requires a surgical redefinition of the cleft condition followed by reconstruction of all tissue elements to their pre-cleft state. Restitution of the original functional matrix is the only manner by which subsequent biologic growth can be normalized.

Release of the dysfunctional matrix from the piriform fossa requires addition of tissue to replace the mucosal deficit. Vermilion paring flaps can be combined with an anteriorly based turbinate flap to accomplish this task 52–53. By virtue of their geometry being at right angles to one other, these flaps achieve a three-dimensional release of the vestibule.

- *Once again, I was dead wrong. Talmant's work disclosed that the problem of the "entrapped" vestibular lining was not that it was small, but rather abnormally stretched by nasalis so that is became mismatched to the overlying soft tissues. Nasalis reassignment and blunt subperiosteal and subperichondrial release eliminate this faux vestibular defect. As a side benefit, the mysterious vestibular "web" disappears.*

Anatomical Evidence (Figs. 18.50, 18.51, 18.52, 18.53, 18.54, 18.55, 18.56, 18.57 and 18.58)

Several lines of experimental data provide evidence that: (1) a specific region in the floor of the nose exists where fusion occurs; (2) growth of the lateral nasal process plays a crucial role in the fusion process; (3) changes in cell proliferation in the LNP (D field) are stage-specific; (4) the actual fusion event is of very short duration; and (5) the final configuration of the lip/nose complex involves a stagespecific mesenchymal translation to the midline. This evidence will be discussed below in sequence.

Where is the Fusion Site?

Scanning electron microscopy demonstrates dramatic changes in the configurations of the LNP, MNP and MxP between Carnegie stages 14 and 18 [45, 46]. As the fields move toward one another and prepare to fuse, a series of morphological and biochemical changes occur; these are most pronounced in the fusion zone—the caudal aspect of the LNP and the MNP. In mouse embryos, epithelial "preparation" at the base of the nasal pit includes loss of microvilli, phenotypic differences and intracellular inclusions indicative of –degeneration in otherwise normal appearing cells [47–55]. Concanavalin A binds selectively at the fusion points of both the LNP and MNP [56].

As the nasal epithelia and mesenchyme differentiate toward fusion (cell death), DNA synthesis decreases. When Wilson and coworkers studied the nasal region of rhesus embryos of advancing developmental age, DNA synthesis decreased successively from the placode stage to the nasal groove to the primary nasal cavity [57]. These findings can be localized to the caudal aspect of the LNP and MNP. DNA synthesis studies of mice [58] and rats [58] show epithelial activity at the fusion site of the nasal groove to be reduced compared with the nonfusion area superior to it. Cell kinetics in the murine fusion region are likewise diminished when labeled with tritiated thymidine [56].

Gui et al. studied the proliferative characteristics of these tissues in mouse embryos using pulse labeling with 5-bromodeoxyuridine to determine epithelial cell proliferation and mesenchymal cell density at differing levels of the

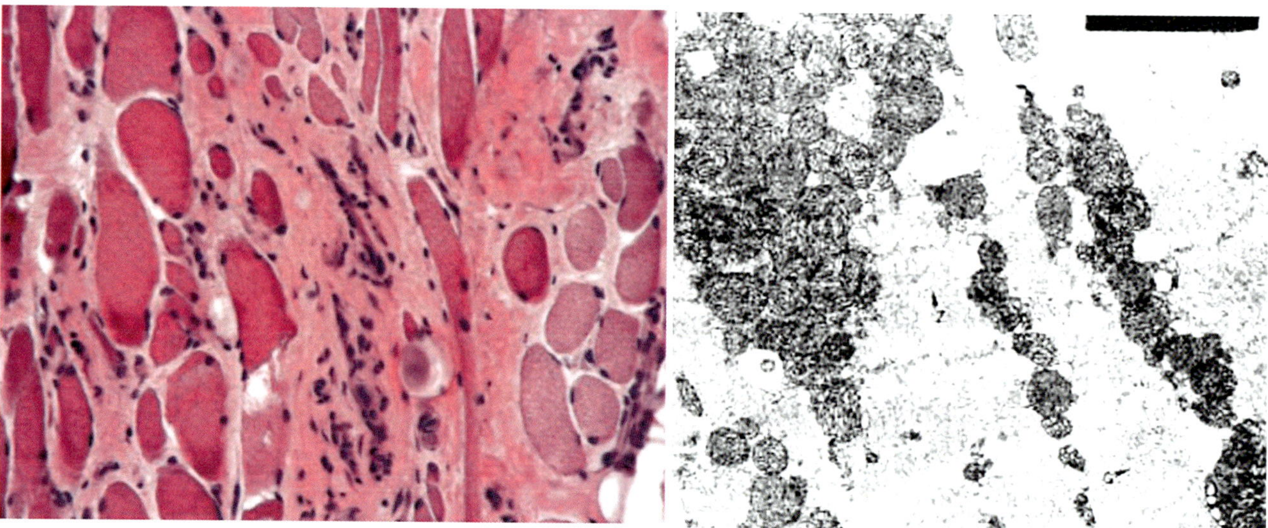

Fig. 18.50 Cleft muscle histology and ultrastructure. *Left*: Muscle at the cleft margin in both cleft lip and cleft palate displays variability of fibers type and size from normal and patterns of disorganization, and fibrosis at the microscopic level both: *Right*: Ultrastructural changes are seen as well, characterized by abnormal clumps of mitochondria with lateration in the cristae. Left: [Reprinted from Lazzeri D. Dystrophic-like alteration characterize orbicularis oris and palatopharyngeus muscles in patients affected by cleft lip and palate. Cleft Palate Craniofac J 2008; 45(6):587–591. With permission from SAGE Publishing.] Right: [Reprinted from Schendel SA, Pearl RM, De'Armond SJ. Pathophysiology of cleft lip muscle. Plast Reconstr Surg 1989; 83(5):777–784. With permission from Wolters Kluwer Health, Inc.]

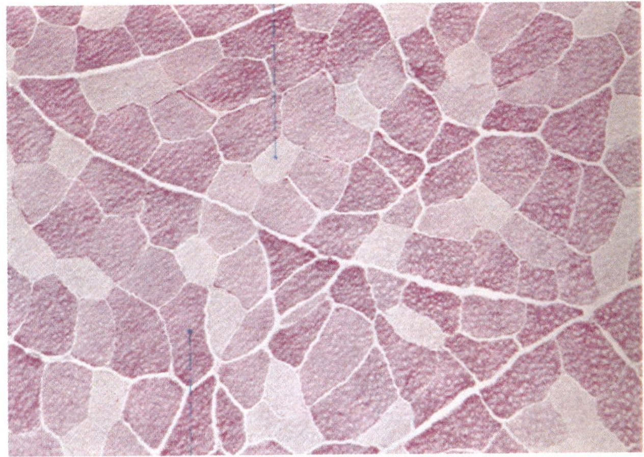

Fig. 18.51 Muscle fiber type. Type 1: Preferentially formed during embryonic phase of myogenesis. The fibers are small, red in color (very vascular), use aerobic respiration to make ATP red color, have lots of mitochondria, and are characterized by slow fatigability for endurance. Type 2A: Preferentially formed during fetal phase of myogenesis. The fibers are larger, white in color (less well perfused), use both aerobic and anaerobic metabolism, have fewer mitochondria, and show easy fatigability for explosive force. Type 2B: same a 2A but uses strictly gycolytic anaerobic metabolism. [Reprinted from Frithjof Hammersen Sobotta's Atlas of Microscopic Anatomy Urban and Scharzenberg, 1985]

nasal groove [59]. These findings were correlated with developmental age, using tail somite number as an indicator. This experimental design permitted three-dimensional variations among four specific regions of the nasal fields. LNP-MNP fusion occurred only at the most caudal (D) level. At tail somite 3, the D-level epithelial labeling index was significantly lower than levels A–C. As the embryo approached TS11, DNA synthesis rapid declined to near zero. The labeling index of the MNP did not reach zero until later at TS13.

Mesenchymal Mass

The mass of the nasal prominences is not equal. In human embryos Warbrick found more rapid mesenchymal expansion on the medial side; these observations were not site-specific [60]. Differences in the LNP versus the MNP were also noted in the mesenchyme. At all tail somite stages (TS3–TS11) of the LNP, labeling was lower at the fusion site (level D). In the MNP, a statistically significant change in labeling between the non-fusion levels and the fusion site did not occur until TS7. Fusion occurred between TS5 and TS7. Gui concluded that *differences in mesenchymal mass* between the—fusion and non-fusion regions of the LNP versus the

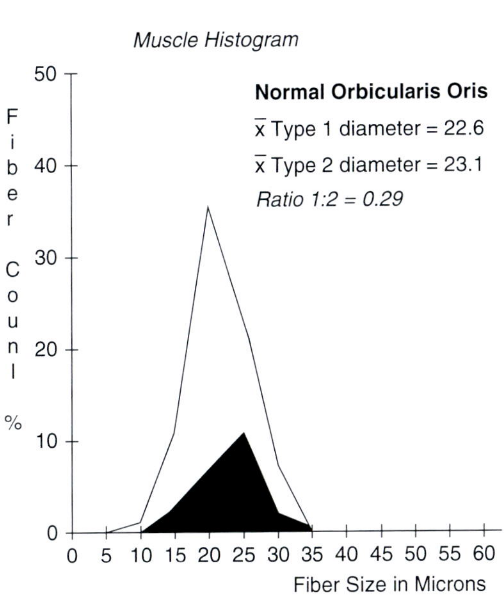

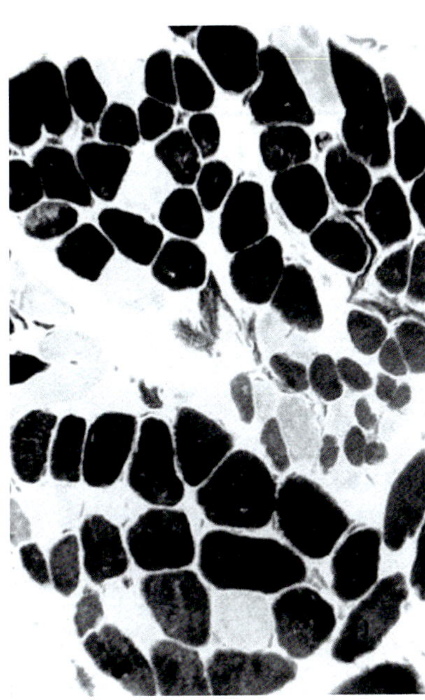

Fig. 18.52 Normal orbicularis muscle from biopsy site unrelated to the cleft margin. Note fiber type with type 1 < type 2, similar fiber diameters, and histology (light staining fibers are type 1 and dark staining fibers are type 2). [Reprinted from de Chalain T, Zuker R, Acklery C. Histologic, histochemical and ultrastructural analysis of soft tissues from cleft and normal lips. Plast Reconstr Surg 2001; 108(3):605–611. With permission from Wolters Kluwer Health, Inc.]

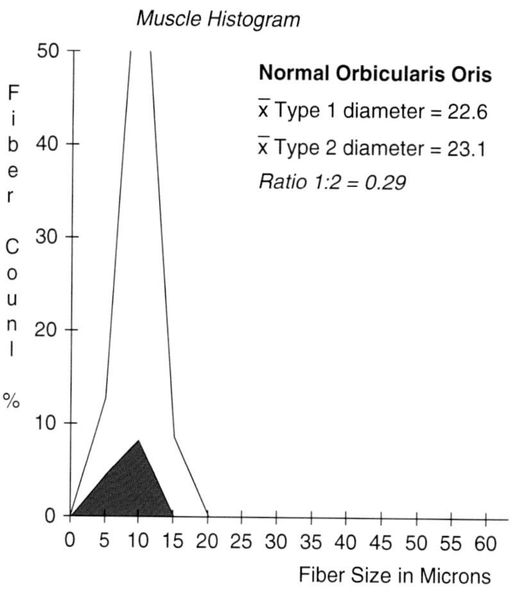

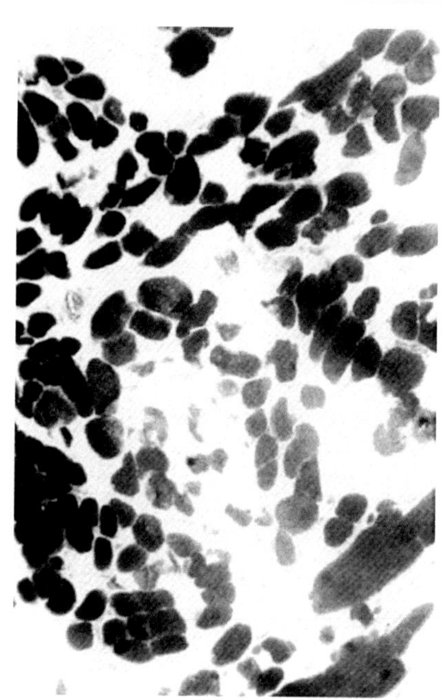

Fig. 18.53 Cleft orbicularis muscle. Type 1 fibers are reduced compared to type 2. Fiber size of type 2 is 50% of normal. Note hypoplasia of the muscle fibers. [Reprinted from de Chalain T, Zuker R, Acklery C. Histologic, histochemical and ultrastructural analysis of soft tissues from cleft and normal lips. Plast Reconstr Surg 2001; 108(3):605–611. With permission from Wolters Kluwer Health, Inc.]

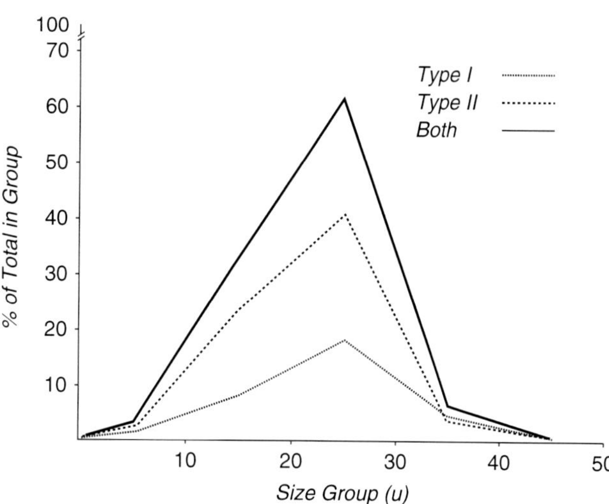

Fig. 18.54 Normal orbicularis muscle histogram Schendel. Type 1 30.3%, Type 2 69.7%. Average type 1 fiber diameter 23.5 μm, Average type 2 fiber diameter 22.4 μm. Note fiber diameters in normal facial muscle are about 50% smaller than for limb muscle. [Reprinted from Schendel SA, Pearl M, De Armond SJ. Pathophysiology of cleft lip muscle. Plast Reconstr Surg 1989;83(3):777–923. With permission from Wolters Kluwer Health, Inc.]

MNP have significance in terms of their relative roles in the fusion process:

"It is believed that the bulk of the MNP in the non-fusing areas expands more rapidly than that of the LNP.... Concerning the presumptive fusion area, we hypothesize conversely that the bulk of the LNP mesenchyme at the side of the nasal groove expands more rapidly than that in the MNP at the other side.... The LNP at the fusion area remarkably protrudes to contact with the MNP. Since DNA synthesis of the LNP epithelium in the fusing area decreases sooner than that of the MNP, it is suggested that *the LNP mesenchyme plays a more important role* in the initial contact between both prominences a t the fusion area, being associated with the epithelial differentiation in the LNP."

Defective genetic programming could alter the coordination of the timetable for field closure. Two genetic loci that cause deficiencies in size of the facial prominences are now associated with cleft lip-producing strains of mice [61]. Alterations in coding or regulation of a growth factor or receptor could impair successful outgrowth or fusion of the LNP and MNP. Once translation of the MxP across the LNP/MNP "bridge" is under way, deficiencies in mesenchymal "flow" may occur. Serial sections of the nasal contact zone in normal and cleft-susceptible mice were measured by Wang and Diewert. In the cleft strains, both the enlargement of the epithelial seam and its "filling out" by a mesenchymal bridge were delayed by tail somite stage compared with normal. Forward growth of the maxillary process was also correlated by Wang with mesenchymal bridge development in both normal and cleft strains.

Fusion itself is of short duration. Surf ace changes in the epithelia are a marker, with disappearance of microvilli seen on scanning electron microscopy at TS6 about 12 h before contact [62]. As the nasal epithelium prepares for fusion between TSS and TS7, alterations in its cellular kinetics occur. The selective decrease in DNA synthesis noted about 6–8 h before fusion further defines the window

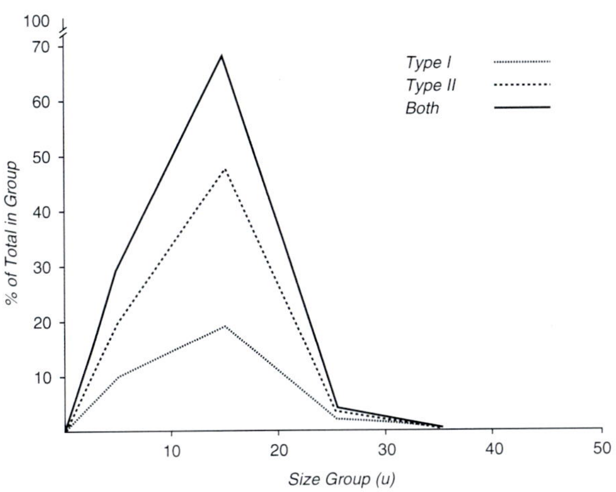

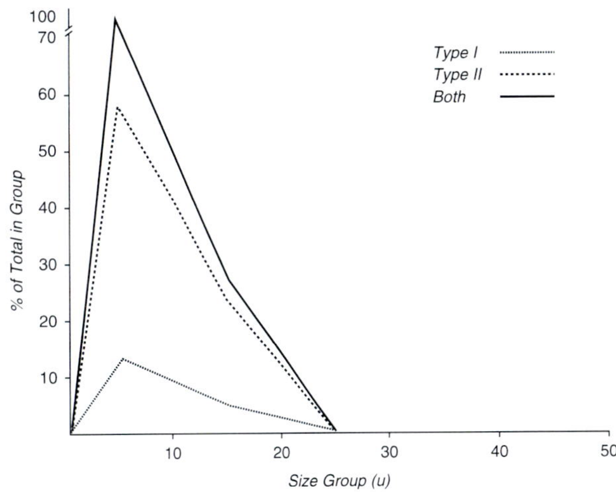

Fig. 18.55 Cleft orbicularis muscle histograms Schendel. Left: Non-cleft side. Type 1, 29.2%; type 2, 70.8%. Average type 1 diameter is 11.1 μm. Average type 2 diameter is 11.8 μm. Right: Cleft side Type 1, 17.4%; type 2, 82.6% Average type 1 diameter is 8.9 μm. Average type 2 diameter is 8.9 μm. [Reprinted from Schendel SA, Pearl M, De Armond SJ. Pathophysiology of cleft lip muscle. Plast Reconstr Surg 1989;83(3):777–923. With permission from Wolters Kluwer Health, Inc.]

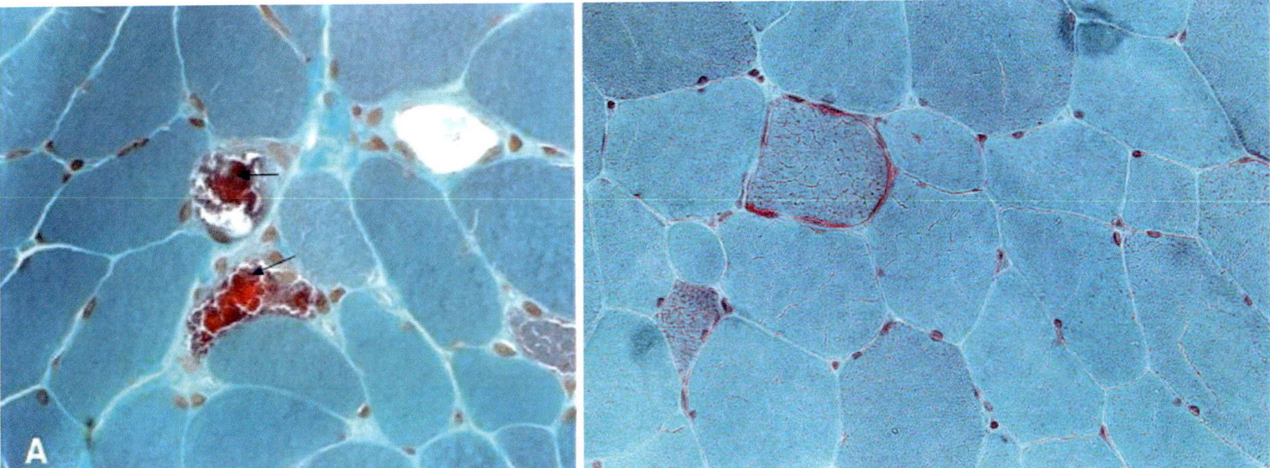

Fig. 18.56 Gomori stain. MERFF is characterized by myopathy with myoclonus, seizures, and cerebellar degeneration. Muscle biopsy in this disease shows ragged red fibers (therefore its name). Diseases of the mitochondria can be caused by defects in nuclear or mitochondrial DNA and result in decreased energy availability for cell processes. When muscle is stained with Gomori Trichrome, characteristic ragged-red fibers are visible under the microscope. This appearance is due to the accumulation of abnormal mitochondria below the plasma membrane of the muscle fiber. These may extend throughout the muscle fiber as the disease severity increases. The mitochondrial aggregates cause the contour of the muscle fiber to become irregular, causing the "ragged" appearance. Besides muscle, what other tissues would you expect to suffer the most damage from a mitochondrial defect? (**a**) MERFF positive biopsy with diseased mitochondria staining red. (**b**) Muscle fiber irregularity clearly seen. (**a**) [Reprinted from Wikimedia. Retrieved from: https://en.wikipedia.org/wiki/Gömöri_trichrome_stain https://commons.wikimedia.org/wiki/File:Ragged_red_fibers_in_MELAS.jpg With permission from Creative Commons License 2.0: https://creativecommons.org/licenses/by/2.0/deed.en.] (**b**) [Reprinted from Wikimedia. Retrieved from https://commons.wikimedia.org/wiki/File:Ragged_red_fibres_-_gtc_-_very_high_mag.jpg. With permission from Creative Commons License 3.0: https://creativecommons.org/licenses/by-sa/3.0/deed.en]

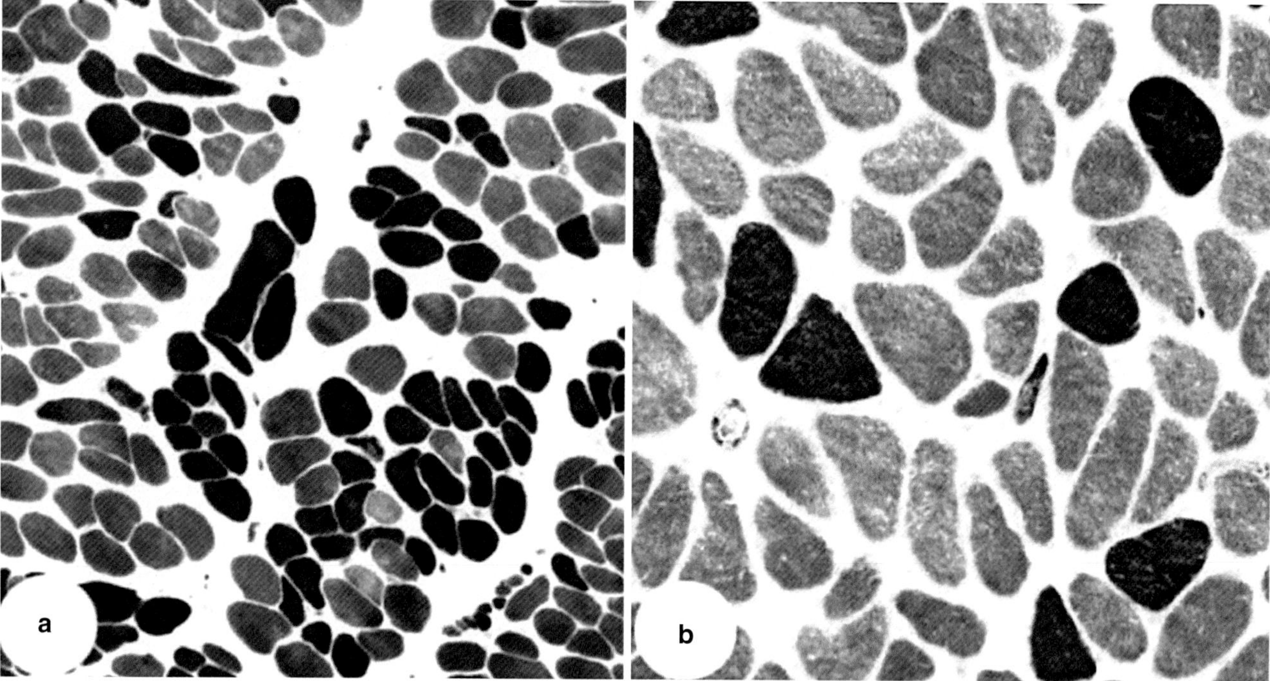

Fig. 18.57 Fiber type in cleft palate third arch Sm7 levator veli palatini. (**a**) LNP in child with cleft Type 1, 36.9% and Type 2, 63.1%. Fiber diameter (mean) 24.9 μm. There is a large proportion of fibrous tissue. (**b**) LVP in normal adult. Type 1, 74.9% and Type 25.1%. Fiber diameter (mean) 46.1 μm. Note: Infant LVP has reversed type 1/type 2 ratio and almost 59% smaller fiber diameter. Fiber diameter relates to degree of use. Light staining fibers = type 1 Dark staining fibers = type 2. [Reprinted from Lindman R, Paulin G, Stål PS. Morphological characterization of the levaator veli palatine muscle in children born with cleft palates. Cleft Palate Craniofac J 2001; 38(5):438–448. With permission from SAGE Publishing]

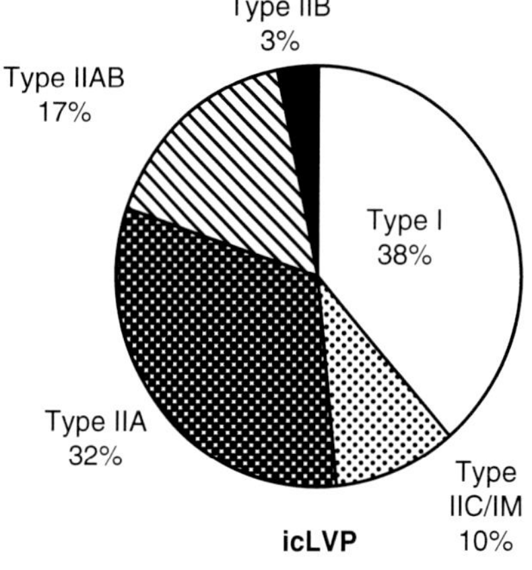

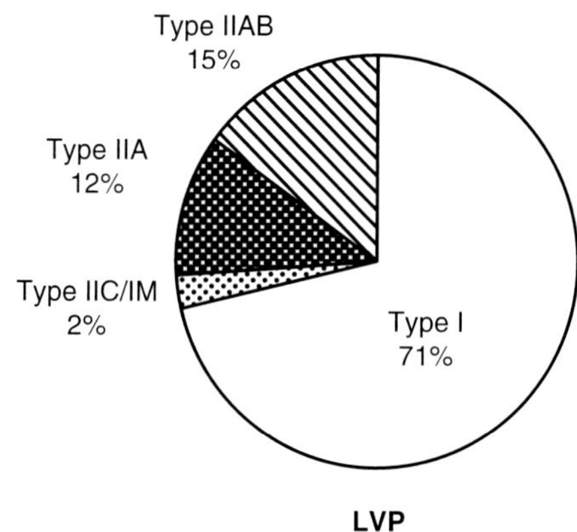

Fig. 18.58 Levator veli palatine in the cleft versus normal state. Cleft state in LVP characterized by 50% less type I fibers. [Reprinted from Lindman R, Paulin G, Stål PS. Morphological characterization of the levator veli palatine muscle in children born with cleft palates. Cleft Palate Craniofac J 2001; 38(5):438–448. With permission from SAGE Publishing]

of prerequisite for later fusion of the lateral lip element to the midline.

Metabolic Insufficiency

Metabolic demands of this rapidly proliferating tissue are undoubtedly high. Immediately before MxP fuses with the unified LNP/MNP complex, the LNP demonstrates a *burst of cell division that renders it vulnerable to insult.* Many possible mechanisms could depress the rate of cell proliferation of the LNP, resulting in the failure to fuse with the MNP; these include: (1) inadequate physical approximation of the processes, (2) Impaired epithelial break-down, and (3) impaired metabolism due to environmental toxicity defective cell programming.

Comparison of cleft formation in three strains of mice provides a useful model to separate out some of the variables. C57Bl/6J embryos have a normal growth pattern of the LNP and MNP, and a negligible incidence of clefts. A/J and CL/Fr embryos have cleft incidences of up to 12% and 36%, respectively. Both strains have alteration in MNP growth that is more parallel to the midline and therefore cannot fuse properly with the LNP. A dramatic *decrease in epithelial activity at the fusion site* is what sets the CL/Fr embryo apart, making it more susceptible to cleft formation than the A/J strain [63].

Cleft may result from *oxygen deprivation* to the nasal fields at this critical juncture. Hypoxia increases the incidence of CL/P in AJ mice. As FiO_2 drops from 21 to 10%, cleft incidence increases from 38 to 90% [47, 49, 50]. Smoking causes perturbations in placental blood flow. Carbon monoxide produced by maternal smoking decreases oxygen release by 42% causing hypoxia which can affect the developing embryo [64]. A Swedish study demonstrated a doubling in the incidence of clefts born to smoking mothers [65]. Hyperoxia can also be protective. CL/Fr mice born to mothers exposed to 50% FiO_2 have only a 10% incidence of clefts. The known susceptibility of A/J mice treated with phenytoin drops from 87% at room air to 25% at FiO_2 [48].

Vascular Insufficiency

Blood supply issues may also come into play. The distal-lateral zone of the A field lies immediately subjacent to the D field and mesial to the internal aspect of the B field. This A–B–D "apex" is a watershed between two vascular systems, between the most distal branches of the internal carotid artery and the most caudal and mesial branches of the external carotid artery. During this early development period, it should be borne in mind that the external carotid system is itself in formation. All facial structures not supplied by the internal carotid artery are initially served by branches of the stapedial artery system. This system must be replaced by means of aortic arch derivatives, selective portions of which coalesce to form the external carotid system [66, 67].

The point of vascular "switchover" leaves the nasal prominences particularly vulnerable to insult for two reasons. First, the transition is itself genetically determined, and errors in this process could occur. Second, extrinsically caused hypoxia could rob whatever "safety margin" the embryo possesses during the switchover. Conversely, the protective effect of hyperoxia in cleftsusceptible mice may apply here. Embryos with a statistically greater likelihood to sustain a vascular insult could be "tided over" by higher O_2 concentrations.

Simonart's Band

What is it that drives the production of the cutaneous Simonart's band? It does not arise from the premaxillary tissues. PMxB is in communication with the caudal tip of LNP which is composed of r2 neural crest. Let's postulate that metabolic activity at the alar base "revs up" the basal LNP to produce this protrusion. In some ways this is analogous to basal MNP when it spits out prolabial process. In any case, the "epithelial experiment" of Simonart's band is short-lived and all-or-nothing. Either it achieves fusion with the medial process or it does not. The success depends upon the physical distance and strength of the biochemical signal.

As we saw, this bridge quickly fills with mesenchyme. If lip development stops here we have a have a very high incomplete cleft. Mucosal fusion (not shown in our diagram) provides the template for DOO migration. The skin envelope fulfills the same function for SOO. Why does SOO stop at the philtral boundary? We can explain this as a question of "like versus unlike." The r2 skin abuts against r1 skin of the philtrum. SOO will along go as far as its template permits. So it simply makes contact with the subcutaneous mesenchyme of the philtrum. Meanwhile, DOO, unimpeded, proceeds blithely along the mucosa until it encounters its contralateral fellow.

In summation, experimental evidence suggests the following

- The closure process of the nose and lip involves a sequence of events, the site of which is located at the caudal aspect of the medial and lateral nasal processes. TS = Theiler stage.
- Physical size (mesenchymal mass) of the processes must put them within the *critical contact distance*.
- Epithelial breakdown along the seams of the processes is the prerequisite for fusion. This requires a diffusible morphogen, BMP4.
- LNP fusion to the MNP is the prerequisite for the migration of MxP mesenchyme medial to eventually unite with prolabium/premaxilla.
- LNP initiates contact with MNP.
  - Both the epithelium and mesenchyme of the LNP at the fusion site are activated *before* those of the MNP.
  - After fusion occurs (TS7), DNA labeling indices continue to decrease in both the LNP and MNP. These reach zero at TS 1 1 for the MNP and TS13 for the MNP.
- Simonart's band: mesenchymal stabilization in these processes correlates with the formation of a mesenchymal bridge at TS12–TS13.

B

Unification of the Lip: Mesenchymal Migration

Lateral lip element migrates along the bridge bringing muscles with it. As more mesenchyme fills the envelope, the fusion process proceeds downward. As BMP4 arrives at the cutaneous border, epithelial breakdown and fusion ensue. Note that all the action is on the lateral lip side as the BMP4 is produced by the premaxillary field.

Intraorally during this time, the medial nasopalatine fields are undergoing posterior-to-anterior fusion. Mesenchyme is added progressively, as discussed in Chap. 14. At precisely the point that this "zipper" mechanism arrives at the incisive foramen, the palatal shelves are lifting up such that contact can be made. At that point medial nasopalatine proceeds forward into the primary palate. Thus, fusion of the secondary palate is anterior–posterior while that of the primary palate is posterioanterior.

- The "flow" of such mesenchyme is lateral to medial. Its timing is short (4 h).
- This period is one of great vulnerability to a number of potential environmental and genetic insults.

Pathophysiology of Cleft Lip/Palate Muscle

We have discussed at great length how developmental field model applies to craniofacial bone fields. Neuroangiosome failure leading to a reduction in mesenchymal volume is manifested as small or absent bone structures that produce smaller amounts of biochemical products, such as diffusible morphogens. We have seen how the feedback loop between bone morphogenetic protein 4 and sonic hedgehog regulates epithelial stability, thereby controlling fusion of neighboring mesenchymal structures, such as palatine shelf to vomer or lateral lip element to philtrum. But the BMP4/Shh mechanism is only one of many that must be operative. In the progressive loss of mesenchyme from PMxF to PMxB the local concentrations of multiple signals and growth factors are affected. In this important section, we shall look at selected aspect of the perioral musculature surrounding the cleft, first, in terms of anatomic organization and function, and second, in terms of histochemistry and ultrastructure. These studies further support the diffusible morphogen model.

- Diffusible morphogens from abnormal field affect multiple tissues, including muscle.
- The reduction in concentrations of these morphogens is highly localized to the submucosal tissues on both sides of the cleft.
- The concentration gradient normalizes distal to the cleft site, with consequent return of microanatomy and histochemistry to normal.
- Non-philtral prolabial tissues are abnormal.
- The fundamental goal of surgical repair is to reassign healthy developmental fields to normal anatomic relationships.

Muscle Anatomy and Function

Traditional anatomic concepts of the orbicularis as a monolithic layer are incorrect. These had led to any surgical publications in which the two muscle layers are dissected as full thickness flaps and sutured either directly or rearranged. Orbicularis is dynamic with two distinct functions: DOO is a sphincter and SOO acts as a dilator. In point of fact, the orbicularis is a layered structure with specific points of insertion, which explains the esthetic "drape" of the lip as well as its prehensile abilities. These layers give the pout of the lip and philtrum. Surgical approaches based on z-plasty muscle flaps and interdigitation can violate the architecture of the nasolabial complex.

The best way to conceptualize this anatomy is to combine the work of Park with that of Delaire, Precious and Nammoun, building on previous studies by Nairn and Nicolau [68–70]. Park's dissections show the orbicularis to possess a deep, sphincteric layer that traverses the entire lip beneath Cupid's bow and permits the lip to pucker. Contraction of DOO causes narrowing of the philtral columns. The deep layer has a flat, peripheral portion hugging the mucosa and an outwardly curving marginal portion: The marginal part does not lie flat but curls outward like the lip of a jug. Its form follows the mucous membrane of the lip as this everts onto the face to become the red lip margin [71] (Figs. 18.11, 18.59, 18.60, 18.61, 18.62 and 18.63).

Superficial to this plane of orbicularis is another superficial layer associated with the muscles of facial expression. *Widening of the philtral columns* results from contraction of this layer. Three-dimensional computed tomographic reconstruction of the lip demonstrate that the superficial orbicularis does not cross the midline but inserts into the lateral margins of the philtrum. Computed tomography could distinguish muscle formation but was unable to detect philtral development until after the second trimester. This discrepancy may reflect the relative insensitivity of the technique. This contradicts the previous work by Latham, who used layered Plexiglas tracings, a methodology which leads to false conclusion that fibers from a single-layer orbicularis oris crossed the philtrum to insert on the opposite side.

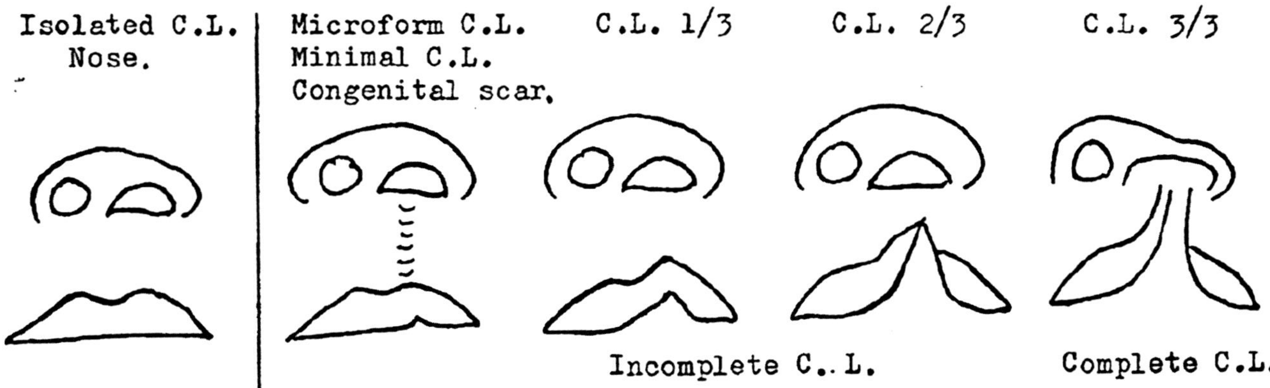

Fig. 18.59 Onizuka classification 1. Onizuka classification of microform clefts is compatible with DFR model. Diffusion of BMP4 as an inhibitor of Sonic hedgehog causes epithelial breakdown. PMx field deficiency results in reduced [BMP4} and therefore resistance of epithelium to fusion. Isolated Cl is defined by Mullein as the "minimicroform" which places emphasis on the lip rather than on the piriform fossa and the nose. [Reprinted from Onizuka T, Ho aka Y, Ayoyama R, Takahama H, Jinnai T, Ursi Y. Operations for microforms of cleft lip. Cleft Palate Craniofac J 1991;28(3): 293. With permission from SAGE Publishing]

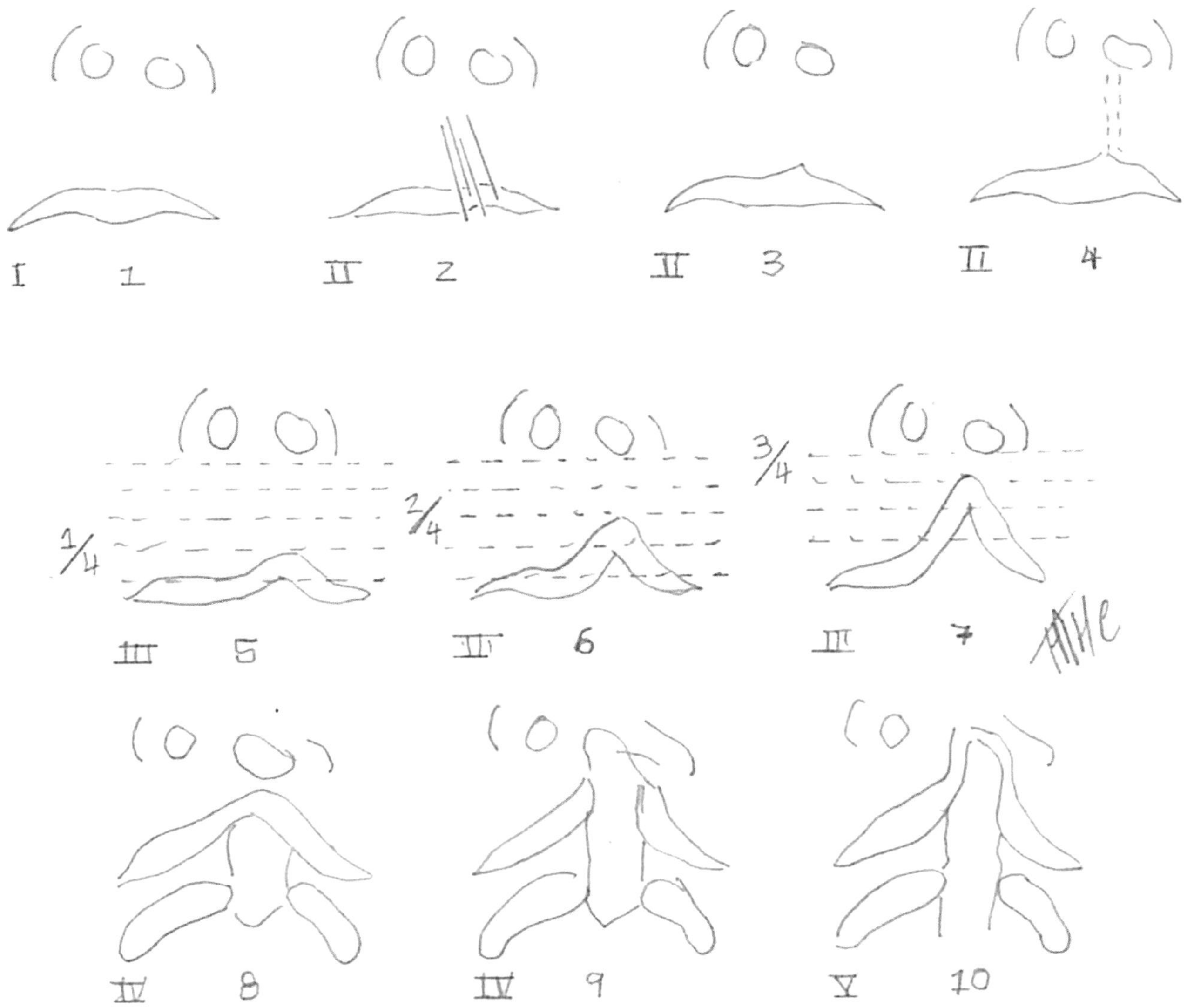

Fig. 18.60 Onizuka classification 2. See text. [Reprinted from Onizuka T, Ho aka Y, Ayoyama R, Takahama H, Jinnai T, Ursi Y. Operations for microforms of cleft lip. Cleft Palate Craniofac J 1991;28(3): 293. With permission from SAGE Publishing]

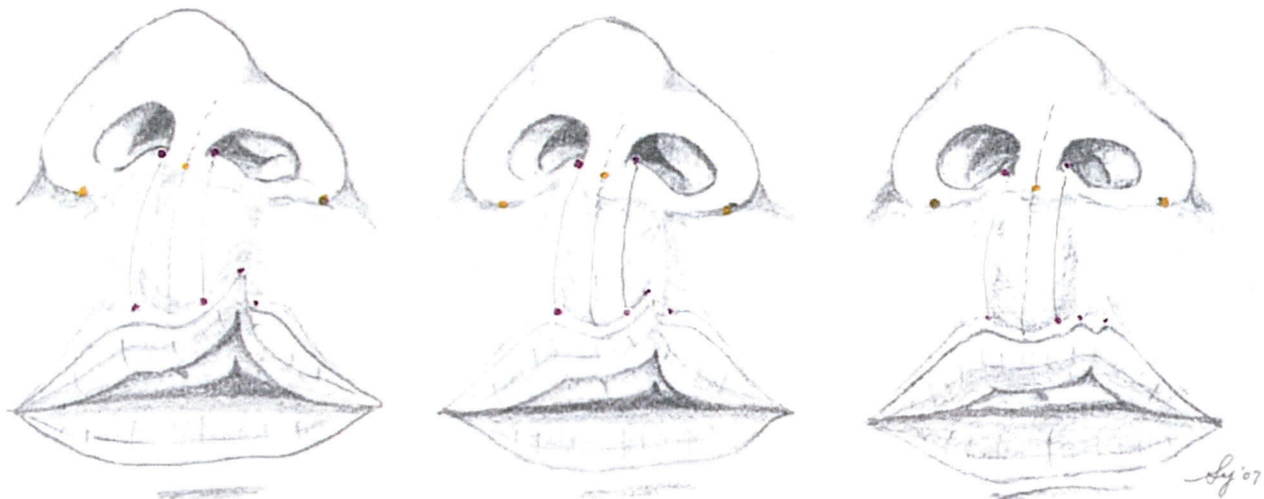

Fig. 18.61 DFR Mulliken classification. Mulliken classification. (Left) incomplete, (center) microform, (right) mini-microform (see text). Red dots (1) Distance from non-cleft side footplate of medial crus to Cupid's bow (13–2); (3) distance from cleft side footplate to Cupid's bow (12–3); maximum height of cleft taken by Mulliken as point 3. Note: DFR defined width of Cupid's bow 2–3 as the territory of the anterior ethmoid neuroangiosomes; it is the distance between the alar footplates (12–13). Orange dots indicate subnasale and the position of the alar bases, on the non-cleft side (4) and on the cleft side (10). Note as alar base displacement becomes more pronounces with increasing severity of the cleft. [Reprinted from Yuzuriha S, Mulliken JB. Minor-form, microform, and mini-microform cleft lip: anatomical features, operative techniques and revisions. Plast Reconstr Surg 2008; 122(5):1489–1493. With permission from Wolters Kluwer Health, Inc.]

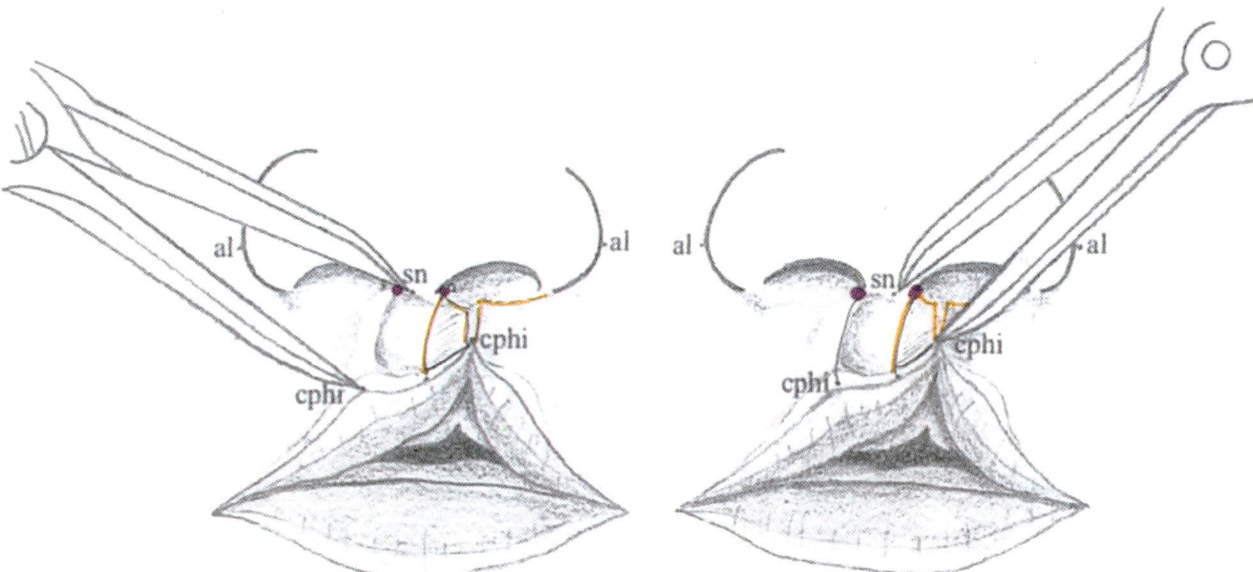

Fig. 18.62 Prolabium paradox. Caliper tip is placed on subnasale and shown to measure a difference in philtral height. Obvious difference between sn-cph non-cleft versus sn-cphi cleft. If caliper tip is placed on the alar footplates 13–2 = 12–3. Non-philtral prolabium, defined by the orange trapezoid, is full-thickness skin and subcutaneous tissue based on terminal branch of medial nasopalatine. It is advance into the columellar release and rotated into in the nasal floor. [Reprinted from Yuzuriha S, Mulliken JB. Minor-form, microform, and mini-microform cleft lip: anatomical features, operative techniques and revisions. Plast Reconstr Surg 2008; 122(5):1489–1493. With permission from Wolters Kluwer Health, Inc.]

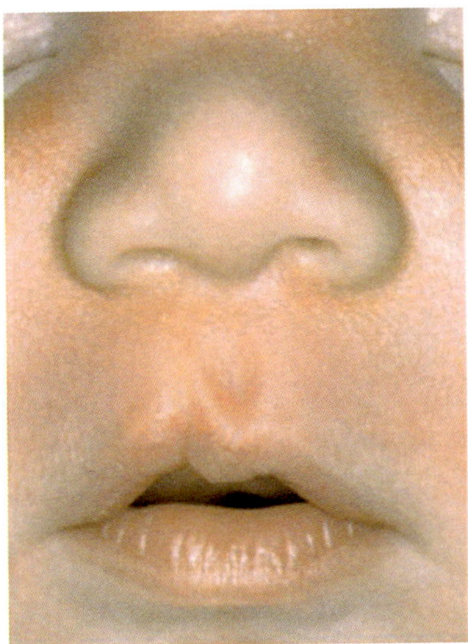

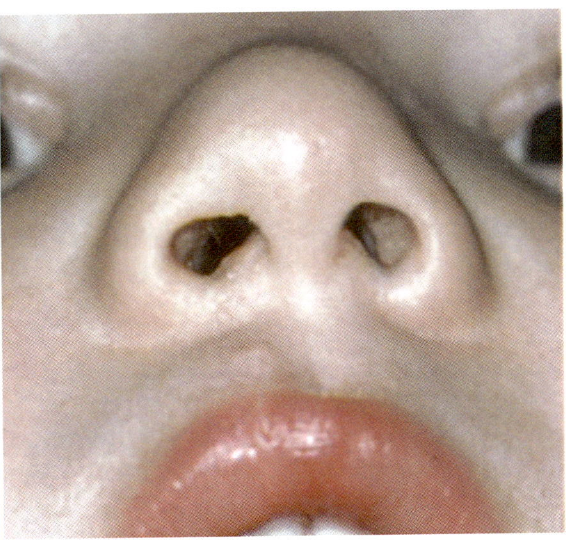

Fig. 18.63 Residual nasal features of right-side microform cleft after repair. Cleft nasal deformity persists post repair: columellar deflection, footplate of medial crus tractioned downward and lies internal to the left side asymmetrical tip, airway reduced in size with compromise of the external valve. Lip shows asymmetry of orbicularis insertion into the right side of columellar base. [Reprinted from Yuzuriha S, Mulliken JB. Minor-form, microform, and mini-microform cleft lip: anatomical features, operative techniques and revisions. Plast Reconstr Surg 2008; 122(5):1489–1493. With permission from Wolters Kluwer Health, Inc.]

The layered orbicularis model provides a functional explanation of the philtrum. It is also compatible with work by Delaire et al. [21] demonstrating that oblique fibers inserted high into the midline near the anterior nasal spine could establish a drape for the lip. These fibers also intermingle with those of the paranasal complex. Joos and Friedburg used magnetic resonance imaging to study the relation of these muscles to the autonomic nervous system (ANS) in normal subjects and in cleft patients both pre- and postoperatively. The lateral muscle, particularly at the level of the spina nasalis, runs in a stepwise fashion dorsally through the nasal floor and is inserted into the nasal septum. This explains why in the cleft state the nasal septum always deviates to the healthy side, where the muscles are inserted in a ring-shaped distribution of the musculature in the anterior facial region. If the muscle ring is not closed ... the musculature of the mid-face contracts outward, and pulls the ala of the nose.

Recognizing the concept of function rings of perioral musculature, Delaire and Markus pioneered in developing the concept of early subperiosteal centralization of the musculature. Cleft repair is thus functional as well as anatomic. Its design should synchronize and facilitate a more normal subsequent facial development (cf Fig. 18.14).

The development and insertion of these muscles follows a strictly coordinated timeline. The overall "game plan" of facial muscle development is from deep to superficial, from lateral to medial [17]. Philtral formation occurs long after the lip has formed due to the late insertion of the SOO [72]. Intrauterine photographs of developing human fetuses demonstrate absence of a philtrum at 10 weeks [72, 73]. Between 12 and 16 weeks, Kraus' specimens demonstrate philtrum formation concomitant with full development of the nostrils and a true nasolabial fold. Delaire's schema of three perioral muscle "rings" develop in an order consistent with facial muscle embryology. The middle ring (the orbicularis oris) is complete at 12 weeks [74, 75]. At 16 weeks, the lip is united to the anterior–inferior septum and the anterior nasal spine; the median cellulous ligament of Delaire is present.

- **37 mm** Orbicularis oris appears but does not encircle the mouth until 58–80 mm.
- **58–80 mm** The levator labii et alequae nasi appears at 58 mm; it is fully inserted at 80 mm.
- **80 mm** The pars alaris of the nasalis is present at 80 mm and subsequently passes deep to the levator labii et alequae nasi to insert bone along the caudal rim of the piriform aperture.
- **142 mm** The depressor septi passes vertically from the caudal septum through the orbicularis to insert on the maxilla. Thus, the most midline and "septal" muscles form last.

A small but extremely important detail must be mentioned here. The proper insertion of the perioral and perinasal fibers is not truly "into" the septum. Rather, the fibers that mingle and make up the septopremaxillary "ligament" insert into the perichondrium of the anterior paraseptal cartilages. The ligament is hemispheric, U-shaped and extends into the interpremaxillary suture. Thus, the septopremaxillary "ligament" is a fusion of fibers that insert into each half of the united mesial leaves of the r1 mesenchyme into which the septum develops [75].

Cadaver studies of nasolabial muscles by Wu in 4 aborted fetuses with cleft lip demonstrate several important points [76]. Muscle fibers on the lateral side of the cleft were thinner and more disorganized. Absent depressor septi nasi on the cleft side explain the traction of the columellar base to the normal side by remaining DSN. Pars transversalis of nasalis was confirmed to insert *lateral to the incisive fossa*, ascending to the dorsum to join the SMAS of the contralateral muscle. Pars alaris of nasalis has its insertion in the center of the incisive fossa. It bypasses the ala to terminate in the dermis above the lateral crus. On the cleft side nasalis fibers were observed to attach into the skin of the ala nasi and into the piriform fossa as well.

Anatomic and Histologic Studies of Cleft Lip Muscle

We begin with the historic work by Fará on cleft muscle using generalized non-selective vascular injection (via the common carotid) and cadaver dissection which were interpreted to show muscle fibers running upward along the cleft site. This was subsequently popularized by Millard and widely accepted, but turned out to be misleading due to fixation artifact. Fará did document in the normal lip anastomosis between columellar arteries and the superior labial artery. An arteriogram is presented showing the exclusive supply of the bilateral cleft prolabium by what we now know to be paired StV1 anterior ethmoids. Because the injections were non-selective, the developmental significance of this finding was missed.

Dado and Kernahan, using electrical stimulation and serial histologic sections, refuted the Fará model finding [77]. At the lateral cleft margin, instead of an upward pull of muscle fibers along the cleft margin (as predicted by Fará), they observed lateral contraction and puckering, indicating insertion into the skin. In histologic examination of a stillborn with complete unilateral cleft lip and palate, the muscle fibers were found to be sparse and hypoplastic with more abundant collagen staining. The immediate submucosal tissue had an abundance of connective tissue.

Heckler writing on 8 cases of microform cleft compared the degree of orbicularis disorganization with the dental defect, the size of the vermilion notch and with the nasal deformity. In each case, a vertical section was taken full-thickness through the scar and submitted for histology. The severity of orbicularis disorganization strongly correlated with the severity of the nasal findings.

De Mey used histologic section from 18 operative cases of unilateral cleft lip. They identified the bilaminar structure of orbicularis. In 4 incomplete clefts DOO, in association with mucosa was simply interrupted. Within the bridge above the cleft they found SOO to be distorted, with the degree of muscle distortion proportional to the degree of the nasal deformity, confirming Heckler's observations.

DOO and SOO were formally identified by Park in normal cadaver specimens. DOO extends from modiolus to modiolus and is continuous with buccinator. Its purpose is to seal the mouth as a constrictor. SOO extends more lateral to modiolus, comingling with adjacent facial muscles.

"When the mouth is pursed up, the vermilion becomes thickened as the marginal portion of the deep orbicularis oris muscle constructs. Simultaneous relaxation of the superficial part (allows) numerous perioral fine wrinkles (to become visible), accentuated the philtral columns and flattens the nasolabial folds. When the mouth is open, in contrast, perioral fine wrinkles and the philtral column are flattened as contraction of the superficial orbicularis oris muscle stretches the perioral skin."

In a stillborn with complete UCL Park found SOO on the non-cleft side to be undistorted and to insert directly into the nasal columella (and philtral column), again on the side of the normal nostril. On the lateral side, it was found to insert into the alar base. Importantly, fibers (attributed to SOO) were observed to be inserted into the *periosteum of the cleft side piriform fossa* (italics mine). These were surely those of the *nasalis complex* coursing along the superior margin of SOO at the level of the alar base fold.

Histochemistry of Muscle in Cleft Lip and Cleft Palate

The margin of soft tissue clefts constitutes a zone of altered development in which muscles from different embryologic origins, second arch orbicularis in cleft lip and third arch levator, palatopharyngeus and uvulus cleft palate show similar patterns of dystrophic change as seen in their histology, ultrastructure and histochemistry. Compared with normal facial muscle biopsies of cleft muscle are markedly different. Routine stains from specimens at the margin show disorganized fibers going in different directions. As the cleft margin is approached muscle fibers are reduced in number and the amount of connective tissue increases (Figs. 18.54 and 18.55).

- We must immediately note work by Solov'ev with the same findings but documenting a *reversion in muscle histology toward normal about* ***5 mm*** *lateral to the cleft margin.*

We now consider two studies by Schendel at Stanford which are absolutely pivotal and deserving of the greatest attention. The first study [78] focuses on the Sm6 facial muscle, orbicularis, comparing muscle from the cleft margin on the cleft side (the prolabium) versus muscle from the margin of the non-cleft side (lateral lip). The second study evaluates a representative Sm7 muscle, the uvulus in two situations: CP, isolated cleft palate (implying a minimal to modest involvement of palatine bone P3 horizontal shelf) and CL/P, cleft palate involving the secondary palate and lip. In the latter case biopsies were able to distinguish the effects of P3 along on muscle development versus more extensive clefting into the maxillary hard palate fields.

- Recall that the secondary hard palate consists of three palatine bone fields, from anterior to posterior, P1–P3.
  - P1–P2 are incorrectly referred to the "maxillary" hard palate but these fields have a separate evolutionary origin. They form the lingual wall of the maxillary alveolus.
  - P3 is commonly considered the horizontal shelf of the palatine bone. Pathology of the lesser palatine neuroangiosome is confined to P3 and the anterior palatine aponeurosis.
  - P4, the ancient ectopterygoid, is no longer part of the mammalian hard palate, being displaced forward to become the medial pterygoid plate.

Terminology: CL, CP and CL/P

CL Cleft lip refers to a deformity of the lip and alveolus from the microform to complete cleft. Why should one include the alveolus? Because the bone stock of PMx is always involved, and, at the very minimum, a bone deficit of the piriform fossa is present, CL implies some degree of alveolar involvement, from a reduction in volume to a frank cleft. A quantitative deficit of the diffusible morphogen, BMP4 from P3 is responsible for failure of soft tissue unification.

CL/P In the literature, this condition implies a concomitant "hit" to both the lip and the secondary palate. With regard to the biochemical effect on the soft palate, the PMx field is too far removed from the soft palate; it is therefore irrelevant. *Any* form of isolated secondary hard palate cleft by virtue of reduced concentrations of BMP4, will affect the soft palate to a degree proportional to the bone defect.

CP This refers to the isolated defect of the soft palate without overt involvement of the hard palate. Recall that in CP, by definition, the specific field defect resides in P3. Once again, BMP4 diffusing backward through the soft tissues, is responsible for the soft tissue cleft in a gradient from posterior to anterior.

Cleft Side Versus Noncleft Side This out-of-date terminology makes a supposition that the cleft pathology somehow resides on the lateral side. To be sure, microform cleft shows only the depression in the lower lateral corner of the piriform fossa due to missing dysplastic or missing PMxF and PMxB. This is an obvious site for decreased BMP4. However, in complete clefts, the BMP4 available to the prolabial cleft margin comes strictly from PMxA. Thus, in terms of muscle development, morphogen deficit exists on both sides of the cleft but the reduction in concentration gradient is less severe on the prolabial side. Thus, it is more appropriate to make reference to the prolabial (noncleft) side and the lateral lip (cleft) side.

Cleft Lip Muscle (Sm6): Orbicularis

Sixty-eight muscle biopsies were taken from *unilateral cleft lip* patients at the time of surgery and submitted for histochemistry and electron microscopy. Fiber type was studied using the following enzyme-specific stains. Although not stated, it is reasonable to assume the DOO layer of orbicularis was biopsied, since SOO on the cleft side does not exist. Here is a summary of the stains used.

- Muscle fiber type
  - ATPase: attached to myosin
    Reaction at pH 9.4 shows type 1 (slow-twitch fibers)
    Reaction at pH 4.6 shows type 2 (fast-twitch fibers): 2a, 2b, 2c
- Mitochondrial location
  - NADH-TR: NADP-tetrazolium reductase is present mitochondria, sarcoplasmic reticulum, and T-tubules
  - SDH: succinic dehydrogenase is exclusive to mitochondria
- Neuromuscular end plates
  - Esterase is specific and indicates degenerating fibers

Table 18.1 Summary of muscle characteristics

	Origin	Fiber type		Fiber diameter µm	
Muscle		Type 1	Type 2	Type 1	Type 2
Extremity	Somites	69.7%	30.3%	23.5	21.4
Orbicularis normal[a]	Sm6	24.5%	75.5%	22.6	23.1
Orbicularis medial	Sm6	29.2%	70.8%	11.1	11.8
Orbicularis lateral	Sm6	17.4%	82.6%	8.9	8.9
LVP normal[b]	Sm7	74.9%	25.1%	46.1 (mean	
LVP CL/P	Sm7	36.9%	63.1%	24.9 (mean)	
Uvulus CP	Sm7	56.7%	43.3%	20.8	20.9
Uvulus CL/P	Sm7	62%	38%	10.16	7.4

[a]Data from [78–80])
[b]Data from [81]

- Z-band identification (can be displaced or degenerate in myelopathies)
  - Gomori trichrome stains collagen

Results from studies by Schendel, de Chalain and Lindman are summarized chart appended below (Table 18.1).

In the normal state, fiber types vary by embryonic origin. The ratio of type 1 compared with type 2 fibers in somitic muscle is 69.7% vs 30.3%. In normal Sm6 muscle the ratio is 24.5% vs 75.5%. In normal Sm7 muscle the ratio is similar to somites, 74.9% vs 25.1%. Recall that the somite system represents the transformation of Sm8 the S1.

- Perhaps the genetic switchover applies to Sm7 as well.

Schendel noted that the orbicularis is abnormal along the cleft margins on both sides compared with the normal. Furthermore, the cleft state has differential effects on the prolabial side versus the lateral lip side. Regarding fiber type, there was no difference on the prolabial size but the lateral side had significantly fewer type 1 fibers. Fiber size (mean) on the prolabial side versus the lateral side 52% and 42% of normal, respectively.

Electron microscopy using Gomori stain showed large numbers of abnormal mitochondria pushing apart the muscle fibers. Instead of being at the periphery of the muscle fibers, the mitochondria were clumped together in the center. They had an unusual shape with abnormal cristae. Glycogen was seen accumulating between the fibrils (Fig. 18.56).

Furthermore, both the Gomori and NADH stains demonstrated large numbers of "ragged red" fibers (RRF). This term refers aggregations of abnormal mitochondria in the sub-sarcolemma of the muscle fibers. RRF is characteristic of myoclonic epilepsy (MERRF, based on the ragged red fibers).

Evidence previously cited regarding metabolism of lateral nasal process tissue on the cleft side in mice indicates *deficient energy production*. A metabolic defect reflected in the mitochondria was hypothesized to be responsible for inadequate muscle migration and development. It should be noted that orbicularis muscle immediately lateral to the cleft site begins to normalize. Solov'ev performed similar studies in Russia and found smaller fibers at the cleft margin with a *reversion in muscle histology toward normal about 4–5 mm lateral to the cleft margin*. This is approximately the width of the non-philtral prolabium.

The model of a biochemical defect leading to changes in cellular metabolism provides a background for findings at the University of Pittsburgh [74] obtained by studying orbicularis anatomy in 29 fetuses, including 9 with cleft lip/palate. Using 3-D CT scanning of section and documented a difference in the speed of development. When compared to the normal time sequence (vide supra) orbicularis oris muscle lagged behind 3.5 weeks, thus affecting the superficial muscle ring and labial muscle ring. They postulated this to be causative for distraction of the premaxillary away from the midline as proposed by Delaire.

- This mitochondrial hypothesis does not explain generalized changes in histology nor does it relate muscle embryogenesis to surrounding bone fields.

Important supportive work by De Chalain evaluated orbicularis biopsies from normal children with lip lacerations versus patients with isolated cleft lip and cleft lip associated with cleft alveolus (a more severe affection of the PMx field).

Sm6 orbicularis in both normal and cleft states was shown to have a fiber type and size profile different from both somitic muscle and from Sm7 palate muscle. Cleft muscle demonstrated hypoplasia, disorganization and clusters of abnormal mitochondria. Ragged red fibers were *not* observed.

- Data for normal orbicularis from this are included with those from Stanford in Table 18.1.

Cleft Palate Muscle (Sm7): Levator Veli Palatini, Palatopharyngeus, Uvulus

Histochemical evaluation of palatal muscle offers similarities and differences. Recall that, with exception of tensor veli palatini from first arch and Sm4, all remaining soft palate muscles originate from Sm7 and therefore come from a different homeotic environment that orbicularis being of second arch origin from Sm6. Findings of fiber type in normal and cleft palate children were reported in Russian in 1988 by Khamidov.

Lindman, working independently from Stanford characterized 11 biopsies of levator veli palatini from children undergoing palatoplasty, using the following histochemical stains: mATPase for fiber classification, NADH-TR to assess mitochondrial oxidation, and Gomori trichrome stain to image cell morphology, membranes and nuclei. Two of 11 were isolated soft palate clefts (P3), and the remainder were clefts involving the secondary palate as well (P2–P3 or P1–P3). On general histology, LVP fibers were loosely packed with types 1 versus 2 at 38% and 72%, respectively. This pattern indicates predominance of fast-twitch response. Interestingly this distribution reverses in adults (Figs. 18.57 and 18.58).

Doménech-Ratto [82] and Cohen et al. [83] documented soft palate morphogenesis in normal fetal specimens. Sm4 tensor veli palatini appears concomitant with the rest of the muscles of mastication with Sm7 muscles maturing later: levator, palatoglossus, palatopharyngeus and uvulus. Bone synthesis followed a similar pattern: mandible, medial pterygoid plate, lateral pterygoid plate and hamulus. Subsequently Cohen et al. [84] noted in 4 cases with cleft palate developmental delay in myogenesis. Interestingly, 3 specimens had a very low amount of myoblasts in Sm7 LVP, while Sm4 TVP remained resistant with myoblast volume comparable with normal.

- An abnormality exists in the early development of soft palate muscles.
- These finding persist throughout fetal life and are present at time of palatoplasty.
- **Note** Similarity in behavior between Sm6 orbicularis in the cleft lip and Sm7 uvulus in cleft palate.

  – In both second arch and third arch muscle fields, a defect in an underlying bone field creates a morphogen deficit which causes a slower rate of development and smaller fiber size.

Why muscle fibers of TVP resistant, whereas those of the levator sling are not? Recall that the bone deficit in P3 follows a gradient inverse to its development. In the cleft state, the rectangular horizontal plate becomes a triangle: bone is missing medially and posteriorly. Laterally P3 at the orbital plate remains normal in volume. BMP4 from P3 diffuses front-to-back along the soft palate edges; i.e. considerably medial to the location of tensor. Additional P3 morphogens critical for muscle development may follow the same pathway.

- Bone deficit of P3 in the medial horizontal plate creates a morphogen deficit which could affect myogenesis in the levator sling.
- The bone volume of P3 is unchanged near TVP thus it is not exposed to a morphogen deficit.

Schendel et al. [80] applied the same methodology used for the cleft lip study to 30 biopsies of the uvulus muscle obtained from 16 patients undergoing primary palatoplasty: 7 cases were isolated soft palate clefts and the remainder were combined with cleft lip in which one side of hard palate was normal and the other was deficient in size. Histochemistry was performed using the same stains as before.

Biopsies of Sm7 muscles showed a different fiber pattern that those from orbicularis, having a much higher percentage of slow fibers. Furthermore, there quantitative differences noted for uvulus specimens from isolated cleft palate versus CL/P. Recall that for soft palate clefts CP there is no "cleft side." But for CL/P there is a cleft side and its anatomy will reflect the great degree of morphogen deficit.

Chapters 14 and 16 presented extensive evidence demonstrating that differences in the biology of soft palate clefts are utterly dependent on pathology in the hard palate platform. Specifically, isolated CP occurs in the context of involvement of the P3 field, i.e., the horizontal plate of the palatine bone. The presentation of isolated CP worsens when the mesenchymal defect advances into the P2 field of the hard palate, even when no obvious bony cleft is visible. Soft palate cleft in the presence of an open hard palate cleft is even more severe. In all these cases, diffusion of BMP4 is responsible for the failure of epithelial fusion. The uvulus is the last site to receive the morphogen and the first to fail when it is inadequate. We shall now consider the Stanford palate data in view of the following principles.

- The greater the degree of palatine bone field involvement, the more severe is the deficit of BMP4 available to the soft palate.
- This universal law of diffusible morphogens applies to all other growth factors produced by the P3 field.
- If development of uvulus muscle is dependent on such morphogens it will be more hypoplastic in cases of CL/P that in isolated CP alone.

Uvulus from isolated CP demonstrates differences from normal orbicularis. There are significantly more type 1 slow-twitch and less type 2 fast-twitch fibers (56.1%/43.3%) compared to normal (29.2%/70.8%). Fiber size of uvulus was much larger than in orbicularis, 20.8 μm versus 11.1/11.8 μm, but comparable to that of skeletal muscle. Interpretation these data must consider differences in origin of the muscles involved.

- Tensor veli palatini, the sole first arch muscle of the soft palate Sm4, arises from Sm4
- Orbicularis is a second arch muscle originating from Sm6
- Palatopharyngeus is a third arch muscle that arises from Sm7

Uvulus specimens from CL/CP when compared with those from CP show similarities and difference. Fiber size for uvulus in the isolated cleft palate is roughly comparable to that of skeletal muscle. It is roughly twofold greater than in cases involving the secondary palate.

- The morphogen effect on the uvulus muscle fiber diameter is inversely proportional to the degree of deficiency in the palatine bone fields.

Finally, mitochondrial stains, both NADH and Gomori trichrome, reveal in the uvulus specimens, of CL/P an abnormal accumulation of mitochondria the center of the muscle, similar to that observed in the orbicularis specimens. These mitochondrial also show the "ragged red" phenotype. In contrast uvulus from CP did *not* show mitochondrial pathology. What could explain this difference? Fiber size is small for both CL and CL/P in both instances whereas CP fibers are much larger. Perhaps a relationship exists between the process that results in reduced fiber size and abnormal mitochondria.

These finding corroborate previous observations by Khamidov in which electron microscopy of CL orbicularis and CL/P palate muscle showed abnormal muscle fibers and mitochondria. Note they did not obtain biopsies from CP patients. Furthermore, patients did not appear to "grow out of" this condition as older children (age 12) showed worse pathology.

Lazzari's group reported similar dystropic changes in Sm6 orbicularis but their study documents changes in Sm7 palatopharyngeus as well. Biopsies were taken at the margins at the time of primary repair cleft lip (7) and cleft palate (20), cleft lip patients (6) presenting for secondary surgery involving takedown of the original repair were also biopsied. Enzymes were the same as Stanford, save SDH and esterase. Orbicularis from both sides of the cleft was taken at unilateral lip repair. Biopsies of superior constrictor and palatopharyngeus were taken at palate repair. Orbicularis did *not* include pathological fibers such ragged red mitochondria but disorganization and fibrosis were documented.

Summary of Histochemical Data

- Second arch Sm6 orbicularis has a fiber diameter similar to somitic skeletal muscle in the normal state.
- Under conditions of cleft (PMx defect) fiber orbicularis diameter becomes 50% smaller on the medial side and 60% smaller on the lateral side.
- Third arch Sm7 uvulus has a fiber type and fiber diameter similar to somitic skeletal muscle in both the normal state and in isolated soft palate cleft (only P3 is affected)
- When hard palate cleft is added (P2 affected), uvulus fibers in the soft palate become 50–67% smaller.

Conclusion

In the cleft state, muscle fibers along both the "cleft-side" lateral lip and along the "non-cleft side" non-philtral prolabium are abnormal for at least 5 mm of distance.

- Lateral lip element muscle should be pared.
- NPP flap, usually about 5 mm wide, should be transferred into the nasal floor.

Clinical Studies of Microform Cleft

Historical Perspectives

Clinical descriptions of minimal cleft are a relatively recent phenomenon in the literature. In 1935, Broderick[101] described hidden radiologic findings in patients with an asymmetric nose and/or a philtral scar. The 1964 case report by Brown of an isolated cleftlip nasal deformity is generally credited as being the first formal description of minimal cleft in the English literature: "The nostril deformity may be an integral part of the cleft-lip syndrome and not secondary to it." This was followed by reports from Stenstrom and Thilander in 1965 and Tulenko in 1968 [85, 86].

Based on previous work by Huffman and Lierle, Brown made a number of important observations [87]. (1) The alar cartilage was held in a faulty position-thereby being forced into distortion; it was *not intrinsically abnormal.* (2) Cleft represented an *asymmetric growth* state. (3) Alar form was somehow linked to the configuration of the *nostril floor.* Reflecting an understandable focus upon the obvious, Brown's explanation focused on a single anatomic variable: the alar cartilage: "The consideration of the case presented and of the embryology of the face suggests that this is another defect of mesoderm, a primary defect of the alar cartilage, and not secondary to developmental abnormality of the lip or alveolus."

That the cleft-lip nose is derived from malformed or deficient cartilage is a *misconception* that has been previously reviewed by this author [88]. In five minimal clefts, Boo-Chai and Tange dissected out the alar cartilages, finding them to be "normal in size and shape, but displaced [10]. It is, therefore unlikely that the condition is due to an intrinsic defect within the cartilage." Stenstrom came to the same conclusion: "We can see no reason to interpret the nasal deformity as a consequence of inherent abnormality of the alar cartilage .... It is more reasonable to regard cleft-lip noses as secondary deformities."

- Nasalis misinsertion as a primary cause was not appreciated for many years. Because the emphasis in the early literature was on appearance, not function, early contributors missed the connection between the normal insertion of nasalis and its relocation to the nostril sill and piriform fossa. All that was missing would have been, in the words of Millard:
  - "Know the beautiful normal."

Dentoalveolar Deficits

Dental considerations in minimal clefting have been noted by many surgeons. Although the palate and arch appeared normal in Brown's case, radiographs revealed an unerupted left lateral incisor (no. 7) and a "minimal" cleft of the alveolus and hard palate. Twenty-one minimal clefts at Columbia-Presbyterian Hospital were studied by Cosman and Crikelair [1]. Dental records from 16 of these cases showed a 100% incidence of incisor anomalies. Radiographs of seven patients demonstrated an alveolar bone defect in six cases despite an apparently intact alveolar arch. All eight case Heckler's series had similar findings.

The orthodontic literature documents patterns of dental anomalies associated with clefts [89, 90]. That these could occur in the absence of an overt cleft was demonstrated in fetuses by Kraus in 1966. Johnson applied these concepts in 4 cases: "The cleft is therefore a symptom of a disorder affecting the whole process of development of the fissural areas and particularly the dental lamina. The presence or absence of a cleft is not the sole criterion for concluding that an error... has taken place."

Stenstrom and Thilander summarized the situation as follows: "It seems as if a developmental disturbance in the lateral incisor region can manifest itself in various ways according to its intensity from a slight disturbance of the tooth germ to its most serious form, a cleft. Since nothing has been forthcoming which can clarify the cause of the defective lateral incisor in an otherwise ideal bite ... it is reasonable to assume that, as it is of the same order as cleft lip nasal deformities, it should be regarded as a *microform* of a cleft manifestation."

- The locus of the cleft is the lateral incisor zone of premaxilla, from which projects the frontal process. The neural crest defect manifests itself not only in the bone but in the dental unit as well.

Orbicularis Abnormalities

The anatomy of the orbicularis oris in minimal clefts has likewise received attention. Orbicularis muscle fiber disorientation is associated with increasing degrees of cleft nasal deformity. Furthermore, the maximum area of disarray was found in those sections taken from the nasal floor and upper lip.

As has been previously discussed, the orbicularis oris in the upper lip is anything but sheet-like. It has deep and superficial components; the oblique fibers of the latter have much to do with the "drape" of the lip. Paranasal levator insertions make important contributions to the geometry of the alar bases, whereas the nasalis insertions determine the curvature of the nostrils. Stenstrom and Oberg recognized that pathologic muscle insertions (or lack thereof) into the nose and septum could play an important role in determining septal, columellar, and alar cartilage configuration in cleft [91]. In two minimal cases, Stenstrom and Thilander [85] noted "obvious defects in the part of the pars alaris of musculus nasalis which holds the nasal wing in against the columella." Restoration of muscle anatomy to the midline is the basis for the Delaire-Joos subperiosteal "functional cheilorhinoplasty [92].

- Attributing a cleft (minimal or otherwise) to a defect in the orbicularis oris (or to any combination of muscles) proved to be an *intellectual dead end*. The biggest drawback in muscle analysis had been the failure to separate out nasalis from orbicularis. When this is inadvertently anchored to the midline, nasalis, a dilator of the airway is converted to a *constrictor*.

Credit for recognition of the nasal floor as "ground zero" for cleft formation surely goes to Cosman and Crikelair, as they correctly identified the locus of the deficiency zone to exist *beneath the plane* of the nostril floor. This zone extends (sagittally) from beneath lip epithelium anteriorly, under nasal floor, posteriorly through the alveolar arch, and terminates at the incisive foramen. What no one anticipated was the existence of the frontal process of premaxilla, which is the primary cause for the misshapen piriform margin.

Heckler observed that the extent of deformity in minimal clefts depends on the severity of the nasal deformity is correct. Greater degrees of premaxillary deficit affect the insertion sites of the nasalis muscle above the canine and lateral incisor causing it to insert pathologically into the nasal floor and piriform fossa. The most striking findings in microform cleft are therefore in the nose. Microform by definition, has an intact lip. What we can see is the quantitative reduction in BMP4 reaching the surface of the skin from below creating a furrow as Cupid's bow.

Onizuka Classification

The concept that microform clefts exist as part of the spectrum of clefting was first advanced by Boo-Chai and Tange: "There may exist a subgroup of the cleft lip deformity in which there is a typical cleft lip nasal deformity with a clinically intact lip. *We believe that they (microform clefts) represent the first link in the progression of severity of the cleft lip syndrome*" [10].

Onizuka expanded on the idea to define the anatomic differences separating various types of clefts (Figs. 18.59 and 18.60). In this classification:

- Minimal (microform) clefts are first degree (isolated nasal deformity) and second degree (upper/lower vermilion border notching, cutaneous striae).
- Incomplete clefts are classified as third degree, having (1) notching of the pars marginalis of the orbicularis oris, which affects both upper and lower vermilion borders; and (2) cutaneous shortening of the ipsilateral philtral column—the degree of the latter being defined as a proportion of total philtral height.

The Onizuka classification should be studied with care for the following reasons: It distinguishes between the variations of minimal cleft presentations. It places minimal clefts into a visible context with other cleft forms. It is visually simple and accurate. It is remarkably compatible with the DFR model.

Class I and Class II: Cleft Nose + Normal Lip

Here, a partial deficit in PMxF affects the insertion sites of nasalis

- Onizuka no. 1 cleft. The most terminal manifestations occur in the development of the most medial and caudal aspect of the SOO. Here the incisal fibers penetrate vertically through the otherwise intact SOO to insert on the vermilion. Note that there is no discontinuity of DOO fibers across the vermilion; the DOO is fully formed by this time.
- Onizuka 2 has a foreshortening of these fibers which causes the vermilion to be elevated; an isolated notch in the lower border of the vermilion results. The white roll is intact.
- Onizuka 3 represents the failure of the caudal margin of the SOO to insert properly into the philtral margin (the lateral border of the A field). The upper border of the vermilion appears notched: white roll discontinuity is present.
- Onizuka 4 occurs when the entire length of the SOO fails to insert at the philtral margin. A cutaneous stria (plural striae) is present; the SOO is seen as a muscle bulge lateral to the pseudophiltrum.

Class III: Cleft Nose + Incomplete Cleft Lip

These represent failure at fusion of lateral lip element with philtral prolabium. Full-thickness notching of the cleft border results. Onizuka subdivides this process into nos. 5, 6, and 7 reflects the fact that the flow of tissue from B to A across the D/C bridge is not an all-or-nothing phenomenon but is a spectrum. The apparently greater "disorientation" of muscle fibers at the cleft margin, seen with more severe degrees of clefting, reflects their development within the confines of a distorted functional matrix at that site. The orbicularis fibers do not "turn to run parallel" to the cleft margin as described by Fará [19]. They simply conform to the preexisting envelope.

Class IV: Cleft Nose + Complete Cleft Lip

As previously discussed, *step* 2 involves failure of the B field to make contact with the D/C "bridge." A complete cleft of the lip, usually (but not necessarily) involving the alveolus, results. The abortive cutaneous D/C bridge, without muscle fibers, is later represented as a cutaneous "Simonart's band." This situation is represented by Onizuka as a no. 8 cleft; this recognizes the importance of Simonart's band as an integral part of the clefting model. Full-fledged *step 1* failure represents the complete separation of the D and C fields, again usually accompanied by an alveolar cleft. This is an Onzuka no. 9 cleft (class IV). A no. 10 cleft includes involvement of the hard palate internal to the alveolus.

Mulliken Classification

Yuzuriha and Mulliken's review of 393 patients with unilateral incomplete cleft lip disclosed 59 with lesser-form variants. These were divided into three categories: incomplete (20), microform (28) and mini-microform (11) the characteristics of which are as follows. Note my addition of features involving the alveolus and nasalis (Figs. 18.61, 18.62 and 18.63).

Incomplete Form

- Defect of the vermilion-cutaneous junction of >3 mm above normal peak of Cupid's bow (on the non-cleft side). [Delta Cupid's bow >3 mm]
  - DOO is discontinuous across the cleft
  - SOO discontinuous below notch, dysplastic or not inserted above the notch
- Vermilion deficient/dysplastic on medial (non-cleft) side
- Muscle depression along philtral column exacerbated by pout
- Cutaneous groove
- Significant nasal deformity
  - Misinsertion of nasalis (both heads)
- Alar base displacement >2 mm
- Alveolar
  - Bone deficit nasal floor significant
  - Alveolar cleft and dental abnormalities are common
- Surgical strategy [93]: modified rotation-advancement with single-limb z-plasty inserted into a slot between the vermilion-skin and vermilion-mucosa, i.e., the wet-dry mucosal border. Alar base repositioning along with lateral lip element

Microform

- Defect of vermillion-cutaneous point <3 mm above normal peak of Cupid's bow (on the non-cleft side); [Delta <3 mm]
  - DOO is continuous across the cleft

  - SOO below notch is discontinuous; above the notch, it is malinserted or dysplastic
- Vermilion deficient/dysplastic on medial (non-cleft) side
- Muscular depression along philtral column but no defect on pucker
- Nasal deformity less: hemicolumella almost equal, the nostril sill has a depression in it
  - Nasalis misinsertion
- Alar base minimally displaced 2 mm
- Alveolus
  - Bone deficit nasal floor
  - Alveolar cleft may be present and/or dental abnormality
- Surgical strategy [93]: single-limb z-plasty at both the wet-dry mucosal border and at the nostril sill. Alar base repositioning is by elliptical excision or v–y advancement

Mini-microform

- Vermilion notch with continuity to the white roll
  - DOO continuous
  - SOO continuous but dysplastic
  - Cupid's bow peaks are equal, vermilion border with skin is discontinuous
- Peaks of Cupid's bow are both level [Delta = 0]
- Free mucosa margin is notched
- Muscle depression seen mostly beneath the nostril sill
- Nasal deformity is subtle: just a depression in the nostril
- Alveolar bone
  - Deficit minimal
  - Alveolus may have a notch and/or dental abnormality
- Surgical strategy [93]: simple excisions in the vermilion and in the nostril sill

Surgical Management of the Incomplete Cleft

As described above, most of literature regarding the microform cleft has emphasized its anatomic features, with par-

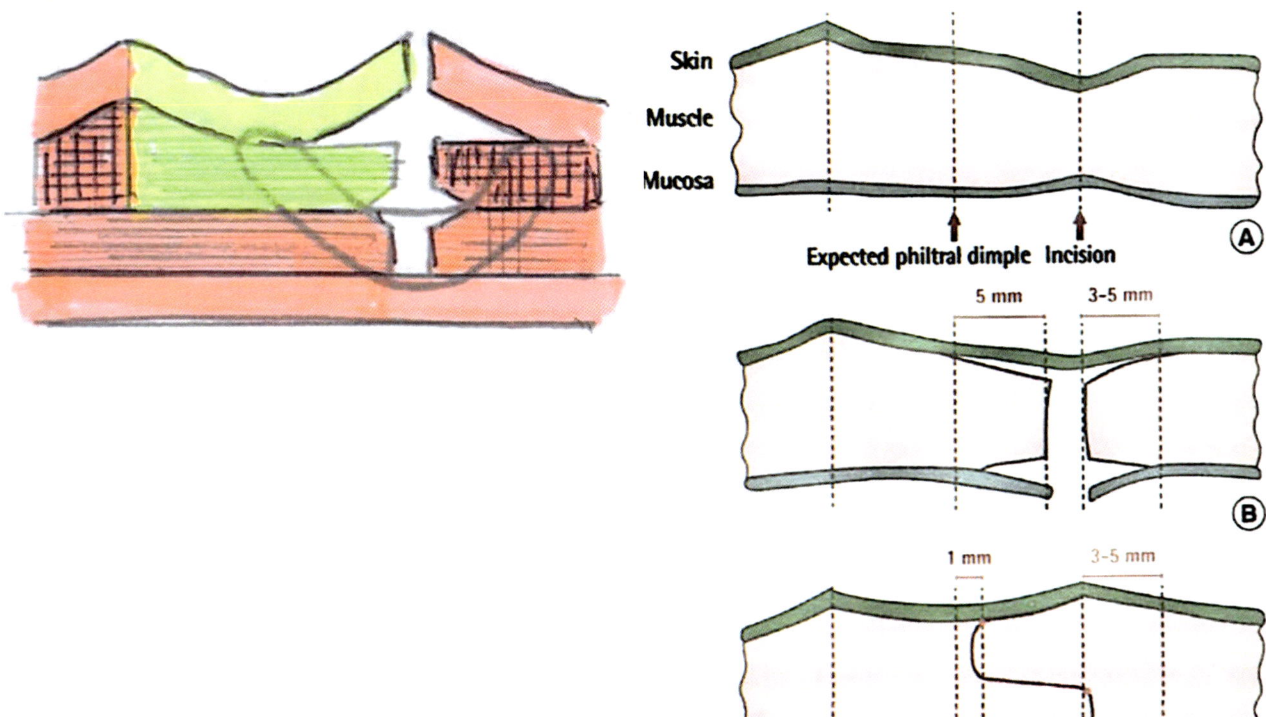

Fig. 18.64 Philtral reconstruction using monolitihic orbicularis as a block. Left: [Reprinted from Kim S, Kwon J, Kwon G-Y, Choi TH. Dynamic reconstruction of the philtrum using coronal muscle splitting technique in microform cleft lip. J Craniofac Surg 2014; 25(3):742–745. With permission from Wolters Kluwer Health, Inc.] Right: [Reprinted from Kim MC, Choi DH, Bae Sg, Cho BC. Correction of minor-form and microform cleft lip using modified muscle overlapping with a minimal skin excision. Archives of Plastic Surgery (Korean Society of Plastic Surgery). 2017; 44(3): 210–216. With permission from Creative Commons License 4.0: http://creativecommons.org/licenses/by-nc/4.0/]

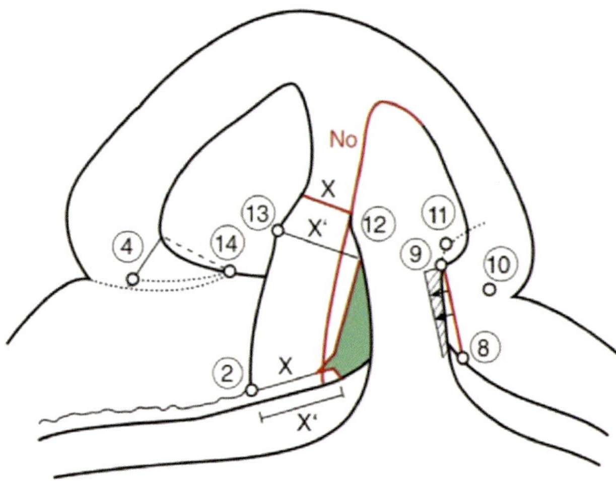

Fig. 18.65 Basic concepts of DFR: markings. See Chap. 19 for details. Width of columella (x); width of philtral prolabium (x') = 12–13; width of Cupid's bow at white roll = 2–3 – 12–13*: height of normal philtral column = 13–2; height of cleft-side philtral column—12–3, height of lateral lip = 8–9. *Difference $x' - x$ is 1–2 mm. This allows for a back-cut for final height adjustment. Normal nostril sill is an isosceles triangle. Base = width of the alar at point 4. Limbs = 4–14. Cleft-side nostril sill is marked the same way beginning at point1 10. Tip of the triangle found internally rotated. When the sill is rotated outward to normal position the space can be fill with NPP or inferior turbinate flap, anteriorly-based. [Courtesy of Michael Carstens, MD]

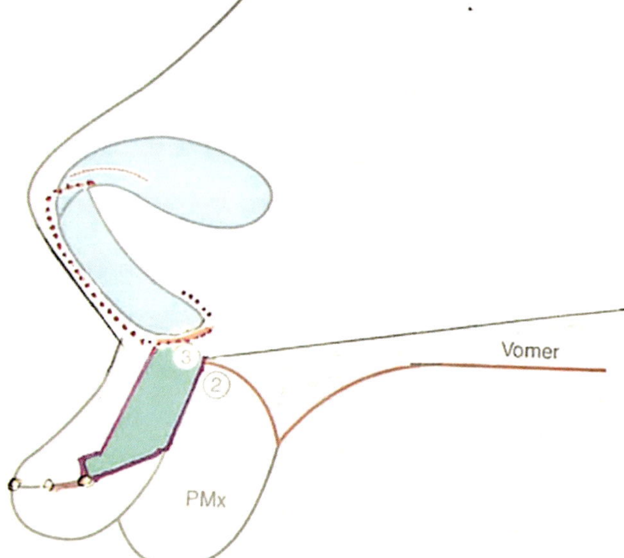

Fig. 18.66 Basic concepts of DFR: medial dissection. Separation of NPP (green) from PP is indicated by red incision lines. Vascular supply to PP from medial nasopalatine is at point 3. Posterior incision ascends to just below the footplate. One then makes a decision depending on how one intends to manage the medial crus and nasal tip. Recall that medial crus is displaced downward and backward as marked by its footplate, point 12 (not shown). Medial crus must be elevated into equality with other side for proper tip support. This can be done in one of two ways:

- Separate skin-cartilage flap, known as *lateral columellar chondrocutaneous flap* (LCC). This requires a skin only incision (orange) just below the footplate. Through this a curved scissors can hug the medial crus all the way to the tip. Spreading at the incision will advance the LCC while leaving the pedicle undisturbed. Spreading around the base at point 3 will give the "shoulder" of PP mobility. At the same time as it is rotated into the nasal floor the "shoulder" can be advanced medially and sutured higher up to fill in the gap created by the upward advance of LCC. Note the should of PP should be mattress-sutured to the other side so it won't pull on LCC
- Lateral columellar incision (red dots) can be brought under the nostril as an "open-closed" rhinoplasty. From point 2 an incision in the membranous septum (red dots) creates a large "bucket handle" flap, the NPP-LCC which can be advance in its entirely upward with PP rotated into the nasal floor. The scar is unnoticeable and exposure to the nasal tip is extensive for suture control of the ipsilateral dome

[Courtesy of Michael Carstens, MD]

ticular emphasis on the nasal deformity. Operative techniques have centered on management of lip height, the creation of a philtral column, muscle repair and repositioning of the alar cartilage. Respiratory function in these patients has not received attention. Of these issues, those specific to the microform cleft are prolabial, symmetry, philtral height and muscle repair. This section is not intended as a comprehensive review (Kim and Cho do a nice job) but to comment on technical details as they conform or at variance with developmental anatomy. Illustrations are focused on general design for the three variants: incomplete, microform and micro-microform. The reader is referred to Chap. 19 for a full presentation of DFR technique (Figs. 18.64, 18.65, 18.66, 18.67, 18.68, 18.69 and 18.70).

The Prolabium

The true philtral prolabium (PP) is the paired neuroangiosome of V1 and StV1 anterior ethmoid arteries. Its subcutaneous mesenchyme is r1 neural crest. The width of PP is equal to that of the columella or (a bit more generously) the distance between the footplates of the medial crura. The optical illusion of the "prolabium" has led to attempts to lower the cleft-side border using rectangular flaps [95], triangular flaps (Tennison), and rotation advancement [6, 96–99]. These techniques are reviewed in Losee and Kirschner. Recently in vogue is complex geometric technique proposed by Fisher [22]. Being unaware of the existence of the non-philtral prolabial field these techniques relay on manipulation of available tissues immediately adjacent to the cleft. What's more, as previously discussed, these marginal tissues are intrinsically affected by the same abnormal biochemical signals (the BMP4/Shh loop) involved in fusion failure as the cleft margin. One does not encounter normal philtral tissue until one reaches the correct neuroangiosome.

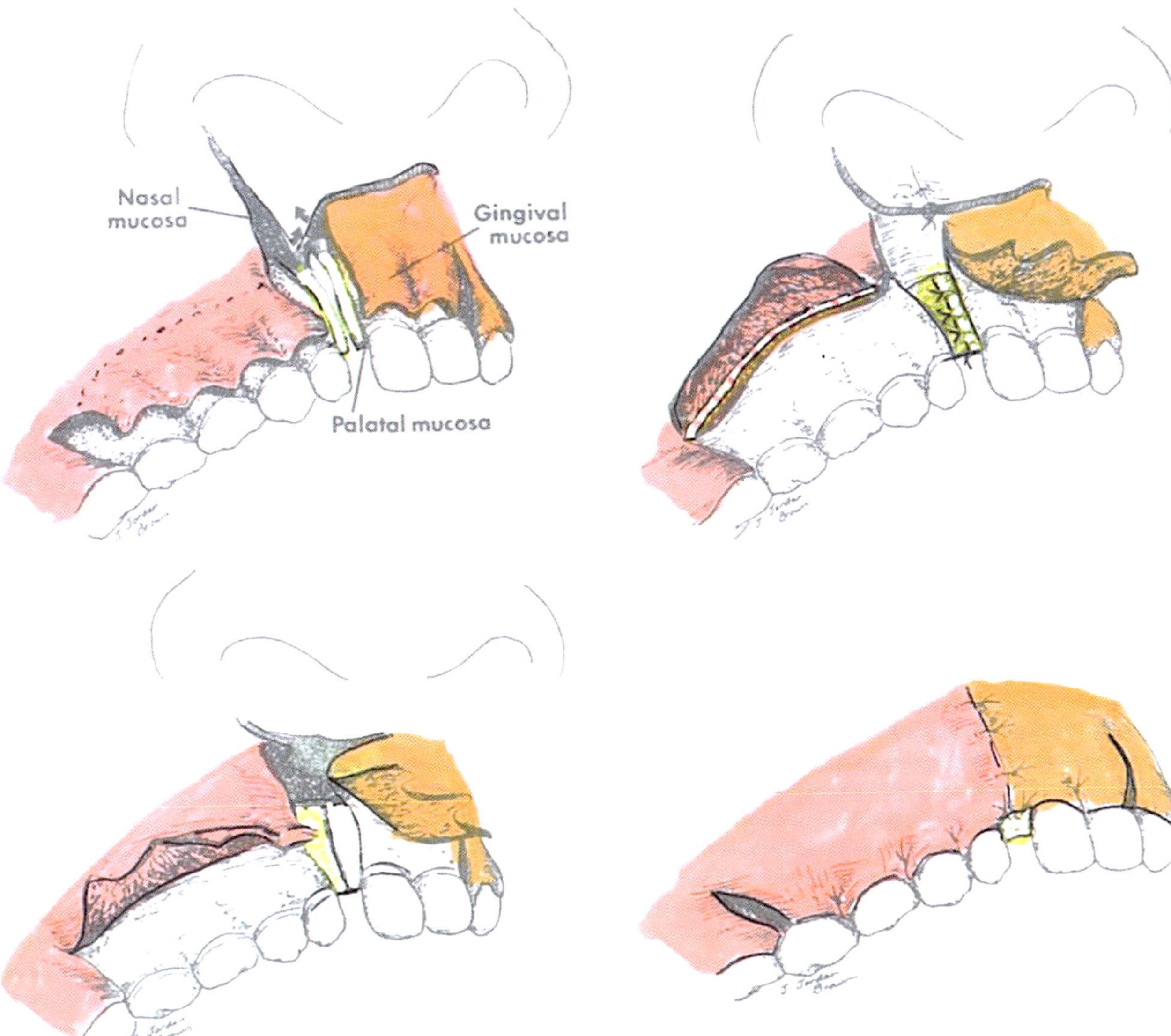

Fig. 18.67 Sliding sulcus mucoperiosteal flaps for closure of alveolar cleft. *Upper left* m S flap, maxilla (pink) is StV2 infraorbital and ECA facial. It is mobilized via pericoronal incision with a back-cut up the buttress. Dotted lines indicated periosteal releasing incision on the other side. S flap premaxilla and contralateral maxilla (orange) has additional supply from StV2 anterior ethmoid. Maxillary alveolar cleft flap (yellow) is greater palatine. Premaxillary alveolar cleft flap (white) is. *Lower left*: nasal floor elevated, flaps elevated. *Upper right*: nasal floor closed, back wall sutured, periosteal releasing incision (white line red dots) mobilizes 1–2 dental units. *Lower right*: closure [94]. [Reprinted from Cohen M, Figueroa AA, Aduus H. The role of gingival mucoperiosteal flaps in the repair of alveolar clefts. Plast Reconstr Surg 1989; 83(5): 812–819. With permission from Wolters Kluwer Health, Inc.]

- **In sum** All conventional prolabial dissections accomplish their purpose but at varying degrees of compromise to normal anatomy.

In 2000, based on vascular injection studies in fetuses, I first recognized the embryonic distinction between the philtrum and its surrounding tissues [3]. At the time, I was unaware of the Padget's work on stapedial system and its implications. Nevertheless, these findings lead to the development of an embryologically based model of the philtrum and prolabium which was translated into the *functional matrix cleft repair*. This concept demonstrated immediate and simplifying clinical results, including the use of the non-philtral prolabial flap for nasal floor reconstruction. It took me another 2 years before I could marry up the basic science with the surgical technique [28]. Neuroangiosome-sparing prolabial dissection has proven applicable across the spectrum of clefts, including incomplete and minimal presentations.

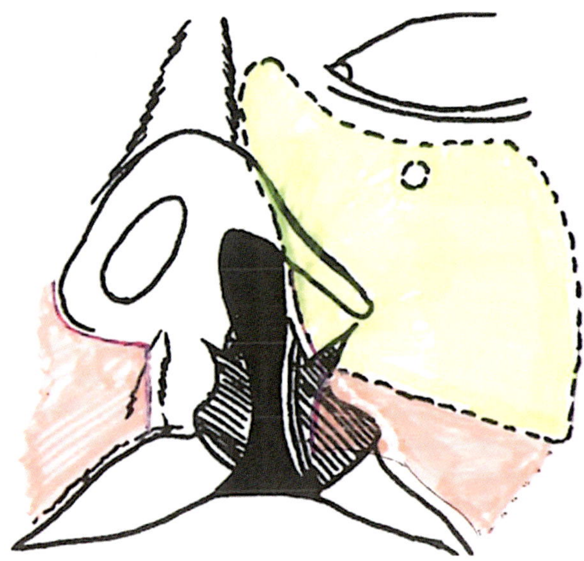

Fig. 18.68 DFR subperiosteal dissection. Delaire dissection plane (yellow) DFR dissection plane (pink) incision line (violet) mucoperiosteum is incised along the alveolar cleft border. This creates a "pocket" above gingival border that extends all the way back to the buttress. At that point an optional full-thickness incision through mucoperiosteum can allow the pocket to further advance as a large bipedicle flap suspended from the orbit above and the gingiva below. DFR undermining is no circumferential around the infraorbital bundle. The neurovascular axis is readily visualized during elevation of the periosteum. The combined incisions sweep upward around the piriform margin to gain access to the nasal bone and the sub-SMAS plane. On the non-cleft side, the mucoperiosteum is incised through a mini-incision just below piriform fossa with lateral elevation as well. Care is taken to sweep the periosteum around the non-cleft piriform margin, elevating the alar base and the nasal floor on that side. [Reprinted from Markus AF, Delaire J. Functional primary closure of cleft lip. British J Oral Maxillofac Surg 1993; 31(5): 281–291. With permission from Elsevier]

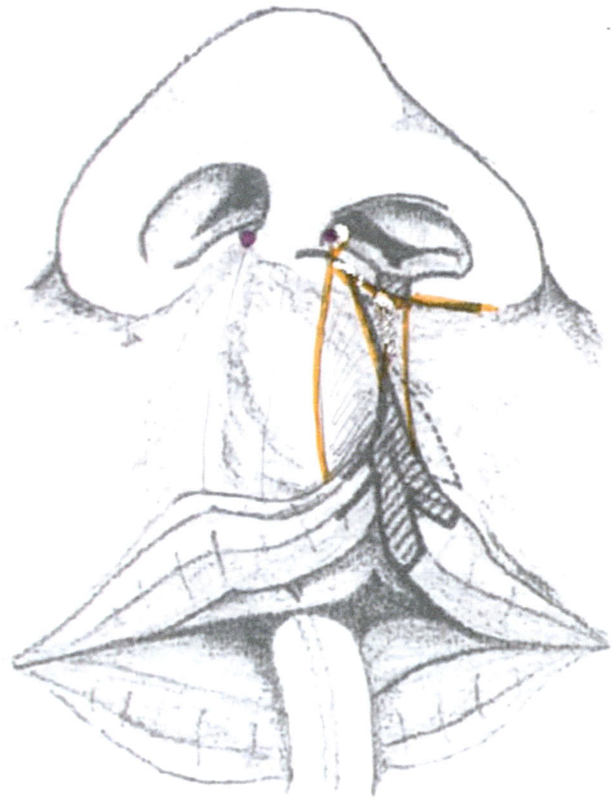

Fig. 18.69 Incomplete cleft DFR. Management consists of DFR subperiosteal mobilization, opening the lip with judicious use of NPP as full thickness flap (which can be thinned), and nasalis dissection. Immediate skin over the scar is discarded. [Adapted from Yuzuriha S, Mulliken JB. Minor-form, microform, and mini-microform cleft lip: anatomical features, operative techniques and revisions. Plast Reconstr Surg 2008; 122(5):1489–1493. With permission from Wolters Kluwer Health, Inc.]

Philtral Column

Creating an aesthetic projection of the ipsilateral philtral column has given rise to a variety of techniques, ranging from skin manipulation, buried dermal grafts and muscle rearrangements. Anatomic traps include (1) failure to recognize the distinct functional components of orbicularis as both sphincter (DOO) and dilator (SOO); (2) misinsertion into dermis rather than r1 mesenchyme; and (3) problems arising from the intrinsically faulty nature of the muscle itself at the cleft margin.

Fascia graft to bulk up the philtrum was reported by Park. The long-term results may be unstable due to absorption.

Reports by Kim and Kwon illustrate the issues of considering the orbicularis to be a single structure. The diagram presents what purports to be bilateral v–y advancement flaps suture together to throw the philtral column into relief. From a mechanical standpoint this works. However, the different embryonic tissues (indicated by colors) show a mismatch. Nonetheless, although not aware of the mesenchyme below the philtral skin, this repair effectively sutures SOO to r1 correctly (Fig. 18.64a).

Cho's group used an intraoral incision excise ½ of the r1 mesenchyme under the philtrum and interpose it with an equal resection of DOO from the cleft side [100]. The two flaps are advanced and overlapped such that SOO now occupies 50% of the territory beneath the philtrum (Fig. 18.64b).

Muscle Repair

Cho reported an intraoral approach to microform with dermal suturing of the lateral flaps [100]. Scar is avoided. However, if nasalis malfunction is present this approach will not provide the exposure needed to dissect out the muscle. The intraoral approach was made further complex by Derosiers who split the orbicularis into three muscle slips from each side and interposed them in a basket-weave manner. Although this thickened the philtral column interdigitation lead to predictable long-term scarring.

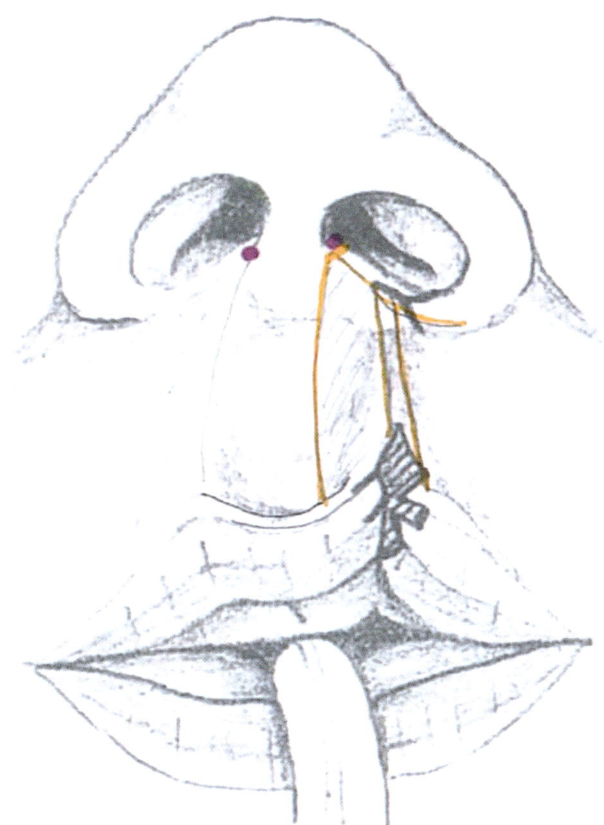

Fig. 18.70 Microform cleft DFR. Management presents variations with the incomplete form. Superiosteal mobilization is the same Lip can be opened completely or just down to DOO. This will give easy access to the nasalis without compromising normal structures. NPP incisions and medial crus elevation are the same but the NPP flap can be thinned considerably. [Adapted from Yuzuriha S, Mulliken JB. Minor-form, microform, and mini-microform cleft lip: anatomical features, operative techniques and revisions. Plast Reconstr Surg 2008; 122(5):1489–1493. With permission from Wolters Kluwer Health, Inc.]

Muscle manipulation via intraoral excision reaches perhaps its most complex in the technique reported by Ningbei Yin which attempt to reconstruct orbicularis according to two lines of tension, the nasal floor and philrum. The flaps are cut full-thickness with no separation of DOO and SOO. Alar part of nasalis is included in the orbicularis insertion into the root of the columella. Functional assessment for nasal breathing was not reported but nasalis was likely converted into a constrictor.

Mulliken Repair Sequence

Yuzuriha and Mulliken describe a logical surgical strategy to deal with the progression of clefts from the most subtle "mini-microform" to the standard microform to the incomplete variant. Using a caliper, they determined the distance from subnasale to the peaks of cupid's bow on the non-cleft and cleft sides. A difference greater than 3 mm is the indication to use rotation advancement. If less than 3 mm, two unilateral z-plasties are used. In both the incomplete and the microform clefts, alar base advancement is described, necessitating at times a v–y advancement. Nasal function is not addressed. The technique is excellent, but from the standpoint of developmental field reassignment, I would make the following points.

- Prolabial dissection errs based the classic misconception of the meaning of the Cupid's bow point. We know well that the philtrum expands with time. Conventional designs simply take the non-philtral prolabial tissue off the table as a reconstructive option for the nasal floor where it is needed most. Cupid's bow width inter-footplate distance is aesthetically pleasing. If short, at the very most a single 1–2 mm interposition flap above the white roll works nicely (Fig. 18.44).
- Subperiosteal dissection makes alar base relocation incredibly easy. v–y advancement simply not needed.
- Elevation of the nasal floor from its depressed condition is not described.
- Elevation and repositioning of the medial crus is not described. This can be done just one passes backwards beneath the footplate of medial crus (vide infra).
- Airway management with nasalis dissection is possible using the Mulliken incisions in all three variants. It had simply not been addressed.
- ***Note***: In Mulliken measurement system, the position of the alar footplates was not considered. By using subnasale as the central point, they missed the fact that the cleft-side columella is retracted downward and inward into nasal introitus.
  - "What you see depends on where you sit."

Developmental Field Reassignment and Incomplete Cleft

Historical Perspective

In the past, the surgical approach to the microform cleft has focused on its skin-muscle presentation (especially the absent philtral column) and on the alar cartilage. Procedures have ranged from skin excision to full-thickness rotation-advancement, to complex muscle flap interdigitation. Various forms of tip rhinoplasty have been reported.

Without the benefit of the multiplanar orbicularis anatomy as currently known, some writers questioned the knee-jerk use of rotation advancement. Curtin queried: "Our personal experience with the microform cleft lip deformity is that the full thickness of the orbicularis muscle is not affected and a through and through incision of the lip in the repair is not necessary. Since modifications of the straight-line repair have been successful in correction of the minimal deft lip deformity, why was it necessary to use the rotation-advancement flap repair in some of the cases?" Millard also reported the use of a simple excisional technique and a straight-line incision in a minimal cleft: "It is interesting to note the correction, which necessitated all other aspects of rotationadvancement with refinements, extensions and adjuncts, still did not require rotation."

That microform clefts have varying manifestations was documented by the extensive series of Tolorova. Onizuka described an individualized operative strategy that addresses the components minimal cleft: (1) alar rim, (2) notching of Cupid's bow, (3) notching of the vermilion free margin, (4) abnormal skin striae and (5) combined deformities.

Superficial rearrangements of skin and vermilion using w- and z-plasties are used for the first three problems. Creation of a philtral column in the fourth problem is achieved by dissection of a vertically oriented "central muscle flap" that is transferred laterally beneath an undermined lateral advancement flap; in so doing, the philtral column is "bulked up." Were this really the case, Millard's categorical critique would be justified: "An attempt at creating a philtral column by undermining skin and gathering it to a fold with sutures tied over stents usually does not maintain great improvement after removal of the sutures." In reality, Onizuka's technique does not depend on "gathering the skin." The success of the procedure lies in the degree to which the differential sutures reconstruct the anatomic insertion of the superficial orbicularis oris to the dermis of the lateral margin of philtrum.

Submental-vertex views from Onizuka's series disclose centralization of the columella and elevation of the deft-sided alar base to achieve symmetry with the noncleft side. His use of muscle suspension sutures is very accurate. But herein lies an ***embryologic error***. The key to alar base repositioning was not subperiosteal release. Instead, restoring communication between the "lateral muscle flap" (pars alaris of the nasalis) and the anterior nasal spine was described as the key objective key. The use of a non-subperiosteal release of the alar base is misleading. An increasing degree of piriform deformity make soft-tissue correction less optimal (and more likely to relapse). Furthermore, as previous stated, *nasalis is not designed to be inserted into the midline*. It requires separation from orbicularis and reassignment over the canine. This detail is crucial for the reconstruction of the cleft airway.

Assessment of the Microform Cleft

Patients presenting with minimal cleft deformities require extra care in assessment. Although they seek attention for a disturbance in their facial features, they may be rather vague as to the nature of their problem. They may report complaints such as: "I have a scar on my lip." "My dentist sent me." "I don't like my nose." "I look funny in photographs" or "I can't pucker up my lips to blow the flute." When the patient is informed as to the true nature of his or her condition, they may react with shock, embarrassment or denial. Being cognizant of this, the surgeon should discuss the matter with tact. Because the anatomic details are subtle, helping the patient to see cause-and-effect of their problem may be difficult. Photographic examples of other minimal clefts may be helpful for the patient at this juncture.

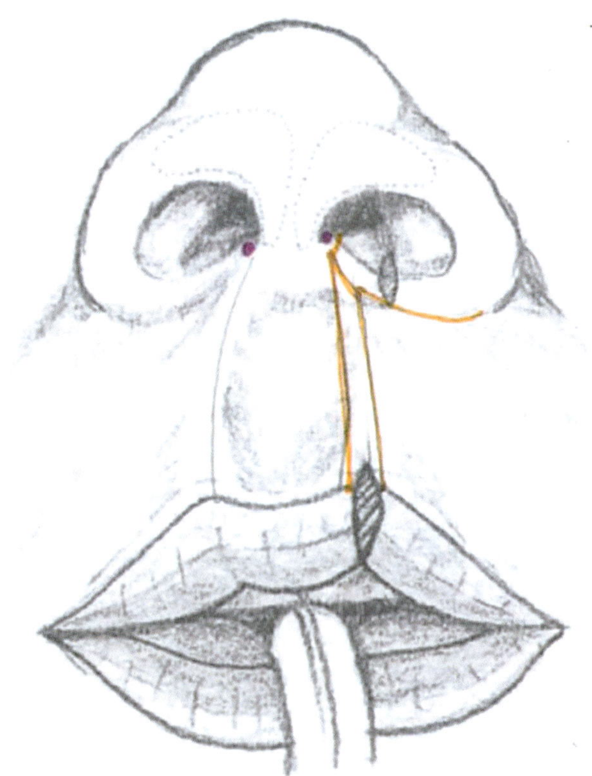

Fig. 18.71 Mini-microform DFR. Subperiosteal mobilization. PP is almost normal with very little NPP. NPP flap cutaneous only. Incision under the footplate only. Separate rim incision as needed for tip control of the ipsilateral dome. If nostril sill deformity is minimal it need not be mobilized and NPP can be discarded. Nasalis dissection is done through the nostril sill incision and the muscle is dropped through that same incision behind orbicularis, retrieved into the mouth by a small buccal sulcus incision, and secured to the periosteum and sulcus over the canine. [Adapted from Yuzuriha S, Mulliken JB. Minor-form, microform, and mini-microform cleft lip: anatomical features, operative techniques and revisions. Plast Reconstr Surg 2008; 122(5):1489–1493. With permission from Wolters Kluwer Health, Inc.]

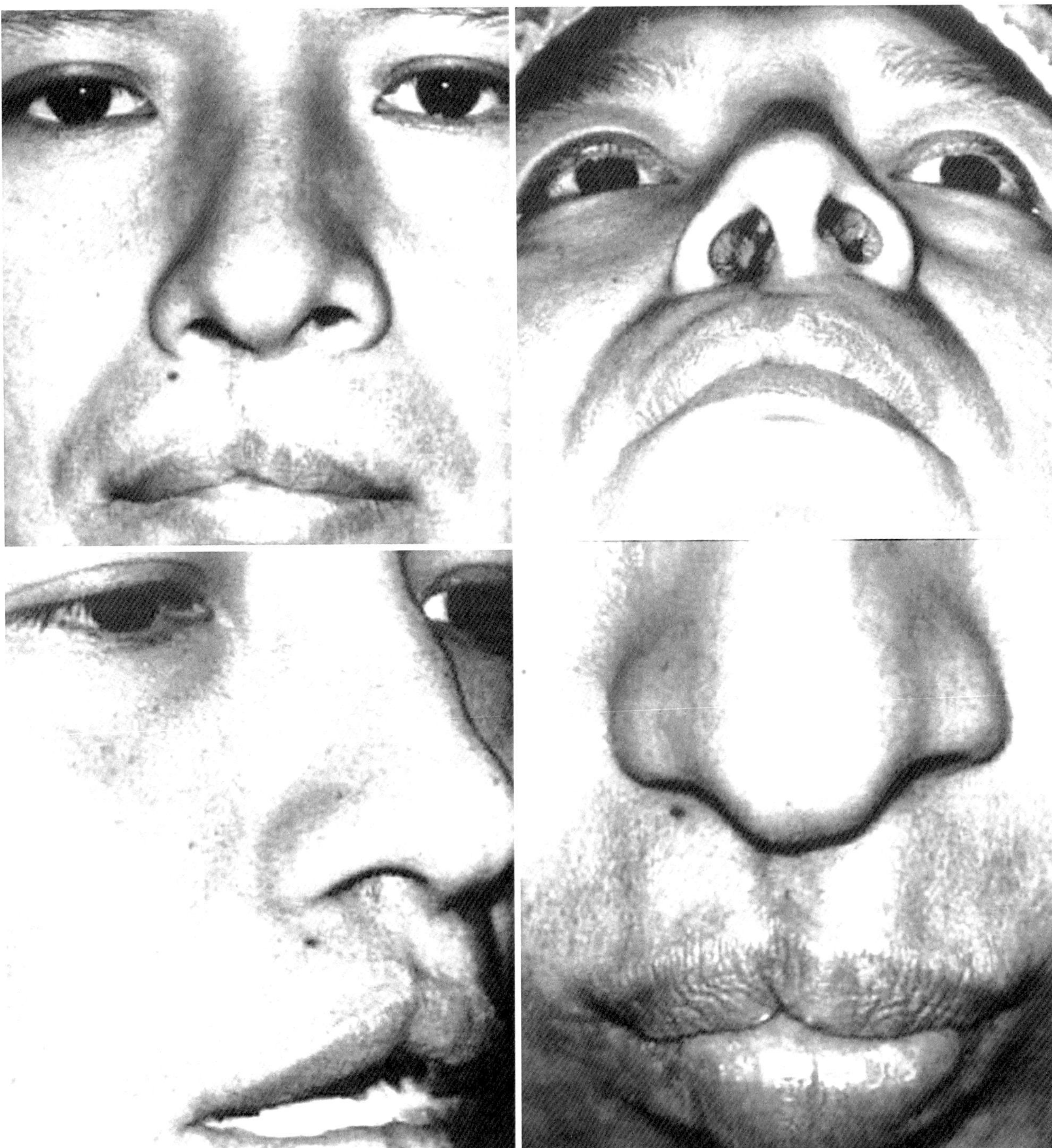

Fig. 18.72 DFR repair of microform cleft 1. In this early case, I did not do a nasalis repair. Medial crus was elevated with a bucket handle incision. The dead-space was filled with a composite chondrocartilage graft taken from the ipsilateral ear with a dermatology punch and the secured to the contralateral side, with a mattress suture. [Courtesy Michael Carstens, MD]

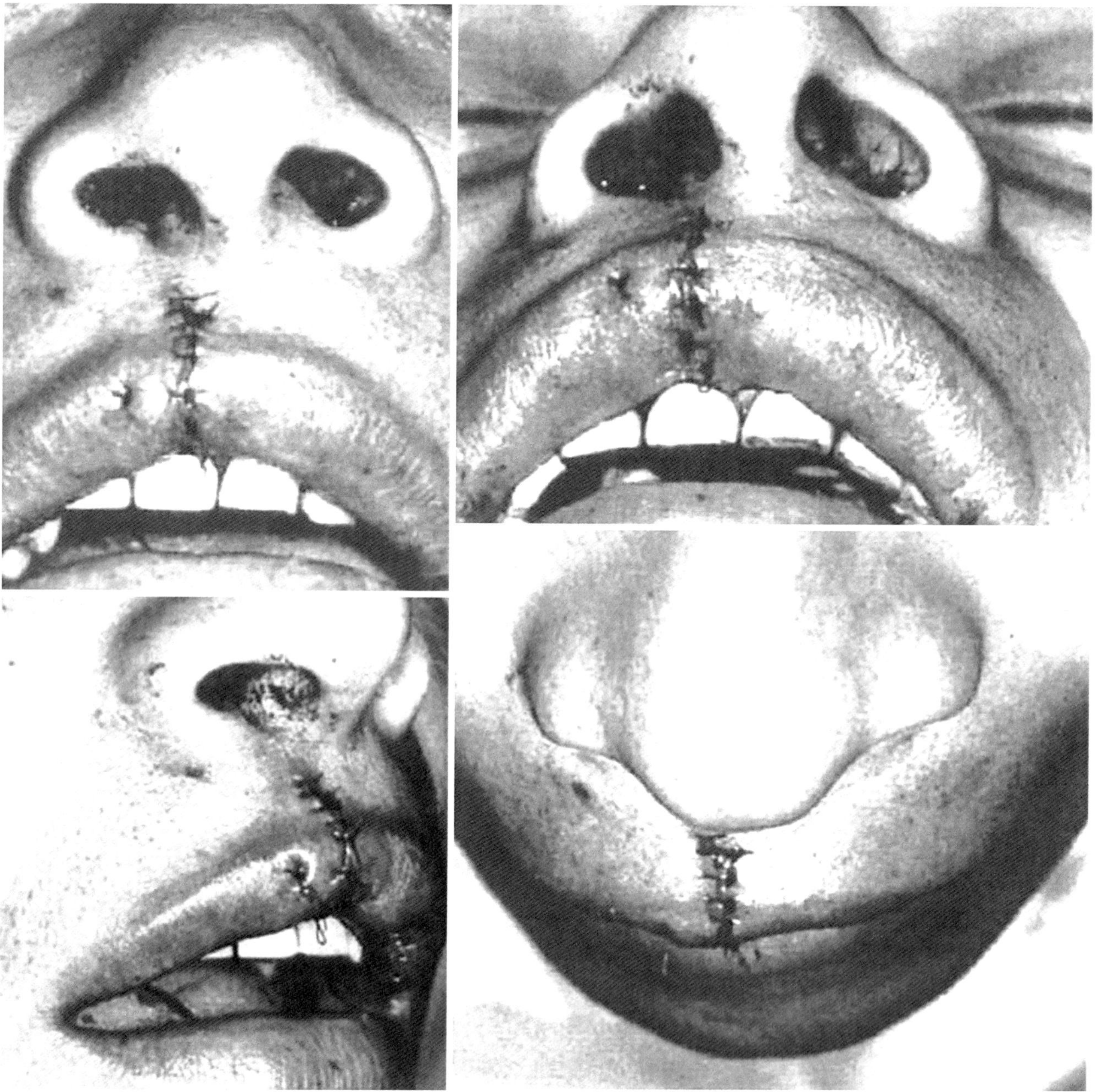

Fig. 18.73 DFR repair of microform cleft 2. Note change in symmetry of the nasal tip, alar base and nostril sill achieved with wide subperiosteal release. [Courtesy Michael Carstens, MDs]

Minimal clefts vary in severity. Some patients may present with a nostril deformity, whereas others possess a vermilion notch, cutaneous groove or muscle bulge. An individualized surgical approach should be designed for each patient according to his or her anatomy and desires. Not every deficit need be addressed. What the surgeon perceives as obvious may be of little significance to the patient. For this reason, the exact anatomic details of the cleft should be clearly understood by the patient so that proper informed consent can be obtained.

Nasalis function should also be tested. Comparison of alar motion can be made standing above with the patient seated and from the submental vertex position.

Problem-Directed Surgical Technique (Figs. 18.65, 18.66, 18.67, 18.68, 18.69, 18.70, 18.71, 18.72 and 18.73)

A depression or notch in the floor of the nose combined with a deformity of the piriform rim are the sine qua non of the *forme fruste* cleft. The nasal mucosa and the alar base have settled into the bony depression and must be released subperiosteally. If no significant insufficiency of the lateral vestibular wall exists, this is all that is required.

The relationship of the alar base to the septum can be redefined with permanent suture placed connecting the base (taking care to stay superficial to the nasalis) to the anterior nasal spine. With adequate superiosteal release, this maneuver is seldom required.

Incision and Subperiosteal Release

The incision design used to accomplish the above depends on the degree of lateral displacement of the alar base. Separation of this variable from the anteroposterior and craniocaudal displacement of the piriform rim can be assessed in the office by placing a finger or cotton swab under the alar base in the sulcus and gently lifting it up. If the elevation achieved in this manner is inadequate, a lateral tethering effect of the ala by the periosteal sleeve over the face of the maxilla exists.

Access to the piriform fossa can be gained by a transverse incision in the mucoperiosteum from the anterior nasal spine beyond the canine fossa. Nasalis muscle will eventually be re-inserted at this point. This produces a broad mucoperiosteal flap that exposes the entire floor of the nose and gives access to the piriform fossa. The periosteum beneath the alar base is elevated. More alar release can be gained using a curvilinear scoring incision through the periosteum around the alar base. This gives additional exposure to the piriform fossa.

Lateral tethering of the alar complex to the anterior face of the maxilla requires a more extensive release but no change in incision design. From the level of the piriform rim, subperiosteal elevation can be directed laterally to the buttress. This elevation stops at the level of the attached gingiva. If further release is needed, an oblique counter-incision is made in the mucoperiosteum along the buttress.

The precision of this approach allows for entry into the fossa (which may drop sharply downward away from the rim itself) without injury to the mucoperiosteum of the nasal floor. Tears at this step may later cause a retethering of the alar complex via scar contracture and are worth avoiding.

Elevation of the Nasal Floor and Volume Restoration

When the alar base is completely released, the true nature of the piriform deficiency becomes apparent. In young children with active bone growth potential, the periosteal elevation will stimulate osteogenesis and support for the alar base may self-correct by the functional matrix reconstruction alone. In adults, actual piriform augmentation may be required. Using the trephine technique (Champy, Walter Lorenz Inc., Jacksonville, FL, U.S.A.), harvest of iliac crest cancellous bone is rapid and relatively noninvasive. A more elegant solution is a small amount of demineralized bone with or without morphogen rhBMP-2, thus eliminating the need for graft may be eliminated altogether.

- In adolescents and adults in whom piriform growth has cease, augmentation beneath the alar base may be required.

Reinsertion of Nasalis Muscle

If a significant amount of alar cartilage tethering is present, the insertion of nasalis should now addressed. This requires dissecting out its two heads. Recall that: (1) lateral head is inserted just beneath the nostril sill; and (2) medial head is inserted into the piriform fossa itself. The two head come together as they wind around the alar base and over the lower lateral cartilage to insert into the SMAS at the nasal dorsum.

Dissection of lateral head entails an external incision in the nostril sill and anterior alar base of 2.5 cm. It does not extend laterally around the alar base. The plane is immediately subcutaneous, above the nasalis. As one proceeds inward, dissection is continued above the *superficial aspect* of the muscle and below the vestibular lining, elevating the latter all the way into the piriform fossa and just beyond the inferior turbinate. The piriform elevation establishes the plane over the *superficial aspect* of medial head of nasalis as well.

In the incomplete cleft, getting below lateral head of nasalis is easy. One can grasp the muscle with a forcep and release it from the superior margin of superficial orbicularis. The two muscles have a different orientation and are easy to identify. One then follows the *deep aspect* of the nasalis around the alar base. At that point the plane is no longer deep to nasalis but *lateral* to it. You proceed upwards to above the alar cartilage. Traction on the lateral head will demonstrate an opening of the airway.

In the microform variant, nasalis dissection is straightforward if the lip is opened completely. If the lip is opened just through the SOO, the plane leading to nasalis is directly visualized as well.

In the mini-microform deformity the dissection is more subtle, as the lip is closed. One must proceed through the existing nostril sill incision, transect the nasalis. And then, putting the muscle on tension with a forcep, dive below it and, with a curved Stevens or Foman scissors, one advance laterally right the way around the alar base. You will then be able to traction the lateral head through the nostril sill to see if your release is adequate.

Dissection of the medial head is now completed through the intraoral incision. Medial head lies with the piriform fossa extending about 2–3 cm inside it. Recall that the vestibular lining has already been reflected away from the *superficial aspect* of head. What one now does is go is release the *deep aspect* of the medial head from the bony wall fossa by means head by proceeding in the superiosteal plane within the fossa. You can thus "fish" the medial head out of the fossa. A short strongly-curved elevator, such a McKenty or a Molt, works well here.

At this point both heads of nasalis are identified. Through the nostril sill incision, spread through the fibers of orbicularis to deliver the lateral head downward into the piriform fossa. Both muscle heads can be treated as a single entity. Place suture of 4–0 vicryl or PDS through the sulcus, taking periosteum as well directly above the canine. Mattress the suture through the nasalis and return it into the mouth. This will re-establish the normal insertion above the canine fossa. Do not tie this suture until completion of the operation. It is just a place-holder. Excessive tension will traction the alar downward.

- *Nota bene*: It the surgeon is reluctant to use an external incision, the medial head in the piriform fossa can be sought and anchored. This may be functionally useful but the normal anatomy is incompletely restored without retrieving the lateral, more superficial head of the muscle.

Nasal Tip Asymmetry

The microform nasal tip may range from essentially normal to a full-blown deformity associated with the complete unilateral cleft. Management principles of the alar cartilage are no different in minimal cleft than in the more severe forms.[7] What should be emphasized in the minimal cleft is that access to the lateral crus of the lower lateral cartilage can be readily gained by the intraoral exposure. One proceeds by subperiosteal dissection along the outside of piriform rim and up over the dorsum: this puts one in the sub-SMAS plane, i.e., below nasalis. A concomitant rim incision puts one above the SMAS. Finally, blunt dissection of vestibular lining (Talmant maneuver) as described in Chap. 18 allows vestibular lining to be mobilized. When the alar complex is elevated from the piriform rim, the plane of the vestibular complete liberation of the lateral crus from the constraining influence of nasalis is an essential component for long-term tension-free repositioning of the tipdefining element. It is therefore important for the surgeon to take advantage of these exposures gained by this technique to accomplish this goal. Various forms of fixation have been described but probably this simplest is long-term stenting with this flexible silicone sheeting.

Vestibular Lining: Deficient or Displaced?

Comment *For some time, I thought that the lining was somehow short and the external vestibular release and addition of tissue using grafts or the inferior turbinate flap would solve the problem. This was a misconception on my part. The vestibular lining is <u>not</u> deficient, it is <u>mismatched</u> to its overlying soft tissues. Depending on the degree of nasal deformity, a Talmant release that bluntly frees the lining from the piriform altogether, may be required.*

- *I include the paragraphs below as examples of ideas* ***<u>not</u>*** *to try.*

"If the deficiency at the cleft site is more severe, superiosteal elevation alone will not be sufficient to release the alar complex. An incision in the lateral wall of the vestibule made in parallel to the nasal floor and terminating at the anterior aspect of the caudal margin of the inferior turbinate will accomplish a full release. The deficit site thus created can be replaced with full-thickness skin graft [101], composite conchal graft deficit [102], or a flap from the inferior turbinate [103–105]. The latter offers distinct advantages: (1) It relates directly to the deficit; (2) The donor site is unobtrusive; and (3) Although the palate is not cleft, intraoral subperiosteal exposure of the piriform fossa makes this flap relatively accessible for harvest."

"If the nasal vestibule exhibits an additional deficiency in the anteroposterior dimension, a releasing incision in parallel with the rim as described by Millard 129 may be required. Once again, the mucosal paring flap can be relied on to fill the deficit. The resultant oronasal fistula is clinically insignificant, acquiring, over time, the dimensions of a pinhole."

Nostril Sill

The nostril sill in a severe unilateral cleft can be readily designed based on the normal side. It is roughly an isosceles triangle, the base of which is equal to the thickness of the alar wall, x (usually about 4 mm), as measured inward from point 4. The anterior limb is 4–14. The posterior limb is drawn from point 14 back to x. After making the marks do not cut the flap: nasalis dissection comes next. If you cut the flap first, the muscle dissection is made more difficult.

The nostril sill is a skin and subcutaneous flap. Just below it lies nasalis. As stated previously, by dissecting subcutaneously all the way into the nasal vestibule as afar as inferior turbinate, one will stay quite nicely on top of nasalis. You can then dive beneath it down to the bone of piriform fossa to bet the remainder of the muscle. Once this is done, go ahead and cut the nostril sill flap.

Outward rotation of the nostril sill flap will put it in place to form the external border of nasal floor. It does not reach all the way to columella. Don't try to force it, as this will tighten the nostril. Instead, place the NPP behind the nostril sill, let them fall into place and suture accordingly.

There is obviously a full-thickness gap created by reassignment of the nostril sill. NPP alone may not reach the way. Keeping mucosa attached to the tip of NPP is not a great idea as it interposes unlike tissue into the lateral nasal floor. Moreover, the blood supply is uncertain. A 100% reliable solution is an anteriorly-based inferior turbinate flap.

The above comments apply to the incomplete cleft but become progressively less important in the microform variant. Here, a simple interposition of the NPP skin flap behind nostril sill is sufficient. In minor degrees of cleft, the nostril sill can be left in situ, undisturbed.

Vermilion Border

Notching is a common presentation of the microform cleft. The issue at hand is the pars marginalis of the DOO as described by Park and Ha. The most caudal muscle curls forward beneath the vermilion, terminating at the white roll. Here it abuts the caudal margin of the SOO, which dead-ends at the white roll as well. The sequence of facial muscle development follows a deep-to-superficial and lateral-to-medial pattern. Three-dimensional computed tomographic reconstruction of the philtrum demonstrates that the pars marginalis of the DOO is continuous across Cupid's bow, whereas the SOO inserts into the r1 FNP mesenchyme of the philtrum. The abutment between r1 and r4 SOO creates the philtral columns. The vermilion notch, located just below the philtral column, represents a failure of the DOO to achieve full union at this site. Correction of the notch is accomplished using a *deepithelialized triangular muscle flap* pared from the lateral margin and inserted into a tunnel beneath the medial margin as described by Guerrero-Santos et al. [106]. This may need to be accompanied by a *zplasty of the vermilion* itself further within the lip. If so, the incision lines should be staggered so as not to create a contraction.

Philtral Column

When the superficial orbicularis oris fails to properly insert into the philtrum notching occurs. This is best appreciated when the patient is asked to pout the lip. When the DOO contracts, the philtral columns are pushed together; discontinuity of the SOO is revealed as a muscle bulge lateral to the pseudo-philtrum. In the same manner, SOO contraction reveals the lack of adequate cutaneous development across the philtrum. When viewed from below, the normal philtral column, albeit flattened, is still discernible. The pseudophiltrum merely flattens out, revealing the lack of support beneath it.

Should the patient desire correction of this problem, no option exists but to open the lip-partially. The thinned-out cutaneous skin of the philtral column is excised. Philtral skin is elevated with a knife from r1 mesenchyme, thus giving one an edge for suturing. The S00 is also undermined about 5 mm. SOO is then tacked to r1 mesenchymal "shelf" beneath the philtrum, and the edges everted. Re-closure of the skin shows "bulked up" philtral column.

- In patients with negligible muscle asymmetries, simple excision and closure of the philtrum column with everting sutures will suffice.

Final Thoughts: How Mechanism Affects the Clinical Applications Developmental Field Reassignment Surgery

- Thou shalt not subdivide or otherwise mutilate a neuroangiosome.
- Replace like with like, whenever possible, by reassigning fields to their correct embryonic relationship.
- Do not include metabolically challenged or embryonically-incomplete tissues in critical reconstruction sites.
  - The non-philtral prolabium must be taken out of the lip and used to support the columella and restore the nasal floor.
- Surgery in of congenital defects in infancy is a four-dimensional equation: growth will occur.
- Mesenchymal stem cell populations are the engine of facial growth.
- A cardinal principle of surgery should be the restoration of stem cell populations to their rightful place in space such that they will produce tissues over time where nature intended them to be.

Pathogenesis of Clefts: The Neuromeric Model

We come now to our final discussion, the heart of this chapter: the developmental sequence leading to the anatomic pathology of the cleft deformity. Let's summarize it up front as a formula; we shall then consider it component parts in sequence.

The Formula of Cleft

Lesion at level of arch formation > neuroangiosome failure > field defect with loss of mesenchymal mass > consequences: (1) physical and (2) biochemical

- Physical/mechanical
  - Critical contact distance between fields is exceeded: fusion failure
  - Effect on surrounding normal fields: distortion
- Biochemical: morphogen failure [BMP4]
  - Epithelial function: fusion failure
  - Muscle binding sites: aberrant insertions
  - Muscle differentiation

Presentations of Cleft Lip and Palate

Cleft lip, in all its forms, is caused by a mesenchymal deficiency in the premaxilla, first affecting the frontal process (PMxF) and then progressing to the lateral incisor zone (PMxB). In rare instances, the central incisor zone is affected as well, either as an isolated entity or in the total loss of the hemi-premaxilla. The cause is insufficient development of the StV2 neuroangiosome, medial nasopalatine artery. The consequences are physical and biochemical. This are best seen in the microform variant, where PMxB remains intact

- The physical effects of deficiency involving PMx and PMxB are as follows;
  - When the volume of PMxB is reduced such that the *critical contact distance* between the premaxillary alveolus and the maxillary alveolus is exceeded, normal fusion cannot take place.
  - All clefts have varying degrees of bone insufficiency. These affect surrounding bone and soft tissue structures. The ipsilateral normal nasal bone can be flattened. Premaxilla can be torqued away from midline. The scooped-out piriform fossa or in the thing of alveolar bone housing the lateral incisor causes the vestibular lining to be pulled down into the cleft site and the alar base to be anchored in a lateralized and posterior position.
- The mesenchymal deficiency also means a *quantitative reduction in diffusible morphogens* which are intrinsic to the process of membranous bone synthesis. These substances move through the overlying soft tissues from above downward and from deep to superficial with varying physiologic effects.
  - BMP4 deficit potentiates the activity of Shh in the surface epithelium, thus preventing fusion from below-upward and from superficial to deep.
  - BMP4, diffusing in the same way, affects the maturation of bone at the insertion sites of nasalis into the incisive fossa and canine fossa, forcing it to relocate to the piriform fossa and the nostril sill.
  - An unidentified morphogen controls development of overlying orbicularis oris muscle. The effect of this deficit is seen along the margins flanking either side of the cleft. As one moves away from the margin (beginning at about 5 mm), normalization of the orbicularis is seen.

Cleft palate, in all its forms, is caused by a mesenchymal deficiency in four bone fields, medially in the vomer and laterally among the three palatine bone fields, P1–P2–P3. In the first case, the cause is a progressive knockout of collateral branches from the medial nasopalatine axis which proceeds with increasing severity from posterior to anterior. In the second case, the cause is also a progressive knockout, first hitting the lesser palatine axis to P3 and subsequently the collateral branches from the greater palatine axis to P2 and P1.

- The physical effect of mesenchymal deficiency is twofold.
  - If the *shelf width* is reduced sufficiently in P1 or further backward in P2 such that the distance between the hard palate and vomer exceeds the critical contact distance, normal fusion cannot occur.
  - When *shelf length* is reduced, as takes place in P3, the palatine aponeurosis is foreshortened, thereby causing the third arch Sm7 muscle sling to be displaced forward into pathologic contact with the posterior shelves of P3.
- The reduction in diffusible morphogens exerts similar effects.
  - BMP4 deficit along the epithelial border prevents normal fusion.
  - An unidentified morphogen affects development of the levator sling but not tensor because the neighboring bone stock of P3 is normal.

Cleft nasal deformity in all its forms, involves (1) a mesenchymal defect in the frontal process of premaxilla and (2) some degree of disturbance in the insertion of nasalis. The latter may even precede the former. Nasal deformity is universal in all premaxillary deficiency states. The spectrum of its presentation is directly related to the lip defect. Every component of the nasal envelope is involved.

- *Columella* is determined by muscle insertions and by the position of the premaxilla.
  - The base of the columella is shifted to the non-cleft (normal) side due to abnormal muscle insertion and the position of the premaxilla. The more complete the cleft, the more deviation occurs. Even though the orbicularis complex is normally inserted in the incomplete form, the force vector is less than on the normal side. Note that in complete CL depressor septi nasa is absent on the affected side.
  - The remainder of columella and nasal tip are angulated toward the cleft side; this is due to the tethering of the nasal skin envelope.
  - In cleft states with a complete alveolar defect, premaxillary may be twisted toward the opposite side; this will flex the caudal septum and columella to the non-cleft side.
- *Septum* has two deflections, making it S-shaped.
  - Anteriorly it is deviated toward the normal nostril anteriorly.
  - Posterior deflection is present about ½ way back is into the cleft nasal passage. The is due to the greater cross-sectional surface area on the cleft side. Hydraulic pressure exerted by fetal breathing of non-compressible amniotic fluid is elevated within the normal airway forcing posterior septum into the cleft.
- *Nasal skin and vestibular lining* is stretched out to fit into the piriform fossa defect. The "program" for lower lateral cartilage is retained. Blunt dissection upward from piriform fossa (Talmant maneuver) will liberated the lining allowing for more stable repositioning of the alar cartilage.
- *Alar cartilage* is normal in size and shape.
  - Lateral crus conforms to a more lateralized position due to mechanical deformation of its lining "program." It is also locked down by the abnormal force vector of overlying nasalis.
  - Medial crus is pulled downward by insufficient bone stock. Its footplate is universally displaced, it can be located about 3 mm down and 3 mm behind the footplate on the normal side.
- *Alar base position* is lateralized due to a combination of muscle insertion and, with complete primary palate cleft, the deflection of maxilla both lateral and posterior. This can reach extremes as in the bilateral clefts.
- *Nasalis* muscle fails to insert into its intended binding sites above the canine and lateral incisor fossae. In the absence nasalis no other muscle attachments are located at those sites. The developmental cause is likely due to lack of morphogen signal from the underlying bone of PMxB which are responsible for activating the binding sites. Nasalis is forced to make other arrangements into the skin of the nostril sill and into the piriform fossa.

These principles are perfectly illustrated by a secondary repair of a right unilateral minimal cleft. The repositioning of the soft tissue is accomplished by developmental field technique combined with wide bilateral subperiosteal elevation of the midface envelope extending downward to include bilateral mucoperiosteal advancement flaps (Figs. 18.72 and 18.73). The result is a centralization of previously displaced soft tissue fields on both sides of the cleft.

Neuromeric Model: *Volume Deficit* Affects Bone and *Morphogen Deficit* Affects Soft Tissues

The spectrum of the *incomplete naso-labial-maxillary clefts*, by definition with an *intact secondary palate*, constitutes a **Rosetta Stone of the embryo** that enables us to decode the secret language of clefts, thus gaining insight into how their anatomic features come about. What's more, the neuromeric model extends far beyond the oral cavity to encompass multiple congenital conditions of the head and neck. So… let's recap the central tenets of what we have learned in sequence, from the general to the specific.

Homeotic Genes: The Universal System of Axis Determination

The first task in the construction of a body plan in all vertebrate embryos consists of axis definition. This is fundamental to the nervous system, the master CPU (central processing unit) to which tissues of the body are "wired." Throughout this text we have seen how homeotic genes, both those of the *original HOX genes* from r3 backwards and *non-HOX homeotic genes* from r2 forward, determine the boundaries and the anatomic contents of all neuromeres. This system is presented in Chaps. 1 and 2. Chapter 5 relates the evolution of the neuromeric system into its final iteration as per Puelles and Rubenstein. Alterations in the homeotic code result in specific tissue defects, i.e., the "knockout" model.

Pharyngeal Arches are Constructed from Paired Rhombomeres

Endodermal, mesodermal and neural crest tissue specific to r2–r11 are assembled into five pharyngeal arches. Each arch thus contains tissues from two distinct homeotic combinations. Specific sectors of the arches are defined by a universal system of Dlx genes.

- The original system of multiple branchial arches in jawless fishes morphed into the 7-arch system of sarcopterygian fishes immediately prior to tetrapods. Over the course of evolution the sixth and seventh arches are represented by neuromeres sp1–sp4.
- In mammals this means that tissue derivatives from c1 to c4 have two fates: between the modern structures of the

neck and primitive derivatives of the arch system, such as sternocleidomastoid and trapezius.
  - It is possible that the esophagus may represents a genetic unit spanning from sp1 to sp8.

Neuroangiosome are Programmed by Genetic Zones in the Arches

Pharyngeal arches are divided into specific zones by the Dlx gene system [107, 108]. Each arch has a designated neurovascular axis that subdivides to supply each Dlx zone or subzone as its unique neuroangiosome. First arch and second arch, because they were reassigned in evolution from respiratory structures to jaws, have a unique vascular composition.

- First arch divides into original mesodermal components, such as branchial chewing muscles related to the oral cavity and "modern" (with placoderms) non-branchial bone fields that produce the maxillary-zygomatic complex and the oro-nasal cavities.
  - Mesodermal (branchial) derivatives are supplied by a single branch of ECA, now known as the internal maxillo-mandibular axis.
  - Neural crest (non-branchial) derivatives are supplied by branches of the stapedial system which are annexed by internal maxillary.
- Second arch has two fates: (1) its anterior zones are fully incorporated into the first arch to form a "sandwich" with (2) remaining myoblasts migrate widely to cover the entire head. Blood supply therefore consists of four distinct branches from ECA.

Neuroangiosomes Have Paired Growth Cones

The neural growth cone produces vascular endothelial growth factor (VEGF) while the vascular growth cone elaborated nerve growth factor (NGF). Both growth cones contain stem cell precursors, pericytes, with differing homeotic origins. Reciprocal interaction between the growth cones takes place within the genetic environment of surrounding mesenchyme.

- Tissue defects (clefts) are caused by failure of a neuroangiosome or of its subsidiary branch. This can occur for two reasons
  - Stem cell failure at the growth cone
  - Abnormal interaction between the growth cones and signal from the surrounding mesenchyme

Neuroangiosome Failure Causes Local Field Defects

Embryonic fields, by definition, is supplied by a specific neuroangiosome. Fields may be homogeneous (e.g. inferior turbinate) in which case the axis has breaks into multiple coequal branches. Other fields, such as premaxilla have several subcomponents which develop in a sequence. In this case the branches of the axis will subdivide to supply each one individually but in a specific order reflecting their temporospatial maturity. Mesenchyme housing the central incisor is laid down first and receive the first-order branches of the axis. Lateral incisor mesenchyme is laid down not next; it receives the second-order branches. The most distal field, the frontal process, arises from PMxB and is supplied last and is therefore vulnerable to axis failure.

Resin casts of vascular injections by Amin in normal mouse palates versus spontaneous cleft palate (both unilateral and bilateral), are very instructive [109]. This report documents in normals a difference in density between the nasal side of the hard palate and the oral side of the hard palate: lateral nasopalatine axis is a smaller circulation joined by transpalatal anastomoses to the greater palatine axis. Furthermore, the medial border of the palatal shelves demonstrated terminal dilatations. Under normal conditions GPA from P1 continues into the premaxilla where it anastomoses with medial nasopalatine.

In the spontaneous cleft group, elevation of the palatal shelves was delayed. Moreover, the extension of the plexus to the medial borders of the shelves on the cleft side showed leakage of resin, indicating *underdeveloped and defective arteries.*

- **In UCL/P, altered development of capillary beds is observed in the nasopalatine region, i.e., the premax-**

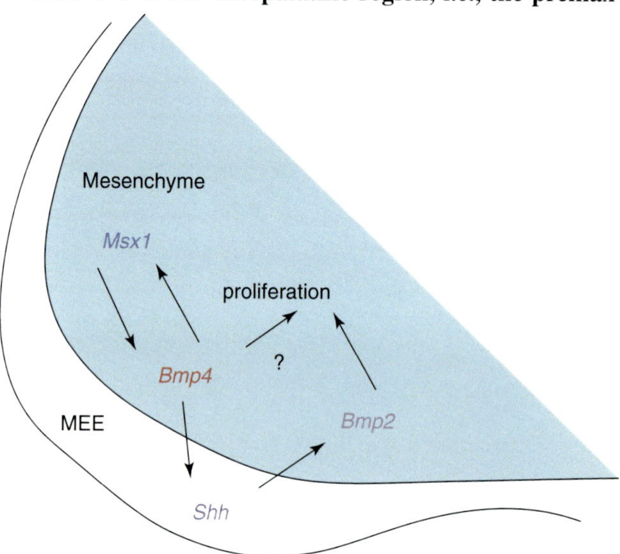

Fig. 18.74 BMP4 and sonic hedgehog (SHH) regulate epithelial fusion. A model for a genetic pathway integrating *Msx1*, *Bmp4*, *Shh*, and *Bmp2* in the epithelial–mesenchymal interactions that regulate mammalian palatogenesis. In this model, it is proposed that in the anterior palatal shelves, mesenchymally expressed *Msx1*, which can be induced by *Bmp4*, is required for *Bmp4* expression in the palatal mesenchyme. Mesenchymally expressed BMP4 maintains *Shh* expression in the MEE and Shh in turn induces *Bmp2* expression in the mesenchyme. BMP2 functions to induce cell proliferation in the palatal mesenchyme, which leads to palatal growth. [Reprinted from Zhang Z, Song Y, Zhao X, et al. Rescue of cleft palate in Msx-1 deficient mice by transgenic Bmp4 reveals a network of BMP4 and Shh signaling in the regulation of mammalian palatogenesis. Development 2002; 129:4135–4146. With permission from Company of Biologists]

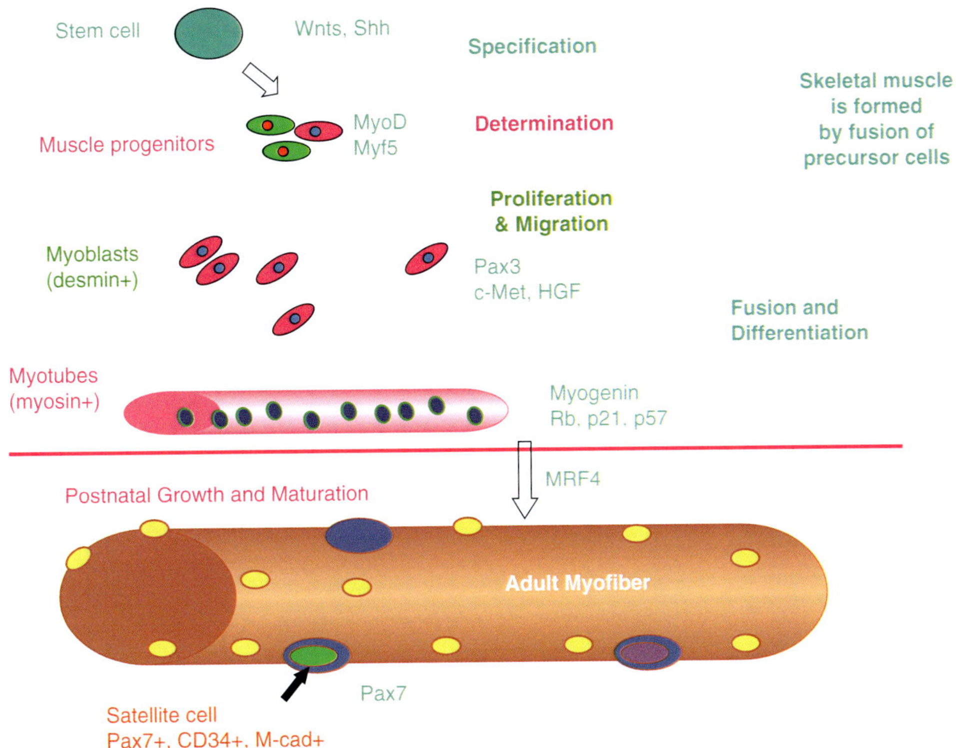

Fig. 18.75 Skeletal muscle embryogenesis 1 BMP4 acts primarily on Myf5. Formation of skeletal muscle tissue involves progressive commitment to the myogenic fate and fusion of myoblasts to form multinucleated myofibers. Specification of myogenic cells from early embryonic stem cells occurs under environmental influences (Wnts, Shh). Subsequently, the expression of MyoD/Myf5 in proliferating muscle progenitors commits them to a muscle fate and expression of muscle-specific proteins such as desmin. Cell cycle arrest, fusion of determined myoblasts into myotubes, and activation of differentiation-specific genes such as myosin are regulated by the MRFs in conjunction with cell cycle regulators. Maturation of myotubes into large myofibers progresses postnatally. Satellite cells (mononucleated muscle precursors) are associated with myofibers and marked by Pax7 expression. [Reprinted from Sachidanandan C, Dhwan J. Skeletal Muscle Progenitor Cells in Development and Regeneration. Proc Indian Nat Sci Acad 2003; 69(5): 719–740. With permission from Springer Nature]

illa. The alternation is ipsilateral in UCLP and affects the entire premaxilla in the BCLP.

Being a "Mesenchymal Midget" has Consequences

When, through neurovascular failure, a developmental field loses one or more of its component sub-field (such as PMxF) or simply experiences a deflation in volume, three things occur: (1) change in shape, (2) reduced production of otherwise normal morphogens, and (3) possible production of abnormal morphogens.

- Shape changes alter the anatomy of otherwise normal partner structure. Plagiocephaly elevates the eyebrow. A scooped-out PMxB lowers the nasal floor or causes a rotation of the lateral incisor.
- Morphogen deficit can lead to cleft formation or failure of muscle binding sites to "turn on the landing lights."
- Abnormal morphogens can cause structural alterations in surrounding tissues. Muscle in the *immediate environment* (5 mm) of the cleft margin shows abnormalities of muscle fibrils and mitochondria. Faulty signal from any of the four zones of parietal bone can cause synostosis in the adjacent suture.

Pathogenesis of Cleft: Molecular Mechanism

A Deficit in [BMP4] is the Common Denominator of Soft Tissue Pathology at the Cleft Margin

From its site of origin in the bone fields of premaxilla BMP4 diffuses through the soft tissues from above downward and from deep to superficial until reaching the margins of prolabium and lateral lip element. Thus, the cleft site is the most distal and the most vulnerable to a reduced or absent concen-

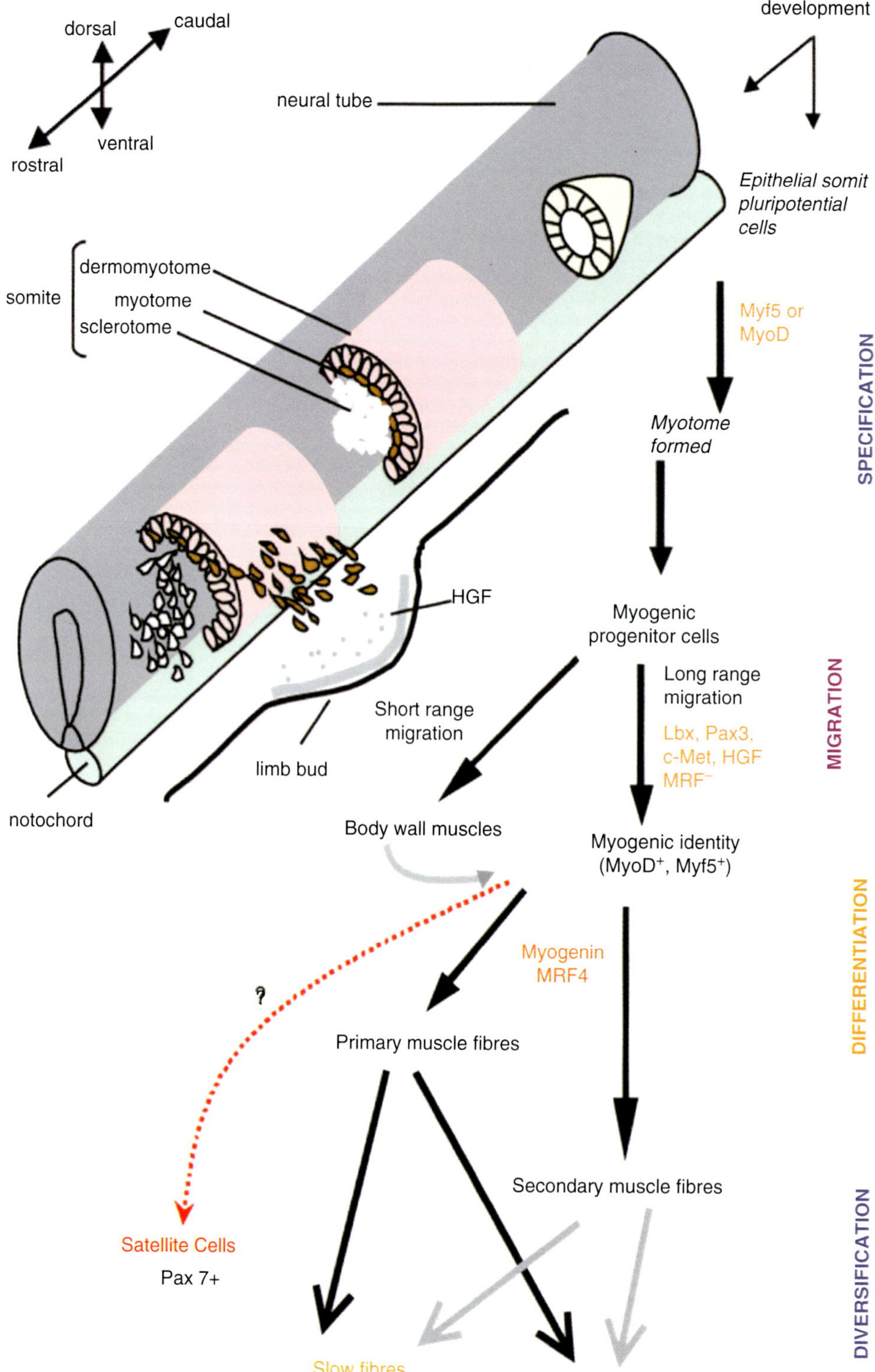

Fig. 18.76 Skeletal muscle embryogenesis 2 BMP4 acts primarily on My5. Schematic representation of somite development indicates the relative location of somites with respect to the axial neural tube and notochord. Although the somite is the source of many cell types (cartilage, muscle, skin), this schematic emphasizes the genes involved in the determination and differentiation of myogenic precursors. Interestingly, muscle progenitors from somites at different axial levels are influenced by different environmental cues. For example, muscle progenitors from somites positioned at the level of the developing limb bud, are induced by HGF to migrate long distances and form the limb muscles. These cells are marked by a distinct set of genes that are not expressed by the precursors of body wall muscles that migrate only short distances. Fusion and differentiation occur in two temporal waves to form primary and secondary myofibers that subsequently mature and diversify into fast and slow fiber types with distinct patterns of myosin expression. Neural signals influence fiber diversification. [Reprinted from Sachidanandan C, Dhwan J. Skeletal Muscle Progenitor Cells in Development and Regeneration. Proc Indian Nat Sci Acad 2003; 69(5): 719–740. With permission from Springer Nature]

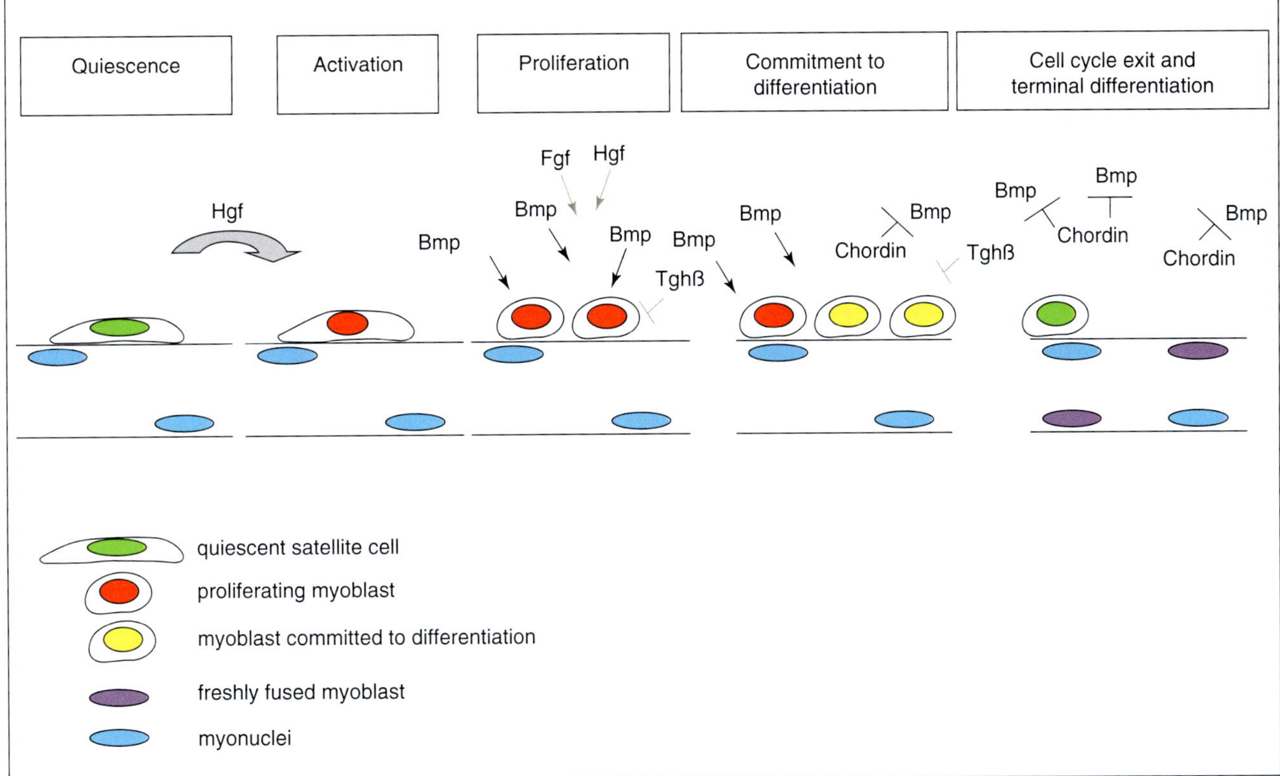

Fig. 18.77 BMP signaling during satellite cell differentiation. Quiescent satellite cells (green) do not respond to BMP signals. Upon activation, satellite cells express MyoD and respond to BMP signals with phosphorylation of p-Smad (red) and reduced differentiation. After satellite cell descendents are committed to differentiate (yellow) they express Myogenin and subsequently the BMP inhibitor Chordin. Upon inhibition of BMP signaling, myoblasts exit the cell cycle, fuse into myotubes and undergo terminal differentiation (purple). Postmitotic myonuclei (blue) irreversibly loose the competence to respond to BMP signals. [Reprinted from Friedrichs M, Wirsdöefer F, Flohé SB, Schneider S, Wuelling M, Vortkamp A. BMP signaling balances proliferation and differentiation of muscle in satellite cell descendents. BMC Cell Biol. 2011; 12:26–43. With permission from Springer Nature]

tration of BMP4. The effects of BMP4 deficit on epithelial surfaces (previously discussed) are highly localized and explain fusion failure at the cleft margins. Also highly localized are muscle abnormalities as documented above. These are similar in both lip and palate muscles: decreased number of muscle fibers, disorganization (failure to properly assemble), abnormal predominance of type II (fast twitch) fibers. BMP4 has a regulatory role at multiple steps in myogenesis (Figs. 18.74, 18.75, 18.76 and 18.77).

- **Conclusion**: *the epithelial and muscle abnormalities at the cleft site evidence the morphogen BMP4 plays central role in the pathogenesis of cleft lip.*
- **Conclusion (see below):** BMP signaling regulated muscle development in the postnatal period by controlling stem (satellite) cell dependent growth of myofibers and the generation of an ongoing reserve of adult muscle stem cells.

How does this work? Let's organize our discussion as follows. First, we will look at the basic properties of BMP4. We then consider the stages of myogenesis and how it is regulated. As we shall see, BMP4 impacts on multiple steps in this process. Satellite cells are stem cells within muscle. They are responsive to BMP4 so we will need to define them. Myogenesis has three phases through development: embryonic, fetal, and perinatal (adult); cleft margin muscle fits into the transition between the first two, beginning around Carnegie stage 20.

BMP4: A Brief Introduction

Bone morphogenetic protein 4, produced by the *BMP4* gene located on chromosome 14q22–q23 and is a member of the transforming growth factor beta superfamily. It is produced at all neuromeric levels of the dorsal notochord and plays a protean role in the formation of the embryonic axis. From the ventral notochord, and opposition to BMP4 is sonic hedgehog.

Later in embryogenesis BMP4 becomes inhibited by dorsal gene expression and assumes a ventralizing role for mesoderm.

BMP4 is responsible for determining the non-neural fate of overlying ectoderm. It affects the development of multiple tissues, including the eye, angiogenesis, the musculoskeletal system and the ureteric bud. Additional roles in craniofacial development include generation of new neurons (stem cell effect) in the subventricular zone of the ventricles; and the control of apoptosis affecting r1 neural crest which is responsible for unification of the nose and centralization of the orbits. See review by Wang et al. [110].

Martin showed histologic changes in "subepithelial" clefts consistent with fibrosis. Ultrasound now demonstrates this as a hypoechoic interruption in an otherwise hyperechoic (black) band corresponding to orbicularis. Demonstrating ultrasound were used.

The Three Phases of Skeletal Muscle Development

Primary myogenesis is embryonic. This population is controlled by *Pax3*. It refers to the formation of the first multinucleate fibers from embryonic precursors the fuse together, creating a scaffold. This phase lasts in mice until E14.5 or approximately Carnegie stages 19–20. Recall that the primary fibers are predominantly type 1.

- Facial skeletal development takes place prior to this time.
- Thus, the effects of decreased [BMP4] are present during embryonic myogenesis.

Secondary myogenesis is fetal. Population control is from *Pax7*. Fetal progenitors add on to the scaffold. They favor type 2 (fast-twitch) fibers. This final differentiation consists of three steps.

- *Exit from the cell cycle*: Myoblasts "decide" to stop proliferating
- *Alignment*: Myoblasts recognized one another and clump together
- *Fusion*: Meltrins, metalloproteinases play a crucial role
  - Primary myoblasts form slow-twitch muscle type 1 fibers
  - Secondary muscle fibers develop with innervation and form fast-twitch type 2 fibers

Tertiary myogenesis is perinatal and stabilizes in the adult. When myogenesis is complete both Pax3 and Pax7 are no longer required thus BMP4 no longer plays a role. In the case of muscle repair after trauma, BMPs again are important for satellite cell activation but via an entirely different mechanism, using chordin as antagonist.

Genetic Regulation of Myogenesis: A Play in 6 Acts

The overall process of myogenesis consists of 6 steps—based on the somite model. Key genes are listed. We shall see them again when discussing where and how BMP4 affects the process. Some familiarity with these genes is worthwhile, trust me! Chai and Pourquíe provide a very comprehensive review.

Note: A homeobox is a sequence of DNA 180 base pairs in length in genes involved in morphogenesis. These genes produce homeodomain proteins have a helix–loop–helix configuration to binds to segments of DNA and unlocks them

- *Delamination from the somite* (PAX3, c-Met) Departure from the dermomyotome is controlled by PAX3 via its induction of c-Met. Failure totally blocks entry into the migration pathway.
  - Control of delamination from somitomeres Sm1 to Sm7 is unclear because PAX3 is not expressed in facial muscles.
- *Migration* (c-Met/HGF, LBX1) Myoblasts move into position in the limbs according to dorsal and ventral signals.
  - c-Met alone is expressed by epithelium whereas c-Met in association with hepatic growth factor is expressed in mesoderm. Cell survival along the migration pathway is dependent on c-Met/HGF.
  - LBX1, ladybird homeobox 1, regulates the actual migration of myoblasts into the limb. It acts unopposed in the hindlimb but the forelimb has separate ventral signals. Thus in the absence of LBX1 flexor muscles develop in the forelimb while the hindlimb is sterile.
  - Control of migration into the orbit and pharyngeal arches from somitomeres is unclear. From levels r8 to r11 tongue muscle migration is distinct from migration of pharyngeal constrictors and laryngeal muscles.
- *Proliferation* (PAX3, c-Met, MOX2, MSX1, Six, My5, MyoD) Myoblasts have to expand in number becoming myocytes.
  - MOX2, mesenchymal homeobox 2 controls expansion of the myoblast pool globally in the hindlimb, causing it to shrivel; in the forelimb only certain muscles are affected. How this gene in the pharyngeal arches is unknown.
  - Myf5, myogenic regulatory factor 5, is a homeobox gene; it is the earliest in a series of MRFs (myogenic regulatory factors). MyoD belongs to the same family and is Myf5 drive myoblasts to commit to skeletal muscle lineage. Under experimental conditions MYF5 can force undifferentiated mesenchymal cells such as

fibroblasts into a muscle. A block in Myf5 causes complete absence of muscle.
  - MyoD, myogenic differentiation factor 1, regulates the committed state of muscle cells. It acts in concert with the more important Myf5. MyoD resides in satellite cells at low levels (vide infra) but gets turned on in response to trauma. MyoD is important for the expression of fast-twitch, type 2 fibers.
  - MSX1, Msx homeobox protein 1, is implicated in cleft palate. Cases of Msx1 deficiency have be "rescued" by BMP4 [111]. This implicates BMP4 in the production of MSX1.
  - SIX1, Sineoculis homeobox homolog 1, is the homolog of the Drosophila eyeless gene. It is involved in the neuromuscular junction. As we shall see, innervation of muscle is necessary for its expansion. SIX1 is required for hypaxial muscles, i.e. muscles of the pharyngeal arches.
- *Determination* (Myf5, MyoD) refers to the final phenotypic expression of skeletal muscle. Assuming the myoblasts have proliferated their conversion into skeletal muscle depends upon proper function of both of these factors.
- *Differentiation* (Myf4, Myf6, MyoD) myocytes now mature.
  - Myf4, myogenin, is required for the fusion of myocytes to either form new fibers or add on to existing one. If permits muscle fibers to grow before the terminal differentiation event.
  - Myf6, herculin, controls differentiation of myotubes in skeletal muscle (not smooth muscle).

Stem Cell Regulation: General Characteristics

Skeletal muscle stem cells ("reserve" cells) have two fates: (1) they serve as precursors of skeletal muscle or (2) they maintain a self-replicating population. Stem cells are found intercalated between the sarcolemma and the basement membrane. Because of their peripheral location they acquired the name "satellite" cells, These cells are small, with few organelles and a high nucleus-to-cytoplasm ratio. Satellite cells can fuse directly to underlying muscle fibers, thus augmenting their nuclei. They can also form new fibers. In the adult stage these stem cells are quiescent but they respond to trauma by first proliferating as myoblasts and then undergoing division.

Satellite cells of non-craniofacial muscle express both PX3 AND PAX6 while craniofacial muscles express only PAX6. In the dormant state, satellite cells express CD34 (typical for baseline mesenchymal stem cells) and Myf5 (indicating stable muscle function). When activated by BMP4, satellite cells initially express the PAX genes but these quickly give way to MyoD and Myf4 (myogenin), indicating ongoing fusion. In general, BMP upregulation *maintains satellite cells in a proliferating state*, thereby expanding the numbers of cells and fibers. Once the cells decide to differentiate, they upregulate Chordin which shuts down BMP as a negative feedback loop to support terminal differentiation [112].

Two experimental situations illustrate this concept

- Noggin has been used to block BMP in utero. Under these conditions, postnatal histology shows decreased satellite cell proliferation and decreased muscle fiber growth (Stantzou 2017).
- *Fibrodysplasia ossificans progressive* is an experiment of nature which shows what happens to muscle stem cells under conditions of abnormally high BMP signaling. The response is quantitative; they are driven to an osteogenic fate thus producing ectopic bone as in Shore et al. [113].
  - **Note**: This is precisely what takes place when rhBMP2 recruits MSCs and converts them into bone (see Chap. 20).

Muscle Stem Cells: Main Population Versus Side Population

Skeletal muscle has more than one type of stem (satellite) cell [114]. A main population (MP) and a side population (SP). SP cells express BMP4 which induces in the MP cells BMP receptor 1a (BMPR1a). The protein Gremlin 1 (Grem1) acts as an antagonist for TG-Beta signaling molecules, specifically blocking BMP2 and BMP4. It is essential for limb bud development. Grem1 is produced by MP cells as a response to the BMP4 stimulus, acting as a negative feedback loop to shut down the process. It has been shown the BMP4+ cells coexist with BMPR1a+ cells in the interstitial cells of muscle.

BMP4 causes cells that are BMPR1a+ to proliferate but has no effect on stem cells that are BMPR1a−. Furthermore, BMP4 inhibits transcription of the gene *MyoD*, which is responsible for causing muscle cells to pull out of the cell cycle and undergo terminal differentiation. As long as BMP4 is active the muscle population expands. Under normal conditions:

- Muscle SP cells cause proliferation of myogenic precursors via BMP4.
- This produces an expansion of the embryonic muscle population to reach its final state prior to shutdown.
- Gremlin stops this process, bringing muscle proliferation to an end.

Cleft Lip Muscle: Why Is It the Way It Is?

Early myoblasts must choose between continued proliferation and differentiation into multi-nucleate myotubules. BMP4 is the potential driver. The myoblast at the margin of the prolabium and maxillary process are the most distal from the "radio signal" of BMP4. They receive the transmission last. Their blood supply is the most distal as well. BMP2, BMP4 are all important for the maintenance of vascular endothelium [115, 116]. What happens to this milieu if the [BMP4] is pathologically reduced?

Muscle fibers within 5–10 mm of the cleft site undergo premature shutdown of proliferation due to morphogen deficit

- Decrease in number of muscle fibers
- Failure of fibers to correctly assemble, leading to disorganization
- Muscle fibers are 50% reduced in diameter
- Interstitial tissue remains in a primitive, fibrotic state
- Reduced number of blood vessels due to BMP4-related endothelial dysfunction

Conclusion

In this chapter, the microform and its variants have served as a Rosetta Stone to help us decipher the secret language of clefts. Using the neuromeric model, we have followed the consequences of a deficit in the premaxilla from local manifestations involving the hard tissue platform to its distal soft tissue effects. In so doing, the stage is now for a final consideration of the complete form of the nasolabial cleft in the chapter that follows.

The soft tissue pathology of the cleft nose is dependent in every way upon the bone platform due to inadequate development of the premaxilla, both in terms of how its soft tissues of the nasal envelope are anchored and how muscles acting on the alar based and lateral nasal wall are anchored. The principles of surgical management are:

- Wide subperiosteal dissection, both within the piriform fossa and nasal vestibule and external all the way to the nasal bones to correctly reassign tissues to their correct relationships
- Correction of the septum, as indicated
- Restoration of the airway by reassigning nasalis to its functionally correct insertion
  - Separate nasalis from orbicularis anchor it to canine, *not* to the midline

The soft tissue pathology of cleft lip takes place in a limited zone of tissue in close proximity to the cleft site characterized by reduction in BMP4 concentration below a critical level required for epithelial stability, normal myogenesis, and microvascular development. The principles of surgical management are as follows:

- Wide subperiosteal mobilization
- Developmental field dissection of the prolabium
- Exclusion of embryologically damaged tissue from the repair site
- Restoration of the nasal floor

Addendum: Alveolar Cleft Closure (with DFR Modification)

Non-DFR flaps are shown in the original illustration with nasal floor closure using nasal mucosa flaps from below as a blind procedure. Under *DFR protocol* the nasal floor is addressed at primary lip repair (6 months) with the backwall and floor created by AEP at 18 months. If the nasal floor has not been closed, prior repair was non-DFR. Secondary lip repair is done with nasalis dissection and water-tight closure.

Step 1 *Upper Left*: Mucoperiosteal flap (r2, pink) gingivoperiosteoplasty flaps (sliding sulcus flap) is mobilized via pericoronal incision with a back-cut up the buttress. The S flap is supplied by medial and lateral branches of StV2 infraorbital and by overlying ECA facial artery. The dotted lines indicate the level of a counter-incision that will be made along the undersurface of the flap. Mucoperiosteum of premaxilla (orange) has also a mixed blood supply from StV1 anterior ethmoid, from StV2 medial nasopalatine, and from StV2 medial branch of infraorbital. Periosteal flap on maxillary side (yellow) is supplied by greater palatine. Periosteal flap on the premaxillary side (white) has a variable blood supply. If the cleft is unilateral, it receives StV2 greater palatine + StV2 medial nasopalatine and back flow from StV1 anterior ethmoid. If the cleft is bilateral, it is perfused by the latter two arteries only. Note the cut shown between lateral incisor and canine on the non-cleft side. In DFR this is usually unnecessary for two reasons: (1) the non-cleft flap can be mobilized in a similar manner. (2) A similar periosteal releasing incision. Bilateral S flaps can cover a total of up to 4 dental units without tension. Dissection time is fast and bloodless, under 5 min per flap.

Step 2 *Upper right*: backwall flaps are sutured, maxillary (yellow) to premaxillary (white). The counter-incision along the undersurface of the S flap is indicated by a white line with red dots. The incision is placed at the level of the buccal sulcus. Although the S flap is mobilized over the entire face of the maxilla, the counter incision frees the alveolar component to translate with ease, with no tethering from above.

Step 3 Mesial translation of maxillary S flap (pink) and premaxillary S flap (orange). Dead space for the flaps is transferred to the first or second molar.

Step 4 Final closure with bone graft/rhBMP2 in place. Incision at premaxillary–maxillary junction on the non-cleft side is not necessary.

Commentary: Karoon Agrawal

The Embryologic Basis of Craniofacial Structure: Developmental Anatomy, Evolutionary Design, and Clinical Applications

Michael Carstens

(Editor in New York: Michelle Tam. Publisher: Springer Verlag)

The world is progressing very fast, and the knowledge is expanding exponentially in breadth and depth; we need to keep pace. This book will be a milestone in the understanding of developmental anatomy, of its evolutionary origin and its clinical applications, especially in reference to the occurrence of facial clefts.

Mr. Michael Carstens is a big name in India. He is a great friend of the Indian Plastic Surgery community and is an honorary member of our Indian Societies. He has visited our country multiple times and has participated in many academic meetings to present his innovative work. He has tirelessly encouraged our scientific community for doing research and making innovations with newer techniques of cleft repair.

I cannot forget my encounter with Mr. Mike for a debate on "Alveolar Extension Palatoplasty" popularized by him. He was speaking in favor, and I was speaking against. In spite of the outcome of the debate, we became great admirers of each other. We kept exchanging our thoughts at regular intervals. The more I interacted with him, the more I realized that this man was born with a mission as he has dedicated his career to solve the jigsaw puzzle of the origins of facial clefts. His concept relates to the stapedial artery not as a single entity, but as a complex system making a relationship between failure of individual stapedial branches as a causation of facial clefts is unique and impressive.

I was surprised by a personal call from Mike for writing a foreword of his book being published by Springer-Verlag, and I am here to do justice to his request. I consider it a great privilege to be asked to write a foreword and express my assessment of this work.

A Note from Dr. Carstens

Karoon Agrawal

For over four decades, Dr. Karoon Agrawal, Professor of Plastic Surgery at the National Heat Institute of New Delhi, India, has been a thought leader with extensive contributions to both cleft lip/palate and hypospadias. A tireless academician, he has also proven himself to be a glutton for punishment on two accounts: first, he is the editor of a six-volume treatise published with Thieme covering the entire specialty; and second, he willingly committed to review this book from cover to cover, contributing careful and in-depth comments. For the time and effort this required, I sincerely hope that his wife will forgive me. Karoon cuts a fine-featured, slender, and scholarly figure, accompanied by a warm smile and restless intellect. He is also a first-class debater, which I discovered to my discomfiture. We were summoned by the Indian Society for Cleft Lip and Palate to debate on the relative merits of alveolar extension palatoplasty versus the traditional concept of cleft palate repair. Right off, I knew that it was a dangerous proposition … no mortal could expect to approximate Karoon's surgical experience. But I had not counted on his sense of humor. There, upon the screen, with his very first slide, in front of the entire audience, was a picture of a crouching King Kong in the posture of a sumo wrestler with my face pasted on top! I suppose it was intended as a backhanded compliment about AEP, but at the time I wanted to crawl in a hole. Anyway, the discussion was lively and the whole audience loved it. Later on, as the journal editor for the ISCLP, Karoon invited me to contribute to a special issue to present the developmental science of palatoplasty. Although it was like pulling teeth, he was patient with me and the result was two articles that eventually formed the blueprint for Chapters 14–17. Karoon's gentle but persistent questioning forced me to focus on mechanism and good writing. I can say, in retrospect, that the content of the chapters are as much his thinking as my own. As both colleague and friend, Dr. Karoon Agrawal represents the very best of Indian plastic surgery, a gift for which I will always be indebted.

This book is a treasure trove for all embryologists, anatomists, cleft clinicians, as well as researchers. The chapters are designed remarkably well, keeping in mind the interests of the readers and beneficiaries. Chapters 1–5 are on the basic concepts; Chaps. 6–13 cover specific anatomic entities (such as bone, fascia and the orbit), Chaps. 14–19 are on the pathologic anatomy and basis of repair of cleft lip and palate and Chap. 20 on BMP is speculative, which I enjoyed the most. The content reflects the unparalleled thought process of a man who has spent all his lifetime thinking, analyzing, and writing about neuroembryology and functional anatomy of the head and neck.

Although the explosion of knowledge in craniofacial surgery coupled with advances in science and technology has improved our clinical outcome, little attention has been given to expanding our understanding of basic anatomic principles. Reading through the chapters of this book, it seems that there are many basics which are still unexplored. This work is an attempt to unravel many such mysteries, not only in clefts, but also in synostosis, orthognathics, and absence or excess of structures in the orbit, nose, neck, and brain. Even the rationale for structures we take for granted, such as the fascial layers, sutures and foramina are explored. All this is clinically relevant. Better understanding of the development of the anomalies will pave the way for better prevention and management strategies.

While going through the chapters, one will be awed by the amount of information that Mike has for all the readers. How he imbibed so much in his lifetime is difficult to fathom. The embryology and pathogenesis of CLP (hard palate, soft palate, and lip/nose complex) are remarkable chapters. At the same time, I do not intend to belittle the relevance of other chapters. No stone is left unturned.

This analysis of principles of development of craniofacial structures is extraordinary. A great amount of research has been done to prepare the manuscript. This is of immense value to every embryologist, anatomist, researcher, and clinician.

Creation of a single-author book of this magnitude is a gigantic task. Mike has spent years working toward the creation of this. Into it he has poured all his knowledge and research, along with his innovative thoughts. I admire his effort. This book is a must-have in the reference section of every institute's medical library and a must-read for every member of the cleft care team.

Karoon Agrawal, MS, MCh

Consultant Plastic Surgeon, National Heart Institute

Member of Smile Train Global Medical Advisory Board

Former Director Professor and Head of Burn, Plastic & Maxillofacial Surgery

VMMC & Safdarjung Hospital, New Delhi, India

Editor: Textbook of Plastic Reconstructive, and Aesthetic Surgery

References

1. Cosman B, Crikelair GF. The shape of the unilateral cleft deformity. Plast Reconstr Surg. 1964;35:484–93.
2. Carstens MH. Correction of the bilateral cleft using the sliding sulcus technique. J Craniofac Surg. 2000a;11(5):137–67.
3. Carstens MH. Functional matrix cleft repair: a common strategy for unilateral and bilateral cleft lip. J Craniofac Surg. 2000b;11(5):437–69.
4. Hinrichsen K. The early development of morphology and pattern of development of the face in the human embryo. Advances in anatomy, embryology and cell biology 98. New York: Springer; 1985.
5. Curtin J. Clinical evaluation of microform cleft lip surgery (discussion of article by Thomson and Delpero). Plast Reconstr Surg. 1984;75:804.
6. Heckler FR, Oesterle LG, Jabaley M. The minimal cleft lip revisited: clinical and anatomic considerations. Cleft Palate J. 1979;16:240.
7. Thomson HG, Del Pero W. Clinical evaluation of microform cleft lip surgery. Plast Reconstr Surg. 1984;75:800–3.
8. Johnson D. Some observations on certain developmental dentoalveolar anomalies and the stigmata of cleft. Dent Pract Dent Res. 1967;17:435.
9. Olin WH. Orthodontics in cleft lip and palate. In: Grabb WC, Rosenstein SW, Bzoch KR, editors. Cleft lip and palate. Boston: Little Brown; 1971. p. 599–615.
10. Boo-Chai K, Tange I. The isolated cleft lip nose: report of five cases in adults. Plast Reconstr Surg. 1968;41(35):28–34.
11. Trott JA, Mohan N. A preliminary report on one-stage open tip rhinoplasty at the time of lip repair in bilateral cleft lip and palate: the Alor Setar experience. Br J Plast Surg. 1993;46:215.
12. Nammoun JD, Hisley GS, Hucthins GN, Van Der Kolk C. Three dimensional reconstruction of the human philtrum. Ann Plast Surg. 1997;38:202–8.
13. Park CG, Ha B. The importance of accurate repair of the orbicularis oris muscle in the correction of unilateral cleft lip. Plast Reconstr Surg. 1995;96:780.
14. Martin RA, Jones KL, Benirschke K. Extension of the cleft lip phenotype: the subepithelial cleft. Am J Med Genet. 1993;47:744–7.
15. Martin RA, Hunter V, Neufeld-Kaiser W, Flodman P, Spence MA, Furnas D, et al. Ultrasonographic detection of orbicularis oris defects in first degree relatives of isolated cleft lip patients. Am J Med Genet. 2000;90:155–61.
16. Marazita ML. Subclinical features in non-syndromic cleft lip with or without cleft palate (CL/P): review of the evidence that subepithelial orbicularis oris muscle defects are part of an expanded phenotype for CL/P. Orthod Craniofac Res. 2007;10:82–7.
17. Gasser RF. The development of the facial muscles in man. Am J Anat. 1967;20:357.
18. Fará M. Anatomy and arteriography of cleft lips in stillborn children. Plast Reconstr Surg. 1968;42:29.
19. Fará M. Anatomy of unilateral and bilateral cleft lip. In: Bardach J, Morris HL, editors. Multidisciplinary management of cleft lip and palate. Philadelphia: WB Saunders; 1990. p. 134–43.
20. Gamba-Garib D, Petrocelli-Rosai J, Sathler R, Okada-Ozaka T. Dual origin of maxillary lateral incisors: clinical implications in patients with cleft lip and palate. Dental Press J Orthod. 2015;20(5):118–25. https://doi.org/10.1590/2177-6709.20.5.118-125.sar.
21. Delaire J, Precious S, Gordeef A. The advantage of wide subperiosteal exposure in primary surgical correction of labial maxillary clefts. Scand J Plast Reconstr Surg. 1993;22:710.
22. Fisher DM. Unilateral cleft lip repair: an anatomical subunit approximation technique. Plast Reconstr Surg. 2011;116:61–71.

23. Stark RB. Development of the center of the face with particular reference to surgical correction of bilateral deft lip. Plast Reconstr Surg. 1958;21:177–92.
24. Peter K. Atlas der Entwicklung der Nase und des Gaumen beim Menschen, rnit Einsdlluss der Entwicklungsstorungen. Jena: Gustav Fischer; 1913.
25. Veau V, Recamier J. Bec-de-Lievre: Formes CliniquesChirugie. Paris: Masson et Cie; 1938. p. 6.
26. Rohrich RJ, Gunter JP, Friedman R. Nasal tip blood supply: an anatomic study validating the safety of the transcolumellar incision in rhinoplasty. Plast Reconstr Surg. 1995;95:795.
27. Song R, Liu C, Zhao Y. A new principle for unilateral complete deft lip repair: the lateral columellar flap method. Plast Reconstr Surg. 1998;102:1848.
28. Carstens MH. Functional matrix cleft repair: principles and techniques. Clin Plast Surg. 2004;31:159–89.
29. Cohen D, Goitain K. Arhinia. Rhinology. 1986;24:287–92.
30. Cohen MM, Sulik KK. Perspectives on holoprosencephaly: Part II. Central nervous system, craniofacial anatomy, syndrome commentary, diagnostic approach, and experimental studies. J Craniofac Genet Dev Biol. 1992;12:196–244.
31. Patel H. Holoprosencephaly with median cleft lip: clinical, pathological, and echoencephalographic study. Am J Dis Child. 1972;214:217–21.
32. Probst FP. The prosencephalies: morphology, neuroradiological appearances, differential diagnosis. New York: Springer; 1979. p. 29–30.
33. O'Rahilly R, Muller F. Developmental stages in human embryo, publication 637. Washington, DC: Carnegie Institution of Washington; 1987. p. 175–302.
34. Streeter GL. Developmental horizons in human embryos: description of age groups XV, XVI, XVII, and XVIII, being the third issue of a survey of the Carnegie collection. Contrib Embryol Carnegie Inst. 1948;32:13–203.
35. Latham RA. The pathogenesis of the skeletal deformity associated with unilateral cleft lip and palate. Cleft Palate J. 1969a;6:404–14.
36. Latham RA. The septopremaxillary ligament and maxillary development. J Anat. 1969b;104:584.
37. Latham RA. The developmental deficiencies of the vertical and antero-posterior dimensions in the unilateral cleft lip and palate deformity. In: Georgiade NG, Hagarty RF, editors. Symposium on the management of cleft lip and cleft palate and associated deformities. St. Louis: CV Mosby; 1974.
38. Latham RA. The structural basis of the philtrum and the contour of the vermilion border: a study of the musculature of the upper lip. J Anat. 1976;121:151.
39. Semb G, Shaw WC. Simonart's band and facial growth in unilateral clefts of the lip and palate. Cleft Palate Craniofac J. 1991;28:40.
40. Enlow DH. Facial growth. 3rd ed. Philadelphia: WB Saunders; 1990.
41. Moss ML. The functional matrix concept and its relationship to temporomandibular joint dysfunction and treatment. Dent Clin N Am. 1983;27(3):445–55.
42. Sasaki A, Takeshita S, Publico AS, Moss ML, Taneka E, Ishino Y, Watanabe M, Tanne K. Finite element growth analysis of the craniofacial skeleton in patients with cleft lip and palate. Med Eng Phys. 2004;26(2):109–18.
43. Avery JK, Happle JD, French HC. Development of the nasal capsule in normal and cleft palate formation. Cleft Palate Bull. 1957;7:8–11.
44. Hairfield WM, Warren OW. Dimensions of the cleft nasal airway in adults: a comparison with subjects without cleft. Cleft Palate J. 1989;26:9–13.
45. Jirasek JE. Atlas of human prenatal morphogenesis. Boston: Kluwer; 1983.
46. Kosaka K, Hama K, Eta K. Light and electron microscopic study of fusion of facial prominences. Anat Embryol (Berl). 1985;173:187–201.
47. Millicovsky G, Johnston MC. Maternal hyperoxia greatly reduces the incidence of phenytoin-induced cleft lip and palate in A/J mice. Science. 1981a;212:671.
48. Millicovsky G, Ambrose JL, Johnston MC. Developmental alterations associated with spontaneous cleft lip and palate in CL/Fr mice. Am J Anat. 1982;164:129.
49. Millicovsky G, Johnston MC. Active role of embryonic facial epithelium: new evidence of cellular events in morphogenesis. J Embryol Exp Morphol. 1981b;63:53–66.
50. Millicovsky G, Johnston MC. Hyperoxia and hypoxia in pregnancy: simple experimental manipulation alters the incidence of cleft lip and palate in CL/Fr mice. Proc Natl Acad Sci U S A. 1981c;9:4723.
51. Poelmann RE, Vermeij-Keers C. Cell degeneration in the mouse embryo: a prerequisite for normal development. In: Muller-Berat N, editor. Progress in differentiation research. Amsterdam: North Holland; 1976. p. 93–102.
52. Gaare JD. Cell degeneration during the fusion of the nasal processes in mice. Anat Rec. 1976;184:407.
53. Gaare JD, Langman J. Fusion of nasal swellings in the mouse embryo: surface coat and initial contact. Am J Anat. 1977a;150:461–475.
54. Gaare JD, Langman J. Fusion of nasal swellings in the mouse embryo: regression of the nasal fin. Am J Anat. 1977b;150:477–499.
55. Gaare JD, Langman J. Fusion of nasal swellings in the mouse embryo. DNA synthesis and histological features. Anat Embryol (Berlin). 1980;159(1):85–99.
56. Figueroa AA, Pratt RM. Autoradiographic study of macromolecular synthesis in the fusion epithelium of the developing rat primary palate in vitro. J Embryol Exp Morphol. 1979;50:145–54.
57. Wilson DB, Hendrickx AG. Quantitative aspects of proliferation in the nasal epithelium of the rhesus monkey embryo. J Embryol Exp Morphol. 1977;38:217–26.
58. Smuts MS. Concanavalin A binding to the epithelial surface of the developing mouse olfactory placode. Anat Rec. 1977;188:29–38.
59. Gui T, Osama-Yamashita N, Eto K. Proliferation of nasal epithelium and mesenchymal cells during primary palate formation. J Craniofac Genet Dev Biol. 1993;13(4):250–8.
60. Warbrick JG. The development of the nasal cavity and upper lip in the human embryo. J Anat. 1960;94:351–62.
61. Juriloff DM. Genetic analysis of the construction of the AEJ. A congenetic strain indicates that nonsyndromic CL(P) in the mouse is caused by two block with epistatic interaction. J Craniofac Genet Dev Biol. 1995;15:l–12.
62. Osumi-Yamashita N, Asada S, Eto K. Distribution of F-actin during mouse facial morphogenesis and its perturbations with cytochalasin D using whole embryo culture. J Craniofac Genet Dev Biol. 1992;12:130–40.
63. Johnston MC, Millicovsky G. Normal and abnormal development of the lip and palate. Clin Plast Surg. 1985;12:521.
64. Longo AD. Carbon monoxide in pregnant mother and fetus and its exchange across the placenta. Ann N Y Acad Sci. 1970;174:131.
65. Ericson A, Kallen B, Westerholm P. Cigarette smoking as an etiologic factor in cleft lip and palate. Am J Obstet. 1979;135:348.
66. Congdon ED. Transformation of the aortic arch system during the development of the human embryo. Carnegie Contrib Embryol. 1922;14:46.
67. Feinberg RN, Scherer GK, Auerbach R. The development of the vascular system. Basel: Karger; 1991.
68. Nairn RJ. The circumoral musculature. Br J Plast Surg. 1975;36:141.
69. Nicolau PJ. The orbicularis oris muscle: a functional approach to its repair in the cleft lip. Br J Plast Surg. 1983;36:141–53.

70. Precious D, Delaire J. Surgical considera tions in patients with cleft deformities. In: Bell WH, editor. Modern practice in reconstructive and orthognathic surgery. Philadelphia: WB Saunders; 1992.
71. Joos U. Muscle reconstruction in primary cleft lip surgery. J Craniomaxillofac Surg. 1989;17:8.
72. Monie IW, Cacciatore A. The development of the philtrum. Plast Reconstr Surg. 1962;30:313–20.
73. England M. Life before birth. Chicago: Year Book Medical Publishers; 1996.
74. Mooney MP, Siegel MI, Kimes KR, et al. Anterior paraseptal cartilage development in normal and cleft lip and palate human fetal specimens. Cleft Palate J. 1994;31:239–45.
75. Ross RB, Semb G, Shaw WC. Simonart's band and facial growth in unilateral clefts of the lip and palate. Cleft Palate Craniofac J. 1991;28:40.
76. Wu J, Yin N. Anatomy research of nasolabial muscle structure in fetus with cleft lip: an iodine staining technique based on microcomputed tomography. J Craniofac Surg. 2014;25(3):1055–62.
77. Kernahan D, Dado DV, Bauer BS. The anatomy of the orbicularis oris muscle in unilateral cleft lip based on a three dimensional histological reconstruction. Plast Reconstr Surg. 1984;73:875–9.
78. Schendel SA, Pearl RM, De'Armond SJ. Pathophysiology of cleft lip muscle. Plast Reconstr Surg. 1989;83(5):777–84.
79. de Chalain T, Zuker R, Acklery C. Histologic, histochemical and ultrastructural analysis of soft tissues from cleft and normal lips. Plast Reconstr Surg. 2001;108(3):605–11.
80. Schendel SA, Cholon A, Delaire J. Histochemical analysis of cleft palate muscle. Plast Reconstr Surg. 1994;94:919–23.
81. Lindman A, Paulin G, Stal PS. Morphological characterization of levator veli palatini muscle in children born with cleft palate. Cleft Palate Craniofac J. 2001;38:438–48.
82. Doménech-Ratto G. Developmental peripheral innervation of the palate muscles. Acta Anat. 1977;97:4–14.
83. Cohen SR, Chen L, Trotman CA, Burdi AR. Soft palate myogenesis: a developmental field paradigm. Cleft Palate Craniofac J. 1993;30:441–6.
84. Cohen SR, Chen LL, Burdi AR, Trotman C-A. Patterns of abnormal myogenesis in human cleft palates. Cleft Palate Craniofac J. 1994;31(5):345–50.
85. Stenstrom SJ, Thilander BL. Cleft lip nasal deformity in absence of cleft lip. Plast Reconstr Surg. 1965;35:160–6.
86. Tulenko JF. Cleft lip nasal deformity in the absence of cleft lip: a case report. Plast Reconstr Surg. 1968;41(1):35–7.
87. Huffman WC, Lierle DM. Studies on the pathologic anatomy of the unilateral hare-lip nose. Plast Reconstr Surg. 1949;4:225–34.
88. Carstens MH. Correction of the unilateral deft lip nasal deformity using the sliding sulcus procedure. J Craniofac Surg. 1999;10:346–64.
89. Dixon D, Newton I. Minimal forms of the cleft syndrome demonstrated by stereophotogrammetric surveys of the face. Br Dent J. 1972;132:183.
90. Ranta R. A review of tooth formation in children with cleft lip/palate. J Orthod Dentofac Orthop. 1986;90:11.
91. Stenstrom SJ, Oberg TRH. The nasal deformity in unilateral cleft lip. Plast Reconstr Surg. 1961;28:295–305.
92. Horswell JBB, Pospisil OA. Nasal symmetry after primary cleft lip repair: comparison between Delaire cheilorhinoplasty and modified rotation-advancement. J Oral Maxillofac Surg. 1995;53:1025.
93. Yuzuriha S, Mulliken JB. Minor form, microform, and minimicroform cleft lip: anatomical features, operative techniques, and revisions. Plast Reconstr Surg. 2008;122(5):1489–93.
94. Demas P, Sotereanos GC. Closure of alveolar cleft with corticocancellous block grafts and bone marrow: a retrospective study. J Oral Maxillofac Surg. 1988;46(8):682–7.
95. Le Mesurier AB. Harelips and their treatment. Baltimore: Williams & Wilkins; 1962. p. 80–100.
96. Lehman JA, Artz JS. The minimal cleft lip. Plast Reconstr Surg. 1976;58:306.
97. Millard DR. Cleft craft. Vol. I: the unilateral deformity. Boston: Little Brown; 1976a. p. 229–377.
98. Millard DR. Cleft craft. Vol I. The unilateral deformity. Boston: Little Brown; 1976b. p. 456–7.
99. Thaller SR, Lee T. Microform cleft lip associated with a complete cleft palate. Cleft Palate J. 1995;32:247.
100. Cho BC. New technique for correction of the microform cleft lip using vertical interdigitatation of the orbicularis oris muscle through the intraoral incision. Plast Reconstr Surg. 2004;114:1032–41.
101. Mcindoe A, Rees TD. Synchronous repair of secondary deformities in cleft lip and nose. Plast Reconstr Surg. 1959;24:150–62.
102. Matsuo K, Hirose T. Secondary correction of the unilateral cleft lip nose using a conchal cartilage graft. Plast Reconstr Surg. 1990;86:991–5.
103. Kimes KK, Siegel MI, Mooney MP, et al. Relative contributions of the nasal septum and airways to total nasal capsule volume in normal and cleft lip and palate fetal specimens. Cleft Palate J. 1988;25:282–7.
104. Noordhoff MS, Chen YR, Chen KT, et al. The surgical technique for the complete unilateral cleft lip-nasal deformity. Oper Tech Plast Reconstr Surg. 1995;2:167–74.
105. Noordhoff MS. The surgical technique for the unilateral cleft lip-nasal deformity. Taipei: Noordhoff Craniofacial Foundation; 1997.
106. Guerrero-Santos J, Ramirez M, Castenda A, et al. Crossed denuded flap as a complement to the Millard technique in the correction of the cleft lip. Plast Reconstr Surg. 1971;48:506–8.
107. Depew MJ, Lufkin T, Rubenstein JL. Specification of jaw subdivisions by DLX genes. Science. 2002;298(5592):381–5. https://doi.org/10.1126/science.1075703.
108. Depew MJ, Simpson CA, Morasso M, Rubenstein JL. Reassessing the DLX code: the genetic regulation of branchial arch skeletal pattern and development. J Anat. 2005;207(5):501–61.
109. Amin N, Ohasi Y, Chiba J, Yoshida S, Takano Y. Alterations in vascular pattern of the developing palate in normal and spontaneous cleft palate mouse embryos. Cleft Palate Craniofac J. 1994;31(5):332–44.
110. Wang RN, Green J, Shi LL, et al. Bone morphogenetic protein (BMP) signaling in development and human diseases. Genes Dis. 2014;1:87–105.
111. Zhang Z, Son Y, Zhao X, Zhang X, Fermin C, Chen Y-P. Rescue of cleft palate in *Msx1*-deficient mice by transgenic BMP4 reveals a network of BMP and SHH signaling in the regulation of mammalian palatogenesis. Development. 2002;129:4135–46.
112. Friedrichs M, Wirsdöerfer F, Fiche SB, Scneier S, Wuelling M, Vortkamp A. BMP signaling balances proliferation and differentiation of muscle satellite cell descendents. Cell Biol. 2011;12:26–43.
113. Shore EM, Xu M, LeMerer M, et al. A recurrent mutation in the BMP type 1 receptor ACVR1 causes inherited and sporadic fibrodysplasia ossificans progressiva. Nat Genet. 2006;38:525–7.
114. Hicks M, Pyle A. The path from pluripotency to skeletal muscle: developmental myogenesis guides the way. Cell Stem Cell. 2015;17(3):255–7.
115. Dyer LA, Pio X, Patterson C. Role of BMPs in endothelial cell function and dysfunction. Trends Endocrinol Metab. 2014;25(9):472–80.
116. Franco C, Gerdardt H. Blood flow boosts BMP signaling to keep vessels in shape. J Cell Biol. 2016;214(7):793.

Further Reading

Bagatain M, Khosh M, Nishoka G, Larrabbe WF Jr. Isolated nasalis reconstruction in secondary unilateral cleft lip nasal reconstruction. Laryngoscope. 1999;109:320–3.

Boddy K, Dawes GS. Fetal breathing. Br Med Bull. 1975;31(1):3–7. https://doi.org/10.1093/oxfordjournals.bmb.a071237.

Boehn A. Dental anomalies in harelip and cleft palate. Acta Odontol Scand. 1963;21(Suppl 38):1–109.

Brown RF. A reappraisal of the cleft-up nose with the report of a case. Br J Plast Surg. 1964;17:168.

Carlson B. Human embryology and developmental biology. St. Louis: CV Mosby; 1999.

Chang J, Pourquíe O. Making muscle: skeletal myogenesis in vivo and in vitro. Development. 2017;144:2104–22.

Chiego D, editor. Avery's essentials of oral histology and embryology: a clinical approach. 4th ed. St. Louis: CV Mosby; 2013.

Cohen M, Figueroa AA, Aduus H. The role of gingival mucoperiosteal flaps in the repair of alveolar clefts. Plast Reconstr Surg. 1989;83(5):812–9.

Conen PE, Erkman B, Metaxotou C. The "D" syndrome: report of four trisomic and one D/D translocation. Am J Dis Child. 1966;111(3):236–47.

Delaire J. The potential role of facial muscles in monitoring maxillary growth and morphogenesis. In: Carlson OS, McNamara Jr JA, editors. Monograph no. 38. Cranial growth series. Ann Arbor: University of Michigan Press; 1978.

Delaire J. Theoretical principles and technique of functional closure of the lip and nasal aperture. J Maxillofac Surg. 1978;6:109.

Delaire J. L'anatomie et la physiologie muscles nasolabiaux chez sujet normal pere d'une fente labiomaxillaire. Acta Orthod. 1980;8:269–85.

Delaire J, Feve R, Chateau JP, et al. Anatomie et physiologie des muscles et du fente de la lievre superieure. Rev Stomatol (Paris). 1977;78:93.

Demas P, Sotereanos GC. Closure of alveolar clefts with corticocancellous block grafts and marrow: a retrospective study. J Oral Maxillofac Surg. 1988;46(6):483–5.

Desrosiers AE, Kawamoto HK, Katchikkan HV, et al. Microform cleft lip repair with intraoral muscle interdigitation. Ann Plast Surg. 2009;62:640.

Fará M, Chlumska A, Hrivnakova J. Musculus orbiculari oris in incomplete hare-lip. Acta Chir Plast. 1965;7:125.

Fischer DM. Unilateral cleft lip repair: An anatomical subunit approximation technique. Plast Reconstr Surg 2005; 116(1): 61–71.

Fox HE. Fetal breathing movements and ultrasound. Am J Dis Child. 1976;130(2):127–9.

Frank N, Kho AT, Schalton T, Gussoni E, et al. Regulation of myogeneic progenitor proliferation in human fetal skeletal muscle by BMP4 and its antagonist gremlin. J Cell Biol. 2006;175(1):99–110.

Gifford GH, Swanson L, MacCollum OW. Congenital absence of the nose and anterior nasopharynx. Plast Reconstr Surg. 1972;50:5.

His W. Anatomie mensdilicher Embryonen. Leipzig-German y: Vogel; 1885.

His W. Die entwerklung der mensdllichen und thierischer Physigonomen. Arch Anat Physiol (Lejpzig). 1892;27:384–434.

His W. Beobachtungen zur Geshichte der Naden und Gaumenbildwlg beim menschlischen Embryonen. Abh Gao Naturw. 1901;27:349–89.

Joos U, Friedburg H. Darstellung des Verlaufs der mirnisclen Muskulatur in der Kemspintomographie. Fortschaft Kiefer Gesichtschir. 1987;32:125.

Kim EK, Khang SK, Lee TL, Kim TG. Clinical features of the microform cleft lip and the ultrastructural characteristics of the orbicularis oris muscle. Cleft Palate Craniofac J. 2010;47(3):297–302.

Kim MC, Choi DH, Bae SG, Cho BC. Correction of minor-form and microform lip using modified muscle overlapping with a minimal skin excision. Arch Plast Surg. 2017;44(3):210–6.

Kraus BS, Jordan RE, Neptune CM. Dental abnormalities associated with cleft lip and/or palate. Cleft Palate J. 1966a;3:439.

Kraus B, Kitamura H, Latham RA. Atlas of developmental anatomy of the face. New York: Harper & Row; 1966b.

Lasseri D, Viacava P, Pollina LE. Dystrophic-like alterations characterize orbicularis oris and palatopharyngeal muscles in patients affected by cleft lip and palate. Cleft Palate Craniofac J. 2008;45(6):587–91.

Losee J, Kirschner R, editors. Comprehensive cleft care. 2nd ed. Stuttgart: Thieme; 2015.

Markus AF, Delaire J. Functional primary closure of cleft lip. Br J Oral Maxillofac Surg. 1978;31:281.

Markus AF, Delaire J, Smith WP. Facial balance in cleft lip and palate. II. Cleft lip and palate and secondary deformities. Br J Oral Maxillofac Surg. 1990;20:296.

Markus AF, Delaire J, Smith WP. Facial balance in cleft lip and palate I. Normal development and cleft palate. Br J Oral Maxillofac Surg. 1992;30:290–5.

Meyer R. Total external and internal reconstruction in arhinia. Plast Reconstr Surg. 1997;99:534–42.

Millard DR, Onizuka T, Hoaka Y, Ayoyama R, et al. Operations for microforms of cleft lip. Cleft Palate Craniofac J. 1991;28:300.

Moore KL, Persaud TV. The developing human. 6th ed. Philadelphia: WB Saunders; 1998. p. 384–9.

Nammoun JD, Siegel MI, Kimes K, et al. Premaxillary development in normal and cleft lip and palate human fetuses using three-dimensional computer reconstruction. Cleft Palate Craniofac J. 1991;28:49–53.

Nishimura H, Okatmoto N. Sequential atlas of human congenital malformations: observations of embryos, fetuses, and newborns. Baltimore: University Park Press; 1976.

Onizuka T. Operative plastic surgery. Tokyo: Takodo; 1982.

Onizuka T, Hosaka Y, Ayoyama R, Takahama H, Jinnai T, Ursi Y. Operations for microforms of cleft lip. Cleft Palate Cranifac J. 1991;28:293.

Park CG. Temporal fascia graft for the correction of the congenital scar of the lip. J Korean Soc Plast Reconstr Surg. 1986;13:67–70.

Patten BM. Olfactory complex. In: Patten BM, editor. Embryology. New York: McGraw-Hill; 1968. p. 289–90.

Sachidanandan C, Dhwan J. Skeletal muscle progenitor cells in development and regeneration. Proc Indian Natl Sci Acad. 2003;69(5):719–40.

Sanvenero-Roselli G. Developmental pathology of the face and the dysraphic syndrome—an essay of interpretation based on experimentally produced congenital defects. Plast Reconstr Surg. 1946;11(1):36–8.

Sperber G. Craniofacial embryology. London: Churchill; 1984.

Sperber G. Embryology of the head and neck. In: Potter's pathology of the fetus and want. 3rd ed. St. Louis: CV Mosby; 1999.

Stantzou A, Schirwis E, Swiss S, et al. BMP signalling regulates satellite cell- dependent post-natal muscle growth. Development 2017; 144(5): 2737–2747

Streeter GL. Developmental horizons in human embryos: description of age group XI, 13 to 20 somites and age group XII, 21 to 29 somites. Contrib Embryol Carnegie Inst. 1942;30:211.

Streeter GL. Developmental horizons in human embryos: description of age group XIII, embryos of 4 or 5 millimeters long, and age group XIV, period of identification of the lens vesicle. Contrib Embryol Carnegie Inst. 1945;31:27.

Suzuki S, Marazita ML, Murray JC, et al. Mutations in BMP4 are associated with subepithelial, microform, and overt cleft lip. Am J Hum Genet. 2009;84(3):406–11. https://doi.org/10.1016/j.ajhg.2009.02.002.

Taylor GI, Palmer JH. The vascular territories (angiosomes) of the body: experimental study and clinical applications. Br J Plast Surg. 1987;46:113.

Taylor GI, Gianoutsos MP, Morris SF. The neurovascular territories of the skin and muscles: anatomic study and clinical implications. Plast Reconstr Surg. 1994;94:1–35.

Tolarova M. Microforms of cleft lip and/or palate. Acta Chir Plast. 1969;1(1):96.

Tolarova H, Havlova Z, Ruzickova J. The distribution of characters considered to be microforms of cleft lip and/or palate in a population of normal 18–21 year-old subjects. Acta Chir Plast. 1967;9:1.

Vissiaronov VA. Correction of a deformity of the upper lip after primary plastic surgery of the bilateral cleft lip. Vestn Khir II Grek. 1988;141(8):86–8.

Vissarionov VA, Tuznaova LI, Blokhina SI. Characteristics of plastic surgery of the central part of the upper lip. Vestn Khir Im II Grek. 1990;145(7):191–2.

Wang MKH. A modified LeMesurier–Tennison technique in unilateral cleft lip repair. Plast Reconstr Surg. 1960;26:190.

Wang K-Y, Diewert VM. A morphometric analysis of craniofacial growth in cleft lip and non-cleft mouse embryos. J Craniofac Genet Dev Biol. 1992;12:141–54.

Wang H, Noulez F, Edom-Vovard F, Le Grand F, Duprez D. Bmp signaling at the tips of skeletal muscle regulates the number of fetal muscle progenitors and satellite cells during development. Dev Cell. 2010;18:643–54.

Weinberg A, Neuman A, Benmeir P, et al. A rare case of arhinia with severe airway obstruction: case report and review of the literature. Plast Reconstr Surg. 1993;91:146–9.

DFR Cheilorhinoplasty: The Role of Developmental Field Reassignment in the Management of Facial Asymmetry and the Airway in the Complete Cleft Deformity

19

Michael H. Carstens

Introductory Remarks

All plastic surgeons involved in the care of cleft-affected children and adults experience first hand both the rewards and the limitations of our craft. This work requires a certain mindset characterized by four qualities: an intense curiosity about cleft biology, a relentless pursuit of good technique, an unflinching assessment of results (both good and bad), and humility in the constant search for better ideas. Of all these characteristics, humility is perhaps the most important. It leads us to constantly search out and appreciate the work of other surgeons. Superior concepts or protocols should be embraced, not rejected.

In this Chap. I shall discuss how my thinking about developmental field reassignment (DFR) for the management of cleft lip and cleft lip nose has morphed over the years. The principle sources of change stem from long-term observations of the outcomes of surgical interventions (my own and those of others) and intellectual contributions of colleagues that made so much biologic sense to me that they demanded to be incorporated. I intend this as a narrative, with personal observations and conjectures. There will be some historical materials as well. Surgical evolution does not take place in a straight line; I have therefore included blind alleys and misconceptions as well, because, in their resolution, value information was gained. This I hope to pass on for your consideration.

We shall cover the following topics

- Evolution of the DFR model
- Embryologic strategy for cleft repair: problems and solutions
- Developmental anatomy of the central lip–nose complex
- Functional lip repair: the 5 As
- DFR: toward a rational protocol
- Technical details of DFR cleft lip–nose repair
- Addendum: Sotereanos alveolar cleft procedure

M. H. Carstens (✉)
Wake Forest Institute of Regenerative Medicine, Wake Forest University, Winston-Salem, NC, USA
e-mail: mcarsten@wakehealth.edu

Developmental Field Reassignment Evolution of a Concept

As we launch into this, our final discussion of cleft management, I would like to share with you a brief explanation of how DFR came into being. What now seems a rational system, based on neuroembryology, neurovascular fields, and biochemical signals started out as a more simplistic model which evolved in fits and starts, in several iterations.

The Process Concept

Developmental field reassignment surgery is a return to basics. The anatomy we see represents a rearrangement of the original building blocks of embryonic tissues. But, unlike the jig saw puzzle on the front of cleft craft, the anatomy is not in two dimensions, rather it is a four-dimensional problem. The event may be as early as early as stage 9, the assembly of the first arch, but its consequences are played out over time until by 8 weeks all the components of the pathology established. Fetal growth merely cements this anatomy into its final form.

As previously recounted, upon returning from Nicaragua, I could not escape the conviction that something was missing in our surgical management of clefts. Why was relapse so common? Why did it always take the same form? Why the need for secondary surgery? It was as if there were processes, unleashed by the cleft event that were not addressed by initial surgery, and that these would lead to an inevitable deterioration of the results (Fig. 19.1).

It occurred to me that a process-oriented cleft repair would have to address four issues.

M. H. Carstens (ed.), *The Embryologic Basis of Craniofacial Structure*, https://doi.org/10.1007/978-3-031-15636-6_19

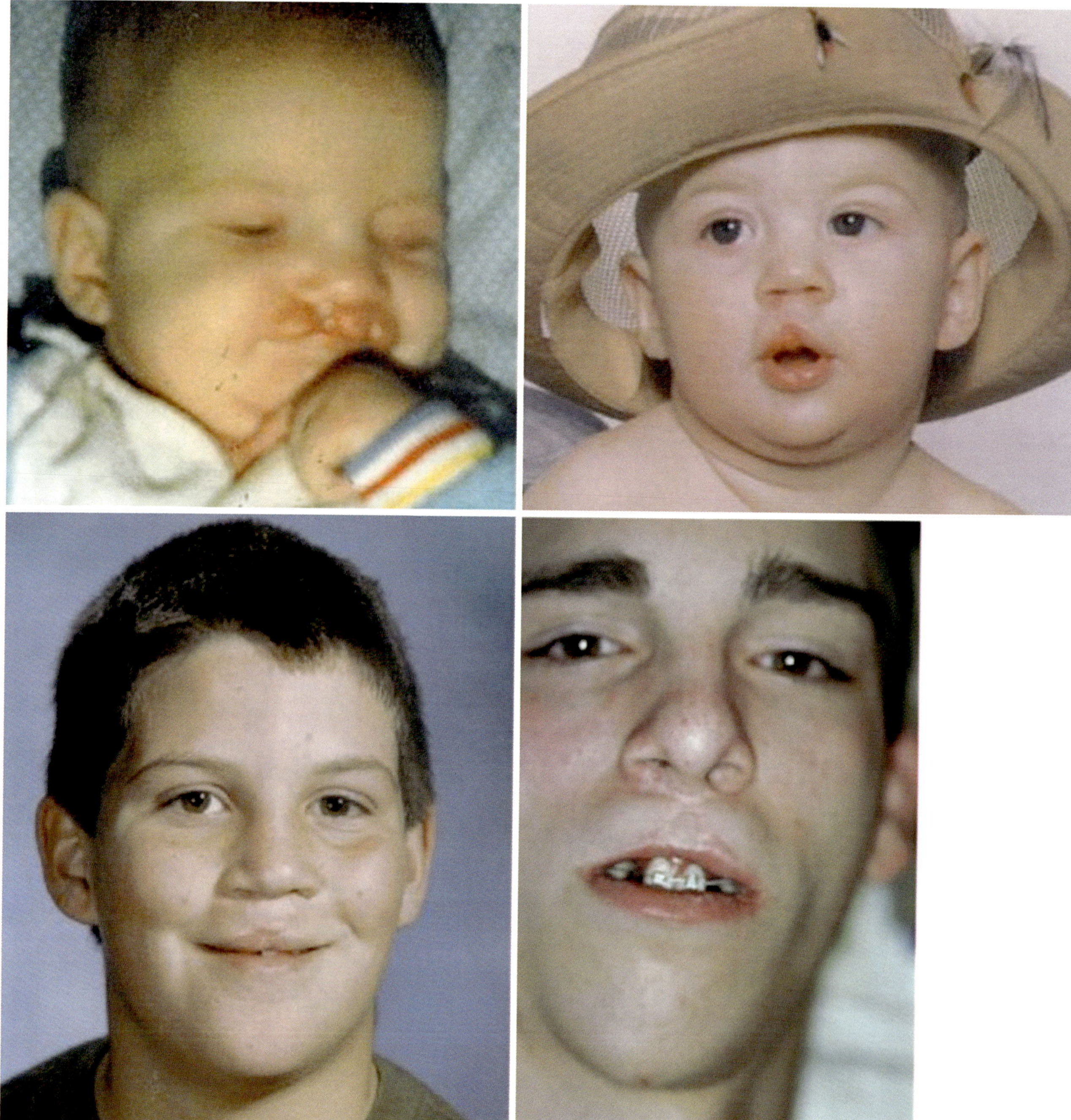

Fig. 19.1 Cleft: a biologic problem in four dimensions. Three processes, a deficiency state leads to a division or maldistribution of tissue. This leads to a displacement of otherwise normal structures. Over time these structures undergo distortion. This case of bilateral cleft, incomplete, right more than left (Top left) demonstrates an acceptable initial result (Top right), albeit with asymmetry of the nose, the right ala being displaced downward (Top right). By late childhood/mixed dentition (Bottom left) the nose is flattening further. (Bottom right): In the late teens (lower right) orbicularis asymmetry is obvious. The nasal airway is compromised. He is a mouth breather. Nose creases over the ala representing nasalis tethering. We shall see him again the end of this photo essay. [Courtesy of Michael Carstens, MD]

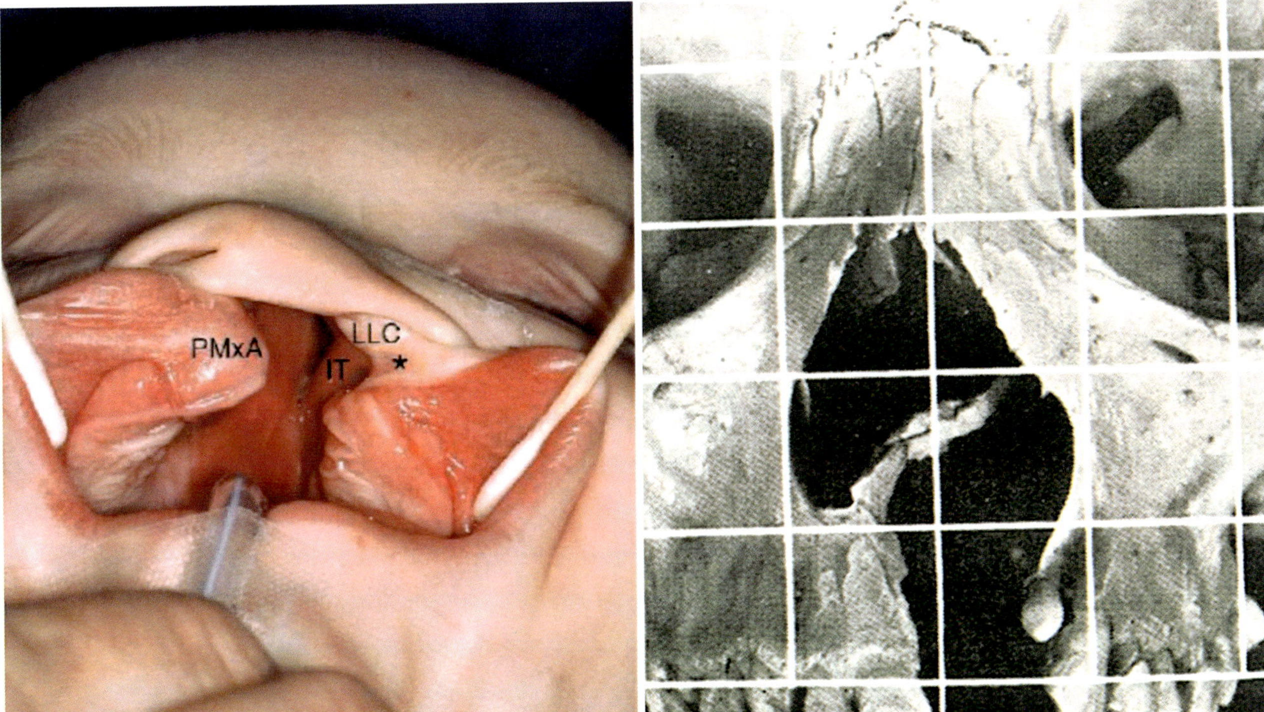

Fig. 19.2 Pathologic anatomy of unilateral cleft. (Left): soft tissues; (Right): skeletal anatomy. Medial prolabium "acquires" displaced tissue from premaxilla

- Philtral prolabium
- Non-philtral prolabium

Medial crus displaced downward

- "Shoulder" of columella flat

Nasal tip distortion

- Foreshortening of nasal floor

Division

- External rotation of premaxilla
- Stem cells distributed asymmetrically in space

Lateral

Pre-turbinate deficiency site > displacement of lateral crus

- Nasal tip distortion
- Inward rotation of nostril sill

Deficient/absent frontal process > misinsertion of nasalis

- "band" in lateral wall
- Dilator becomes constrictor

Division of force vectors

[Courtesy of Michael Carstens, MD]

Deficiency of tissue at some specified site (I was not sure, but it seemed to be in the nasal floor). Somehow this was related to a d*ivision* or cleft ranging in severity from microform to complete. The false insertion of muscles and the consequent imbalance [1] would lead to *displacement*, as in the alar base, columella, and premaxilla. Finally, over time, the sum of these processes would worsen, causing *distortion* of otherwise normal tissues. I set out to see how these processes could be undone so that relapse could be avoided (Figs. 19.2 and 19.3).

While a resident and fellow at the University of Pittsburgh, I saw the ability of the wide subperiosteal approach designed by Sotereanos to close alveolar clefts of any dimension. At the same time, it gave an *aesthetic correction* of the midface, a centralization of the lip and nose which remained *stable over time*. The sliding sulcus procedure was merely an application of the concepts of Soteranos and Delaire to primary cleft repair. As initially reported, the lip dissection remained rotation-advancement. Although primary gingivoperiosteoplasty (GPP) using a gingival incision proved to be impractical the biologic lessons stuck with me (Figs. 19.4, 19.5, 19.6, 19.7 and 19.8).

- Avoid a buccal sulcus incision to protect osteogenic periosteal cells from injury
- Transfer those cells into correct position to make membranous bone where desired during the period of ensuing rapid facial growth
- Freeing the alar base
- Correction of the septum

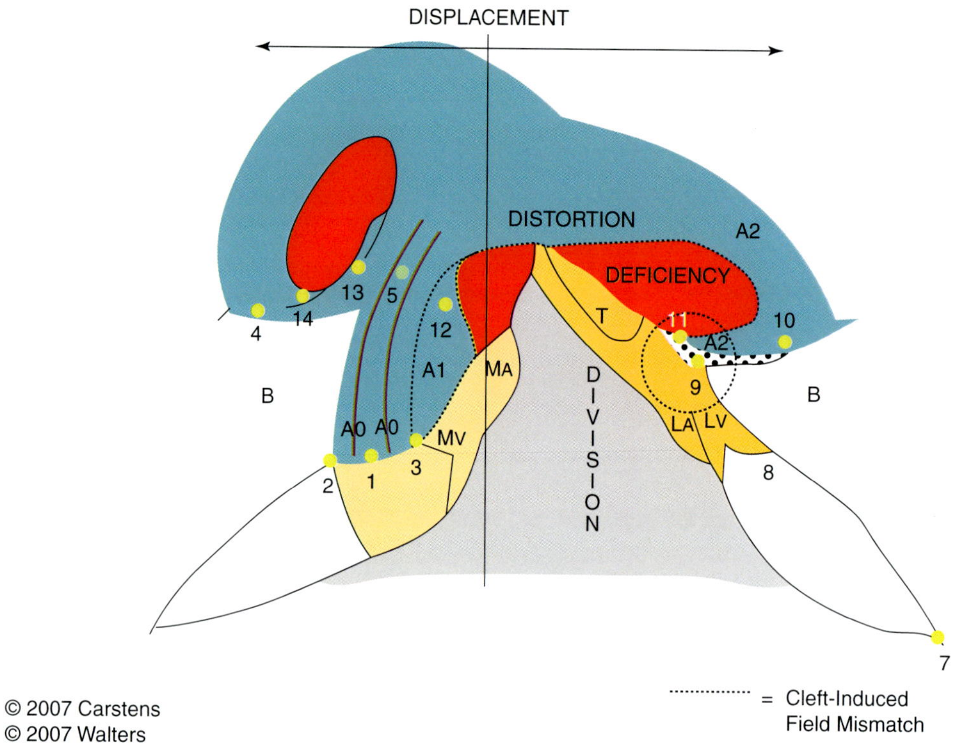

Fig. 19.3 4 Ds of cleft repair (developmental model for cleft prior to injection studies). **D**eficiency > **D**ivision > **D**isplacement > **D**istortion. Key: Blue = Frontonasal skin with r1 neural crest and StV1 ethmoids. Red = vestibular lining with r1 neural crest and StV1 ethmoids. Tan = r2 medial. This older drawing does not appreciate the invasion of first and second arch elements into the lateral nasal process (thereby changing its color). [Reprinted from Carstens MH. Functional matrix cleft repair: Principles and techniques. Clin Plast Surg 2004; 31: 159–189. With permission from Elsevier.]

The Functional Matrix Concept: Enter Neuroembryology

In 1999–2001 five events came together to jolt my thinking into a higher level. From 1997 to 1999 I prepared a series of papers regarding applications of process theory and subperiosteal repair for unilateral, bilateral, and cleft nose deformities. This process required extensive review of the literature, including turning all three volumes of cleft craft into an underlined and annotated shambles. So many papers seemed to be variations on design with little embryology to back them up. But I was impressed by the *functional matrix* concept of Moss and Opitz, the idea, born out of orthodontics and genetics, that tissues were organized into blocks with individual behaviors. I was not aware at the time of Taylor's work with angiosomes. To understand this better I devoured standard embryology texts but found them utterly lacking regarding craniofacial development. In particular, I wanted to understand better the blood supply to the central face, nose, columella, and lip. Somehow there must be functional matrices involved in the pathology of clefts.

About this time, I had the good fortune to meet and work with fetal pathologist, Dr. Geoffrey Machin at Kaiser Hospital in Oakland. Geoff and I procured fetal specimens, some with normal facies, some with variants of holoprosencephaly and a right unilateral cleft lip and alveolus with an intact secondary palate. We injected them with results published previously [2] and discussed in Chap. 18 (Figs. 19.9 and 19.10). In so doing we discovered the anatomic boundaries of what I thought (falsely) to be the internal carotid/ophthalmic supply to the midline. This first developmental field map was of neurovascular origin; although it lacked a developmental rationale it was immediately practical, as is suggested a different way to dissect the lip–nose complex, initially termed **Functional Matrix Repair**. The functional matrix concept gave much better facial symmetry and nasal projection but did not give a better airway (Figs. 19.11 and 19.12).

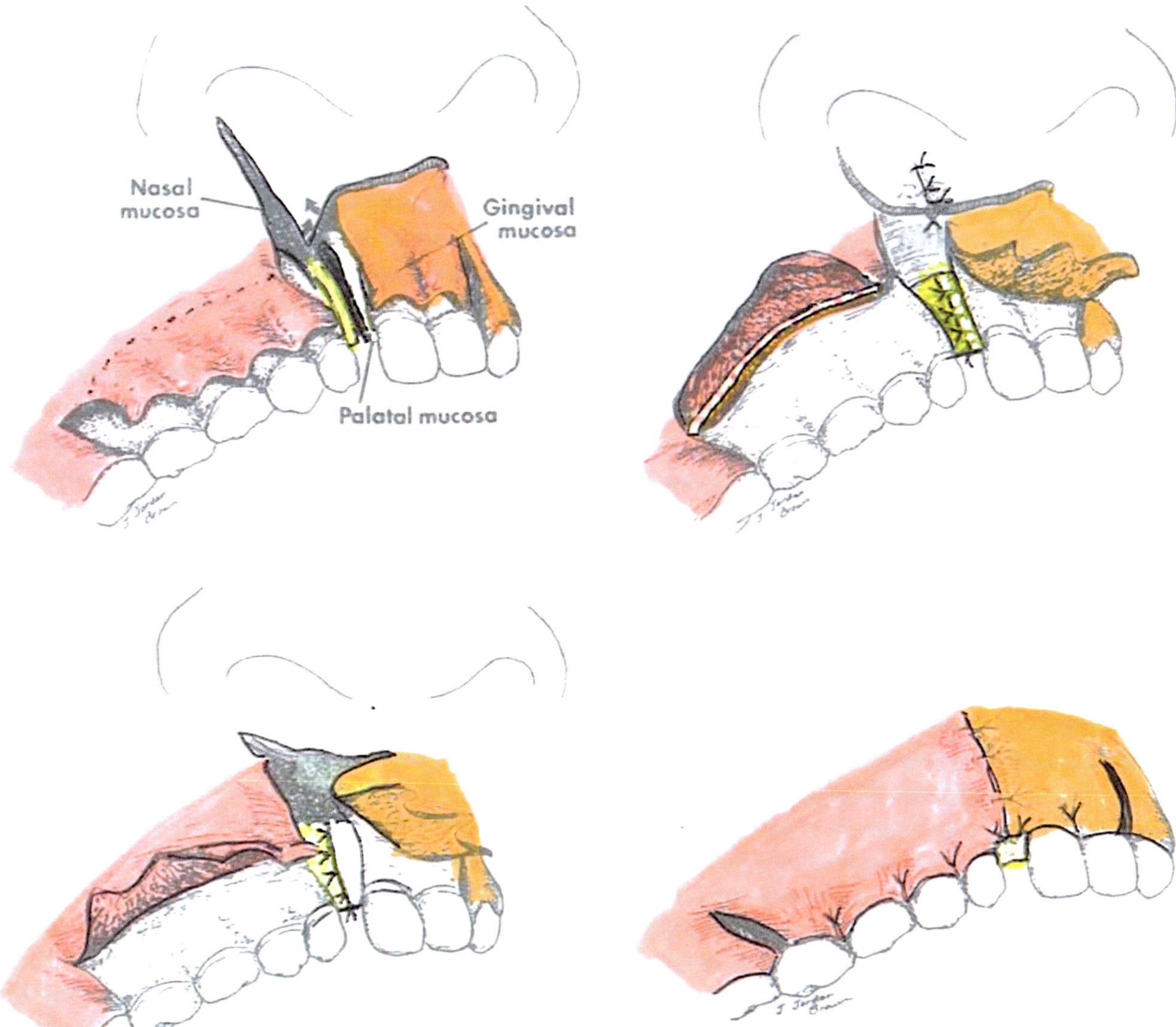

Fig. 19.4 Sotereanos muco-gingivoperiosteoplasty (GPP). Developmental field reassignment originated from experience in alveolar cleft bone grafting and in observations in multiple secondary cases, all showing the same pattern of relapse. Sliding sulcus flap (S) transfers two dental units of tissue per side. Bilateral flaps can close fistulas of almost any size. The S flaps are accompanied by a wide periosteal dissection (Delaire). This results in a tension-free centralization of the entire midface soft tissue envelope with normalized projection on the cleft side. In bilateral cases the effect is striking. What also occurs is a reassignment of bioactive stem cell populations from the periphery to the center. On the cleft side, left to their own devices, stem cells will correctly synthesize bone in the wrong position in space. In 1997 stem cells per se were not on my radar screen but my thinking underwent in shock treatment in 2000 when Martin Chin and I translated Boyne's work with rhBMP2 in orthopedics to morphogen-driven craniofacial bone reconstruction, **in situ osteogenesis** (ISO). See Chap. 20. (Top left): Full-thickness incisions in oral side mucoperiosteum of cleft-side maxillary (pink) and non-cleft side premaxillary. (Bottom left): nasal side mucoperiosteal flaps (yellow on both sides) reflected backward into the cleft. (Top right): counter-incisions (transverse and vertical) made in maxillary mucoperiosteal advancement. (Bottom right): Maxillary flap translocated mesially two dental units. [Reprinted from Cohen M, Figueroa A, Aduss H. The role of gingiva mucoperiosteal flaps in the repair of alveolar clefts. Plast Reconstr Surg 1989; 83(5): 812–816. With permission from Wolters Kluwer Health, Inc.]

The discovery of the anatomic basis of developmental fields I attribute directly to Dr. Machin. In the process of our dissections Geoff told me about a new theory that he found intriguing: the existence of homeobox genes and how that related to the body axis. At Machin's prompting, I discovered the 1989 work on rhombomeres and cranial nerves by Lumsden and Keynes [4]. This had instant application to the Tessier system classification system of craniofacial clefts. Neuroanatomy had always been one of my obsessions. When I first learned about the Tessier system in 1987, it was apparent these rare clefts had some relationship to the sensory distribution of the trigeminal nerve (Figs. 19.13 and 19.14). What's more, rhombomeric compartments seemed to give a neuromeric explanation for the pharyngeal arches which was

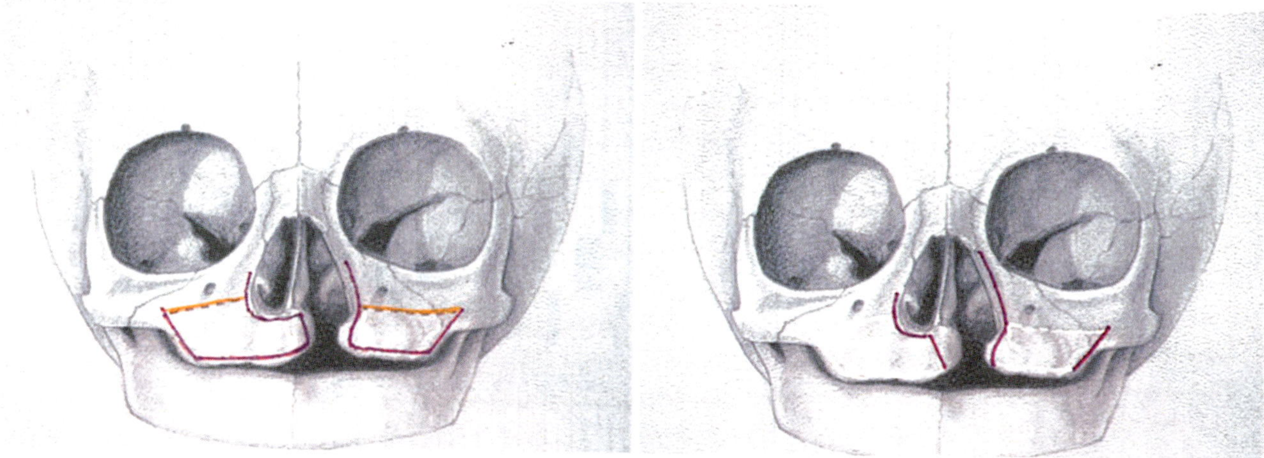

Fig. 19.5 The sliding sulcus procedure. Initial attempt to address facial asymmetry, both as perceived, and with future growth, was based on subperiosteal centralization of the soft tissue envelope. This had the effect of reassigning the membranous bone-forming "factory" into correct position such that future osteogenesis would take place centrically and not in a displaced position. If the car factory belongs in Detroit, don't make your Fords in Flatbush. [Reprinted from Carstens MH. Functional matrix cleft repair: Principles and techniques. Clin Plast Surg 2004; 31: 159–189. With permission from Elsevier.]

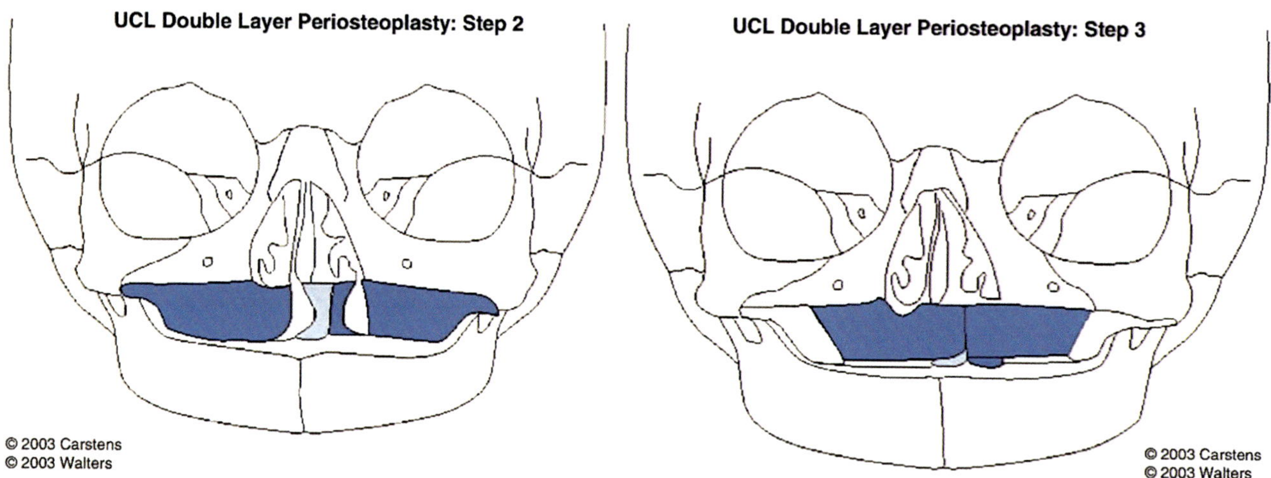

Fig. 19.6 Subperiosteal dissection with sliding sulcus flaps. (Left): Original construct of sliding sulcus repair used a gingivoperiosteoplasty (GPP) with incision along the margin of the alveolus as in the sotereanos alveolar cleft procedure (shown at the end of this essay). Bone production was not adequate and did not prevent lateral segment from collapsing but forward growth was stimulated. (Right): The same mobilization can be obtained without the gingival incision, reserving the S flaps for later use to close the alveolar cleft at age 4 (ideally) or later. [Reprinted from Carstens MH. Functional matrix cleft repair: Principles and techniques. Clin Plast Surg 2004; 31: 159–189. With permission from Elsevier.]

directly relevant to understanding the development of the face and the cleft condition.

Very murky, however, was the basis for development and vascularization of the upper face. Why the reliance of a supposedly internal carotid source of supply for tissues unrelated to the forebrain? Or, more profoundly, were they actually related? Perhaps it was not that "the face predicts the brain" but rather that "the brain predicts the face." In the meantime, I heard that John Rubenstein's group at UCSF was doing work with hox genes. So it was, that one foggy morning, I traipsed over to his lab on Parnassus hill. There I discovered the prosomeric system. Luis Puelles and Rubenstein had successfully linked the development of the entire embryonic neuraxis to the homeotic system [5]. I knew that a functional matrix cleft repair could be designed to make use of a neuromeric tissue map.

Functional matrix cleft repair abandoned rotation-advancement forever. The identity and anatomy of the non-philtral prolabium (NPP) was clear from injection, consistent with r2 mesenchyme and supported by a known neu-

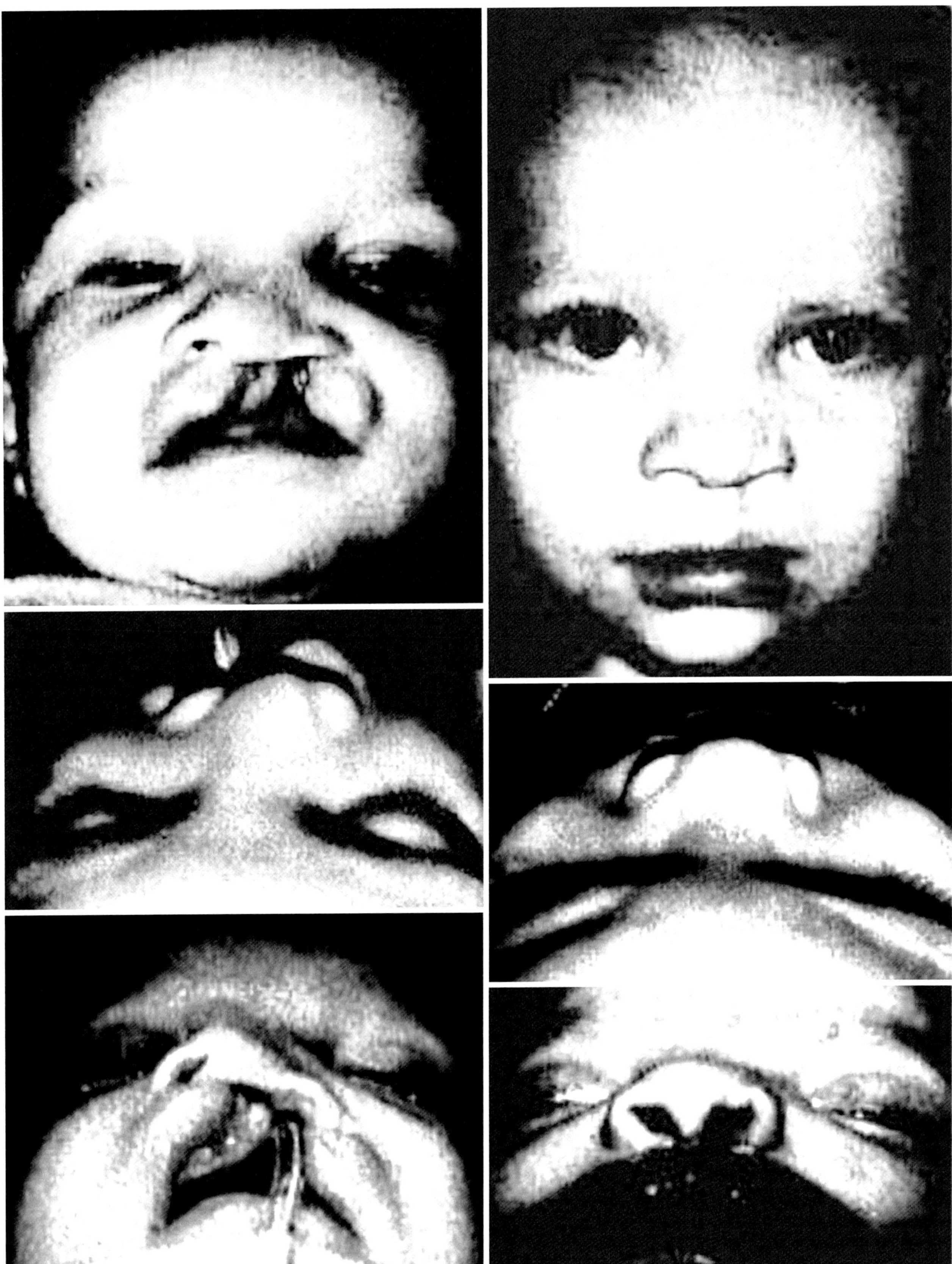

Fig. 19.7 Original case of sliding sulcus subperiosteal repair with gingivoperiosteoplasty. Midline symmetry is striking and position of the left alar base is maintained at 3 years. Unfortunately, would be many years before I recognized the importance of nasalis reassignment restoration of the airway. It also correcting the alar base, which is here asymmetrical. [Reprinted from Carstens MH. The sliding sulcus procedure: simultaneous repair of unilateral clefts of the lip and primary palate—a new technique. J Craniofac Surg 1999; 10(5): 415–429. With permission from Wolters Kluwer Health, Inc.]

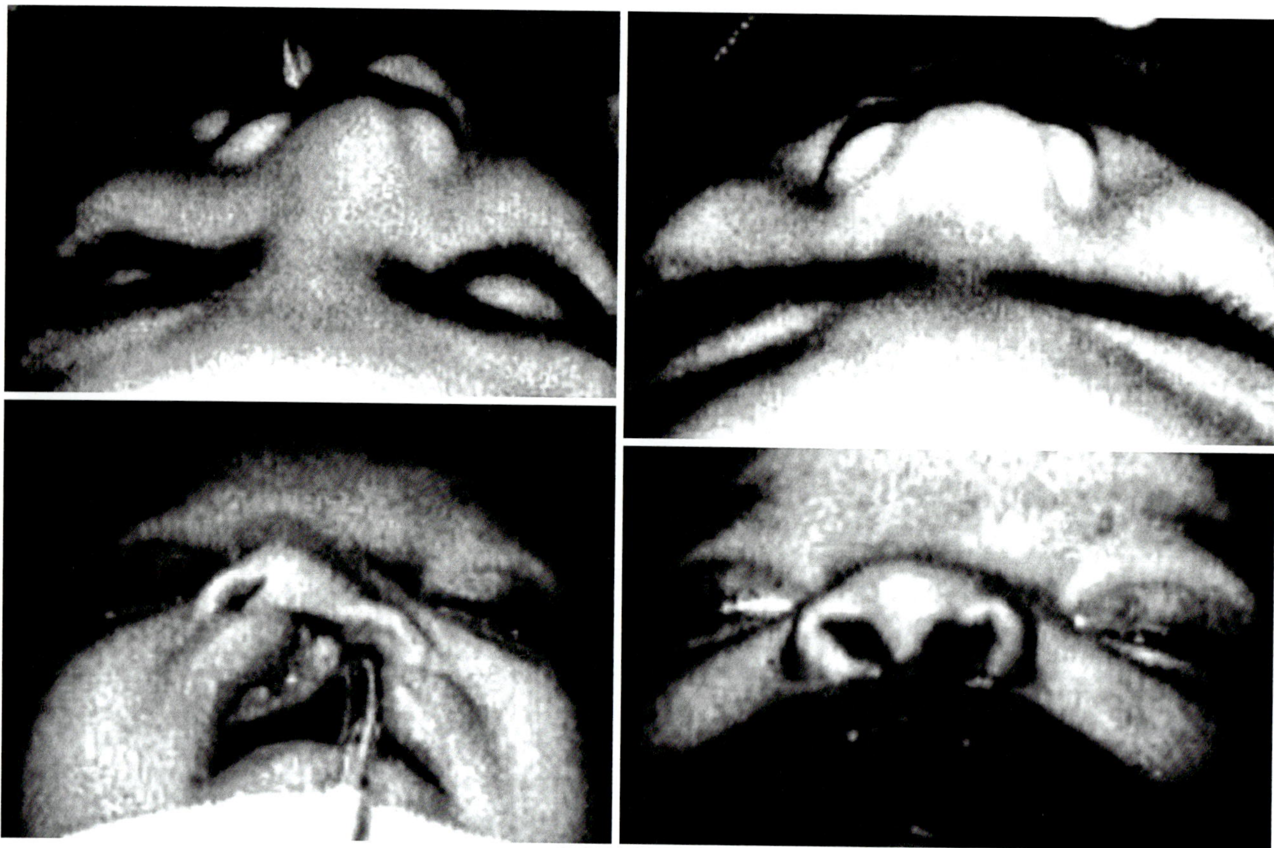

Fig. 19.8 An original subperiosteal repair with gingivomucoperiosteoplasty. The photos are very old but what is so striking in the incredible degree of centralization and the complete forward re-positioning of the left ala. [Reprinted from Carstens MH. The sliding sulcus procedure: simultaneous repair of unilateral clefts of the lip and primary palate—a new technique. J Craniofac Surg 1999; 10(5): 415–429. With permission from Wolters Kluwer Health, Inc.]

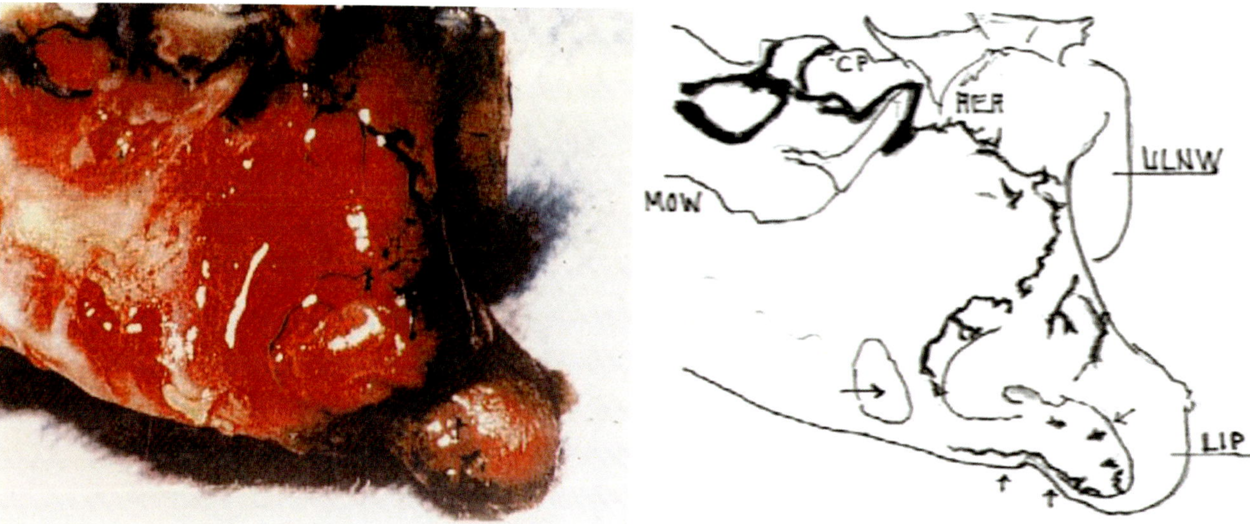

Fig. 19.9 Blood supply to the prolabium: a change in perception. Injection study showing dye from ethmoid circulation descending the columella and entering the prolabium, which is shown here in hemisection to reveal the vessels. [Courtesy of Michael Carstens, MD]

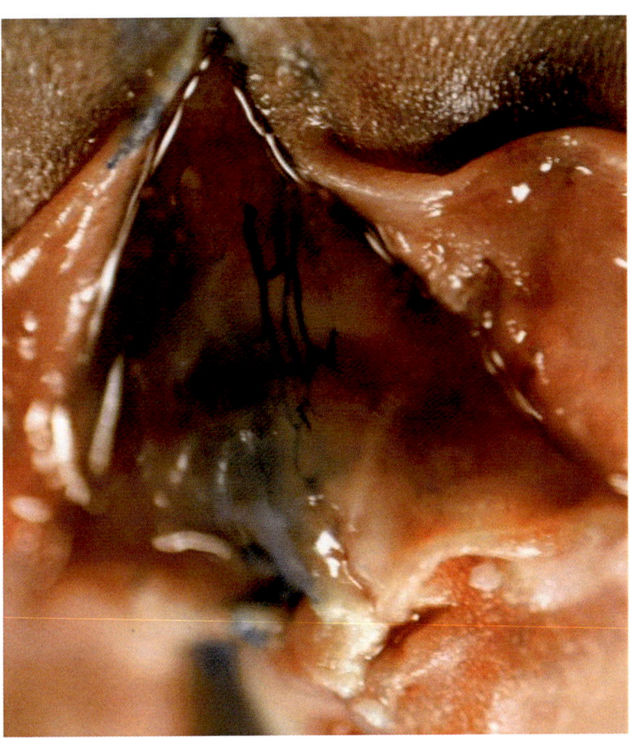

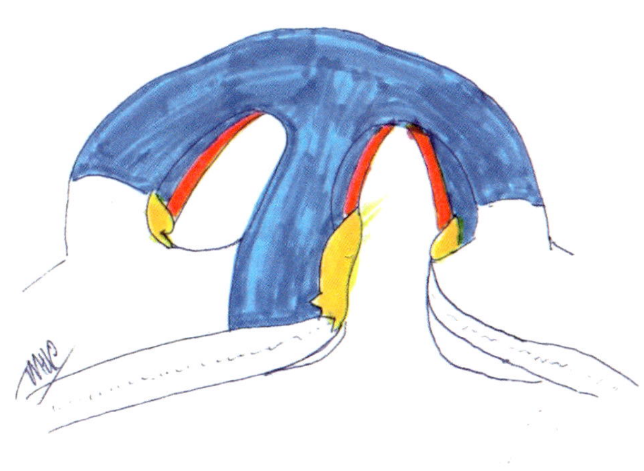

Fig. 19.10 Field separation: AP. Fetal cadaver nasal dorsum showing paired StV1 anterior ethmoid arteries descending over the nasal tip for the columella 0. Midbrain neural crest: nasal skin (hp2 neural fold epithelium), vestibular skin (hp2 placode epithelium). Hindbrain neural crest: vomer, premaxilla (r2). **Cleft prolabium** = philtral prolabium + non-philtral prolabium. [Courtesy of Michael Carstens, MD]

roangiosome, the medial nasopalatine artery. Functional matrix dissection applied equally well to both unilateral and bilateral clefts, eliminating the cognitive dissonance shared by all surgeons doing two very different procedures for the very same pathology, something that never made sense to me. The rationale and design of the functional matrix repair were published in 2000 and updated in 2004 with better understanding of the nasal tissues.

But in 2000 along came Flor, a little girl born with a severe left-sided lateral facial cleft involving structures on both sides of the midline. I sat by the bedside and made a drawing of what I saw. Suddenly, an entire map of the face appeared with upper deck structures from r2 and r4 matching up with lower deck structures of r3 and r5 neural crest. This fit the Dlx system I had learned about from Michael DePew at Rubenstein's lab. It was obvious that all pharyngeal arches had to have the same mapping system to create compartments that would become developmental fields. We shall meet Flor again in our final chapter (Chap. 20) as she posed a reconstructive challenge that lead to the first application of recombinant human bone morphogenetic protein-2 in craniofacial surgery: it was the birth of in situ osteogenesis.

Developmental Field Reassignment: The Impact of Bruce Carlson and Dorcas Padget

Let's fast forward to 2006. The functional matrix was working nicely. I was working with Martin Chin on application of rhBMP2 to the alveolar cleft and closing it up at the primary surgery. [As noted before, I have now come to the conclusion that secondary repair at age 4 (a Talmant concept), again, ideally with rhBMP2, is the best management for the dental arch.] Blood supply issues between alveolar extension palatoplasty (AEP) (Chap. 17) and the lip made a lot of sense. By this time with texts of molecular embryology by Bruce Carlson and of developmental biology Scott Gilbert incorporated into neuromeric theory things were falling into place. But I found myself lost at sea when it came to the vascular system. The arterial system was the door to understanding the developmental map of the face. But where was the key?

References are a good thing. Hinrichsen's invaluable SEM studies of facial development (which I recommend to all those interested) were just such a footnote. So too, as I

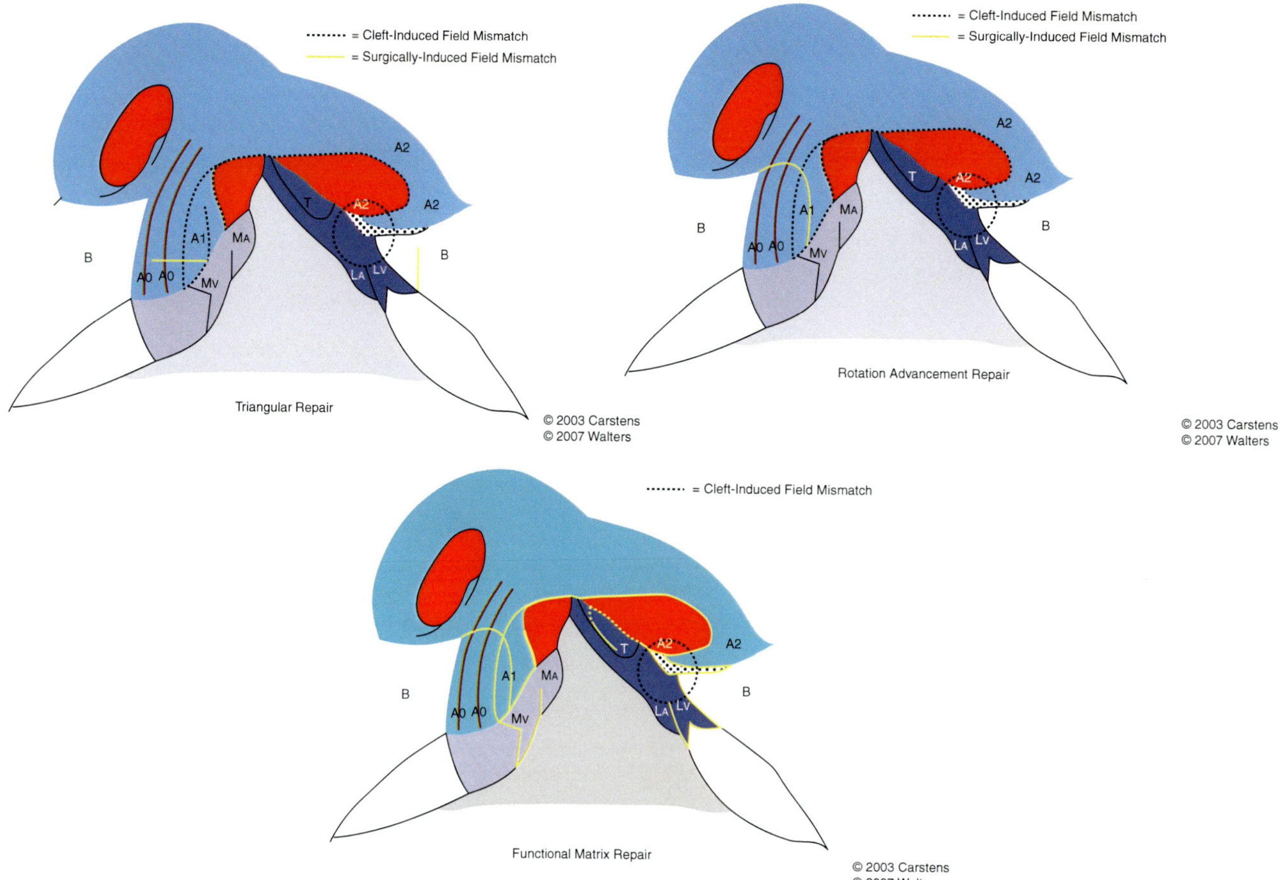

Fig. 19.11 Contemporary surgical designs violate embryonic fields. Incisions (yellow) in both triangular (Tennison-Randall, Fisher) and rotation-advancement transect the embryonic supply to the philtrum. They do not give direct access to the nose. Key: light blue = PNC skin and StV1 ethmoid mesenchyme; red = placodal vestibular skin and StV1 ethmoid; blue-gray = hindbrain mucosa and r2 medial nasopalatine axis; dark blue = hindbrain mucosa and r2 lateral nasopalatine axis [3]. [Reprinted from Carstens MH. Functional matrix cleft repair: Principles and techniques. Clin Plast Surg 2004; 31: 159–189. With permission from Elsevier.]

looked for clues to the vascular system, I found Carlson's text a reference research by a woman artist working at the Carnegie Institution on the arterial supply to the head in staged embryos. Her fascinating story, recounted elsewhere, was an inspiration and her drawing superlative. In this book I have attempted to bring them alive with color coding and simplified concepts. Nevertheless, Padget showed me the unifying concept of blood supply as multiple iterations through development with the stapedial system as the crowning innovation explaining those structures of the face that are never considered as part of the first arch yet form the interface between maxilla and brain.

Once the neurovascular map was complete, it seemed appropriate to rename the conceptual basis for cleft repair as DFR. The DFR concept applies to many craniofacial problems, both as a better means to understand their pathogenesis and promoting more rational surgical strategies. In the future others will use field mapping to explore old problems in a new way. As for the remainder of this chapter, we shall see how DFR technique was further modified, in particular, in service to a functional aspect of clefts that has previously defined my best efforts and those of many others: the nasal airway (Fig. 19.15).

And herein, at risk of being a bit informal, I'd like to include some personal details as to how these innovations have emerged. No craniofacial surgeon works in a vacuum. The contributions of long-term work by David Matthews, Jean Claude Talmant, Jean-Pierre Lumineau, and Luis Monasterio are essential components in the evolution of the developmental field model. What is remarkable is that each of us, coming from different experiences, has arrived at a very similar set of priorities and techniques. My purpose here is to bring these perspectives together into a single unitary philosophy based on the developmental field model of cleft formation. As Victor Veau put it so well "the surgery of clefts is merely experimental embryology." This chapter intends to prove him right.

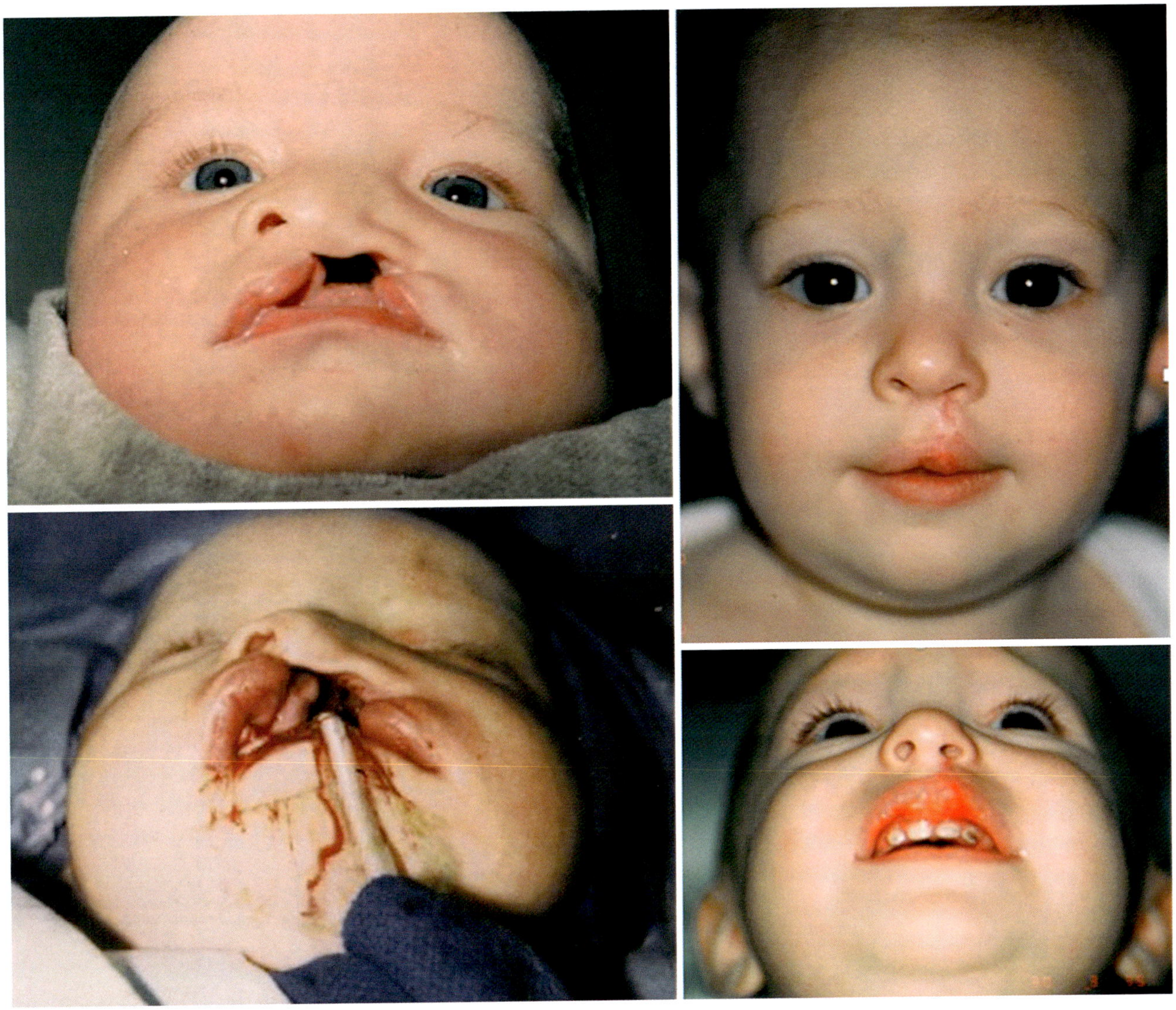

Fig. 19.12 Primary subperiosteal repair with gingivoperiosteoplasty. Symmetry is maintained at 2 years. Good growth of the ipsilateral maxilla which was recessed pre-op. The left nostril is slightly smaller (no nasalis repair). Note normal eruption of lateral incisor into the cleft site. There is discoloration of the enamel. [Courtesy of Michael Carstens, MD]

Toward an Embryologic Strategy for the Surgical Management of Clefts

The organization of this chapter is built around five sets of issues that I have found most problematic. We shall consider each of these in turn.

Medial Wall Dissection

Certain aspects of the DFR design produced results that were not ideal or did not make sense. (1) In some cases, I observed *flattening of the cupid's bow* at the intersection of the white roll and the cleft-side philtral column. (2) Since 2003 I have been advancing the cleft-side medial crus into the nasal tip using an *anterolateral columellar incision*. Although this design worked well (with excellent scars) I had the nagging suspicion that a simpler design would accomplish the same goals. (3) I realized that the intranasal extension of the prolabial incision was an *embryologic challenge* because it would have to conform to the neurovascular field map of the medial nasal wall. What were the precise boundaries between skin of forebrain neural crest origin and skin originating from the hindbrain neural crest?

Lateral Wall Dissection

Nasal airway expansion, a top priority in DFR surgery, was difficult to maintain, despite the near-perfect fit of the NPP flap into the lateral wall releasing incision. Why should the cleft-side nasal airway arrive at its particular shape? What cause or causes could explain the functional limitations of breathing on the cleft side? There seemed to be a missing piece to the puzzle. What was it?

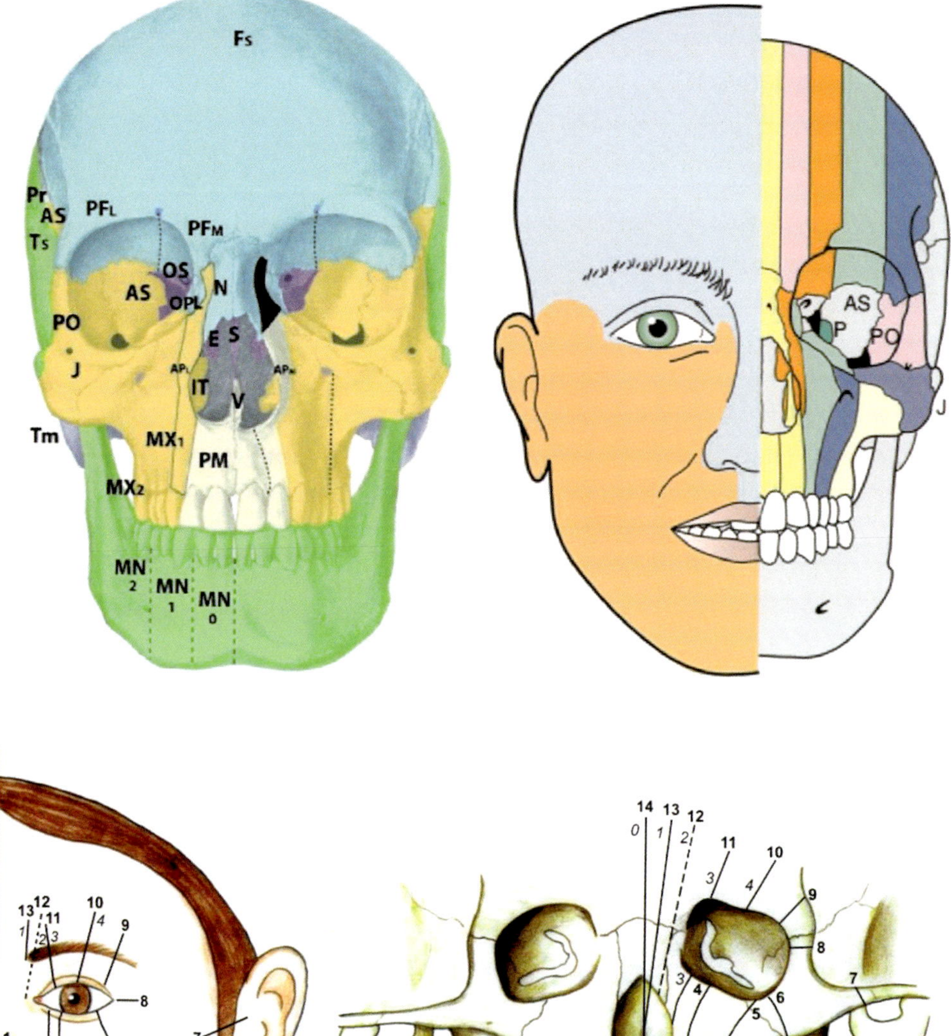

Fig. 19.13 Tessier cleft classification system. From the get-go I realized there had to be a neuroanatomic explanation for the cleft zones because they were concordant with the trigeminal system. This perception in 1987 would eventually morph into neurovascular fields based on neuromeres. Left: [Courtesy of Michael Carstens, MD]. Right: [Reprinted from Carstens MH. Developmental field reassignment cleft surgery: reassessment and refinements. In: Bennun R, Harfin J, Sandor G (ed). Cleft Lip and Palate Management: A Comprehensive Atlas. Hoboken, NJ: John Wiley & Sons; 2016:83–111. With permission from John Wiley & Sons.]

Fig. 19.14 Tessier cleft zones. Skull bones on the left are color coded for the origin of the neural crest mesenchyme: prosomeric (blue), mesomeric (white), rhombomeric (tan). On the right, individual cleft zones are depicted. The exact biologic relationship between orbital and maxillary cleft zones could be due to similarities in the homeotic code (or a different gene code) specifying their respective neurovascular axes. The following fields are all in register with rhombomere r2 and supplied by StV2 branches of the stapedial from the sphenopalatine fossa: *fluorescent* = medial nasopalatine axis, *tan/flesh* = lateral nasopalatine axis, *light green* = descending palatine axis (greater and lesser), *green* = medial infraorbital axis, *blue* = lateral infraorbital axis, *gold* = superior alveolar axis and *purple* = zygomaticofacial axis, and *pink* = zygomaticotemporal axis. Fields supplied by the non-stapedial external carotid system are colored *orange*. Fields supplied by StV1 branches are *light blue* for hp2 frontonasal skin and *red* for hp2 placodal skin. [Reprinted with permission from Sari, E. Tessier Number 30 Facial Cleft: A Rare Maxillofacial Anomaly. Turkish Journal of Plastic Surgery, 2018; 26(1): 12–19.]

Dental Arch Management

Coming to clarity regarding the surgical-orthodontic sequence with respect to the alveolar arch was a very frustrating problem for me. The long-standing debate regarding presurgical orthopedics has been characterized by loudly stated opinions and a near-total lack of developmental biology. Understanding the cleft maxilla and its reconstruction requires the juxtaposition of new input from developmental field biology, with a surgical and orthodontic protocol in functional agreement with basic science. Above all, a biologic protocol should be backed up by long-term results.

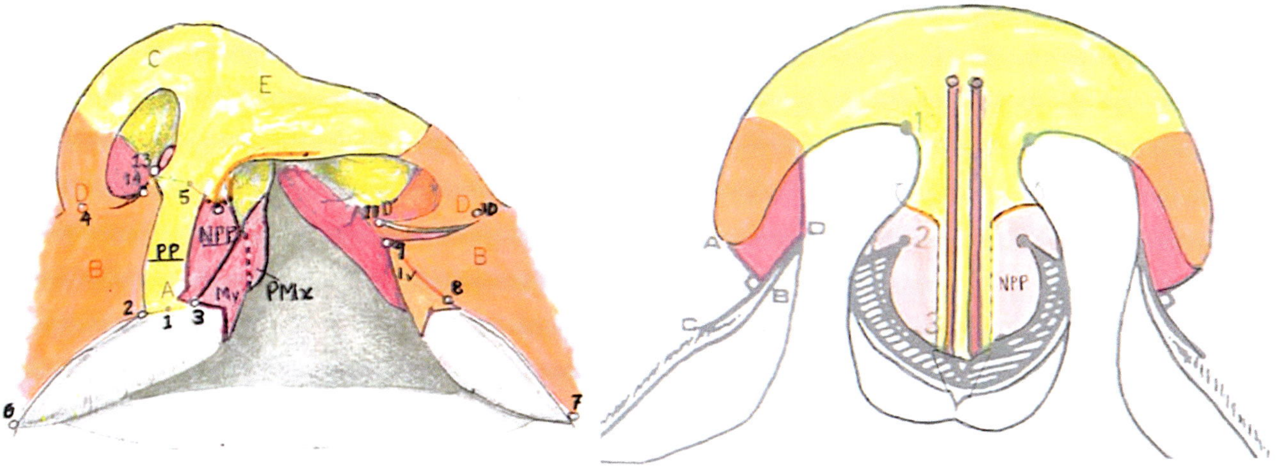

Fig. 19.15 Developmental field map: UCL and BCL. (*Left*): Prolabium consists of two pairs of fields, StV1 (yellow) and StV2 (pink). Normal embryogenesis assigns StV2 as medial nasal floor above the central and lateral incisor fields, PMxA and PMx B. In the cleft state tissues from r1 remain abnormally fused with r1. This results in an island of medial nasopalatine non-philtral prolabium (pink) located on one of both sides of the true anterior ethmoid philtrum. (*Right*): Bilateral clefts have symmetrical fields. Key: yellow = StV1 anterior ethmoid; pink = StV2 medial nasopalatine; magenta = StV2 lateral nasopalatine; orange = StV2 infraorbital + ECA facial Gray vermilion flaps = ECA. *Note:* DFR flaps. Diagram shows NPP elevated along with LCC, but these can be separated (as per Matthews). Lateral wall mucoperiosteum is elevated from what was formerly PMxF is now Mx F. Subvestibular mobilization with elevator continues under the periosteum inside the nostril until reaching nasal bone. Mucosal flaps are dropped as "diapers" without committing them until the end of the case. M can be sutured to the lateral lip for coverage. It can also be based proximally and rotated into the sulcus release site under the ala which gave access to the bone. M can provide coverage to the lip or turned up into the vestibule. [Courtesy of Michael Carstens, MD]

Microform Cleft

It has long been observed that a small number of patients have a nasal deformity characteristic of cleft lip but a minimal to absent affectation of the lip itself. The literature on this subject is limited but very consistent. All such patients have asymmetries of the nasal floor/alar base; some degree of septal deflection is always present. The microform cleft is truly the "*form fruste*" in the problem. It demonstrates most elegantly that the source of the pathology lies within the piriform fossa, not the lip. In addition, the microform cleft demonstrates how differences in the fluid mechanics of fetal ventilation cause deformation of the septum. Of late, I have been able to find more satisfactory answers to these questions.

The Airway in Facial Clefts

Breathing begins in utero. As we shall see, the biologic problems of facial clefts begin with the airway itself, as the deficiency state of the premaxillary field, no matter how slight, causes deformation of the piriform fossa, the nasalis muscle complex, and the vestibular lining of the lateral nasal wall. All other manifestations of cleft in the nose, as seen in the nasal tip, proceed from this triad. What's more, if the palate is also cleft the nasal airway is compromised by two sites of deformation involving the septum. The achievement of correct nasal function is the Sahara Desert of cleft surgery, a harsh and uncompromising terrain, poorly understood and clinically neglected. No matter how much attention we place on prolabial designs and soft tissue flaps, at the end of the day, the majority of our patients are left with life-long breathing dysfunction on the cleft side. As we shall see, it turns out that basic features of the embryologic features of the cleft nose involving the septum, medial footplate, and the lateral crus are hugely significant. They point the way to understanding the pathology and correcting it completely. Understanding the pathology of the cleft airway and addressing it as an integral part of the initial repair are a major focus of DFR.

Developmental Anatomy of the Midline Lip–Nose Complex

Understanding the development of the premaxilla and its soft tissue coverage, the prolabium is fundamental for cleft lip surgery. Before we take on the issues raised previously, let's review how the midline structures of the nasal envelope premaxillary platform, the prolabium, and the nose are organized.

Nasal Skin

Let's return to Chap. 4. Recall that frontonasal skin flows forward and downward in development, forming a bilaminar envelope. Its outer surface is epidermis formed from the anterior prosencephalic folds of hypothalamic prosomeres hp1 and hp2. Its dermis originates from posterior prosencephalic neural crest from prosomeres p1–p3 (probably the latter). The inner lining develops from nasal placodes, in register with hp2. These sink into the mass on either side. In so doing they create two tunnels, the future nasal cavities. The unique vestibular epithelium of the nasal airways is derived from placodal ectoderm, the medial half of which is dedicated to the pheromonal accessory olfactory system and GnRH neurons of nervus terminalis, while the lateral half forms olfactory system. Growing forward into this mass we find extensions of r1 neural crest which are interposed in between the frontonasal layers like cheese in a sandwich. These are responsible for making nasal cartilages.

The nasal chambers wind up as U-shaped structures supplied by posterior and anterior internal nasal and septal branches of the posterior and anterior ethmoid arteries. Its innervation is V1. The nasal floor is constructed from r2 neural crest from the first arch, which is innervated by V2 and supplied by corresponding Stapedial artery related to cranial nerve V2 (StV2) branches from the pterygopalatine fossa. Tessier cleft zones 13–12 are represented within the nasal chamber.

The external nasal skin paired zones of 13–12 along the dorsum and zone 11 along the side walls. These are supplied, respectively, by external nasal branches of anterior ethmoids and by the dorsal nasal artery from the infratrochlear stem.

Columella

Although manipulated and or incised in most cleft repairs the columella has suffered from anatomic neglect. Over the years a vast literature has accrued regarding the components of the nasal tip and their support structures [6] but with little emphasis on the design of the columella. Previously, this author [2] demonstrated the presence of paired anterior ethmoid arteries that transmitted dye into the philtrum but not beyond the philtral columns (Figs. 19.9 and 19.10). As is expected, these vessels are surrounded by r1 neural crest fat filling the space between the medial crura all the way down to the footplates. The combination of fat, medial crura, and skin abutting the septum is the membranous septum. U-shaped incisions allow the upward mobilization of the lateral crus which remains perfused independently from the central columellar. The reason for this is the abundant blood supply reaching the membranous septum from vessels of the internal nasal chamber and rim. Furthermore, the entire philtrum-columella-membranous septum can be readily elevated in a plane anterior to the septum thus elevating the nasal tip, as seen in reports of bilateral cleft report from Trott (which we will discuss later on) (Fig. 19.16).

Blood supply to the skin over the medial crura is quite independent of the columella. The lateral columellar incision described in the technical section an incision below the footplate and the parallel incision in the membranous septum creates a lateral columellar chondrocutaneous (LCC) flap that can be repositioned upward into the nasal tip. It is supplied by a network of StV1 vessels of the dorsum combine by ECA collaterals ascending around the rim of the nostril. This robust blood supply prompted Song to report the use of the columella flap rotated downward into the prolabium in primary unilateral repair. The design transects the columellar arteries to philtrum from above, a sort of Millard-in-reverse. But the technique is worth knowing because it demonstrates the abundant blood supply to the lateral columella (Fig. 19.17).

Congenital absence of the columella is a rare occurrence but quite significant. It can occur as part of holoprosencephaly or as an isolated field defect. This condition was first reported by Jacobs [7]. It comes in two forms, total columella loss without and with caudal septum involvement. All cases report absent medial crura. A familial tendency has been documented by both Lewin and Mavili [8, 9] (Figs. 19.18 and 19.19). These cases reveal preservation of two mounds of tissue just overlying the footplates of the medial crura: this is referred to as the so-called "shoulder" of the columella. This tissue originates from r2 and is supplied by a terminal branch of medial nasopalatine artery. The philtrum is preserved. No case of unilateral columellar loss has been reported. This is because the anterior ethmoid arteries are paired into the philtrum; loss of one would be compensated by the other.

The medial crura are lost in columellar aplasia. This makes an important point: *Nasal tip cartilage, although seemingly a continuous structure, consists of distinct developmental subunits*. Medial crus belongs to zone 13 and lateral crus to zone 12. They are connected by intermediate crus. For this reason, the number 13 cleft presents as medial alar notch and the number 12 cleft as a defect in the lateral crus.

The *philtrum remains intact* in columellar aplasia. This may occur because the PNC mesenchyme has reached the level of the lip and is vascularized by collateral circulation from the surrounding lateral lip elements. In the meantime, vascular failure eliminates the columella. I cannot find an example of congenital columellar aplasia in the presence of cleft lip. For example, case should show an abnormal or absent philtrum.

Prolabium

The prolabium refers to the non-mucosal soft tissue coverage of the premaxillae. In normal development it consists of a caudal extension of columella which includes frontonasal skin, anterior ethmoid arteries (usually paired) about 2–4 mm apart, and r1 neural crest fatty-fibrous tissue. This midline structure is known as the philtrum. Its innervation is V1 (Fig. 19.15).

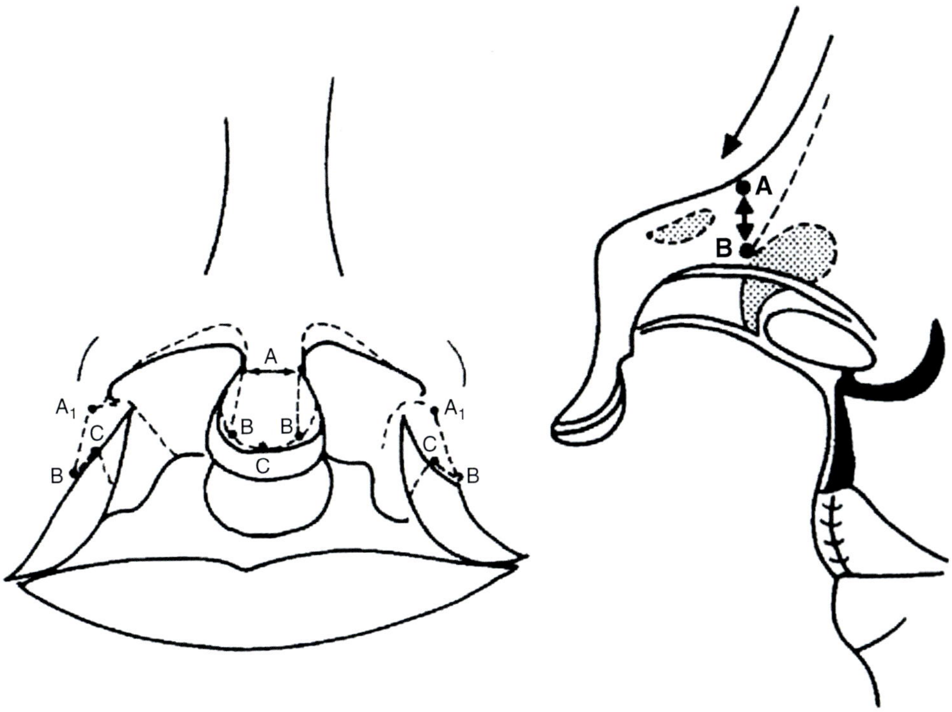

Fig. 19.16 Trott–Mohan BCL design proved the vascular independence of columella and philtral prolabium. They elevated the entire PP and columella to gain exposure to the nasal tip. This is an extension of the Rethi and Harashina procedures. In general enough elasticity is present in DFR that elevating PP is not necessary. A specific indication for this however is in the insufficient prolabium. In this case PP becomes the columella and the Cupid's bow is constructed with a graft. This can be later replaced with a composite skin-cartilage graft from the cymba of the ear the curvature of which will mimic the depth of Cupid's bow. [Reprinted from Trott JA, Mohan N. A preliminary report on one-stage open tip rhinoplasty at the time of lip repair in bilateral cleft lip and palate: the alor Setar experience. Br J Plast Surg 1993; 46;215–222. With permission from Elsevier.]

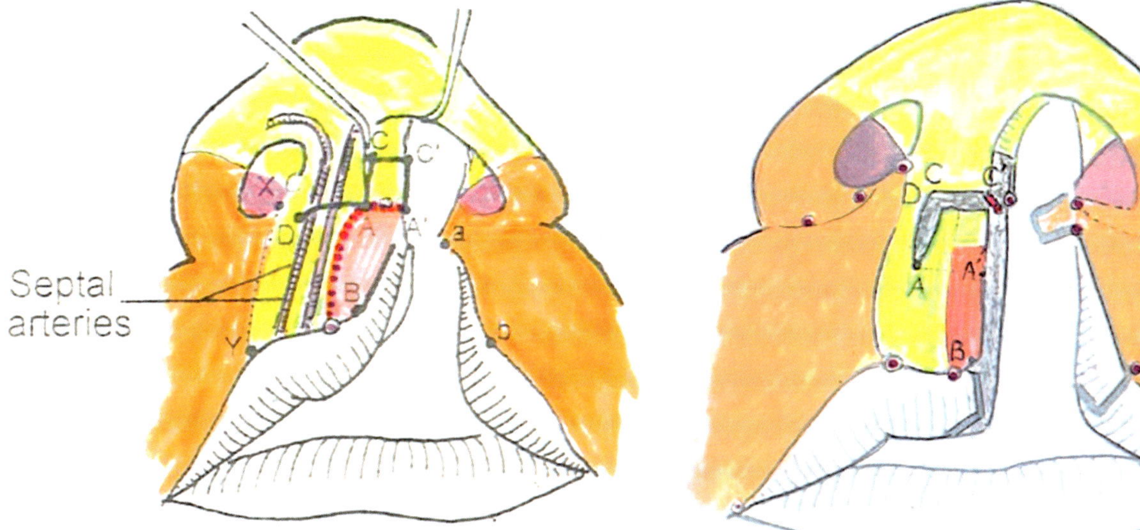

Fig. 19.17 Lessons of the lateral columellar flap. Song's variation on the original Reichert procedure (described by Honigmann) violates columellar fields but does illustrate alternative blood supply to the NPP. Skin flap extracted from lateral columella is a random extension of NPP, an idea that originated from Reichert. Both arteries to the philtrum are cut. Blood supply is reverse from the labials. NPP is separated from its vascular axis, becoming random based on philtral prolabium, the blood supply to which is now reduced by 50%. Compare this design to DFR. Pink, medial nasopalatine, magenta, lateral nasopalatine, lemon, anterior ethmoid, orange, ECA facial. [Adapted from Song R, Liu C, Zhao Y. A new principle for unilateral complete deft lip repair: The lateral columellar flap method. Plast Reconstr Surg 1998; 102:1848. With permission from Wolters Kluwer Health, Inc.]

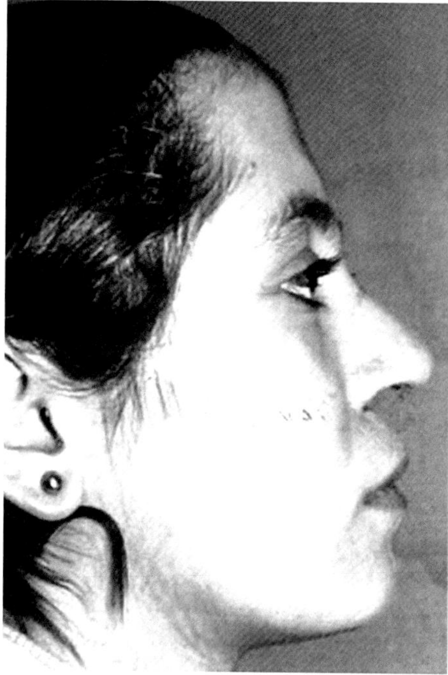

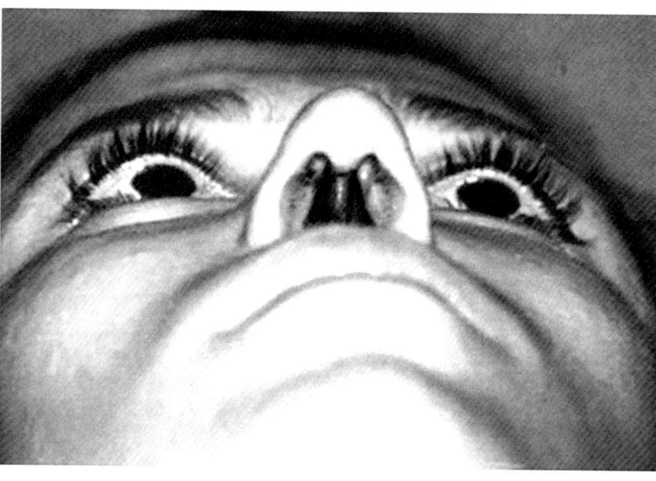

Fig. 19.18 Congenital absence of columella 1. Note absence of medial crura but otherwise normal nasal tip with normal projection. Medial crura are *not* necessary for tip support. Orbicularis is normally inserted into both r1 columella and into the r2 fields below the footplates. In this case the normal sweep upward into columella is lost but lip "drape" remains. Nostril sills are hiding in the nasal floor. They normally blend into the r2 "shoulder" of columella. Columella represents paired independent developmental fields. Medial crus is programmed by columellar skin. Familial tendency (seen in twins). Proves *vascular independence* from the septum. [Reprinted from Mavalli ME, Akyúrek M. Congenital isolated absence of the nasal columella: Reconstruction with an internal nasal vestibular skin flap and bilateral labial mucosa flaps. Plast Reconstr Surg 2000; 106(2):393–399. With permission from Wolters Kluwer Health, Inc.]

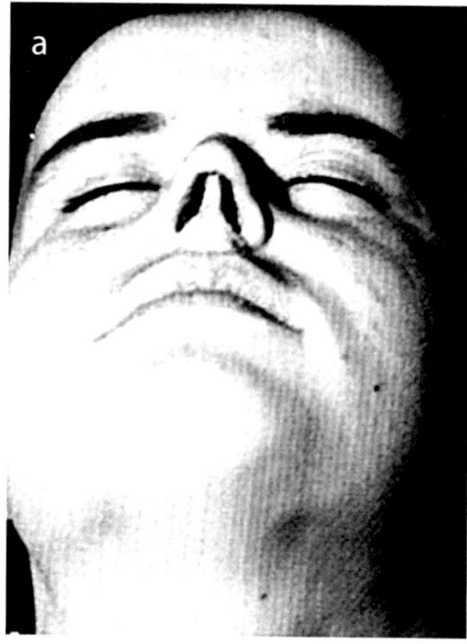

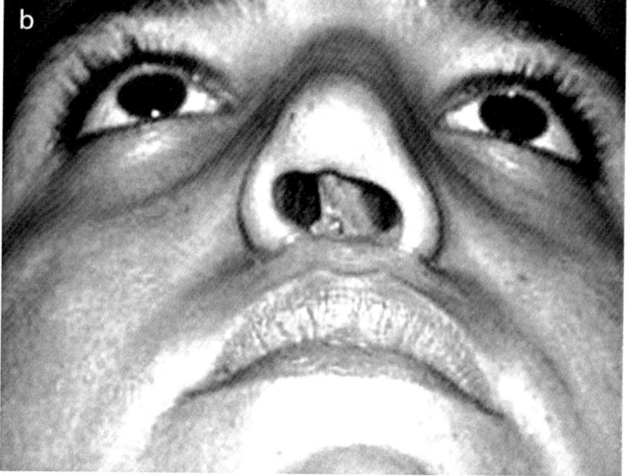

Fig. 19.19 Congenital absence of the columella. (**a**) Loss of columellar tissue in the center but with residual normal "flow" of columellar tissue into the "shoulders" at the base. (**b**) a more severe defect in which both midline and basal tissues are absent. Left: [Reprinted from Lewin ML. Congenital absence of columella. Cleft Palate Journal 1985; 25(1):58–63. With permission from SAGE Publications.] Right: [Reprinted from Ayhan M, Sevin A, Aytug Z, Gorgu M, Erdogan B, Reconstruction of congenital and acquired columella defects: Clinical review of 38 patients. J Craniofac Surg 2007; 18(6):1500–1503. With permission from Wolters Kluwer Health, Inc.]

Prolabial anatomy is "more than meets the eye" because the prolabium is a *composite of tissues* that originate from *different locations in the embryo*. Under normal conditions the original prolabium *fuses* with the lateral lip elements. Its lateral embryonic components become tucked inside the nose as the "shoulder" of the columella that gives soft tissue coverage of the lateral incisor. We tend to forget that they exist. Thus, the best way to study the embryology of the prolabium is in its "native" state: the complete bilateral cleft. We will then discuss how the prolabium in unilateral cleft patients contains extra tissues belonging to the first and second pharyngeal arches.

Prior to fusion with the lateral lip elements, the original prolabium looks just like that of the bilateral cleft. It has central PNC elements supplied by the Stapedial artery related to cranial nerve V1 (StV1) anterior ethmoids and lateral elements (the "shoulders") supplied by StV2 facial. Note that in a bilateral cleft the entire prolabial complex can be elevated and the split into three component parts. In conventional techniques, this elevation proceeds up to the anterior nasal spine because of vascular concerns. In DFR, as a lesson from the Trott and Mohan design, the entire columella and membranous septum (philtral prolabium) can be elevated upward, leaving the columellar shoulders (NPP) in place. This is accomplished by means of a simple incision straight back underneath the alar footplates.

- Philtral prolabium is PNC
- Non-philtral prolabium is RNC
- The prolabium in the embryonic state acquires muscle by fusion with lateral lip elements
- Bilateral cleft lip repeats the embryonic state and is devoid of muscle

Lateral Lip Elements

Recall that in development first arch and second arch fuse to form a sandwich. This is clearly demonstrated by the structural layers of the lips. Consider the upper lip as a saddlebag suspended from the maxillary complex. Both inner and outer surfaces are r2 ectoderm: externally it forms skin and internally it forms mucosa. Muscles of mastication are dedicated to suspending the jaws and they do not extend into the lip. Thus, there is no layer of DIF fascia in the lips. Into this r2 "envelope" projects second arch myoblasts enclosed within SMAS (SIF) fascia. Deep orbicularis (DOO) is programmed by the mucosa, while superficial orbicularis (SOO) is programmed by the skin. DOO curls upward at the vermilion border where it comes into edge-to-edge contact with SOO. This transition is marked by the white roll (Fig. 19.20).

All second arch muscles, both those of mastication and those of facial expression, arise from Sm6. The former are associated with deep investing fascia (DIF), while the latter make a primary insertion into the SMAS and then pursue three options (Figs. 19.21, 19.22, 19.23 and 19.24)

- Facial muscles can remain within the SMAS, inserting into a contralateral counterpart from the other side, and thereby form a sphincter as in orbicularis oculi and orbicularis oris. Frontalis has a secondary insertion into the muscle-free zone of central galea aponeruotica which is stationary, acting like a tendon that permits frontalis to lift the forehead.
- SMAS muscles can drop downward to insert into bone as in zygomaticus major and minor, nasalis, and depressor septi nasi (DSN). Note that the SMAS over the vertex of skull is devoid of muscle but acts like a tendon, into which insert both frontalis and occipitalis.
- SMAS muscles can be directed upward to insert into skin, as in corrugator superciliaris.

Lower lip development precedes that of the upper lip. Sm6 buccinator from mandible inserts into the modiolus first followed by that from the maxilla. This criss-cross of fibers is important to recreate in cases of lateral clefting in which the vest-over-pants commissure must be reconstructed.

Under normal conditions the lateral lip elements fuse to the PNC mesenchyme overlying premaxilla, bringing three sets of muscles into position, each of which has a distinct fate.

- SOO stops at the sidewall of the prolabium. It is incapable of penetrating frontonasal mesenchyme. The interface between the mesenchyme beneath philtral skin and SOO results in the formation of a philtral column. Reports in the literature of the philtral columns arising by criss-crossing interdigitation are incorrect. This concept, propagated by Latham, resulted as an artifact of the methodology used. These fibers are not seen in the dissection of unilateral cleft. Once one has pared the cleft side, one sees skin, a fibrous mesenchyme devoid of muscle, and distinct layer of DOO.
- DOO, on the other hand, follows the r2 mucosa as it covers over the premaxilla, making a sulcus. DOO flows in a plane deep to sub-philtral mesenchyme, achieving continuity with itself across the midline. Note that the branches of facial artery run in the plane between DOO and SOO.
- DSN is a small muscle associated with DOO, that moves into the midline at anterior nasal spine [10]. The fibers diverge to insert into the preseptal ligament and into the alar bases. Cadaver dissections of DSN show the fibers descending from SMAS into the periosteum below the ANS [11, 12] (Fig. 19.25). This small muscle is important because it can pull down the nasal tip or, on smiling, give the appearance of a transverse crease, the "tight lip." Clinical approaches to DSN are reviewed by Sinno et al. [13].
  - I am not aware of any reports documenting the presence of DSN in unilateral clefts, but it must surely be so. Barbosa documents an asymmetrical insertion but I am not aware of the circumstances.

Premaxilla and Vomer

The neuroembryologic model of cleft remains the same and has been well described elsewhere [14–16] and *pari passu* in this book. In broad brush strokes, the spectrum of unilateral

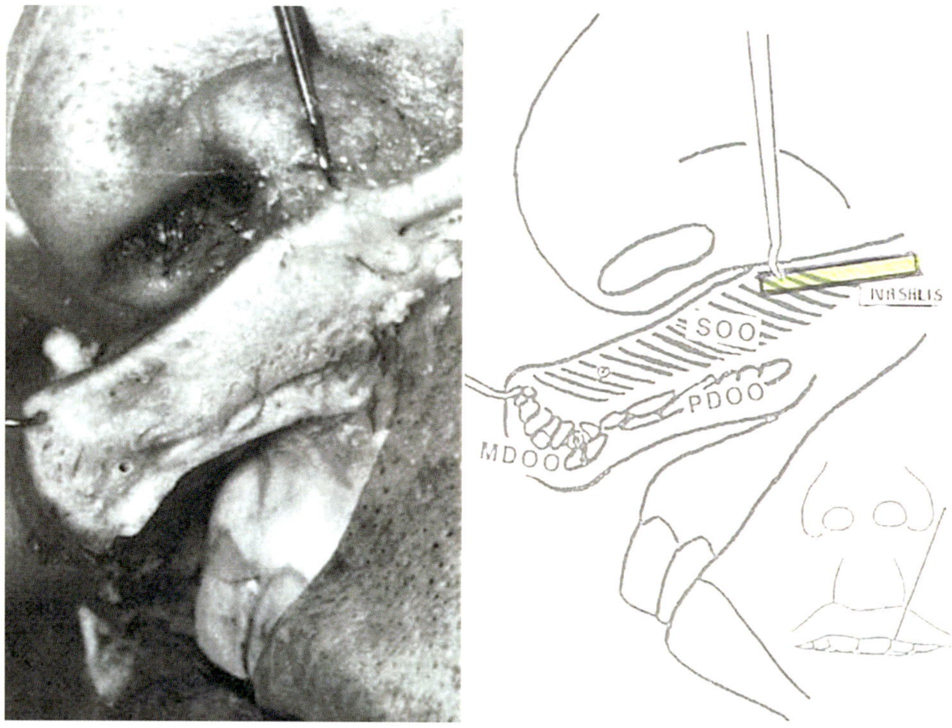

Fig. 19.20 Lateral lip element contains three muscles. DOO follows the wet mucosa around in the form of the letter "J" with marginal fibers MDOO ascending beneath the dry mucosa. Boundary be between dry mucosa and skin is the white roll. SOO stops short of the white roll. Pit ends proximally with nasalis fibers at the level of the alar base and nostril sill. Nasalis (yellow) in cleft state is located subcutaneously from the nostril sill into the vestibule as far in an inferior turbinate. [Reprinted from Park C, Ha B. The Importance of Accurate Repair of the Orbicularis Oris Muscle in the Correction of Unilateral Cleft Lip. Plast Reconstr Surg 1995; 96(4): 780–8. With permission from Wolters Kluwer Health, Inc.]

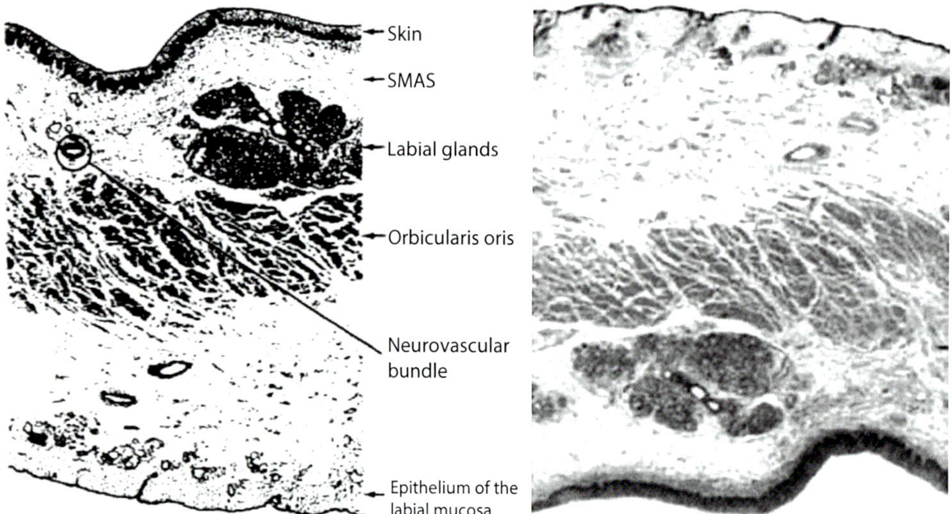

Fig. 19.21 Superficial musculoaponeurotic system upper lip. Lip and nose anatomy both demonstrate SMAS which conforms to the five criteria of Tessier (see text). Left: [Reprinted from Pensler JM, Ward JW, Parry SW. The superficial musculoaponeurotic system in the upper lip: an anatomic study in cadavers. Plast Reconstr Surg 1985; 75(4):488–492. With permission from Wolters Kluwer Health, Inc.] Right: [Reprinted from Letourneau A, Daniel RK. The superficial musculo-aponeurotic system of the nose. Plast Reconstr Surg 1988; 82(1): 48–57. With permission from Wolters Kluwer Health, Inc.]

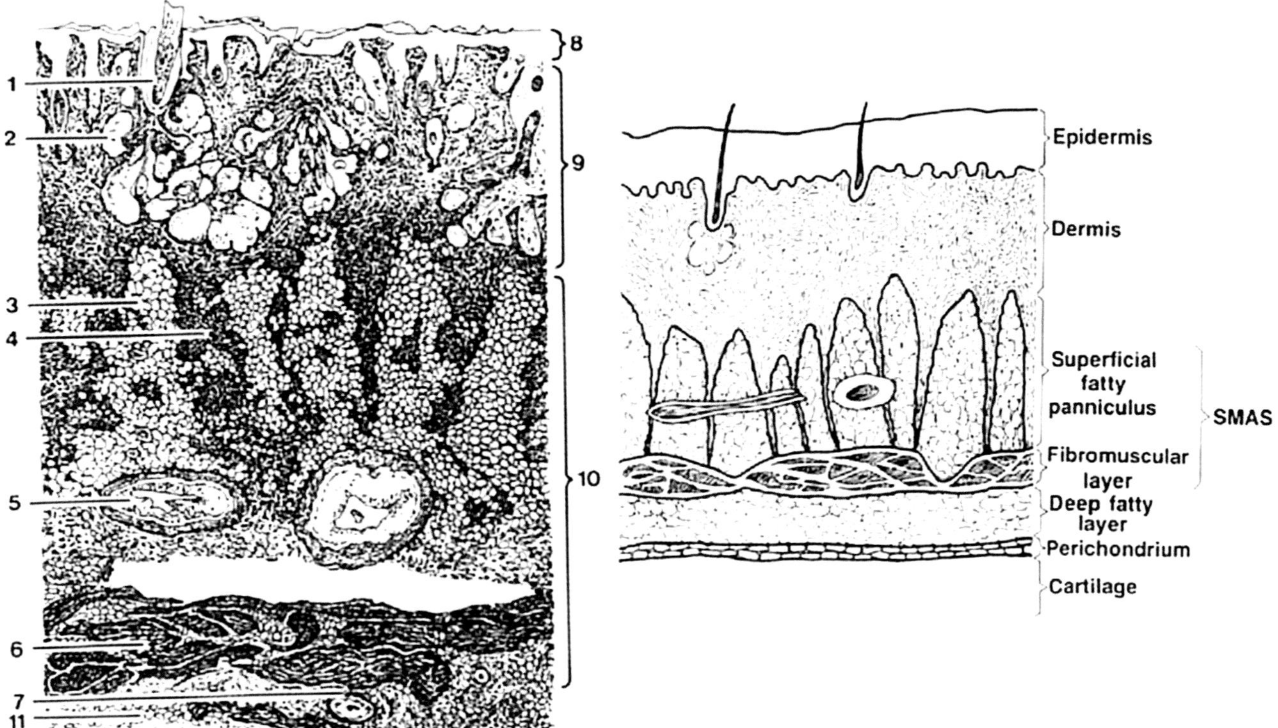

Fig. 19.22 Nasal SMAS histology. Key: Mallory's triple stain: 1, hair follicle; 2, sebaceous gland; 3, superficial fatty layer; 4, fibrous septae; 5, anterior ethmoid artery, terminal branch; 6, fibromuscular layer (i.e., procerus muscle); 7. arteriole; 8, epidermis; 9, dermis; 10 SMAS; 11, deep fatty layer; 12, perichondrium. [Reprinted from Letourneau A, Daniel RK. The superficial musculoaponeurotic system of the nose. Plast Reconstr Surg 1988; 82(1): 48–57. With permission from Wolters Kluwer Health, Inc.]

cleft lip alone or in combination with cleft palate results from defects along the axis of the *medial nasopalatine neuroangiosome (MNP)*, the most medial of all the branches emanating from the sphenopalatine fossa. This neurovascular axis supplies two r2 bone fields: the premaxilla and the vomer. Both bones are paired and for the sake of simplicity, I am going to refer to them in the singular (Figs. 19.26 and 19.27).

Each premaxilla consists of three distinct zones (from oldest to newest): central incisor, lateral incisor, and a frontal process stretching upward from the lateral incisor all the way to the frontal bone. *When pathology strikes the premaxilla, the frontal process is the first to be affected* (causing the scooping out of the piriform rim). If the pathology is more extensive, the lateral incisor zone takes the hit. Very rarely, the entire field may be wiped out [17].

Note that the number of incisors in basal mammals is *three* (in dinosaurs, it was *four*) (Fig. 19.28). For this reason, it is entirely normal for the medial branch of anterior superior alveolar artery to supply, via a separate branch to the frontal process of maxilla, a third incisor, with its main vessel supplying the canine.

Vomerine bones are triangular in shape. Because the vomer sits under the septum, its vertical height anteriorly is very small but, as one proceeds posteriorly, the height increases. Development of the vomer is (1) anterior-to-posterior and (2) dorsal-to-ventral. *When pathology strikes the vomer, the posterior height is affected first.* The deficient sector of vomer will fail to reach the plane of the palatal shelves.

Defects of the medial nasopalatine neurovascular axis can affect the premaxilla, the vomer, or both, as we shall see below.

The vascular anatomy of medial nasopalatine artery was discussed in Chaps. 14 and 15 but let's recap the important point. The axis descends along the septum with side branches distributed to mesenchyme such that deficiencies hits the posterior parts of the field first. At incisive foramen the artery flows laterally with the distal zones affected first. These facts explain the spatial sequence of cleft lip and of midline cleft palate involving the vomer.

Premaxillary deficiency always causes a contour deformity of the piriform fossa (the cleft lip nose). Depending upon its severity, isolated cleft lip or cleft lip plus alveolar defect can occur. Vomerine deficiency always affects the hard palate. When associated with premaxillary deficiency, the combination of cleft lip and cleft palate is observed.

The construction of the hard palate involves multiple neuroangiosomes. The *intranasal anterior ethmoid* and *poste-*

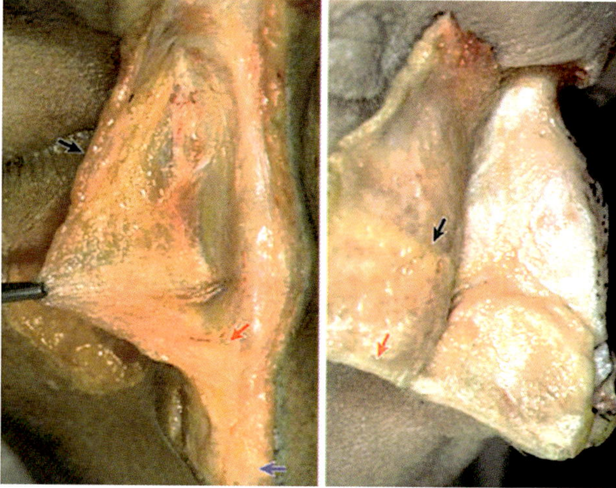

Fig. 19.23 Superficial musculoaponeurotic system (SMAS) of the nose. The SMAS contains the Sm6 muscles of the nose. These have a primary insertion into the fascia and a secondary insertion into soft tissue structures of the nasal envelope. SMAS transmits forces in differential ways due to its flexibility. *(Left)*: Superficial layer of SMAS. The nasal SMAS before its division (black arrow); note its insertion into the internal nasal valve, which appears yellowish (red arrow). The medial superficial layer covers the lower lateral cartilage (blue arrow) and inserts on to the skin of the alar margin. *(Right)*: Deep aspect of nasal superficial musculoaponeurotic system (SMAS). SMAS insertions have been separated at the level of the internal nasal valve (black arrow) and the margin of the nostrils (red arrow). [Reprinted from Saba Y, Amodeo CA, Hammou J-C, Polselli R. An anatomical study of the nasal superficial musculoaponeurotic system: surgical applications in rhinoplasty. Arch Facial Plast Surg 2008; 10(2):115. With permission from Mary Ann Liebert, Inc. Publishers.]

rior ethmoid supply the perpendicular plate and septum. *The medial nasopalatine* supplies the vomer and premaxilla. The *greater palatine* supplies the oral surface of the secondary hard palate: the palatine bone fields P1 and P2 with what we term the palatine bone P3 supplied by the *lesser palatine* axis. The *lateral nasopalatine* supplies the nasal surface of the secondary hard palate and the inferior turbinate. Thus, the spectrum of cleft palate is more complex than that of cleft lip alone. Embryologic classification of cleft palate is a subject unto itself and was discussed in Chap. 14. In this chapter, we shall place our emphasis strictly upon the medial nasopalatine axis: the premaxilla and prolabium.

The connection between underlying bone field pathology and soft tissues is as follows. Whenever membranous bones are synthesized, BMP-4 is released. This protein diffuses upward through overlying soft tissues until it reaches the cleft margin. En route, it fulfills 3 functions (Fig. 19.29).

- BMP4 signals the attachment of nasalis muscle over the lateral incisor and canine.
  - If the signal if absent, nasalis inserts into nostril sill and piriform fossa.
- BMP4 is required for orbicularis muscle development in late embryogenesis (stages 18–23). It holds the myoblasts in a proliferative state until the proper number is achieved and the system can be pass into terminal differentiation.
  - If BMP4 is reduced, muscle fibers in the distal 5 mm of the cleft margin commit prematurely while, in the

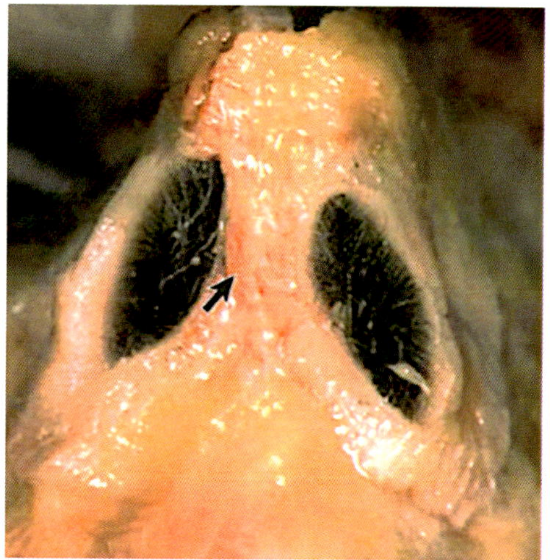

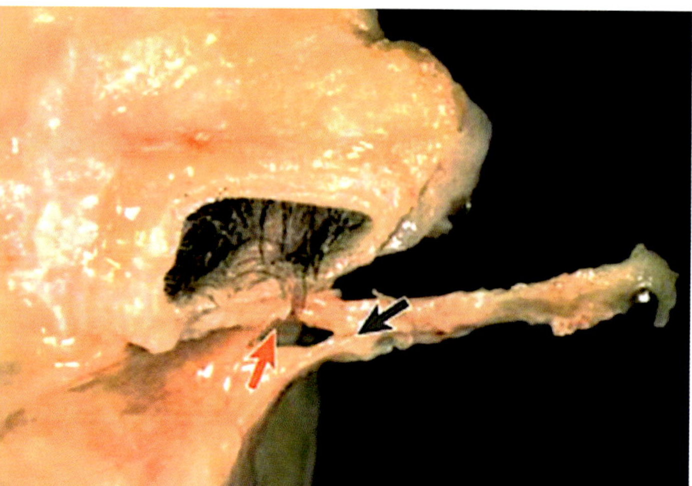

Fig. 19.24 Superficial musculoaponeurotic system (SMAS) of the nose. *(Left)*: Basal view of the nasal SMAS. This layer is penetrated by vascular elements lying immediately beneath the dermis (black arrow). *(Right)*: Superficial medial layer of the nasal SMAS running into the columella. Note the columellar arteries (black arrow) and the relationship between this layer and the fibers of the depressor septi nasi muscle (red arrow). [Reprinted from Saba Y, Amodeo CA, Hammou J-C, Polselli R. An anatomical study of the nasal superficial musculoaponeurotic system: surgical applications in rhinoplasty. Arch Facial Plast Surg 2008; 10(2):115. With permission from Mary Ann Liebert, Inc. Publishers.]

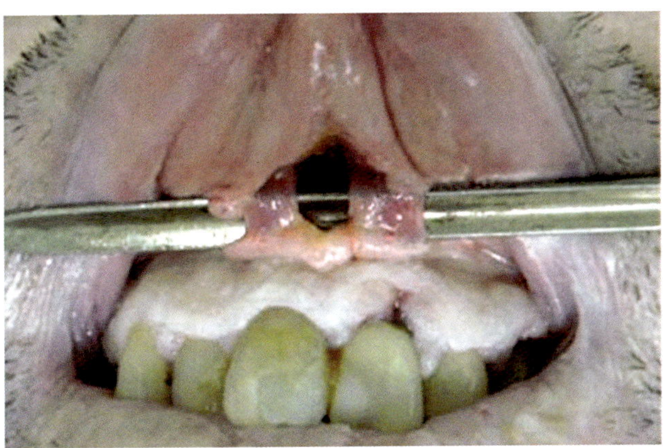

Fig. 19.25 Depressor septi nasi has primary insertion into fossa above the central incisor. These fibers are often muddled together with those of nasalis in which the triangular muscle complex bears the obscurantist name *myrtiformis*. [Reprinted from Barbosa MV, Nahas FX, Fereria LM. Anatomy of the depressor septi nasi. J Plast Surg Hand Surg. 2013 Apr;47(2):102–5. With permission from Taylor & Francis]

embryonic state, forming fewer fibers in a disorganized array.

- BMP4 acts on the epithelium where it blocks sonic hedgehog (SHH), a stabilizer of the epithelium. This block leads to epithelial breakdown and mesenchyme fusion between lateral lip and prolabium.
  - Insufficient [BMP4] leads to fusion failure. This mechanism is quantitative and directional. Defects in [BMP-4] affect the extent of downward diffusion. Thus, a minor reduction creates a cleft of the vermillion. The greater the reduction in [BMP-4], the higher the soft tissue cleft ascends [18–20].

Neuroangiosomes are the functional basis of embryology and of facial clefts. Before we proceed onward with mapping out the prolabium and premaxilla let us get one concept straight. *Sensory nerves induce arteries.* As the face develops, various families of arteries arise, reorganize,

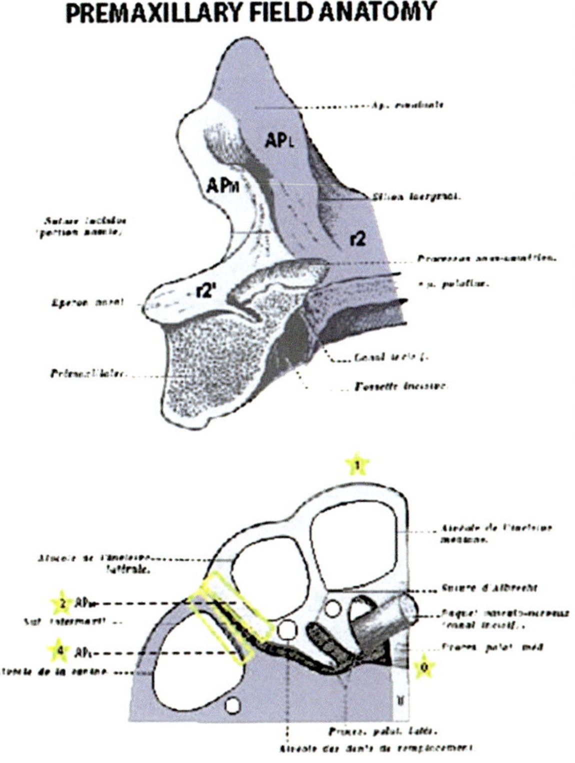

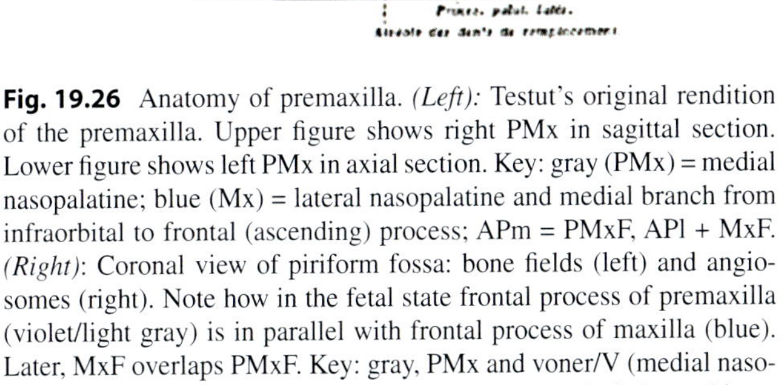

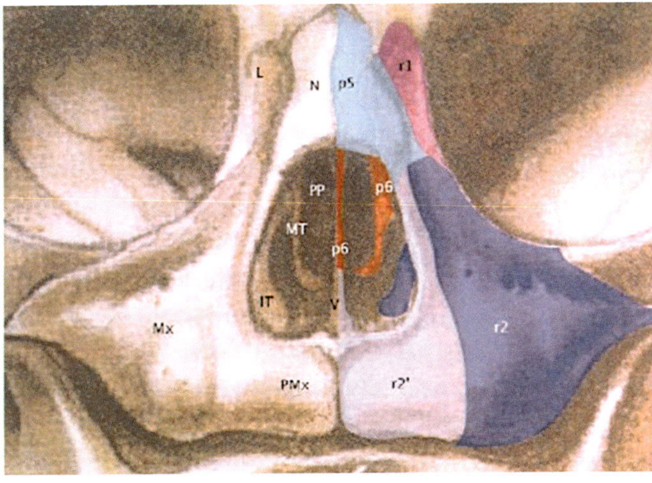

Fig. 19.26 Anatomy of premaxilla. *(Left)*: Testut's original rendition of the premaxilla. Upper figure shows right PMx in sagittal section. Lower figure shows left PMx in axial section. Key: gray (PMx) = medial nasopalatine; blue (Mx) = lateral nasopalatine and medial branch from infraorbital to frontal (ascending) process; APm = PMxF, APl + MxF. *(Right)*: Coronal view of piriform fossa: bone fields (left) and angiosomes (right). Note how in the fetal state frontal process of premaxilla (violet/light gray) is in parallel with frontal process of maxilla (blue). Later, MxF overlaps PMxF. Key: gray, PMx and voner/V (medial nasopalatine), dark blue, Mx (infraorbital); lighter blue, inferior turbinate (lateral nasopalatine); red, middle turbinate (ethmoid); aqua, nasal bone (ethmoid and infratrochlear; magenta, lacrimal (infratrochlear). **Key (French)** *Above*; premaxiliaire, canal incisive, fossette, incisive, processus naso-frontalis, silion lachrymal, eperon nasal. *Below*: alviole de l'incisive laterale, alviole de la canine, alveole de dents de remplacement, paquet vaselo-nerveus dans canal incisive, procesus palatine median, procesus palatine laterale, suinure de Albrecht. Left: [Reprinted from Testut L. Traite d'anatomie humaine: anatomie descriptive, histologie, developpement [French]. Paris, France: Gustave Doin, 1899.] Right: [Courtesy of Michael Carstens, MD]

Fig. 19.27 Growth of premaxillary fields. Observe that frontal process grows upward out of the PMxb (here labeled i2). The two frontal processes, MxF and PMxF, are initially separate but fuse. [Reprinted from Barteczo K, Jacob M. A re-evaluation of the premaxillary bone in humans. Anat Embryol (Berlin) 2004; 207(6):417–437. With permission from Springer Nature]

Premaxilla:

Origin: 2nd rhombomere **N/V supply:** medial sphenopalatine

Sub-fields: :central incisor > lateral incisor > frontal process

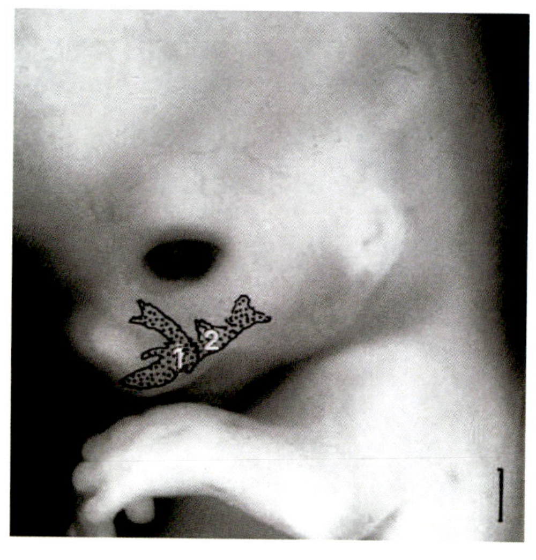

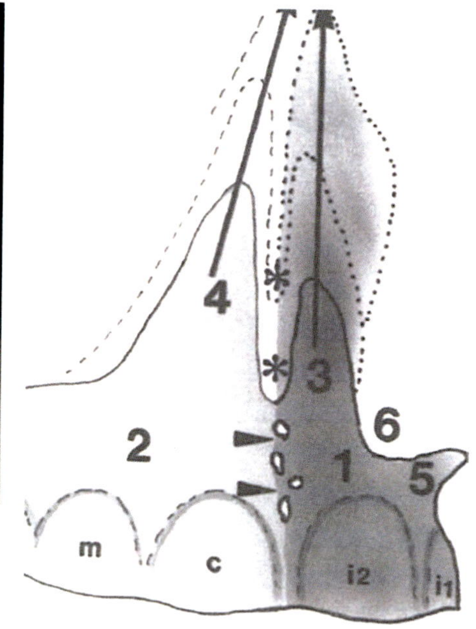

Fig. 19.28 Premaxilla in evolution. (**a**) Premammal tetrapods had 4 peg-like teeth, as in *Tyrannosaurus Rex*. Mammals had specialization into incisors. Placental mammals have basal formula of 11 teeth per quadrant with three incisors: I3, C1, P4, M3. (**b**) Key: i1, central incisor (PMxA); i2, lateral incisor (PMxB); 3, frontal process of premaxilla (PMxF); 4, frontal process of maxilla (MxF); 5, canine eminence (lateral head of nasalia); 6, lateral incisor eminence (medial head of nasalis). (**a**) [Reprinted from Barteczo K, Jacob M. A re-evaluation of the premaxillary bone in humans. Anat Embryol (Berlin) 2004; 207(6):417–437. With permission from Springer Nature]. (**b**) [Courtesy Michael Carstens, MD]

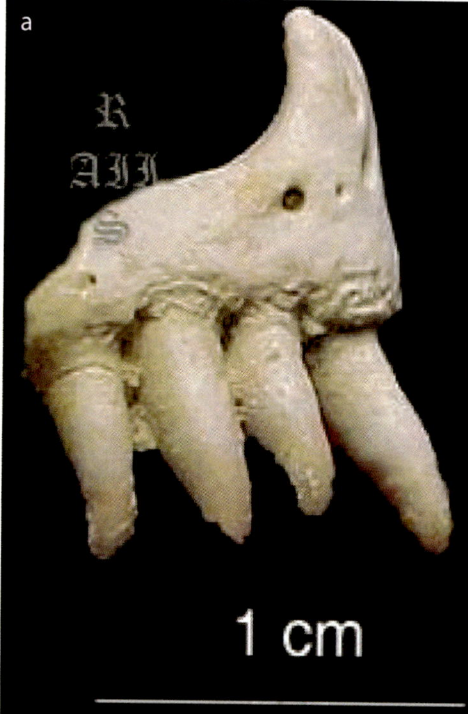

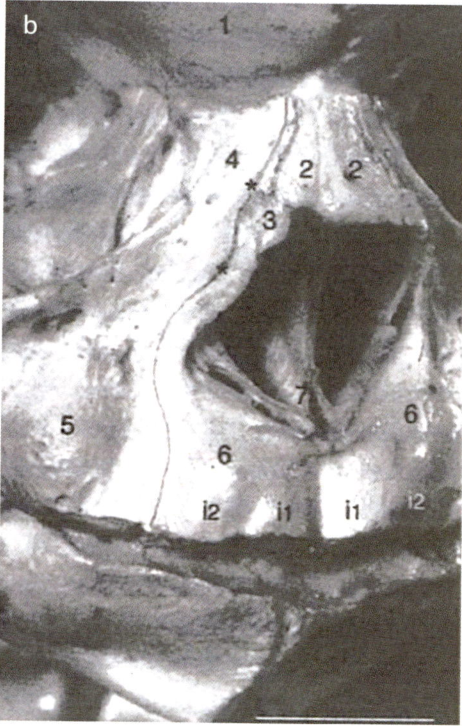

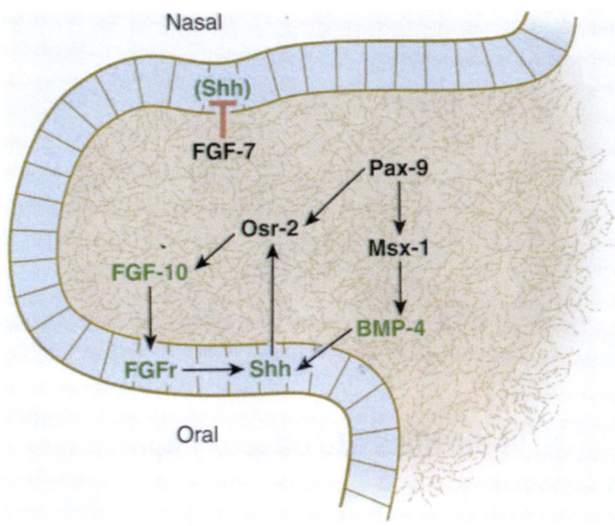

Fig. 19.29 Mesenchymal deficiency = morphogen deficiency. Signaling interactions in the developmental palatal shelves. *BMP* bone morphogenetic protein, *FGF* fibroblast growth factor, *FGFr* fibroblast growth factor receptor, *Shh* sonic hedgehog. [Reprinted from Carlson BM. Human Embryology and Developmental Biology, sixth edition. St. Louis, MO: Elsevier; 2019. With permission from Elsevier.]

and interact. How this takes place is a fascinating story covered elsewhere.

As we have seen, in the middle stages of development, just at the time of emergence of the cranial nerves, a new *stapedial arterial system* appears. Its stem is the dorsal remnant of the defunct second aortic arch artery. It traverses the middle ear, where it divides, one part entering the skull and the other part exiting into the face. All branches of the stapedial system, both intracranial and extracranial, are programmed by sensory branches of cranial nerves [21–23].

The consequences of this neuroembryology were intuitively grasped by Paul Tessier when he developed his classification system of rare craniofacial clefts. *Orbital clefts* (zones 10–13) represent individual "knock-outs" of branches from the V1-induced supraorbital stapedial. *Maxillary clefts* (zones 1–8) represent individual "knock-outs" of branches from the V2 and V3-induced infraorbital stapedial. Clefts in zone 9 are considered the most rare because of its dual blood supply from StV1 lacrimal and StV2 recurrent meningeal. Recall that trigonocephaly may arise in this zone so it is not so obscure as originally thought. Zone 14 does not exist: it merely represents the failure of normal tissue involution (apoptosis) required to approximate the facial midline. Such patients have normal brains but large ethmoid complexes and widened interorbital dimensions [24, 25] (Fig. 19.11).

Having waded through the deep waters of vascular development, we are now in a position to "map out" craniofacial developmental fields using the tools of neuroembryology. Remember, arteries represent inductions from sensory nerves. Thus, *prolabium in the bilateral cleft can be "mapped" into four distinct embryonic zones.*

- The ***philtral prolabium*** In the center, paired *extranasal anterior ethmoid nerves and arteries* from the V1 stapedial system run down the columella about 4 mm apart. The anterior ethmoid fields make up the philtrum, that is, the Cupid's bow. The width of the philtrum = the width of the columella (as defined by the footplates of the inferior crura).
- The ***non-philtral prolabium*** Laterally, additional tissues flank the philtrum. These are supplied by the *MNPs*, from the V2 stapedial system [26, 27].

The skin and subcutaneous tissues of the philtral prolabium (PP) are unique: ***ectoderm and mesoderm are not present***. Recall that the mature embryonic forebrain is divided into five developmental zones: diencephalon develops from *prosomeres*, p1–p3, and secondary prosencephalon develops from two hypothalamic prosomeres, hp1 and hp2. Orbitofrontonasal skin has its *epidermis* from the non-neural ectoderm (not neural crest) of hp1 and hp2. Its *dermis* arises from p1 to p3 neural crest. Nasal placed contributes a unique form of lining. Take a look inside your nose: the color difference between epithelial nasal skin and placodal vestibular lining is obvious. The boundary is marked by nasal vibrissae.

- All the remaining facial skin consists of r2–r3 ectoderm and neural crest dermis.
- One does not encounter true mesodermal dermis until dermatome level C2.

Embryologic Cleft Surgery: Core

Principles of Medial Dissection: Prolabium, Nasal Tip, and Medial Nasal Wall (Figs. 19.16, 19.17, 19.20, 19.30, 19.31, 19.32, 19.33, 19.34, 19.35, 19.36, 19.37, 19.38, 19.39 and 19.40)

Under normal conditions, unification between the premaxilla and the maxilla involves two sets of structures. The frontal process of the premaxilla fuses with its its counterpart, the frontal process of the maxilla. The lateral incisor zone of the premaxilla fuses with the canine zone of maxilla. This unites the MNP with the lateral nasopalatine neuroangiosome. In this way, the soft tissues covering the premaxilla and those of the nostril sill become internalized within the floor of the nose.

Under normal conditions of lip fusion, the two layers of orbicularis oris do not "migrate," instead they passively accompany the epithelium with which they are associated. Mucosa of the lateral lip fuses with its tr2 counterpart lying

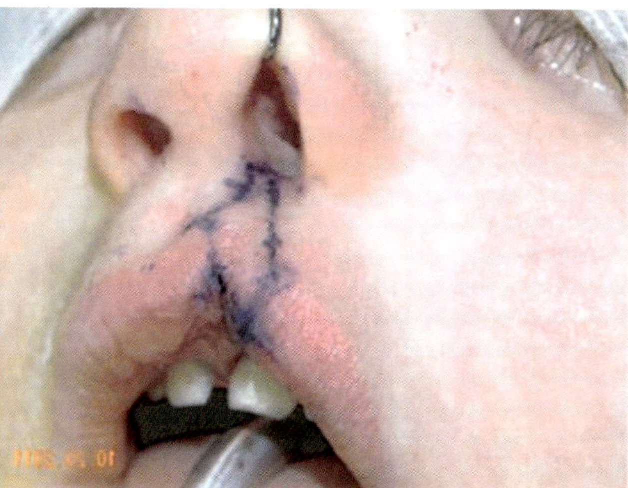

Fig. 19.30 David Matthews innovations
improved prolabial incision
- Marking the PP flap wider based on the footplates allows for back-cut directly above point 3. This allows downward displacement of 1–2 mm which is filled with a half z-plasty flap from the lateral lip. Enables a final "cut-as-you-go" strategy for DFR

Non-invasive elevation of medial crus
- Medial crus can be elevated using a "stealth" incision under the alar footplate separates NPP from LCC, making them two independent flaps

NPP–LCC elevated in continuity with vomer flap (V). Septum (S), gold color, accessed from vomer, yellow. Circle denote pedicle to NPP at level of nasopalatine duct. M flap can pared from the NPP or kept with it [Courtesy of Michael Carstens, MD]

beneath prolabium and covering premaxilla. Like a train following a track, DOO migrates medially beneath philtrum. SOO follows different "rules." It develops in association with r2 skin of the upper lateral lip element (medial infraorbital neuroangiosome). This skin fuses with the frontonasal skin of the philtrum (external anterior ethmoid neuroangiosome). These structures are biologically incompatible. SOO will not penetrate prosomeric mesenchyme. SOO-containing skin is inherently thicker than prolabial skin. The philtral column results from this discrepancy.

The central theme of medial dissection is the separation of embryonic fields with preservation of angiosomes. The philtrum as seen in the unilateral complete cleft philtrum is the ideal model. Bilateral cases are simply a variation on this theme (Figs. 19.15, 19.16, 19.34 and 19.35).

- *Non-cleft side* consists of the lateral lip element, a normal philtral column, the r1 columella and its r2 "shoulder" which lies in the nasal floor, and the PP. Note that SOO stops at Cupid's bow and DOO continues all the way to the cleft side.
- *Cleft side* consists of a skin envelope derived from r2: the "shoulder" of columella (which is not truly part of columella), and the NPP.

The *non-philtral prolabium* develops from hindbrain neural crest. It represents a distal soft tissue extension of the medial sphenopalatine artery, originating as the terminal branch of the internal maxillary system in the sphenopalatine fossa. *Under normal conditions, NP represents the skin tissue coverage of the medial the nasal floor*. Tucked inside the nostril sill, it is difficult to appreciate. The reason NPP is displaced into the nasal floor is because the lateral lip element is inserted between PP and NPP.

- In cleft lip patients, NPP is externalized and readily seen, being a lateral "add-on" to the true philtrum. The abnormal fusion is due to failure of the lateral lip element to reach the midline. Because NPP remains connected to the nasal floor, it draws the cleft side philtrum upward, giving the impression of "shortness."

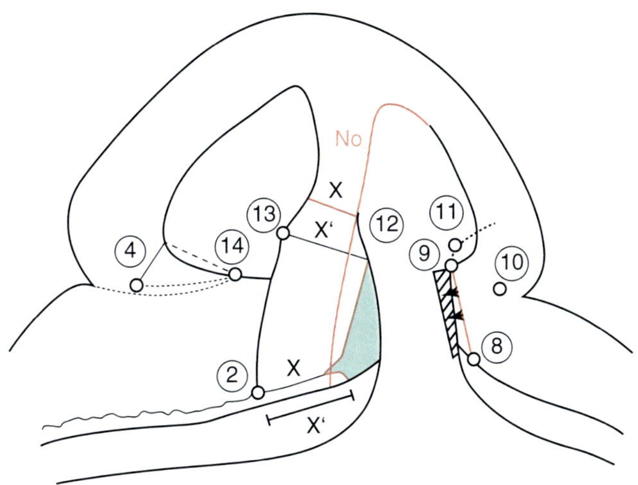

Fig. 19.31 DFR markings. Numbering is based on the original numerical sequence popularized by Millard. The width of the "true" philtral prolabium (P) is the width of the columella at the level of the alar footplates at the tips of the medial crura. Let us call this distance *x*. Point 2 is the normal/non-cleft PMx philtral column at the white roll. The new philtral column on the cleft side, point 3, is 2-*x*. Point 1 in the Millard system, the visual "center" of the cupid's bow, is therefore irrelevant. Points 4 and 10 are the centers of the alar bases on the non-cleft and cleft sides, respectively, as defined by the light reflex. Point 5, the Millard back-cut, is irrelevant. Points 6 and 7 are the commissures. Point 8 is the tentative location of the cleft-side philtral column on the lateral lip element at the white roll. This can be adjusted. Measuring distances 2–6 and 7–8 are rough guides to equality but not terribly useful. The height of the Cupid's bow is measured from the alar footplates, with point 13 on the non-cleft side and point 12 on the cleft side. Distance 13–2 is the true height of the lip and will equal 12–3 with the addition of the Matthew's triangle (discussed below). 13–2 will equal 8–9. Point 9 can be marked as the highest point on the skin margin of the lateral lip element. The alar base on the cleft-side is rotated inward and this translates the nostril sill internally. The nostril sill is a triangle defined on the non-cleft side by 4–14, with point 14 being the terminus of the sill. This is usually 3–4 mm. You can take the compass and measure across the sill from point 4 into the nose and find the other leg of the triangle. In similar fashion, the nostril sill on the cleft side can be marked out from point 10 based on the measurements on the normal side. The tip of nostril sill flap, point 11, is inserted at the base of point 12 to re-establish the normal width of the nostril floor. Note: The **NPP field** (green) is powered by a designated branch of medial nasopalatine located just at the septopremaxillary junction, i.e., at incisive foramen. It is below the footplate of the medial crus. [Reprinted from Carstens MH. Developmental field reassignment cleft surgery: reassessment and refinements. In: Bennun R, Harfin J, Sandor G (ed). Cleft Lip and Palate Management: A Comprehensive Atlas. Hoboken, NJ: John Wiley & Sons; 2016:83–111. With permission from John Wiley & Sons.]

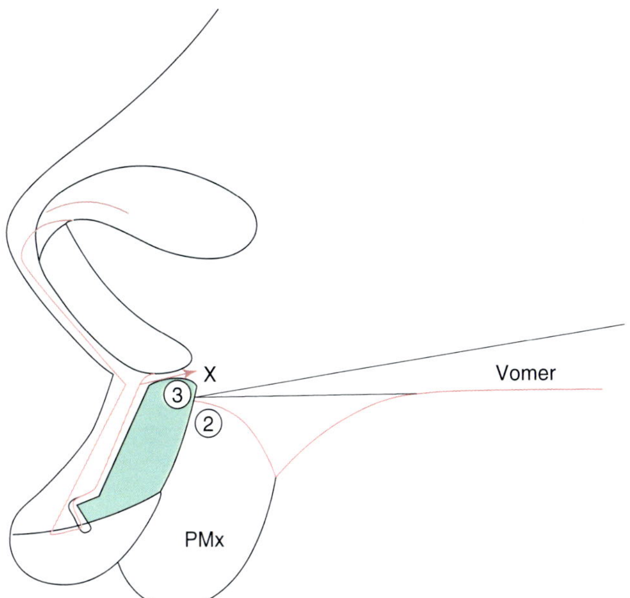

Fig. 19.32 Medial dissection of NPP flap: independent flap vs. composite with LCC. Markings of the both versions of DFR are show here. Here the incision separating the non-prolabium (NP) from the philtral prolabium (PP) sweeps upward along the lateral margin of the columella: **open-closed rhinoplasty**, a modification of Trott a small counter-incision in the membranous septum helps lift up the medial crus like a "boot strap." This will leave a 3–4 mm gap. One can replace it by backing up the "shoulder" of the NPP into the deficit at the same time as it is repositioned into the nasal floor. This may require some stretching around the pedicle to gain mobility. Another alternative is to simply rotate NPP on its existing axis and fill the gap with a composite punch graft from the ear. The lower paring incision of the NP flap is carried over the alveolus. If the alveolar cleft is complete, it can reach to the vomer to elevate an optional vomer flap and thus achieve closure of the nasal floor. A lateral nasal incision elevates tissues in front of the inferior turbinate. With the Talmant dissection of the nasalis complex, this lateral design is not useful. There may be situations in which the access gained by the lateral columellar incision justifies its use. The resulting scar behaves well and is inconspicuous. Matthew's modification brings the incision underneath the alar footplate. It can stop at point 1. Placing curved scissors through the incision beneath the footplate places the surgeon immediately beneath the medial crus. The dissection extends readily up to the tip, where it becomes superficial to the intermediate crus and to the lateral crus. Option 2 follows the alveolus backwards to the vomer permitting elevation of a vomer flap to close the floor of the nose. Complete closure of the nasal floor over an alveolar cleft is critical to avoid an iatrogenic fistula and for successful grafting at age 4. In the Talmant protocol, soft palate closure is achieved at the first surgery but nothing is done with the hard palate. Certain surgical situations, where recall is uncertain, may require closure on the nasal side because it will never be more readily accessible. The extension of the NPP flap incision can be readily carried backward on the vomer to accomplish this goal, leaving completion of the hard palate with mucoperiosteal flap mobilization to a later stage in the sequence. Note that the lateral columellar incision lifts up the entire medial crus and re-sets it into position *vis-a-vis* the normal side. This maneuver proved highly effective with exceedingly good scar but was superseded by Matthew's innovation. [Reprinted from Carstens MH. Developmental field reassignment cleft surgery: reassessment and refinements. In: Bennun R, Harfin J, Sandor G (ed). Cleft Lip and Palate Management: A Comprehensive Atlas. Hoboken, NJ: John Wiley & Sons; 2016:83–111. With permission from John Wiley & Sons.]

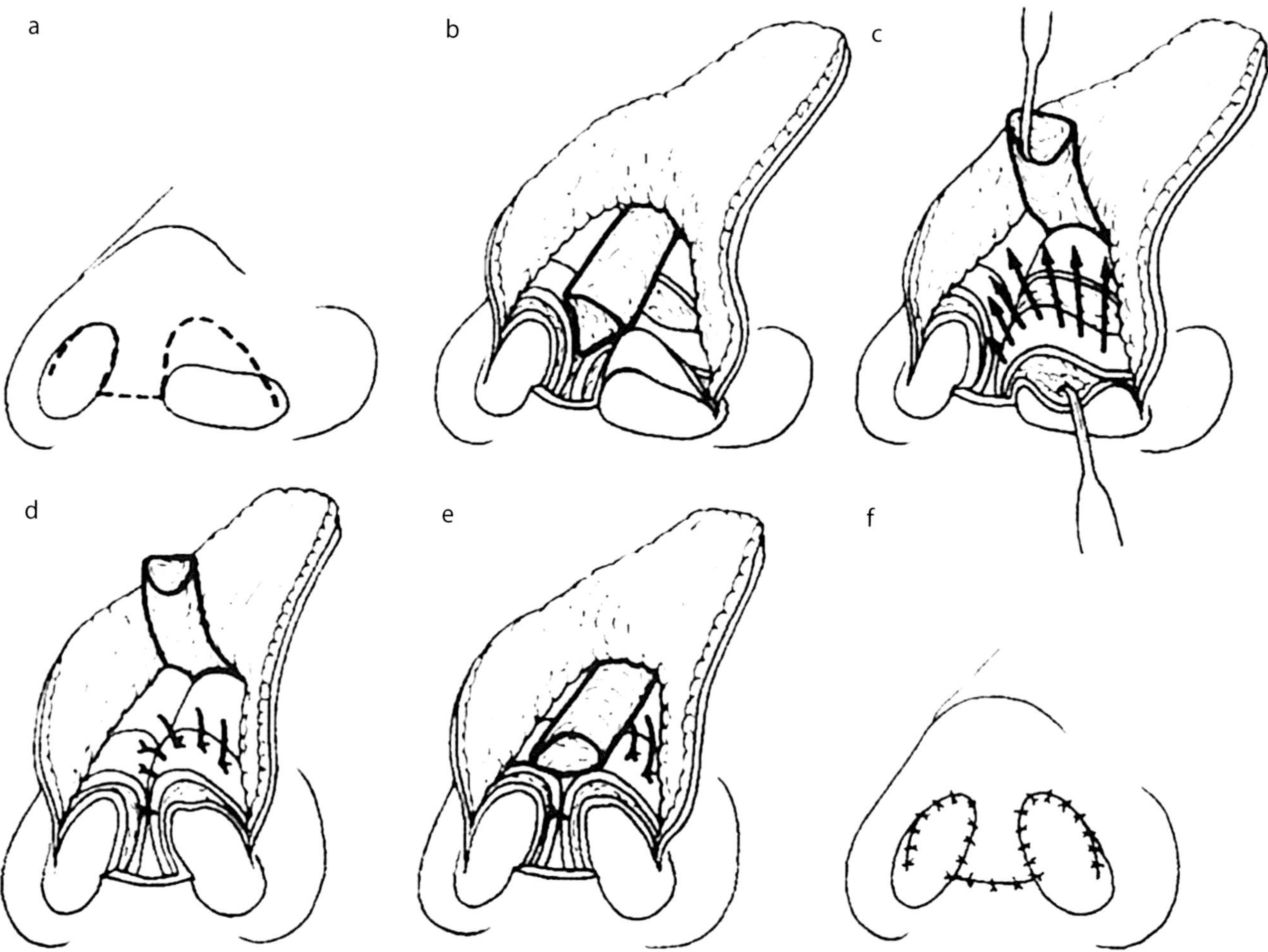

Fig. 19.33 Access to the nose via anterior columellar incision. Harashina showed a variation of the 1919 Rethi incision which opens the lateral columella in front of the medial crura and includes part of the central columella, transecting it, as in an open-rhinoplasty. Had he proceeded further down into the lip and philtrum he would have isolated the entire StV1 field. Trott and Mohan extended this idea. Perfusion of remaining lateral columella is profuse vessels of the rim and septum. (**a**) Transcolumellar incision with extension along the rim of the non-cleft side and, on the cleft side, along the caudal border of the alar cartilage. (**b**) Skin flaps elevated and fibro-fatty tissue elevated away from the nasal tip. (**c**) Suture fixation of the lower lateral cartilage and cleft-side alar cartilage is advanced. (**d**) Narrowing of tip with interdomal sutures. (**e**) Fibro-fatty tissue replaced back over tip. (**f**) Closure. [Reprinted from Harashina T. Open reverse U incision technique for secondary correction of unilateral cleft lip deformity. Br J Plast Surg 1990; 86:872–881. With permission from Elsevier.]

- Surgical separation of non-philtrum from the Cupid's bow makes use of this embryonic fusion plane. NPP is reassigned into the nasal floor and medial nasal wall nasal floor. It joins with tissues from the alar to recreate the missing nostril floor anterior to that of the mucoperiosteum. NPP reassignment releases the medial crus and expands the airway. Simultaneously, the ipsilateral philtrum is dropped into position. Modifications of the DFR incisions based on the contributions of Matthews and Talmant make the surgical design embryologically accurate and (not surprisingly) give significant functional improvement.

Medial Dissection: Modifications

Ten years ago, I met David Matthews at the ASPS in Orland, Florida. Over the years we have developed a close professional cooperation. We found that, from a technical standpoint, our dissections were very similar. Several years ago, Dr. Matthews began to incorporate the NPP flap into the nasal floor. However, when we compared long-term outcomes my cases demonstrated a flattening of the Cupid's bow at the junction of the white roll and the cleft-side philtral column, whereas my colleague's patients maintained eversion all along the Cupid's bow. Clearly, there was something

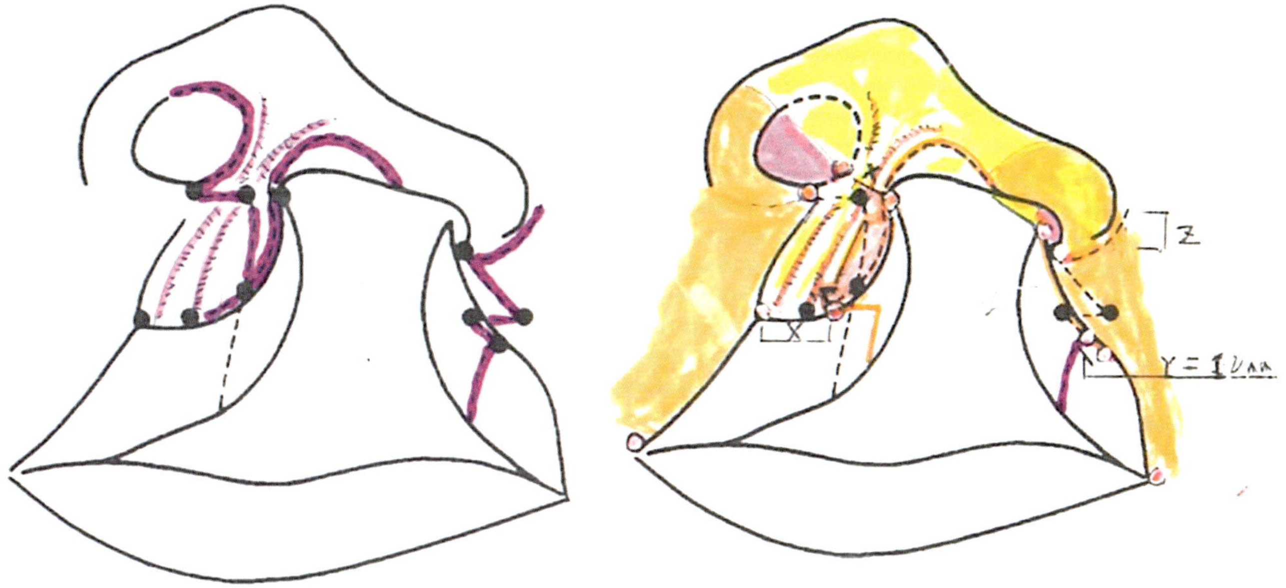

Fig. 19.34 Trott and Mohan came very close to the open-closed rhinoplasty. This design for UCL demonstrates typical tension with BCL where developmental fields are obvious in their bilateral design the NPP was staring them in the face. If they had only carried over these observations to the UCL repair they might have been able to use a full NPP. Unfortunately the dogma of the Millard "c flap" proved to be a stumbling block to innovation. Note my superimposed DFR incision marks (orange) [28–30]. [Reprinted from Song R, Liu C, Zhao Y. A new principle for unilateral complete cleft lip repair, the lateral columellar flap method. Plast Reconstr Surg 1998; 102(6):1848–1852. With permission from Wolters Kluwer Health, Inc.]

amiss with the design of my straight-line incision as it approached the caudal margin of the Cupid's bow (Figs. 19.30, 19.31 and 19.32).

- ***Matthews #1***: The answer was the *preservation of the uppermost fibers of the deep orbicularis oris* by means of a small back-cut incision that drops the muscle into proper position with respect to that of the non-cleft side. Technical details of this modification will be described below.
- ***Matthews #2***: I also observed that the proximal limb of Matthews' NP incision could be nicely carried into the nasal floor just beneath the footplate. The trick was to incise *just the skin*, leaving the subcutaneous pedicle powered by intact MSP artery branches. The resulting NP flap rotates perfectly into the lateral nasal incision. Furthermore, the incision *permits access to the internal surface of the medial crus*. One simply inserts the curved scissors inward, locating the internal surface of the medial crus, and elevates. From the very moment I saw his approach it was obvious to me that *the misplaced medial crus could be corrected perfectly well without an external incision*. Simple is always better.

The embryologic implication of Matthews' incision is that the true width of the PP was not (as I had previously thought) merely the width of the columella alone. Instead, *PP equals the distance between the alar footplates*. This **adds 2–3 mm** to the width of the Cupid's bow along the white roll. As we shall see, the terminal point of Matthews' back-cut is aligned with the old definition for PP.

Medial Dissection: Embryologic Lessons, Access to the Nose

A side benefit of Matthews' critique of the DFR incision design (not the concept) was that it forced me to rethink the embryologic boundaries between the columella, the nasal floor, and the lip. Landmarks were needed to define the junction between frontonasal skin (of forebrain origin) and facial skin (of hindbrain origin). To illustrate this point, we must discuss the embryologic origins of body skin, facial skin, and frontonasal skin. All three arise in a radically different manner. Here is a quick and dirty summary. Armed with this information, we can understand how the developmental fields of the lip and nose are laid out.

Body skin arises from two tissue sources. Epidermis comes from ectoderm, whereas dermis is mesoderm. During gastrulation (the creation of a three-layer embryo) the mesoderm becomes segmented into somites. Each somite is supplied by a sensory nerve, the nucleus of which lies within the corresponding developmental unit of the CNS: the neuromere. These tissues are in genetic register with each other. T3-innervated dermis is genetically connected with the third

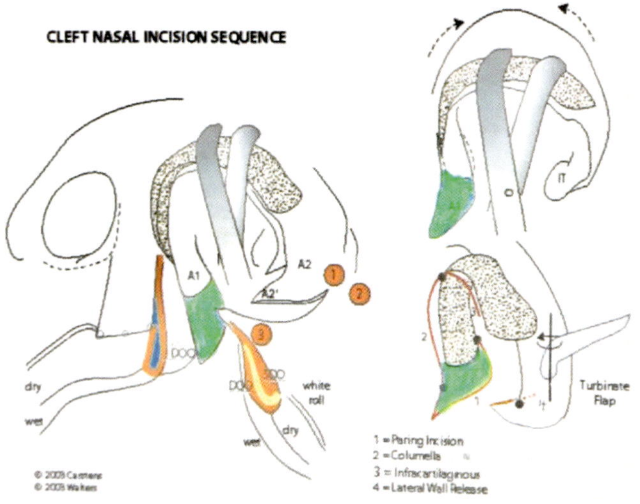

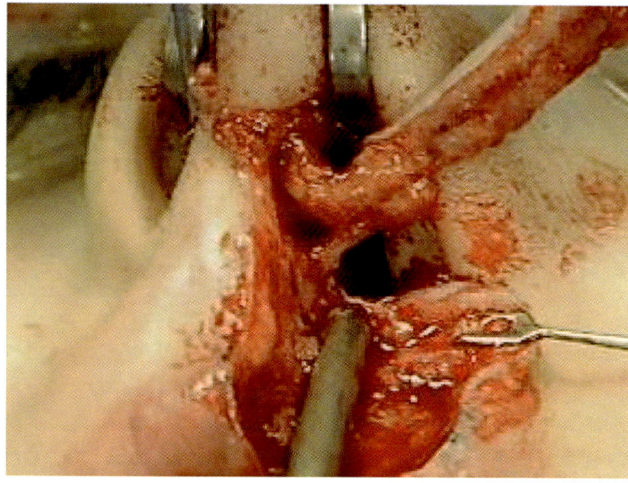

Fig. 19.35 Composite NPP–LCC flap and "open-closed" rhinoplasty. Left: NPP–LCC is a small flap with a long name: non-philtral prolabium-lateral columellar chondrocutaneous flap. Right: NPP flap (green) is in continuity with LCC. Vascular pedicle is outlined by black tape. Footplate of the medial crus is in the raw tissue just above the skin of the flap. Flap includes subcutaneous scar which can be trimmed. Note that the NPP flap has been dissected using a lateral columellar incision. The blue tape marks the pedicle. NPP flap is a skin flap. Note here the mucosal extension to the NPP flap. Note: In the cleft state, the prolabium is wider than normal. This is because non-philtral tissue that should be assigned to the premaxilla becomes included. I have therefore considered the cleft "prolabium" to be a composite structure consisting of a "true" philtral component (supplied by the terminal branches of the StV1 anterior ethmoid axis) and a "false" non-philtral component (supplied by the StV2 medial nasopalatine axis). Thus, the incision that runs upward from the lip at point 3 separates the *P* philtral prolabium, from the *NPP* non-philtral prolabium. Note: The true position of the NPP is to provide soft tissue coverage over the lateral incisor, that is, the introitus of the nasal floor. The vessels supplying the NPP flap come from the vomer–premaxillary junction. These can be readily visualized, but blunt dissection in the subperiosteal plane is protective. Note here (right) that the NPP flap has been dissected using a lateral columellar incision. The blue tape marks the pedicle. NPP flap is a skin flap. Note here the mucosal extension to the NPP flap. Left: [Reprinted from Carstens MH. Developmental field reassignment cleft surgery: reassessment and refinements. In: Bennun R, Harfin J, Sandor G (ed). Cleft Lip and Palate Management: A Comprehensive Atlas. Hoboken, NJ: John Wiley & Sons; 2016:83–111. With permission from John Wiley & Sons.] Right: [Reprinted from Carstens MH. Functional matrix cleft repair: Principles and techniques. Clin Plast Surg 2004; 31: 159–189. With permission from Elsevier.]

thoracic neuromere. Each "swatch" of mesoderm is "assigned" to a zone of overlying ectoderm; together they share a common genetic definition. This skin unit is what textbooks refer to as a "dermatome" (sic). Precisely where does body skin begin? The alert reader may ask, "Why is there no C1 dermatome?" All somites contain dermatomes, but those of the first four occipital somites and the first cervical somite are *unstable* and degenerate. Thus, the body skin (dermatome-derived dermis) does not appear until the second cervical somite.

Facial skin arises from two different sources as well. The epidermis comes from ectoderm but the dermis arises from neural crest. Once again, neuroanatomy comes to our rescue. Facial skin is innervated by V2 and V3, whereas frontonasal skin is innervated by V1. Since the V2 nucleus resides within the second rhombomere, it makes sense that the neural crest dermis of maxillary division skin arises from the neural fold just above r2. These tissues share a common genetic "signature." Mandibular dermis arises from the r3 neural crest.

Frontonasal skin is utterly different: it arises from a single source. Here, specialized non-neural ectoderm of the neural folds overlying the secondary prosencephalon (hp1 and hyp2 brain) gives rise to all frontonasal epidermis except vestibular lining of the nose. This specialized epithelium contains neurons and arises from the nasal placode which lives within the anterior forebrain neural fold. The source of dermis for frontonasal skin if neural crest from prosomeres p1–p3 encoding the diencephalon.

For readers new to the subject of neuromeric mapping, the above description probably seems abstract. Let's convert this to anatomic terms understandable to all surgeons, that is, to *neuroangiosomes*. Vestibular skin (hp2 placode) and fronto-orbital-nasal skin (hp1–hp2) are innervated by V1. Arterial supply to both these regions is from branches of the V1 stapedial branches of ophthalmic axis. Vestibular "skin" consists of septal mucoperichondrium and lateral nasal wall mucoperiosteum. These tissues are supplied by the posterior and anterior ethmoid arteries, both of which send out nasal branches to their respective targets. External nasal envelope skin is supplied by terminal branches of the anterior ethmoid. These exist from beneath the nasal bones and run downward to supply the distal (non-vestibular) internal nasal skin and the columella. The skin lying immediately beneath the alar footplates belongs to a separate developmental field, the r2 NPP, supplied by the medial nasopalatine axis.

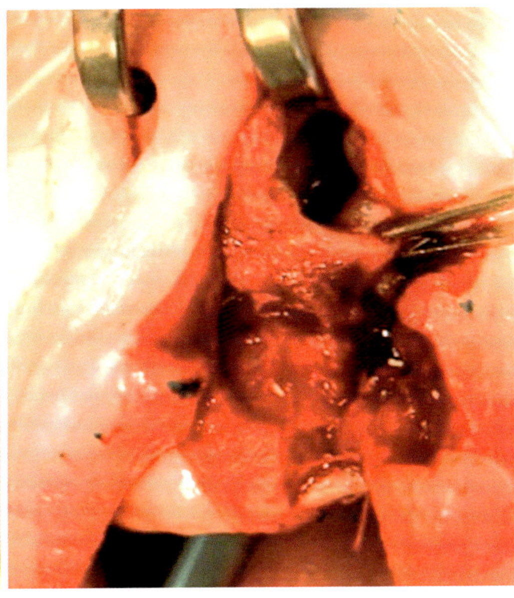

Fig. 19.36 Creation of composite flap NPP–LCC. Original design of NPP–LCC had two incisions: (1) anterior columella into the nose, and (2) membranous septum 50%. Results was a U-shaped "bucket handle" flap. Here we see the separation of non-philtral prolabium (NPP) from philtral prolabium (PP). Note how incision flares around columella until just before the footplate. It then ascends up *the side* of lateral columella *in front of the medial crus* and into the nose as far as intermediate crus. To elevate the flap, I used to make a counter-incision in the membranous septum the medial crus. Mobilization around the nasopalatine foramen is blunt to avoid injury to the pedicle. The medial crus is elevated into the nasal tip, bringing NPP with it. At this point, I don't think the counter-incision is necessary, just good mobilization. [Courtesy of Michael Carstens, MD]

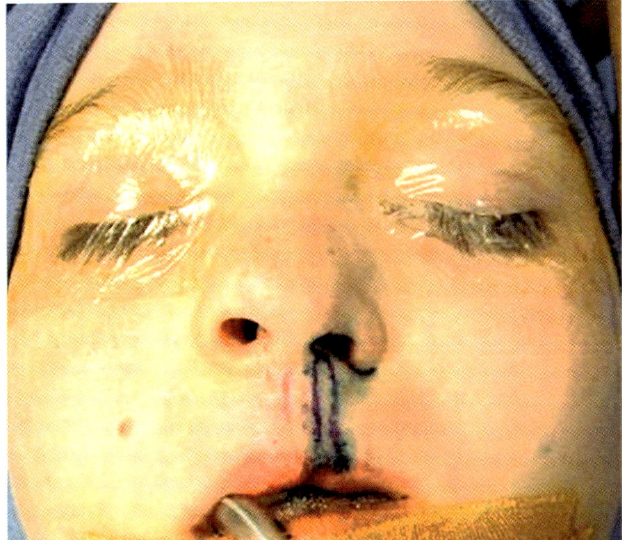

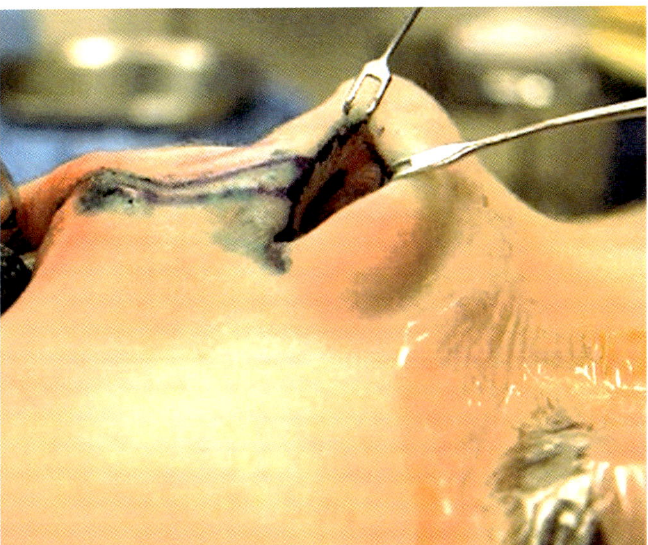

Fig. 19.37 Vissiaronov flap. In cleft surgeries that leave behind the NPP, subcutaneous tissues (buried) are still capable of supporting a full-thickness skin flap harvested from the scar but widening below to include blood supply (now random). This was the basis of the flap described (in Russian) by Vissiaronov. We are indebted to Millard for preserving his work. Lateral columella chondrocutaneous flap (LCC) is elevated as a "bucket handle" with two parallel incisions connected beneath the footplate. But the surprising finding with the NPP scar flap (recall that some NPP tissue may have been discarded at the previous surgery is much more than the skin. Tracing subcutaneous tissue (including the scar) will lead you right up to the pedicle just beneath the footplate. When LCC advances, the "shoulder" of NPP can be "backed up" into the defect, with the remainder of the flap rotated into the nasal floor and trimmed to fit the defect (as there is always an excess of tissue). [Courtesy of Michael Carstens, MD]

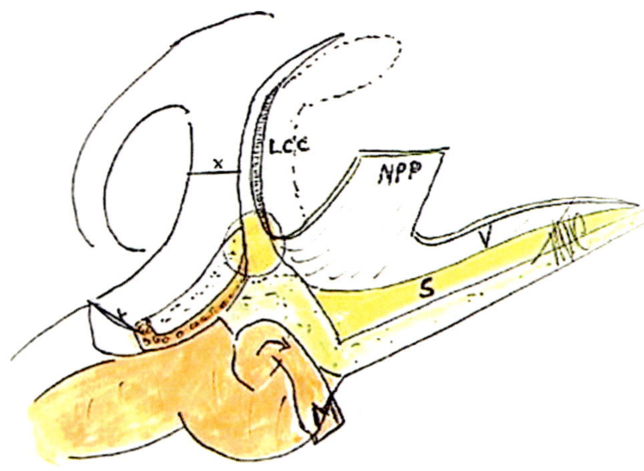

Fig. 19.38 Closure of the nasal floor. NPP–LCC elevated in continuity with vomer flap (V). Septum (S), gold color, accessed from vomer, yellow. Circle denote pedicle to NPP at level of nasopalatine duct. M flap can pared from the NPP or kept with it.

- Advantage: open-closed rhinoplasty
- Advantage: no trans-columellar scar as in Harashina; lateral columella scar well tolerated
- Advantage: counter incision in membranous septum behind the medial crus is not required
- Advantage: direct access for mobilization of medial nasal vestibular lining (Talmant 1 maneuver)
- Disadvantage: additional closure time (15 min)

[Courtesy of Michael Carstens, MD]

Cartilages of the nose develop when neural crest mesenchyme comes in contact with, and is "instructed" by an epithelial program. The size and shape of the upper lateral (triangular) cartilages are *vestibular lining* of zone 12, whereas the contours of the lower lateral (alar) cartilages are fixed by interaction with zone 12 skin. The medial crura are programmed from nasal skin of zone 13.

- The footplates of the medial crura are the landmark for where skin ends *and where r2 upper lip skin begins*.

Freeing the Medial Crus: Lateral Columellar Incision

The whole idea of the DFR incision is field separation. In the original iteration of DFR I took the prolabial separation incision right up the side of the columella and combined two unrelated neuroangiosomes: r1 frontonasal + r2 medial nasopalatine. The flap was large: NPP skin (medial nasopalatine) in continuity with the entire ipsilateral columella (anterior ethmoid). It merited an equally awkward term: NPP–LCC (lateral columella chondrocutaneous) flap. This design did not achieve embryologic field separation, but it seemed to work. I rationalized this compromise by convincing myself that the downward displaced, to be repositioned correctly, the cleft-side medial crus needed a surgical release (Figs. 19.35, 19.36, 19.37 and 19.38).

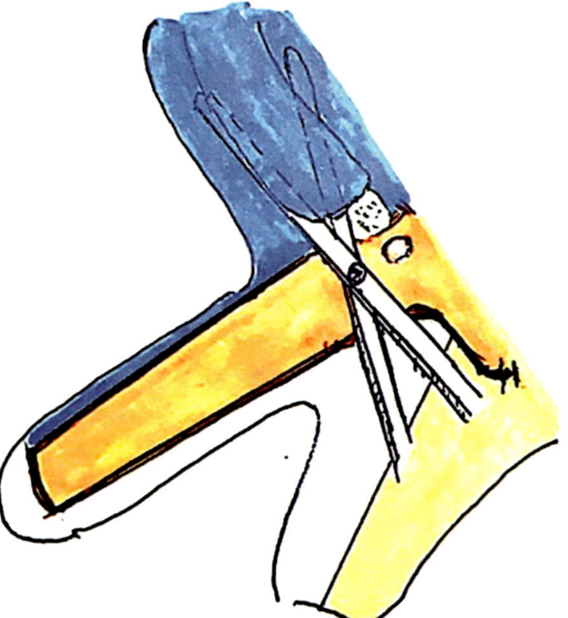

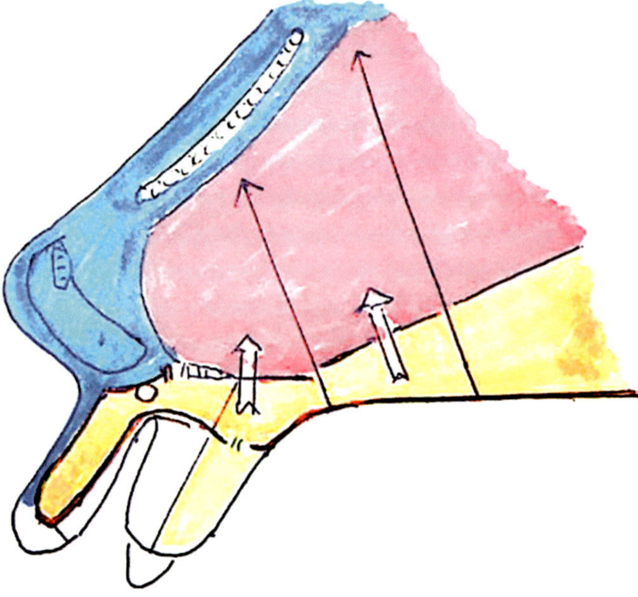

Fig. 19.39 NPP flap alternative incision design (Matthews). Lateral columella incision and membranous septum incisions are *not* done. NPP extends proximally in curvilinear fashion just below footplate of medial crus. It puts the access incision to columella *above* the NPP flap. From here, once again, the NPP flap can be kept continuity with vomer… preserving the pedicle… and used to close the nasal floor above the alveolus by suturing NPP-V to the mucoperiosteum elevated from nasal side of the secondary palate

- Advantage: Talmant 1 elevation of medial nasal vestibular lining can be done
- Disadvantage: open-closed rhinoplasty not available

Key: blue, columella with anterior ethmoid; pink, septum with anterior/posterior septals from ethmoid; orange, medial nasopalatine angiosome, lemon mixture of greater palatine (if continuity with one maxilla) with medial NP

[Courtesy of Michael Carstens, MD]

Freeing the Medial Crus: the Infra-Footplate Incision

This concept proved to be unnecessary. Correct separation of PNC and r2 RNC skin takes place *just below the footplates*. This permits elevation of the media crus from below just as effectively as the lateral columella incision with shorter time of closure. The incision, just through the skin, permits gentle spreading of areolar tissue. It automatically puts the scissors inside the columella; by hugging the medial crus one can advance right the way up into the nasal tip with virtually no bleeding. Finally, the infra-footplate incision puts one safe above the medial nasopalatine pedicle permitting safe dissection of the NPP flap (Figs. 19.39 and 19.40).

Principles of Lateral Dissection: Lip–Nose Muscles, Nasal Dorsum, and Lateral Nasal Wall (Figs. 19.41, 19.42, 19.43, 19.44, 19.45, 19.46, 19.47, 19.48, 19.49, 19.50, 19.51, 19.52, 19.53, 19.54, 19.55, 19.56, 19.57, 19.58, 19.59, 19.60, 19.61, 19.62, 19.63, 19.64, 19.65, 19.66 and 19.67)

Wide Subperiosteal Release

This is one of the first principles I learned that comes directly from work of Delaire and Soterianos [31]. What I did not understand 20 years ago was that this was really a case of functional stem cell transfer. We shall recount its history and literature.

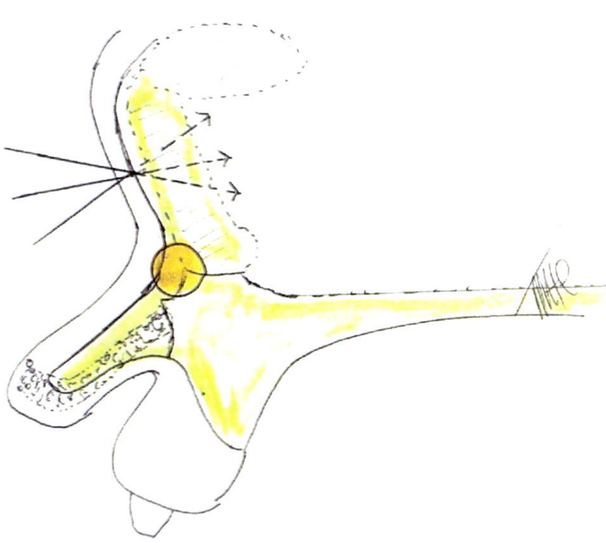

Fig. 19.40 Lateral columellar incision vs. infra-footplate incision. Subperiosteal dissection of NPP from premaxilla takes you safely right to the pedicle (circle)

- Lateral columella incision: very easy to elevate medial cruse, direct access to septum
- Infra-footplate incision accesses the septum. Elevation of medial crus is blind

NPP-vomer flap = makes use of entire medial sphenopalatine neuroangiosome

- Access the septum from below
- Elevation of medial crus less facile

[Courtesy of Michael Carstens, MD]

Separating the Orbicularis Layers in the Lateral Lip Element

DOO and SOO insert differently. DOO needs to extend further forward to the midline to joint its opposite just under anterior nasal spine. SOO, on the other hand, has a different vector and is sutured to the mesenchyme of the PP. Paring of the mucosal border exposes these two layers and the artery that runs downward between them to the vermilion border. Naturally, the vascular axis is surrounded by adipose tissue which can be gently spread in the upper half of the lateral lip to gain mobility between the layers. The vertical incision running upward along the anterior border of the alveolar cleft provides access to the subperiosteal plane for wide mobilization. A counter-incision through the mucoperiosteum running up the buttress can be used as needed to convert the entire lip–cheek unit into a giant bipedicle flap.

Fig. 19.41 Jean-Claude Talmant: cleft surgery must restore airway. His brother demonstrated movements of the nasal cartilages with fetal breathing. The nasal chamber expanded and contracts with the flow of amniotic fluid. Distortions of airway diameter affect development of nasal envelope and, in particular, the shape of the septum. [Courtesy Dr. Jean Claude Talmant]

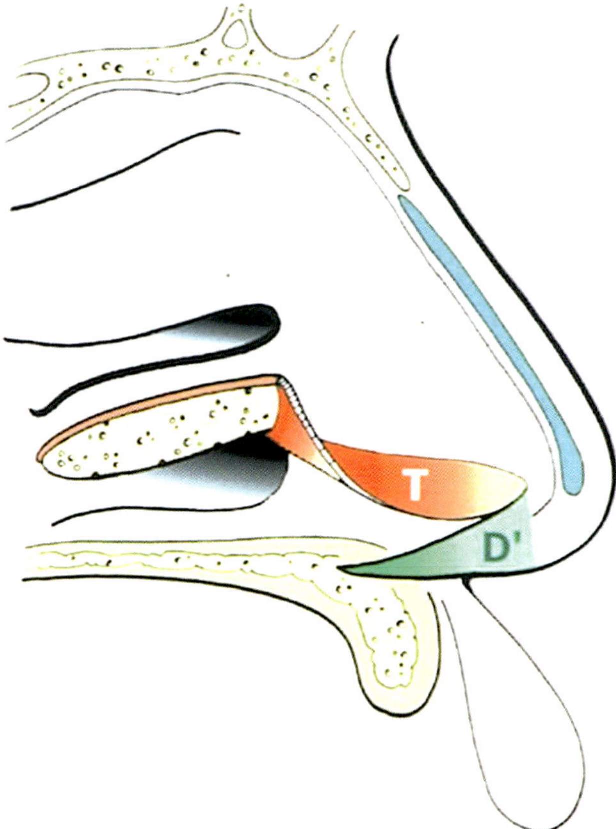

Fig. 19.42 Inferior turbinate flap (after Noordhoff). Very useful "patch" for lateral nasal wall. I did not recognize extent of mobilization until the Talmant 2 maneuver. Simple excision, spreading, and turbinate flap did not give reliable advance of the alar cartilage. At the present time, with the extensive mobilization, the deficit created sometimes not reliably covered with NPP. Important not to tighten the airway. Thus, turbinate flap becomes relevant again. [Courtesy of Michael Carstens, MD]

Management of the Lateral Nasal Wall: Trial and Error

The Vestibular "tightness": A Case of Misunderstood Mismatch.

In normal patients, the lateral nasal wall just in front of the inferior turbinate contains the soft tissues and bone of the premaxillary frontal process. In virtually all cleft patients this zone appears contracted. A prominent vertical fold or web is seen running downward from the distal margin of the lateral crus and terminating at the palatal shelf. My initial concept of the cleft nasal deformity was that the vestibular lining was somehow deficient. This was very much influenced by the work of Markus, Delaire, and Precious (more on this later). The deficiency site could be readily "patched" with an inferior turbinate flap, rotated 180° forward. This became part of what I considered the "process-oriented" cleft rhinoplasty [32]. Alternatively, I tried filling in the deficiency site with composite skin-cartilage graft from the ear. The cartilage was intended to prevent contraction and relapse. Despite stenting, I was disappointed in the results. As it turned out, the model was wrong and I had not yet stumbled on the developmental field concept.

Upon discovery of the PMx subfields I thought I has found a better solution. Absence of the PMx subfield seemed to explain the entrapment of the lateral crus. After working out the neurovascular anatomy of the prolabium in 2000, it seemed logical that this problem could be addressed by wide subperiosteal release, mucosal release incision, and addition of the NPP flap into the defect. As an adjunct, the nasal airway was supported by postoperative stenting for up to 3 months.

By 2010 I was convinced that this idea was insufficient. Although the initial release seemed to create a nearly perfect airway, in many patients the expansion was unstable. The nostril sill was better but airflow in these patients continued to be reduced compared to the normal side. Furthermore, on physical examination, I started to note a dynamic component to the cleft airway. Watching unilateral patients breathe from below and from above demonstrated an asymmetry of alar movement. Despite the addition of adequate soft tissues, forceful inspiration would be accompanied by an *inward contraction*. It was time to return to the drawing board: Gray's Anatomy and the work of Jean Claude Talmant (Fig. 19.41).

Reassignment of the Nasalis Muscle and Vestibular Lining

I was fortunate to meet Dr. Talmant in 2005 as invited faculty for the Indian Society for Cleft Lip, Palate and Craniofacial Anomalies in Guwahati, Assam. His emphasis on the nasal airway as a key to maxillary growth made biologic sense and his results were impressive. His concepts are well summarized in "Evolution of the functional repair concept for cleft lip and palate patients," published in *IJPS* 2006; 39(2):197–209. This article should be required reading for all cleft surgeons. This being said, I must confess that my initial review of Talmant's work was too superficial. I underestimated the physiologic significance of the nasalis muscle and I missed its embryologic relationship to the frontal process of the premaxilla [33].

Four years later, Dr. Talmant's work came to my attention once again via David Matthews. After visiting Talmant in Nantes he brought back three key ideas: (1) reassignment of the nasalis muscle to its correct position; (2) subperiosteal release and the internal nasal lining; and (3) effective techniques to correctly reassign the nasal lining using packing and a much better technique for airway stenting. These three maneuvers made instant biologic sense. Cleft surgery without them is inconceivable.

These maneuvers describe below are based on three embryologic principles

- First, the concept of functional muscle repair, as per Delaire–Markus, is of great value but contains a funda-

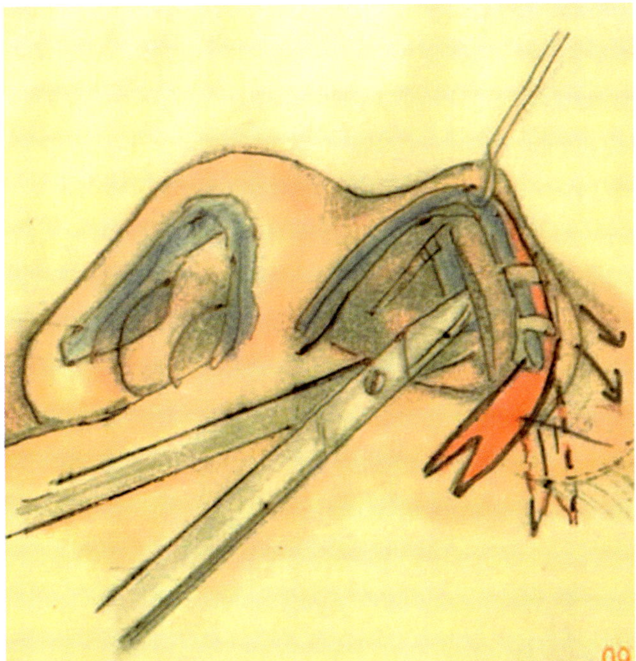

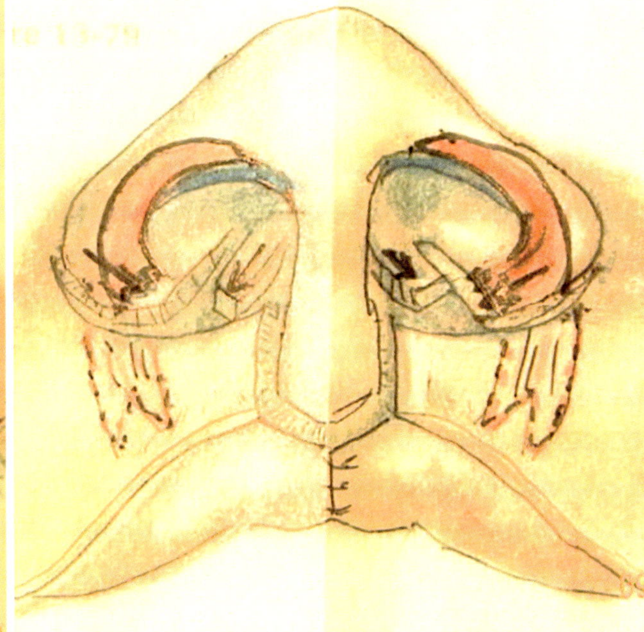

Fig. 19.43 Consequences of conventional surgery: two problems. Traditional approaches to cleft rhinoplasty include lateral approaches to the lateral crus via the ala, via vestibular incisions. These can be combined with medial approaches, such as Rethi incision. Here we see two parallel incisions, pre-cartilage and intercartilage to "free" the alar cartilage to reposition in into the tip. Recall that the cartilage is "programmed" by the vestibular lining. Thus, freeing the vestibular lining for overlying soft tissues (Talmant) accomplishes lasting alar cartilage stability. Failure to do so results in

- Tip definition compromised
- Relapse of the lateral nasal wall
- Improper nasal valve function

The nasalis complex (red) is not reassigned to its normal insertion site (dotted lines) but transferred to the midline where it acts as a constrictor of the airway, as times seen as paradoxical motion of the ala with forced inspiration

- Constriction of the airway
- Vestibular "web"

[Courtesy of Michael Carstens, MD]

mental embryologic misconception about the nasalis muscle. By assuming that it inserts into the midline at or near the anterior nasal spine, and by failing dissect it out as a separate structure, all standard lip repairs, be the supraperiosteal or subperiosteal, anchor nasalis into the midline, thereby condemning it to function as a constrictor, instead of as a dilator.
  – Virtually 100% of complete unilateral and bilateral cleft repairs having the nasalis centralized will have paradoxical alar movement during respiration and impaired nasal breathing.
- Second, the surface area of vestibular lining is normal. It is simply mismatched with respect to the overlying skin envelope. The layers can be separated bluntly with an elevator and the internal lining rotated upward toward the midline. The distal anchorage point is release as well during the process of bone dissection into the alveolar cleft and piriform fossa.
- Third, low-grade stenting in the postoperative period is very effective and, depending on the technology used, quite well tolerated.

Talmant #1 Like all muscles innervated by the facial nerve, the nasalis muscle originates from the paraxial mesoderm of somitomere 6. It flows forward within the substance of the SMAS along the trajectory of the buccal branch of VII. Upon reaching its destination it inserts at two distinct locations. Under normal conditions, the muscle has two proximal sites of attachment, both of which are into the mucoperiosteum of the canine fossa. A lateral "head" lies over the root of the canine while the medial "head" is located above the lateral incisor. These proximal attachments require the correction position of the embryologic fields making up the piriform rim. First to develop is the frontal process of the maxilla, arising from just above the canine. Second to develop is the

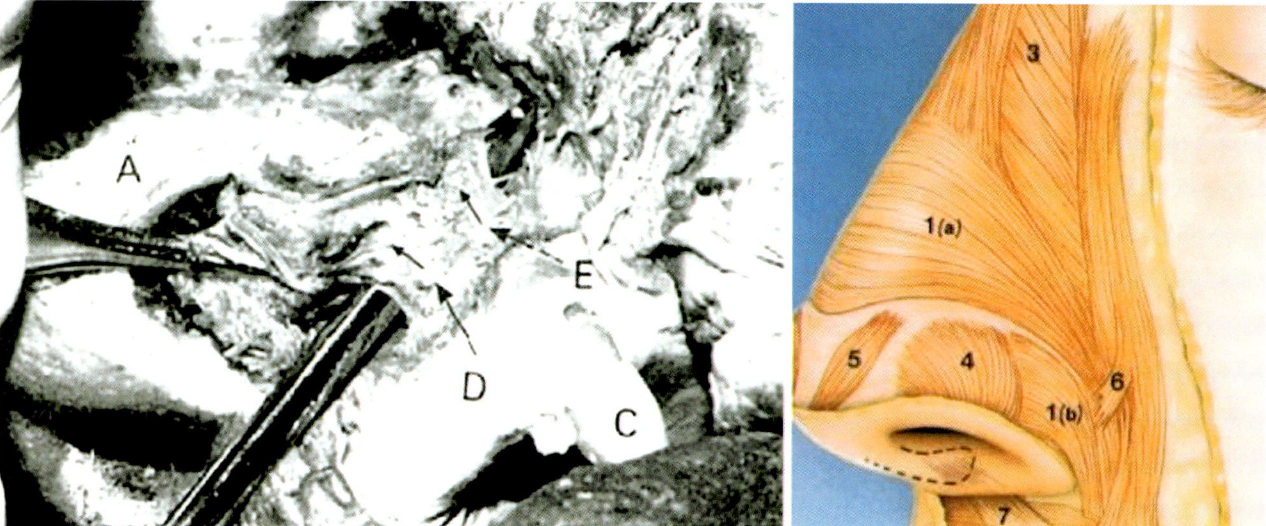

Fig. 19.44 Alar anatomy: cleft surgery must restore airway. Soft triangle defined by gap between compressor and pars alaris. Nostril sill is that segment that connects ala with "shoulder" of columella. It corresponds to Simonart's band between LNP and MNP. 1a, nasalis pars alaris (dilator) inserts above lateral incisor; 1b, nasalis pars transversus inserts into canine eminence; 3, anomalous nasi (dilator) is part of nasalis; 4, anterior dilator nasi is functionally part of pars alaris; 5, compressor nasi; 6, levator labii et alaque nasi); 7, oblique fibers of Delaire from SOO are superficial to insertions of nasalis and; 8, depressor septi nasi (no numbered) is deep so SOO, inserts over central incisor, and is shown inserting into the columella. Note pars transversus of nasalis is often labeled as a compressor but, acting as a chain attached to accessory cartilage and thence to the canine, is really a dilator. When surgically mal-inserted into the midline, it acts as a constrictor and a powerful constraint on the lateral crus. [Reprinted from Oneal RM, Beil J. Surgical anatomy of the nose. Clin Plast Surg 2010; 37(2): 191–211. With permission from Elsevier.]

frontal process of the premaxilla arising from just above the lateral incisor. The proximal heads of nasalis form their insertions in the same sequence. The SMAS then migrates upward, tracking along the vestibular lining, carrying the remaining nasalis myoblasts upward into the nose. At the lateral crus, the cartilage-vestibular attachment forces the SMAS over the dorsal surface of the alar cartilage. Here it forms a distal insertion into the perichondrium of the lateral crus.

In cleft patients the frontal process of the premaxilla is reduced or missing. The proximal muscle mass is pathologically displaced into the piriform fossa. It is attached to the mucoperiosteum immediately in front of the inferior turbinate, *the exact location of the missing frontal process of the premaxilla*. Distal nasalis anatomy remains unaffected. At operation, the muscle is readily encountered, using the same type of releasing incision as for DFR. The proximal muscle mass is substantial, one simply has to look for it. It is detached from two points: (1) from the mucoperiosteal lining, using sharp dissection; and (2) from the internal surface of the piriform margin using the subperiosteal plane. The distal muscle mass may be optionally released from the lateral crus using the subperichondrial plane. Reassignment of the displaced nasalis is accomplished by placing a mattress suture into the distal muscle mass and then suturing it to the alveolar mucoperiosteum just above the canine. This maneuver instantly corrects the pathologic vector of muscle contraction; it now acts functionally to *open the airway*, not to constrict it.

Talmant #2 Redistribution of the nasal lining is simple and elegant. It involves four maneuvers: (1) elevation of the septal mucoperichondrium all the way backward and upward to the nasal bone; (2) subperiosteal dissection from the internal piriform margin all the way backward and upward to the nasal bone; (3) subperiosteal dissection of the maxilla, external piriform margin and nasal bones; and (4) optional subperichondrial dissection via a rim dissection. Note that (3) and (4) involve the sub-SMAS plane.

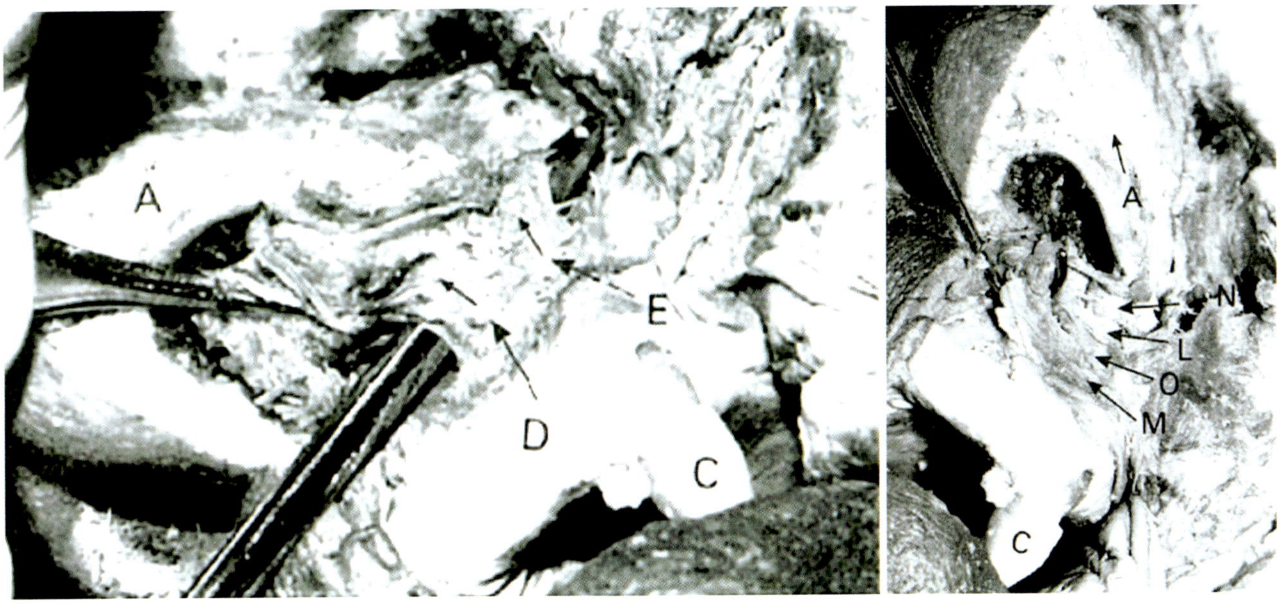

Fig. 19.45 Myrtiformis. Key, *(left)*: *A* alar margin, *C* canine, *D* inner fibers of myrtiformis (act as a depressor), *E* outer fibers of *myrtiformis* (resemble myrtle, its berries or leaves). Myrtiformis act as both (1) a depressor/stabilizer of the columella and ala and (2) via SMAS as nasalis, (both pars alaris and pars dorsalis) as a dilator of the airway. Key, *(right)*: *A* alar margin, *N* nasalis, *L* labial portion of levator labii et alaque nasi; *O* orbicularis, *M* inner fibers of myrtiformis (oblique fibers of Delaire from SOO?) [Reprinted from Figallo EE, Acosta JA. Nose muscular dynamics: the tip trigonum. Plast Reconstr Surg 2001;108(5):118–1126. With permission from Wolters Kluwer Health, Inc.]

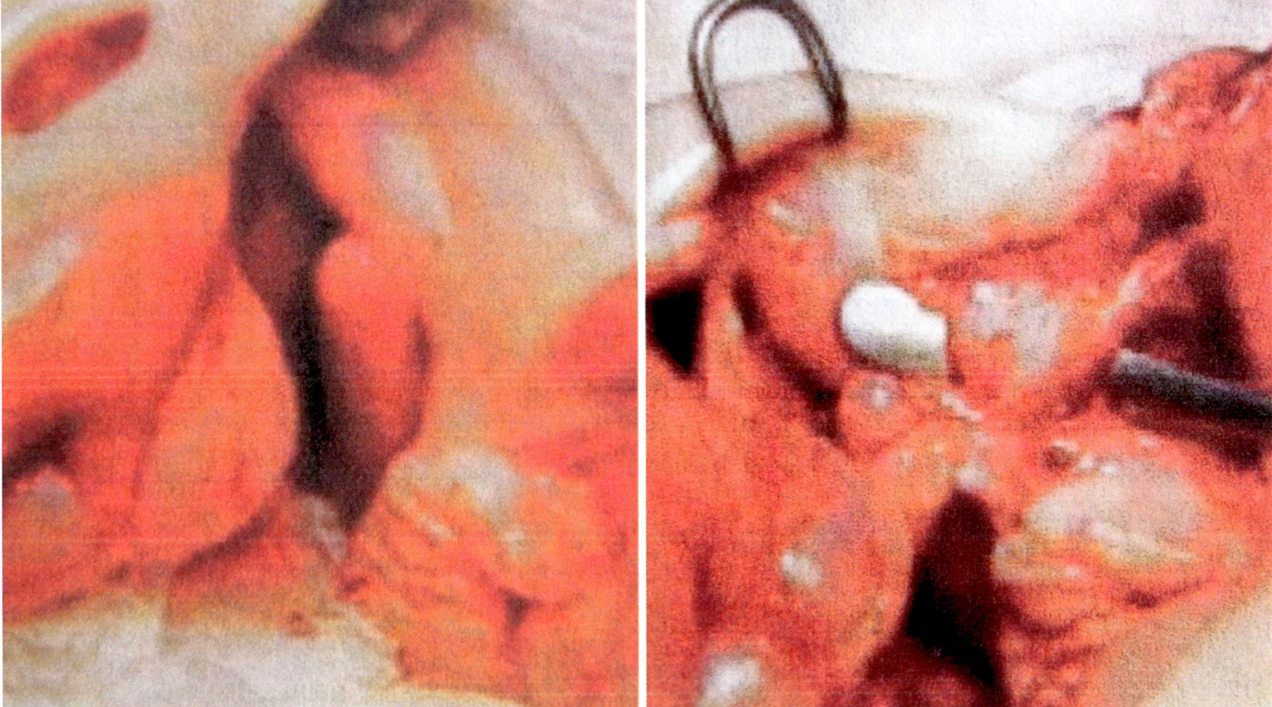

Fig. 19.46 Vestibular web is due to nasalis. *Left*: The mal-inserted nasalis exerts a "bowstring" effect. *Right*: Insertion into piriform fossa of deep head, pars transversus, is shown by the cotton tip applicator. *Note*: If you look under the skin of the nostril sill, just where it flows into the lateral lip, you will see fibers of pars alaris. They are the same thickness as SOO (0.5 mm) and are 1–1.5 cm wide. After dissecting the superficial plane of pars alaris, one dives beneath the muscle (while staying on top of pars dorsalis). This plane takes you directly to bone. Slip the elevator inside the piriform fossa and elevated pars dorsalis. It can be "delivered" from the fossa. The two heads are sutured as one unit above the canine. With release of nasalis, the "web" goes away. [Courtesy of Michael Carstens, MD]

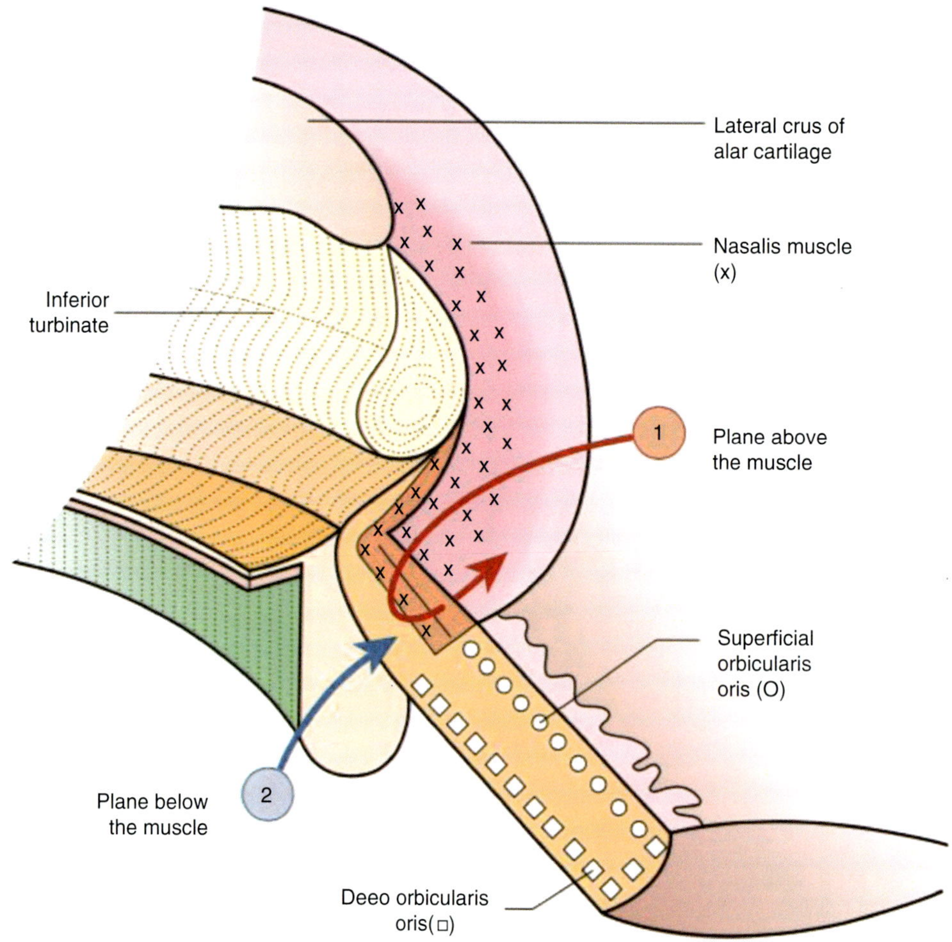

Fig. 19.47 Lateral dissection 1. Lateral dissection opens the lip and separates: (1) the skin from the superficial orbicularis—about 5 mm; (2) the superficial orbicularis (SOO) from the deep orbicularis (DOO); and (3) the nasalis from both SOO and DOO. NB: actual separation of the nasalis is most easily accomplished after the muscle has been completely dissected out because it can be tensed away from the lateral lip element. Recall that the nasalis has two heads, both of which become displaced in the cleft condition. The superficial head should insert over the lateral incisor but instead becomes stranded beneath the nostril sill. The deep head should insert over the canine but instead it falls inside the piriform rim and attaches low on the wall. Dissection of the nasalis proceeds in two planes. First, the skin of the nostril sill is elevated off the muscle. Dissection is then extended up into the nose undermining the vestibular mucosa off the muscle, proceeding in front of the inferior turbinate, and extending all the way up to the edge of the lateral crus. Next, while grasping the muscle, one proceeds deep to it, separating it from the orbicularis and going directly to the rim of the piriform fossa. One then proceeds inward and, hugging the inner wall of the piriform rim, one elevates the deep head off the floor of the piriform. A McKenty or a Molt #9 elevator can prove useful as one "fishes out" the muscle from the piriform fish tank. Note here that the superficial plane of the nasalis (pink arrow) is dissected by proceeding just beneath the skin of the nostril sill and thence upward beneath the vestibular lining until one achieves the edge of the lateral crus. The deep plane of the nasalis (blue arrow) is achieved by following the mucosa down to the piriform fossa where one proceeds in the subperiosteal plane along both its internal aspect and upward to the nasal bones. [Reprinted from Carstens MH. Developmental field reassignment cleft surgery: reassessment and refinements. In: Bennun R, Harfin J, Sandor G (ed). Cleft Lip and Palate Management: A Comprehensive Atlas. Hoboken, NJ: John Wiley & Sons; 2016:83–111. With permission from John Wiley & Sons.]

Nasalis detachment requires entry into the piriform fossa. Using a sweeping motion, the mucoperiosteal sleeve of the lateral nasal wall is elevated off the bone completely, all the way up to the nasal bone. Here, the lateral dissection becomes continuous with the medial dissection of the septal mucoperichondrium. The result of these maneuvers is the *complete liberation of the nasal lining*; the lateral wall rotates internally. Embryologic anchorage of the lateral crus to the underlying vestibular skin (its "program") ensures that it will be dragged into a correct position. At this juncture, the tip defining point is assessed. If additional medicalization is needed, the lateral crus can be freed up from its overlying nasalis attachment, using the *subperichondrial* plane. This requires a rim incision.

Talmant #3 Once separation of the nasal lining from the SMAS is achieved, how can one effectively reposition the lower lateral and upper lateral cartilages? A roll of thin sili-

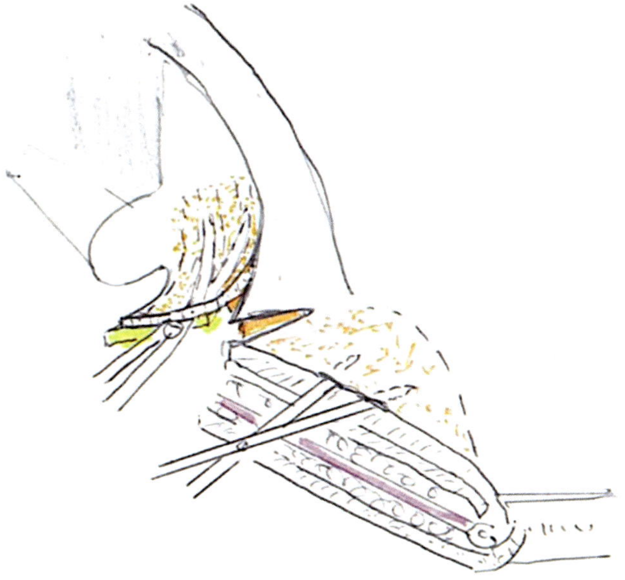

Fig. 19.48 Lateral dissection 2. Field separation, sagittal first step. Height of the lateral lip (8–9) = cleft-side philtral column (3–12) = non-cleft-side philtral column (2–13). The lateral lip incision ascends from point 8 to point 9 which is tucked just inside the nostril rim. Alar base, point 10 and tips of nostril sill flap, point 11 are clearly seen. From 9 the incision then darts lateral behind and in parallel to the nostril sill. It then ascends for about a cm just in front of the inferior turbinate. The triangular nostril sill flap, the apex of which is point 11, is of the same dimensions as that of the non-cleft nostril floor (or slightly greater for overcorrection)

- Superficial orbicularis (SOO) and deep orbicularis (DOO) both in green are separated in the fatty vascular plane (magenta). The two heads of nasalis (yellow and tan) are released from orbicularis by
- Dis-insert *pars alaris* (yellow) from vestibular lining
  - Pars alaris is intermingled with the upper fibers of orbicularis for about 1 cm
  - Just transect OO away from ala
- Dis-insert *pars dorsalis/transversus* (orange) from internal surface of piriform fossa
  - Note: pars transversus is positioned behind pars alaris. To get at it you must first reflect up the more superficial muscle

[Courtesy of Michael Carstens, MD]

cone sheeting is placed into each nostril and secured with transfixion sutures. During this maneuver, the now mobilized lining can be physically manipulated and the position of the alar cartilage observed. The silicone stents are then packed with vaseline gauze. This expands the nasal cavity and effectively pushes the upper lateral cartilage into a new relationship with the nasal bone and with the septum.

Lateral Wall Dissection: Embryologic Implications

The shape of the piriform fossa dramatically affects the appearance of the nasal soft tissues. Recall that this structure is bilaminar. It is composed, externally, by the frontal process of the maxilla and, internally, by the frontal process of the premaxilla. The alar base is pinioned to the external rim. The vestibular lining contains the "program" of the lateral cartilages, determining their size, shape, and position. In cleft patients, a deficiency or absence of the premaxillary frontal process causes the rim to be posteriorly recessed and inferiorly scooped out. The canine fossa is ablated. This causes pathologic displacement of the distal nasalis. The medial head is transferred from above lateral incisor to within the piriform fossa. The lateral head shifts from above the canine to the skin of the nostril sill. A lateral vestibular "web" is the result. The internal nasal skin envelope contracts into the piriform fossa "sink." This further flattens the lateral nasal.

Talmant's first contribution is the recognition of the physiologic role of the nasalis to maintain a patent airway. By demonstrating over two decades the consistent displacement of the nasalis muscle inside the piriform fossa, his work explains observable anatomy long misunderstood, that is, the nasal "web." Moreover, it confirms the role of the frontal process in constructing the normal nasal vault.

Talmant's second contribution is supported by two important embryologic points. (1) Soft tissues produce membranous bone. Biosynthesis of the internal piriform fossa depends upon stem cells within an internal lining of adequate dimensions. When the lining is contracted (as in clefts), over time bone deposition and resorption of the piriform fossa cannot take place normally. *Freeing up the internal lining reassigns stem cells into a correct functional position*. Thus, growth over time creates a normal nasal cavity. The end result of Talmant's mucoperiosteal dissection is a *redistribution of the osteosynthetic envelope into a developmentally correct position*. (2) The frontal process of the premaxilla is the missing Lego® piece of the piriform fossa. Its absence results from an absolute lack of premaxillary soft tissue immediately in front of the inferior turbinate. This deficit causes the "scooped out" appearance of the piriform fossa in cleft patients. The soft tissues of the NPP represent the "housing" of the missing lateral incisor and frontal process fields. The NPP flap (based on the medial nasopalatine artery) can be separated from the PP and transferred into the floor of the nose and the lateral nasal wall. It fits into the releasing incision required for nasalis dissection.

- By replacing the missing frontal process field, NPP reassignment accomplishes two important goals: (1) it potentiates Talmant's lining release and (2) it increases the surface area at the external valve.

Nasal breathing plays a vital role in shaping the nasal air passage. It determines, in no small measure, the physical dimensions of the nasopharynx and, by extension, of the maxilla itself. This is vital for speech because the functional position of the soft palate depends upon the bony platform to which it is attached.

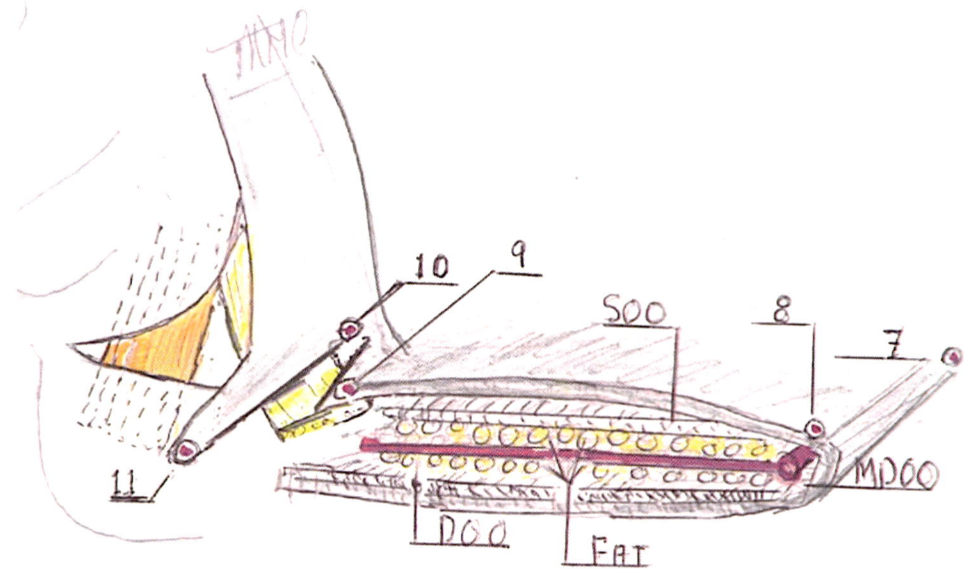

Fig. 19.49 Lateral dissection 3. Field separation, sagittal, second step
- Dis-insert *pars* alaris (yellow) from skin of nostril sill and extend up under vestibular lining past the inferior turbinate
- Dis-insert *pars transversus* (orange) from internal surface of piriform fossa, above half-way

Key: The skin/mucosa envelope of lateral lip element are shaded. DOO is thicker than SOO. Muscles separated by fat plane with axial vessel running the length of the flap. PP are in gray. SOO ends about 5 mm below the alar crease. Nostril sill flap (point 11) is separated from lateral lip margin (point 10)
Pars alaris (yellow) required two planes of dissection. In the superficial plane it is separated from upper margin of SOO and followed up hard under the skin and vestibular lining. The deep plan goes between SOO and DOO down to bone. Holding the muscle with a forcep one follows it right the way around the outside of piriform, hugging the bone. One will encounter the previous subperiosteal plane
Pars transversus is elevated from the inner surface of piriform. It can readily be grasped and extracted. The two heads are gathered together and secured to the sulcus at the canine with a 4–0 vicryl or PDS mattress suture
[Courtesy of Michael Carstens, MD]

Cleft patients have impaired nasal breathing for the following reasons
- Deficit of the soft tissues responsible for synthesis of the frontal process of premaxilla reduces the surface area of the vestibular lining, dragging the lateral crus downward.
- In the absence of the frontal process, the proximal attachments of nasalis are malpositioned. Muscle contraction restricts the airway rather than opening it.
- Contraction of the vestibular lining means that ongoing osteosynthesis of the piriform fossa cannot be normal. Its physical dimensions become abnormal.
- The lower and upper lateral crura provide support for the vestibular lining. In particular, correction of the upper lateral cartilage has been largely ignored in cleft surgery.

The sum of these four factors is *increased turbulence at the external valve* and *reduced airflow*. Talmant's concepts provide an embryologically sound answer to these issues.

Microform Cleft: Premaxillary Deficiency and Poiseuille's law $\Delta P = 8uLQ/\pi r^4$

The underlying problem in the microform or "minimal" cleft is no different than in more explicit manifestations of cleft. The stigmata arise from a mesenchymal deficiency in the premaxilla. The frontal process zone deficit causes a warping of the piriform fossa. A deficit in the lateral incisor zone causes the nasal floor to be "scooped out." To the degree that the bone volume is reduced, the strength of the BMP-4 signal will also be diminished. SHH, the protein within lip tissue that stabilizes the epithelium, will be less inhibited [34].

In sum, the spectrum of minimal clefting is absolutely quantitative, with the degree of soft tissue involvement proportional to the reduction in bone stock. Premaxillary deficits always affect the insertion of the nasalis muscle with consequent depression of the alar cartilage. Nasalis misinsertion, when significant, manifests itself as a "band" in the lateral nasal wall. That these factors should also cause a deviation of the anterior nasal septum is not intuitively obvious.

Talmant deserves credit for drawing our attention to the effect of hydraulic forces resulting from fetal ventilation of amniotic fluid upon the shape of the nasal fossae. As previously mentioned, sagittal real-time ultrasound studies demonstrate how the alar cartilages and septum respond to the influx and efflux of fluid through the nares. In the case of cleft lip associated with cleft palate, Talmant postulates a difference in pressure between the two sides, the non-cleft nostril having higher pressures than those within the

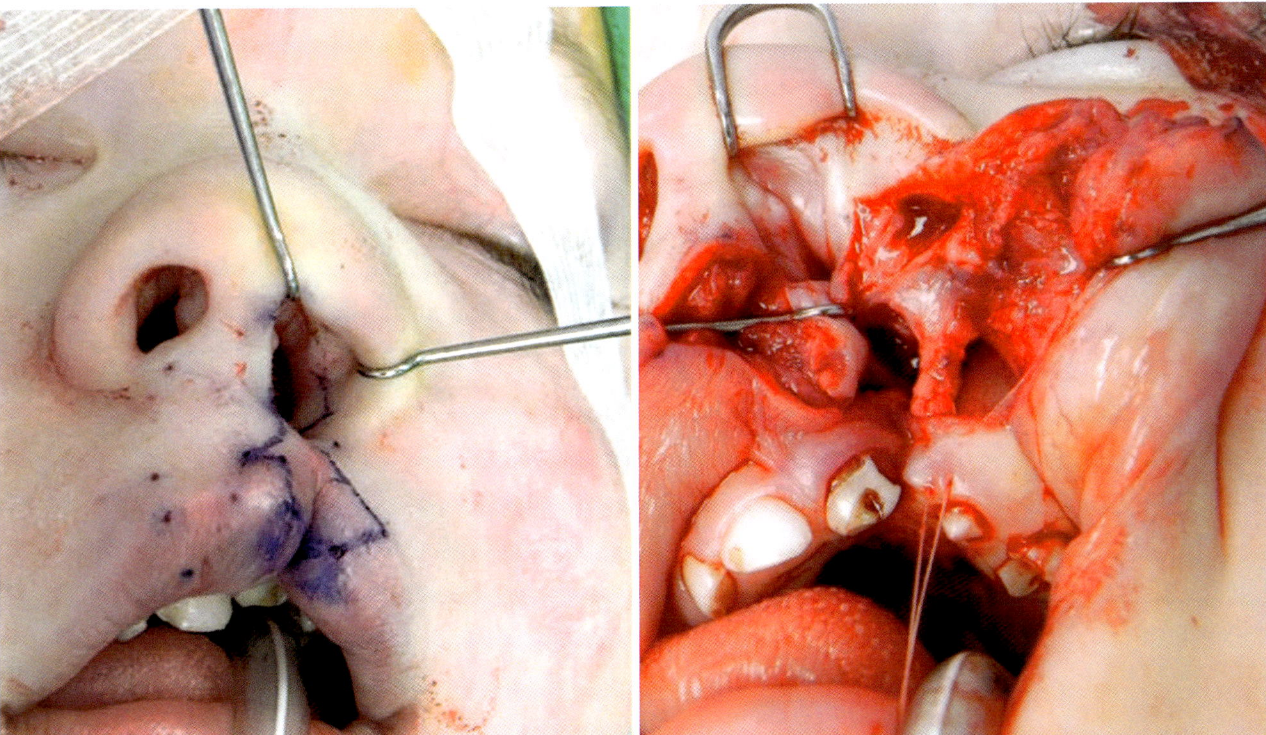

Fig. 19.50 Nasalis complete cleft: dissection and reposition
- primary cases, NPP in continuity with vomer flap
- secondary cases NPP developed separately including scar

[Courtesy of Michael Carstens, MD]

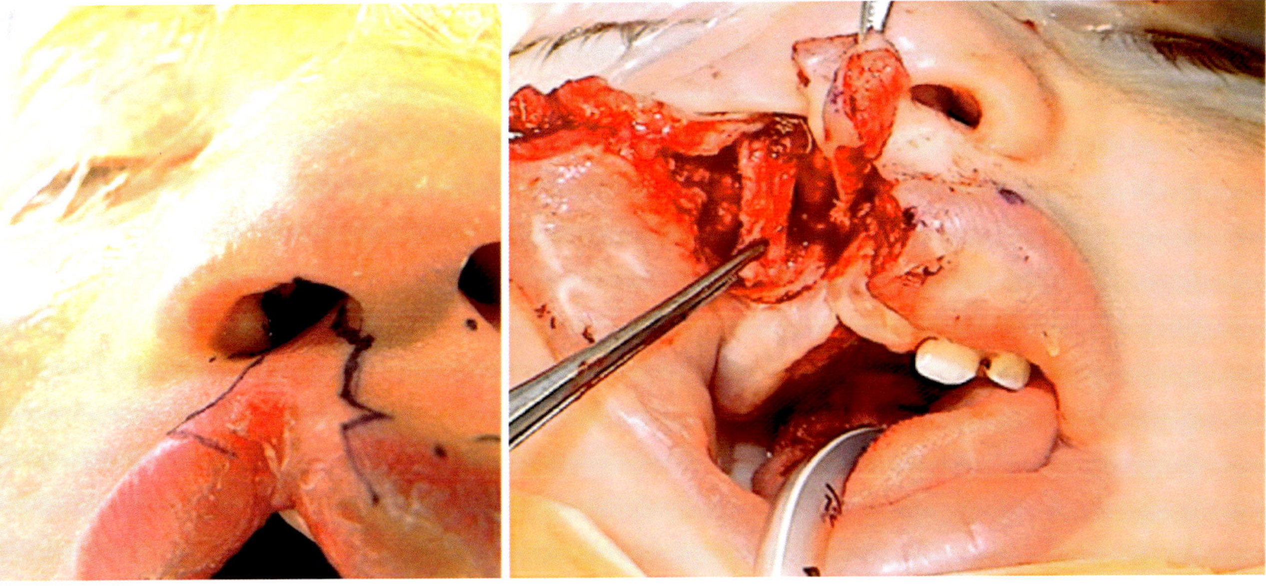

Fig. 19.51 Nasalis, incomplete cleft. *DFR ± Talmant technique:* In this case of primary CP only, nasal floor scooped up from depressed bone with optional alveolar augmentation using ICBG, Osteocel®, rhBMP2. *(Left)*: Lip incision shown curving around footplate and descending to just above white roll where Matthews back-cut is planned. If height adjustment is needed It will to receive a *half z-plasty* at the conclusion of the procedure. Lateral incision from 8 to 9 is seen turning lateral behind the nostril sill. *(Right)*: Forcep (left) tractions well defined nasalis. Undermining of skin for 5 mm demonstrates SOO, the fee border of which is hanging down. NPP is suspended from its pedicle. [Courtesy of Michael Carstens, MD]

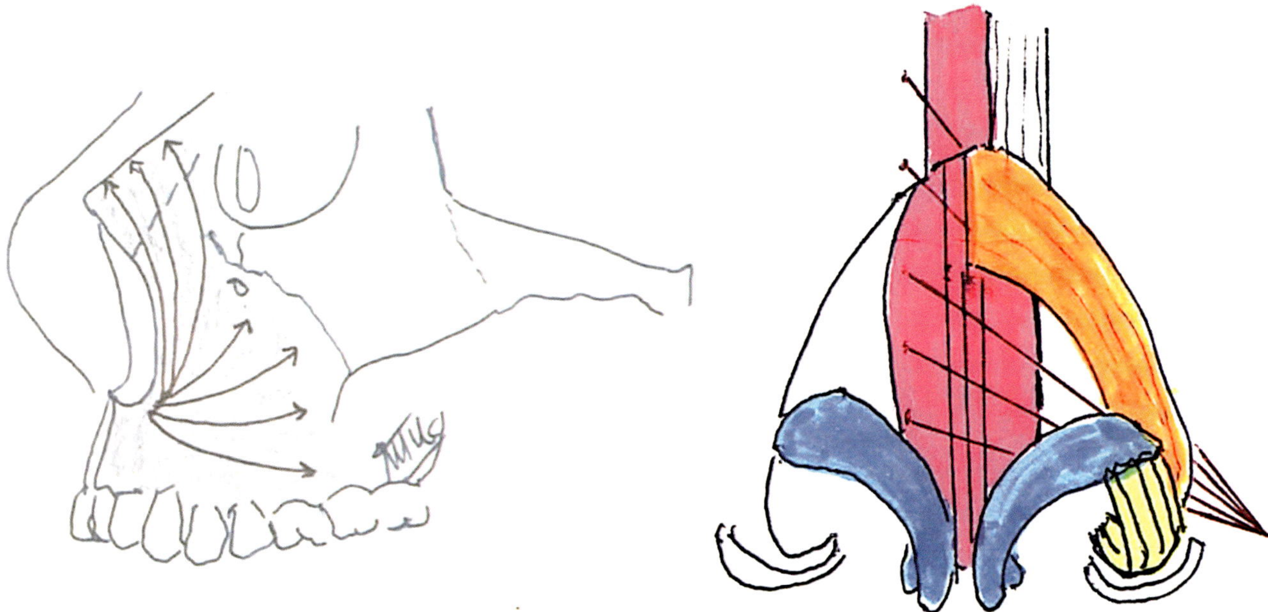

Fig. 19.52 Subperiosteal dissection *external* to the piriform fossa (Delaire). *Left*: Wide subperiosteal dissection of the maxilla and nasal bone (black arrows). Dissection is carried out immediately after the *superficial plane dissection* of nasal skin via the rim incisions. *Right*: The next step, *deep plane dissection*, takes one directly down to the bone of the piriform rim. From that point the maxilla is widely freed up, sometimes if more mobility is needed, with a counter-incision, going up the buttress. Subsequently the elevator is brought along the external surface of the maxillary frontal process, all the way to the nasal bone. It is then brought downward and forward, passing bluntly into the nasal tip, as described by Tessier. Note that the plane passes *below* pars dorsalis of the SMAS (orange). [Courtesy of Michael Carstens, MD]

Fig. 19.53 Subperiosteal dissection *internal* to the piriform fossa (Talmant). *Left*: Talmant dissection inside the piriform is *bidirectional* (orange arrows). It frees the entire internal surface of frontal process of maxilla. It sweeps upward to the nasal bones and then pulls downward and forward, stretching the lining. This causes (1) passive repositioning of the cartilages and (2) expansion of the airway. *Right*: Vestibular lining (pink, StV2 and white, StV1) is completely mobilized with Talmant 1 freeing the medial wall and Talmant 2 freeing the lateral wall. [Courtesy of Michael Carstens, MD]

Fig. 19.54 Lateral nasal wall subperiosteal dissection: lengthening the nasal lining as elevator descends it dislodges "scroll" of upper lateral cartilage from inferior margin of nasal bone. Orange nasalis pars transversus yellow nasalis pars alaris. [Courtesy of Michael Carstens, MD]

labiomaxillary cleft. It is easy to conceive how the anterior septum could be warped.

The "fly in the ointment" of this argument is the minimal cleft. Here, the septal deformity is exactly the same yet the nasal floor remains intact. The answer, once again, lies in the reduction in bone volume within the frontal process and lateral incisor fields of the premaxilla. Poiseuille's law predicts that minor increases in radius will significantly reduce the pressure drop (ΔP) in a linear tube of length L. Let us consider the nostrils as two tubes of equal length in parallel sharing a common wall (the septum). Fluid viscosity (u) and fluid flow velocity (Q) upon entry into the tubes stays constant. Although the piriform fossae are not circular, the premaxillary deficit increases the overall perimeter on the cleft side. This translates into a non-traditional "radius" that exceeds that of the non-cleft piriform fossa. Unequal pressures within the nostrils result in the difference being maximal at the level

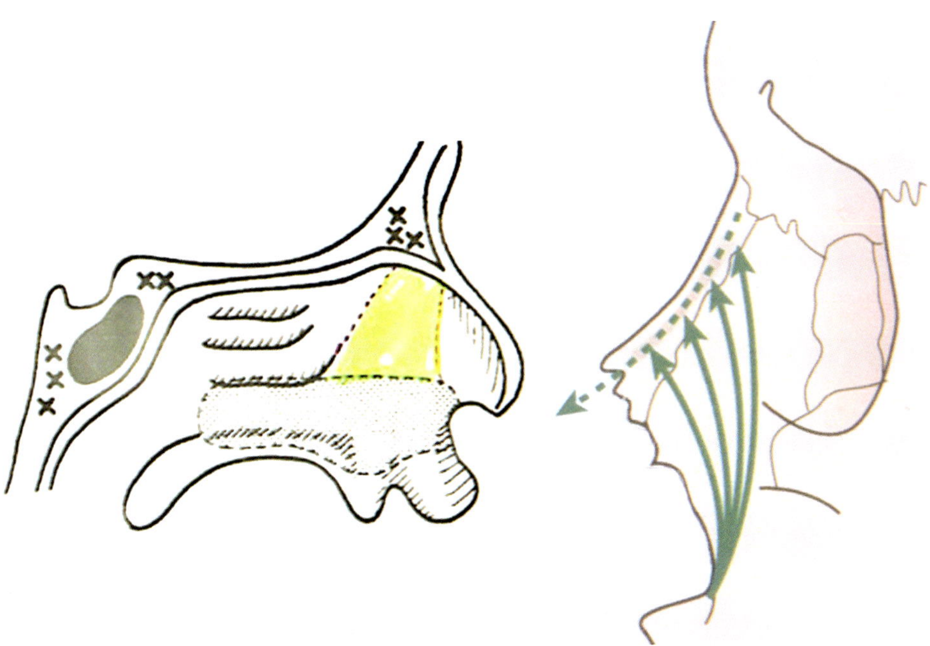

Fig. 19.55 Lateral nasal wall dissection: variations (Trott and Talmant). *(Left)*: Trott and Mohan design for mobilization of the lateral nasal mucosa (indicated in gray stipple). Talmant elevation is in yellow. *(Right)*: I could never get the lateral nasal wall right until I became aware of Talmant's findings. Formerly, I had tried lateral crus elevation using a V–Y incision, broad subperiosteal freeing, and so on, all to no avail. Nasalis dissection is key but another maneuver worked out by Talmant is very helpful. Using a long curved elevator, one proceeds inside the piriform rim and frees it all the way to the nasal bones. Then you sweep downward—always in the subperiosteal plane—until you reach the nasal mucosa underlying the alar cartilage. Now, you do exactly the same maneuver outside the piriform fossa—again in the subperiosteal plane—until you reach the nasal bones and then proceed downward bluntly separating the SMAS muscle layer from the underlying upper lateral (triangular) cartilage and thence to the lower lateral (alar) cartilage. The alar cartilage will be nicely mobilized. You will also fall into the dissection plane you previously created by the medial dissection with scissors up the columella. Recall Tessier's concept of blunt tissue dissection following embryonic planes. These maneuvers are simple and quick but a nicely sharpened elevator is a must. Left: [Courtesy of Michael Carstens, MD]. Right: [Reprinted from Carstens MH. Developmental field reassignment cleft surgery: reassessment and refinements. In: Bennun R, Harfin J, Sandor G (ed). Cleft Lip and Palate Management: A Comprehensive Atlas. Hoboken, NJ: John Wiley & Sons; 2016:83–111. With permission from John Wiley & Sons.]

Fig. 19.56 Open-closed rhinoplasty: elevation of the nasal tip
Step 1: mobilization of the medial and lateral crura
Step 2: Elevation of the medial crus by 5–0 PDS sutures to the contralateral side. Alar cartilage repositioning
Step 3: Achieving definition of the nasal tip

- Intercrural suture at the inflection between the medial and intermediate crura
- Interdomal suture to align the dome on the cleft side with the non-cleft side
- Suspension sutures placed to septum or upper lateral cartilage, depending on the in situ geometry. NPP is now ready to be placed into the nasal floor. If it has been raised with anterior vomer flap the closure is more extensive

Note: In the case of complete cleft palate, the width of the cleft may be considerable. The use at primary lip–nose repair of a concomitant soft palate closure will result in an inevitable narrowing of the secondary palate defect, making vomer flap closure easier. At definitive palatoplasty, access to the nasal mucoperiosteal flap when the lip is closed remains quite simple. A small incision in the vestibular wall in front of inferior turbinate will allow one to slip in a narrow periosteal elevator. Proceed straight backward, hugging the bone. Now sweep medially until one reaches the edge.
Closure of the nasal floor is essential to block oronasal fistula
[Courtesy of Michael Carstens, MD]

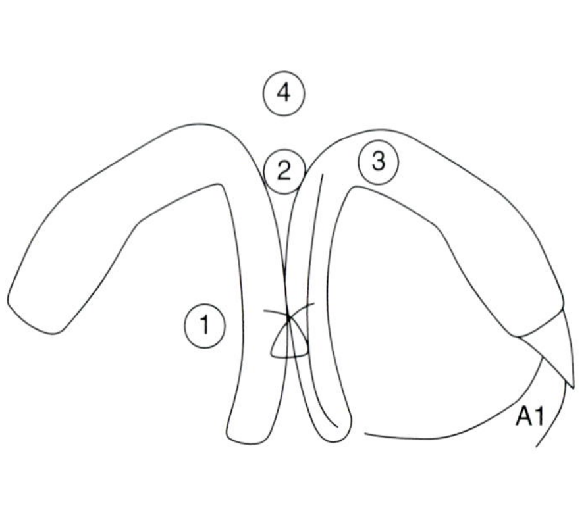

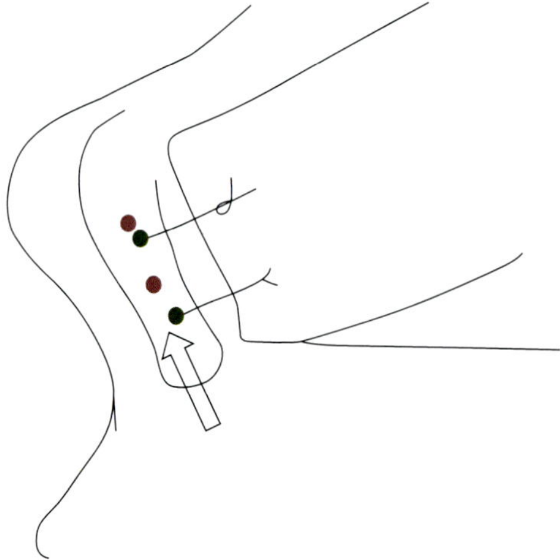

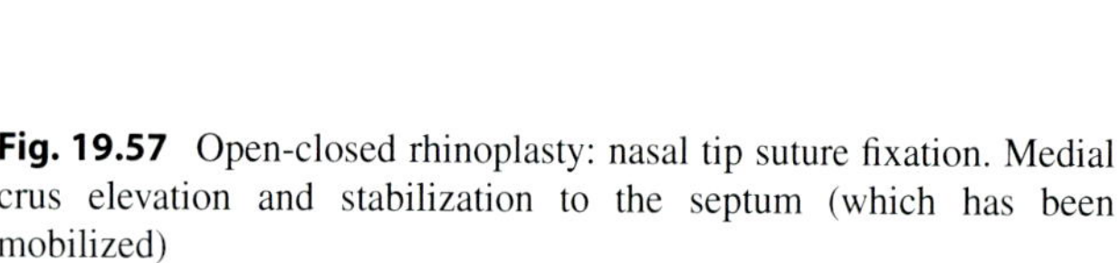

Fig. 19.57 Open-closed rhinoplasty: nasal tip suture fixation. Medial crus elevation and stabilization to the septum (which has been mobilized)

- Medial crura joined (red sutures) and fixed to septum (green sutures)
- Oblique head SOO attached to base of columella (arrow) > fixation of septum

Nasal tip control: interdomal suture to narrow the angle of domal divergence (2), intercrural fixation to elevate cleft-side tip defining point (3), and suspension of lateral crus from the contralateral upper lateral cartilage or septum (4)
Restoration of the nasal floor
[Courtesy of Michael Carstens, MD]

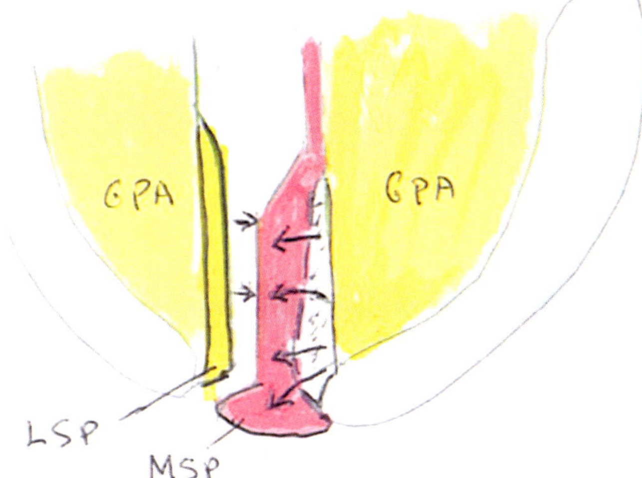

Fig. 19.58 NPP flap *augments* nasal floor; can be combined with vomer flap. NPP flap *augments* nasal floor; can be combined with vomer flap. Optional nasal floor closure if NPP combined with vomer flap. NPP-V flap (pink) has two parts NPP flap is rotated 90° and "fits into" the vomer flap

- Liberate mucoperiosteum on nasal side of the secondary palate using *anterior approach* in front of vomer
- Unification of neuroangiosomes = medial SP (NPP and vomer) + lateral SP (inf turbinate)
- MSP (pink), LSP (gold), greater palatine (yellow), infraorbital (dark orange), anterior ethmoid (light orange)

Left: Dorsal view of the lip–nose complex showing un-elevated NPP flap in situ
Right: Oral view showing anatomic continuity between NPP and the anterior vomer flap
[Courtesy of Michael Carstens, MD]

of the external nasal valve. Growth of the nasomaxillary complex and of the septum takes place in the fetal period. Unequal intranasal pressures during this time account for warping of the septum as seen in the microform cleft and in all other variants.

Finally, displacement of the lower and upper lateral cartilages in the microform cleft demonstrates the importance of epithelial–mesenchymal programming, a concept we have invoked previously. Nasalis in the microform is minimally displaced but the piriform fossa deformity remains. The finite deficiency of nasal skin from the defective premaxilla creates an insufficiency state within the lateral nasal wall. The vestibular lining, displaced downward, synthesizes a perfectly normal cartilage in the wrong place. A similar effect can be observed, on occasion, in microform clefts in which one notes an asymmetry of the nasal bones, with slight flattening on the side of the cleft. The effect can only be explained by an abnormally positioned epithelial program.

The surgical consequence of repairing the microform cleft using a Talmant intranasal subperiosteal dissection is that the biologically active vestibular "bone factory" will be repositioned into its normal state. As membranous bone synthesis continues over time one can expect progressive improvement in the shape and volume of the piriform fossa.

Dental Arch Management

The impact of cleft surgery on maxillary growth after surgery has been intensively studied. Blame in such cases is usually attributed to surgical intervention versus maxillary "hypoplasia." Nonetheless, multiple studies of *unoperated* CL(P) patients demonstrate two patterns of fundamental scientific importance. (1) The cleft-side maxilla has an *abnormal piriform fossa*; this can exert distal effects on the overall shape of the anterior maxillary wall. (2) The cleft-side alveolar process has *normal growth potential*. We have already discussed the first phenomenon so we shall concentrate on the second. What is at the root of all this confusion?

Prior to the era of molecular embryology and stem cell biology, the mechanism of membranous bone growth was poorly appreciated. Textbooks would describe the maxilla as developing in membrane from a single ossification center. This model, based upon nineteenth-century descriptive embryology, is completely out of date.

We now know that osteoblasts arise from stem cells residing in the cambium layer of periosteum. The relationship between membranous bone and soft tissue is like butter and bread: the bread synthesizes the butter. More precisely, membranous bone develops when a unique population of

Fig. 19.59 Nasal floor closure: how to harvest the lateral mucoperiosteal flap. This brings me to a final detail regarding the lateral wall that is important if one wants to achieve a hermetic closure of the nasal mucoperiosteum. As we all know, elevating the lining away from the nasal/dorsal aspect of the palatal shelf can be challenging. Two maneuvers make this easier and safer. First, after making the incision along the edge of the palatal shelf, a small dental amalgam packer can be used to elevate off all fibers of periosteum. The head of this instrument is angulated and delicate. It can nicely lift up the edge without tearing it. Second, the incision to elevate the triangular nostril sill flap places one directly in front of the inferior turbinate—in the subperiosteal plane. Using a periosteal elevator or the amalgam packer, one proceeds straight backward beneath the turbinate, elevating the mucoperiosteum away from the vertical wall of bone. Then, at the horizontal palatal shelf, one simply proceeds in the same plane, elevating lateral to medial until one gains the cut edge of the palatal shelf. Hard palate mucoperiosteum

- Nasal side + inferior turbinate = medial sphenopalatine neuroangiosome (yellow)
- Oral side = greater palatine branch of descending palatine (lemon)

[Courtesy of Michael Carstens, MD]

Fig. 19.60 Nasal floor closure: anterior approach to the nasal layer of palatal mucoperiosteum. Anterior sub-turbinate approach elevates mucoperiosteum from nasal side of P1 (). NPP flap with extension to the vomer (pink) reaches easily. Note the L-shaped closure as seen in Fig. 19.58b. Maxillary lamina of alveolus and the face of maxilla (orange) are supplied by infraorbital. Dedicated branch from medial IO toe zone 4 supplies frontal process of maxilla (not seen here). [Courtesy of Michael Carstens, MD]

neural crest cells organized as a *neuroangiosome* comes in contact with an epithelial surface (mucosa, skin, meninges) from which it receives a "program" determining the size and shape of the bone product.

Some membranous bones are *unilaminar*. The nasal bone is synthesized from forebrain neural crest (sixth prosomere). Its "program" is nasal mucoperiosteum supplied by the V1 anterior ethmoid neuroangiosome. Others membranous bones are *bilaminar*, with two sources of programming. Such bones are characterized by a *separation plane* occupied by *sinuses or bone marrow*. The prepalatine bone (Sic, the horizontal plate of the maxilla) is synthesized from hindbrain neural crest (second rhombomere). It has two "programs." The upper (nasal) layer is the V2 lateral nasopalatine neuroangiosome. The lower (oral) layer is the V2 greater palatine neuroangiosome (GPA). Sinuses in the palatine bone fields of the maxilla (P1 and P2) are a well-described anatomic variant. Membranous bones of the calvarium, such as the parietal bone, arise from dual sources of mesenchyme (dermis and dura); these are programmed by skin and neuroepithelium. The interspace of the parietal bone contains marrow.

Let us now consider the membranous bones of the dental arch. The purpose of the alveolus is to house the dental apparatus. Two factors determine the size and shape of the arch: the *number of dental units present* and the *effective biosynthesis of its bony walls*. Both the upper and the lower dental arches are constructed in exactly the same way. They are "sandwiches." In the center is a dental field arising from neural crest cartilage. It has its own intrinsic neurovascular supply. The lingual wall is composed of four distinct bone fields: premaxilla, P1–P3. The buccolingual wall is composed of an apparently single bone, the maxilla proper, divided into three developmental zones supplied by branches of the superior alveolar arterial system. Stem cells of neural crest origin within these various neuroangiosomes lay down membranous bone on either side of the dental anlage, much like armor plating.

The following table lists the neurovascular territories of the three layers (see Table 8.1). Here are the main take-home points: (1) The alveolar bones of both jaws are complex structures composed of multiple neuroangiosomes. (2) Alveolar osteogenesis is utterly dependent upon the integrity of these fields. (3) Malformations of the dental arch arise from abnormalities in the number or size of teeth, deficits in the alveolar housing, or both.

The nasomaxillary complex is a series of neural crest bones, each of which is innervated by a specific branch of V2. Using a VEGF mechanism each nerve induces an accompanying artery from the stapedial system (the internal maxillary). All these neuroangiosomes have a common point of origin: the pterygopalatine fossa. The maxilla per se has a partial role in the alveolus: it is responsible for the dental

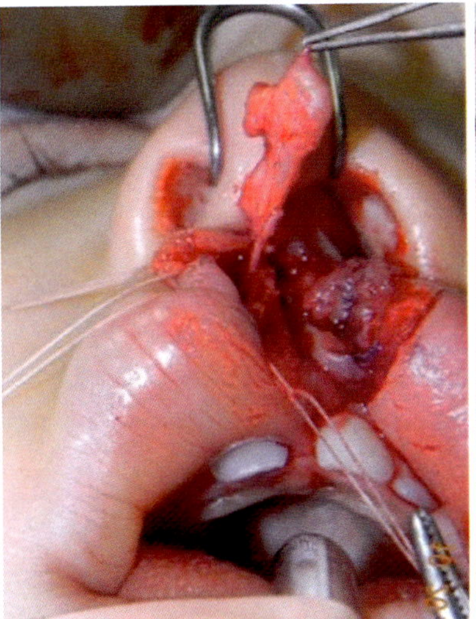
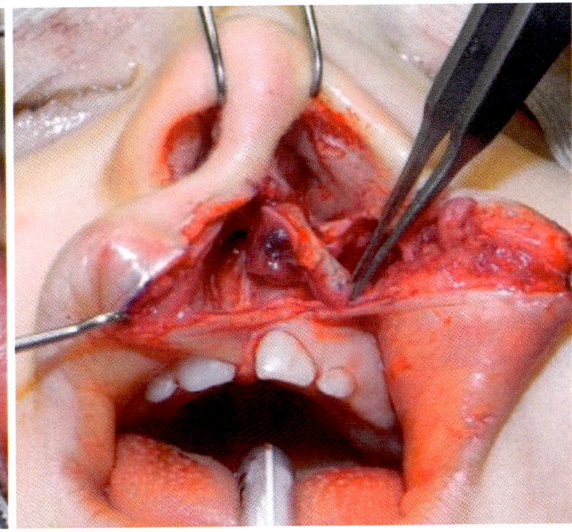

Fig. 19.61 NPP inset into lateral wall. (Left): NPP flap is lifted up on its pedicle. It is 3 mm wide. Left nostril sill has not been rotated outward. (Right): nostril flap out-rotated, leaving tissue deficit behind it. NPP flap (forcep) fits into the slot, permitting nostril sill to be exteriorized. [Courtesy of Michael Carstens, MD]

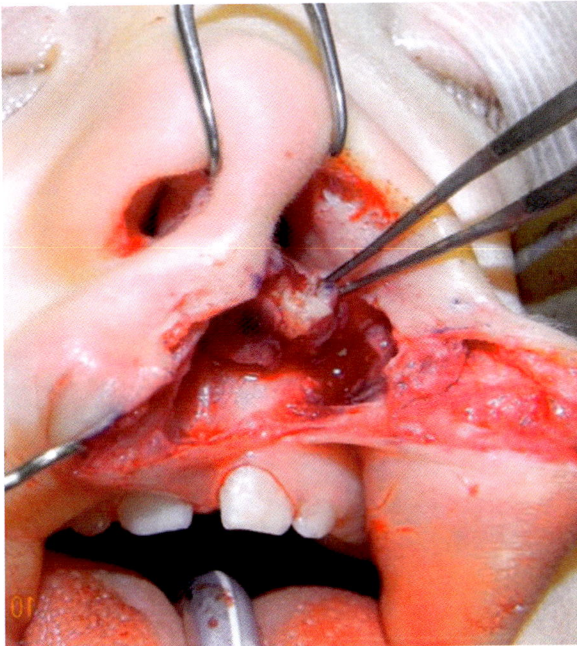
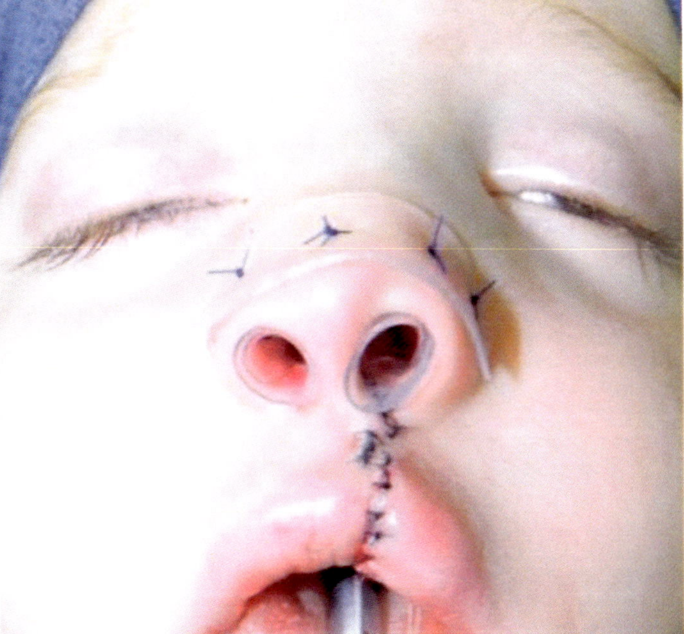

Fig. 19.62 NPP inset and nasalis. Another case shows NPP flap fitting into the release of nostril sill. (*Left*): Nasalis seen below nostril sill. It is clearly separated from SOO. PDS or Vicryl suture about to be placed through sulcus at the level of the canine, being passed in mattress fashion through the bulk of the nasalis and then returned to the oral cavity, where it is tied at the base of the nose at the end (literally) of the case. (*Right*): roomy nostril and good nostril sill seen at closure around the Talmant stent. Note cleft-side nostril overcorrected. [Courtesy of Michael Carstens, MD]

field and the external lamina. The prepalatine and palatine bones form the internal lamina. Although the premaxilla is an autonomous bone, under normal conditions, its external lamina receives mesenchyme of the maxilla.

The mandible is equally complex. In its primitive tetrapod form it consisted of multiple bones. The original dental lamina arose from Meckel's cartilage and consisted of a *dentary bone* plus two to three *coronoid bones*. The body of the mandible supporting the teeth had two splenial bones. The primitive ramus had four bones: pre-articular and articular were lingual. The fates of these bones are listed. The mammalian mandible bears muscle insertions corresponding to the genetic "idea" of these ancient bone fields.

Response of the Dental Arch in Clefts to Surgery

Clefts are experiments of nature. They present us with clues as to the mechanisms behind craniofacial embryology.

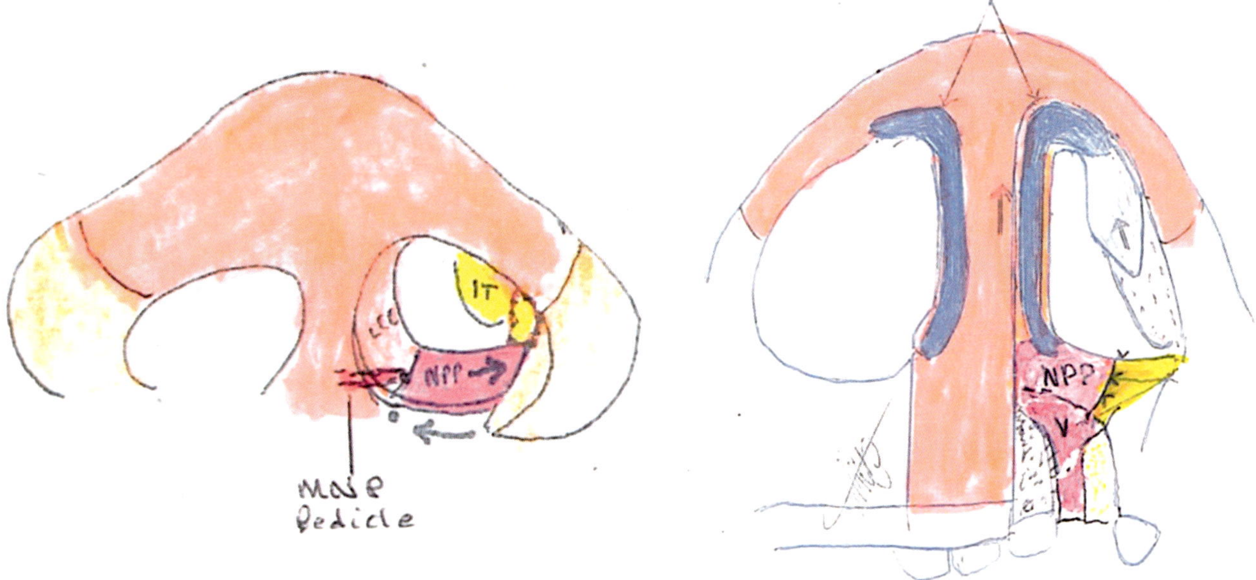

Fig. 19.63 Final closure of the nasal floor +/– back wall of alveolar cleft. On the left NPP (pink) is shown achieving closure behind the release incision of the nostril sill, medial branch of infraorbital (orange). This is frequently not possible, necessitating mobilization of LNP nasal flap (gold) seen on right from the opposite side. If one wants to close back wall of the alveolar cleft another option is to include the anterior sector of the vomer flap as a mucoperiosteal extension of NPP. After advancement of the vestibular lining, with proper mobilization (using Talmant technique) no residual raw surface remains (as seen on right). If a raw surface is present it can be patched using a turbinate flap or a skin graft (full thickness or composite) from the back of the ear. [Courtesy of Michael Carstens, MD]

Fig. 19.64 Muscle closure: DOO and SOO. *Left*: NPP flap placed into the floor of the nose to close the upper margin of the "box." *Right*: DOO (brown) is sutured first to the non-cleft side at its highest point just below anterior nasal spine. This suspends the lip. SOO is sutured to the mesenchyme of the prolabium. Recall that in unilateral clefts contralateral SOO does not extend past the philtral column. Sm6 muscle cannot penetrate r1 neural crest mesenchyme. [Courtesy of Michael Carstens, MD]

Nowhere is this more apparent than in the completely divided palate. The cleft-side maxilla is dynamic; it responds to surgical intervention in two primary ways: *retroposition* (growth restriction) and *collapse* (cross-bite). We shall deal with these problems in sequence.

Retroposition

At birth, the cleft-side maxilla appears smaller and/or retropositioned when compared with its normal counterpart. This situation likely results from disruption of force vectors in utero. Forward, centric growth of the ethmoid, vomerine, and premaxillary fields is disconnected from the cleft-side maxilla (in bilateral clefts, from both maxillae). Nonetheless, multiple studies of *unoperated* CL(P) patients demonstrate two patterns of fundamental scientific importance. (1) The cleft-side maxilla has an *abnormal piriform fossa*; this can exert distal effects on the overall shape of the anterior maxillary wall. (2) The cleft-side alveolar process has *normal growth potential*. We have already discussed the first phenomenon so we shall concentrate on the second. Why does

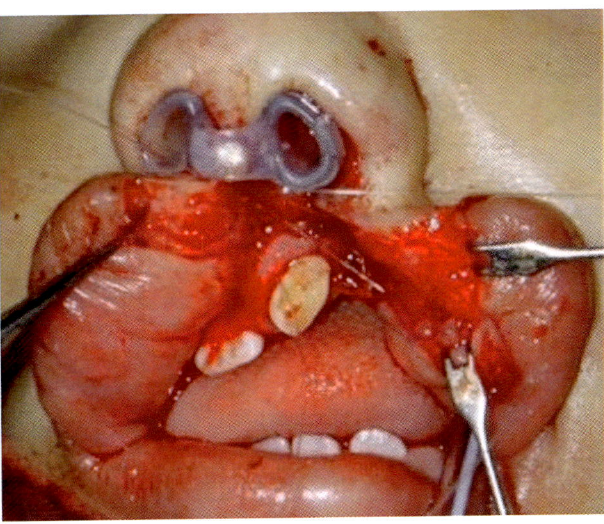

Fig. 19.65 Orbicularis repair: what *not* to do. Muscle repair is nuanced as per the individual surgeon. I think the key factor is anchoring DOO to its counterpart right under the nasal spine. This serves to set the height of the lateral lip flap. SOO is repaired to the subdermal tissue high in the flap. The upper corner (fibers of Delaire) can be sutured to the base of the columella to see if an improvement in the "aesthetic drape" of the lip is obtained. Remember that SOO is superficial to nasalis. This case was performed before I learned about nasalis re-insertion. I include it here because it shows a very common MISTAKE of lip repair. *(Left):* Nasal stents not yet placed. The gingival flaps are being turned inward from the cleft margin. NPP flap is sutured into place in the nasal floor. Just under the skin of the lateral lip element you can see muscle. The uppermost 1–1.5 cm is nasalis (here left in situ with orbicularis. The subcutaneous plane for freeing nasalis from the nostril sill is obvious. *(Right)*: Here you can see the fibers under the nostril see sutured to the columellar base. Do not do this. Suture nasalis to just above the canine with a mattress that catches the periosteum and the sulcus. The knot is in the mouth. Do not tighten this or you will depress the alar base. Just gently approximate it and let nature do the rest. This is the **final step** in the DFR procedure. [Courtesy of Michael Carstens, MD]

Fig. 19.66 Oblique fibers of Delaire. This final suture can help the aesthetic "drape" of the lip. It is placed as a mattress into the *upper margin* of SOO through in incision to free nostril sill about 1–1.5 cm back from the leading edge of the muscle. It can be mattressed into the base of the columella. 5–0 PDS glides nicely through the tissue permitting you to see if it accomplishes an improved suspension. If not, simply remove it. [Courtesy of Michael Carstens, MD]

the dental arch in cleft patients present as it does? Why does it self-correct over time (in the absence of surgery)? How does surgical intervention affect this process?

Recall that the *primary determinant of alveolar size is the number of teeth*. Assuming that the cleft-side alveolus has all six dental units, one can expect that it will eventually achieve normal proportions. Dental development involves the interaction between the epithelial program and the underlying mesenchyme. Although this process is delayed on the cleft-side, it eventually takes place. Assuming a normal mucoperiosteal envelope, the alveolar walls of the cleft maxilla will respond to accommodate the dentition. For this reason, patients with unoperated complete cleft palate can present with a dental arch of normal dimensions.

Restriction in maxillary growth after surgery has been intensively studied. Blame in such cases is usually attributed to surgical intervention versus hypoplasia intrinsic to the maxilla. Studies by Bardach (in the *supraperiosteal* plane) showed that increasing degrees of dissection were accompanied by increased growth reduction. At the same time, long-term work by Tessier and Delaire demonstrates that extensive *subperiosteal* dissection does not result in growth. What is at the root of all this confusion?

As previously discussed, biosynthesis of membranous bones is wholly dependent upon the vascular integrity of the neuroangiosomes from which they are constructed. These

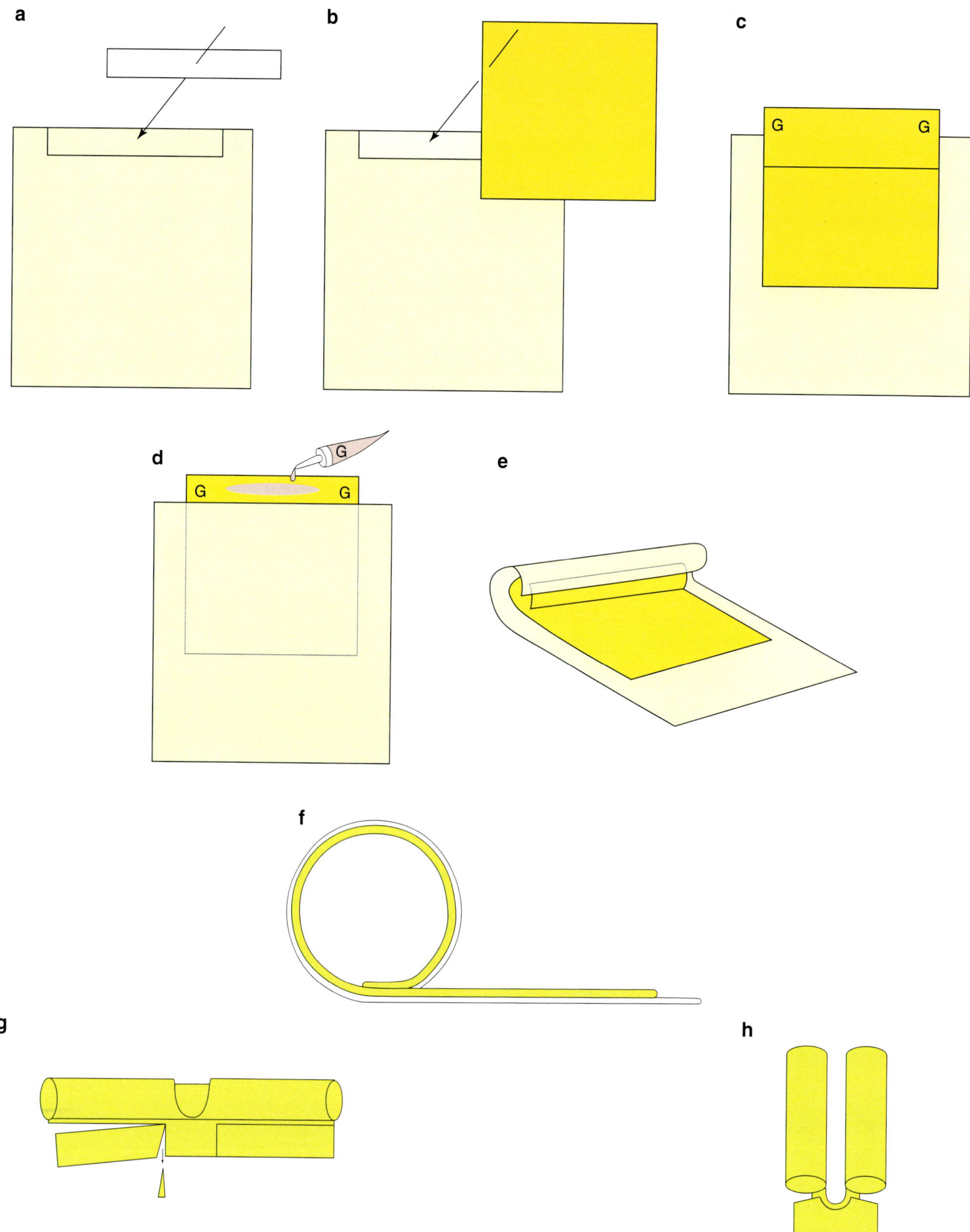

Fig. 19.67 Talmant nasal stent. Use 0.5 mm silicon sheeting and appropriate glue. Let dry overnight. These are also commercially available. Note: Cases 1–6 are primary DFR repairs using followed for 5 years. Case 7 is a primary repair using rotation-advancement and revised with DFR with results shown at 1-year post-revision. Case 0 began this chapter and is followed demonstrating DFR correction of multiple secondary problems at 1-year post-revision. [Courtesy of Michael Carstens, MD]

contain stem cells, located in the cambium layer of the periosteum. *Any surgical intervention that invades or falsely subdivides a neuroangiosome will potentially compromise the blood supply to the stem cell population* and negatively affect osteogenesis.

Traditional techniques of lip repair have mobilized the lateral lip element using the readily accessible *supraperiosteal plane*. As a consequence, cleft lip surgery disrupts blood supply to the stem cell rich cambium layer. Restriction in growth of the maxillary wall (Tessier zones 4, 5, and 6) can lead to malposition of the alveolus per se. Blood supply to the buccolabial mucoperiosteum can also be reduced. Bardach's findings in beagles, using the supraperiosteal plane, were originally interpreted as a caution against wide dissection [35–37]. In point of fact, these papers demonstrate just the opposite: the use of the supraperiosteal plane violates the principles of embryology by devascularizing developmental fields. Because membranous bone is just a by-product of its soft tissue envelope, preservation of the resident stem cell population in the cambium layer of the periosteum requires the use of the subperiosteal plane. Applications of this concept to cleft surgery were described in detail with a detailed literature review by this author [31, 38, 39].

In cleft palate repair, oral mucoperiosteal flaps are commonly elevated using an incision placed at the junction between the lingual alveolar wall and the horizontal palatal shelf. This maneuver disrupts blood vessels directly laterally from the greater palatine axis to the lingual mucoperiosteum of the alveolus. As a consequence, blood supply to the lingual mucoperiosteum becomes random, based upon collateral flow from the opposing buccolabial mucoperiosteum and upon the underlying alveolar bone.

Just as the whole is greater than the sum of its parts, the most negative effects on alveolar development are observed in patients having combined surgical interventions for cleft lip and cleft palate surgery. The resulting process, *sequential vascular isolation*, is the *primary cause of maxillary growth arrest after cleft surgery* [32].

Collapse: A Reversible Phenomenon

Unoperated patients with complete clefts of the lip and palate demonstrate a gap corresponding to the absence of the two premaxillary fields (lateral incisor and frontal process). In the growing child, lip repair re-establishes force vectors across the alveolar cleft that are unopposed by bone: collapse ensues. This situation is dynamic; it is easily reversed by orthodontic expansion. The Achilles' heel of this approach to cleft management is access to orthodontic care, a resource not readily available in many countries with a high-volume cleft population.

In theory, primary alveolar stabilization would prevent collapse and reduce or simplify subsequent orthodontic management. Inclusion of soft palate repair with the primary lip intervention ameliorates, but does not prevent, arch collapse. Long-term splinting and delayed repair of the hard palate lead to good size and position of the arch but poor speech. Over the years, various attempts have been made to address this problem using soft tissue procedures with or without some form of grafting. Early work suffered from soft tissue considerations stemming from use of the supraperiosteal plane. Observation of maxillary growth restriction led to a consensus that primary grafting should be abandoned in favor of secondary grafting at ages 6–9 (eruption of the canine).

This dogma was successfully challenged by Kernahan and Rosenstein at the University of Illinois [40, 41]. Using extremely limited dissection and rib grafts, stabilization was achieved. Long-term growth studies (up to 30 years) in these patients demonstrated arch development without maxillary restriction. Of note, the grafts were placed as struts high up over the alveolar cleft, leaving the actual walls of the cleft site largely untouched. Rosenstein's final report is of great scientific importance for several reasons. First, it shows that primary arch stabilization is achievable. Second, and more importantly, the relationship between grafting and maxillary growth restriction reported in previous studies is an artifact of the surgical technique resulting from entry into the supraperiosteal plane.

More recently, GPP using local flaps at the alveolar cleft site has been reported with the hope that autologous bone would be generated at the cleft site (an idea dating back to Skoog) [42]. This approach is limited by the dimensions of the cleft site (wide clefts are not amenable). The dissection is technically demanding: flaps are taken directly over alveolar bone that is extremely delicate. The amount of bone generated by this technique has been disappointing. Elevation of the flaps may compromise the stem cell population within them. Furthermore, scar formation at the site may have consequences for dentition, such as tipping or rotation. Such problems are certainly amenable to orthodontic correction, but the limited benefits of GPP and its technical nuances do not recommend it.

Contributions of Jean Delaire

Delaire's work demonstrates the surgical importance of the subperiosteal plane in cleft surgery [1, 43]. Elevation and medicalization of such flaps from the face of the maxilla represents the transfer of bioactive stem cell-bearing mucoperiosteum into a correct midline position. This concept is of fundamental importance for DFR cleft repair because it corrects the displacement of soft tissues that is present on both sides of the cleft. Such displacement is asymmetrical in the unilateral cleft and symmetrical in the bilateral cleft.

As previously mentioned, an unintended side effect of the Delaire's dissection is that it draws the malpositioned nasalis into the midline along with the rest of the lateral lip musculature. This changes the function of the nasalis and constricts the airway. Talmant, having inherited many of Delaire's patients, recognized this problem and worked out how the nasalis could be repositioned to its natural attachment at the canine fossa. This simple dissection frees both heads of the muscle and passes them below the orbicularis complex. Technical details are provided in the addendum.

Using the Delaire dissection plus alveolar cleft closure as described by Demas and Sotereanos, in 1996 this author described use of an extended alveolar flap that can span a distance of two tooth units [38]. This size of this flap permitted abundant tissue to be mobilized backwards from the alveolar margins for a secure soft tissue closure in larger clefts. The *sliding sulcus* technique was therefore more versatile than the GPP. It worked well for patients 2 years of age and up. In younger patients, the sliding sulcus had the same technical drawbacks of soft tissue manipulation at the alveolar cleft site. In 2004–2006 the problem of inadequate stem cell response seen in the above techniques was addressed by placing rhBMP2 in the alveolar cleft site. CT scans demonstrated bone formation that filled nearly the entire vertical dimensions of the cleft site. At the end of 10 years of work and observation, I came to the conclusion that *gaining alveolar cleft closure at the primary surgery was doable but not worth the effort*. A simpler approach was required. As we shall see, the search for this better solution led eventually to the Talmant protocol.

The bottom line is this: *Only nature can make a dental arch ... and only the orthodontist can set it right*. In the ungrafted complete cleft palate the dental arch is unstable; all forms of lip repair cause alveolar collapse. Fortunately, such perturbation is only temporary. DFR transfers stem cell populations into correct, centric positions such that they produce bone where desired. DFR is designed to assist the orthodontist in all other aspects of cleft palate development: airway, soft palate positioning, and achieving complete closure of the anterior hard palate.

Functional Lip and Palate Repair: The Five As

Let's put the dental arch into the overall scheme of the cleft problem. Patients with complete clefts of the lip and palate require a prioritized, functional approach characterized by the "five **A**s": ***a**irway*, ***a**rticulation*, ***a**nterior fistula avoidance*, ***a**rch*, and ***a**esthetics*. What is the rationale behind this order? There is no question that intelligible speech is the most important social benefit for these patients, so why prioritize the airway before articulation?

Airway

The anatomic and physiologic relationships between the hard and soft palate can be described by the *pinball machine* model. The soft palate is like the "*flipper*" in a pinball machine (a moveable lever that prevents the ball from escaping and keeps it in play). The hard palate is the *platform* to which the flipper is attached. The function of the soft palate flipper is to reach backward and upward to make contact with the pharynx; thereby closing off the nasal airway during phonation. The ability of the soft palate to accomplish this goal depends on: (1) normal motor innervation from cranial nerves V3 and IX; (2) normal muscle mesenchymal mass from somitomeres 4 (first arch) and 7 (third arch); and (3) normal length of the bony platform to which the soft palate musculature is attached. In other words, one must be able to control the flipper. It must be of normal length and not be floppy. Above all, the flipper mechanism must be positioned on the platform in such a way that it closes off the escape of the ball.

The size and shape of the hard palate bones (prepalatine and palatine) determine the positioning of the soft palate. When these bones are deficient, the soft palate musculature cannot make contact with the pharynx. In some patients, however, an apparently normal hard and soft palate complex is inadequate, causing velopharyngeal incompetence. What is the problem here? The secondary hard palate is not an isolated entity. It is part of a larger bone complex, the nasal chamber. The prepalatine and palatine bones are in anatomic contact with the frontal process of the maxilla, the inferior turbinate, and the sphenoid. Thus, *the overall dimensions of the nasal airway determine the spatial relationship between the soft palate and the posterior pharynx* and consequently its function.

Surgical implications of the above are as follows: First, the vestibular lining of the cleft nose is malpositioned due to deficiency/absence of the premaxillary frontal process. If left in this state, the ongoing process of osteogenesis during the first years of life will produce a permanent deformity in the bony dimensions of nasal chamber. Although the primary determinants of soft palate position are the prepalatine and palatine bones, these in turn are affected by the spatial dimensions of the surrounding bones, that is, the nasal airway. Correction of the airway by reassignment of the internal lining field (and therefore the cartilages) and of the nasalis is required to achieve optimal hard palate dimensions.

When should nasal airway correction be undertaken? Talmant believes that if nasal breathing patterns are not established in the first 6 months of life, oral breathing patterns will develop. Mouth breathing has negative effects that are difficult to reverse. These include head positioning, body posture, and maxillary shape. This logic is indisputable but it is unclear to me how rigid one should be on this issue. Children can present late, with wide clefts or with complications. The good news is that airway correction can be easily done secondarily, either concomitant with lip revision or as an isolated procedure. It seems thus reasonable to conclude that nasal breathing should be accomplished in all children by age 2.

A caveat to the reader here: technical details regarding intraoperative nasal packing and postoperative nasal stenting are of the utmost importance for airway correction. These accomplish the following: (1) a subtle, but very precise, control of alar cartilage position; (2) upper lateral cartilage repositioning; and (3) adequate diameter of the external nasal valve. These maneuvers will be described in the technical section that follows ... stay tuned.

Articulation

Timing of palate closure is yet another area of cleft management that is rife with controversy. As we have previously seen, palate clefts can be classified by embryologic mechanism, depending on which components are affected. Cleft palate arises from faulty mesenchyme involving bones (neural crest from rhombomeres r1, r2, r3), muscles (paraxial mesoderm from somitomeres 4 and/or 7), or a combination of both. In the absence of such classification, studies relating surgical technique to speech are inherently unreliable; apples cannot be compared with oranges.

Nonetheless, some general concepts of cleft palate repair make sense. First, soft palate closure mechanically positions the tongue, helping prevent the acquisition of motor patterns that are difficult to correct later. This can be done as early as 6 months, as in the Talmant protocol (see below). Soft palate repair establishes normal force vectors across the posterior midline. These balance against force vectors created by lip repair or lip adhesion. Any form of early lip closure exerts force vectors across the alveolar cleft leading to arch collapse and cross-bite. Thus, lip repair and soft palate repair are a rational combination at 6 months.

A quick review of the embryology of the soft palate musculature is required. Five sets of muscles are involved. First pharyngeal arch muscles arise from paraxial mesoderm of the fourth somitomere. The sole Sm4 palatal muscle is tensor veli palatini.

Third pharyngeal arch muscles arise from the seventh somitomere. These are levator veli palatini, palatoglossus, palatopharyngeus, uvulus, and superior constrictor.

These muscles can readily be tested using a Peña® muscle stimulator (Integra Life Sciences, Plainsboro, New Jersey). Normal function is graded as follows: normal = 2, hypotonic = 1, and no response = 0. Differences frequently exist between the left and right sides. Such information is valuable for classification of the palate cleft and for diagnosis of speech problems, if these should arise.

Soft palate muscles exist in two layers. *Nasal mucosa* provides the "program" for the tensor and levator. *Oral mucosa* is the "program" for the palatopharyngeus and palatoglossus. Uvulus is programmed by both mucosal surfaces. The *upper pharyngeal mucosa* contains the "program" for the superior constrictor.

Only the muscles of the nasal layer (tensor and levator) are responsible for palatal elevation/elongation. Only these muscles are directly attached to the palatine bone. These are the one that need to be set back. The dissection plane between the nasal and oral layers is readily defined. *Extensive three-layer intervelar veloplasty is not embryologically rational and causes unnecessary scarring*. Soft palate repair is best viewed as a "first-pass" event. If the functional result is good, nothing more is required. If VPI is present, and the *elevation pattern is V-shaped*, secondary lengthening with a double-opposing z-plasty will readily resolve the issue.

Anterior Fistula Avoidance

The GPA supplies two bone fields: the horizontal shelf of the palatine bone and the oral lamina of the prepalatine bone. It runs all the way forward to the junction of the canine and lateral incisor. There it makes an anastomosis with the MNP. In traditional cleft palate surgery, hard palate incisions located at the junction of the palatal shelf and lingual alveolar wall constitute an *embryologically incorrect subdivision* of the greater palatine fields. Not only does this incision *interrupt blood supply* to the alveolar mucoperiosteum; it also creates *short flaps*. Troublesome anterior fistulae result from this design.

In the unoperated state, the length of the GPA field is designed to reach the premaxilla. Achieving a water-tight closure is simply a matter of using the entire neuroangiosome, using a pericoronal incision (or one just lingual to the teeth). The term AEP refers to the additional tissue gained by this flap. The anatomy and developmental biology of this procedure have been previously described. The AEP procedure is also useful in secondary cases. When enough time has elapsed after the primary palate repair, collateral flow is re-established between the palatal shelf and alveolar mucoperiosteum. One can therefore use the AEP incision design to re-elevate the entire mucoperiosteum of the hard palate and lingual alveolus as a single flap. In unilateral clefts, the AEP flap from the non-cleft side always projects beyond the alveolar cleft, while that from the cleft side reaches the back wall of the alveolar cleft. These flaps permit closure of anterior fistulae in virtually all situations.

Arch

Surgeons and orthodontists get along best when goals of each specialty are respected

- Early lip closure (at 6 months) will cause the alveolar cleft to narrow down and may create a cross-bite of the primary dentition, but this is readily correctible.

  - ***Orthodontist to surgeon****: Alveolar soft tissues over un-erupted teeth are delicate ...* ***leave them alone****.*
- Achieving a perfect closure of the entire nasal floor is a major goal of the primary surgical sequence. The most technically difficult site for this is the anterior floor, just over the alveolar cleft, exactly where the collapse has taken place. When nasal palatine mucoperiosteum is joined with that of the vomer, the *floor will always be short* in the anterior–posterior dimension. The NPP flap corrects this deficit: it is dissected out of the lip, transferred into the nasal floor, and sutured to the vomer-maxilla closure.
  - ***Surgeon to orthodontist****: The anterior nasal floor is tight and tricky ... can you* ***expand the alveolar cleft*** *prior to hard palate repair and then* ***keep it open*** *until I graft it?* What is required here is a means of palatal expansion that can be maintained over time. This also favors proper positioning of the lateral nasal wall. Such a *device cannot be fixed*; it must be capable of adapting to changes in the arch. As detailed below, the modified quad helix as designed by Lumineau fulfills these requirements and produces, often by age 4, a dental arch ready for bone grafting.

Developmental Field Reassignment

At the beginning of the chapter, we set out to define a set of biologic goals for cleft repair. At this point, the protocol that comes closest to achieving them is that published by the Nantes group. Their results have been gained over 20 plus years. Credit should be given to Jean Delaire, from whose work Talmant's concepts have evolved. In any case, the outcomes of the airway and dental arch using this protocol not only are the best I have seen but also have a sound basis in developmental biology (Figs. 19.68, 19.69, 19.70, 19.71, 19.72, 19.73, 19.74, 19.75, 19.76, 19.77, 19.78, 19.79, 19.80, 19.81, 19.82, 19.83, 19.84, 19.85, 19.86, 19.87 and 19.88 reference DFR cases done with 5-year follow-up.)

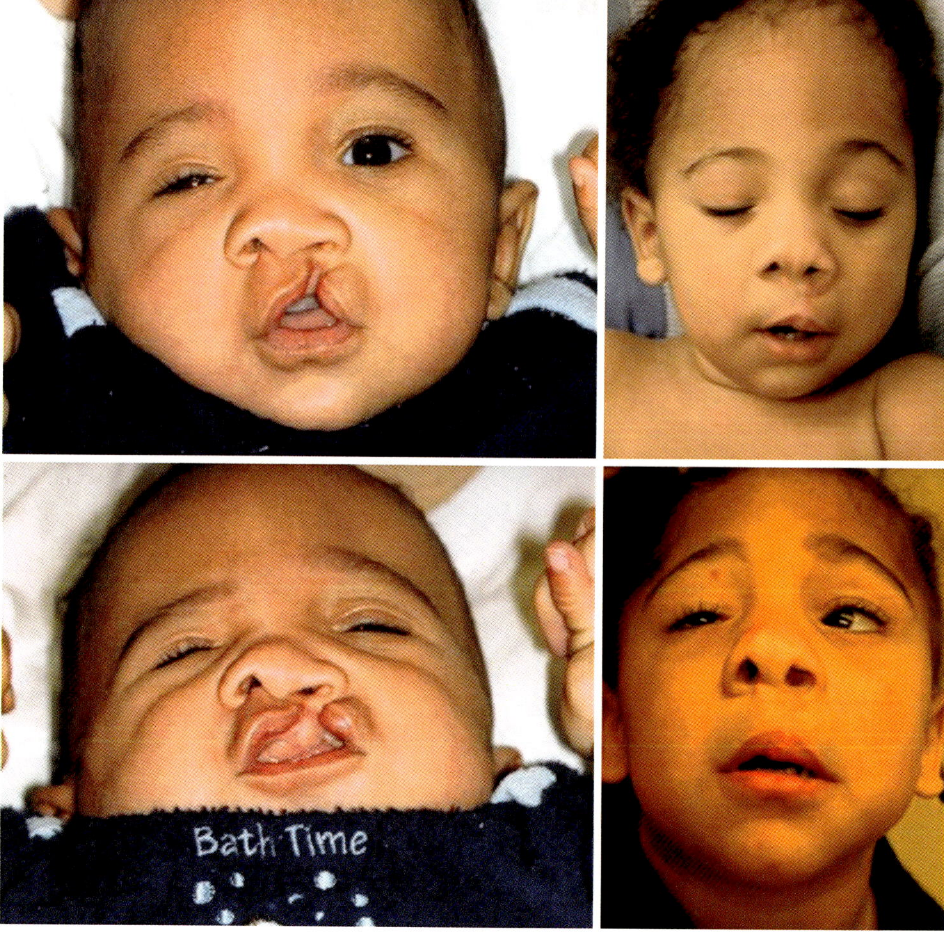

Fig. 19.68 Case 1 5-year result: left UCL, incomplete with complete cleft of primary and secondary palate. Patient has microphthalmia and ptosis. Nasal projection maintained. No nasalis was done so nasal tip is asymmetric and nostril axis is oblique versus normal vertical. Top/bottom left pre-op; top/bottom right 5 years post-op. [Courtesy of Michael Carstens, MD]

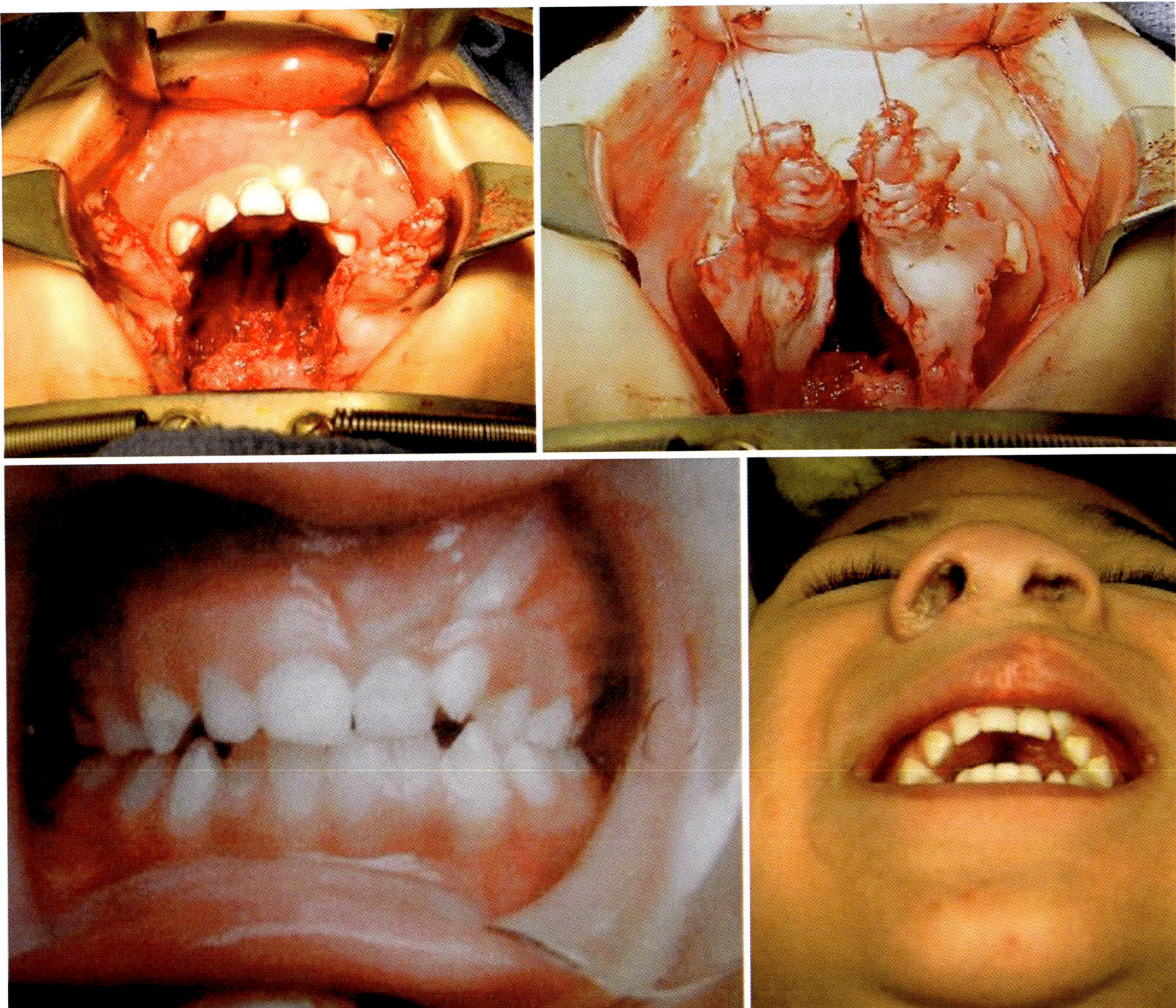

Fig. 19.69 Case 1 Alveolar extension palatoplasty with rhBMP2. Top left: surgical defect. Top right: flaps elevated. Note their length; they exceed the boundaries of the alveolar cleft. Bottom left/right: At 5 years arch is preserved and lateral incisor has erupted. [Courtesy of Michael Carstens, MD]

Talmant–Lumineau Protocol

- Six months: soft palate repair plus full lip/nose repair. Wide bilateral subperiosteal dissection is carried out, both extrapiriform and intrapiriform. *In some countries, OR and anesthesia conditions may require that the soft palate repair be carried out separately.*
- Step 2. As soon a primary dentition permits, a *modified quad helix* is placed. The cleft site is expanded over a period of approximately 1 month and then maintained. Unlike traditional quad-helix designs, it is *not* fixed to the molars. Instead, it is secured with three wire ligatures. The expander has four "eyelets."
  - These are 360° loops placed posteriorly at the level of the second molars (*E*) and anteriorly at the level of the canines. The expander is attached to each molar, with the "pull" of each wire loop placed in opposite directions.
  - A third wire is passed through the posterior "eyelet" and over the second wire. The purpose of this wire is to prevent backward slippage of the expander. The third wire can be tightened to keep the device in place. In this way, the quad helix remains in situ but can be advanced forward to accommodate dental arch growth. Expansion is maintained until alveolar bone grafting is completed.
- Two years: hard palate repair.
- Four years: alveolar bone grafting. The timing for the procedure follows the extent of distraction: *the maxillary*

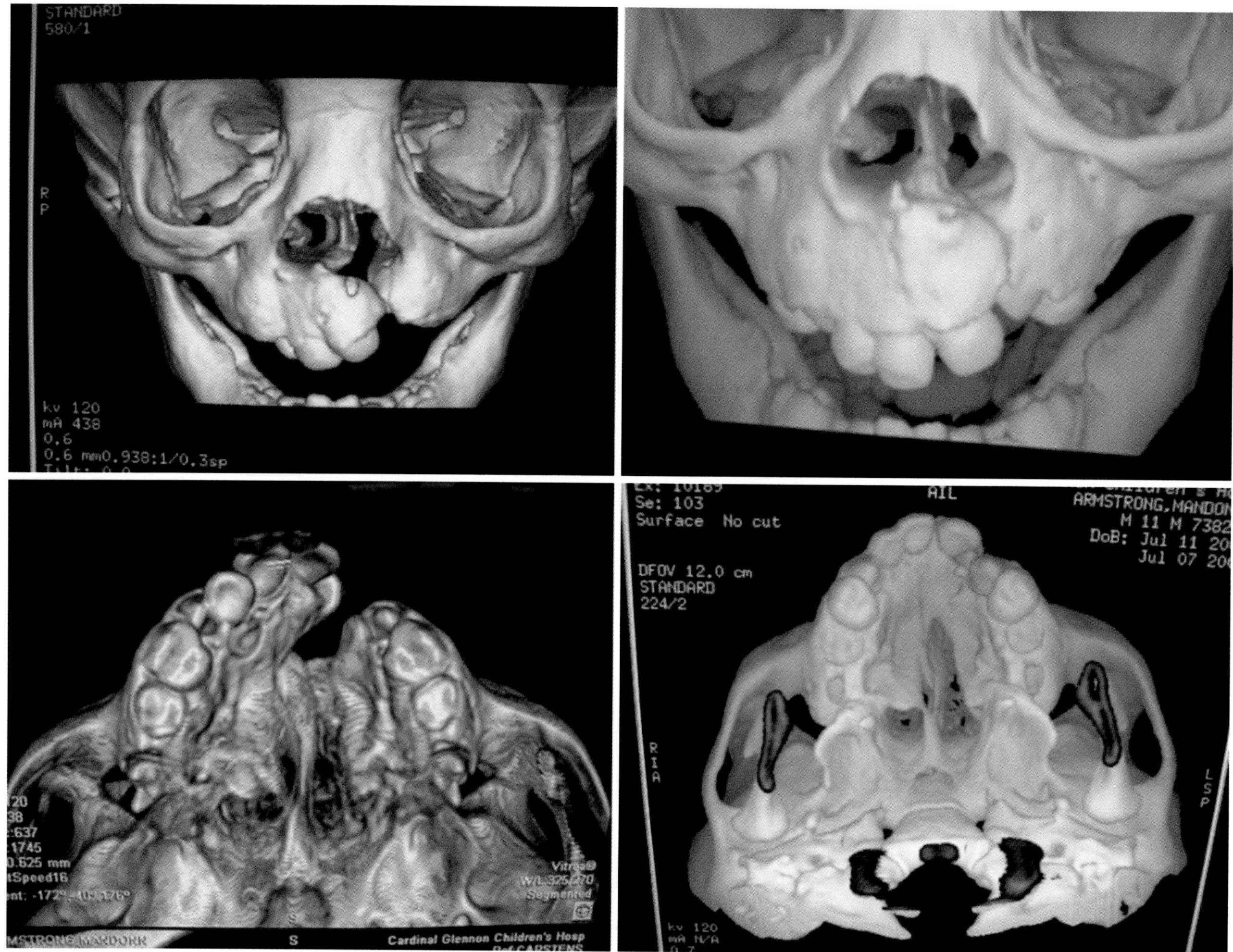

Fig. 19.70 Case 1 3-D CT scan AP and palatal views pre-op versus 5 years showing complete filling of the primary cleft with good height of bone maintained at 5 years. [Courtesy of Michael Carstens, MD]

intercanine distance must exceed the mandibular intercanine distance by 6 mm.

- 4.5 years later: remove quad helix.

Comments

The Talmant lip repair accomplishes the following goals. (1) It establishes an adequate nasal airway in both the short and the long term. (2) Subperiosteal elevation of the internal nasal lining elevates nasal cartilages into correct position, providing immediate support for the external nasal valve. (3) Reassignment of nasalis into an anatomic insertion changes the force vectors of breathing, both immediately and in the long term. (4) Lining release reassigns biosynthetic mucoperiosteum into a correct relationship with the developing bony cavity.

Talmant's lip/nose repair *does not touch the nasal mucosa of the hard palate*; furthermore, it *does not make use of a vomer flap*. The reader should be aware that some schools of thought (particularly in Europe) maintain that vomer flap closure to the nasal mucoperiosteum is to be avoided. Certainly, repair of the anterior nasal floor is not required to accomplish the four goals listed above. However, the decision to not unite the nasal floor completely, once and for all, may be technically disadvantageous.

Concomitant with the lip repair, soft palate repair helps with early tongue posturing, narrows the posterior palate cleft, and offsets the anterior forces created by the cleft lip repair. Talmant wants the hard palate cleft to narrow down in anticipation of palatoplasty at 2 years. He also knows that Lumineau's quad helix will compensate for collapse.

Hard palate repair at 18 months has no particular drawbacks. As we shall see (below) the eruption of dentition and the quality of alveolar bone make *AEP* flaps (when needed) simple to design and execute.

At age 4, the mucoperiosteum lining the (now fully expanded) cleft is easy to elevate. The alveolar bone is strong as well, protecting unerupted permanent dentition.

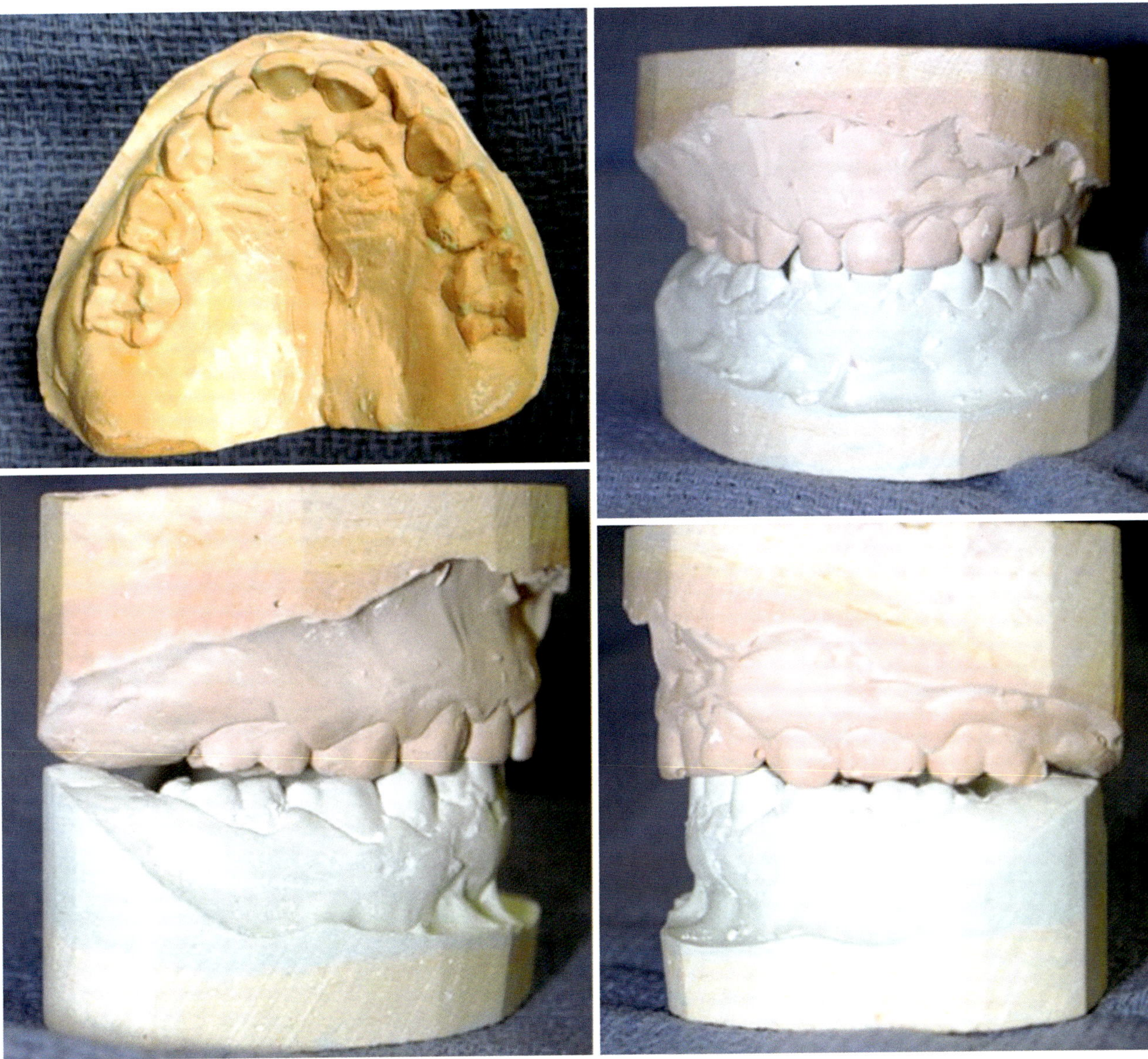

Fig. 19.71 Case 1 Dental models taken at age 5. Occlusion is Class 1. Note the left lateral incisor is in small degree of cross-bite. [Courtesy of Michael Carstens, MD]

Intervention at this age is rational because it does not pose a risk to the stem cell layer nor to the tooth buds.

DFR-Modified Talmant–Lumineau Protocol

- Six months: DFR including complete nasal repair + soft palate with buccinator interposition palatoplasty.
  - Soft palate repair can be performed separately as soon as conditions permit.
- At eruption of dentition: Lumineau adjustable quad helix.
- 18–24 months: hard palate repair.
  - If lip adhesion was done, revise with DFR. Steps 4 and 5 as per Talmant–Lumineau.
- Four years: alveolar bone grafting. The timing for the procedure follows the extent of distraction: *the maxillary intercanine distance must exceed the mandibular intercanine distance by 6 mm.*
- 4.5 years later: remove quad helix.

Comments

Doing a full lip repair in isolation from the hard palate presents theoretical problems directly over the alveolar cleft site. (1) Prolabial tissue is brought into the nasal floor and remains isolated from hard palate mucoperiosteum. The NPP flap, to be most effective, brings with it periosteum from the lateral "shoulder" of the premaxilla. NPP is sutured under direct vision of the nasal palate closure, periosteum to periosteum.

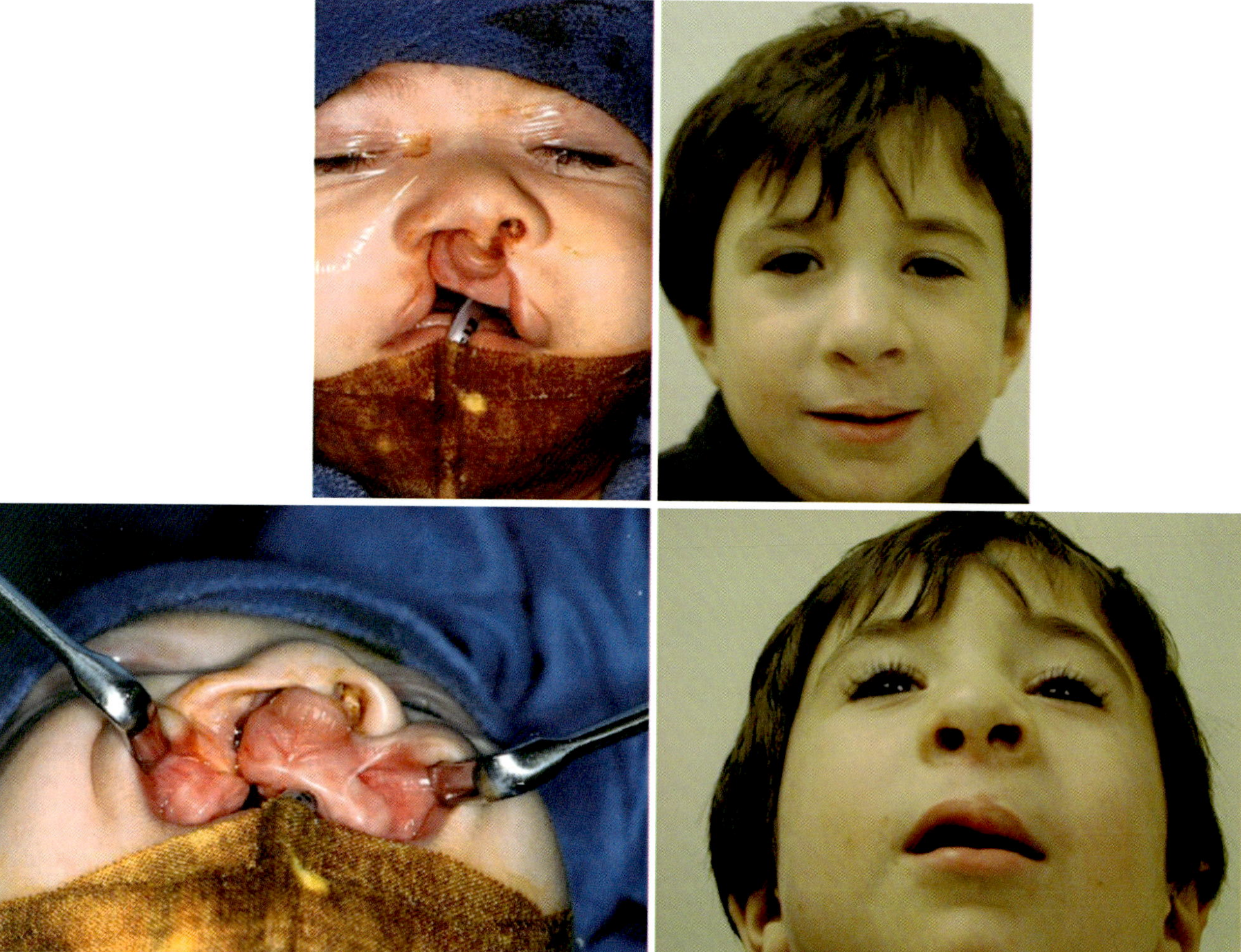

Fig. 19.72 Case 2 Primary BCL/P, complete on right and incomplete on the left. Palate cleft was complete, unilateral on the right. AP and submental vertex views show that airway symmetry at 5 years is maintained. [Courtesy of Michael Carstens, MD]

This type of continuity may be difficult to achieve at a second surgery. (2) These authors do not subscribe to a "no touch" approach to the vomer. From a developmental standpoint, the nasal side of the floor of the nose is supplied by two neuroangiosomes: medial nasopalatine *from the vomer* and the lateral nasopalatine coming *from the lateral nasal wall below the inferior turbinate*. The oral side of the nasal floor is supplied by the greater palatine. Thus, in normal embryologic conditions, each vomer is united to the maxillary crest of the ipsilateral P1–P2 shelf.

Achieving definitive nasal repair at 2 years versus 6 months represents an obvious trade-off. Do the requirements of cerebral "programming" demand nasal surgery at 6 months? Management of nasal stents is undoubtedly simpler in younger patients. Lip–nose surgery performed as a secondary procedure offers technical simplicity (the structures are larger and less delicate, have secure tissue planes, and a readily accessible continuity between the mucoperiosteum of the premaxilla, vomer, and septum). The length of the AEP flaps is such that the entire floor of the alveolar cleft will be covered. No oronasal fistula is produced. The technical design of doing a simultaneous lip and hard palate closure closes five of the six sides of the alveolar "box": roof, floor, sides, and back wall. Thus, soft tissue closure of the alveolar cleft is possible at this stage. The relative merits of this are unknown.

Embryologic Basis of Occlusion and the Lateral Facial Cleft

The maxilla and mandible develop within the confines of the first pharyngeal arch. In its initial state, PA1 hangs downward from the axis of the embryo like a saddlebag. Pharyngeal arches are almost exclusively composed of neural crest. Running through the core of each arch is a mesodermal structure, known as an aortic arch. Let us explore how the first aortic arch develops.

Recall that the primitive embryonic vascular system consists of paired *dorsal aortae* running the entire length of the

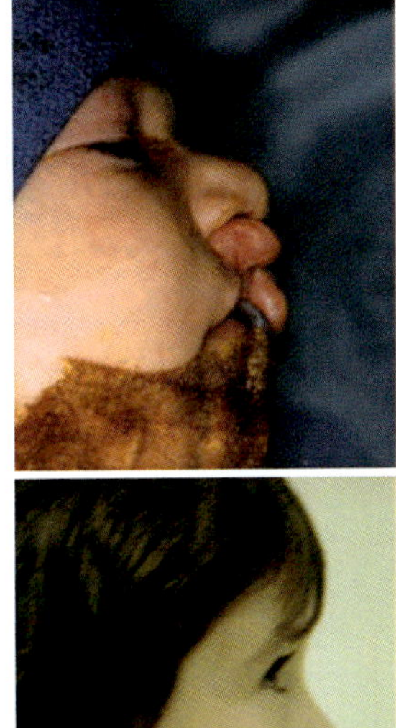

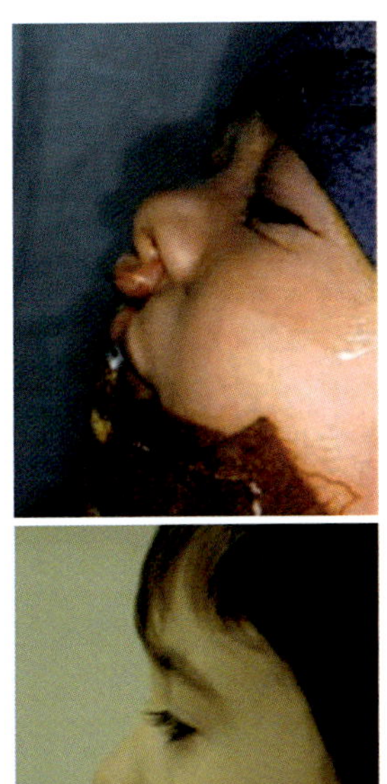

Elevation of nasal time/
Airways patent

Fig. 19.73 Case 2 Subperiosteal release reassigns soft tissues without tension despite projection of premaxilla. Upper panel show projection premaxilla. Lower panel at 5 years shows symmetrical facial growth. Note equality of nasal creases. Good nasal tip projection. [Courtesy of Michael Carstens, MD]

embryo. In the most anterior region of the stage-8 embryo (anterior to the brain) the dorsal aortae form a U-shaped loop. Here, the dorsal aortae take a different name, *primitive heart tubes*; these fuse to form the embryonic heart. The most distal (the most anterior) segment of the primitive heart is the atria, being connected to the vitelline veins. Stage 9 is characterized by three major events: embryonic folding, neural crest migration, and the appearance of the first pharyngeal arch. Brain growth is the driving force behind folding. The heart turns 180°, brining it *ventral to the embryonic face*. The functional components of the heart also undergo a 180° reversal. As a result, the atria now are positioned posterior to the future ventricular outflow tract. The two vessels connecting the cardiac outflow tract with the dorsal aortae become surrounded by the neural crest mesenchyme migrating downward from rhombomeres 1–3 of the hindbrain. The connecting arteries acquire a new name: the first aortic arch arteries.

From an evolutionary and embryologic standpoint these bones are not single entities; maxillary bone fields arise from neural crest of the second rhombomere, whereas those of the mandible arise from neural crest of the third rhombomere. In its initial state, the first arch hangs down like a saddlebag. Its rostral half is all r2, while the caudal half is made from r3. Thus, maxillary and mandibular bone fields sit directly across from one another along the "neuromeric fault line." As we shall see, this situation explains their eventual occlusal relationship.

Oral mucosa is an *ectodermal* tissue all the way back to the buccopharyngeal membrane (represented in the fetal state by Waldeyer's ring). Beyond that point, pharyngeal mucosa is endodermal in origin. Oral mucosa contains the "program" for each tooth. Neural crest mesenchyme simply responds, producing the appropriate dental units. For this reason, ectodermal dysplasias can result in malformed or absent teeth. The premaxilla and the dentary bone (being formed from neural crest) merely "present" potential tooth germs to the epithelium.

In any case, all five pharyngeal arches are divided into quadrants by a set of *distal-less* genes (Dlx) genes: Dlx-1, Dlx-2, and so on. Recall that the first and second pharyngeal arches fuse together very early, probably by the time the third pharyngeal arch makes its appearance (Carnegie stage 11). It can be postulated that a nested set of genes related to, or controlled by, the Dlx system exists along the midline of each arch. This would ensure that a potential zone of apoptosis exists in the first arch running along the longitudinal midline from distal to proximal. Epithelium invading the apoptotic zone contains "mirror image" genes specifying opposing maxillary and mandibular dentition.

Under normal conditions, the apoptotic event is confined to the deep-lying bone fields of the first arch, not the skin. However, if the signals responsible for this division are also located in the second arch and/or in the overlying skin, a lateral facial cleft will be present. Such a cleft extends progressively proximal from the oral commissure to the external

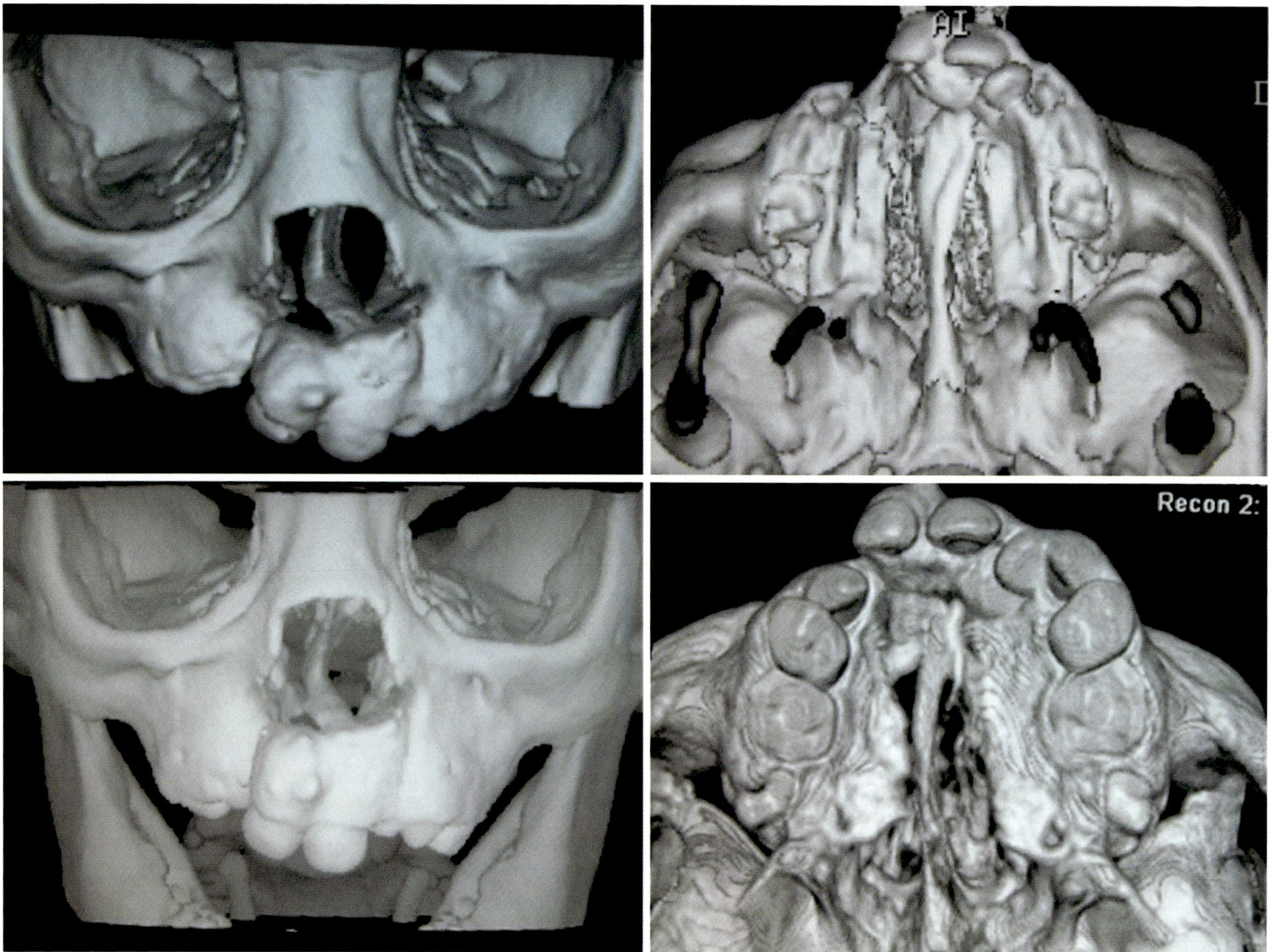

Fig. 19.74 Case 2 Comparison of 3D CT scans pre-op versus 5 years showing that graft with rhBMP2 has stabilized the arch but anteriorly the cleft is only about 50% filled. Will need more graft for implant placement if desired. [Courtesy of Michael Carstens, MD]

auditory canal, as successive Dlx fields are involved. Thus, the cleft described as number 7 has nothing whatsoever to do with Tessier cleft zone 7, that is, the jugal bone and overlying malar soft tissues.

Technical Details of Developmental Field Reassignment Surgery

Children with CL(P) affecting the *primary hard palate* only (intact arch) are brought to the OR at 6 months of age for definitive repair [44].

When CL(P) affects both the *primary palate and secondary hard palate*, that is, when the cleft is complete (through the secondary palate) and narrow, DFR/Talmant lip–nose repair is done at 6 months.

In some cases, cleft lip accompanied by a wide complete palate cleft may benefit from a passive splint or from a Dynacleft® device prior to surgery. Lip adhesion should not be considered a "failure." This procedure has a definite role; the resulting scar is not an issue because it will be relocated within the nose along with the non-philtral flap.

Embryologic Definition of the Philtral Prolabium

As previously discussed, the lip incision in DFR cleft surgery divides the prolabium into two distinct embryologic components: a "true" PP and a "false" NPP.

In normal patients, the central lip element (Cupid's bow) and the PP are synonymous. The true PP is the terminal extension of columella and nasofrontal skin, all of which (epidermis, dermis, and subdermal tissue) originate from p5 forebrain neural crest. The philtrum consists of two paired fields; both are supplied by a V1 sensory nerve and a terminal branch of the anterior ethmoid artery. The two AEA vessels run about 2–4 mm apart. The developmental field of the prolabium is therefore in continuity with the *entire columella, including the lateral walls* all the way to the pink p6

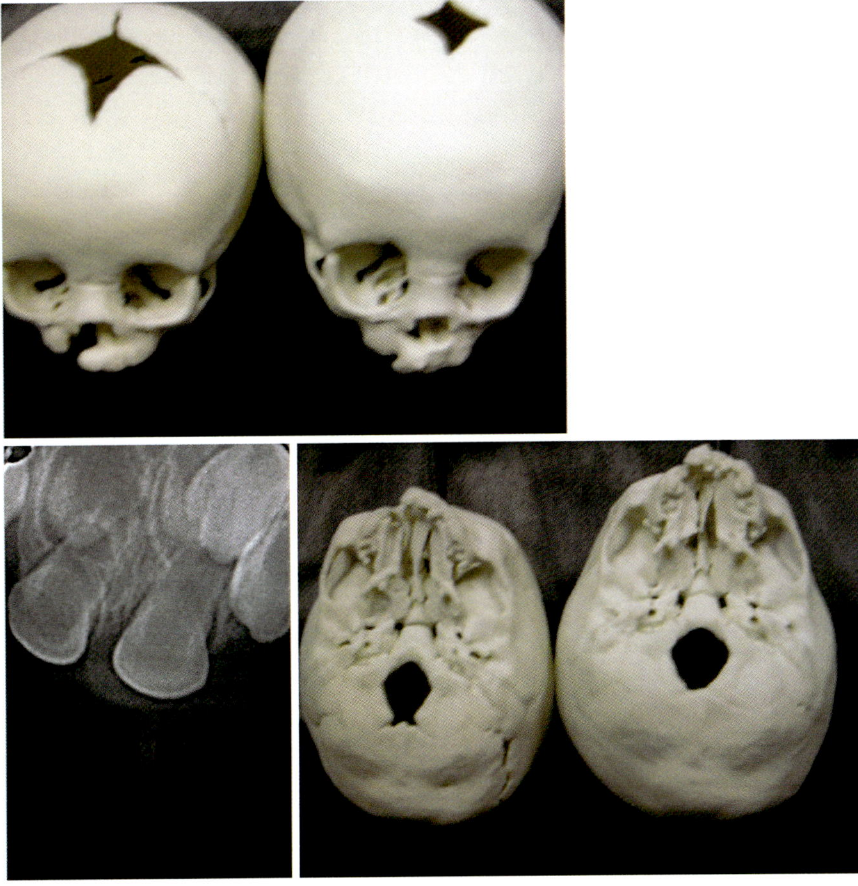

Fig. 19.75 Case 2 Skull models made from the 3-D CT scan: pre-op and 5 years. Courtesy of Medical Modeling, Inc. Vertex submental views (upper left) and submental vertex views (lower right) show symmetrical midface growth at 5 years. Lower left: Apical dental film shows collar of bone around the erupted canine in the cleft. [Courtesy of Michael Carstens, MD]

septal mucoperiosteum. Contained within the philtral–columellar fields are the medial crura and footplates of the lower lateral cartilages. In the Matthews model, the *philtral columns ascend from the white roll to the footplates*. In this interaction, the width of the PP is no longer considered just that of the columella; it is defined by the *transverse distance between the footplates*.

In cleft lip patients the prolabium contains additional tissue, the NPP. This tissue is in continuity with the mucoperiosteum of the underlying premaxilla and vomer. The mesenchyme of the non-philtrum originates from r2 hindbrain neural crest. NP, premaxilla, and vomer share a common neurovascular axis: the medial sphenopalatine (nasopalatine) artery. Note that the mucoperiosteum of vomer and septum appears to be continuous, but these are actually embryologically separate fields. The soft tissue walls within which the septum is synthesized are derived from p6 forebrain neural crest. They are therefore innervated by branches of V1 and supplied by anterior and posterior nasal branches from the anterior ethmoid axis.

In normal patients, NP is never seen. It lies within the nasal floor as the skin cover for two premaxillary sub-fields: the lateral incisor field and the frontal process field. In the cleft situation, however, the neural crest bone elements of lateral alveolus and frontal process are gone. NPP becomes "shipwrecked," a lonely mass of mesenchyme is cast ashore alongside the pre-existing PP. In bilateral clefts the prolabium will therefore have four separate fields: two PP fields in the center and one NPP field on either side.

In sum, the normal prolabium consists of a tissue complex with four layers, the *skin* and the underlying *non-muscle-bearing mesenchyme*. Both arise from forebrain neural crest (prosomere 5). These are supplied by the terminal branches of the anterior ethmoid artery. All remaining layers arise from the hindbrain. An *intervening layer of fat* conveys branches of the facial artery to the fascia and muscle of the *DOO oris*. The fascia is neural crest from rhombomere 5, while the muscle arises from somitomere 6. *The facial nerve* supplying the DOO has its nucleus in r5 as well. The *mucosa* is a neural crest structure arising from rhombomere 2. It is supplied by branches from the medial infraorbital artery.

Medial Dissection: the Non-Philtrum Flap, Septum, the Medial Nasal Fossa, and Nasal Tip

The DFR incision is designed to separate out tissues that are embryologically distinct. Recall that the purposes of the non-philtral flap are (1) to add length to the nasal floor, that is replace the missing "housing" of the lateral incisor; (2) to

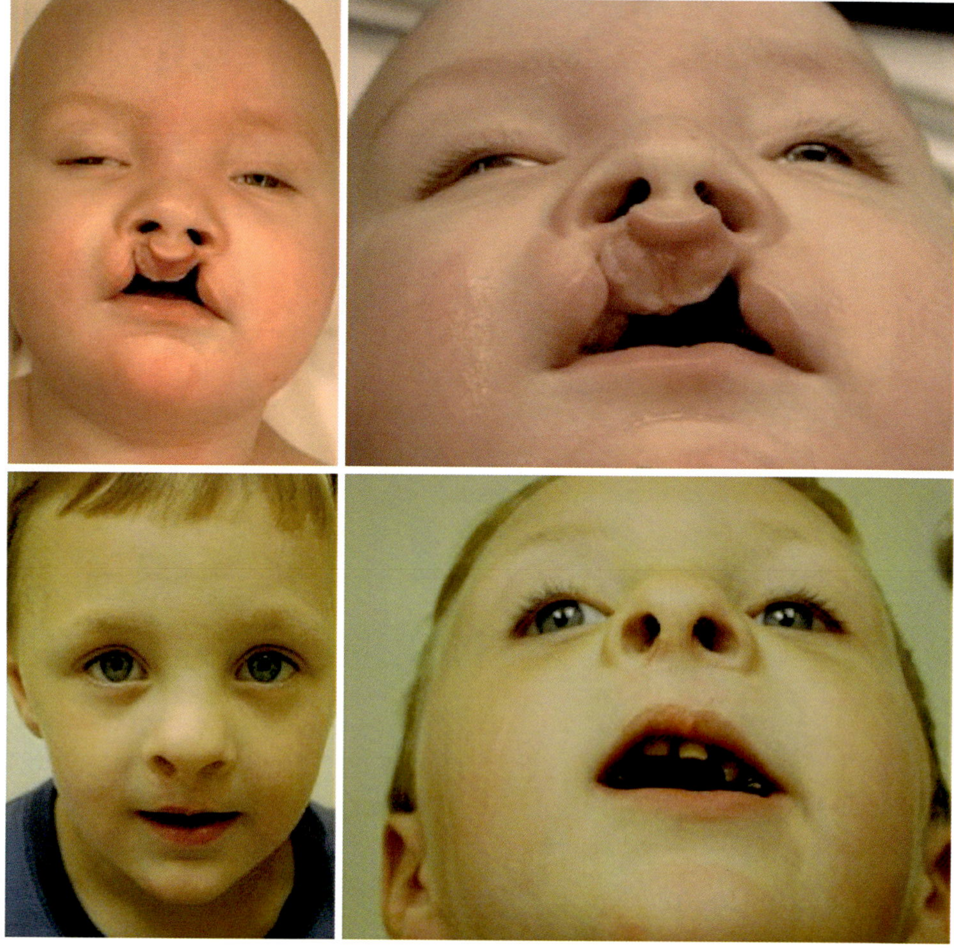

Fig. 19.76 Case 3 BCL incomplete with bilateral CP, incomplete on right, and complete on the left. Upper panel demonstrates a small philtrum. Lower panel shows 5-year aesthetics with good nasal tip projection. Despite the initial size of the philtrum DFR achieved good repositioning of the columella without need for NAM. [Courtesy of Michael Carstens, MD]

replace tissue deficit resulting from release of the lateral nasal wall from the piriform fossa, that is, reconstruct the missing frontal process of the premaxilla; (3) to release the alar footplate and permit advancement of the lower lateral cartilage into the nasal tip; and (4) to create direct access to the septum from below.

In harvesting the NP flap, knowing the width of the PP is critical. Previously, this author had described the transverse dimensions of the true philtrum as being equal to those of columella proper. Matthews' definition above is broader and better. *The width of the true philtrum is equal to the transverse distance between the alar footplates.* This has two implications: (1) the resulting Cupid's bow is embryologically more accurate; and (2) surgical dissection using these landmarks has technical advantages. Both these points will be discussed later (Figs. 19.30, 19.31 and 19.32).

The prolabial incision we are describing has implications for lip height and lip aesthetics. Using the former (narrower) definition of philtral width (in which 2–3 equals columellar width) only lip height is almost uniformly equal. That is, the height of the philtral column on the non-cleft side (13–2) equals the distance on the cleft side (12–3) equals the height of the lateral lip element (8–9). *The aesthetic problem with this incision resides at the white roll at the cleft-side philtral column.* The normal upper lip has natural pucker (a slight eversion) all along the white roll. This aesthetically important feature is caused by the presence of marginal fibers of the DOO. Recall that anatomic territory of DOO is biologically "programmed" oral mucosa. Thus, the distal margin of DOO follows the vermillion, curving upward in the form of a fishhook. It terminates at the mucosa–skin interface, that is, at the white roll. In a non-cleft situation, these terminal fibers extend right across the philtral column. By lending bulk to the vermillion just below the while-roll, the terminal fibers are responsible for eversion.

Matthews has pointed out that that when a DFR incision is brought straight down based upon columellar width alone, *eversion at the white roll is obliterated.* How can this be avoided and, at the same time, preserve lip height? Three simple steps will do the trick. (1) The definition of Cupid's bow width is expanded, marking it out using the transverse dimensions of the footplates and normal mucosal landmarks. (2) The DOO fibers should be conserved all across the philtrum. (3) A measured back-cut above the "ridge" of the DOO will preserve lip height. Over time this incision becomes a

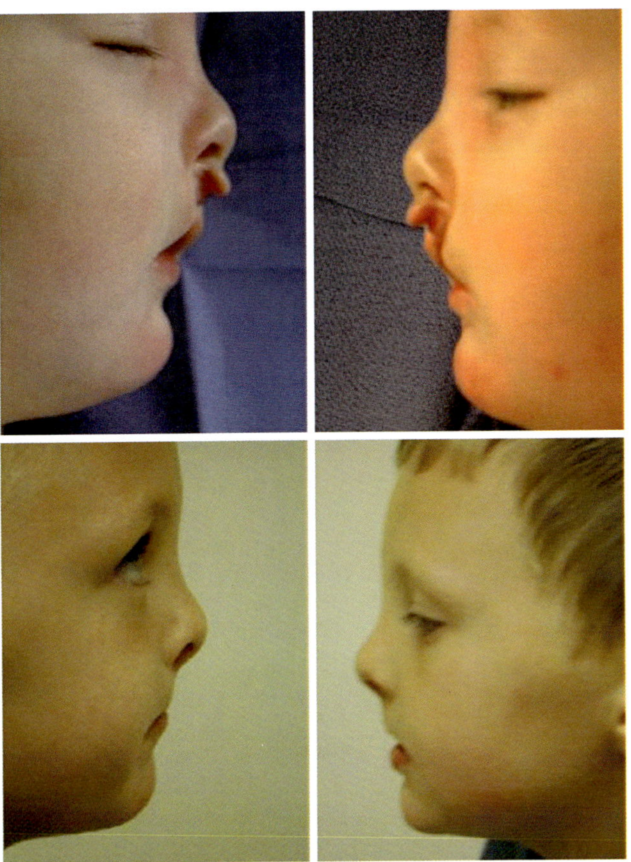

Fig. 19.77 Case 3 Nasal tip anatomy and projection. Upper panel compared with lower panel shows good aesthetics at 5 years. DFR achieves good repositioning of the columella. [Courtesy of Michael Carstens, MD]

straight line and *conserves the aesthetically important eversion* of the lip all the way across the repair.

In the Matthews modification of DFR, the PP is mapped out as follows. Point 1 (the center of PP) is located *directly above the frenulum*. Point 2 is the non-cleft philtral column. Point 3 (the cleft-side philtral column) is marked out with distance 1–2 equal to 1–3. Remember that PP and columella share a common source of mesenchyme (forebrain neural crest) and a common neurovascular supply (V1 and the terminal branches of anterior ethmoid arteries). The width of these fields is the transverse distance between the footplates. In a unilateral cleft the non-cleft footplate is point 13 and the cleft-side footplate is point 12. Matthews' prolabial markings create a philtrum of the same width. Thus, **2–3 = 13–12**. These medial markings will match those of the lateral lip element **8–9**.

The eversion (pucker) of the lip is marked all the way across the philtrum. Because the DOO follows the mucosa, this will be observed to curve slightly upward. With our new definition, cleft-side lip height will fall short by 3–4 mm. One compensates for this by lengthening PP using a back-cut made *above the natural roll of the muscle along the margin of the Cupid's bow*. The back-cut will dart inward about 2 mm and then back outward by the same amount at a 45° angle, where it joints the original curve of the DFR incision. This gives two benefits: (1) parity of lip height and (2) a natural fullness of muscle at the base of the new philtral column.

The proximal component of the prolabium incision ascends toward the columella. It circles around the base of the columella, passing beneath the footplate of the medial crus. The incision terminates at the transition between the columellar skin and the septal mucosa. Initially, the depth of this incision is *skin only* (Figs. 19.39 and 19.40).

The depth of the prolabial incision is important. Initially, it is entirely cutaneous. This permits elevation of the philtral skin from the underlying non-muscle-bearing mesenchyme for about 5 mm. Once the initial incision and undermining are accomplished, the prolabial incision is deepened from the white roll upward to the base of the columella (the underlying mucosa is spared). At the columellar base the incision remains cutaneous only. This is to preserve the underlying mesenchyme within which is contained the pedicle from the medial nasopalatine artery. This tissue is left intact. The cutaneous incision beneath the alar footplate permits entry into the columella with curved scissors. The tips of the scissors follow the medial aspect of the medial crus all the way into the nasal tip (Fig. 19.39).

A paring incision is now done from the white roll up into the nose. Again, at the base of the columella, care is taken to make this incision cutaneous only. The non-philtrum flap is now elevated with a generous subcutaneous base. The flap is lifted off the premaxilla in the subperiosteal plane up to, but not beyond, the junction of the premaxilla with the vomer. This ensures mobility of the flap without compromise of the pedicle. In sum, the non-philtrum flap is a skin island supported by subcutaneous tissue supplied by the nasopalatine artery.

The disposition of the medial mucosa is at the discretion of the surgeon. The M flap can be dropped downward like a baby's diaper. It can be included with the non-philtrum flap for greater epithelial width. Finally, it can be inferiorly based and rotated. I prefer option 1 because it leaves option 3 open. The width of the non-philtrum flap is not strictly determined by the skin paddle. Subcutaneous tissue will readily epithelialize.

Septal mobilization is carried out next. This can be done by extending the incision like a hockey stick upward at the anterior septal border. A dental amalgam packer is very useful because its curved tip is flat and cross-hatched. It can rasp through the mucoperichondrium to the correct plane with great delicacy. The septum is completely dislocated.

Herein we encounter part 1 of Talmant's nasal fossa reconstruction: the dissection is extended upward to the nasal bone and the vestibular lining is freed in the subperios-

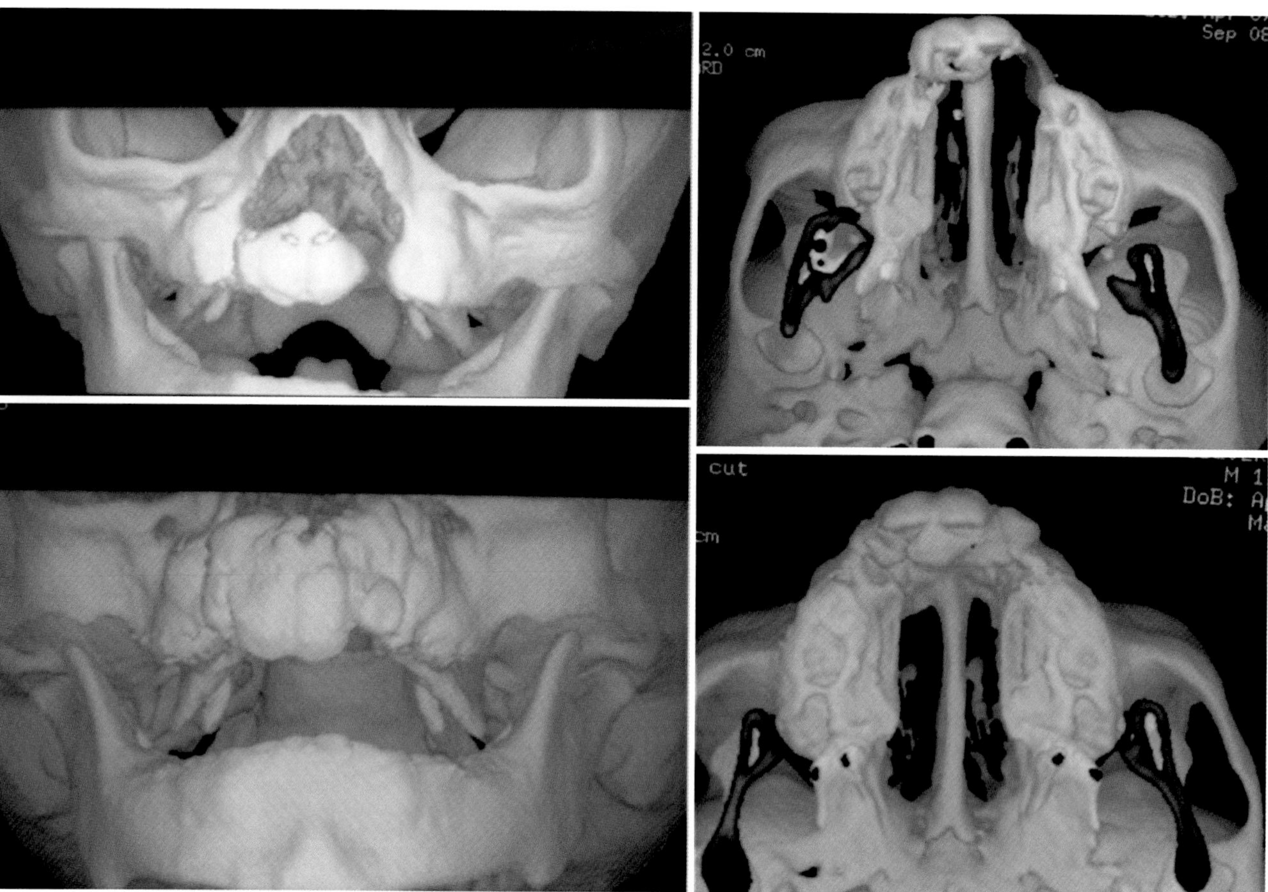

Fig. 19.78 Case 3 At 5 years, unification of arch with rhBMP2 at palatoplasty has resulted in stability, class 1 and a nicely rounded, projection maxilla. [Courtesy of Michael Carstens, MD]

teal plane from beneath the nasal bone. This frees up the upper lateral cartilage from beneath the nasal bone. Recall that the abnormal anatomy of the nasalis predisposes to a flattening of the nasal bone and entrapment of the upper lateral cartilage. This maneuver also lengthens the dimensions of the nasal fossa [45].

We have now completed the medial dissection. It is now time to ensure that the goals of our nasal dissection are accomplished. The nasalis complex needs to be freed in the subperiosteal plane from the underlying upper and lower lateral cartilages. A small infracartilaginous incision provides access to the nasal dorsum. It is a good idea to visualize the alar cartilage (once again, the amalgam packer proves helpful). A McComb dissection is carried out. One should drop downward to encounter the bone along the piriform fossa. Spatial limitations of the nasal incision limit what one can accomplish. We must await the lateral dissection to ensure that our subperiosteal nasal dissection is complete.

Lateral Dissection: Muscle Separation, Nasalis Transposition, and the Lateral Nasal Fossa

Paring incision proceeds from point 8 all the way into the nasal cavity. Recall that distances 8–9 = 13–2 = 12–3. Also recall that the width of the keratinized mucosa at point 8 should equal that at point 3. With the lateral lip element stretched, one immediately undermines the skin from the SOO oris for a distance of about half a fingertip. This separation is continued upward to the nasal skin, involving both the lower alar base and the nostril sill. The superficial and DOO muscles are separated by a fat pad analogous to that found in the prolabium. SOO and DOO are separated proximally all the way to the termination of the deep layer (Figs. 19.20, 19.47, 19.48 and 19.49).

Nasalis and SOO are still confluent. Our next task is to separate them and gain control of the nasalis. This is most easily done using the concept of two planes: subcutaneous

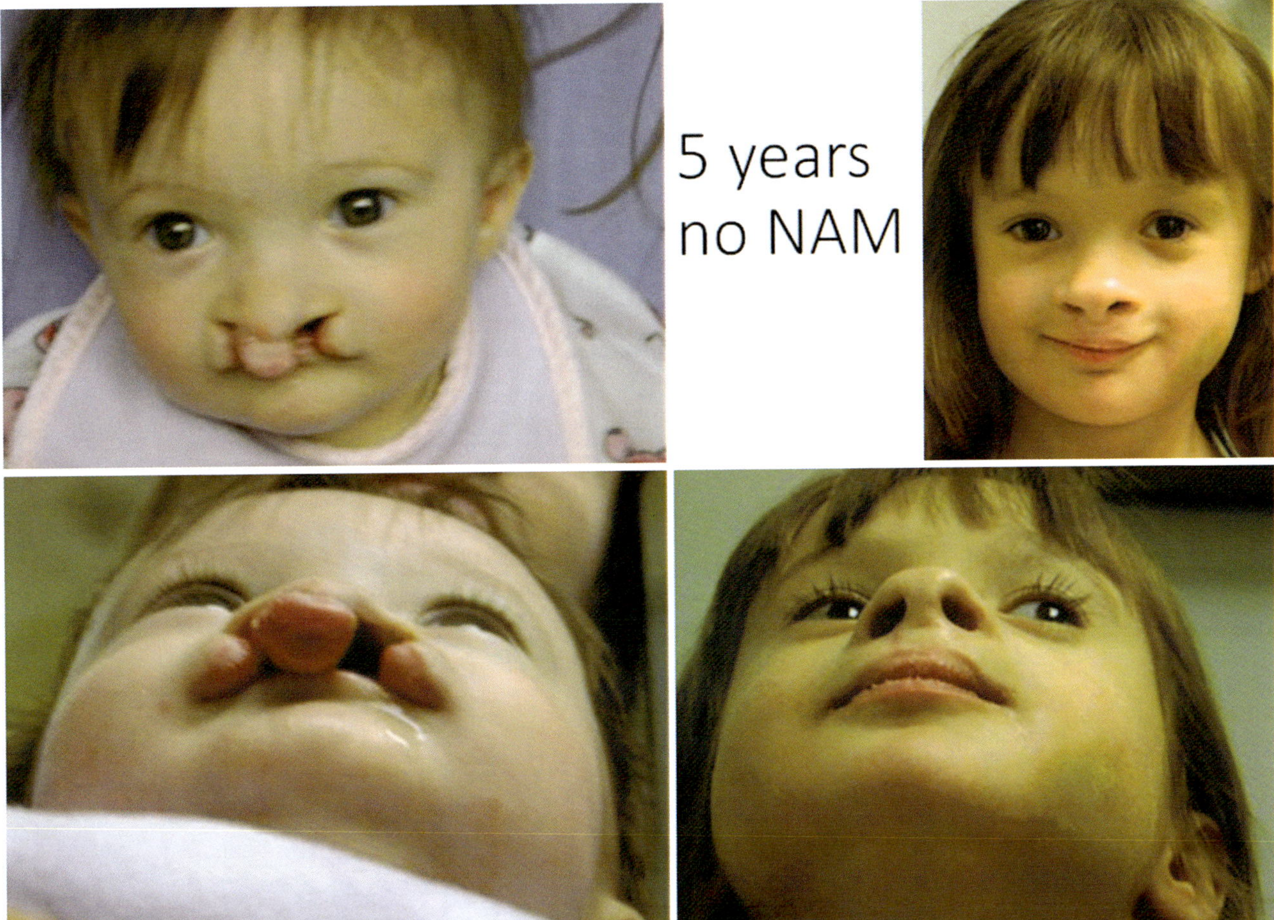

Fig. 19.79 Case 4 Primary BCL with long-term growth of midface, stability of nasal projection, and protection of the airway. Note at 5 years (upper right and left) the nostrils are widely patent. [Courtesy of Michael Carstens, MD]

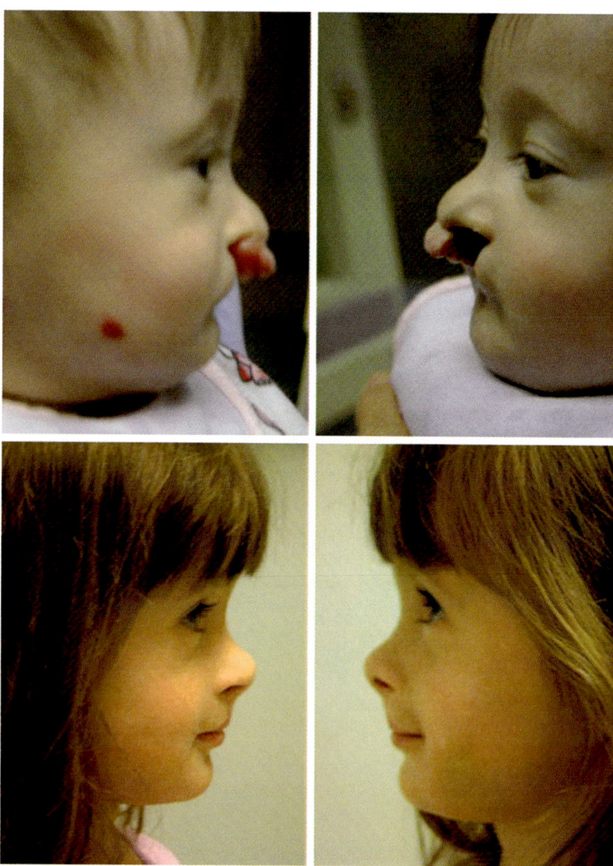

Fig. 19.80 Case 4 5-year growth of the midface with stem cell reassignment has brought the premaxillae and maxillae into a normal relationship. Lower panel demonstrates ability of DFR to bring soft tissue envelope forward without compromising nasal projection. [Courtesy of Michael Carstens, MD]

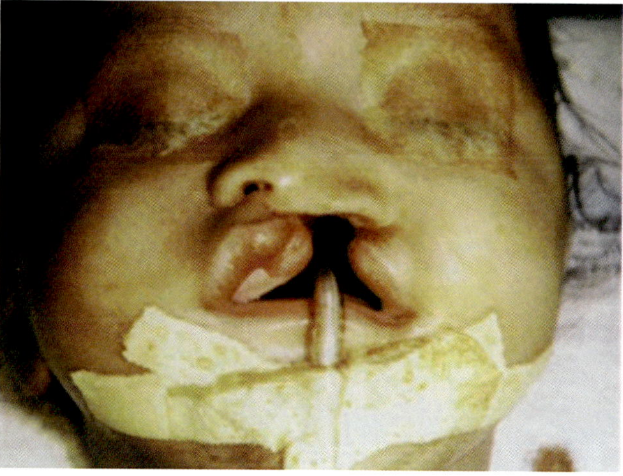

Fig. 19.81 Case 5 Personal comparison of rotation-advancement and DFR: same patient, same surgeon, different concepts, better biology. Complete UCL/P (wide) with primary repair using rotation-advancement presents with typical secondary deformities. Shown here is the on-table pre-op appearance at primary repair for which I was the attending surgeon. On-table results were nice but she represented at age 9 with the complete panoply of secondary deformities. [Courtesy of Michael Carstens, MD]

and subperiosteal. Recall that the medial head of the nasalis is inserted into the vestibular lining, while the lateral head of nasalis resides in the piriform fossa, extending halfway up its vertical extent. We shall start with the medial head. Extension of the skin under-mining into the nasal cavity now continues in a plane superficial to the medial head of the nasalis. The dissection should extend in front of the inferior turbinate and all the way up to the lower lateral cartilage. One now proceeds in the subperiosteal plane beginning at the entrance into the piriform fossa, that is, in the triangular zone corresponding to the nostril sill. One is now beneath the medial head of nasalis. The muscle is grasped and sectioned away from the bone. With the lateral lip once again on tension, the SOO is divided away from medial nasalis at the level of the alar base. Tension placed on the nasalis at this point will transmit to the lateral nasal wall but will not be free.

The lateral rim of the piriform fossa is swept clean upward to the halfway point. One has now gained access to the lateral margin of the fossa. An elevator is passed upward along the lateral wall of the nose until it achieves the nasal dorsum. At this point one should switch to an elevator that is sharp and strong, with a broad and curved tip, such as a McKenty. A customized Molt elevator works well. The elevator is swept downward underneath the SMAS (including the nasalis) all the way to the nasal tip. The result of this *McComb dissection* should be a complete liberation of the dorsal nasal skin and SMAS from the underlying upper and lower lateral cartilages (Fig. 19.52).

Recall that all unilateral cleft noses have an overall deviation of the soft tissue envelope on the non-cleft side away from the midline. This is an opportune moment to correct this problem and to centralize the entire midface envelope. Wide subperiosteal dissection is carried out over the face of the maxilla on both sides. Care should be taken to free up the non-cleft alar base as well. Because the mucoperiosteal envelope contains the stem cells required for future membranous bone synthesis, one has now "centralized the biosynthetic envelope" such that the external dimensions of piriform fossa (and the maxillae in general) will continue to autocorrect over time.

We now turn our attention to inside the piriform fossa to complete the dissection of the lateral head of nasalis. Using a curved elevator the nasalis muscle fibers lying within piriform are literally "scooped out" of the fossa. At this point the two heads are evaluated. They may appear distinct or as a single mass. The functionality of the muscle dissection is now tested. Traction is placed on the nasalis; motion along the lower lateral cartilages will be seen. The range is between 5 and 10 mm. If restriction is encountered, one can release the muscle further by first re-entering the piriform fossa to take down any residual attachments. One can then proceed along the external aspect of the muscle about halfway up, spreading it away from the overlying skin. Excessive exter-

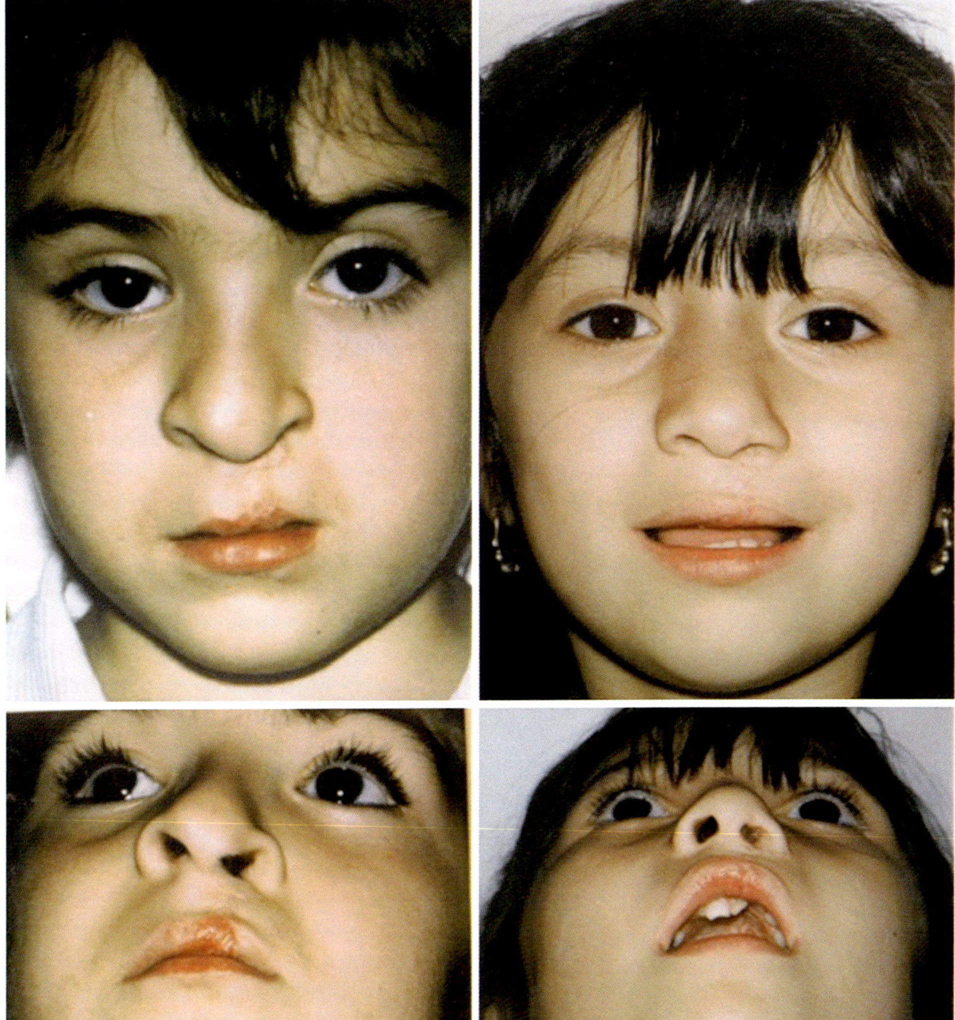

Fig. 19.82 Case 5 Shown here (left) at 9 years post-rotation-advancement the patient has a lip notch, muscle asymmetry, nasal deformity and mouth breathing. Right panels show after secondary repair with DFR (Talmant) restoration of symmetry and a functional airway. [Courtesy of Michael Carstens, MD]

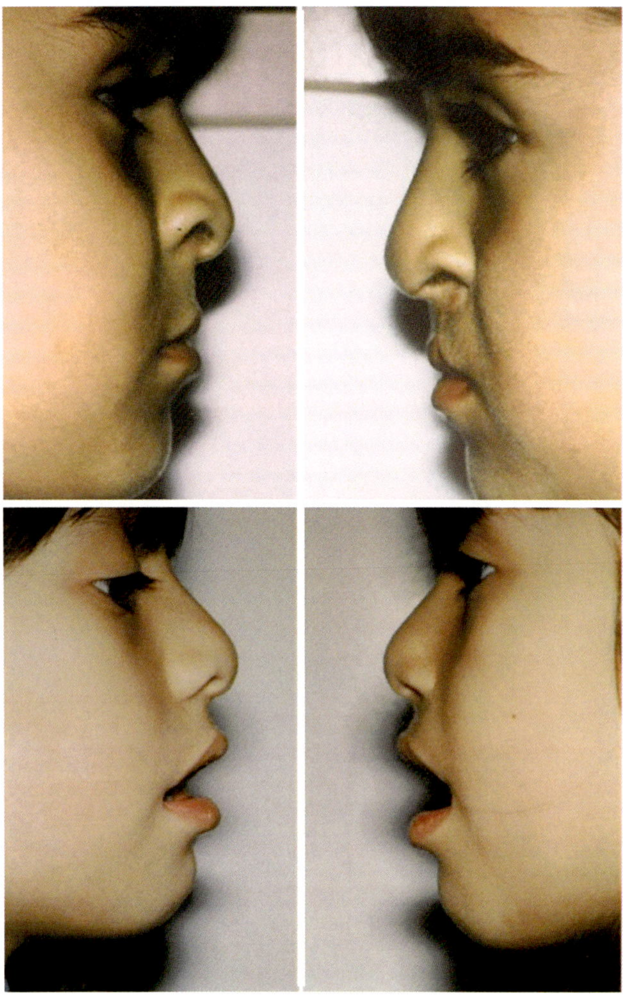

Fig. 19.83 Case 6 Lateral views showing midface and nasal profiles. Note on the left panels aesthetic improvement of projection in what is apparently the "normal" right side. [Courtesy of Michael Carstens, MD]

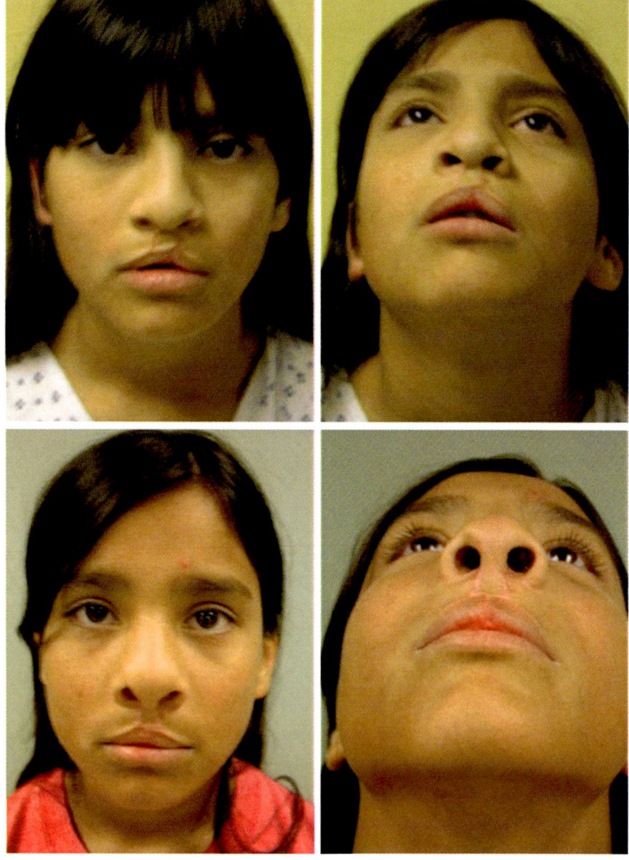

Fig. 19.84 Case 7 Secondary BCL with nasal airway restoration. Upper panel shows typical compression of the nose and compromise of the nasal airways. Lower panel demonstrates functional and aesthetic improvements after DFR. [Courtesy of Michael Carstens, MD]

nal dissection will encounter blood supply to the muscle from the facial arcade.

Repositioning of the nasalis is done by anchoring the muscle to the mucoperiosteum of the canine fossa and the sulcus using a mattress suture of 4–0 vicryl. The suture is passed from the buccal sulcus upward behind the orbicularis. It loops through the muscle and is returned to the mouth where the suture is placed on a clamp. It will be approximated as the final maneuver of the surgery (Figs. 19.51 and 19.52).

Having cleaned out the piriform fossa we are now in position to complete part 2 of Talmant's nasal fossa reconstruction. A curved elevator is passed anterior to the inferior turbinate and is then directed backward and upward until the nasal bone is reached. Using this combination of subperichondrial and subperiosteal dissection, a complete freeing of the mucoperiosteal lining of the nasal fossa is achieved. Once again, the biologic advantage of this maneuver is to reposition the stem cell envelope correctly such that osteosynthesis within the piriform fossa proceeds in a normal manner from 6 months onwards. The goal is normalization of the internal dimensions of the piriform fossa over the course of time.

Closure and Nasal Splinting

At this juncture, the dissection is complete. All cleft surgeons have their preferred sequence of steps to achieve closure. I will simply make note of a few maneuvers that I have found helpful over time. The closure sequence begins inside the nose. Access incisions to the septum and the nasal tip are addressed first. The non-philtrum flap is sutured to the release incision in the lateral nasal wall. If a vomer flap has been raised, the NP flap is sutured on its back side to the vomer flap. Otherwise, the nostril sill flap created by the releasing

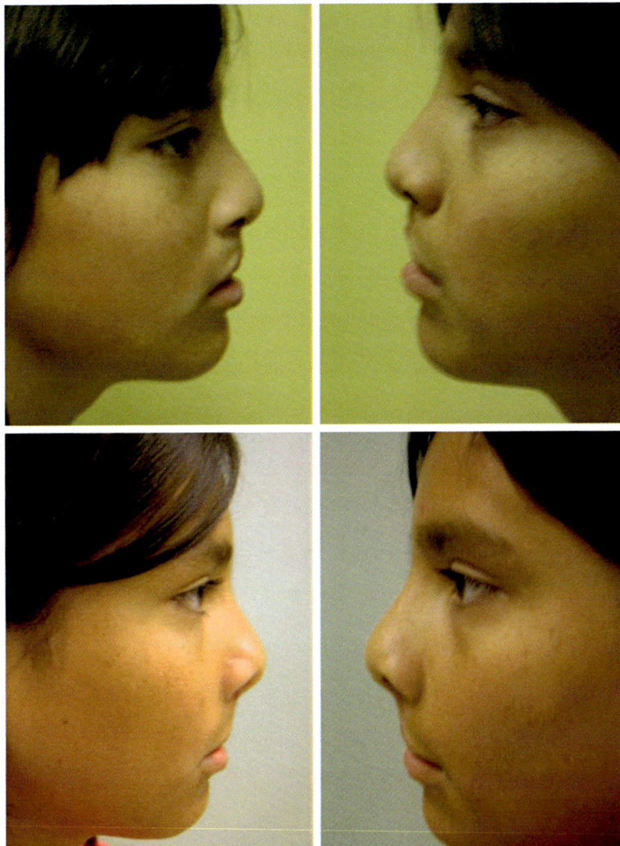

Fig. 19.85 Case 7 Bilateral secondary repair: lateral views show nasal projection and forward drape of the upper lip. Lower panel demonstrates increased tip projection and angulation of the alar plane. [Courtesy of Michael Carstens, MD]

incision previously described is brought along the front side of the NP flap and sutured with 5–0 vicryl sutures taking care to maximize the "fit" between the two flaps. A double hook is then placed in the nose and gently lifted. The posterior margin of the NP flap is sutured to the footplate of the alar cartilage. In this way one strives to re-establish the elusive "shoulder" of the columella (Figs. 19.61, 19.62, 19.63 and 19.64).

Lateral lip height is first established by suturing the extreme tip of the DOO to its counterpart at the anterior nasal spine. Alignment of the white roll is carried out using 4–0 or 5–0 vicryl in the DOO where it curls upward toward the white roll. A confirmatory suture of 5–0 monocryl is placed at the dermal–epidermal junction of the white roll. An additional vicryl suture is placed at the wet–dry mucosal junction. With the lip on traction, the remainder of the mucosa and DOO is closed as a single layer. Final decisions regarding the M and L flaps are taken at this time. The use of the L flap as a "diaper" to cover the raw area of the medial lip vermillion often makes sense.

The extreme tip of SOO (sometimes referred to as the oblique head of the SOO) is attached by a mattress suture through the columella just below the footplate of the lateral crus on the non-cleft side. The suture is tightened to check the position but not tied. The SOO is then sutured to the non-muscle-bearing mesenchyme of the prolabium with three to four vicryl sutures. This helps achieve some bulkiness beneath the new philtral column. Several sutures of 6–0 monocryl (or nylon, if nothing else is available) carefully placed at the dermal–epidermal junction seal up the skin. A running 6–0 vicryl suture completes the closure. Ophthalmic 6–0 is particularly good because of its dissolution characteristics.

Suspension of the nasal cartilages is not a part of the Talmant technique. Nonetheless, I think that tip projection is improved by suturing the medial crura together in their new position with two sutures of 5–0 vicryl/monocryl. The unified crura can then be elevated and sutured to the septum with 4–0 vicryl/monocryl. Luis Monasterio has a very ingenious tip graft which is shaped like the number 1 with the upper limb pointing backwards. The *number 1 graft* is placed through a lateral columellar incision exactly the way I used to do if for the non-philtrum flap. The incision falls just anterior to the medial crus. The graft is harvested from the septum. Closure of the incision with 6–0 chromic is virtually unnoticeable. For simplicity and versatility, the number 1 graft is my first choice for secondary rhinoplasty or in adult cleft cases. The graft is fixed in place using the same suture sequence: medial crural unification followed by fixation to the septum.

Silicon stents cut from 0.5 mm sheeting are then placed (Fig. 19.67). The initial two stents are only temporary. They are curled up like snails inside each nasal cavity. An additional sheet of silicon is placed over the dorsum and the tubes are secured with two sutures of 4–0 prolene, beginning with the non-cleft side. When one places these sutures on the cleft side, a very important phenomenon takes place. On the second pass from the nasal cavity back to the dorsum, the tip of the needle is engaged with the silicon sheeting. Moving the sheeting further repositions the alar cartilage. The needle is then thrust upward through the sheeting, fixing the cartilage in its final position.

Vaseline gauze or Xeroform® gauze is then used to pack the nose. This further expands the nasal airway. During the packing phase, antibiotics are given. At 1 week, both the packing and the temporary silicone stents are removed. Nasal stenting is then continued for 4 months using the design as illustrated. Once again 0.5 mm silicon sheeting is used. In countries where this is not available, intravenous solution bags can be cut up instead.

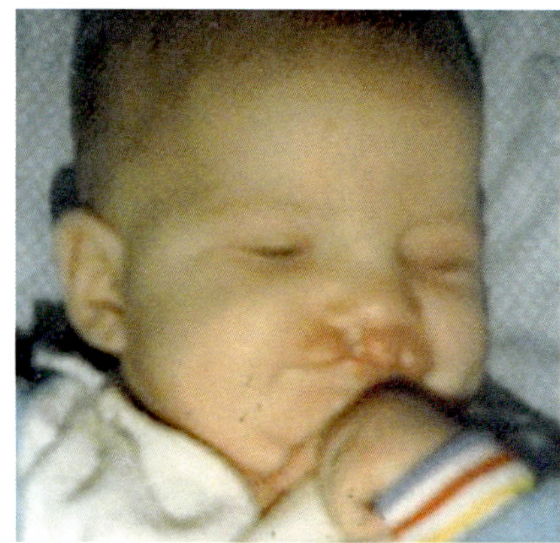

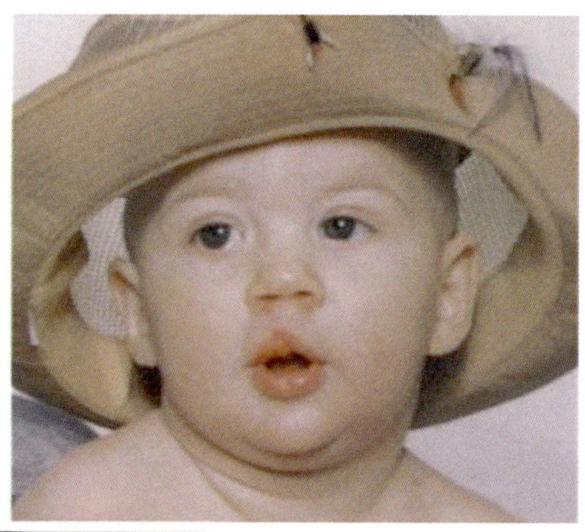

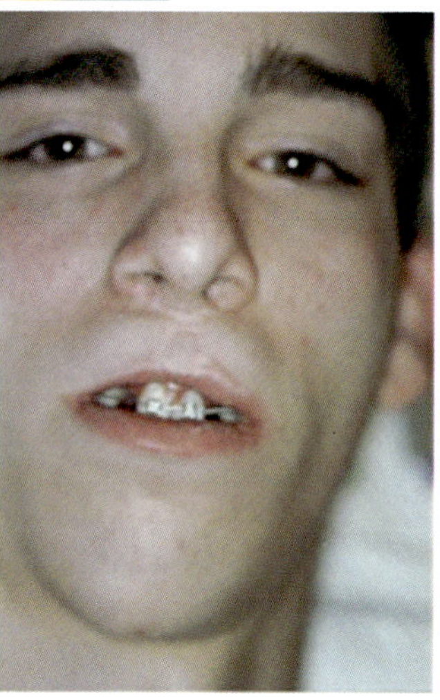

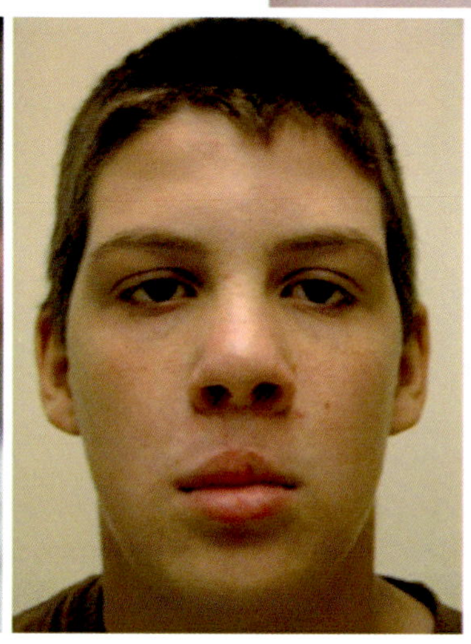

Fig. 19.86 Case 0 From the beginning, we return to this story of relapse, now treated with DFR and rhBMP graft. Original pathology was complete UCL/P on the right with incomplete/microform on the left. At 1-year post-op from DFR cleft lip–nose revision (lower right) a striking correction of the "recessed" maxillary soft tissue envelope has been achieved. [Courtesy of Michael Carstens, MD]

ADDENDUM: Sotereanos Muco-Gingivoperiosteoplasty (see Fig. 19.4)

Alveolar cleft closure (with DFR modification) is extremely useful as part of secondary cleft repair or as a stand-alone procedure. The following comments are intended to simplify the steps.

Non-DFR flaps are shown in the original illustration with nasal floor closure using nasal mucosa flaps from below as a blind procedure. Under *DFR protocol* the nasal floor is addressed at primary lip repair (6 months) with the back wall and floor created by AEP at 18 months. If the nasal floor has not been closed, prior repair was non-DFR. Secondary lip

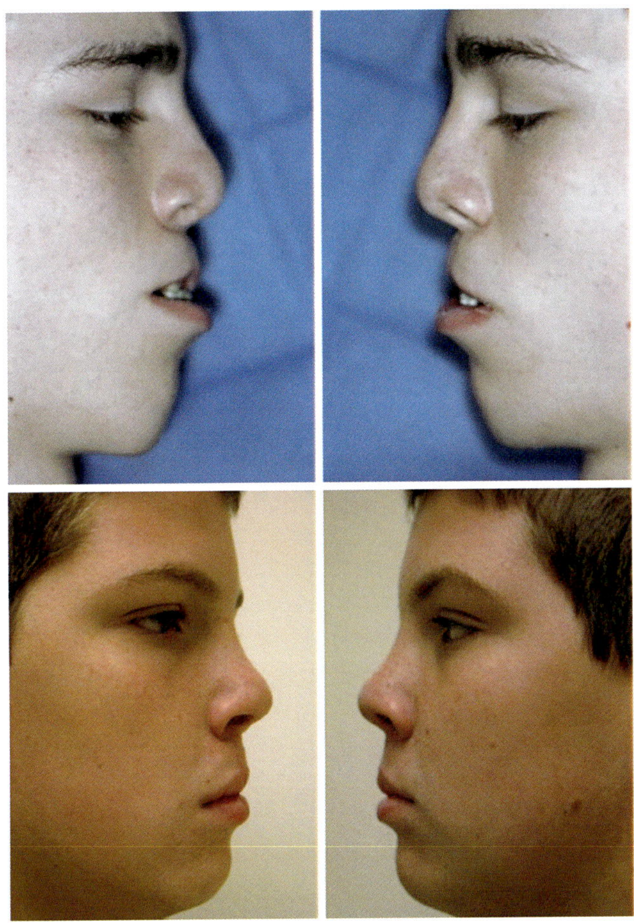

Fig. 19.87 Case 0 Lower panel shows fullness of the midface and normalization of upper–lower lip aesthetics and significant changes in alar angulation, nasal projection, and normalization of the nose–lip relationship. [Courtesy of Michael Carstens, MD]

repair is done with nasalis dissection and water-tight closure.

Step 1 Upper Left Mucoperiosteal flap (r2, pink) GPP flaps (sliding sulcus flap) is mobilized via pericoronal incision with a back-cut up the buttress. The S flap is supplied by medial and lateral branches of StV2 infraorbital and by overlying ECA facial artery. The dotted lines indicate the level of a counter-incision that will be made along the undersurface of the flap. Mucoperiosteum of premaxilla (orange) has also a mixed blood supply from StV1 anterior ethmoid, from StV2 medial nasopalatine, and from StV2 medial branch of infraorbital. Periosteal flap on maxillary side (yellow) is supplied by greater palatine. Periosteal flap on the premaxillary side (white) has a variable blood supply. If the cleft is unilateral, it receives StV2 greater palatine + StV2 medial nasopalatine and back flow from StV1 anterior ethmoid. If the cleft is bilateral, it is perfused by the latter two arteries only. Note the cut shown between lateral incisor and canine on the non-cleft side. In DFR this is usually unnecessary for two reasons: (1) the non-cleft flap can be mobilized in a similar manner. (2) A similar periosteal releasing incision. Bilateral S flaps can cover a total of up to four dental units without tension. Dissection time is fast and bloodless, under 5 min per flap.

Step 2 Upper Right Backwall flaps are sutured, maxillary (yellow) to premaxillary (white). The counter-incision along the undersurface of the S flap is indicated by a white line with red dots. The incision is placed at the level of the buccal sulcus. Although the S flap is mobilized over the entire face of the maxilla, the counter incision frees the alveolar component to translate with ease, with no tethering from above.

Step 3 Mesial translation of maxillary S flap (pink) and premaxillary S flap (orange). Dead space for the flaps is transferred to the first and second molar.

Step 4 Final closure with bone graft/rhBMP2 in place. Incision at premaxillary–maxillary junction on the non-cleft side is not necessary.

A closing quote, re-stated: "Cleft surgery is applied embryology." (Victor Veau) Better outcomes for the treatment of facial anomalies will inevitably be achieved when empiricism yields to an understanding of the normal mechanisms of development (Figs. 19.89 and 19.90).

Acknowledgements The author wishes to credit the following individuals for their intellectual and technical contributions to developmental field reassignment cleft surgery: Paul Tessier, M.D. (deceased)—for his pioneering work in facial cleft classification; Jean Delaire, M.D., D.M.D.—for his demonstration of wide subperiosteal dissection in cleft surgery; David Matthews, M.D.—for his refinements in the design of the non-philtral prolabial flap; Jean Claude Talmant, M.D.—for his discovery of the nasalis muscle and the importance of correct nasal airway restoration; Jean-Pierre Lumineau, M.D., D.M.D.—for his unique technical advances in the management of the alveolar cleft; and Luis Monasterio, M.D.—for demonstrating the importance of the developmental field approach to cleft palate repair using the alveolar extension palatoplasty to treat or prevent fistulae and maintain dental development.

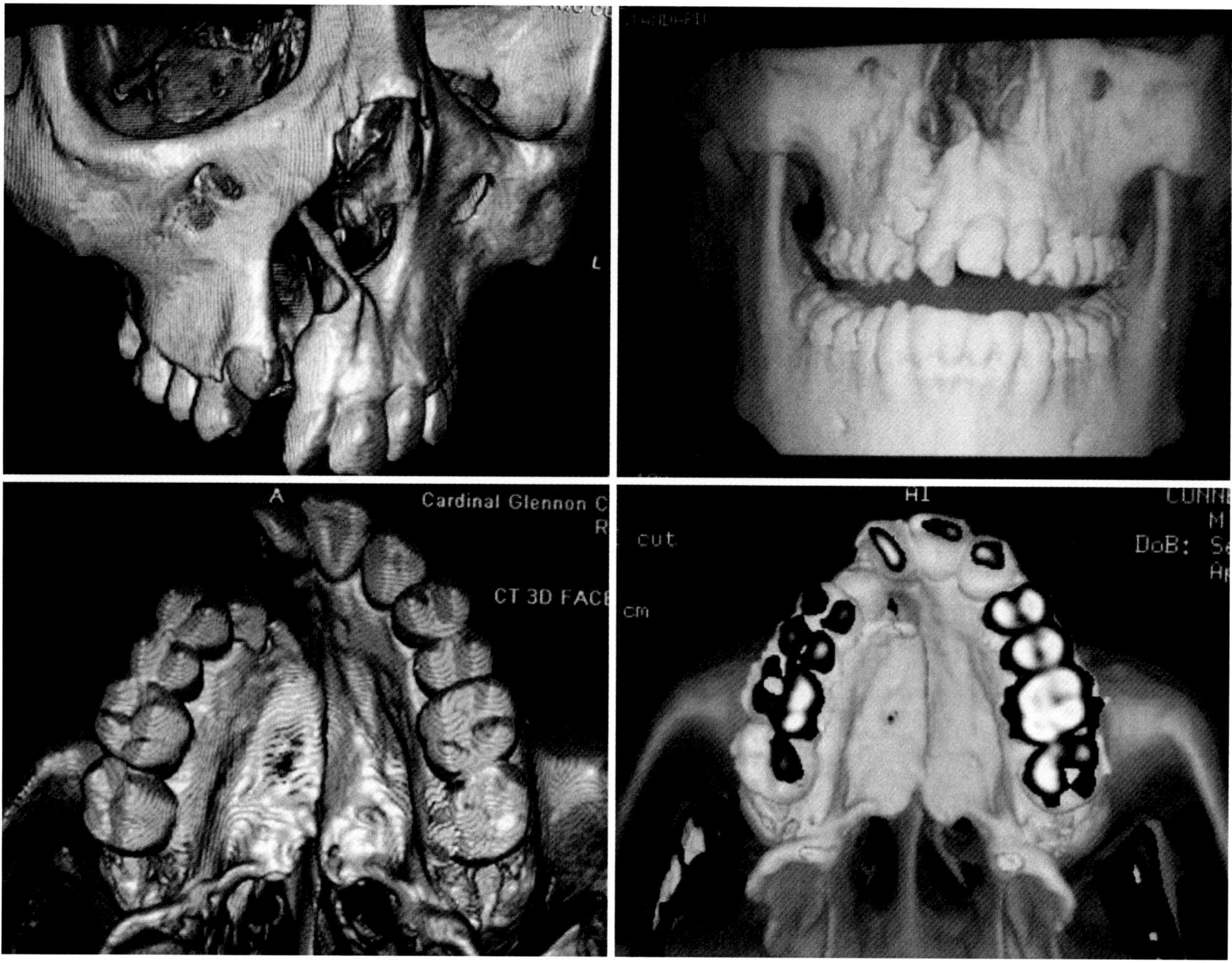

Fig. 19.88 Case 0 CT demonstrates ability of rhBMP2 to achieve a complete fill of this large alveolar defect, extending back to incisive foramen. Orthodontics closed down the arc giving him good aesthetics and bite. [Courtesy of Michael Carstens, MD]

Fig. 19.89 Pioneers in molecular embryology. *(Top left)* Luis Puelles, Universidad de Murcia, Spain. *(Top right)* John Rubenstein, University of California San Francisco. *(Bottom left)* Neural crest mapping: Nicole LeDouarin. *(Bottom right)* Mesodermal mapping: Drew Noden, Cornell University. [Top left: Courtesy of Luis Puelles, MD]. [Top right: Courtesy of John Rubenstein, University of California, San Francisco]. [Bottom left: Reprinted from Flickr. Retrieved from https://www.flickr.com/photos/ministere-enseignementsup-recherche/11046611646/in/album-72157638039342903/. With permission from Creative Commons License 2.0: https://creativecommons.org/licenses/by-sa/2.0/.] [Bottom right: Courtesy of Drew M. Noden PhD]

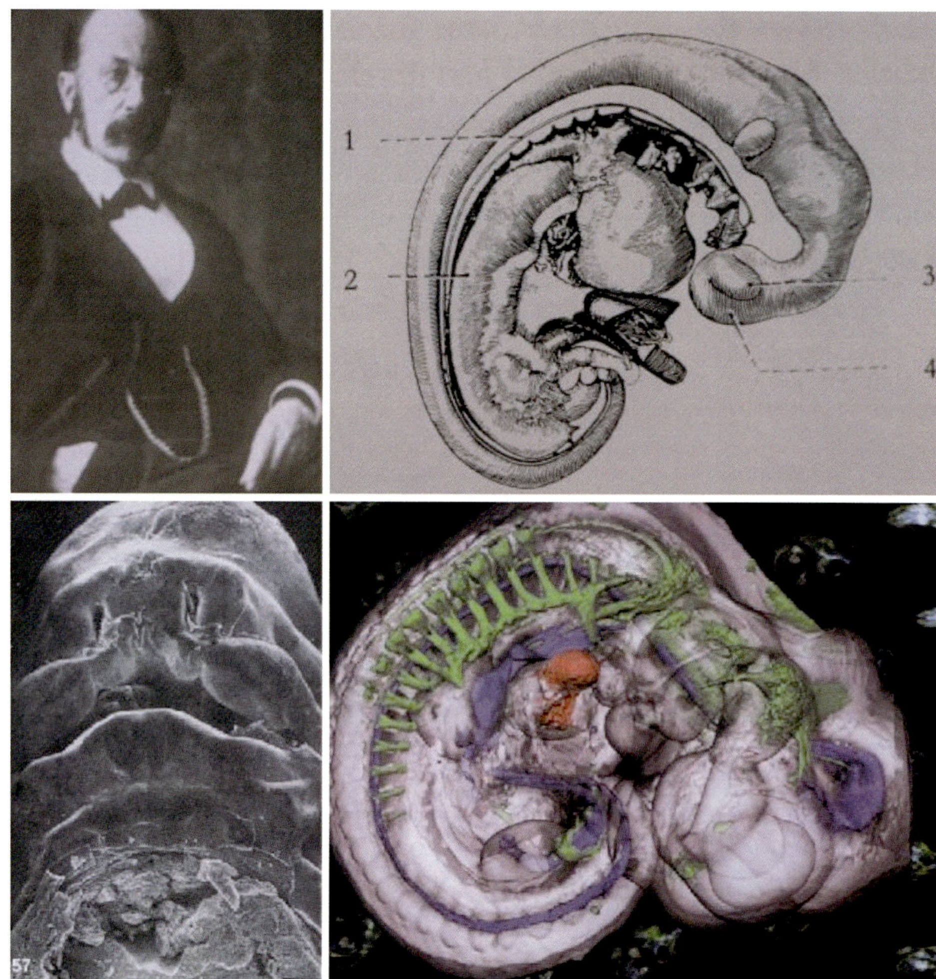

Fig. 19.90 Developmental biology: engine of change for craniofacial surgery. Swiss anatomist Wilhelm His invented the microtome for histologic sections and became the father of descriptive embryology. To his right is an original drawing by His from his embryo collection. Scanning electron microscopy enabled Hinrichsen to carry out his monumental study of facial development. And now, with gene tracers and dynamic imaging, the development of the face can be understood as never before. Top left: [Reprinted from Wikimedia. Retrieved from https://commons.wikimedia.org/wiki/File:Wilhelm_His.jpeg.]. Top right: [Wilhelm His drawing, circa 1889]. Bottom left: [Courtesy of Prof. Kathleen K. Sulik, University of North Carolina.] Bottom right: [Reprinted from Gilbert SF, Barresi MJK (ed). Developmental Biology, 11th ed. Sinauer Associates; 2016. With permission from Oxford Publishing Limited.]

Commentary: Jean-Claude Talmant

Writing the afterword of this multidisciplinary work on the treatment of cleft lip and palate is a delicate mission and a great responsibility. With good reason, our patients always expect more from us. May this contribution remind us that anatomy is nothing but the support of function and that we must understand it and respect it to order to convert their hopes into progress.

If I had to translate my thought in a few lines, these aphorisms would be my interpreters:

"Diagnose before you treat." Harold Gillies

"Repair must be applied embryology." Victor Veau

"When planning the restoration, the first consideration should be function." Harold Gillies

History has left us with a confusing situation in which, with consistency, most cleft teams, regardless of country, treat lip/palate clefts according to their own protocol, the one they were taught, one which is reassuring and comfortable. They feel legitimate since their patients have the expected acceptable result. As Janus Bardach [46], with rare lucidity, suggested in the preface to his book, even the most respected of protocols are not evaluated over the long term for the simple reason that their less-than-optimal results deter them from doing so. It is as if we have resigned ourselves to a certain mediocrity, to the point that our primary goal has become to help patients accept their handicap by attenuating their deformity and preparing their families and society to welcome them well [47]. This renunciation traps many therapists who spend their lives following others, complacent in their routine. For them, difficulty is not an opportunity. They know. This knowledge has been passed on to them. You must not challenge it. But then, how to do better?

Should One Prioritize the Judicious Choice of Technique or to the Magic of Its Execution?

Twenty years of daily management of a generation of patients are barely enough for the surgeon to evaluate his protocol. It will take yet another generation to usefully compare that protocol with others and to identify if progress has (or has not) been made. Understanding is a matter of patience. Senior surgeons finally have the experience and the time to reflect on the consistency of their choices... if only they take the risk to share their regrets. Their doubts are born of their imprecise perceptions about cleft biology itself, and of their sense that a hierarchy of techniques actually exists; and that these techniques can somehow be put into play in an attempt to restore functional anatomy without compromising growth. Surgery is, in any case, iatrogenic. However, there exist studies that maintain that some surgeons are less iatrogenic than others [48]. These assessments are not innocent. Sneakily, they add to the confusion, attributing results to talent. *Whatever the dexterity or the visual intelligence of a surgeon might be, what really underlies a successful outcome is the reasoned choice of a protocol, and of those procedures required for its execution.*

Our understanding of facial growth mechanisms has not seen any major advances in recent decades. The quality of dialog required to achieve progress in the field of cleft management continues to be hindered by an uneven level of culture and knowledge of basic anatomy, combined with a lack of curiosity and a resistance to open-mindedness. Certainly, technical concepts differ from one school to another, but the extent of our misunderstanding of basic science is appalling. In 2011, a renowned chairman, commenting on my presentation on the treatment of bilateral cleft lip and palate, threw out this comment to the audience: "But, Jean-Claude, why do you think that nasal breathing might play a role in facial growth?" This question left me speechless for a moment... So we are still there!

What is the Nature of the Cleft?

Craniofacial embryology, focused on molecular biochemistry, has clarified many questions but has remained without translation into surgical practice. Nevertheless, Michael Carstens, seeking the similarities between the vascular supply of the maxilla during the last stages of embryology and the clinical classification of facial clefts established by Paul Tessier, demonstrated their perfect correspondence [49]. Thus, the very essence of the labio-maxillary-palatal clefts finds its genesis at the level in a specific development field, that of the premaxilla. A dysfunction in the forward advance of the median naso-palatine vascular pedicle which supplies this field, on or before the sixth week of uterine life, changes the course of things. The earlier this occurs, the more it affects the entire bone structure, where, according to an immutable cephalo-caudal and medio-lateral spatio-temporal chronology, there exist 3 subfields: the central incisor subfield is the most primitive, followed by that of the lateral incisor, and, finally, that of its distal extension, the frontal apophysis of the premaxilla. This latter structure doubles the thickness of the outline of the piriform orifice as it underlies the ascending apophysis of the maxilla. A deficiency state of the apophysis is present in all clefts, even in their most subtle form fruste. Most often, the entire embryological field of the lateral incisor is affected with a wide variety of expressions of the deficit of its mesenchyme, especially as an additional vascular supply, coming from the lateral naso-palatine artery and from a branch of the greater palatine artery, joins the canine sector of the lateral maxillary segment. This model enables us to better understand the frequency and the considerable variety of anomalies of the lateral incisor, from its duplication in a good number of labio-alveolar clefts, to an isolated precanine, passing through its pure and simple agenesis in more than 50% of cases of complete clefts.

This direct link between cleft lip and palate and agenesis of the lateral incisor is perfectly consistent with the chronology of embryology and the clinical experience of clefts. It is the illustration of Kundera's aphorism: "It is at the moment of its genesis that the essence of a phenomenon is revealed." If the volume of the mesenchyme is sufficient, the joining of the frontonasal, external nostril, and maxillary buds forms the epithelial wall whose disappearance conditions their fusion, ensuring the continuity of the facial envelope, the buccal orifice, and the tip of the nose. The diffusion of BMP-4 from the mesenchyme is necessary to inhibit the action of Sonic hedgehog (Shh) which stabilizes the epithelial wall. In the event of a cleft, the BMP-4 deficiency linked to that of the mesenchyme leaves the epithelial wall the opportunity to oppose the fusion of the buds.

Influence of a Missing Embryonic Territory on Facial Muscular Organization

In a complete cleft, the distribution of the muscular network of the facial envelope and the beginning of nasal ventilation take place posterior in time to the development of the lateral incisor field. This territory is a strategic site, the keystone, both of the narrowest section of the respiratory tract, the nasal valve, and of the intercanine distance governing the centric and symmetrical mastication useful for facial growth. It should be emphasized that this is agenesis of the lateral incisor in the context of a cleft in the maxillary structure, on its two sides (nasal and oral) and not agenesis within an arch not split with diminished consequences. The alveolo-dental and maxillary impact of clefts must therefore be approached in a much more active and efficient way. In a nutshell, *early orthopedic action is needed in the cleft space to restore the width useful for canine function.*

Self-organization of Muscles in Space and Time

The image of the cutaway with its radiant arrangement of the facial skin muscles centered on the orbicular muscles is in no way different from the diagram of the forces acting within the facial envelope. A rupture in this system necessarily modifies this distribution. From the 7th week of uterine life, the premuscular cells amassed in the second pharyngeal arch invade the facial envelope to join the midline, eight weeks later, thereby constructing the nasalis muscle, the cupid's arch, and the philtrum of the lip. As soon as the premyoblasts become contractile, around the 9th week, they polarize and orient themselves where they are, according to the axis of the forces which reign within the envelope. They then come together in bundles like iron filings in an electric field [50]. If the skin muscle bundles have, by definition, a maxillary or zygomatic bone insertion initiated by the power of attraction exerted by BMP-4, their orientation responds to a simple, self-organizing rule of physics, without genetic sophistication. This concept is largely demonstrated by advances in the new physics of soft matter and active matter. Beginning in the years 2000–2010, studies in this novel field have explored the mechanisms of mobility for both intra- or extra-cellular living structures as well as their orientation [51].

With the disappearance of the embryological field of the lateral incisor, the ruptured lateral facial envelope loses its support on the midline. The new distribution of forces within it changes the muscular organization. Held back by the buccinator, the envelope finds a new anterior anchorage on the end of the small maxillary fragment. Nasalis and myrtiformis deprived of their common extensions in the nostril sill merge into a single muscular body, the lower head of which is the myrtiformis. Powerful, because very stressed by its new function, the muscle slides on the lateral crus of the alar cartilage and constrains it to a caudal rotation. Their superposition raises a vertical webbing in the nostril vestibule, the correction of which requires *complete separation between myrtiformis and alar cartilage by extensive dissection, which is best performed in primary surgery*. Orthopedics can only act on the width of the cleft. It has no effect on the abnormal relationship between cartilage and muscle.

Influence of the Missing Element on Fetal Ventilation

Fetal ventilation, by mobilizing the incompressible amniotic fluid 50 times/minute in the nasal fossae, has a considerable morphogenic role from the 11th week onward. Opening the nasal cavity into the oral cavity causes the expiratory pressure to drop on that side, so that the relative hyperpressure widens the opposite nasal cavity and *pushes the septum and vomer back into the cleft nostril* with *dislocation of the septal cartilage out of its vomer groove*. On the cleft side, all the structures retract, and the weak cyclic stimulation explains the quantifiable growth defect at the level of the premaxilla from the 14th week, whereas the cleft appeared between the 6th and 7th weeks and the ventilation starts at the 11th week [52]. This chronologic correlation reinforces the credibility of the hypothesis which recognizes the primordial role of fetal ventilation. The pressure states induced to fetal ventilation also explain why, in the unilateral cleft, the median vomero-septal axis is disrupted, whereas, in the bilateral cleft, the axis remains aligned. Only surgery can correct these nasal cartilaginous deformities; they are inaccessible to orthopedics. Their magnitude is such that the vomer in some total unilateral clefts can actually horizontalize, thus seeming to be integrated into the palatal vault. Exploiting this deformity by closing the hard palate with the vomerian flap does not reconstruct a normal anatomy and is not without harmful consequences, as we will see.

The Concept of Facial Envelope

In utero, the combined actions of ventilation and the facial envelope shape and transform the entire nasal capsule and upper lip into a sophisticated anatomy called upon to manage future air ventilation. The facial envelope is all the soft tissues of the face; its physical characteristics exceed the sum of those of its constituents and its mechanics become so different that it is necessary to master this concept to analyze the dysfunctions and better treat them. As for the morphogenesis of the nasal and perinasal structures, it is the work of ventilation. Nothing escapes it, from the prow of the philtrum to the concha of the turbinates, passing through the individualization of the tip of the nose, the telescoping of the plica nasi, the conformation of the nasal valve and its takeover by the commissural pillars of the facial envelope. With this in mind, one can decipher, after birth, both the role of the facial envelope in the rest posture for the control of ventilation (its only vital function), and that of swallowing, which is subordinate to it (ventilation) although synchronized.

We can read influence of the facial envelope on the more or less brachyfacial or dolichofacial evolution of each of us according to our anatomical particularities and its performance. We no longer wonder if nasal ventilation influences facial growth... We can see it! The bases of essential knowledge, concerning the facial envelope, are gathered in the report of the 76th scientific meeting of the SFODF on: "The vertical dimension" by J. Talmant et al, published in "L'Orthodontie Française," 2003; 74, 2: pp 137–313.

Respect the Coherence of the Theory of Open Systems

The trampling of our knowledge and the confusion of protocols show that Descartes' theory is no longer sufficient to understand the complex interactions linking ventilation and facial growth. It is necessary to appeal to the systems theory of von Bertalanffy [53]. The ventilatory system which constantly adjusts the balance between nasal and oral ventilation on demand is a particularly complex open system: It constantly exchanges and interacts with the environment [54, 55]. It is itself composed of subsystems which interact and evolve over time and with external factors, such as the change of fluid before and after birth, or the change of dentition. The ventilatory system, because it is the only one to be vital for life, is the great organizer of all the subsystems which interact with it, such as those of mastication and phonation. It is therefore not coherent for the practitioner, nor healthy for the patient, to align the teeth by extraction or close spaces in the dental arch resulting from missing teeth. This represents an optimization of the masticatory subsystem at the expense of the system as a whole, with ventilation being primordial.

Optimal ventilation is nasal, at rest and in sleep. It alone ensures the oral posture (associating lingual posture and posture of the facial envelope in charge of labial occlusion and regulation of the nasal valve) *that is necessary for normal and balanced facial growth* [56]. Hairfield and Warren have clearly shown that the ventilation of almost 75% of children operated on for a complete uni- or bilateral cleft is more oral than nasal. Overall on inspiration, the smallest section of the nasal passages of an adult which corresponds to that of the nasal valve is reduced by 40% on inspiration in the event of uni- or bilateral total cleft [57]. By the Hagen–Poiseuille equation, a reduction by half of the diameter of the nasal valve reduces its flow 16 times; thus we can measure the functional influence of the closure of the missing territory of the lateral incisor.

Precocious Closure of the Lateral Incisor Space: Its Consequences

After birth, the widths of the nasal valve and the piriform orifice remain essential. They are governed by the dimensions of premaxilla, and therefore by its dental contents. If the lateral incisor is missing in a total cleft, preserving its space is essential.

The Latham and Millard Protocol

We can verify the consequences of an early closure of this space sought by the supporters of preoperative active orthopedics. As soon as the two edges come into contact, gingivoperiosteoplasty leads to bone fusion, the base of which is supposed to facilitate future labial-nasal repair. Millard and Latham had developed this protocol in 1977, for the difficult treatment of bilateral clefts [58]. After the initial enthusiasm, they became disillusioned, while refraining from publishing their results. Berkowitz took charge of this, emphasizing the very harmful impact of this protocol on facial growth [59]. At that time, the retired Millard had incidentally dedicated his photo to Tony Wolfe, his former student and partner, writing on the silver print: “Tony, plastic surgery is not as easy as it seems”, thus testifying to his awareness of this unfortunate choice.

The Latham appliance took effect in 3–4 weeks, the two premaxillae separating to pass on each side of the vomer in a telescoping movement. After stopping the compression, the premaxilla was expected to resume its growth, but the vomero-premaxillary suture had been abused too much. The brutality of this form of dental orthopedics has been called into question. Nonetheless this protocol has a lesson to teach. Closure of the space was a tacit recognition of a previously existing field defect. Once fusion occurred, the deficit could no longer be reproduced naturally. Maintaining the space for the defect and then reconstructing it at the proper time are logical alternatives.

Nasoalveolar Molding by Grayson and Cutting (NAM)

Grayson has developed a "soft" version of this closure of the cleft, in 3–4 months, without forcing the vomero-premaxillary suture, with a Hotz plate on which the addition of nasal extenders simultaneously ensures the expansion of the columella and the projection of the wing dome. This nasoalveolar molding (NAM) remains quite in vogue. Like their predecessors, Grayson and Cutting were tight-lipped about their long-term results. After careful assessment, Berkowitz denounced the frequency of dental crowding and the major problems of maxillary growth; this led Cutting to announce in 2011, during a congress in Perth, in which I participated, that he had abandoned this technique in unilateral cleft. In this book, the chapter by Aurélie Majourau, begins with the case of a young patient, born in New York and treated according to the NAM protocol [60]. Figure 19.91 details the evolution, revealing the terrible repercussions of the first bone fusion of the cleft. Closing and sealing this space in order to then try to reopen it with frenzied techniques are illogical and do little for the patient other than to obligate him to undergo unnecessary and heavy procedures. Obviously, these practitioners have given no credence to the key role of the lateral incisor field. It is also true that nasal ventilation is not their cup of tea.

The Case of the Vomer Flap

Due to its simplicity and the opportunity, it offers to treat the entire cleft at one time, the most universal technique for closing the hard palate is the one-plane vomer flap. The face of the flap left raw heals by secondary epithelialization, known for the persistence of its intense retractile power. The vomerian flap is an integral part of the Malek protocol, which has been widely used in France and in French-speaking countries in recent decades [61]. We have repeatedly observed its drawbacks by surgically exploring its labial, nasal, and alveolar sequelae. Early scar retraction closes the space of the lateral incisor, damage increased in case of agenesis. It

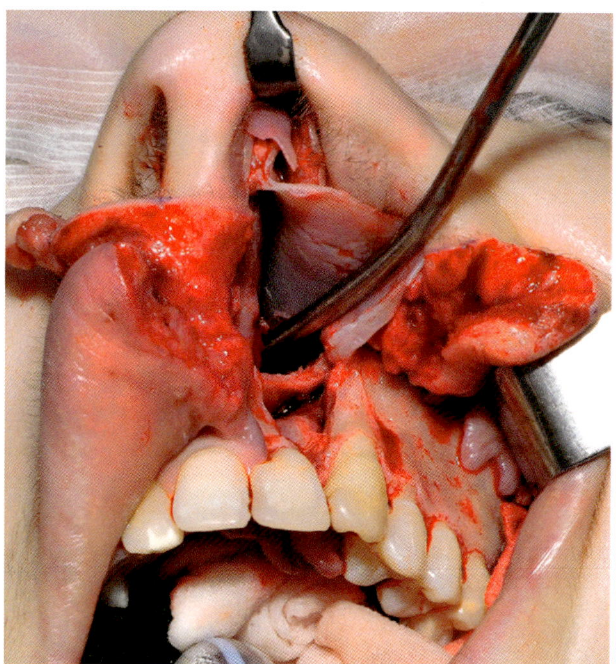

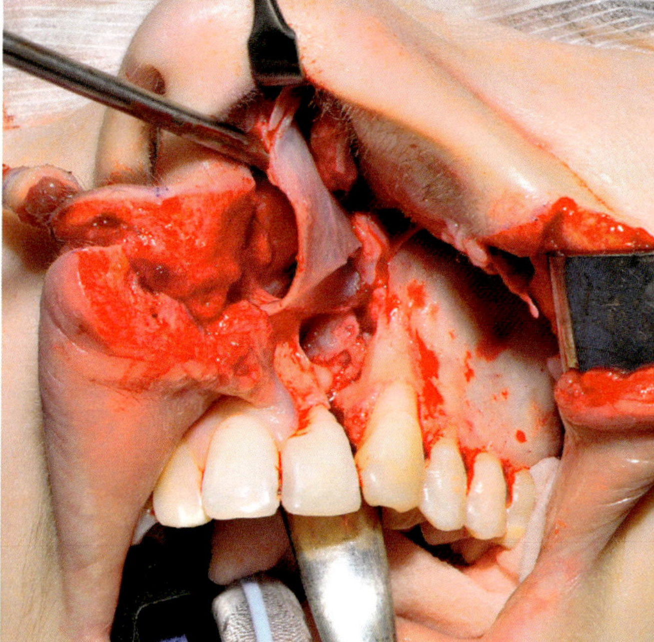

Fig. 19.91 Revision of a left cleft operated according to Malek. Canine and central incisor are in contact. The space of the missing lateral incisor was closed early with the one-plane vomer flap. The septum touches the lateral nasal wall. The vomer lies on the nasal floor; between the two the inferior meatus is excluded from any ventilation. The nasal fossa, amputated by the width of the lateral incisor, is obstructed

touches the vault of the palate, an eminently functional structure which has all its growth to do. This retractile tissue installed early in the cleft is a nuisance that is not limited to the arch and the anterior palate but hits the nasal valve, the piriform orifice, and the entire nasal cavity hard. The inferior meatus, between the vomer and floor, is excluded from any ventilation while the residual space is obstructed by the septal deformity (Fig. 19.91).

Malek preferred the vomerine flap to large maxillary and palatal subperiosteal dissections, which he accused of being the cause of growth defects. Many European publications of the last decade compare protocols which, in fact, all use the vomerian flap, simple and universal. Can we be surprised by their conclusions which show that these protocols are equal and that none really stands out? [62]. The harmfulness of the vomerian flap is the most probable cause of the leveling of the results whose mediocrity, accepted with resignation, is wrongly attributed to the cleft.

It is clear that 50 years of statistics have not been enough to provide useful information. We probably didn't ask the right questions! Rare are the publications showing with the most perfect photographic rigor, the results at the end of growth (20 years of follow-up) of all the first consecutive cases operated according to a precise protocol of a unilateral and bilateral complete cleft lip and palate. The first ten cases of a protocol are binding objectives, indeed difficult to fulfill by the surgeon as well as the patient [63], but so rich in teaching! I know of only two such attempts. The first is that of Harold McComb of Perth, who became a very dear friend 30 years ago, who encouraged me, given my liberal status, to build my credibility in this way. At the end of 2018, I thus published the first 5 consecutive cases of total unilateral cleft lip and palate treated with my protocol since 1997, and also did so for bilateral clefts in the *"Encyclopédie médico-chirurgicale"* thanks to digital possibilities [64].

Abandoning subperiosteal dissections during labial-narinal repair limits the projection of the soft tissues in the naso-labial angle. This contributes to the "cleft look," a defect that Georges Herzog, in his chapter on the closure of the lateral incisor space, regrets in his patients after Lefort I osteotomy [65]. He is not responsible for it. He undergoes the situation created by the surgery. His chapter is very informative, with a good iconography of the palatal vault and the evolution of occlusion as is rarely shown. Only the incidences that allow the nasal sequelae to be seen are lacking. The protocol implemented in these observations is that of Malek.

Most of my students came from this protocol and adopted mine because I showed my results, and they wanted to improve theirs or those of their master. Comparative studies by them in a variety of fields do justify their choices [66–75]. The Lausanne Protocol has a Lefort I osteotomy rate close to

60% cases in complete unilateral clefts. This situates its harmfulness on growth. The 3 cases of complete unilateral clefts presented by Georges Herzog reveal the consequences of the choices made during primary surgeries that we will analyze below.

Preoperative Passive Orthopedics

It is clear that the palatal orthosis does not have the expected effects on the protection of the palatal sutures, the conformation of the arch and maxillary growth. This confirms the work of teams who have compared series of patients treated with or without passive orthopedics. They could not demonstrate any lasting advantage in any area.

The Vomerian Flap: Its Flaws

The major problem is the vomer flap itself. We are not going to insist, again, on the constant nasal obstruction of this protocol. In the first case, the vomer itself is largely integrated into the palatal vault. The resulting rigidity does not facilitate recentering of the midpoint. In another context, orthopedics in deciduous dentition on a normalized palate could have solved this problem and avoided this adventure of a Lefort I osteotomy at 9 years old, when the maxillary arch had good proportions. This resulted in the extractions of 14 and 24, a second osteotomy, the persistence of a fistula and a labial "cleft look."

The vomer mucosa does not have the qualities required to replace the palatal fibromucosa. Observations 1 and 3 show recurrent bucco-nasal fistulas, very wide in the 3rd case. These fistulas disappeared in our two-stage palatal closure protocol. Putting the fibromucosa of the banks in its place according to Victor Veau's principle solves this problem definitively from the primary stage. The vomer flap does not facilitate the restoration of the alveolar cleft: it contracts it. In the absence of bone graft, the *quality of the periodontium at the edge of the cleft is mediocre*, especially for the lateral if it is present. In our protocol, any tooth, even dysmorphic, is respected until the end of the treatment—with very rare exceptions—as a precious bone capital to be preserved in total clefts. Here, it is the dental capital which is adapted to the narrowness of the arch contracted by the vomer flap. The missing part of the embryological field of the lateral incisor that can be assessed with precision by reestablishing the maxillary intercanine distance in relation to its mandibular reference is quite simply obliterated—both compacted in the patient and erased in the mind of the therapist. Should we really continue onward this path?

The Logic: Open the Space of the Lateral Incisor

This situation can be reversed, subject to action. A labial-nasal, septal, and alveolar revision is possible at any age (the ideal is between 5 and 6 years old) and with the usual precautions (post-operative nostril conformation of 4 months). This secondary surgery is technically demanding but effective. Prepared by an anterior maxillary expansion for an iliac cancellous bone graft, it restores its effectiveness in orthodontics. Any procrastination is a waste of luck while the growth potential that is known to be good in the cleft becomes exhausted.

Aurélie Majourau's sub-chapter on maintaining lateral incisor space provides many answers [60]. Read it and reread it. It is solid, didactic, ambitious, full of hope for today's patients, addressing all the questions of a multidisciplinary practice where surgeons and orthodontists must play the same score with a repair that respects the structures, modeled on the embryology. Cleft orthodontics must be concerned with the need for efficient orthopedics, therefore early and active in the deciduous dentition and at the beginning of the mixed dentition. With the progress made, the prosthetic solution of the implant which requires an excellent periodontium and will stabilize the final result is the logical conclusion which awaits only a generalized implementation of the protocol and the procedures which are described, undoubtedly still perfectible, but whose broad outlines should be confirmed with time.

Nothing is easy in this pathology, but inaction is not defensible when oral ventilation can "make the bed" for obstructive sleep apnea and its consequences (See the good sub-chapter on OSAS by Elisabeth Ruppert and Delphine Wagner [76]). Orthognathic surgery is not an end in itself. It lacks the harmony of normal growth and it is not uncommon for it to fail in cleft lip and palate where recurrences are feared. It would be preferable for the quality of the primary surgery to preserve us from this scenario, but, if an orthognathic procedure becomes necessary, the primary repair surgery can create conditions for a favorable result: a reduced negative overjet, a single median scar on the palate, no fistula, concordant arches, good bone volume in the alveolar cleft, normal intercanine width with preservation of the lateral incisor space, and a functional nostril. Of course, a precanine lateral incisor of sufficient size, with a quality periodontium, must be preserved. After coronoplasty, this living tooth will be preferable to an implant.

Finally, there is a dysfunction whose considerable influence we all underestimate, especially in the clefts: the persistence of primary swallowing. This is the most probable cause of our current therapeutic limits. This subject must concern us and become a priority from the age of 5–6 years.

Retrospective study. Plast Reconstr Surg. 2022;150:613e–24e. https://doi.org/10.1097/PRS0000000000009477.

76. Ruppert E, Wagner D. Prise en charge pluridisciplinaire autour des fentes labio-palatines, autres que l'odontologie et la chirurgie: Le syndrome d'apnées hypopnées obstructives du sommeil. In: Grollemund B, Etienne O, editors. Fentes labio-palatines: approches thérapeutiques chirurgicales, orthodontiques et prothétiques. Editions CDP; 2022.

Further Reading

Ayhan M, Sevin A, Aytüg Z, Gorgu M, Erdogan B. Reconstruction of congenital and acquired columellar defects: clinical review of 38 patients. J Craniofac Surg. 2007;18(6):1503.

Bennun R, Harfin J, Sandor G, editors. Cleft lip and plate management: a comprehensive atlas. New York: Wiley; 2015.

Bilkay U, Tokat C, Ozek C, Erdem O, Cagdas A. Reconstruction of congential absent columella. J Craniofac Surg. 2004;15(1):60–3.

Carstens MH. Functional matrix cleft repair: principles and techniques. Clin Plast Surg. 2004;31:159–89.

Carstens MH. Neural tube programming and the pathogenesis of craniofacial clefts, part 2: mesenchyme, pharyngeal arches, developmental fields, and the assembly of the human face. Handb Clin Neurol. 2008c;87:277–339.

Carstens MH. Developmental field reassignment cleft surgery: reassessment and refinements. In: Bennun R, Harfin J, Sandor G, editors. Cleft lip and palate management: a comprehensive atlas. New York: Wiley; 2015.

Demas PD, Sotereanos GC. Closure of alveolar clefts with corticocancellous block graft and bone marrow. J Oral Maxillofac Surg. 1988;46(8):682–7.

Etchevers HC, Couly G, Le Douarin NM. Morphogenesis of the branchial vascular sector. Trends Cardiovasc Med. 2002;12(7):299–304.

Fisher DM. Unilateral cleft lip repair: an anatomical subunit approximation technique. Plast Reconstr Surg. 2005;116:61–71.

Fisher DM, Sommerlad BC. Cleft lip, cleft palate, and velopharyngeal insufficiency. Plast Reconstr Surg. 2011;128(4):342e–60e.

Flores-Sarnat L, Sarnat HB, Hahn JS, et al. Axes and gradients of the neural tube form a genetic classification of nervous system malformations. Handb Clin Neurol. 2007;83:3–11.

Harashina T. Open revers-U incision technique for secondary correction of unilateral cleft lip deformity. Br J Plast Surg. 1990;86:872–81.

Honigmann K. Experiences with the Reichert procedure in closure of unilateral cleft lips. Plast Reconstr Surg. 1980;65(2):164–8.

Le Douarin NM, Kalchiem C. The neural crest. 2nd ed. Cambridge: Cambridge University Press; 1999.

Lee MR, Malafa M, Roostaeian J, Rohrich RJ. Soft tissue composition of the columella and potential relevance to rhinoplasty. Plast Reconstr Surb. 2014;134(4):621–5.

Letourneau A, Daniel RK. The superficial musculoaponeurotic system of the nose. Plast Reconstr Surg. 1988;82:48–57.

May H. The Rethi ncision in rhinoplasty. Plast Reconstr Surg. 1951;8(2):123–31.

Mohler LR. Unilateral cleft lip repair. Plast Reconstr Surg. 1987;80:511–7.

Oneal RM, Beil J. Surgical anatomy of the nose. Clin Plast Surg. 2010;37(2):191–211.

Oneal RM, Izenberg PH, Schlesinger J. Surgical anatomy of the nose. In: Daniel RK, editor. Aesthetic plastic surgery rhinoplasty. Boston: Little, Brown and Company; 1993. p. 3–37.

Onizuka T. A new method for the primary repair of unilateral cleft lip. Ann Plast Surg. 1980;4:516–22.

Pensler JM, Ward JW, Parry SW. The superficial musculoaponeurotic system in the upper lip: an anatomic study in cadavers. Plast Reconstr Surg. 1985;75(4):488–92.

Ribatti D, Nico B, Crivellato E. The role of pericytes in angiogenesis. Int J Dev Biol. 2011;55:261–8.

Reichert K. Die Formung des Philtrums beim pimären Verschluss einseitiger Lipp-und Lipp-Liefer-Spalten. Forts Kiefer Gesichtschi. 1973;16:72.

Saba Y, Amodeo CA, Hammou J-C, Poiselli R. An anatomic study of the nasal superficial musculoaponeurotic system: surgical applications in rhinoplasty. Arch Facial Plast Surg. 2008;10(2):115.

Salyer K, Bardach J. Atlas of cleft and craniofacial surgery. Philadelphia: William & Wilkins Lippencott; 1999.

Sari E. Tessier number 30 facial cleft: a rare maxillofacial anomaly. Turk J Plast Surg. 2018;26(1):12.

Song R, Liu C, Zhao Y. A new principle for unilateral complete deft lip repair: the lateral columellar flap method. Plast Reconstr Surg. 1998;102:1848–52.

Talmant JC, Talmant JC. Cleft rhinoplasty, from primary to secondary surgery. Ann Chir Plast Esthet. 2014;59(6):555–84. https://doi.org/10.1016/j.anplas.2014.08.004.

Talmant JC, Talmant JC, Lumineau JP. Secondary treatment of cleft lip and palate. Ann Chir Plast Esthet. 2016;61(5):360–70. https://doi.org/10.1016/j.anplas.2016.06.012.

Talmant JC, Talmant JC, Lumineau JP. Primary treatment of cleft lip and palate. Its fundamental principles. Ann Chir Plast Esthet. 2016;61(5):348–59. https://doi.org/10.1016/j.anplas.2016.06.007.

Testut L, Latarjet A, Latarjet M. Tratado de Anatomía Humana. Bacelona: Salvat Editores S.A; 1988.

Thomas C. Primary rhinoplasty by open approach with repair of the bilateral complete cleft lip. J Craniofac Surg. 2009;20(Suppl 2):1715–8.

Thomas C. Primary rhinoplasty by open approach with repair of the unilateral complete cleft lip. J Craniofac Surg. 2009;20(Suppl 2):1711–4.

Trott JA, Mohan N. A preliminary report on open tip rhinoplasty at the time of lip repair in bilateral cleft lip and palate: the Alor Setar experience. Br J Plast Surg. 1993;46(3):215–22.

Trott JA, Mohan N. A preliminary report on open tip rhinoplasty at the time of lip repair in unilateral cleft lip and palate: the Alor Setar experience. Br J Plast Surg. 1993;46(5):363–70.

Biologics in Craniofacial Reconstruction: Morphogens and Stem Cells

20

Michael H. Carstens

This chapter recounts a story of how clinical observations, basic science discoveries and technical innovations melded together over time to produce an important clinical breakthrough: the bioengineering of bone, in situ osteogenesis (ISO). Since this author has been fortunate to witness this process in real time, the story is anecdotal as well, a sort of surgical *bildungsroman*, if you will. Our approach here will be historical, as craniofacial bone reconstruction has evolved through successive iterations that reflect an improved understanding of embryogenesis and development. This progression is seen in a progressive simplification of reconstructive options from the tedious "walking" of tube flaps up from the chest to the face, to improvements in bone grafting, to plate fixation, to microsurgical tissue transfer, and finally, to biologic reconstructions that are able to inducing its own blood supply, thus simplifying the necessary surgery, while reducing the biologic "cost" to the patient.

Let's begin with a simple equation and consider its components, one by one. Over time, its variable have been filled in by progressively more sophisticated biomaterials. Although some substitutions will evolve over time, the fundamentals will remain unchanged. Let's consider its components, one by one.

Bone = resident stem cells + mesenchymal signaling cells + morphogen + lattice + template + blood supply

- *Resident stem cells* capable of transformation into osteoblasts are specific for the neuromere(s), angiosome, and tissue type. Sources include periosteum, dura, and skeletal muscle. Under normal anatomic circumstances resident stem cells are a constant in this equation, but under conditions of trauma, tumor ablation, or congenital deficiency, the number of resident stem cells in situ can be reduced or altogether absent. Resident stem cells respond to rhBMP-2.
- *Mesenchymal signaling cells* (in common parlance mesenchymal stem cells, MSCs) are pericytes. MSCs inhabit a perivascular niche and are less differentiated. These are readily obtained from adipose tissue in great numbers or from bone marrow, albeit in lesser amounts requiring ex vivo expansion. When stimulated, MSCs not only proliferate but also activate resident stem cells. Acting alone, MSCs can be driven to differentiate into osteoblasts. Under conditions of hypoxia, they are intensely angiogenic, via production of VEGF [1]. MSCs respond to rhBMP-2.
- *Morphogens* induce specific transformation to the osteogenic line in both resident stem cells and MSCs. Bone morphogenetic proteins (BMPs) (14 in number) are members of the transforming growth factor family and have varying degrees of osteogenic activity. In recombinant form rhBMP-2 and (to a lesser degree) rhBMP-7 have been shown to be clinically effective. TGF-$B1$ per se has an additive effect for osseointegration.
- *Lattice or scaffold* substrate is required for (1) cell adhesion and the formation of bone structure and (2) binding of morphogen. Scaffold materials take various forms: semi-solid in the form of granules (tricalcium phosphate, cancellous bone), a poloxamer-based gel, bovine bone blocks, fibrin glue, and absorbable collagen sponge, and demineralized bone [2–5].
- *Template* refers to a surrounding membrane or structure which determines the final shape of the synthesized bone and/or provides necessary fixation to surrounding structures. It may be permanent (titanium cage or mesh) or absorbable (PGA. DMB putty); the physical characteristics of templates can be expected to be progressively more biocompatible in the future.

M. H. Carstens (✉)
Wake Forest Institute of Regenerative Medicine, Wake Forest University, Winston-Salem, NC, USA
e-mail: mcarsten@wakehealth.edu

M. H. Carstens (ed.), *The Embryologic Basis of Craniofacial Structure*, https://doi.org/10.1007/978-3-031-15636-6_20

note, in the NYU series were those patients whose alveolar clefts could not be closed. This was ascribed to "incomplete approximation" (the flaps were not large enough to provide coverage). The technique required approximation of the segments of about 1 mm.

- The "closing down" of the alveolar cleft dimensions in the NYU protocol was a fundamentally non-embryologic strategy since it could only produce the amount of bone needed to bridge the gap, rather than the volume of bone equivalent to the original defect (Fig. 20.5).

The advantage of the Sotereanos technique is its provision of extensive amounts of soft tissue for alveolar cleft closure. This makes the procedure less critically dependent on the degree of presurgical orthodontic success.

Badran reported a series of 14 patients treated over a 25-year period using marginal gingivomucoperiosteal flaps reflected mesially and covered by a mucosal flap from the sulcus of the lateral lip element [39]. The average cleft width was, 9.6 mm; the largest was 14 mm. A normal alveolar arch configuration was achieved in all patients. Biopsies were acquired from three alveolar clefts showing osteoid tissue up to 20 months. Occlusion was good if only the primary palate was involved; secondary palatal clefts showed crossbite at the cleft site only.

In Sum GPP represents a scientific advance because it demonstrated the innate ability of MSCs resident within the periosteal envelope to produce bone under stimulation by factors in the blood clot contained within the GPP envelope.

- This suggests that conventional iliac crest bone graft works by enhanced stimulation of the MSCs by morphogens (BMP-2, BMP-4) native to the bone marrow.
- Implants using rhBMP-2 at supraphysiologic dose of morphogen to periosteum.

Muscle Considerations

Muscle anatomy in clefts has been extensively studied [40–43]. Seminal work by Delaire [44, 45] proposed that muscle reconstitution is required to overcome the pathological forces on cleft margins exerted by these by the abnormally inserted muscles. Reestablishment of the muscle sling is facilitated by the exposure created by the subperiosteal approach [46]. By separating out periosteal islands corresponding to the paranasal and orbicularis muscles, these structures can be identified and mobilized differentially. Anatomic reconstitution of these muscles affects the drape of the lip, its motion, and its relation to the ala. When the paranasal sleeve is brought into the anterior nasal spine, the perialar groove is created. Maintaining the continuity of alar, cheek, and lip tissue is preferable to the unnatural effect created by perialar incisions.

- Over the years, I have not seen any benefit to "separate mobilization" of muscles
- At the same time nasalis must be separated from the orbicularis and relocated behind it

Buccal Sulcus Incision: A Non-embryologic Strategy

The subperiosteal mobilization in DFR addresses several issues arising from the buccal sulcus incision. The first is sulcus depth. At the level of the buttress, dense soft tissue attachments suspend the sulcus a full finger-breadth above the molar. These are potentially disrupted by supraperiosteal mobilization. Transfer of the lip-sulcus-mucosal aesthetic unit maintains the height. A second problem with the buccal sulcus incision is the aesthetic mismatch of vermilion it creates. When the lip is stretched it behaves as an "elastic flap," the properties of which were described by Goldstein. The principle refers to the inherent stretch of certain tissues when tethered. In buccal sulcus procedures the lip vermilion thins disproportionately to the skin. Greater tension is present at the site of attachment, adding to lateral thinning. The lip is not stretched linearly; it is forced to move around the arc of the maxilla. Rather than retain a rectangular shape, the elastic tissue seeks the position of minimum tension (i.e., a circle). The result is that the nonkeratinized mucosa rolls under, resulting in a "thin lip." These findings can be confirmed on any postoperative patient by unrolling the lip and then watching it recoil to its contracted state. The elastic flap effect is avoided by two periosteal releases, distally up to the buttress and transversely at the level of the piriform fossa. The "curtain" of the lip–cheek unit can be moved mesially, without unpleating (Fig. 20.6).

The Legacy of the Sliding Sulcus Procedure for DFR Primary Cleft Repair

Wide subperiosteal dissection is central for DFR. Its original iteration, in 1997, the sliding sulcus procedure, was an application of Sotereanos flaps to the infant maxilla combined with used a Millard prolabial dissection. The S flaps released by means of a gingival incision were very generous. There was no alveolar gap that could not be spanned. At the same time, the nasal floor was closed. Based on the literature, I was hopeful that this operation would accomplish the following:

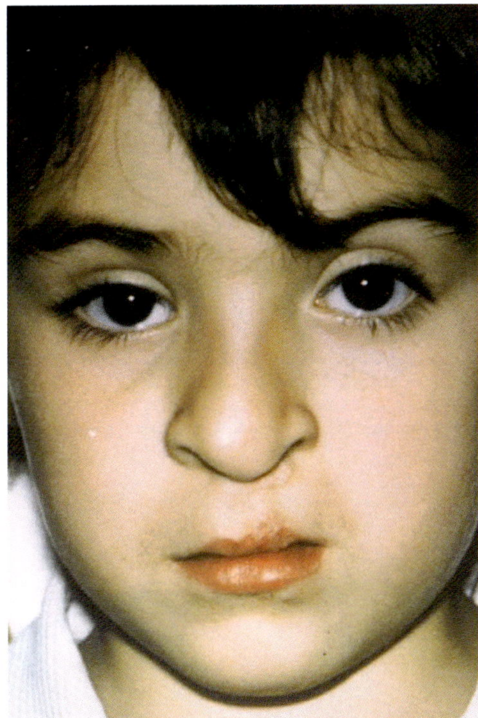
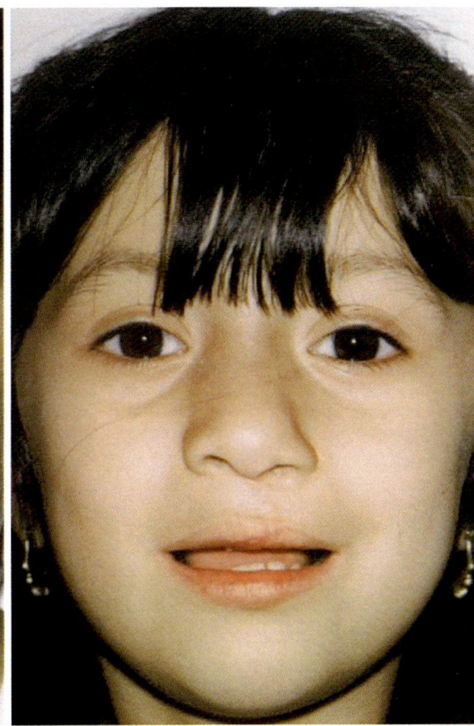

Fig. 20.6 My own patient previously corrected in infancy with Millard rotation-advancement using supraperiosteal mobilization. At age 7 she presents with facial asymmetry which I corrected using DFR lip/nose design with wide subperiosteal dissection S flap and iliac crest bone graft to the alveolus. [Courtesy of Michael Carstens, MD]

- Correct the asymmetries of rotation-advancement.
- S flaps were large enough to cover any size gap (compared to NYU).
- Seal the nasal floor.
- Protect facial growth.
- Prevent lesser segment collapse by strapping the maxilla to the premaxilla.
- Possible formation of bone within the GPP chamber.

Gingivoperiosteoplasty proved disappointing but offered valuable lessons which why we are considering it now. As pointed out by Cutting, the dissection from the alveolus is tricky and I quickly switched to operating at 6 months to work with better bone stock. Nonetheless, as reported, it was effective for soft tissue closure; but, to what end? Bone formation in the GPP was not as robust as in iliac crest bone grafting. The stem cells in the periosteum was either not sufficient in number or not activated fully. But the fact that bone formed made an important point which becomes important with later work using rhBMP-2.

- ICBG is primarily a lattice. Its cells are rendered anoxic. ICBG succeeds the morphogens that the envelope. For this reason, graft survival, the so-called "creeping substitution," occurs from the periphery to the center as the BMP-stimulated MSCs in the vascularized envelope respond.

This biology was proven by the use of rhBMP-2 (which we shall discuss anon) as it stimulated the GPP envelope with migrating MSCs that filled up the alveolar cleft, the process of ISO. The combination of primary GPP and BMP-2 from 2000 to about 2006 was impressive in its ability to fill up the palatal apical portion of the graft, to achieve good root coverage. Dental eruption took place through the graft, but the effects on the local tissues into mixed dentition were of concern.

Technical advances in DFR demonstrated that the gingival incision was not necessary for mobilization of the lip/nose complex. Addition of AEP meant that alveolar cleft closure could be accomplished at palatoplasty using generous local flaps, not the smaller gingival flaps I originally used. It was obvious that rhBMP-2 obviated the need for ICBG entirely, eliminating concerns about donor sites issues from the hip graft. Secondary alveolar bone grafting using morphogen-based ISO is the realization of Skoog's prediction: the "boneless bone graft" will inevitably replace the trephine. For these reasons, I discarded GPP altogether. Alveolar cleft management with rhBMP-2 should be done later, either at palatoplasty or, preferably at age 4, as per the protocol of Talmant–Lumineau.

- SABG is done with the alveolar "box" closed on 5 sides by the previous DFR and AEP procedures. The S flap constitutes "front door" of the box. Graft: rhBMP-2 + (collagen sponge and/or demineralized bone).

Over the last 20 years, the consequences of primary GPP has become clear. The Chang Gung unit compared GPP performed at 6 months with SABG in non-syndromic patients with UCL/P [47]. The success of the procedures was defined by the presence of a bone bridge greater than or equal to 75% of the bone height and a less that 25% residual bone defect as seen on occlusal X-rays [48]. The success rate of GPP was 72% versus 96% for SABG. GPP is particularly deficient in making bone in the palatal sector.

- The palatal apical sector is covered with local tissue which does not have the same high MSC content of as that of the alveolar periosteum.

Šmahel's group in Prague followed facial growth of UCL/P patients between ages 10 and 15 years in 48 primary palatoplasty patients and 18 patients with SABG. Initially (age 10) few differences were noted between GPP and SAGB, but over time several trends were documented. First, the position of the alveolus (both maxillary and mandibular) with respect to facial plane was better in SABG patients. Second, the overlying soft tissue envelope was more convex (especially in the nasal area), that of GPP being flatter. Third, maxillary inclination was better. Fourth, the vertical intermaxillary relationships worsened with both groups but SABG was more favorable.

- Note that in the protocols, lip repairs were supraperiosteal and palatoplasties subdivided the GPA angiosome, thereby ensuring some degree of SVI.
  - Abnormal vertical dental relationship results from SVI.

Principles of DFR Cleft Repair

- The treatment of cleft defects requires a sequenced restoration of soft tissue, skeletal, and dental anatomy with emphasis on the functional relationships between muscle and bone.
- The relationship between the maxilla and its soft tissue envelope must be altered so that centralization and continuity of tissues across the cleft is achieved.
- Extensive subperiosteal elevation will wrest tension from the midline, allow proper alar base positioning, and reestablish the muscle sling emphasizing suspension of the orbicularis in two layers and reassignment of nasalis to its correct insertion.
- Periosteoplasty with S flaps at the primary surgery does not make bone effectively.
- Incisions for the S flaps at primary repair are located along the bony borders of the cleft. Gingival incision should be reserved for alveolar cleft grafting as a secondary procedure.
- Elimination of fistula is a priority. This should be done by correct nasal floor closure.
- The sulcus and the alveolar mucoperiosteum constitute composite developmental fields supplied by the medial and lateral branches of StV2 infraorbital with collateral support from the facial.
- Dissection of the prolabium in DFR respects the integrity of the philtral prolabium (StV1 anterior ethmoid) and of the non-philtral prolabium (StV2 medial nasopalatine).
- Alar base access is subperiosteal with the incision beneath the nostril sill only. It reaches point 10 and does not curve around the ala, a natural aesthetic unit.
- Nasal airway restoration involves nasalis correction, vestibular lining release, and alar cartilage control.
- Alveolar extension palatoplasty creates the receptor site for bone graft. Cancellous bone or rhBMP-2 can be placed at age 4 with anterior mucoperiosteal flaps closing the 6th (anterior) site of the "box."

Conclusions

DFR, AEP, and ISO accomplish the following

- These procedures conserve angiosomes and respect developmental fields.
- Subperiosteal repair restores centric soft tissue anatomy (Fig. 20.5).
- Flap mobilization is a form of stem cell transfer in which biologically active MSC populations are reassigned to their correct site to make bone where desired.
- Bone produced is from neural crest, native to the site.
- The use of morphogen-driven ISO spares patients considerable suffering from graft harvest and is capable of generating alveolar bone in large quantities.
- Pericoronal gingival incision for alveolar cleft closure is delayed.

We shall now turn our attention forward to the evolution of ISO (in situ osteogenesis) and DISO (distraction-assisted in situ osteogenesis).

In Situ Osteogenesis: Regeneration of Membranous Bone Using rhBMP-2 Graft

Introduction

The Birth of the Field Concept

In 2000 Martin Chin and I were called to evaluate an infant girl with multiple left-sided congenital anomalies of her facial skeleton and soft tissues (Fig. 20.7). A complete cleft of the lip and palate was present in combination with a lateral facial cleft extending from the left commissure through the ear. On either side, the muscles of facial expression were normal. Temporalis muscle was present and functional, but

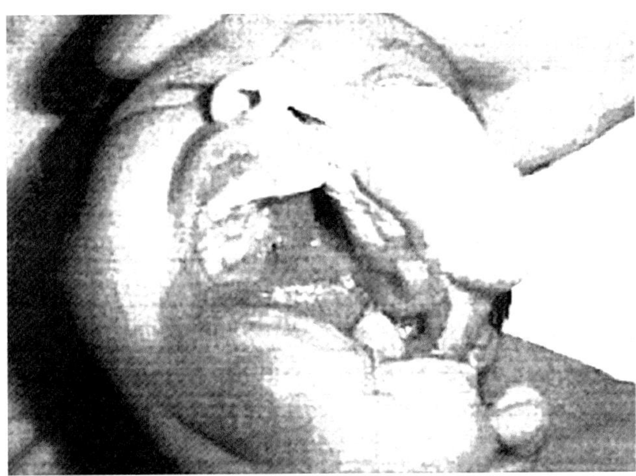

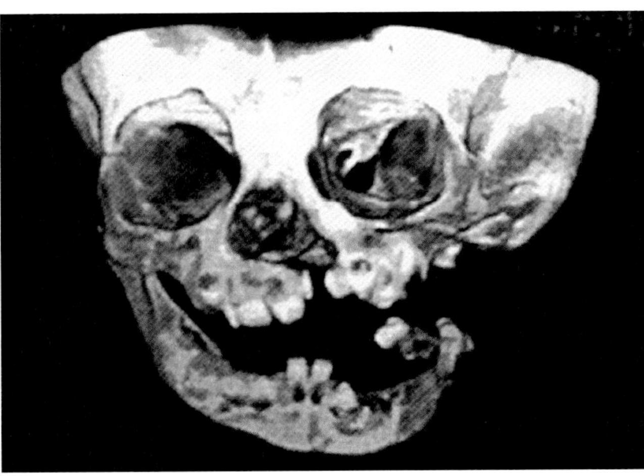

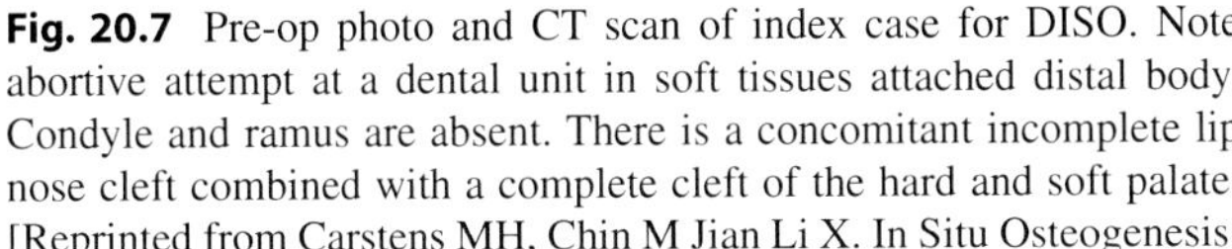

Fig. 20.7 Pre-op photo and CT scan of index case for DISO. Note abortive attempt at a dental unit in soft tissues attached distal body. Condyle and ramus are absent. There is a concomitant incomplete lip nose cleft combined with a complete cleft of the hard and soft palate. [Reprinted from Carstens MH, Chin M Jian Li X. In Situ Osteogenesis: Regeneration of 10-cm Mandibular Defect in Porcine Model Using Recombinant Human Bone Morphogenetic Protein-2 (rhBMP-2) and Helistat Absorbable Collagen Sponge. J Craniofac Surg 2005 16(6): 1033–1042. With permission from Wolters Kluwer Health, Inc.]

masseter and pterygoids were absent. Soft tissue bulk over the region of the posterior mandible was reduced.

The ear cleft was oblique, extending from anterior to posterior in a caudal to cranial manner. The upper portion of the ear contained skin and cartilage; the lower portion contained a dysplastic external auditory canal, the remainder of the pinna, and the lobule. The anterior structures above the cleft were absent. Below the cleft were a deformed tragus and abortive external auditory canal. Posterior structures included an aberrant antitragus and earlobe.

As in all labiomaxillary clefts, a deficiency state existed in the anterolateral piriform fossa. Continuity between mucoperiosteum lining the alveolar cleft and the vestibular lining of the nose resulted in caudal displacement of the alar base and distortion of the alar cartilage. Curiously, the position of the maxillary alveolus with respect to the premaxilla was not displaced, as would normally be the case (probably to disconnection from the lateral cheek tissues). The ipsilateral palatal shelf was smaller than that of the right side. Soft palate volume was also diminished but not atrophic. The cleft extended posteriorly around the tuberosity, becoming continuous with the lateral cleft. The eustachian tube was open at its lateral aspect; the tensor veli palatini was absent (thus explaining the change in soft palate volume).

Spiral computed tomography (CT) scan revealed all bones on the maxillary side of the cleft to be present (Fig. 20.2). The zygoma and alisphenoid (greater wing) were normal in volume, but the maxilla and lateral pterygoid process (of alisphenoid) were smaller than those on the right. Medial pterygoid process (of presphenoid) was normal, as were the squamous, petrous, and mastoid temporal bone fields.

The mandible and associated structures were remarkably affected. The body was foreshortened. The ramus, coronoid process, and condyle were absent. A separate bony segment projected superiorly from the proximal end of the residual mandible. Masseter muscle attached to the ramus was missing. There was no evidence of masseter along the zygomatic arch. The volume of temporalis muscle arising from squamous temporal bone was normal, despite the lack of coronoid and condylar processes.

We thought that treatment should be predicated on embryogenesis using developmental field theory. The existence of individual soft tissue bone units having a common embryologic basis was first proposed by Moss as the functional matrix [49]. Opitz in 1998 considered these units, termed **developmental fields**, to be genetically related to developmental units of the neural tube [50]. The resultant embryologic model, termed **neuromeric model**, based on neuroembryology, is presented in Chaps. 1 and 2 and throughout this book. The pathologic anatomy of this case involved developmental fields of both the maxillary and the mandibular sectors of the first pharyngeal arch. The cleft represented an actual biologic "seam" between these fields in the embryonic state. A "gradient of severity" was also present, with caudal and posterior fields being the more severely affected. Thus, *posterior mandible was the epicenter of the disturbance.*

Chin was actively working on distraction osteogenesis (DO). Aware of Phil Boyne's research at Loma Linda University, he commuted down from San Francisco to work with Boyne on the use of rhBMP-2 for distraction of the edentulous mandible, work they presented in Paris in 2001 [51]. Chin's experience with rhBMP-2 immediately suggested its potential to activate the resident stem cells in the Sotereanos flaps, thereby filling the alveolar cleft. Just then, as luck would have it, patient FP came into our lives. The newly hatched developmental field model seemed applicable; that all her deficits (both bone and muscle) could be seen as arising a common field defect, one that we could somehow reconstruct by making a ramus in situ and transferring existing muscle to it. Martin envisioned using a distractor to make a three-dimensional space within the periosteal envelope, one which we could subsequently fill with rhBMP-2. But would this work?

In situ osteogenesis was experimentally tested using a critical-size mandibular resection in a pig model.

Filling a distraction chamber expanded to match the mandibular defect using rhBMP-2 made sense both scientifically and from a patient-care standpoint. The ability of rhBMP-2 to stimulate local angiogenesis was very attractive avoid graft loss. We were looking for a relatively non-invasive way to give Flor the graft materials she needed, avoiding, if possible, painful and disfiguring donor sites. We also perceived that ISO was potentially modular. Additional graft could be added and the regenerate could be further expanded using distraction.

So it was that we embarked on securing from the FDA a compassion use protocol for the case. But first it was necessary to prove the feasibility of ISO for large defects. A porcine jaw model was chosen as it far exceeded the dimensions of patient's defect. Moreover, the model was well known in plastic surgery for its replication of human wound biology. The results of this work were irrefutable, gaining the support of the FAD and of the Children's Hospital Oakland IRB. It was published later in 2005 *Journal of Craniofacial Surgery*.

In preparing this chapter, the relevance of this paper is striking, as it represents a clear discussion of how ISO works with nice histology and good seminal references. Moreover, it has historical interest as it represents the groundwork needed to carry out the first application of morphogen-directed ISO in craniofacial surgery, which will be discussed in the subsequent section.

Osteoconduction vs. Osteoinduction (ISO)

Bone replacement in reconstructive surgery remains a challenge. Vascularized bone flaps, such as rib, scapula, fibula, or iliac crest, leave painful and deforming donor sites. Free grafts are limited by blood supply issues at the recipient site. The ideal technique would be one that would not depend upon microvascular transfer, would recruit cells at the donor site, and would be size independent. Bone substitutes, such as hydroxyapatite, are not helpful because they depend upon time-consuming processes of **osteoconduction** and creeping substitution. An ideal approach would involve ***osteoinduction*** using the implantation of inert, absorbable materials accompanied by powerful molecular signals capable of inducing bone formation from surrounding mesenchymal cells. The authors term such a process ISO.

Recombinant human bone morphogenetic protein-2 (rhBMP-2) is now commercially available in a purified form of standardized concentration. Application of rhBMP-2 to a Helistat (Integra Life Sciences, Plainsboro, NJ) absorbable collagen sponge (ACS) in a standardized manner creates an implant into which cells can migrate, proliferate, and differentiate. The dosage of BMP-2 is highly species specific. Humans require the highest concentration, 1.5 mg/mL. Thus, an rhBMP-2/ACS implant will function reliably in other, less demanding animal systems, including dogs, pigs, rodents, and primates.

The bulk of preclinical work with BMP-2 was concentrated on bones of interest to orthopedic surgeons, i.e., the axial skeleton below the skull and the appendicular skeleton of the extremities. The axial skeleton below the skull originates from the paraxial mesoderm of somites, whereas the appendicular skeleton of the extremities originates from non-somitic lateral plate mesoderm (LPM). These mesodermal bones develop via *endochondral ossification* [52]. Note that PAM and LPM are both in register with the neuromeric system [53].

The membranous bones of the facial skeleton are a different story. They originate from neural crest cells, and develop within a periosteal environment directly via *membranous ossification* [54]. In some instances (the less wing of sphenoid) neural crest mesenchyme condenses first to cartilage and is then transformed into membranous bone. This little-known process is called *chondroid ossification* [55]. Histologic studies of bone created by DO demonstrate the presence of chondroid ossification in the distraction chamber [56]. We shall return to this important concept later.

The differentiation of neural fold cells into neural crest cells occurs as a response to BMP-4 signals from ectoderm outlying the neural tube [57]. All neural crest cells consequently possess membrane-bound BMP receptors [58, 59]. Thus, it is logical that stimulation of periosteum with rhBMP-2 will result in bone formation.

- Bone grafts containing marrow contain BMPs as well. When surrounded by a periosteal source containing MSCs, it is likely that the bone graft "take" in alveolar cleft reconstruction, for example, is due primarily to the influence of native morphogens on the soft tissue envelope, rather than to the graft, which acts primarily as a scaffold.

Preclinical work testing the applications of ISO to oral surgery was carried out by Boyne, who documented the ability of rhBMP-2 to augment the alveolar ridge, fill the maxillary sinus, and replace surgically created mandibular defects. Using similar methodology he demonstrated that an rhBMP-2 implant was capable of repairing simulated cleft palate defects [60–63].

Conventional bone grafts provide a "scaffold" that is eventually overtaken by native cells. This process is termed *osteoconduction*. Because free bone grafts initially have no blood supply, some degree of graft loss occurs. This places finite limits on how large a defect may be successfully grafted. In critical-size defects, the current approach to this dilemma relies on microsurgical transfer of vascularized bone. Free flap procedures are inevitably complex, frequently lengthy, require intensive postoperative monitoring, and have a small, but finite, incident of flap loss. For these reasons, reconstruction of relatively common maxillofacial defects may lie beyond the capabilities of district hospitals, despite the technical abilities of the staff.

The mechanism of ISO is biologically distinct; it takes place via osteoinduction. When implanted into a periosteal chamber, rhBMP-2/ACS stimulates all areas of the environment equally.

The mechanism of ISO is biologically distinct; it takes place via osteoinduction. When implanted into a periosteal chamber, rhBMP-2/ACS stimulates all areas of the environment equally. Osteogenesis does not need to "build" up in any particular direction. The entire cambium layer is a cell source [64]. Resident MSCs have their own autogenous blood supply. Thus, osteoblast activity and subsequent bone formation take place independently from the dimensions of the defect. ISO can potentially change the clinical management of critical-size bone deficits.

This communication describes the regeneration of a surgically created mandibular defect of critical size using rhBMP-2-mediated ISO. Because of its long-standing usage in wound healing, a porcine model is used. The clinical behavior of the rhBMP-2/ACS implant is documented. The histopathologic findings of ISO are described in detail. Discussion of the case centers on how the mechanism of action of rhBMP-2 is particularly relevant for facial bone reconstruction. Potential clinical applications of ISO are presented (Figs. 20.8 and 20.9).

Materials and Methods

Given the experimental protocol using current standards of animal safety was reviewed and accepted by the animal rights committee at Covance Surgical Laboratory (Berkeley, CA). An adolescent Yorkshire pig was chosen because of its known growth characteristics. With the pig under general anesthesia, the right mandible was exposed using a modified Risdon incision. A reconstruction plate was shaped, applied,

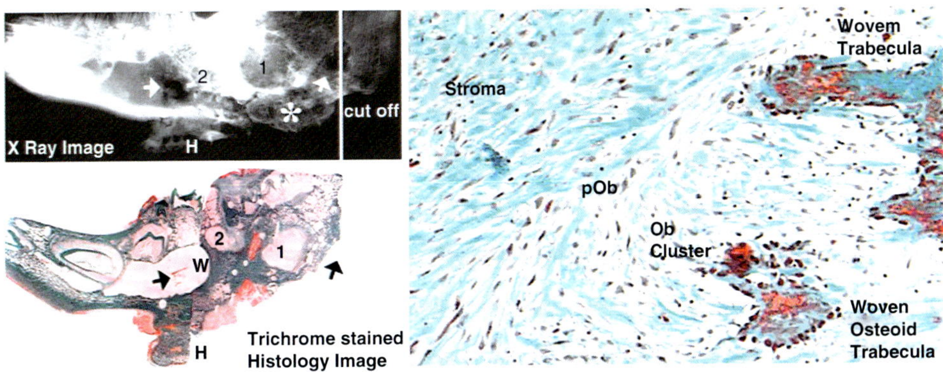

Fig. 20.8 Porcine model for reconstruction of critical-size bone defect in mandible with rhBMP-2. (Left): X-ray image (top) and trichrome-stained histologic image (bottom) of the right mandible are shown here at a matched scale. Note that the posterior portion of mandible shown in X-ray was cut off to reduce its size for histologic processing. At both sides of the defect, arrows indicate screw holes. "1" and "2" signify the two incompletely filled regions within the defect. Heterotopic bone formation (H) is observed on the periosteal surface under the anterior screw hole. The anterior screw hole locates near the posterior wall of mandibular foramen (W), where the dense cortical bone is formed. (Right): Residual ACS is located in the central region (left), surrounding it is a stromal zone (stroma) that is concentrated with dense mesenchymal stromal cells MSC. These MSC differentiate into preosteoblasts or osteoblastic precursors (pOb), which further proliferate to increase in number and consolidate into an osteoblastic cluster, where they synthesize type I collagen matrix, which is later mineralized into woven bone trabecula. When more woven bone spicules are generated, they connect to each other and form a woven trabecular bone network (WTb, mid-portion of the image), which, replaced by lamellar trabecular, becomes a partial lamellar trabecular bone network (LTb) on the right of the image. Under polarized light, collagen fiber of woven trabecular bone appears disorganized, whereas that of partial lamellar trabecular bone shows organized fiber orientation on the surface of the trabeculae. The pOb cells have proliferated and consolidated into an osteoblastic cluster (Ob cluster). In the Ob cluster, the consolidated osteoblasts are just about to form a woven osteoid trabecula by synthesis of type I collagen matrix. Woven osteoid trabecula later becomes mineralized into woven trabecula. [Reprinted from Carstens MH, Chin M Jian Li X. In Situ Osteogenesis: Regeneration of 10-cm Mandibular Defect in Porcine Model Using Recombinant Human Bone Morphogenetic Protein-2 (rhBMP-2) and Helistat Absorbable Collagen Sponge. J Craniofac Surg 2005 16(6): 1033–1042. With permission from Wolters Kluwer Health, Inc.]

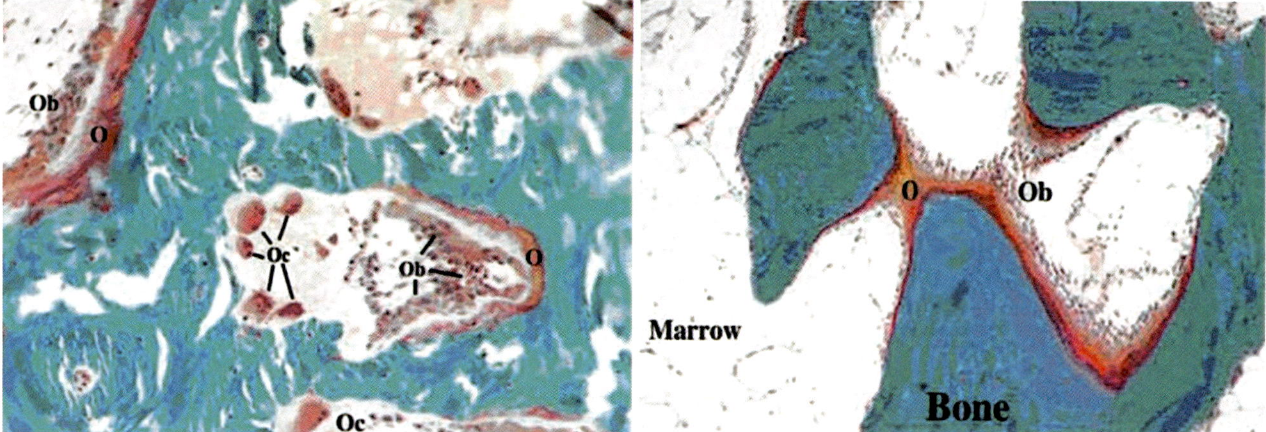

Fig. 20.9 (Left): Microphotograph represents the process of bone remodeling. In the center there are a group of cells working together to reshape a piece of bone. Originally the entire bone was solid woven cortical bone. Osteoclasts (Oc) start to remove the bone from right to left resulting in a resorption cavity. After resorption osteoblasts refill the cavity with osteoid (O) which later mineralizes into a lamellar Haversian system. (Right): Greater than 50% of the trabecular surface is activated from quiescent to formation stage by local increased concentration of BMP-2. Normally, less than 10% of a trabecular surface is actively involved in forming new bone (via remodeling). This BMP-2-induced effect increases trabecular thickness. Most marrow space (M) is occupied by adipocytes in host bone, while it is occupied by MSC in defect region. This 20 × image reveals a thick layer of non-mineralized osteoid (o) covering all actively forming trabecular surfaces. A layer of osteoblasts (Ob) is seen aligned along the surface of the osteoid. In this particular area, three neighboring trabecular surfaces are forming toward each other, bridging the marrow spaces between, forming an increasingly connected trabecular network structure. [Reprinted from Carstens MH, Chin M Jian Li X. In Situ Osteogenesis: Regeneration of 10-cm Mandibular Defect in Porcine Model Using Recombinant Human Bone Morphogenetic Protein-2 (rhBMP-2) and Helistat® Absorbable Collagen Sponge. J Craniofac Surg 2005 16(6): 1033–1042. With permission from Wolters Kluwer Health, Inc.]

and removed. Subperiosteal osteotomy was then performed, creating a trapezoidal defect 10 cm in length from the mandibular body. Continuity of the jaw was re-established with the reconstruction plate. Implantation of rhBMP-2 into the periosteal chamber was then performed.

Purified rhBMP-2 (Genetics Institute, Cambridge, MA; now from Medtronic) was reconstituted as two vials of 8.4 mL each. At a concentration of 1.5 mg/mL, the total dose was 12.6 mg. The protein was then applied uniformly to two Helistat® ACSs 4 cm x 3 cm in dimension. Binding time was 20 min. The rhBMP-2/ACS implants were placed into the periosteal pocket at the osteotomy site. Using 2-mm absorbable screws, the implant was affixed by securing the reconstruction plate through the pilot holes previously drilled on mandible body at both sides of the trapezoidal defect. The periosteum was subsequently advanced over the reconstruction plate and sutured with 4–0 Vicryl. Soft tissue closure was carried out using 3–0 and 4–0 to the investing fascia and superficial fascia, respectively; 4–0 PDS was placed in running subcuticular fashion, followed by 4–0 nylon interrupted to the external surface.

- Given the unusual pattern of wound healing that was exhibited by the neck incision, these details may be of some importance.

In addition to the mandibular reconstructive procedure we also wanted to determine the osteogenic effect of an rhBMP-2/ACS graft in a pure mesodermal environment. To accomplish this, stainless steel mesh was fashioned into a trapezoid based on the dimensions of the bone specimen. It was subsequently filled with rhBMP-2/ACS and implanted into the right quadriceps muscle. We surmised that this could be readily studied radiographically and histologically.

After an uneventful convalescence, the animal demonstrated normal function and jaw growth. Radiographs were taken of the jaw at monthly intervals. Bone formation was rapid, with the operative site apparently filled at 2 months. The muscle implant behaved in a similar manner.

At 3 months, the animal was euthanized by intravenous injection. The right mandible was removed, cleaned of soft tissue to the level of the periosteum, fixed in 70% ethanol, and sent to Wyeth Bone Research Laboratory in Cambridge, Massachusetts.

Before histological processing, X-ray imaging was performed on the pig mandible. Because of the huge size (25 cm × 12 cm × 3.5 cm) of the pig mandible, multiple holes (2 mm diameter) were drilled into bone marrow to allow solution penetration to obtain the best quality during histologic processing. The entire right mandible was dehydrated in ascending gradient of alcohol (cleared in xylene) under vacuum and gentle agitation in the Fisher LX 120 Automatic Tissue Processor. Under 4 °C, the mandible was immersed in liquid methyl methacrylate for infiltration and polymerization during a 1-month period. After completion of polymerization, the plastic block embedded with mandible was sectioned using a Riechert Jung Polycut. Eight-micron-thick sections were cut at multiple levels, with 500 microns between levels. Sections were stretched and pressed on gelatin–chrome alum slides and placed in an oven at 45–50 °C overnight to ensure tissue adherence. Sections were then de-plasticized, stained with Goldner's trichrome, and mounted with a coverslip. Under different magnifications of objective lenses, qualitative evaluation of the sections was performed.

Results

At the time of surgical exposure, complete consolidation of the defect was noted. The newly formed bone expected the confines of the periosteal envelope. No ectopic bone was found external to the periosteum. The periosteal layer was incised and reflected. ISO was complete across the defect; the periosteal chamber was completely filled with no "skip areas." Bone formation extended through the holes in the plate to reach lateral confines of the periosteal pocket. The plate itself was buried (i.e., the interface zone between the lateral margin of the plate and the periosteum had also formed bone).

Radiographs demonstrated a progression of ossification within the surgical defect. At 30 days, a faint haziness was noted. By 60 days, definite, but non-uniform, ossification was present. At 90 days, complete ossification was present.

X-ray image and Goldner's trichrome stained histologic image (Fig. 20.8) of the mandible both reveal that the 10-cm trapezoidal defect between the two screw holes (arrows) was filled completely with radiopaque bony tissues, with the exception of two small regions (regions 1 and 2). Heterotopic bone formation (H) occurred on the periosteal surface where rhBMP-2 leaked from one of the screw holes. In areas between and below regions 1 and 2, newly formed bone is consolidating into a solid cortical region to rebuild the posterior wall of the mandibular foramen (W). In area posterior to the region 1, newly formed bone spicules are connected to form a trabecular network, regenerating a new metaphyseal region of the posterior mandibular body. Regions 1 and 2 appear to be incompletely filled defect areas, within which residual ACS and mesenchymal stroma are observed in the center and ingrown osteoblastic new bone formation is visible on the periphery to completely fill the regions eventually.

Under the polarized lens, all bone tissue within the defect region appears to be immature woven bone compared with mature lamellar bone observed outside of the defect; as such, the arrangement of interwoven coarse collagen fibers and the distribution of osteocytes are both in a random fashion. In the periphery of the defect, where bone was regenerated earlier, bone tissue appears to be increasingly mature, as indicated by the replacement of woven bone by lamellar bone through the remodeling process. In these regions, lamellar bone packets are observed on the trabecular surface (Fig. 20.9), and a Haversian system (H) with multiple layers of concentric lamellas is observed in the cortical region.

In regions 1 and 2, a typical process of membranous ossification can be clearly visualized step by step under the transmitted microscope (Figs. 20.8 and 20.9). Because of the rhBMP-2-induced chemotactic effects, concentrated mesenchymal stromal cells (MSC) are observed around the residual ACS. These stromal cells differentiated into spindle-shaped preosteoblasts (or osteoblastic precursors; pOb), proliferated to increase cell number, consolidated into a preosteoblastic cluster, completely differentiated into osteoblasts, synthesized and released osteoid, and formed many non-mineralized trabeculae (Tb). This BMP-2, implant-induced cascade of ISO processes maintains is its momentum until the entire defect is filled with woven bone. Soon after the defect is filled completely by woven bone, the remodeling process occurs to remove the woven bone and replace it with lamellar bone in both the trabecular and cortical regions.

In regions outside of the defect, native bone tissue appears to be mature lamellar bone under the polarized lens. Lamellar bone demonstrates alternating light–dark layers, representing the parallel arrangement for each layer and the perpendicular arrangement between neighboring layers of collagen fibers. In this region, most bone surface is actively forming appositional bone because of the BMP-2-induced osteoconductive effect. Thus, a thick layer of osteoid is visible in most of the trabecular surface (Fig. 20.6).

Discussion

Chemistry and Mechanism of Action of rfhBMP-2

Recombinant human bone morphogenetic protein-18 is a glycosylated dimer. It binds onto specific membrane-bound receptors of mesenchymal cells causing them to differentiate into cartilage-forming, bone-forming, or even fat-forming cells [65]. The choice of which pathway a given MSC will take depends upon the presence of cofactors in the surrounding environment. For example, insulin, TGB 1, and ascorbic acid signal work together with BMP-2 to produce a chondrocytic lineage [66].

Bone morphogenetic proteins belong to the transforming growth factor beta superfamily. At least 20 human BMPs have been identified. BMPs are involved in embryogenesis and skeletal formation. BMPs are osteogenic proteins; that is, they have the ability to induce/initiate bone formation by attracting and binding to MSCs. When MSCs are thus stimulated, they become bone-forming cells (osteoblasts). Bone may be formed via three pathways: (1) directly (*intra-*

membranous); (2) directly via a transient cartilage intermediate (*chondroid-membranous*); and (3) indirectly via cartilage (*endochondral*).

The amount and density of bone induced within a target site depends upon the initial concentration of rhBMP-2 at the site and the residence time the rhBMP-2 remains present at the site. Preclinical studies show that when rhBMP-2 is administered intravenously, it is cleared from the systemic circulation in a matter of minutes, being redistributed to the liver and reticuloendothelial system, and subsequently being excreted in the urine within 24 h [64]. For this reason, in order for rhBMP-2 to have an appropriate residence time at the target site, it must be applied to an appropriate biodegradable carrier.

Helistat®, an implantable collagen sponge long in use as a hemostatic device for dental extraction sites, binds rhBMP-2 within 15–20 min [67]. These authors recommend 45 min of binding. Implantation should be carried out within 2 h of reconstitution.

Optimal bone induction depends upon the concentration of rhBMP-2. This follows a species-specific gradient (rodents < dogs < non-human primates < humans). The optimal human dose is 1.5 mg/mL. MSCs from the surrounding environment invade the matrix. Degradation of the matrix occurs concomitant with formation of cartilage or woven bone (or both). Intense vascular ingrowth also occurs. With time, bone formation extends from the periphery of the implant toward its center until the implant is completely replaced by trabecular bone. Subsequent remodeling of the newly formed bone takes the place according to the physiology of the site. Thus, osteoinduction takes place at the surface of the implant. In conclusion, the local concentration of rhBMP-2 (expressed as mg rhBMP-2 per unit volume of matrix) is a more relevant concept than that of total local dosage.

The mechanism of rhBMP-2/ACS involves six steps. These are as follows:

- *Implantation*: the surgical creation and/or modification of an environment containing MSCs. This involves contact with periosteum, muscle, or bone marrow;
- *Chemotaxis*: MSCs from as much as 2 inches away are attracted by rhBMP-2 to the implantation site;
- *Proliferation*: rhBMP-2/ACS implant provides a local environment where MSCs can multiply before differentiation;
- *Differentiation*: rhBMP-2 binds to specific receptors on MSC cell surface causing them to change into osteoblasts;
- *Bone formation and angiogenesis*: osteoblasts respond to local mechanical forces to produce osteoid. Intense new blood vessel formation is observed; and
- *Remodeling*: bone remodels in response to local environment and mechanical forces.

Clinical performance of rhBMP-2/ACS is relatively predictable. The amount of bone produced relates to the size of the regenerate chamber. Complete filling of the regeneration chamber with the sponge is important. In spinal fusion, physical stability is provided by the transverse processes. The presence of an external container, either of absorbable PGA/PGL mesh or titanium crib/cage, has proven useful in providing a shape to the regenerate bone. Serial CT scan provides evidence of the time course of ISO. At 6 weeks, bone formation is clearly seen. By 12 weeks, consolidation is present. Critical-size defects in porcine calvaria demonstrate identical histologic findings and tensile strength compared with the control side at 12 weeks [68]. Humans may lag behind, but evidence for this remains preliminary. Final remodeling can be considered complete at 24 weeks. This seems to an appropriate time for plate removal or additional manipulations, such as osteotomies for distraction.

The clinical behavior of soft tissues in the presence of an rhBMP-2 implant varies considerably from other surgical models. The intense vascular influx causes swelling to peak more slowly (2 days) and disappear more slowly (5 days) than the normal course of events. Wound healing is affected in an unknown manner. Scar formation above an implantation site seems to bypass some of the typical early stages. Scars at 1 month may have the clinical appearance of scars 6–12 months old.

The ISO Regeneration Chamber

The amount of rhBMP/ACS required for a given defect is volumetric. The implant should replace the defect milliliter-for-milliliter. In the experimental protocol, we prepared the ACS by rolling it up into a tube and suturing it with 5–0 chromic. Two such "sushi rolls" fit the defect nicely. Now available commercially as Infuse® bone graft (Medtronic Sofamor Danek, Memphis, TN), rhBMP-2/ACS is distributed in three sizes. Implant size can be augmented using absorbable granules of 85% hydroxyapatite and 15% calcium triphosphate (Mastergraft®, Medtronic). Because the dose of rhBMP-2 is supraphysiologic, volumetric expansion of the implant probably is both rational and efficient.

Mesenchymal stem cells undergo considerable migration to populate the implant scaffolding. Cellular migration within the regeneration chamber is the key element. The lattice-like structure of the implant facilitates this process. Physical abutment of a stain less steel reconstruction plate against the periosteal membrane was not sufficient to prevent bone formation at the interface.

The biologic processes involved in ISO for neural crest bone of periosteal origin differ in several key aspects for ISO of the long bones. Periosteal neural crest cells differentiate into membranous bone readily under rhBMP-2 stimulus.

Formation of chondral bone in the extremities is a more lengthy process. Bone produced in this manner has a slower onset. Radiologic evidence mandibular mineralization is present at 10 days, whereas initial visualization in tibia occurs at 25 days or later [69, 70]. The histologic environment of endochondral bone is distinct from that of membranous bone. Periosteum surrounding long bones is not a "template." Instead, cells must be recruited from surrounding muscle and soft tissues not in direct apposition to bone. Proliferation is also slower because it starts with fewer cells. Differentiation also requires a much longer time.

Membranous bone formation correlates with vascular density. Because of an abundant blood supply, facial soft tissues are very sensitive to BMP-2. ISO from this substrate occurs rapidly; this favors membranous over chondral bone formation. The effective dosage of rhBMP-2 may also influence membranous ossification. Low doses produce cartilage that is subsequently turned into bone, whereas larger doses lead to earlier ossification. At high concentration, BMP-2 induces direct (membranous) ossification [71]. Thus, the relative amount of endochondral versus membranous ossification is associated with the biology of the implantation site and the concentration of BMP-2 implanted.

Embryogenesis of Bone: Which Type to Make?

Bone develops from three mesenchymal cell sources: neural crest, PAM via somitomeres and somites, and LPM. The axial skeleton is formed as follows. Craniofacial membranous bone is exclusively derived from neural crest. Paraxial mesoderm forming just outside the neural tube also participates in skull formation. Beginning at the level of the notochord, PAM is organized into distinct segmental units called somitomeres (Sm), each one of which is associated with a unique corresponding developmental unit of the nervous system (neuromere). The first seven somitomeres do not form somites. Nonetheless, they give rise to the parietal bone from Sm 2 and Sm 3 and to the temporal bone from Sm 6 and Sm 7 [58, 72]. Paraxial mesoderm at the level of the 8th–11th somitomes becomes transformed into the four occipital somites, the sclerotomes of which undergo a complex process of fusion to form the basioccipital, exoccipital, and supraoccipital bones [73].

At the foramen magnum, the axial skeleton takes on a new organizational pattern, in which individual vertebrae arise as a fusion of two sclerotomes. This process is known as parasegmentation. Thus, the caudal fourth occipital somite combines with the cranial half of the first cervical somite (somite 5) to produce the proatlas. This ancient structure is incorporated into the skull as the occipitodental ligament attaches the occiput to the dens of the atlas. In mouse embryos, application of retinoic acid causes a "reshuffling" of the parasegmentation process, in which the mysterious proatlas appears; the atlas and axis are converted to "standard" cervical vertebrae; and the total count of cervical vertebrae is eight [74].

The appendicular skeleton consists of the tubular bones of the extremities and their support platforms (the shoulder girdle and pelvis). These bones are not derived not from somitic sclerotomes. Instead, they arise as limb buds at junction zones of the somite system. In all embryos, limb buds develop at the cranial and caudal limits of the trunk. LPM is the cell source for these tubular bones. Parasegmentation is not involved. Instead, limb bones develop from LPM at multiple segmental levels. Coding of a limb bud is revealed by its sensory neuroanatomy. At each developmental level of the spinal cord (neuromere), a unique set of homeobox genes is expressed. This gene pattern is also expressed along the axis of the limb bud. Thus, the upper limb is assembled from LPM corresponding to neuromeric levels c5 to t1. This type of neuromeric "coding" system has also been applied to craniofacial bones by this author.

The "decision" as to which type of bone an embryo will form may reduce down to a series of interactions between a mesenchymal substrate and the local environment. Mesenchymal cell membranes may differ in the density or sensitivity of BMP receptors. The relative concentration of BMP signals from surrounding tissues may also have a role. This, in turn, could be related to variations of local vascularity. Neural crest cells (bearing BMP receptors) also play a role in tissue vascularization because of elaboration of vascular endothelial growth factor [75].

Histology of the ISO Regenerate: Distraction Osteogenesis Explained?

Despite a large literature dedicated to its technical aspects, the biologic mechanisms of DO remain poorly understood [76–78]. DO initially was applied to chondral bones of the extra cranial skeleton and later to membranous craniofacial bone. In the current study, histologic sections from the center of the specimen demonstrate two findings of relevance to the mechanism of DO. First, intact, unresorbed ACS at the center of the specimen was surrounded by osteoblasts producing osteoid. Thus, BMP-2 activity may be surprisingly long lived, depending upon the integrity of the carrier. Wherever the BMP-2 signal is present, osteoblast activation will occur. The leading edge of the distractor exerts a local mechanical force on the periosteal membrane containing potentially responsive neural crest cells. *Biologic activity of BMP-2 may thus "follow" the leading edge of the distractor.*

Also present in the specimen were two distinct forms of ossification, membranous and chondroid. In normal craniofacial development, certain neural crest populations form intermediate cartilaginous structures that are subsequently transformed into membranous bone. The lesser wing of the sphenoid forms by chondroid ossification via the orbital cartilage [55]. Chondroid ossification is a common factor linking DO with ISO. Histologic studies of bone produced by DO demonstrate the presence of cartilage [56]. Distraction regenerates have enhanced vascularity [76, 79]. Human osteoblasts respond to mechanical strain in a number of ways [80]. Production of bone matrix protein occurs independent of hormonal regulation [81]. Type I BMP receptors have been noted [82]. Expression of BMP mRNA in distraction-produced regenerates has been confirmed [83, 84].

- Thus, we hypothesize that the biologic basis of DO and ISO involves a common molecular mechanism, using BMPs as signaling intermediates.

The Clinical Potential of ISO in Craniofacial Surgery

ISO can be put to immediate use in many aspects of craniofacial surgery. First, it may be used to fill defects secondary to ablation or trauma. Children requiring calvarial bone can avoid donor site morbidity with rhBMP-2/ACS. In situations in which pre-shaped, fully formed bone is required, full-thickness calvarial bone may be harvested for grafts, put into place, and the full-thickness donor site reconstructed with confidence using rhBMP-2/ACS. This can be considered a "**BMP switch**." Mandibular and maxillary defects resulting from clefts or resection can be reconstructed with the local periosteum as the cell source. In the case of clefts, wide subperiosteal undermining will permit transfer of adjacent periosteum into the cleft site. Mucoperiosteal transfer is the basis of the functional matrix cleft operation using the sliding sulcus technique.

Blast injuries or resections resulting in extensive periosteal loss necessitate cell source transfer from a distant source. Vascularized periosteum in extensive amounts can be harvested as a galeal–subgaleal fascia flap based on the superficial temporal artery. Access to the oral cavity can be achieved by tunneling it beneath the zygomatic arch and through the fat pad of Bichat. The flap remains viable even in the face of external carotid artery (ECA) ligation because of back fill from the internal carotid via the middle meningeal artery.

GSF-periosteum + rhBMP-2 could provide a new form of reconstruction after cancer. Preclinical testing in all strains of tumor-bearing mice has failed to demonstrate any tumorigenic behavior of rhBMP-2. The safety of rhBMP-2 has been tested in 13 clinical studies involving 1000 enrolled patients with no increased risk of malignancy. Antibody formation to the type I bovine collagen of the carrier was not noted [64]. For example, hemimandibular defects could be treated using a reconstruction plate, importation of cell source, and osteoinductive agent implantation. Such a system could function in the face of postoperative irradiation. Radiotherapy can be initiated 6 weeks after surgery. Small-vessel damage takes effect approximately 6 weeks later. Within 12 weeks, mandibular resynthesis via ISO will be virtually complete. In selected patients at risk for extensive free-flap procedures, such as smokers or the elderly, ISO may prove a valuable option to reduce morbidity and hospital stays.

Existence of a genetically based segmentation system in the embryonic neural tube has brought a new understanding to head and neck development. Assembly of many common structures can be understood as craniofacial developmental fields. The physical confines of many of these fields may coincide with craniofacial bones. This model provides a highly effective surgical strategy for cleft reconstruction (see Chap. 18). Reconstruction of congenitally absent bone fields also is possible. Using distraction technology, an existing osseous structure can be used to expand surrounding soft tissues in a predetermined direction. A resulting periosteal "pocket" can be filled with rhBMP-2 to create a new bone structure. This would, in turn, maintain the newly expanded soft tissues in a stable configuration. An example of this would be the de novo synthesis of a missing mandibular ramus via expansion of a segment from the distal mandibular body. This method of reconstruction would logically be termed distraction-assisted in situ osteogenesis (DISO).

Summary

Neural crest MSCs form the membranous bones of the craniofacial skeleton. Residing in the periosteum, these cells possess BMP receptors. Implantation of rhBMP-2/ACS into a periosteal environment causes these cells to migrate, proliferate, and differentiate into osteoblasts. This results in a faithful reproduction of bone native to that specific environment. This process, ISO, occurs by osteoinduction, rather than osteoconduction. Because the rhBMP-2 implant causes a massive influx of vasculature, ISO does not follow the constraints of traditional free-grafting techniques. The shape and size of bone produced by ISO are dictated strictly by the biology of the envelope. Reconstruction using ISO is rapid, versatile, and biologically sound. New structures could be potentially synthesized using DISO.

Distraction-Assisted In Situ Osteogenesis

Introduction

The porcine trial provided convincing evidence that rhBMP-2, when placed within a periosteal envelope, could produce membranous bone almost unlimited amounts. The only limitation was the size of the envelope. Our clinical problem was how translate this to the laterofacial cleft, commonly (but erroneously) referred to as #7. Flor's defect, extended from the left oral commissure through the horizontal embryologic axis between upper deck maxillary fields and lower deck mandibular fields, reaches the external auditory canal. It was accompanied by an ipsilateral cleft of the lip, alveolus, and palate. The ramus, coronoid, and condyle were missing, and the distal (posterior) aspect of the mandibular body was foreshortened. Masseter and pterygoid muscles were absent, but temporalis was present (albeit uninserted). Her periosteal envelope was therefore foreshortened and stopped 3 cm short of the ear. How could we create in this space a neo-ramus?

Distraction is a form of tissue expansion. Could we not use a distractor to enlarge the periosteal envelope such that we could accomplish within it a morphogen-driven ISO? In this model we would osteotomize the existing mandible and, by distracting the proximal segment backward, create an expanded periosteal chamber to receive the BMP-2 implant. We termed this process **DISO**.

This 2005 paper presents the first clinical application of DISO to the human craniofacial skeleton. Herein we present in detail the clinical anatomy of the case, the rationale of the reconstructive sequence, and the surgical techniques involved. Discussion involves the following topics: (1) mechanism of action of rhBMP-2/ACS; (2) clinical performance of rhBMP-2 within the expansion chamber; (3) implications of DISO as to the biology DO; (4) the concept of craniofacial developmental fields; and (5) reconstruction of field deficits using DISO (Figs. 20.10, 20.11, 20.12, 20.13, 20.14 and 20.15).

ISO and DISO: Two New Concepts

Congenital deficiency or absence of the mandibular ramus, coronoid process, and condyle are characteristic of Treacher-Collins syndrome, craniofacial microsomia, and the #7 cleft as described by Paul Tessier. Note that with the neuromeric system, this is known as a laterofacial cleft. It represents the developmental fusion plane in the maxillary process between the "upper deck" zygomaticomaxillary fields of r2 and r4 and the "lower deck" mandibular fields of r3 and r5.

Surgical management of missing bone in these conditions has involved autogenous bone grafting and, more recently, DO. The biologic behavior of surrounding soft tissue frequently constitutes a limiting factor in such cases. Development of new reconstructive strategies suffers from an inadequate theoretical model with regard to their embryology.

Recent availability of morphogens, such as recombinant human bone morphogenetic protein (rhBMP-2), offers new possibilities for mandibular reconstruction. When an implant of rhBMP-2 in an appropriate carrier is introduced into a soft tissue environment, MSCs native to that environment differentiate into osteoblasts; new bone forms. This process is known as ISO.

Common sources for MSCs are periosteum, bone marrow, and muscle. Most bones of the craniofacial skeleton develop in a periosteal envelope via membranous ossifica-

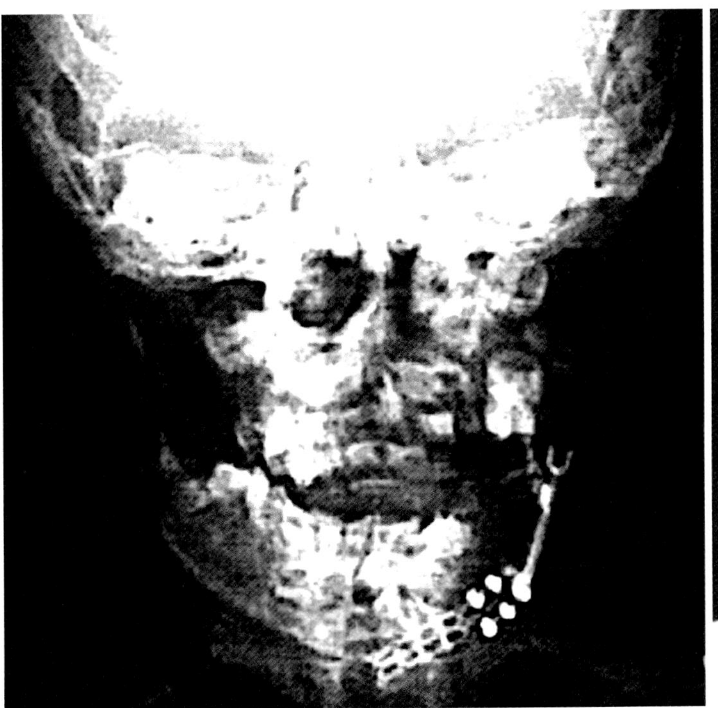

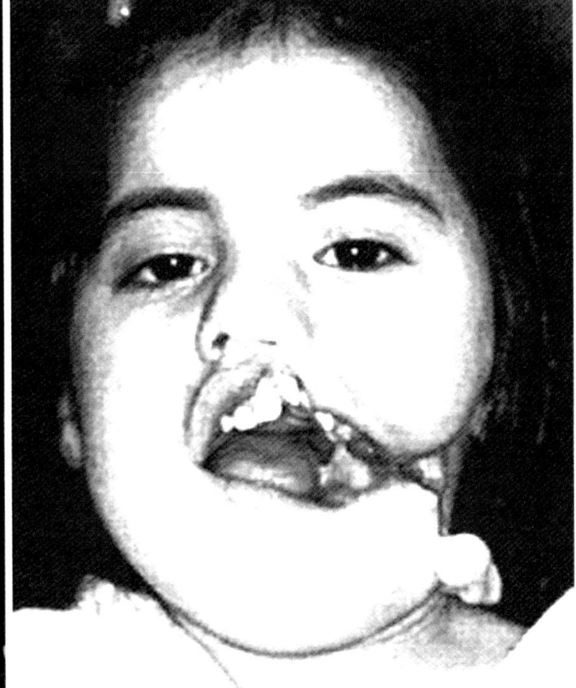

Fig. 20.10 (Left): 2 years of age showing distractor immediately before rhBMP-2/ACS placement. Extension bracket used at second stage of distraction to gain additional length can be seen mesially. (Right): 3 months after implantation. [Reprinted from Carstens MH, Chin M, Ng T, Tom WK. Reconstruction of a #7 facial cleft with distraction-assisted in situ osteogenesis (DISO): Role of recombinant human bone morphogenetic protein-2 with Helistat-activated collagen implant. J Craniofac Surg 2005 16(6)1023–1032. With permission from Wolters Kluwer Health, Inc.]

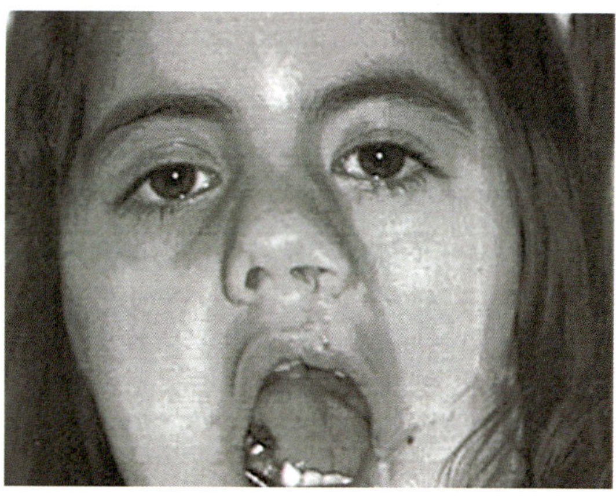
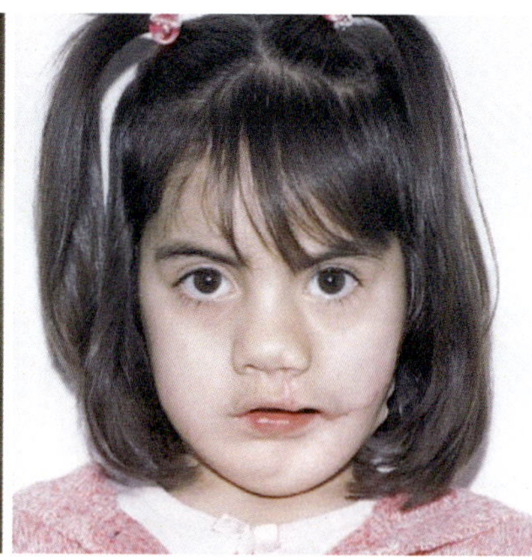

Fig. 20.15 Case 2: (Left): at 4.5 years, the patient has a corrected septum, level alar bases, and level lip line. Residual canting of mandibular occlusal plane, diminished soft tissue volume of left cheek, widened left commissure, and hypertrophic lip scar remain, requiring orthodontics, fat grafting, scar revision, and the potential need for additional distraction. Horizontal volume and vertical height of the left cheek have been substantially corrected by DISO field expansion, resulting in improved overall symmetry. Centralization of osteosynthetic periosteal matrix of maxilla preserved with functional matrix cleft repair. Level position of alar bases and midline nasal symmetry demonstrate ongoing maxillary growth. Note: Patient able to achieve centric opening of the mouth through the pseudoarthrosis. (Right): DFR cleft repair and lateral cleft repair give good symmetry in the midline with anatomic muscle repair. Intraorial opening is near normal. She will require scar revision during mixed dentition prior to adolescence or in the post-adolescent period. Left: [Reprinted from Chin M, Ng T, Tom WK, Carstens MH. Repair of Alveolar Clefts with Recombinant Human Bone Morphogenetic Protein (rhBMP-2) in Patients with Clefts. J Craniofac Surg 2005; 16(5):778–789. With permission from Wolters Kluwer Health, Inc.]. Right: [Reprinted from Carstens MH, Chin M, Ng T, Tom WK. Reconstruction of #7 facial cleft with distraction-assisted in situ osteogenesis (DISO): Role of recombinant human bone morphogenetic protein-2 With Helistat-activated collagen implant. J Craniofac Surg 2006; 16(6): 1023–1032. With permission from Wolters Kluwer Health, Inc.]

rhBMP-2/ACS implant can be placed. Vector control of the expansion process can theoretically produce a chamber of desired size and direction. Production of bone by rhBMP-2/ACS within such a chamber is called DISO.

Materials and Methods

Clinical Case

The patient presentation has been previously discussed (vide supra). In sum the case involved concomitant presentation of a lateral facial cleft with a complete cleft of the ipsilateral lip, alveolus, and hard/soft palate. The mandibular defect was Pruzansky III with a foreshortened body, absent ramus, and absent masseter. Taking advantage of the developmental field model, reconstruction of the osseous defect was undertaken using an expansion of the autologous periosteum as a source of stem cells (Figs. 20.9, 20.10, 20.11, 20.12, 20.13 and 20.14).

The surgical strategy involved both short-term and long-term objectives

- securing nutrition using obturator for oral competence
- promotion of speech patterns with palate closure
- creation of a neo-ramus/condyle using DISO
- transfer of neo-ramus to skull base as a pseudoarthrosis with temporalis transfer for motor control
- closure of lateral cleft
- reconstruction of cleft lip, alveolus, and nose
- revision of commissure
- reconstruction of the left ear

We thought that this reconstruction should address the underlying embryologic mechanism responsible for the condition. Clearly, the developmental field forming ramus, coronoid, condyle, and masseter muscle was defective. Although no means to reconstitute masseter exists, we knew that transfer of temporalis could be effective, provided the patient could gain an appropriate lever. Absence of the temporomandibular joint (TMJ) could be overcome with a pseudoarthrosis. Creation of such a neo-ramus/condyle was the key. Such a bone unit would expand the deficient soft tissue matrix and hold it rigidly in place.

- Placement of conventional graft from rib or hip would not produce the soft tissue expansion required. Furthermore, these bones are chondral, being derived from paraxial mesoderm. Bone native to the site is membranous and is derived from neural crest. ISO within a periosteal sleeve using an rhBMP-2/ACS had been shown effective in preclinical animal models. The key was to produce a periosteal sleeve of the desired dimensions. We decided to make use of distraction technology to achieve just such a distraction periosteogenesis (DP). Regenerate bone within such a chamber would result from the induction of MSCs in situ by rhBMP-2/ACS. Such bone would be membranous. Its physical dimensions would retain the physical changes in soft tissues wrought by the distraction. Moreover, with time, the child eventually would "outgrow" her reconstruction. The DISO created neo-ramus would be distracted in the future to re-expand the deficient field.

Preoperative Preparation

Because of the unusual circumstances of the case, rhBMP-2 was acquired from Genetics Institute (Cambridge, MA) under a compassionate use release from the Food and Drug Administration.

- Preoperative testing of rhBMP-2 in the porcine model was carried out and demonstrated to produce 10 cm of histologically normal bone.
- Appropriate consent for the surgical protocol was approved by the Institutional Review Board of Children's Hospital Oakland. The parents were fully informed and signed Spanish language consents for the surgical sequence.
- Note: rhBMP-2 (Infuse®), collagen sponge (Helistat®), and tricalcium phosphate scaffold (Mastergraft®) are now commercially available (Medtronic Sofamor Danek, Memphis, TN).

Operative Sequence

9 Months: Alveolar Extension Palatoplasty

The surgical sequence was begun when impressions for facial moulage were taken and alveolar extension palatoplasty (AEP) was performed. Despite the large size of the lip cleft, the alveolus was not displaced laterally as expected. We hypothesized this to be a consequence of the lateral cleft because, with growth, the divided lateral fields would exert less of a traction effect on the ipsilateral maxillary alveolus. The high alveolar incision used in AEP technique is designed to ensure preservation of blood supply to the entire osteogenic cambium layer of mucoperiosteum. The objective of AEP is to ensure biologic activity of lingual functional matrix, thus promoting subsequent alveolar growth.

25 Months: Osteotomy of the Proximal Mandibular Segment and Placement of Distractor

It should be noted that when the distraction device was placed (Fig. 20.4), *bone production was not the objective*. Rather, we sought a distraction of the periosteal sleeve in a proximal and superior direction. Such a *DP* would create a recipient chamber lined with MSCs that, when subsequently filled with an rhBMP-2/ACS implant, would create bone in the desired dimensions and shape.

- A 15-mm KLS Martin (Jacksonville, FL) distraction device was modified to fit the contours of the mandible. The external portion was placed through the posterior margin of the cleft. The footplates were fixed with 1-mm diameter screws 7 mm in length with center drive heads. Four were used on the proximal/lesser fragment (i.e., the future condyle), and five were placed on the mesial/greater fragment. The Cardanic extension for activation was brought out from the posterior aspect of the lateral cleft.
- After a 5-day latency period, distraction was carried out at 2 mm per day.

26 Months: Modification of the Distractor

At this time 20 mm had been achieved. To gain additional length of the chamber we modified the existing device.

- The screws anchoring the greater fragment footplate were removed. Counter-clockwise rotation of the activating rod transported the footplates toward the proximal (posterior) end. Examination of the distraction chamber noted it to be essentially *devoid of regenerate bone*. The resultant gap between the fragments was bridged using a section of titanium mesh. The distal mesh was secured with 2-mm diameter screws to the greater fragment. The device was then reactivated, and the footplate was transported 4 mm to ensure a satisfactory vector. Wound closure was carried out in a manner similar to that of the first surgery.

27 Months: rhBMP-2/ACS Implant

3 weeks after the previous surgery, the patient again underwent surgery, and the device was exposed (Fig. 20.5). The gap between the proximal and distal fragments was now 25 mm. The discontinuity itself was without soft tissue or osseous content. When the chamber was probed, it was noted that *a degree of bone-like tissue was present along the medial wall of the chamber*. No sign of bone formation was present in the superior or lateral aspect of the chamber. The cardanic activating device was removed.

- Preparation of the implant was carried out as dictated by the instructions. The protein powder was reconstituted with sterile saline provided. This resulted in a final concentration of 1.5 mg/mL. The solution was then applied using a drop-by-drop technique over the Helistat-activated (Integra Life Sciences, Plainsboro, NJ) collagen sponge, the goal being to achieve a uniform degree of saturation over the entire surface of the ACS. The device was then rolled up and sectioned into five segments. Each segment was individually placed into the chamber such that the three-dimensional volume of the discontinuity was completely occupied by the implant. The entire collagen sponge was used.

29 Months: Temporalis Transfer

8 weeks later, suspension of the jaw to the cranial base and muscle transfer were undertaken (Fig. 20.6). The temporalis muscle fascia was followed beneath the zygomatic arch until reaching the upper cleft margin. Opening of the posterior cleft margins left a tunnel through which a tendon graft could be passed. At the margin *no muscle was noted*; instead, *poorly developed fatty tissue* was present.

- The presence of fatty tissue instead of muscle was an indication that inadequate concentration of BMP-4 due to absent bone formation was the reason that the mesenchyme in situ failed to differentiate (see Chap. 18).
- *Aberrant dental units* were extracted just anterior to the rhBMP-2/ACS reconstruction site. We thought these might represent: (1) an abortive attempt to form a coronoid process or a portion of proximal maxilla; (2) in the primitive pre-tetrapod state, dental units are found widely distributed in the bones of the oral cavity and even on the branchial arches.

Fascia lata was harvested from the ipsilateral thigh. The tendon graft was woven into the temporalis muscle fascia and then passed down through the tunnel. The graft was then looped around the neo-ramus and passed back beneath the zygomatic arch to the temporal fossa.

- Tension of the fascial sling around the ramus was adjusted in the temporalis fossa against passive opening of the mandible.

We brought the neo-condyle into position just anterior to tympanic bone. Suspension was carried out with 3–0 nylon. The lateral cleft was then closed in three layers. Commissure reconstruction was deferred until after subsequent lip repair. Care was taken to excise the epithelial lining of the blind external auditory canal.

32 Months: DFR Cheilo-Rhinoplasty [85]

Wide subperiosteal mobilization was carried out all the way to the buttress; thus, the entire maxillary functional matrix was advanced forward into its correct position in space. This follows the principles of Delaire, Precious, and Soteranos and colleagues. Care was taken to suture the two planes of deep and superficial orbicularis in two planes as described by Park and Ha. The deep orbicularis oris was repaired in continuity between both sides, whereas the superficial orbicularis oris was sutured to the mesenchyme below philtral margin. Complete release of nasal vestibular lining from the ipsilateral piriform fossa and from the overlying nasal envelope was achieved. The resultant defect in the lateral nasal wall was replaced by the paring tissue non-philtral prolabium (NPP) in continuity with the medial crural chondrocutaneous flap.

- Note: I was not aware of Talmant's work at the time and did not reassign the nasalis to its correct insertion.

36 Months: Osteotomy and Distraction of the BMP-2 Regenerate, Lateral Cleft Repair

At this time, the neo-ramus consisted of 25 mm of fully consolidated bone. However, due to continued facial development, the patient had outgrown these dimensions and we were resolved to see if more could be achieved using the morphogen-based bone using a conventional distraction protocol. Using the distractor as a guide the central portion of the regenerate was determined and an osteotomy was done. A new Zurich distraction device (KLS Martin) was placed on the lateral side of the mandible. Distraction was completed in the usual manner. An additional 10 mm was achieved.

- Bone biopsy taken from the osteotomy site demonstrated normal histology.

Initial revision of the initial lateral repair was undertaken to "bulk up" the cleft edges and perform initial commissure repair. Repositioning of the alar base was also done.

- A curious finding was noted at revision surgery. The same submandibular incision used in the three previous operations had an appearance far different from expected. It was soft and not inflamed. It seemed more like the mature scar of a year's duration than a fresh postoperative scar. This was reminiscent of a similar reaction observed in the porcine model of ISO used in preparations for this case.

40 Months: Removal of Distractor

At this time the regenerated was verified to be mechanically stable.

Results

At 48 months the patient has achieved normal body growth and developmental milestones. She speaks Spanish and English equally well. Oral competence is good; no oral-nasal fistula is present. Her left TMJ pseudoarthrosis functions well with excellent interincisal opening. The mandibular neo-ramus is 3.5 cm long and provides good soft tissue contour and projection. CT scan discloses a lingual cortex and a buccal cortex separated by an intervening space consistent with marrow (Figs. 20.7 and 20.8).

Dental development is proceeding normally in unaffected maxillary and mandibular fields. Maxillary growth is almost symmetrical; residual hypoplasia is noted at the tuberosity. The previously deficient lateral pterygoid plate is growing proportionately, albeit smaller than on the right side. Mandibular growth continues to be reduced on the left side. Although no outright deficiency state exists in the right mandible, overall growth is also reduced. Thus, both sides of the mandible have been affected, but all fields continue to develop at their own intrinsic pace (no one site is diverging away from the others). Soft tissue development is proceeding well. The lip remains level and the left airway remains patent. The nasal dorsum is straight; alar bases are level and equally projected in space. Mild asymmetry of commissures persists.

At 48 months of age, the patient continues to have good nasomaxillary symmetry (Fig. 20.9). Her oral opening is excellent with very little deviation from the midline. All phases of initial reconstruction are complete with exception of microtia repair. Additional distraction is anticipated as growth proceeds. The hypertrophic lip scar will be revised with two cycles of adjunctive postoperative radiation to the surgical site. In contrast, the neck incision is not hypertrophic, despite the multiple entries at the site. Soft tissue augmentation with fat injection will be used to further refine her facial contours.

Discussion

Bone morphogenetic protein-2 is one of at least 15 BMP molecules, all of which belong to the transforming growth factor superfamily [3]. BMP-2, 6, and 9 have the most osteogenic activity [64]. The mechanism of rhBMP-2/ACS involves six steps [71]:

- *Implantation*: The surgical creation or modification of an environment containing MSCs. This involves contact with periosteum, muscle, or bone marrow.
- *Chemotaxis*: MSCs from up to 2 inches away are attracted by rhBMP-2 to the implantation site.
- *Proliferation*: The rhBMP-2/ACS implant provides local environment where MSCs can multiply prior to differentiation.
- *Differentiation*: Binding of rhBMP-2 to specific receptors on the MSC cell surface causing them to change into osteoblasts.
- *Bone formation* and *angiogenesis*: Osteoblasts respond to local mechanical forces to produce osteoid. Intense new blood vessel formation is observed.
- *Remodeling*: Bone remodels in response to local environment and mechanical forces.

Clinical performance of rhBMP-2/ACS is relatively predictable [64]. The amount of bone produced relates to the size of the regenerate chamber. Complete filling of the regeneration chamber with sponge is important. In spinal fusion, physical stability is provided by the transverse processes. The presence of an external container, either of absorbable PGA/PGL mesh or of stainless steel, has proven useful in providing a shape to the regenerate bone. Serial CT scan provides evidence of the time course of ISO. At 6 weeks, bone formation is clearly seen. By 12 weeks, consolidation is present. Critical-size defects in porcine calvaria demonstrated identical histologic findings and tensile strength compared with the control side at 12 weeks [68]. Humans may lag behind, but evidence for this remains preliminary. Final remodeling can be considered complete at 24 weeks. This seems to an appropriate time for plate removal or additional manipulations, such as osteotomies for distraction.

The clinical behavior of soft tissues in the presence of an rhBMP-2 implant varies considerably from other surgical models. The intense vascular influx causes swelling to peak more slowly (2 days) and disappear more slowly (5 days) than the normal course of events. Wound healing is affected in a currently unknown manner. Scar formation above an implantation site seems to bypass some of the typical early stages. Scars at 1 month have the clinical appearance of scars 6–12 months old.

Maxillofacial Applications of rhBMP-2: Preclinical and Clinical Studies

Boyne [60–62] documented the ability of rhBMP-2 to augment the alveolar ridge, fill the maxillary sinus, and replace surgically created mandibular defects. Using similar methodology, he demonstrated the ability of rhBMP-2 to repair simulated cleft palate defects [63]. Following long-standing work with alveolar distraction, Chin and colleagues [51, 86–88] reported successfully application of rhBMP-2 in human distraction cases. In preparation for the clinical management of the case presented here, Carstens and Chin tested the ability of rhBMP-2/ACS to produce ISO in a critical-size, 10-cm mandibular defect in 2001 (reported in 2003). Fully formed trabecular bone was observed at 3 months. The success of this work led these workers to postulate a surgical sequence in which rapid DP would be followed by rhBMP-2/ACS into the implantation chamber. Successful implementation of DISO technique in the human mandible was first reported by Chin et al. [88].

- Subsequently, I found DISO to be very useful for infant mandibular distraction in severe cases of Pierre Robin requiring intubation, reducing the time of airway support to 24 h [89].

Clinical Significance of the #7 Cleft: *Developmental Fields Exist*

Descriptive embryology as a science began from pioneering observations by Wilhelm His in the 1870s using light microscopic analysis and histological staining. The approach was morphologic, rather than cellular. His was a staunch opponent of the idea that genetic material might reside within the nucleus. His descriptive terminology (e.g., "lateral nasal process") can be found in all textbooks. But what is the anatomic nature of a "process?" What are its constituent parts and from where in the embryo do they come? All surgeons are well aware that clefts, for example, occur in an orderly spectrum of presentations. Unfortunately, concepts such as "failure of fusion" or "failure of mesoderm penetration" are incapable of providing a rational explanation for this varying degree of pathology. Embryology seems a jumble of terms with no clinical relevance. But the fact remains: without a detailed understanding of the developmental anatomy of the face based on modern developmental biology, genetics, comparative anatomy, and neuroembryology, pediatric plastic surgery is a collection of techniques in search of a science.

Based on clinical observations of secondary cleft patterns, this author arrived at the following hypotheses: (1) unidentified **developmental fields** might constitute the "building blocks" of the face; (2) a deficiency in such a field would be characterized by an inadequate osteosynthetic capability; (3) the pattern of osseous deficiency in a cleft might be a clue to its pathogenesis; (4) in common labiomaxillary clefts, a bony insufficiency exists in the inferolateral piriform fossa/lateral nasal wall; and (5) such a functional matrix deficiency state might account for the relapse pattern observed in patients with cleft after primary repair.

To test these ideas, it seemed logical as a first approximation to study the relative contributions of the internal carotid artery (ICA) and ECA circulations to the skin and epithelium of the nasal fossa. Contrast injections into isolated internal carotid arteries in a series of aborted fetuses were performed. The author found, to his surprise, that the upper border of the inferior turbinate combined with the skin/mucosa junction of the inferolateral piriform fossa just anterior to the inferior turbinate constituted a potential field interface zone. At this site three distinct biologic systems (vascularization, innervation, and genetic programming) functioned in precisely the same manner:

- The internal carotid supplied the mucosa (but not skin) of the lateral nasal wall but only as far as the upper border of the inferior turbinate. The mucosa beneath the turbinate and the skin margin along the infracartilaginous nostril remained un-perfused;
- The sensory innervation followed the exact same distribution pattern. The epithelium supplied by the ICA corresponded to sensory supply from the 1 branch of the trigeminal, whereas that supplied by the ECA was innervated by V2;
- The inferolateral piriform fossa also represents an interface zone between three entirely different developmental zones of the embryonic neural crest.

Comment We discussed the fetal injection study in Chap. 17. What shows in this 2005 paper is an incipient model of neurovascular fields immediately prior to discovery of Padget's work and the stapedial system leading to final neuromeric mapping of craniofacial circulation (Chaps. 6 and 7). Also by this time I had been able to map out the three zones of premaxilla. This gave rise to the ideas of cleft pathology being the result of a progressively worsening vascular "hit" affecting PMxF and progressing backwards to PMxB. Knowledge of the BMP-4-SHH loop had to wait until the 2007 and the discovery of Zhang's 2002 paper describing how BMP-4 could "rescue" cases of Msx-1-deficient mice from forming cleft palate.

This case convinced me of the existence of a **developmental field map**, with many fields conveniently defined by the bone content. I was later to learn that the fields described are actually derived from the Dlx gene system which divides up all the pharyngeal arches into geometric sectors. But here in 2005 is how the map looked like simple set of quadrants.

The mammalian mandible has four distinct developmental fields. The first three are defined by the dentition and their sensory supply. Mn0 has the incisors, Mn1 contains the canines and premolars, and Mn2 houses the molars. Three distinct sensory nerves serving each of the above are enveloped within a common inferior alveolar canal [90, 90]. The ramus MnR constitutes a fourth separate field. It has two outgrowth structures, the coronoid and the condyle. Masseter has is primary insertion into the ramus. It therefore develops in the same zone of first arch as the ramus. Defects in MnR would be expected to progressively involve the condyle, the ramus itself, and the masseter muscle.

On the cranial side of the occlusal plane are four "partner fields" corresponding to those of the mandible. Thus, the incisors belong to the premaxilla, PMx; the lateral maxillary incisor, canine and premolars to Mx2; and the molars to Mx3 Just opposite MnR lie the palatine bone field (Pl) and the lateral pterygoid process (LPt). It is of interest to note that reports of Tessier #7 clefts document the presence of a *crease in the back wall of the alveolar tuberosity*, occasional *absence of a molar*, and *reduced palatine size*. The case described here has several of these features, including a significant reduction in size of LPt.

The clinical significance of this case was the restoration of the missing MnR field, some features of which were amenable to synthesis and some of which were not. The absent masseter muscle could not be reduplicated. The mechanical concept was to create appropriate bony platform (via DISO) of appropriate size and shape to which a muscle could be attached. Tendon transfer from the temporalis to the neoramus could recreate a mastication system. In sum, field reconstruction can be achieved using DISO as a means to expand a soft tissue envelope. Although not all components of a field can be recreated, traditional plastic surgical techniques (such as tendon transfer) may give to such a field an acceptable degree of function.

Reconstruction of the Developmental Field Map: *The Role of DISO*

As has been stated previously, the cell membranes of neural crest cells uniformly possess receptors for BMPs [70]. Osteoblasts respond to rhBMP-2 signals by production of osteoid [81]. Mechanical strain, when applied to a distraction chamber, causes a similar production by osteoblasts of bone matrix protein [91]. If the mechanism of DO proceeds via induction of local osteoblasts, why not create such a chamber rapidly without regard to bone formation? The local environment merely requires that its cellular constituents receive appropriate BMP signals; osteogenesis will ensue. If a uniform concentration of rhBMP-2 can be provided to a chamber via an ACS implant, it does not matter if the chamber is created slowly or rapidly. All surfaces of the chamber are equally stimulated simultaneously.

The most efficient system would be one in which the geometrical dimensions of the chamber are rapidly achieved followed by secondary introduction of the desired rhBMP-2 dose to stimulate the MSC population residing in the walls of the chamber. Such a strategy is entirely feasible using rapid expansion in a specific, planned direction. When this "distraction periosteogenesis" is complete, neighboring soft tissues also will be expanded. Synthesis of bone in the chamber by DISO will create a permanent "internal strut," of desired size and shape, that will maintain the expanded state of the soft tissue surround.

An rhBMP-2/ACS implant placed into a distraction chamber contains 1.5 mg/mL of protein. This represents a supraphysiologic dose, more than 200,000 times the level found in normal demineralized human bone. The ability of ISO to fill a 10-cm chamber in 120 days invites comparison between the clinical performance of DISO and that of conventional distraction [92]. At a rate of 1 mm per day, the conventional DO would require more than 100 days to reach the desired length without a uniform time for maturation of the distraction chamber. In summation, knowledge of the developmental theory can enable surgeons to use the principles of DISO to recreate missing tissue units by planned vector-controlled tissue expansion.

Summary

We began our exploration of morphogen-based bone reconstruction with a challenging case of multiple facial clefts, using anatomical analysis based on the neuromeric model of developmental fields. The embryonic defect was localized to the MnR field, involving neural crest arising from the third rhombomere and, to a less degree, neural crest from the second rhombomere forming the palatine and lateral pterygoid plate fields. Reconstructive strategy was predicated on de novo synthesis of a mandibular ramus and muscle transfer from the normal, ipsilateral temporalis. Deliberate, spatially directed expansion of the defective field using periosteal expansion followed by implantation of rhBMP-2 proved a highly effective strategy.

What Did We Learn?

Craniofacial abnormalities involving deficient or absent bone structures, each one unique, present many challenges. Any surgeries performed upon these structures can undergo significant scarring or breakdown if too much tension/pressure is placed upon the tissues. The underlying bony structure may not have the available blood supply to grow. Multiple surgeries may be required in the same anatomic site as these patients continue to grow throughout their lives. Due to different training philosophies, treatment approaches may be varied yet the goal of every surgeon remains the same. The mission is the conversion of an abnormal defect, either soft or hard tissue, to normal appearance and function.

Distraction osteogenesis has the advantage of regenerating soft and hard tissues at the same time as well as enlarging the anatomical complex. The use of distraction does not generally require the harvesting of bone. This is a significant feature of this surgical technique, as many children do not have a large supply of available bone by which bone can be harvested. Distraction will lessen the morbidity to the patient and decrease the length of care that the patient requires in terms of hospitalization or postoperative care. Unfortunately, DO does not always produce the desired amount of bone necessary for anatomic correction. In some cases, the deformity is so large that the potential for consolidation and union of the distracted bone is not expected because of compromise in tissue quality (scar), vasculature, or available cells.

The use of morphogens such as BMP-2 allows the patient to form their own bone de novo. The advantage of this molecular therapy is that it is truly osteoinductive. The rhBMP will promote chemotaxis of the MSCs and direct the differentiation of the stem cells to become osteoblast. The recombinant technology will allow for amplification of the cellular differentiation. The advantage to the patient is faster consolidation of the bone growth in the deficient area. Newly generated bone has been shown to be as mechanically stable as the patient's existing bone. In our case, once the bone was consolidated, it behaved like normal bone. Upon cutting the cortex and entrance into the marrow portion of the bone, there was normal bleeding. The placement of a distractor combined with additional rhBMP produced a second engineered ossicle via ISO. The bone produced was mechanically stable clinically and demonstrated normal histology. The use of ISO allowed the soft tissue to be enlarged while the bone was consolidating. The risk of wound breakdown is lessened by this technique as well as obviating the need for bone harvest. As the child continues to mature, it is proposed that more surgery will be necessary and it is anticipated that the bone regenerate will respond as appropriately as before to the surgical procedure.

It should be emphasized that a previously underappreciated aspect of morphogen-driven ISO is the ability to induce local blood supply. It was thought initially that BMP-2 was the agent producing VEGF, but recent experience with MSCs, particularly in the form of adipose-derived stromal vascular fraction (SVF), indicates that the MSCs per se secrete VEGF. This is a form of in situ angiogenesis (ISA). Treatment of peripheral vascular insufficiency with SVF has demonstrated the ability of ISA to revascularize tissue previously rendered ischemic due to diabetes or arterisclerotic disease. Thus, the morphogenetic effect of BMP-2 on the target MSC population includes not only proliferation and differentiation but also the immediate production of local blood supply to the support the bone graft.

Conclusions from Experience with DISO

1. Distraction of the bone segment constructed with rhBMP-2 resulted in rapid consolidation with viable bone. In contrast, prior distraction of native bone in this same patient resulted in no bone formation, i.e., non-union.
   - This was likely due to intrinsic mesenchymal deficiency in the bone stock: not enough available cells.
2. Once the original morphogen-based regenerate matured, the repeat application of BMP-2 showed a similar clinical and histologic response.
   - Since DO is essential a form of stem cell activation, DISO gives more predictable results due to the high concentration of morphogen.
3. Tissue-engineered bone with BMP-2 is physiologic: it supports dental eruption, periodontal regeneration, and orthodontic manipulation.

Post-script: Autologous MSCs, the Next Dimension for DO

The future direction of ISO and DISO will be altered by the addition of SVF or similar stem cell preparations to the implant for the following reasons. At the conclusion of this chapter we shall explore the following topics in greater detail.

- SVF exerts an anti-fibrotic and trophic effect on tissues which directly addresses issues, such as fibrosis and poor-quality interstitial tissues
- SVF is anti-inflammatory, thereby improving the wound healing response to surgical trauma and reducing the reaction to both the distractor and the morphogen implant
- SVF is highly angiogenic, acting immediately with production of blood vessels as early as 72 h. It can thus provide the necessary bridge for survival for bone. Furthermore, the space occupied by the implant does not require a high degree of vascularity initially. It is quickly invaded by new blood vessels and migrating MSCs
- SVF overcomes the limitations of cell supply presented by periosteum and local tissues by providing an unlimited supply of autologous MSCs available for transformation by the morphogen

Alveolar Cleft Reconstruction

By 2004 we knew that rhBMP-2 in the presence of a periosteal source of stem cells produced membranous bone of normal histology, responded physiologically to DO. The immediate and obvious translation of this work was in the management of alveolar clefts. We set out to assess the performance of morphogen grafts in secondary cases (age six and later) and as a component of primary cleft management, either a lip repair (6 months) or palate repair (18 months).

BMP-2 implantation represents a superior form of treatment compared to iliac crest bone grafting as it would address important liabilities of conventional surgery. First, hip graft donor sites are painful and graft harvest has its own attendant complications. Second, ICBG reconstruction often results in uncertain production of bone within the cleft chamber. Recall that the Achilles heel recipient site is the soft tissue closure leading to the floor of the nose, the "roof" of the alveolar cleft "box." Under conventional conditions the floor of the nose is closed off with mucosal tissue. This technique has the following drawbacks. Fistulas here are common, leading to graft loss. Moreover, non-mucoperiosteal tissues are not bone forming in nature. They many possess an inadequate number of stem cells. What's more, conventional ICBG provides insufficient stimulus to existing responder cells such that *root coverage high in the cleft may be inadequate*. BMP-2, on the other hand, is capable of mobilizing MSCs peripheral to the site. These cells actually migrate into the graft, thus compensating for any deficiency in number or biologic responsiveness in the local progenitor population. Finally, the use of BMP-2 grafting is modular; additional graft can be placed when and where needed.

This section discusses the clinical applications and outcomes of rhBMP-2 for alveolar cleft repair, both secondary and primary. We shall first discuss **secondary BMP-2 cases** because this algorithm, using Sotereanos S flap technique, has the best overall performance. We shall then detail our experience with **primary BMP-2 grafting**, both as part of lip–nose repair at 6 months and, in a later group, at palate repair using AEP. Although our current recommendation is for BMP-2 grafting at 4 years as per the Talmant-Lumineau protocol, this earlier experience is valuable to review at it provides important insights into the biology of ISO in cleft patients.

Secondary Protocol

Introduction

Staged reconstruction of cleft lip/palate cases includes alveolar cleft repair and closure of the associated oronasal fistula. In 1974, Boyne reviewed his experience with repair of the alveolar discontinuity using autogenous marrow from the iliac crest. Since that time, the use of autogenous iliac crest grafts for alveolar cleft repair has become the standard treatment. The outcomes are generally good but morbidity associated with harvesting bone from the iliac crest is substantial. Potential complications associated with the iliac crest harvesting include significant pain and potential paresthesia over the lateral thigh. Obtaining bone from the iliac crest typically involves general anesthesia. In an effort to avoid the adverse effects associated with autogenous iliac crest harvesting, a variety of *graft substitutes* have been tried. These include human banked bone, anorganic bovine bone mineral, and alloplast [22, 93]. Alternative autogenous donor sites have also been used. These include the calvarium, tibia, and mandible.

In general, management of the soft tissue is similar, regardless of the type of graft material used. Typical surgical repair involves elevation and rotation of mucosa flaps to establish a closed chamber at the alveolar cleft site. The use of S flaps as per Demas and Sotereanos greatly facilitates anterior closure.

The osseous discontinuity of the alveolar cleft is filled with cancellous bone harvested from the patient's iliac crest. Timing of surgical repair, relative age, and stage of dental development vary between treatment centers. All of the grafting material options have been associated with some limitations. Banked bone is primarily osteoconductive and requires viable bone at the periphery. Alloplastic materials, such as hydroxyapatite, may impede tooth eruption and orthodontic movement. Autogenous bone has been associated with root resorption when placed in contact with exposed tooth roots. To date, the use of autogenous iliac crest remains the standard treatment. Our purpose here is to present the use of recombinant human bone morphogenetic protein (rhBMP-2) as an alternative to autogenous iliac crest bone as a graft source.

Before we embark on this discussion, it is worthwhile to point out a paradox: *the central problem posed for successful alveolar cleft grafting: primary cleft lip surgery*. Patients with a wide-open lip and palate defect present an anatomy that virtually begs for an integral solution, one that ensures a hermetic seal of the nasal floor. As described in the previous chapter, this can be readily achieved by proper inset of the NPP flap plus or minus a vomer flap combined with contributions the nasal mucoperiosteum of the palatal shelf. Raw space left behind by the release of the nostril sill flap and advance of the lateral vestibular mucosa can be replaced with an anterior-based turbinate flap and/or a proximally based mucosal L flap harvested from the lateral lip.

Recombinant Human Bone Morphogenetic Protein-2

Bone morphogenetic proteins are a group of endogenous proteins that are a part of the transforming growth factor beta superfamily. These proteins are involved in embryologic development and formation of the skeleton. Minute quantities of these proteins are contained in the mature skeleton and may be involved in bone maintenance and fracture repair. Marshall Urist, an orthopedic surgeon conducting basic research on the nature of bone mineralization, first conceptualized the existence and role of these agents. His research demonstrated that when demineralized, lyophilized segments of allogeneic bone matrix were placed into muscle pockets in rabbits, bone formed. This phenomenon led him to postulate that agents present in the bone matrix possessed the capability of inducing the formation of bone-forming cells [94]. The concept that undiscovered agents, contained in bone matrix, had the capability to cause bone formation without the introduction of bone cells was revolutionary. In 1971, Urist and Strates introduced the term *BMP* to identify these agents. As a result of extensive purification of bones, minute quantities of BMPs were isolated [95]. In 1988 Wozney sequenced rhBMP-2 and cloned it.

Recombinant technology now allows production of large, pure quantities of rhBMP-2 that can be used clinically and in the laboratory. Until the recent past, the use of these materials was limited to the research environment. It is now commercially available as Infuse® bone graft (Medtronic Sofamor Danke, Memphis, TN). Clinical studies in maxillofacial surgery included construction of bone in the maxillary sinus for placement of dental implants. Orthopedic research demonstrated the efficacy and safety of using rhBMP-2 for spine fusion and management of long bone fracture non-unions. Current clinical indications for Infuse® approved by the Food and Drug Administration are as follows: 2018 spine fusion application, sinus lift, and alveolar ridge augmentation. This material and others are now available for human use.

When placed in the proper environment, rhBMP-2 can cause bone formation de novo. *The introduction of bone-forming cells is not necessary to initiate the process*. Instead, rhBMP-2 acts in situ to concentrate host stem cells at the site and then to influence their differentiation into osteogenic cells. This process, termed *morphogenesis*, is unique to this new class of agents. Introduction of exogenous stem cells is not needed, even in the mature human. To achieve a clinical effect, administration of super-physiologic doses is needed. Following the current manufacturer's recommendations for spine fusion, for example, the device is prepared at 1.5 mg of rhBMP-2 per milliliter. This represents approximately 200,000 times the estimated physiologic concentration of natural BMP-2 found in bone. Extensive preclinical and clinical studies have demonstrated the efficacy and safety of rhBMP-2. In a prior report, Chin et al. [50] showed that rhBMP-2 enhanced the healing of alveolar distraction sites in non-human primates. Chin demonstrated the use of rhBMP-2 for enhancement of alveolar distraction in humans. Chin and Carstens showed that rhBMP-2 was effective in establishing osseous union across discontinuity defects found in major facial clefts. In the following study the clinical and radiologic consequences of alveolar cleft grafting with rhBMP-2 are described.

Materials and Methods

Patients

This series consists of 43 children undergoing repair of 50 cleft sites with rhBMP-2. Cleft types included unilateral cleft lip and palate (n = 30), bilateral cleft lip and palate (n = 7), midline maxillary cleft (n = 4), and lateral facial cleft (n = 2). Ages ranged from 6 to 14 years. Follow-up was 6–25 months. Two children were being treated for failed autogenous bone grafts.

Surgical Protocol

Group 1: Moderate Alveolar Clefts

Children presenting with typical unilateral alveolar discontinuity related to cleft lip and palate underwent surgery involving closure of the oronasal fistula and creation of a graft chamber by advancement of mucosa flaps. The technique for management of the soft tissue followed that described by Boyne [96] and Betts and Fonseca (2). Nasal and palatal mucosa flaps were elevated and rotated to achieve approximation without tension. Interrupted sutures stabilized the closure. Labial flaps were advanced with the aid of periosteum releasing incisions, a posterior, vertical incision up the buttress, and, in some cases, a releasing incision in the periosteum along the undersurface of the flap. This latter incision, just through the periosteal membrane, extended from the piriform fossa to the buttress was chosen to provide additional medial translation of the lower have of the flap [97]. Enough labial flap advancement was performed to allow for tension-free closure and adequate keratinized gingiva at the cleft site.

Twenty minutes before placement of the graft material, the bone graft device was prepared. The device used in this series was the Infuse® Bone Graft (Sofamor-Danek, Memphis, TN). The two components of this system, rhBMP-2 and an ACS, are combined to form the rhBMP-2/ACS device. Following the manufacturer's recommendation, the rhBMP-2 was reconstituted with sterile water to achieve a 1.5 mg/mL concentration. The solution of rhBMP-2 was then used to saturate the ACS supplied as part of the infuse device. The saturated ACS was allowed to soak for 20 min in the rhBMP-2 solution. The rhBMP-2/ACS device was then packed into the site. Closure of the labial mucosa flaps completed the repair. All patients were started on antibiotics at the beginning of surgery, cefazolin 25 mg/kg. The patients were maintained on oral antibiotics, penicillin v potassium 25 mg/kg/day in four divided doses, for 5 days. Radiographs show consolidation of bone at the alveolar cleft site and eruption of teeth through the generated bone (Figs. 20.16, 20.17 and 20.18).

Children presenting with bilateral cleft lip and palate were managed in the similar manner. These patients often present with limited mucosa overlying the premaxilla. The lack of soft tissue complicates creation of an enclosed chamber of ideal volume into which graft material can be placed. The osseous discontinuities in the alveolar process were grafted with rhBMP-2/ACS devices and labial mucosa flaps advanced from the posterior segments. Fifteen months after grafting with rhBMP-2, radiographs show osseous continuity and eruption of permanent teeth into constructed alveolus. The teeth respond normally to orthodontic treatment (Figs. 20.19, 20.20 and 20.21).

> **Note**: Palatoplasties done previously for these patients were *not* with the AEP technique. We have found in all patients treated subsequent to this 2005 report, AEP flaps are generous in length and provide a complete seal behind the premaxilla. This has made soft tissue coverage of graft material much easier.

Group 2: Large Alveolar Clefts

In this group of patients, the volume of the defect and quality of the soft tissue would not allow repair with the single-stage method described for typical alveolar clefts. These patients were managed in multiple stages. The first stage of treatment involved narrowing the cleft with *horizontal distraction osteogenesis*. In the second stage, after the cleft margins were in closer proximity, rhBMP-2 was used to consolidate the docking site.

A 12-year-old presented with a wide midline maxillary cleft (Figs. 20.22, 20.23 and 20.24). A previously placed rib

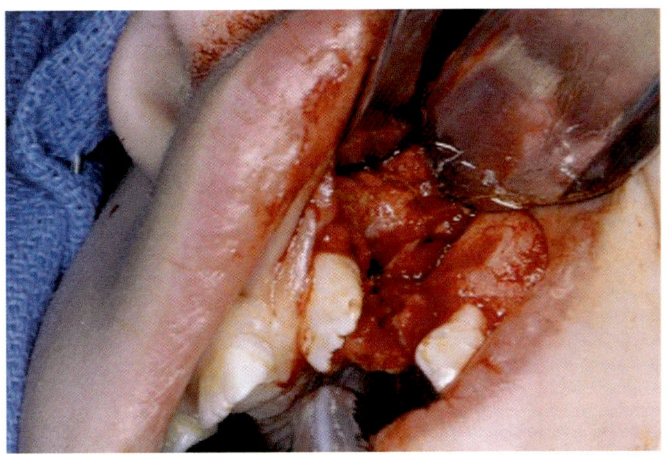
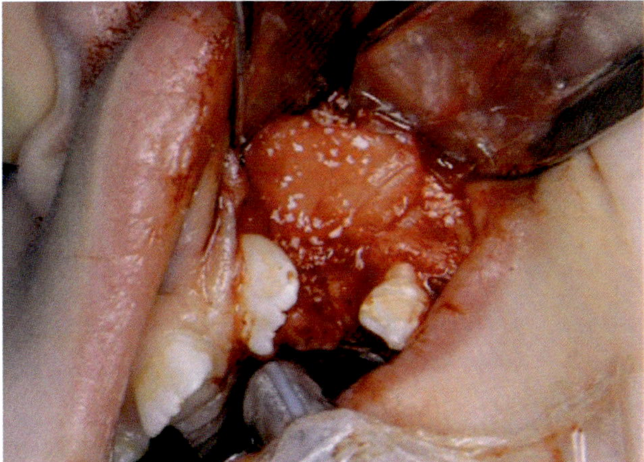

Fig. 20.16 Case 2: Alveolar cleft prior to implantation (left) showing closure of the nasal floor. Right image shows rhBM-2/ACS graft filling the space prior to closure with an S flap. [Reprinted from Chin M, Ng T, Tom WK, Carstens MH. Repair of alveolar clefts with recombinant human none morphogenetic protein (rhBMP-2) in patients with clefts. J Craniofac Surg 2005 16(5):778–789. With permission from Wolters Kluwer Health, Inc.]

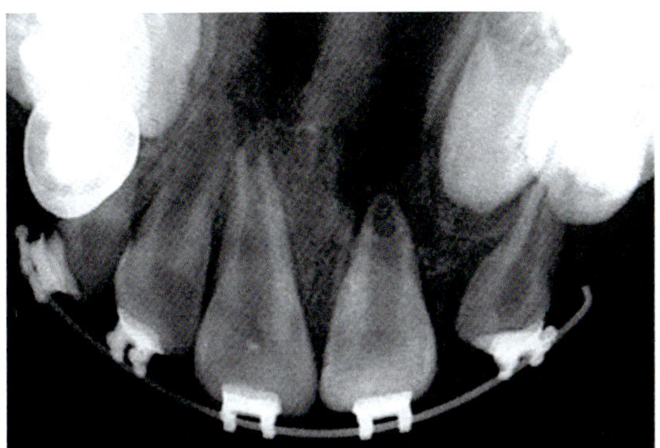
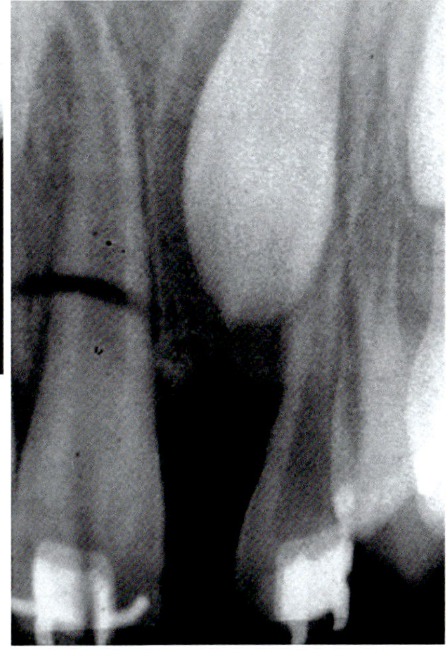

Fig. 20.17 Case 2: Alveolar cleft with bone formation at 12 months postop (left) and at 17 months postop (right). Note canine erupting through the ISO-created bone. [Reprinted from Chin M, Ng T, Tom WK, Carstens MH. Repair of alveolar clefts with recombinant human none morphogenetic protein (rhBMP-2) in patients with clefts. J Craniofac Surg 2005 16(5):778–789. With permission from Wolters Kluwer Health, Inc.]

graft constructed piriform rims, but a large defect in the alveolar process remained. Maxillary retrusion made construction of a functional alveolar process even more challenging. Because of the combination of the large volume defect, limited soft tissue, and underlying class III skeletal relationship, conventional grafting was likely to fail. This patient was treated in multiple stages combining DO and rhBMP-2 grafting. In a first stage, Le Fort I distraction advanced the maxilla. The result was to improve the position of the dentulous maxillary segments, but the procedure did not improve the midline defect or quality of the soft tissue. The cleft was then narrowed using bifocal, horizontal distraction of alveolar segments. The method and device described by Liou and Chen were used [98]. Segments were transported until they docked in the midline.

As expected, the docking site did not spontaneously form an osseous union. After distraction segment transport was complete, the result was a narrow alveolar cleft in the midline of the maxilla with an oronasal fistula. This residual, midline cleft was managed by an approach similar to typical alveolar clefts. The site was opened and mucosa flaps rotated to establish a closed chamber. At the time of surgery, it was found that one tooth at the docking site was totally without bone support, and it was extracted. The remaining tooth adjacent to the docking site had root exposure involving 75% of the root. This root was covered with an rhBMP-2/ACS device

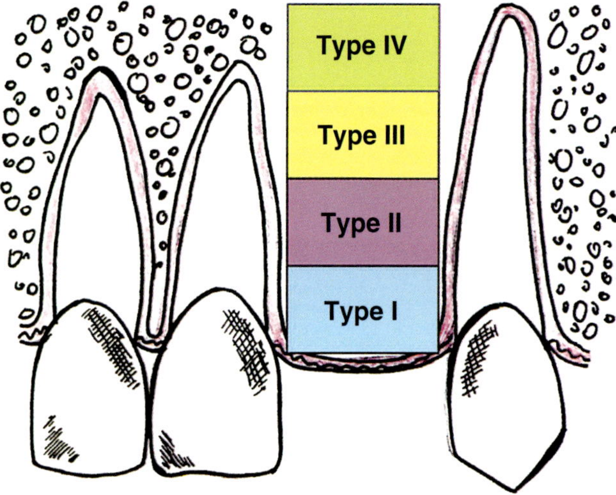

Fig. 20.18 Case 2: Abyholm/Bergland classification of alveolar cleft graft success. Note bone synthesis filling the cleft to the level of the alveolar ridge (level 1). [Reprinted from Kim Y-E, Han J, Baek R-M, Ki, B-K, Alveolar bone grafting with simultaneous cleft lip rhinoplasty. JPRAS 2016; 69(11): 1544–1550. With permission from Elsevier.]

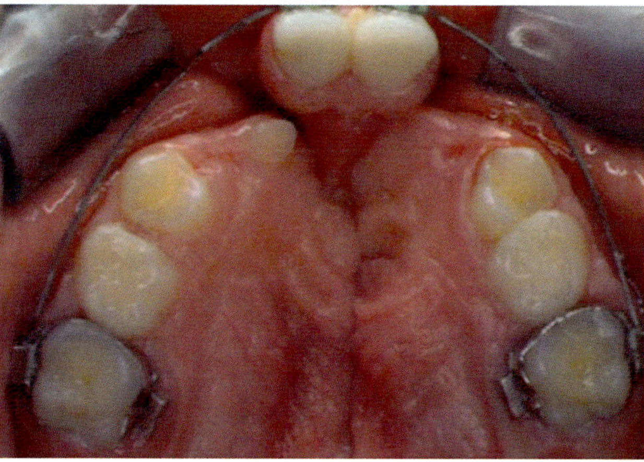

Fig. 20.19 Case 3: Bilateral cleft palate with classic collapse of the lateral segments behind the premaxilla. Soft tissue coverage in this situation is difficult. Posteriorly AEP flaps can be advanced forward. Anteriorly Sotereanos S flaps (sliding sulcus) will bring forward 2 dental units of tissue. Premaxillary mucoperiosteum is split in the lateral midline but the posteriorly based flap can be wider. Decision to use AEP flaps depends on the dimensions of the premaxillary flaps. [Reprinted from Chin M, Ng T, Tom WK, Carstens MH. Repair of alveolar clefts with recombinant human none morphogenetic protein (rhBMP-2) in patients with clefts. J Craniofac Surg 2005 16(5):778–789. With permission from Wolters Kluwer Health, Inc.]

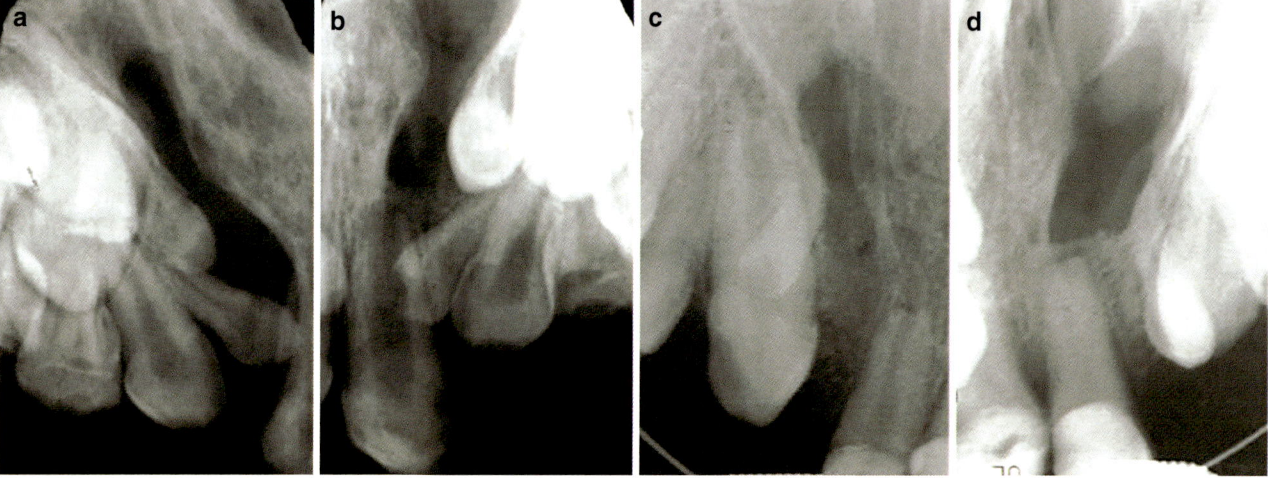

Fig. 20.20 Rh BMP-2 grafting of bilateral alveolar clefts. Bilateral alveolar cleft reconstruction with rhBMP-2 showing graft extending down to the alveolar ridge (Abyholm stage 1). (**a**) Right alveolar cleft pre-op, (**b**) left alveolar cleft pre-op, (**c**) right alveolar cleft post-graft, (**d**) left alveolar cleft post-graft. [Reprinted from Chin M, Ng T, Tom WK, Carstens MH. Repair of alveolar clefts with recombinant human none morphogenetic protein (rhBMP-2) in patients with clefts. J Craniofac Surg 2005 16(5):778–789. With permission from Wolters Kluwer Health, Inc.]

(1.5 mg/mL). The remaining chamber including the dental extraction site was filled with rhBMP-2/ACS. Follow-up radiographs confirm ossification. *The tooth with root exposure showed no sign of resorption.* It was clinically stable and was successfully moved with orthodontic appliance. The volume and position of the constructed alveolar process was satisfactory to place dental implants, and there was marked improvement in the facial balance.

Group 3: Severe Deficiency

In these patients the bone and soft tissue deficiency exceeds the ability of conventional DO to generate enough bone to meet the objectives of reconstruction. Because of the large magnitude of bone transport necessary, *non-union is the likely outcome*. The strategy for this group of patients is to transport bone segments to promote soft tissue regeneration but to accept that a non-union will likely result. In this situa-

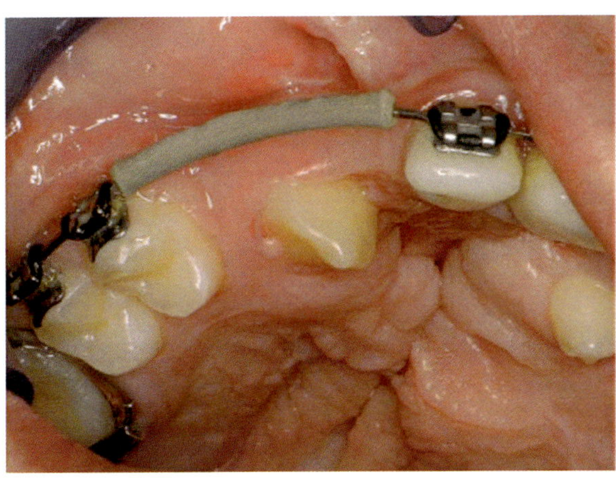

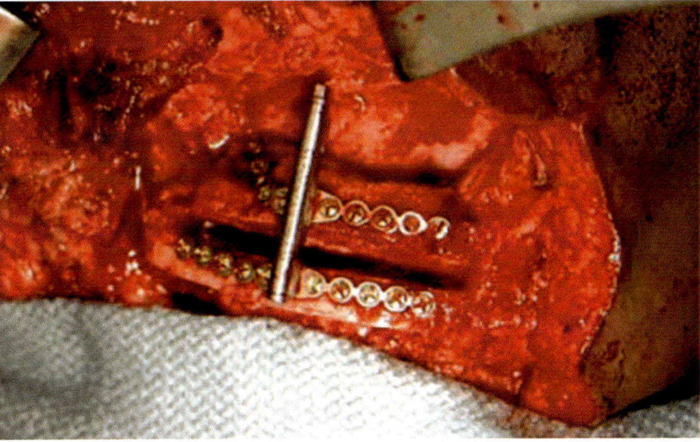

Fig. 20.21 Case 3: Intraoral photographs of postoperative results using rhBMP-2 to graft the bilateral alveolar clefts from 20.20. The graft extends down to the alveolar ridge [Abyholm stage. (Left): Right alveolar cleft graft site; (right): Left alveolar cleft graft site]. [Reprinted from Chin M, Ng T, Tom WK, Carstens MH. Repair of alveolar clefts with recombinant human none morphogenetic protein (rhBMP-2) in patients with clefts. J Craniofac Surg 2005 16(5):778–789. With permission from Wolters Kluwer Health, Inc.]

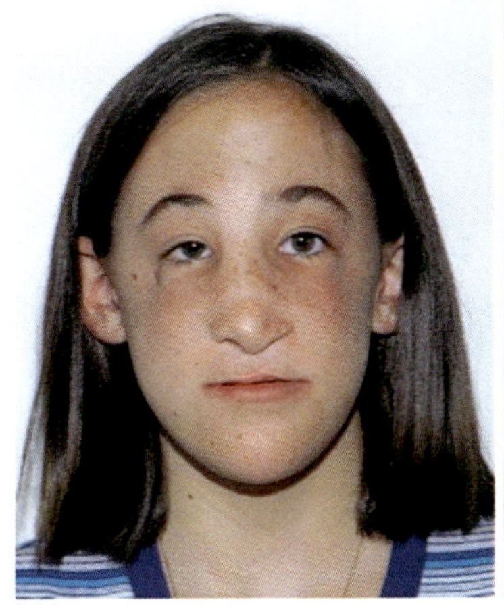

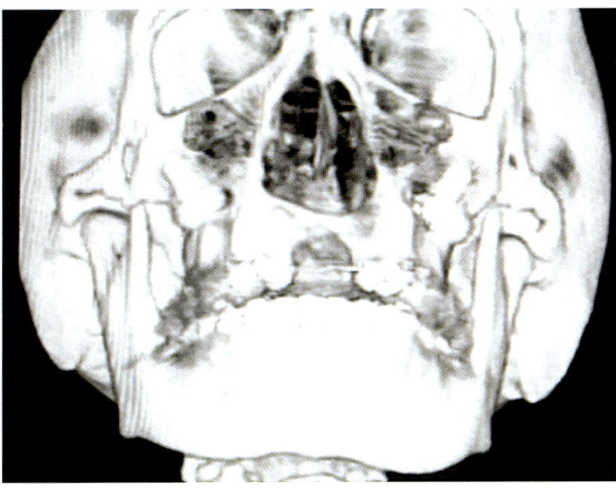

Fig. 20.22 Case 4: Late effects of rib grafting. *Left*: Midline maxillary cleft bridged by a previous rib graft seen. LeFort I advancement was performed. *Right*: CT scan shows reconstructed bony arch but with need for additional bone for stability of the alveolar arch. [Reprinted from Chin M, Ng T, Tom WK, Carstens MH. Repair of alveolar clefts with recombinant human none morphogenetic protein (rhBMP-2) in patients with clefts. J Craniofac Surg 2005 16(5):778–789. With permission from Wolters Kluwer Health, Inc.]

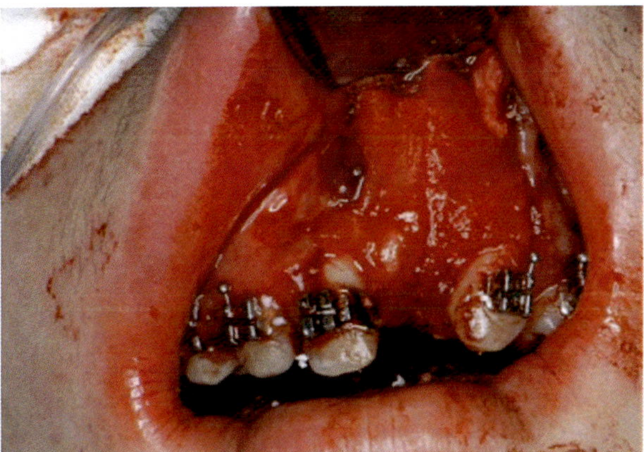

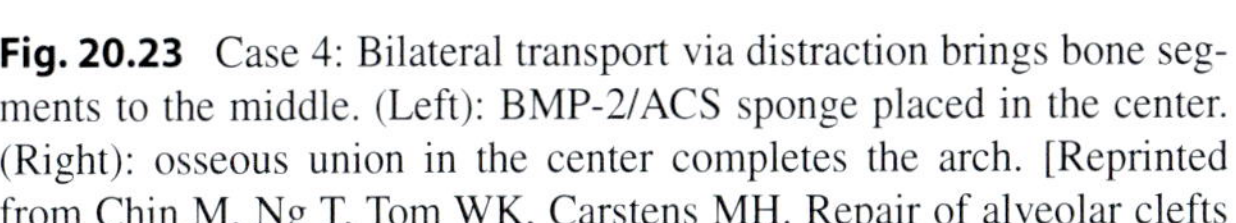

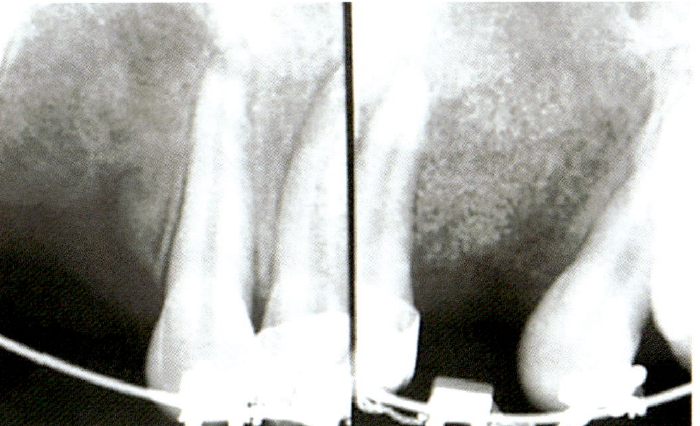

Fig. 20.23 Case 4: Bilateral transport via distraction brings bone segments to the middle. (Left): BMP-2/ACS sponge placed in the center. (Right): osseous union in the center completes the arch. [Reprinted from Chin M, Ng T, Tom WK, Carstens MH. Repair of alveolar clefts with recombinant human none morphogenetic protein (rhBMP-2) in patients with clefts. J Craniofac Surg 2005 16(5):778–789. With permission from Wolters Kluwer Health, Inc.]

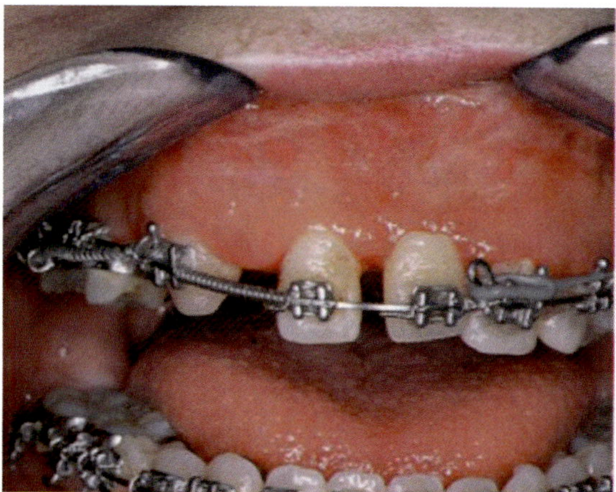
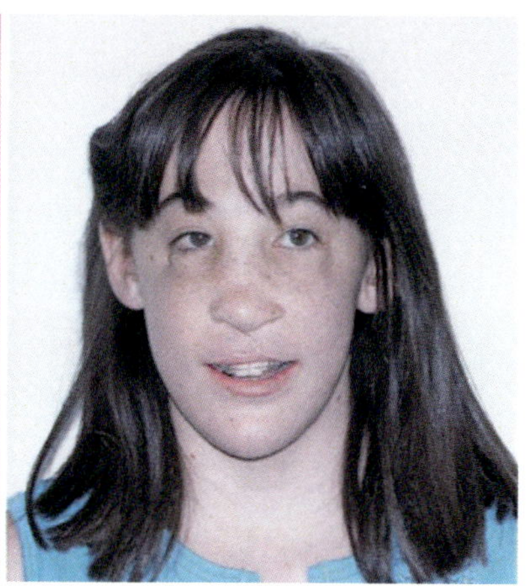

Fig. 20.24 Case 4: Bone formation by paramedian distraction with good fill at 18 months postop. Excellent postop symmetry and class 1 occlusion. [Reprinted from Chin M, Ng T, Tom WK, Carstens MH. Repair of alveolar clefts with recombinant human none morphogenetic protein (rhBMP-2) in patients with clefts. J Craniofac Surg 2005 16(5):778–789. With permission from Wolters Kluwer Health, Inc.]

tion, rhBMP-2 is introduced into the regeneration chamber to *convert the intentional non-union into a structural, anatomic bone field*. This case is exemplified by our treatment plan for FP, whose lateral facial cleft was previously described.

Results

Of the 50 alveolar and facial clefts repaired, 49 had successful consolidation demonstrated by radiographs and resolution of the oronasal fistula. One alveolar cleft site failed to consolidate with bone. In the patient who failed to ossify the rhBMP-2/ACS graft, it was felt that the extensive exposure of the central incisor root into the cleft site contributed to the poor response to the procedure. No systemic adverse effects of the rhBMP-2 were found in any patient. Follow-up radiographs showed no evidence of root resorption in any patient. Clinical and radiologic examination showed ossification across the alveolar cleft site only. No case exhibited bone formation outside of the site prepared for construction. Sites that were re-entered for additional surgery showed no sign of ectopic bone formation or extension of the generated bone beyond the contours of normal anatomy. There was no sign that the rhBMP-2 had an adverse effect on adjacent tissue.

> Note: In point of fact, morphogen-driven release of VEGF from target MSCs resulted in significant improvement in vascular supply, decreased inflammation, and faster wound healing.

Discussion

The child who went underwent repair of the midline cleft after horizontal distraction went on to full consolidation. Placement of the rhBMP-2/ACS device directly on the exposed root surface did not have an apparent adverse effect on the tooth. Teeth in the region responded to orthodontic movement into the constructed site. During repair of the conventional alveolar clefts, many of the teeth adjacent the cleft site had root exposure. These teeth were not removed unless there was no bone support at all. Many demonstrated significant mobility. To date, none of these teeth have shown resorption, and none have been lost.

One of the concerns about clinical use of this material has been the effect of displaced rhBMP-2 on adjacent structures. During the early studies on BMPs, Urist [99], recognized that these materials had the capability of forming bone at ectopic sites. Confining the material to a specific site is not possible with the current carrier system. When the wound is closed, pressure on the rhBMP-2/ACS device results in some displacement of the rhBMP-2 into the adjacent tissue. Despite obvious leakage of rhBMP-2 into adjacent soft tissue, no ectopic bone formation was observed. The child undergoing reconstruction of the lateral facial cleft with DO and rhBMP-2 had surgical exposure of the mandible by a submandibular incision. As part of the dissection, the facial artery and vein and branches of the facial nerve were exposed. During placement of the rhBMP-2/ACS device and closure of the soft wound, obvious displacement of graft material occurred onto the vessels and nerves. Contact with rhBMP-2 did not have any apparent adverse effect on these structures.

Conclusion

There is a clear, obvious advantage of rhBMP-2 use over iliac crest bone from the standpoint of postoperative recovery. Sparing children iliac harvesting is, in itself, a compelling reason to use rhBMP-2. The efficacy of the material in establishing an osseous union was clear. Failure of the graft in one patient in this series implies that successful healing is multifactorial. Although it is possible that the patient was able to respond to rhBMP-2, it is more likely causes were *failure of the mucosal closure* or *periodontal compromise of the exposed root*.

Because the surgical procedure of alveolar cleft repair using this material involves only oral mucosa flap manipulation, it is possible that some patients would tolerate the surgery *without general anesthesia*, using a complete trigeminal block at the pterygopalatine fossa with sedation only.

This feature may make alveolar repairs available in situations where access to general anesthesia is limited or not desirable. In many centers, there is a trend toward repairing craniofacial anomalies at younger ages. Because these deformities typically involve significant deficiencies of bone, there is often the need for substantial amounts of tissue. In young patients, 3–5 years of age, the iliac crest is often cartilaginous and offers *limited volume of bone* for harvesting without incurring major morbidity. Using rhBMP-2 may be the best method to achieve major bone volume augmentation in these patients. Because craniofacial children often require multiple, additional surgeries, sparing autogenous donor sites reserves more options for future treatment. Limited displacement of rhBMP-2 into adjacent tissue did not result in ectopic bone formation. Sensory and motor nerves coming into contact with rhBMP-2 did not exhibit altered function.

Future Directions

The surgical protocol used in this series was adopted from the clinical trials of this device. These studies involved use of the rhBMP-2/ACS in spine fusion, long bone non-union repair, and maxillary sinus elevation for dental implants. Material handling, concentration, and collagen sponge carrier use were taken from the spine fusion recommendations. Protocols to optimize efficacy in pediatric craniofacial grafting will require additional study. Dosing and use of the ACS as a carrier may require modification when used for alveolar grafting. Construction of the alveolus differs significantly from those procedures investigated in the clinical trials. Preparation of a site for placement of an alveolar graft involves rotation of mucosa flaps. The fate of the rhBMP-2 graft is partially dependent on the effective handling of the soft tissue closure and the patient's ability to heal the mucosa. If there is breakdown of the soft tissue closure, rhBMP-2 cannot be expected to result in bone formation. Impact on teeth adjacent to the surgical exposure, graft material, and developing bone is also important. All materials used previously for alveolar cleft grafting have been associated with occasional adverse effects on adjacent teeth. Autogenous bone, when placed against tooth roots, may result in root resorption or ankylosis. When used for alveolar cleft reconstruction, non-resorbable ceramics, such as Interpore 200 (Interpore International), may interfere with the eruption and orthodontic movement of teeth [100].

Management of roots exposed into the cleft site remains an open question. One option is to remove teeth adjacent to the cleft that have exposed roots to improve the probability of successful graft healing. The alternative is to attempt to repair the periodontal defect with rhBMP-2. In this short series, both successful and unsuccessful outcomes occurred when teeth were retained. The effect of rhBMP-2 on different types of periodontal defects is largely unknown. In this series, no adverse effects were found on teeth that came into direct contact with the rhBMP-2/ACS devices.

Primary Protocol

Introduction

In cleft surgery, primary alveolar cleft bone grafting has been desirable in concept but elusive in execution. Replacement of "like with like" makes eminent sense to control dental arch stability, as well as to provide the proper environment for tooth eruption.

Past attempts to achieve this have proven difficult because of conceptual blocks. (1) Cleft pathogenesis is poorly understood; the biologic basis of the bone biology dictates how it must be reconstructed. (2) Misunderstanding of periosteal anatomy and function underlies dissection strategies that violate principles of maxillary embryogenesis. The fundamental problem in cleft surgery is a developmental field defect involving the premaxilla, PMx. Craniofacial development can be thought of as a process in which distinct biologic units (fields) arising from disparate sites of the embryo are assembled like Lego® pieces in a strict spatiotemporal order to produce the final anatomic design. Deficiency or absence of any Lego piece will affect the final structure, particularly as surrounding normal fields grow. Such a system is inherently unstable. Surgical interventions that fail to reconstruct the missing piece will, over time, tend toward relapse.

Cleft Pathogenesis

The premaxilla arises from neural crest-derived stem cells that produce mesenchyme. After the bone forms, MSCs, in the craniofacial skeleton (excluding the skull base), reside in the cambium layer of periosteum and in the dura. Premaxilla is innervated exclusively by cranial nerve V2, the nucleus of which reside in the second rhombomere. Therefore, PMx neural crest cells arises from the neural folds in register with r2. PMx is supplied by the terminal branch of medial sphenopalatine artery. It has 3 subunits: a medial incisor zone (PMxA), a lateral incisor zone (PMxB), and an ascending frontal process (PMxF) extending vertically upward from PMxB to articulate with the frontal bone. The piriform fossa is a bilaminar structure, with PMxF lying internal to the frontal process of the maxilla. Note that MxF has its own blood supply from a medial branch of the anterior superior alveolar axis. PMxF and MxF form a bicortical buttress that proves useful in facial fracture management.

The components of each developmental field are laid down in a specific sequence. Eruption of the central incisor precedes that of the lateral incisor. Therefore, PMxA is biologically "older" than PMxB. PMxF is the "newest" subfield. Cleft lip pathology is a distal-to-proximal failure of the vascular axis (see Chap. 18). This explains why a full-blown cleft nasal deformity can coexist in the presence of a minimal cleft or normal lip. Increasing severity of cleft affects more proximal zones of the premaxilla. Deficiency in PMxB results in a gradient of alveolar cleft pathology from a simple notch to loss of the lateral incisor. Hits to the oldest zone, PMxA, are rare but can result in a diminutive premaxilla.

The space in the dental arch created by an alveolar cleft represents a mesenchymal deficit, the dimensions of which must be *conserved* and *reconstructed*, ***not collapsed.***

- In virtually all alveolar clefts, the skin and soft tissues of the nasal floor lying above PMxB are not missing; they can be found instead in the prolabium. Thus, in the unilateral cleft, the prolabium has 3 soft tissue zones.
- The 2 zones of the *philtral prolabium* (PP) contain mesenchyme supplied by paired anterior ethmoid arteries and innervated by V1. The width of philtral prolabium is exactly that of the columella.
- The remaining tissue, the NPP, contains mesenchyme supplied by the medial sphenopalatine artery and innervated by V2.
- Dissection of the prolabium along this neuroembryologic "fault line" is a key component of DFR cleft surgery.
- When NPP is elevated in continuity with the lateral mucoperiosteum of the pPMx and the ipsilateral vomer, the nasal floor can be hermetically sealed. This sets the stage for a complete enclosure of the alveolar cleft using stem cell-bearing periosteal flaps.

Periosteum as a Bone-Forming Tissue

The second conceptual block concerns the biologic relationship between soft tissues and the periosteum that belongs to them and the underlying membranous bone product. The DFR sequence involves the reassignment of these tissues into a proper centric location and morphogen-based activation, either using native iliac crest bone graft or rhBMP-2 bone graft.

Developmental field reassignment is performed at 6 months of age followed by alveolar extension palatoplasty at 18 months [84]. DFR closes the alveolar AEP on the *nasal aspect*. It fuses a pericoronal incision along the lingual dentition to create long (extended) mucoperiosteal flaps. This design avoids an anatomically false subdivision of the greater palatine angiosome. It also permits closure of the alveolar cleft from the *lingual aspect*. The combination of DFR and alveolar extension palatoplasty results in a 5-sided mucoperiosteal box. The addition of an extended mucoperiosteal S flap covers the *labiobuccal aspect*, the sixth side of the box.

In this model 5 of the 6 six sides of the box contain MSCs with receptors for BMP embedded in their membranes (Carstens and Chin 2005a, b). Tissue coverage of the 6th side of the box consists of NPP tissue. **NB:** *If the NPP flap is harvested with a periosteal window from the premaxilla, it too will have osteogenic MSCs*. The stage is now set for morphogen-mediated ISO, which achieves a definitive reconstruction of the missing Lego piece, PMxB. Of all the advantages of ISO over conventional graft procedures, perhaps the most significant is the morphogen-based induction of local blood supply. This is an important factor in the prevention of graft resorption.

Overall success rate of alveolar grafting varies between 88 and 95%, but the amount of bone produced is variable. Loss of volume in alveolar cleft reconstructions using iliac crest spongiosa was stratified by Abyholm stratified into four-point scale: grad I = 0–25% bone loss, grade II = 25–50% bone loss, grade III = 50–75% bone loss, and grade IV = 75–100% bone loss [101] (Fig. 20.18). Radiologic evaluation demonstrates a typical pattern with grade I noted in 69%, grade II in 19%, grade III in 10%, and grade IV in 1% of cases. In the Schultze-Mosgau study success was defined as having resorption rates <50% of alveolar height.

Variables impacting success are (1) the state of canine eruption and (2) the width of the cleft. For this reason, many treatment algorithms seek to un-roof unerupted canines in order to guide them orthodontically and/or to close down the gap of the cleft. In an idea world neither are optimal as the goal should be to maintain normal dimension to have a full-mouth smile. The issue at hand really has to do with a misunderstanding of the nature of the graft. Cancellous bone graft interacts with clot at the cleft site to primarily provide morphogens (in small quantities) to the MSCs overlying the site. Based on what we know, these MSCs migrate into the cleft site, populate the lattice, multiply, and differentiate. In conventional bone grafting procedures these biologic processes are uncontrolled. We cannot ensure an adequate dose of morphogens to the periosteal cover. Neither can we control vascularization. We are at the mercy of Ingrowth of blood vessels to supply the cells of the graft and thereby maintain its volume. Finally, conventional grafting does not affect the number of embryologically appropriate MSCs available for repopulation of the graft.

The twin lynchpins of biological alveolar cleft repair are the provision of an adequate amount of morphogen and a healthy periosteal envelope. As we have seen, rhBMP-2 is perfectly capable of attracting large numbers of MSCs into the scaffold, be it of collagen or demineralized bone. Thus, rhBMP-2 shows that the actual need for ICBG is nil.

We have from time immemorial be laboring under the false assumption that the corticocancellous material was indeed a "graft," providing cells that would become bone. In actual fact, it merely served as a source of morphogens and acted as a mechanical lattice for repopulation.

Recall that maxillary periosteal MSCs are of neural crest origin, whereas bone marrow MSCs are derived from paraxial mesoderm. Cleft patients with ectomesenchymal grafts from mandibular symphysis demonstrated better incorporation with less resorption than those received conventional mesenchymal grafts from iliac crest [102]. Thus, it is quite possible that ICBG simply does not reliably produce enough morphogens to support the migration of neural crest MSCs, whereas corticocancellous bone graft from r3 mandible is a better biologic "match" to stimulate the MSCs in the overlying soft tissue envelope.

We therefore postulated that rhBMP-2 could induce membranous bone formation of quality and durability that was equal or superior to that achieved via conventional means. The present study's aim was to provide radiographic evidence of de novo synthesis of bone using rhBMP-2 in the surgically designed DFR "box" [103] (Figs. 20.25, 20.26, 20.27 and 20.28).

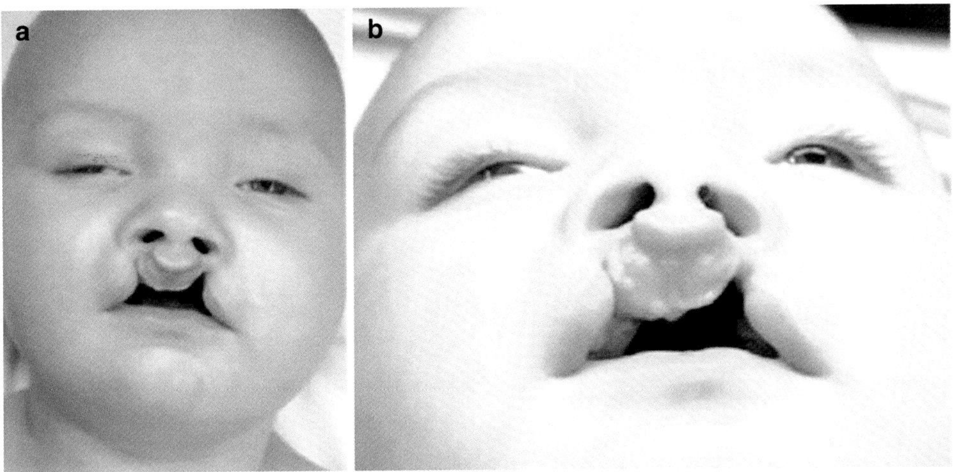

Fig. 20.25 Case 5: Complete bilateral cleft palate with bilateral incomplete cleft lip (Simonart's bands). (**a**) Frontal view demonstrates "short" prolabium. Note the Simonart's bands spanning from the premaxilla to the lateral lip elements. (**b**) Intraoral view (**b**) shows alveolar bone clefts separating the premaxilla. These bands act to retain the premaxilla in place. [Reprinted from Falluco M, Carstens MH. Primary reconstruction of alveolar clefts using recombinant human bone morphogenetic protein-2: clinical and radiographic outcomes. J Craniofac Surg 2009; Suppl 2:1759–1764. With permission from Wolters Kluwer Health, Inc.]

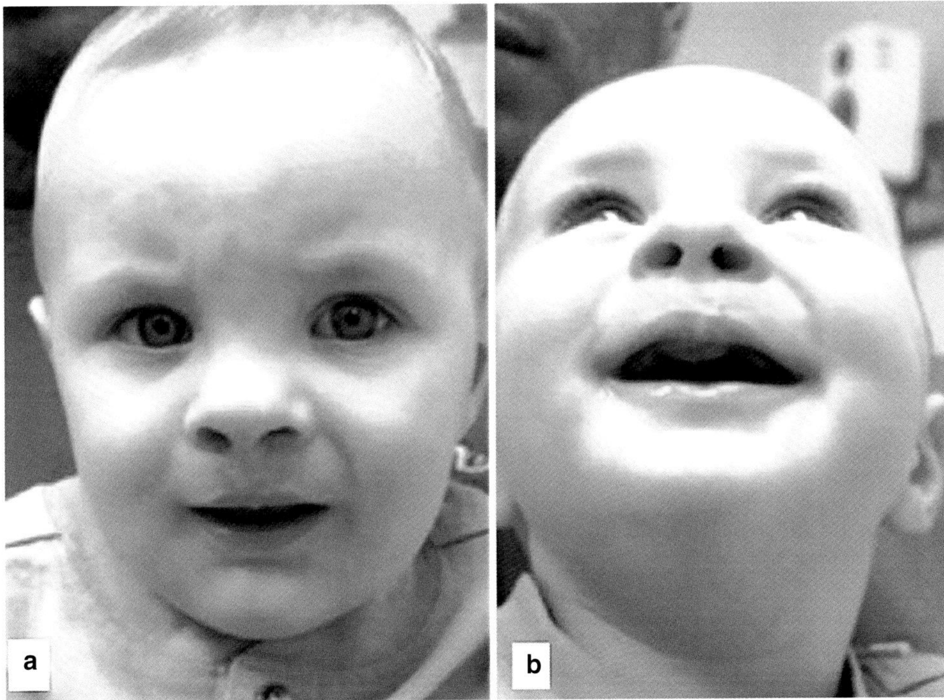

Fig. 20.26 Case 5: Postop facial symmetry nasal projection and patent airways. Postop showing facial symmetry nasal projection and patent airways. (**a**) Vertical height of the prolabium is adequate. (**b**) Airways patent and nasal tip not depressed. [Reprinted from Falluco M, Carstens MH. Primary reconstruction of alveolar clefts using recombinant human bone morphogenetic protein-2: clinical and radiographic outcomes. J Craniofac Surg 2009; Suppl 2:1759–1764. With permission from Wolters Kluwer Health, Inc.]

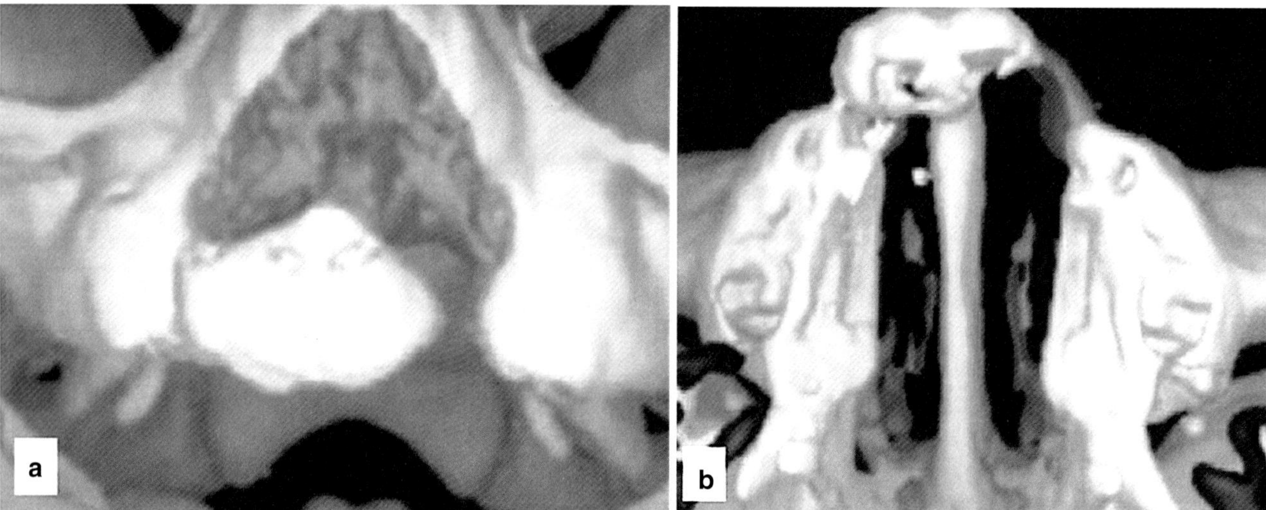

Fig. 20.27 Case 5: Pre-op before placement of rhBMP-2. (**a**) Right side is incomplete and left side is complete. (**b**) Force vectors are disrupted on the left side of the cleft permitting right-sided rotation of the premaxilla. [Reprinted from Falluco M, Carstens MH. Primary reconstruction of alveolar clefts using recombinant human bone morphogenetic protein-2: clinical and radiographic outcomes. J Craniofac Surg 2009; Suppl 2:1759–1764. With permission from Wolters Kluwer Health, Inc.]

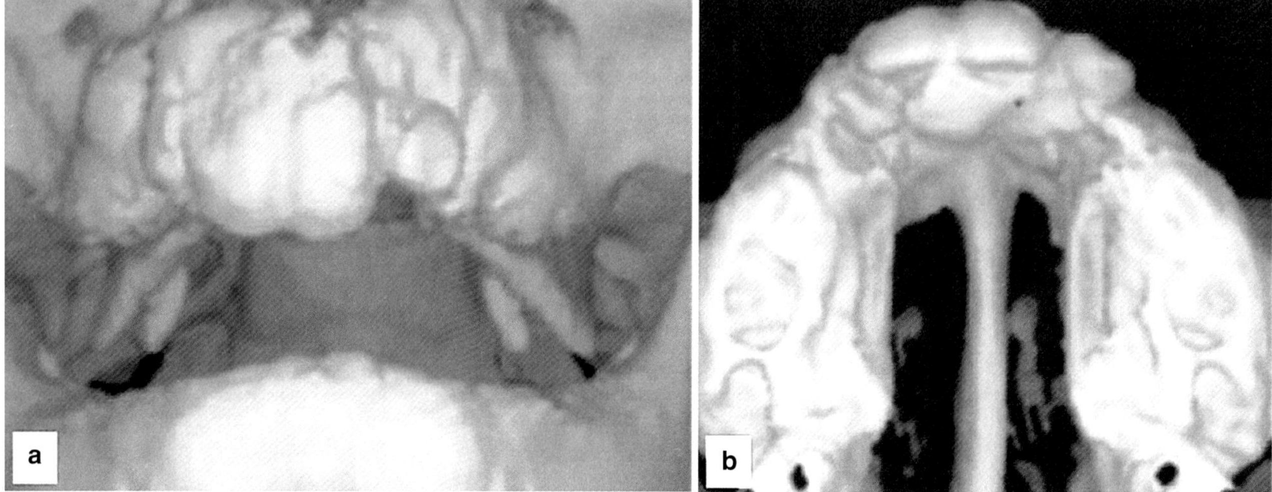

Fig. 20.28 Case 5: Postop bone formation with rhBMP-2 fills the alveolar clefts. (**a**) Vertical height of the cleft filled with rhBMP-2. (**b**) Volume of the arch restored on either side of the premaxilla. [Reprinted from Falluco M, Carstens MH. Primary reconstruction of alveolar clefts using recombinant human bone morphogenetic protein-2: clinical and radiographic outcomes. J Craniofac Surg 2009; Suppl 2:1759–1764. With permission from Wolters Kluwer Health, Inc.]

Materials and Methods

An institutional review board-approved class IV study retrospectively evaluated primary alveolar cleft patients from 2004 to 2006. Subjects chose an off-label application of rhBMP-2 impregnated on an ACS carrier to reconstruct alveolar clefts, all greater than 3 mm in width. There were 22 males and 11 females who underwent primary cleft reconstruction involving the lip, nose, and anterior palate. From these patients, the resulting 42 alveolar cleft sites were distributed as follows: 16 unilateral right sided, 21 unilateral left sided, and 5 bilateral. The rhBMP-2/ACS implant used was a small (2.6 mL) kit for unilateral and a medium (5.8 mL) kit for bilateral repairs. In 2 cleft sites, acellular dermal matrix (Alloderm; LifeCell Corporation, Branchburg, NJ) was used to reinforce the soft tissue repair; it was not part of the implant itself.

Patient evaluation through the study's protocol included a preoperative and postoperative CT scan. An initial group of patients (25 cleft sites) were assessed at 3 months postoperative; rapid and effective alveolar bone formation was observed. Because of the theoretical possibility of prolonged rhBMP-2 activity, the timing of postoperative CT was changed to a mean of 6 months to assess the bone at its greatest possible density. This second group (17 cleft sites) is the basis of the current study [103].

To assess radiographic outcomes, low-dose spiral CT with 1-mm cuts limited to the maxilla was used. Two radiologists independently calculated the Hounsfield units (HUs) in the region of interest (ROI) from the CT scans (Table 20.1). Both the 1-mm-thin cut CT scans and the same area for ROI, 11 mm^2, were used to minimize disparities in volume averaging of the HU calculation. In cases where there was a difference in HU for the ROI that showed bone versus no bone formation, a third radiologist, blinded to the other 2 radiologists' results, was consulted to mediate. Dental arch fusion was defined as continuous filling of the previous air- or mesenchymal-filled alveolar cleft with radiographic HU density in the ROI greater than 226.

Surgical Technique

Careful preparation of the alveolar cleft site is required for successful implantation of rhBMP-2. First, a soft tissue pocket surrounding the recipient osseous defect is created, and hemostasis is ensured for a dry implant site. Mucoperiosteal flaps containing stem cells are elevated. Careful elevation from bone surrounding the tooth buds in the cleft is essential. Key to the design is the embryologically driven dissection of the lip through DFR, details of which have been previously described [1]. Back-table preparation of rhBMP-2 is completed as described by the package insert.

Table 20.1 Clinical characteristics of patients with chronic limb ischemia

Case	Age/sex	Dx	Status	Category	Wound
1†	57F	DM	Rest pain	6	5 cm × 4 cm
2†	64F	AS	Claudication	3	
3	71F	AS	Claudication	6	3 cm × 3 cm
4	72F	AS	Rest pain	4	
5	75M	DM	Rest pain/WC	6	6 cm × 4 cm
6	76F	DM	Claudication	6	4.5 cm × 3 cm
7	77F	DM/AS	Claudication	3	
8	78F	DM/AS	Rest pain/WC	6	7 cm × 5 cm
9	78F	DM/AS	Claudication	3	3 cm × 2.5 cm
10†	85F	AS/RF	Rest pain	6	3 cm × 2.5 cm

Chemical binding of the protein to the collagen sponge requires a minimum of 15-min contact time. The implant is then placed into the surgical site, and the MSC-lined mucoperiosteal box is closed.

Defect size determines the implant requirement. If the defect were flat, such as in the membranous skull, surface area alone would dictate choice of kit size. This, in turn, depends on the surface area of the ACS. Currently, rhBMP-2 comes in 3 sizes: 25-cm^2 (2.6 mL), 50-cm^2 (5.8 mL), and 75 cm^2 (8.4 mL) defects. It should be noted that the volume of the sponge is similar to that of Integra without the silicone backing. Therefore, to achieve significant fill in the third dimension, a bulking agent is sometimes required. In primary clefts, a 2.6-mL implant is sufficient for unilateral defect; a bilateral defect requires a 5.8-mL implant.

Results

Radiologic Outcomes

The 17 cleft sites followed radiographically by the 3 radiologists had baseline preoperative alveolar cleft mean tissue density measurements of −38, −j83, and −129 HU. The negative values demonstrated in the preoperative alveolar cleft mean HU values reflect the standard HU value for air of **−1000 HU**, Table 20.1. The maximum HUs in the ROI for all 17 alveolar clefts were all below the threshold for trabecular bone density.

Clinical Outcomes

One of the 17 cleft sites repaired with rhBMP-2/ACS failed to meet radiographic criteria of dental arch fusion; however, the aesthetic result was pleasing, and possibility for subsequent grafting exists. Failure in this particular case was attributed to excessive irrigation of the implant site before final closure, leading to a dilution effect of protein from the sponge.

Discussion

A 35-year review of early autogenous bone grafting reported by Rosenstein et al. in 2003 constitutes a direct challenge to accepted dogma on the subject dating back to Skoog. The successful outcomes demonstrated in this report were the direct result of the technique used by Kernahan, which deliberately avoided the type of supraperiosteal mobilization in vogue at that time. This important study demonstrates that there are no intrinsic deleterious effects of bone grafting on midfacial growth. Design of the surgical pocket is all important.

Gingivoperiosteoplasty has been carefully followed by Grayson and Cutting, with 60–73% success rate (no secondary alveolar bone grafting) and 0% fistula rate.

An important drawback of the NYU technique was its requirement for a 1-mm approximation of the segments. In preparation for GPP, the technique of closing down the cleft site is a non-embryologic strategy because *the end result cannot produce a volume of bone equivalent to the original defect*. Furthermore, in 27–40% of cases, GPP is unsuccessful. Despite these drawbacks, the GPP experience was an important scientific advance because it emphasizes the innate ability of MSCs within a periosteal environment to produce bone.

Recreation of the embryonic defect would require full restoration of the lateral incisor zone; *dental arch dimensions should be conserved*. If the spatial gap exceeds the ability of MSCs to migrate and fill in the defect, failure of GPP is guaranteed. On the other hand, if cleft closure is undertaken using an osteoinductive cytokine plus an appropriate resorbable matrix for cell migration, volumetric reconstitution of the defect can be achieved. Comparing osteoinduction (rhBMP-2/collagen sponge) versus traditional osteoconductive (iliac crest) bone grafting, a UCLA study demonstrated increased bone volume (95%) vs. 63% volume via osteoinduction in skeletally mature patients [104].

We propose that rhBMP-2 combined with the initial lip repair decreases the amount of secondary surgeries, stabilizes the dental arch and anterior nasal floor, and avoids donor site morbidity. ISO using rhBMP-2 generates bone in virtually unlimited quantities that is histologically similar and mechanically superior to native bone without donor site morbidity (P. J. Boyne, personal communication, June 2007).

This series demonstrates a marked increase in radiographic tissue density within the rhBMP-2/ACS-implanted alveolar clefts consistent with the de novo synthesis of bone (Figs. 20.2 and 20.3). Optimal volumetric outcome of ISO in cleft depends upon the complete reconstruction of the original embryonic defect. DFR technique reestablishes anatomic boundaries and reimports stem cells into the cleft environment. Current work with DFR at our institution indicates that primary grafting at palatoplasty (18 months) (1) is technically easier, (2) produces larger volumes of graft, and (3) is more anatomically precise.

Non-cleft Craniofacial Reconstructive Applications

Hemimandibular Reconstruction: Sequential ISO and DISO

Juvenile active ossifying fibroma (JAOF) is a benign fibro-osseous lesion with a tendency to occur in children and adolescents. Its clinical behavior is notable for locally aggressive growth within the craniofacial bones and a significant recurrence rate, varying between 30 and 58%. Although this is a rare lesion, it can result in significant facial distortion in a young patient. Controversy exists regarding the treatment of JAOF. There are proponents for both enucleation and complete resection. When a tumor reaches a size that mandates resection of the involved bone, the focus should be on achieving the best functional and aesthetic reconstruction for the child. The authors present a case of recurrent, locally aggressive tumor requiring hemimandibulectomy and a new technique for mandible reconstruction involving ISO with recombinant bone morphogenetic protein (rhBMP-2) (Figs. 20.27, 20.28, 20.29, 20.30 and 20.31).

Clinical Report

In January 2002, a 5-year-old boy presented to an outside hospital with a painless, rapidly growing right mandibular mass. X-ray examination revealed a radiolucent expansile lesion arising within the body of the right mandible with marked cortical thinning. Intraoperatively, the intracortical lesion appeared spongy, pale, and did not involve tooth elements. It was removed in piecemeal fashion by curettage; the operating surgeon thought the lesion had been extirpated in its entirety. The pathologic diagnosis of this 4.5 × 3 × 0.8 cm lesion was *juvenile psammomatoid ossifying fibroma*. The child was lost to postoperative follow-up because of social circumstances.

In September 2003, the patient was referred to our craniofacial clinic with recurrence of his right mandibular mass. The mass involved the body of right mandible, was firm and painless to palpation, and had extraoral distortion of his face. There was no palpable lymphadenopathy. CT showed a 2.8 × 2.5 × 3.75 cm rounded expansile lesion involving the mandibular body extending to the angle. The lesion surrounded adjacent teeth and demonstrated internal heterogeneity with a central area of hypodensity. A thin rim of bone surrounded the lesion. The child was not seen again until September 2004. Follow-up CT showed interval expansion of this lesion to 4.7 × 4.7 × 3.75 cm; it now involved the body, angle, and ramus.

Because of the size of this lesion and its involvement of the right hemimandible, an incisional biopsy of the lesion submitted for review to the Armed Forces Institute of Pathology (AFIP) showed a highly cellular stroma composed of spindle shaped cells and interspace depositions of non-calcified osteoid rimmed by polygonal osteoblasts. A diagnosis of *JAOF* was made, with the recommendation for complete resection, with margins (Figs. 20.29, 20.30, 20.31, 20.32 and 20.33).

Given the size of the lesion and the recurrent nature of JAOF, the required segmental resection would generate a defect extending from the right angle to the contralateral canine.

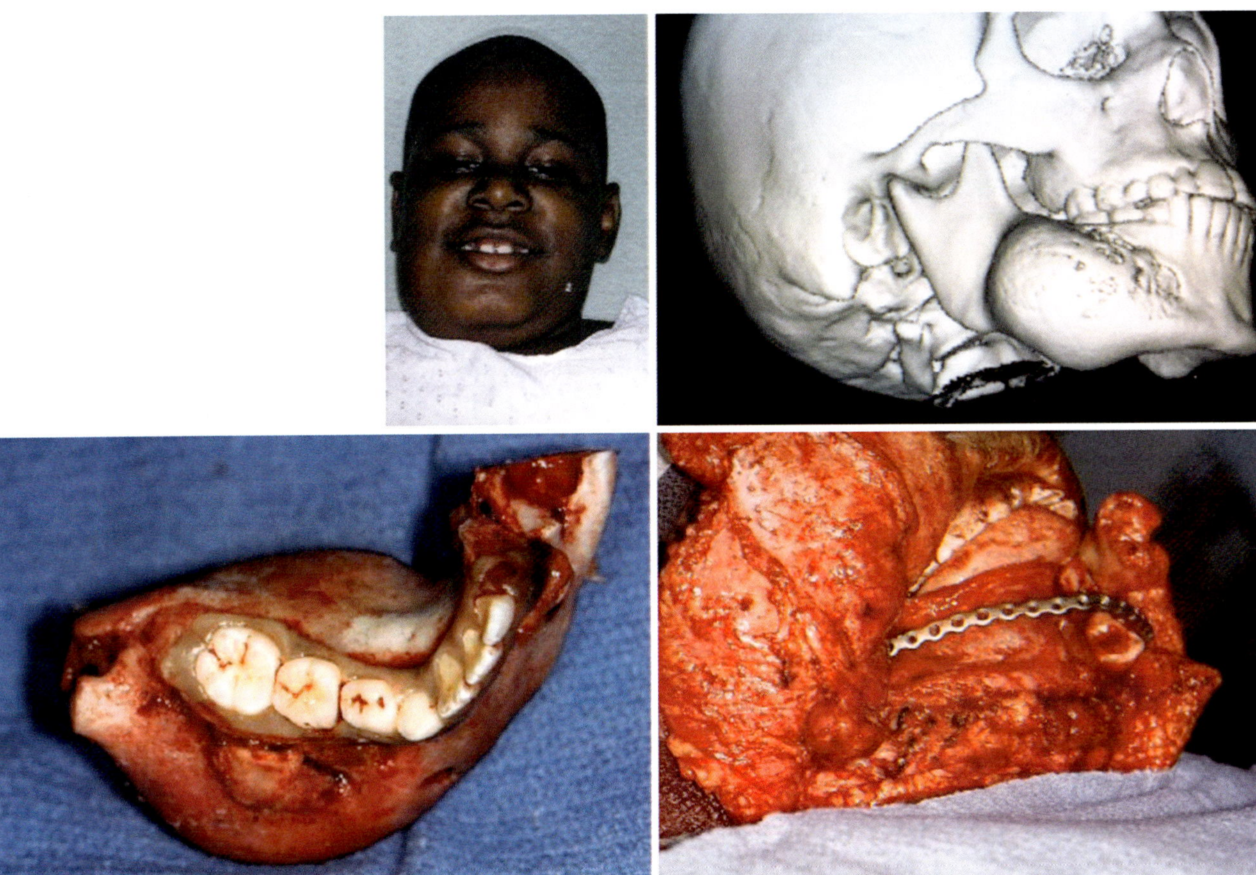

Fig. 20.29 Case 6: A 7-year-old boy with large juvenile psammomatoid ossifying fibroma. Model of the CT scan was very accurate in predicting the spread of the lesion. Reconstruction plate keeps the occlusion and periosteal flaps are elevated to receive the rhBMP-2/ACS graft. [Reprinted from Chao M, Donovan T, Sotelo C, Carstens MH. In situ osteogenesis of hemimandible with rhbmp-2 in a 9-year-old boy: osteoinduction via stem cell concentration. J Craniofac Surg 2006; 17(3):405–412. With permission from Wolters Kluwer Health, Inc.]

Because a subperiosteal plane would be used for surgical resection, neural crest-derived stem cells residing within the cambium layer of the periosteum would be left undisturbed. Because these cells bear membrane-bound BMP receptors implantation of rhBMP-2 within the resulting periosteal chamber would induce migration of these stem cells into the implant, with subsequent proliferation and differentiation into osteoblasts. Creation of bone within a defined space using an osteoinductive agent, such as rhBMP-2, is known as ISO.

Preoperative planning involved design of a reconstruction plate to match the anticipated defect. This was done using a three-dimensional (3-D) skull (Medical Modeling, Golden, CO). To our surprise, *the model disclosed the extent of tumor invasion within the substance of the mandible itself with greater precision than did the original CT scan*. This enabled us to plan a resection sparing the contralateral incisor. A mirror-image right hemimandible based on the geometry of normal left side was used to pre-shape a titanium reconstruction plate (KLS Martin, Jacksonville, FL).

Surgical Treatment

At surgery, hemimandibulectomy was performed from the right ramus to the left incisor foramen via a Risdon incision. Subperiosteal dissection permitted preservation of surrounding soft tissue elements. The resultant defect measured 12 cm. The mandibular segments were stabilized with the plate using 2-mm screws. The two leaves of elevated periosteum were sutured together to form an envelope in the shape of the original defect, the plate remaining external to the ISO chamber. Two large kits of Infuse® rhBMP-2 (Medtronic Sofamor Danek, Memphis, TN) were used, each one containing 8.4 mg of rhBMP-2 in powder form. Each vial of powder was reconstituted as per protocol and applied to a Helistat collagen sponge (Integra Life Sciences, Plainsboro, NJ). After 40 min of binding time (minimum binding time is 15 min), the sponges were wrapped around granules of 85% tricalcium phosphate and 15% hydroxyapatite (Mastergraft®, Medtronic Sofamor Danek, Memphis, TN) which were included to bulk up the ISO chamber. Positioning of the

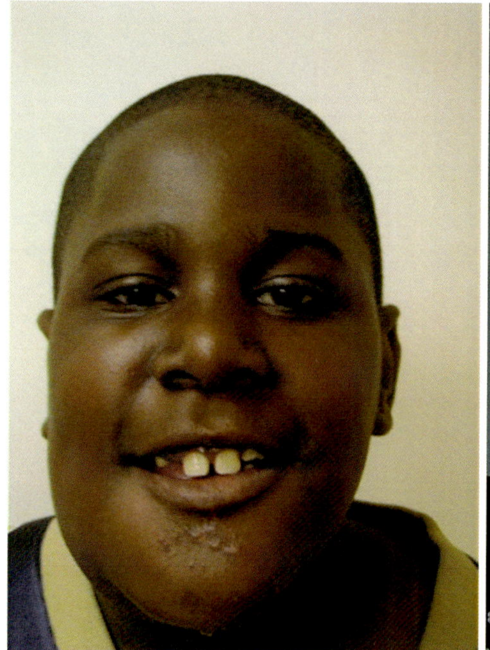

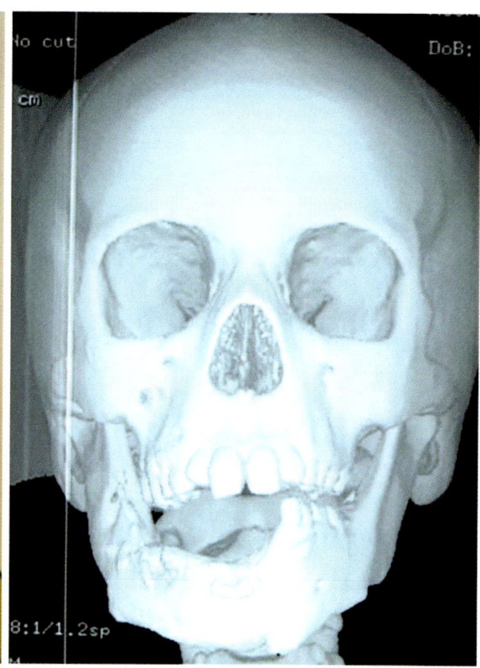

Fig. 20.32 Case 6: Good facial symmetry achieved. Alveolar bone height sufficient to support dental implants. [Reprinted from Chao M, Donovan T, Sotelo C, Carstens MH. In situ osteogenesis of hemimandible with rhbmp-2 in a 9-year-old boy: osteoinduction via stem cell concentration. J Craniofac Surg 2006; 17(3):405–412. With permission from Wolters Kluwer Health, Inc.]

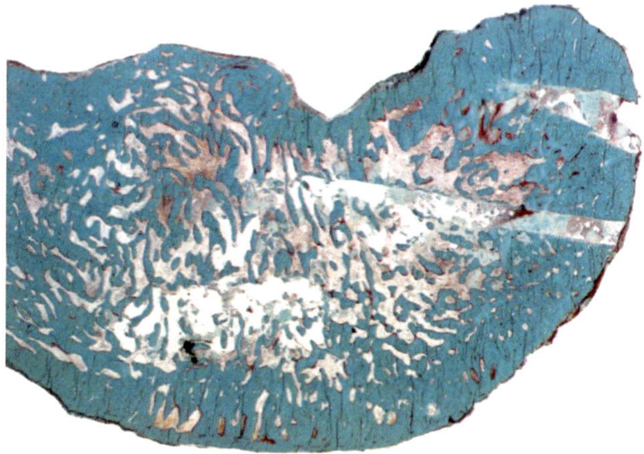

Fig. 20.33 Histology of ISO bone. Islands of woven bone entrapped within trabecular bone are characteristic of rapid osteoinduction via rhBMP-2. [Reprinted from Chao M, Donovan T, Sotelo C, Carstens MH. In situ osteogenesis of hemimandible with rhbmp-2 in a 9-year-old boy: osteoinduction via stem cell concentration. J Craniofac Surg 2006; 17(3):405–412. With permission from Wolters Kluwer Health, Inc.]

posed of a cell-rich fibrous tissue containing bands of cellular osteoid without osteoblasts together with trabeculae of more typical woven bone [105–109]. Small foci of giant cells may be present, and in some parts, there may be abundant osteoclasts related to the woven bone. Usually no fibrous capsule can be demonstrated, but the lesion is well demarcated from the surrounding bone. These lesions demonstrate variations in their histopathology. Most medical literature reports recognize the existence of a psammomatoid variant (juvenile psammomatoid ossifying fibroma [JPOF]) with similar clinical course [110]. JPOF is sometimes referred to as cemento-ossifying. The histology of the psammomatoid variant generally consists of highly cellular fibrous stroma with closely packed spherical ossicles resembling psammoma bodies [111, 112]. To date, there continues to be diagnostic uncertainty between these variants, as demonstrated in our case in which the outside institution surgical pathologist diagnosed the lesion as JPOF, whereas the AFIP diagnosis was JAOF.

Management

Similar controversy exists regarding surgical treatment of JAOF [113, 114]. Because this tumor is benign, many reports exist in the literature recommending curettage or limited resection. However, treatment of JAOF by these methods has a high failure rate, especially in children. Johnson found that 90% of all recurrences occurred in patients younger than 10 years. Therefore, we agree with the AFIP regarding complete excision of lesions for the following indications: (1) children younger than 10 years, (2) history of recurrence; and (3) large tumors that affect the stability of the mandible. The patient in question demonstrated all three of these indications.

Current mandible reconstructive options include alloplastic materials, non-vascular bone grafts, and osseous free flaps. The advantages of autologous bone far outweigh the technical issues raised with its harvest. However, these methods are not without their intrinsic drawbacks. Free bone grafts placed into a mandibular defect function primarily as biologic struts. They are eventually replaced by ingrowth of autologous bone from the environment. This process is known as *osteoconduction*. Take of such grafts is entirely dependent upon the vascular characteristics of the recipient site. The process of osteoconduction involves a finite amount of cellular death. The ability of the environment to respond is also variable, depending upon age and comorbid factors, such as disease, smoking, and radiation.

Microvascular transfer of the fibula by experienced teams has been highly successful in addressing these issues [67]. Children heal significant mandibular defects using free graft with remarkable ease. Unfortunately, harvest of large grafts from the iliac crest is painful. Successful use of fibular free flaps in children is well known. The donor site defect is significant, leaving the lower extremity *destabilized for active participation in sports*. Furthermore, fibula is a poor substitute for the mandible. It is tubular bone develops from LPM, not neural crest. It lacks the volume and shape to adequately serve as bone stock for dental implants. **Note**: We shall subsequently discuss will a case of *fibular augmentation* using ISO.

The ideal replacement for a mandibular defect would be re-synthesis of the mandible itself. Stem cells native to the environment constitute a readily available source of mesenchyme for ISO to occur. Neural crest stem cells respond to a BMP stimulus. In point of fact, the specific signal directing cells along to neural ridge to become neural crest cells in the first place is BMP-4 from the outlying ectoderm. Neural crest cells respond in a dose-dependent manner. *Low-dose BMP-2 induces cartilage formation, whereas high-dose BMP-2 will produce bone*.

Reconstruction of continuity defects using ISO has theoretical and practical advantages. (1) Correct application of rhBMP-2 to the collagen carrier produces a concentration/unit volume that is 200,000 × physiologic. This resulting chemoattraction brings in neural crest cells from as far away as 2 inches. ISO is thus a form of *stem cell concentration*. (2) ISO is a means to force the cellular environment surrounding the defect to reproduce membranous bone nati*ve to that environment*. Such bone is more likely to respond appropriately to secondary surgical manipulation, such as DO, than is chondral bone, such rib or fibula. (3) The Infuse® implant is acellular. ISO occurs *independent of blood supply* because it attracts living cells into the implantation site and stimulates vascular ingrowth from the environment. (4) Operative time is reduced. (5) Donor site morbidity is avoided. (6) ISO is *modular*. Additional amounts of rhBMP-2 can be added to a site for augmentation purposes with relative ease.

The case presented here illustrates a number of technical points regarding ISO. The presence of periosteum does not determine the ultimate shape of the regenerated bone. This comes about from all the associated attachments on the external surface of the periosteal sleeve. However, in the presence of a plate, the sleeve needs to be suspended in space from the plate to prevent sagging of the regenerate. The volume of bone generated depends upon a precise filling of the defect. Bone will not be produced to any great degree beyond the confines of the sponge. The need to add additional Infuse at the secondary surgery resulted from inadequate filling at the original surgery. Vascularization as stimulated by rhBMP-2 may occur with great speed.

Functional restoration of the mandible is best carried out using a membranous bone stock resembling, as closely as possible, the native mandibular bone. The case in point supports the potential application of rhBMP-2 as a form of mandibular ridge augmentation. In situations in which ISO is performed first, the timing of implant placement probably should follow the same rules as that for distraction. Cases involving an intact edentulous mandible may lend themselves to simultaneous implantation and augmentation, but these applications will follow from additional experimental and clinical work.

What Does the Implant Do?

Implants, such as demineralized bone (DMB) and collagen sponge (Helisat®), serve several functions. First, they occupy volume within a surgically created space, thereby allowing bone formation to be distributed uniformly within the chamber. Second, they prevent leakage of BMP-2. Morphogens, being soluble, cannot simply be injected into tissues. By adsorbing BMP-2 the implant localizes the signal source. Third, the implant itself doesn't cause stem cells to congregate; it merely provides a scaffold within which the stem cells will become activated and convert to osteoblasts. As such, the implant must be biocompatible with the developing bone so that it is eventually absorbed into the bone matrix.

What If the Periosteum Is Gone?

Cases in which periosteal loss has been incurred via trauma, resection, or in congenital absence (agnathia) constitute a challenge as the local source of MSNCs is non-existent. The Infuse/Mastergraft "fajita" can be wrapped using a free periosteal graft from the skull with reasonable expectation of success. Some situations could be addressed by importation of vascularized periosteum using a superficial temporal artery-based flap.

- **Note**: This particular scenario would eventually lead to the concept of stem cell grafting using MSCs in the form of SVF transplanted into the desired site.

Dose–Response of rhBMP-2

The dose–response curve of rhBMP-2 is species specific. Humans and high primates require 1.5 mg/cc. ISO continues in humans for as long as 6 months. For these reasons, although CT evidence of bone is readily visualized at 3 months, secondary procedures performed on the ISO regenerate (such as distraction) are probably best deferred until 6 months after implantation.

A Look to the Future: ISO, DISO, and Stem Cell Transplantation

Stem cell concentration within a periosteal chamber is abundant. In a primate model, ISO takes place even at advanced age. Manipulation of a periosteal chamber using distraction technology has permitted a creation of a mandibular ramus with of specific size and shape in a patient with a number 7 lateral facial cleft and a Pruzansky III mandibular defect. Such an approach is termed **DISO**. As additional progress occurs with the developmental field model of craniofacial embryogenesis, deficiency or outright absence of specific fields (mandibular ramus, zygoma) will become feasible using ISO/DISO methodology. The problem of an inadequate responder population can potentially be addressed by transplantation of either SVF MSCs or expanded bone marrow MSCs.

Conclusion

Reconstruction of a critical-size 12-cm mandibular defect resulting from subperiosteal resection of a JAOF is presented using ISO with an rhBMP-2/activated collagen sponge carrier implant. Clinical findings and basic science concepts of ISO are reviewed. Chemoattractive signals from the implant produced a critical concentration of stem cells within a predetermined environment; the resulting membranous bone was histologically identical with that of the original bone. Serial CT scans demonstrated ISO to be a rapid process, with bony continuity established as early as 6 weeks after surgery. At 24 weeks, the regenerate was osteotomized and tolerated distraction successfully.

ISO provides a novel means to reconstruct large membranous defects without recourse to microvascular tissue transfer. Because ISO is osteoinductive, critical issues involving local blood supply are potentially circumvented. Fabrication of an rhBMP-2/ACS implant is rapid and simple. The resultant savings in time, morbidity, and intraoperative technology make ISO an attractive means for re-establishing continuity in the human mandible and in other membranous bones of the human craniofacial skeleton.

Pierre Robin Sequence: Rapid Expansion DISO as Treatment for the Infant Airway

Robin syndrome (PRS) is a relatively uncommon condition defined by the triad of micrognathia, glossoptosis, and cleft palate. The incidence of PRS has been estimated to be between 1:2000 and 1:30,000 live births [115]. Neonates with PRS can develop respiratory distress due to an inadequate airway related to microretrognathia and glossoptosis. A wide array of potential treatment modalities, ranging from simple non-surgical interventions to invasive surgical procedures, has been described depending on the extent to which the patient's airway is compromised. Previous studies have shown that 12–40% of neonates with respiratory distress due to PRS require endotracheal intubation or tracheostomy [116–118]. These methods are not ideal because they are only temporizing measures that are associated with significant morbidity and mortality [119].

Distraction osteogenesis was first described by Codivilla in 1905 where he reported its use in lengthening a femur by axial distraction [120]. Subsequently, in 1992, McCarthy introduced mandibular distraction osteogenesis (MDO) as a means of correcting craniofacial abnormalities without the need for bone grafting. Although this was an incredible advance in craniofacial surgery, it has not been without complications and it requires prolonged treatment time with multiple surgeries. MDO has the unique advantage of simultaneous bone and soft tissue distraction. DO, however, is a lengthy process that is variable in duration but most commonly involves a 3–7-day latency period, a distraction phase that proceeds at approximately 1 mm/day, and then a consolidation phase lasting 4–12 weeks.

Complications of mandibular DO include infection at pin or osteotomy sites, inferior alveolar nerve dysfunction, and resorption and ankylosis of the temporomandibular joint. Dauria and Marsh demonstrated that DO could prevent the need for tracheostomy in patients with PRS [121]. Monasterio also demonstrated a successful treatment of respiratory problems in 15/15 patients with PRS using DO [122].

The aim of this retrospective analysis was to evaluate the safety and effectiveness of BMP-2 in enabling a rapid protocol DO in neonates with PRS.

Clinical Cases

A review of all patients treated in the department between February 2003 and February 2008 was performed. Three patients with the Pierre Robin syndrome who underwent DO with rhBMP-2 were identified. Inpatient and outpatient charts were reviewed for time to completion of distraction, age at distraction, need for tracheostomy, and complications of the mandibular distraction (Figs. 20.34 and 20.35).

Case 1

An infant (patient A; Table 20.1) with PRS and respiratory distress was evaluated in the neonatal intensive care unit (NICU). The patient was intubated because she was unable to protect her airway secondary to micrognathia and glossoptosis (Fig. 20.1). The patient was taken to the operating room at 10 days of age where she had bilateral osteotomies of the mandibular body and placement of a distractor on each side. Then, rapid protocol distraction was performed on both

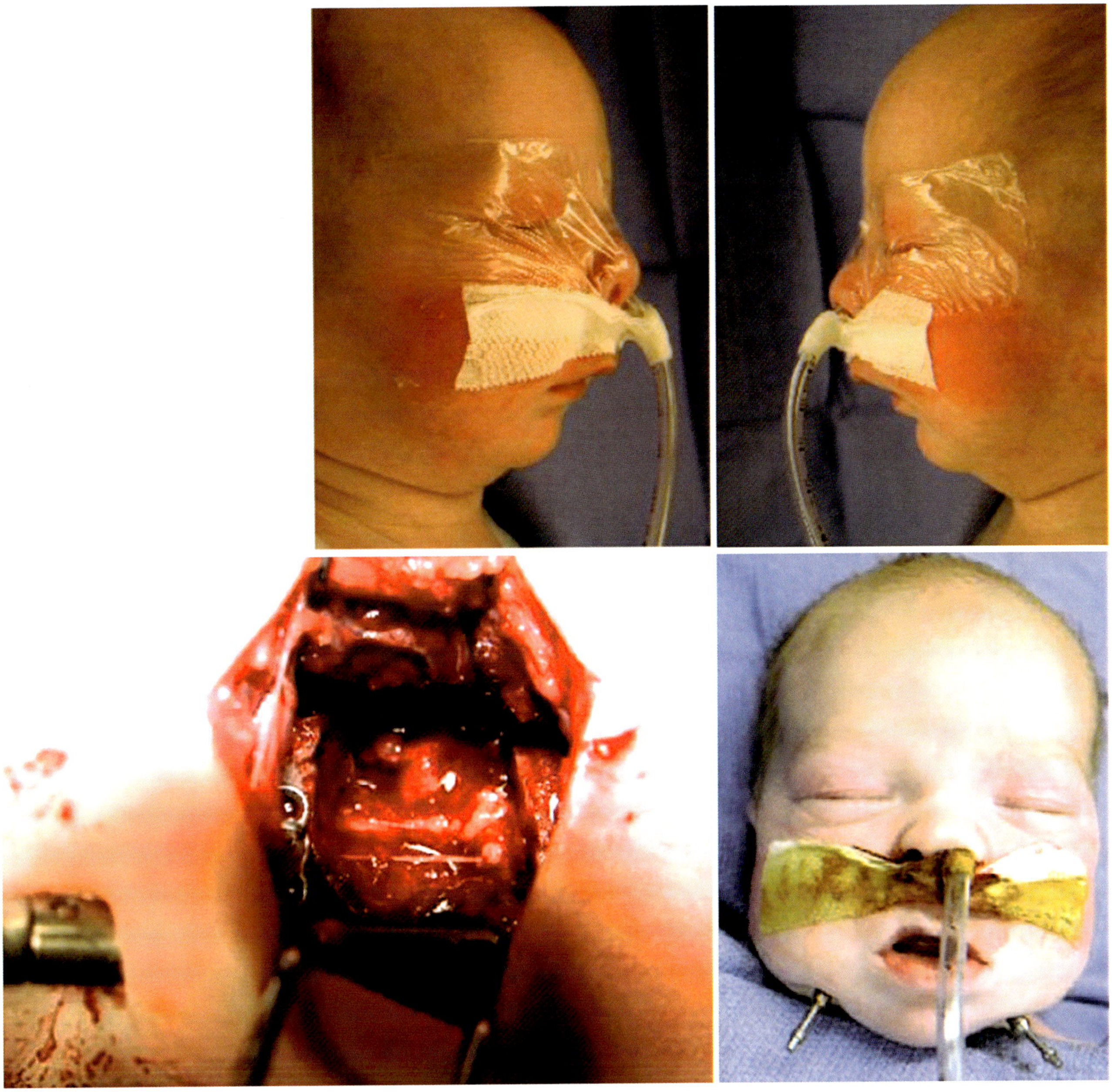

Fig. 20.34 Neonatal distraction osteogenesis for mandibular deficiency secondary to Pierre Robin sequence. The inferior alveolar nerve is easy to identify and preserve. Cuts are made circumferentially with the final opening using a 2–3-mm straight osteotome. Nerve stretches easily with on-table distraction. Note projection of the chin at conclusion of DISO. [Reprinted from Franco J, Coppage J, Carstens MH. Mandibular distraction using bone morphogenic protein and rapid distraction in neonates with Pierre Robin syndrome. J Craniofac Surg 2010; 21(4):1158–1161. With permission from Wolters Kluwer Health, Inc.]

hemimandibles. The right side was distracted 8.1 mm, and the left was distracted 9 mm intraoperatively. rhBMP-2 was placed in the resulting defects. This distraction placed the infant in slight class III occlusion (Fig. 20.2). The patient returned to the NICU intubated and was *extubated on postoperative day 2*. Ten weeks later, she was brought back to the operating room for removal of both of her mandibular distraction devices (Figs. 20.34 and 20.35).

Case 2

The second patient (patient B; Table 20.1) to undergo rapid distraction was in the NICU for 3 weeks with intermittent respiratory compromise. After discussion with the intensive care team, it was decided that the patient might benefit from mandibular distraction to resolve her respiratory issues. The patient underwent rapid distraction protocol as described previously, with both sides of her mandible being distracted

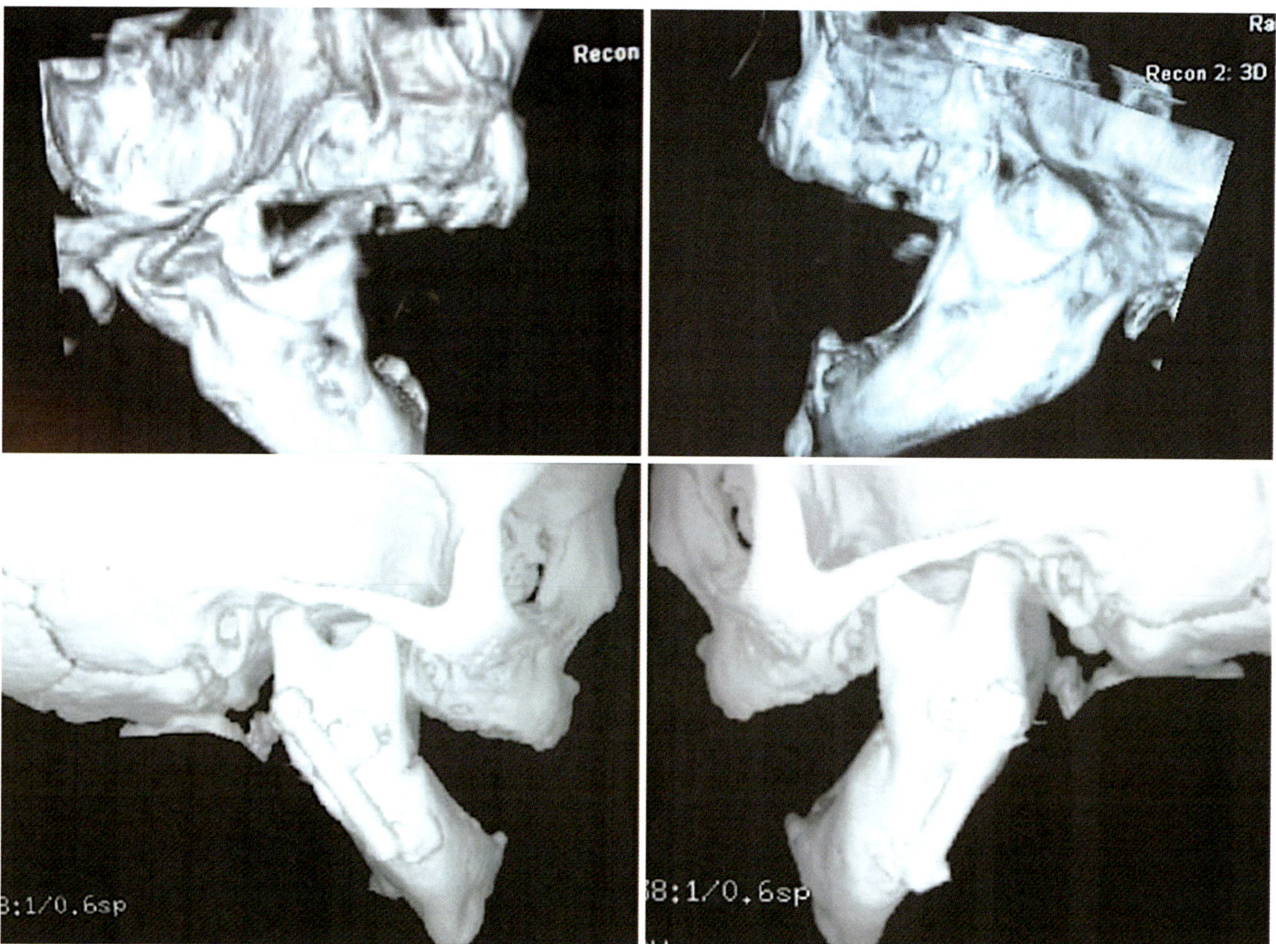

Fig. 20.35 DISO leads to early extubation. The DISO regenerate is shown with the distractor in place prior to removal. [Reprinted from Franco J, Coppage J, Carstens MH. Mandibular distraction using bone morphogenic protein and rapid distraction in neonates with Pierre Robin syndrome. J Craniofac Surg 2010; 21(4):1158–1161. With permission from Wolters Kluwer Health, Inc.]

9.0 mm with placement of rhBMP-2 in the osseous gap created by the distraction. This placed the neonate in class III occlusion (Fig. 20.4). She was able to *avoid tracheostomy* and was discharged home several days after placement of the distractors. Fifteen weeks after placement of the distractors, the patient returned and had both distractors removed. She experienced minor complication of a pin tract infection.

Case 3

The last patient (patient C; Table 20.1) in this series is a 20-day-old neonate who required intubation in the NICU secondary to severe respiratory distress. The patient underwent rapid protocol distraction with placement of rhBMP-2. Both hemimandibles were distracted at 15 mm, bringing the mandible into class III occlusion. He was returned to the NICU and was able to be *extubated on postoperative day 2*. He had 2 complications during his treatment course, 1 of which was a pin site infection that was treated as an outpatient with oral antibiotics. The second complication was a non-union of the left hemimandible, which required a second operation. A rib graft was placed at the site of non-union, which has subsequently healed.

Results

Three patients (6 hemimandibles) with Pierre Robin syndrome underwent rapid protocol distraction with rhBMP-2. Postoperative airway support was discontinued in 48 h in two patients and avoided altogether in the third. Mean age at initial distraction was 17.3 days. Mean time from device placement to removal was 89.3 days. A single case of non-union was managed successfully with an additional graft. No patient required tracheostomy.

Discussion

Distraction osteogenesis has continued to evolve for the last 100 years since it was first described by Codivilla. Its use in the treatment of mandibles and other craniofacial skeletal issues was solidified by the work of McCarthy et al. on the subject in 1992 [123]. Distraction has not been without its problems since this time because it is a multistage process and it has been associated with *complication rates in excess of 40%*. The high complication rate is often attributed to the prolonged treatment period and complexity of distraction. As the science and understanding of distraction and bone morphology continues to grow, so do the techniques and instruments that are used to facilitate the process. An example of this is the use of absorbable plates for distraction as described by Burstein [124]. The continual development of new techniques and instruments for distraction will lead to a decrease in the morbidity associated with the current distraction techniques.

Pierre Robin syndrome is unique in that neonates with this problem directly benefit from efficient distraction without a latency period and lengthy distraction interval. The children in this review all experienced acute respiratory distress that required intubation or were in imminent danger of needing intubation. Before the studies by Dauria and others, these neonates often required tracheostomy until they could receive definitive treatment for their micrognathia and glossoptosis. Tracheostomy in the pediatric population carries its own morbidity and mortality. Carron reviewed 204 pediatric patients needing tracheostomy and reported a complication rate of 44% and a 3.6% mortality directly related to the tracheostomy [125].

The use of *rapid protocol distraction*, in which patients are distracted to the desired length at the time of distractor placement and with the use of rhBMP-2, has several advantages over traditional treatment methods. This method, like previous distraction treatments for Pierre Robin syndrome, prevents the need for tracheostomy. The benefit of this method over traditional distraction methods is the time in which the neonate's mandible is distracted to its final length. The mandibles of the 3 neonates discussed in this review were all distracted to their final length at the time of distractor placement. This *avoids the latency period and the distraction interval*. If the mandible of the third patient in this series was to be distracted by the traditional method, the patient would have undergone a 5-day latency period, and then at 1 mm/day, it would have taken an additional 15 days to distract his mandible to its final length. The traditional method, although improved over prolonged intubation and tracheostomy, would have required the patient to be intubated and hospitalized for an additional 20 days. The additional time on the ventilator leads to increased morbidity associated with intubation and mechanical ventilation such as pneumonia. This gradual distraction process also leads to a prolonged hospitalization. *All of the patients in this series were extubated within 2 postoperative days.*

Among the benefits of distraction is the simultaneous stretching of the accompanying soft tissue. In this series, there were no complications due to relapse; however, this may be due to the extremely pliable soft tissues in this age group. This small series did have 2 minor complications and 1 major complication. The major complication of non-union was seen in the patient whose mandible was distracted 15 mm; whether this represents a limit to the length of distraction that is possible by this method or an isolated case can not be determined by this small series.

Scar Formation Over rhBMP-2 Implantation Sites

An interesting observation was made regarding scar formation in the soft tissue overlying the BMP-2 implantation site. A previously described the morphogen acts in a paracrine fashion resulting in the local production of VEGF. The quality of the scar over the mandibular access site was faint in color, razor thin, and flat. Diffusion of morphogen products into the tissues is postulated to be the mechanism of this extraordinary outcome.

Conclusion

Rapid protocol distraction with rhBMP-2 was used effectively in 5 of 6 hemimandibles (the sixth side required an additional graft). This method, coupled with other innovative techniques such as the use of absorbable plates, will continue to decrease the morbidity associated with mandibular distraction. The method of rapid protocol distraction was effective in relieving the respiratory distress from which all 3 of the patients were experiencing. Although larger studies and prospective trials are needed to further define the role of rapid protocol distraction, the technique offers obvious advantages and demonstrates the ability of the responder population to produce membranous bone with an expanded periosteal envelope, more rapidly than was previously thought possible.

Calvarial Reconstruction

Our focus so far has been on the interaction of facial bone periosteum with morphogen-based ISO to produce single-lamina bone. Reconstruction of the skull follows the same principles, the only difference being the presence of dura as a source of an MSC responder population.

Cranioplasty

A 13-year-old girl sustained closed head trauma when she struck a tree with her forehead while riding her bicycle at high speed. A frontal bone flap, removed for purposes of decompression, was inadvertently rendered non-viable during the storage procedure leaving the entire frontal lobe without protection albeit with viable dura. The defect was reconstructed with using four kits of Infuse® combined with tricalcium phosphate as an onlay over the dura. Recover was uneventful. She was returned to the operating room 4 months later for an additional only procedure for a local zone of asymmetry, again without sequelae. The final aesthetic result was acceptable and she continued with routine sports in the following school year (Figs. 20.36, 20.37 and 20.38).

Calvarial "Switch" Procedure

An 11-year-old boy presented with supraorbital deficiency status post reconstruction of the forehead for X using titanium mesh. He had simply outgrown the dimensions of the original mesh. The aesthetic defect was particularly noticeable in the supraorbital region, where he lacked adequate protection laterally for his globes. At surgery, the mesh was found to be inextricably involved with the underlying dura

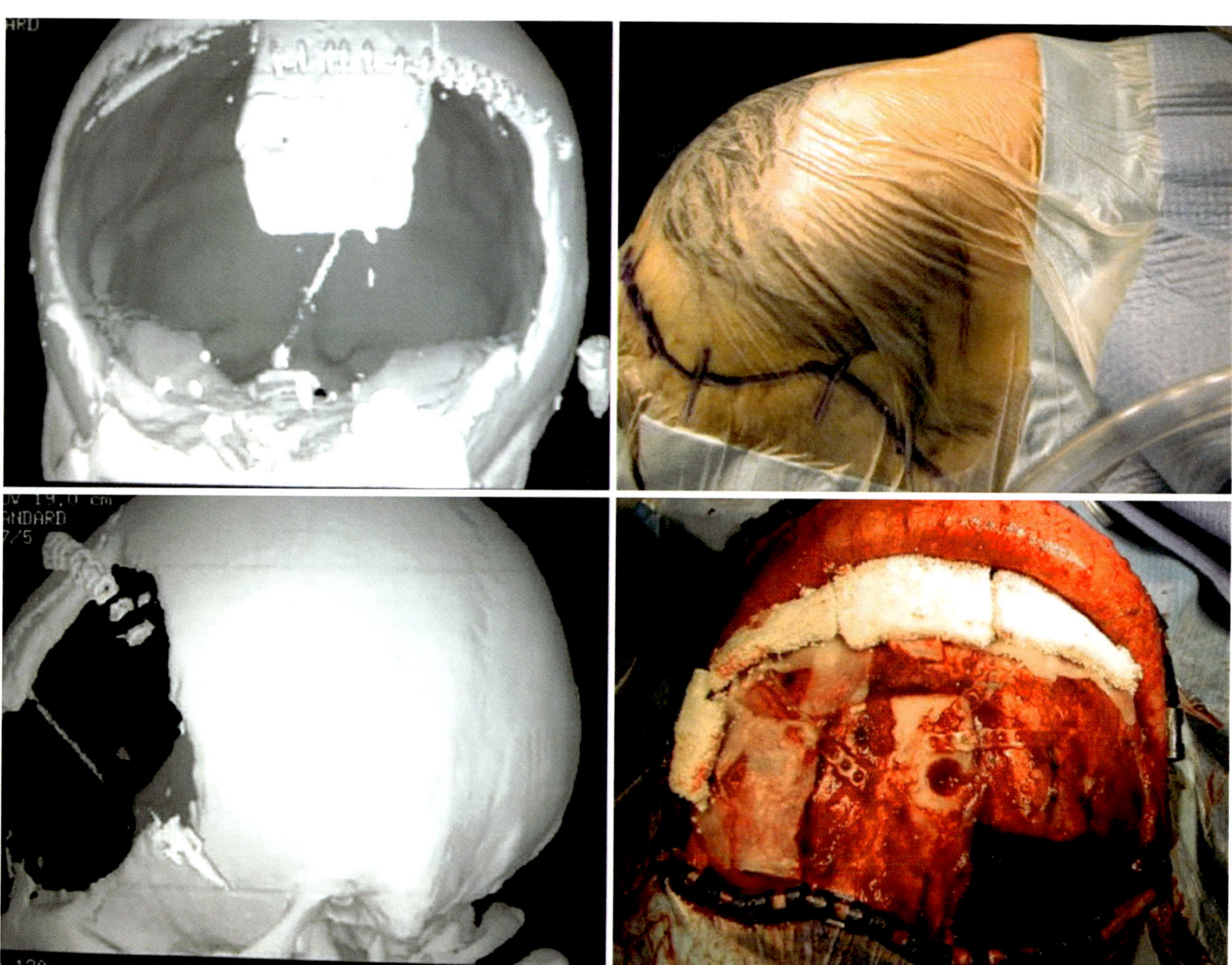

Fig. 20.36 Case 8: Post-traumatic loss of virtually entire frontal bone with preservation of dura. Reconstruction with rhBMP-2/ACS + Mastergraft® TCP. [Courtesy of Michael Carstens, MD]

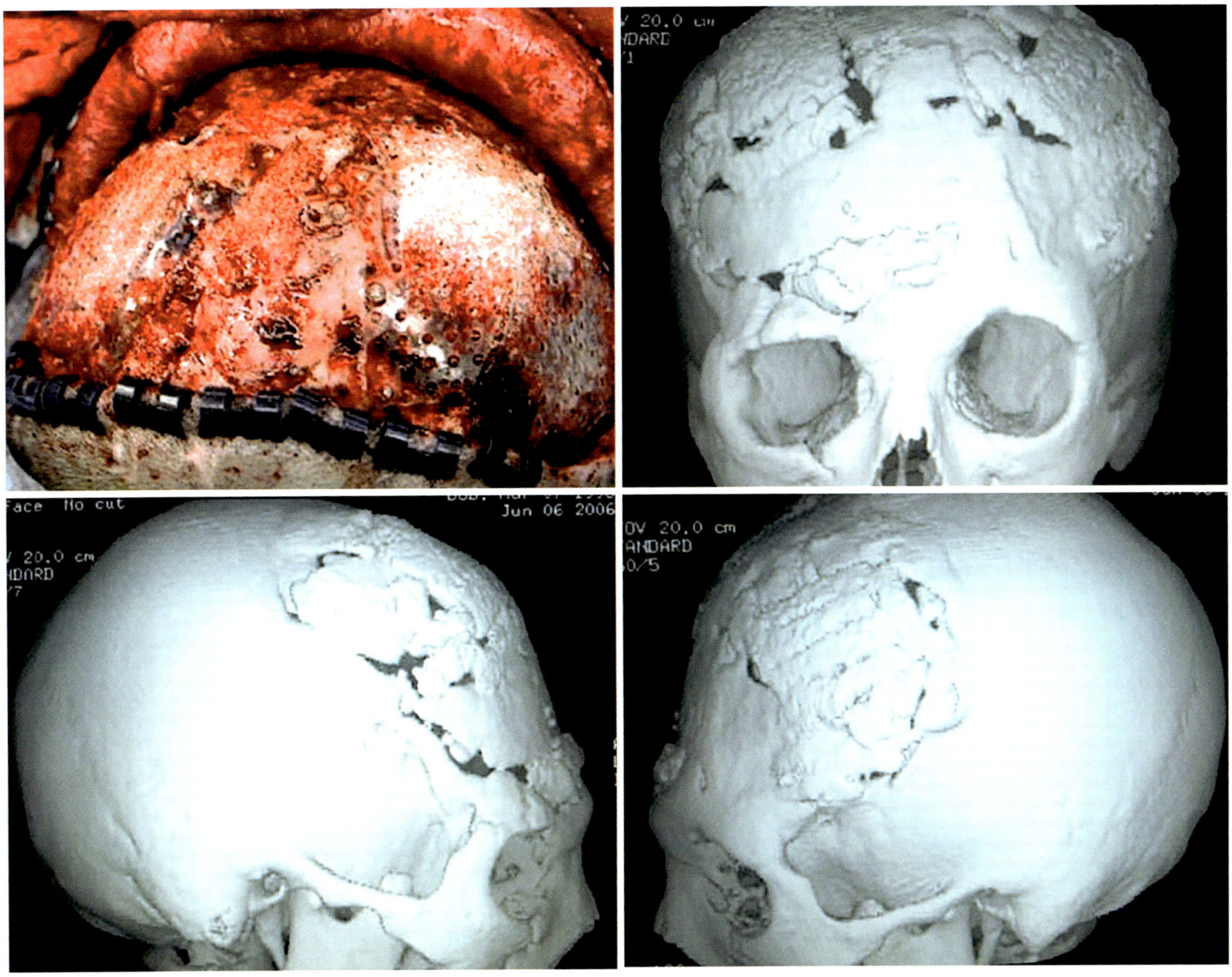

Fig. 20.37 Case 8: Second-look cranioplasty demonstrating the fast action of previous ISO reconstruction. Consolidated frontal bone from the previous operation was augmented with placement of additional rhBMP-2 and TCP to create a smooth supraorbital contour. Top left: Intraoperative view from above (patient supine) shows a residual supraorbital volume deficit. Bottom left: Lateral CT scan shows visible recession of the frontal bone above the supraorbital bar. The goal of the second-look procedure was to replace this deficit. Top right: AP view post graft shows good transverse dimensions and elimination of temporal hollowing. Bottom right: Forehead volume restored with normal frontal vertical projection (Bottom left and bottom right). [Courtesy of Michael Carstens, MD]

such that its removal would lead to multiple tears and dural defects. The overlying pericranium was undamaged and constituted a single-layer source of MSCs for reconstitution of the defect. Once again rhBMP-2 and TCP were applied, three kits being required. The defect was slightly over corrected. Postoperative appearance was acceptable and scars demonstrated a significant supraorbital advance due to the only osteoplasty (Figs. 20.39, 20.40, 20.41, 20.42 and 20.43).

BMP "Switch Procedure"

A 5-year-old boy was attacked by a dog, sustaining an avulsive injury to the left temporal bone, with loss of dura and partial loss of cerebral cortex. After debridement and soft tissue coverage using scalp expansion his neurologic recovery was remarkably normal. He remained without protection for virtually the entire hemicalvarium. A switch procedure was executed based on the intact dura of the contralateral size. Harvest of a temporoparietal bone flap to match the defect was done and the flap secured into place with absorbable plates. The intact dura on the right side was reconstructed with three kits of Infuse® and TCP. Symmetry was achieved with consolidation complete at 3 months.

Augmentation Osteoplasty

A 14-year-old girl presented with left facial asymmetry after resection at age of the entire mandibular body for an aggressive ameloblastoma. She was reconstructed elsewhere with a fibular free flap. In mid-adolescence facial growth has led to a loss of bulk (Figs. 20.44 and 20.45).

Fig. 20.38 Case 8: Postoperative appearance 6 months after second grafting showing contour restoration. Top left and top right: Residual supraorbital volume deficit with visible recession of the frontal bone above the supraorbital bar. Bottom left and bottom right: Lateral views show good frontal bone contour and aesthetic reconstruction of the naso-frontal angle. Projection of the nasal dorsum can be enhanced with future only bone grafting. [Courtesy of Michael Carstens, MD]

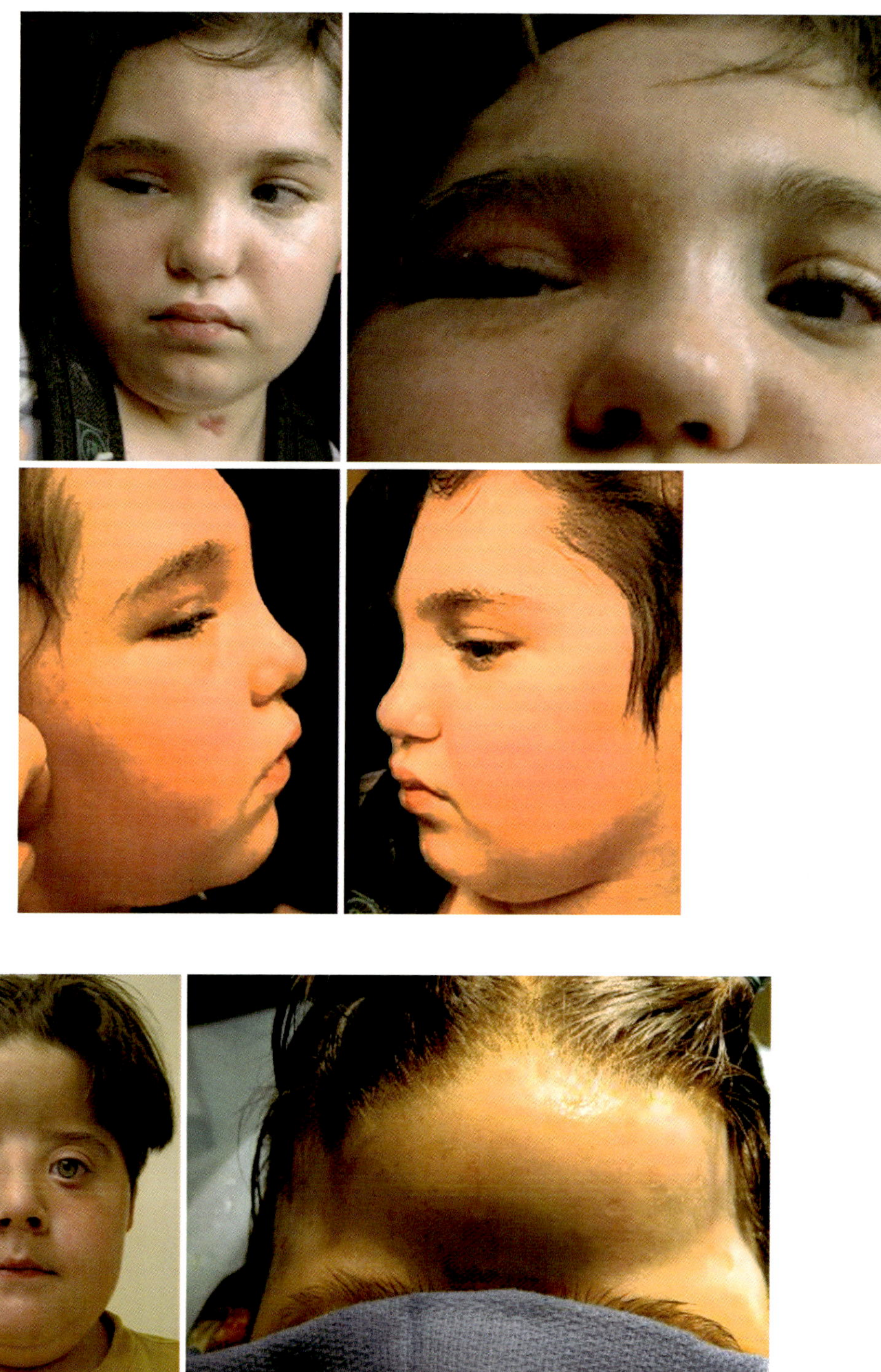

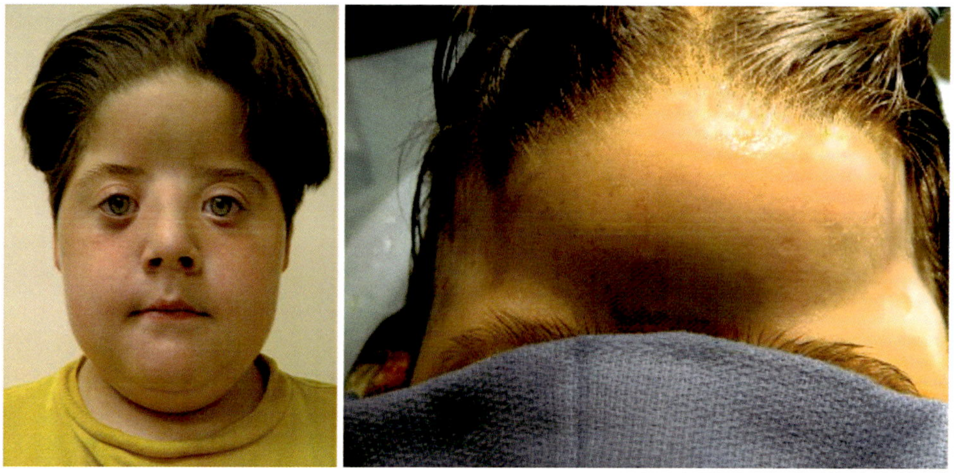

Fig. 20.39 Case 9: Augmentation osteoplasty: Saethre-Chotzen with orbital roof deficiency state. Lateral globes unprotected for sports rim. Previous titanium mesh cranioplasty was encased by dura throughout, making resection of the mesh dangerous. Forehead osteoplasty carried out a U-shaped bi-parietal bone graft to create a "pseudo-bandeau." Parietal defects filled in with BMP-2/ACS + Mastergraft in horizontal strips over the temporoparietal dura. [Courtesy of Michael Carstens, MD]

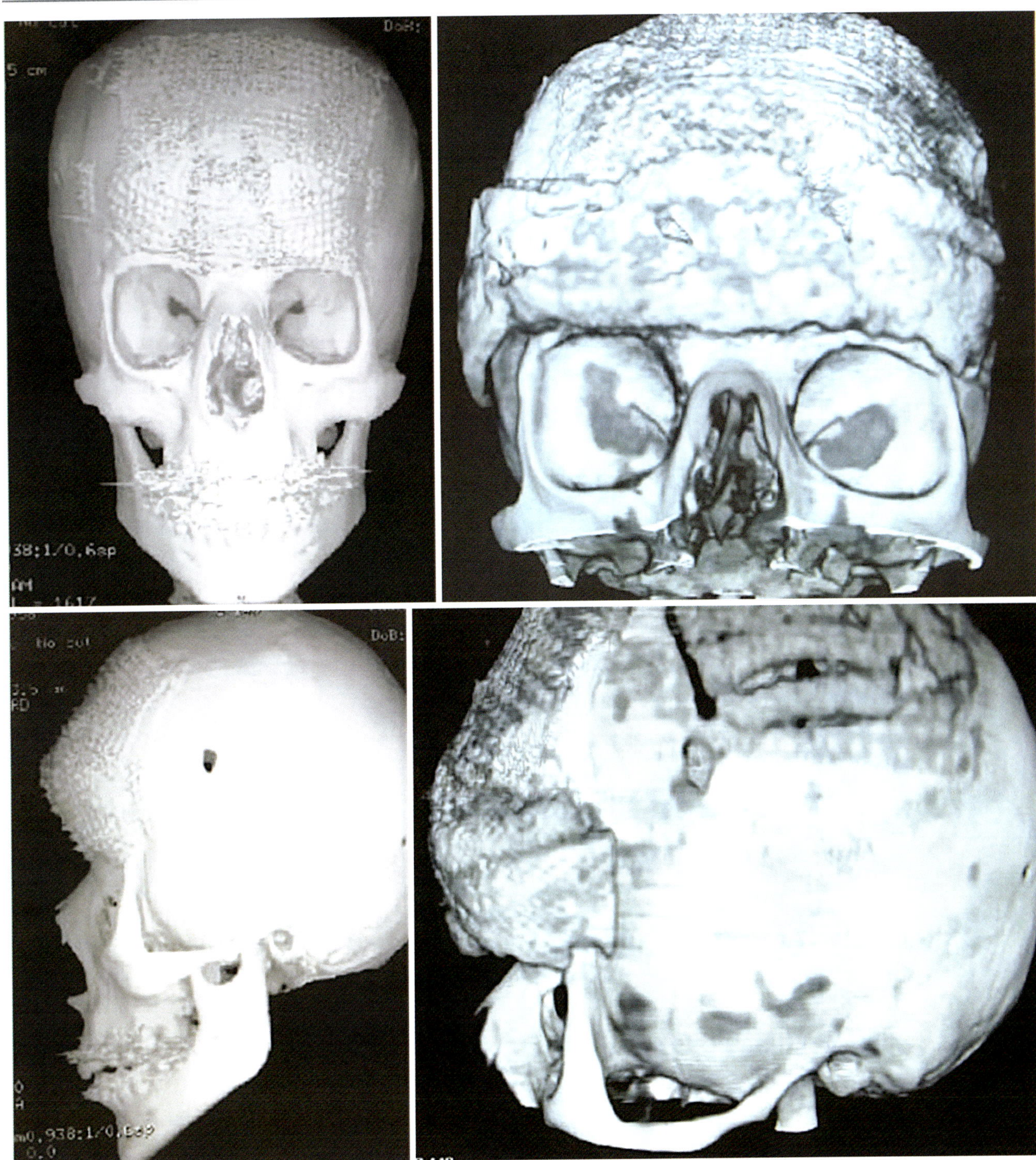

Fig. 20.40 Case 9: Previous titanium mesh cranioplasty encased by dura throughout making resection of the mesh dangerous. Forehead osteoplasty carried out a U-shaped bi-parietal bone graft to create a "pseudo-bandeau." Parietal defects filled in with BMP-2/ACS + Mastergraft in horizontal strips over the temporoparietal dura. [Courtesy of Michael Carstens, MD]

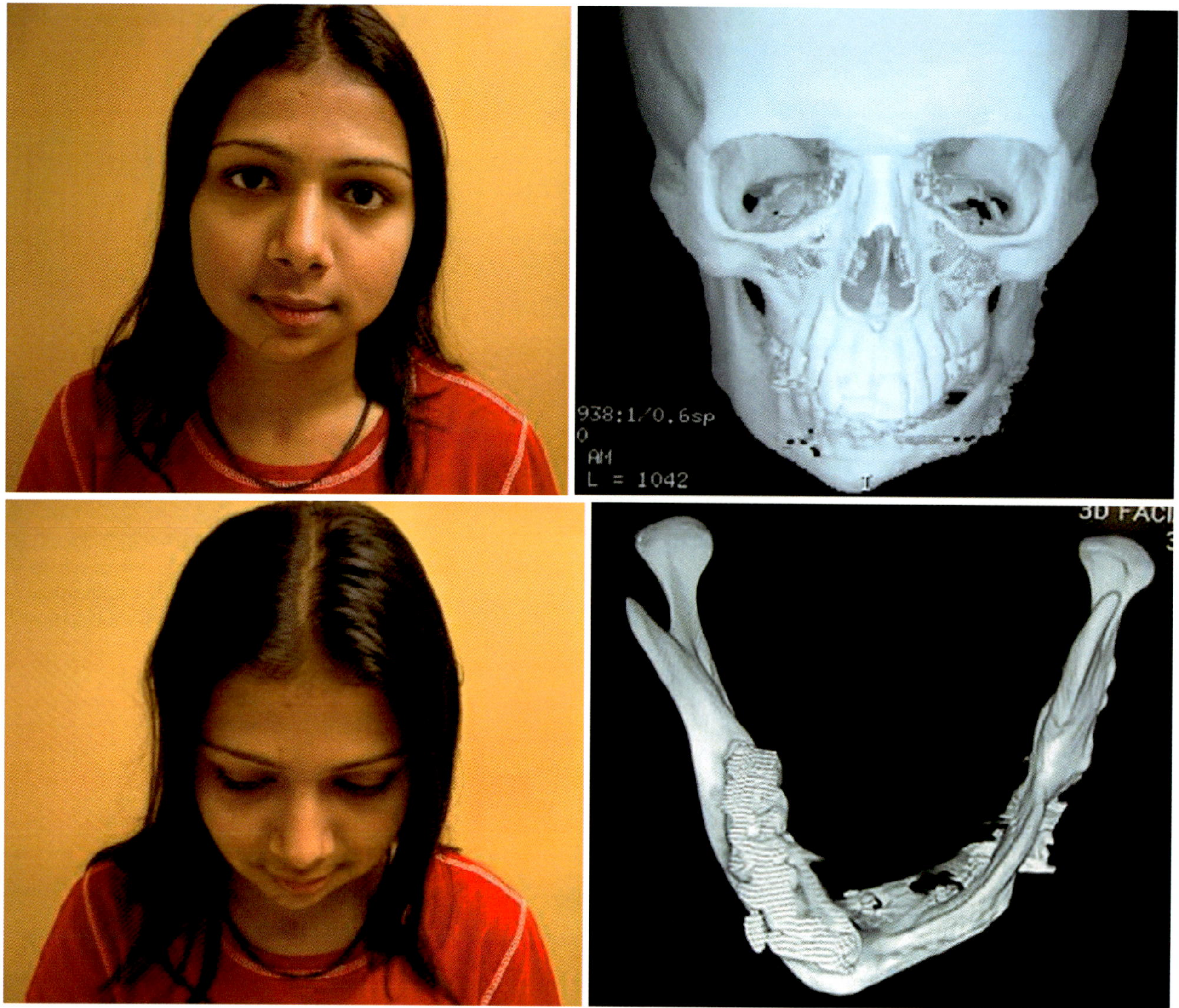

Fig. 20.44 Case 10: A 14-year-old girl with left facial asymmetry after resection of tumor left mandibular body followed and reconstruction with fibular free flap. The inadequate mandibular height and volume preclude dental implants. Preservation of the pedicle to the bone graft is a priority. Contour restoration planned with BMP-2 onlay osteoplasty designed to provide new bone in a previously irradiated wound bed and preserve the pedicle to the bone flap. [Courtesy of Michael Carstens, MD]

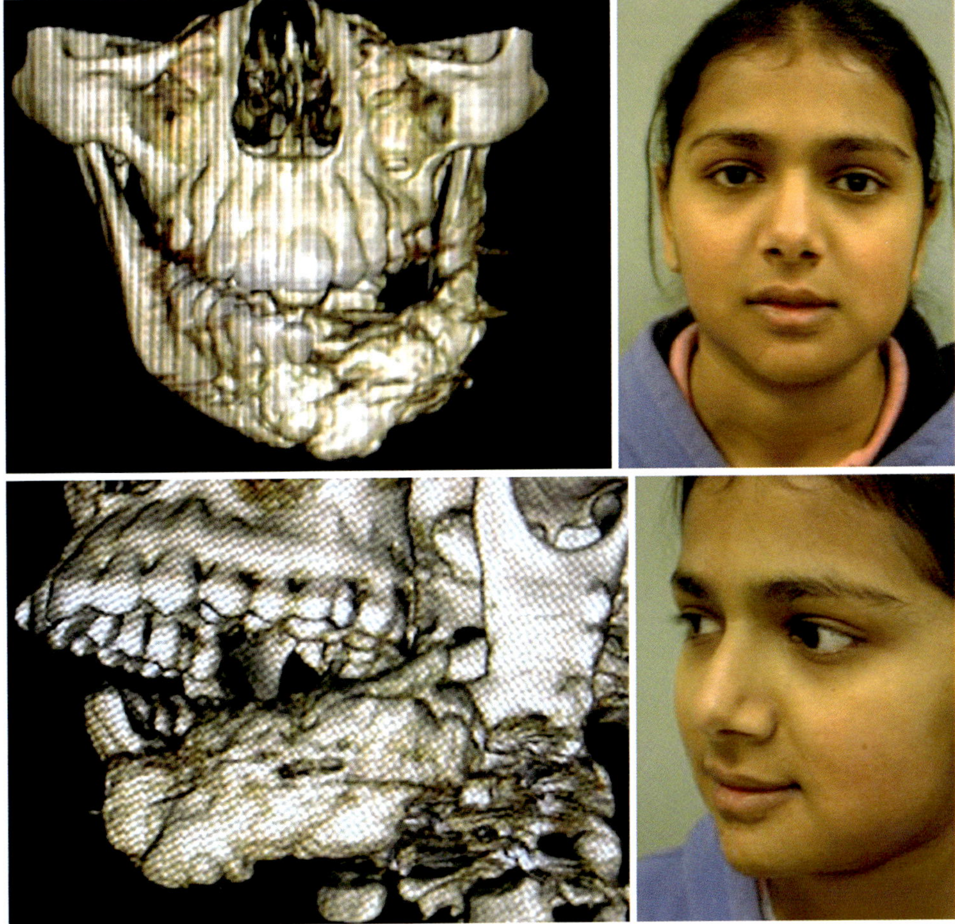

Fig. 20.45 Case 10: Onlay osteoplasty showing adequate mandibular height for implants. Residual prominence on the left side of the chin was taken down with a burr in a subsequent procedure using intraoral access. [Courtesy of Michael Carstens, MD]

Reconstruction of Stem Cell Responder Population Defects

In situ osteogenesis is an equation in which involving five variables: (1) resident stem cells, (2) mesenchymal signaling cells, (3) a morphogen, (4) a scaffold upon which the bone is constructed, (5) a lattice representing a surrounding membrane or structure which determines the final shape of the synthesized bone, and (6) blood supply. It should be noted that the biomechanics of the scaffold, if sufficiently sturdy can fulfill the fourth requirement. Let's look more closely as variables 6 and 3.

No bone reconstruction can survive without rapid vascular support. As we have seen, the ability of BMP-2 to stimulate resident stem cells also resulted in an intense degree of vascular support through VEGF produced by those cells. Recall that resident stem cells are already vascularized. They do not need to be supported. As they multiply, the new supply ensured the survival of the ISO-created bone. But what if the resident population is inadequate or missing? Surgeons are commonly faced with situations in which, through trauma or tumor resection, resident MSCs are simply gone. Surrounding soft tissues of the face, even in the presence of muscle, may not be sufficient to provide an adequate response, despite implantation of a powerful morphogen. Previously, this scenario had only one solution: a bone-bearing flap with all its attendant advantages and drawbacks. In this case, the bone graft (in whatever shape) survives based on its own blood supply which is either *pedicled*, permitting rotation into position, or *free*, with implantation based on a microvascular anastomosis.

This point was driven home to me in one case, circa 2009, in which I attempted to reconstruct a mandibular defect in the absence of periosteum, thinking that resident MSCs of mylohyoid and platysma would suffice. I was dead wrong, of course… the case resulted in failure. It was a bitter lesson but left me convinced that a source of exogenous autologous MSCs would resolve the dilemma. It would be another 5 years, thousands of miles, and an apparent re-direction of research before the answer would come in the form of a visit from Bill Futrell, Gertrudis's diabetic wound, and the unrecognized work of a Canadian maxillofacial surgeon working in Finland.

Stromal Vascular Fraction: A New Kid on the Block

The concept of ISO based on mesenchymal signaling cells from adipose tissue (ASCs) represents an inflection point in biotechnology. This final section then is the story of how adipose-derived MSCs, in the form of the SVF combined with the morphogen, rhBMP-2, would prove to be a paradigm shift for craniofacial reconstruction. To tell the story, and drive home the science, I'd like to pull the following disparate threads together.

In 1999 Bill Futrell had tasked his residents, Adam Katz and Ramón Llull, to work on what of value might be found in lipoaspirate, the material routinely thrown away after aesthetic liposuction cases. They discovered the existence of MSCs is fat in numbers far greater than any other tissue (>500 × more than in bone marrow). The results, first published in 2001, indicated that MSCs, with the extensive clinical potential, could be readily harvested and could assume multiple lineages into mesodermal derivatives. Over the ensuing decade, an effective process to collect and concentrate these cells as the SVF was worked out.

In this way, from 2013 onward, the Pittsburgh technology was approved and a series of trials were drawn up to investigate safety and efficacy in proof-of-concept studies approved by the ethics committee at the National University of Nicaragua and filed with the Ministry of Health (MINSA) of that country. These included work in burn rehabilitation, post-traumatic fibrosis, osteoarthritis of the knee, rheumatoid arthritis of the hand, end-stage peripheral vascular disease (PVD), and finally the intra-arterial administration of SVF for Mesoamerican nephropathy (MeN).

As these studies and their references demonstrate, SVF is a multicellular mixture of pericytes (the precursors of MSCs), stem cells proper, endothelial cells, and monocytes. Inflammatory components, such as red cells and leukocytes, are washed away. SVF has multiple trophic effects, including intense angiogenesis, anti-fibrosis through action on matrix metalloproteinases, reduction of inflammation through production of hepatic growth factor (HGF) in opposition to transforming growth factor-beta 1 (TGF-$B1$), anti-apoptosis, and the support of regenerative cells in situ. Furthermore, SVF cells home in to anatomic sites of inflammation [126].

> To make our bone equation work we need to demonstrate two things. (1) the adipose-derived signaling cells in SVF should be capable of providing vascular support for themselves by **producing VEGF** *independent of a morphogen stimulus*. (2) ASCs should show **morphogen-driven differentiation** to osteogenic line. In this section we will look at the evidence that, under ischemic conditions, ASCs respond by producing neo-angiogenesis.

Non-reconstructable Critical Limb Ischemia Treated with SVF Cells: A 4-Year Study: SVF Makes VEGF Under Hypoxic Conditions and Independent of a Morphogen

Introduction

Peripheral vascular disease due to arterial insufficiency is a common problem, affecting up to 6.9% of individuals >45 years of age [127]. The most common causes in the U.S. population are macrovascular lesions due to arteriosclerosis (AS) (70–80%), and microvascular lesions due to diabetes mellitus (DM) (20–30%) [128], with a large proportion of patients presenting concomitant infrapopliteal arteriosclerotic disease [129]. Local and regional ischemia follow a clinical progression, starting with claudication, referred as muscle pain distal to the occlusion site brought on by predictable levels of exertion. Later, critical limb ischemia (CLI) presents with rest pain typically at the level of the metatarsal foot, unrelenting, and relieved by dangling the foot, and when blood flow falls below a critical perfusion pressure, ulceration or frank ischemic necrosis ensues. This symptomatic progression can be classified using the criteria of Fontaine and Rutherford [130, 131].

The diagnosis is confirmed by vascular examination including pulse Doppler and the ankle/brachial index (ABI) calculation (the ratio between the systolic pressures at the ankle and the mid-arm). Normal ABI or stage I is ≥1.0. Stage II occurs at an ABI of 0.9, while rest pain at stage III is characterized by an ABI of 0.5. Additional imaging-based modalities, such as computerized tomography (CT) angiography and/or magnetic resonance (MRI) angiography, localize sites of occlusion and help determine if surgical bypass is feasible. 3-D Doppler ultrasound is very sensitive for vascular imaging of the foot.

Thirty percent (30%) of patients with CLI present with non-reconstructable disease, a situation particularly critical for diabetic patients, due to calcific and fibrocalcific disease in the distal vasculature [132–134]. Furthermore, 40% of diabetic patients diagnosed with CLI will progress to tissue necrosis and/or gangrene, versus 9% of patients with AS disease [135]. Current treatment options percutaneous transluminal angioplasty (PTA) have a variable efficacy, with patients resulting in partial revascularization and restenosis, especially in infrapopliteal arteries. In addition, endovascular revascularization procedures are precluded in about 30% of patients with CLI due to high operative risks and/or unpropitious vascular anatomy [136]. Given these numbers, the need exists for novel cost-effective treatments for CLI.

Gene therapy was introduced as a way to provide angiogenic growth factors (e.g., VEGF, HIF1a, HGF, FGF1), based on their known inductive capacity to form vascular structures [137], a fact initially appreciated in cardiac ischemia [138]. Cell-based therapies have also been proposed to fabricate new blood vessels in ischemic areas. For instance, bone marrow-derived and peripheral blood circulating (with or without previous mobilization) mononuclear cells (MNCs) containing a population of endothelial progenitor cells (EPCs) have been used in clinical trials for vascular insufficiency with variable degrees of success (reviewed in [139]). In those studies, *intramuscular (IM) injections* of unselected or selected ($CD34^+$) cells have been shown to ameliorate the symptoms associated with poor distal blood supply.

On the other hand, transplantation of adult MSCs has great promise. These cells secrete a wide range of growth factors and cytokines acting in a paracrine fashion. Many of these factors are proangiogenic (e.g., VEGF) while others assist in the repair process of injured tissues [126, 140]. MSCs can be obtained from a variety of tissues, including bone marrow and adipose tissue, the latter being particularly rich in MSCs [141]. MSCs are readily obtained, as a multicellular SVF, after enzymatic digestion and centrifugation of lipoaspirate. SVF is a heterogeneous population of MNCs that includes adipose-derived stem cells (ADSCs) of mesenchymal phenotype (analogous to MSCs), EPCs, hemopoietic progenitors, monocytes, leukocytes, and pericytes [142, 143]. Pericytes represent the perivascular phenotype of native MSCs [144–147] and constitute a key cell component of SVF during angiogenesis, as they stabilize nascent blood vessels [148, 149].

Animal models of CLI, including the hindlimb ischemia model after femoral artery ligation in rats [150] and rabbits [151], have been proven useful to assess the effects of various cell types and to study potential mechanisms. For instance, IM injections of culture-expanded ADSCs increased flow and induced a higher systemic presence of EPCs [152]. Murakami and Tanaka [153] demonstrated the superior angiogenic potential of bone marrow-derived MSCs over MNCs; the former were able to differentiate into both endothelial cells and vascular smooth muscle cells. Finally, Hao et al. [151] reported the neovascularization effect of both ADSCs and bone marrow-derived MNCs. In these and other studies, ADSCs came to be recognized as a source for angiogenic factors acting through a paracrine mechanism, and in concert with other cellular players (e.g., EPCs and macrophages) [154–156].

Preclinical data have prompted multiple groups to explore the feasibility, safety, and efficacy of bone marrow-derived cell-based therapy for PVD, through the design and execution of small clinical trials (summarized in [139, 157–159]). Powell et al. [160] reported the interim results of the RESTORE-CLI trial, where IM injections of tissue repair cells (analogous to a MNC mixture) proved no serious adverse effects, with increased amputation-free survival of patients and improved wound healing. In addition to the IM route, intra-arterial (IA) administration (through a femoral artery catheter) has been also safely and efficaciously used to inject allogeneic, expanded, bone marrow-derived MSCs [161].

In sum, adult stem cell transplantation constitutes a paradigm shift in the treatment of chronic limb ischemia (CLI), especially for diabetic patients [162–164]. The safety and efficacy of culture-expanded ADSCs derived from SVF for CLI have been documented [7, 165] and reviewed [166], although further studies with more rigorous designs, including randomization, standard-of-care, or placebo controls, are still needed. However, to the best of our knowledge, *no report has been made so far with fresh, non-fractionated, uncultured, point-of-care administered SVF in CLI.* Therefore, an open label, non-randomized study to assess the safety and efficacy of non-culture-expanded adipose-derived SVF cells administered IM and subcutaneously to ten patients with non-reconstructable CLI was designed, approved, and executed at the National Autonomous University of Nicaragua in León (Figs. 20.44, 20.45 and 20.46).

Materials and Methods

Medical Ethics

This study (not registered in clinicaltrials.gov) was approved by the Medical Ethics Committee of UNAN-Leon and by the Ministry of Health of Nicaragua (MINSA). The procedures followed were in accordance with the ethical standards of the responsible committee on human experimentation (institutional and national, Universidad Nacional Autónoma de Nicaragua-León) and the Helsinki Declaration of 1975, as revised in 2000. Informed consent was obtained from all participants in accordance with standards of MINSA and the

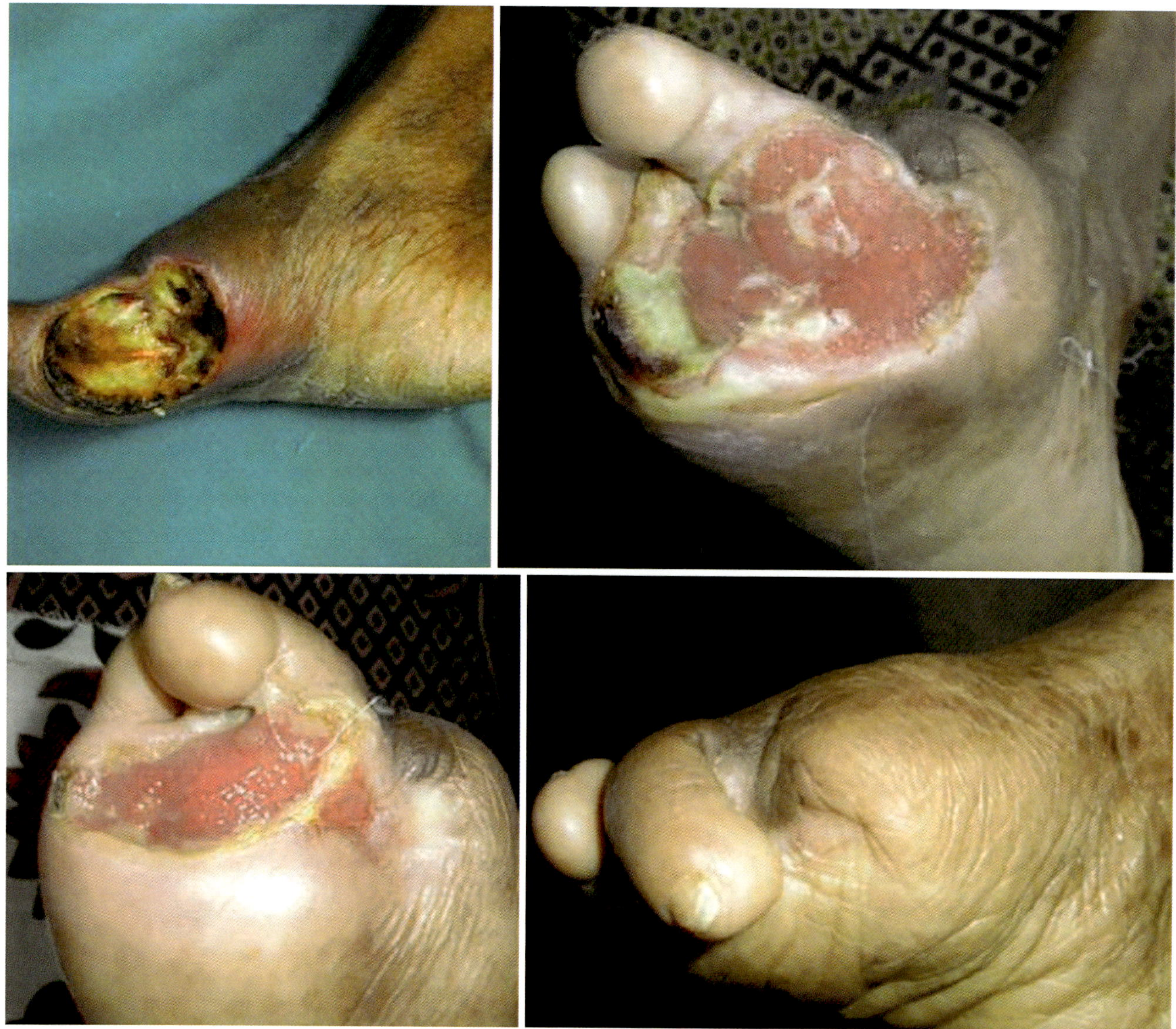

Fig. 20.46 Case 11: 57 females with diabetes. Rutherford 6 and amputation of the right first and second toes ABI: 1.4 > normal due to stiff arteries. Photo series shows wound pre-op, 3 months, 6 months, and 9 months after SVF implantation. ABI measured at 9 months = 1.9. [Reprinted from Carstens MH, et al. Non-reconstructable peripheral vascular disease of the lower extremity in ten patients treated with adipose-derived stromal vascular fraction cells. Stem Cell Res 2017; 18: 14–21. With permission from Elsevier.]

World Health Organization and included consent to publish this study in all formats.

The original criteria for admission were rest pain or claudication at least of one half block or less, non-healing ulcer or non-healing surgical amputation site for greater than 3 months, and inoperable PVD due to medical reasons. All patients at this advanced stage of disease were amputation candidates for amputation. Criteria for exclusion were as follows: age 40 years, unstable cardiovascular disease at the moment of enrollment, smoking and/or the presence of chronic pulmonary disease, ongoing infection and/or sepsis, and uncontrolled diabetes. At time 7 of the original group are alive: diabetes (DM) only (2), diabetes with arteriosclerotic disease (3), and arteriosclerosis only (2).

Surgical Procedure and SVF Processing

Details of this procedure are provided in our original paper. The surgical sequence consisted of the following steps: (1) lipoaspiration into a sterile cannister (GID SVF-1, Louisville, CO, USA); (2) serial washing to remove leukocytes and erythrocytes (thus producing "dry fat"); (3) enzymatic digestion with GMP-grade collagenase (GIDzyme, GID, Louisville, CO, USA) at 37 °C in a warmer-shaker cabinet (MaxQ 4450, Fisher Scientific) for 40 min; (4) neutralization 2.5% solution of human albumin; and (5) centrifugation for 10 min at 800 g (Sorvall ST40, Fisher Scientific) and resuspension. Samples were assayed by an image cytometer (ADAM MC, Portsmouth, NH, USA) to determine the final viable cells count.

Outcomes Assessment

The purpose of this study was to evaluate the safety and feasibility of an SVF cell-based therapy for advanced PVD in ten patients. Safety was evaluated by the absence of adverse effects after the procedure, related to inflammation, infection, or local necrosis of the injection site. Feasibility was evaluated using clinical and imaging criteria. Clinical data were both subjective (pain assessment and ulcer healing when present) and objective [ankle-brachial index (ABI)]. The ulcers, when present, were photographed and dimensions measured recording their two main axes (in cm). Patients were evaluated for pulses and the ABI calculated by averaging the ratios of three consecutive measurements of systolic blood pressures of the ankle to mid-arm. Imaging data involved a pre- and postoperative MRI-based angiography. These parameters were collected before the procedure and after at 2, 4, 6–8 and 10 and 18 months (pain and ulcers), and 4 months (ABI and angiography).

Statistics

To establish a potential correlation between age, lipoaspirate volume, viable cell injected (i.e., cell dose), and the changes in ABI (delta ABI), the Pearson correlation coefficient (r), along with a 95% confidence interval and a P value were calculated and graphed using Prism 6 for Mac (GraphPad Software, Inc.).

Results

Patient Characteristics

18-month data have been previously published. In follow-up all patients were examined and photographed to document the healing of previous wounds and the absence of recurrence. ABIs were measured. A Toshiba Doppler ultrasound was used to image both the treated and untreated ankles and feet.

Statistical Considerations

Data in our previous study regarding age, lipoaspirate volume, the cell "dose" injected, and ABI were compared with a Pearson correlation coefficient (r), with a 95% confidence interval and a P value using Prism 6 for MAC (GraphPad Software, Inc.). The disease process in the treated and untreated extremities is considered to be asymmetrical (the treated extremity being the more symptomatic) but serves as a qualitative backdrop. Comparison between ABIs pre-op, at 18 months, and at 4 years is presented.

Results

Patient Characteristics

In the original group, at the time of treatment, ages ranged from 57 to 85 (average 72 years). Demographics are shown in Table 20.1. Note that in the interim, three patients (1, 2, and 10) have expired, all due to cardiac causes.

Diagnoses are as follows: DM alone (2), DM combined with AS (3), and AS alone (2). All diabetic patients remain under insulin treatment and dietary restriction. One patient (#8) continues with anti-hypertensive therapy and has received a pacemaker. One patient (#9) had an amputation of the untreated extremity 6 months after her SVF procedure (Table 20.2).

Follow-Up

- **Pain**: Five patients started with rest pain (#5 and #8 on a wheelchair), while the remaining five had claudication triggered at walking for half a block or less. After the treatment, two out of the five patients with rest pain (#1 and #4) were able to ambulate pain free for at least 200 m. Wheelchair-bound patients had reversal of rest pain. Case #10 expired prior to final documentation. The five patients with claudication were able to walk from 300 m to several blocks without claudication symptoms.
  - At an 18-month follow-up these results remained unchanged.
  - A 4-year follow-up for the remaining 7 patients disclosed continued pain relief.
- **Ulcers**: Four out of six patients with ulcers achieved closure between 8 and 10 months post-op. Case #8 under-

Table 20.2 Cell counts and clinical results

Case	Cells	%	Pre	18 mo	48 mo	48 Ctl	Amb	Wound
1†	30.8	81	1.4	1.8			200 m	Closed 8 mo
2†	53.2	75.0	0.3	0.8			400 m	
3	80.4	88.5	0.7	1.5	0.9	0.6	8 blks	Closed 9 mo
4	34.0	86.0	0.3	0.8	0.5	0.8	200 m	
5	126.0	82.0	1.4	2.0	0.6	0.9	WCh	Closed 9 mo
6	42.5	83.5	0.4	0.8			15 blks	Closed 9 mo
7	158	73.0	0.4	0.7			3 blks	
8	19.1	83.0	0.3	0.8			WCh	STSG 5 mo
9	44.5	85.5	0.4	0.6	0.8	Amp	300 m	Closed 6 mo
10†	130.0	88.0	0.3	NA			NA	Closed 4 mo

Dagger indicates patient deceased

went successful skin grafting at 5 months post-procedure. Early wound healing was present in case #10 at the time of her demise.

- **ABI ratios**: An increase in the calculated ABI ratios was documented in all patients. The increments ranged between 1.29 and 2.67 times (Post/Pre) with an average increase of 1.92 times. Of note, two out of six patients with DM had "normal" preoperative ABI ratios (1.4), which in diabetic patients with small-vessel vasculopathies may be a normal finding (ABI ratios ≥1.0). Nevertheless, those two patients also experienced an increase in their postoperative ABI ratios (1.8 and 2.0, respectively).
  - A 4-year follow-up showed progression of disease with a decline in the ABI score compared with 18 months but the ABIs in all treated extremities remained improved compared to pre-op.
- **Angiography**: MRI-based angiography was performed at 6 months in 6/9 patients. 3/9 patients had medical contra-indications (e.g., compromised kidney function). The tenth patient was deceased at 4 months. Of the 6 studies successfully completed, 5 demonstrated neovascularization in the leg, foot, or both. One patient presented no change.
- **3-D Doppler ultrasonography** at 4 years demonstrated in 13 limbs: (1) consistent increase in diameter of dorsalis pedis and tibialis posterior; (2) better flow with return of triphasic signal in some cases; and (3) improved quality of subcutaneous tissue with virtually no edema.

Discussion

This study describes clinical outcomes obtained when fresh, non-fractionated, non-culture-expanded, adipose derived SVF cells were used to treat patients with CLI. The rationale for using SVF instead of expanded ADSCs was twofold: (1) to test the feasibility of a point-of-care administration of a cell-based therapy approach and (2) to take advantage of the presence within SVF off additional regenerative cells types (e.g., endothelial cells, monocytes, and pericytes) known to have angiogenic and blood vessel-stabilizing properties.

Despite a wide range of cell doses (viable cells injected), all patients at 18-month completion demonstrated positive clinical responses in all assessments in terms of pain control and in the ability to ambulate (when possible) without claudication (when possible) (in the seven patients in which the evaluation was feasible). ABI ratios were increased in all patients at 18 months. In the diabetics with elevated ABIs we considered this to be due to increased inflow upstream. Over time the seven survivors demonstrated a decline of the ABIs consistent with their disease. All ulcers remained healed with no recurrences at 4 years. Angiographies showed evidence of neovascularization in 5/6 patients, but the particular pattern was noted either by localization (e.g., proximal, middle, or distal 1/3 of the leg), or by responding artery. Most importantly, the degree of revascularization in the leg was not as striking as the overall clinical response (Figs. 20.47 and 20.48).

We recognize the limitations of angiography in these patients are technically difficult [167]. Our impression was that angiography in cases of severe distal disease involving small vessels is of *limited clinical value* for assessment. Doppler ultrasound studies at 48 months were easier to obtain and gave more information.

These results are comparable with other two previously reported clinical series, using cultured adipose-derived ADSCs with controlled cell doses. The first reported trial with ADSCs for CLI, enrolled 15 subjects [12 with thromboangiitis obliterans (TAO) and 3 with diabetic foot] exhibiting rest pain with/without non-healing ulcers or tissue necrosis. Treatment consisted of IM injections of culture-expanded ADSCs [165]. Clinical improvement, based on pain scores, claudication distances, and ulcer healing, was documented in 66.7% of the patients. Digital subtraction angiography (DSA) at 6 months revealed a significant improvement in the collateral vessel formation score in eight out of ten TAO patients and in two out of three of the diabetic patients.

Notwithstanding the observed clinical and imaging similarities between our study and the ones reported with cultured-expanded ADSCs, a major difference relates to the *effective cell dose* used. The studies reported by Lee and Bura used a fixed number of ADSCs, significantly higher than in our study. Two considerations need to be taken into account for this contrast.

- First, the reported percentage of SVF constitutive cell populations is variable. For instance, ADSCs can be present in a range between 10 and 30%, while EPCs and pericytes can comprise 10–20% and 3–5%, respectively [142]. Therefore, the absolute number of ADSCs administered in our study was significantly lower than the one used in the other two studies.
- Second, DM can exert an effect over different progenitor cell populations within the adipose tissue, which can alter even more the percentage distribution of cells within SVF, and more importantly their functional effects [142, 143]. For instance, Rennert et al. [168] described that DM impairs the angiogenic potential of ADSCs (both in vitro and in vivo) by selectively depleting specific cellular sub-populations.

These findings can serve as the basis to hypothesize a potential therapeutic benefit of SVF when compared with

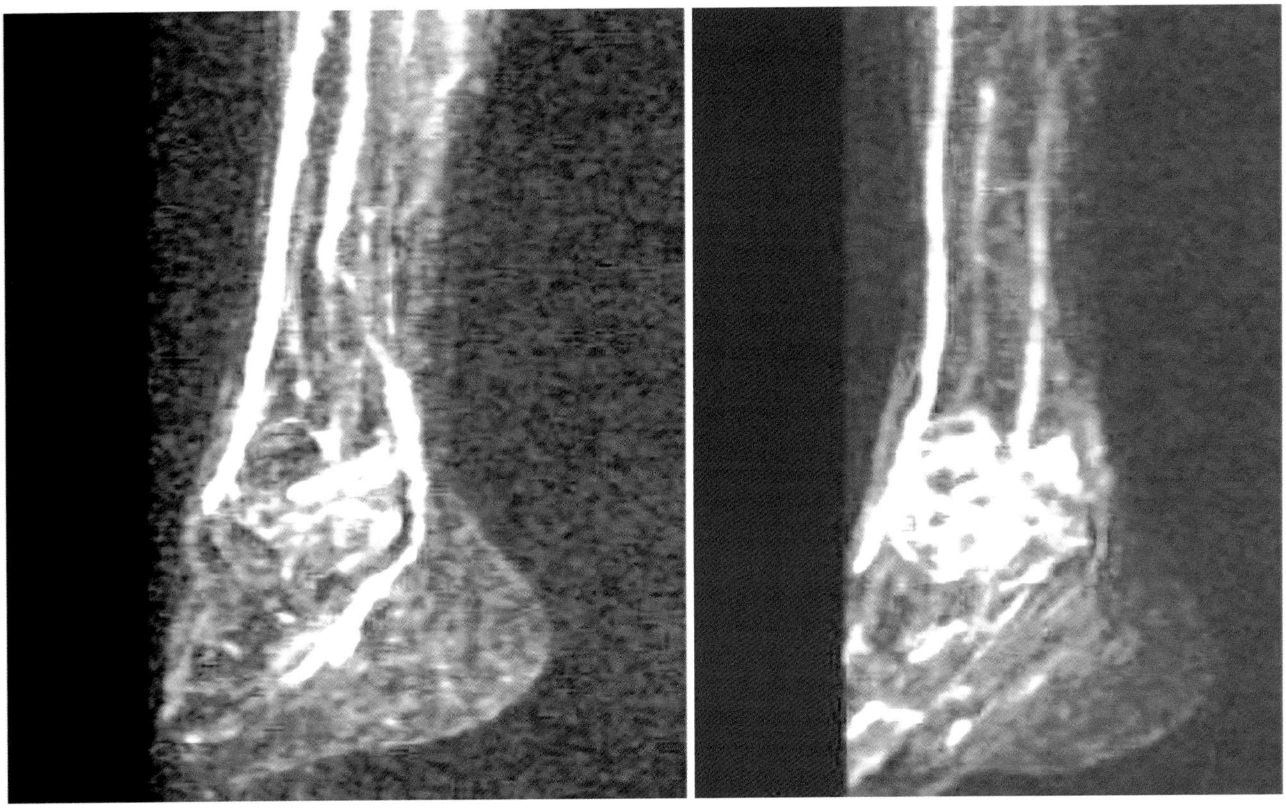

Fig. 20.47 Case 11: Angiogram of the right foot at 6 months shows multiple anastomoses with dorsalis pedis and tibialis posterior with arteries extending down into the foot. [Reprinted from Carstens MH, et al. Non-reconstructable peripheral vascular disease of the lower extremity in ten patients treated with adipose-derived stromal vascular fraction cells. Stem Cell Res 2017; 18: 14–21. With permission from Elsevier.]

culture-expanded ADSCs alone, again, based on the presence of additional cells with angiogenic capabilities. In fact, Zimmerlin et al. [169] identified within the SVF mix *two populations of endothelial cells* (one mature and one progenitor, discriminated by the expression of CD34 and CD90), and *two populations of perivascular stabilizing cells* (pericytes and supra-adventitial cells discriminated by the expression of CD146, α-SMA, and CD34). Collectively, these cell populations along with ADSCs may have a stronger and faster angiogenic response than ADSCs alone and may help explain the slightly better results with non-fractionated SVF than with culture-expanded ADSCs.

What About "Dose"?

Remarkably, we could not find any statistical correlation between the lipoaspirate volume obtained and parameters, such as age and the resulting viable cell yield. Moreover, the cell dose administered (viable cells injected) was not correlated with the degree of clinical improvement. For instance, the patient with the highest cell dose (#7) was the one with no demonstrable angiographic changes. On the other hand, two of the patients that received significantly lower doses (#1 and #6) exhibited impressive wound healing results and angiographic changes.

- Of note, the patient who received the lowest dose of all (#8) corresponds to the one who underwent successful skin grafting, suggesting a *potential minimum cell "dose"* required for a therapeutic result, or that intrinsic factors of the individual patient's tissues determine the wound response.

Since all patients received the same injection sequence into the gastrocnemius muscles, they theoretically received a relatively uniform cell dose distributed down the course of the leg. However, some areas showed more constrast than others reflecting local actors such as a zone of critical stenosis. It is possible then that the improvement in such cases took place in the microcirculation, beyond the resolution of the scans.

- *In sum*, a dose–response correlation is difficult to ascertain.

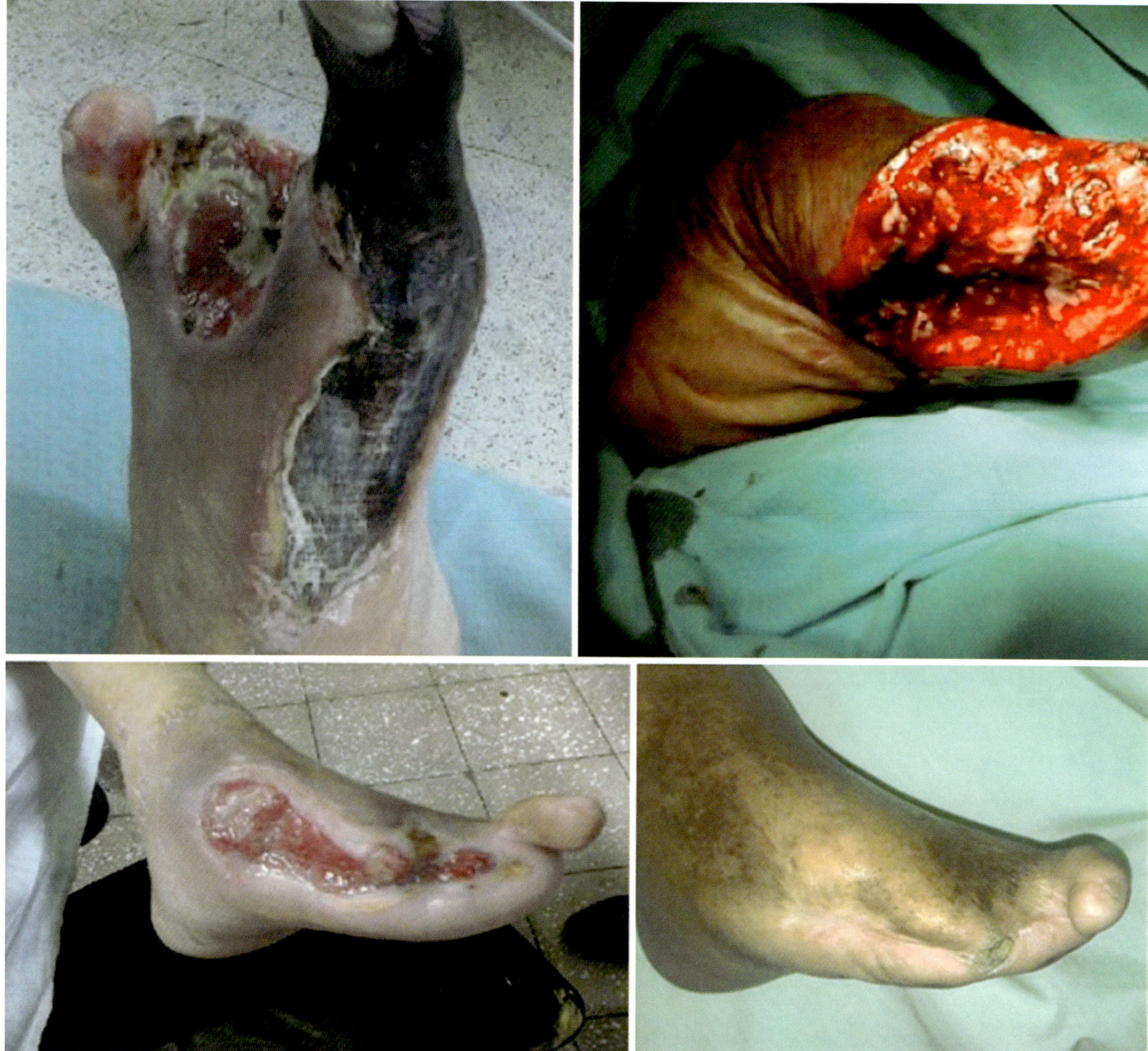

Fig. 20.48 Case 12: A 75-year-old diabetic male with 4-year follow-up. Top left: Preoperative condition prior to amputation of the great toe and debridement. SVF cells were transplanted at a separate procedure 1 week later. Top right: 3 months after SVF the vascularity of the wound is increased. Bottom left: 6 months after SVF the wound is closing. Bottom right: 9 months after SVF the wound was healed without surgery. This photograph shows the foot at 4-year follow-up without evidence of further tissue breakdown. Note, as a diabetic, the pre-op ABI was artificially elevated to 1.4 and, due to increased inflow, was further increased to 2.0 at 18 months. At 4 years postop, due to anti-fibrotic action of SVF ABI has declined into the normal range. [Reprinted from Carstens MH, et al. Non-reconstructable peripheral vascular disease of the lower extremity in ten patients treated with adipose-derived stromal vascular fraction cells. Stem Cell Res 2017; 18: 14–21. With permission from Elsevier.]

Conclusion: Is Angiogenic Under Ischemic Condition Without a Morphogen

The injection of autologous adipose-derived SVF cells represents a novel alternative for a point-of-care cell-based therapy for symptomatic patients with non-reconstructable PVD and CLI that offers the possibility of relief from ischemic pain, improved healing of chronic wounds, and the possibility of tissue preservation.

- Cells localize over time to areas of maximum ischemia. These are angiosome specific.
- Application of cells can be directly into the tissues, intramuscular, or intra-arterial.
- Based on our experience with ASC injection into the renal artery ($n = 38$) without complications suggests that the safety profile of femoral artery infusion be established and the downstream effects of this cell infusion be confirmed.

- Long-term anatomic changes in the pedal arteries (size, waveform) suggest that cell therapy could be complementary to angioplasty, perhaps prolonging its effectiveness.
- The effects of SVF infiltration on peripheral neuropathy should be studies and quantitative measurements pursued.

Composite In Situ Osteogenesis: ASCs Should Show Morphogen-Driven Differentiation to the Osteogenic Line

In this chapter we have been considered the development of morphogen-based reconstructions, followed by the introduction of mesenchymal signaling cells in the form of ASCs. Consider once again our equation:

Bone = resident stem cells + mesenchymal signaling cells + morphogen + lattice + template + blood supply

This chapter has consisted of two quite distinct themes: the development of morphogen-based reconstructions, followed by the introduction of mesenchymal signaling cells in the form of ASCs. To this point these themes has so far been kept separate. We have not yet thrown MSCs in the mix. We shall now consider three forms of ISO using ADSCs: ASCs plus a biomaterial alone. ASCs with rhBMP-7 and ASCs with rhBMP-2.

In 2004 Lendeckel, citing the reports Zuk and Futrell of multilineage differentiation of human adipose tissue, reported application of 295 × 10^6 MNCs obtained from enzymatic digestion of 42 g of autologous fat with an expected yield of 2–3% giving a dose of 8.85 × 10^6 "stem cells" which were subsequently spread out over a bed of 120 cancellous bone chips from the iliac crest to reconstruct a 120 cm^2 post-traumatic calvarial defect. To keep the cells in place, autologous fibrin glue was added. 3-D CT scan at 3 months demonstrated the defect to be completely filled. The entire procedure was carried out in the operating room. This report is as far as I know the first human application of ASCs for a cranial defect.

Subsequently Thesloff and Suuronen at Tampere University, Finland (2011) performed 4 cranioplasties for defects averaging 7.8 × 6.1 cm. In these cases the adipose tissue was first minced and digested with a recombinant collagenase; the cells were then expanded over 22 days. A total of 15 × 10^6 cells were then admixed with 60 mL of beta-TCP granules 48 h prior to surgery. These critical-size defects were closed at 3 months. CT scan at 1-year post-op showed Hounsfield units of bone density equivalent to surrounding bone.

In 1998, Clokie and Sandor at the University of Toronto, in cooperation with orthopedic surgeon Marshall Urist at UCLA, reconstructed a 6-cm mandibular body defect resulting from resection for ameloblastoma in a patient previously treated with chemotherapy and high dose radiation for Hodgkin's lymphoma using 200 mg of "native human BMP," origin and process not specified. Recall that the recombinant form of BMP-2 was not available from Genetics Institute. Martin Chin and I obtained it via a compassionate use protocol with the FDA in 2001.

Urist was the true godfather of BMP-2 but his work was carried into maxillofacial surgery by Phil Boyne at Loma Linda. It was Boyne's pioneering experimental work, including in primates that influenced George Sandor and Martin Chin. Clokie [170] had worked directly with Urist on a binding vehicle for BMP, Poloxamer. When BMP-7 became available as OP-1, Clokie and Sandor [171] developed a protocol for mandibular reconstruction mixing rhBMP-7 with a demineralized bone matrix (Dyngraft® putty) which hardens into a shape at body temperature. They suspended the DBM in a reverse-phase copolymer medium using poloxamer 407. They documented 10 cases of post-resection mandibular defects using this construction (Figs. 20.49, 20.50 and 20.51). Clokie remained at Toronto while Sandor continued his interest with ISO and BMPs at the University of Tampere where he collaborates with Suuronen who had in 2007 independently combined rhBMP-2 with ASCs.

Sandor's work at Tampere continued using cultured ASCs with rhBMP-2 (InductOs, Wyeth Europe) in 12 mg batches using variety of scaffold material applied to a large mandibular defect permitting the eventual placement of implants [172]. Subsequently a series was reported with consisting of 13 craniomaxillofacial defects [173] resulted in success with 10/13 cases, the remainder requiring re-operation (successful) for problems derived from the resorbable meshes employed. A perspective on this work is available in Bennun's *Cleft Lip and Palate Management: A Comprehensive Atlas*, Wiley 2015 (Figs. 20.49, 20.50, 20.51, 20.52 and 20.53).

First, Sandor's work of great value demonstrates conclusively the possibility of successful bone reconstruction using ISO and ASCs with a BMP-2 morphogen. Second, the work is diverse and well documented. It readily translates to SVF. Of note, DBM putty can be of great value in shaping a reconstruction.

The literature regarding applications of ASCs and BMP-2 is expanding steadily as all variables of our equation appear to be filled. What remains to be defined is the potential contribution of SVF as the vehicle containing ASCs. The characteristics of SVF have been summarized [142, 143]. The technology of its production renders high cell counts with rapid processing times and cell harvests as low as 100–150 cc of lipoaspirate.

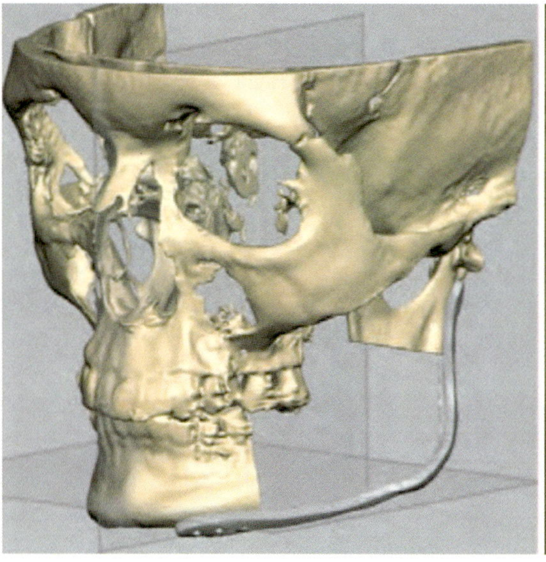

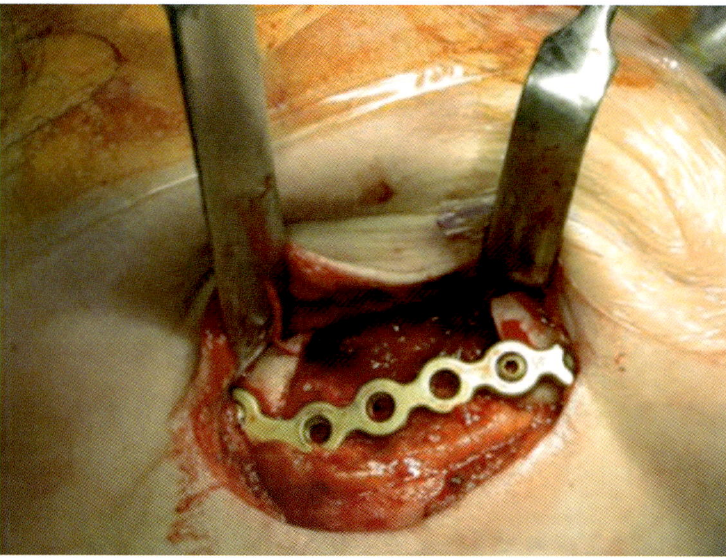

Fig. 20.49 Case 13: (Left): Mandibular reconstruction planned for resection of recurrent ameloblastoma. Adipose tissue was resected minced and digested with recombinant collagenase NB6 (InVitrogen Paisley Scotland) and then expanded. After 3 weeks cells numbering approximately 10 × 106 were combined with B-TCP granules and 12 mg of rhBMP-2 (InductOs Wyeth Europa). Careful analysis of the ASCS was performed including viability and immunophenotyping. (Right): resection defect seen spanned with a prefabricated reconstruction plate. Defects ranging in size from 6 to 10 cm were successfully bridged. [Reprinted from Wolff J, Sándor GK, Miettinen A, et al. GMP-level adipose stem cells combined with computer-aided manufacturing to reconstruct mandibular ameloblastoma defects: Experience with three cases. Ann Maxillofac Surg 2013; 3(2):114–125. With permission from Wolters Kluwer Health.]

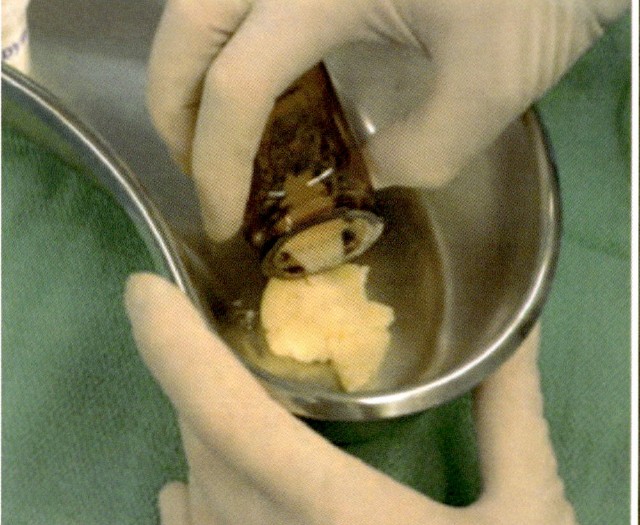

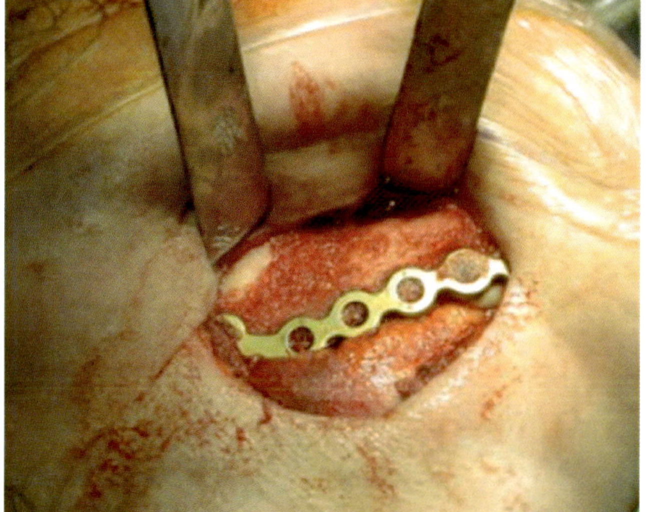

Fig. 20.50 Case 13: DBM putty being mixed and placed into the defect. rhBMP-7 + beta TCP expanded ASCs. [Reprinted from Wolff J, Sándor GK, Miettinen A, et al. GMP-level adipose stem cells combined with computer-aided manufacturing to reconstruct mandibular ameloblastoma defects: Experience with three cases. Ann Maxillofac Surg 2013; 3(2):114–125. With permission from Wolters Kluwer Health.]

- Quick turn-around time in the operating room and high cell counts make SVF a logical "go-to" source of ASCs for single-stage craniofacial reconstructions

And so it is that we come to the end of this brief foray into the role of biologics in craniofacial bone reconstruction. Obviously, what is presented here is just a sampling of the current research. Between the lines you can see a story of how concepts and techniques have morphed over time. But evolution is a continuous process. All clinicians and basic scientists contribute to the whole. And so it is for you, the reader, and those with whom you work, to write the next chapter in this story to expand our understanding of craniofacial structure, its evolution, its pathologies and thereby provide better solutions for our patients.

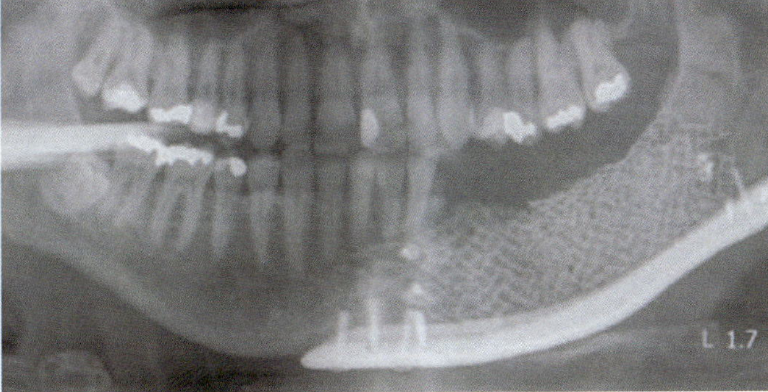

Fig. 20.51 Case 13: Bone graft in place consisting of 1000 g ASCs 12 mg rhBMP-2 (InductOS® Wyeth Europa) 40-mL Beta-TCP (Chronos® Synthes). Seen on panorex at 3 months with bone fill readily appreciated. [Reprinted from Wolff J, Sándor GK, Miettinen A, et al. GMP-level adipose stem cells combined with computer-aided manufacturing to reconstruct mandibular ameloblastoma defects: Experience with three cases. Ann Maxillofac Surg 2013; 3(2):114–125. With permission from Wolters Kluwer Health.]

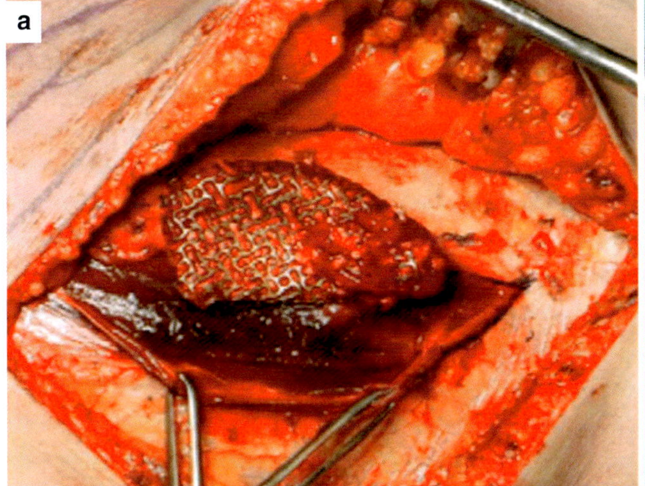

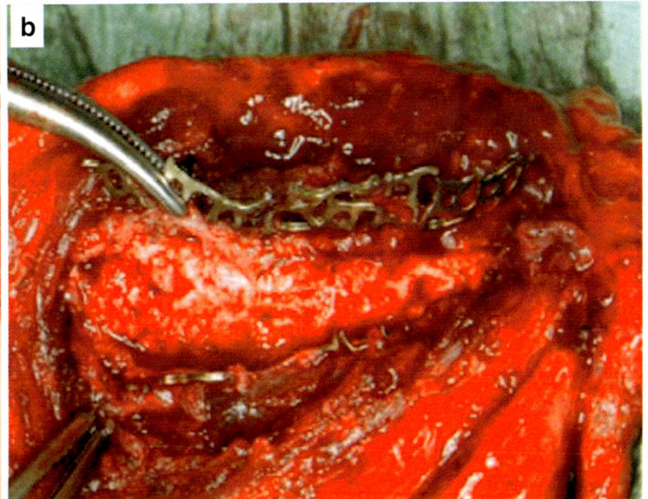

Fig. 20.52 Case 14: Prefabricating a rectus abdominis free flap for reconstruction of the right maxilla using 12 mg rhBMP-2 (InductOs® Wyeth Europa). (**a**) Muscle is implanted with a titanium cage filled with adipose stem cells and B-TCP. (**b**) Rectus free flap raised and the graft is exposed. Note the bleeding bone. In addition to the ASCs, muscle provided a source of resident MSCs. [Reprinted from Mesimaki K, Türnwall J, Suuronen R, et al. Novel maxillary reconstruction wity ectopic bone formation by BMP adipose stem cells. Int J Oral Maxillofac Surg 2009; 38(3):201–209. With permission from Elsevier.]

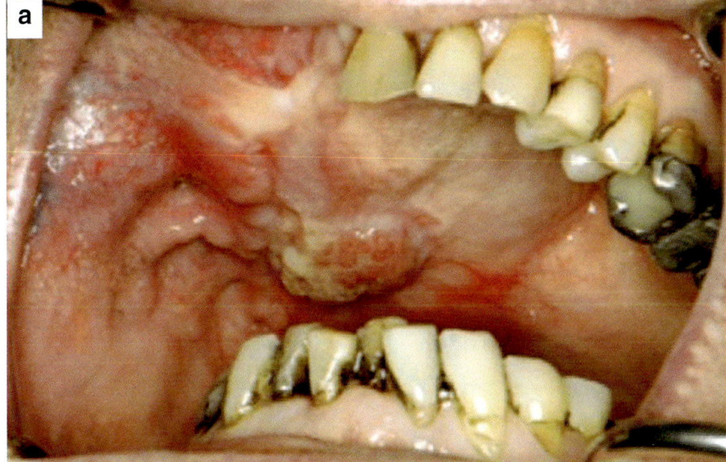

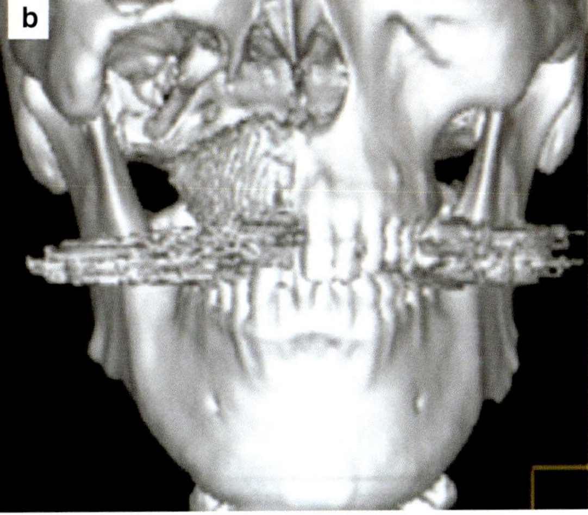

Fig. 20.53 Case 15: Right maxilla and hard palate reconstructed. (**a**) Showing ectopic bone formation by BMP adipose stem cells. (**b**) CT scan showing restoration. [Reprinted from Mesimaki K, Türnwall J, Suuronen R, et al. Novel maxillary reconstruction with ectopic bone formation by BMP adipose stem cells. Int J Oral Maxillofac Surg 2009; 38(3):201–209. With permission from Elsevier.]

Commentary: S. M. Balaji

This book is the culmination of a few decades' long experience as well as the lifelong accumulated knowledge of the author—Prof. Michael Carstens. It is an enthusiastic celebration of many innovations, modifications, and discoveries that the author has done in the field of facial cleft rehabilitation, reconstruction, and regenerative medicine. He relies on and uses tools that are either of natural origin or biomimetics of human proteins in function and most importantly biological relevance. This deep understanding of biological mechanisms is presented in the earlier part of the book where the process of embryo formation and face development is meticulously described. Michael and I have worked together in complex craniofacial cases in India and several parts of the world. During these times, I have closely observed the way he would translate basic knowledge into innovative surgical results. His unique and deep understanding of the complex anatomy, physiology, and molecular biology would thrill me. I always looked forward to operating with him. In this process, we mutually learnt a lot while our patients stood benefitted.

This textbook—**Craniofacial Structure: Developmental Anatomy, Evolutionary Design, and Clinical Applications**—is seminal and is also a fitting unique tribute to the many forerunners in the field. The abundance of colorful illustrations—clinical as well as infographics—adds unique value to this compilation of experiences. Michael Carstens' ability to interweave clinical situations with basic molecular biology in a very appealing way is a standing testimony to his focus and mastery of the subject. Although the emphasis of this book is predominantly clinical, the overall effect is that such a treatment approach will revolutionize the way healthcare delivery is done. The ultimate benefactor of this newly emerging scientific field will be society. After all, the destiny of science and well-being of society are highly linked. Every chapter abounds with the enthusiasm of the author and his collaborators. The book is the result of an intellectual fascination with human facial development, which has motivated to understand exactly how this process works. The work is unique in its profundity and breadth. It represents a passionate search for underlying mechanisms as the fundamental requirement for the treatment of facial clefts. The results of this quest are told as a tightly knit story in an intriguing and visually appealing way. I sincerely believe that this book will become a cornerstone of craniofacial education. This will also take craniofacial cleft reconstruction beyond the science of mere observation. Such a developmental approach based on neuroembryology represents a new method of inquiry leading toward a better understanding of the complexity of human facial development.

Donning several hats—clinician, surgeon, academician, researcher, entrepreneur, and scientist—Prof. Carstens has condensed his skills, experience, and knowledge into a single textbook, with a noble intention to pass on this hard-earned knowledge to the next generation of surgeons. The approach is a unique and forward-thinking approach; in point of fact, it represents a challenge, a throwing-down of the gauntlet as it is the obvious intention of the author to motivate the reader to think, probe, and act. He has been quite successful in this noble goal. The book succeeds in captivating the readers and takes them through various stages of face formation right from the embryo. Later, it branches into the neuromeric theory and the neurovascular embryology behind how various clefts are formed. In my opinion, this knowledge is directly relevant for the diagnosis and treatment of a wide variety of craniofacial anomalies. This book serves as a reminder that medical-surgical science will continue to accomplish unimaginable feats so long as people like Prof. Michael Carstens continue to look beneath the surface and challenge the dogma of the status quo, in the search for better results in the well-being of patients.

As a colleague and friend, I wish Prof. Michael Carstens all success and all readers a fruitful reading and learning process.

S.M. Balaji

Balaji Dental and Craniofacial Hospital, Teynampet, Chennai, India

A Note from Dr. Carstens

S.M. Balaji

From the vantage point of the Balaji Dental and Craniofacial Hospital in Chennai (formerly Madras), Tamil Nadu, India, Prof. Dr. S.M. Balaji cuts an imposing figure with an impossibly long list of accomplishments. As his full name is also impossibly long, I shall refer to him as Bala, and he is a true force of nature. A relentless perfectionist, academician, optimist, and general overachiever, his energy (seemingly boundless) is somehow contained within a sizeable and powerful frame. However, behind the booming voice and irrepressible smile is a restless and innovative mind, rather unencumbered by dogma and quick to seek clinical implementation for concepts he considers of value. Given this constellation of attributes, it is not surprising that Bala has been the driving force behind the introduction of rhBMP-2 to reconstructive surgery for the Indian subcontinent and Southeast Asia. We started out doing cleft cases together, but he quickly found a way innovate with bone grafts with an eye toward regenerative applications in the future. In the operating theatre, I found him to be a master technician and a never-ending source of new ideas. Away from the hospital, Bala and his wife, Sachin, were the most gracious of hosts. But our cooperation did not stop there, for Bala is a tireless organizer and connector—his conferences in the Seychelles and Maldives gathered together like minds with the results that were eye-opening. Bala showed me the promise of a much larger world. I am convinced that innovation in medicine for the twenty-first century will come about as clinicians and scientists from India harness the incredible power of their clinical experience and produce studies that will change the direction of our thinking and techniques. As this story unfolds, I am sure that Dr. S.M. Balaji will be in the forefront; I hope to follow along to see its denouement.

References

1. Carstens MH, Gomez AF, Cortes R, Turner E, Perez C, Ocon M, Correa D. Non-reconstructable peripheral vascular disease of the lower extremity in ten patients treated with adipose-derived stromal vascular fraction cells. Stem Cell Res. 2017;18:14–21.
2. Carstens MH, Chin M. Regeneration of 10 cm mandibular defect in pig using recombinant human bone morphogenetic protein-2 and Helistat® absorbable collagen sponge. J Craniofac Surg. 2005;16(6):1033–42.
3. Chen B, Lin H, Wang J, Zhao Y, Wang B, Zhao W, Sun W, Dai J. Homogenous osteogenesis and bone regeneration by demineralized bone matrix loading with collagen-targeting bone morphogenetic protein-2. Biomaterials. 2007;28:1027–35.
4. Leindeckel S, Jodicke A, Heidinger K, et al. Autologous stem cells (adipose) and fibrin glue used to treat widespread calvarial defects: case report. J Craniomaxillofac Surg. 2004;32:370–3.
5. Warnke PH, Springer IN, Witfang J, et al. Growth and transplantation of a custom vascularized bone graft in a man. Lancet. 2004;364:766–70.
6. Markus AF, Delaire J, Smith WP. Facial balance in cleft lip and palate. I. Normal development and cleft palate. Br J Oral Maxillofac Surg. 1992;30:290.
7. Bura A, Planat-Benard V, Bourin P, Silvestre J-S, Gross F, Grolleau J-L, Saint-Lebese B, Peyrafitte J-A, Fleury S, Gadelorge M, Taurand M, Dupuis-Coronas S, Leobon B, Casteilla L. Phase I trial: the use of autologous cultured adipose-derived stroma/stem cells to treat patients with non-revascularizable critical limb ischemia. Cytotherapy. 2014;16:245–57.
8. Walker JC Jr, Collito MB, Mancuso-Ungaro A, Meijer R. Physiologic considerations in cleft lip closure. The C-W technique. Plast Reconstr Surg. 1966;37:552
9. Bardach J, Kelly KM. The influence of lip repair with and without soft tissue undermining on facial growth in beagles. Plast Reconstr Surg. 1988;82:747.Leipziger LS, Schnapp DS, Haworth RD, et al. Facial growth after timed soft-tissue undermining. Plast Reconstr Surg. 1992;89:809.
10. Ross RB. Treatment variable affecting facial growth in unilateral cleft lip and palate: 3. Alveolus repair and bone grafting. Cleft Palate J. 1987;24:33.
11. Bishara SE, Jakobsen JR, Krause JC, Sos-Martinez R. Cephalometric comparisons of individuals from India and Mexico with unoperated cleft lip and palate. Cleft Palate J. 1986;23:116.
12. Ortiz-Monasterio F, Rebeil AF, Valderrama M, Cruz R. Cephalometric measurements on adult patients with non-operated cleft palates. Plast Reconstr Surg. 1959;24:53.
13. Ortiz-Monasterio F, Serrano A, Barrera G, RodriguezHoffman H. A study of untreated adult cleft palate patients. Plast Reconstr Surg. 1966;38:36.
14. Kapucu MR, Guler Gursu K, Enacar A, Aras S. Effect of cleft lip repair on maxillary morphology in patients with unilateral complete cleft lip and palate. Plast Reconstr Surg. 1996;97:1371.
15. Siebert JW, Angrigiani C, McCarthy JG, Longaker MT. Blood supply of the LeFort I maxillary segment: an anatomic study. Plast Reconstr Surg. 1997;100:8434.
16. De Lacure MD. Physiology of bone healing and bone grafts. Otolaryngol Clin N Am. 1994;27:859.
17. Muir IFK. Repair of the cleft alveolus. Br J Plast Surg. 1966;19:30.
18. Backdahl M, Nordin KE. Replacement of the maxillary bone defect in cleft palate. A new procedure. Acta Chir Scand. 1961;122:131.
19. Johnsson G, Stenstrom S, Thilander B. The use of a vomer flap covered by an autogenous skin graft as part of the palatal repair in children with unilateral cleft lip and palate. Arch dimensions and occlusions up to the age of five. Scand J Plast Reconstr Surg. 1980;14:13.
20. Ritsala V, Alhuporo S, Gylling U, Rintala A. The use of free periosteum for bone formation in congenital clefts of the maxilla. Scand J Plast Reconstr Surg. 1972;6:57.
21. El Deeb ME, Hinrichs JE, Waite DE, et al. Repair of alveolar cleft defects with autogenous bone grafting. Periodontal evaluation. Cleft Palate J. 1986;23:126.
22. Vanarsdell RL, Corn H. Soft tissue management of labially positioned unerupted teeth. Am J Orthod. 1977;72:53.
23. Steedle SR, Profitt WR. The pattern and control of eruptive tooth movements. Am J Orthod. 1985;87:56.
24. Wolfe SA, Price GW, Stuzin JM, Berkowitz S. Alveolar and anterior palatal clefts. In: McCarthy JG, Wood RJ, Grayson BH, Cutting CB, editors. Gingivoperiosteoplasty and midfacial growth, vol. 34. New York: Springer; 1997. p. 17.
25. Rintala AE, Ranta R. Periosteal flaps and grafts in primary cleft repair: a follow up study. Plast Reconstr Surg. 1989;83:17.
26. Šmahel Z, Mullerova Z. Effects of primary periosteoplasty on facial growth in unilateral lip and palate: a 10-year follow-up. Cleft Palate J. 1988;25(4):356–61.
27. Šmahel Z, Mullerova Z. Facial growth in unilateral cleft lip and palate prior to the eruption of permanent incisors after primary bone grafting and periosteoplasty. Acta Chir Plast. 1996;38(1):30–6.
28. Cohen M, Polley JW, Figueroa AA. Secondary (intermediate) alveolar bone grafting. Clin Plast Surg. 1993;20:691.
29. Cohen SR, Figueroa AA, Aduss H. The role of gingival mucoperiosteal flaps in the repair of alveolar clefts. Plast Reconstr Surg. 1989;83:812.
30. Cohen SR, Kalinowski J, LaRossa D, Randall P. Cleft palate fistula: a multivariate statistical analysis of prevalence, etiology, and surgical management. Plast Reconstr Surg. 1991;87:1041.
31. Brusati R, Mannucci N. The early gingivoalveoloplasty. Preliminary results. Scand J Plast Reconstr Surg Hand Surg. 1992;26:65.
32. Dixit UB, Kelly KM, Squier MA, Bardach J. Periosteum in regeneration of palatal defects. Cleft Palate J. 1995;32:228.

33. Massei A. Reconstruction of the cleft maxilla with periosteoplasty. Scand J Plast Reconstr. 1986;20(1):41–4.
34. Santoni-Rugiu P. Periosteal flaps and grafts in primary cleft repair; a follow-up study. Plast Reconstr Surg. 1989;83:23.
35. Friede H, Johansen B. A follow-up study of cleft children treated with primary bone grafting. Scand J Plast Reconstr Surg. 1974;8:88.
36. Lehman JA, Curtin P, Haas DG. Closure of anterior palatal fistulae. Cleft Palate J. 1978;15:33.
37. Lehman JA, Douglans BK, Ho WC, Husami TW. One-stage closure of the entire primary palate. Plast Reconstr Surg. 1990;86:675.
38. Badran HA, Maher H, El Barbary A. Experience with the use of primary periosteoplasty in the repair of alveolar clefts. Presented at the PSEF international symposium on surgical techniques in cleft lip and palate: long-term results. San Francisco, CA. September 19, 1997
39. Dado DV, Kerhahan DA. Anatomy of the orbicularis oris muscle in incomplete unilateral cleft lip based on histological examination. Ann Plast Surg. 1984;15:90.
40. Fara M. Anatomy and arteriography of cleft lips in still-born children. Plast Reconstr Surg. 1968;42:29.
41. Kernahan DA, Dado DV, Bauer BS. Functional cleft lip repair: a sequential, layered closure with orbicularis muscle realignment. Plast Reconstr Surg. 1983;72:459.
42. Kernahan DA, Dado DV, Bauer BS. The anatomy of the orbicularis oris muscle in unilateral cleft lip based on a threedimensional histologic reconstruction. Plast Reconstr Surg. 1984;73:875.
43. Delaire J, Precious DS. Influence of the nasal septum on maxillonasal growth in patients with congenital labiomaxillary cleft. Cleft Palate J. 1986;23:270.
44. Delaire J. Theoretical principles and technique of functional closure of the lip and nasal aperture. J Maxillofac Surg. 1978;6:109.
45. Precious DA, Delaire J. Surgical considerations in patients with cleft deformities. In: Bell WH, editor. Modern practice in orthognathic and reconstructive surgery. Philadelphia: Saunders; 1992. p. 390–425.
46. Wang Y-C, Liao Y-F, Chen PKT. Comparative outcomes of primary gingivoperiosteoplasty and secondary alveolar bone grafting in patients with unilateral cleft lip and palate. Plast Reconstr Surg. 2016;137:a218–27.
47. Bergland O, Semb G, Abyholm FE. Elimination of the residual alveolar cleft by secondary bone grafting and subsequent orthodontic treatment. Cleft Palate J. 1986;23:175.
48. Moss ML, Vilman H, Das Gupta G, Salak R. Craniofacial growth in space-time. In: Carlson BR, editor. Craniofacial biology. Ann Arbor: University of Michigan; 1981.
49. Opitz JM. Errors of morphogenesis and developmental field theory. Am J Med Genet. 1998;76:291–6.
50. Chin M, Boyne P, et al. Distraction osteogenesis with bone morphogenetic protein enhancement in the extension of edentulous bone. In: Proceedings of the 3rd international congress on cranial and facial distraction processes, Paris, France, 2001, pp. 19–22.
51. Gilbert SF. Developmental biology. 11th ed. Sunderland: Sinauer; 2016.
52. Carstens MH. Functional matrix cleft repair: principles and techniques. Clin Plast Surg. 2004a;31:159–89.
53. Morriss-Kay GM. Derivation of the mammalian skull vault. Development. 2001;199:143–51.
54. Barghusen HR, Hopson JA. The endoskeleton: comparative anatomy of the skull and visceral skeleton. In: Wake MH, editor. Hyman's comparative vertebrate anatomy. 3rd ed. Chicago: University of Chicago; 1979.
55. Yasui N, Sato M, Ochi T, et al. Three modes of ossification during distraction osteogenesis in the rat. J Bone Jt Surg. 1997;79B:824–30.
56. Liem KF, Tremmal C, Roelink H, et al. Dorsal differentiation of neural plate cells induced by BMP-mediated signals from epidermal ectoderm. Cell. 1995;82:969–79.
57. Le Douarin NM, Kalcheim C. The neural crest. 2nd ed. Cambridge: Cambridge University Press; 1999.
58. Maschoff KL, Baldwin HS. Molecular determinants of neural crest migration. Am J Med Genet. 2000;97:280–2888.
59. Boyne PJ. A feasibility study evaluating rhBMP-2/absorbable collagen sponge for maxillary sinus floor augmentation. Int J Periodontic Restorative Dent. 1997;17:11–25.
60. Boyne PJ. Animal studies of application of rhBMP-2 in maxillofacial reconstruction. Bone. 1996;19(Suppl):83S–92S.
61. Boyne PJ. Application of bone morphogenetic proteins in the treatment of clinical oral and maxillofacial osseous defects. J Bone Jt Surg. 2001;83A(Suppl 1):S146–50.
62. Boyne PJ, Nath R, Nakamura A. Human recombinant BMP-2 in osseous reconstruction of simulated cleft palate defects. Br J Oral Maxillofac Surg. 1998;36:84–90.
63. Valentin-Opran WJ, Wozney J, Csiima C, Lilly L, Reidel GE. Clinical evaluation of recombinant human bone morphogenetic protein-2. Clin Orthop Relat Res. 2002;395:110–20.
64. Majumdar MK, Wang E, Moris EA. BMP-2 and PMP-9 promote chondrogenic differentiation of human multipotential mesenchymal cells and overcome the inhibitory effect of IL-1. J Cell Physiol. 2001;189:275–84.
65. zur Nieden NI, Kempka G, Rancourt DE, Ahr H-J. Induction of chondro-, osteo-, and adipogenesis in embryonic stem cells by bone morphogenetic protein-2: effect of cofactors on differentiating lineages. BMC Dev Biol. 2005;5:1–15.
66. Seeherman H, Wozney J, Li R. Bone morphogenetic protein delivery systems. Spine. 2002;27(165):516–23.
67. Chang SC, Wei FC, Chuang H, et al. Ex vivo gene therapy in autologous critical-size craniofacial bone regeneration. Plast Reconstr Surg. 2003;112:141–50.
68. Toriumi D. Mandibular reconstruction with a recombinant bone morphogenetic protein (rhBMP-2). Arch Otolaryngol Head Neck Surg. 1991;117:1101–12.
69. Sciadini MF, Johnson KD. Evaluation of rhBMP-2 as a bone-graft substitute in a canine segmental model. J Orthop Res. 2000;18:289–302.
70. Ebara S, Nakayama K. Mechanism for the action of bone morphogenetic proteins and regulation of their activities. Spine. 2002;27(16S):S10–5.
71. Noden DM. Interactions and fates of avian craniofacial mesenchyme. Development. 1988;103(Suppl):121–40.
72. Muller F, O'Rahilly R. Segmentation in staged human embryos: the occipitocervical region revisited. J Anat. 2003;203:297–315.
73. Conlon RA. Retinoic acid and pattern formation in vertebrates. Trends Genet. 1995;11:314–9.
74. Ruberte J, Carretero A, Navarro M, et al. Morphogenesis of blood vessels in the head muscles of avian embryos: spatial, temporal, and VEGF expression analysis. Dev Dyn. 2003;227:470–83.
75. Irianov YM. Peculiarities of angiogenesis in distraction regenerates. Genji Oropedii. 1996;2:132.
76. Samchukov ML, Cherkashin AM, Cope JB. Distraction osteogenesis: origins and evolution. In: McNamara Jr JA, Trotman CA, editors. Distraction osteogenesis and tissue Eng. Ann Arbor: University of Michigan; 1998. p. 32.
77. Samchukov ML, Cope JB, Cherkashin AM, editors. Craniofacial distraction osteogenesis. CV Mosby: St. Louis; 2001.
78. Sawaki Y, Heggie ACC. The vascular change during and after mandibular distraction. In: Diner PA, Vasquez MP, editors. 2nd International congress on cranial and facial bone distraction processes. Paris: Monduzzi Editore; 1999.

79. Jones DB, Nolte H, Scholobbers JG. Biomechanical signal transduction of mechanical strain in osteoblast-like cells. Biomaterials. 1991;12(2):101–10.
80. Harter LV, Hruska KA, Duncan RL. Human osteoblast-like cells respond to mechanical strain with increased bone matrix protein production independent of hormonal regulation. Endocrinology. 1995;136:528–35.
81. Ishidou Y, Katajima I, Obama H. Enhanced expression of type 1 receptor for bone morphogenetic proteins during bone formation. J Bone Miner Res. 1995;10:1651–9.
82. Sato M, Ochi T, Nakase T. Mechanical tension-stress induces expression of bone morphogenetic protein BMP-2 and BMP-4, but not BMP-6, or BMP-7. J Bone Miner Res. 1999;14(7):84–95.
83. Sato M, Yasui N, Nakase T. Expression of bone morphogenetic protein mRNA during distraction osteogenesis. J Bone Miner Res. 1998;13:1221.
84. Carstens MH. Developmental field reassignment: reassessment and refinements. In: Bennun R, Sandor GKB, editors. Cleft lip and palate management: a comprehensive atlas. New York: Wiley-Blackwell; 2015.
85. Chin M. Bone morphogenetic protein enhancement of alveolar distraction in humans. In: From the proceedings of the 4th international congress on cranial and facial distraction processes, Paris, 2003, pp. 49–51.
86. Chin M. Distraction osteogenesis in maxillofacial surgery. In: Lynch SE, Genco RJ, Marx RE, editors. Tissue engineering: applications in maxillofacial surgery and periodontics. Carol Stream: Quintessence Publishers; 1998.
87. Chin M, Carstens M, et al. Distraction osteogenesis with bone morphogenetic protein enhancement: facial cleft repair in humans. In: From the proceedings of the 4th international congress on cranial and facial distraction processes, Paris, 2003, pp. 197–200.
88. Franco J, Carstens MH. Mandibular distraction using bone morphogenetic protein and rapid distraction in neonates with Pierre Robin syndrome. J Craniofac Surg. 2010;21(4):1158–61.
89. Kjaer I. Etiology-based craniofacial and dental diagnostics. New York: Wiley-Blackwell; 2016.
90. Kjaer I, Keeling JW, Fisher-Hansen B. The prenatal human cranium. New York: Munksgaard/Wiley; 1999.
91. Sato Y, Grayson BH, Garfinkle JS, et al. Success rate of gingivoperiosteoplasty with and without secondary bone grafts compared with secondary alveolar bone grafts alone. Plast Reconstr Surg. 2008;121:1356Y1367.
92. Chao M, Donovan T, Sotelo C, Carstens MH. In situ osteogenesis of Hemimandible with rhBMP-2 in a 9-year-old boy: osteoinduction via stem cell concentration. J Craniofac Surg. 2006;17(3):405–12.
93. Betts N, Fonseca R. Allogeneic grafting of dentoalveolar clefts. Oral Maxillofac Surg Clin North Am. 1991;3:617–24.
94. Urist MR. Bone: formation by autoinduction. Science. 1965;150:893–9.
95. Urist MR, Sato K, et al. Purification of bovine bone morphogenetic protein by hydroxyapatite chromatography. Proc Natl Acad Sci U S A. 1984;81:371–5.
96. Boyne PJ. Use of marrow cancellous bone grafts in maxillary alveolar and palatal clefts. J Dent Res. 1974;43:821–4.
97. Demas PN, Sotereanos GC. Closure of alveolar clefts with corticocancellous block grafts and marrow: a retrospective study. J Oral Maxillofac Surg. 1988;46:682.
98. Liou E, Chen PK, et al. Interdental distraction osteogenesis and rapid orthodontic tooth movement: novel approach to approximate a wide alveolar cleft or bony defect. Plastic Reconstr Surg. 2000;105:1262–72.
99. Urist MR, Strates BS. Bone morphogenetic protein. J Dent Res. 1971;50:1392–406.
100. El Deeb M, Wolford L. Utilization of alloplastic ceramics in repair of alveolar clefts and correction of skeletofacial deformities in patients with cleft palate. Oral Maxillofac Surg Clin North Am. 1991;3:625–40.
101. Abyholm PE, Bergland O, Semb G. Secondary bone grafting of alveolar clefts. Scand J Plast Surg. 1981;15:127.
102. Koole R. Ectomesenchymal mandibular symphysis bone graft: an improvement in alveolar cleft grafting? Cleft Palate Craniofac J. 1994;31(3):217–23.
103. Falluco M, Carstens MH. Primary reconstruction of alveolar clefts using recombinant human bone morphogenetic protein-2: clinical and radiographic outcomes. J Craniofac Surg. 2009;Suppl 2:1759–64.
104. Dickson BP, Ashley RK, Wasson KL, et al. Reduced morbidity and improved healing with bone morphogenic protein-2 in older patients with alveolar cleft defects. Plast Reconstr Surg. 2008;121:209–17.
105. Alawi F. Benign fibro-osseous diseases of the maxillofacial bones: a review and differential diagnosis. Am J Clin Pathol. 2002;118(Suppl):550–70.
106. Leimolo-Vertanenen L, Vahatelo K, Syrjanen S. Juvenile ossifying fibroma of the mandible: report of 2 cases. J Oral Maxillofac Surg. 2001;54:439–44.
107. Rinaggio J, Land M, Cleveland DB. Juvenile ossifying fibroma of the mandible: a review. J Pediatr Surg. 2003;38(4):648–50.
108. Rosenberg M, Moktari H, Slootweg PJ. The natural course of an ossifying fibroma: a case report. Int J Oral Maxillofac Surg. 1999;26(6):454–6.
109. Williams HK, Mangham C, Speight PM. Juvenile ossifying fibroma: an analysis of eight cases and a comparison with other fibro-osseous lesions. J Oral Pathol Med. 2000;29(1):13–8.
110. Brannon RB, Fowler CB. Benign fibro-osseous lesions: a review of current concepts. Adv Anat Pathol. 2001;8(3):126–43.
111. Johnson LC, Yousefi M, Vinh TN, et al. Juvenile active ossifying fibroma: its nature, dynamics and origin. Acta Otolaryngol Suppl. 1991;488:1–40.
112. Slootweg PJ, Panders AK, Kootmans R. Juvenile ossifying fibroma: an analysis of 33 cases with emphasis on histopathologic aspects. J Oral Pathol Med. 1994;23(9):385–8.
113. Gurol M, Uckan S, Guler N, Yatmax PI. Surgical and reconstructive treatment of a large ossifying fibroma of the mandible in a retrognathic patient. J Oral Maxillofac Surg. 2001;59(9):1097–100.
114. Zama M, Gallo S, Santecchia L, et al. Juvenile ossifying fibroma with massive involvement of the mandible. Plast Reconstr Surg. 2004;113(3):970–4.
115. Bush PG, Williams AJ. The incidence of the Robin Anomalad (Pierrre Robin syndrome). Br J Plast Surg (now JPRS). 1983;36(4):434–7.
116. Benjamin B, Walker P. Management of airway obstruction in the Pierre Robin sequence. Int J Pediatr Otolaryngol. 1991;22:29–37.
117. Caouette-Laberge L, Bayet B, Larocque Y. The Pierre Robin sequence: review of 125 cases and evolution of treatment modalities. Plast Reconstr Surg. 1994;93:934.
118. Tomaski SM, Zalzal GH, Saal HM. Airway obstruction in the Pierre Robin sequence. Laryngoscope. 1995;105:111–4.
119. Zeitouni A, Manoukian J. Tracheostomy in the first year of life. J Otolaryngol. 1993;22:431–4.
120. Codivilla A. On the means of lengthening in the lower limbs, the muscles of and tissues which are shortened through deformity. Am J Orthop Surg. 1905;2:353–7.
121. Dauria D, Marsh JL. Mandibular distraction osteogenesis for Pierre Robin sequence: what percentage of neonates need it? J Craniofac Surg. 2008;19:1237–43.

122. Monasterio-Alfaro L. Alveolar extension palatoplasty demonstrates reduction in palatal fistulae. Orlando: American Cleft Palate Association; 2008.
123. McCarthy JG, Schreiber J, Karp N, et al. Lengthening the human mandible by gradual distraction. Plast Reconstr Surg. 1992;89:1–8.
124. Burstein F. Resorbable distraction of the mandible: technical evolution and clinical experience. J Craniofac Surg. 2008;19:637–43.
125. Carron JD, Derkay CS, Strope GL, et al. Pediatric tracheotomies: changing indications and outcomes. Laryngoscope. 2000;110:1099–104.
126. Caplan AI, Correa D. The MSC: an injury drugstore. Cell Stem Cell. 2011;9:11–5.
127. Stoffers HE, Rinkens PE, Kester AD, Kaiser V, Knottnerus JA. The prevalence of asymptomatic and unrecognized peripheral arterial occlusive disease. Int J Epidemiol. 1996;25:282–90.
128. Marso SP, Hiatt WR. Peripheral arterial disease in patients with diabetes. J Am Coll Cardiol. 2006;47(5):921–9.
129. Chen Q, Smith CY, Bailey KR, Wennberg PW, Kullo IJ. Disease location is associated with survival in patients with peripheral arterial disease. J Am Heart Assoc. 2013;2:e000304.
130. Nehler MR, McDermott MM, Treat-Jacobson D, Chetter I, Regensteiner JG. Functional outcomes and quality of life in peripheral arterial disease: current status. Vasc Med. 2003;8:115–26.
131. Rutherford RB, Baker JD, Ernst C, Johnston KW, Porter JM, Ahn S, Jones DN. Recommended standards for reports dealing with lower extremity ischemia: revised version. J Vasc Surg. 1997;26:517–38.
132. Bishop PD, Feiten LE, Ouriel K, Nassoiy SP, Pavkov ML, Clair DG, Kashyap VS. Arterial calcification increases in distal arteries in patients with peripheral arterial disease. Ann Vasc Surg. 2008;22:799–805.
133. Dormandy JA, Rutherford RB, TASC Working Group. TransAtlantic Inter-Society Consensus (TASC). Management of peripheral arterial disease (PAD). J Vasc Surg. 2000;31:S1–S296.
134. Norgren L, Hiatt WR, Dormandy JA, Nehler MR, Harris KA, Fowkes FGR, Rutherford RB, TASC II Working Group. Inter-society consensus for the management of peripheral arterial disease. Int Angiol. 2007;26(2):81–157.
135. Lévigne D, Tobalem M, Modarressi A, Pittet-Cuénod B. Hyperglycemia increases susceptibility to ischemic necrosis. Biomed Res Int. 2013;2013:490964–5.
136. Mamidi MK, Pal R, Dey S, Bin Abdullah BJJ, Zakaria Z, Rao MS, Das AK. Cell therapy in critical limb ischemia: current developments and future progress. Cytotherapy. 2012;14:902–16.
137. Sedighiani F, Nikol S. Gene therapy in vascular disease. Surgeon. 2011;9:326–35.
138. Isner JM, Asahara T. Angiogenesis and vasculogenesis as therapeutic strategies for postnatal neovascularization. J Clin Invest. 1999;103:1231–6.
139. Raval Z, Losordo DW. Cell therapy of peripheral arterial disease: from experimental findings to clinical trials. Circ Res. 2013;112:1288–302.
140. Caplan AI, Dennis JE. Mesenchymal stem cells as trophic mediators. J Cell Biochem. 2006;98:1076–84.
141. Zuk PA, Zhu M, Mizuno H, Huang J, Futrell JW, Katz AJ, Benhaim P, Lorenz HP, Hedrick MH. Multilineage cells from human adipose tissue: implications for cell-based therapies. Tissue Eng. 2001;7:211–28.
142. Guo J, Nguyen A, Widgerowe AD. Stomal vascular fraction: a regenerative reality? Part 2. Mechanisms of regenerative action. JPRAS. 2016;69:180–8.
143. Nguyen A, Guo J, Widgerow AD. Stromal vascular fraction: a regenerative reality? Part 1. Current concepts and review of the literature. JPRAS. 2016;69:170–9.
144. Crisan M, Yap S, Casteilla L, Chen C-W, Corselli M, Park TS, Andriolo G, Sun B, Zheng B, Zhang L, Norotte C, Teng P-N, Traas J, Schugar R, Deasy BM, Badylak S, Buhring H-J, Giacobino J-P, Lazzari L, Huard J, Péault B. A perivascular origin for mesenchymal stem cells in multiple human organs. Cell Stem Cell. 2008;3:301–13.
145. da Silva Meirelles L, Chagastelles PC, Nardi NB. Mesenchymal stem cells reside in virtually all post-natal organs and tissues. J Cell Sci. 2006;119:2204–13.
146. da Silva Meirelles L, Caplan AI, Nardi NB. In search of the in vivo identity of mesenchymal stem cells. Stem Cells. 2008;26:2287–99.
147. Sacchetti B, Funari A, Michienzi S, Di Cesare S, Piersanti S, Saggio I, Tagliafico E, Ferrari S, Robey PG, Riminucci M, Bianco P. Self-renewing osteoprogenitors in bone marrow sinusoids can organize a hematopoietic microenvironment. Cell. 2007;131:324–36.
148. Armulik A, Genove G, Betsholtz C. Pericytes: developmental, physiological, and pathological perspectives, problems, and promises. Dev Cell. 2011;21:193–215.
149. von Tell D, Armulik A, Betsholtz C. Pericytes and vascular stability. Exp Cell Res. 2006;312:623–9.
150. Rochester JR, Brown NJ, Reed MW. Characterisation of an experimental model of chronic lower limb ischaemia in the anaesthetised rat. Int J Microcirc Clin Exp. 1994;14:27–33.
151. Hao C, Shintani S, Shimizu Y, Kondo K, Ishii M, Wu H, Murohara T. Therapeutic angiogenesis by autologous adipose-derived regenerative cells: comparison with bone marrow mononuclear cells. Am J Physiol Heart Circ Physiol. 2014;307:H869–79.
152. Kondo K, Shintani S, Shibata R, Murakami H, Murakami R, Imaizumi M, Kitagawa Y, Murohara T. Implantation of adipose-derived regenerative cells enhances ischemia-enhanced angiogenesis. Arterioscler Thromb Vasc Biol. 2009;29(1):61–6.
153. Murakami Y, Tanaka M. Evolution of motor innervation to vertebrate fins and limbs. Dev Biol. 2011;355(1):164–72.
154. Nakagami H, Maeda K, Morishita R, Iguchi S, Nishikawa T, Takami Y, Kikuchi Y, Saito Y, Tamai K, Ogihara T, Kaneda Y. Novel autologous cell therapy in ischemic limb disease through growth factor secretion by cultured adipose tissue-derived stromal cells. Arterioscler Thromb Vasc Biol. 2005;25:2542–7.
155. Rehman J, Traktuev D, Li J, Merfeld-Clauss S, Temm-Grove CJ, Bovenkerk JE, Pell CL, Johnstone BH, Considine RV, March KL. Secretion of angiogenic and antiapoptotic factors by human adipose stromal cells. Circulation. 2004;109:1292–8.
156. Sumi M, Sata M, Toya N, Yanaga K, Ohki T, Nagai R. Transplantation of adipose stromal cells, but not mature adipocytes, augments ischemia-induced angiogenesis. Life Sci. 2007;80:559–65.
157. Lawall H, Bramlage P, Amann B. Stem cell and progenitor cell therapy in peripheral artery disease. Thromb Haemost. 2010;103:696–709.
158. Lawall H, Bramlage P, Amann B. Treatment of peripheral arterial disease using stem and progenitor cell therapy. J Vasc Surg. 2011;53:445–53.
159. Liew A, O'brien T. Therapeutic potential for mesenchymal stem cell transplantation in critical limb ischemia. Stem Cell Res Ther. 2012;3:28.
160. Powell RJ, Comerota AJ, Berceli SA, Guzman R, Henry TD, Tzeng E, Velazquez O, Marston WA, Bartel RL, Longcore A, Stern T, Watling S. Interim analysis results from the RESTORE-CLI, a randomized, double-blind multicenter phase II trial com-

paring expanded autologous bone marrow-derived tissue repair cells and placebo in patients with critical limb ischemia. J Vasc Surg. 2011;54:1032–41.

161. Das AK, Bin Abdullah BJJ, Dhillon SS, Vijanari A, Anoop CH, Gupta PK. Intra- arterial allogeneic mesenchymal stem cells for critical limb ischemia are safe and efficacious: report of a phase I study. World J Surg. 2013;37:915–22.
162. O'Neill CL, O'Doherty MT, Wilson SE, Rana AA, Hirst CE, Stitt AW, Medina RJ. Therapeutic revascularisation of ischaemic tissue: the opportunities and challenges for therapy using vascular stem/progenitor cells. Stem Cell Res Ther. 2012;2012(3):31.
163. Powell RJ. Update on clinical trials evaluating the effect of biologic therapy in patients with critical limb ischemia. J Vasc Surg. 2012;56:264–6.
164. Weck M, Slesaczeck T, Rietzsch H, Münch D, Nanning T, Paetzold H, Florek H-J, Barthel A, Weiss N, Bornstein S. Noninvasive management of the diabetic foot with critical limb ischemia: current options and future perspectives. Ther Adv Endocrinol Metab. 2011;2:247–55.
165. Lee HC, An SG, Lee HW, Park J-S, Cha KS, Hong TJ, Park JH, Lee S-Y, Kim S-P, Kim YD, Chung SW, Bae YC, Shin YB, Kim JI, Jung JS. Safety and effect of adipose tissue-derived stem cell implantation in patients with critical limb ischemia: a pilot study. Circ J. 2012;76:1750–60.
166. Zhi K, Gao Z, Bai J, Wu Y, Zhou S, Li M, Qu L. Application of adipose-derived stem cells in critical limb ischemia. Front Biosci. 2014;19:768–76.
167. Gates J, Hartnell GG. Optimized diagnostic angiography in high-risk patients with severe peripheral vascular disease. Radiographics. 2000;20:121–33.
168. Rennert RC, Sorkin M, Januszyk M, Duscher D, Kosaraju R, Chung MT, Lennon J, Radiya-Dixit A, Raghvendra S, Maan ZN, Hu MS, Rajadas J, Rodrigues M, Gurtner GC. Diabetes impairs the angiogenic potential of adipose-derived stem cells by selectively depleting cellular subpopulations. Stem Cell Res Ther. 2014;5:79.
169. Zimmerlin L, Donnenberg VS, Pfeifer ME, Meyer EM, Péault B, Rubin JP, Donnenberg AD. Stromal vascular progenitors in adult human adipose tissue. Cytometry A. 2010;77:22–30.
170. Clokie CML, Urist MR. Bone morphogenetic protein excipients: with observations on Poloxamer. Plast Reconstr Surg. 2000;105:628–37.
171. Clokie CML, Sandor GKB. Reconstruction of 10 major mandibular defects using bioimplants containing BMP7. J Can Dental Assoc. 2008;74:67–72.
172. Sandor GK, Tuovinen VJ, Wolff J, et al. Adipose stem cell tissue-engineered construct used to treat large anterior mandibular defect: a case report and review of the clinical application of good manufacturing practice-level adipose stem cells for bone regeneration. J Oral Maxillofac Surg. 2013;71:938–50.
173. Sandor GK, Numminen J, Wolff J, et al. Adipose stem cells used to reconstruct 13 cases with cranio-maxillofacial hard-tissue defects. Stem Cells Trans Med. 2014;3:530–40.

Further Reading

Anderson DJ. Cellular and molecular biology of neural crest migration. Trends Genet. 1997;13:267–80.

Bardach J, Mooney M, Giedrojc-Juraha ZL. A comparative study of facial growth with and without soft-tissue undermining: an experimental study in rabbits. Plast Reconstr Surg. 1982;69:745.

Barro WB, Latham RA. Palatal periosteal response to surgical trauma. Plast Reconstr Surg. 1981;67:6.

Birgfield CB, Roberts S, Wang Y-C, Liao Y-F, Chen PKT. Comparative outcomes of primary gingivoperiosteoplasty and secondary alveolar bone grafting in patients with unilateral cleft lip and palate. Plast Reconstr Surg. 2016;137(1):228–9.

Boo CK. The unoperated adult bilateral cleft of the lip and palate. Br J Plast Surg. 1971;24:250.

Boyne PJ, Sands NR. Secondary bone grafting of residual alveolar and palatal clefts. J Oral Surg. 1972;30:87–01.

Cagáňová V, Borsky J, Šmahel Z, Veleminská J. Facial growth and development in unilateral cleft lip and palate. Comparison between secondary alveolar bone grafting and primary periosteoplasty. Cleft Palate Craniofac J. 2014;51(1):15–22.

Carstens MH. Correction of unilateral cleft lip nasal deformity using the sliding sulcus procedure. J Craniofac Surg. 1999;10:346–64.

Carstens MH. The sliding sulcus procedure: simultaneous repair of unilateral clefts of the lip and primary palate-a new technique. J Craniofac Surg. 1999;10:415–29.

Carstens MH. Functional matrix cleft repair: a common strategy for unilateral and bilateral clefts. J Craniofac Surg. 2000;11:437–69.

Carstens MH. Internal carotid artery distribution to the face: evidence for fusion of paired olfactory fields. PSEF basic science presentation. Los Angeles: American Society of Plastic Surgeons; 2000.

Carstens MH. Development of the facial midline. J Craniofac Surg. 2002;13:129–87.

Carstens MH. Neuromeric programming and craniofacial cleft formation. 1. The neuromeric organization of the head and neck. Eur J Ped Neurol. 2004b;8:181–210.

Carstens M, Chin M, Ng T, Tom WK. Reconstruction of #7 facial cleft with distraction-assisted in situ osteogenesis (DISO): role of recombinant human bone morphogenetic protein-2 with Helistat-activated collagen implant. J Craniofac Surg. 2005;16(6):1023–32.

Carlson BR. Human embryology and developmental biology. 5th ed. Philadelphia: Mosby; 2016. p. 185–90.

Cestero HJ, Salyer KE. Regenerative potential of bone and periosteum. Surg Forum. 1975;26:555.

Cestero HJ, Salyer KE, Johns DF. The periosteum and craniofacial growth. Surg Forum. 1976;27:556.

Cheng H, Jiang W, Phillips FM, et al. Osteogenic activity of the 14 types of human bone morphogenetic proteins (BMPs). J Bone Jt Surg. 2001;85A:1544–11551.

Chin M. Alveolar process reconstruction using distraction osteogenesis. In: Diner PA, Vasquez MP, editors. Tissue engineering and facial bone distraction processes. Bologna: Monduzzi Editore; 1997.

Chin M, Ng T, Tom WK, Carstens M. Repair of alveolar clefts with recombinant human bone morphogenetic protein (rhBMP-2) in patients with clefts. J Craniofac Surg. 2005;16(5):40–58.

da Silva Meirelles L, Fontes AM, Covas DT, Caplan AI. Mechanisms involved in the therapeutic properties of mesenchymal stem cells. Cytokine Growth Factor Rev. 2009;20:419–27.

Delaire J. Surgical considerations in patients with cleft deformities. In: Bell WH, editor. Modern practice in orthognathic and reconstructive surgery. Philadelphia: WB Saunders; 1992.

Delaire J, Precious DS, Gordeef A. The advantage of wide subperiosteal exposure in primary correction of labial maxillary clefts. Scand J Plast Reconstr Surg. 1988;22:147–51.

Duboule D. Patterning in the vertebrate limb. Curr Opin Genet Dev. 1991;1(2):21–6.

Georgiade NC. Anterior palatal-alveolar closure by means of interpolated flaps. Plast Reconstr Surg. 1967;39:162.

Georgiade NR. Repair of anterior palatal-alveolar clefts in the cleft lip and palate patient. In: Georgiade NR, Hagarty RF, editors. Symposium on the management of cleft lip and palate and associated deformities. St. Louis: CV Mosby; 1974. p. 62–6.

Goldstein MH. A tissue expanding vermilion myocutaneous flap for lip repair. Plast Reconstr Surg. 1984;73:768.

Goldstein MH. The elastic flap for lip repair. Plast Reconstr Surg. 1990;85:446.

Hellquist R, Ponten B. The influence of infant periosteoplasty on facial growth and dental occlusion from 5 to 8 years of age in cases of complete unilateral cleft lip and palate. Scand J Plast Reconstr Surg. 1979;13(2):305–12.

Hellquist R, Svardstrom K, Ponten B. A longitudinal study of delayed periosteoplasty to the cleft alveolus. Cleft Palate. 1983;J20:277.

His W. Beobachtungen zur Gesichte und Gamenbildung beim menschlishen embryo. Kgl Akad Wis. 1901;27:349–89.

Hrivnakova J. Comparison of the upper jaw development in facial clefts following primary osteoplasty and bridging the cleft with a periosteal flap. Rozhl Chir. 1982;61:805.

Hrivnakova J, Fara M, Mullerova Z. The use of periosteal flaps for bridging maxillary defects in facial clefts. Acta Chir Plast. 1981;23:130.

Januszyk M, Sorkin M, Glotzbach JP, Vial IN, Maan ZN, Rennert RC, Duscher D, Thangarajah H, Longaker MT, Butte AJ, Gurtner GC. Diabetes irreversibly depletes bone marrow-derived mesenchymal progenitor cell subpopulations. Diabetes. 2014;63:3047–56.

Joos U. Muscle reconstruction in primary cleft lip surgery. J Craniomaxillofac Surg. 1989;17(Suppl 1):8–10.

Kardong KV. Vertebrates: comparative anatomy, function, and evolution. 7th ed. New York: McGraw Hill; 2015.

Katagiri T, Yamaguchi A, Komaki M, et al. Bone morphogenic protein-2 converts the differentiation pathway of C2C12 myoblasts into the osteoblast lineage. J Cell Biol. 1994;127:1755–66.

Kitagawa Y, Murohara T. Implantation of adipose-derived regenerative cells enhances ischemia-induced angiogenesis. Arterioscler Thromb Vasc Biol. 2009;29:61–6.

Kitamura S. Comparison of angiogenic potency between mesenchymal stem cells and mononuclear cells in a rat model of hindlimb ischemia. Cardiovasc Res. 2005;66:543–51.

Lee CT, Grayson BH, Cutting CB, Brecht LE, Lin WY. Prepubertal midface growth in unilateral cleft lip and palate following alveolar molding and gingivoperiosteoplasty. Cleft Palate Craniofac J. 2004;41:375–80.

Matsuo K, Hirose T. Secondary correction of the unilateral cleft lip nose using a conchal cartilage graft. Plast Reconstr Surg. 1990;86:991.

May JW, Little JW, editors. Plastic surgery, vol. 4. Philadelphia: WB Saunders; 1990.

Millard DR. Cleft craft: the evolution of its surgery. The unilateral deformity, vol. I. Boston: Little, Brown; 1976.

Moon MH, Ludwig R, Halbhurner S, Burshe K, Stoll T. Human adipose tissue derived mesenchymal stem cells improve postnatal neovascularization in a mouse model of hindlimb ischemia. Cell Physiol Biochem. 2006;17:17279–90.

Mooney MP, Siegel MI, Kimes KR, Todhunter J. Development of the orbicularis oris muscle in normal and cleft lip and palate human fetuses using three-dimensional computer reconstruction. Plast Reconstr Surg. 1988;81:336.

Noden DM. Origins and patterning of craniofacial mesenchymal tissues. J Craniofac Genet Dev Biol Suppl. 1986;2(15–31):33.

Noordhoff MS. Reconstruction of vermilion in unilateral and bilateral cleft lips. Plast Reconstr Surg. 1984;73:52.

Noordhoff MS, Chen Y, Chen K, et al. The surgical technique for the complete unilateral cleft lip-nasal deformity. Oper Tech Plast Reconstr Surg. 1995;2:167.

Ortiz-Monasterio F, Drucker M, Molina F, et al. Distraction osteogenesis in Pierre Robin sequence and related respiratory problems in children. J Craniofac Surg. 2002;13:79–83.

Ow AT, Cheung LK. Meta-analysis of mandibular distraction osteogenesis: clinical applications and functional outcomes. Plast Reconstr Surg. 2008;121:e54–69.

Park CG, Ha B. The importance of accurate repair of the orbicularis oris muscle in the correction of unilateral cleft lip. Plast Reconstr Surg. 1995;96:780–8.

Precious D. Alveolar bone grafting. Oral Maxillofac Surg Clin North Am. 2000;12:501–13.

Precious DS, Delaire J. Balanced facial growth: a schematic interpretation. Oral Surg Oral Med Oral Pathol. 1987;63:637.

Roselli D. Bilateral labiopalatoschisis-early closing of the osseous fissure through free graft of periosteum in one stage. Ann Plast Surg. 1982;9:18.

Rubenstein JLR, Puelles L. Development of the nervous system. In: Epstein CJ, Erickson RP, Wynshaw-Boris A, editors. Inborn errors of metabolism: the molecular basis of clinical disorders of morphogenesis. Oxford: Oxford University Press; 2004. p. 75–88.

Sandor GKB. Tissue engineering and regenerative medicine: evolving applications toward cleft lip and palate surgery. In: Bennun R, Sandor GK, Harfin J, editors. Cleft lip and palate management: a comprehensive atlas. New York: Wiley; 2015.

Schendel SA, Delaire J. Facial muscles: form and function in dentofacial deformities. In: Bell WH, editor. Surgical correction of dentofacial deformities: new frontiers. Philadelphia: WB Saunders; 1985.

Schendel S, Pearl RM, De'Armond SJ. Pathophysiology of cleft lip muscle. Plast Reconstr Surg. 1989;83:777.

Schultz RC. Free periosteal graft repair of maxillary clefts in adolescents. Plast Reconstr Surg. 1984;73:556.

Semb G. Secondary bone grafting and orthodontic treatment in patients with bilateral complete clefts of the lip and palate. Ann Plast Surg. 1986;17:460.

Singer NG, Caplan AI. Mesenchymal stem cells: mechanisms of inflammation. Annu Rev Pathol. 2011;6:457–78.

Sitzmann F. The alveolar flap for the repair of the cleft alveolus related to the development of the upper jaw. J Maxillofac Surg. 1979;7:81.

Skoog T. The use of periosteal flaps in the repair of clefts of the primary palate. Cleft Palate J. 1965;2:332.

Skoog T. The use of periosteum and surgical for bone restoration in congenital clefts of the maxilla. A clinical and experimental investigation. Scand J Plast Reconstr Surg. 1967;1:113.

Skoog T. Skoog's method of repair of unilateral and bilateral cleft lip. In: Crabb WC, Rosenstein SW, Bzoch KR, editors. Cleft lip and palate. Boston: Little, Brown; 1971.

Smahel Z, Müllerovná Z. Facial growth and development in unilateral cleft lip and palate during the period of puberty: comparison of the development after periosteoplasty and after primary bone grafting. Cleft Palate Craniofac J. 1994;31:10.

Tessier P. Anatomical classification of facial, cranio-facial, and latero-facial clefts. J Maxillofac Surg. 1976;4:70–92.

Tessier P, Tulasne JF, Delaire J, Resche F. Therapeutic aspects of maxillonasal dystosis (Binder syndrome). Head Neck Surg. 1981;3:207.

Veau V. Bec-de-lievre. Paris: Masson; 1938.

Walker JC Jr, Collito MB, Mancuso-Ungaro A, Meijer R. Physiologic considerations in cleft lip closure. The C-W technique. Plast Reconstr Surg. 1966;37:552.

Wozney J. Overview of bone morphogenetic proteins. Spine. 2002;27(165):52–8.

Wozney JM, Rosen V, et al. Novel regulators of bone formation: molecular clones and activities. Science. 1988;242:1528–34.

Yamaguchi A, Katagiri T, Ikeda T, et al. Recombinant human bone morphogenic protein-2 stimulates osteoblast maturation and inhibits myogenic differentiation in vitro. J Cell Biol. 1991;113:681–7.

You ZH. The study of vascular communication between jaw bones and their surrounding tissues by SEM of resin casts. West Chin J Stomatol. 1990;8:235.

You ZH, Zhang ZK, Wang Y, et al. Distribution of minor nutrient foramina on the bone surfaces of the maxilla. In: Presented at the third conference of Chinese oral and maxillofacial surgeons. Xi'an, China. 1990.

You ZH, Zhang ZK, Zhia JL. A study of maxillary and mandibular vasculature in relation to orthognathic surgery. Chin J Stomatal. 1991;26:263.

Yoshimura K, Shigeura T, Matsumoto D, Sato T, Takaki Y, Aiba-Kojima E, Sato K, Inoue K, Nagase T, Koshima I, Gonda K. Characterization of freshly isolated and cultured cells derived from the fatty and fluid portions of liposuction aspirates. J Cell Physiol. 2006;208:64–76.

Index

M. H. Carstens (ed.), *The Embryologic Basis of Craniofacial Structure*, https://doi.org/10.1007/978-3-031-15636-6

M

N

O

P

R

S

T

V